Environment *an*

The SAGE
Handbook *of*
Environment *and* Society

Edited by
Jules Pretty, Andrew S. Ball,
Ted Benton, Julia S. Guivant,
David R. Lee, David Orr,
Max J. Pfeffer and Hugh Ward

SAGE Publications
Los Angeles • London • New Delhi • Singapore

SAGE Publications Ltd
1 Oliver's Yard
55 City Road
London EC1Y 1SP

SAGE Publications Inc.
2455 Teller Road
Thousand Oaks
California 91320

SAGE Publications India Pvt Ltd
B 1/1 1 Mohan Cooperative Industrial Area
Mathura Road, Post Bag 7
New Delhi 110 044

SAGE Publications Asia-Pacific Pte Ltd
33 Pekin Street #02-01
Far East Square
Singapore 048763

Library of Congress Control Number 2007922921

British Library Cataloguing in Publication data
A catalogue record for this book is available from the British Library

ISBN 978-1-4129-1843-5

Typeset by Cepha Imaging Pvt. Ltd., Bangalore, India
Printed in Great Britain by the Cromwell Press Ltd, Trowbridge, Wiltshire
Printed on paper from sustainable resources

Contents

List of Contributors

Andrew S. Ball
School of Biological Sciences
Flinders University
GPO Box 2100
Adelaide 5001, South Australia

Ian J. Bateman
Centre for Social and Economic Research on the
    Global Environment (CSERGE)
School of Environmental Sciences
University of East Anglia
Norwich NR4 7TJ, UK

Richard Bawden
Community, Agriculture Recreation and
    Resource Studies (CARRS)
330 Natural Resources
Michigan State University
East Lansing
Michigan 48824, USA

Ulrich Beck
Institute for Sociology
Ludwig-Maximilians University - Munich
Konradstr. 6
Munich, 80801
Germany

Charles Benjamin
Center for Environmental Studies
Williams College
Kellogg House
P.O. Box 632
Williamstown, MA 01267, USA

Ted Benton
The Department of Sociology
University of Essex
Wivenhoe Park
Colchester, CO4 3SQ, UK

Steven R. Brechin
Department of Sociology
Syracuse University
302 Maxwell Hall
Syracuse, NY 13244, USA

Henry Buller
Department of Geography
University of Exeter
The Queen's Drive
Exeter, Devon
UK EX4 4QJ

Stuart W. Bunting
Department of Biological Sciences
University of Essex
Wivenhoe Park
Colchester
Essex CO4 3SQ, UK

Chris Cocklin
Faculty of Science, Engineering and IT
James Cook University
Townsville, QLD 4811
Australia

Ian Colbeck
Department of Biological Sciences
University of Essex
Wivenhoe Park
Colchester
Essex CO4 3SQ, UK

Johan Colding
The Beijer Institute of
    Ecological Economics
The Royal Swedish Academy of
    Sciences
Box 50005
SE-10405 Stockholm
Sweden

Thomas D. Crocker
Department of Economics and Finance
University of Wyoming
P.O. Box 3985
Laramie, WY 82071-3985
USA

Leanne Cullen
Coral Reef Research Unit
Department of Biological Sciences
University of Essex
Wivenhoe Park
Colchester, CO4 3SQ, UK

Carl Folke
Stockholm Resilience Centre: Research for
    Governance of Social-Ecological Systems
Stockholm University
SE-10691 Stockholm
Sweden

The Beijer Institute of
    Ecological Economics
The Royal Swedish Academy of Sciences
Box 50005
SE-10405 Stockholm
Sweden

Warwick Fox
Centre for Professional Ethics
University of Central Lancashire
Preston PR1 2HE, UK

Howard Frumkin
Centers for Disease Control
    and Prevention
1600 Clifton Rd
Atlanta, GA 30333
USA

Madhav Gadgil
Agharkar Research Institute
Agarkar Road
Pune 411004
India

Steven Griggs
Social Sciences, Government
    and Politics
University of Birmingham
Edgbaston
Birmingham B15 2TT, UK

Julia S. Guivant
Department of Sociology and Political Science
Universidade Federal de Santa Catarina
Campus Universitário
Florianópolis SC 88040-900
Brazil

Thomas Hahn
Stockholm Resilience Centre:
    Research for Governance of
    Social-Ecological Systems
Stockholm University
SE-10691 Stockholm
Sweden

Jonathan Hastie
Department of Government
University of Essex
Wivenhoe Park
Colchester CO4 3SQ, UK

Ronald J. Herring
Department of Government
Cornell University
313 White Hall
Ithaca, NY 14853, USA

David R. Howarth
Department of Government
University of Essex
Wivenhoe Park
Colchester CO4 3SQ, UK

Ron Johnston
School of Geographical Sciences
University of Bristol
University Road
Bristol BS8 1SS, UK

Gideon Kossoff
Schumacher College
The Old Postern
Dartington
Devon
TQ9 6EA, UK

Randall A. Kramer
Nicholas School of the Environment
Duke University
Durham
North Carolina 27708
USA

Cordula Kropp
Institute for Sociology
Ludwig-Maximilians University - Munich
Konradstr. 6
Munich 80801
Germany

David R. Lee
Department of Economics and Management
441 Warren Hall
Cornell University
Ithaca, NY 14853, USA

Laura Little
School of Geography and Environmental
  Science
Building 11
Monash University
Vic 3800
Australia

Amory Lovins
Rocky Mountain Institute
1739 Snowmass Creek Road
Snowmass CO 81654-9199, USA

Luisa Maffi
Terralingua
217 Baker Road
Salt Spring Island
British Columbia
Canada V8K 2N6

Mary Mellor
Sociology and Criminology
School of Arts and Social Sciences
Northumbria University
Newcastle upon Tyne
NE1 8ST, England, UK

Carol Morris
Department of Geography
University of Nottingham
Nottingham,
NG7 2RD, UK

Joe Morris
Department of Natural Resources
School of Applied Sciences
Cranfield University
Bedfordshire, MK43 0AL
England, UK

Grant Murray
Institute for Coastal Research
Malaspina University-College
900 5th Street
Nanaimo
British Columbia, V9R 5S5
Canada

Harini Nagendra
Center for the Study of Institutions, Population,
  and Environmental Change (CIPEC)
Indiana University
408 North Indiana
Bloomington, Indiana 47408
USA

Ashoka Trust for Research in Ecology and the
  Environment (ATREE)
659, 5th A Main, Hebbal
Bangalore 560 024
India

Tim O'Riordan
School of Environmental Sciences,
University of East Anglia
Norwich
NR4 7TJ, UK

Per Olsson
Stockholm Resilience Centre: Research for
  Governance of Social-Ecological Systems
Stockholm University
SE-10691 Stockholm
Sweden

Peter Oosterveer
Environmental Policy Department
Wageningen University
Postbus 8130
6700 EW Wageningen
The Netherlands

David W. Orr
Room 210
Environmental Studies Program
Adam Joseph Lewis Center
Oberlin College
122 Elm Street
Oberlin, OH 44074, USA

Elinor Ostrom
Center for the Study of Institutions, Population,
    and Environmental Change (CIPEC)
Indiana University
408 North Indiana
Bloomington, Indiana 47408, USA

Workshop in Political Theory and Policy
    Analysis
Indiana University
513 North Park Avenue
Bloomington, Indiana 47408, USA

Center for the Study of Institutional Diversity
School of Human Evolution & Social Change
Arizona State University
Tempe, AZ 85287-2402, USA

Christina Page
Director, Energy and Climate
Yahoo! Inc.
701 First Avenue
Sunnyvale, CA 94089

Max J. Pfeffer
Development Sociology Department
133 Warren Hall
Cornell University
Ithaca, NY 14853-7801, USA

Sarah Pilgrim
Department of Biological Sciences
University of Essex
Wivenhoe Park
Colchester
Essex CO4 3SQ, UK

Val Plumwood
School of Environmental Studies
The Australian National University
Canberra ACT 0200
Australia

Jules Pretty
Department of Biological Sciences and Centre
    for Environment and Society
University of Essex
Wivenhoe Park
Colchester CO4 3SQ, UK

David J. Rapport
Ecohealth Consulting
217 Baker Road
Salt Spring Island
British Columbia
Canada V8K 2N6

Institute for Applied Ecology
Chinese Academy of Sciences
Shenyang, China

Paty Romero-Lankao
Institute for the Study of Society and
    Environment
National Center for Atmospheric
    Research
PO Box 3000
Boulder
Colorado 80307, USA

Alan Rudy
Department of Sociology, Anthropology and
    Social Work
Central Michigan University
124 Anspach Hall
Mount Pleasant
Michigan 48859
USA

David Smith
Coral Reef Research Unit
Department of Biological Sciences
University of Essex
Wivenhoe Park
Colchester CO4 3SQ, UK

Gert Spaargaren
Environmental Policy Department
Wageningen University
Postbus 8130
6700 EW Wageningen
The Netherlands

Linda P. Wagenet
Development Sociology Department
Cornell University
322 Warren Hall
Ithaca, NY 14853-7801, USA

Hugh Ward
Department of Government
University of Essex
Wivenhoe Park
Colchester CO4 3SQ, UK

Albert Weale
Department of Government
University of Essex
Wivenhoe Park
Colchester CO4 3SQ, UK

Damian F. White
Department of Sociology and Anthropology
James Madison University
Sheldon Hall
Harrisonburg, VA 22807, USA

Thomas J. Wilbanks
Oak Ridge National Laboratory
P.O. Box 2008
Oak Ridge, TN 37831, USA

Chris Wilbert
Department of Geography and Tourism
Anglia Ruskin University
Bishop Hall Lane
Chelmsford CM1 1SQ, UK

1

Introduction to Environment and Society

Jules Pretty, Andrew S. Ball, Ted Benton,
Julia S. Guivant, David R. Lee, David Orr,
Max J. Pfeffer and Hugh Ward

PERSPECTIVES ON SUSTAINABILITY

It is only in recent decades that the concepts associated with sustainability have come into more common use. Environmental concerns began to develop in the 1960s, and were particularly driven by Rachel Carson's book *Silent Spring* and the publicity surrounding it (Carson, 1963). Like other popular and scientific studies at the time, it focused on the environmental harm caused by one economic sector, in this case agriculture. In the 1970s, the Club of Rome identified the problems that societies would face when environmental resources were overused, depleted or harmed, and pointed towards the need for different types of policies to maintain and generate economic growth. In the 1980s, the World Commission on Environment and Development, chaired by Gro Harlem Brundtland, published *Our Common Future*, the first serious attempt to link poverty to natural resource management and the state of the environment. Sustainable development was defined as 'meeting the needs of the present without compromising the ability of future generations to meet their own needs'. The concept implied both limits to growth, and the idea of different patterns of growth, as well as introducing questions of intergenerational justice (WCED, 1987).

In 1992, the UN Conference on Environment and Development was held in Rio de Janeiro, taking forward many themes prefigured at the UN Conference on the Human Environment held in Stockholm in 1972. The main agreement was Agenda 21, a forty-one chapter document setting out priorities and practices for all economic and social sectors, and how these should relate to the environment. The principles of sustainable forms of development that encouraged minimizing harm to the environment and human health were agreed. However, progress has not been good, as Agenda 21 was not a binding treaty on national governments, and all are free to choose whether they adopt or ignore such principles (Pretty and Koohafkan, 2002). The Rio Summit was followed by some international successes, including the signing of the Convention on Biodiversity in 1995, the Kyoto Protocol in 1998 and the Stockholm Convention on Persistent Organic Pollutants in 2001. The ten years after the Rio World Summit on Sustainable Development was then held in Johannesburg in 2002, again raising the profile of sustainability, but also failing to tie governments to clear actions and timetables.

Over time, the concept of sustainability has grown from an initial focus on environmental aspects to include first economic and then broader social and political dimensions:

• *Environmental or ecological* – the core concerns are to reduce negative environmental and health externalities, to enhance and use local ecosystem resources, and preserve biodiversity. More recent concerns include broader recognition of the potential for positive environmental externalities from some economic sectors (including carbon capture in soils and flood protection).

- *Economic* – economic perspectives recognize that many environmental services are not priced by markets and that, because of this, it may be economically rational to use the environment in unsustainable ways and to undersupply environmental public goods. In response to this, some seek to assign value to environmental goods and services, and also to include a longer time frame in economic analysis. They also highlight subsidies that promote the depletion of resources or unfair competition with other production systems.
- *Social and political* – there are many concerns about the equity of technological change. At the local level, sustainability is associated with participation, group action and promotion of local institutions and culture (Ostrom, 1990; Pretty and Ward, 2001; Grafton and Knowles, 2004). At the higher level, the concern is for enabling policies that target preservation of nature and its vital goods and services. Many believe that liberal democracies are more likely to give rise to such policies than are autocracies, as part of generally better governance (United Nations Development Programme, 2003), but the empirical evidence for this is ambiguous (Midlarsky, 1998; Barrett and Graddy, 2000; Fredriksson *et al.*, 2005). Partly because of this some argue that the liberal democratic state needs to be transcended by adding in representation of other species, other generations and other nations (Eckersley, 2004) and by enhancing the potential for open deliberation about the issues, to bring together the knowledge that different groups and communities have and to reduce the corrosive impact of narrow self-interest (cf. Saward, 1993; Dryzek, 1996).

SOCIAL PERSPECTIVES ON ENVIRONMENT AND SOCIETY

An important feature of this Handbook centres on how social organization constrains humans' relationships with nature, but also how social organizations are shaped by nature. Perhaps the most distinctive feature of such an approach is that it rejects the notion that any form of social organization or structured human action is ideal or given by nature. While much human action is constrained by social structures (e.g. market behaviour), it is assumed that those structures are socially constructed and subject to change. This stance implies that human behaviour in relation to nature can be redirected if social structures change. Furthermore, changes in nature may force changes in social structure which in turn lead to changes in human behaviour.

Social scientists have long striven to develop an understanding of the relationship between the natural environment and society, but until the 1970s treatment by sociologists of this relationship remained more implicit than explicit. At this time, sociologists began to consider the nature–society nexus, and contemporary environmental sociology became a reaction to growing social activism for environmental protection. This activism reflected discontent with the dominant pro-technology and pro-growth economic policies following World War II. During the Cold War era, these policies might have tended to be either market- or state-centred, but regardless of ideological orientation economic growth driven by technological innovation was the overarching approach to economic development. This dominant worldview held that human domination of nature was unproblematic from a practical standpoint and was morally justified as well. But this point of view came to be challenged on both practical and moral grounds (Catton and Dunlap, 1978; Buttel, 1987; Beck, 1992a,b; Seippel, 2002).

From a practical standpoint, environmental deterioration became visible to the untrained eye. Air and water pollution became public issues of great concern (Buttel, 1997; Mertig *et al.*, 2002). Although the scientific community had been the foundation of technological development, critics of various technologies began to emerge from within it as well. Perhaps the most celebrated scientist to mount a sustained critique of the environmental impacts of technology was Rachel Carson. Many observers claim that the publication of her book, *Silent Spring* (Carson, 1963), marked the rise of contemporary environmentalism in the USA but there is clear evidence that concern about environmental destruction had already been stirring throughout the industrial world (Rootes, 1997; Mertig *et al.*, 2002). The rise of the environmental movement in the USA, for example, led to the enactment of a variety of unprecedented environmental legislation.

Sociologists were somewhat taken by surprise by the environmental movement, and struggled to understand it. Its substantive focus as well as the composition of its adherents appeared to be somewhat different from the other social movements of the day. The movement's adherents were initially thought to be more middle class and perhaps more mainstream than the anti-war and civil rights activists of the time. Substantively, the movement seemed to be charting a new course that was not rooted in the dominant socialist or capitalist ideologies. For this reason some sociologists began to suspect that environmentalists were advocating an entirely new paradigm – one that politically was neither left nor right, but entirely different. For this reason some initial thinking by sociologists was that an entirely new theoretical underpinning would need to be formulated

(Catton and Dunlap, 1978; Dunlap and Catton, 1994; Dunlap, 1997).

Initially, existing social theories were largely rejected on the assumption that they had been deficient in considering the active part played by the natural environment in societal development and had considered the impact of society on nature as inconsequential. Without a clear theory to guide the development of an alternative sociology of the environment, early efforts moved in a variety of directions that steered environmental sociology away from established theories of society.

Environmental sociologists initially criticized existing social theories for their hubris in assuming that humans through science and technology could dominate nature without significant impacts on the natural world or society. This paradigm was labelled 'human exemptionalism' (the assumption that human society is exempt from the biophysical law that control other species) (Catton and Dunlap, 1978; Dunlap and Catton, 1994). It was immediately clear that any sociology of the environment would need to focus on the relationship between that natural environment and society. A more careful treatment of this issue would challenge many assumptions in sociology. For example, sociology had assumed that all social structures could be explained by human agency. From this point of view, the physical and biological worlds were passive objects in the human construction of the social world (Murphy, 1994). But environmentalists' concerns about the destruction of nature and its consequences for society led to a reconsideration of how nature shapes society. Some claimed that what was distinctive about environmental sociology was its emphasis on the mutual constitution of nature and society (Freudenburg *et al.*, 1995; Norgaard, 1997). From this perspective, some sort of unidirectional and exclusively human construction of the life world is impossible.

So, what shapes the relationship between society and the environment? Some early attempts to apply sociological theory to the understanding of nature–society relationships drew on Marxist political economy. Political economists focused on the nature of the capitalist organization of production and how the functional demands of this system defined the use of nature. Some of the early thought in this area emphasized how capitalism's requirement for the continuous expansion of production into new areas would inevitably lead to the destruction of nature (Schnaiberg, 1980; Schnaiberg and Gould, 1994; Buttel, 1997). More recently, there has been greater emphasis on how capitalism is constrained by the biological and physical limits imposed by the natural world (Benton, 1989, 1998; Dickens, 1996, 1997).

The implications of the dominant system of market capitalism for nature–society relationships are a point of considerable contention in sociology. Some would argue that the capitalist economy is fundamentally destructive of the environment and for this reason is unsustainable in the long run. From this point of view, environmental destruction is the 'Achilles heel' of capitalism. This approach is deeply suspicious of claims that science and technology can always produce adequate substitutes for depleted natural resources (O'Connor, 1998). Recently, a decidedly more optimistic theory of ecological modernization has come into play. From this point of view, environmental destruction reflects a lack of investment in modern technologies and this deficit can be remedied with state policies that prohibit production practices wasteful or destructive of the environment. Ecological modernization is not just about technology, though. It is as much about bringing ecological considerations into market decision making through appropriate pricing of environmental services. In this theory the state plays a prominent role, with little real significance attached to abstractions like the 'free' market. The state constrains markets through policies that establish incentives to channel market behaviour in environmentally sound directions (Simonis, 1989; Mol, 1996, 2001; Mol and Spaargaren, 2000; Spaargaren *et al.*, 2000).

These opposing viewpoints on the environmental impacts of market economies point to the distinctiveness of this approach to understanding nature–society relations. Regardless of their theoretical orientation, sociologists consider organizational forms to be social constructs that are subject to change. This assumption implies that human behaviour is not inherent or given, but moulded by the social structures in place at any time in history. Thus, sociologists emphasize the distinctiveness of processes of societal rationalization, or the elaboration of a historically specific logic that structures the interaction between nature and society. Any particular rationalization is not 'natural' but has a distinctive form that constrains options for human interactions with nature (Murphy, 1994).

Since sociologists assume that social organization does not take some sort of 'ideal' form, the organization of human interactions with nature is a subject of particular interest to sociologists. Given an infinite number of possible forms of organization, why are similar forms of organization widely dispersed across a wide range of social and natural environments? This question has become especially salient with the emergence of the processes of globalization (Yearley, 1996). Economic, environmental and social organization displays some striking similarities in far-flung parts of the world. This organizational isomorphism is of growing interest to sociologists (Buttel, 1997; Frank, 2002; Frank *et al.*, 2000;

Schelhas and Pfeffer, 2005; Pfeffer *et al.*, 2006). But just as interesting to sociologists are some of the distinctive ways that these organizations are refashioned by local interests and the local natural resource base (Pfeffer *et al.*, 2001, 2005).

ENVIRONMENTAL ASSETS AND EXTERNALITIES

Many economic sectors directly affect many of the very assets on which they rely for success. Economic systems at all levels rely on the value of services flowing from the total stock of assets that they influence and control, and five types of asset, natural, social, human, physical and financial capital, are now recognized as being important. There are, though, some advantages and misgivings with the use of the term capital. On the one hand, capital implies an asset, and assets should be cared for, protected and accumulated over long and intergenerational periods. On the other, capital can imply easy measurability and transferability. Because the value of something can be assigned a monetary value, then it can appear not to matter if it is lost, as the required money could simply be allocated to purchase another asset, or to transfer it from elsewhere. But nature and its wider values is not so easily replaceable as a commodity (Coleman, 1988; Ostrom, 1990; Putnam, 1993; Flora and Flora, 1996; Benton, 1998; Uphoff, 1998, 2002; Costanza *et al.*, 1997; Pretty and Ward, 2001; Pretty, 2003; MEA, 2005).

Nonetheless, as terms, natural, social and human capital have become widespread in helping to shape concepts around basic questions about the potential sustainability of natural and human systems. The five capitals have been defined in the following ways:

1   *Natural capital* produces environmental goods and services, and is the source of food (both farmed and harvested or caught from the wild), wood and fibre; water supply and regulation; treatment, assimilation and decomposition of wastes; nutrient cycling and fixation; soil formation; biological control of pests; climate regulation; wildlife habitats; storm protection and flood control; carbon sequestration; pollination; and recreation and leisure.
2   *Social capital* yields a flow of mutually beneficial collective action, contributing to the cohesiveness of people in their societies. The social assets comprising social capital include norms, values and attitudes that predispose people to cooperate; relations of trust, reciprocity and obligations; and common rules and sanctions mutually agreed or handed down. These are connected and structured in networks and groups.
3   *Human capital* is the total capability residing in individuals, based on their stock of knowledge skills, health and nutrition. It is enhanced by access to services that provide these, such as schools, medical services and adult training. People's productivity is increased by their capacity to interact with productive technologies and with other people. Leadership and organizational skills are particularly important in making other resources more valuable.
4   *Physical capital* is the store of human-made material resources, and comprises buildings, such as housing and factories, market infrastructure, irrigation works, roads and bridges, tools and tractors, communications, and energy and transportation systems, that make labour more productive.
5   *Financial capital* is more of an accounting concept, as it serves in a facilitating role rather than as a source of productivity in and of itself. It represents accumulated claims on goods and services, built up through financial systems that gather savings and issue credit, such as pensions, remittances, welfare payments, grants and subsidies.

As economic systems shape the very assets on which they rely for inputs, there are feedback loops from outcomes to inputs. For instance, some economists emphasize the way that markets respond to resource scarcity is by pushing up prices, encouraging substitution and searching for technical change (Beckerman, 1996). However, such market feedbacks cannot work properly if environmental assets come for free. Thus, while sustainable systems will have a positive effect on natural, social and human capital, unsustainable ones feed back to deplete these assets, leaving fewer for future generations. For example, an agricultural system that erodes soil whilst producing food externalizes costs that others must bear. But one that sequesters carbon in soils through organic matter accumulation helps to mediate climate change. Similarly, a diverse system that enhances on-farm wildlife for pest control contributes to wider stocks of biodiversity, whilst simplified modernized systems that eliminate wildlife do not. Agricultural systems that offer labour-absorption opportunities, through resource improvements or value-added activities, can boost local economies and help to reverse rural-to-urban migration patterns (Carney, 1998; Dasgupta and Serageldin, 1998; Ellis, 2000; Morison *et al.*, 2005; Pretty *et al.*, 2006).

Any activities that lead to improvements in these renewable capital assets thus make a contribution towards sustainability. However, the idea of sustainability does not suggest that all assets are improved at the same time. One system that contributes more to these capital assets than another can be said to be more sustainable, but there may still be trade-offs with one asset increasing as another falls, though some environmental assets are essentially irreplaceable and vital, so they cannot be substituted – see the discussion of the idea of sustainability below. In practice, though, there are usually strong links between changes in natural, social and human capital, with systems having many potential effects on all three.

Many economic systems are, therefore, fundamentally multifunctional. They jointly produce many environmental goods and services. Clearly, a key policy challenge, for both industrialized and developing countries, is to find ways to maintain and enhance economic productivity. But a key question is: can this be done whilst seeking both to improve the positive side effects and to eliminate the negative ones? It will not be easy, as modern patterns of development have tended to ignore the considerable external costs of harm to the environment.

VALUING THE ENVIRONMENT

The idea that the environment and the services it provides can be valued strikes some as antithetical to the intrinsic values of environmental resources and the role that these resources play in society, history and culture. How can we possibly assign an economic or monetary value, it might be asked, to unique biodiversity such as the bald eagle or the snow leopard, to views of the Alps or the Rocky Mountains, or to water resources that are essential to life and that many societies consider to be an inherent human right? If economic/monetary values of these and similar resources can be estimated, how can they possibly be accurate if underlying conditions of scarcity change, as they inevitably will, leading to changes in associated scarcity values? And, if economic/monetary values are assigned to resources, whatever those values may be, does this valuation in and of itself inevitably lead to political trade-offs that may degrade those resources in the interests of economic development or other goals?

For these and many other reasons, the valuation of environmental resources is often fraught with contention, both conceptually and certainly in practice, where many empirical estimation and measurement issues arise. Yet, as mentioned above, the treatment of environmental assets as natural capital and associated exercises in measurement, valuation and evaluation are increasingly common in both academic analysis and policymaking. This is for several reasons. First, without such valuations, society has done a remarkably poor job in managing its stewardship of environmental resources; surely, any mechanism that can help improve on society's past dubious record in environmental policy is an advance. Second, since at least the 1960s and 1970s, the environmental impacts of economic development and human interventions in the landscape have been central to policy debates as society has increasingly been concerned with both the direct effects and opportunity costs of those interventions – e.g. what is lost when development proceeds. Third, in the two decades since the publication of the Brundtland Report (WCED, 1987), issues of sustainability have achieved much higher prominence in public debate in many countries, highlighting the needs of future generations in decisions made today about resource use. This has increased interest in how to trade off current versus future demands on the environment and how to deal with associated intergenerational equity concerns, which, in turn, has increased interest in mechanisms, like economic valuation, that permit these intertemporal comparisons.

In addition to these general factors stimulating interest in environmental valuation, efforts at economic and monetary valuation of the environment have flourished over the past several decades because they address several additional specific needs that are increasingly evident in environmental policymaking. First, the importance of the divergence between social valuation of resources and their incomplete (or non-existent) valuation in the market is increasingly apparent. How can we begin to address the problem of global warming, for example, if the externalities of industrial pollution are so poorly measured and understood, and consequently devalued in the policy arena, compared with the measurable jobs and income that are created? Second, as the human population expands and many formerly abundant resources are increasingly scarce – clean water and clean air, wilderness, open space, even silence – accounting for, and valuing, the public good dimensions of these resources has become increasingly important in prioritizing their survival in policy debates. How else, outside of moral suasion, will the scarcity value of public goods be understood and taken into account? Third, as the demand for economic valuation has expanded since the 1960s and 1970s, specific valuation methods and estimation procedures have also improved significantly, permitting a more accurate – though still frequently problematic – estimation of economic and monetary values of environmental resources and associated services.

An additional factor has to do with the response to policymaking itself. The limitations of 'command and control' and 'fences and fines' approaches to environmental policymaking have become increasingly evident, both in industrialized countries, where the institutions are often in place to deal effectively with at least some environmental problems, and certainly in developing countries, where such institutions are often non-existent, irrelevant or functionally powerless. Yet, 'command and control' policy and regulatory approaches often generate responses by private decision-makers that are, at best, socially inefficient and wasteful of resources, and, at worst, stimulate rent-seeking behaviour and strategic decision-making that yield perverse outcomes. Is it not preferable to develop policies and regulatory frameworks that are compatible with private incentives and that, in fact, employ these incentives and knowledge of human behaviour in innovative ways to lead to socially desired outcomes? Much of the recent interest in environmental valuation has been concerned with precisely these questions, specifically, the development of incentive-compatible policies and regulatory approaches that yield desired outcomes in ways that may be less costly and more socially efficient. Hence, the interest in tradable emissions permits, carbon-trading schemes, the pricing of heretofore free water resources, valuation, compensatory and payment transfer mechanisms for environmental services, and other such innovations.

Although alternative typologies exist, one common framework for organizing our thinking about resource valuation distinguishes four types of ecosystem values (Pearce and Turner, 1990): (1) *direct use values*, due to the direct utilization of resources and ecosystem services; (2) *indirect use values*, attributable to the externalities of ecosystem services; (3) *option value*, due to preserving the option for future use of the resource (also directly addressing sustainability criteria); and (4) *non-use values*, which are attributable to a variety of intrinsic ecosystem characteristics. This nomenclature aside, perhaps inevitably, much of the attention in environmental valuation has focused on specific methodologies and analytical approaches to assigning economic and monetary values to resources, especially those resources that have typically been outside the formal market (Hanley and Spash, 1993; Freeman, 2003).

Accordingly, as discussed further in several chapters in Section II, these approaches are commonly divided into 'expressed (or 'stated') preference' approaches and 'revealed preference' approaches. The former approaches ask consumers and other private agents to assign resource values and rankings directly; these approaches include 'contingent valuation' methodologies in which people are asked for their 'willingness-to-pay' to pay for environmental benefits, for example. The latter approaches indirectly elicit consumer valuations through methods such as the 'travel cost' approach and 'hedonic pricing', which estimate resource values through statistical analysis of

factors underlying human behaviour and the preferences (e.g. values) that are thus revealed. All of these methods have acknowledged strengths and deficiencies (also discussed in Section II). Yet, they have achieved wide acceptance because they continue to be at least partially successful in giving policy analysts and policymakers useful mechanisms and standards for achieving a better understanding of the values of environmental resources, thus enabling them to make better decisions regarding resource management, including the conservation and preservation of environmental resources in the face of competing uses.

THE CONSUMPTION TREADMILL

Since the World Commission on Environment and Development began deliberating on the links between environment and economy, there have been at least a couple of hundred further definitions of sustainability, and the term has now entered our common language. But where are we now with this sustainability idea? Does it offer some new hope for the world, or has it just hidden a much greater problem? The biggest challenge to sustainable development is now the consumption treadmill. The figures are worrying. People in North America now consume 430 litres of water per day; in developing countries, 23% have no water. In North America, 308 kg of paper are consumed by each person annually; in Europe 125 kg, in China 34 kg, and in India and Africa just 4 kg. In North America, there are 75 motor vehicles per 100 people, in Japan 57, in Europe 24, and in China, India and Africa just six to nine (see Table 1.1). Worldwide, some 400,000 hectares of cropland are paved per year for roads and parking lots (the USA's 16 million hectares of land under asphalt will soon reach the total area under wheat). The world motor-vehicle fleet grows alarmingly, as the nearly wealthy look to other parts of our global community for guidance as to what to buy. By almost every measure of resource consumption or proxy for waste production, the USA and Europe lead the way. And what model is being held up as the one to aspire to? There are now few people in the world who do not now aspire to the same levels of consumption as North America, which is, after all, presented as the pinnacle of economic achievement.

This consumer boom is already happening (see Meadows *et al.*, 1972; Bell, 2004; see also Frank, 1999; Kasser, 2002; Schwartz, 2004; Nettle, 2005). The new consumers (Myers and Kent, 2003, 2004) have already entered the global economy, and are aspiring to have lifestyles currently enjoyed by the richest. A number of formerly poor countries are seeing the growing influence of

Table 1.1 Indicators of consumption from different countries and regions of the world (data from 2004–2005)

	USA	Europe	China	India	Asia	Africa	Latin & Central America	World
Passenger cars per 1000 people	750	240	7	6	20	9	56	91
Annual petrol and diesel consumption (litres per person)	1624	286	33	9	47	36	169	174
Annual energy consumption per person (kg oil equivalent)	8520	3546	896	515	892	580	1190	1640
Annnual carbon dioxide emissions (tonnes per person)	20.3	8–12	2.7	0.99	<1	<1	<1	3.85
Annual paper and board consumption (kg per person)	308	125	34	4	29	4	38	52
Annual meat consumption (kg per person)	125	74	52	5	28	13	58	40
Daily water consumption (cubic metres per person)	4.6	1.59	1.35	1.74	1.72	0.47	1.47	1.73
Population (million, 2005)	293	730	1306	1080	3667	887	518	6500
Children born per woman	2.08	1.56	1.72	2.78	3.1	4.82	2.75	2.55

Sources: Pretty (2007), using Brown (2004); Myers and Kent (2004); WRI (2006)

affluence, as the middle classes of China, India, Indonesia, Pakistan, Philippines, South Korea, Thailand, Argentina, Brazil, Colombia and Mexico engage in greater conspicuous consumption. The side effects are already being felt – the average car in Bangkok spends 44 days a year stuck in traffic. But there is still a long way to go. The car fleet of the whole of India is still smaller than that of Chicago, and that of China is half the number of cars in greater Los Angeles. At the same time as a consumer boom is occurring among newly affluent urban elites, poor people in such countries as India and China lack access to the basics such as clean water and health care.

This is now the concern: the idea of sustainable economic development seems to imply that the world can be improved, or even saved, by bringing everyone up to the same levels of consumption as those in the industrialized countries. We can, it is said, grow out of many kinds of economic trouble. This cannot be done, as we would need six worlds at European and eight to nine at North American levels and patterns of consumption (Rees *et al.*, 1996; Rees, 2002, 2003). How much, we might wonder, would be enough (see Suzuki, 1997)?

The currently dominant idea about the inevitable benefits of progress would appear to be a modern invention. Indigenous peoples do not believe that their current community is any better than those in the past. To them, past and future are the same as current time. Their ancestors, and those of animals too, constantly remind them to be humble as they move about their landscapes. But the myth of progress permits the losses of both species and special places, as it is believed that

losses can be offset by doing something else that is better. The myth permits a belief in technological fixes, which are indeed effective in many ways, but rarely seem to make everyone happier, even if some of them contribute to human longevity and reduce suffering. Environmental problems are, after all, human problems. New technologies will make improvements, but possibly not fast enough to save us. They also bring some new risks, possibly rendering society more vulnerable. To come soon will be fabulous electronic memory, a genomics revolution, renewable energy, and human brains augmented by computers, though as Rees (2002) puts it, 'a super-intelligent machine could be the last invention humans ever make'. Rees recounts the 1937 efforts by the US National Academy of Sciences to predict breakthroughs for the rest of the last century. They made a good stab at agriculture, rubber and oil, but completely missed nuclear energy, antibiotics, jet aircraft, space travel and computers (see also Gray, 2002, 2004).

It is now clear from a variety of studies of people in the USA and Europe that people were happier in the 1950s compared with today. We can only guess more about earlier times, as the data do not exist in comparable form. But it does seem that our programmed happiness is about striving for, not actually increasing, happiness (Frank, 1999; Kasser, 2002; Schwartz, 2004; Nettle, 2005). One reason is that we compare our consumption with others around us, and we do not necessarily feel better off or happier if others' consumption is also increasing. There is always a nagging gap between present levels of contentment and how it could be. We believe we will be happier in the future, but seldom are. We also are constantly worrying about how future life events affect our happiness. As Bell (2004) has pointed out, we could work four hours per day, or just for about half a year, if we consumed at 1940s levels, yet be equally happy. But would anyone choose this option if they could?

EMERGING PERSPECTIVES ON POPULATION AND THE ENVIRONMENT

Population will continue to grow in many countries at least until mid-century, posing considerable problems in relation to providing for basic needs and dealing with environmental damage in some. Yet population is already declining in some rich countries, and others' population can be expected to stabilize then to decline, as the age structure of the population shifts and social practices change. A psychological problem yet to be faced is the consequence of coming population decline. Thomas Malthus (1798) argued that human population growth would always outstrip resources. 'Population, when unchecked', he said, 'increases in a geometrical ratio. Subsistence increases only in an arithmetic ratio. A slight acquaintance with numbers will shew the immensity of the first power in comparison with the second'. Since then, most policies and practice regarding natural resources and food have been shaped by concerns about our growing numbers. Humans are, after all, an extraordinarily successful species. When agriculture emerged, some 10,000 years ago, there were probably five million people worldwide. To the mid-19th century, world population then doubled eight times. Since then it has doubled four more times, and will continue to grow to probably eight and a half billion people by the middle of the 21st century. It will then stabilize for a while, and subsequently fall. Not because of wars, climate change or infectious diseases (though they may contribute to greater declines), but because of changing fertility patterns. More choices about contraception and decreasing poverty reduces the need to have so many children, and changing lifestyles among the rich delay child-bearing ages. When one generation produces fewer daughters, and fewer daughters are produced by them, then the replacement rate soon falls below the 2.1 needed to maintain population stability.

Today, the average woman in industrialized countries has fewer than 1.6 children, in the least developed countries 5 children, and in the other developing countries 2.6. The lowest fertility rates are now in southern Europe, at 1.1 children per woman. In the mid-1970s, the average Bangladeshi woman had six children; today she has about three; in Iran, fertility has fallen from more than five children in the late 1980s to just over two today. The worldwide annual gain is still 76 million people (down from 100 million in 1990), but this is expected to fall to zero by 2050 as the number of children falls from today's average of 2.55 to 2.0. Life expectancy at birth was 47 years in 1950–1955, rose to 65 years by 2000–2005, and will rise again to 75 years worldwide by 2045–2050. By then, the number of people over 60 will have tripled to 1.9 billion, and the number over 80 will have risen from today's 86 million to 395 million. Of course, these changes will not be evenly spread. Some countries are predicted to triple their numbers by 2050: these include Afghanistan, Burkina Faso, Burundi, Chad, Congo, DR Congo, DR Timor-Leste, Guinea-Bissau, Liberia, Mali, Niger and Uganda. But the populations of 51 countries will fall, including Germany, Italy, Japan and most of the former USSR (UN, 2004, 2005a).

What will happen after this peak, less than two generations away from us now? The United Nations (2005b) has made population predictions for the next 300 years, and uncertain though these must be, the medium fertility estimates suggest at least a levelling of world population for 250 more years at 8.5 to 9 billion. At low fertility (at the kind of levels we are already seeing today – after all, 93 out of 222 countries already have fewer than 2.1 children per women, and 37 have less than 1.5), world population declines to 5.5 billion by the end of this century, to 3.9 billion by 2150, and down to 2.3 billion by 2300. Which track we end up on depends entirely on early changes in fertility. Demographers cannot, of course, agree on the probability of stability or decline. But any kind of fall will bring huge changes. In 2000, people on average retired two weeks before mean life expectancy (at 65 years); by 2300, people will retire more than 30 years short of life expectancy (unless age of retirement changes), when on average women will live to 97 and men to 95 years. This does not take account of potentially revolutionary changes to human longevity that new medical technologies might bring.

Caldwell (2004) says that 'the low scenario is by no means implausible', and that the low projections 'would probably portend to many the fear of human extinction'. Governments would try to raise fertility levels, but it could be very difficult to achieve, as people do not always do the bidding of their governments. What, then, will happen to all those settlements we do not need? What of the fields and farms that become surplus to requirements? What of the wild animals – will we see their return to places where they had long since been eliminated (not the extinct species, of course, as they are gone forever)? Or might the vision be quite different – of spreading urban wastelands, of forgotten linkages to nature, of the nightmare of decivilization (a term coined by Timothy Garton Ash, in Porritt, 2005)?

DUALISM, SEPARATION AND CONNECTIONS

In recent years, with growing concerns for sustainability, the environment and biodiversity, many different typologies have been developed to categorize shades of deep to shallow green thinking. Arne Naess sees shallow ecology, for example, as an approach centred on efficiency of resource use, whereas deep ecology transcends conservation in favour of biocentric values. Other typologies include Donald Worster's imperial and arcadian ecology (Worster, 1993) and the resource and holistic schools of conservation. For some, there is an even more fundamental schism – whether nature exists independently of us, or whether it is characterized as post-modern or as part of a post-modern condition. Nature to scientific ecologists exists. To some post-modernist perspectives, though, it is mostly a cultural construction. The truth is, surely, that nature does exist, but that we socially construct its meaning to us. Such meanings and values change over time, and between different groups of people.

There are many dangers in the persistent dualism that separates humans from nature. It appears to suggest that we can be objective and independent observers – rather than part of the system and inevitably bound up in it. Everything we know about the world we know because we interact with it, or it with us. Thus, if each of our views is unique, we should listen to the accounts of others and observe carefully their actions. Another problem is that nature is seen as having boundaries – the edges of parks or protected areas. At the landscape level, this creates difficulties, as the whole is always more important than each part, and diversity is an important outcome (Foreman, 1997; Klijn and Vos, 2000).

This can lead to the idea of enclaves – social enclaves such as reservations, barrios or Chinatowns, and natural enclaves like national parks, wildernesses, sites of special scientific interest, protected areas or zoos. Enclave thinking can lead us away from accepting the connectivity of nature and people, though it has the advantage of creating niches for specialization. One consequence is that biodiversity and conservation can be considered to be in one place, and productive agricultural activities in another (Cronon et al., 1992; Deutsch, 1992; Brunkhorst et al., 1997; Pretty, 2002). It is no longer acceptable

to cause damage in some natural landscapes, provided we leave some areas protected. Enclaves also act as a sop to those with a conscience – the wider destruction can be justified if we fashion a small space for natural history to persist.

By continuing to separate humans and nature, the dualism also appears to suggest that technologies can always intervene to reverse damage caused by this very dualism. The greater vision, and the more difficult to define, involves looking at the whole, and seeking ways to redesign it. Cartesian dualism that puts humans outside nature remains a strange concept to many human cultures. It is only modernist thinking that has separated humans from nature in the first place, putting us up as distant controllers. Most peoples do not externalize nature in this way. From the Ashéninha of Peru to the forest dwellers of former Zaire, people see themselves as just one part of a larger whole, as do many people who adhere to major modern religions – even Christians who are often accused of treating nature as something to be plundered. Their relationships with nature are holistic, based on 'both/with' rather than 'either/or' (Benton, 1998; Gray, 1999). Recent research on the biophilia hypothesis of E. O. Wilson is indicating that natural or green places are good for mental health, irrespective of social context (Kellett and Wilson, 1993; Pretty, 2004; Pretty et al., 2005).

The idea of the wilderness struck a chord in the mid-19th century, with the influential writers Henry David Thoreau and John Muir setting out a new philosophy for our relations with nature. This grew out of a recognition of the value of wildlands for people's well-being. Without them, we are nothing; with them, we have life. Thoreau famously said in 1851, 'in wildness is the preservation of the world'. Muir in turn indicated that: 'wildness is a necessity; and mountain parks and reservations are useful not only as fountains of timber and irrigating rivers, but as fountains of life.' But as Roderick Nash, Max Oelschlaeger, Simon Schama and many other recent commentators have pointed out, these concerns for wilderness represented much more than a defence of unencroached lands. (For the Thoreau quote, see Nash, 1973, p. 84 – quoted in turn from a speech by Thoreau on April 23rd, 1851, to the Concord Lyceum. For the Muir quote, see Oelschlaeger, 1991. See also Nash, 1973; Schama, 1996; and Vandergeest and DuPuis, 1996.) It involved the construction of a deeper idea, which proved to be hugely successful in reawakening in North American and European consciences the fundamental value of nature.

Debates have since raged over whether 'discovered' landscapes were 'virgin' lands or 'widowed' ones, left behind after the death of indigenous peoples. Did wildernesses exist, or did we create them? Donald Worster, environmental historian, points out for North America that 'neither adjective will quite do, for the continent was far too big and diverse to be so simply gendered and personalised' (Worster, 1993). In other words, just because they constructed this idea does not mean to say it was an error. Nonetheless, they were wrong to imply that the wildernesses in, say, Yosemite were untouched by human hand, as these landscapes and habitats had been deliberately constructed by Ahwahneechee and other native Americans and their management practices to enhance valued fauna and flora.

Henry David Thoreau developed his idea of people and their cultures as being intricately embedded in nature as a fundamental critique of mechanical ideas that had separated nature from its observers. His was an organic view of the connections between people and nature (For a good review of Thoreau, see Oelschlaeger, 1991, pp. 133–171). In his *Natural History*, Thoreau celebrates learning by 'direct intercourse and sympathy' and advocates a scientific wisdom that arises from local knowledge accumulated from experience combined with the science of induction and deduction. But he still invokes the core idea of wilderness untouched by humans – even though his Massachusetts had been colonized just two centuries earlier and had a long history of 'taming' both nature and local native Americans.

The question, 'is a landscape wild, or is it managed', are perhaps the wrong ones to ask, as it encourages unnecessary and lengthy argument. What is more important is the notion of human intervention in a nature of which we are part. Sometimes such intervention means doing nothing at all, so leaving a whole landscape in a 'wild' state, or perhaps it means just protecting the last remaining tree in an urban neighbourhood or hedgerow on a field boundary. Preferably, intervention should mean sensitive management, with a light touch on the landscape. Or it may mean heavy reshaping of the land, for the good or the bad.

So it does not matter whether untouched and pristine wildernesses actually exist. Nature exists without us; and with us is shaped and reshaped. Most of what exists today does so because it has been influenced explicitly or implicitly by the hands of humans, mainly because our reach has spread as our numbers have grown, and as the effects of our consumption patterns have compounded the effect. But there are still places that seem truly wild, and these exist at very different scales and touch us in different ways. Some are on a continental scale, such as the Antarctic. Others are entirely local, a woodland amidst farmed fields, a saltmarsh along an estuary, a mysterious urban garden, all touched with private and special meanings.

In all of these situations, we are a part, connected, and so affecting nature and land, and being affected by it. This is a fundamentally different position to one which suggests that wilderness is untouched, pristine, and so somehow better because it is separated from humans – who, irony of ironies, promptly want to go there in large numbers precisely because it appears separate. But an historical understanding of what has happened to produce the landscape or nature we see before us matters enormously when we use an idea to form a vision that clashes with the truth. An idea that this place is wild, and so these local people should be removed. Another idea that this place is ripe for development, and so a group of people should be dispossessed. The term wilderness has come to mean many things, usually implying an absence of people and presence of wild animals, but also containing something to do with the feelings and emotions provoked in people. Roderick Nash (1973) takes a particularly Eurocentric perspective in saying, 'any place in which a person feels stripped of guidance, lost and perplexed may be called a wilderness', though this definition may also be true of some harsh urban landscapes. The important thing is not defining what it really is, but what we think it is, and then telling stories about it.

SOCIOLOGY AND THE ENVIRONMENT

The classical approaches to understanding the structure of society shared two basic features. One was the ambition to provide ways of conceptualizing the large-scale structural features of whole societies, and to situate them in the context of long-term historical change and in relation to the alternative social forms and historical tendencies in the rest of the world. The other was the insistence that human social and historical life was a distinct order of reality in its own right, not to be explained away in terms of the biological sciences of the day: industrial development, social inequality, crime, suicide rates, gender divisions and the like were to be understood in terms of social and cultural causes, not racial inheritance, genetic endowment or physiological constitution. This second feature was the basis for a process of 'separate development', through which the life and social sciences proceeded in ignorance of one another: 'nature' and 'culture' were distinct and contrasting realms, knowing and needing to know nothing of one another (Benton, 1996, 2001).

A common feature of the classics was their insistence on human social and cultural life as an order of reality in its own right, irreducible to the biological realm. Through most of the 20th century, this was taken to be an unquestioned assumption: social processes were to be explained in terms of social causes. This resistance to biological explanation was strongly reinforced by widespread revulsion at the consequences of Nazi doctrines of racial superiority, and the racist underpinning of much European imperial domination of non-Western peoples. With the rise of new social movements from the 1960s onwards, challenging established inequalities and social exclusions based on gender difference and sexual orientation, the terms 'nature' and 'natural' came to be viewed with suspicion. Sociologists sympathetic to the struggle for women's emancipation or gay rights critically exposed the way dominant ideologies justified oppression in the name of a distinction between what was 'natural' and what was 'unnatural', and therefore pathological. In this way, the strong links between sociology and progressive social movements reinforced the assumption already built into the main sociological traditions that biology, the 'natural', should be held at arms length and viewed with suspicion. It became a standard procedure for sociologists, and especially those who identified with a critical stance towards established society, to call into question all authoritative claims to knowledge of 'nature' or 'reality' (Soper, 1995; MacNaughten and Urry, 1998).

Then, from the late 1970s onwards, developments in linguistics and cultural theory became very influential, and approaches which (following Weber and Simmel, among the classics) focused on symbolic meaning and the role of language in shaping our experience of the world flourished. Questions about the material reality of nature and our relation to it now became excluded as a matter of methodological principle: all experience of the world is to be understood as mediated by language and culture. But there is no way anyone can stand outside the available language and culture to see reality in itself: we are left with the task of characterizing the role played in social life by various different and often conflicting linguistic and cultural 'constructions' of reality. It is important to remain neutral and agnostic about which, if any, of these 'constructions' is true. Critical sociology can aid emancipatory social struggles by exposing the 'constructed' character of the prevailing oppressive accounts of what is 'natural', thus 'deconstructing' them and challenging their authoritative hold over peoples' lives. These are the core insights of the approaches called 'constructionist'.

The sensitivity of sociology to the social and cultural movements and issues in the wider world outside the academy now presented it with a deep challenge: from the early 1960s the progressive social movements with which many sociologists had become identified also included a burgeoning radical environmental movement whose intellectual leaders (often dissident natural scientists) raised public alarm about the growing threat posed by our affluent, growth-oriented throwaway society to its own planetary life-support systems. Here was a new and powerful basis for a radical, critical politics, but one which celebrated nature, and claimed authoritative knowledge of the terrible destruction of it unleashed by contemporary society. This was a deep challenge in two ways. First, it was an intellectual challenge. Sociology had established its right to exist as a distinct discipline by a radical separation of the realms of nature and culture, but now faced pressing questions about the consequences of the mutual interconnection, the shared fate, of natural processes and social life. The second challenge was rooted in the normative commitment of critical sociologists and was particularly strongly felt by those who sympathized with such emancipatory movements as anti-racist, gay rights and women's liberation activism but were also drawn to the emergent green politics with its passionate defence of 'nature'.

There emerged two very broad, and to some extent conflicting, ways of addressing the new environmental agenda. One, typically 'constructionist', and deriving from the 'modest' tradition, tended to avoid large-scale theorizing. The great strength of this tradition has been its detailed case studies of particular environmental issues, social movements, campaigns and episodes of conflict. Rather than use the new environmental agenda as an occasion for questioning the basic inherited assumptions of the discipline, this sort of approach has concentrated on treating environmental issues as a new field in which to demonstrate the value of already-established sociological concepts and styles of argument. For instance, this approach has debunked many myths about the environmental movement. It was found that while parts of the movement retain a radical and progressive edge, many had evolved into highly professional lobby organizations seeking insider status in government decision making through moderating their demands (Dalton, 1994). At the same time, many members do little or nothing beyond giving an annual donation, having little or no direct involvement in local politics and living remarkably standard middle-class lives. The key standard concepts turn out to be institutionalization as a consequence of the problem of resource mobilization, which in turn derives from rational choice by individuals to do little, as captured in the Prisoner's Dilemma metaphor and other models of collective action failure (Jordan and Maloney, 1997).

For sociologists of science, the natural sciences are thoroughly social in character, their conceptual organization, research priorities and methodological procedures all shaped by social interests and cultural values – generally subservient to the dominant group or elite interests. Similarly with technologies – these are designed to serve powerful interests and cannot be fully understood independently of the social practices and relationships which their use either maintains or transforms. This way of understanding scientific and technical innovation has much to offer in the environmental field. It opens up the possibility of analysis of the kinds of pressures, power relationships, forms of regulation, etc., which promote environmentally damaging technologies, and also suggests the sorts of social and economic change which might encourage more benign forms of technical change.

Contructionist approaches have also produced valuable research in the field of environmental social movement mobilization and organization. The key insight which informs their approach is recognition that there is no one-to-one correspondence between the existence of, say, air pollution, or biodiversity loss, on the one hand, and the emergence of a social movement which identifies it as an unacceptable condition and campaigns for change, on the other. A leading constructionist environmental sociologist, John Hannigan (1995), provides an illuminating set of concepts for analysing the social and cultural processes involved in 'constructing' an environmental problem. First, a problem-claim has to be 'assembled': evidence, including scientific evidence, has to be collected and put together in such a way as to show that the state of affairs is significant enough to justify public concern and action. Next, it has to be 'presented': the problem has to be characterized in ways which will attract attention, and provoke the desired public concern. Since the media are now so central to communication to wider publics, this will also involve ways of engaging with the media in such a way as to ensure not only their attention, but also that media representations coincide with the movement's own 'framing' of the issue. The case of Greenpeace's use of dramatic film footage of whaling is a good example. The visual images were irresistible material for the electronic media, and their vivid portrayal of the violent death of great and beautiful creatures had more impact on public conscience than a thousand books. Perhaps, too, the constructionists might argue that the ethical and aesthetic power of these images far outweighed the influence of detailed scientific studies of the population dynamics and risk of extinction of the different whale species.

Finally, Hannigan notes that success on the part of social movements in making their 'problem-claim' is not the end of the matter. In each case, interests will be threatened by the raising of an issue – in the case of whaling, for example, the industry itself, and spin-off processing and retail interests, as well as consumer cultures in certain countries and indigenous people for whom whaling is central to their whole way of life. So, the raising of an issue will generally be met with counter-arguments, and competition for media framing and public acceptance. Hannigan calls this the 'contestation' of movement claims. Social movement theory also has developed concepts for analysing the processes involved in establishing, maintaining and coordinating social movement activity, for studying the culture of such movements and how they shape the identities of individuals who participate in them (Eyerman and Jamison, 1991; Yearley, 1991, 1994; Munck, 1995).

It is increasingly common for constructionists to defend a more limited or 'contextual' constructionism, which does not deny either the reality or the importance of actual environmental change. So, there is a convergence between constructionist and the alternative 'realist' approaches in their underlying philosophy. Even so, the rhetoric of a more radical constructionism is often retained, and a lack of analysis of the crucial ambiguities of concepts such as 'construction' itself can give the impression that an account of the cultural *construction* of an environmental change as an 'issue' somehow also explains the socio-economic *causes* of the change itself. In other words, the constructionist approaches may be true to the 'nature-sceptical' critical traditions, but they do not, in the end, address the need to revise those traditions to reconnect our understanding of society with its material basis in nature.

The four basic types of approach are, first, the 'new environmental paradigm' advanced in the late 1970s, second, 'reflexive modernization', as advocated by Giddens, Beck and others, third, a more recent cluster of approaches referred to as 'ecological modernization', and, finally, a range of approaches deriving from the Marxist, or historical materialist, tradition in various combinations with green, feminist and anti-racist ideas. These latter approaches can be collectively referred to as 'radical political economy'. The pioneers of the first approach were the US sociologists, R. E. Dunlap and W. R. Catton. In a series of articles from the late 1970s onwards (see Catton and Dunlap, 1978, 1980; Dunlap and Catton, 1994) they criticized mainstream sociology for working with a 'human exemptionalist' paradigm: that is, sociologists had tried to understand human societies in abstraction from their interdependence with the rest of nature, as if we were 'exempt' from the laws of nature which apply to all other beings. Instead, they proposed a 'new environmental paradigm' which would locate human societies within the wider web of environmental interactions.

Clearly, their proposal was for an ecology-inspired radical revision of the whole sociological tradition. Very much in line with these original proposals is an influential approach which attempts to measure the scale of materials and energy taken up by and emitted by particular societies at different historical periods. Key concepts in materials and energy flow accounting are the 'metabolism' between societies and nature, and 'colonisation' of nature and natural resources by social processes (Fischer-Kowalski and Haberl, 1993; Foster, 1999; Schandl and Schulz, 2000). This approach offers a means of quantitative measure of the extent of 'ecological modernization' over time in the industrialized countries (Adrianne *et al.*, 1997; Matthews *et al.*, 2000).

Whereas the focus of the radical political economy analyses is modern capitalism, its expansionary tendencies and political implications, the key concept for these other approaches is 'modernity' and the key process 'modernization'. The shift from 'modern' as an adjective, to the idea of 'modernity' as a way of characterizing a whole society or historical period (Craib, 1992, 1997; Stones, 1998) is associated with a tradition known as 'functionalism'. This approach assumed an evolutionary development in the history of societies toward more complex and functionally differentiated societies. The Western societies represented the highest developmental stage, and models were devised to foster 'development' in the rest of the world, on the assumption that it would follow the model already achieved in the west. This process was 'modernization'. Its outcome would be capitalist and liberal-democratic. 'Modernity' was the state we in the West had already attained, and, by implication, one to which everyone else would, or should, aspire. In this early version the notion included three aspects: modernity was the destiny of the whole world, the West was leading the way, and this was a good thing. Initially influential as a cold-war ideology, this assumed more triumphalist forms with the fall of Soviet and East European state-centralist regimes at the end of the 1980s (Fukuyama, 1992). This period also marked a revival in the use of the terms 'modernity' and 'modernization' by sociologists, often as a way of avoiding the more politically contentious term 'capitalism'.

Most relevant to our theme have been two theoretical approaches which have linked 'modernity' and 'modernization' with ecological change and environmental social movements: ecological modernization theory and the notion of 'reflexive modernization' associated with Ulrich Beck and Anthony Giddens. Both approaches see 'modernity' as a phase in historical development as well as a type of society, and both subdivide modernity itself into successive developmental phases.

In this respect, reflexive modernization theorists, especially, incorporate some of the themes of post-modernism as characterizing a significant transition within modernity. Beyond this, the two traditions diverge quite radically.

Ecological modernization theory had its origins in the 1970s. Its earliest advocates shared the optimistic evolutionary/developmental perspective of the American functionalist versions. They distinguished an early phase of modernity, a phase of industrial 'construction', in which increased production was won at the cost of increased environmental degradation, from a more recent phase of 'reconstruction'. In this latter phase, industrial production and consumption were increasingly governed by a new, 'ecological rationality'. Scientific and technical innovation was increasingly devoted to adapting the industrial society to environmental constraints (Murphy, 2000).

The idea of reflexive modernization, too, has a two-phase model of development within 'modernity'. Modernity itself is defined (in Giddens's version) as a combination of four distinct institutional dimensions: a liberal democratic state, concerned with surveillance, a military establishment which monopolizes the legitimate use of force, an economic system, characterized by private property and market, and industrial technologies as the mode of appropriation of nature. However, in recent decades, this model of 'simple' modernity is rendered increasingly inappropriate by three interrelated social processes. Globalization, which, for Giddens, is primarily a matter of increased international flows of communication and information, opens up all closed communities and stable traditions to the existence of alternatives. A new cosmopolitanism emerges in which it is impossible to maintain traditions 'in the traditional way'. So, along with globalization comes 'de-traditionalization'. Freed from the constraints of localism and traditionalism, both individuals and institutions become more 'reflexive': more self-conscious, and consequently more open to revising their practices and identities. Instead of a life whose main outlines are determined by the contingencies of birth – class, sex, locality – we are increasingly required to turn our lives into a 'reflexive' process of flexibly inventing and re-inventing our identities. Traditional forms of gender relation and family forms, established authority relations and norms of conduct and especially the traditional political divisions of left and right, rooted in traditional class identities, are expected to dissolve in the acid of reflexivity (Giddens, 1994; for commentary, see O'Brien *et al.*, 1999; Benton, 2000).

Beck and Giddens concur in their expectation that the mass politics of left and right, like the class identities which that expressed, will fade away, to be replaced by a new politics 'beyond left and right'. Giddens speaks of this as a politics of life-style and voluntary activity, whilst Beck's hope is for a 'new modernity' in which non-institutionalized social activism will demand democratic accountability from technocrats and politicians in the way science and technology are developed and introduced.

In the face of such analyses from the reflexive modernizers, and from the developments of the radical political economy approach, the contemporary advocates of ecological modernization have significantly reworked their inherited theory. Writers such as A. Weale, G. Spaargaren, A. P. J. Mol, M. A. Hajer and others have acknowledged that the advance of their hoped-for 'ecological rationality' is more problematic than earlier writers such as Huber and Janicke had supposed (Weale, 1992; Hajer, 1995; Mol and Sonnenfeld, 2000). Recent work in this tradition differs from the earlier in three main respects. First, the earlier emphasis on technology is broadened to include the importance of accompanying changes in culture, consumer behaviour, organization and governmental intervention and regulation as fostering environmental adaptation. Along with this is a shift away from the functionalism of the earlier version in favour of recognition of the role of social agency in bringing about change, and, finally, the recognition that ecological modernization is a 'project', facing resistance, obstacles and reverses, not an inherent, smoothly operating tendency inherent in the historical development of 'modernity'.

The ecological modernizers remain, however, significantly more optimistic about the environmental prospects of 'modernity' than either the analysts of 'risk society' (Beck, 1992) or the radical political economists (for more on these approaches, see Chapter 6). In favour of the ecological modernizers is the evidence that the 'advanced' industrial societies have made significant progress in environmental regulation and 'green' taxation, most have ministries devoted to environmental policy, significant gains have been made in combating important sources of air, water and soil pollution, recycling, materials substitution, and increasingly energy and resource-efficient technologies have been developed and employed. Evidence on materials and energy flow over a twenty-year period for some of the industrialized countries does indeed show the looked-for 'decoupling' of economic growth measured in financial terms from measurable environmental impact: industry in these countries does appear to be increasingly ecologically efficient per unit of economic value produced.

In the domain of environmental politics, the green movements in most advanced industrial societies have changed their role from a marginal, oppositional and 'outsider' status, to insiders, collaborators with business, government and technocrats in setting mainstream policy objectives. In the international sphere, the EU has gained democratic legitimacy for its vigorous espousal of environmental issues, both in relation to the wider global scene and in relation to the record of member states. At the global level, a series of conferences leading up to Rio in 1992 have provided an overarching concept of sustainable development embracing both social justice and long-term environmental protection, as well as international agreements on, among other important issues, trade in endangered species, climate change, ozone depletion and conservation of biological diversity.

There is a major debate about how effective these agreements are, and indeed about what effectiveness means (Underdal, 1992; Young and Levy, 1999; Sprinz and Helm, 2000). International agreements do not operate in isolation from each other and they frequently have negative side effects on other environmental problems (Ward et al., 2004). For instance, some substitutes for CFCs are powerful greenhouse gases, so the Montreal Convention on ozone-depleting substances has side effects on the Kyoto Protocol, eventually leading to international action. Because of such interconnections and side effects the real issue is whether the *system* of international environmental agreements promotes *sustainability* on balance (Ward et al., 2004). There is evidence that it does do so, pushing countries beyond what they would otherwise have done to promote sustainability (Ward, 2006).

As the constructionists clearly demonstrate, the formation and transformation of environmental issues as an agenda for public attention and policy-making depends on complex interactions between social movement activists, researchers, media communicators, policy networks and communities, industrial lobbies, government departments, international organizations and many other sorts of actors. In the realist approaches favoured here, recognition of the roles played by these heterogeneous and often conflicting social actors has to be complemented by acknowledgement of the active causal role played by non-human beings, relations and forces: both those purposively mobilized in the course of technologically mediated human social interaction with nature, and those unintentionally and often unexpectedly 'striking back'. Scholars have an important place in the effort to understand the systemic connections between the social, economic, political and biophysical dimensions of our increasingly problematic 'metabolism' with non-human nature. The intellectual demands of such an enterprise, and the great divisions of interest and of value judgement at stake in it suggest that it will always be a thoroughly contested enterprise. The less encouraging aspect of our situation, however, is that the socio-ecological processes of destruction and degradation escalate as we argue.

SECTION I: ENVIRONMENTAL THOUGHT – PAST AND PRESENT

In the first chapter of this section (Chapter 2) on the enlightenment and its legacy, Ted Benton sketches some of the historical background to our contemporary debates about the relationship between human society and the rest of nature. This chapter begins with the influence of the 17th century scientific revolution on the thinkers of the Enlightenment. The views of Hobbes, Locke and Rousseau are compared, to illustrate the great diversity of thought within the Enlightenment. Rousseau, especially, is introduced as a precursor of the Romantic movement, which challenged the prevalent view of nature as merely a set of resources to be utilized for human purposes. Instead, the Romantics offered views of our relationship to nature as one in which aesthetic appreciation – even awe and wonder at nature's magnificence – were essential to full human flourishing.

Benton goes on to note the importance of the legacy of romanticism for Darwin's revolutionary understanding of the historical character of evolving nature, and for his sense of wonder at the immense diversity of life. Despite Darwin's own initial reluctance to elaborate on the implications of evolution for our understanding of human nature and prospects, he was soon drawn into the intense debates about these questions that followed the publication of his *Origin of Species*. Here, Benton attempts to show that the influence of Darwin's ideas on social thought were much more diverse than is often recognized.

Damian White and Gideon Kossoff then assess the history of anti-authoritarian thought in anarchism, libertarianism and environmentalism in the second chapter (Chapter 3). They trace the diverse connections between anarchism, the broader libertarian tradition, environmentalism and scientific ecology. Anarchists maintain that it is the very coercive ideologies, practices and institutions of modernity that are the source of the disorder and social chaos they are designed to prevent. The authors demonstrate that the resistance many contemporary forms of ecological politics holds for conventional leadership patterns, individualism and division of labour has a long pedigree. At the same time, social anarchist, left libertarian and ecological anarchist currents have all influenced thinking about social–nature relations. It is apparent that many politics going under the loose term ecology continue to find these traditions invaluable

sources of ideas and innovation. The search for self-organizing societies continues, as does concern for the establishment of sustainable cities and other settlements.

In the third chapter (Chapter 4) of this section, Mary Mellor analyses the development of thinking around ecofeminism, gender and ecology (see also Mellor, 1992). Ecofeminism is based on the claim that there is a connection between exploitation and degradation of the natural world and the subordination and oppression of women. It also takes the view of the natural world as interconnected and interdependent, with humanity systematically gendered in ways that subordinates, exploits and oppresses women. Unlike some other writers, Mellor does not make a claim that women have a superior vision, or higher moral authority, but indicates that an ethics that does not take account of the gendered nature of society is doomed to failure, as it will not confront the structure of society and how that structure impacts on the material relationship between humanity and nature.

The problem, of course, is how political change can occur. Should it be driven from the top, or does political agency need to come from people and groups who are exploited, marginalized and excluded by the existing social and ecological structures? Mellor indicates that building coalitions and coordinated political action are essential. The basis for this position is that knowledge about the natural world will always be partial, and so awareness of pervasive uncertainty should be the starting point of all other knowledge. Humanity is part of a dynamic iterative ecological process where the whole is always more than the sum of the parts. Far from being a restriction on feminism, ecofeminism offers analyses that show how exploitative and ecologically unsustainable systems have emerged through the gendering of human society. Such an analysis demands radical change.

In the fourth chapter (Chapter 5) on deep ecology, Ted Benton suggests that the orientations to nature expressed in the art and literature of the Romantic movement (Chapter 2) find more systematic philosophical and political expression in the stream of modern environmentalism known as 'deep ecology'. Benton presents an outline of the thought of the Norwegian philosopher, Arne Naess, who is generally recognized as the 'founding figure' of the deep ecology movement. Naess made a sharp contrast between 'shallow ecology', which seeks mainly to manage resources for human purposes, and his own, 'deep ecological' perspective, which understands humans and nature as bound together in a single indivisible totality, every part of which is (in principle) equally valuable. Not surprisingly,

Naess's distinction itself, as well as the implications of his deep ecological alternative to 'mainstream' environmentalism have been very controversial. Benton goes on to present some of the main arguments of the critics of deep ecology, and the replies offered by the deep ecologists and other 'ecocentrics'. The debate is presented as open-ended, and as having much to offer to our current practice of environmental politics.

In the fifth chapter (Chapter 6) on greening the left, Ted Benton explores some of the historical background to the present tendency for social justice (a traditional concern of the political left) to be linked closely to the demand for environmental protection (a central concern of the green movement). He suggests that there has been a long history of the intertwining of these two sets of concerns in the thought and practice of some of the traditions of the left. Beginning with Marx and Engels's ways of analysing the different historical forms of human society and historical change in terms of their 'metabolism' with the rest of nature, he suggests that they have valuable insights to offer to today's environmental movements – this despite the dreadful environmental record of many of the regimes established in Marx's name (see also Benton, 1989, 1996; Foster, 1999).

With the re-emergence of radical environmental politics in the 1960s, some of the radical thinkers of the left responded by drawing on and developing the legacy of the earlier socialist traditions. Their aim was to address what they saw as the close connections between the social and ecological crises of our own times. The work of the late 19th century designer, artist, craftsman, environmentalist and socialist, William Morris, has been an important inspiration. Benton also discusses the more recent ideas of Andre Gorz and the American eco-Marxist, James O'Connor, going on to introduce an approach called 'World System Theory'. This is an attempt to understand the causes of continuing inequalities in the global economy and in the relations between different nation states. Benton suggests that this approach has much to offer in explaining global ecological degradation and the current lack of success in tackling its causes.

The sixth chapter (Chapter 7) of this section contains an exposition by Warwick Fox on the problems that need to be addressed by a theory of general ethics. Old ethics has generally occurred in a closed moral universe, whilst new ethics, that conducted in a whole earth, or Gaian, context seeks to work in an expansive moral universe. There are problems, though, with new ethics. If biodiversity is important to preserve, what do we make of introduced (or alien) species that are ecologically destructive? Should they be removed, even if they increase net biodiversity? What if they

are sentient themselves? The consideration of the holistic integrity of ecosystems is further considered, along with the difficulties of being both comprehensive and consistent. In this article, eighteen problems as they relate to interhuman ethics, animal welfare ethics, life-based ethics, ecosystem integrity ethics, and the ethics of human-constructed environments are discussed and analysed. This effectively sets out a map of the ethical terrain for those addressing environmental and society-related issues and the likely dilemmas they will encounter.

The final chapter of this section (Chapter 8) is by Damian White, Chris Wilbert and Alan Rudy, and addresses the contemporary and growing problem of anti-environmentalism. The emergence of the Lomborg controversy was seen by some as a new phase of criticism of environmentalism, by some even a unique critique. Yet there were many antecedents, arising from left, right and technocratic sources to post-war environmentalism, then to the global environmentalism of the 1990s (after the Rio conference and as a result of the efforts to establish international treaties) and then the modern contrarians exemplified by Lomborg and others. There remains a fundamentalist form of contrarianism that is at the centre of greenwash attempts by anti-environmental industry. Yet framing of debates as primarily being between contrarians and radical ecologists misses many important developments in both thinking and action. There are, for example, distinct tendencies of green optimism in industrial ecology, sustainable architecture and sustainable agriculture. At the same time, there are others who frame arguments in technologically pessimistic terms.

SECTION II: VALUING THE ENVIRONMENT

In the first chapter of this section (Chapter 9), Thomas Crocker examines the basic economic questions underlying the social choice of environmental management instruments and institutions. The author argues that, at its root, this social choice is motivated by competing 'deontological' versus 'individualistic' visions and their associated management options. Neither vision, in its extreme, is seen as an accurate or realistic basis for environmental management. Rather, the author suggests that environmental management is based on discovering 'collective procedural rationality', not to be confused with the 'limited elemental rationality' of the individual. To the extent that exchange institutions – markets and other incentive-compatible environmental policies and instruments – accurately reflect available information and options, they can help create collectively rational mechanisms which 'guide people to their own interests'. This process is based on market prices which provide incentives for collectively rational behaviour, but that are themselves subject to a variety of limitations which interfere with achieving efficient outcomes: incomplete information, non-zero transactions costs, misdirected incentives, and undefined, non-transparent or illegitimate initial distribution of rights over assets. The conclusion is that top–down decision-making of environmental authorities regarding the selection of control instruments and effort spent on monitoring and compliance to mandate 'what to do and how to do it' is obsolete, assumes scarce or incomplete information, and is expensive due to strategically interdependent decisions of the authority and users. But neither will the total privatization of environmental decision-making by individuals typically be collectively rational.

The best, then, that authorities can often do is help guide asset owners and resource users to make private decisions which lead to environmental outcomes that are compatible with collectively rational mechanisms. In the past 20 years, this principle has been extended to numerous examples, including effluent charges, tradable permits and liability standards. The chapter offers an extended example of the use of tradable permits to address biodiversity conservation, specifically the wildlife habitat requirements mandated by the US Endangered Species Act. Achieving a lower-cost, lower-risk incentive-compatible outcome is shown to be dependent on a clear definition of the habitat units to be traded, the baseline distribution of units, and a carefully defined institutional framework for exchange. In this and other similar examples, public goods constraints are a further obstacle to least-cost collectively rational outcomes and also must be considered. In general, collectively rational institutions for environmental management require three things: the credible commitments of economic agents, transparent market or shadow prices, and effective arbitrage opportunities. In the end, for these instruments to work and represent an effective alternative to command-and-control policies, careful initial attention must be given to institutional design based on a fully informed understanding of the use and users of the natural asset.

The next chapter (Chapter 10) by Ian Bateman provides a comprehensive review of three central questions related to the valuation of environmental impacts. The first is comparison and contrast of the two principal approaches used in the evaluation of environmental impacts: cost–benefit analysis (CBA) and environmental impact analysis (EIA). The author states that CBA assumes an anthropocentric approach, growing out of

economic analysis, and typically focuses on the precise measurement and evaluation of multiple impacts, discounted to the present. In execution, it is highly quantitative and attempts to incorporate multiple impacts into a single money value numeraire, with the attendant pro's and con's. It is not good, however, at addressing the distribution of costs and benefits among different groups nor in assessing sustainability dimensions. By contrast, EIA does not attempt to assess monetary impacts comprehensively but focuses on evaluating diverse physical environmental impacts, both quantitatively and qualitatively. Its wide variety of impact assessment measures is a positive feature of this approach, enabling long-term sustainability impacts to be more easily be incorporated than in CBA. However, by failing to incorporate the assessment of multiple impacts into a single measure, it becomes more difficult to compare projects and interpret their results using compatible criteria.

The author also discusses and summarizes a number of important conceptual and empirical distinctions that arise in valuing environmental impacts. To begin with, prices do not equate with values for either private or public goods (due to non-zero consumer surplus). The chapter outlines the basic distinction between private and public goods, with the key result that open-access resources may be highly valued even though private prices may be wholly absent. The broader concept of total economic value (TEV) (also discussed in Chapter 11) comprises both use values (option and bequest values, for example) and non-use values (existence and non-human values). These lead to complications in valuation which are reinforced by the existence of complex trade-offs and the multidimensional valuation criteria used by different individuals. In theory as well as practice, 'willingness to pay' measures (to obtain a gain or avoid a loss) very often differ from 'willingness to accept' measures (to forgo a gain or suffer a loss). Context specificity, loss aversion and 'part-whole' problems further complicate environmental valuation in practice.

Economists have developed a wide array of alternative approaches to conduct empirical monetary valuation of environmental public goods. These are often differentiated as 'pricing' approaches and 'valuation' approaches; each are briefly summarized in Chapter 11 (and discussed separately in Chapter 12). The former includes approaches which employ estimates of: opportunity costs, costs of alternatives, mitigation costs, shadow project costs, government (subsidy) costs or dose–response value estimates. All of these approaches suffer from the flaw that the 'prices' that are estimated may differ from true economic valuation. The latter set of 'valuation' approaches include two categories of methods. Expressed or stated preference methods such as contingent valuation, preference ranking and conjoint analysis all involve explicit, direct valuation (or ranking) of environmental goods by respondents. Revealed preference approaches – specifically, the travel cost method and hedonic pricing – assess environmental values by measuring respondent's actual market behaviour and statistically estimating the resultant 'revealed' environmental values. Overall, the author argues that while valuation methods are more cumbersome in their application, they have wider applicability and address the difference between prices and values. It should be noted that the Bateman chapter only reviews the relevant environmental economics research through the late 1990s; this is an active area of ongoing research in the field.

Chapter 11 by Randall Kramer reviews many of the basic economic valuation concepts covered in the previous chapter – use values, option values, non-use values – and applies them to the valuation of a particularly important environmental resource: water. The author summarizes some of the recent research regarding the non-market valuation of environmental services, and the advantages and limitations of alternative valuation methods. Several empirical examples employing standard non-market environmental valuation concepts and methods are introduced and discussed: impacts of lake pollution on water recreation (using the travel-cost method); impacts of water quality on residential land prices (contingent valuation); and estimation of the value of water quality protection (contingent valuation estimates subsequently used in a cost–benefit analysis). By focusing on the valuation of environmental services (specifically the value of water quality) the author emphasizes the fact that the 'true value of nature' termed by the Millennium Ecosystem Assessment (MEA) (2005) is best assessed by highlighting the scarcity value of the services provided by environmental resources.

Chapter 12 by David R. Lee treats the topic of environmental tradeoffs addressed elsewhere in this section, but with an explicit focus on developing countries which are home to many of the most vexing environmental management and policy problems. Given the severe resource constraints facing many developing countries, achieving environmental management goals typically must occur in the context of simultaneously realizing food security, economic growth and improved livelihood objectives. But doing so is more often characterized by tradeoffs among these goals than by synergistic relationships. The author argues that significant insights into understanding these relationships lie in the empirical evidence at both macro- and micro-levels. At the macro-level,

much of the discussion over the past decade has centred around the 'Environmental Kuznets Curve' (inverted 'U') hypothesis and empirical evidence supporting or contradicting it. Overall, the evidence is distinctly mixed for most indicators, suggesting that a country's ability to 'grow its way' out of environmental degradation problems is not a generalizable policy result.

At the micro or household level, the evidence is also limited, for different reasons. Comparing the results of household- and village-level studies is difficult due to the use of non-comparable analytical methodologies, the lack of results estimated over time (which would demonstrate the sustainability of production and livelihood systems), and the use of different empirical measures for key economic, production and environmental indicators. Several case study examples which surmount these obstacles and in which positive environmental outcomes are shown to be achieved alongside other social objectives are discussed. The factors which generally condition the achievement of sustainable environmental outcomes in the context of jointly realizing production, food security and economic livelihood objectives are identified and discussed.

In the final chapter of this section (Chapter 13), Joe Morris applies economic concepts and analytical tools to the analysis of the Water Framework Directive (WFD) of the European Union. The WFD seeks to prevent the deterioration of surface and groundwater sources and aquatic ecosystems in the EU and provide for good surface and ground water quality by 2015. It operates through a dual approach; first, of 'command and control' regulatory methods to establish environmental quality standards and control pollution discharges, and second, employs various economic measures and incentive pricing mechanisms to achieve targeted outcomes. As the author indicates, the setting of water quality standards by regulatory fiat means that cost-effectiveness, rather than economic efficiency, is the standard by which delivery mechanisms are evaluated. However, the WFD does treat water as an 'economic commodity' and employs economic analytical and policy tools widely.

The chapter describes in considerable detail the scope for using economic analytical tools in estimating water demand and the values of water's multiple uses, and in evaluating alternative measures to improve water quality. Specifically, these tools are used in estimating: a range of user benefits stemming from alternative water demands and uses; the external uncompensated costs of water supply; the cost-effectiveness of alternative delivery mechanisms; and the impacts of incentive pricing on consumers of water. While there are still considerable practical and methodological challenges involved in implementing economic-based mechanisms in the WFD, the application of these analytical tools in identifying cost-effective measures for water quality management can be expected to be broadened to address issues of non-point source pollution and agricultural land management.

SECTION III: KNOWLEDGES AND KNOWING

What do we know about environment and society links, and what do we need to know to escape from the emerging environmental crises? Perhaps more importantly, we will need to ask how we can develop systems of knowing about the world that are transformative. Knowledge on its own is not a sufficient condition for change. What is needed is ways of knowing that change the way people see the world, interact with one another, and bring their views to bear over critical challenges for a complex and contested world.

In the first chapter in this section of the Handbook (Chapter 14), David Orr sets out the components and principles of ecological design and education. In spite of nearly a century of substantial economic growth, a large proportion of the world is either on the edge of starvation in absolute poverty, or is suffering the consequences of over-consumption in their worlds of traffic jams, bad diets, addictions, boredom and mental ill-health. These two worlds may appear to some to be diverging, but may actually be on a collision course. The inability to solve ecological and social problems points to deeper flaws in a faith in human capability to solve all the problems we bring on ourselves.

Ecological designers know one big thing – everything is hitched to everything else. This suggests a need for a blending of nature with human crafted space, a bringing together of arts, crafts, science and architecture. But this is easy to say, and hard to achieve. We will need to spend more time thinking about how we see the world, and how we learn from it. A number of key principles are set out for a new type of design that recalibrates education with ecology. Nature is not something to be mastered, but a potential tutor and mentor for human actions. But ecological design is deeper than mimicry. It should encourage us to ask what will nature permit us to do? Another key principle is that humans are not infinitely plastic. There are biological and evolutionary constraints that shape our interactions with the world. All design is, of course, inherently political, as it is about both provision of goods and services, but also the distribution of risks, costs and benefits. Ecological design implies robust economics, an honest assessment of human capabilities, a capacity to understand the lessons of history and past civilizations, and above all offers opportunities of healing. Designers are story-tellers that aim to speak to the human spirit, and this is where education must mimic, and tell better stories about the world.

Richard Bawden then develops the theme of knowing systems and the environment in the second chapter (Chapter 15). Once again, the problem lies in how we have come to risk the world on the back of such great achievements in economic and technological development. The chapter focuses on systems, both hard and soft, and on coming to know. Our quest, says Bawden, in seeking to come to terms with sustainability, must start with learning. What we think we mean when we use terms like development and sustainability. We have made the world as it is, and so it is up to us collectively to make meaning through our learning. In a state of denial, about how bad circumstances are, we are going to need to devise different ways to think, interact and act very quickly.

An important contrast centres on how we conceptualize systems' ideas, and thus bring some cognitive coherence to bear on a complex world. Earlier pioneers of systems' thinking focused on cybernetic regulative processes that maintained steady states, and many ideas about resilience and adaptation have since been developed. But strangely, systems ideas in the social sciences have seen declining support in recent decades. Another conceptualization, however, centres less on systems in the world, and more on systems of cognition, in which inquiry about the world is the soft system that can be both revealing and transformative. In this way, learning becomes less about the acquisition of knowledge and more about the transformation of experience, whereby knowledge is fluid, being created, recreated and used by individuals as they seek to make sense of the world. The quest for sustainability focuses on new types of engagement between people with their different worldviews and paradigms, and the world about us.

Max J. Pfeffer and Linda Wagenet show in Chapter 16 how such new ways of knowing are playing out in the environmental volunteering sector in the USA. Volunteer environmental monitoring offers the possibility of directly involving citizens in environmental decision-making. It may also reinforce public confidence in science-based decision-making, and offer the means to increase more direct interactions with the environment and its resources. Such volunteering is likely to be important where there are already extensive environmental regulations and clear compliance standards, and where concerned citizens have the time and resources to participate. Existing literature contains no comprehensive review of volunteer environmental monitoring, and this chapter reviews its importance over more than a century in tracking weather data, bird ringing (or banding), game fish tagging, water quality monitoring, wastewater plant monitoring, and the Christmas bird survey. All of these represent important types of citizen science in action.

More than half of Americans are engaged in some kind of local volunteer activity, however minimal, and forms of civic environmentalism have become common and indeed effective. It is widely known that a lack of meaningful public involvement can lead to the emergence of barriers to environmental management, and since the 1990s there has been growing uptake by federal agencies, particularly for watershed management. Some authors are confident that this represents the potential for positive outcomes for both human and ecosystem well-being; others are cynical, characterizing the interactions as no more than the scientifically illiterate versus the politically clueless. Nonetheless, such community science does have transformative potential, not only for individuals but also for groups who coalesce to act together. The chapter addresses three key questions: do volunteers generate data that meet acceptable scientific standards? Are such data then used by agencies engaged in environmental management? And finally, does this activity reduce the gap between environmental science and the lay public?

In the fourth chapter in this section (Chapter 17) on environmental ethics, Val Plumwood goes on to draw some of these themes together by asking do only human lives and humans count, as we relentlessly drive other species from the planet? How we think about these kinds of questions determines partly how humans act in this world. Plumwood explores a series of perspectives on value, including instrumentalism, utility and intrinsic value, and teases apart common default settings that are often ignored in environmental narratives. Interspecies relationships may be the key task of environmental ethics, but such an ethic will also need to challenge conventional concepts of human identity too. The problem with instrumentalism is that it is seen to draw the life, meaning and wonder from the world, as we progressively commodify relationships with nature and its goods and services.

Instrumentalism also suggests a human apartness from nature, which echoes concerns about intrahuman dominance, especially on the grounds of gender and race. Non-humans are taken to be naturally inferior, and lacking qualities that are supposed to matter, such as mind, rationality and individuality. A human-centred (or anthropocentric) worldview and its misunderstandings of human nature pose risks to both human and non-human survival. Commodities become taken for granted, and nature is starved of resource for its own maintenance. Sustainability is a project aimed at countering the exhaustion of the planet's resources for life, and Plumwood indicates why we should recognize human and non-human needs as part of this concept. The chapter concludes with a perspective on counter-hegemonic structures and communicative ethics, and includes how processes of knowing and coming to know can break down discontinuities between humans and nature, reconstruct human identity, dehomogenize nature and human categories, and acknowledge difference.

In the final chapter of this section (Chapter 18), Luisa Maffi analyses the concept of biocultural diversity and how it relates to current concerns about both ecological and cultural sustainability. Biocultural diversity draws on anthropological, ethnobiological and ethnoecological insights about the relationships between human language, knowledge and practices with the environment. Evidence now indicates that the idea of the existence of pristine environments unaffected by humans is erroneous. Humans have maintained, enhanced, and even created biodiversity through culturally diverse practices over many thousands of generations. There are some suggestions that biodiversity and cultural diversity in the form of linguistic differences are associated, though at the local level these relationships do not always stand scrutiny. But the role of language is nonetheless critical as a vehicle for communicating and transmitting cultural values, traditional knowledges and practices, and thus for mediating human–environment interactions.

Landscapes can be networks of knowledge and wisdom, conveyed by the language of local people. But the problem is that many languages are under threat. There are some 5000–7000 languages spoken today, of which 32% are in Asia, 30% in Africa, 19% in the Pacific, 15% in the Americas and 3% in Europe. Yet only half of these languages are each spoken by more than 10,000 speakers. Some 550 are spoken by fewer than 100 people, and 1100 by between 100 and 1000 people. A small group of less than 300 languages is spoken by communities of one-million speakers or more. Some 90% of all the world's languages may disappear in the course of this century – yet these very languages are tied to the creation, transmission and perpetuation of local knowledge and cultural behaviour. As language disappears, so does people's ability to understand and talk about their worlds. Natural and cultural continuity are thus connected. The phenomenon of loss has been called the extinction of experience – and the loss of traditional languages and cultures may be hastened by environmental degradation.

Yet in many parts of the world, both in developing and industrialized countries, such traditional ecological knowledge (or ecological literacy) is declining and under threat of extinction. As humans coevolved with their local environments, and have now come to be disconnected, so knowledges that coded stories, binding people to place, have become less valued. New efforts to analyse biocultural diversity on a country-by-country basis are reviewed, and despite some important progress in the international sphere, such as in the Convention on Biodiversity, the most fundamental changes must come from ground-up actions. In this way, the field of biocultural diversity has embraced strong ethics and human rights components.

SECTION IV: POLITICAL ECONOMY OF ENVIRONMENTAL CHANGE

The fourth section of this Handbook explores questions of distribution, risks, winners and losers in the quest for representation and access to resources. Environmental change occurs at a different pace at different human scales, and affects different groups of people in different ways. As a result, incentives and inclinations to act differ greatly, even though all humans are part of the same world system. These differences raise contradictions, complexities and conflicts, and positive social outcomes for some may mean negative environmental outcomes for others.

In the first chapter (Chapter 19), Ron Johnston explores questions of representative democracy and the solution of environmental problems that require collective action at different scales. Many environmental problems have three common characteristics. They are produced by individual actions, but their intensity may be more than the sum of individual contributions. Most problems affect others, and these spatial overspills require that all those (or at least most) must reduce or end their contributions. And third, individual contributors can gain advantage over others by declining to participate in efforts designed to solve the problem. In small-scale situations, generally trust and enforcing agreements are possible, and indeed have been very effective in many parts of the world, but at higher scales, efforts have to centre on either privatizing the commons or on external regulation by bodies with the power to ensure compliance.

There is always a range of scientific and political challenges to be overcome. An issue has to be identified, recognized that there is an associated problem, a postulated cause accepted, and then acceptance that the problem can be tackled or remedied. But tackling a problem requires commitment of resources (and thus always in short supply), which have to be obtained from citizens. For a solution to be implemented, there must therefore be both political and public support. The challenges of environmental problems thus play out in different ways according to whether they are confined within individual states, are shared by two or more states, and confined within their boundaries, or involving interactions with large numbers of states. Most governments have short-time horizons, and this adds further complications to the need to address pressing current and future problems.

In the second chapter (Chapter 20), Ronald J. Herring analyses how the genomics revolution in biology seems to be creating novel analytical and policy questions for political ecology. Such politics reinforce the centrality of science to all political ecology, which in turn presents new challenges to the way interests in nature are understood by citizens and political classes that control states. Much indeterminacy of interests in nature is knowledge based, and so radically different levels of ecological knowledge occur amongst mass publics, political actors and administrative managers over time. There are many contradictory positions. There are global conflicts over transgenic organisms that focus, at least in part, on ecological threats arising in agriculture (even though modern agriculture is itself quite destructive of nature), yet transgenic pharmaceuticals seem to be quite immune to protest. There are, of course, many political reasons for this selectivity – miracle drugs save lives and are ineffective targets for opposition.

At the same time, it is clear that public goods and bads are not objectively perceived, but rather are embedded in normative logic and cultural norms. A swamp was once seen to be unhealthy and thus gladly drained (except for the people living there); but wetlands now purify water and are for preserving. In the contested politics surrounding such normative spectrums, new and unpredictable relationships emerge. In the genomics revolution itself, new values are created in natural landscapes, as yesterday's obscure species becomes an object for bioprospecting and biopiracy. Whatever regulators may seek to do, there will be circumstances in which the practice of individuals forces further change. The seed sharing amongst farmers in India and Brazil is an example where states had to follow what farmers themselves preferred to do. In the end, though, the science of ecology frustrates policy, as unexpected interconnections amongst parts of systems keep being discovered. Honest science is always incomplete at the frontier, and yet such uncertainty is the most powerful weapon of opposition movements.

In the third chapter (Chapter 21), Steven Griggs and David Howarth explore protest movements, environmental activism and environmentalism in the UK, using examples of struggles against road building and airport expansion. There are not many people who are in favour of fewer pollution controls, more greenhouse gases and greater species extinction, and the public goods struggled for by many environmental movements and organizations are goods desired by large numbers of people. Yet, despite the appeal of many of these environmental demands, the translation of such popularity into a populist form of politics has not been straightforward or even successful. Populist discourses appeal to a collective subject, such as the people or a nation. They are grounded on the construction of an underdog versus establishment frontier, the latter being seen as the enemy or adversary of the people. And they are centred on an appeal to all the people in a space or a domain – there are, after all, universal concerns, it is commonly claimed.

The authors explore three phases in environmental politics – from early conservation environmentalism to mid-late 20th century ecological environmentalism, and the later emergence of radical environmentalism. Over time, membership of some environmental groups has grown remarkably, and their size and scope has caused them to become institutionalized, thus blunting the radical aspirations of some people or members. As some have become larger (and more effective in certain spheres), so others have moved away from such insider routes to set up alternative movements. Some of these have resulted in direct action against roads and airports, and indeed have led to the melding of unlikely social groups, such as radical protestors and middle-class residents. Tactics are often different, but new alliances have had some influence on how national politics frames environmental problems and solutions.

In Chapter 22, Tim O'Riordan explores the many faces of the sustainability transition by suggesting that the phrase sustainable development has become so universal that it now means everything and so is in danger of meaning nothing. Sustainable development binds a range of movements – peace, democracy, development and environment, and yet current economic development patterns are widening the gap from wealthy to poor and destroying the natural resources and life-support systems daily, and so are rapidly moving away from sustainable development by the day. Despite concerns, though, about reaching global tipping points arising from the huge collective human influence on the world, the chapter suggests that localism offers real opportunities to create sustainable communities. People can form communities where safety, security and sustainability can flourish and form livelihoods that offer hope for all involved.

We do not, however, know enough about the changing state of planetary support systems. Forecasts remain uncertain, and so there is great difficulty in making predictions about how political systems and their leaders will respond. At the same time, of course, we are not good at delivering well-being for both people and nature. The UK has a very good sustainable development strategy, but as yet there is little or no capacity in the UK government for a change of direction. We will need to build from below, and seek to find ways to leap to sustainability in one generation. Several zones of sustainability engagement show promise, including some change in businesses, consumer behaviour and use of purchasing power, in that tipping points are beginning to be noticed, and in that well-being is appearing on the political agenda.

The final chapter (Chapter 23) of this section takes forward one of these themes, as Christina Page and Amory Lovins explore whether businesses can be greened, and whether the very idea represents an opportunity or contradiction. Businesses have recently begun to move beyond command and control environmentalism towards the mindset where pursuit of sustainability is seen as a competitive advantage. Private businesses and companies can be a source of innovation and invention, and so can create novel solutions to some social and environmental challenges. Assets in socially responsible investing have grown faster than all other professionally managed assets, and this too is causing a rethink. The authors set the scene for a natural capitalism framework. Industrialized capitalism liquidates rather than values important forms of natural, social and human capital, yet sustainability calls for ways to protect and invest in these assets over the long term.

What can businesses themselves do? They can seek to increase radically the productivity of resources – do more, better, with far less and for longer. This is easy to say, but there are indeed compelling examples of where this is working. They can practice biomimicry, by designing individual systems with closed loops, no waste and no toxicity. They can shift from a product-based economy to a solutions-based one, and finally, they can reinvest in natural capital. Progress, however, may bring unintended consequences, as successful enterprises that use less per unit of product may see demand so increase that at the aggregate level an increase in negative impact on the environment may occur. The path forward suggests the need to think in whole systems and to adopt full cost accounting to capture the problem of externalities. But there will still be a need for civil society, shareholders and government to apply more than a little pressure to help in the transition.

SECTION V: ENVIRONMENTAL TECHNOLOGIES

The fifth section of this Handbook explores key questions around environmental technologies, the history of pollution, the scales at which environmental problems are manifested, and some potential options for intervention that could solve hitherto intractable problems. In the first chapter (Chapter 24), Thomas Wilbanks and Patricia Romero-Lankao analyse the human dimensions of global environmental change, a term that covers a wide range of processes and phenomena. There are three major categories: human driving forces that lead to environmental change, human impacts of environmental change and human responses to environmental change. To these has recently been added human decision support, which links information about driving forces and impacts with decisions that can moderate driving forces or reduce impacts. The range of key drivers include industrialization, world population demographics, technological change that encourages greater consumption, and institutional change. Vulnerability to environmental change is related to exposure and sensitivity to changes, and the capacity to cope. Human responses then centre on mitigation or adaptation, and when impacts are negative, there are many types of adaptive responses.

The chapter details three specific cases to explore these issues: human settlements and carbon footprints, economic growth and development, and governance and society. The human dimensions of global environmental change have the potential to be profoundly important for the fundamental challenges of sustainability, equity and peace. Human societies, economies and responses to these impacts, and concerns about the risks of them, in turn shape further changes. We will need to improve our understanding of these dynamics if sustainable futures for both nature and society are to be discovered.

Howard Frumkin then discusses the concept of environmental health in the second chapter (Chapter 25). The human impact of environmental exposures, it is suggested, should be considered broadly. The environment affects people along many dimensions, including medical status, psychological well-being and spirituality. While the focus of much scientific attention has been on environmental exposures that are toxic, it is clear that other exposures can also be health promoting. These differences have shaped the evolving definitions of environmental health over time. The chapter explores the ancient origins of environmental

health, the industrial awakenings, combined with the emergence of new analytical methods, and the modern era.

A range of themes have developed in the modern era, beginning with the recognition of chemical hazards, and the linkages to ill-health, supported by advances in toxicology and epidemiology. A new development was in environmental psychology, founded on E. O. Wilson's theory of biophilia (Wilson, 1984). Further developments included the continued integration of ecology with human health, and the expansion of clinical services related to environmental exposure. Environmental health policy has continued to emerge, at both national and international levels. A new theme, though, has been a growing focus on environmental justice, born of a fusion of environmentalism, public health and civil rights. Environmental justice is one example of a broader trend, a focus on susceptible groups rather than whole populations.

In the third chapter (Chapter 26), Ian Colbeck explores the history of actions and effectiveness of change in influencing air pollution and its impacts. Despite some technological and policy advances, air pollution continues to impose a heavy burden on the health of populations in many parts of the world, particularly now in urban areas of developing countries. In the European Union, though, particulate matter claims an average of 8.6 months from the life of every person. Other key problems arise from ozone, sulphur dioxide and oxides of nitrogen. Air pollution also has other key effects on the environment, including on forests, lakes, agriculture, wildlife and buildings.

There has been a long history of the recognition of the problems of air pollution, dating back at least to ancient Rome. But it was the industrial revolution that substantially increased the burden of pollutants in the air, leading to many combined efforts by civil society and policy makers, which in turn did affect attitudes amongst the public. Single large events had a significant effect on change, such as from the anticyclone that covered London in December 1952. The smoke-laden fog caused the deaths of at least 4000 people (possibly nearer 12,000 according to recent assessments), the asphyxiation of cattle, the suspension of public transport, and even the suspension of an opera performance when smog in the auditorium made conditions intolerable for the audience and performers. The Clean Air Act of 1956 was considered a success, and has been followed by a number of examples of helpful policy interventions. In general, though, there has been a change from permissive to mandatory legislation with the setting of specific air quality standards.

The fourth chapter (Chapter 27) by Andrew Ball addresses terrestrial environments and the potential arising from bioremediation to solve difficult environmental problems. In soils, decomposition of organic compounds returns them to their inorganic form, thus making them available to plants for uptake. But what if compounds in the soil are not naturally present? Will they be broken down, or persist? If they are toxic, what effect will they have on the biotic community? There exists in the natural soil community the potential for the breakdown and recycling of a wide range of compounds by microbes. If this potential can be identified, new technological options may follow.

Bioremediation is the use of microorganisms to return an object or area to a condition which is not harmful to plant or animal life. One of the advantages of bioremediation centres on the possibility of treating a polluted soil without having to remove it elsewhere, thus reducing the cost of treatment. A range of options are available, including biodegradation, biostimulation, bioaugmentation and biorestoration. The increasing sophistication of chemical industries, combined with a growth in complexity of waste materials, means that the opportunities for bioremediation are large. However, efforts have to be paid to questions of social acceptability of these methods, as a failure to anticipate public concerns can derail potentially beneficial technologies. Key dimensions centre on types of dialogue, attention to constituents affected and interested, and the details of the physical, social and institutional context. There are many areas of land heavily contaminated, and bioremediation is a growing and relatively cheap and effective set of technologies.

The fifth chapter (Chapter 28) by Stuart Bunting contains an analysis of the environmental problems brought about by aquaculture systems, and offers guidance for their reconfiguration to make them productive, environmentally sensitive and equitable. Aquaculture has emerged in recent decades as an important food production sector, now worth some $60 billion annually. However, aquaculture appropriates a wide range of environmental goods and services, and where demand exceeds carrying capacity, then adverse impacts can occur. The consequences of such negative environmental impacts include self-pollution, restricted amenity, reduced functionality, and impacts on option and non-use values. In some locations, social tensions and conflicts have arisen, especially where traditional access rights and resource-use patterns are disrupted.

There are, however, a range of regeneration strategies and policies that can be employed. These include using resources more efficiently, especially for neighbouring production systems, horizontally integrated production, again to better use of wastes, and efforts to increase the sustainability of both feed and seed supplies.

Community-based management is a crucial option in many locations, yet many past efforts have ignored the involvement of local people and their institutions. There are relatively few helpful policies, yet these could help to reduce negative impacts and improve access to benefits. A wide range of institutions need to be involved, including national and local government authorities, extension agents, development practitioners, education establishments and communities themselves.

The final chapter in this section (Chapter 29), by Peter Oosterveer, Julia S. Guivant and Gert Spaargaren, addresses the emergent issue of sustainable food consumption, one of the key features in the green consumption trends. Starting in the 1990s, this trend has been consolidated through the role of a new global actor: the supermarkets. Recent data show how countries where most organic products are sold via supermarket chains tend to be the countries where the organic market shares are the highest as well. But what is the role of supermarkets in possible transitions to more sustainable food systems? This is a topic still not significantly recognized in social sciences in its relevance for the transformation of the horizon of the provision of green food-products and also the changing profile of the consumer. The authors take this challenge and elaborate an original theoretical framework in dialogue with the current perspectives on the sociology of consumption and the ecological modernization theory.

The retail outlet is considered as a special example of the meeting point of different rationalities (production, distribution and consumption) and as the 'locale' constitutive for their interaction. The transitions are characterized in a non-essentialistic way, opening the analysis to identify new developments within the global network society. The authors also identify plural and complex profiles of sustainable consumers, suggesting four dimensions that are not mutually exclusive: naturalness, food safety, animal welfare and environmental related. Examples are presented from different countries and special incidences discussed, such as food scares, and this global approach allows the authors to translate their theoretical proposal into an outlook of variables that could be part of a future research agenda.

SECTION VI: REDESIGNING NATURES

If things have become bad in many environments across the world, what are the prospects for making improvements? Are there options for redesign of sectors and relationships? The fact that some environments can be rehabilitated does not justify their damage in the first place, nor does it suggest that complacency is acceptable when environments are further threatened. However, given our current knowledge about harm to all types of environments across the world, combined with the losses of key environmental goods and services, then redesign is a crucial challenge.

This section begins with a chapter by David Rapport (Chapter 30) on the evolving paradigm of healthy ecosystems. The chapter reviews the evolution of the concept of ecosystem health and its potential to motivate and guide the politics of the environment. The timetable is, of course, pressing, as harm to the world's environment may soon be a challenge to humanity's future. The concepts of health and illness offer new perspectives, and these lead to the development of diagnostic indicators to aid assessment of the state of the environment. The term health is, however, somewhat enigmatic, and many argue it is too subjective a term to provide real utility. On the other hand, it does aid the identification of stressors on systems and their capacities to self-regulate and function.

For ecosystem health, there are three key measures: vitality (or productivity), organization and resilience. All of these involve analyses of the connections between social and biological aspects, and therefore must transcend the boundaries of single disciplines. This further suggests the need to understand the interfaces between human health and ecosystem health, between cultural health and ecosystem health, and between governance and ecosystem health. Design for regional eco-cultural health will have to be proactive if there is to be a lighter human footprint on the planet.

The second chapter (Chapter 31) is by Laura Little and Chris Cocklin, and addresses the question of environment and human security. While consensus over definitions remains elusive, many discourses on sustainable development have shared a greater recognition and understanding of the interdependence of human societies and the natural environment. This chapter asks specifically what can the viewing of environmental issues through the lens of security contribute, both to the understanding of the current relationship between human societies and the environment, and to recognizing what must be done to shape future transformations. Definitions of security vary, depending on what activities are trying to be made secure, and what are defined as threats to security.

The authors indicate that the security discourse is a powerful political tool to channel energy and resources in particular new directions. Environmental degradation can clearly be seen as a threat to human security, either in terms of welfare or development, or to survival itself. A number of perspectives are relevant, including the military and security, national economic interest and security (played out on both domestic and international arenas), and the links between security and sustainable development. A human security perspective focuses specifically on the interconnections between environmental and social, political and cultural issues. Thus, environmental concerns are human social and political problems as much as scientific and economic ones.

The third chapter (Chapter 32) by Jules Pretty addresses key questions of redesign in agricultural and food systems. Concerns about sustainability in agricultural systems centre on the need to develop technologies and practices that do not have adverse effects on environmental goods and services, that are accessible to and effective for farmers, and that lead to improvements in food productivity. Despite great progress in agricultural productivity in the past half-century, with crop and livestock productivity strongly driven by increased use of fertilizers, irrigation water, agricultural machinery, pesticides and land, it would be over-optimistic to assume that these relationships will remain linear in the future. New approaches are needed that will integrate biological and ecological processes into food production; minimize the use of those non-renewable inputs that cause harm to the environment or to the health of farmers and consumers; make productive use of the knowledge and skills of farmers, so substituting human capital for costly external inputs; and make productive use of people's collective capacities to work together to solve common agricultural and natural resource problems, such as for pest, watershed, irrigation, forest and credit management.

These principles help to build important capital assets for agricultural systems: natural, social, human, physical and financial capital. Improving natural capital is a central aim, and dividends can come from making the best use of the genotypes of crops and animals and the ecological conditions under which they are grown or raised. Agricultural sustainability suggests a focus on both genotype improvements through the full range of modern biological approaches, as well as improved understanding of the benefits of ecological and agronomic management, manipulation and redesign. The ecological management of agroecosystems that addresses energy flows, nutrient cycling, population regulating mechanisms and system resilience can lead to the redesign of agriculture at a landscape scale. Sustainable agriculture outcomes can be positive for food productivity, for reduced pesticide use and for carbon balances. Significant challenges, however, remain to develop national and international policies to support the wider emergence of more sustainable forms of agricultural production across both industrialized and developing countries.

In the fourth chapter (Chapter 33), Henry Buller and Carol Morris explore questions relating to animals and society. Animals and humans are rarely wholly apart, even thought the spaces they occupy are increasingly differentiated. They share common origins, common biologies, and a long history of interaction and interdependence, yet in modern industrialized settings are increasingly disconnected, continuing the lengthy process of anthropocentric disassociation from nature and the wild. Animals and humans are usually studied separately, yet an ethnoethology would bring together contemporary approaches to help understand relations between human and non-human animal society. The article explores a variety of issues. Humans are animals, and much of human social organization and behaviour to non-human animals can be explained by this human animality. Although the otherness of animals is still commonly evoked, there is a need to develop less anthropocentric conceptualizations of the non-human world.

The modernist legacy has been separation, yet this chapter analyses recent and less dualistic approaches to human–animal relations by assessing humans as animals, animals as others, and human–animal hybrids. The common theme is that interactions, such as use, enjoyment, observation, killing and eating of animals, are so unavoidable and so universal that they have been central constituents of human society from it origins. These relationships do not break down clearly into binary categories, and so it is better to think of them as part of a network, permitting perhaps the intermixing of humans and non-humans in practice and thought, and perhaps too ways to link social, natural, constructed and realist conceptions of the living world.

Madhav Gadgil explores questions of social change and conservation in the fourth chapter (Chapter 34). Human society has been both prudent and harmful to the natural environment over thousands of generations, and Gadgil uses the concepts of ecosystem people and biosphere people to explore the continuum between those who rely mostly on local resources and those that have exploitative access to additional sources of energy and resources from outside. Ecosystems people in many parts of the world continue to exhibit a variety of cultural traditions of conservation practices, in spite of widespread loss of control over the resource base. However, there are now very few examples of entirely autonomous people, fully in control of their local ecosystems with very light human demands.

But when control is lost, local communities can easily lose their motivation for sustainable use, together with their local institutions that arrange rules, sanctions and behavioural norms. Political and economic subjugation, combined with market forces, have made it progressively more difficult for local communities to continue practices that may have been sustainable over many generations. As a result, the costs of conservation can increase, even to the point where the state intervenes, feeling it can do better. Ultimately, options for ecodevelopment will have to arise from below, but will need new forms of external support if they are to succeed in providing both livelihood options and protection for the natural environment.

The final part of this section (Chapter 35) moves from the terrestrial domain to the highly biodiverse and now threatened environments of coral reefs, in which David Smith, Sarah Pilgrim and Leanne Cullen address a range of issues relating to human pressures, valuation and management. Coral reefs represent one of the largest natural structures on the planet, and are home to more species than any other marine system. They are also important for the welfare of millions of people, providing a range of vital environmental goods and services. However, the majority of coral reefs worldwide are now overexploited, and 60% show severe signs of decline. During the course of the next century, pressures are likely to increase, with some estimates suggesting that 70% of coral reefs could be completely lost by 2050.

Despite the value of coral reefs to local communities, and their long-term dependence on them, it has become clear that efforts to govern and sustain reef fisheries have frequently failed. Yet many self-management systems have been very successful at maintaining resource levels over long periods. Local knowledge of species and ecological interactions, combined with institutions to set norms and rules, have been successful in many parts of the world. But centralized conservation, where ownership changes hands, or responsibility towards local resources is lost or abandoned, does not always work. Government-imposed authority frequently backfires, even if it is originally driven by a desire to protect resources sustainably. The dynamics of reefs systems can never be fully understood by those external to it, and thus co-management options need to be developed and implemented.

SECTION VII: INSTITUTIONS AND POLICIES FOR INFLUENCING THE ENVIRONMENT

The final section of the Handbook explores how institutions from local to national level shape and influence environmental outcomes. What are the best options for those with different types of knowledge? How do social–ecological systems develop over time, and what are the best approaches for community-based natural resource management? At the national level, how do questions of precaution affect policy development, and finally, in what form do environmental risks manifest themselves in the configuration of society?

In the first chapter (Chapter 36), Jonathan Hastie assesses the role of science and scientists in environmental policy, and shows how there is no straightforward relationship between science and politics. There are four institutional norms characteristic of science: organized scepticism (judgement is suspended until evidence is convincing), universalism (knowledge claims are tested with universal criteria), disinterestedness (scientists support ideas on the basis of merit, not self-interest) and communism (findings are shared in order for knowledge to progress). Scientists, of course, have differing opinions and hypotheses, yet where scientists disagree, so policy makers and interest groups may take advantage, using only those findings that support their pre-existing preferences. Sometimes, political interests use the products of science after their generation, on other occasions they seek to intervene during the assessment or funding process. In a variety of ways, therefore, science does not linearly produce evidence that policymakers simply then adopt. Scientific knowledge can be exploited, influenced or even ignored.

Scientists themselves may, too, become actively involved in political struggles, seeking to promote certain policies, either as individuals or groups. Today, appointed scientific advisors themselves have great power. Epistemic communities theory accepts the notion that scientists are far from disinterested, and examines how they build consensus to gain authority. In a similar way, discourse coalitions can focus around sets of shared ideas and principles. In this way, a constructivist (compared with a positivist) model of science in society sees scientific knowledge as constructed within a social process. In observing environmental policy, it is important therefore to study science, policy and the shifting boundary between the two with equal intensity.

The second chapter (Chapter 37) by Carl Folke Johan Colding, Per Olsson, and Thomas Hahn analyses the characteristics of social–ecological systems. They seek to provide a rich understanding of not just human–environment interactions but of how the world we live in actually works and the implications it has for current policies and governance. The chapter emphasizes that the social landscape should be approached as carefully as the ecological in order to clarify features that contribute to the resilience of social–ecological systems. In this context, Pretty and Ward (2001) find that relations of trust, reciprocity, common rules, norms and sanctions, and connectedness in institutions are critical. Folke *et al*. have similar findings that include vision, leadership and trust; enabling legislation that creates social space for ecosystem management; funds for responding to environmental change and for remedial action; capacity for monitoring and responding to environmental feedback; information flow through social networks; the combination of various sources of information and knowledge; and sensemaking and arenas of collaborative learning for ecosystem management. Their work illustrates that the interplay between individuals (e.g. leadership, teams, actor groups), the emergence of nested organizational structures, institutional dynamics and power relations tied together in dynamic social networks are examples of features that seem critical in adaptive governance which allows for ecosystem management and for responding to environmental feedback across scales.

An important lesson from the research is that it is not enough to create arenas for dialogue and collaboration, nor is it enough to develop networks to deal with issues at a landscape level. Further investigation of the interplay between key individuals, actor groups, social networks, organizations and institutions in multilevel social–ecological systems in relation to adaptive capacity, cross-scale interactions and enhancement of resilience is needed. We have to understand, support and perhaps even learn how actively to navigate the underlying social structures and processes in the face of change. There will be inevitable and possibly large-scale environmental changes, and preparedness has to be built to enhance the social–ecological capacity to respond, adapt to and shape our common future and make use of creative capacity to find ways to transform into pathways of improved development. They conclude that the existence of transformative capacity is essential in order to create social–ecological systems with the capability to manage ecosystems sustainably for human well-being. Adaptive capacity will be needed to strengthen and sustain such systems in the face of external drivers and events.

In the third chapter (Chapter 38), Stephen Brechin, Grant Murray and Charles Benjamin analyse the current challenges and opportunities in community-based natural resource management. The article links four bodies of work. The first concentrates on the social and political issues related to demarcated land-based conservation initiatives, particularly focusing on management issues involving local people. The second addresses similar issues in marine protected areas. The third addresses questions of state-centred devolution of responsibilities that are redefining community-based efforts, and the last reviews the social promises and pitfalls of ecotourism. The evidence clearly now shows that the future of bio-diversity conservation rests on finding more effec-tive and connected ways of integrating local people and communities into the conservation process, and not in their greater separation.

There are many questions, though, on how to ensure greater social justice, how to address the specific needs of indigenous people (who some commentators have called the 'danger within'), the rise of private parks, the growth of big interna-tional NGOs (BINGOs), and the challenges of developing processes that are effective across whole landscapes. Community-based conserva-tion is increasing in relevance and importance, partly through decentralization, and partly because of the emergence of strong evidence to show its effectiveness when the social, ecological and polit-ical conditions are right. The future of biodiversity conservation must rest largely on working together with people and communities, both in developing and industrialized countries.

Harini Nagendra and Elinor Ostrom explore a range of institutional and collective action ques-tions in the fourth chapter (Chapter 39) of this sec-tion of the Handbook. Until recently, the dominant theory predicted that individual users of common pool resources would always overuse and/or underinvest in the resources unless these were owned privately or by government. In this chapter, the theoretical perspectives are first reviewed, and the central principles of alternative positions sum-marized: with the right institutions, rules and boundary conditions, it is possible for communi-ties to manage common pool resources over very long periods. There is, however, a need for flexi-ble rather than blueprint thinking, a recognition of the importance of differing contextual variables, an understanding of how financial benefits can serve as incentives for effective management, and an acknowledgement that heterogeneity can be positively associated with successful collective action.

The case of Nepal is analysed in detail, and the particular problem of blueprint thinking identi-fied. A consequence of the growing appreciation of the value of community-based efforts for forest conservation has resulted in their increased pro-motion by government, with over 8500 forest user groups now formed in the hills and plains. But where models are applied from above rather than developed iteratively from below, then successful management may be threatened. At the same time, financial benefits are rarely evenly shared between communities, especially those in buffer zones of parks bringing in substantial ecotourism revenue. In conclusion, scholars interested in environmental policies will need to pay more attention to the need for adaptive development of institutions to fit the ecological system of interest.

The fifth chapter (Chapter 40) is by Albert Weale, and contains a clear analysis of the precau-tionary principle in environmental politics. There is an interesting conflict in environmental policy – on the one hand, there is widespread agreement on the need to act to protect biodiversity and encour-age sustainable development. On the other, there remains controversy as to what to do to attain these apparently consensual goals. Uncertainty is a central element of contemporary environmental policy, with many key questions on the frontier of scientific knowledge and understanding. Sometimes uncertainty seems to suggest taking no action, and on other occasions it appears to com-mend immediate action. The precautionary princi-ple has received widespread attention in many policy instruments, and again has been invoked in many different ways. Thus, governments dispute its formulation and contest its applications, and policy commentators and activists are divided on whether it is useful or not.

Discussions of the precautionary principle centre on three interrelated questions. How is the principle defined and what claims are being asserted? How should policymakers deal with inevitable uncertainties about cause and effect? How do the values protected by the application of the principle of precaution stand relative to other values? The varying conceptions of precaution suggest that there is not one precautionary princi-ple, rather a precautionary attitude, characterized by a willingness to act on threats, even when the risk is unclear or unlikely, and to the differing degrees to which threats and costs are evaluated. Proponents of a strong conception will act with less evidence than those who hold to a weaker conception of the principle. The bigger question, however, centres on whether it is possible to democratize decision on precaution.

The final chapter of this section and of the Handbook (Chapter 41) is by Ulrich Beck and Cordula Kropp, and explores issues relating to environmental risks and public perceptions. The backdrop is Beck's concept of the world risk soci-ety. Global approaches to problems can work, but face three problems: relevant (both lay and expert) knowledge is rarely clear about global hazards, global definition of environmental problems can be seen itself as a kind of ecological imperial-ism, and the very idea of nature conservation can be perverted into a new kind of world manage-ment. Underpinning these questions are issues of uncertainty – existing ones and self-generated manufactured ones. Can risks be brought under control, or will they always escape, leading per-haps to ecological flashpoints? In the world risk society, therefore, industrial projects become political ventures. Thus, what is required is global action from above, such as through international treaties and institutions, and globalization from below, such as through new transnational actors operating beyond the system of parliamentary politics and challenging established political organizations and interest groups.

In the crisis of global interdependence are global financial risks, the threats from terrorist networks and ecological risks. All three have the potential to cause cross-border conflicts, though environmental ones have particular features, such as having long periods of latency, the need to pass scientific, media and public attention to come into existence, and the difficulty of individualizing risks which generally spread over and under national borders. Global environmental risks are potentially transformative, especially where the desire for sustainability has eclipsed or displaced the long-held notions of economic and technical progress.

REFERENCES

Adrianne, A. (1997) *Resource Flows: The Material Basis of Industrial Economies.* Washington: W.R.I.

Barrett, S. and Graddy K. (2000) Freedom, growth and the environment. *Environment and Development Economics* 5(4):433–456.

Beck, U. (1992a) From industrial society to risk society: questions of survival, social structure and ecological enlightenment. *Theory, Culture and Society* 9:97–123.

Beck, U. (1992b) *Risk Society: Towards a New Modernity.* London, Newbury Park and New Delhi: Sage.

Beckerman, W. (1996) *Through Green Coloured Glasses: Environmentalism Reconsidered.* Washington DC: Cato Institute.

Bell, M. M. (2004) *An Invitation to Environmental Sociology*, 2nd edn. Thousand Oaks: Pine Forge Press.

Benton, T. (1989) Marxism and natural limits: an ecological critique of reconstruction. *New Left Review* 178:51–86.

Benton, T. (ed.) (1996) *The Greening of Marxism.* New York: Guilford Press.

Benton, T. (1998) Sustainable development and the accumulation of capital: reconciling the irreconcilable? In Dobson, A. (ed.) *Fairness and Futurity.* Oxford: Oxford University Press.

Benton, T. (2000) Reflexive modernization. In Browning G., Halcli, A. and Webster, F. (eds) *Understanding Contemporary Society.* London: Sage.

Benton, T. (2001) Environmental sociology: controversy and continuity. *Sosiologisk tidsskrift (Journal of Sociology).* 9:4–48.

Brown, L. (2004) *Outgrowing the Earth.* New York: W. W. Norton.

Brunkhorst, D., Bridgewater, P. and Parker, P. (1997) The UNESCO biosphere reserve program comes of age: learning by doing; landscape models for sustainable conservation and resource use. In Hale, P. and Lamb, D. (eds) *Conservation*

Outside Nature Areas. University of Queensland, pp. 176–182.

Buttel, F. H. (1987) New directions in environmental sociology. *Annual Review of Sociology* 13:465–488.

Buttel, F. H. (1997) Social institutions and environmental change. In Redclift M. and Woodgate G. (eds.) *The International Handbook of Environmental Sociology.* Cheltenham, UK: Edward Elgar Publishing, pp. 40–54.

Carney, D. (1998) *Sustainable Rural Livelihoods.* London: Department For International Development.

Carson, R. (1963) *Silent Spring.* Boston: Houghton Miflin.

Catton, W. R. and Dunlap, R E. (1978) Environmental sociology: a new paradigm. *The American Sociologist* 13:41–49.

Catton, W. R. and Dunlap, R E. (1980) A new ecological for post-exuberant sociology. *American Behavioural Scientist* 24:15–47.

Coleman, J. (1988) Social capital and the creation of human capital. *American Journal of Sociology* 94, suppl. S95–S120.

Costanza, R., d'Arge, R., de Groot, R., Farber, S., Grasso, M., Hannon, B., Limburg, K., Naeem, S., O'Neil, R. V., Paruelo, J., Raskin, R. G., Sutton, P. and van den Belt, M. (1997) The value of the world's ecosystem services and natural capital. *Nature* 387:253–260.

Craib, I. (1992) *Modern Social Theory.* Hemel Hempstead: Harvester Wheatsheaf.

Craib, I. (1997) *Classical Social Theory.* Oxford: Oxford University.

Cronon, W., Miles, G. and Gitlin, J. (eds) (1992) *Under an Open Sky. Rethinking America's Western Past.* New York: W. W. Norton.

Dalton, R. J. (1994) *The Green Rainbow: Environmental Groups in Western European Politics.* New Haven: Yale University Press.

Dasgupta, P. and Serageldin, I. (eds) (1998) *Social Capital: A Multiperspective Approach.* Washington, DC: World Bank.

Deutsch, S. (1992) Landscape of enclaves. In Cronon, W., Miles, G. and Gitlin, J. (eds) (1992) *Under an Open Sky. Rethinking America's Western Past.* New York: W. W. Norton.

Dickens, P. (1996) *Reconstructing Nature: Alienation, Emancipation and the Division of Labour.* London and New York: Routledge.

Dickens, P. (1997) Beyond sociology: Marxism and the environment. In Redclift M. and Woodgate G. (eds.) *The International Handbook of Environmental Sociology.* Northampton: Edward Elgar Publishing, pp. 179–194.

Dryzek, J. S. (1996) Strategies of ecological democratization. In Lafferty, W. M. and Meadowcroft, J. (eds) *Democracy and the Environment.* Cheltenham: Edward Elgar Publishing.

Dunlap, R. E. (1997) The evolution of environmental sociology: a brief history and assessment of the American experience. In Redclift M. and Woodgate G. (eds.) *The International Handbook of Environmental Sociology.* Northampton: Edward Elgar Publishing, pp. 21–39.

Dunlap, R. E. and Catton W. R. (1994) Struggling with human exemptionalism: the rise, decline and revitalization of environmental sociology. *The American Sociologist* 25(1):5–30.

Eckersley R. (2004) *The Green State: Rethinking Democracy and Sovereignty.* Cambridge, MA: MIT Press.

Ellis, F. (2000) *Rural Livelihoods and Diversity in Developing Countries*. Oxford: Oxford University Press.

Eyerman, R. and Jamison, A. (1991) *Social Movements: A Cognitive Approach*. Cambridge: Polity.

Fischer-Kowalski, M. and Haberl, H. (1993) Metabolism and colonisation: modes of production and the physical exchange between societies and nature. *Innovation in Social Science Research* 6(4):415–442.

Flora, C. B. and Flora, J. L. (1996) Creating social capital. In Vitek, W. and Jackson, W. (eds) *Rooted in the Land: Essays on Community and Place*. New Haven and London: Yale University Press, pp. 217–225.

Foreman, R. T. (1997) *Land Mosaics. The Ecology of Landscapes and Regions*. Cambridge: Cambridge University Press.

Foster, J. B. (1999) Marx's theory of metabolic rift: classical foundations for environmental sociology. *American Journal of Sociology* 105(2):406–454.

Frank, D. J. (2002) The origins question: building global institutions to protect nature. *Organizations, Policy, and the Natural Environment*. Stanford: Stanford University Press, pp. 41–56.

Frank, D. J., Hironaka, A. and Schofer, E. (2000) The nation-state and the environment over the twentieth century. *American Sociological Review* 65(1):96–116.

Frank, R. H. (1999) *Luxury Fever*. Princeton: Princeton University Press.

Fredriksson, P. G., Neumayer, E., Damania, R. and Gates, S. (2005) Environmentalism, democracy, and pollution control. *Journal of Environmental Economics and Management* 49(4):343–365.

Freeman, A. M. (2003) *The Measurement of Environmental and Resource Value: Theory and Methods*, 2nd edn. Washington, DC: Resources for the Future.

Freudenburg, W. R., Frickel, S. and Gramling, R. (1995) Beyond the nature/society divide: learning to think about a mountain. *Sociological Forum* 10(3):361–392.

Fukuyama, F. (1992) *The End of History and the Last Man*. London: Hamish Hamilton.

Giddens, A. (1994) *Beyond Left and Right*. Cambridge: Polity.

Grafton, R. Q. and Knowles, S. (2004) Social capital and national environmental performance: a cross sectional analysis. *Journal of Environment and Development* 13(4): 336–370.

Gray, A. (1999) Indigenous peoples, their environments and territories. In Posey, D. (ed.) *Cultural and Spiritual Values of Biodiversity*. London: IT Publications and UNEP.

Gray, J. (2002) *Straw Dogs*. London: Granta Books.

Gray, J. (2004) *Heresies. Against Progress and Other Illusions*. London: Granta.

Hajer, M. A. (1995) *The Politics of Environmental Discourse: Ecological Modernisation and the Policy Process*. Oxford: Clarendon.

Hanley, H. D. and Spash, C. (1993) *Cost-Benefit Analysis and the Environment*. Aldershot: Edward Elgar Publishing.

Hannigan, J. A. (1995) *Environmental Sociology: A Social Constructionist Approach*. London and New York: Routledge.

Jordan, G. and Maloney, W. A. (1997) *The Protest Business: Mobilizing Campaign Groups*. Manchester: Manchester University Press.

Kasser, T. (2002) *The High Price of Materialism*. Cambridge, MA: MIT Press.

Kellert, S. R. and Wilson, E. O. (1993) (eds) *The Biophilia Hypothesis*. Washington, DC: Island Press.

Klijn, J. and Vos, W. (eds) (2000) *From Landscape Ecology to Landscape Science*. Dordrecht: Kluwer Academic.

MacNaughten, P. and Urry, J. (1998) *Contested Natures*. London: Sage.

Malthus, T. (1798) *An Essay on the Principles of Population*.

Matthews, E., Amann, C., Bringezu, S. and Fischer-Kawalski, M. (2000) *The Weight of Nations*. Washington DC: WRI.

Meadows, D. H., Meadows, D. L. and Behrens, W. III (1972) *The Limits to Growth*. New York: Universe.

Mellor, M. (1992) *Breaking the Boundaries*. London: Virago.

Mertig, A., Dunlap, R. E. and Morrison, D. E. (2002) The environmental movement in the United States. In Dunlap, R. E. and Michelson, W. (eds) *Handbook of Environmental Sociology*. London: Greenwood Press, pp. 448–481.

Midlarsky, M. (1998) Democracy and the environment: an empirical assessment. *Journal of Peace Research* 35(3): 341–361.

Millennium Ecosystem Assessment (MEA) (2005) *Ecosystems and Human Well-Being*. Washington, DC: Island Press.

Mol, A. P. J. (1996) Ecological modernisation and institutional reflexivity: environmental reform in the Late Modern Age. *Environmental Politics* 5(2):302–23.

Mol, A. P. J. (2001) *Globalization and Environmental Reform: The Ecological Modernization of the Global Economy*. Boston: MIT Press.

Mol, A. P. J. and Sonnenfeld, D. A. (eds) (2000) *Ecological Modernisation Around the World: Perspectives and Critical Debates*. [Special issue of *Environmental Politics* 9(1)].

Mol, A. P. J. and Spaargaren, G. (2000) Ecological modernization theory in debate: a review. *Environmental Politics* 9(1):17–49.

Morison, J., Hine, R. and Pretty, J. (2005) Survey and analysis of labour on organic farms in the UK and Republic of Ireland. *Int. J. Agric. Sustainability* 3(1):24–43.

Munck, G. L. (1995) Actor formation, social co-ordination, and political strategy: some conceptual problems in the study of social movements. *Sociology* 29(4):667–685.

Murphy, J. (2000) Ecological modernization. *Geoforum* 31:1–8.

Murphy, R. (1994) *Rationality and Nature: A Sociological Inquiry into a Changing Relationship*. Boulder: Westview.

Myers, N. and Kent, J. (2003) New consumers: the influence of affluence on the environment. *Proc. Natl. Acad. Sci.* 100, 4963–4968.

Myers, N. and Kent, J. (2004) *The New Consumers*. Washington DC: Island Press.

Naess, A. (1990) *Ecology, Community and Lifestyle: Outline of an Ecosophy* (transl. D. Rothenberg). Cambridge: Cambridge University Press.

Nash, R. (1973) *Wilderness and the American Mind*. New Haven: Yale University Press.

Nettle, D. (2005) *Happiness*. Oxford: Oxford University Press.

Norgaard, R. B. (1997) A coevolutionary environmental sociology. In Redclift, M. and Woodgate, G. (eds) *The International Handbook of Environmental Sociology*. Northampton: Edward Elgar Publishing, pp. 158–168.

O'Brien, M., Penna, S. and Hay, C. (eds) (1999) *Theorising Modernity*. London and New York: Longman.

O'Connor, J. (1998) *Natural Causes: Essays in Ecological Modernization*. New York: Guilford Press.

Oelschlaeger, M. (1991) *The Idea of Wilderness*. New Haven: Yale University Press.

Ostrom, E. (1990) *Governing the Commons: The Evolution of Institutions for Collective Action*. New York: Cambridge University Press.

Pearce, D. W. and Turner, R. K. (1990) *The Economics of Natural Resources and the Environment*. Hemel Hempstead: Harvester Wheatsheaf.

Pfeffer, M. J., Schelhas, J. and Day, L. A. (2001) Forest conservation, value conflict, and interest formation in a Honduran National Park. *Rural Sociology* 66(3):382–402.

Pfeffer, M. J., Schelhas, J. W., DeGloria, S. and Gomez, J. (2005) Population, conservation, and land use change in Honduras. *Agriculture, Ecosystems and the Environment* 110:14–28.

Pfeffer, M. J., Schelhas, J. W. and Meola, C. (2006) Environmental globalization, organizational form and expected benefits from protected areas in Central America. *Rural Sociology* 71(3):429–450.

Porritt, J. (2005) *Capitalism as if the World Matters*. London: Earthscan.

Pretty, J. (2002) *Agri-Culture. Reconnecting People, Land and Nature*. London: Earthscan.

Pretty, J. (2003) Social capital and the collective management of resources. *Science* 302:1912–1915.

Pretty, J. (2004) How nature contributes to mental and physical health. *Spirituality & Health International* 5(2):68–78.

Pretty, J. (2007) Agricultural sustainability: concepts, principles and evidence. *Phil. Trans. R. Soc. Lond. B.* (in press).

Pretty, J. N. and Koohafkan, P. (2002) *Land and Agriculture: From UNCED Rio to WSSD Johannesburg*. Rome: FAO.

Pretty, J. and Ward, H. (2001) Social capital and the environment. *World Development* 29(2):209–227.

Pretty, J., Peacock, J., Sellens, M. and Griffin, M. (2005) The mental and physical health outcomes of green exercise. *International Journal of Environmental Health Research* 15:319–337.

Pretty J., Noble, A. D., Bossio, D., Dixon, J., Hine, R. E., Penning de Vries, F. W. T. and Morison, J. I. L. (2006) Resource-conserving agriculture increases yields in developing countries. *Environmental Science & Technology* 40(4):1114–1119.

Putnam, R. D., with Leonardi, R. and Nanetti, R. Y. (1993) *Making Democracy Work: Civic Traditions in Modern Italy*. Princeton, NJ: Princeton University Press.

Rees, M. (2003) *Our Final Century*. London: Arrow Books.

Rees, W. (2002) Ecological footprints. *Nature* 421:898.

Rees, W., Wackernagel, M. and Testemae, P. (1996) *Our Ecological Footprint. Reducing Human Impact on the Earth*. Gabriola, Canada: New Society Publ.

Rootes, C. A. (1997) Environmental movements and green parties in Western and Eastern Europe. In Redclift, M. and Woodgate, G. (eds) *The International Handbook of Environmental Sociology*. Northampton: Edward Elgar Publishing, pp. 319–348.

Saward, M. (1993) Green democracy? In Dobson, A. and Lucardie, P. (eds) *The Politics of Nature: Explorations in Green Political Theory*. London: Routledge, pp. 63–80.

Schama, S. (1996) *Landscape and Memory*. London: Fontana Press.

Schandl, H. and Schulz, N. (2000) *Using Material Flow Accounting to Operationalize the Concept of Society's Metabolism*. Colchester: University of Essex.

Schelhas, J. and Pfeffer, M. J. (2005) Forest values of national park neighbours in Costa Rica. *Human Organization* 64(4):385–397.

Schnaiberg, A. (1980) *The Environment, From Surplus Scarcity*. New York: Oxford University Press.

Schnaiberg, A. and Gould, K. A. (1994) *Environment and Society: The Enduring Conflict*. New York: St. Martin's Press.

Schwartz B. (2004) *The Paradox of Choice*. New York: Harper.

Seippel, O. (2002) Modernity, politics, and the environment: a theoretical perspective. In Dunlap, R. E., Buttel, F. H., Dickens, P., and Gijswijt, A. (eds) *Sociological Theory and the Environment: Classical Foundations, Contemporary Insights*. New York: Rowman and Litttlefield.

Simonis, U. E. (1989) Ecological modernization of industrial society: three strategic elements. *International Social Science Journal* 121:347–361.

Soper, K. (1995) *What is Nature?* Oxford: Blackwell.

Spaargaren, G., Mol, A. P. J. and Buttell, F. H. (eds) (2000) *Environment and Global Modernity*. London: Sage.

Sprinz, D. F. and Helm, C. (2000) Measuring the effectiveness of international environmental regimes *Journal of Conflict Resolution* 45(5):630–652.

Stones, R. (1998) *Key Sociological Thinkers*. Basingstoke and London: Macmillan.

Suzuki, D. (1997) *Sacred Balance*. London: Bantam Books.

UN. (2004) *World Population to 2300*. New York

UN. (2005a) *Long-range World Population Projections: Based on the 1998 Revision*. New York: UN Population Division.

UN. (2005b) *World Population Prospects. The 2004 Revision*. UN Population Division. UN. 2004. *World Population to 2300*. UN Population Division.

Underdal, A. (1992) The concept of regime 'effectiveness'. *Cooperation and Conflict* 27(3):227–240.

Uphoff, N. (1998) Understanding social capital: learning from the analysis and experience of participation. In Dasgupta, P. and Serageldin, I. (eds) *Social Capital: A Multiperspective Approach*. Washington, DC: World Bank.

Uphoff, N. (ed.) (2002) *Agroecological Innovations*. London: Earthscan.

Vandergeest, P. and DuPuis, E. M. (1996) Introduction. In DuPuis, E. M. and Vandergeest, P. (eds) *Creating Countryside. The Politics of Rural and Environmental Discourse*. Philadelphia: Temple University Press.

Ward, H. (2006) International linkages and environmental sustainability: the effectiveness of the regime and IGO networks. *Journal of Peace Research* 43:149–166.

Ward, H., Grundig, F. and Zorick, E. (2004) Formal theory and regime effectiveness: rational players, irrational regimes. In Underdal, A. and Young, O. (eds) *Regime Consequences: Methodological Challenges and Research Strategies*. Dordrecht: Kluwer Academic.

WCED, (1987) Our Common Future. World Commission on Environment and Devleopment. Oxford: Oxford University Press.

Weale, A. (1992) *The New Politics of Pollution*. Manchester: Manchester University Press.

Wilson, E. O. (1984) *Biophilia – The Human Bond with other Species*. Cambridge, MA: Harvard University Press.

Worster, D. (1993) *The Wealth of Nature: Environmental History and the Ecological Imagination*. New York: Oxford University Press.

WRI (2006) World Resources Institute Statistics. At www.wri.org

Yearley, S. (1991) *The Green Case: A Sociology of Environmental Issues, Arguments and Politics*. London: HarperCollins.

Yearley, S. (1994) Social movements and environmental change. In Redclift, M. and Benton, T. (eds) *Social Theory and the Global Environment*. London: Routledge, pp. 150–168.

Yearley, S. (1996) *Sociology, Environmentalism, Globalization: Reinventing the Globe*. Thousand Oaks: Sage.

Young, O. R. and Levy, M. A. (1999) The effectiveness of international environmental regimes. In Young, O. R. (ed.) *The Effectiveness of International Environmental Regimes: Causal Connections and Behavioural Mechanisms*. Cambridge, MA: MIT Press, pp. 1–32.

Environmental Thought: Past and Present

2

Humans and Nature: From Locke and Rousseau to Darwin and Wallace

Ted Benton

INTRODUCTION

This chapter gives an overview of some strands of Western thinking about the place of humans in nature from the emergence of modern physical science during the 17th century, to the evolutionary debates of the late 19th century. We often find in today's accounts of the rise of modern science and technology, and the philosophy of 'enlightenment' that followed it a rather simplified and one-sided view. A 'modern project' is sometimes held to involve elevating humans above the rest of nature, seeking to realise human potential by subordinating and controlling the rest of the natural world. This picture does represent one very influential 'grand narrative' of the modern period, but it is also important to remind ourselves that there were always alternative views: sometimes present as explicit opposition to the dominant view, but also often present as a 'sub-text', intertwined with the dominant view in the thought of a single writer. Although it is, of course, necessarily very selective, and also contains many oversimplifications, the following account is intended to give some illustrations of the complexity and ambiguity present in the thought of this period.

FROM THE MEDIEVAL VIEW TO THE NEW MECHANICAL SCIENCE

The official view of the universe that dominated the European middle ages represented it as a set of concentric spheres, the outer ones moving with perfect circular motion, carrying with them the heavenly bodies. At the centre was the earth, a region of change and decay, and the home for human kind. This view was, in essence, derived from the physics of Aristotle (384–322 BC), but with the heaven and hell of Christian doctrine, and the hand of God as creator added on. For the great majority of the rural population through this period, everyday life was lived in close interaction with 'nature'. Living quarters were often shared with domesticated animals, and the untamed forces of nature, the passage of the seasons and recurrent famines and epidemics favoured a common-sense acceptance of an 'organic' connection between humans and nature.

Carolyn Merchant (1982) has traced the variety of medieval images of nature as an organism, specifically as a nurturing mother, through to the 16th century. But the expansion of more intrusive forms of human practical relationship to nature, most especially mining, and, eventually, the coming of industrial production, led to a shift of perceptions according to which female nature came to be seen as disorderly and threatening, in need of taming and domination. From the middle of the 16th through to the end of the 17th century what is now recognised as the revolution of modern science began a process of replacing the earlier organic and integrated views of humans' place in nature in favour of a mechanical view. In Merchant's argument, the earlier organic view had imposed ethical constraints on the way nature could be treated. But these constraints were lifted as a new 'instrumental' view of nature as a mere mechanism became dominant. As well as representing a consolidation of male power (see Chapter 4), the scientific revolution, and the later industrial revolution that it made possible, transformed the relation between humans and nature. It also offered a profound challenge to the power and moral authority of the Church, for which the Aristotelian hierarchical view had provided the foundations.

The new mechanical science of Copernicus, Galileo, Descartes and Newton displaced the earth, and so humans, from the centre of the universe. In Galileo's metaphor, the world is a great book, open to the gaze of all, but to read it we need more than mere vision. The book is written in the language of mathematics, without which we cannot understand a single word. Systematic observation, measurement and mathematical reasoning are the methods to be employed for a proper understanding of the universe as a colossal mechanical system, acting according to a small number of universal, mathematical laws. In Newton's celebrated synthesis, three laws of motion and the law of universal gravitation unified in a single system the motions of the planets around the Sun, the Earth's rotation, free fall close to the Earth's surface, the paths of projectiles and other mechanical interactions. Some free thinkers drew sceptical conclusions from this new view of the universe and attracted condemnation and punishment from the Catholic Inquisition. But for many of the new 'natural philosophers', including Newton, their scientific method was another route to a knowledge of God through His creation. For these scientists, there was no contradiction between their religious faith and the new methods of discovery, and they struggled to reconcile their new account of the universe with scriptural interpretation.

Despite this, the new science remained unsettling, and an alternative source of legitimacy was sought in strong claims for the utility of science in the service of human welfare. The earlier utopia of Francis Bacon (1561–1626) saw 'active science' as a means of transforming and managing all of nature in the enlargement of 'the bounds of Human Empire'. This advocacy of science and technology applied to industry and agricultural improvement was a powerful means of persuading ruling elites of the value of freedom of scientific thought. In Britain it soon found approval from the most respectable circles in the foundation of the Royal Society of London in 1660–1662. As we shall see, this proved to be a development with the profoundest practical consequences for the future, as modern industrial production transformed the relationships between human society and its nature-given conditions of existence.

THE ENLIGHTENMENT: STATE OF NATURE AND CIVIL SOCIETY

However, in the short term, it was the philosophical and political implications of the new science that were perhaps the most apparent. The achievements of the Galilean/ Newtonian method inspired the thinkers of the 17th and 18th centuries to set morality, politics and human social life itself on new, rational foundations, with Newton's vision of a law-governed nature as their model. The political philosophers of the period, now loosely referred to as the 'Enlightenment', posed fundamental questions such as the justifications for the obedience of citizens to their rulers, the conditions under which they had a right to rebel, the basis for the right to property, and the status of inequalities among people. Though these thinkers commonly quoted biblical texts to support their arguments, they also justified their claims by an appeal to reason. These secular, rational arguments have since come to provide the main justification for the institutions of Western liberal democracies, as well as the principles underlying international law, and global institutions such as the United Nations.

The contrast between a supposed original 'state of nature' and 'civil society' was the framework for much of this early modern social and political thinking. The method was to take the social or political institutions or principles that were to be considered, and to imagine a state of human existence prior to their establishment. It could then be asked, why would people (usually 'men') in such a state choose to enter into agreements to be bound by obedience to government, law, property, or whatever? For our purposes, these speculations (and often they quite explicitly admitted their accounts were, indeed, purely speculative) about human existence in a 'state of nature' prior to society or to law and government provide interesting insights into how the thinkers concerned conceptualised our relationships to other species and to non-human nature in general. The contrasting speculations of just two of these thinkers – John Locke (1632–1704) and Jean-Jacques Rousseau (1712–1778) – illustrate both the diversity of Enlightenment thought, and also the origins of later alternative views of nature.

John Locke: Reason, Labour and Property

Famously, it was Thomas Hobbes (1588–1697) who painted the most bleak picture of human life in the state of nature as 'solitary, poor, nasty, brutish and short', a state in which each individual is at war with the others. For John Locke, however, the state of nature is already a condition of social coexistence, in which there is family life, property and even trade, in the form of direct exchange. Locke's state of nature is one in which 'men' are free and equal, and live according to rational principles: 'natural law'. In such a state men live together 'according to reason without a common superior on earth'. It is a state of 'peace, good-will, mutual assistance, and preservation'. In the absence of agreement to establish a sovereign power, each individual has a right to exact reparation and punish transgressors against the law of nature, as such individuals have put themselves into a state of war. But there are 'inconveniences' of the state of nature and reason dictates each individual should consent to handing over to a common authority their right to be judge and jury in their own case, if agreement is possible [Locke, 1971 (1690)].

So, what of natural man's relationship to the rest of nature? Locke starts with three connected principles: that men have a natural right to self-preservation; that God gave men the world in common; and that the world was given 'for the support and comfort of men'. So, the Earth and 'all the fruits it naturally produces, and beasts it feeds, belong to mankind in common, as they are produced by the spontaneous hand of nature'. But it would make no sense for God to have given us the Earth for our benefit, and made us dependent on the fruits of the Earth for our subsistence, if there were no way of individually appropriating 'meat and drink and such other things as nature affords'. Fortunately, in Locke's view, God gave us reason and property in our own person. This both enables and entitles us to take from the common property of humankind what we need for our 'best advantage and convenience'. What entitles each individual to take what they need from the common store is the application of labour to the earth or its products. But this conversion of common property into private property is subject to three restraints or conditions:

1　That enough is left for others;
2　That what is left is of as good quality as what is taken;
3　That only so much is taken as can be used without it spoiling or being destroyed.

The earth itself, as well as its 'spontaneous' productions, may also be taken from the common store, subject to the above conditions, by the application of labour to it, in the form of 'tilling, planting, improving, cultivating'. So, where the population is sparse and land is plentiful, as in America, those who are 'industrious and rational' are entitled to take land for cultivation without seeking agreement from others, as the above three conditions are met. Thus, the lands of indigenous people can be appropriated directly by the 'industrious and rational' colonists. Only in a country such as England must consent be sought for the privatisation of common land, as taking some would not leave as much or as good for others.

Locke's way of thinking about the relationship of 'natural man' and the rest of nature is quite revealing. He has what we would now call an instrumental, anthropocentric account of this relationship, both at the level of the individual and our species as a whole. Other animals, lacking rationality, live outside the scope of natural law, as our inferiors, and they are subject to our purposes. Indeed, tigers, lions and other 'wild savage beasts' are, like humans who breach natural law and threaten our lives, in a state of war with us and may justifiably be killed.

Also of note is Locke's emphasis on the value-creating role of labour. Mixing one's labour with nature in cultivating virgin ground not only converts it to your own property, but also 'improves' it: renders it more productive of 'civilised goods', as against the meagre spontaneous productions of 'unassisted nature'. Consequently 'land that is left wholly to nature that hath no improvement of pasturage, tillage or planting, is called, as it is, waste: and we shall find the benefit of it amount to little more than nothing'. And, as to the products of the earth, 'as they come to our use', nine tenths is 'on the account of labour', only the remainder 'purely owing to nature'.

Locke thus appears to acknowledge no dimension of value in nature other than its capacity to serve our purposes. Even here, its capacity to do so is very limited unless subjected to human industriousness, rationality, and the allocation of property rights.

However, Locke's understanding of the 'law of nature' does impose some interesting restraints on the appropriation of nature as private property: the first two conditions impose a requirement of justice and the third a requirement of conservation – together they could even be seen as foreshadowing our contemporary notion of sustainability. But, for Locke, what changes all this is the presumption of general consent to the use of durable materials, gold and silver, as measures of wealth. The introduction of money allows indefinite accumulation of wealth without risk of spoiling or decay, and, along with that, unlimited appropriation from nature, the creation of scarcity and inequality among humans in access to the useful products of nature.

Jean-Jacques Rousseau: Natural Man and the Origins of Inequality

Rousseau was, like Locke, also a critic of Hobbes's bleak view of the state of nature, but he goes much further than Locke, seeing the passage from a state of nature to that of civil society as a process of loss as well as of compensating gains. The life of man in the state of nature is, for Rousseau, as it is for Hobbes, a solitary one. But, in contrast to Hobbes's view, it is also one in which the hazards and challenges of life cause men to 'form a robust and almost indestructible constitution' [Rousseau, 1974 (1754), p. 147]. In places, the state of nature is described almost lyrically:

> I see an animal less strong than some, less agile than others, but on the whole, the most advantageously constituted of all. I see him sitting under an oak tree, quenching his thirst at the nearest stream, finding his bed at the foot of the same tree that supplied him with his meal; and thus all his needs are satisfied. The earth, left to its natural fertility and covered with immense forests that have never been mutilated by an axe, offers abundant food and shelter to animals of every species. Men, scattered among them, observe and imitate their industry, and thereby attain the instincts of beasts, with the advantage that, whereas each of the other species has only its own instincts, man, who may never have had any peculiar to himself, appropriates all of them, eats most of the different foods that the other animals divide among themselves, and consequently finds his sustenance more easily than any of them (*ibid.,* pp. 146–147).

Against Hobbes, Rousseau argues that the state of nature is a condition in which individual efforts at self-preservation are least prejudicial to the well-being of others, so it is the state most conducive to peace and 'advantageous to mankind'. Hobbes, comments Rousseau, mistakenly includes in his account of the state of nature 'passions' (presumably for power and glory) that are in reality only engendered by the social state. This is why Hobbes mistakenly sees the state of nature as necessarily a state of war. It is only the passions evoked by life in society that necessarily result in antagonism and conflict. In Rousseau's account, natural man lives a simple and regular life, needing only physical satisfactions. Lacking either virtue or vice, language or sustained social bonds, the life of natural man is not 'miserable', as it is in Hobbes's account. On the contrary:

> I would like someone to explain to me what kind of misery could afflict a free man whose heart is at peace and whose body is in good health. I ask whether it is social life or natural life that is more likely to become unbearable to those who live it. We see around us hardly anyone who does not complain about his life; some even put an end to it... (*ibid.,* p. 162).

Rousseau's natural man lives a life of comparative ease, gaining sustenance, along with the other animals, from an abundant nature. Even the ferocious species that Locke thought were at war with man, are much less of a problem for Rousseau's 'savage': '... it seems that no animal naturally attacks man, except in cases of self-defence or extreme hunger, or shows toward him that violent antipathy which appears to indicate that nature intends one species to be food for another' (*ibid.,* p. 148). More than this, we are at one with other animals in our bodily needs for food, drink and sex. And this commonality extends also to our mental lives: '... every animal has ideas, since it has senses; it even combines its ideas to some extent' (*ibid.,* p. 153). So, in these respects we differ from other animals in degree only. Like animals we have a natural inclination, or passion for self-preservation, but this is moderated by natural compassion. This latter is defined by Rousseau as a 'principle prior to reason' which 'gives us a natural repugnance to seeing any sentient creature, especially our fellow man, perish or suffer'. The combination of these two passions is the source of all the rules of natural right. But for Rousseau these rules extend beyond the boundaries of the human species. Since other animals are sentient beings, if we do not resist our inner impulse for compassion, we will not harm any sentient creature except when self-preservation requires it. Animals have a right not to be needlessly mistreated by humans.

But humans do differ from animals in two respects. Animals behave in ways dictated by their nature, whereas, though humans may feel the same inclinations, they are aware of their freedom to acquiesce or resist them. This power of choosing is a specifically spiritual attribute, not physically explicable, and not shared with other animals. Second, and as a consequence of their power of choosing, humans have a capacity for 'self-improvement'. However, this remains latent for so long as the state of nature persists. Indeed, the state of nature is so favourable to human life, Rousseau speculates, that it must have persisted for many centuries, and was only abandoned in favour of social existence as a consequence of accumulating 'accidents'. Unlike Hobbes, for whom the state of nature is so terrible that subservience to a sovereign ruler is seen as a necessity, and unlike Locke, for whom civil government is a welcome solution for various 'inconveniences', Rousseau sees the advent of civil society as accidental, bringing with it profound losses, as well as benefits:

> Having shown that perfectibility, the social virtues, and the other faculties that natural man possessed in latent form could never have developed of

themselves, that they required the fortuitous concurrence of a number of extraneous causes which might never have arisen and without which man would have remained eternally in his original condition, I must next consider and correlate the various accidents that may have improved human reason while deteriorating the species, made man malicious while making him sociable, and, from that remote beginning, brought him and the world to the point where we see them now (*ibid.*, pp. 171–172).

Among the baleful consequences of abandonment of his natural state is that man eventually becomes 'a tyrant over himself and nature' (*ibid.*, p. 154). The private enclosure of land, justified by Locke in terms of the contrast between the 'waste' of uncultivated nature and the utility of land 'improved' by human labour, is denounced by Rousseau as the source of a train of evils:

The first person who, having enclosed a piece of land, took it into his head to say, 'this is mine,' and found people simple enough to believe him, was the true founder of civil society. The human race would have been spared endless crimes, wars, murders, and horrors if someone had pulled up the stakes or filled in the ditch and cried out to his fellow men, 'Do not listen to this impostor! You are lost if you forget that the fruits of the earth belong to everyone, and the earth to no one!' (*ibid.*, p. 173).

In Rousseau's account, the immensely long historical journey of human kind passes through the invention of tools and weapons, clothing and dwellings, the establishment of familial bonds and human settlements, the development of language and the emergence of metallurgy and agriculture, together with their associated divisions of labour. With these historical transitions come social bonds and luxuries, but, at the same time, envy, dissatisfaction, competitive pursuit of honour and power, and, finally, despotism. In all these developments, the institution of property is the one that brings most misfortunes in its wake:

But as soon as one man needed another's help, as soon as one man realised that it was useful to have enough provisions for two, equality disappeared, property came into being, work became necessary, and vast forests were changed into smiling fields which man had to water with his sweat, and in which slavery and poverty soon germinated with the crops (*ibid.*, p. 180).

So, by contrast to Locke's celebration of the rational and industrious 'improvement' of the land, and justification of the property rights that flow from it, Rousseau traces the evils, inequalities,

oppressions and dependencies of modern society back to this original act of privatisation of the earth and its fruits. These ideas were widely seen as playing their part in the great French Revolution of 1789, but they also pointed in the direction of a new artistic and cultural sensibility that came to be known as Romanticism.

THE ROMANTICS: NATURE, SELF AND SENTIMENT

Romanticism is commonly understood as a reaction that set in during the latter part of the 18th and early 19th centuries, against the Enlightenment's elevation of reason over sentiment, against the scientific representation of nature as a colourless and law-governed mathematical system, and against the destruction of nature's beauty by the encroachment of industry and urbanism. Rousseau's philosophical writings, but perhaps more than these, his autobiographical *Confessions*, were sources of inspiration for key figures of the Romantic movement – for example, the English Romantic poet and revolutionary, Percy Bysshe Shelley (1792–1822), declared Rousseau's name to be 'sacred'. Romanticism more often found its expression in the creative arts – especially in painting, music and poetry. It is not surprising, therefore, that there is no clear unity of doctrine or formal belief shared by the Romantics. There were very different national traditions, and the Terror that followed the French revolution caused deep divisions politically. We can, however, identify common tendencies of thought and expression. In their portrayal of human nature the Romantics, like Rousseau, celebrated the natural, undomesticated free spirit against the individual as subdued, domesticated, a slave to convention and artifice. As in Rousseau's thought, the passions are seen as more fundamental than reason, even contrasted to reason as the preferred animator of our activity in life.

Just as the natural is preferred to the domesticated in the human world, so Romanticism celebrates wild, untamed nature. The contrast between the Romantic view and Locke's sober, utilitarian view of nature as given 'for the support and comfort of men', and in its uncultivated state being mere 'waste' the benefit of which 'amounts to almost nothing' is profound. In Europe the contrast is shown in the difference between the formal gardens and orderly, landscaped parklands of the gentry, and the poetic celebration of the awesome vastness and wildness of the Alps. In Britain, the Lake District and the mountains of Wales and Scotland were the favoured environments of the Romantic poets and painters. Nature is no longer thought of as a mere bundle of potential resources at the service of human labour, made ever-more productive by the application of science to industry. On the contrary, the new poetic exploration of the inner world of the self finds its complement in the awe felt by the solitary individual ('wandering lonely as a cloud') in a landscape of towering mountains and rushing torrents.

Many Romantics came from aristocratic families, and were hostile to the new wealth that came from industry and commerce, seeing it as vulgar and degraded by its commercial values. For many of them, mere economic, monetary value, and the utilitarian measure of nature as a means to human comfort and benefit were demeaning to both humans and nature. Instead, they promoted an aesthetic valuation of nature as the authentic source of beauty, but it was not the beauty of a symmetrically ordered, cultivated and productive landscape. Rather the Romantic painters, poets and composers depicted the beauty of nature as expressed in its awesome, even terrifying, scale and majesty, but also in the perfection of its tiniest creatures, in its destructive power as much as in its richness, diversity and fertility. For some, too, the awe they felt in the contemplation of nature suggested something beyond the sensory experience of beauty. For them, the ultimate experience of nature was something sublime, a religious or spiritual sense of connectedness to a grander unity of the world. The poets Shelley and Wordsworth gave voice to just such a philosophy:

Spirit of Nature! Here!
In this interminable wilderness
Of worlds, at whose immensity
Even soaring fancy staggers,
Here is thy fitting temple.
Yet not the lightest leaf
That quivers to the passing breeze
Is less instinct with thee:
Yet not the meanest worm
That lurks in graves and fattens on the dead
Less shares thy eternal breath.
(Shelley, from *Queen Mab*)

Black drizzling crags that spake by the wayside
As if a voice were in them – the sick sight
And giddy prospect of the raving stream,
The unfettered clouds and region of the heavens,
Tumult and peace, the darkness and the light,
Were all like workings of one mind, the features
Of the same face, blossoms upon one tree,
Characters of the great apocalypse,
The type and symbols of eternity,
Of first, and last, and midst, and without end.
[Wordsworth, from *Prelude*, p. 218 (book vi)]

In some versions, Romanticism, especially in the wake of political disillusionment, took a backward-looking form, celebrating an imagined medieval past of simple rural communities, resurrecting, or inventing their myths and legends. Some historians have seen this development of Romanticism as one source of a later, reactionary and racist appeal to 'blood and soil' that fed into the ideology of European Nazi and Fascist movements. But this identification with a real or imagined rural past could also inspire alternative visions of a future reconciliation of humans with one another and with nature. The Romantic poets' celebration of wild nature came just as, in England, the landowning class turned increasingly to a commercial agriculture that dictated enclosure of the commons, grubbing out of woodlands, draining of the marshes, and in all, driving both wild nature and the poorer classes of humans from their ancient homes. The English 'peasant poet' John Clare began his working life as a rural labourer – a ploughboy, reaper and thresher and jobbing gardener. The intensity of his love of nature is grounded in this practical and sensory dwelling within it, as is the intensity of his hatred of its economically motivated destruction:

Now this sweet vision of my boyish hours
Free as spring clouds and wild as summer flowers
Is faded all – a hope that blossomed free,
And hath been once, no more shall ever be
Inclosure came and trampled on the grave
Of labour's rights and left the poor a slave.

But Clare's sympathy is not solely with the rural labourer. It is shared with the plight of the non-human denizens of the woods and heaths:

Each little tyrant with his little sign
Shows where man claims earth glows no more divine
But paths to freedom and to childhood dear
A board sticks up to notice 'no road here'
And on the tree with ivy overhung
The hated sign by vulgar taste is hung
As tho' the very birds should learn to know
When they go there they must no further go
Thus, with the poor, scared freedom bade goodbye
And much they feel it in the smothered sigh
And birds and trees and flowers without a name
All sighed when lawless law's enclosure came.
(from *The Mores*)

THE 19TH CENTURY: SCIENCE, INDUSTRY AND EVOLUTION

Alongside the changes in the countryside wrought by enclosure and commercial agriculture came expansion of the towns and cities. Formerly trading and administrative centres, the growing urban areas were increasingly centres of manufacturing industry. A new urban industrial class of wage labourers was in process of formation, drawing in much of the rural population displaced by the transformations of the countryside. During the latter part of the 18th century, the methods of the new mechanical science of Galileo and Newton had been extended to new fields of enquiry, with scientific discovery and industrial application often going hand in hand. This period had seen great advances in the design of the steam engine, the development of chemistry, especially the chemistry of combustion, by Priestley, Lavoisier and others, greater understanding of electricity and magnetism, as well as geology and astronomy. These discoveries were, during the late 18th and early 19th centuries in Britain, later in other European countries, to transform manufacturing industry and with that, the lives of both urban and rural populations in what came to be known as the 'industrial revolution'. In terms of the 'metabolism' between human society and nature this revolution inaugurated a new form of civilisation, wholly dependent on the combustion of fossil fuels.

Although Francis Bacon's practical vision of science in the service of a growing human mastery of the forces of nature seemed about to be fulfilled, it is important to keep in mind another aspect of science's place in society. From the 1790s onwards, there was a great increase in the circulation of printed literature – newspapers, journals, books – throughout much of Europe. An increasingly literate public – including significant numbers of the newly forming industrial working class – were eager to share in the new scientific knowledge, and many scientists were equally keen to communicate their discoveries to a wide public readership. Progress in science was steadily promoting new ways of understanding the world, and the place of humans in it, among a lay public.

The Economy of Nature

Perhaps the most pervasive image of nature in the 18th century is captured in the idea of an 'economy of nature'. This idea had numerous variant forms in the work of major natural philosophers and observers, such as the Swedish botanist Linnaeus, the English naturalist John Ray and the theologian William Paley. But a unifying theme is the notion of a hierarchy of beings, each initially created by God, as a distinct and unchanging type, and each with its proper place and purpose in creation. The lower orders in nature and society exist to serve the higher, with humans at the apex of the natural series, midway between the earthly chain of beings and the supernatural hierarchy of angels,

all of them expressions of the purposes of the Supreme being, God. Within this 'economy of nature', human dominion over the natural order was assured, and, as we saw in Locke's version, this entailed a right – perhaps even an obligation – to bend inferior nature to human purposes through cultivation, 'improvement' and the application of rational enquiry.

But the limitations of this essentially static view of nature were gradually being exposed by new discoveries. Of particular significance were the new discoveries and ideas of the geologists. The exposure of underlying rock formations by mining, quarrying and building work gave a special impetus to geology, giving evidence of past transformations of land forms, and exposing a vast array of fossil remains. Of course, all this could be interpreted in different ways, but the view gained ground that the earth was of great antiquity, that it had undergone great upheavals in the course of its long history, and that past epochs had favoured life forms quite different from those currently seen on earth. In Germany, where Romanticism had strongly influenced science and philosophy as well as the arts, and in France, as well as Britain, these new insights into the historical character of the Earth and its inhabitants led to evolutionary speculations: to the thought that the living forms of today, including ourselves, had descended, through some form of transformative influence, from those of the deep past, as revealed in the fossil record.

Darwin: History, Transformation and Diversity in Nature

To orthodox religion, of course, such speculations were anathema: God had created the earth and its inhabitants as we see them today, in a mere six days, and that only some 6000 years ago (many still believe this today, amazingly enough!). The early decades of the 19th century were a period of great upheaval and transformation – enclosure, urbanisation, industrialisation and the spread of conflict between opposed social classes. The new ideas inspired by science were a further source of instability, challenging the settled authority of the Christian churches as sources of knowledge and moral certainty. This is one explanation of the reluctance of Charles Darwin (1809–1822) to release his intellectual time-bomb into the public arena. Darwin had been deeply by impressed by Paley's version of the theological argument for God's creation of living species. The astonishingly complex and seemingly perfect adaptation of living forms to their conditions of life could surely only be explained by the hypothesis of creative design.

However, Darwin's early studies and field-work in geology and botany, and his long voyage on the

Beagle (1831–1836) had provided him with strong grounds for scepticism. By the time he embarked on his famous voyage, Darwin had been deeply impressed by his reading of the German writer, Alexander von Humboldt's *Personal Narrative of Travels*. Humboldt, in turn, was strongly influenced by the great German Romantic, Goethe, but he attempted to combine a profoundly emotional response to nature with a systematic, observational science of the interconnections and mutual dependencies between plants, animals and climate in different regions of the world. But perhaps even greater than Humboldt's influence was that exerted on the thought of the young Darwin by the first two volumes of Charles Lyell's path-breaking *Principles of Geology*. These two works, combined with Darwin's own emotional and intellectual responses to his experiences on his Beagle adventures, impressed upon him several aspects of the natural world that no longer seemed compatible with the harmonious order of a specially created 'economy of nature'.

Both Humboldt's narrative and his own observations on the Galapagos Islands illustrated the geographical diversity in associations of animals and plants: there was no *single* economy of nature, but, rather, a multitude of locally accidental economies, in which taxonomically quite different species might be adapted to play similar roles in different localities. But Lyell's geology, and Darwin's own fossil-hunting had revealed the historical character of nature: that vast transformations had occurred in the long history of the Earth's structure and climate, and that along with these changes, successions of different life forms – many of them long-since extinct – had flourished. In yet another way, Lyell's geology and Darwin's own experience of both nature and human society on his travels called into question the harmonious economy of nature: everywhere he saw conflict, violence and bloodshed, seeming to confirm Tennyson's vision of 'nature red in tooth and claw'. But for Darwin, both aspects – association, adaptation and mutual dependence, as well as predation, conflict and war – were held in tense combination with one another.

Darwin (unlike Lyell) had come hesitantly, and unwillingly to the conclusion that current life forms must be descendants from those of the past, and far from having been created as they are now, they must have undergone great transformations during those past epochs. Lyell's 'uniformitarian' approach to geology (that an accumulation of gradual, small-scale changes, brought about by still familiar natural forces, could, over vast time-scales, eventually yield massive transformations of land forms) prepared Darwin for the thought that the gradual accumulation of small-scale changes in *living* forms over similar time-scales

might also lead to great transformations in them. However, the mechanisms responsible for such modifications remained unknown. This was one reason why earlier attempts at evolutionary theory had failed to gain scientific assent.

Very soon after his return from the Beagle voyage, Darwin began his notebooks on the transformation of species, jotting down snippets of information, thoughts, observations and speculations that might have a bearing on this problem of mechanism of organic change. The notion of nature's economy continues to play a significant role in his thinking, but in Darwin's own version, as an immensely complex web of relationships (interdependent as well as conflictual) among local associations of individuals and species of animals and plants and their various physical conditions of life. But, for him, what has to be explained is not just the presumed fact of modification from generation to generation, but also the tendency of that modification in the direction of ever-closer *adaptation* of the members of a species to the demands of their organic and inorganic conditions of life. Only an explanation of this would serve to address the problem of the appearance of 'design' in nature.

Two further aspects of his thinking at this point are also very evident in the notebooks. One is his interest in the modification brought about in domesticated animals and plants by selective breeding and hybridisation. The comparison of this with the supposedly much more powerful and long-lasting effects of nature's 'selection' of living forms is already present in the notebooks. So, also, is a growing awareness of similarities between humans and other animals, not just in their physical forms, but also in their mental life. Mental life itself is increasingly recognised as a function of the brain and nervous system: 'Oh! You materialist!' he admonishes himself. His encounter with 'Jenny', an orang-utan, at London Zoo resulted in this notebook comment:

> Let man visit Ourang-outang in domestication, hear expressive whine, see its intelligence when spoken; … & then let him dare to boast of his proud pre-eminence.
> [Darwin, 1987 (1838), p. 64, C79]

This is a key moment in the formation of Darwin's evolutionism: theological notions of the special status of humans, their 'pre-eminence' in the order of nature, their unique possession of reason and their consequent entitlement to subject the rest of nature to their mastery are shattered at once. As he put it '… whole fabric totters and falls' (*ibid.*, p. 273, C76). Once the kinship of humans with other animals is entertained, a great theoretical obstacle to evolutionary thought is removed, and, at the same time, humans are re-introduced as belonging to nature, as bound together with the rest of life, suffering its vicissitudes, and sharing with it a common history and ancestry. Further still, in one brief, startling moment in the note-books, Darwin draws the most radical of ethical conclusions from this thought: just as racial distinctions are used to justify slavery, so, perhaps, the distinction between humans and animals is no more than a device to justify our enslavement of them:

> Animals – whom we have made our slaves we do not like to consider our equals. Do not slave-holders wish to make the black man other kind? … the soul by consent of all is superadded, animals not got it, not look forward if we choose to let conjecture run wild then animals our fellow brethren in pain, disease, death, & suffering & famine; our slaves in the most laborious work, our companion in our amusements. They may partake, from our origin in one common ancestor we may be all netted together [*ibid.* (1837): pp. 228–9, B231–2].

However, this move, in which Darwin recovers something of the radical Romantic philosophy of nature, is not fully sustained in his later writings. His encounter with the 'savage' people of Tierra del Fuego, during the voyage of the Beagle, convinced him of the closeness of humans to animals in the opposite direction: the old hierarchy of humans over animals could be displaced by one of the civilised over the savage, the latter remaining as evidence of our kinship with our animal ancestors. As we shall see, these two radically different ways of placing humans back into nature had profoundly opposed political implications for subsequent evolutionists.

Darwin, Malthus and Natural Selection

But still the question of the mechanism of organic change-as-adaptation had not been fully answered. This problem was in his mind when, 'for amusement', in late 1838, he read Thomas Malthus' *Essay on the Principle of Population* (by now in its sixth edition). Malthus (1766–1834) was a parson and political economist who wrote the first version of his essay as a refutation of the radical ideas of William Godwin (father-in-law of the poet Shelley). Since his ideas have been an important influence on some environmentalists in our own time, as well as on Darwin, it will be worth giving a brief account of them. In its original version, Malthus' argument was very simple. In humans, as for animals, there is a tendency for the population to grow, as each pair can bring more offspring into the world than would be required simply to replace them. But this growth tendency has a particular mathematical character: a constant rate of growth produces an ever-escalating growth in actual numbers (like compound interest on savings).

Malthus calls this 'geometric' increase (in our own time, the term 'exponential' is used to make the same point). At the same time, he argues that the growth of the food supply is limited by the extent of land that can be brought into cultivation, and the growth of agricultural productivity. At best, he claims, this can expand only 'arithmetically' – i.e. by constant increments year by year. So, the growth of population must necessarily tend to outstrip the availability of food to feed it. Disease and starvation are an unavoidable predicament of humans, as of animals, and well-meaning attempts to improve the lot of the poor will only encourage them to breed more, and so make the problem still worse. In later editions of his *Essay*, Malthus 'softened' his argument, suggesting that 'moral restraint' might reduce family size, whilst the threat of starvation would motivate the poor to industry and self-help, so reducing the role of starvation and disease in limiting the human population. However, the 1834 poor-law reform, modelled on Malthus' ideas, established the dreaded workhouse system, and was certainly not experienced as a 'softening'. It provoked riots and generally contributed to the heightened social and political conflict of the period.

However, Darwin's reading was informed by quite different questions, and he drew quite different conclusions from Malthus' 'law'. Though Malthus had advanced his law as applying quite generally to both humans and animals, Darwin immediately realised that the law must apply with far greater force to non-human animals and plants. They are unable to 'soften' the effect of the law by sexual restraint, producing fewer offspring, and neither are they able to apply agricultural improvements to enhance their food supply. It seems likely that it was Malthus' mathematical representation of the sheer scale of selective pressures on populations that enabled Darwin to envisage them as a power sufficient to produce the immense diversity of living beings. Being prepared by his familiarity with the small-scale individual variations within each species, together with the metaphor of 'selection', Darwin drew the conclusion that if some individuals possessed a (heritable) feature that gave them an advantage in the competition for survival, then they would be more likely to survive and pass that character on to the next generation. The long-term result would be a gradual modification of subsequent generations away from the original stock in the direction of improved adaptation to the challenges posed by their organic and inorganic conditions of life. So, in contrast to Malthus' concern with the purely *quantitative* outcome of his 'law', Darwin drew a *qualitative* conclusion: differential 'selection' exerted by forces of nature on surplus offspring will eventually lead to qualitative change in the population, and the emergence of new organic forms.

But, as we have seen, Darwin's reading and observations had brought him to a different and more complex view of the economy of nature than either that of the 18th century theologian–naturalists or that of Malthus. For Darwin, the economy of nature is an immensely complex web of relationships, varying from place to place, but also being transformed both in its constituents and in its overall shape by the power of natural selection. In particular, this means that Malthus' simplified model of population growth in relation to food supply is already surpassed by Darwin's more complex grasp of the forces that bear in on an organism throughout its life, and determine its chances of surviving to reproduce itself:

> (T)ake Europe on an average, every species must have same number killed, year with year, by hawks. by. cold & c. – even one species of hawk decreasing in number must effect instantaneously all the rest. One may say there is a force like a hundred thousand wedges trying force every kind of adapted structure into the gaps in the economy of Nature, or rather forming gaps by thrusting out weaker ones. The final cause of all this wedging, must be to sort out proper structure & adapt it to change [*ibid.* (1838): pp. 375–6, D135e].

The forces that continually shape the conditions under which each organism struggles to survive and reproduce itself are so complex that we remain ignorant of them even in the best-understood cases, but they are generally so finely balanced that 'the face of nature remains for long periods of time uniform' [Darwin, 1882 (1859) p. 57].

Darwin is now universally recognised for his discovery of the mechanism, 'natural selection', by which organic change and adaptation occurs. With this concept it now becomes possible to make sense of a very wide range of phenomena – the fossil record, the structural similarities of whole groups of organisms, similarities in the embryonic stages of different species, the patterns of geographical distribution of organic forms, the presence of seemingly functionless 'vestigeal' organs, and many others. Above all, we have a way of thinking about the history and geography of life as the source of its own proliferating diversity. The economy of nature as Darwin conceives of it is immensely complex, locally specific, dynamic, largely unknown to us, and not subject to divine or human purposes. By implication, humans are simply part of this evolutionary scene,

doubly 'netted together' with other species: sharing with them descent from a common ancestor in some distant past, but also forming part of complex ecological webs of competitive interdependence.

In Darwin's breakthrough there is opened up the possibility of a view of nature that is both secular and genuinely non-anthropocentric. Organisms literally or metaphorically 'struggle' for their own ends in whatever conditions of life they are thrust into, and the age-long, unwilled, unpredictable outcome of their myriad activities and accidental modifications is the teeming diversity of life that now covers the surface of the planet. Humans are just one (possibly transitory) outcome of all this, among all the rest, with no special place, or superior standing.

Alfred Russel Wallace: In Darwin's Shadow

But if this seems to be an implication of Darwin's hypothesis, and one that Darwin himself occasionally recognised, it proved hard to sustain as Darwin's ideas made their way into the wider cultural world of Victorian England. Darwin had seen, early on, the radically subversive implications of the conclusion to which he had been drawn. For some historians, his anxiety on this score is sufficient explanation of the recurrent illnesses he suffered, and it certainly seems likely that it explained his great reluctance to share his revolutionary ideas with any but his closest circle of scientific acquaintances (Desmond and Moore, 1992). What eventually – some 20 years after his great discovery – prompted him to go public was receipt of a letter from a relatively unknown fellow naturalist. The letter came from the Malay archipelago, and its author was Alfred Russel Wallace (1823–1913). Wallace had independently hit on the mechanism of organic change, and Darwin was persuaded to allow his and Wallace's papers to be read at a meeting of the Linnaean Society in 1858. A much trimmed statement of his views in book form was hurried to publication in 1859 as *The Origin of Species.*

Wallace's early life could hardly have been more different from that of Darwin. He left home and school at 13 to join his brother, an apprentice in the London building industry. He mixed with manual workers, and attended the 'Hall of Science' off Tottenham Court Road. In these circles he gained an education in Owenite socialism and the radical, sceptical writings of Thomas Paine and others. He subsequently joined another brother as a land surveyor, from which he gained a strong interest in the geology and the flora and fauna of several parts of Britain, together with an abiding love of the open countryside. By the early

1840s he was already a committed sceptic in religious matters and had acquired a considerable scientific education. During a short spell as a teacher in Leicester he made active use of the town library, and met up with another young naturalist, Henry Walter Bates. From their correspondence it is clear that by the late 1840s Wallace was acquainted with current evolutionary writing, and was already convinced by some version of organic evolution. In 1848 he and Bates set out on a collecting expedition to the Amazon, from which Wallace returned in 1852. His collections and reports earned him a reputation among the leading naturalists of the day, and he was soon able to set out on a further adventure: this time to the Malay archipelago. He set off for Singapore early in 1854, travelling from there to Sarawak and Borneo, where he stayed for 14 months.

During the Asian expedition, Wallace continued to correspond with other naturalists, including Darwin, and composed two major scientific papers. The first of these noted that each new species appeared closely in time and space to allied species – strongly suggesting, without actually stating, that new species emerged by transformation from earlier, closely similar ones. It was his next paper, 'On the Tendency of Varieties to Depart Indefinitely from the Original Type', sent to Darwin in 1858, that reported Wallace's independent discovery of the mechanism of organic change. Interestingly, despite great differences in Darwin's and Wallace's social background and economic circumstances, there were some common preconditions of their discovery: both were 'uniformitarian' in their view of change in nature, and both had been impressed by the patterns of geographical distribution of animals and plants, witnessed on their travels.

Two other encounters immediately preceded the breakthrough in both cases: with the orangutan and with Thomas Malthus. On Wallace's own account, it was a reading of Malthus' *Essay* that led to his own breakthrough, just as it had for Darwin. However, Wallace's encounter with the orang was rather more intimate than Darwin's. There is some evidence that Wallace's choice of the Malay archipelago for his second great adventure was influenced by the thought that he might study the orang-utan at close quarters. He appears to have already been convinced of our close kinship with the great apes. However, in his accounts of his encounters with the orang there is a deep contradiction that sheds some light on his later differences with Darwin over the great question that dominated the evolutionary debates of the latter half of the 19th century: 'man's place in nature'.

Wallace's published account of the orang in Borneo focuses on his unremitting attempts to track down and kill as many orangs as he and his native helpers can locate. Each successful kill is accompanied by detailed measurements and there are general comments on the creature's great physical strength, nest building, geographical distribution, arboreal skill and fierceness when attacked. There is little to indicate a special evolutionary interest in this species, and no indication of remorse or ethical scruples about his daily slaughter. However, one incident betrays a wholly different relationship to the 'man-like' ape. He took pity on the tiny offspring of an adult female he had shot, and attempted to rear it. A lengthy account of the experience is contained in a letter sent home during his travels. In it he compares the infant to a human baby, notes its entertaining ways and its emotional expressions. He is unusually frank about his great affection for his 'dear little duck of a darling of a little brown hairy baby' and his sadness when it dies.

How can we explain this apparent contradiction in Wallace's view of the orang? One possibility is the moral and political outlook Wallace took with him on his travels. Already deeply critical of the inequality, exploitation and what he saw as moral degeneracy in his own society, his response to the indigenous peoples of the Amazon was quite different from Darwin's feelings about the 'savages' of Tierra del Fuego. In his autobiography he lists his encounter with the indigenous Amazonians as one of the three great experiences of his adventures:

> …the third and most unexpected sensation of surprise and delight was my first meeting and living with man in a state of nature – with absolute uncontaminated savages!… they walked with the free step of the independent forest-dweller …. In every detail they were original and self-sustaining as are the wild animals of the forests, absolutely independent of civilization, and who could and did live their lives in their own way, as they had done for countless generations before America was discovered … The true denizen of the Amazonian forests, like the forest itself, is unique and not to be forgotten (Wallace, 1908, p. 151).

In this combination of admiration for humans in their natural state with disgust at the inequity and corruption of civilised society, Wallace reminds us of Rousseau's elevation of the 'noble savage'. However, in Wallace's case the state of nature is already a social state. In fact, in its moral aspect it approaches the perfection of the future socialist society for which Wallace hoped:

> Now it is very remarkable that among people in a very low stage of civilization we find some approach to such a perfect social state … There are

none of those wide distinctions, of education and ignorance, wealth and poverty, master and servant, which are the product of our civilization … [Wallace, 1962 (1869), p. 456].

Like Darwin, Wallace was committed to the uniformitarian view of change as taking place slowly and gradually, by many small steps. But in the case of human evolution, this presents some problems. Wallace was, as a socialist and humanist, committed to a universalistic morality of human equality. Darwin, when he later confronted the question of human origins, could compare the gulf between the mental and moral state of the lowest 'savages' and that of the civilised races with that between the lowest invertebrates and higher primates as a way of making believable the gradual transition from one level to the next [Darwin, 1874 (1871)]. This move was not available to Wallace, given both his political values and his direct experience of the moral order of indigenous society: 'The more I see of uncivilized people, the better I think of human nature on the whole, and the essential differences between civilized and savage man seem to disappear' (Wallace, 1908, p. 178).

Eventually Wallace's uniformitarianism lost out to his strong progressive humanitarian values, in favour of a dualistic interpretation of human nature and origins. From the early 1860s onwards, he remained convinced that humans had descended from primate ancestors, but that, once a certain stage of social and intellectual development had been reached, these distinctively human capacities became the primary object of selective pressures. The result would have been a very rapid elevation of the human species above the rest of nature as 'a new and distinct order of being'. Increasingly, Wallace emphasised distinctively human traits – a sense of humour, love of music, religious and metaphysical concerns, even the capacity for advanced mathematics – that could not be explained in terms of natural selection. This took him in the direction of a spiritualist belief in a supernatural force directing evolutionary change in a progressive direction.

Darwin also acknowledged human distinctiveness, even, implicitly, recognising that natural selection was insufficient to account for it. But in his *Descent of Man* Darwin retained both his consistent scientific materialism and his uniformitarianism. He was able to maintain continuity between humans and other animals by a 'pincer movement': emphasising the complex psychological attributes of the 'higher' animals, at the same time as reminding his reader of the low mental and moral state of the Fuegian 'savages' that he had encountered in his Beagle adventure. As to the explanation of change, Darwin included 'use-inheritance', sexual selection, the acquisition of a 'social instinct' and other mechanisms alongside natural selection to account for the origins of human distinctiveness.

USES AND ABUSES OF THE EVOLUTIONARY IDEA

Despite their differences in the explanation of human origins and distinctive attributes, both Darwin and Wallace converged on a version of evolutionary history that confirmed the dominant Victorian ideology of 'progress'. For both of them, the initial conception of evolutionary change as a process of radiating adaptation of organisms to their local environmental conditions of existence had been silently submerged in favour of evolution as a grand narrative of progressive development, with humans as its ultimate outcome and 'highest' expression. But the *content* of the idea of 'progress' continued to divide and polarise the many political uses of evolutionary thought. For Wallace and some other socialist evolutionists, progress would consist in a moral development of humanity towards new forms of human solidarity and compassion. Darwin himself continued to assert the depth of the gulf between 'savage' and 'civilised' humanity, but he, like Wallace, nevertheless also held to a vision of human moral progress as involving sympathetic concern for the suffering of others. Savages were to be educated and delivered of the benefits of Christian civilisation, but not enslaved or brutalised.

However, this milder cultural imperialism did not prevent others using Darwin's exposure of nature as an arena of unremitting war, conquest and extinction to justify the greatest excesses of European imperial domination and extermination of other peoples. For example, Ernst Haeckel (1834–1899), the leading German evolutionist of the latter part of the 19th century, and an acknowledged pioneer of 'ecology' as a distinct discipline, found in evolutionary ideas an ethical justification for genocide. He distinguished ten 'species' of men, of which:

The first, primitive man, is dead this long time past. Of the nine others, the next four will pass in a shorter or longer time… Even now these four races are diminishing day by day. They are fading away ever more swiftly before the o'er-mastering white invaders. Melancholy as is the battle of the different races of man, much as we may sorrow at the fact that might rides at all points over right, a lofty consolation is still ours in the thought that, on the whole, it is the more perfect, the nobler man that triumphs over his fellows… [Haeckel, 1883 (1865), p. 85].

This 'social Darwinist' extension of the Darwinian struggle for existence in nature to the relationships between human races, and to the competitive struggle between individuals, provided justifications for a ruthless 'free market' capitalism, in which reforms aimed at improving the condition of the poor were seen as running counter to nature. If the weak, poor and inferior examples of humanity were weeded out by the intensity of competition with their fellows, then, sad as this might seem, the outcome could only be an improvement in the quality of the survivors and their progeny.

But Darwin's great propagandist, T. H. Huxley (1825–1895), drew quite contrary conclusions. Having demonstrated to a broad lay readership the scientific case for inclusion of humans as one species of primate among others, descended from animal ancestors by the same natural mechanism as governed the emergence of all new species, he went on to insist on the 'vastness of the gulf between civilized man and the brutes' (Huxley, 1895a, 153). For him, 'intelligent speech', a uniquely human attribute, makes possible the accumulation and organisation of experience from generation to generation. The emergence of the bonds of sympathy and cooperation in human society give that social life a great advantage in the struggle for existence with the rest of nature. No longer subject to environmental conditions of life, humans become capable of altering those conditions to favour animals and plants that satisfy human needs and wants. In other words, the great law of the struggle for existence ceases to operate *within* human society, but is redirected into a struggle between humans and the rest of nature. Huxley expresses this in terms of a struggle on the part of the 'state of art' against the 'state of nature', and likens the process to the creation, then extension, of a walled garden.

This is certainly what would now be seen as an 'anthropocentric' view of the relationship between humans and nature, and one in which the forces of nature are seen as antagonistic to human practical and moral progress. Nevertheless, the purposes for which the forces of nature are to be resisted are not solely utilitarian or exploitative. The horticultural work of protecting and nurturing aims 'to bring about the survival of those forms which most nearly approach the standard of the useful, or the beautiful, which he has in mind' (Huxley, 1895b, p. 14).

Wallace, too, denied the relevance of the 'struggle for existence' to the relations among humans. To him, the social Darwinists were wrong to see ruthless competition and elimination of the weak as mere expressions of human nature. On the contrary, their opposition to progressive social reform was itself a denial of a central aspect of evolved human nature: the development of strong social bonds of mutual sympathy and compassion. Like many social reformers of his time, Wallace took from evolutionism a strong sense of the importance of environmental conditions in shaping human character and development. Since, for him, there was no *essential* difference between 'savage' and 'civilised' man, their different physical and social environments must have been at work in creating their different characters. Applied to his own contemporary society, this meant systematic exposition of the degrading and debilitating conditions under which the great majority of industrial workers, men, women and children, lived, worked and died (e.g. Wallace, 1913). For Wallace, this illustrated the great imbalance between the strides that had been made in scientific and technical mastery of the forces of nature, on the one hand, and the moral degradation that had accompanied it, on the other.

In this respect, Wallace had much in common with other progressive and socialist environmentalists, demanding greater protection from hazards at work, measures of public health, restrictions on working hours, enhanced educational opportunity and so on. However, there is little evidence in his writing of resistance to the project of mastery of nature itself. His arguments for the nationalisation of land, for example, concentrate on the ways cooperative enterprise would allow for a more equitable sharing of the benefits, and reduction of the human costs, of material progress. However, something more than this is indicated by his proposals for the future management of Epping Forest, on the outskirts of London. Following a determined popular struggle against illegal enclosures, the forest was, in 1877, subject of an act of parliament guaranteeing the 'preservation of its natural aspect' for public enjoyment and recreation. Wallace's (unsuccessful) proposal was to plant up areas that had already been denuded of trees with examples taken from the other northern temperate forests. But, alongside this piece of self-indulgence on the part of the founder of the discipline of plant geography, were insightful proposals for the management of the remaining unspoilt parts of the forest. These proposals testify to a deep understanding of the value of contact with nature to human well-being, as well as an advanced ecological approach to conservation management.

After bemoaning the enclosure of land as private property, and the injustice that excludes the people from enjoyment of the 'beautiful scenery of their native land', Wallace goes on to enthuse about the securing of the forest for the public: 'Here at length every one will have a right to roam unmolested, and to enjoy the beauties which nature so lavishly spreads around when left to her own wild luxuriance' [Wallace, 1900 (1878), p. 75].

When he turns to the matter of how to manage the 'native' forest, he sounds a warning note on the new powers vested in the conservators to drain wet areas of the forest. He insists that boggy areas, swamps and damp hollows are essential to the 'natural aspect' of any forest:

> Every lover of nature finds them interesting and enjoyable. Here the wanderer from the great city may perchance find such lovely flowers as the fringed buck-bean, the delicate bog pimpernell and creeping campanula .... These and many other choice plants would be exterminated if, by too severe drainage, all such wet places were made dry; the marsh birds and rare insects which haunted them would disappear, and thus a chief source of recreation and enjoyment to that numerous and yearly-increasing class who delight in wild flowers and birds, and insects, would be seriously inter-fered with (*ibid.*, p. 93).

Wallace adds to these considerations a comment on the role of wet areas in the forest as 'natural reservoirs' in a region of low rainfall, essential to the preservation of a local climate favourable to the vegetation of the forest as a whole.

Whilst it is true that both Wallace's arguments against excessive drainage of the forest relate to the human interest in the 'natural aspect', they still exemplify a grasp of the complex interdependencies of the forest ecosystem, and go beyond a mere instrumental view of nature. The delight of nature lovers from the great city has unmistakable links to the Romantic exaltation of nature for its own sake.

But what of the radically egalitarian ethic of universal kinship that Darwin, as we saw, expressed in his early notebooks? Although this never again surfaced in quite such clarity, it never quite disappeared, either. Though the Darwin of the *Descent of Man* had conceded to the Victorian ideology of 'progress', and used a hierarchical view of the human races to support the doctrine of human descent from 'lower' animals, there was another side to the same argument. This was the elevation of non-human animals to kinship with humans, and an insistence on their possession of the same range of emotional and psychological attributes – differing from humans in degree only. This leads Darwin to a view of moral progress as a bit-by-bit extension of the scope of sympathy and benevolence toward others: first, to others in one's family, then tribe, nation and to the human race as a whole. But even this is not the culmination of civilised morality:

> Sympathy beyond the confines of man, that is, humanity to the lower animals, seems to be one of the latest moral acquisitions .... This virtue, one of

the noblest with which man is endowed, seems to arise incidentally from our sympathies becoming more tender and more widely diffused, until they are extended to all sentient beings [Darwin, 1874 (1871), p. 123].

But it was left to others to render explicit and directly political this 'subterranean' aspect of Darwinian thought. Radical figures such as William Morris, Edward Carpenter and Henry Salt saw essential connections between reform of society and a transformed relationship between humanity and nature, a vision most clearly expressed in this extract from Henry Salt's autobiography:

> Humanity and science between them have exploded the time-honoured idea of a hard-and-fast line between white man and black man, rich man and poor man, educated man and unedu-cated man, good man and bad man; equally impossible to maintain, in the light of newer knowledge, is the idea that there is any difference in kind, and not in degree only, between human and non-human intelligence. The emancipation of men from cruelty and injustice will bring with it in due course the emancipation of animals also. The two reforms are inseparably connected, and neither can be fully realized alone (Salt, 1921).

REFERENCES AND BIBLIOGRAPHY

Darwin, C. (1874) [1871] *The Descent of Man and Selection in Relation to Sex* (2nd edn.) London: John Murray.

Darwin, C. (1882) [1859] *The Origin of Species* (6th edn.) London: John Murray.

Darwin, C. (1987) [1838] In P. H. Barrett, P. J. Gautrey, S. Herbert, D. Kohn and S. Smith (eds) *Charles Darwin's Notebooks 1836–1846*. British Museum (Natural History)/Cambridge University.

Desmond, A. and Moore, J. (1992) *Darwin*. London: Penguin.

Haeckel, E. (1883) [1865] (tr. E. Aveling) *The Pedigree of Man*. London: Freethought.

Huxley, T. H. (1895a) *Man's Place in Nature and other Anthropological Essays*. London: Macmillan.

Huxley, T. H. (1895b) *Evolution and Ethics and other Essays*. London: Macmillan.

Locke, J. (1971) [1690] An essay concerning the true original, extent and end of civil government. In E. Barker (ed.) *Social Contract: Essays by Locke, Hume and Rousseau*. Oxford: Oxford University Press.

Malthus, T. R. (1798) *An Essay on the Priniciple of Population*. London: J. Johnson.

Merchant, C. (1982) *The Death of Nature*. London: Wildwood House.

Rousseau, J. -J. (1974) [1754] Discourse on the origin and basis of inequality among men. In M. Josephson (ed.) *The Essential Rousseau*. New York: New American Library.

Salt, H. (1921) *70 Years Among Savages*. London: unpublished autobiography.

Thomas, K. (1983) *Man and the Natural World*. Harmondsworth: Penguin.

Wallace, A. R. (1900) [1878] Epping Forest and temperate forest regions. In A. R. Wallace (ed.) *Studies Scientific and Social*, vol. II. London: Macmillan.

Wallace, A. R. (1908) *My Life*. London: Chapman & Hall.

Wallace, A. R. (1913) *Social Environment and Moral Progress*. London: Cassell.

Wallace, A. R. (1962) [1869] *The Malay Archipelago*. New York: Dover.

Worster, D. (1994) *Nature's Economy: A History of Ecological Ideas*. Cambridge: Cambridge University Press.

3

Anarchism, Libertarianism and Environmentalism: Anti-Authoritarian Thought and the Search for Self-Organizing Societies

Damian Finbar White and Gideon Kossoff

INTRODUCTION

Few intellectual currents have played as influential a role in the development and shaping of modern environmentalism as the anarchist and libertarian tradition of social and political thought. Generalizations about common ideological roots to a politics as diverse and internally divided as environmentalism are of course hazardous. Yet, when we consider some of the currents that run through much of the radical green worldview: philosophical naturalism, advocacy of economic, political and technological decentralization or the desire to ground a sustainable society in participatory institutions, the spirit of the classic anarchists clearly looms over much of this conversation. Indeed, it could be noted that at one time or another in the last two centuries many of the organizing ideas of the more radical currents of contemporary ecological politics have been initiated and developed by people who would have called themselves 'anarchists' or 'libertarians'.

In this chapter we seek to trace the diverse connections that can be found between anarchism, the broader libertarian tradition, environmentalism and scientific ecology. We begin by establishing the historical context of anti-authoritarian thought. Since the Enlightenment, anarchists and libertarians from Godwin to Proudhon have advanced the idea that social order is generated through the voluntary association of human beings. As such, this tradition stands in sharp contrast to the mainstream of social and political theory which has maintained that social order is generated by the external imposition of authority. Indeed, anarchists have maintained that it is the very coercive ideologies, practices and institutions of modernity that are the source of the disorder and social chaos they are designed to prevent. We elaborate on this worldview in the first section of this chapter and argue that the resistance that many contemporary forms of ecological politics demonstrate for conventional leadership structures, and the advanced division of labour has a long pedigree.

In the second part of this chapter, we focus more specifically on the impact that social anarchist, left libertarian and more recent ecological anarchist currents have had on the development of thinking about society–nature relations. The dominant figures here are Peter Kropotkin and Murray Bookchin. In these thinkers we can find a range of important contributions being made to eco-philosophy and environmental ethics, from attempts to cultivate a metaphysics of nature and develop naturalistic ethics, to reflections on scientific ecology and evolutionary biology.

In the third part of this chapter, we go on to consider the broader impact that anarchist and libertarian thinkers have had on debates about the 'built environment'. Anarchist thought has often been presented by its critics as simply advocating a pastoral vision of the future. Such readings though ignore the extent to which, as Peter Hall has observed '… the anarchist fathers had a magnificent vision of the possibilities of urban civilisation …' (Hall, 2002). In the work of Geddes and Howard, Bookchin and Ward, a stream of thought can be recovered which moves from advocacy of garden cities and city gardens to championing the virtues of allotments, participatory planning, ecological technology and urban direct democracy. This 'urban ecological' strand of anarchist and libertarian thought is doubly important. Not only does it suggest that the 'lived practice' of much green social movement activity could be viewed as 'anarchism in action' but also the claim of anarchist urbanists, that the optimal human environment would be the self-organized human-scaled city that is carefully integrated into the region and the broader natural environment, potentially make important contributions to contemporary discussions of the importance of 'sustainable cities'.

Finally, we conclude by considering critical evaluations of anarchist and libertarian social and political theory and the future relationship between anarchism, libertarian thought, environmentalism and ecology. Serious intellectual challenges have been posed to the coherence of this tradition as social and political theory and as providing a basis for green social and political theory. However, we argue that libertarian and anarchist themes continue to work their way into the environmental debate from a surprising range of areas.

APPROACHES TO ANARCHISM

The word anarchism is derived from two ancient Greek words *an* and *arkhê*. It literally means the absence of authority (Guérin, 1970, p. 11) or the condition of being without a ruler (Marshall, 1992, p. x). As Peter Marshall notes, from the beginning the term has been associated with both 'the negative sense of unruliness which leads to disorder and chaos, and the positive sense of a free society in which rule is no longer necessary' (Marshall, 1992a, p. 3). Refining the central commitments of the anarchist and libertarian

tradition further, though, is not an easy task. Guérin, for example, has claimed 'anarchists reject society as a whole' (1970, p. 13). Many anarchist rhetorics give the impression that they reject government. At the same time though, many self-identified anarchists and libertarians have championed 'society' and been enthusiastic supporters of radically democratic and communal governing structures.[1]

Further complexities emerge from the manner in which anarchist and libertarian discourses can originate from both radically individualist and radically collectivist social philosophies. Furthermore, the terms 'libertarianism' and 'anarchism' are sometimes used interchangeably in the literature and sometimes given different meanings. Some clarification of definitions is therefore necessary.

In this chapter, we suggest that anarchism and libertarianism are best treated as a common anti-authoritarian tradition. By definition, this family of social and political thought can be identified by its central unifying desire to criticize the view that authority should be the organizing principle of social life. Yet, this critique has invariably been expanded to oppose all institutional and psychological forms of domination, hierarchy and authoritarianism. Much anarchist and libertarian advocacy, then, takes the form of social philosophies that explain how such a constellation of repressive forces and structures came into being. It offers political philosophies which argue for the transcendence of such structures and suggest alternative social, political, economic and technological forms that would maximize the realm of freedom, autonomy and self-management.

There are, however, tensions and differences within the anti-authoritarian tradition. Tensions exist between communitarians and individualists; between those that view social solidarity as a precondition for the free society (social anarchists) and those who argue that primacy should be given to individual sovereignty and private judgement (anarcho-individualists). The tradition is also marked by notable tensions between scientific rationalists and romantics; between those that see capitalism as antithetical to a free society and those who view markets as the most efficient coordinating mechanism for decentralized societies. Additional differences emerge from the fact that anti-authoritarians who self-identify as 'anarchists' tend to hold to the view that the free society must necessarily be stateless. In contrast, self-identified 'libertarians' are more likely to tolerate minimal state forms for the foreseeable future or pragmatically aspiring, like Buber, to 'substitute society for the State to the greatest degree possible' (Buber, 1947, p. 80). Nevertheless, despite these differences it is a shared hostility towards the 'specific form of government which emerged in post-renaissance Europe' (Miller, 1984, p. 5), that is, the modern state, that brings together libertarians and anarchists of assorted persuasions.

ANARCHISM, SOCIAL ORDER AND FREEDOM

The words 'law' and 'order' are often paired as if they were indissolubly united. According to the modern mind there is no 'order' without 'law', where 'law' is understood as a body of rules devised and imposed from without. This law is enforced by the authority of the courts, the police, the army and ultimately, the government. Social order, according to this worldview, emanates from the institutions of the modern state. This view was central to Thomas Hobbes' justification of the modern state and was developed by a range of social-contract theorists in the 17th and 18th century. William Godwin (1756–1836) was the first of a long line of thinkers, who we now look upon as the classic anarchists, including, for example, Michael Bakunin (1814–1876), Pierre-Joseph Proudhon (1809–1865) and Peter Kropotkin (1842–1921) who resisted such arguments and the spread of state centralization more generally.

Synthesizing and developing currents of French 18th century liberal thought with indigenous traditions of English dissenting radicalism (Woodcock, 1986) Godwin is frequently referred to as the father of the libertarian tradition. In his 'Enquiry Concerning Political Justice' (1793), a text which immediately followed the French Revolution, Godwin argued that the power of one man over another is only achieved by conquest or coercion. By nature, he maintained, we are all equal. In the infancy of society men associate for mutual assistance. It is 'the errors and perverseness of the few' that leads to calls for restraint in the form of government (Marshall, 1992a, p. 19). Government, then, is initially intended to suppress injustice. However, turning to consider government in the form of the modern state, Godwin argued that it is increasingly clear that government only perpetuates injustice. It has dangerously concentrated the force of the community and aggregated the power of inequality (Marshall, 1992a, p. 19).

From Godwin onwards many anarchists and libertarians have made a distinction between government and society, the former being seen as an 'artificial social form', the latter a 'natural' form. It is maintained that authentic social order, or social harmony, cannot be imposed from outside a community by an external authority. Government is therefore seen to be an imposition on society, and the order generated by state power is seen to be 'inauthentic'. An authentic social order, on the other hand, is created through the interpersonal relationships of those who live within the ambit of a community.

Social order is, therefore, seen to come about spontaneously through the daily interactions of people living in proximity to one another, in their work, in their families, in their friendships and in the economy and culture. Out of the personalized relationships of everyday life, it is argued, communities develop the ability to manage their own needs and affairs. In other words, human societies have the capacity to become self-governing. The defence and cultivation of social spontaneity, which is the barometer of societal health, reveals the typical anarchist view of human nature – that we are fundamentally social beings.

It needs to be noted here that whilst the classic anarchists identified the 'authority principle' as the source of social disorder, they refer to externally imposed, controlling authority: authority per se is not usually rejected. Typical, for example, is Godwin's distinction between three kinds of authority: 'the authority of reason', the authority given to a person worthy of 'reverence and esteem', and authority which is buttressed by sanction and therefore dependent on force. It is the latter which is 'the species of authority that properly connects itself with the idea of government' (Godwin, 1986, p. 104) and is therefore to be rejected.

Anarchists following Godwin have argued that state power is inseparable from domination. It is maintained that those in power (whether the government is representative or despotic) are the privileged minority of the age (be it, to paraphrase Bakunin, priestly, aristocratic, bourgeois or bureaucratic) who profess to understand the interests of the majority better than the majority can itself.

Much classic anarchist advocacy has sought to contest this claim by turning to history and anthropology to consider the social, cultural and political arrangements that are to be found in pre-modern societies and contemporary people living in small-scale societies.

MUTUAL AID AND THE 'UNNATURALNESS' OF THE STATE

Peter Kropotkin's *Mutual Aid* (1902) constitutes one of the most serious attempts in social and political theory to undermine the notion that the modern state's legitimacy is based on a social contract or that social life can be explained in the competitive and aggressive terms of social Darwinism. Kropotkin asked why people in 'the state of nature' should have consented to be ruled if, in the absence of government, their communities were already cohesive? Directly repudiating Hobbes's *Leviathan*, Kropotkin argues

> It is utterly false to represent primitive man as a disorderly agglomeration of individuals, who only obey their individual passions, and take advantage of their personal force and cunningness against all other representatives of the species. Unbridled individualism is a modern growth, but it is not a characteristic of primitive mankind (Kropotkin, 1987, p. 82).

He goes on to suggest that if we view the historical and anthropological record in all its breadth, we can gain a much more diverse sense of the social institutions that human beings have created across time.

In *Mutual Aid* it is reasoned that if humanity had been mutually antagonistic we would not have been able to create enduring human communities and would, as relatively slow and weak creatures, have passed from the evolutionary scene aeons ago. Reflecting on the historical record as well as drawing from ethnographies of tribal peoples such as the Buryat and Kabyle, Kropotkin argues that mutual aid, not competition, is the norm of social organization and that, therefore, mutually beneficial collaborative and cooperative networks have been a persistent feature of human society.

Whilst by no means seeking to undermine the gains of 'civilization', Kropotkin suggests we have to recognize the 'creative genius' of early humans (Kropotkin, 1987). It is argued early human 'clan societies' shared food and other necessities, held property in common, developed communal care for children, and assisted the weak. This practice of mutual aid was extended, Kropotkin argues, as village communities emerged. Thus, Kropotkin argues we can see the emergence of village cultures that held and worked land in common and collectively cleared marshes, drained forests and built roads, bridges and defences. Such communities also developed systems of customary law backed not by coercion but by the moral authority of the folkmote (an ancient assembly of the people gathered to discuss matters of the common wealth). What these examples demonstrate, Kropotkin reasons, is that the principle of voluntary and direct association between people has historically provided the basis of a robust and creative social fabric.

Kropotkin maintains that mutual aid did not simply pass from the scene with the rise of feudalism. Whilst we can find an authoritarian tradition consolidating itself around the monarchy and the barons, equally from the 12th century, and for several hundred years thereafter Europe saw an important 'counter-trend' in terms of a 'communalist revolution'. From the 12th century onwards one can point to the slow rise of hundreds of cities seeking to emancipate themselves from their feudal lords to become self-governing entities. Kropotkin maintains these 'free cities' not only experimented with a highly decentralized form of neighbourhood organization, but also developed a new form of mutual aid, the guilds, which provided fraternal and egalitarian associations for each of the trades, arts and crafts.

According to Kropotkin the free cities liberated huge intellectual and creative forces across Europe. Because of this, within three or four hundred years Europe was covered with 'beautiful sumptuous buildings, expressing the genius of freed unions of free men' (Kropotkin, 1987). This demonstrates, he says, that 'authority simply hinders men from giving free expression to their inherent social tendencies' (Kropotkin, 1987, p. 8).

How can we explain the triumph of the 'authority principle'? Kropotkin admitted that alongside mutual aid, one can find an instinct of self-assertion which could take the form of a will to dominate and exploit others (Miller, 1984, p. 73). Authoritarian institutions can appeal to such a spirit. This is what happened to the free cities from the end of the 15th century when the rising centralized State took advantage of divisions that had arisen both within the cities and between them and the countryside, and proceeded to rip them apart. It was the state that

> weeded out all institutions in which the mutual aid tendency had formerly found expression .... Folkmotes .... courts and independent administration .... lands were confiscated ... guilds spoliated of their possessions .... Cities divested of their sovereignty ... the elected justices and administration, the sovereign parish and parish guild ... were annihilated; the State's functionary took possession of every link of what formerly was an organic whole ... [absorbing] all its social functions ... [it saw] ... in the communal lands a means for gratifying its supporters .... They [the States] have broken all bonds between men (Kropotkin, 1987, p. 11).

The 19th century anarchists' case against the state, with its edifice of bureaucratic, legal, militaristic, educational and religious structures, was summed up by Pierre-Joseph Proudhon (1809–1865):

> To be governed is to be at every operation, at every transaction, watched over, inspected, spied on, directed, legislated at, regulated, docketed, indoctrinated, preached at, controlled, assessed,

weighed, censored, ordered about, by men who have neither the right nor the knowledge nor the virtue. To be governed is to be, on the pretext of the general interest, taxed, drilled, held to ransom, exploited, monopolised, extorted, squeezed, hoaxed, robbed, then at the least resistance, at the first word of complaint, to be repressed, fined, abused, annoyed, followed, bullied, beaten, disarmed, garrotted, imprisoned, machine gunned, judged, condemned, deported, flayed, sold, betrayed, and finally mocked, ridiculed, insulted, dishonoured. That's government, that's its justice, that's its morality (Proudhon, in Miller, 1984, p. 6).

ALTERNATIVES? FIELDS, FACTORIES AND WORKSHOPS

What, then, is the alternative to the modern state? The classic anarchists generally concurred that it was necessary to defend and expand the instances of mutual aid, voluntary association and self-organization that survived and lingered in capitalist societies. But, in addition to salvaging these practices, it was held that credible political resistance to the 'authoritarian principle' involved developing political projects and movements that would seek to locate power at the administrative unit that is closest to the people. Thus, attention was focused on the commune or the municipality. Anarchists also, became some of the principle advocates of confederalism. As such, it was envisaged that confederated networks of communes or municipalities, free cities and free regions could eventually replace the State. Kropotkin in *Fields, Factories and Workshops*, for example, envisaged the state being substituted by 'an interwoven network, composed of an infinite variety of groups and federations of all sizes and degree, local regional national and international – temporary or more or less permanent ...' (Kropotkin, 1993, p. 7).

Yet, what to do with the economy? From its inception, the classic anarchists stood alongside emerging socialist and Marxist currents in their protests against industrial capitalism.

Arguing that labour is the source of economic value, not capital, currency or land Godwin maintained that 'the faculty of disposing of the produce of another man's industry' (Marshall, 1992a, p. 211) was unacceptable. Similarly, Kropotkin argued in *The Conquest of Bread* that, since the heritage of humanity is a collective one in which it is impossible to measure the individual contribution of any one person, this heritage must be enjoyed collectively (Woodcock, 1986, p. 169). In contrast, however, to their Marxist contemporaries, anarchists maintained that the fundamental problem with capitalist industrialism and agriculture

was not simply the social relations which underpinned these, but their gigantic scale, their centralization and their reliance on an increasingly advanced division of labour. Unlike many Marxists then, who welcomed centralizing institutions as marking a further 'progressive rationalization' of the mode of production (aiming merely, according to Bakunin, at turning society into a 'barrack' where 'regimented working men and women will sleep, wake, work and live to the beat of a drum' (Miller, 1984, p. 11), the classic anarchists viewed many elements of capitalist rationalization as socially and culturally regressive, leading to the expansion of the authority principle of uniformity and homogenization in social life, and to the undermining of autonomy, skill, craft and self-organizing tendencies in the work place.

Kropotkin, in his *Fields, Factories and Workshops*, took this line of thought further than had other anarchists, arguing that industry could and should be decentralized and integrated with agriculture – not simply for the reasons stated above but because of the opportunities it opened for a more balanced and healthy life:

> The scattering of industry over the country – so as to bring the factory amidst the fields, to make agriculture derive all those profits which it always finds in being combined with industry and to produce a combination of agricultural with industrial work – is surely the next step to be taken … This step is imposed by the necessity for each healthy man and woman to spend a part of their lives in manual work in the free air; and it will be rendered the more necessary when the great social movements, which have now become unavoidable, come to disturb the present international trade and compel each nation to revert to her own resources for her own maintenance.

Significant differences emerge, however, as to what follows from this. Kropotkin, for example, advocated an anarcho-communist and market-free vision of the future significantly influenced by the socialist and communist movements of the time (albeit free of the state). In contrast, Proudhon may well have declared at one point 'all property is theft' but, along with Benjamin Tucker, he also argued that limited property ownership was necessary to secure and protect individual liberty. As such, Proudhon advocated a more 'market' friendly system of mutualism. Proudhon's mutualism consisted of a pluralist and confederal mixed economy organized around market exchange between producers, production being carried out by self-employed artisans and farmers, small producers' cooperatives, consumers' cooperatives and worker-controlled enterprises. He argued that all this should be underpinned by a currency controlled by a democratically elected people's bank.

While Proudhon presented his mutualism as striking a balance between extreme collectivism and extreme individualism, it is certainly the case that the role that market's 'private judgement' and individualism should play in any libertarian future has become a key point of division as the anarchist tradition has developed across the 20th century. In the post-war period in particular one can see an increasingly sharp division between self-styled anarcho-capitalists such as Murray Rothbards or 'libertarians' such as Ayn Rand, who, following the disaster of State Socialism, concluded that any form of collectivist economics is incompatible with an anti-authoritarian politics. In contrast, 'left' libertarians have continued to argue that corporate capitalism is the central issue that needs to be addressed.

Over the last two decades, anarcho-capitalists and individualist libertarians have argued that the free market and the minimal state provides the political and economic form which optimizes decentralization, individual autonomy and private judgement. Their argument has unquestionably made major gains in academic and political circles, particularly in the USA. Establishing themselves as the radical wing of the US conservative movement, such figures have frequently directly impacted on US policy discussions and have virtually annexed the term 'libertarian' so that it is almost solely associated with free-market radicalism.

Such manifestations of the individualist traditions of anarchism, however, have had little to say in relation to environmental concerns and little impact on environmental movements. In contrast, it is social anarchists, left libertarians and assorted other utopian fellow travellers that have had the most sympathy with environmental issues in general and radical ecology in particular. We go on to explore their influences in the next section.

SOCIAL ANARCHISM AND SOCIETY–NATURE RELATIONS

The relationship between social anarchism, left libertarianism and society–nature relations is complex. It may well be the case that Henry David Thoreau (1817–1862) retreated to the woods of Massachusetts to be on more intimate terms with the natural world, and Proudhon demonstrated deep sympathies with the peasant farmer and lamented the commodification of the land (Marshall, 1992a, p. 237). On the other hand, the social anarchists of the 19th century were indeed very much products of the Enlightenment. And as such they firmly sought to contest forms of naturalistic reductionism present in such thinkers as Malthus and in social Darwinism. Godwin, Kropotkin and Reclus firmly rejected

Malthus' claim that 'natural limits' reached through overpopulation would provide a 'natural check' on any progressive project. Godwin saw distinct gains in the technologies of the industrial revolution, and saw no virtue in unpleasant toil (Marshall, 1992a, p. 215). Kropotkin, advocating 'the Conquest of Nature', believed that the stock of energy in nature was 'potentially infinite' and that as population became more dense, means of food cultivation would improve, which would circumvent population pressures (see Marshall, 1992a, p. 331). Reclus similarly believed that advanced technology would help increase production and improve life for all (Marshall, 1992a, p. 342). It would therefore be quite incorrect to read the classic social anarchists as proto-radical ecologists. Indeed, many of the classic social anarchists would have looked in horror at the technophobic pronouncements of contemporary neo-Malthusians and 'primitivists' (e.g. Zerzan, 1994).

Nevertheless, the enormous debt that much late 20th century environmentalism and ecologism owes to the social philosophy of 19th century social anarchism has been much commented on in the literature on green politics (O'Riordan, 1981; Dobson, 1990; Eckersley, 1992; Pepper, 1993; deGeus, 1999) and the lines of connection should already be apparent. For example, Kropotkin's utopian visions, his sympathetic reading of the histories of pre-modern and small-scale societies and his advocacy of decentralization have clearly influenced much green utopian thinking in the 20th century. The social anarchist critique of the state and authoritarianism clearly informs the preference for loosely knit network forms of organization and loose social movements structures that can be found amongst many radical environmental and ecological groups and attempts by various Green Parties in the 1980s to experiment with various 'anti-party, party' structures. The differences between Kropotkin and Proudhon on economic organization mirror ongoing debates in environmentalism and radical ecology concerning the role that markets should play in the development of decentralized sustainable societies.

We can identify three further and overlapping areas where social anarchists such as Kropotkin and more contemporary thinkers such as Murray Bookchin have contributed to debates about society-nature relations. First, from Kropotkin onwards, social anarchists have been drawn to ethical naturalism and social organicist modes of thinking, that is, the desire to look to 'nature in the large' to provide guidelines for the organization of a liberated society. Second, social anarchists have made direct interventions in scientific debates in biology, evolutionary theory and latterly ecology.

Third, more recent social anarchists and libertarians have broken with Kropotkin's commitment to the 'conquest of nature', raising concerns about the Enlightenment tradition's commitment to the idea of the 'domination of nature'.

ETHICAL NATURALISM AND THE CLASSIC SOCIAL ANARCHISTS

The classic social anarchist tradition, as Peter Marshall (1992a) has observed, is deeply infused by a kind of 'cosmic optimism' a sense that anarchism is somehow an expression of the natural way of things. A deep-seated ethical naturalism which views order, reason, creativity and, ultimately, meaning as woven in some way into the fabric of the natural world, pervades much 19th century anarchism. Godwin, for example, argued that the rational, deterministic order of the universe could potentially be translated into the rational and benign order of society. The implicit message was that social systems based on power, authority and control somehow go against the grain of both human and non-human nature.

Many social anarchist and libertarian currents after Godwin have been attracted to organismic as opposed to mechanistic metaphors. Charles Fourier (1772–1837), for example, took this metaphor to its extreme in his suggestion that the universe was not a Newtonian machine but a vast living organism that pulsated with life: everything it contained was governed by the principle of 'passionate attraction'. Fourier believed that given appropriate social forms, this force of passionate attraction could be released in social life. Once again, though, of the 19th century thinkers it is Kropotkin who emerges as the key figure in developing the naturalistic side of anarchist advocacy.

One of the more interesting and controversial aspects of Kropotkin's thinking is his critique of the notion that somehow the principle of authority lurks in nature. As a practising naturalist, geographer and indeed evolutionary theorist, field work he conducted in Siberia as a young man led him to believe that mutual aid was at work in the natural world as well as in society at large. His position was in stark contrast to many contemporary 19th century social theorists such as T.H. Huxley and Herbert Spencer, who appropriated Darwin's theory of evolution to buttress the ideology of *laissez-faire* economics.

Kropotkin recounts how, whilst he was doing field research in Siberia, 'under the fresh impression of *Origin* … [he and his colleague] vainly looked for the keen competition between animals of the same species which the reading of Darwin's works had led us to expect' (Kropotkin, 1989). He did not deny in *Mutual Aid* that there was struggle in nature, and particularly struggles between species, but proposed that the 'fittest' in this struggle were those given to cooperation within their own species. 'Who are the fittest: those who are continually at war with each other, or those who support one another? We at once see that those animals which acquire habits of mutual aid are undoubtedly the fittest' (Kropotkin, 1987, p. 24) Competition in a harsh natural world is seen as a waste of energy and resources: cooperation, on the other hand, enables animals to secure food, to protect themselves from predators, and to rear their offspring. Thus mutual aid, not mutual antagonism, association, not competition, became the most significant agent of natural evolution, or, as Kropotkin says 'the most efficacious weapon in the struggle for existence' (Kropotkin, 1992, p. 45). This applied especially to the weakest, the slowest, and therefore the most physically vulnerable animals, amongst whom were human beings.

Kropotkin cites many examples of this phenomenon, including insects, land-crabs, bees, birds of prey, migratory and nesting birds, and nearly all mammals: the higher up the scale of evolution the more conscious such association becomes, until in human beings, it becomes a reasoned process. Indeed it is this process, he says, which takes animals up the evolutionary scale, and has been responsible for the development of their longevity and intelligence. In human and non-human nature, 'life in societies', he concludes, 'is the most powerful weapon in the struggle for life' (Kropotkin, 1989).

Kropotkin concurred with Darwin in the *Descent of Man*, that sociality helped animals survive and in the higher animals, such as humans, it engendered mutual sympathy, compassion and ultimately love, and also gave birth to notions of equity and justice. Thus, both the source of ethical behaviour, 'the rudiments of moral relations' and the basis for anarchism could be found in the natural world, specifically in the sociality that had pre-human origins but had been most highly developed in humans. Kropotkin contended that 'Nature has to be recognized as the first ethical teacher of man. The social instinct … is the origin of all ethical conceptions and all the subsequent development of morality … the moral feelings of man … are further developments of the feelings of sociality which existed amongst his remotest pre-human ancestors' (1992, p. 45).

This naturalistic argument challenged theologians and philosophers who assigned ethics a supernatural or metaphysical origin.

Perhaps the most important libertarian thinker to take up Kropotkin's ontological and ethical legacy in the 20th century has been Murray Bookchin (1922–2006).

BOOKCHIN'S SOCIAL ECOLOGY

Since his 1965 essay *Ecology and Revolutionary Thought,* Bookchin has sought to integrate ecology and the libertarian tradition into a grand synthesis he has termed 'social ecology'. Bookchin's work diverges from Kropotkin's in rejecting the idea that the domination of nature is a necessary and inevitable feature of the human condition.

Early societies, Bookchin says, lacked concepts of domination and therefore could not develop the concept of the domination of nature. But the hierarchical sensibility that later emerged and which 'conceptually equipped humanity to transfer its social antagonisms to the natural world' (Bookchin, 1982, p. 82) was projected on to nature and the idea of dominating nature was born: nature became 'a taskmaster – either to be controlled or obeyed' (Bookchin, 1989, p. 33). Thus, the very idea that humanity must dominate nature is intimately related to the rise of hierarchy in human societies. The ecological crisis, therefore, has social roots: since it arises out of the domination of human by human it will only be resolved not simply by dismantling state institutions but, more generally, all hierarchical social forms and ideologies.

Bookchin's argument reverses the position, held since Classical times, that the domination of human by human, the cruel treatment and exploitation of one economic class by another, is justified by the myth of 'a blind, mute, cruel, stingy and competitive nature' (Bookchin, 1995, p. 39). Bookchin maintains [t]his ideology holds that wealth can only be created through the treatment of nature as a resource, and the need to force wealth from such a stingy nature became an apologetic for the 'stingy' behaviour of ruling elites and providing the utilitarian underpinning for modern ideologies such as liberalism and Marxism.

Whilst acknowledging the existence of scarcity, Bookchin argues that ruling elites have often exaggerated nature's stinginess, and indeed have often artificially induced scarcity (Bookchin, 1995, p. 99). The dominant classes have maintained that authoritarian institutions are needed to protect people from the struggle that would ensue as a result of scarcity in the natural world. Or, as Kroptkin earlier put it, the belief 'that without authority men would eat one another' (1992, p. 49). All this has led to the instrumental treatment of the natural world as a collection of resources, a set of raw materials.

Like Kroptkin, Bookchin turns to modern developments in scientific ecology and evolutionary theory. However, Bookchin attempts to integrate such insights with what are viewed as complementary broader insights about nature that can be found in the Western tradition of process philosophy and dialectic thought: from Aristotle and Schelling to Fichte and Hegel.

Seeking to dispel 'the marketplace image of nature' Bookchin (1982) has suggested that the post-war science of ecology and evolutionary theory provides us with a very different vision of nature from that of Malthus, Marx or Adam Smith. Far from being competitive, Bookchin says, scientific ecology reveals nature as characterized by interactive and participatory relationships; far from being stingy it is fecund; far from being blind it is creative and directive; and far from being necessitarian it provides the grounding for an ethics of freedom.

Bookchin's 'dialectical naturalism' (1990, p. 16) as he calls it, suggests that the most appropriate way to understand socio-ecological relations is not to focus on the development of individual species in isolation from other species (he argues the tendency to do so is a reflection of our culture's entrepreneurial bias) but to focus on their interdependent development within the context of ever changing ecocommunities. Ecocommunities are best viewed as interactive and integrated (but also evolving and unfinished) wholes which, he argues, are at the foreground of evolution. They can be characterized by the principle of 'dynamic unity in diversity' (1982, p. 24) providing the context for the differentiation, or evolution, of species and individuals.

'The thrust of evolution', Bookchin maintains 'is toward the increasing diversification of species and their interlocking into highly complex, basically mutualistic relationships' (1995, p. 41). The diversity within an ecocommunity is not only the source of its stability but is also responsible for its evolution, its increasing differentiation, and this becomes a source of 'nascent freedom'. According to Bookchin, diversity, as it develops in the course of evolution provides 'varying degrees of choice, self-directiveness, and participation by life-forms in their own development ... the increase in diversity in the biosphere opens new evolutionary pathways, indeed, alternative evolutionary directions, in which species play an active role in their own survival and change'. He further contends that this 'dim choice' emergent subjectivity, and capacity to select its own environment, increases as species 'become structurally, physiologically, and above all neurologically more complex' (1995, p. 44). In other words, as a species becomes more advanced it participates to an ever greater extent in its own evolution.

As ecocommunities become more complex in their diversity in the course of evolution, new evolutionary pathways open up and new kinds of interactions become possible and the avenues for this participatory process become more various.

Evolution has, therefore, not only a mutualistic but also a participatory dimension. This view of life as being actively, relationally and creatively engaged in its own evolutionary development is of course at variance with the conventional view that in their own evolution species are but passive objects of 'exogenous' forces (1995, p. 44). This, Bookchin argues, is simply a modern expression of the idea of nature as necessitarian, or deterministic.

Thus, not only human society, as Kropotkin argued, but human will, subjectivity, choice, intentionality, reason and therefore freedom all exist as latent potentialities within the natural world, which have unfolded, or graded, in the course of evolution. These capacities have emerged *out* of nature, not *in spite* of nature as Western civilization has usually held, therefore nature can no longer be seen as a blind, uncreative object. The desire to formulate a rational, libertarian ethics, then, need no longer be haunted by the fear of relativism or be premised on a sharp dualism between society and nature, for we can see:

> Mutualism, freedom and subjectivity are not strictly human values and concerns. They appear, however germinally, in larger cosmic and organic processes that requires no Aristotelian God to motivate them, no Hegelian spirit to vitalise them (Bookchin, 1982, p. 365).

According to Bookchin, dialectical naturalism gives rise to an active ecological humanism. Human beings emerge from 'first nature' (the natural world) to construct second nature (society). Yet, we need to view ourselves as an expression of an inherent striving towards consciousness present in first nature: 'We have been constituted to intervene actively, consciously and purposely into first nature with unparelled effectiveness and alter it on a planetary scale' (Bookchin, 1990, p. 42). The relevant question for Bookchin is whether we are increasing or diminishing social and ecological diversity and complexity?

SOCIAL ANARCHISM, NATURE AND THE BUILT ENVIRONMENT: FROM GARDEN CITIES TO ECOLOGICAL URBANISM?

What might be the social forms that facilitate the development of social and ecological complexity and diversity? At various points over the last two-hundred years, beginning with Fourier's rural 'phalanstries', many groups influenced by anarchism and libertarianism have maintained that cooperative and self-sufficient rural communes embedded in local ecologies, mixing small-scale agriculture and craft production, could provide a desirable alternative to industrial capitalism. Such currents have clearly played a major influence on the diverse forms of eco-monasticism and green communitarianism that have ebbed and flowed over the last forty years (see Eckersley, 1992). More extreme still, over more recent times has been the rise of various forms of 'anarcho-primitivism'. Premised on an apocalyptic vision of ecological crisis coupled with a romanticized view of the ecological virtues of hunter-gatherers and a desire to recover 'wildness', such currents have argued a sustainable or ecological society must involve a wholesale rejection of modernity, urbanism, cities and indeed, 'civilization' (e.g. Zerzan, 1994). Such currents have clearly influenced many of the more regressive and backward looking manifestations of contemporary green politics. Yet, the assumption that this is the only contribution that the anarchist and libertarian tradition has made to the contemporary environmental debate is problematic in the least.

As Graham Purchase has noted, many of the classic 19th anarchists indeed firmly rejected the idea that a return to a pre-industrial world constituted any kind of viable solution. The anarchist geographer Elisee Reclus for example 'firmly rejected the notion that small scale experimental communities … provided anything approaching an adequate solution to the problem of human co-existence' and instead championed 'the autonomous and eco-regionally integrated city' (Purchase, 1997, p. 16). Kropotkin equally envisaged urban communes which would be 'large autonomous and self sustaining agro-industrial agglomerates the largest of which might be the size of Paris' (Purchase, 1997, p. 20) and rural worlds supported by the spread of advanced technologies.

Amongst 20th century libertarians and anarchists such as Patrick Geddes and Ebenezer Howard, Martin Buber, Murray Bookchin and Colin Ward, it is the idea of the 'human scaled city' which time and again emerges as the object of fascination and study and is posited as the site of potential liberation for society and nature. One can find in all these thinkers the view that the source of the socio-ecological dilemmas of contemporary societies is not urbanism (or indeed technology) per se but the development of forms of urbanism that are inherently poor in structure and that undermine the potential of the classic city. Martin Buber, for example, maintained that a rich social structure is one made up of dense and overlapping forms of community association, and that it was the tendency of industrial-capitalism to destroy such forms of association (Buber, 1949). Twentieth-century anarchist urbanists, then, have generally argued that the social and ecological contradictions of the age will only be resolved through rebuilding complex ecological and urban social structures. Recovering the urban has often been viewed as the first step in reorganizing social and ecological life more generally. This would entail combining existing currents of mutual aid with a project to create consciously designed, living environments that rework civic, democratic, communal, technological and ecological materials to facilitate the rise of self-organized societies.

For example, Ebenezer Howard (1850–1928) and Patrick Geddes (1854–1932), in their desire to design healthy and democratic urban spaces, were hugely influenced by Kropotkin. Yet both take Kropotkin's thinking much further. Reacting to the unplanned mess of Victorian slums, Howard advocated radical changes in private and public land ownership to develop carefully planned and aesthetically designed cities that would at once maximize freedom of choice and community and enable people to live in a more harmonious relationship with nature (deGeus, 1999). Recognizing that neither the contemporary city nor the countryside allowed for a full human life, Howard proposed 'garden cities', human-scaled cities that could combine the best of 'town' and 'country': beautiful gardens and rich cultural institutions, spacious boulevards and public parks, advanced workplaces and public transport systems, clean production centres and good sanitation. It was envisaged such cities would strike a balance between society and nature, culture and ecology. They would be rationally planned and surrounded by dense green belts that would allow nature to flourish. Indeed, they would be close to works of art. As Howard states:

> … it is essential, as we have said, that there should be unity of design and purpose – that the town should be planned as a whole, and not left to grow up in a chaotic manner as has been the case with all English towns, and more or less so with the towns of all countries. A town like a flower or a tree, or an animal, should at each stage of its growth, possess unity, symmetry completeness, and the effort of growth should never be to destroy that unity, but to give it greater purpose, not to mar that symmetry, but to make it more symmetrical; while the completeness of the early structure should be merged in the yet greater completeness of the later development (Howard, *Garden Cities of Tomorrow*, cited in deGeus, 1999, p. 121).

Moving beyond bureaucratic collectivism and Victorian capitalism Howard envisaged that such garden cites would create a decentralized but rationally planned, confederated and cooperative commonwealth, within which there would be a combination of private and municipally owned property.

Geddes, in *City Development* and *Cities in Evolution,* advocated reshaping the environments of the town, the city and the countryside so as to allow their inhabitants to become engaged in a popular activity of 'civic planning'. As Colin Ward has observed 'The direct expression of ordinary citizens' aspirations in the reshaping of the town or city is the message that comes from so many of Geddes' environmental perceptions' (Ward, 1991, p. 110). A central theme of Geddes' work is:

> … the idea that the average citizen has something positive to contribute towards the improvement of his environment. Geddes was convinced that each generation has the right to build their own aspirations into the fabric of their town (Ward, 1991, p. 110).

Geddes believed that in order to achieve this, a basic level of civic understanding had to be created through education. As such, he canvassed schools, societies and associations and attempted to draw them into making surveys and plans of their locality; creating play spaces, planting trees and painting buildings. He seized on any vehicle to expose people to situations in which they had to make judgments (Ward, 1991, p. 110).

Moreover, Geddes is of central importance for arguing that cities should be imagined in more organic and holistic terms, for his sensuous appreciation of their broader environments and for recognizing that cities belong to regions. Like Howard and the anarchists before him, Geddes believed that such human-scale urban communities should be not only be exquisitely tailored to the ecologies they find themselves in but also form confederations of autonomous regions that would replace nation states with a more benign collective commonwealth.

Both Geddes and Howard played a central role in developing the progressive traditions of town planning in the UK in the pre-war period. Two of the central post-war figures that have sought to keep the vision of a libertarian and ecological urbanism alive have been Murray Bookchin and Colin Ward.

Bookchin's vision of the ecological future is significantly informed by an urbanist sensibility. His vision of the restoration of the city is informed by classical notions of human scale, citizenship and direct democracy. This, Bookchin argues, will permit the re-establishment of the balance between 'town and country' that is central to establishing the ecological society.

Bookchin's first full-length publications *Our Synthetic Environment* (1962) and *Crisis in Our Cities* (1965) are concerned with the notion of an unfolding 'urban crisis'. Suggesting that life in the modern post-war American 'megalopolis' is 'breaking down', attention is paid to the demographic shifts occurring from the cities into the suburbs. And while he notes that much energy has been spent by critics ridiculing this exodus, Bookchin argues in these texts that the impulse to escape from the bloated sprawl of the post-war urbanization is perfectly rational. In an attempt to escape the *reification* at the heart of modern life, the 'average American' is seen as 'making an attempt however confusedly, to reduce his environment to a human scale'. He is 'trying to re-create a world that he can cope with as an individual'. At root, this reflects 'a need to function within an intelligible, manipulable, and individually creative sphere of human activity' (Bookchin, 1962, p. 238).

Bookchin argues in these texts that there is a need to recover 'the normal, balanced, and manageable rhythms of human life – that is an environment which meets our requirements as individual and biological beings' (Bookchin, 1962, p. 240). While present trends towards the development of formless urban agglomerates are seen as profoundly undesirable, equally as problematic is the notion that we can return to some pre-industrial rural past. Rejecting any kind of arcadian or primitivist alternative (the use of farm machinery after all does not necessarily conflict with sound agricultural practices, nor is industry and agriculture incompatible with a more natural environment) Bookchin argues that we need a 'new type of human community,' a community which constitutes 'neither a complete return to the past nor a suburban accomodation to the present' (Bookchin, 1962, p. 242).

The ecological project should not, therefore, reject urbanism but should reconsider the city in all its historical diversity (Bookchin, 1962, 1965). Inspired in part by Kropotkin, Geddes, Howard and others, Bookchin argues that there is a need to integrate some of the virtues of modernity with those of the urban forms which provided the well springs of Western civilization, such as the Athenian polis, the early Roman Republic and the free cities of the Renaissance:

> It is no longer fanciful to think of man's future environment in terms of a decentralised, moderate sized city that combines industry with agriculture, not only in the same civic entity but in the occupational activities of the same person (Bookchin, 1962, p. 242).

Bookchin maintains that the problem that environmentalists and ecologists should recognize is not that of urban life per se. Rather, the problem is urbanization *under capitalism*, or the way in which capitalism generates 'urbanization without cities'. Capitalist forms of urbanization, he argues, undermine and hollow out any real sense of civic life and commitment, community or active citizenship. They impose ecologically irrational burdens on the surrounding environment and create 'grossly unbalanced' (1965, p. 173) societies populated by 'nervous excitable individuals'. What is needed, then, is that we develop our environment '… more selectively, more subtly, and more rationally' to bring forth 'a new synthesis of man and nature, nation and region, town and country' (1962, p. 244).

Bookchin's vision of an ecological urbanism is interesting not simply because of its championing of the moderate-sized city as potentially the site for an ecological politics, but for its attempt to integrate this urban environmental project with a politics of ecological technology and a politics of participation and citizenship.

In Bookchin's 1965 essay *Towards a Liberatory Technology*, he notes that, with the advent of Stalinism and the cold war, the case for a simple and direct correlation between technological advance and social progress has been shattered. Modern attitudes have become 'schizoid, divided into a gnawing fear of nuclear extinction on the one hand and a yearning for material abundance, leisure and security on the other' (1986, p. 107). However, tendencies to resolve these tensions by presenting technology as 'imbued with a sinister life of its own', resulting in its blanket rejection, are seen as simplistic as the optimism that prevailed in earlier decades. If we are not to be paralysed by this 'new form of social fatalism', a fatalism attributed to social theorists of technology such as Jacques Elul and Friedrich Juenger, it is argued that 'a balance must be struck' (1986, p. 108). Concerning where exactly the balance should lie, much of Bookchin's earliest writings argue that we need to recover a sense of the *liberatory* possibilites of new technology.

Bookchin argues that a radically decentralized society is not only compatible with many aspects of the modern technological world but is potentially *facilitated* by new developments. For example, he argues that technological innovations may have actually made the need for huge concentrations of people in a few urban areas *less* important as the expansion of mass communications and transportation have ensured that 'the obstacles created by space and time are essentially gone' (1986, p. 241). Concerning the viability of industrial decentralization, it is suggested that new developments in miniaturization, computing and engineering have ensured that small-scale alternatives to many of the giant facilities that dominated industrial societies are now increasingly viable.

Bookchin speculates that the sheer scale of labour-saving possibilities created by automation makes a toilless future imaginable, perhaps for the first time in history. He argues that virtually all the utopias and revolutionary programmes of the early 19th century faced problems of work and want. Indeed, much socialist thinking was so effected by such imagery, lasting well into the 20th century, that there emerged on the left a virtually puritanical work ethic, a fetishization of toil and a view of socialism as the industrious society of full employment. The technological developments in the post-war era, Bookchin argues, hold the potential for replacing this 'realm of necessity' with a 'realm of freedom'. The critical issue, however, is not whether technology can liberate humanity from want but the extent to which it can contribute to humanizing society and human–nature relations.

A future society should, therefore, be based on ecological technologies which are both 'restorative of the environment and perhaps, more significantly, of personal and communal autonomy' (Bookchin, 1980, p. 130). Eco-technology should not only 'reawaken man's sense of dependence on the environment' (1986, p. 136) but restore selfhood and competence to a 'client citizenry' (1980, p. 130). There may well be logistical or technical reasons why 'small is beautiful', but for Bookchin attention to the human scale is important since it renders society *comprehensible* and, hence, *controllable* by all. What is needed

> ... is not a wholesale discarding of advanced technologies then, but indeed a shifting, indeed a further development of technology along ecological principles (1982, p. 37).

Bookchin has an Aristotelean preference for balanced communities, for 'the well rounded individual' and for politics as a domain of ethics and participation. He argues that direct democracy, rather than representative democracy, would be central to the functioning of an ecological society. Such a society would see every individual as capable of participating directly in the formulation of social policy which would 'instantly invalidate social hierarchy and social domination' (1982, p. 340), This political culture, Bookchin argues, would create the conditions for decisively undercutting the idea that humanity needs to 'dominate nature' and invite the widest possible participation, permitting the recovery of human beings not as 'taxpayers', 'constituents' or 'consumers' but as citizens.

An ecological society needs, therefore, to be populated by libertarian political institutions, that is, institutions structured around direct, face-to-face relationships and based on participation, involvement, 'and a sense of citizenship that stresses activity', not based on 'the delegation of power and spectatorial politics' (Bookchin, 1982, p. 336). It would be committed to the cardinal principle that 'all mature individuals can be expected to manage social affairs directly – just as we expect them to manage their private affairs' (1982, p. 336).

The British anarchist, environmentalist and urbanist Colin Ward is another notable successor of Kropotkin and Buber. Ward's hugely underestimated intellectual project over the last 50 years could be described as the excavation of the enduring forms of mutual aid that persist even in the most capitalist of cities. In contrast to much anarchist advocacy, Ward's central message has been that anarchism is not simply some far off rationalist utopia but a persistent social practice. He argues that an anarchist society, a society which organizes itself without authority, is always in existence, 'like a seed beneath the snow' (Ward, 1988, p. 14). Today, it can be found wherever there is communal voluntary action and bottom-up self-organization: from allotments to free schools, self-build housing to city gardens and community-supported agriculture.

Ward argues that we should not see the anti-authoritarian tradition as an ideology demanding total social transformation. Rather, it is more usefully viewed as advocating certain types of social practice: 'The choice between libertarian and authoritarian solutions is not a once-and-for- all cataclysmic struggle, it is a series of running engagements' (Ward, 1988, p. 136).

SCEPTICS AND CRITICS: THE LIMITS OF ANARCHISM AND ECO-ANARCHISM

At many points over the last two centuries anarchists and libertarians have been subject to severe criticism. They have been dismissed as holding to either a hopelessly naïve, nostalgic or romantic view of the past and/or proposing a utopian, unworkable or indeed dangerous view of the future. Critics have argued that the anti-authoritarian tradition is premised on inordinately optimistic assumptions about human nature and human benevolence. Leninists, for example, have dismissed the anarchist and libertarian critique of the state, authority and centralization as symptomatic of an 'infantile disorder' (Lenin, 1985). Many Marxists more generally have maintained that anarchist advocacy demonstrates a persistent tendency to idealize pre-capitalist social relations and persistently fails to grasp the progressive dimensions of the capitalist rationalization process.

Liberals have argued that a society without the rule of law would look more like Hobbes's state of nature than Kropotkin's utopia, and emphasized that the rights of individuals and the democratic community will frequently come into conflict. Social democrats have maintained that the absence of the state as a political and economic coordinating mechanism could lead to the re-emergence of significant regional, national and international inequalities. Recently, post-structuralists have argued that 'organic communities' are parochial and stifling places, frequently intolerant of difference, multiculturalism and individuality.

With the re-emergence of anarchist and libertarian tendencies in the environmental debate such discussions have been restaged. As such, recent currents in green political theory and environmental sociology have argued that some of the greatest weaknesses of modern environmentalism (particularly its radical wing) stem from its predilection for anarchist intellectual assumptions and political strategies (Barry, 1998). Doubts have been raised at the extent to which transnational environmental problems can be credibly dealt with by radical decentralist solutions. Ecological modernizers have argued that ecological anarchisms informed by 'counter productivity positions' in particular fail to grasp the electoral unpopularity of such strategies and underestimate the extent to which liberal democracies have been open to certain degrees of ecological reforms or the importance that centralized bodies (whether states or super-state forms) currently play in brokering and enforcing environmental agreements (Mol, 2003). More recent defenders of 'the green state' have severely critiqued the anarchist ambivalence to the state that pervades many green movements (Eckersley, 1992; Barry, 1998; Monbiot, 2003). Indeed, some commentators have argued that in an age of neo-liberal globalization, where unaccountable private institutions wield extraordinary power, we should be building global state-like institutions as a bulwark against global capitalism, rather than undermining the state (Monbiot, 2003).

In response to such claims, modern eco-anarchists have been keen to point out that it was the anarchist tradition, more than any other body of ideas, that anticipated the last great attempt to 'force people to be free' via deploying all the power of vanguards, the centralized state and the disciplined party cadre. And as such, given the disaster of the 'party state' constructed by Marxism–Leninism, it has been argued that contemporary ecologists and environmentalists would do well to reflect more extensively on the pros and cons of wielding of state power to resolve environmental problems (Bookchin, 1971). Eco-anarchists have continued to argue that the capitalist rationalization process is clearly beset with numerous irrationalities, not least of which are environmental problems, affluenza and ennuie in the north and a lack of the basic means of life in the south. They have noted how the modern liberal democratic state continues to expand its capacities for surveillance and control, and how capitalism continues to discipline its 'subject', albeit now more by 'work and spend' ideologies and regimented leisure cultures than direct coercion. Figures such as Bookchin have continued to insist doggedly that confederal arrangements could indeed provide communities with perfectly viable coordinating mechanisms to deal with environmental and other social problems (1987, 1989). Moreover, on the feasibility of anarchist and libertarian utopias, perhaps it is worth citing the near forgotten figure of Paul Goodman, who once responded to such a criticism in the following fashion:

> the issue is not whether people are 'good enough' for a particular type of society; rather it is a matter of developing the kind of social institutions that are most conducive to expanding the potentialities we have for intelligence, grace, sociability and freedom (Goodman, 1964).

ANARCHISM, LIBERTARIANISM AND ENVIRONMENTALISM: ANTI-AUTHORITARIAN THOUGHT AND THE SEARCH FOR SELF-ORGANIZING SUSTAINABLE SOCIETIES

How then can we think about green anarchism, green libertarian currents and the search for plausible sustainable futures? Many environmentalists and radical ecologists will argue that however far off the project may be, it is the anarchist solution to the environmental problematic, the vision that runs through the work of Kropotkin, Geddes, Howard and Ward, Buber and Bookchin, that ultimately underwrites the imaginative and ethical horizons of green politics.

For sceptics, perhaps a set of different response is necessary. At a bare minimum, it could be argued that the anti-authoritarian tradition as social philosophy is perennially important in posing a set of important challenges to any serious body of transformative thought (ecological or otherwise). How should a project for social transformation balance the desire for human equality with the reality of human diversity? Is there a danger that centralized institutional power structures, however 'democratic', rarely appreciate their own limits? Does human emancipation presuppose the ideology of the 'domination of nature' and the subordination of all other species to industrial modes of production? Is the continual development of an advanced division of labour and the search for ever more complex and regionally specific economies of scale compatible with a humane, diverse and ecological mode of production? Indeed, is the regimented workplace (capitalist, socialist or other) or the disciplined political party the best place for developing a character structure and social sensibility that nurtures and values self-organization, autonomy and active engagement? How is the aspiration to an ecological society to be made compatible with a society which still values spontaneity, playfulness, hedonism and craft? How can societies so committed to standardization and efficiency allow the qualitative features of the human condition to flourish?

The anarchist and libertarian traditions are of importance to the environmental debate because, minimally, no other tradition of social and political theory poses these questions in such a direct fashion. Yet, more generally still and despite its marginality in the academy, it would also have to be recognized that as a social philosophy of self-organization, anarchist and libertarian advocacy has had an extraordinary capacity to continue to influence a diverse range of debates in the environmental social sciences.

For example, if we return to the 'built environment', since the late 1960s anarchist and libertarian sensibilities have played a significant role in inspiring debates in urban sociology and urban studies. Anarchism has influenced experiments in participatory urbanism, urban design, neighbourhood governance, popular planning and so on. Even now, some of the more progressive contemporary discussions of 'sustainable' and 'green' cities echo much of the thinking of Kropotkin and Geddes, Howard, Ward, Bookchin and others who stand in this tradition (see Wheeler and Beatley, 2003).

Over the last two decades, organizational theory has challenged the merits of the centralized, hierarchically managed firm and championed the virtues of regional economies, decentralized flexible production systems and flexible networks (Piore and Sabel, 1984; Hirst, 1994; Castells, 1996). Of course, the primary motivation for much managerial 'experimentation' is the desire to reduce costs. Yet, as Paul Hirst has argued in his advocacy of associational democracy (directly inspired by Proudhon at times) or Robin Murray (Murray, 2002) in his investigations of decentralized craft based 'post fordist' production, such developments need not be taken in this fashion but, used imaginatively, have the potential to open up new horizons for social and ecological reorganization.

Even anarchistic naturalism continues to influence debates in the environmental sciences and technology. Many features of Kropotkin and Bookchin's advocacy have entered into the mainstream of debate about scientific, technological and agricultural innovation. For example, the increasingly serious attempts to envisage the contours of a low-carbon, zero-waste economy, drawing from developments in modern 'industrial ecology', clean production, ecological architecture and design and renewable energy, can be understood as attempts to realize Bookchin's infrastructural vision (Hawkins *et al.,* 1999; Milani, 2002; Murray, 2002). Likewise, the notion that industrial agriculture should decentralize and become more pluralistic has many articulate modern champions (Pretty, 2002). Indeed, where Kropotkin and Bookchin looked to participation, self-organization and emergent subjectivity to ground their radical ecological ethic, it is interesting to note the extent to which recent developments in 'complexity science' seem to reiterate such themes. Complexity theory increasingly argues that self-organization, a theme at the heart of much anarchist theory, is a more or less ubiquitous feature of the natural world (Capra, 1996). Thus, out of the internal dynamics of organizationally closed physical, chemical and biological systems, it is suggested that pattern, order and form emerge spontaneously. Indeed, self-organization operates at every level of biological systems – from the cell, through to ecosystems and even, in Gaia theory, to the planet as a whole (Lovelock, 1987).

Such influences suggest that while the death of anarchism and libertarian thought has long been proclaimed, the search for self-organizing sustainable societies will continue to return to anarchist and libertarian social philosophy for inspiration.

NOTES

1 As Marshall (1992a) and Miller (1984) have noted part of the problem here is that there has been a certain looseness of terminology between 'the state' and 'government' in anarchist writings. Some anarchists have spoken of the state and government as synonymous—notably Godwin and various anarcho-individualists. Others, such as Bookchin, make a clear distinction between states and governing institutions. Proudhon, as Marshall observes, reflects such inconsistencies when he argues that 'government of man by man is servitude' yet he goes on to define anarchy as the absence of a sovereign or ruler as being a 'form of government' (1992a, p. 19). It has to be recognized, though, that extreme anarcho-individualists such as Max Stirner would indeed view anarchism as defined by a rejection of both government and society.

REFERENCES AND BIBLIOGRAPHY

Audi, Robert (ed.) (1995) *The Cambridge Dictionary of Philosophy*. Cambridge: Cambridge University Press.

Barry, John (1999) *Rethinking Green Politics – Nature, Virtue, Progress*. London: Sage.

Biehl, Janet (1997) *The Murray Bookchin Reader*. London: Cassell.

Briggs, Asa (ed.) (1962) *William Morris, Selected Writings and Designs*. Middlesex: Penguin Books.

Bookchin, Murray (1962) *Our Synthetic Environment*. New York: Knopf. (Initially published under the pseudonym Lewis Herber.)

Bookchin, Murray (1965) Ecology and Revolutionary thought. *Post scarcity Anarchism*. Berkeley, CA: Ramparts Press.

Bookchin, Murray (1965) *Crisis in Our Cities*. New York and Englewood Cliffs, NJ: Prentice Hall. (Initially published under the pseudonym Lewis Herber.)

Bookchin, Murray (1971) *Post Scarcity Anarchism*, 1st edn. Berkeley: Ramparts Press.

Bookchin, Murray (1980) *Toward an Ecological Society*. Montreal: Black Rose Books.

Bookchin, Murray (1982) *The Ecology of Freedom*. Palo Alto: Cheshire Books.

Bookchin, Murray (1986a) *Post-Scarcity Anarchism*. Montreal: Black Rose Books (Second Edition).

Bookchin, Murray (1986b) *The Modern Crisis*. Philadelphia: New Society Publishers.

Bookchin, Murray (1987) *The Rise of Urbanization and the Decline of Citizenship*. San Francisco: Sierra Club Books.

Bookchin, Murray (1989) *Remaking Society*. Montreal: Black Rose Books.

Bookchin, Murray (1990) *The Philosophy of Social Ecology*. Montreal: Black Rose Books.

Bookchin, Murray (1995) *Social Anarchism or Lifestyle Anarchism: An Unbridgeable Chasm*. Edinburgh: AK Press.

Buber, Martin (1949) *Paths in Utopia*. New York: Collier Books.

Capra, Fritjof (1996) *The Web of Life*. London: Harper Collins.

Clark, John (1984) *The Anarchist Moment: Reflections on Culture, Nature and Power*. Montreal: Black Rose Books.

Clark, John (ed.) (1990) *Renewing the Earth: A Celebration of the Work of Murray Bookchin*. London: Green Print.

Clark, John P. and Martin, Camille (2004) *Anarchy, Geography, Modernity*. Maryland: Lexington Books.

deGeus, Marius (1999) *Ecological Utopias: Envisaging the Sustainable society*. Utrecht: International Books.

Dobson, Andrew (1990) *Green Political Thought*. London: Unwin Hyman.

Eckersley, Robin (1992) *Environmentalism and Political Theory*. London: UCL.

Fourier, Charles (1971) In Poster, M. (ed.) *Harmonian Man*. Mark Poster, editor. New York: Anchor Books.

Godwin, William (1986) In Marshall, P. *The Anarchist Writings of William Godwin*. London: Freedom Press.

Goodman, Paul (1964) *Utopian Essays and Practical Proposals*. New York: Vintage Books.

Goodman, Percival and Goodman, Paul (1947) *Communitas: Ways of Livelihood and Mean of Life*. New York: Columbia University Press.

Guerin, Daniel (1970) *Anarchism*. New York: Monthly Review Press.

Hall, Peter (2002) *Cities of Tomorrow. An Intellectual History of Urban Planning and Design in the twentieth century*. London: Blackwell.

Hawkins, Paul, Lovins, Amory and Lovins, Hunter (1999) *Natural Capitalism*. London: Earthscan.

Hirst, Paul (1994) *Associative Democracy*. Cambridge: Polity.

Krimerman, Leonard I., and Perry, Lewis (ed.) (1966) *Patterns of Anarchy*. New York: Anchor Books.

Kropotkin (1902) *Mutual Aid*.

Kropotkin, Peter (1974) In Ward, C. (ed.) *Fields, Factories and Workshops Tomorrow*. London: Freedom Press.

Kropotkin, Peter (1987a) In Walter, N. (ed.) *Two Essays: Anarchism and Anarchist Communism*. London: Freedom Press.

Kropotkin, Peter (1987b) *The State*. In Walter, N. (ed.) London: Freedom Press.

Kropotkin, Peter (1987c) *Mutual Aid: A Factor of Evolution*. London: Freedom Press.

Kropotkin, Peter (1992) *Ethics: Origin and Development*. Montreal: Black Rose Books.

Kropotkin, Peter (1993) *Anarchism and Anarchist Communism: Its Basis and Principles*. London: Freedom Press.

Kropotkin, Peter (1995) *Evolution and Environment*. Montreal: Black Rose Books.

Lenin, V. I. (1985) *Left Wing Communism: An Infantile Disorder*. London: International Publishers.

Lovelock, J. E. (1987) *Gaia: A New Look at Life on Earth*. Oxford: Oxford University Press.

Marshall, Peter (1992a) *Demanding the Impossible: A History of Anarchism*. London: Harper Collins.

Marshall, Peter (1992b) *Nature's Web: An Exploration of Ecological Thinking*. London: Simon & Schuster.

Maximoff, G. P. (1953) *The Political Philosophy of Bakunin*. New York: The Free Press.

Milani, Brain (2000) *Designing the Green Economy*. New York: Rowman and Littlefield.

Miller, David (1984) *Anarchism*, London and Melbourne: J. M. Dent and sons.

Mol, A. (2003) *Globalization and Environmental Reform*. Boston: MIT Press.

Monbiot, George (2003) *The Age of Consent*. London: Flamingo.

Mumford, Lewis (1938) *The Culture of Cities*. London: Secker & Warburg.

Mumford, Lewis (1967) *The Myth of the Machine*. London: Secker & Warburg.

Mumford, Lewis (1986) *The Future of Technics and Civilization*. London: Freedom Press.

Murray, Robin (2002) *Zero Waste*. London: Greenpeace Trust.

O'Riordan, Tim (1981) *Environmentalism*. London: Pion.

Pepper, David (1993) *Ecosocialism: From Deep Ecology to Social Justice*. London: Routledge.

Piore, Michael, J. and Sabel, Charles (1984) *The Second Industrial Divide – Possibilities for Prosperity*. London: Basic Books.

Pretty J. (2002) *Agri-Culture*. London: Earthscan.

Purchase, Graham (1997) *Anarchism and Ecology*. Montreal: Black Rose Books.

Susser, Bernard (1981) *Existence and Utopia: The Social and Political Thought of Martin Buber*. New Jersey: Associated University Press.

Ward, Colin (1988) *Anarchy in Action*. London: Freedom Press.

Ward, Colin (1993) *Influences: Voices of Creative Dissent*. Devon: Green Books.

Wheeler, Stephen and Beatley, Tim (2003) *The Sustainable Urban Development Reader*. London: Routledge.

Woodcock, George (1986) *Anarchism: A History of Libertarian Ideas and Movements*. London: Penguin Books.

Zerzan, John (1994) *Future Primitive and Other Essays*. New York: Autonomedia.

4

Ecofeminism: Linking Gender and Ecology

Mary Mellor

INTRODUCTION

Ecofeminism emerged alongside second-wave feminism in the 1970s asserting a link between gender and ecology, or rather women and nature (Ruether, 1975; Griffin, 1978). Ecofeminism is based on the claim that there is a connection between the exploitation and degradation of the natural world and the subordination and oppression of women. With greens, ecofeminism takes a view of the natural world (which includes humanity) as interconnected and interdependent and shares with feminism a view of humanity as systematically gendered in ways that subordinates, exploits and oppresses women. Ecofeminists see gender inequality as resulting in a destructive relationship between humanity and the rest of the natural world.

From the start ecofeminists have spoken in a range of voices, making their point in many ways. These voices have been protesting, spiritual, poetic, academic, political, often all at the same time. Alongside the early analysis of women and nature by theologians and poets such as Rosemary Radford Ruether and Susan Griffin, environmental activists like Wangari Maathai, founder of the Kenyan Green Belt Movement, were engaged directly in environmental action (Jones and Maathai, 1983). In the 1980s Vandana Shiva saw the Himalayan Chipko (tree-hugging) movement as emblematic of ecofeminism, although whether it can be accurately described as such has been disputed (Agarwal, 1992). In 1980, six-hundred women gathered in Amherst (USA) under the banner 'Women and Life on Earth: A conference on Ecofeminism in the '80s'. Ecofeminism was strongly linked with peace and anti-nuclear campaigns such as Women's Pentagon Action in 1980/81 and the marches against Cruise Missiles in Europe that led to women's peace camps such as Greenham Common (Liddington, 1998). Ecofeminist ideas were also linked with many environmental and anti-technology campaigns such as toxic waste or against nuclear weapons testing on native American land (LaDuke, 1993).

While the assertion of a link between the subordination of women and the degradation of nature was the core of ecofeminism, explaining that connection proved contentious. The titles of UK and US based anthologies that capture the early years of ecofeminism reflect a vision of women as saviours of the planet: 'Reclaim the Earth' (Caldecott and Leland, 1983); 'Healing the Wounds' (Plant, 1989); 'Reweaving the World' (Diamond and Orenstein, 1990). However, even within those anthologies there was a variety of perspectives including socialist ecofeminism (Merchant, 1990) and social ecofeminism (linked to Murray Bookchin's social ecology) (King, 1989, 1990) as well as ecofeminist spirituality (Spretnak, 1990). As Eaton and Lorentzen point out: 'Ecofeminism encompasses a variety of theoretical, practical and critical efforts to understand and resist the interrelated dominations of women and nature' (2004, p. 1). Summarising the concerns of three decades of ecofeminism they conclude:

Analysis of critical links between militarism, sexism, classism, racism and environmental destruction became central to ecofeminist thought and action. Ecofeminism now reflects the concerned efforts of women trying to integrate their personal, ecological and socio-political concerns (Eaton and Lorentzen, 2004, p. 2).

As with feminism in its early days, there was a desire to have an all-embracing ecofeminist movement. Speaking of the specific experience of the United States, Noel Sturgeon points out that ecofeminism faced very quickly the issue of women's diversity (Sturgeon, 1997) as did the wider feminist movement (Coote and Campbell, 1982). Ecofeminism also found itself at odds with some aspects of feminism, particularly the liberal feminist agenda of equal opportunities. As the American social ecofeminist Ynestra King put it, 'what is the point of partaking equally in a system that is killing us all' (1990, p. 106).

While all ecofeminists see a connection between feminist issues and environmental questions, this is expressed in many different ways. For some early ecofeminists a direct affinity appeared to be claimed. Women 'understand' nature through their physiological functions (birthing, menstrual cycles) or some deep element of their personalities (life-oriented, nourishing/caring values). This version of ecofeminism draws heavily on radical/cultural feminism in promoting the particular and exclusive interests/values of women (Collard, 1988). This led to the political charge that ecofeminism romanticised women's mothering and nurturing roles, thereby feeding reactionary patriarchal politics (Davion, 1994).

As Kate Soper points out, it was understandable that feminists would resist any perspective that seemed to argue for the 'naturalness' of nature because of the danger of endorsing 'the naturalisation of sexual hierarchy' (1995, p. 121). This creates a dilemma for ecofeminism where the de-naturalising impulse of feminism is at odds with the nature-valorising approach of ecology which risks opening up 'the potentially reactionary dimensions of ecological naturalism' (Soper, 1995, p. 124). However, as Soper acknowledges, 'despite the pervasive resistance of feminism to any naturalization of gender relations, there has been an equally widespread sense that there is an overall affinity and convergence of feminist and ecological political aims' (1995, p. 121). For Soper, the feminist celebration of relational ethics chimes with the green notion of interdependency. Greens reject the view of nature as bestial and something which must be controlled, while feminists reject the view of women as being naturally inferior. Both see rational and technocratic values as destructive: for greens the fault lies with the scientific revolution and technological modernisation per se, while for ecofeminists the problem rests more specifically in its gendered roots (Merchant, 1983).

Within academic ecofeminism, two broad and sometimes overlapping approaches have emerged to explain the subordination of women and nature in contemporary societies. One is based on a critique of western culture from a philosophical and/or spiritual perspective and the other rests on a socio-economic critique.

DUALISM: THE LOGIC OF DOMINATION

Dualism is a major focus of environmental philosophers such as Val Plumwood (1993, 2002) and Karen Warren (1994, 1996, 1997, 2000). For Plumwood:

> Western culture has treated the human/nature relation as a dualism and ... this explains many of the problematic features of the west's treatment of nature which underlie the environmental crisis, especially the western construction of human identity as 'outside' nature (1993, p. 2).

Both Plumwood and Warren see dualism as representing a cultural institutionalisation of power relations. This is described by Warren as a 'logic of domination' and by Plumwood as a 'logic of colonisation'. Such a logic not only divides categories of thought and life but also prioritises one over the other. Plumwood refers to this as hyperseparation or radical exclusion (1993, p. 49) calling instead for a 'critical ecological feminism' that will resolve these dualisms by exposing the assumptions that underpin them. Plumwood, like Ruether (1975), sees the origin of dualism as lying with the Greeks, particularly Plato, who separated the sphere of ideas from the rest of human existence. For Plumwood this has led to a 'hyperseparated conception of the human' which forms the 'master identity' of Western culture (1993, p. 72). The master culture emphasises rationality, freedom and the transcendence of nature. This has led to an existential homelessness in Western thought based on 'an alienated account of human identity in which humans are essentially apart from or outside of nature, having no true home in it or allegiance to it' (1993, p. 71). What is needed Plumwood argues, is an 'ecological identity' based on connection with the natural world. However, masculinist culture is not moving in this direction. Instead, it is struggling to take control of the world of nature and the body through 'progressive' science and technology while still denying the materiality of human existence. This 'backgrounding' of nature has had destructive consequences in the way, for example, that nature has been 'externalised' from economic calculations and concern.

For Plumwood, Platonic dualism is reinforced by Descartes 'stripping out' of the human mind from body/nature leaving only disembodied consciousness 'I'. This is the autonomous human-centred form of consciousness that both deep ecologists and ecofeminists see as lying at the heart of the destructiveness of Western thinking. All agency is denied to the natural world so that nature has no 'originative power' within itself :

> *Consciousness* now divides the universe completely in a total cleavage between the thinking being and mindless nature, and between the thinking substance and 'its' body, which becomes the division between consciousness and clockwork. Gone is the teleological and organic in biological explanation. Mind is defining of and confined to human knowers, and nature is merely alien (1993, p. 116, italics in original).

More specifically, consciousness is seen as resting with the male knower as women are identified with the natural. Plumwood argues that this gendered dualism cannot be challenged by an 'uncritical reversal' based on an assertion of the superiority of 'woman's culture' (1993, p. 31). To celebrate womanhood is to celebrate something which has been created by inequality, as other feminists have pointed out (Biehl, 1991). The challenge must be to the dualism itself such that 'we reconceive ourselves as more animal and embodied, more "natural", and that we reconceive nature as more mindlike than in the Cartesian conception' (Plumwood, 1993, p. 124). The Greek/Enlightenment concept of 'humanity' needs to be challenged from a feminist perspective since it has been constructed in a 'framework of exclusion, denial and denigration of the feminine sphere, the natural sphere and the sphere associated with subsistence' (1993, p. 22). It is this bodily association that links women with other oppressed groups and with non-human nature:

> The body is 'feminine-associated' but it is even more clearly associated with other oppressed groups, such as 'primitives', animals, slaves and those who labour with their bodies (1993, p. 116).

Plumwood identifies two possibilities for radical change. One is a cultural challenge to the master culture which asks 'active and intentional subjects ... to recognise and eject the master identity in culture, in ourselves, and in political and economic structures' and instead to develop 'forms of rationality which encourage mutually sustaining relationships between humans and the earth'. Such alternative frameworks would be based on examples of 'care, friendship and love' ... 'radical democracy, co-operation, mutuality' (1993, pp. 195–6). The other is the possibility that the 'master story' will falter through inherent material contradictions:

> After much destruction, mastery will fail, because the master denies dependency on the sustaining other; he misunderstands the conditions of his own existence and lacks sensitivity to limits and to the ultimate points of earthian existence (1993, p. 195).

Karen Warren also argues that patriarchal systems are unhealthy because the patriarchal conceptual framework is based on 'faulty beliefs' (2000, p. 207). Such beliefs lead to impaired thinking and the language of domination, which in turn leads to behaviours of domination. Such behaviours make life unmanageable which then feeds back into faulty beliefs. Warren sees part of the problem as lying in traditional philosophical theories about absolute and universal rights, duties and rules and rejects an ethics based on abstract, ahistoric, transcendental, essentialist principles based on reason (2000, p. 105). Instead she develops an ecofeminist ethics based in particularities. 'Ethics is not about what detached, impersonal, objective, rational agents engaged in grand theorising deduce. Rather, ethics is and should be about what imperfect human beings living in particular historical socioeconomic contexts can and should do, given those contexts' (2000, p. 114).

For Warren, ecofeminist ethics are interweaved and emergent (care-sensitive ethics) rather than abstract, rational and detached (the justice ethic). However, Warren wants to distinguish her concept of care-sensitive ethics from the notion of an ethic of care. That is, it is not about the specific case for caring, but for care as needing to be central to all ethics. Warren's ecofeminist philosophy sees a care-sensitive ethic as emergent from women's situated experience of care practices: '*care practices* are practices that either maintain, promote, or enhance the health (well-being, flourishing) of relevant parties, or at least do not cause unnecessary harm to the health (well-being, flourishing) of relevant parties' (2000, p. 115).

Recently, Warren and Plumwood have both stressed the importance of spirituality in achieving cultural change. This is an interesting development given that Chris Cuomo as late as 1998 was praising Warren and Plumwood for developing a 'non-spiritualist ecological feminist theory' (1998, p. 139). Theology and spirituality have had strong links with ecofeminism since its earliest days with publications by women such as Rosemary Radford Ruether (1975), Starhawk (1982), Charlene Spretnak (1982) and Riane Eisler (1987) variously addressing feminist theology or celebrating Earth-based consciousness, women's spirituality or the cult of the Goddess. An emphasis on spirituality is still strong; in 2002 the self-named 'Wicca' practitioner Starhawk called for an Earth-based spirituality to underpin the global struggle against structures such as the WTO.

Plumwood and Warren, like many other ecofeminists and greens, draw strength from the traditional spirituality of indigenous peoples. Warren stresses the importance of what she describes as place-oriented peoples, and particularly the women of the Chipko and native American movements. She sees place-based spirituality as playing a role in developing alternatives to the destructiveness of contemporary society (2000). Plumwood also speaks favourably of indigenous philosophers such as Carol Lee Sanchez who 'have mounted powerful critiques of the human-centredness of western spirituality and its denial of ecological inclusion and connection' (2002, p. 225). For Plumwood indigenous spirituality is 'a spirituality that both locates human beings within nature and acknowledges ethical, personal and narrative relationships as extending to the more-than-human world' (2002, p. 224).

She believes that most ecofeminists are 'calling for specifically ecological kinds of spirituality that are nondualist and immanent in orientation rather than transcendent and rationalist' (Plumwood, 2002, pp. 218–9). This would be 'a materialist spirituality which recognises that spirit is not a hyper-separated extra ingredient but a certain mode of organisation of a material body, unable to exist separately from it' (Plumwood, 2002, p. 223).

This spirituality is materialist because it does not seek to transcend death and because it is related to a materialist spirituality of place. This is necessary as 'both critical and spiritual projects require that we acknowledge caring relationships to multiple places as does any achievable goal of recognising the "ecological footprint" of daily living' (2002: 233). Plumwood recognises that any place-based approach risks the danger of a 'blood and soil' identity with land (2002, p. 219). She therefore argues that in contemporary society the most important thing is to be 'place-sensitive' (2002, p. 232). However, she goes on to argue that 'a language of the land requires a deep acquaintance with some place, or perhaps a group of places' (2002, p. 231). For both Warren and Plumwood their personal experience of wild nature is of great importance in developing their perspectives and philosophy. For Plumwood it is 'the experience of twilight communication in the southern Gondwanic rainforest' (2002, p. 231). For Warren it is rock-climbing (2000, pp. 102–3) and swimming with dolphins (2000, pp. 20–1).

While philosophers such as Warren and Plumwood are embracing more spiritual approaches, some ecofeminist theologians are questioning the relevance of feminist theology or ecofeminist spirituality to the material struggles around women and environment. In a recent book Heather Eaton recounts how as a feminist theologian she has found ecofeminist religious

discourses as fallible in the face of corporate global rule: 'We have myriad theories and ideals of relationality, respect for diversity and ethics of mutuality. We can understand empowerment and earth ethics. We can name the "Changing of the Gods" … but … where does all of this intersect with globalization?' (2004, p. 3). While acknowledging that within ecofeminist philosophy 'the emphases on methodology, and in particular epistemology, have strengthened feminist discourses', Eaton argues that 'there are losses, such as a lack of attending to the material world'. Ecofeminist emancipatory visions and strategies 'appear weak in the context of today's world' (Eaton, 2004, pp. 2–3).

Eaton quotes with favour Rosemary Radford Ruether's preface to the same book that warns of the dangers of a middle-class spiritual ecofeminism becoming disconnected from a socio-economic critique. If ecofeminism became preoccupied with 'self-cultivation' it might become a 'leisure class "spirituality" unconnected to poverty' (Ruether, 2004, p. x). Although herself coming from a theological background, it has always been Ruether's view that the devastation of the Earth has to be seen in a global economic context (1996). This would seem to call for more emphasis on the material analysis of the socio-economic structures of contemporary society.

TRANSCENDENCE, IMMANENCE AND THE EXPLOITATION OF WOMEN'S WORK

Ecofeminists who stress a socio-economic or materialist perspective start from the argument that the sexual division of labour, and the gendering of work in general, results in economic systems that systematically degrade both women and the natural world (Mellor, 1997a; Perkins, 1997; Peitila, 1997; Salleh, 1997; Mies, 1998). They see the present global market economy as subordinating women and nature through an interconnected web of cultural and economic valuing that is based on hierarchical dualisms that are materially exploitive and destructive. Men are socially and economically valued to the material detriment of women. Women are socially and economically marginalised and excluded and their labour exploited. Much of their domestic work is not valued at all. Market systems which favour men are valued and rewarded as against the subsistence systems on which many women rely for their survival. While resources are valued in market exchange their role in local eco-systems or subsistence systems are ignored. Individual wealth is valued over systems of social reciprocity. Traditional experience and knowledge, often associated with women, are equally devalued as against the science and technology of the cities and the 'advanced' economies unless they become marketable as in the current rush to patent indigenous resource, knowledge or products through TRIPS (Trade Related Intellectual Property Rights).

In these hierarchical and exploitative dualisms the dominant socio-economic position of men is directly related to the subordination and exploitation of women and nature (Salleh, 1997). Dominant men could not maintain their power without the free or underpaid work of women, a case made by many feminist economists (Himmelweit, 2000). Equally, the natural world is used as a free resource; common lands, amenity value and biodiversity are not recognised unless they can be sold. Women in subsistence societies are particularly vulnerable to the loss of common resources such as land, water and trees (Shiva, 1989; Braidotti et al., 1994; Harcourt, 1994).

The claim of an emerging ecofeminist political economy is that the link between women's subordination and the degradation of the natural world lies in women's centrality to the support economies of unpaid domestic work and social reciprocity, that is, the home and the community (Pietila, 1997). It develops a materialist ecofeminism which argues that the roots of patriarchy go much deeper than the present globalised capitalist market economy – it is a reflection of the way human activities are structured in all societies that devalue women (Salleh, 1997). Ecofeminist political economy is not just concerned with the unfairness of women's lives within the globalised economy, but their position at the boundaries of economic systems and the particular role of 'women's work' (Mellor, 1992, 1997a).

The essence of 'women's work' (which is not always done by women or even by all women) is that it is embodied in humanity's physical existence and embedded in the local (and global) environment. It is the work of the human body and its basic needs, the maintenance and sustenance of the human body and psyche through the cycle of the day, and the cycle of life (birth to death). Unlike labour in the profit-driven economy or tax-dependent sectors, this work has to be done in sickness and in health, in youth and in age, paid or unpaid. Inevitably such work is very responsive to environmental factors, particularly the health of the local ecosystem. 'Women's work' represents care work, child care, sick care, aged care, animal care, community care (volunteering, relationship building), family care (listening, cuddling, sexual nurturing, esteem building) and environmental care. Women's work tends to be embedded, it is of necessity local, communal, close to home (Shiva, 1994). In subsistence economies it is critically embedded in the local ecosystem. Women's work

is time-consuming because it is often routine and repetitive: cooking, cleaning, fetching and carrying, weeding. Often it involves just being there, watching, waiting, available, dependable, being on call. The more time an activity takes and the more limited it is spatially, the more likely it is to be associated with women.

For the materialist perspective of ecofeminist political economy, the history of gender relations has seen the creation of economic systems in which dominant men (and some women) claim social space and time while subordinate women and excluded social groups have remained engaged in the routine and necessary labours of life, close to home, domestic responsibilities and the necessities of existence (Mellor, 2000). This subordinated, marginalised and exploited work is the basic work that makes socially valued and economically rewarded activity possible. The material and materialist argument that ecofeminist political economy makes is that dominant men can transcend the limitations of human embodiment and embeddedness by using the labour of others, particularly women (Salleh, 1994; Mies, 1998).

The exploitation of this work is the mechanism by which unsustainable economic systems emerge from the embedded nature of human existence. The gendered nature of human society means that male-dominated socio-economic systems have not acknowledged the embodied and embedded nature of human existence. As a result 'women's work' and the 'work' of the ecosystem lies at the margins or beyond the boundaries of dominant cultures and economies (Mellor, 1992). The money-based economy does not recognise or internalise the holistic nature of human existence (Perkins, 1997). Patriarchal economic systems (capitalist or otherwise) have therefore been erected on a false sense of transcendence (Mellor, 1997b). They appear to have transcended human material conditions because they largely ignore them; such economic systems are therefore both exploitative and ecologically unsustainable (Hutchinson et al., 2002).

Ecofeminist political economists have paid particular attention to the exploitation of women and the natural world in the process of globalisation. Christa Wichterich goes so far as to argue that women are the fuel of the process of globalisation (2000). Like Maria Mies (1998) she sees 'globalised woman' as central to the accumulation of capital on a global scale. For Maria Mies, whose analysis was first published in 1986, the key process was 'housewification' where the women of the 'North' became cast in the role of consumers while the women of the 'South' were forced by necessity into the role of producers as their economic base in the subsistence economy was destroyed.

Mies has been accused of romanticising women's position in pre-market societies (Agarwal, 1992; Jackson, 1994), but her analysis of the role of patriarchy in the exploitation of women on a global scale is acute. As key providers for their families, and often as heads of households, women have had to bear the impact of trading and structural adjustment policies as governments try to compete their way to the bottom of the wages, conditions and public services league to gain 'comparative advantage'. Mies points to the acute degree of exploitation experienced by women where subsistence economies are penetrated by the capitalist market. Wages paid to both male and female labourers are calculated on the assumption that their basic subsistence is being met by their own subsistence economy. Men are not given wages sufficient to sustain themselves, let alone their families, and women are paid even less. This is the process that Maria Mies has described as super-exploitation:

> I define their exploitation as super-exploitation because it is not based on the appropriation (by the capitalist) of the time and labour over and above the necessary labour time, the surplus labour but of the time and labour necessary for people's own survival or subsistence production. It is not compensated for by a wage, the size of which is calculated on the 'necessary' reproduction costs of the labourer, but is mainly determined by force or coercive institutions. This is the main reason for the growing poverty and starvation of third world producers (1998, p. 48).

As the work of Mies and Wichterich shows, the position of women within globalisation can only be understood as an interrelationship between capitalism and patriarchy, or as Mies has called it patriarchal capitalism.

ECOFEMINISM, DEVELOPMENT AND GLOBALISATION

In the mid-1980s a global movement began to emerge that questioned the whole 'development' process that was shifting the focus of many countries from primarily rural indigenous economies to export-oriented commercialisation. In particular, it focused on the hardship women faced as the introduction of the global market economy, and particularly the commercialisation of agriculture, undermined subsistence economies (Sen and Grown, 1987; Dankelman and Davidson, 1988; Shiva, 1989; Sachs, 1996). Traditionally, much of women's access to resources was secured by usufruct, i.e. rights to use common or family land and resources without individual ownership.

The commercialisation of agriculture and forestry led to the 'privatization' or 'statization' (state ownership/control) of land (Agarwal, 1997). Commercialisation of agriculture and forestry also meant that there was less attention paid to subsistence needs, traditionally women's area of agricultural and domestic work. Lack of ready access to land and natural resources, such as water, led to women becoming increasingly impoverished and vulnerable. As Rocheleau *et al.* (1996, p. 291) note 'access to resources … proves to be an important environmental issue for women virtually everywhere'. Evidence grew that women's agricultural work (paid and unpaid) was being substantially under-reported and this was affecting their economic claims (Waring, 1989).

In 1984 a network of women from developing countries formed DAWN (Development Alternatives with Women for a New Era). DAWN challenged the prevailing claims that women's involvement in the development process would enhance their quality of life. Founder member Devaki Jain, summed up their disillusionment:

> Economic development, that magic formula, devised sincerely to move poor nations out of poverty, has become women's worst enemy. Roads bring machine-made ersatz goods, take away young girls and food and traditional art and culture; technologies replace women, leaving families even further impoverished. Manufacturing cuts into natural resources (especially trees), pushing fuel and fodder resources further away, bringing home-destroying floods or life-destroying drought, and adding all the time to women's work burdens (quoted in Pietila and Vickers, 1990, p. 35).

In 1987 DAWN published their alternative to the Western-driven development process 'Development Crises and Alternative Visions' (Sen and Grown, 1987). Arguing that women in subsistence economies were 'managers of human welfare' (1987, p. 18), they called for 'women's values of nurturance and solidarity' to characterise human relationships and that 'basic needs should become basic rights' (1987, p. 80).

This approach challenged the prevailing approach to women's marginalisation in development, Women in Development (WID), that advocated women's inclusion in the development process. Increasingly the WID approach was criticised for uncritically endorsing development (Braidotti *et al.* 1994; Kabeer, 1994). It was also argued that women were being used in an instrumental way by development agencies to secure the implementation of their programmes rather than to further the interests of women themselves (Jackson, 1994). DAWN and supportive ecofeminists promoted the notion of WED

(Women, Environment and Development) (Harcourt, 1994). Ecofeminists became involved in a range of issues including opposition to population-control policies and reproductive technologies generally (Sen, 1994). Ecofeminists played a key role in the formation of FINNRAGE (Feminist International Network of Resistance to Reproductive and Genetic Engineering) in 1985.

The link between gender and ecology began to emerge within global politics. One of the most influential books was Marilyn Waring's work on the invisibility of women and the natural world in international measures of economic wealth (1989). As a New Zealand MP she found that world decision-making systems rendered women, women's work and the natural environment invisible. At that time the United Nations System of National Accounts (UNSNA) explicitly excluded women's work from its definition of production. Subsistence work defined as 'primary work' was held to have no value 'since primary production and the consumption of their own produce by non-primary producers is of little or no importance' (Waring, 1989, p. 78). Waring saw this as reflecting the 'blindness and arrogance of patriarchy' which enshrined 'the invisibility and enslavement of women in the economic process' (1989, p. 78). She argued that not taking account of women's subsistence work had profound consequences for countries with large subsistence economies as it underestimated the real productive capacity of the country and led aid programmes to ignore women's needs for such fundamental things as access to clean water or sanitation.

The global linking of women's interests and environmental issues directly challenged the international programme of development sponsored by the industrialised nations and the United Nations. The UN responded to the growing movement by acknowledging the need for a specific women's input to the 1992 'Earth Summit' held in Rio de Janeiro, Brazil. Two conferences were hurriedly called in Miami in 1991. The first 'Global Assembly of Women for a Healthy Planet' brought over two-hundred women from all over the world to present their experiences of managing and protecting the environment to five-hundred invited delegates from development organisations. The second called fifteen-hundred women together from eighty-three countries to prepare a Women's Action Agenda for the 'Earth Summit' with the result that:

> for the first time ever women across political/geographical, class, race, professional and institutional divides came up with a critique of development and a collective position on the environmental crisis, arrived at in a participatory and democratic process (Braidotti et al., 1994, p. 103)

In parallel to the Summit itself in June 1992 the Brazilian Women's Coalition organised a women's conference, Planeta Femea, at the NGO forum in Rio de Janeiro. The result of these meetings was an input to Agenda 21 in which the position of women was addressed, specifically in Chapter 24, where the need for the active involvement of women in economic and political decision-making was acknowledged. The relative lack of importance of women, however, can be shown by the fact that a calculation of the costs of implementing Agenda 21 was $600 billion for the whole programme but only $40 million for programmes relating to women. Even the Planeta Femea group was not without its problems. As Sabine Hausler points out the remarkable unity among all the delegates around women and the environment led to a lack of sensitivity to divisions and inequalities between women, and the problems of racism and lack of representation were raised (1994, p. 150).

For Christa Wichterich the most important outcome of global capitalism is that while it has exploited and marginalised women it has also mobilised them (2000). While there are a legion of injuries to women that have sharply worsened their lives and livelihoods, the global process has given them the means to build global networks. The hope for Wichterich lies in South–South networking through organisations such as DAWN and the UN women's world conferences. Wichterich is not rosy-spectacled about these events and senses a conference weariness as women trudge from forum to forum seeing few positive outcomes. However, reflecting on the 1995 Women's Conference in Beijing, Wichterich saw the global feminist agenda as being led by women from the South: 'women from the North were often there as listeners and learners, as the main issues were overwhelmingly introduced by women from the South' (2000, p. 146). The way forward for Wichterich is not global feminism but an alliance of differentiated local feminisms. She sees differences and diversity as part of the strength of the women's movement, but argues that this does not mean there are no mutual experiences to be shared. DAWN also argue that transformation through participation is the way forward and that despite the contradictions of globalisation, change will not be achieved without engagement with it. Solidarity around issues that transcend cultural difference will provide the basis of emancipatory struggles. However, in recent years some ecofeminists have argued that emancipation can only lie in a rejection of the whole programme of globalisation and development, not engagement with it.

REJECTING DEVELOPMENT

Maria Mies and Vandana Shiva have been particularly vocal in claiming that Western-driven technological changes reflect 'mal-development' (1993, p. 297). Instead they advocate a withdrawal from the market economy in favour of a 'subsistence perspective'. This approach is being developed in a number of centres. Maria Mies works closely with Veronika Bennholdt-Thomsen, Director of the Institute of the Theory and Practice of Subsistence at Bielefeld, Germany (Bennholdt-Thomsen and Maria Mies, 1999; Bennholdt-Thomsen *et al.,* 2001). Research centres for alternative development have also been founded elsewhere. Vandana Shiva is Director of the Research Foundation for Science Technology and Ecology in New Delhi, and in Dhaka Policy Research for Alternative Development (UBINIG) is led by Farida Akhter.

In their case for a 'subsistence perspective' Veronica Bennholdt-Thomsen and Maria Mies reject the whole premise and promise of modernity (1999). They reject the aim of moving all human communities towards modernity and the market and money-based industrial form. They oppose the enclosure of resources that were once held in common, particularly land and the segregation of the public world of politics and economics from the private world of the oikos (households) and non-commodified work. They reject the separation of people from place, of humanity from nature. For those who embrace the subsistence perspective the claims of modernity have been a sham. They see 'development' as based on the ideological notion that human communities are moving from 'primitive' forms of society and economy to the maturity of industrialisation and the global market economy through the application of rationalised thinking. Such an approach argues that it is in the interest of all human societies to give up their collective resources and autonomous means of subsistence in order to embrace the market and 'development'. For Bennholdt-Thomsen and Mies this is a total misrepresentation of the true situation. What is in fact happening is a violent attack on the rights of women and indigenous peoples.

To critics who say that the subsistence perspective is romanticised and apolitical, depending on small-scale initiatives and personal change, Bennholdt-Thomsen and Mies argue that it represents an alternative form of politics and economics that is much needed if there is to be an alternative to global capitalism. They base their case on the experience of indigenous, peasant and communal economies in Africa, Latin America and the South Pacific as well as Europe (Bennholdt-Thomsen *et al.,* 2001). Those who support the resurgence of subsistence argue that it should not be read as a wholesale return to traditional farming systems (although this should be preserved where possible) but the issue is the ownership and control of the means of consumption and production at the local level. Food security, in the broadest sense. Where local subsistence systems have already been destroyed the aim would be to rebuild local systems of production and circulation through new structures such as eco-villages, cooperatives or LETS (Local Exchange Trading Systems) schemes.

The subsistence perspective has raised a number of questions for ecofeminists. The first is whether it is any longer explicitly ecofeminist as against being a more general celebration of the value of indigenous communities and traditional farming societies. To the extent that the particular role of women in indigenous subsistence systems is praised, the reality of women's experience has been questioned. Agarwal (1992), Braidotti *et al.* (1994), Carlassare (1994), Jackson (1995), Sturgeon (1997) and Sandilands (1999) have all expressed concern that there is a tendency to romanticise indigenous societies and in describing women as being guardians of the ecology of traditional subsistence-based farming, the repressive and exploitative cultural framework within which those systems operate is being ignored.

There is also evidence that women in indigenous communities do not necessarily reject Western values or embrace environmental issues. Rosemary Radford Ruether found in her interviews with women in indigenous societies that they were not likely to idealise their indigenous heritage or set their own traditions against those of the West. Instead they were concerned with the practicalities of life (1996). A study of the Mijikenda people of Kenya by Celia Nyamweru found that women did not behave in a more sustainable way than men in relation to their sacred kaya forests (2004). They appeared to have no greater concern than men about ecological risks. However, she argued that this was partly due to the heavily patriarchal and gerontocratic nature of the society. In her study of the Mexican Chiapas, Lois Lorentzen found that women were not culturally identified with nature and even when women found a politicial voice they did not identify ecology as an issue: 'It is striking that in the Women's Revolutionary Laws formed by Zapatista women and affirmed at the Indigenous Women's Congress in Chiapas in 1997, no mention was made of nature, of women's connection to nature, or of the impact of environmental deterioration specifically on women' (Lorentzen, 2004, p. 61). As Lorentzen points out it would be wrong to assume that indigenous equals traditional lifeways (2004, p. 65).

It was also important not to ignore indigenous inequalities:

> Overemphasizing the West's economic system's culpability in harming indigenous women and nature, great as it may be, leaves us without analytic tools to uncover class, property and gender relations within a particular nation, society or group (2004, p. 65).

For Cecile Jackson there is a danger to the whole feminist agenda:

> Ecofeminist prescriptions are for women to reject transcendence, embrace the body, bond to our mothers, remain embedded in our local ecosystems, abandon the goals of freedom and autonomy, rely on and care for our kin and community and remain in subsistence production. Such conservatism can hardly claim empowerment for women (1995, p. 129)

Sabine Hausler has also expressed concern that women, far from being seen as disadvantaged within the global process are coming to be seen as the solution:

> The prevailing image of women as agents fighting the effects of the global ecological crisis casts them as *the* answer to the crisis: women as privileged knowers of natural processes, resourceful and 'naturally' suited to provide the 'alternative' (1994, p. 149, italics in original)

This raises the critical issue for ecofeminism. Is there a particularity of (all?) women in relation to the dysfunctional relationship between human societies and the non-human environment?

ECOFEMINISM AND THE PARTICULARITY OF WOMEN

The starting point of ecofeminism is that women share with the natural world subordination and exploitation at the hands of male-dominated societies. For those who stress cultural, philosophical or spiritual approaches, the solution lies in recognising the distortions of current ways of thinking and believing and developing and embracing new visions. For those ecofeminists who stress the more material relations of the sexual division of labour, gendered structures of economic and social power will need to be dismantled so that all humanity lives within its ecological means. Male-dominated economic systems will no longer be able to function as if the labour of daily life and the renewal of ecosystems do not exist. The problem is how this change

will come about and what particular role women will play.

Some ecofeminists have asserted that there is a direct link between women's roles in society and their attitudes to the natural world. It is argued that since women are primarily responsible for life-giving and sustaining nurturance work, they exhibit caring values and behaviours that can be incorporated into an ecofeminist politics (Collard, 1988). Another version of the link between the social location of women and their worldview is the claim that the gendered relationship between humanity and nature leads women to have a particular view or knowledge of the world. Implicitly or explicitly this version of ecofeminism has a theory of knowledge that privileges women; Ariel Salleh, for example, identifies 'an epistemology from below'. She argues that women's experience in capitalist patriarchal societies has given them a materially grounded base which privileges them 'temporarily as historical agents par excellence'. Women are in a contradictory position in that they are both inside and outside of patriarchal systems. From this position whether they are 'dominated' or 'empowered' women are well placed to 'take up the case for "other" living beings' (1994, p. 120).

The view of women as privileged 'knowers' has given rise to criticisms of essentialism (Evans, 1993; Seager, 1993) and the spiritual and bodily linkages between women and nature that were being made in the name of ecofeminism led some writers to adopt a different terminology. Bina Agarwal put forward the concept of feminist environmentalism and from a structuralist perspective called for a transformative struggle around the woman–nature relationship rather than a celebration of it (1992). Chris Cuomo (1998) and Karen Warren (1997) have used the term ecological feminism. Critics of essentialism in ecofeminism have generally been concerned that the reassertion of women's association with the natural world, whether through their bodies, their caring role as mothers or nurturers or their traditional forms of subsistence work, far from being an agent for change in society, could become a reaffirmation of women's present position.

Chris Cuomo argues that concepts like 'women', 'femininity', 'motherhood' 'diversity' and 'unity' have been used far too simplistically and although ecofeminism has produced 'truly insightful, moving and carefully argued work' it has also 'spawned some very bad theory' (1998, pp. 139–40). In particular:

> proponents of an unmodified reclamation of ... ethics based on care and compassion may be drawing from a conception of female rooted not in the actual material and social conditions of

women living anywhere today, but from mythical ideals of femininity and historical woman (1998, p. 130).

Catriona Sandilands also rejects ecofeminism as a politics of identity based on women's particularity. From a post-modern perspective she argues for ecofeminism as an aspect of radical democracy (1999, p. 49) linking feminist with ecological struggles.

Karen Warren sees women's care-sensitive experience as an important aspect of what she calls situated universalism. 'Situated universals are "situated" in that they grow out of and reflect historically particular, real-life experiences and practices: they are "universal" in that they express generalisations common to, and reflective of, lives of diverse peoples situated in different historical circumstances' (2000, p. 114). Women are, therefore, part of a network of experiences that express the commonality of the human predicament.

Similarly, Ynestra King, a social ecofeminist, does not assert a 'natural' affinity between women and the natural world, arguing instead that the socially constructed identity of women and nature that has been imposed upon women, could be consciously used as a 'vantage point' for:

creating a different kind of culture and politics that would integrate intuitive, spiritual and rational forms of knowledge, embracing both science and magic insofar as they enable us to transform the nature–culture distinction and to envision and create a free, ecological society (1989, p. 23).

Other ecofeminists have asserted the importance of not distancing ecofeminism from the implications of an embodied view of women (Carlassare, 1994; Cook, 1998). Cook sees ecofeminism as reclaiming the politics of the body:

Women are personally and collectively revaluing and redefining denigrated and naturalized female experiences, including childbirth, breastfeeding, menstruation and menopause (1998, p. 246).

Ariel Salleh advocates a 'materially embodied standpoint' but denies that this is an essentialist position (1997). She argues that it is not women's bodies that locate them politically, but their position within the sexual division of labour. Also, women do not spontaneously recognise this contradiction, they have to develop an 'ecofeminist consciousness'. Salleh therefore follows the Marxist model of distinguishing between a sex/gender in itself and a conscious sex/gender 'for itself', or rather 'for nature'.

Kate Soper also acknowledges that the danger of essentialism may always be present in ecofeminism simply because it is engaged with such fundamental questions about the relationship between humanity and non-human nature. 'Human beings, like all other living creatures, are determined by biology in the sense that they are embodied mortal entities' (Soper, 1995, pp. 125–6) … 'there is an extra-discursive and biologically differentiated body upon which culture goes to work and inscribes its specific and mutable gender text' (1995, p. 133). In this context to speak of nature is 'to speak of those material structures and processes that are independent of human activity (in the sense that they are not a humanly created product) and whose forces and causal powers are the necessary condition of every human practice, and determine the possible forms it can take' (Soper, 1995, pp. 132–3). Ecofeminism's role is, therefore, to start from this material reality and address the future of the relationship between humanity and the rest of the natural world. At the same time Soper rejects a cultural revaluation of the feminine:

an ecofeminist politics that calls on us to celebrate previously derided 'feminine' values, or that would look to that feminine 'difference' which culture has hitherto excluded, as the site of renewal, does not necessarily go very far in de-gendering the implicitly masculinist conception of humanity that has gone together with the feminization of nature (1995, p. 127).

Noel Sturgeon has argued that in a patriarchal society it may be necessary for ecofeminists to adopt a 'strategic essentialism' in order to address the questions that have been ignored (1997). Women may need to unite around their female identity for political reasons, but this does not mean that they do not acknowledge differences between women or assert an essential commonality.

One of the earliest exponents of ecofeminism, Rosemary Radford Ruether restated in 2004 that ecofeminism cannot 'treat women as a univocal category' (2004, p. vii). Nor can it restrict itself to a consideration of women alone 'the complexity of gender within class and race, reveal[s] why ecofeminism must interconnect with the movements against environment racism and for ecojustice' (2004, p. viii). Ruether sees ecofeminist analysis as starting from a critique of the connection between sexism and ecology and linking this to all the other categories of oppression, in particular class exploitation and racism: 'Global ecofeminism shows how … patterns of impoverishment of nature and of emiserated humans are interconnected in a worldwide economic system skewed to the benefit of the rich beneficiaries of the market economy' (2004, p. viii).

Far from being a restriction on feminism, ecofeminism offers an analysis that shows the way in which exploitative and ecologically unsustainable systems have been constructed through the gendering of human society. Such an analysis demands radical change. As Ruether argues, feminism should be 'about converting the relations of patriarchal domination for both women and men into new relationships of mutuality ... bringing men into the work of care for the household and earth that is now borne disproportionately by women' (2004, p. ix).

REFERENCES

Agarwal, Bina (1992) The Gender and Environment Debate: Lessons from India. *Feminist Studies* 18(1), pp. 119–158.

Agarwal, Bina (1997) Re-sounding the Alert – Gender, Resources and Community Action. *World Development* 25(9), pp. 1373–1380.

Bennholdt-Thomsen Veronika and Maria Mies (1999) *The Subsistence Perspective*. London: Zed Press.

Bennholdt-Thomsen Veronika, Nicholas Faraclans and Claudia von Werlhof (2001) *There is an Alternative: Subsistence and Worldwide Resistance to Corporate Globalization*. London: Zed Press.

Biehl, Janet (1991) *Finding Our Way: Rethinking Ecofeminist Politics*. Montreal: Black Rose Books.

Braidotti, Rosa, Ewa Charkiewicz, Sabine Hausler and Saskia Wieringa (1994) *Women, the Environment and Sustainable Development*. London: Zed Press.

Caldecott, Leonie and Stephanie Leland (Eds) (1983) *Reclaim the Earth*. London: The Women's Press.

Carlassare, Elizabeth (1994) Destabilizing the Criticism of Essentialism in Ecofeminist Discourse. *Capitalism, Nature, Socialism* 5 (3), Issue 19, pp. 50–66.

Collard, Andree, with Joyce Contrucci (1988) *Rape of the Wild*. London: The Women's Press.

Cook, J. (1998) The Philosophical Colonization of Ecofeminism. *Environmental Ethics* 20 (3), pp. 227–246.

Coote, Anna and Beatrix Campbell (1982) *Sweet Freedom*. London: Picador.

Cuomo, Chris J. (1998) *Feminism and Ecological Communities: An Ethic of Flourishing*. London: Routledge.

Dankelman, Irene and Joan Davidson (1988) *Women and Environment in the Third World*. London: Earthscan.

Davion, Victoria (1994) Is Ecofeminism Feminist? (in) Karen Warren (Ed.) *Ecological Feminism*. London: Routledge.

Diamond, Irene and Gloria Feman Orenstein (Eds) (1990) *Reweaving the World*. San Francisco: Sierra Club Books.

Eaton, Heather (2004) Can Ecofeminism Withstand Corporate Globalization? (in) Heather Eaton and Lois Ann Lorentzen (Eds) *Ecofeminism and Globalization: Exploring Culture, Context and Religion*. Rowman and Littlefield. MD: Lanham.

Eaton, Heather and Lois Ann Lorentzen (2004) *Ecofeminism and Globalization: Exploring Culture, Context and Religion*. Lanham MD: Rowman and Littlefield.

Eisler, Riane (1987) *The Chalice and the Blade*. London: Unwin.

Evans, Judy (1993) Ecofeminism and the Politics of the Gendered Self (in) Andrew Dobson and Paul Lucardie (Eds) *The Politics of Nature*. London: Routledge.

Griffin, Susan (1978) *Woman and Nature: The Roaring Inside Her*. London: The Women's Press.

Harcourt, Wendy (Ed.) (1994) *Feminist Perspectives on Sustainable Development*. London: Zed Press.

Hausler, Sabine (1994) Women and the Politics of Sustainable Development (in) Wendy Harcourt (Ed.) *Feminist Perspectives on Sustainable Development*. London: Zed Press.

Himmelweit, Susan (Ed.) (2000) *Inside the Household: From Labour to Care*. London: Macmillan.

Hutchinson, Frances, Mary Mellor and Wendy Olsen (2002) *The Politics of Money: Towards Sustainability and Economic Democracy*. London: Pluto.

Jackson, Cecile (1994) Gender Analysis and Environmentalisms (in) Michael Redclift and Ted Benton (Eds) *Social Theory and the Global Environment*. London: Routledge.

Jackson, Cecile (1995) Radical Environmental Myths: A Gender Perspective. *New Left Review* No. 210, March/April, pp. 124–140.

Jones, Maggie and Wangari Maathai (1983) Greening the Desert: Women of Kenya Reclaim the Land (in) Leonie Caldecott and Stephanie Leland (Eds) *Reclaim the Earth*. London: The Women's Press.

Kabeer, Naila (1994) *Reversed Realities*. London: Verso.

King, Ynestra (1989) Toward an Ecological Feminism and a Feminist Ecology (in) Judith Plant (Ed.) *Healing the Wounds: The Promise of Ecofeminism*. London: Green Print.

King, Ynestra (1990) Healing the Wounds: Feminism, Ecology and Nature/Culture Dualism (in) Irene Diamond and Gloria Feman Orenstein (Eds) *Reweaving the World*. San Francisco: Sierra Club Books.

LaDuke, Winona (1993) A Society Based on Conquest Cannot Be Sustained: Native Peoples and the Environmental Crisis (in) Richard Hofrichter (Ed.) *Toxic Struggles*. Philadelphia: New Society Publishers.

Liddington, Jill (1989) *The Long Road to Greenham*. London: Virago.

Lorentzen, Lois Ann (2004) Indigenous Feet: Ecofeminism, Globalization, and the Case of Chiapas (in) Heather Eaton and Lois Ann Lorentzen (Eds) *Ecofeminism and Globalization: Exploring Culture, Context and Religion*. Lanham, MD: Rowman and Littlefield.

Mellor, Mary (1992) *Breaking the Boundaries*. London: Virago.

Mellor, Mary (1997a) Women, Nature and the Social Construction of 'Economic Man'. *Ecological Economics*, (20) 2, pp. 129–140.

Mellor, Mary (1997b) *Feminism and Ecology*. Cambridge: Polity.

Mellor, Mary (2000) Challenging the New World (Dis) order: Feminist Green Socialism (in) Susan Himmelweit (Ed.) *Inside the Household: From Labour to Care*. London: Macmillan.

Merchant, Carolyn (1983) *The Death of Nature*. New York: Harper & Row.

Merchant Carolyn (1990) Ecofeminism and Feminist Theory (in) Irene Diamond and Gloria Feman Orenstein (Ed) *Reweaving the World*. San Francisco: Sierra Club Books.

Mies, Maria (1998) *Patriarchy and Accumulation on a World Scale*. London: Zed Press.

Mies, Maria and Vandana Shiva (1993) *Ecofeminism*. London: Zed Press.

Nyamweru, Celia (2004) Women and Sacred Groves in Coastal Kenya: A Contribution to the Ecofeminist Debate (in) Heather Eaton and Lois Ann Lorentzen (Eds) *Ecofeminism and Globalization: Exploring Culture, Context and Religion*. Lanham MD: Rowman and Littlefield.

Perkins, Ellie (Ed.) (1997) Special Issue: Women, Ecology and Economics. *Ecological Economics* 20(2).

Pietila, Hilkka (1997) The Triangle of the Human Economy: Household – Cultivation – Industrial Production. An Attempt at Making Visible the Human Economy *in toto*. *Ecological Economics* 20(2), pp. 113–128.

Pietila, Hilkka and Jeanne, Vickers (1990) *Making Women Matter*. London: Zed Books.

Plant, Judith (Ed.) (1989) *Healing the Wounds: The Promise of Ecofeminism*. London: Green Print.

Plumwood, Val (1993) *Feminism and the Mastery of Nature*. London: Routledge.

Plumwood, Val (2002) *Environmental Culture: The Ecological Crisis of Reason*. London: Routledge.

Rocheleau, Diane, Barbara Thomas-Slayter and Ester Wangari (Eds) (1996) *Feminist Political Ecology*. London: Routledge.

Ruether, Rosemary Radford (1975) *New Woman, New Earth*. New York: The Seabury Press.

Ruether, Rosemary Radford (1996) *Women Healing Earth: Third World Women on Ecology, Feminism and Religion*. New York: Orbis Books.

Reuter, Rosemary Radford (2004) Preface (in) Heather Eaton and Lois Lorentzen (Eds) *Ecofeminism and Globalisation: Exploring Culture, Context and Religion*. Lanham, MD: Rowman and Littlefield, pp. vii–vi.

Sachs, Carolyn (1996) *Gendered Fields*. Boulder, CO: Westview Press.

Salleh, Ariel (1994) Nature, Woman, Labour, Capital: Living the Deepest Contradiction (in) Martin O'Connor (Ed.) *Is Capitalism Sustainable*. New York: Guilford.

Salleh, Ariel (1997) *Ecofeminism as Politics*. London: Zed Press.

Sandilands, Catriona (1999) *The Good-Natured Ecofeminist*. Minneapolis: University of Minnesota Press.

Seager, Joni (1993) *Earth Follies*. London: Earthscan.

Sen, Gita (1994) Women, Poverty and Population: Issues for the Concerned Environmentalist (in) Wendy Harcourt (Ed.) *Feminist Perspectives on Sustainable Development*. London: Zed Press.

Sen, Gita and Caren Grown (1987) *Development, Crises and Alternative Visions*. New York: Monthly Review.

Shiva, Vandana (1989) *Staying Alive*. London: Zed Press.

Shiva, Vandana (Ed.) (1994) *Close to Home*. Philadelphia: New Society Publishers.

Soper, Kate (1995) *What is Nature?* Oxford: Blackwell.

Spretnak, Charlene (1982) *The Politics of Women's Spirituality*. New York: Doubleday.

Spretnak, Charlene (1990) (in) Irene Diamond and Gloria Feman Orenstein (Ed.) *Reweaving the World*. San Francisco: Sierra Club Books.

Starhawk (Ed.) (1982) *Dreaming the Dark*. Boston: Beacon Press.

Starhawk (2002) *Webs of Power: Notes from the Global Uprising*. Philadelphia: New Society Publishers.

Sturgeon, Noel (1997) *Ecofeminist Natures*. London: Routledge.

Waring, Marilyn (1989) *If Women Counted*. London: Macmillan.

Warren, Karen J. (Ed.) (1994) *Ecological Feminism*. London: Routledge.

Warren, Karen J. (Ed.) (1996) *Ecological Feminist Philosophies*. Bloomington: Indiana University Press.

Warren, Karen J. (Ed.) (1997) *Ecofeminism*. Bloomington: Indiana University Press.

Warren, Karen (2000) *Ecofeminist Philosophy*. Lanham: Rowman and Littlefield.

Wichterich, Christa (2000) *The Globalised Woman*. London: Zed Press.

5

Deep Ecology

Ted Benton

INTRODUCTION

Deep ecology is a diverse social movement, taking its inspiration from a set of radical philosophical ideas about nature and the relationship of humans to it. The key thinkers of the deep ecology movement start with a view of humans as inseparably connected to the rest of nature. So deep is this connection and interdependence that, for them, it makes no sense to see humans as uniquely valuable. The dominant traditions of modern Western philosophy (as well as common-sense thinking) have assigned a special status to humans in virtue of our supposedly unique attributes such as rationality, capacity for language, possession of self-consciousness or freedom of choice. Because of this, human individuals have been seen as worthy of moral consideration in their own right, with other living and non-living beings having value only in so far as they affect the well-being of humans. As we saw in Chapter 2, Darwin, as well as some utilitarian thinkers in the 19th century, recognised that some other animal species can experience pleasure and pain, fear, loyalty and other emotions, and to this extent they were able to extend the circle of direct moral concern beyond the boundaries of our own species. However, the thinking of the deep ecologists takes an even more radical departure from the 'mainstream' than this.

The American Romantic environmentalists, H. D. Thoreau and John Muir, are often cited as predecessors of the deep ecologists. The American forester–naturalist Aldo Leopold is another important inspiration. His 'land ethic' proposed to include in the moral community not just humans, animals and plants, but the land itself: 'A thing is right when it tends to preserve the integrity, stability, and beauty of the biotic community. It is wrong when it tends otherwise' [Leopold (1949), 1977]. Another way of understanding their sources is to think back to the values and sentiments of the Romantic poets we discussed in Chapter 2. As we saw, they tended to avoid systematic theory building, and although some had strong political views, there was no Romantic political movement as such. However, the deep respect and wonder at wild nature that is such a feature of much Romantic art and literature is something that the Romantics share with the deep ecologists. However, the latter do attempt to give a more systematic philosophical grounding for their outlook, and they draw activist political conclusions from it.

THE PHILOSOPHY OF ARNE NAESS

The Norwegian philosopher Arne Naess is widely accepted as the founding theorist of the deep ecology movement. He spent much of his life as an academic philosopher, concerned with questions of meaning and communication in language, as well as making a special study of the 17th century rationalist philosopher, Benedict de Spinoza (1632–1677). He also took a particular interest in the non-violent political philosophy and activism of Mahatma Gandhi. Naess's own approach shows the influence of each of these aspects of his academic studies. The interest in language is one source for Naess's emphasis on networks of relationships in systems, as against analytical approaches that consider things (words, people, objects, etc.) as isolated 'atoms'. From Spinoza he takes a holistic conception of the world as a unified totality, and an ethic of 'self-realisation' as identification of the self with this greater unity of the world (see de Jonge, 2004). Finally, and more directly, he takes from Gandhi's non-violent liberation struggle against British imperialism as his model for the activism of the Green movement.

Naess introduced the term 'deep ecology' in 1973 in the context of a rather stark opposition between what he saw as two very different streams within the environmental movement. 'Shallow' ecology, the dominant tradition, was primarily concerned with issues of pollution and resource depletion, and its main objectives were health and high living standards in the developed countries. By contrast, 'deep' ecology, a minority tradition, was characterised by two main features: first, an 'ontological' view (i.e. an account of the nature of the world) and, second, a value perspective that is taken to follow from it. The ontological view rejects what Naess calls the 'man-in-environment' image (typical of shallow ecology) in favour of a 'relational, total-field image'.

There are two main differences between these contrasted 'images' of reality. The 'man-in-environment' image puts humans at the centre, and sees other beings and relationships as forming the context of human life (in other words it is 'anthropocentric'). By contrast, the 'total-field' image refuses to make humans the central focus. But the two images also differ in the way they think about things (objects, people, animals, etc.) and the relationships between them. In every-day common sense, as well as most scientific thinking, we distinguish between the characteristics that are intrinsic to things, and those that they have only in virtue of their relationships to other things. So, we might distinguish between personal characteristics such as being two metres tall, having red hair and brown eyes, as against 'relational' characteristics, like being a mother or father, a student, or an asylum seeker. We might think that the latter set of characteristics are not 'essential' to our identity, because they are true of us only because we have a certain alterable position in society, or relationship to another person or set of social institutions. But Naess uses several lines of argument, some drawn from an analogy with the meaning of words in language, others from the psychology of perception, to suggest that a deeper analysis would show all characteristics are really relational ones. So, being two metres tall is only true of me in virtue of a conventional method of measurement, whilst brown eyes and red hair will both look very different depending on the light conditions and the perceptual apparatus of the observer. So Naess extends the emphasis in ecological thought on the relationships between populations and communities of living things and their non-living conditions. He extends it into a general view of the world-as-a-whole in which there are ultimately no 'things' but only networks of relationships.

To this ontological view, Naess adds a distinctive value perspective: 'Biospherical egalitarianism – in principle'. He sees it as an intuitively obvious 'value axiom' for those who have a deep understanding and connection with the rest of nature. The example he uses is the 'ecological field worker', but he and other deep ecologists also refer often to their experiences of uncultivated wild nature, mountain landscapes and 'wilderness'. In this respect it is probably worth noting that the leading thinkers of the deep ecology movement have often come from countries – Scandinavia, the USA and Australia – that still have vast areas of uncultivated land, much of it, of course, the contested home of indigenous people. However, Naess still thinks that the moral sentiments that lead him and others to accept 'biospherical egalitarianism' are potentially universal among humans. To deny our need for close partnership with non-human life is to be alienated from our own nature. For Naess, full human development involves establishing a sense of self-identity that goes so far beyond 'egoism' as to include our unity with the rest of life. This is what he terms 'Self-realisation'.

POLITICS: THE PLATFORM OF DEEP ECOLOGY

But these rather abstract philosophical ideas are still some distance from the political programme of a social movement, and here Naess recognises the need for a much more open 'platform' to express the common ground among deep ecologists, many of whom will not share his own particular philosophical views. So, during the 1980s Naess worked with an American deep ecologist, George Sessions, to draw up just such a platform. This is worth representing in full, but it needs quite a lot of further explanation (Table 5.1).

According to Naess, there may be many philosophical or religious outlooks ('ecosophies', he calls them), including his own ('ecosophy T') which might lead people to agree with these eight principles. This allows for pluralism in the movement, so long as there is a consensus around this core platform. Of course, there are numerous vague terms in this statement – 'quality of life', 'vital needs', 'intrinsic value' and so on. Naess admits this, but in part the vagueness is a deliberate attempt to make possible common ground, where greater precision would narrow down the possible areas of agreement. But Naess also notes the tendency on the part of a science-dominated culture to devalue as 'vague' anything that is not strictly quantifiable and measurable.

Table 5.1 A platform of the deep ecology movement

1. The flourishing of human and non-human life on Earth has intrinsic value. The value of non-human life forms is independent of the usefulness these may have for narrow human purposes.
2. Richness and diversity of life forms are values in themselves and contribute to the flourishing of human and non-human life on Earth.
3. Humans have no right to reduce this richness and diversity except to satisfy vital needs.
4. Present human interference with the non-human world is excessive, and this situation is rapidly worsening.
5. The flourishing of human life and cultures is compatible with a substantial decrease of the human population. The flourishing of non-human life requires such a decrease.
6. Significant change of life conditions for the better requires change in policies. These affect basic economic, technological and ideological structures.
7. The ideological change is mainly that of appreciating *life quality* (dwelling in situations of intrinsic value) rather than adhering to a high standard of living. There will be a profound awareness of the difference between big and great.
8. Those who subscribe to the foregoing points have an obligation directly or indirectly to participate in the attempts to implement the necessary changes.

Elaborations

The first three principles are effectively restatements of Naess's own axiom of 'biospherical egalitarianism in principle'. The emphasis on 'flourishing' relates to the concept of 'life quality' (principle 7). The concern here is not for mere survival, nor even for higher material living standards, but rather with the full realisation of potentials, with life lived according to ultimate values. In this respect the deep ecological view overlaps with a humanist insistence on the importance of all-round development and self-realisation, but this is extended to all other life forms. These, too, have an 'equal right to live and blossom'. The 'richness and diversity' of life forms includes what is now understood as 'biodiversity', but it also involves such things as rivers, landscapes, cultures and ecosystems that would not ordinarily be understood as life forms. The deep ecological position is, therefore, more accurately described as 'ecocentric', rather than (more narrowly) biocentric. The term 'richness' refers to populations of life forms: mere survival as minimum sustainable populations would not do as an adequate conservation strategy. For deep ecology, flourishing of non-human life forms has to include sufficiently large populations for biological evolution to proceed. This, in turn, partly explains their policy emphasis on the preservation of large areas of 'wilderness', free from significant modification by human activity.

Allowing for the flourishing of non-human life entails strong restrictions on human activity in relation to nature. However, since humans are themselves life forms that, no less than the others, need to draw on nature to meet their needs, some 'killing, exploitation and suppression' is unavoidable. The concept of satisfying 'vital needs' is the key idea for setting normative limits on what humans are entitled to inflict on the rest of nature.

People will differ, of course, over what will count as a 'vital need': access to clean water, unpolluted air to breathe, sufficient nutritious food, and shelter, however, are likely to be uncontroversial. At the other end of the spectrum, the 'need' of urban dwellers for four-wheel drive cars, for numerous holiday jet flights, for the latest fashions in clothing and household gadgets and so on might be more questionable. The point here is not to arrive at some definitive answer, but, rather, to provoke serious reflection on how much of the taken-for-granted consumption of the affluent is essential to our quality of life, taking into account both its impact on non-human nature as well as impoverished human populations.

Point 4 in the platform claims that the human impact on non-human nature is already too great, and becoming worse. Environmentalists have two rather different ways of approaching this issue. One is to focus on population: there are just too many of us. The other is to stress the huge *per capita* environmental impact of affluent, mostly Western people, compared with that of the poor in 'developing' countries, who barely manage to meet their 'vital needs'. The platform combines both approaches: reduction in the human population is necessary for the flourishing of non-human nature, and can be achieved without lessening cultural diversity or human fulfilment. But also the platform calls for deep-seated changes in economic, political and technical organisation as well as in cultural values.

However, the platform is silent on the question of the *content* of the socio-economic changes that are needed, and on how they might be achieved, while the cultural shift is identified as a move away from aspirations for increased (material) living standards in favour of quality of life. The emphasis on a cultural shift is important for two main reasons. First, in liberal democracies, a direct appeal to decision-makers on behalf of the full deep ecological programme could have no hope of overcoming the power of vested interests unless it had the active support of the wider population. Second, the emphasis on cultural change offers a strategy for avoiding the narrowly 'survivalist' or 'hair-shirt' responses of some environmentalists to the perception of impending ecological disaster: we don't need to see the demand to use less of the Earth's resources as a privation, requiring author- itarian measures to curtail freedom. Once we see that we can have more fulfilled and meaningful lives while damaging our environment less, we will not need to be coerced into changing our behaviour by authoritarian state measures.

Technology

Naess himself has provided some more detail on the technical, political and economic changes implicit in the realisation of the deep ecological platform. On technology, Naess effectively dis- misses a common stereotype of green thinking as backward looking and anti-technology. On the con- trary, as technologies are central to our interaction with the rest of nature, he argues that the issue of what technologies are used is of key importance. He criticises the very widespread ideology of 'tech- nological determinism': a set of assumptions shared by 'modernisers' as well as by some ver- sions of Marxism (see Chapter 6). Technological determinists tend to see scientific and technical innovation as an autonomous process that drives 'progress', and to which the rest of society must adapt. The most 'advanced' technologies, on this view, are the ones based on the most recent scien- tific developments, and often the ones enabling the most profitable commercial exploitation. Also asso- ciated with technological determinism is a ten- dency to rely on technological innovation to solve social and ecological problems – the 'technical fix'. Naess argues for a 'systems' approach (consistent with ecological systems-thinking), according to which technologies have to be understood in rela- tion to other aspects of society and culture. The adoption of a new technology is a social choice, although we do not experience it as such because we are excluded from the narrow technocratic cir- cles who currently make it on our behalf.

Against the technological determinist approach, Naess argues for a process of evaluation of proposed new technologies in terms of their potential contribution to long-term values and goals of the culture, and reduced interference with nature. Technologies should be considered 'advanced' to the extent that they make such a contribution. The implication is that such deci- sions should be decentralised and devolved to local communities, so that the technologies chosen will sustain autonomy and reflect the cultural diversity of those communities. This has important implications for the policy of technology transfer to the poorer countries of the 'Third World'. Drawing especially on his admiration for Gandhi, Naess criticises transfer of 'hard' technologies (large-scale energy generation, armaments, agribusiness) developed in the rich countries to foster 'development' in the Third World. Such technologies always bring with them social and cultural changes that erode the independence and distinctiveness of local cultures.

Instead, 'soft' and 'intermediate' technologies (recycling, renewable energy, organic agriculture, the bicycle) offer gains in reducing waste and risks to health, addressing the needs of the poor as a priority, giving individuals and communities more control over their technologies, and reducing hierarchies of power and dependency.

Economics

Naess's comments on the need for economic change focus on the concept of GNP as a measure of economic progress, and on well-meaning attempts by economists to revise it or provide alternative measures. His critique of GNP is one that is widely shared among green thinkers: it measures the satisfaction of wants rather than needs, it includes on the positive side negativities such as environmental clean-up and health expen- diture, it takes no account of distribution of bene- fits among different groups in the population, and of necessity it fails to measure unquantifiable and non-marketed contributions to welfare. Also, the categories of mainstream (neo-classical) econom- ics do not take into account the possible external limits imposed by resource scarcity or the capacity of planetary life-support systems, and so implicitly assume that material growth of economies can go on indefinitely. Naess cites the work of economists such as Georgescu-Roegen and Fritz Schumacher as offering an important, ecologically informed alternative, but is critical of attempts to revise the mainstream approach by, for example, assigning money values to environmental assets ('shadow pricing'). In his view, these ultimately depend on introducing alternative norms and values into eco- nomic analysis. This would, of course, be a good thing, but in Naess's view the better strategy is to render explicit the values of mainstream economics and conduct the whole debate about economic policy within an alternative framework of discussion of long-term goals and values.

Politics

Despite the seemingly very radical implications of the deep ecological outlook, Naess is committed to a broadly based, non-sectarian approach to environmental politics. The Gandhian practice of non-violence implies not just using peaceful means for advancing the cause. Political 'enemies' must be treated with respect and understanding, rather than dismissed with derogatory stereotypes. The aim should be constructive dialogue in the hope of reaching an eventual consensus. Within the deep ecological movement itself, Naess argues that those who give priority to changing their own lifestyle and consciousness, as well as those who focus rather on political action at community or societal level, all have a place. These should be seen as mutually complementary aspects of a single movement, rather than contradictory strategies.

In Naess's view, the politics of deep ecology shares many of the concerns of the 'shallow' movement – resource depletion, pollution, population, global inequalities and so on – but deep ecologists tend to ask more questions and press for more fundamental changes. On resources and pollution, for example, deep ecologists seek to address the fundamental structures of production and consumption that are the underlying causes of the problems. This is because they seek long-term solutions, rather than short-term palliatives. On population, it is clear that the current and projected world population could not be sustained on the assumption of equalised living standards comparable to those of western industrialised societies. Since there is little hope of short-term social and economic transformations necessary to reduce *per capita* environmental impact, human population reduction must be a priority. This can be done consistently with increasing both the quality of life for humans, and allowing more space for 'free nature'. On global inequalities and free-trading regimes, Naess argues that many governmentally organised projects for aid and development of Third World countries, as well as international free trade are ecologically damaging, and also undermine local communities and their cultural distinctiveness. As far as is feasible, needs should be met by local production, with international trade judged by how far it meets culturally valued objectives, such as self-realisation and autonomy. An alternative way of addressing global inequalities would be through dialogues conducted between local communities, leading to a process of mutual aid, as far as possible by-passing big institutions.

Naess has an open view on the variety of political tactics and strategies available to the deep ecology movement. Consistently with the influence of Gandhi on his thinking, Naess favours non-violent direct action. So far as is possible this should be conducted within the law. But whether lawful or not, action should always be non-violent, always carefully explained and justified, and have some democratic legitimacy: that is, the expectation

should be that a fully informed public opinion would support the action. The relationship between green movement activism and the formal system of political parties is a vexed issue. Naess acknowledges the current tendency for Green political parties in Europe to form alliances with parties of the left, and he recognises shared values between greens and the left: opposition to aggressive individualism, production for use rather than profit, opposition to capitalism and big differentials of wealth and income, and so on. However, he also notes some socialist 'slogans' not acceptable to greens: centralisation, high consumption, economic growth and 'materialism'. He also sees support for individual enterprise and hostility to big government and bureaucracy as positive aspects of conservative political values. The greens, in Naess's view, are not just another political tendency between red and blue, but are, rather, on a different dimension. Their role should be to ecologise other political parties, with independent Green parties only a transitional measure.

The upshot of these comments on aspects of policy is the outline of the sort of society towards which the deep ecologists are working. The main aim is to decentralise key decision-making to local communities. These, ideally, should be small enough for it to be possible for the members to be directly acquainted with one another. Decisions should be made by a system of direct democracy. Production should be mainly organised locally, and make use of 'soft' technologies, allowing for a high degree of participation and local self-reliance. Differentials of income and wealth should be small enough for all to share ways of life. The geographical scale should be small enough for modes of transport such as walking and cycling to be feasible as the main means of getting about. Informal 'friendly' mechanisms should be the main means of maintaining order. However, Naess also accepts that that some larger-scale institutions, including some international bodies to maintain environmental standards, will still be needed.

ROBYN ECKERSLEY AND ECOCENTRIC PHILOSOPHY

Despite Naess's professed commitment to a non-sectarian and broadly based environmental movement, it could be claimed that his own distinction between 'shallow' and 'deep' ecology implies a negative view of alternative approaches – few activists would be happy to refer to their own positions as 'shallow'. Also, the rather polarised classification of environmental movements overlooks something of the 'richness and diversity' of environmental culture and politics. A very useful corrective is the work of Robyn Eckersley (1992). She offers an alternative classification of environmentalisms, locating them on a spectrum from narrowly instrumental approaches to nature through to fully non-anthropocentric ones. In her account (followed here), 'deep ecology' figures as one major variant within a wider category of 'ecocentric' philosophies.

Eckersley distinguishes five main currents of environmentalist thought and action: resource conservation, human welfare ecology, preservationism, animal rights and ecocentrism. Resource conservation involves a recognition of the material dependence of humans on nature, and seeks to use natural resources in a rational and efficient way so as to meet human needs. This is a fully anthropocentric approach, that limits intrinsic value to humans, and sees non-human nature as valuable only in so far as it is actually or potentially useful to humans. Human welfare ecology also rests on an anthropocentric value-system, but recognises a much wider range of human needs in relation to nature. Human welfare ecologists understand that full human flourishing involves not just access to food, fresh water, clothing and shelter, but for aesthetic delight and what Marx called 'spiritual nourishment' in a many-sided relationship to non-human nature.

Both preservationism and animal rights advocacy come closer to the full ecocentric recognition of intrinsic value in non-human nature. In the case of preservationism, priority is given to untouched 'wild nature', or wilderness, to be respected, even regarded as sacred, independently of any instrumental human interest in it. Eckersley has two reservations about it. First, the tendency is to preserve those environments that fit the Romantic values of great beauty or sublime grandeur, at the expense of less striking but often ecologically diverse and significant habitats such as wetlands, traditional farmland, roadside verges and ex-industrial sites. Second, preservationism tends to reinforce a dualistic opposition between humanly inhabited nature and wild, untouched nature. This overlooks, sometimes with catastrophic political consequences, the indigenous human inhabitants of 'wilderness', as well as the continuing coexistence of humans with non-human nature even in highly modified human settlements. Finally, the philosophy of animal rights does recognise intrinsic value in non-human beings, but does so in a more restrictive way than fully fledged ecocentrism. Intrinsic value is limited to sentient beings only, not to all living beings, ecosystems and even non-living beings as in the ecocentric perspectives. Not only this, but there is an implicit anthropocentrism at one remove, in that the non-human animals singled out for the attribution of rights are selected on the basis of characteristics (sentience, psychological complexity, a sense of self, and so on) that they share with humans.

Eckersley's account of ecocentrism coincides with the main principles of Naess's 'deep ecology' in most respects. Like him, she is committed to a philosophy of internal relations, and to seeing humans as just one species within a wider network of relations with other species, ecosystems and the rest of nature. She is also committed to recognising intrinsic value in non-human beings, and to an egalitarianism that extends beyond the boundaries of human society. Instead of Naess's 'biospherical egalitarianism in principle', she speaks of 'a *prima facie* orientation of nonfavouritism' in our dealings with non-humans. However, she, like Naess, argues that humans no less than other species have a right to live and blossom. This necessarily entails some interference, killing and harm to other beings and their habitats. The ecocentric approach simply favours choices that minimise the harm to non-human nature consistently with maintaining human flourishing.

However, where Naess's 'ecosophy T' seems simply to assert as a fundamental moral axiom that non-human nature has intrinsic value, Eckersley notes other ecocentric value theories. 'Autopoietic intrinsic value theory' attributes value to those beings that 'strive' to maintain and reproduce their organisational structure. This is a much wider category of beings than are allowed value in the animal rights philosophy, and seems to include all living beings as 'ends in themselves'. It also may be held (more controversially) to extend to ecosystems and even the biosphere (as in the claims made by Lovelock in his notion of 'Gaia'). An alternative approach is a development of Naess's own notion of 'Self-realisation' by Warwick Fox (Fox, 1990). This version of ecocentric philosophy has much in common with some non-Western philosophies including Buddhism, Taoism and some indigenous Australian and American belief systems, and is called 'transpersonal ecology'. Fox's view rests on the ecocentric conception of the world as a network of internal relations. For him, full personal development involves a growing ability to identify with and respect beings beyond the narrow egotistical self. Eventually, the fully developed capacity for indentification will extend to all of humanity and beyond to the whole of non-human nature, with the recognition of mutual interdependence of humans and other beings in nature. This sort of approach is held to have the advantage that with a transformed subjectivity in relation to nature, humans will spontaneously respect and care for it, rather than being disciplined into proper behaviour by an authoritative set of rules or laws, imposed from above. Eckersley also discusses ecofeminism as an alternative value basis for an ecocentric orientation (see Chapter 4 of this book).

Initially, deep ecology informed radical ecological political movements and organisations as part of a broader 'emancipatory' stream of activism. Deep ecologists, like social ecologists (see Chapter 3) and others identified strongly with aspects of non-human nature, and saw a need for profound transformations of existing Western industrial capitalist societies. They also tended to share a concern for the poorest and most threatened humans both in Western societies and also in the 'Third World', seeing connections between the creation of a more just social and economic order and the emergence of a new, more respectful and harmonious relationship between humans and the rest of nature. However, through the 1980s, and especially in the USA, divisions within the movement began to appear, and to intensify. The most prominent activist movement inspired by deep ecology – *Earth First!* – attracted to itself some individuals strongly committed to direct action to defend 'wilderness' against the encroachment of logging and mineral extraction (Foreman and Hellenbach, 1989; Taylor, 1995). Out of this came a tendency to hostility towards not just those industries, or even the economic system of which they were a part, but to humans as such. The most extreme expressions of this tendency were notorious proclamations of welcome to Third World famines and epidemics such as AIDS that would reduce the human population. This led to accusations of misanthropy, even ecological fascism, from the more socially oriented parts of the radical ecology movement. Meanwhile, the social ecologists were accused, in their turn, of reverting to a narrowly anthropocentric concern with human welfare only.

SOME CRITICISMS AND RESPONSES

The leading social ecologist, Murray Bookchin (see Chapter 3) brought matters to a head with his attack on the deep ecology movement at the US National Green Gathering at Amhurst in 1987. Fiery exchanges followed, but there was at least a partial reconciliation when Bookchin and Dave Foreman (one of the most outspoken of the wilderness campaigners) were brought together for a debate in 1989 and 1990 (Bookchin and Foreman, 1991). The details of this debate need not detain us, but during this episode and since a number of common criticisms of deep ecological/ecocentric philosophies and politics have emerged. These concern: the philosophical basis of deep ecology; the moral and political priorities of the movement; and the practical feasibility of its political programme. We will consider each of these in turn.

Philosophy

The deep ecological view of nature as a vastly complex network of relationships is open to some philosophical objections, and is also only doubtfully consistent with deep ecological values. The 'relational field image' is introduced as an alternative to the atomistic philosophy of classical mechanical science as well as individualistic social and political theories. Often the cosmology of modern physics and scientific ecology are cited as support for the relational view of the world (e.g. Capra, 1977). This can be justified to some extent, in that ecology, for example, does focus on the relationships between living organisms, either as individuals or as local populations, and the other organisms and non-living beings and conditions they interact with. However, this does not imply that everything ultimately resolves into relations. On the contrary, for the idea of relationships to make sense there have to be substantive things, beings, communities, etc., that are related. In other words, as well as relational properties, there must be some attributes that things have independently of the relationships they happen to have with other things. For example, being a student is a relational property of a person, because they can only be a student by virtue of belonging to an educational institution of some kind, or having a relationship with a particular body of knowledge or thought. However, to be a student also implies being a living person, having a certain set of mental abilities, a brain and central nervous system and so on. In a basic sense, this person can be identified as 'the same' person before, during and after the period of time she was a student. This notion of intrinsic identities of beings, including humans, is what Naess and other deep ecologists seem to be denying.

This view of reality is very effective in undermining the atomistic and utilitarian–individualist ways of thinking that have been the dominant Western ideologies in modern times, and have sustained anthropocentric approaches to nature. However, the relational view arguably goes too far in the opposite direction, threatening to 'drown' any notion of autonomous identity or agency in a sea of ever-shifting relationships. This is not only ultimately incoherent, but it also threatens to undermine the very set of ecological values that the deep ecologists advocate. The principle of 'biospherical egalitarianism in principle' (or 'non-favouritism') implies a commitment to identifiable individuals whose interests can be discovered and compared. Indeed, it is hard to see how full recognition of the intrinsic value of another being can be sustained if we think of all beings as having no core identity, and as ultimately resolvable into relationships within some larger totality. This may be one possible source of a tendency in *some* deep-ecological thinking to accept the sacrifice of individuals for the sake of the greater good. Also, if we think of the activist dimension of deep ecological thought, it is clear that a high level of personal integrity and ability to think and act independently – even against the overwhelming domination of anti-ecological forces – is required. The ecological activist needs to combine a strong sense of personal identity and autonomy with a capacity for sensitive awareness of the interconnectedness of each individual life with all those with whom she coexists. Arguably, a philosophy of 'beings-in-relationships' would be better able to sustain this than either 'atomistic' individualism or thoroughgoing relationism.

Further, as Eckersley points out, the relational view of the world, whilst it undermines the basis for anthropocentric thought, does not automatically bring with it a justification for an alternative ecological value perspective. This opens up the hotly debated issue of how to justify attributing 'intrinsic value' to beings other than humans – even to non-living beings and systems. This is the core value commitment of ecocentrism. The advocates of animal rights have a strong argument for extending moral concern to individuals of other species on the basis of consistency. If we accept that humans have intrinsic value, then how can we consistently deny moral status to other living beings that are like us in morally relevant respects? But the problem for ecocentrics is how to extend that circle of morally significant beings beyond the rather narrow category of conscious, sentient animals. 'Autopoietic intrinsic value theory' is one way of doing this. However, it, like the animal rights approach, still limits intrinsic value to living beings and (more controversially) to some forms of association among them. Also, like the animal rights perspective, it can be accused of 'anthropocentrism at one remove', because it relies on the notion of beings as 'ends in themselves', as having their own inner goals or purposes (like humans). So, apart from the analogy with humans, why should we find it convincing that intrinsic value should be attributed on the basis of a being's striving to maintain its own organisational structures?

The problem is more readily solved if one has a religious outlook: whether one thinks of the world as the creation of a transcendent deity (as in Christianity) or whether the deity is thought of as inherent in the world itself (as in pantheism), the intrinsic value, even sacredness, of nature seems to follow directly. Paradoxically, however, this has been very much a minority tradition within Christianity, given the Genesis account of non-human nature as created for human use (but see Linzey, 1987).

But this is no help to those of us who look for a secular solution to the problem of value. In the absence of belief in the existence of supernatural beings, it seems to make no sense to talk of 'value' in abstraction from beings who are capable of assigning value to things. As it happens, humans seem to be the only beings currently on this planet who can assign value, in the fullest sense. Certainly, other animals, especially those close to us in their social or psychological natures, have preferences, clearly enjoy some activities more than others and so on. Some of the extravagant outcomes of evolutionary sexual selection, the plumage of the birds of paradise, the elaborate songs of many species of warbler, for example, signal something that must resemble our aesthetic sense in the females that select the best performance. However, the 'values' expressed in the behaviour of other species are attached to very specific displays, performances or other objects; they do not have the universality required of human values, including those that belong to the core beliefs of ecocentric philosophy.

So, short of religious or spiritual beliefs, it seems that the doctrine of 'intrinsic value' has to be taken as an axiom that stands without any independent justification. Indeed, it seems difficult, from a secular perspective, to make sense of the idea of values independently of some conscious being capable of assigning value. We can make sense of the existence and activities of, for example, dinosaurs, which existed prior to the evolution of hominids. But can we make sense of the idea that they had 'intrinsic value' in the absence of anyone capable of valuing them? If we follow this thought where it leads, then the conditions for something to have value turn out to include the existence of sentient beings that can express preferences, respond emotionally to elements in their environment, and so on. More than this, and particularly if we consider the sort of valuing practices that are favoured by deep ecologists, it can be argued that socialisation into a culture that sustains and nurtures the various practical engagements, skills and sensibilities that develop our ability to value is also necessary. So, for example, the practical and literary engagements with nature associated with the Romantic movement (Chapter 2) may be favourable to a certain sort of non-instrumental respect and appreciation of nature, whilst people brought up in cultures in which Romanticism never took root might find this more difficult and alien.

The difficulty in this line of reasoning, from the standpoint of ecocentric philosophy, is it makes the value of non-human nature appear to be a purely subjective matter: either you value it or you don't. There is no basis for an ethical or a cognitive requirement to *recognise* value in nature.

This is why the insistence on 'intrinsic' value is so important. So, are there any moves toward the deep ecological position that can be made from within a secular perspective? I think there are. We can move from pure subjectivism in two distinct stages. First, we learn to value nature (or each other, or art objects, or ...) as a result of participation in a culture that gives us the resources to do so (through education, free enquiry, public access to the valued objects of the culture, enough time free of basic toil to develop our ability to appreciate them, and so on). Since learning, skill, understanding, sensibility and so on are needed for us to value things properly, we can no longer think of valuing as a purely subjective matter. The store of wisdom, thoughtful judgement and critical understanding that the culture has accumulated are objective resources that we draw on as a necessary condition of our own development. This does not imply, of course, that there can be no disagreement, or that each of us merely takes on trust a certain canon of past critical judgements. In this context 'objectivity' simply means an independently existing context of established views, theories, judgements and arguments against which our own experiences and developing skills can be tested.

But this doesn't get us far enough for the deep ecologist. Though cultures may provide some basis for 'objectivity' in our valuing, including the ways we value non-human nature, cultures themselves vary. What are we to say of cultures that simply may not provide the sorts of experiences and aesthetic or spiritual traditions that are necessary to the formation of an ecological sensibility? The step we have made so far avoids pure subjectivism, but lands us with the problem of cultural relativism. Who is to say that some sorts of cultural values are better, truer, than others? A second move can be made as a response to this. We need to press the notion of 'culture' a bit further. Thinking of the Romantic poets, it is inconceivable that the sentiments that inspired their creative work could have arisen without their direct experience of and practical engagement with the natural objects they wrote about. Something parallel is also true of the philosophers of deep ecology; for example, Naess's work is full of references to his own direct experiences of the Scandinavian mountains. But we can also think of more mundane activities, particularly ones that need craft skills, such as cooking, woodwork or playing a musical instrument. In each case more is required than merely learning a set of principles or theoretical knowledge. The material with which we work has its own properties, it offers resistance to the things we want to do, and so it imposes a kind of discipline on us: don't blow too hard or the reed will squeak!, saw the wood at this angle, or it will

split!, etc. The idea here is that creative work on nature, including craft work, mountaineering, scientific research or writing poetry is something not well conceptualised as merely 'human activity'. Rather, the outcome, including the criteria of cultural valuation themselves, emerges out of conscious but also material interaction between embodied humans and the objects of their activity. So, when a culture offers resources to recognise value in non-human nature, these resources bear the imprint of the independent activity of nature itself in shaping our responses to it.

Now, since all cultures must of necessity sustain practices of conscious material interaction with non-human nature, then all cultures contain at least the potentiality to develop the sorts of understanding, respect and sensibility that are at the core of an ecocentric approach to the world. This suggests that there will be the basis in elementary experiences of people brought up within any culture for ecocentric values to emerge. The wide variety of scientific, religious and philosophical traditions, both Eastern and Western, that offer ways into a deeper appreciation of human embedding in a wider natural world, and an associated respect for and wonder at non-human nature, is consistent with this.

But in bringing in human material and cultural practices as essential to a notion of value in non-human nature, are not we still far short of the deep ecological insistence on 'intrinsic' value? This may well be the view of many ecocentric thinkers, but it does provide us with a secular basis for sustaining much of the deep ecological outlook, and also gives us a new way of making the key distinction between anthropocentric and ecocentric values. Whilst recognising that ways of valuing non-human beings arise from the forms of engagement humans have with them, we can still distinguish between different sorts of human engagement with the world, and between the different sorts of value orientation that are associated with them. We can, in other words, distinguish between value systems that value non-human nature only for its usefulness to humans and those which value non-human nature for its own sake – for what it is in itself. We can readily understand what it is to love another person for who they are, without wanting to change them, possess them or subordinate them to our will. There is no fundamental reason why such an attitude of love as respect for the independent qualities, and 'otherness' of another should not be directed also to the non-humans with whom we share the planet. It is an attitude that many of us find develops spontaneously with the depth of our practical engagement with or understanding of aspects of the rest of nature. Evelyn Fox Keller, a feminist philosopher of science writes of the developmental biologist, Barbara McClintock:

> McClintock can risk the suspension of boundaries between subject and object without jeopardy to science precisely because, to her, science is not premised on that division. Indeed the intimacy she experiences with the objects she studies – intimacy born of a lifetime of cultivated attentiveness – is a wellspring of her powers as a scientist (Keller, 1985, p. 164).

Morals and politics

Anti-humanism?

Deep ecologists have much in common with other wings of the environmental movement in the sorts of issues that concern them: pollution, loss of bio-diversity, degradation of landscapes and so on, but Naess and Eckersley both single out two issues in particular as distinctive emphases of ecocentrism: the preservation of wilderness and reduction of the human population. Both issues can be, and have been, taken up in ways that are rightly criticised as 'misanthropic', Deep ecologists have become increasingly careful about the way they present the population issue, taking into account both the huge disparities between rich and poor countries in their *per capita* environmental impact, and the ethical implications of different ways of securing a reduction in the population. Early formulations (such as that given in the 'platform') make no mention of restraints on acceptable ways of reducing the population, beyond saying that population could be reduced consistently with the 'flourishing of human life and cultures'. The objectionable attitudes to epidemic disease and Third World famine attributed to some early Earth First! activists might well claim some legitimacy from loose formulations such as this. This is particularly so as the 'platform' implies a contradiction between current human population size and the flourishing of the rest of nature.

Against this, it is clear that the broader ecocentric philosophy of internal relations and associated commitment to the flourishing of all beings implies that 'humans are just as entitled to live and blossom as any other species' (Eckersley, 1992, p. 57). Far from being misanthropic, when properly understood, ecocentric environmentalism favours human flourishing – but in ways that minimally interfere with the flourishing of other species. It is an unfortunate fact that religions and philosophies are frequently taken up by their self-professed followers in ways not foreseen and certainly not desired by their originators – Marxism, Christianity and other beliefs have all been turned into their opposites by would-be 'followers'.

However, originators can never quite renounce responsibility for what others do in their name: a much clearer set of policy and ethical guidelines is needed for the ecocentric commitment to population reduction to be protected from misinterpretation.

Some similar considerations apply to the emphasis on wilderness preservation. Again, exclusion of humans from 'wilderness' areas in the name of conservation has occasionally led to violent mass expulsion of indigenous people from their land – sometimes with the complicity of 'conservation' organisations. Usually, advocates of ecocentric approaches fully acknowledge the rights of indigenous people, as well as their role in ecosystem maintenance. Nevertheless, the term 'wilderness' does have connotations that suggest an absence of human habitation. As with the issue of population, a clearer and more prominent advocacy of indigenous rights is needed to offset mis-representations.

Bad priorities?

Short of actual misanthropy, ecocentrism is often criticised for a distorted view of moral and political priorities. The doctrines of non-favouritism, or 'biospherical egalistarianism', seem to suggest that we must count the life of a mosquito – or even a virus – as equivalent in importance to that of a human. Moreover, the emphasis on assigning value to the flourishing of non-human life has been seen as showing insufficient sensitivity to inequality, oppression and suffering within the human species.

On the first point, the 'platform' is clear that humans are entitled to meet their 'vital needs', and it is acknowledged that this will involve causing suffering and loss to some non-human beings, whilst Eckersley argues that 'non-favouritism' merely implies that the interests of other beings should not be *a priori* ruled out of account simply because of their status as non-humans. So, we should always take account of the interests of other beings in our decision-making, but we may sometimes override them if it is necessary to our own flourishing. This makes the ecocentric position seem less contrary to prevailing value priorities, but it still does not provide us with clear ethical guidelines. Eckersley does take us further, suggesting that a range of criteria, such as the degree of sentience of a being, whether it belongs to an endangered species, or whether its population is crucial for the maintenance of an ecosystem, might guide ethical choice in particular contexts. There is no attempt to provide fully developed systematic value theory here, so we do not know, for example, how much weight should be attached to the different criteria. But this may be desirable – too much emphasis in rigorous moral theorising can lead to insensitive dogmatism and tell against wise judgement in the light of particular circumstances.

On the second point, the allegation of insufficient concern for inequalities and suffering among humans, the main thinkers of deep ecology have a strong defence. It will be remembered that Naess's initial distinction between deep and shallow ecology criticised the latter (quite unfairly!) as limiting its concern to health and living standards of populations in the developed countries. Eckersley, too, insists that ecocentrism is committed to 'emancipation writ large': that is to say, to justice both within the human species, and between humans and others. From an ecocentric standpoint, the criticism that their concern for the flourishing of non-human beings must imply lack of concern for justice among humans derives from a mistaken assumption: that, in principle, human interests can only be pursued at the expense of non-human flourishing. This they would deny. On the contrary, as Naess puts it, the experience of 'free nature' is essential to human well-being.

PRACTICAL POLITICS AND FEASIBLE UTOPIAS

Ecocentric thinkers tend to focus on philosophical and general ethical issues, rather than on matters of specific policy. This is appropriate enough, given that ecocentrism is a broad and quite diverse strand of environmentalism, rather than a coherent, organised political movement. However, the 'platform', as we have seen, does call for changes in basic economic, technological and ideological structures, and Naess has quite a lot to say about appropriate tactics and strategies for the movement. There is also a fairly wide consensus about the sort of organisation of social life that would most fully realise ecocentric values.

Naess notes with regret that deep ecologists are often ill-equipped to participate in economic debates, and this may be a symptom of the tendency of ecocentric thinkers and activists to emphasise the importance of changes in consciousness. Both Naess's notion of 'Self-realisation' and Warwick Fox's 'transpersonal ecology' reflect this tendency. Similarly, Robyn Eckersley, noting some of the difficulties in extending liberal notions of rights to non-humans, contrasts that strategy with an ecocentric one that: 'emphasize(s) the importance of a general change in consciousness and suggest(s) that a gradual cultural, educational and social revolution involving a reorientation of our sense of place in the evolutionary drama is likely to provide a better long term protection of the interests of the nonhuman world' (Eckersley, 1992, p. 59). This emphasis on consciousness change has its justification in the radical difference between fully-fledged ecocentrism and the anthropocentric, consumption-oriented values that are currently dominant. Spectacular but non-violent direct action, as well as 'prefigurative' attempts at setting up small eco-communities can be seen as having a role in changing public consciousness. From changing public consciousness, then, might come pressure for alternative policies on technological innovation, and economic organisation and decentralisation of decision-making power.

This raises the familiar 'chicken and egg' problem. As Naess and others often point out, attitudes of respect for non-human nature grow out of appropriate forms of experience and engagement with nature. More broadly, personal and social identities are formed as individuals grow through and within forms of social and cultural organisation: households, schools, places of work, modes of material and cultural consumption, leisure organisations and so on. What is significant here is not just the explicit values and beliefs that prevail in these contexts, but also the tacit understandings, practical constraints and opportunities they impose on us. So, while schools may have an explicit ethos of non-selfishness and cooperation, the pressures to succeed in tests and exams cultivate and reward competitive and individualist personality traits. In later working life, most people have little control over the environmental or social policies of the firms or public bodies they work for, and often have little choice over what job they take. In short, the prevailing social and economic framework in which most of us live presents powerful obstacles to the formation of the sorts of identity favoured by deep ecology, and often punishes those who attempt to live by its values.

Certainly, Naess accepts this, and argues that more personal concerns with changing consciousness and lifestyle are complementary to organised attempts to change the policies of governments and other large organisations. The difficulty here, of course, is that access to direct influence on policy-making, even in the most open of liberal democracies, is very selective. The attempt to become an 'insider' to the process of policy formation entails strong pressures to compromise with the prevailing ways of framing the issues and the associated terms of discussion. So, for example, although Naess notes the harmful effects of 'free market' thinking on the independence and cultural integrity of Third World countries, a large number of established environmental organisations now speak the language of markets as their passport to big business and government. In other words, short of a deep crisis in currently dominant power relations, the strategy of attempting to have direct influence on policy-making necessarily involves environmentalists in a shift to the 'shallow' end of the environmental spectrum. This is clearly a great dilemma – but it is one shared by all strands within the movement.

Finally, a few words need to be said about the long-term vision of deep ecology. Their preference for small-scale, self-sufficient communities, with direct democratic participation in decision-making, and the use of 'soft', appropriate technologies is quite widely shared with other radical 'alternative' movements; utopian socialists and anarchists as well as greens. It seems reasonable to expect that such a community, being dependent for most of its needs on the local environment, would behave responsibly in protecting its long-term quality. Participatory forms of democracy should ensure that the interests of the whole community, not just those of an elite minority, were served. Such a community might also be one in which the various forms of necessary labour involved in producing food and clothing, building and repair of homes and implements, caring for the children, the sick and the elderly and so on could be shared, strengthening social bonds and a sense of connectedness to nature.

However, there are several sorts of problem that could be envisaged. The first is that there would be huge differences between communities in the character of their local environments; some may be situated in dry, inhospitable terrain, with low soil fertility, while others might be situated in fertile river valleys, blessed with warm and favourable climates. In the absence of wider institutional frameworks that could organise redistribution, equality within each community could coexist with inequalities between communities as great as those we see in our world as presently constituted. Moreover, it seems unlikely that such inequalities could persist indefinitely without the emergence of hostilities and conflicts between communities.

Another, related, problem derives from the interconnectedness of ecosystems. While one community manages its local environment to its best long-term advantage, it may well deprive an adjacent community of the means to do the same: up-stream river-bank communities limiting the flow of water to down-stream environments might be an example. Of course, dialogue and mutual aid between communities is always possible, and favoured by Naess, as an alternative to the involvement of large institutions in resolving disputes of this sort. Against this, however, it could be argued that a shared sense of the justness of a particular solution would have to emerge and made convincing to each community. The sorts of localised 'particularistic' identities that are often found in small-scale communities now might constitute strong obstacles to this. By contrast, the frameworks of abstract and universalistic conceptions of justice that have grown up in the context of modern liberal democratic legal and political systems might turn out to be more favourable (or, perhaps, we should say 'less unfavourable') to peaceful settling of disputes.

A quite different sort of problem with the ideal of small, self-sufficient communities has to do with whether they would provide an adequate context for 'human flourishing' in aspects of life less directly connected with ecological sustainability. Notwithstanding the widely recognised 'ecological footprints' of big urban centres, they do offer opportunities for valued experiences not obviously available in a society wholly devolved into small, predominantly rural communes. The intimacy of daily interaction in a small community has its attractions, but for many people can be experienced as an oppressive loss of privacy and personal autonomy. This might be intensified if the eco-communities of the future were as narrowly inward looking and intolerant of diversity as many of those of the past and present. Indeed, it might be argued that city life presents a more favourable context for fostering the richness and diversity of human cultures than does small, self-sufficient community life. Perhaps in the light of such considerations, Robyn Eckersley makes a case for the value of urban civilisation, while Naess himself recognises the continuing need for some overarching institutions.

A final difficulty has to do with how to get from here to there. The large-scale hierarchical institutions and consumerist values that currently prevail are associated with intractable concentrations of military, economic and political power, as well as deep psychological and cultural obstacles to change. At the personal and cultural levels, the challenge faced by deep ecology is to persuade or demonstrate that a simpler, less affluent life is consistent with a happier, more fulfilled and joyful one. At the level of military, economic and political power structures the challenge is two-fold: to mobilise a sufficient breadth and depth of popular resistance, and to popularise an attractive and viable vision of a future sustainable world.

REFERENCES AND BIBLIOGRAPHY

Bookchin, M. and Foreman, D. (1991) *Defending the Earth*. Montreal and New York: Black Rose.

Capra, F. (1977) *The Tao of Physics*. London: Fontana.

Dobson, A. (2000) *Green Political Thought*. London: Routledge.

Eckersley, R. (1992) *Environmentalism and Political Theory*. London: University College.

Foreman, D. and Hellenbach, T. O. (1989) *Ecodefence: a Field Guide to Monkeywrenching*. Tucson: Ned Ludd.

Fox, W. (1990) *Toward a Transpersonal Ecology*. Boston: Shambhala.

Jonge, E. de (2004) *Spinoza and Deep Ecology*. Aldershot: Ashgate.

Keller, E. Fox (1985) *Reflections on Gender and Science*. New Haven: Yale University.

Keulartz, J. (1998) *Struggle for Nature: A Critique of Radical Ecology*. London: Routledge.

Leopold, A. [1949] (1977) *A Sand County Almanac*. Oxford: Oxford University Press.

Linzey, A. (1987) *Christianity and the Rights of Animals*. New York: Crossroad.

Lovelock, J. E. (1979) *Gaia: A New Look at Life on Earth*. Oxford: Oxford University.

Lovelock, J. E. (2006) *The Revenge of Gaia*. London: Allen Lane.

Luke, T. (1997) *Ecocritique*. Minneapolis: University of Minnesota.

Naess, A. (1989) *Ecology, Community and Lifestyle* (tr. and ed. D. Rothenberg). Cambridge: Cambridge University.

Sessions, G. (ed.) (1995) *Deep Ecology for the 21st Century*. Boston and London: Shambhala.

Sylvan, R. (1985) A Critique of Deep Ecology (parts 1 and 2) *Radical Philosophy* 40: 2–12; 41: 10–22.

Taylor, B. R. (1995) Earth First! and global narratives of popular ecological resisitance. In Taylor, B. R. (ed.) *Ecological Resistance Movements*. Albany: State University of New York.

6

Greening the Left? From Marx to World-System Theory

Ted Benton

INTRODUCTION

Many writers and activists in the Green movement have thought of the new politics as somehow 'beyond left and right'. Unfortunately, this is a difficult view to sustain in the face of the practical choices that any serious Green politics has to make. For one thing, political thought and practice must begin with an understanding of the nature of the society and political system, and a diagnosis of the problems and issues they produce. Left and right have strikingly different understandings to offer. Second, the policy options chosen to address the ecological issues which are at the core of Green politics necessarily have other impacts. So, for example, a strategy of seeking to conserve water or energy by using the price mechanism will usually involve privatisation of the relevant utilities (popular with the political right), but is likely to deprive (and has deprived) the poorer sections of society from access to a basic necessity – something likely to be objected to from the left.

Increasingly, Green politics has come to recognise a need to combine together the left's traditional concern for social justice with action to defend the environment. But this process did not happen overnight, and it corresponds to a parallel development within parts of the left, increasingly recognising that social justice was unachievable without action to defend the environment. This chapter is an introduction to some of the main traditions of thought on the left, showing the relevance of many of their ideas to an adequate understanding of our contemporary ecological issues.

FROM THE ROMANTICS TO MARX

In Germany, Romanticism inspired scientific work, as well as painting, poetry and the other creative arts. We already saw in Chapter 2 the influence of Humboldt's writings on Darwin. Goethe (1749–1832), too, was a towering influence in both literature and science in Germany. His evolutionary thinking long predated that of Darwin, and so the notions of historicity in nature, and of humans as part of that history, continuous with the rest of nature, was already familiar in German-speaking cultures. The most pervasive German intellectual influence following Goethe was the philosopher G.W.F. Hegel (1770–1831). His 'dialectical' view of the world as in process of historical transformation toward some future state of full 'self-realisation' proposed a view of humans as interconnected with the rest of nature, and evolving historically along with it. However, the reunification of conscious humanity with material nature, in Hegel's philosophy, was to be actualised in the domain of spiritual, or 'ideal' existence.

It was one of Hegel's younger and radical followers, Ludwig Feuerbach (1804–1872), who first proposed a materialist 'inversion' of the Hegelian philosophy. The notions of historical development and interconnectedness of humans and nature were to be retained, but the 'self-realisation' towards which history was moving was now to be understood as full *human* self-realisation; the full development of humans as real, embodied beings. According to Feuerbach, religious beliefs, as well as the Idealism of Hegel, were to be understood as forms of 'self-alienation': we imagine ourselves to be creations of God, whereas, in fact, the reverse is true; it is we humans that invent 'Gods', or notions such as 'Absolute Spirit'. We do so, Feuerbach argued, because our lives in the here-and-now are stunted, unfulfilled, and so we invent imaginary beings and attribute to them the powers of fully developed humans. The implication is that to realise full human development on Earth would make religion unnecessary.

These ideas immediately attracted the young Karl Marx (1818–1883) and his subsequent close collaborator, Frederick Engels (1820–1895). For Marx, Feuerbach's ideas were the starting point for a radically new materialist view of human nature and history, contained in a set of notes, 'The Economic and Philosophical Manuscripts', which were written in 1844 but not published until long after Marx's death. Meanwhile, Engels, who was the son of a Manchester industrialist, was busy writing a much more (literally) down-to-earth critical exposure of the dreadful living conditions of the working population of the English industrial cities. Both texts have much to offer, Marx's as a significant contribution to environmental philosophy, Engels as a pioneering work of environmental socialism.

THE YOUNG MARX: HUMANS AND NATURE

Marx's 'Manuscripts' are incomplete and difficult to understand, but profound. Their main theme is one of humans as a species developing their potentials through practical interaction with nature, towards some ultimate stage of full self-realisation. Like Feuerbach, Marx saw the phases of the historical process prior to full human self-realisation as ones in which humans are 'alienated' – that is, as separated from their own true nature or potentials. This was true of the current stage of historical development – early industrial capitalism – so full human self-realisation, and thus the overcoming of 'alienation', could only be achieved by the overthrow of capitalism and the creation of a new form of society: socialism or communism. This would be a form of society in which full self-realisation of humans would involve reconciliation among humans but also a reunification of humans with the rest of nature:

> History itself is a *real* part of *natural history* – of nature developing into man. Natural science will in time incorporate into itself the science of man, just as the science of man will incorporate into itself natural science: there will be *one* science (Marx and Engels, 1975, p. 303–304).

Whereas Feuerbach had been mainly concerned with religion as a form of self-alienation, Marx goes on to ask the further question: what is it that stunts human development, prevents people from realising their potentials? Consistent with his materialism, Marx concluded that the source of self-alienation must be the material conditions under which people lived. The most obvious source of understanding of those material conditions was, for him, the most advanced economic thought of the time; especially the work of the Edinburgh philosopher and economist, Adam Smith, and subsequent political economists, notably Ricardo and James Mill.

Marx takes from the political economists some general characterisations of capitalism and its tendencies: the growth of industry and the towns at the expense of rural landowners and small farmers; the growing social division between industrial wage workers and the industrial capitalists; the reduction of diverse qualities and values of people and things into a single measure, money value; and the tendency for competition to lead to the elimination of small producers and the emergence of monopoly, and so on. At this stage, Marx takes much of this on trust, but criticises the political economists for simply *describing* economic relations. They fail, he argues, to 'comprehend', or to explain them. Whereas Smith, for example, sees market exchange as an expression of universal human nature, Marx sees it as a (temporary) phase in the long historical process of human development.

So, we need to have some idea about how Marx understood that process. At the centre of his account is human labour, understood as creative, embodied interaction with nature:

> The worker can create nothing without nature, without the sensuous external world. It is the material on which his labour is realised, in which it is active, from which and by means of which it produces (Marx and Engels, 1975, p. 273).

Labour in this sense is the source not just of means of subsistence, or material wealth, but it is also the process by which humans create themselves, both as individuals and as a species. Labour in this sense includes a great diversity of different sorts of engagement with nature – art, literature, craft production, industry and aesthetic contemplation, not only production merely to meet 'immediate physical need'. Indeed, Marx denies that the latter is genuinely *human* production at all. All these forms of productive activity are understood by Marx as social in character; even a solitary painter or sculptor is engaged in a social activity, as the aesthetic sense itself emerges from the social process:

> Only through the objectively unfolded richness of man's essential being is the richness of subjective

human sensibility (a musical ear, an eye for beauty of form – in short, sense capable of human gratification, senses affirming themselves as essential powers of *man*) either cultivated or brought into being ... The *forming* of the five senses is a labour of the entire history of the world down to the present. (Marx and Engels, 1975, pp. 301–302).

But the social character of labour is at first small scale and localised, as is the domain of nature upon which it acts. In Marx's philosophical view, human history is a long-run process in which social cooperation expands its reach, drawing in ever-wider human populations, and at the same time extending the scale and scope of its transformative engagement with nature. Certainly, Marx saw the technological development of modern industry as an expression of this long-run historical process. However, he did not share the widespread 'modernist' equation of scientific and technical domination of nature with 'progress'. For him, capitalist industrial development was a distorted and perverse expression of the development of human cooperative labour. The future communist overcoming of the capitalist form of this development would enable the emergence of a quite new relationship of humans to nature.

The key ideas here are humans as a 'species being' and the notion of 'humanisation of nature'. Humans are a 'species being' in the sense that they can take the whole species as a matter of their conscious concern, and also in the sense that they have the potential – to be realised through the historical process – to coordinate their activity on a species-wide basis. That is, Marx anticipated what we now call 'globalisation', though, of course, for him it was to be a quite different *sort* of globalisation. Marx's vision of full human self-realisation was, rather, one in which humans would freely participate in a great creative project of 'humanisation of nature'. Through their creative work on nature they would finally live within a natural environment in which they felt at home, and which would in some sense reflect their own fully developed humanity. At first reading this seems close to the vision advocated rather later in the century by Darwin's propagandist, T. H. Huxley (see Chapter 2). The view seems to be that after a long history in which humans struggle to wrest their means of subsistence from nature, their antagonistic relation to non-human nature is overcome as a result of their transforming the whole of nature in their own image, or to suit their own aesthetic values. We can think of this metaphorically as a 'landscape gardening' view of the reunification of humans with nature. Looked at from a deep ecological point of view it is clearly anthropocentric,

but it is certainly not narrowly instrumental or exploitative.

But, is this an adequate interpretation of Marx's view? There are some strong indications that it is not the full story. Perhaps the most well-known concept in Marx's early work is one we have already mentioned: the idea of 'alienation' or 'estrangement'. This idea figures very commonly in critical writings in the 19th and early 20th centuries, but in Marx's early work it is applied specifically to human labour prior to the overcoming of private property. Under feudal and capitalist private property, human labour is alienated; it is labour in which the person who works is separated from what he/she produces, in the sense that it belongs to another. But there are other dimensions to this state of 'separation': workers are separated from the very activity of labour itself, as it is specified in advance by someone else what is to be produced, how, and under what conditions. There is, in other words no free, creative expression of the workers identity – as Marx puts it, work loses all its 'charm' for the worker, and becomes mere drudgery. Since labour (in Marx's very wide sense of this term) is essential to what it is to be human, to be alienated from one's work is to be alienated from oneself, from other humans and from the species.

These are the aspects of alienation as a set of social divisions that have been emphasised by most commentators on Marx's early writings, but if we keep in mind what Marx says about the dependence of labour on nature (see the above quotation – Marx and Engels, 1975, p. 273), then it is clear that for him alienated labour is, first and foremost, labour that is alienated from its own source, means and conditions; that is to say, from nature. Far from an arrogant vision of humans remaking nature in their own image, there is a strong sense of the dependence of humans, in all aspects of their well-being, on their continued interaction with the rest of nature. This is very clearly stated in the following passage:

Just as plants, animals, stones, air, light, etc., constitute theoretically a part of human consciousness, partly as objects of natural science, partly as objects of art – his spiritual inorganic nature, spiritual nourishment which he must first prepare to make palatable and digestible – so also in the realm of practice they constitute a part of human life and human activity. Physically man lives only on these products of nature, whether they appear in the form of food, heating, clothes, a dwelling, etc. The universality of man appears in practice precisely in the universality which makes all nature his *inorganic* body – both inasmuch equalise as nature is (1) his direct means of life, and (2) the material, the object, and the instrument of his life activity.

Nature is man's *inorganic body* – nature, that is, insofar as it is not itself human body. Man *lives* on nature – means that nature is his *body*, with which he must remain in continuous interchange if he is not to die. That man's physical and spiritual life is linked to nature means simply that nature is linked to itself, for man is a part of nature (Marx and Engels, 1975, pp. 275–276).

This metaphor of nature as 'man's' body (why 'inorganic' remains a mystery!), with the associated idea of 'continuous interchange' or 'metabolism' between socially organised human labour and its naturally given conditions, is a lasting contribution to the understanding of human ecology (see, e.g. Fischer-Kowalski and Haberl, 1993; Matthews *et al.*, 2000; Schandl and Schulz, 2000). But it is also important to note what Marx has to say about nature as 'spiritual nourishment' and the part played by nature in human spiritual life, as object of both artistic and scientific activity. There is some ambiguity in this: the metaphor of 'spiritual nourishment' certainly implies that the 'food' of artistic and scientific activity exists independently of the activity, and so is not a merely 'subjective' construct. However, that 'spiritual inorganic nature' must first be prepared for us to digest it seems to take us back in the direction of the 'landscape gardening' vision.

This interpretation is possible, but there is good evidence in the text for an intriguing alternative. Contrasting human and animal 'production', Marx says that 'man knows how to produce in accordance with the standard of every species, and knows how to apply everywhere the inherent standard of the object. Man therefore also forms objects in accordance with the laws of beauty' (Marx and Engels, 1975, p. 277). Later, he says:

Private property has made us so stupid and one-sided that an object is only *ours* when we have it – when it exists for us as capital, or when it is directly possessed, eaten, drunk, worn, inhabited, etc. – in short, when it is *used* by us (Marx and Engels, 1975, p. 301).

If we take these passages together with Marx's historical–cultural understanding of the development of human sensibility, they can be seen as pointing to a vision of a future society in which our aesthetic, spiritual and scientific appreciations of the rest of nature coincide with the 'inherent' reality of their objects, independently of their utility for us:

... the dealer in minerals sees only the commercial value but not the beauty and the specific character of the mineral. He has no mineralogical sense. Thus, the objectification of the human essence,

both in its theoretical and practical aspects, is required to make man's *sense human*, as well as to create *human sense* corresponding to the entire wealth of human and natural substance (Marx and Engels, 1975, p. 302).

ENGELS AND THE CONDITION OF THE WORKING CLASS

Meanwhile, Friedrich Engels, who was to become Marx's lifelong collaborator, was conducting a practical survey and subsequent denunciation of the living conditions of the working class in the industrial districts of England. The resulting work, which drew on the official reports of the time as well as his own first-hand experience, can be seen as a founding work of ecological socialism (Benton, 1996b). Engels's *Condition of the Working Class in England* begins with a historical account of the origins of the Industrial Revolution, but focuses on the concentration of the industrial working population in the great towns and cities, notably Manchester and industrial Lancashire. What distinguishes his approach from much orthodox Marxism is his emphasis on the importance of class relations not just in peoples' direct working lives, but also in their living conditions in the working-class districts, their degraded food, lack of clean water and sanitation and associated risks to health. The outcome is a remarkable synthesis of environmental concerns with historical–social analysis.

Engels offers vivid descriptions of the conditions of life of the slum dwellers, paying attention to the physical organisation and spatial distribution of their districts. Not only is there a residential segregation of the classes, but building around the main thoroughfares is so disposed as to 'conceal from the eyes of the wealthy men and women of strong stomachs and weak nerves the misery and grime that form the complement of their wealth' (Engels, 1969, p. 60). Within the working-class districts there are internal differentiations of condition. Engels uses diagrams to show how contractors maximise the profitable use of space in the design of terraces that can command different levels of rent according to stratifications among the working class families.

In the very poorest districts are concentrated the lodging houses. As noted in a contemporary report:

They are nearly all disgustingly filthy and ill-smelling, the refuge of beggars, thieves, tramps and prostitutes, who eat, drink, smoke, and sleep here without the slightest regard to comfort or decency in an atmosphere endurable to these degraded beings only (cited in Engels, 1969, p. 70).

But, despite these differentiations, there are characteristics common to all the working-class districts:

> The streets are generally unpaved, rough, dirty, filled with vegetable and animal refuse, without sewers or gutters, but supplied with foul, stagnant pools instead. Moreover, ventilation is impeded by the bad, confused method of building the whole quarter, and since many human beings here live crowded into a small space, the atmosphere that prevails in these working-men's quarters may readily be imagined (Engels, 1969, p. 60).

The working-class districts lack sewers, gutters or 'privies', and domestic animals, notably pigs, are kept in the courts between the cottages. The result is piles of rotting refuse in the immediate vicinity of dwellings, and pollution of local rivers and streams, which are the sources of water for washing and drinking. These external conditions combine with poor construction, overcrowding and filth within the dwellings to take an immense toll in terms of ill-health and mortality. Here, Engels follows a recent report on the sanitary condition of the labouring poor by Edwin Chadwick:

> All putrefying vegetable and animal substance give off gases decidedly injurious to health, and if these gases have no free way of escape, they inevitably poison the atmosphere. The filth and stagnant pools of the working people's quarters in the great cities have, therefore, the worst effect upon the public health, because they produce precisely those gases which engender disease; so, too, the exhalations from contaminated streams (Engels, 1969, p. 128).

But Engels is not content simply to outline the effects of these environmental conditions on the bodily health of working-class families. He also echoes the concerns of the more respectable social reformers for the mental and moral character of people forced to live out their lives in such squalor. How can family life and personal decency be maintained in these conditions? Should anyone be surprised if theft, prostitution, drunkenness and 'sexual excess' prevail?

Engels goes on to detail the insufficient nutrition, but also poisoned and adulterated food to which the working-class families are subjected, linking this, together with their general environmental conditions of life, with a systematic analysis of industrial hazards and diseases, occupation by occupation. He gives particular attention to the maldevelopment of children that results from this combination of both squalid living conditions and hazardous working environments.

Of course, it may well be argued that, while Engels's denunciation of the links between environmental degradation and class divisions were appropriate enough in his day, our own world is very different. It is true that many of the extremes of environmental degradation that Engels described were addressed by subsequent waves of environmental reform, public health measures and industrial regulation (Engels himself commented on these in his preface to the 1892 English edition). However, if we take the international perspective encouraged by Marx and Engels themselves, it becomes clear that new forms of international fluidity of capital, modes of regulation of trade and investment and technological change have produced new global distributions of the environmental damage and hazards associated with industrial capitalism.

On a global scale we now witness a scene broadly comparable with that exposed by Engels in his own day: socio-economic deprivation tends to be strongly associated with environmental degradation. Conditions comparable to those described by Engels persist in many cities and 'industrial zones' in 'third world' countries. Even in the 'core' countries of the 'north', as well as in the former 'communist' states of eastern Europe, there are deteriorating conditions of life in many urban complexes, where endemic high levels of unemployment combine with infrastructural decay, social fragmentation and homelessness. A vigorous 'environmental justice' movement in the USA testifies to the continuing pattern of racial and class disadvantage associated with environmental degradation (see Faber, 1998).

But there is another continuity between Engels's analysis and our contemporary world. As he made clear, for all the attempts to hide the human cost of industrial wealth creation, this cost could not be confined to the poor alone:

> I have already referred to the unusual activity which the sanitary police manifested during the cholera visitation. When the epidemic was approaching, a universal terror seized the bourgeoisie of the city. People remembered the unwholesome dwellings of the poor, and trembled before the certainty that each of these slums would become a centre for the plague, whence it would spread desolation in all directions through the houses of the propertied class (Engels, 1969, p. 97).

The toxic combination of endemic poverty, disease and environmental damage suffered by so much of the urban and rural populations of South America, Africa and south Asia stands not just as an ethical reprimand to the still-rich countries. The present threat of global pandemics, of impending local and regional resource wars, the rise of international terrorism and the endemic failure of international policy-making in the face of these global issues are all linked to deepening socio-ecological inequalities.

MARX'S LATER THOUGHT: HISTORICAL MATERIALISM

Returning to Marx, it seems that he soon became dissatisfied with the philosophical work of his early years, and turned more to detailed historical and journalistic studies, while continuing his thorough critical review of the economic theories of his time. From these studies there eventually emerged a new materialist conceptual framework for analysing historical processes, as well as a vast, systematic and critical account of the nature of modern capitalism.

The materialist approach to historical understanding (subsequently termed 'historical materialism') had, as its central concept, the idea of 'modes of production'. In Marx's view, the key to understanding how a society works, its long-term tendencies, inherent patterns of social conflict, and liability to change is the pattern of social relationships through which people act on nature to meet their needs. We need to look at the kinds of work people do, the technologies they employ, the way cooperation between them is organised, as well as the rules that govern how the results of all this work are distributed through the population. Marx's concept of modes of production is his way of classifying the different ways human society has organised its social relationship to nature – as well as giving the outlines of future possibilities. So, Marx distinguishes between the classical civilisations with their 'Ancient' mode of production; the dominant system of the European middle ages – the 'Feudal' mode of production; an oriental system – the 'Asiatic' mode; and, finally, the modern 'Capitalist' mode. Marx's expectation was that the capitalist mode would eventually be overthrown in favour of a socialist or communist mode, in which productive property would be held in common, and the human relation to nature would be rationally regulated, rather than left to market forces.

Although ecological issues were not central to Marx's own thinking, the idea of modes of production has much to offer for environmentalists. It enables us to get away from misleading simplifications in explaining the causes of ecological problems. Instead of abstract questions about what 'Man' has done to nature, or unthinking explanations in terms of population growth, belief systems or 'life-styles', Marx's approach allows us to investigate *historically specific* forms of socio-economic interaction with natural forces and conditions. In our own times, as in Marx's, these forms of interaction with nature are predominantly capitalist. Hence, the continuing relevance of Marx's monumental study of the way capitalism works if we want to understand the ultimate causes of ecological change and its impact on different groups in society.

According to Marx, modern capitalism could only be established when peasants and rural labourers were driven off the land, giving rise to a floating population of people with no way of meeting their basic needs other than by selling their labour to another. This is one part of what Marx meant when he argued that capitalism was based on 'alienation' from nature. The 'others' to whom the newly 'freed' population were to sell their labour were the owners of manufacturing workshops, precursors of the modern industrial factory system. In Marx's view the key social division in the new society built on the basis of industrial production was the opposition between the wage workers, on the one hand, and, on the other, the owners of the factories, machinery and raw materials in and with which the workers were paid to labour.

In exchange for their work, the labourers are paid a money wage, with which they buy such items as they need to maintain themselves and their families. Meanwhile, the factory owners, the capitalists, see the wages as one cost, among others, that they have to pay in order to make a profit from selling the items the workers produce. But since other factory owners are also attempting to sell their produce, each capitalist is set into a competitive relationship with all the others in his branch of production. To survive in competition he has to reduce costs, by paying his workers less, making them work harder, inventing more productive technologies, or increasing the size of his operations to get economies of scale – perhaps taking over or driving out of business his competitors. So, on this analysis, the capitalist faces a structural imperative (whatever his personal feelings) to strive for more and more profit, to invest, grow and beat the competition. This is the beginnings of an explanation of the way the never-ending pursuit of growth is an intrinsic part of capitalism; not something that people could just choose to stop doing.

The importance of monetary calculation in capitalist society is a consequence of the fact that productive activity is not simply organised around the meeting of needs, but, rather, to make a profit from the sale of goods and services. But, of course, at least some needs have to be met most of the time for social life to continue at all. Taking account of this, Marx developed a dual analysis of modern capitalism: it is at the same time *both* a process of producing goods with a value in exchange on the market ('exchange values') *and* a process whereby, through the circulation of goods in the market, diverse specific wants are met. So, for Marx, every commodity has both an exchange value (which can be expressed in monetary terms) and a 'use value'. We can analyse the whole process of production in terms of the bringing together of many qualitatively diverse materials, transformed by workers with different skills and technologies to produce many different kinds of goods, designed to meet a similarly great variety of human purposes. However, what is specific to capitalism is that all of these *qualitatively* different aspects of the process are subjected to a single, *quantitative* calculation; they enter into the considerations of the capitalist, and so enter into investment decisions, purely in terms of their market value ('exchange value'). They figure simply as monetary costs of potential sources of income and profit.

Marx emphasised some of the consequences of this for social relations under capitalism. One of these is that the overall pattern of investment in labour, materials, etc., is determined by an impersonal force – the market – which we experience as outside our control, almost as a force of nature ('you can't buck the market'). So, there is nothing in the system that guarantees investment will flow to the production of things or services ('use values') that are really needed. True, investment may follow signals of market demand for goods, but there is no necessary correspondence between social need and market demand. The latter depends on ability to pay, whereas the former is often greatest among those who cannot afford to pay. Hence, the common experience of extreme poverty and unmet need alongside immense wealth and luxury consumption.

MARX'S ECOLOGY

Just as the diversity of human needs can remain unmet if they do not issue in market demand, so the naturally given and ecological environmental conditions that contribute to production (and social life generally) remain outside the sphere of monetary calculation. Environmental pollution, degradation of living conditions, extermination of living species and destruction of their habitats, exhaustion of raw materials and so on do not come up for consideration unless they can somehow be made to figure as monetary costs. In some cases (e.g. many raw materials) these conditions do figure as monetary costs, and scarcity may lead to rising costs, and consequent increases in efficiency, or new technologies to avoid using them. However, many of these conditions do not have a monetary value, and so will continue to be degraded and destroyed in the absence of some external legal or political intervention to regulate capitalist development.

The tendency of capitalist production to undermine its own environmental or ecological–material conditions was given less emphasis in Marx's writings than was his concern for the degradation of its human victims, the wage workers. Nevertheless, the makings of a sophisticated ecological critique of capitalism were certainly present. The idea of the dependence of humans on continuous interaction with the rest of nature was, as we have seen, a key theme of Marx's earlier philosophical writings. However, by the 1860s, as he worked on his major critical work, *Capital*, he was increasingly interested in the question of declining soil fertility. The development of large-scale commercial agriculture in the early decades of the 19th century was giving rise to increasing demand for alternative sources of plant nutrients as these became depleted in the soil. A great international trade in guano and other manures developed, but was inadequate to the demand.

The crisis gave rise to the development of the science of soil chemistry, and foremost in developing this science was Justus von Liebig. Marx studied Liebig's work closely, and followed him in an increasingly pessimistic view of the role of modern industrialised agriculture. Accepting Liebig's central concept of 'metabolism', as characterising the cyclical flow of nutrients from soil through to agricultural produce and back to the soil, Marx identified a contradiction in the relation between capitalist agriculture and modern industry. This refers back, interestingly, to Edwin Chadwick's report on the sanitary condition of the poor that had been so influential for Engels.

Essentially, in drawing the former agricultural population into the great cities, modern industry produced an immense problem of disposal of polluting human and animal waste, whilst at the same time failing to return to the soil the nutrients drawn from it to feed the urban population. Marx refers to this as the 'metabolic rift', endemic to the combination of modern industrial capitalism with large-scale industrial agriculture:

> Capitalist production collects the population together in great centres, and causes the urban population to achieve an ever-growing preponderance. This has two results. On the one hand, it concentrates the historical motive force of society; on the other hand, it disturbs the metabolic interaction between man and the earth, i.e. it prevents the return to the soil of its constituent elements consumed by man in the form of food and clothing; hence it hinders the operation of the eternal natural condition for the lasting fertility of the soil ... all progress in capitalist agriculture is progress in the art, not only of robbing the worker, but of robbing the soil; all progress in increasing the fertility of the soil for a given time is a progress toward

ruining the more long-lasting sources of that fertility (*Capital* 1, pp. 637–638, cited in Foster, 2000, p. 156).

SCIENCE AND TECHNOLOGY: THE 'DEVELOPMENT OF THE FORCES OF PRODUCTION'

The application of soil chemistry to agricultural production was but one example of the increasing integration of natural scientific research with the priorities of capitalist industry, to the extent that Marx referred to science itself as a 'force of production'. In Marx's analysis, the application of science to promote technological innovation was another intrinsic feature of modern capitalism. In his account, new technologies had the dual role of replacing human labour, increasing productivity and so reducing costs per unit output, and also achieving more control over the labour process. So, the twin stimuli of capitalist competition and worker resistance made for rapid technological innovation. Modern industrial capitalism, then, looked set fair to realise the dream of Francis Bacon and the founders of the Royal Society (see Chapter 2), of an ever-growing human technical mastery over the forces of nature:

> The bourgeoisie cannot exist without constantly revolutionising the instruments of production, and thereby the relations of production, and with them the whole relations of society ... The bourgeoisie, during its rule of scarce one hundred years, has created more massive and more colossal productive forces than have all preceding generations together. Subjection of Nature's forces to man, machinery, application of chemistry to industry and agriculture, steam-navigation, railways, electric telegraphs, clearing of whole continents for cultivation, canalisation of rivers, whole populations conjured out of the ground – what earlier century had even a presentiment that such productive forces slumbered in the lap of social labour? (*Communist Manifesto* in Marx and Engels, 1976, pp. 487–489).

Many subsequent readers of Marx, including those claiming to be his followers, interpreted such paragraphs as unequivocally endorsing the human project to master the forces of nature through technological development. This was, indeed, the way Marx was interpreted by Stalin and subsequent leaders of the former Soviet Union, with their ruthless drive for industrial development no matter what the human and environmental cost.

However, as we have seen, this is not the only, or even the most plausible, understanding of Marx's view. Marx was well aware of the environmental

damage implicit in capitalist industrialisation, and seriously posed the question of the sustainability of capitalist agriculture. His vision of a socialist future was, rather, one in which the associated producers would regulate their metabolism with nature according to rational principles:

> Freedom in this area can only consist in socialised man, the associated producers, rationally regulating, bringing under their common control, their interchange with nature, instead of being ruled by it as by a blind power (*Capital* 3, ch. 48, sect. 3; tr. David McNeill).

And it seems that at least one relevant principle was to be sustainability:

> Even an entire society, a nation, or all simultaneously existing societies taken together, are not owners of the earth. They are simply its possessors, its beneficiaries, and have to bequeath it in an improved state to succeeding generations as *boni patres familias* (good heads of the household) (*Capital* 3, p. 911, cited in Foster, 2000, p. 164).

However, Marx does seem to have envisioned the future socialist society as one in which the great productive power unleashed by technological change under capitalism would be the basis for a material abundance shared by the whole society, rather than appropriated by the minority of rich business men and their allies. He also seems to have thought of technological advance primarily in terms of reducing the labour time necessary to produce a given quantity of goods. As we shall see, some later green thinkers in the socialist tradition were critical of both these aspects of Marx's thought; perhaps the destructiveness of capitalist industrialism might be at least partly due to the technology itself, independently of the property relations under which it was employed? Perhaps we should see the development of technology under capitalism, not as value neutral, but as reflecting the requirements of capital. And is not the vision of a future society in which we are liberated from the need to work a serious retreat from Marx's early vision of fulfilment through cooperative and creative work?

CAPITALISM AND CONSUMPTION

But Marx was impatient of attempts to say much in advance about the nature of the future society. For him and Engels, the key concern was with the contradictory and unstable character of contemporary capitalism, and the prospects for its overthrow by the increasingly organised and rebellious European working class. Although a competitive scramble for profit and the accumulation of money wealth was endemic to capitalism, Marx believed the system to be contradictory, in the sense that it constantly generated obstacles to its own dynamic of growth. We have seen that one of these was its tendency to undermine its own environmental conditions (a theme to be developed later), but Marx himself concentrated on several 'crisis-tendencies' that he thought were built into the structure of capitalism, preventing it from ever stabilising itself. One of these has to do with the mismatch between the immense productive power unleashed by capitalism and the availability of purchasing power in the economy. As we saw, each capitalist seeks to reduce his labour costs to cheapen the product and so gain an advantage over his competitors. However, to make a profit, he must actually sell the goods he produces so cheaply. So, it is in his interests that other employers pay their workers well, so they have the purchasing power to buy his goods (except that his workers would soon notice this and demand more, or go to another employer!). So, one cause of recurrent economic crises is that low wages result in insufficient purchasing power in the economy, cutting the capitalists' return on their investment, with consequent mass unemployment and bankruptcy.

This internal problem of capitalism has great importance for understanding the way capitalism developed in the West during the 20th century. From the 1920s onwards, and especially after World War II, capitalism underwent important reforms. Under pressure from mass labour movements and socialist political parties, it was recognised that profitability could be maintained while allowing a sizable section of the working class to enjoy job security and improved living standards, so long as their relatively high wages were used to maintain high levels of consumer spending. The new system is sometimes called 'Fordism', after the US car manufacturer who has come to symbolise this important shift in the nature of capitalist society (see Gramsci, 1971). Instead of a life of privation, workers in the 'core' manufacturing industries were encouraged by sophisticated advertising to desire ever-increasing levels of material consumption. The capacity of the system to 'deliver the goods' was to serve both as a means of avoiding destabilising crises, and also as a way of giving legitimacy to the continuation of rule by big business. To a considerable extent, the 'cold war' between the Soviet Union and the West came to be fought in terms of the rival claims of the two systems to satisfy consumer demand. The new framework of modern capitalism also involved a much increased role for the state. The 'welfare state' offered guarantees of security in the face of ill-health and unemployment, whilst enabling the state to offset impending economic crises by public spending. Perhaps the most insightful commentators on this new form of capitalism were the thinkers of the Frankfurt School of critical theory, notably Adorno, Horkheimer and Marcuse (Held, 1980, remains an excellent introduction).

With the demise of the Soviet Union, there have been significant shifts away from this 'Fordist' model in the West, but what remains as an absolute imperative is the system-requirement for ever-higher levels of consumption (see Harvey, 1990, 1996). Governments are judged by their success in maintaining economic growth, and in the face of the terrorist threat, enjoin their subjects to consume still more as a patriotic duty. Of course, high levels of economic growth and consumption do not translate directly into environmental degradation. Many defenders of the prevailing social system pin their hopes on future technologies that will allow for continuing increases in material living standards, not just in the already 'developed' societies, but across the 'developing' world, whilst maintaining the world's ecological life-support systems intact (see e.g., Mol, 1996). Currently, the evidence points in the opposite direction. Greenhouse gas emissions continue to rise, along with rapidly expanding use of the private car and escalating air transport, whilst destruction of habitat and extinction of species proceeds unabated. However, the key point here is that on a Marxian analysis, ever-increasing levels of consumption and associated economic growth are *system imperatives* for modern capitalism, not mere matters of personal lifestyle choice. Any large-scale decline in consumer demand triggers economic crisis with mass unemployment and political instability. Serious change to a less ecologically destructive pattern of living could only be achieved alongside deep changes in the whole social organisation of production and distribution.

THE FUTURE SOCIETY: MORRIS AND NEWS FROM NOWHERE

So far we have seen that a broadly Marxian account of the way capitalism works can give valuable insights into the fundamental causes and dynamics of environmental change in modern society. However, this is still some way from a plausible account of the possibility of a future, non-capitalist society that might enable us to live both happily and sustainably with the rest of nature. Marx and Engels took the view that it was not the role of intellectuals to provide blueprints for future utopia – the new society would be built democratically by those who brought it into being. However, other thinkers in the Marxian tradition have seen the importance of utopian thinking as necessary for opening up the political imagination to alternative possibilities.

One such thinker was William Morris (1834–1896). A contemporary of Marx, Morris is now best known for his wallpaper designs, but in his day he was an influential figure in radical political circles, as well as inspirational leader of the arts and crafts movement. He became dissatisfied with his work as an interior designer servicing the well-to-do households, and, increasingly drawn to Marxist socialism, insisted that beautiful surroundings were an entitlement for all, as was a share in fulfilling and creative work. He denounced the polluting and degrading effects of modern industrial production, and championed freely creative work in the craft traditions:

> ... civilization has reduced the workman to such a skinny and pitiful existence, that he scarcely knows how to frame a desire for any life much better than that which he now endures perforce. It is the province of art to set the true ideal of a full and reasonable life before him, a life to which the perception and creation of beauty, the enjoyment of real pleasure that is, shall be felt to be as necessary to man as his daily bread ... ('How I became a socialist', *Justice*, 16th June 1894, in Briggs, 1973, p. 37).

One of Morris's most influential writings is his utopian novel, *News from Nowhere*. He imagines waking up in the 21st century, after a revolutionary uprising that has dispensed with the factory system and wage labour. In this new world, people work for the joy of the activity itself and for the pleasure of helping their neighbours. Spared the debilitating effects of poverty and unwholesome working and living conditions, the people are physically handsome and long lived. His guide introduces him to once-familiar parts of London, that have long since been converted into meadows, orchards, woodland and gardens, interspersed with buildings. Where once stood East End 'slums', mere 'stews for breeding and rearing men and women in such degradation that ... torture should seem to them mere ordinary and natural life', were now pleasant meadows and gardens, with buildings 'mostly built of red brick and roofed with tiles, [that] looked, above all, comfortable, and as if they were ... alive and sympathetic with the life of the dwellers in them'. Some of the older 'pre-commercial' buildings have been retained, but persist in radically changed surroundings and uses. The Houses of Parliament have been preserved, due to the efforts of an antiquarian society, but now they are used as 'a kind of subsidiary market, and a storage place for manure' (Briggs, 1973, p. 209). An elderly relative of his guide, who still retains some knowledge of the previous order of society,

explains the change in the organisation of work. In the old order:

> ... men had got into a vicious circle in the matter of production of wares. They had reached a wonderful facility of production, and in order to make the most of that facility they had gradually created (or allowed to grow, rather) a most elaborate system of buying and selling, which has been called the World Market; and that World Market, once set a-going, forced them to go on making more and more of these wares whether they needed them or not ... By all this they burdened themselves with a prodigious mass of work merely for the sake of keeping their wretched system going ... To this 'cheapening of production', as it was called, everything was sacrificed: the happiness of the workman at his work, nay, his most elementary comfort and bare health, his food, his clothes, his dwelling, his leisure, his amusement, his education – his life, in short – did not weigh a grain of sand in the balance against this dire necessity of 'cheap production' of things, a great part of which were not worth producing at all (Briggs, 1973, pp. 263–264).

So far, Morris's denunciation of the compulsive growth dynamic of capitalism, its separation of the exchange value of goods from their real utility, and subordination of all values to the accumulation of money capital follows the main argument of Marx's *Capital*. However, Morris's vision of an alternative world of work and consumption seems to echo the earlier, philosophical Marx. The Marx of *Capital* seems to have given up on the possibility of a society of freely creative producers, deriving fulfilment from their work, in favour of a future in which advanced technology reduces necessary labour time to a minimum and gives us time to pursue a variety of leisure activities. In Morris' utopia work has become, again, a source of pleasure. The factories have disappeared, and been replaced by 'banded workshops' where people collect together to do 'handwork in which working together is either necessary or convenient; such work is often very pleasant':

> As to the crafts, throwing the clay must be jolly work: the glass-blowing is rather a sweltering job; but some folk like it very much indeed; and I don't much wonder: there is such a sense of power, when you have got deft in it, in dealing with the hot metal (Briggs, 1973, p. 222).

Freed from the demand to mass produce cheap and often unnecessary goods for the market, the link between production and the meeting of needs is restored. Quality goods, enjoyable to make, fit for use, and only so many as are needed are the new rational principles governing work:

> The wares which we make are made because they are needed: men make for their neighbours' use as if they were making for themselves, not for a vague market of which they know nothing, and over which they have no control ... So that whatever is made is good, and thoroughly fit for its purpose. Nothing can be made except for genuine use; therefore no inferior goods are made ... and as we are not driven to make a vast quantity of useless things, we have time and resources enough to consider our pleasure in making them (Briggs, 1973, p. 267).

But Morris acknowledges that not all work can be done by hand with pleasure, and such 'irksome' work is done by 'immensely improved machinery'.

The novel ranges over many questions, including the position of women in society (where Morris's vision seems decidedly traditionalist by the standards of modern feminism), the education of children, the keeping of order, government, architecture and domestic furnishing. His descriptions of the new physical landscapes suggest a vision in which the opposition between town and countryside is overcome, with open pastures, meadows and orchards where once was dense urban housing. This would be consistent with Marx's critique of the 'metabolic rift', but in Morris' case probably motivated more by aesthetic and social considerations – populating and rendering more cosmopolitan the rural areas, while rendering more beautiful the urban scene. However, the landscapes of Morris' utopia, whilst beautiful, are productive and shaped by human cultivation. Indeed, the land is described as a 'garden'. Nevertheless, Morris' traveller does see the edge of the 'Middlesex and Essex forest', and asks if such things have any place in a garden. His informant replies:

> My friend, we like these pieces of wild nature, and can afford them, so we have them; let alone that as to the forests, we need a great deal of timber, and suppose that our sons and son's sons will do the like ... I assure you that some of the natural rockeries of our garden are worth seeing. Go north this summer and look at the Cumberland and Westmorland ones – where by the way, you will see some sheep feeding, so that they are not so wasteful as you think; not so wasteful as forcing-grounds for fruit out of season (Briggs, 1973 p. 247).

There is, then, a place in utopia for wild nature, but this is not a wilderness from which humans are excluded. Natural environments can retain their status as such, while still offering some human utility. The reference to the 'Middlesex and Essex forest' is a piece of contemporary polemic. After the successful struggle to save Epping Forest from enclosure there was a considerable debate about how it should be managed (see Chapter 2). Morris railed against the prospect that the forest would become victim of the landscape gardeners, or even be converted to 'golf grounds'.

SOVIET ECOLOGY AND ITS FATE

The puzzle remains why, with this theoretical and imaginative background in the Marxist tradition, the environmental record of the first great attempt at socialist construction should have turned out so disastrously. Fascinating historical research – see especially Weiner (1988) and Gare (1993, 1996) – made possible by the opening up of Soviet culture by Gorbachev, and by the subsequent collapse of the regime, begins to answer this question. In the period immediately following the 1917 revolution there was a great flowering of Soviet culture. Influenced by the philosophy of A.A. Bogdanov, who had developed a version of systems theory based on the flows of energy through nature, ecology came to occupy an important place in Soviet science and culture. Bogdanov was a leading figure in the 'Proletarian Cultural and Educational Organisations' (*Proletkul't*), later replaced by the sinister-sounding 'Commissariat for Enlightenment'. However, under the leadership of A.V. Lunacharsky, a supporter of Bogdanov's ideas, the Commissariat promoted Soviet science, and, in particular, the science of ecology. In pre-revolutionary Russia plant ecology was already well advanced, and pioneers of this discipline were arguing that natural ecological communities were a model of efficiency and productivity that should be studied and emulated in agriculture.

This necessitated the setting aside of protected nature reserves for study. Lunacharsky was a strong supporter of these ideas, and persuaded Lenin to hand over responsibility for creating and managing reserves to the Commissariat, to prevent them being threatened by short-term economic pressures. By 1929 sixty-one reserves had been created, covering almost four million hectares. Subsequently, ecology was increasingly included in university and school curricula, and by the late 1920s ecologists were making a bid to influence state policy. One of the most original soviet ecologists, V. V. Stanchinskii, for example, was arguing for studies of energy flows through ecological communities as a basis for economic activity and for the use of biological pest control instead of pesticides. Under his influence the 1929

All-Russia Congress for the Conservation of Nature resolved that 'the distinction and tempo of economic growth can be correctly determined only after the detailed study of the environment and the evaluation of its productive capacities with the aim of its conservation, development and enrichment'. On this basis, ecologists should play a major part in the formulation and monitoring of the Five-Year Plan.

However, none of these ideas achieved practical shape. By 1930 Stalin had consolidated his rule and began to implement a ruthless vision of total subordination of nature to human purposes. A sustained drive against ecology resulted in the closure of Stanchinskii's research station, and his arrest in 1934. With a few remarkable exceptions, after the 1930s the various traditions of Marxism abandoned the earlier concerns with ecology. In the poorer countries Marxists tended to favour rapid urbanisation and industrialisation to rid themselves of dependency on the Western imperial powers, and to advance the living standards of the workers. Meanwhile, in the Western countries, there developed a variety of 'Western Marxisms' that generally concerned themselves with social and cultural issues that had been left undeveloped by the economic focus of classical Marxism. Where, as in some Western European countries, there remained powerful mass Communist parties, these generally formed alliances with other parties of the left with an agenda of progressive parliamentary reform.

THE 1960s AND GREEN–LEFT REVIVAL

The next great upheaval came in the late 1960s and 1970s, with the combined challenge of a revitalised feminism and the emergence of an independent, radical Green political movement. Among other things, feminism forced Marxists to rethink the contribution made by human labour outside the industrial workplace, mostly by women, unwaged, in the household. Not only had the gendered character of the division of labour in capitalist society been neglected, but so had the existence of other forms of oppression, exclusion and exploitation not directly attributable to wage labour. The resulting explosion of political struggles on these issues, as well as theoretical discussions of 'domestic labour' led to the (re)posing of 'new' questions about the nature of the future society. In particular, some of the more radical feminists were insisting on the importance of emotional and evaluative dimensions of work, advocating an ethos of care, nurturance and concern for future generations. The connection of these values with the new ecological politics was also increasingly clear. As the case of the Soviet

Union made only too obvious, mere state ownership of production did not of itself guarantee liberation of the people, or a new era of ecologically sustainable production and consumption.

These challenges provoked a ferment of creative work, debate and intellectual renewal among those modern Marxists prepared to take the risk of addressing them. There is space here to mention only a small selection of these developments, and the continuing debate they have engendered.

Andre Gorz and the dual economy

Andre Gorz, a leading French radical writer, authored a series of articles during the 1970s, collected together with additions, and in translation published as *Ecology as Politics* in 1980. Gorz reworks the classical Marxian analysis of capitalism to show how ecological destruction obstructs the accumulation of wealth and thus affects the rate of profit. A part of his analysis (strongly influenced by the ideas of Ivan Illych) is a critique of the broadly optimistic view of technological development still common among the 'traditional' left. Instead, he offers a view of technology as both a product of the prevailing system of power, and a nexus through which that system sustains and reproduces itself: a new kind of society would need quite different sorts of technology.

> ... the techniques on which the economic system is based are not neutral. In fact they reflect and determine the relations of the producers to their products, of the workers to their work, of the individual to the group and the society, of the people to the environment. Technology is the matrix in which the distribution of power, the social relations of production, and the hierarchical division of labour are embedded ... [C]apitalism develops only those technologies which correspond to its logic and which are compatible with its continued domination. It eliminates those technologies which do not strengthen prevailing social relations, even where they are more rational with respect to stated objectives. Capitalist relations of production and exchange are already inscribed in the technologies which capitalism bequeaths to us (Gorz, 1980, p. 19).

This view of the way existing power relations and socio-economic priorities are built into technologies clearly has very important implications for environmental policy. So-called 'technology transfer' to poorer countries is not so innocently benevolent as it might seem, and the strategy of addressing ecological problems with 'technological fix' solutions can be seen as a heavily loaded political choice.

Gorz's vision of an alternative future of voluntary cooperation, with free and self-determining individuals and communities, would set quite different criteria for the development of technologies and production methods. Acceptable technologies and methods of production:

- can be used and controlled at the level of the neighbourhood or community;
- are capable of generating increased economic autonomy for local and regional collectivities;
- are not harmful to the environment;
- are compatible with the exercise of joint control by producers over products and production processes.

It is instructive to apply these criteria to the current debate over energy generation between green proposals for micro-generation and intensive special-interest lobbying for nuclear power.

Gorz is also one of the few thinkers (R-GSG, 1995, and Martin Ryle, 1989, are two other examples) to have followed Morris' example in imagining a future 'feasible utopia'. In fact, his vision ('A possible utopia' in Gorz, 1980, pp. 43–50) has much in common with that of Morris. Like Morris, he remains committed to the idea of pleasure and fulfilment in work, but also recognises that some work may be intrinsically irksome. This forms the basis of an envisioned dual economy. In a formal sector, devoted to production of necessary goods, efficient technologies will be employed to produce 'a restricted number of standardized products of equal quality and in sufficient amounts, to satisfy the needs of all'. Production would use methods consistent with the criteria just listed. The aim would be to so organise the production process that goods necessary to meet essential needs could be produced, whilst reducing the working week to not more than 24 hours. There would be a standard income unrelated to one's employment, and opportunities for people to exchange jobs, to get some variety in their work experience. The provision of a basic income would allow of the possibility for voluntary organisations to provide neighbourhood services such as caring for children, helping the elderly or sick and teaching skills. As more free time came available, with the reduction of hours spent in the formal economy, people could do freely creative work of their own choosing in well-equipped studios and workshops that would be located in each neighbourhood or even apartment block.

James O'Connor and the second contradiction of capitalism

A number of writers based in the USA have also contributed significantly to the development of theoretical and practical integration of green ideas with a reworking of the Marxian heritage. A good deal of this writing is carried in the journal *Capitalism, Nature, Socialism* whose major influence has been the work of J. O'Connor. His seminal work has been the postulation and development of the idea of a 'second contradiction' of capitalism (see O'Connor, 1996, 1998). As we saw above, Marx identified a central contradiction in capitalism in the form of growing incompatibility between the development of an increasingly cooperative production process and the continuation of private accumulation of wealth. This contradiction, according to Marx, was the structural basis for the social conflict between organised labour and the employing class. O'Connor argues that, as well as this 'first contradiction', another, 'second contradiction' is endemic to capitalism. This is a contradiction between the forces (technology, etc.) and relations (property relations) of capitalism, on the one hand, and the conditions of production, on the other. This is really a systematic development of the notion we have already encountered in Marx's own analysis: the idea that capitalist development tends to undermine its own conditions. What O'Connor adds is a more thorough theorisation of the conditions, and an explanation of capitalist production's tendency to degrade them.

O'Connor distinguishes three broad categories of conditions of production:

1 'Communal general conditions of production', including means of communication, transport infrastructures, sewage and waste-disposal systems.
2 'Labour power', including the mental and physical health of the work force, skills, training and both socialisation and biological reproduction.
3 'External physical conditions', including ecological life-support systems, the quality of air, soil and water, reserves of raw material and so on.

His general explanatory hypothesis is that these conditions tend to be over-exploited and so degraded because capitalists treat them as commodities, whereas they are not produced or reproduced as commodities. The clearest example is that of the reproduction of the labour force. The 'production cycle' here is very prolonged, and takes place wholly outside the framework of capitalist relations, in the household, and is conducted on the basis of values and motivations quite other than production of children for purposes of exchange on the market!! So, the expectation of a rough correlation between supply and demand that applies to other commodities certainly does not apply to the generation-by-generation 'production' of new workers. So far as the first category of conditions is concerned, these may, of course, be provided by capitalist firms for profit, but the rate at which they are provided, their location, effectiveness and so on cannot be left to the vagaries of the market, and state planning and part-financing cannot be avoided to compensate for the irrationalities of market allocation. As to the third category, ecological life-support systems and general ecological/material conditions, these are often not treated by capital as commodities. Rather, they tend to be treated as free goods and consequently over-exploited and degraded. Only with state intervention to 'internalise' these environmental costs in the shape of green taxes or other financial instruments do they come to be treated as commodities. But, again, environmental conditions are not produced as commodities – indeed, they are not produced at all, but given by nature.

O'Connor's thesis, then, is that, perhaps for rather different reasons, the three main categories of conditions of production tend to be over-exploited and degraded by capitalist development, with the consequence not only that the living and working conditions, health and quality of life of the wider population suffers, but also that the further accumulation of capital is itself threatened. The general experience of environmental degradation fosters the development of an environmental movement, which, in turn, puts pressure on the political system to repair and restructure the conditions of production. There is a general parallel here with the way the labour movement acts to put pressure on the political system to ameliorate and regulate the exploitation of labour, introduce welfare reforms and so on. In both cases, O'Connor argues, wider social movements have to act politically so as to save capitalist economic activity from its own destructive effects. The political outcome, therefore, is a potential alliance between the labour movement and the environmental movement to represent the general interest of society against the particular interest of capital, with a resultant broadening of the popular basis for socialism.

Globalisation and the environment: world-system theory

Whilst this aspect of O'Connor's thought focuses on the relationships between capitalism, the environment and social movements at the level of the nation state, a great deal of 20th century Marxism has been concerned with analysing the structure of economy and society at the global level. As capitalism has become the overwhelming form of economic organisation worldwide, this implies the need for a general account of the geographical spread of capitalism and of the radically unequal development experienced by different regions of the world.

One of the more successful approaches to this task is called 'World System Theory' (WST) or, more loosely, the world system 'perspective'. The approach owes much to earlier writers in the Marxist tradition, such as the analyses of imperialism by V.I. Lenin, Rosa Luxemburg and others, and more recently to theorists in the 'dependency' school who attempted to explain the continued poverty in 'Third World' countries in terms of the legacy of Western political and economic dominance. However, WST does depart from orthodox Marxism in several respects, for example, in having a wider definition of capitalism, including as 'capitalist' enterprises that use forced labour, as distinct from the Marxian emphasis on the wage relation. The generally accepted originator of the approach is Emmanuel Wallerstein, but a very useful introduction is Shannon (1996).

According to WST, the economic structure of a country, and the condition of its people and environment need to be understood in terms of its past history, and, in turn, that has to be understood in terms of the positioning of that country in the wider world. The 'world system' within which each country is located has two closely related components: a broad pattern of 'economic zones' and the system of nation states, with their various international relationships. WST distinguishes three main economic zones: the 'core', characterised by capital-intensive, high-technology industries, and high levels of capital accumulation. Geographically the core includes Western Europe, the USA and Japan. At the opposite extreme are the 'peripheral zones'. Traditionally, these countries, often former colonies of the Western powers, derive their income from exporting primary products or low-technology manufactured goods to the 'core'. Many of the poorer countries of Africa and South America can be included in this category. A third zone is termed the 'semi-periphery'. Here, there are moderate levels of industrialisation, often including industrial sectors that are in decline and so relocated from core economies. The semi-periphery supplies primary and some manufactured goods to the core, but also exports manufactured goods to the periphery. Examples include Mexico, India and some of the countries of south-east Asia.

The position of nation states within each of these zones is rooted in their historical past, very often linked to past relationships of colonisation. As Roberts and Grimes (2002) put it: 'These historic links to the world economy and polity shape many of the types of products a nation makes, which commodities are traded with whom and on what terms, the conditions for both capital and labour, as well as the nation's global power *vis-à-vis* other nations'. This continuing legacy of trading and investment relationships, with its consequences for the economic, military and diplomatic power of nation states within the world system explains why many poorer countries remain trapped with very limited options for breaking out of their subordinate position. There is no single view of the main mechanisms keeping the system in place. Some WST theorists emphasise inequality in trading relationships, with peripheral countries having to sell primary goods cheaply on the world market, while core countries can export their high-technology goods at higher market prices, with the net effect of continuing transfers of wealth from periphery to core. But other mechanisms include core transnational corporations investing in the periphery to take advantage of cheap labour and low environmental regulation, with profits flowing back to the core. Current institutions regulating international trade, investment and development aid (WTO, World Bank, IMF) are also seen to represent the interests of the more powerful core economies and transnational corporations. Covert subversion and outright military invasion by the powerful 'core' states are, of course, reserve mechanisms to 'discipline' recalcitrant nations.

Until quite recently few world systems theorists included environmental issues into their analysis, but there is increasing interest and research into this dimension. The approach is particularly useful in explaining global patterns of inequality in ecological degradation alongside inequalities of wealth. For example, the pressure to compete in the global economy, given the combined effects of neo-liberal WTO rules, debt repayments and IMF structural readjustment policies, results in many countries of the periphery imposing unsustainable levels of exploitation both of labour and of their natural environments and resources. Exports of timber, cash crops, scarce minerals and other commodities are examples that have been researched in detail by followers of the WST approach (see literature cited in Roberts and Grimes, 2002). Another area of environmental concern well analysed by WST is the relocation of declining industries in the semi-periphery. These industries often use outdated, polluting technologies, and to maintain incentives for foreign investment are frequently located in special zones with low environmental and labour standards. This problem is exacerbated by determined efforts on the part of some of the more powerful states in the semi-periphery to acquire 'core' status. China and India, in particular, are industrialising with great rapidity, but at enormous cost to their own environments and the health of their working population and poor. These recent changes may herald structural change in the global economy and inter-state system, representing a major challenge to the global dominance of the USA and its allies. In terms of both environmental pollution on a global scale and increasing conflict over scarce resources, notably oil, natural gas and water, we may be facing a period of extreme socio-ecological disruption, authoritarian measures and increased military conflict.

PROSPECTS FOR CHANGE?

However, central to the Marxian and neo-Marxian traditions has been a commitment to recognising the significance of oppositional social movements. As we saw, the early Marx and some later writers in the tradition held out a vision of a future society in which wealth would be held in common, and a harmonious relationship with the rest of nature restored through an approach to production that relinked it closely to the needs of consumers and to the pleasure of the producers. Whether or not they refer to the Marxian tradition, many of the activists in today's global coalitions of social movements and NGOs share these values of sustainability, social justice and production for need. Particularly in South America, we are beginning to see national policies strongly influenced by these grass-roots movements, and offering resistance to the dominant pattern of global economic regulation.

An important case is that of Cuba (see Gott, 2004). A former Soviet 'satellite', the Cubans faced disaster with the collapse of the Soviet Union, and with that the loss of their main export markets. Combined with the trade and investment boycott enforced by the USA this took Cuba close to economic collapse in the early 1990s. However, emergency measures put in place to develop a self-sufficient economy have led to an unprecedented 'greening' of the Cuban economy, with a shift to agricultural cooperatives and large-scale conversion from intensive to organic food production. Out of the increasingly chaotic and conflict-ridden global disorder of the present decade, it is at least possible that alternative visions of a greener and more just world order may eventually gain a new prominence.

REFERENCES

Benton, T. (ed.) (1996a) *The Greening of Marxism*. New York and London: Guilford.

Benton, T. (1996b) 'Engels and the politics of nature'. In C. J. Arthur (ed.) *Engels Today: A Centenary Appreciation*. Basingstoke: Macmillan, pp. 67–94.

Briggs, A. (ed.) (1973) *William Morris: Selected Writings and Designs*. Harmondsworth: Penguin.

Dunlap, R. E., Buttel, F. H., Dickens, P. and Gijwijt, A. (eds) (2002) *Sociological Theory and the Environment*. Lanham, Boulder, New York and Oxford: Rowman & Littlefield.

Engels, F. (1969) *The Condition of the Working Class in England*. London: Panther.

Faber, D. (ed.) (1998) *The Struggle for Ecological Democracy*. New York: Guilford.

Fischer-Kowalski, M. and Haberl, H. (1993) 'Metabolism and colonisation: modes of production and the physical exchange between societies and nature'. *Innovation in Social Science Research* 6(4): 415–442.

Foster, J. B. (2000) *Marx's Ecology*. New York: Monthly Review.

Gare, A. E. (1993) *Beyond European Civilization: Marxism, Process Philosophy and the Environment*. Bungendore, Australia: Eco-logical Press.

Gare, A. (1996) 'Soviet environmentalism: the path not taken'. In T. Benton (ed.) *The Greening of Marxism*. New York and London: Guilford, pp. 111-128.

Gorz, A. (1980) *Ecology as Politics*. London: Pluto.

Gott, R. (2004) *Cuba: A New History*. New York: Yale University.

Gramsci (1971) *Selections from the Prison Notebooks*. London: Lawrence & Wishart.

Harvey, D. (1990) *The Condition of Postmodernity*. Oxford: Blackwell.

Harvey, D. (1996) *Justice, Nature and the Geography of Difference*. Cambridge: Blackwell.

Held, D. (1980) *Introduction to Critical Theory*. London: Hutchinson.

Marx and Engels (1975) *Collected Works. Vol. 3: Marx and Engels 1843–1844*. London: Lawrence & Wishart.

Marx and Engels (1976) *Collected Works. Vol. 6: Marx and Engels 1845–1848*. London: Lawrence & Wishart.

Matthews,, E., Amann, C., Bringezu, S., Fischer-Kowalski, M.,, Huttler, W., Kleijn, R., Moriguchi, Y., Ottke, C., Rodenburg, E., Rogich, D., Schandl, H., Schutz, N., van der Voet, E. and Weisz, H. (2000) *The Weight of Nations: Material Outputs from Industrial Economies*. Washington: WRI.

Mol, A. P. J. (1996) 'Ecological modernisation and institutional reflexivity.' *Environmental Politics* 5(2): 302–323.

O' Connor, J. (1996) Chapter 9 in Benton (ed.) *The Greening of Marxism*. New York and London: Guilford, pp. 187–239.

O'Connor, J. (1998) *Natural Causes: Essays in Ecological Marxism*. New York and London: Guilford.

Red–Green Study Group (R-GSG) (1995) *What on Earth is to be Done?* Manchester (www.redgreenstudygroup.org.uk)

Roberts, J. T. and Grimes, P. E. (2002) 'World-system theory and the environment: towards a new synthesis.' In R. E. Dunlap *et al.* (eds) *Sociological Theory and the Environment*. Lanham, Boulder, New York and Oxford: Rowman & Littlefield. pp. 167–194.

Ryle, M. (1988) *Ecology and Socialism*. London: Radius.

Schandl, H. and Schulz, N. (2000) 'Using material flow accounting to operationalize the concept of society's metabolism'. Colchester: University of Essex ISER Working Paper.

Shannon, T. R. (1996) *An Introduction to the World-System Perspective*. Boulder and Oxford: Westview.

Weiner, D. R. (1988) *Models of Nature: Ecology, Conservation, and Cultural Revolution in Soviet Russia*. Bloomington: Indiana University.

7

Human Relationships, Nature, and the Built Environment: Problems that Any General Ethics Must Be Able to Address

Warwick Fox

In my book *A Theory of General Ethics* (Fox, 2006) I coin and define the term 'General Ethics' as referring to a single, integrated approach to ethics that encompasses the realms of interhuman ethics, the ethics of the natural environment, and the ethics of the human-constructed (or built) environment. A truly General Ethics would therefore constitute a 'Theory of Everything' in the domain of ethics. In this chapter, I want to outline no less than eighteen problems that confront any attempt to construct a General Ethics, which is also to say the range of ethical problems and that anyone seriously interested in environment and society related issues must be able to address. I will outline each of these problems according to the main approaches they relate to – interhuman ethics, animal welfare ethics, life-based ethics, ecosystem integrity ethics, and the ethics of the human-constructed environment. My own approach to these problems – my own General Ethics – can be found in *A Theory of General Ethics*. What you have here is, if you like, a map of the ethical terrain that those dealing with environment and society related issues are liable to encounter.

CENTRAL PROBLEMS RELATING TO INTERHUMAN ETHICS

Problem 1: 'Why are humans valuable?'

Although we tend to take it for granted that humans are extremely valuable when considered in their own right – as opposed to being merely 'human resources' that we can *use* – it is nevertheless important for any ethical theory to be able to give an adequate answer to this question. This is especially so given that the traditional secular and religious answers to this question have come under fire from a number of quarters. For example, the idea that humans are essentially and uniquely rational has been called into question by what we have learned not only from Freud and the panoply of developments in clinical psychiatry and psychology since Freud but also from human cognitive psychology, comparative psychology, and cognitive ethology. These studies suggest, in short, that humans are not as rational as we have traditionally liked to think and that nonhuman animals are not as irrational as we have traditionally liked to think. Similarly, the idea that humans are uniquely endowed with a soul has become entirely contentious. Moreover, insofar as the idea of a soul is linked to human consciousness, neuroscientific evidence clearly points to the conclusion that consciousness is entirely dependent on neural processes and thus ceases to exist when these neural processes cease.

As if this weren't enough, a number of thinkers – and most notably the animal welfare ethicist Peter Singer – have followed the lead of Jeremy Bentham (1748–1832), the founding father of utilitarianism, and asked what being rational or having a soul has to do with being deserving of moral consideration in the first place. That is, they have suggested that the traditional reasons that have been given for bestowing moral consideration uniquely upon humans are irrelevant! For these thinkers, the reason that we should not, say, torture a six-month-old baby is not because it is rational (or the sort of being that will become rational) and not because it has a soul (if it does) but because it would *suffer*, here and now. Thus, for these thinkers, as for Bentham, the essential moral question is not 'Can this being reason (or, for that matter, does it have a soul)?' but 'Can it suffer?' However, if that question is accepted as the litmus test of which beings are deserving of moral consideration, then we are into a whole new ethical ball game – one that extends to all sentient beings and not just humans.

We can see, then, that giving a decent answer to the 'Why are humans valuable?' question is not a simple matter, and much can rest on it in terms of further ethical implications, both within the human sphere (e.g. how should we respond to the issues of abortion and euthanasia?) and beyond the human sphere (e.g. should we stop eating other animals and all be vegetarian?).

Problem 2: Abortion; Problem 3: Euthanasia

In view of what I have just said in regard to problem 1, it ought to be fairly clear that how we answer the 'Why are humans valuable?' problem is of the first importance to how we should approach these two problems, which concern the beginnings and ends of human life, respectively. For example, if one regards any form of human life as sacred to God and considers that it is therefore a sin against God to take such life – irrespective of the quality of life that a fetus might go on to have or the quality of life that a patient with a terminal illness might now have – then clearly one has a straightforward answer to both of these problems, namely, one of implacable opposition to both abortion and euthanasia. However, given that there are many other views on these issues and that a great many people do in fact want and, where they can, exercise their freedom

to make quite different choices in regard to them, it is easy to see why these issues have been sources of considerable contention – especially when considered in the context of the medical means that are now available to us both to support and terminate human life. Clearly, any adequate General Ethics must be able to address these two problems in a sensible and defensible way.

Problem 4: 'What are our obligations to other people?'

This question is one of the central questions for any form of interhuman ethics. What kinds of obligations do I have to others? Am I obliged simply not to harm others (and if so, why?) or are my obligations more extensive than that? For example, am I also obliged to offer what we might call 'saving help' to others even though I might be in no way responsible for the harm or distress that has befallen them? Am I obliged to go even further and offer other people ongoing supportive help rather than just limited forms of saving help? What about completely bonus forms of help? Where do my obligations to other people begin and end? And am I supposed to extend these obligations equally to all other people? That is, do I owe just as much to strangers as I do to my own nearest and dearest? For example, am I just as obliged – or perhaps even more obliged – to relieve suffering by donating to famine relief as I am to funding my children's education, or to taking my family on a holiday? The influential utilitarian philosopher Peter Singer argued in a famous and much reprinted 1972 paper entitled 'Famine, Affluence, and Morality' that we are so obliged, for 'if it is in our power to prevent something bad from happening, without thereby sacrificing anything of comparable moral importance, we ought, morally, to do it' (Singer, 2002, p. 573). And, in Singer's view, this applies regardless of 'proximity or distance': 'If we accept any principle of impartiality, universalizability, equality, or whatever, we cannot discriminate against someone merely because he is far away from us' (Singer, 2002, p. 573). Thus, we are, in principle, just as obliged to help strangers as we are to help those to whom we are closest.

Any adequate General Ethics must be able to offer a sensible and defensible approach to all of these kinds of questions regarding our obligations to others. [However, it needn't be anything like Singer's approach to these particular questions. It is therefore worth noting in this regard that the approach I develop in *A Theory of General Ethics* (Fox, 2006) is far more sensitive to the various questions I have raised above than Singer's one-size-fits-all approach to ethics. Moreover, I argue that my approach is a far more defensible one than

Singer's because it takes into account central features that pertain to any moral problem situation that Singer's approach fails to take into account.]

Problem 5: 'What is the best structural form of politics?'

It is easy to see from the above discussion that questions regarding our personal obligations to others can easily spread into overtly political questions. For example, to what extent do I *personally* owe saving help to faraway strangers who are experiencing famine or being persecuted in a war (and Singer, as we saw, believes that we owe a great deal at a *personal* level in these contexts) and to what extent might it be more morally reasonable – and not simply a 'cop-out' – to say that my *nation-state* has certain obligations in these contexts such that my personal obligations in respect of the above problems actually kick in at the level of my being obliged to support the kind of political system and government that will live up to these obligations (and draw on my taxes to do so)? Clearly, a truly General Ethics must be able to give us some guidance here.

However, the first form of guidance that we want from a General Ethics in regard to politics is guidance with respect to the *structural* form of politics that we ought to support. By this I mean that a General Ethics ought to be broad enough to endorse – and be able to explain why it endorses – one or more kinds of political structure (such as dictatorship, monarchy, aristocracy, plutocracy, oligarchy, democracy, or anarchy) over others. Now, as with the 'Why are humans valuable?' question, it is easy for most of us who live in democracies simply to assume the answer here for it seems just as obvious to most of us that democracy is the best structural form of politics as it does that people are extremely valuable. But a General Ethics really needs to provide an explicit answer here, just as it does for the previous questions.

Problem 6: 'What is the best "flavor" of politics?'

Essentially the same structural forms of politics can nevertheless take on very different 'flavors.' For example, a dictatorship (or any system in which power is overwhelmingly concentrated in the hands of a few and from whom that power cannot easily be removed) can, in theory, be brutal, benign, or benevolent; anarchy can, in theory, consist of 'mutual aid' or a 'war of all against all' that proceeds in the absence of any rule of law whatsoever; democracies can and typically are distinguished in terms of the extent to which they are socially oriented (and so taxed accordingly in order to fund socially oriented programs, including all the state administrative apparatus that these programs entail) as opposed to individualistically oriented (and so taxed accordingly in order to fund a more minimal state apparatus, including more minimal administrative and social services). Thus, whatever our answer to the 'What is the best structural form of politics?' question, we still want to know what 'flavor' this structural form of politics ought to have since (political) structure, by itself, does not determine (political) content. Indeed, this is precisely why we vote *within* a democratic *structure*: to determine the 'flavor' – or, in other terms, the *content* – we want that democratic structure to have (at least for the next few years!). Ideally, then, we want a truly General Ethics to provide an explicit answer to the question not only of the kind of political structure that we ought to endorse but also of the kind of 'flavor' that that political structure ought to have.

The six questions that I have outlined here – the 'Why are humans valuable?' question, the abortion question, the euthanasia question, the 'What are our obligations to other people?' question, the 'What is the best structural form of politics?' question, and the 'What is the best "flavor" of politics?' question – arguably represent the six most central questions in interhuman ethics. A General Ethics needs to be able directly to address them all, to offer sensible and defensible answers to each of them, and also to address a wide range of ethical questions that run far, far beyond these questions, as we will see in what follows.

CENTRAL PROBLEMS RELATING TO ANIMAL WELFARE ETHICS

Problem 7: 'Why are sentient beings valuable?'

The best known and most influential answers to this question have been advanced by the utilitarian ethicist Peter Singer [1990 (1975)] and the rights-based ethicist Tom Regan (1983, 2003) under the names of 'animal liberation' and 'animal rights,' respectively. Both of these approaches turn, in their different ways, on the basic idea that sentient beings in general (which, for Singer, includes anything more complex than mollusks) or some more specialized subset of sentient beings (such as mammals and birds in Regan's more recent expositions) have an experiential welfare that ought to be respected. Singer adopts the utilitarian approach of arguing that we ought to take the interests of other sentient beings into account in our actions by weighing these interests impartially against the interests of other sentient beings (including our own). Regan adopts the rights-based approach of arguing that the possession of an experiential welfare – at least an experiential welfare of a certain order, such that these beings constitute what Regan refers to (quite unclearly in my view) as 'subjects-of-a-life' – makes a being sufficiently 'inherently valuable' as to possess 'rights' to life and liberty. The possession of such 'rights' means that the being's interests in continued life and liberty cannot be 'traded off' against the interests of others, as in the utilitarian approach. Singer and Regan both argue that it is *speciesist* to recognize only the interests or rights to life and liberty of members of one's own species – and the parallel with the ideas of sexism and racism is quite deliberate here. [The term *speciesism* was coined by Richard Ryder who has himself more recently developed an as yet not very well known but nevertheless significant partial synthesis of Singer's and Regan's views, which he refers to as *painism* (Ryder, 2000, 2001). Basically, Ryder rejects Regan's emphasis on being the subject-of-a-life in favor of Singer's more straightforward emphasis on the moral importance of pain while also rejecting Singer's utilitarian preparedness to aggregate pleasures and pains across different beings in favor of Regan's rights-based opposition to such aggregation.]

In order to refer to the animal liberation and animal rights (and, for that matter, painism) approaches collectively – and without privileging one of these names over the other(s) – a number of commentators, including myself, find it convenient to refer to them as the *animal welfare approach* (or animal welfare approaches, depending on the degree of specificity intended) since these approaches proceed from some version of the idea that sentient beings are valuable because they have an experiential welfare such that they can fare better or worse. I have used this *animal welfare* terminology earlier in this chapter, as well as in the heading of this section, and will continue to do so as appropriate in what follows.

Although I would want to take issue with the details of Singer's and Regan's (and, by implication, Ryder's) basic arguments for their approaches in a longer discussion, it is sufficient for now to note the following. These approaches assign essentially the same level of moral status to all sentient beings or subjects-of-a-life – including humans. They consider that, other things being equal (such as the level of comfort or distress that a being is experiencing), our obligations in respect of nonhuman sentient beings are just as strong as our obligations in respect of other humans. Considered from the other side, these approaches flatly deny that we have any direct obligations in respect of *nonsentient* living things. Thus, there would be nothing wrong in principle in destroying

nonsentient living things such as plants and trees simply because it was our pleasure to do so. As Singer says:

> If a being suffers, there can be no moral justification for disregarding that suffering, or for refusing to count it equally with the like suffering of any other being. But the converse of this is also true. If a being is not capable of suffering, or of enjoyment, there is nothing to take into account (1990, p. 171).

Now any General Ethics obviously needs to address the important question of why sentient beings are valuable – and to what extent they are valuable (e.g. even if we set anthropocentric prejudices aside, is it actually rationally defensible to assign the same general level of moral status to nonhuman sentient beings – or subjects-of-a-life – as to humans?). However, to the extent that we think that it is sensible to ask questions about the values we should live by – that is, ethical questions – in respect of a great many things that are not sentient, such as plants, trees, ecosystems, and buildings, and to the extent that we think that the proper answers to these questions cannot simply be reduced to the interests of sentient beings, then a General Ethics will clearly need directly to address a great many more issues than those addressed by the animal welfare approaches. Not only that, but any adequate General Ethics will need to address even the above 'Why are sentient beings valuable?' question within the context of a far more comprehensive theoretical framework than that offered by the animal welfare approaches that I have referred to here. The reasons for this can be seen from considering the problems that I will outline as problems 8 through 13 below.

Problem 8: Predation

The animal welfare approaches cannot adequately explain why we should, on the one hand, stop the suffering or rights violations of other animals in terms of our (human) predation upon them, but, on the other hand, not attempt to intervene to stop the suffering or rights violations of other animals in terms of their predation upon each other. The problem here, of course, is that from the point of view of the animal being torn apart, it does not necessarily make any difference whether it is a human or a nonhuman animal that is causing its suffering or violating its rights. (Indeed, a bullet through the brain might well be 'preferable' to being torn apart.) Why, then, stop at opposing human predation alone?

As Mark Sagoff (2001) asked in an influential paper, which carried the revealing title 'Animal

Liberation and Environmental Ethics: Bad Marriage, Quick Divorce': if we accept any of the main versions of the animal welfare argument, then

> Where should society concentrate its efforts to provide for the basic welfare – the security and subsistence – of animals? Plainly, where animals most lack this security, when their basic rights, needs, or interests are most thwarted and where their suffering is most intense. Alas, this is in nature (2001, p. 91).

Arguing that animals typically die violently in nature through predation, starvation, disease, parasitism, and cold, that most do not live to maturity, and that very few die of old age, Sagoff (2001, p. 92) proceeds, with deliberately provocative intent, to suggest that if wild animals could themselves understand the conditions into which they are born, then they 'might reasonably prefer to be raised on a farm, where the chances of survival for a year or more would be good, and to escape from the wild, where they are negligible.' Thus, 'One may modestly propose the conversion of national wilderness areas, especially national parks, into farms in order to replace violent wild areas with more humane and managed environments' (Sagoff, 2001, p. 92).

Why not reduce suffering and rights violations by doing this? That way, prey could be killed humanely and fed to predators. Alternatively, we could follow the equally provocatively intended suggestion advanced by the influential ecocentric ethicist J. Baird Callicott in a devastating review of Tom Regan's *The Case for Animal Rights* and simply humanely eliminate all predators. Callicott (1985) argues that because Regan makes it clear that all subjects-of-a-life possess equally strong rights, demanding equally strong degrees of respect, it must follow that:

> If we ought to protect humans' rights not to be preyed on by both human and animal predators, then we ought to protect animals' rights not to be preyed upon by both human and animal predators. In short, then, Regan's theory of animal rights implies a policy of humane predator extermination, since predators, however innocently, violate the rights of their victims (1985, p. 371).

Singer and Regan have both attempted to resist these kinds of conclusions by arguing that we should not interfere with nature in these ways because there is a big difference between human predation upon nonhuman animals and nonhuman animals' predation upon each other: specifically, humans are moral agents and so can assess the rights and wrongs of their actions, whereas nonhuman animals are not moral agents and so cannot assess the rights and wrongs of their actions. But it just will not do to dismiss the problem of nonhuman predation by saying that nonhuman animals 'don't know any better,' therefore cannot be blamed for their actions, and *therefore* should be allowed to carry on with these actions. This is an entirely misplaced argument at best and an egregiously sophistical argument at worst: moral agents (i.e. normal mature humans) can reasonably be held responsible for *allowing* nonmoral agents (such as nonhuman predators) to cause harm or violate the rights of others. As both Steve Sapontzis (1998) and J. Baird Callicott (1985) have pointed out, respectively, we might not hold a young child who 'doesn't know any better' to be morally responsible for tormenting a rabbit, nor might we hold a brain-damaged sadist to be morally responsible for torturing a child, but this does nothing to lessen our responsibility *as* moral agents to stop the young child or the brain-damaged sadist from doing these things. As Sapontzis says:

> Young children cannot recognize moral rights and obligations; nonetheless, it is still wrong for them to torment and kill rabbits. Adults who see what the children are doing should step in to protect rabbits from being killed by the children. Similarly, humans can have an obligation to protect rabbits from being killed by foxes, even though the foxes cannot understand moral concepts (1998, p. 276).

Callicott drives the point home this way:

> Imagine the authorities explaining to the parents of a small child tortured and killed by a certifiably brain-damaged sadist that, even though he had a history of this sort of thing, he is not properly a moral agent and so can violate no-one's rights, and therefore has to be allowed to remain at large pursuing a course of action to which he is impelled by drives he cannot control (1985, p. 370).

Thus, Singer's and Regan's concern with the question of whether or not nonhuman predators are moral agents misses their own morally relevant point: the morally relevant question is not whether these *predators* are *moral agents* but whether their *prey* are *moral patients* (i.e. beings that we, as moral agents, have an obligation to protect from harm). And, according to both Singer's and Regan's versions of the animal welfare approach, there is a broad class of prey animals that fall into this category; thus, it must follow that their views imply that we should intervene where doing so is likely to lessen the overall amount of pain and suffering in the world or, if we adopt a rights-based approach, to stop the violation of the rights of prey (regardless of utilitarian considerations regarding the total amount of pain and pleasure in the world).

If the animal welfarists conceded their shaky ground here and decided to prosecute a worldwide campaign to stop predation in nature generally, then (setting aside the likelihood of ecological meltdown for the sake of the argument!) they would effectively end up domesticating or otherwise taming what remains of wild nature. The animal welfare ethicists *say* that we should not do this, but the problem here is that they are not rationally *entitled* to say this in terms of the theoretical approaches to which they are committed. This means that while accusing human predators of applying a double standard, these ethicists are elsewhere applying a double standard of their own. On the one hand, they charge that human (meat-eating) predators think that we should not cause suffering or violate each other's rights by eating each other (i.e. engaging in cannibalism), but that it's OK to cause suffering or violate the rights of nonhumans by eating them. However, on the other hand, the animal welfarists are themselves saying that humans in general should not cause suffering or violate the rights of any sentient or rights-holding animals, but that it's OK for any other animals to cause suffering or violate the rights of any sentient or rights-holding animals. Thus, Tyler Cowen's (2003, p. 170) damaging, but I think correct, observation: 'Through casual conversation I have found that many believers in animal rights reject policing [of other animals with respect to predation] out of hand, though for no firm reasons, other than thinking that it does not sound right.'

Suppose, however, that animal welfarists agreed to apply their own arguments consistently, even though that would mean policing nature to the extent of totally domesticating or taming it. This raises the question 'What would be wrong with that in any case?,' which brings us to the next point.

(Note that from here on I will just refer to sentient animals, but you can substitute subjects-of-a-life/rights-holding animals as you wish, depending on your preferred version of the animal welfare approach. I will also take it as read – and so will not keep stating explicitly – that any General Ethics needs to be able to offer sensible and defensible responses not only to the predation problem but also to each of the following problems in regard to the animal welfare approach.)

Problem 9: The wild/domesticated problem

Because of their thoroughly individualistic foci, the animal welfare approaches imply that a *wild* sentient animal or a population of wild sentient animals is no more valuable or deserving of moral consideration than a *domesticated* sentient animal or a population of the same number of domesticated sentient animals of the same average level of sentience. (This is because, in both cases, one has just as many sentient animals with just as many total 'units of sentience'; or, substituting for the main alternative animal welfare view, just as many rights-holding animals.) This runs against the sense, shared by many reflective people, that – if we set the special case of companion animals (or 'pets') to one side – there is, somehow, 'something' that is ultimately more valuable about a wild sentient animal or a population of wild sentient animals than a domesticated sentient animal or a population of the same number of domesticated sentient animals of the same average level of sentience. As I have implied, people might well disagree with this statement if it is taken to include the special case of their companion animals, which can come to be seen as members of the household, with many of the status privileges – and even, to some extent, responsibilities – that being a member of the household brings with it. But this potential point of disagreement speaks of the special value of these animals *to us*; it does not speak to the value of these animals in more general, less obviously self-interested terms. If we therefore set the special case of companion animals to one side and consider the issue in terms of those domesticated animals with which we have no special relationship (such as the sheep, cows, pigs, chickens, and so on that we keep for instrumental reasons and that constitute the vast bulk of the domesticated animal population even if they are largely hidden from us), then we can get to the heart of the question being asked here: Are these domesticated-animals-in-general as valuable as wild animals? The animal welfare approaches are theoretically committed to saying that, in principle, they are. This, in turn, implies that a world of totally domesticated animals would, other things being equal, be just as good as a world of wild animals or a world containing a mixture of the two. In that case, then, why not domesticate the planet completely if it suits our purposes to do so?

Not only do the animal welfare approaches invite this question, but there are grounds for thinking that the advocates of these approaches ought to be enthusiastic about realizing such a world. After all, it would help us to sort out the previously discussed problem of nonhuman predation, for we could police nature much more effectively in a totally domesticated world. It would, for example, be much easier to exterminate all predators humanely or, alternatively, kill their prey humanely and then present it to the recalcitrant predators at feeding time. My cat – a skillful wildlife predator when left to her own devices – seems quite happy with this arrangement, especially around 5:30 p.m. each evening when she gets fed what are, in fact, parts of another dead animal, out of a tin. Why wouldn't every other animal be happy with this arrangement?

Problem 10: Indigenous/introduced problem

Because of their thoroughly individualistic foci, the animal welfare approaches similarly imply that an *indigenous* sentient animal or a population of indigenous sentient animals is no more valuable or deserving of moral consideration than an *introduced* sentient animal or a population of the same number of introduced sentient animals of the same average level of sentience. This runs against the sense, shared by many reflective people (and certainly most nature reserve and wildlife management agencies), that there is, somehow, 'something' that is ultimately more valuable about an indigenous sentient animal or a population of indigenous sentient animals than an introduced – especially an *invasive* – sentient animal or a population of the same number of introduced sentient animals of the same average level of sentience. Yet the animal welfare approaches invite the question: Why not populate the world with whatever cute and fluffy, colorful, or otherwise interesting introduced sentient animals we like, even if this leads to a loss of biodiversity overall [which is exactly what it does since a certain percentage of introduced species will turn out to be invasive – although we often do not know which ones in advance – and invasive species represent, after habitat alteration, the second leading cause of loss of global biodiversity (Holmes, 1998; Bright, 1999)]? Why should it matter if a sentient animal isn't indigenous to a particular region? After all, who really cares about the standardization of our fauna and flora through the processes of ecological globalization? Home gardeners 'mix 'n' match' all the time, using the world's flora as their palette to make pleasing gardens. Why shouldn't we do this to get whatever mix of sentient animals happens to please us?

Problem 11: Local diversity/monoculture

Because of their thoroughly individualistic foci, the animal welfare approaches likewise imply that a *diversity* of sentient animals is no more valuable or deserving of moral consideration than a *monoculture* (or something approaching a monoculture) of the same number of sentient animals of the same average level of sentience. Again, this runs against the sense, shared by many reflective

people, that there is, somehow, 'something' that is more valuable about a diversity of sentient animals than a monoculture (or something approaching a monoculture) of the same number of sentient animals of the same average level of sentience. Yet the animal welfare approaches invite the question: Why not populate the world with monocultures of sentient animals, especially if it suits our purposes to do so? (It might seem unusual to think of non-human animals – rather than plants – in terms of monocultures, but that is effectively what, for example, vast herds of cattle are when the distinction is applied to sentient species.)

This problem can be taken as posing the question of diversity and monoculture on a case-by-case basis without reference to the overall amount of biodiversity in the world. This follows from the fact that we could at least imagine a world in which there are many, many small monocultures (or near monocultures) but monocultures that are sufficiently different from each other to add up to a world in which the overall diversity is just as great as another world in which there are mixtures of considerable (but not always dissimilar) diversity everywhere (and thus no monocultures at all). This means that the issue of diversity/monoculture at any given local level is conceptually distinct from the issue of biodiversity (or the preservation of a wide range of species) at a global level even if there is a strong relationship between the two at a practical level. With this in mind, we can now turn to consider the conceptually distinct but practically related question of the overall diversity of species globally.

Problem 12: Species (or global biodiversity)

Because of their thoroughly individualistic foci, the animal welfare approaches imply that the last remnants of a population of sentient animals are no more valuable or deserving of moral consideration than the same number of sentient animals of the same average level of sentience drawn at random from a population that exists in plague proportions. This also runs against the sense, shared by many reflective people, that there is, somehow, 'something' that is valuable about the preservation of a species as such, even though a species as such cannot feel and so has no 'experiential welfare' to be concerned about (only the individual flesh-and-blood *members* of a species can feel and thus possess an experiential welfare; a species as such is just an abstract category; it just refers to a *type* of entity not to token instances of that entity). The animal welfare approaches therefore invite the question: Why care about biodiversity at all? Why not populate the world with equal numbers of a relatively small range of those plants and nonhuman animals that are most useful to us or that simply most take our fancy?

Problem 13: Ecosystem integrity/preservation in zoos and farms

Because of their thoroughly individualistic foci, the animal welfare approaches imply that free-ranging sentient animals that actively participate in rich networks of ecosystemic processes, including food webs, are no more valuable or deserving of moral consideration than the same number of sentient animals of the same average level of sentience and experiencing the same average level of experiential satisfaction confined in a zoo or on a farm. This similarly runs against the sense, shared by many reflective people, that there is, somehow, 'something' that is more valuable about the former animals than the latter – or at least about the former *situation* than the latter. I add this rider because there is perhaps a sense in which we can generally agree that the value of a tiger considered in its own right, which is to say 'in isolation' from everything else, is whatever it is regardless of whether it is in the wild or in a zoo. However, the fact is that *nothing exists in isolation*. What we ultimately need to consider, then, is the overall value of the two situations: tiger in the wild and tiger in the zoo.

The problem for the animal welfare approaches, however, is that their thoroughly individualistic foci mean that they cannot 'see' contextual issues. All they are concerned about is the value of sentient animals as such (or, as I noted at the end of my discussion of problem 8, you can substitute 'rights-holding animals' here as you prefer). The very best they can do in accounting for contextual issues is to consider them in a second-order, derivative fashion and argue, for example, that a wild animal would be, say, happier in the wild, and that this would be a reason for preferring this situation to a zoo or a farm. But this argument is quickly countered: we can easily think of examples in which it is plausible to argue that an animal would have a longer and less stressed life living in some reasonable form of captivity than, as it were, taking its ecosystemic chances. [In this connection, recall Mark Sagoff's (2001, p. 92) sober assessment that animals typically die violently in nature through predation, starvation, disease, parasitism, and cold; that most do not live to maturity and that very few die of old age; and that many might 'reasonably prefer to be raised on a farm (or, we might add in this context, a good zoo), where the chances of survival for a year or more would be good, and to escape the wild, where they are negligible.'] In these cases, animal welfarists should see the zoo or farm scenario as preferable

to that of the animal being left to the not-so-tender mercies of nature.

At the very least, the fact that animal welfare approaches are blind to contextual matters in anything other than a second-order, derivative way, means that they have no ultimate grounds for preferring happy or miserable animals in zoos to equally happy or miserable animals in nature. All that these approaches are equipped to 'see' are the sentient (or, to repeat, the rights-holding) animals in nature; they cannot 'see' or place value upon the more abstract, ecosystemic processes of nature that ultimately connect these animals. It is as if their moral vision allows them to see the individual sentient dots in the picture, but not to join them up. Thus, the long and the short of the ecosystem integrity problem for the animal welfare approaches is that their individualistic foci mean that they place no value on ecosystem integrity per se. Its value is purely derivative. Many reflective people think that that is not good enough. The difficult question remains, however, of explaining why it isn't good enough.

CENTRAL PROBLEMS RELATING TO LIFE-BASED ETHICS

Problem 14: 'Why is life valuable?'

The standard argument that has been advanced by the main life-based ethicists – such as Albert Schweitzer (see Warren, 2000), Kenneth Goodpaster (2001), Robin Attfield (2002), Paul Taylor (1986), and Gary Varner (1998, 2002) – for the value of all living things, whether sentient or not, is that even a nonsentient living thing can be thought of as in some sense embodying a biologically based (but, of course, nonconscious) 'will to live' (Schweitzer), 'interests' (Goodpaster and Attfield), 'needs' (Varner and Attfield), or 'good of its own' (Taylor). But, alas, this general form of argument turns out to be seriously flawed in at least two respects. First, we simply cannot make proper sense of the argument that nonsentient living things can be said (literally rather than metaphorically) to have *wills, interests, needs,* or *goods of their own* – of any kind. Singer, a staunch defender of the view that the criterion of sentience is 'the only defensible boundary of concern for the interests of others' (Singer, 1990, p. 9), puts the point succinctly when he argues that the problem with the standard defenses offered by life-based ethicists is that

> [T]hey use language metaphorically and then argue as if what they had said was literally true. We may often talk about plants 'seeking' water or light so that they can survive, and this way of

thinking about plants makes it easier to accept talk of their 'will to live,' or of them 'pursuing' their own good. But once we stop to reflect on the fact that plants are not conscious and cannot engage in any intentional 'behaviour', it is clear that all this language is metaphorical; one might just as well say that a river is pursuing its own good and striving to reach the sea, or that the 'good' of a guided missile is to blow itself up along with its target. ... [In fact, however,] it is possible to give a purely physical explanation of what is happening; and in the absence of consciousness, there is no good reason why we should have greater respect for the physical processes that govern the growth and decay of living things than we have for those that govern non-living things (1993, p. 279).

We can easily *attribute* wills, interests, needs, and goods of their own to nonsentient living things, but we are doing so entirely from our own point of view, from our own ways of thinking about things in terms of ascribing intentions to them. We should not kid ourselves, however, that we can seriously – or, as Singer says, literally as opposed to metaphorically – claim that these features exist from the point of view of the nonsentient living thing under consideration, because *a nonsentient living thing doesn't have a point of view*. It is not *like* anything to be a nonsentient living thing; if it were, then, by definition, that thing would be sentient rather than nonsentient. Thus, it is quite misleading of Paul Taylor (1986, p. 63) to suggest, repeatedly, in respect of nonsentient living things that 'Things that happen to them can be judged, *from their standpoint*, to be favorable or unfavorable to them' (my emphasis), for we can no more judge benefits or harms 'from the standpoint' of a plant or a tree than we can judge these things 'from the standpoint' of a rock – and for the same reason. We can easily make these judgments in respect of plants or trees from *our* standpoint or point of view (and note here that *standpoint* literally refers to 'a physical or mental position from which things are viewed,' i.e. a point of view), but it is not literally possible to make such judgments from *their* standpoint or point of view because they do not have one. The attribution of nonconscious wills, interests, needs, or goods of their own to nonsentient living things is, in the final analysis, incoherent.

The second problem with the rational foundations of the standard argument for the life-based approach – and one that I am not aware of having been raised before – is that it is circular. Consider: it is simply not the case that every desire, interest, need, or good of one's own is automatically valuable; for example, someone might feel that they have an interest in, or a need to, or that it might

further their own good to see someone dead, or have sex with someone by force if necessary, or lie badly to someone, and so on. It therefore becomes quite important to specify more precisely which interests, needs, or goods of their own are deemed to be valuable and which are not. For life-based ethicists, the interests, needs, or goods of their own that are deemed to be valuable are clearly those that are directed toward the maintenance of essential life processes, that is, those interests, needs, or goods of their own that make an entity an *autopoietic* system (literally, a *self-making* and, by extension, *self-remaking*, or *self-renewing* system). But in that case we can ask: 'Well, why do you think that these essential life processes – autopoietic processes – are valuable?' The answer that we will then get from the life-based ethicists is in terms of living processes being valuable because they embody (nonconscious) interests, needs, or goods of their own! And so the circle continues:

1   The standard life-based argument: living things are valuable because they embody (nonconscious) interests, needs, or goods of their own.
2   Critical question: but since not all interests, needs, or goods of their own are valuable (e.g., murder, rape, serious lying), what is it that makes these interests, needs, or goods of their own valuable?
3   Answer: the fact that they are directed toward the maintenance of living things.
4   Question: so what? What is so important about the maintenance of living things?
5   Answer: return to 1.

And so it goes. But circular reasoning offers no substantial reasons at all; it just chases its own tail instead of giving a solid answer to a problem.

We can note here that whatever the other strengths and weaknesses of the standard answers to the 'Why are humans valuable?' question and the 'Why are sentient beings valuable?' question, they are not circular. The standard kinds of answers we will get from the supporters of these approaches are answers like 'Because humans are rational,' or 'Because humans have a soul,' and 'Because sentient beings are capable of feeling and so can be benefited or harmed from their own point of view.' If we then ask, 'Well, are these features valuable in themselves?', the supporters of these approaches can easily say 'Yes' and proceed to tell us why in a noncircular way. For example, they can tell us that the possession of these features is what makes the possessor's life valuable *to them* – and then expound further why these beings should be respected on that account (just as we wish to be). But suppose we ask a life-based ethicist 'Why are even nonsentient living things valuable?' and they say 'Because nonsentient living things embody biologically based (but, of course, nonconscious) wills to live, interests, or goods of their own that are directed toward their own survival.' If we then ask, 'Well are these things valuable in themselves?', the supporters of this approach cannot give the same kind of answer as those we have just considered; that is, they cannot say 'Of course they are – these capacities are the very things that make the lives of nonsentient living things valuable *to them*' because it is not like anything to be a nonsentient living thing; nothing is valuable *to them*. Life-based ethicists must therefore reach for another answer, but, unfortunately for them, that answer is the circular answer outlined above.

In view of these problems, I would suggest that Gary Varner is on safer – albeit extremely vague – ground when he offers a second, nonstandard argument for a life-based approach to ethics. In this argument he asks us to imagine two worlds – one that is rich in nonsentient life-forms and one that is not. Then he asks us which world we think is more valuable. In answering his own question, he drops considerations relating to biologically based needs and so on altogether and simply appeals to our intuitive sense that 'the mere existence of nonconscious life adds *something* to the goodness of the world' (Varner, 2002, p. 114). Many of us would agree with that, as far as it goes, but the problem remains that Varner fails to tell us what this special 'something' is – and I am not aware of any other contributors to this approach who have been any more forthcoming; indeed, most do not even mention this second, more intuitively based argument. Even Varner (2002, p. 113) admits to deliberately omitting this argument from an earlier book because he 'doubted that it would be persuasive to anyone not already essentially convinced.' However, despite this, Varner (2002, p. 113) nevertheless thinks that 'this second argument expresses very clearly the most basic value assumption of the biocentric individualist [i.e. people who believe that all individual living things are valuable in their own right].'

But what Varner fails to see here is that this second argument – which serves to highlight an intuition rather than provide a detailed set of reasons for a conclusion – can be applied just as well to other comparisons. For example, imagine these two worlds: one that is rich in nonsentient life forms that are arranged in botanical gardens attended by robots and one that is rich in the same number of nonsentient life forms that exist in natural, ecosystemic arrangements; or imagine these two worlds: neither has any life forms at all, but one consists of nothing more than barren rock whereas the other is an abandoned world in which all life has died, but which still retains ruins of buildings and sculptures that would rival the finest you've ever seen. Could we not equally well argue that the second of the comparisons in each case is the intuitively preferable one, that 'the mere existence of ecosystems in the first example, or the mere existence of such highly organized architectural and sculptural complexity in the second example, adds *something* to the goodness of the world in both cases?' Yet if this is reasonable, then Varner's own form of argument undercuts his own biocentric individualist position. Varner does not wish to say that anything other than individual living things are valuable in their own right, yet his own intuitively based argument can easily be adapted to suggest that *holistic* systems (in this case ecosystems), rather than what he thinks of as *individual* living things, add something to the value of the world, and that certain formations of *nonliving* things can add something to the value of the world as well. Where, then, do these extensions of his own argument leave his biocentric individualist view that only *individual living* things can add something to the value of the world? Thus, it seems that even this second, nonstandard form of argument cannot be used to sustain a strictly biocentric individualist position.

[It is worth noting here that the approach that I develop in *A Theory of General Ethics* (Fox, 2006) tells us exactly what the mysterious 'something' is that is added to the goodness of the world in each of the 'two world' comparisons discussed above – Varner's and mine. However, although the approach I develop in *A Theory of General Ethics* embraces and explains what is right in Varner's intuitive demonstration, it is not limited to and can in no way be summed up as being simply or even primarily a 'life-based approach' to ethics.]

The life-based approach and Problems 8 through 13 revisited

Beyond these problems with its rational foundations, the individualistically focused life-based approach recapitulates the same range of problems that afflicts the animal welfare approaches on the basis of their individualistic foci of interest. That is, the life-based approach suffers from *the wild/domesticated problem* (after all, wild and domesticated plants are just as alive and, therefore, just as valuable as each other); *the indigenous/introduced problem* (similarly, indigenous and introduced plants are just as alive and, therefore, just as valuable as each other); *the local diversity/monoculture problem* (considered at the local level and without reference to overall global biodiversity, we can have just as many living things and, therefore, just as much value whether the living things in question are extremely diverse or all much the same); *the species (or global biodiversity) problem* (the same reasoning applies with respect to overall global biodiversity: we can have just as many living things and, therefore, just as much value whether the living things in question are extremely diverse or all much the same); and *the ecosystem integrity/preservation in zoos and farms problem*, which we can now expand to read: *the ecosystem integrity/preservation in botanical gardens and zoos and farms problem* (e.g. we can put all the plants in an ecosystem into a botanical garden, look after them really well, and have just as much life and, therefore, just as much value in both cases).

The predation problem is a separate kind of problem to those linked to the individualistic foci of the life-based approaches as such; however, it also recapitulates the formulation of this problem for the sentience-based approach – with a vengeance. In this context, the problem is this: Does the recognition of the value of nonsentient living things mean that moral agents should not destroy (and that includes eat) any living things? If so, how are we to live? Moreover, does the recognition of the value of nonsentient living things mean that moral agents should intervene to stop other living things destroying (and that includes eating) any other living things? (We saw that the argument for doing this is much stronger than the argument against it with respect to the animal welfare approaches.) The life-based approach is clearly untenable at a practical level unless it is made compatible with some sensible kind of hierarchy of value that explains why the value of nonsentient living things can be trumped by the value of other living things – and especially other sentient animals – maintaining their own existence. The main life-based ethicists recognize this problem and have generally attempted to develop either a set of priority rules that kick in under different circumstances (e.g. by distinguishing between basic and nonbasic needs) or else explicit hierarchies of value, such that while the idea of something being valuable in its own right kicks in at the level of individual living things (and not before), the value hierarchy goes on to ascribe greater value to more complex kinds of living things, such as sentient beings in general (or certain of their interests) and humans in particular (or certain of their interests).

The fact that the predation question alone more or less forces life-based ethicists to develop priority rules or explicit hierarchies of value in order to give us a workable theory – one that allows us to eat, for a start – might lead some of us to reflect back on the animal welfare approaches and ask: If some sensible hierarchy of value is the only way to make any practical sense out of the life-based approach, then why should the hierarchy of value

flatten out at the level of sentient animals? Is there not a sensible hierarchy of value to be found there too? But, if so, what are the implications of this for the best known animal welfare approaches?

A final point that I will mention in regard to the life-based approach is that it is, obviously enough, a *pro-life* approach, with all that that entails. Any argument for the value of nonsentient individual living things is, therefore, clearly a *prima facie* argument against both abortion and euthanasia. Life-based ethicists might be quite happy with this – or they might wish to call on their various priority rules or hierarchies of value in order to allow abortion and euthanasia in various circumstances. The problem is, they don't say. Even though their approach is a pro-life approach, we will generally search in vain for any mention, let alone real discussion, of abortion and euthanasia in their arguments. It is as if they haven't made the connection. Clearly, however, any life-based approach needs to address these issues since this kind of approach invites their discussion. And I have already noted earlier in this chapter that any General Ethics – by virtue of being a *general* ethics – must do the same (see my discussion of problems 2 and 3).

CENTRAL PROBLEMS RELATING TO ECOSYSTEM INTEGRITY ETHICS

This brings us to the ecosystem integrity approaches. These seem to solve a number of the problems that confront the previous, individualistically focused approaches. For example, ecosystem integrity approaches dissolve *the predation problem* created by both the animal welfare approaches and (if adopted sufficiently zealously) the life-based approaches in that predation, in a great many forms at least, is seen as part and parcel *of* the maintenance of ecosystem integrity. Similarly, *the wild/domesticated problem* is cashed out primarily in terms of what contributes to or disrupts ecosystem integrity. I say 'primarily' here because ecosystem integrity is not the only kind of value under consideration if we are talking about an inclusive ecosystem integrity approach, that is, one that also recognizes the value of individual living things. However, I also say 'primarily' here because I take it that the point of an ecosystem integrity approach is to favor ecosystem integrity over the value of individual living things when and where these values come into conflict. Thus, the right balance with respect to *the wild/domesticated problem* is to be found in terms of humans meeting their own needs for domestication within the context of preserving ecosystem integrity. *The indigenous/introduced problem* is also cashed out in terms of this understanding of what contributes to or disrupts ecosystem integrity. This means favoring indigenous living things (whether wild or not) over introduced living things and especially over invasive living things. The same understanding applies to *the local diversity/monoculture problem*: the right balance here is that which maintains ecosystem integrity and thus the approach that favors characteristic diversity over an increase in diversity for its own sake or a reduction in that diversity. *The species (or global biodiversity) problem* is just the diversity/monoculture problem at the global or ecospheric level as opposed to the local ecosystemic level. Essentially the same answer therefore applies: the right balance is that which maintains ecospheric integrity overall and thus the approach that favors characteristic ecospheric diversity over an increase in diversity for its own sake or a significant reduction in that diversity. Finally, this approach cannot remotely be accused of being blind to *the ecosystem integrity problem* because a concern for ecosystem integrity is its *raison d'être*.

But all is not as rosy as it seems with the ecosystem integrity approach. In particular, this approach adds the following new problems to the list of problems that any General Ethics must be able to address.

Problem 15: 'Why is ecosystem integrity valuable?'

The prototype of an ecosystem integrity ethics was first advanced by the American forester, wildlife ecologist, and conservationist Aldo Leopold (1887–1948) in a now famous essay entitled 'The Land Ethic,' which forms the concluding section of his classic *A Sand County Almanac*, first published in 1949 (a date that is remarkably early relative to the development of environmental ethics as a formal field of inquiry only since the mid-to-late 1970s). Leopold (1981, pp. 224–225) famously asserted that we should expand our traditional notions of ethics to include the 'biotic community' by adding the following principle to our existing moral codes: 'A thing is right when it tends to preserve the integrity, stability, and beauty of the biotic community. It is wrong when it tends otherwise.' Unfortunately, Leopold named this principle the 'Land Ethic,' which is quite unhelpful given that it can be applied just as much to ecosystemic relationships in riverine, estuarine, marine, and, presumably, atmospheric environments as to terrestrial environments. Alas, the misleading 'Land Ethic' label has stuck, but we can think of his proposal as a – even *the* – prototypical form of ecosystem integrity ethics.

Another unfortunate aspect of Leopold's central maxim concerns his use of the term 'beauty.' Partly because this term can mean different things

to different people and partly because Leopold provided no independent elaboration of and defense for his inclusion of this term, later commentators have either ignored this aspect of Leopold's formulation or rendered it in terms of 'ecological integrity' – or some roughly equivalent formulation. Thus, for example, James Heffernan (1982, p. 237) says that (i) 'The characteristic structure of an ecosystem seems to be what Leopold means by its integrity,' and then suggests (ii) that we can equate the idea of the 'objective beauty' of an ecosystem with its characteristic structure [which, from (i), also means its integrity]. For Heffernan (1982, p. 237), then, 'when Leopold talks of preserving the "integrity, stability, and beauty of the biotic community" he is referring to preserving the characteristic structure of an ecosystem and its capacity to withstand change or stress.' In consequence, Heffernan (1982, p. 247) drops any explicit reference to 'beauty' in his own suggested reformulation of Leopold's Land Ethic: 'A thing is right when it tends to preserve the characteristic diversity and stability of an ecosystem (or the biosphere). It is wrong when it tends otherwise.' J. Baird Callicott (1996, p. 372) likewise drops any reference to 'beauty' in his suggested reformulation of Leopold's Land Ethic: 'A thing is right when it tends to disturb the biotic community only at normal spatial and temporal scales. It is wrong when it tends otherwise.' Thus, Leopold's Land Ethic is generally understood to refer to matters concerning ecological integrity not to what we might ordinarily understand as aesthetic matters as such. This is probably just as well because one person's idea of a 'beautiful' landscape can be an ecologist's idea of a 'disaster area' – a landscape overrun with invasive species and so on; similarly, one person's idea of an ugly or uninteresting landscape – like a 'swamp' – can be an ecologist's idea of a precious 'wetland.'

Leopold's proposal bears a similar relationship to contemporary ecosystem integrity ethics (*à la* Heffernan and Callicott) as Schweitzer's prototypical 'reverence for life' approach does to contemporary life-based ethics. Specifically, neither Leopold nor Schweitzer were professionally trained philosophers, and this shows in the relative looseness of their arguments (as well as Leopold's name for his approach and even his formulation of the Land Ethic), but they did both pioneer ethical directions that later philosophers have been inspired by and have attempted to develop in more detailed and rigorous ways. Thus, notwithstanding the relative fame of Leopold's Land Ethic, the essay in which he advances this ethic, although pregnant with significant ideas, offers little in the way of anything that philosophers would recognize as a rigorously reasoned argument. Even so, we can discern the basic structure of an argument in Leopold's essay if we dig deep enough, and this is what it looks like: Leopold argues that ethics are not a fixed and firm thing but rather a 'product of social evolution'; that 'All ethics so far evolved rest upon a single premise: that the individual is a member of a community of interdependent parts'; that there are now both theoretical and practical reasons for extending our conception of what our community is – and, thus, what our ethical concerns should cover – from the human level to the ecological level; and that the 'mechanism of operation [for this social evolutionary development] is the same for any ethic: social approbation for right actions: social disapproval for wrong actions' (Leopold, 1981, pp. 225, 203, 225, respectively). What Leopold is suggesting, then, is that the next stage in the social evolution of our ethics needs to be one in which we collectively embrace the wider ecological context of which we are a part as part of our extended community and, thus, as falling within the scope of our moral concerns and sympathies.

J. Baird Callicott (1987, 1989, 1999) has, through a sustained and influential output over many years, done much to draw out the Humean and Darwinian roots of this kind of argument and to develop it further. These Humean and Darwinian roots are essentially as follows. David Hume (1978, pp. 575, 618) argued in his masterpiece *A Treatise of Human Nature* (first published in 1739–40) that ethics is grounded in the sympathies and antipathies that are part and parcel of human nature: 'The minds of all men are similar in their feeling and operations' and 'sympathy is the chief source of moral distinctions.' Thus, rather than our reasoning about ethics driving our feelings (or passions), our reasoning about ethics is, ought, and can only be, as Hume (1978, p. 415) famously said, 'the slave of the passions.' [Or should that be: 'as Hume infamously said?' Given that the general thrust of Western philosophical ethics has been and remains very much concerned with using human reason to channel and curb the acting out of our passions in various ways, we can see why Robert Arrington (1998, p. 234) describes Hume's claim as 'one of the most notorious claims in the literature of moral philosophy.'] For Hume, then, we express our moral sentiments when we express approval or disapproval for those things and actions that we find useful or agreeable to ourselves and others. Darwinian evolutionary thought, in turn, informs our understanding of human nature and, thus, how our natural sympathies and antipathies got to be the way they are. It also informs our understanding of ecology and, thus, the interdependent relationship we have with the rest of the natural world. Callicott argues that Leopold draws upon and contributes to this line of intellectual and moral development in suggesting that we as a human community should now learn to extend our concerns and sympathies to the wider ecological community of which we are a part.

But where does this get us? If we pursue this line of thinking, then it seems obvious, to me at least, that the natural sympathies we share on the basis of our evolutionary inheritance are such that we do and will continue to feel most strongly for our immediate kin and kith, followed by whatever we take our most immediate wider group to be, and then perhaps outward to our own species and so on, but that the wider 'natural world' or 'ecological community' will inevitably, when weighed in this kind of balance, remain a relatively distant concern in terms of our evolutionary endowed sympathies, passions, or just plain old gut feelings. It therefore seems 'natural' that people will keep clearing land or fishing their seas and lakes not only in order to feed their families in some subsistence sense but even, on grander scales, in order to allow their families to live in luxury – and this even when they are endangering or extirpating the remaining members of a particular species. Thus, Hume's moral sentiments, as honed by Darwinian evolution, would not appear to provide us with a sufficient degree of motivation to move to Leopold's proposed next stage of social evolution in anything beyond a token sense; that is, we effectively say: 'Sure, we are members of a wider ecological community, but the evolutionary distant members or aspects of this community matter much less to me than my immediate kin, kith, and kind.'

However, we also know that it doesn't have to be this way, at least not entirely. We know that, contra Hume, we can channel and curb our sentiments – including the natural priorities of these sentiments as they run from kin, kith, and kind to our wider ecological context – if we are given a sufficiently good reason to do so. But what would constitute a sufficiently good reason? Two obvious kinds of reasons suggest themselves. The first is that we should value our ecological contexts much more than we do because doing so is crucial to our own survival and well-being as well as that of our own kin, kith, and kind. This is completely compatible with a Humean and Darwinian account of value, but it pays the price of collapsing Leopold's celebrated Land Ethic into an *instrumental* (or *use*) value approach to the value of ecological integrity. Moreover, this argument is vulnerable to the charge that if our concerns with ecological integrity boil down to its usefulness to us, then these concerns will have to take their place alongside other self-interest-based arguments regarding the possible alternative uses of various natural areas for such things as dams for generating hydroelectricity, housing, farming, logging, mining, and so on. No change there then.

But all this would seem to be a far cry from what it appears that Leopold wanted to say (and is typically taken as saying) because Leopold (1981, p. 223) suggested in his essay that the Land Ethic involved valuing land 'in the philosophical sense,' which most philosophers have taken to mean on the basis of its *intrinsic* value, its value in its own right. This, then, brings us to the second kind of reason that might persuade us to override the natural priorities bestowed by our evolutionary endowed sentiments and value our ecological contexts much more than we do: specifically, we might accept that we should regard ecosystem integrity as valuable, not only because of its usefulness to us but also because it is valuable in its own right and that 'such and such is the reason why.' This orientation to the problem leads us away from the kind of subjectivist approach to value that runs through the Hume–Darwin–Leopold line of thinking that Callicott endorses (an approach that locates the basis of our evaluations, as Hume [1978, p. 468] says, 'in [ourselves], not in the object [toward which these evaluations are directed]') and toward an objectively based reason that would explain why ecosystem integrity is valuable in its own right. But what might such a reason be?

The main objectively based reason that has been advanced in regard to the value of ecosystem integrity is that ecosystems are alive and living things are valuable in their own right. Leopold himself toyed with this idea in an essay written much earlier than his famous 'Land Ethic' but only published in 1979, and this approach has since been taken up by James Heffernan (1982) in his objectivist, distinctly non-Callicottian, interpretation of Leopold. But the problem with this kind of ecosystem integrity argument is that it just reduces to an expanded version of the life-based argument. Indeed, it is an even more controversial version of this argument than those that I have already considered above because, whereas we can at least all agree that, say, individual plants and trees are alive, it is simply not clear that the ecosphere is alive in anything like the same sense (although there is at least a sensible argument to be had here in terms of formal definitions of life and so on). This in turn means that this objectivist approach to the value of ecosystem integrity is just as flawed as the standard argument for the life-based approach because it is just the standard argument for the life-based approach extended to include ecosystems as living things, which just adds another shaky layer to an already incoherent and circular argument.

Perhaps we would do better simply to say, in the style of Varner, that 'the mere existence of ecosystem integrity – of longstanding, self-sustaining, complex webs of relationships between individual living things themselves and between them and their physical environments – adds *something* to the goodness of the world.' But what *is* that something? [As I have already indicated in my discussion of problem 14 – the 'Why is life valuable?' problem – the approach that I develop to General Ethics in *A Theory of General Ethics* (Fox, 2006) tells us exactly what that 'something' is.]

Problem 16: The subtraction and addition of ecologically benign species

This might be regarded as more of a worry than a serious problem, but then again .... The worry, or problem, is this: although the ecosystem integrity approach appears to give the right answers with respect to questions about diversity – that is, it supports the maintenance of characteristic diversity over an increase in diversity for its own sake or a significant reduction in that diversity – it is not at all clear that this approach genuinely entitles its advocates to object to *the subtraction and addition of ecologically benign species*. This is because this approach is concerned primarily with the maintenance of the ecological integrity – or self-sustaining capacity – of an ecosystem, and 'ecologically benign species' refers, by definition, to species whose loss or addition does not significantly disrupt this integrity or self-sustaining capacity. Of course there are two immediate points to be made here both for and against this concern. The pro-point is that it is just not the case that every species is vital to, or even has any great impact upon, the self-sustaining capacity of an ecosystem. Not every species is a *keystone* species – or anything like it. Ecosystems are not like a rug that unravels if a single thread is removed – unless of course it happens to be a 'keystone thread' (to thoroughly mix architectural and weaving metaphors). Neither will an ecosystem necessarily unravel if one more species is woven into it. The contra-point, however, is that the relationships in ecosystems are so complex that we often cannot know with any certainty what might happen if we do subtract or add a species that we think is ecologically benign. This, then, gives an advocate of the ecosystem integrity approach a *practical* way of responding to *the subtraction and addition of ecologically benign species problem*. They can simply say that we should not attempt to add or subtract species to or from an ecosystem, no matter how ecologically benign we think our actions are, because we can never be sure. We should therefore adopt the maintenance of characteristic diversity, which has been tried and tested through evolutionary processes, as our default position.

But consider, for the sake of the argument, the following rejoinder: 'Oh, I see, you think we should adopt the maintenance of characteristic diversity as our default position because you *aren't sure* what would happen to ecosystem integrity if we didn't. Well, have I got news for you! Through a complex procedure known as quantum–relativistic informational time tunneling, I've been able to download a program from an intergalactic civilization far in advance of ours that enables us simply to scan a geographical area (using the well-known Zooly–Mischoff scanning procedure) and then be able to tell *exactly* what will happen if we add or subtract any given species to that area. Now, c'mon, be honest, wouldn't you like to be able to add a bit more ecological diversity around here if it were ecologically benign? It'd make things more interesting, right? And wouldn't you like to be able to remove the odd species – especially those that get in your way one way or the other – if you knew that it wouldn't have any other ill-effects?'

You can say that this response is fanciful, but the point at issue is a serious one: if the ecological integrity approach is concerned primarily with the maintenance of the ecological integrity – or self-sustaining capacity – of an ecosystem, then this approach provides no grounds to object *in principle* to the subtraction or addition of ecologically benign species precisely because, by definition, this subtraction or addition makes no significant difference to the ecological integrity of the ecosystem. This means that advocates of this approach who want to object to the subtraction or addition of ecologically benign species have to fall back on 'What *might* happen if ...' kinds of arguments. Yet many informed judges in this area feel that there ought to be a way of objecting to the subtraction or addition of ecologically benign species in principle. But is there any good argument for this?

Problem 17: The (catastrophic) way evolution works

A question also arises regarding the relationship of ecosystem integrity to evolutionary processes. Given that we now understand evolutionary processes to include the odd catastrophic cosmic collision between an asteroid and the Earth – and that such collisions have constituted a major structuring agent of the biosphere in which we ourselves have evolved; indeed, that they may even be responsible for our existence through seeing off the dinosaurs and allowing the spread and rise of mammals – then there would seem to be a tension between our normal understanding of ecosystem integrity, on the one hand, and evolutionary processes, broadly understood, on the other.

How are we to reconcile this tension? If we lean too far in the direction of trying to maintain ecosystem integrity in the absence of evolutionary processes, then we are in danger of deep-freezing ecosystems and regarding any new evolutionary developments as bad. If we lean too far in the direction of embracing any and all evolutionary processes, then our response to the prospect of a catastrophic cosmic collision will be 'bring it on.'

Callicott (1996, p. 372) has suggested a middle way in his own reformulation of Aldo Leopold's Land Ethic. For Callicott, 'A thing is right when it tends to disturb the biotic community only at normal spatial and temporal scales. It is wrong when it tends otherwise.' But, even here, the tension remains, for it has in fact been 'normal' for cosmic collisions or other equally catastrophic factors to declare 'Game Over' for a tremendous number of species every hundred million years or so. This *is* the 'normal temporal scale' for catastrophic disturbances of the biotic community, and we owe our existence to it. So any ecological integrity approach that wants to embrace evolutionary processes at 'normal spatial and temporal scales' has to accept this normal temporal scale of catastrophe. These considerations therefore raise a significant question: Is there any way in which we can *consistently* embrace the more gradual kinds of evolutionary processes that we usually think in terms of (and, thus, avoid committing ourselves to 'deep-freezing' ecological processes) while also rejecting – and acting in whatever ways we can to prevent – catastrophic forms of evolutionary restructuring?

THE CENTRAL PROBLEM RELATING TO THE ETHICS OF THE HUMAN-CONSTRUCTED ENVIRONMENT

Problem 18: The human-constructed environment problem or comprehensiveness problem

Consider the following example: suppose we have two buildings, one of which, when considered purely at the level of design (i.e. when considered purely at the level of its built *form*), is contextually fitting with its natural environment and one of which is not. In other words, one of these buildings seems to blend in beautifully with its landscape while the other 'sticks out like a sore thumb,' is a 'blot upon the landscape,' and so on. But suppose also that neither building disrupts ecological integrity any more than the other (i.e. they are, for example, equally energy efficient and nonpolluting, or, for that matter, equally energy inefficient and polluting). We might personally *prefer* one building to the other on aesthetic grounds but the fact remains that an *ethics* that is limited to concerns regarding ecological integrity is unable to offer any support for the view that, *prima facie*, it is wrong *in principle* to build in the contextually ill-fitting way (since neither building disrupts ecological integrity any more than the other). Moreover, it is not even clear that people's overall preferences in instances like this will necessarily follow the response that I am trying to motivate here. For example, many people might come to prefer or at least 'not really mind' the contextually *ill-fitting* building 'all things considered' because, whatever its faults, it is 'just so convenient,' or offers easier parking, or has stores that offer cheaper prices. (A look at the human-constructed environment around you might serve to confirm this suspicion.) If these reasons get sufficiently mixed together with whatever preferences people might (or, alas, might not) have in terms of architectural design, then the users of the contextually ill-fitting building can come to see it as not being particularly ugly – or perhaps just come not to *see* it in various ways, such as in terms of any wider contextual understanding.

Thus, if we are to address the question of the ethics of the human-constructed environment *directly* – at the level of *principle* – rather than *indirectly* via either human preferences (which might not go in the direction that we think they 'ought' to) or concerns about ecological integrity (which, again, might not go in the direction that we think they 'ought' to, since a contextually ill-fitting building can, for example, be just as energy efficient and nonpolluting – or even more energy efficient and nonpolluting – than a contextually fitting building), then we clearly need an ethics that can directly address concerns at the relatively intangible level of design. The problem is, however, that we do not presently have such an ethics. [For steps in the direction of developing such an ethics see my edited collection *Ethics and the Built Environment* (Fox, 2000). My introduction to that book also contains references to the few previous contributions that have been made in this direction.] If we did have some kind of ethics that was directly concerned with the human-constructed environment, then it is possible that the discussion of this ethical approach would have generated a range of problems that would enable me to list, say, four or six of the main problems in this area, much as I have done in regard to the interhuman, animal welfare, life-based, and ecosystem integrity ethical approaches discussed above. Instead, we simply have one big problem in regard to the ethics of the human-constructed environment, namely, the fact that there presently isn't one!

This lack of an ethics in respect of the human-constructed environment represents the lack of an ethics in respect of what we might think of as the third main realm of our existence, that is, the realm of material culture (which includes all the 'stuff' that humans intentionally make) as opposed to the biophysical realm (which includes ecosystems and the plants and animals that live in them) or the realm of symbolic culture (which is constituted by language-using human moral agents). It follows from this observation that any ethics that cannot directly address problems in this 'third realm' is not even a candidate for a truly General Ethics. Thus, we can think of this last, human-constructed environment problem as a test for the *comprehensiveness* of any approach that is already able to address problems in respect of the biophysical and symbolic cultural realms. For this reason, it is convenient to refer to this problem not only as *the human-constructed environment problem* but also as *the comprehensiveness problem*, which is what I have done in the heading for this section. environment and society related issues.

WHAT NOW?

The foregoing eighteen problems represent a survey of the central ethical problems that anyone seriously interested in the full range of environment and society related issues must be able to address. Moreover, the fact that complex interactive effects arise between many of these problems means that it is important that we eschew partial approaches to these problems (such as those that are pitched primarily or exclusively at the level of concerns regarding humans, animals in general, living things in general, or ecosystem integrity) and work toward the development of a single, integrated approach that can not only directly address each of the problems I have outlined but also provide compelling reasons for prioritizing our recommendations when value conflicts occur (e.g. between concerns regarding animal welfare and ecosystem integrity, or human preferences and contextual fit). If we can do this, then we will have a General Ethics that is truly worthy of the name. (For the detailed development of such an ethics see Fox, 2006.)

REFERENCES

Arrington, Robert (1998) *Western Ethics: An Historical Introduction.* Malden, MA, and Oxford: Blackwell.

Attfield, Robin (2002) "The Good of Trees." In David Schmidtz and Elizabeth Willott (eds) *Environmental Ethics: What Really Matters, What Really Works.* New York and Oxford: Oxford University Press, pp. 58–71.

Bright, Chris (1999) *Life Out of Bounds: Bioinvasion in a Borderless World.* London: Earthscan.

Callicott, J. Baird (1985) Review of Tom Regan, 'The Case for Animal Rights'. *Environmental Ethics* 7: 365–372.

Callicott, J. Baird (ed.) (1987) *Companion to 'A Sand County Almanac': Interpretive and Critical Essays.* Madison, WI: University of Wisconsin Press.

Callicott, J. Baird (1989) *In Defense of the Land Ethic: Essays in Environmental Philosophy.* Albany, NY: State University of New York Press.

Callicott, J. Baird (1996) "Do Deconstructive Ecology and Sociobiology Undermine Leopold's Land Ethic?" *Environmental Ethics* 18: 353–372.

Callicott, J. Baird (1999) *Beyond the Land Ethic: More Essays in Environmental Philosophy.* Albany, NY: State University of New York Press.

Cowen, Tyler (2003) "Policing Nature." *Environmental Ethics* 25: 169–182.

Fox, Warwick, ed. (2000) *Ethics and the Built Environment.* London and New York: Routledge.

Fox, Warwick (2006) *A Theory of General Ethics: Human Relationships, Nature, and the Built Environment.* Cambridge, MA: The MIT Press.

Goodpaster, Kenneth (2001) "On Being Morally Considerable." In Michael Zimmerman (gen. ed.) *Environmental Philosophy: From Animal Rights to Radical Ecology,* 3rd edn. Upper Saddle River, NJ: Prentice Hall, pp. 56–70.

Heffernan, James (1982) "The Land Ethic: A Critical Reappraisal." *Environmental Ethics* 4: 235–247.

Holmes, Bob (1998) "Day of the Sparrow." *New Scientist,* 27 June, 32–35.

Hume, David (1978) [first pub. 1739-40]. In L. A. Selby-Bigge (ed.) 2nd edn. *A Treatise of Human Nature,* rev. P. H. Nidditch. Oxford: Oxford University Press.

Leopold, Aldo (1979) "Some Fundamentals of Conservation in the Southwest." *Environmental Ethics* 1: 131–141.

Leopold, Aldo (1981) [first pub. 1949] *A Sand County Almanac.* Oxford: Oxford University Press.

Regan, Tom (1983) *The Case for Animal Rights.* Berkeley: University of California Press.

Regan, Tom (2003) *Animal Rights, Human Wrongs: An Introduction to Moral Philosophy.* Lanham, MD: Rowman & Littlefield.

Ryder, Richard (2000) *Animal Revolution: Changing Attitudes towards Speciesism.* Oxford and New York: Berg.

Ryder, Richard (2001) *Painism: A Modern Morality.* London: Centaur Press.

Sagoff, Mark (2001) "Animal Liberation and Environmental Ethics: Bad Marriage, Quick Divorce." In Michael Zimmerman (gen. ed.) *Environmental Philosophy: From Animal Rights to Radical Ecology,* 3rd edn. Upper Saddle River, NJ: Prentice Hall, pp. 87–96.

Sapontzis, Steve (1998) "Predation." In Marc Bekoff (ed.) *Encyclopedia of Animal Rights and Animal Welfare.* London and Chicago: Fitzroy Dearborn, pp. 275–277.

Singer, Peter (1990) [first pub. 1975] *Animal Liberation,* 2nd edn. London: Jonathan Cape.

Singer, Peter (1993) *Practical Ethics,* 2nd edn. Cambridge: Cambridge University Press.

Singer, Peter (2002) "Famine, Affluence, and Morality." In Hugh LaFollette (ed.) *Ethics in Practice: An Anthology,* 2nd edn. Malden, MA, and Oxford: Blackwell, pp. 572–581.

Taylor, Paul (1986) *Respect for Nature: A Theory of Environmental Ethics.* Princeton, NJ: Princeton University Press.

Varner, Gary (1998) *In Nature's Interests?: Interests, Animal Rights, and Environmental Ethics.* New York and Oxford: Oxford University Press.

Varner, Gary (2002) "Biocentric Individualism." In David Schmidtz and Elizabeth Willott (eds) *Environmental Ethics: What Really Matters, What Really Works.* New York and Oxford: Oxford University Press, pp. 108–120.

Warren, Mary Anne (2000) *Moral Status: Obligations to Persons and Other Living Things.* Oxford and New York: Oxford University Press.

Reprinted (with minor modifications) from Chapter 2 of Warwick Fox, A Theory of General Ethics: Human Relationships, Nature, and the Built Environment (Cambridge, MA.: The MIT Press, 2006) by kind permission of The MIT Press.

8

Anti-Environmentalism: Prometheans, Contrarians and Beyond

Damian Finbar White, Alan P. Rudy and Chris Wilbert

INTRODUCTION

In 2001, the Danish political scientist and statistician, Bjorn Lomborg, published a text entitled *The Skeptical Environmentalist*. Across some five hundred pages, Lomborg argued that if the statistics on environmental trends and human welfare are surveyed at the global level, the numbers look rather better than is often assumed. He went on to suggest that one of the central reasons for the wide-scale (mis)perception amongst publics and policy makers in the Organization for Economic Cooperation and Development (OECD) – that we face rapid global environmental decline or even a potential 'global environmental crisis' – is because environmental groups and non-government organizations (NGOs) have been guilty of systematically exaggerating the scale of environmental problems. Committed to proclaiming a standard 'litany' of environmental doom at every opportunity, Lomborg argued that environmental NGOs pay very little attention to the costs and benefits of dealing with environmental problems. More damning still, he argued, environmentalists spend much of their time campaigning about environmental problems that are either bogus or highly uncertain (Lomborg, 2001).

The response to the claims of *The Skeptical Environmentalist* was striking. Lomborg was courted by influential media such as *The Economist*, *The Washington Post* and other media outlets in Europe and the USA as a new, original, lone voice

willing to 'face down' the Green movement. Yet, reviews of *The Skeptical Environmentalist* in the scientific press (with a few exceptions) read Lomborg's skepticism skeptically. Moderate critics in the environmental science community suggested that, while Lomborg's reading of upward trends in human welfare was broadly correct, his reading of the direction of environmental trends was based on questionable readings of the data. More severe critics argued that Lomborg had fundamentally misunderstood key tenets of environmental science. Environmental campaigners more generally argued that it was Lomborg himself that was the charlatan.[1]

The emergence of Bjorn Lomborg and what became known as 'the Lomborg controversy' has been treated by many as constituting a unique and original critique of the environmental movement. From a broader perspective though, many environmental social scientists have observed that this controversy is best thought of as the public high point of an increasingly organized wave of anti-environmental thought and organization that had been building, particularly in the USA, since the early 1990s. This political phenomenon has distinct resonances with an earlier 'Promethian' or 'cornucopian' critique of environmentalism that has been around since the initial emergence of post-war environmentalism in the late 1960s.

In this chapter, we seek to get to grips with these developments. As will become immediately clear, we suggest mapping the terrain between environmentalism and its critics is more complex than the above narrative suggests. This is because standard narratives of the history of the environmental debate, which counter-poses 'survivalists' against 'Promethians' or cornucopians, fail to capture the multiplicity of the thinkers and currents that have positioned themselves in a critical relationship to environmental discourses. Critiques of both mainstream and radical environmentalism/ ecologism have taken very different forms at different historical periods. This is true as much because of the success of environmental movements as it is due to the long and fierce disagreements internal to these movements over ideological and political issues. As such, many issues presently associated with contemporary contrarian currents surfaced, or even originated, within the environmental movement itself.

To make this argument concrete, we suggest that if these currents are analyzed in their intellectual, political and sociological specificity, we can see a much more complex debate unfold. We begin this chapter by mapping the diverse critical discussions that emerged from left, right and technocratic sources to the first wave of post-war environmentalism.

In the second part of the chapter, we go on to consider the explosion of a 'second wave' of contrarian thought that followed the apparent consolidation of 'global environmentalism' at Rio de Janeiro and the Republican takeover of the US Congress in 1994. In this second wave, we suggest that it is certainly the case that right/libertarian think tanks and industry groups played a central role as instigators, propagators and coordinators of anti-environmental thinking (Beder, 1999a, 1999b; Monbiot, 2006). However, once again, it is evident that the diverse criticisms that have been made of environmental movements and discourses cannot be simply reduced to such developments.

In the third, and main part of this chapter, we go on to critically appraise the strengths and weaknesses of the most familiar forms of modern contrarian thinking. We consider the effects such currents have had on environmental movements and the coherence of environmentalism as an ideology. Finally, we consider some emerging suggestions that the debate between mainstream environmentalism and contrarian discourses might be in the process of being transcended.

THE POPULATION/RESOURCES DEBATES OF THE 1960s AND 1970s

Environmental discursive practices have long attracted critics. Indeed the central issues of contention running through the modern environmental debate can be traced back to some of the disputes that occurred between Godwin, Malthus, Ricardo, Jevons, Mill and Marx one hundred and fifty years ago (Benton, 1989, 1996; Foster, 1999). In any event, the first wave of post-war environmental political discourse was dominated by neo-Malthusians (Vogt, 1948; Osborn, 1954; Ehrlich, 1968; Meadows *et al.*, 1972). As is now well known, this current can be identified by its attempt to frame the environmental debate around the issue of 'exponential growth in a finite system.' Particular emphasis was placed by such currents on the dangers of rapid human population growth, in particular generating scarcity of geological resources.

Following these conceptual principles, it was argued by the population biologist Paul Ehrlich that if population pressures remained at 3.3 billion at current levels of demand lead would run out in 1983, platinum in 1984, uranium in 1990 and oil in 2000 (Ehrlich, 1968). Similarly, using the World2 computer simulation, the *Limits to Growth* model (Meadows *et al.*, 1972) argued that if existing exponential growth trends continued in human population, agricultural and industrial production, resource depletion and pollution, it could be 'scientifically' demonstrated that there would a fairly sudden global decline in industrial and population capacity by the mid 21st century.

Broadly speaking, we can locate four central discourses that sought to critically respond to such claims: (i) neo-Classical economists; (ii) skeptics of modeling and futurology; (iii) promethean Marxism; and (iv) new Left 'post-scarcity' thinkers.

Neo-classical critics

There is no doubt that the primary discourse that led the way in critiquing the first wave of environmental concern emerged from neo-classical economists. John Dryzek (1997, pp. 46–47) suggests that the basic contours of the post-war neo-Classical critique of environmentalism can be found in Barnett and Morse's *Scarcity and Growth* (1963). Responding to some of the earliest post-war stirrings of neo-Malthusian thought (e.g. Vogt, 1948; Osborne, 1954), Barnett and Morse argued that the fear that developed nations might be running out of resources was mistaken. They suggest that not only is it the case that the amount of resources in the Earth's crust probably exceeded present known resources by many thousand or million fold but if 20th century trends in extraction were examined carefully the cost of extracting materials had fallen significantly over the 20th century with the exception of forestry and possibly copper.

The emerging central claim then of neo-Classical economists was that price should be seen as the central measure of scarcity (Dryzek, 1997, p. 46). It is reasoned that if prices are falling (as they had been historically throughout the 20th century) then this indicates resources are increasing. Second, neo-Classical economists have gone on to argue that neo-Malthusians persistently underestimate the role that scientific and technological innovation play in further expanding the realm of known resources, the manner in which price increases can additionally make more economically viable previously uncompetitive sources of exploitation and allow for substitution of resources. Rather than being static objects, it is argued that resources expand to meet demand. The economist Julian Simon, popularized such arguments in *The Ultimate Resource* (1981). Expanding these arguments beyond resource pricing to include further indicators of human well-being such as food supply per capita, life expectancy and so on, Simon argued the fact that life expectancy is on an upward curve globally clearly suggests the overall effects of pollution are dropping (Dryzek, 1997, p. 48).

Technocrats and systems modeling

A second current of skepticism to *The Limits to Growth* report could be found amongst social scientists that focused on the limitations of using computer modeling to understand the interplay between social and natural systems at a planetary level. Specifically, it was argued that the Limits to Growth model simply did not adequately take into account feedback processes and the effects of human decision-making (Cole *et al.*, 1973). It was argued that the whole strategy of projecting exponential growth rates into the future whilst keeping technology and innovation fixed offered limited insight. Macrae for example observed, if we projected the trends of the 1880s, it would have shown that cities would be buried in horse manure by the mid-1920s (cited in Sandbach, 1978, p. 499). Concerns were additionally raised with the managerial and technocratic tendencies of the *Limits to Growth* analysis given the emphasis placed in this report on the need to 'control and regulate consumption.' It was argued by some that much of this modeling carried authoritarian and technocratic tendencies (Sandbach, 1978).

The Marxist critique

It is interesting to note, however, that in the first wave of criticism of post-war environmentalism, social scientists associated with the political left, and particularly Marxism, were equally involved. The Marxist and socialist tradition has long had a rather ambivalent relationship with the environmental movement (Benton, 1996). With its intellectual legacies of Promethianism and productivism, declarations of a looming 'global environmental crisis' initially generated a highly ambivalent response. Beyond initial charges that environmentalism was primarily a preoccupation of the middle classes in common with neo-Classical economists, a good deal of critical scrutiny was given to the essentially Malthusian theoretical assumptions underpinning this work with its stress on technological pessimism and natural limits' ideologies (Harvey, 1974).

Marxist critiques of neo-Malthusian environmentalism in the 1970s focused specifically on the naturalistic reductionism of much environmental discourse. Equally, some Marxists criticized the manner in which 'limits' and 'no-growth' oriented discourses appeared to be insensitive to the dispossessed 'at home' or the material aspirations of the developing world (Enzenberger, 1974). Specifically, it was argued the natural limits arguments of Ehrlich and others naturalized problems specific to capitalism.

Notably, it was argued there was a distinct tendency in green understandings of 'natural limits' to obscure the fact that poverty is a product of the maldistribution of resources and not due to *natural limits* in producing the resources themselves. Additionally, it was argued that, under capitalism, there is a failure to see that scarcity is a social and not a natural (Commoner, 1971; Harvey, 1974; Enzenberger, 1974). As Benton notes, for many Marxists (who often reflected the mood in more general critical circles) 'the new politics of ecology was reactionary in content and elitist in terms of the interests it represented' (1996, p. 7).

Social ecology and the New Left

A further part of this story which is frequently missed and complicates matters still, is that we can find a range of 'new left' inspired thinkers, who critiqued neo-Malthusian arguments in the 1970s yet nevertheless sought to defend a different type of social environmentalism (Bookchin, 1971; Commoner, 1971; Leiss, 1972; Gorz, 1975; Stretton, 1976; Schnaiberg, 1975, 1980). Perhaps one of the most sophisticated figures in this current of New Left environmentalism was Murray Bookchin.

Arguing for a non-Malthusian and post-scarcity 'social ecology', Bookchin's work effectively sought to turn the neo-Malthusian/cornucopian debate upside down. In essays dating as far back as the early 1960s, Bookchin suggested that the emerging obsession amongst some environmentalists with overpopulation simply distracted attention from the much greater environmental problems generated by the US economy (Bookchin, 1964, pp. 85–86). He went on to argue that in eliding consideration of the complex cultural, political and historical factors involved in population booms, neo-Malthusian demography obscured the eminently *social factors* which have produced hunger and famine. More broadly, he suggested particular attention needed to be given to declarations of overpopulation, racism and imperialism. As Bookchin noted:

> We must pause to look more carefully into the population problem, touted so widely by the white races of North America and Europe – races that have wantonly exploited the peoples of Asia, Africa, Latin America, and the South Pacific. The exploited have delicately advised their exploiters that what they need are not contraceptive devices, armed 'liberators', and Prof. Paul R. Ehrlich to resolve their population problems; rather, what they need is a fair return on the immense resources that were plundered from their lands by North America and Europe. (1980, p. 37).

Concerning the energy and resource depletion arguments of Ehrlich and the *Limits to Growth*, Bookchin dismissed such claims as 'a media myth' (1980, p. 305). Following Marxist critics (and Julian Simon), he argued that historically even the most extravagant estimates of petroleum reserves and mineral resources have proved to be hugely underestimated. Second, Bookchin argued that many 'shortages' were the outcome of commercially created interests and oligopolistic market manipulation – rather than being statements of the (unknown) realities of the oil or other resources of the world. Third, it was suggested that such arguments and fears of shortage, more generally, 'serve the interests of price fixing operations, not to mention crassly imperialist policies' (1980, p. 306). The central point made here – against any simple endorsement of the 'Limits to Growth' thesis – was simply that '"scarcity" is a social problem not merely a "natural" one …' (1980, p. 306).

According to Bookchin's 'social ecology', the central problem with the Club of Rome report was that it failed to recognize that 'the greatest danger these practices raise is *not depletion* but *simplification* …', that is '… the limits to capitalist expansion are *ecological* not *geological*' (1980, p. 306). Entirely rejecting the prospect of imminent geological scarcity or Malthusian understandings of 'overpopulation', it is quite a different series of problems which are presented as the central components of 'ecological crisis' in social ecology. Notably from 1952 to 1965, when Bookchin's work was devoted specifically to

analyzing ecological problems, we can see concerns raised with:

1 excessive use of pesticides and insecticides in farming;
2 water and air pollution;
3 the proliferation of toxic chemicals, radioactive isotopes and lead;
4 industrial pollution;
5 waste generation;
6 the debilitating lifestyles that accompany a sedentary, congested, stressful, urbanized world.

It was additionally speculated in Bookchin's 1964 essay *Ecology and Revolutionary Thought* that a longer-term problem may result from the changing proportion of carbon dioxide to other atmospheric gases through the burning of fossil fuels (Bookchin, 1952, 1962, 1964, 1965). Later writings never return to surveying environmental issues in any detail. We can find in these works though a growing stress on the overall deterioration of 'basic planetary cycles.' Major concerns were identified as:

1 increase in the ratio of carbon dioxide to oxygen in the atmosphere;
2 widespread deforestation and soil erosion;
3 the role that chlorofluorocarbons played in thinning out the ozone layer;
4 simplification of wildlife and plant biodiversity (Bookchin, 1982, 1990).

A complex debate

As we can see, marking out a clear terrain of engagement between 'environmentalists' and 'cornucopians' and Prometheans is not a straightforward operation. Even at the beginnings of the first wave of the post war environmental debate, we can see a complex range of positions and debates emerging. To be sure there is a neo-Malthusian–neo-Classical/Promethian axis at some level in this debate. Yet, it is also evident that one can find a range of figures from incipient political ecologists such as Barry Commoner (1971), Hans Magnus Enzenberger (1974) and Francis Moore Lappe (Lappé and Collins, 1977; Lappe and Schurman, 1990; Lappé *et al.*, 1998) to post-scarcity, and in some respects cornucopian social ecologists such as Murray Bookchin, who muddy the waters considerably.

Such currents suggest that the politics of the environment has been, and is, much more messy and potentially embedded in questions of power, capital, race and other issues than the conventional 'survivalists versus promethians' discourse allows. Regardless of this, however, it would seem to be the case that by the early 1980s this first wave of the debate had died down considerably. To an extent in the USA, this can be traced to certain political victories on the part of the environmental movement. The Nixon White House responded to environmental pressure in enacting the biggest wave of legislative reform in post-war history. At the same time neo-Malthusian thinking became part of US foreign policy. Radical environmental movements went into abeyance and a popular impression was left that it was the conceptual resources of the politics of ecology that had run out.

CONTRARIANS AGAINST GLOBAL ENVIRONMENTALISM

The second act of the environmental debate is usually traced to the rise of the global environmental agenda from the late 1980s onwards (Mol, 2003). Following in the wake of the UN-sponsored 'Brundtland' report of 1987 – which rejuvenated environmentalism from the deep-freeze of the Reagan/Thatcher era – we saw a major broadening of the environmental debate well beyond the population/resources debate of the 1970s. New concerns with global warming, biodiversity loss, and stratospheric ozone depletion galvanized a new wave of global environmentalism. Whilst this second wave of the discussion did not entirely displace the population/resources debate of old, we can see a shift from concerns about resource scarcity to a more general concern emerge about the hazardous effects of ecological simplification on the health of ecosystems. The new environmental agenda became more diversified and yet also global in its range. The adage 'contrarian' is a term that increasingly came into circulation in the mid-1990s to identify an increasingly vocal and well-organized series of critics of this new global agenda.

From promethean to contrarian

In some respects, it is interesting to note how the use of the term itself 'contrarian' – contrary to the norm – tells us a lot about the major inroads that environmentalism had made into the mainstream of scientific and policy debate since the mid-1970s. As John Dryzek has observed the very term 'contrarian' itself clearly suggests that to some degree the mainstream of the debate in Western liberal democracies had largely inculcated environmental values or significantly accommodated to many of the central themes of the environmental movement (Dryzek, 1997). As Dryzek notes, if we consider that many of the central assumptions of modern contrarian thought were

majoritarian assumptions underpinning the modern West, perhaps the really interesting irony is to consider how such 'majoritarians' became contrarians, how the dominant discourse of the West had to be 'articulated and defended, rather than just taken for granted' (Dryzek, 1997, p. 46). What is striking about the second wave of anti-environmental thought is the leading role that libertarian think tanks, 'wise use' groups and corporate funders played in its development (Boston, 1999; Luke, 2000; McCright and Dunlap, 2000a,b; Monbiot, 2006).

New right and wise-use contrarians

The heart and soul of much of the new economic right's contrarianism comes from a neo-Classical foundation: a commitment to property rights over civil rights – in the neo-Marxist parlance of the 1970s, a commitment to accumulation over legitimation. While it is fairly clear that the Wise Use movement is more the product of corporate sponsorship than grass roots activism, the historical brand of American populism provided fertile ground for arguments as to the encroachment of 'big government' on 'the little guy's' property rights (Echerrevia and Eby, 1994; McCarthy, 1998, 2002). Furthermore, the agrarian and backwoods localism that inheres in populism's small + local = good vs. big + national/global = bad theory of politics resonates strongly with suggestions that property owners know their land, resources, crops and animals and how to manage them far better than some urban, laboratory scientist who has never got his or her boots muddy, hands bloody or face sweaty. Onto this Romantic culturally conservative terrain, the Wise Use 'movement' and its funders has layered a discourse of 'environmental takings' that fits well with Neoliberal economic conservatism – as each are deeply committed to bureaucratic deregulation and legislative devolution.

It is interesting to note odd resonances, though that can be found between ecosceptics in the Wise Use movement that focus on property mechanisms and conservative conservation groups like The Nature Conservancy, who have sought to buy ecologically 'important' landscapes and/or the development rights to historically modified and working agrarian, forest, and wetland properties (Luke, 1997). Furthermore, as will be developed later, market mechanisms have been encouraged by other strands of neoliberal ecoscepticism resonant with Malthusian ethics – over and above regulation based on biological and ecological science – as the proper means of conserving resources and allocating/distributing pollution. That these techniques are not wholly different from certain pragmatic implementations of eco-logical modernization theories is another moment in the ironic complexity of the relation between environmentalism and its sceptics. In all, the question of whether or not the object of concern is really nature, ecology, or science – as opposed to concerns with economic, political, or social reproduction – comes increasingly to the fore.

The rise of fundamentalist contrarianism

Our goal is to destroy, to eradicate the environmental movement. We're mad as hell. We're not going to take it anymore. We're dead serious – we're going to destroy them. Environmentalism is the new paganism. Trees are worshipped and humans sacrificed at it's altar. It is evil. And we intend to destroy it. No one was aware that environmentalism was a problem until we came along (Arnold, 1987).

Scholars of environmental politics have long been familiar with the notion that green political discourse can be marked by varying distinctions between 'reformist' and 'radical' currents, between 'realo' and 'fundi' branches and so on (Mayer and Ely, 1998). With the evolution and complexification of contrarian politics in the 1990s, it has become increasingly apparent that contrarian politics similarly requires a differentiating typology. Contrarian politics would seem to recommend itself to two different discourses.

First, in terms of understanding the problem, it needs to be seen that there are clear differences between contrarian discourses which claim certain sympathies with environmentalism yet wish to contest environmental apocalypticism (see Easterbrook, 1995; Lewis, 1995) and forms of contrarian discourse which maintain that all environmental discourses are essentially 'little green lies' and wish to see environmentalism eradicated (Arnold, 1987). There are also clearly considerable differences between contrarian currents that proceed through a commitment to reasoned argument, theoretical and empirical engagement and in essence relying on 'the strength of the better argument' to carry the day, and campaigning forms of contrarianism, especially those at the furthest extreme of fundamentalist contrarianism that involve media manipulation, deception and perhaps even violence.

However, differences between such perspectives have been made ever-more cloudy for a variety of reasons. First, mainstream environmentalists themselves paint with too broad a brush, labeling all of their critics as contrarians, thus stifling debate which the movement needs and indeed should thrive upon. Second, we have seen a rise in policy think tanks and corporate-sponsored front groups that have insinuated themselves into the center of policy-making arenas, taking advantage of opaque political decision-making structures in Western liberal democracies (Beder, 1999a, 1999b; Monbiot, 2006). Such think tanks and front groups have presented themselves as independent centers offering the kinds of expertise that mass-media organizations tend to favor (Monbiot, 2006).

The techniques used by many of these new right think tanks, as well as corporate-backed PR companies and front organizations such as the Global Climate Information Project in the USA, have also fed into wider cultural transformations in terms of public attitudes toward the sciences and politics more generally regarding trust and risk (Giddens, 1991; Beck, 1992). These techniques have mainly been focused on undermining the science-based discourses of environmental politics or confusing the public and policy-makers on particular environmental issues – especially regarding global climate change. As Sharon Beder argues:

Think tanks have sought to spread confusion about the scientific basis of environmental problems, to oppose environmental regulations and promote free market remedies to those problems such as privatization, deregulation and the expanded use of property rights. Corporations that wish to portray themselves in public as environmentally concerned often fund such think tanks, whom they are not readily identified with, to oppose environmental reforms (1999a).

Most controversially for many environmentalists in this regard has been the role of the oil company Exxon in financing 'front groups' and research that casts doubt on the notion of anthropogenic-induced climate change – though they are only one of many corporations involved in such activities (Monbiot, 2006). It must also be said that mass-media forms of reporting stories has played into the hands of such media savvy contrarian groups in the media desire for supposedly 'balanced' stories that present one side with an antagonistic other (often a corporate funded front group, or a dissident scientist) thus keeping a pretence of impartiality that the viewer is then supposed to determine rationally for themselves. In the process, of course, an impression is given of undue uncertainty over the issue.[2]

Contrarianism thinking and Republican politics

If we consider the influence of contrarian currents in North Atlantic rim societies, there have been notable points of commonality between the modern Republican Party in the USA and contrarian currents. As Dryzek notes, Promethian discourses played a particularly important role in the Reagan administration (1980–1988) with Reagan's Secretary of the Interior James Watt and Anne Gorsuch – the Administrator of the EPA – being highly influenced by such discourses (see Dryzek, 1997, p. 54–57). After a modest drop off in influence in the late 1980s and early 1990s contrarian currents became particularly powerful in the 'backlash' to the Clinton administration and the subsequent Republican renaissance in Congress. In 1994, a growing group of Republican senators were elected with links to the anti-environmental movement who explicitly declared their aim to be to undermine much of the environmental legislation passed in the 1970s such as the Clean Air and Clean Water Acts, and the Endangered Species Act (Tokar, 1995). Contrarian currents have additionally expanded their influence in the Republican Party as figures such as Pat Robertson in the Christian Coalition started to view environmentalists at the evil priests of a new paganism. The administration of George W. Bush has have continued this close relationship between the Republican Party and contrarian views on environmental policy. The Bush administration has pursued a much more systematic attempt to roll back US environmental legislation achieved from the 1970s onwards. President Bush for example appointed Gale Norton, a member of the Wise Use Movement, to head the department of the Interior. Indeed, contrarian currents have gone on to play a major role in impeding further progress of the Kyoto protocol.

THE INSTITUTE OF IDEAS/SPIKED ONLINE GROUP

The spread of contrarian ideas in the UK has principally occurred through a rather mysterious group of journalists and intellectuals around the think tank 'Institute of Ideas' and the website 'Spiked Online.' This group has its roots in an orthodox Marxist – Leninist group – the Revolutionary Communist Party – that developed from the 1970s and was founded by the sociologist Frank Furedi. However, from the early 1990s onwards the group increasingly moved away from socialist politics whilst maintaining strong commitments to the vangardism, scientism and technological determinism of orthodox Marxism–Leninism. The end result has generated an unusual fusion for anti-environmentalism currents. Declaring themselves 'beyond left and right' (Furedi, 2005), the Institute of Ideas/Spiked Online Group argue from an ideological position that draws together radical anti-environmental

contrarianism, with the aspects of Leninism concerned with deference to authority and leadership structures. This is combined with a Nietzschian celebration of the unencumbered individual and an Ayn Rand style defense of elitism and the unrestricted free market.

Furedi has increasingly argued that a concern with environmental questions in any form is merely a manifestation of a ubiquitous 'Culture of Fear' that has gripped Western societies with the collapse of Communism. It is argued that Western societies are now characterized by constant waves of moral panics, cultural relativism and a refusal of hierarchy and leadership, with environmentalists and leftists more generally playing the leading role as 'problem mongers.' Claiming that the whole agenda of 'sustainable development' is merely a symptom of societies' 'culture of low expectations', the 'Institute of Ideas' and 'Spiked' have emerged as central conduits channeling the thinking of US libertarian right think tanks into the UK media. More generally, they have become libertarian promoters of all kinds of contrarian ideas about everyday life and the sciences: from disputing the relationship between HIV and AIDS to denying the existence of global warming and the Rwandan genocide.

Explaining the continued influence of contrarian currents

So, what might explain the continued influence of contrarian currents? Five factors would seem to come into play here. First, the environmental sciences are clearly characterized by high degrees of uncertainty and complexity that lend themselves to high degrees of expert disputation. Second, in media-saturated societies where the popular press constantly searches for new angles to generate controversy 'contrarian currents' serve a useful function in social distraction. Third, it would have to be recognized that the ongoing influence of contrarianism is linked to the manner in which the institutionalization of environmental legislation in OECD countries and the political maneuverings of the environmental movement has resulted in mainstream environmentalism generating a range of enemies over the years. Fourth (as we will argue later), many environmentalist arguments themselves have been characterized by such a high degree of economic, political, cultural and ecological blindnesses that they have generated legitimate criticism from both the left and the right. Finally, it surely would seem to be the case that contrarians maintain their high status because as John Dryzek has noted particularly in the USA such discourses 'resonate with the interests of both capitalist market zealots and Christian conservatives, not to mention miners, loggers and ranchers accustomed to subsidized access to resources' (Dryzek, 1997, p. 57).

CRITICAL EVALUATION AND DISCUSSION: CONSIDERING SOME OF THE STRENGTHS AND WEAKNESSES OF CONTRARIAN DISCOURSE

Whilst it is very difficult to find any positive virtues in the more extreme forms of contrarian politics, it is also difficult to know what to do with the type of medieval rhetoric espoused by figures such as Ron Arnold (1987) who have simply declared all environmentalism as 'evil.' Moreover, other radical forms of contrarianism seem so fanatically tied to an extreme anti-government right-libertarian agenda that it seems very difficult to engage seriously with such currents. However, given that the critique of environmentalism moves on a sliding scale, there are reasons to feel that moderate contrarians, such as Beckerman (1995) and 'eco-optimists' like Greg Easterbrook (1995), or even rather inconsistent figures such as Lomborg (2001) have made an academic and political impact and posed some important questions for the environmental movement. In particular, such currents have set out some very important challenges particularly in relation to neo-Malthusian, ecocentric and anti-humanist forms of radical ecology. We suggest that there are six central challenges.

The critique of eco-apocalyptic and neo-Malthusian thought

First, it would seem to be the case that the contrarian critique of environmentalism is at its strongest when critiquing the 'apocalyptic' frames that have dominated much of late 20th century environmentalism. Here, many environmentalists (Ehrlich and Ehrlich, 1996, pp. 45–46) as well as academics generally sympathetic to the movement (Harvey, 1996; Katz, 1998) have noted that the regular 'doomsday predictions' of politically diverse environmental currents across many fields have not withstood the test of time. The neo-Malthusian and survivalist arguments of the 1970s on population and resources look rather embarrassing – given their premises in static understandings of scientific and technological development and ahistorical understandings of natural limits. And in this respect, Simon's critique of 'the population bomb' may be seen, in retrospect, as a less problematic piece of argumentation that is generally acknowledged.[3]

It is also notable that the strong statements found in *The Limits to Growth* and *The Population Bomb* no longer receive support, even

from their authors. Such a concession has indeed largely been accepted by the Ehrlichs (1996). Thus, the simple argument that the physical scarcity of specific local resources is generating impending environmental social crisis is now downplayed:

> ... this study may have under-rated (and we once did) the amount of technological innovation and substitution that can be called forth in the short term by prices driven up by scarcity ... (Ehrlich and Ehrlich, 1996, p. 95).

And even concerning their emphasis on population, they have noted:

> ... since people in both industrialized and middle income nations are almost all better fed and paying less in relation to incomes for food than they were in 1968, our projects were inaccurate (Ehrlich and Ehrlich, 1996, p. 34).

The reduction in the apocalypticism of the Ehrlichs reached the point in the mid-1990s when New Right efforts to systematically unravel landmark environmental legislation passed in the 1960s and 1970s finally led to the defense of such legislation as the source of real, ecologically meaningful improvements in water quality, air pollution, urban sewage and industrial effluents (Ehrlich and Ehrlich, 1996, p. 46).

Misanthropy and racism in environmentalism

Second, a further and genuine contribution that contrarians have made is out lining the close relationship between neo-Malthusian thought, eugenics, misanthropy and racism (Ross, 1998). There remain those within environmentalism who claim that the exaggerations of 1960s neo-Malthusianism are forgivable due to the role they played in 'raising the ecological alarm.' Part of the problem with this form of analysis is that it simply avoids the manner in which population control rhetoric has long had an underlying classist, racist, and sexist quality.[4] Indeed, there are good reasons to feel that the discussion of human population growth in terms of alarmist metaphors like 'bombs going off' did indeed contribute to the rise of coercive population control polices from the 1960s onwards. Julian Simon was correct to note the manner in which population bomb rhetoric frequently fed into broader eugenicist projects which diversely have sought to reduce the fertility of the poor, African Americans and Africans 'for their own good' (see Simon, 1981: 8; additionally see Ross, 1998).

It also needs to be recognized that despite the discrediting of many neo-Malthusian arguments general neo-Malthusian ideological assumptions continue to resonate in respectable environmental institutions like the Worldwatch Institute or the academic disciplines of demography, ecology, or conservation biology (Luke, 1997). Lester Brown, the leader of the Worldwatch Institute has been warning of an imminent collapse in agriculture for some 20 years. Lomborg (2001) is on strong ground here when he demonstrates how Brown has consistently overstated the case.

Anti-humanism

Third, moderate contrarians and eco-optimists legitimately take up further pathologies of radical environmentalism particularly in the latter's tendencies toward anti-humanist, ecocentric, and deep ecological manifestations most prevalent in the 1980s in North America. Some of the latter currents are well known, having been the subject of much criticism from social ecologists (e.g. Bookchin, 1987; Bradford, 1989) and ecofeminsts (Salleh, 1984, 1993) within the ambit of environmentalism. However, once again contrarians have been quite correct to argue that a strain of green thinking has not only been attracted to controversial and contradictory biocentric leveling arguments, but has also on occasion explicitly embraced misanthrophic politics, and even forms of eco-fascism (Biehl and Staudenmaier, 1995; Ross, 1998).

The questionable priorities of northern environmentalists

Fourth, contrarians have also raised reasonable questions about the priorities which have dominated the agendas of many currents of Northern environmentalism. Easterbrook is correct to argue that at times environmentalists have been fixated on attempting to 'contain small conjectural risks' (Easterbrook, 1995, p. 30). The focus of many environmental protection campaigns does reflect class and racial bias. It seems increasingly evident for example that the manner in which many environmental currents in the 1960s and 1970s (particularly in the USA and Australia) embraced the cult of wilderness has generated many social pathologies. Not only did such an emphasis directly lead to a marginalization of the environmental concerns of everyday working people (Gottlieb, 1993) but 'wilderness ideology' also encouraged an extremely problematic romanticization of the ecological practises and environmental histories of native peoples and indeed a

dehistorization of landscape more generally. Contrarians are also correct to note that questions of priorities have also generated real problems between 'northern environmental NGOs and development campaigners in the Global South.' For example, questions could be asked about the wisdom of campaigns to continue the ban on DDT spraying when, according to Hollander and others, scientific studies have not substantiated the claim by Rachel Carson that DDT is a human carcinogen (Hollander, 2003, p. 8) and many alternative approaches to malaria prevention are ineffective. Environmentalists have given insufficient attention to indoor air pollution. Conservation is clearly a lower priority for many people who do not have clean drinking water (Harvey, 1996).

In fact, it is in the context of the environmental movements discursive commitments to 'nature' in the face of its regulatory successes around the basic need for clean air, water, and residence that the environmental justice movement's concerns with the built environment – rural and urban – comes critically to the fore (Gottlieb, 1993; Harvey, 1996). How to characterize these groups in the environmentalist–contrarian formulation is difficult. They tend to be very critical of the naturalism – and historically implicit racism – of traditional environmentalism on the one hand and have a fraught relationship – grounded in the use of science to stifle normative debates – with the traditionally underdemocratic progressive scientism of the environmental state on the other. There are indeed legitimate grounds for criticism of the activities of conservation organizations that focus on species conservation, not least in their problematic attitudes toward local people in developing countries where they often work (Luke, 1997; Chapin, 2004).

Naturalistic reductionism

Fifth, interesting issues are also raised by some contrarian writings that chart certain cultural pathologies of some forms of environmentalism. Many currents of green politics have been informed by tendencies to see the 'natural' as 'good', human-made as 'bad'; nature left 'alone' good, intervention bad; the natural/organic good, the artificial bad. Such arguments can be found in a range of environmental currents, with the anti-ecological restoration arguments of Elliot (1997) being just one manifestation. There is clearly a real problem with such valorizations. As the works of Latour (1993) and Haraway (1985) have argued, critical scrutiny of such dualisms reveals they are hard to sustain. More generally, it could be observed that by colluding with a sense of alienation that some citizens of late modernity may feel in terms of the direction of contemporary society, such dualisms can end up alienating humanity from itself and developing a kind of contempt for the 'human-made,' as well as the historical struggles people in all walks of life have had with the 'natural' world. This of course also links into previously mentioned discussions on deep ecology and primitivism where anti-humanism became an element in some groups in the 1980s. Perhaps most importantly, it is the undifferentiated character of homogenized humanity and trans-historical nature – much less the diversity of relations, both material and semiotic, between the two – that generates much in the way of the ecological blindnesses, policy failures, and social contradictions of traditional/conservationist/naturalist environmentalism.

Dynamic and resilient nature

Finally, some forms of contrarianism raise challenging issues about conceptualizing 'natures.' For example, rather than taking the view of nature as 'fragile', 'delicate' and easily damaged, contrarians tend to stress qualities such as tenacity, resilience, and regenerative powers. Whilst still problematic in many ways, not least in implying a separation of nature and society, such views tend to stress more dynamic views of nature and of society's complex entanglements with nature. Of course, the resilience of nature can be overemphasized – leading to something of a see-saw argument. Indeed this was to be seen regarding recent discussions of oil spillages, with some environmentalists stressing disaster, and other contrarians arguing nature can recover quickly from accidental spillages (Kaplinsky, 2002). Here, this contrarian strand arises as the evil twin to traditional environmental transhistoricism. For both, nature is always a thing (or set of things) and is always the same. It is just that one side sees it, nature, as a fragile thing or set and the other sees it as a robust and resilient thing or set. In neither case is that materiality dynamics of environmental history or the social situatedness of human–nature relations taken into account.

THE LIMITATIONS OF THE CONTRARIAN CRITIQUE

We argue then that there are reasons to believe that environmentalism and the environmental social sciences have some things to learn from *some* contrarian critiques. At the same time though, such

comments should not detract from the fact that contrarian currents seem to suffer from some fairly substantive weaknesses of their own. Nine points will be made here.

Misrepresenting the complexity of the social–environmental debate

Perhaps one of the most striking features of the 'contrarian case' is the extent that *most contrarian discourses tend to misrepresent or ignore the complexity of environmental politics, debates currently occurring in the environmental social sciences and the complexities of the environmental debate*. Most notably, the standard maneuver of contrarian arguments involves presenting a generic 'green' or 'environmentalist' position in terms that simply evoke neo-Malthusian, anti-humanist and technophobic currents. As such, there is no recognition amongst contrarians that humanist eco-socialists such as Barry Commoner and social ecologists such as Murray Bookchin – from the 1970s onwards – were the first figures to counter overpopulation arguments, critique ecocentric thought for its technophobia, and, more generally, mainstream environmentalism for its class and racially questionable positions.

As we have argued throughout this chapter, all the environmental debates get muddy quickly when they are closely explored. Contemporary contrarian discourses without exception rewrite the history of the environmental debate to avoid such complexities and to achieve the kind of rhetorical closure and claims of originality that authors such as Easterbrook or Lomborg or writers from on-line sources such as Tech-Central and Spiked Online regularly rely on. As such, contrarian narratives contain no recognition of the messy internal critique of the environmental mainstream that has been conducted by diverse 'red greens,' political ecologists, social ecologist in the 1970s and 1980s, and critical eco-feminists and environmental justice advocates in the 1990s. Additionally, it could be noted that contrarian discourses have paid very little attention to how developments in the environmental social sciences over the last ten years – from environmental history to political ecology, from industrial ecology, the sociology of ecological modernization to the production of social nature debate – have radically altered the whole context of the environmental debate.

Ecological constraints/enablements

Whilst it is the case that most sensible forms of contemporary ecocritique concede to early contrarian thought (and Bookchin's social ecology) in rejecting the idea that the 'limits to growth' are simply geological, more sophisticated currents argue it is better to talk about historically and culturally specific material enablements/constraints (Benton, 1994; Harvey, 1996) than simple limits and to recognize growth may have different trajectories; Promethian/contrarian arguments continue to have some major blind spots. Notably, on 'geological matters' and 'technological fixes,' ecological economists and environmental sociologists have made various counter arguments:

1   Not all resources are substituted for in a time period sufficient for alternative modes of growth/ development/reproduction. Indeed, societies which are characterized by poor ecological monitoring systems and a high degrees of contrarian skepticism are likely to be characterized by low adaptive capacities to achieve adequate substitution. For example, it could be observed that at the moment it is the alliance of contrarian intellectuals and policy-makers, eco-skeptic oil and car manufacturers that are the principle social forces retarding the possibility of achieving a smooth transition to a post-carbon economy.
2   Technological 'fixes' are not cost free but involve diversion of labour time, physical material and land from other purposes not involving human welfare to offset a self-induced deterioration (Benton, 1993); some substitutions can have real effects and create new problems.
3   The availability of resources is affected not simply by technical and economic but by political and social factors. It may well be the case that the Middle East has significant reserves of oil. Whether it makes geopolitical sense to be so reliant on such sited resources is another matter.
4   Ecological economists further argue that there is clearly a complex set of enablements/constraints present at an ecological level or the 'ecosystem' or 'eco-regulatory' services that are provided by the planet. It is very difficult to see how such services can be easily/quickly replaced. Moreover, it is argued that a massive increased throughput of materials on the planet and an increased human impact may well be having an effect on these central 'eco-regulatory' features of the planet, on agriculture soils, ground water, biodiversity (including fishstocks, etc), the broader recycling and absorption capacities of the planet (Benton, 1989; Costanza, 1997).
5   It could be additionally noted that the claim by contrarians that technological innovation and substitution under existing capitalist social relations will endlessly allow exponential improvements for all social groups is actually premised on a similar logic to the neo-Malthusian notion

of exponential growth. Eco-Marxists such as Benton have argued that it is not entirely logical to believe that efficiencies will improve endlessly or that innovations will avoid perverse or unwanted side-effects particularly when such processes are dominated by the priorities of capitalist social relations. For example, the well-worn case of the Green Revolution of the 1960s did see major improvements in yields through technologies. Such improvements occurred through a much more intense use of mean energy inputs and brought with them ecological and social problems which impacted on different social groups in very different ways (Sandbach, 1978, p. 500). A world transition from fossil fuel to nuclear power to limit greenhouse gas omissions could generate profound problems of nuclear proliferation.

Panglossian optimism and philosophical idealism

If the key pathology of green discourse is to default into armageddonist arguments, it would seem apparent that the key default of contrarian arguments is frequently to move from technologically optimistic worldviews to super-optimistic Panglossian futurology. Julian Simon was probably the first to set this trend. Not happy with attempting to unravel the problems behind neo-Malthusian analysis, Simon ended up claiming that we will have enough resources to last 10 billion years. Gregg Easterbrook's *A Moment on the Earth* (1995) begins with a cautious nod to the real achievements environmental legislation has achieved in the USA over the last thirty years alongside a critique of the tendency of environmentalists to illegitimately extrapolate trends of decline over the long duration. We are told in a sober fashion: '… there is little chance Western life can be sustained exactly in its current manifestation. Fossil fuels, for one, are sure to become scarce eventually. It may take a decade, it may take a century; it will happen' (Easterbrook, 1995, p. 17). Yet, this is then followed by a narrative which itself ends up extrapolating certain positive trends without any care.[5] Easterbrook goes on to declare the end of world pollution in our lifetime; that the most feared environmental disasters are almost certain to be avoided; and that 'once rational decision making becomes the rule in world affairs, the pace of progress will accelerate.' A similar naive optimism can be found in Lomborg's analysis in *The Skeptical Environmentalist* when speculative technologies such as nuclear fusion are announced as the solution to the energy problem or the waste problem is dismissed because all the world's waste could merely be buried in the Arizona desert in a mile square hole!

It could be observed that an underlying philosophical problem here is that many contrarian discourses find it very difficult to deal with issues of materialism and the recalcitrance of objects and processes. The legitimate desire to reject the environmental determinism of neo-Malthusian thinking frequently ensures that many contrarian currents are pushed towards radically idealist ontologies and forms of super-social constructionism.

Binary argumentation and false dichotomies

Almost all contrarian discourses rely on binary forms of argumentation. The strategic maneuvers of contrarian discourse (ironically mirroring neo-Malthusian discourses) invariably involve counter-posing 'jobs' versus 'nature', 'people' versus 'planet', 'growth' versus the 'environment' and so on. What follows from this is that virtually all contrarian arguments involve simplistic reductions of complex scenarios, as if several differing goals cannot be seen as complementary, or that there are good and complex reasons why diverse ecologies are good things for people to have as well as clean, cheap water.

Lomborg provides many examples of this style of posing binary arguments and false dichotomies. Lomborg for example accepts the existence of anthropogenic global warming yet argues on a cost–benefit basis 'the world as a whole would benefit more from investing in tackling problems of poverty in the developing world and in research and development of renewable energy than in policies focused on climate change' (2001, p. 259), the cost of which would be prohibitively expensive. Beyond the fact that Lomborg's calculation of the cost of addressing climate change is widely disputed by economists (Stern, 2007), more generally though, one could wonder – why do we need to choose between these two priorities? Why not pursue poverty reduction, renewables and appropriate climate change policy based on the best evidence?

Lomborg's response appears to rely on making recourse to fiscal responsibility – there is only enough money to do so many things. Yet, if we are looking around for cuts in the budgets of OECD nations at present that start the process of funding a major technological shift to a solar/hydrogen economy it could be argued why don't we start with military budgets, then we could move on to considering reversing tax cuts for the wealthy and the various other corporate subsidies and kickbacks that corporate America and the nuclear and fossil fuel sector have been receiving for decades. In a world where funds are found at the drop of a hat for 'star wars,' nuclear weapons, and war in Iraq, Lomborg's invocation of fiscal responsibility seem to lack critical depth.

No space and (inhuman) time

Invariably, contrarian discourses find it very difficult to deal consistently with issues of space and time concerning questions of socio-ecological change and transformation. For example, contrarians have quite correctly argued that environmentalists invariably ignore improvement in environmental indicators in the OECD. However, contrarian discourses themselves flatten the spatial variations that underline socio-ecological improvement. Thus, a standard contrarian argument highlights how countries with higher incomes correlate with greater environmental sustainability. However, a growing array of empirical literatures emerging out of the sociology of environmental justice and world systems theory suggest that contrarian currents ignore the extent to which such societies clean-up by displacement across space, time and other media within and between countries, much less the extent to which clean-up in the OECD depends on imported resources for both productive and reproductive purposes (see Bunker, 1996; Frey, 2001). Perverse subsidies and the hazardous waste stream in the global world-system do ensure that contrarians are almost entirely uninterested in the socio-ecological dynamics of core/periphery/ semi-periphery relations within the capitalist world system. Such issues additionally emphasize the manner in which contrarian arguments do not engage with the manner in which market prices do not fully reflect the full social and ecological cost of resources, harvesting and consumption.

On a related issue, it is equally striking how contrarian arguments are often framed around a defense of humanism against the incoherence of ecocentric thought. Yet they frequently deploy timescales to undermine environmentalist arguments which are entirely irrelevant to human well-being. For example, it is regularly argued by contrarians that current concerns with biodiversity loss is irrelevant because at previous points in the Earth's history, the Earth has experienced catastrophic changes. As is well known, in the Permian extinction, 96% of species became extinct, e.g. 99% of all species which have come into being are now extinct. However, as Stephen J. Gould has argued many of these geological arguments are completely fallacious when related to the social–environmental debate. This is because it may be the case that the Earth has been struck by asteroids at different periods and survived yet this does not tell us what is good for us as *Homo sapiens* or the kind of world we may flourish in, prosper and wish to inhabit. It is interesting to note then that extreme contrarians here fold into a type of anti-humanist thinking which is every bit as problematic as ecocentric anti-humanism. In stressing the resilience of planet Earth over the long run, they seem to factor out the matter that over the long run (to paraphrase Keynes) we are all dead.

Markets and the non-contradictory nature of contemporary capitalism

Directly linked to the rather general sunny optimism of many contrarians is a tendency to hold to extraordinarily optimistic accounts of the power of the market and the non-contradictory nature of contemporary capitalism. Even where contrarians are prepared to recognize the existence of substantive environmental problems (e.g. global warming), the shift from conventional capitalism to eco-capitalism is frequently viewed as entirely unproblematic. Gregg Easterbrook tells us that 'western use of petroleum has begun shifting markedly in the direction of conservation. Such changes are partly driven by prices and government policy. But partly they will be seen as organic self adaptation – society reacting just as nature would to self correcting a resource imbalance' (1995, p. 18).

Yet again though there is no sense here that not all sectors of capital share the same interest. It is maintained by Hollander (2003) and Lomborg (2001) that new environmental technologies and renewables energy sources are simply going to come on line and receive extensive global proliferation without any difficulties. It seems striking how such contrarians' arguments seem to ignore how different sections of capital are already demonstrating very different attitudes to the possibility/desirability of producing ecological friendly technologies.

The class and race bias of contrarian thought

Such a line of thinking takes us to questions of class and race. As we have seen there are undoubtedly distinct class and racial overtones that can be found in mainstream environmentalism. What is often missed though is the extent to which many radical contrarian discourses can equally be viewed as marked by a distinct class and race bias. This can be seen in numerous ways. We have seen already that contrarian discourses for example show no interest in the empirical information provided by the environmental justice movement that workers and minority groups in the OECD are disproportionately likely to live in conditions of environmental degradation and they are disinterested in how environmental improvement is not a generalized phenomenon. Yet, one can see

arguments emerging that are every bit as problematic as Paul Ehrlich's being made when contrarians celebrate the toxic waste export industry or export of nuclear repossessing plants to the Third World as really potential for industrial development, or even when they resolutely declare that the developing world will simply have to put up with a dirty industrial transition eventually to achieve the gains of modernity. It is striking here how northern, white and mostly middle class contrarian writers are willing to tolerate social and ecological conditions of existence for people living in the Global South that they would not dream of defending for their own children.

TINA and crisis rhetoric

Finally, it could be observed that many radical contrarian narratives, despite certain superficial surface appearances, are every bit as tied to crisis imagery and a denial of human agency as their neo-Malthusian adversaries. Indeed, it is striking to note for example that whilst neo-Malthusians stress looming natural apocalypse if uninformed southern families and northern consumers do not change their ways, many radical contrarian narratives (e.g. Furedi, 1996, 2005) are addicted to mirror image narratives; yet here, the emphasis is on social and cultural (but not natural) apocalypse if environmentalists and romantic spiritualists are not politically emasculated, and the traditional norms of positivism, hierarchy and a passive public that follows a knowledgable scientific establishment are not put back in their place. Contrarians such as the Institute of Ideas in the UK or the Cato Institute in the USA regularly work with an apocalyptic discourse of environmentalists, leading a more generalized 'culture of fear' that has gripped and threatens to undermine Western civilization, its standards, its scientific institutions and all we hold dear.

It could be observed here that this narrative, with its imagery of an uninformed irrational rabble pressing up against the prepolitical world of science, is every bit as apocalyptic about contemporary social life as that of the Ehrlichs of the 1960s. Moreover, such contrarian 'cultural declinist' narratives, in a similar fashion to neo-Malthusians, are marked by profound pessimism about human agency. The whole notion that *the* natural sciences could profit in any way from being more open to the agency of a reflexive public (through deliberative forums, consensus conferences) or that the priorities and processes of scientific and technological innovation could and should be subject to democratic check is indeed treated as *the* 'demob-cracy bomb' as dangerous and to be feared as the 'population bomb' of previous decades. Finally, it could be noted that the manner in which both currents further frame their whole understandings in terms of 'binary arguments' does ensure that ultimately both neo-Malthusains and radical contrarians, are profoundly committed (yet in very different ways) to the idea that there is no alternative. Neither current can conceive of the idea that there might be radically different ways to organize the social metabolization of nature, science and technology and social affairs. Neither discourse can entertain the possibility that perhaps we could reorganize social affairs so as to ensure we address both the challenge of climate change and clean water, abundance, ecological integrity and social justice.

THE ENVIRONMENTAL DEBATE IN THE 21ST CENTURY: BEYOND CONTRARIANS AND RADICAL ECOLOGISTS?

In this chapter, we have argued that while we can certainly identify a certain articulation of environmental politics that stands in direct conflict with a form of contrarian thought which is largely neo-liberal and Promethean in orientation, we certainly do have a fundamentalist form of contrarianism that is at the center of an anti-environmental industry involved in constant 'greenwash'; the attempt to frame the whole of the environmental debate in this context is far from adequate. Indeed, it might be argued that radical contrarianism and perhaps neo-Malthusian/radical ecology currents have mutually problematized each other and that the real debate now is slowly being transcended. Why might this be the case? We will conclude this chapter by focusing on three developments which may lead to the transcending of the environmental debate being framed primarily in terms of 'contrarians' versus 'radical ecologists.'

Revisionism in the environmental social sciences and the environmental movement

First, within the environmental movement and the environmental social sciences more generally, there would seem to be incipient signs that there is a desire in many quarters to move away from the kind of catastrophe versus cornucopia framing of the environmental debate. Many factors have influenced this. The environmental movement has seen notable changes over the last decade and has been challenged not simply by the external critique of contrarians but internally by the rise of environmental justice movements, civil environmental movements and movements focusing on green cities and sustainable urbanism. The last decade in the environmental social sciences has also

seen a number of intellectual developments such as the rise of environmental history, political ecology, environmental sociology and 'new ecology.'

Such currents have ensured that environmentalism is slowly being forced into a more nuanced engagement with environmental science and one can denote general tendencies away from arguments based on plenary claims (Castree, 2002) towards greater focus on complexity, and greater reflection on the status of environmental science as post-normal science. More attention is being paid to the power dynamics and winners and losers in environmental change and in environmental legislation. This revisionist mood can be overestimated and the continued grip of neo-Malthusian, romantic and catastrophist modes of thinking is still very prevalent in the environmental movement. Yet, even in the work of the Ehrlich's we can now see much greater emphasis being placed on uncertainty as a key feature of the environmental sciences.

Revisionist currents in the contrarian movement

At the same time, it could be observed that there are few academic contrarian who presently support the type of extreme contrarian currents that seek to argue environmental problems can all simply be viewed as lies or symptomatic of a 'culture of fear.' For example, if we turn to Lomborg's *Skeptical Environmentalist*, while the sales pitch surrounding this text claimed that it demonstrated things are getting 'better and better,' in many respects far from bolstering the views of extreme contrarians, in its more sober moments *The Skeptical Environmentalist* actually marked a significant repositioning and moderating of the contrarian case. For example, even if we take Lomborg's empirical data on its own terms it actually demonstrates we presently have some extremely worrying problems and are storing up real trouble for the future.

Thus, on climate change, Lomborg accepts the reality of man-made global warming (2001, p. 259) and believes the dramatic increase of carbon dioxide in the atmosphere is a serious problem. While contesting the scale of deforestation and biodiversity loss claimed by some environmental NGOs, Lomborg's own conservative revisions still concede that tropical deforestation (home to the largest mass of plants and animals on the planet) is running at 0.5% per year (Lomborg, 2001, p. 159) and biodiversity loss is running at a rate 'about 1,500 times higher than the natural background extinction' (2001, p. 235). Air pollution through small particles is estimated to kill over three times the number of people killed in road accidents in the USA. Lomborg notes that Beijing, New Delhi and Mexico City all have estimated particle pollution levels eight times that of the USA. In terms of top soil loss Lomborg estimates that the USA lost 12 tons per hectare in 1974 (p.105). He agrees with UNEP (United Nations Environmental Program) figures that argue over 17% of land is degraded to some extent, and, regarding overfishing, it is conceded that a third of fish are taken from stocks showing decline (p.107). If we turn to the question of long-term solutions to such problems it is interesting to note that in a balanced chapter on energy production Lomborg largely agrees with arguments that environmentalists have been making for the last four decades. The potential for ecotechnologies and renewable energy is recognized to be possibly enormous and that over the long term, a transition from fossil fuels to a solar/hydrogen economy is necessary and desirable.

Who are the defenders of progress? Qualitative development and the green industrial revolution

Further evidence of a debate in transition comes from the manner in which the claim that the environmental debate could be cleaved in a straightforward way between contrarian technological optimists or defenders of 'progress' and green technological pessimists have become increasingly difficult to sustain. The work of Amory Lovins (Lovins, 1977; Hawkins *et al.*, 1999) and the post-scarcity politics of Murray Bookchin have long stood as a significant problem to this notion. However, over the last decade there has been a notable technological turn amongst greens and environmentalists (see White, 2002). For example, whether it is Lovins' work on natural capitalism, which argues for the possibility of win–win environmental solutions or Brian Milani's (2000) advocacy of the possibilities of a post-industrial eco-socialism, a distinct green technological optimism can be seen to be emerging in discourses on industrial ecology, sustainable technological innovation and sustainable architecture. It is interesting over recent times to take one example – it has been green currents in the USA that are now the principle proponents of the notion that the car industry should be forced to embark on a shift towards hybrid or even fuel-cell technologies. In contrast, it is interesting how the default position of contrarians seems to be to defend the internal combustion engine. We might wonder here – who is the defender of progress in this scenario?

NOTES

1 For a range of reviews of The Skeptical Environmentalist see: 'Doomsday postponed' *The*

Economist, September 6, 2001; 'Greener Than You Think,' by Denis Dutton in *The Washington Post*, October 21, 2001; Chris Lavers: 'You've never had it so good,' *The Guardian* September 1, 2001; 'Nine things journalists should know about The Skeptical Environmentalist,' World Resources Institute; 'The Skeptical Environmentalist: A Case Study in the Manufacture of News,' Committee for the Scientific Investigation of Claims of the Paranormal, January 23, 2003; 'Something Is Rotten in the State of Denmark,' *Grist Magazine*, December 12, 2001; 'No need to worry about the future,' by Stuart Pimm and Jeff Harvey in *Nature*, 8 November, 2001. Stephen Schneider, John P. Holdren, John Bongaarts, Thomas Lovejoy: 'Misleading Math about the Earth.' *Scientific American*, January, 2002. For an account of the effect that Lomborg has had on Danish environmental policy, see Jamison (2004).

2 In saying this we do not claim that no debate should be had about environmental issues. We agree they should. However, the way the media deals with questions such as global warming, along with the activities of think tanks and so on, leaves the debate in a state of is it happening or not, when more crucial questions need to be addressed around the implications of climate change.

3 This is not to say that it is not problematic, however. Simon's analysis has its own tendencies to under-analyze history and misunderstand technoscience. Yet equally, he needs to be commended for outlining in *The Ultimate Resource* how population bomb rhetoric was directly feeding eugenicist polices, particularly in the USA.

4 This is of course an argument that has been made by anarchists, socialists and feminists since the 19th century.

5 See: L. Haimson and B. Goodman (eds) (1995) 'A Moment of Truth: Correcting the Scientific Errors' in Gregg Easterbrook's *A Moment of Truth.'* Available at the Earth Defense Fund web site: http://www. environmentaldefense.org

REFERENCES AND BIBLIOGRAPHY

Arnold, R. (1987) *Ecology Wars: Environmentalism as if People Mattered*. Bellevue, WA: The Free Enterprise Press.

Austin, A. (2002) 'Advancing Accumulation and Managing its Discontents: The US Anti Environmental Countermovement,' *Sociological Spectrum* 2: 71–105.

Bailey, R. (1993) *Ecoscam: The False Prophets of Ecological Apocalypse*. New York: St. Martin's Press.

Barnett, H. J. and C. Morse (1963) *Scarcity and Growth: The Economics of Natural Resource*. Baltimore: John Hopkins Press.

Beckermann, W. (1995) *Small is Stupid: Blowing the Whistle on the Greens*. London: Duckworth.

Beder, S. (1999a) 'The Intellectual Sorcery of Think Tanks.' *Arena Magazine* 41, June/July: 30–32.

Beder, S. (1999b) 'Corporate Hijacking of the Greenhouse Debate.' *The Ecologist*, March/April: 119–122.

Benton, T. (1989) 'Marxism and Natural Limits: An Ecological Critique and Reconstruction.' *New Left Review* 178: 51–86.

Benton, T. (1994) 'Biology and Social Theory in the Environmental Debate'. In *Social Theory and the Global Environment*, ed. M. Redclift and T. Benton. London: Routledge, pp.

Benton, T. (1996) *The Greening of Marxism*. New York: Guilford Press.

Biehl, J. and P. Staudenmaier (1995) *Ecofascism: Lessons From The German Experience*. San Francisco: AK Press.

Bird, E. A. R. (1987) 'The Social Construction of Nature: Theoretical Approaches to the History of Environmental Problems.' *Environmental Review* 11(4): 255–264.

Bookchin, M. (1952) 'The problem of Chemicals in Food,' *Contemporary Issues*, 3. no. (12) (Jun–Aug). (Written under pseudonym of Herber, Lewis.)

Bookchin, M. (1962) *Our Synthetic Environment*. New York: Alfred. A. Knofi. (Written under pseudonym of Herber, Lewis.)

Bookchin, M. (1965) *Crisis in Our Cities*. Englewood Cliffs, Prentice Hall (Written under pseudonym of Herber, Lewis.)

Bookchin, M. (1964) Ecology and Revolutionary Thought'. In *Post Scarcity Anarchism*. Berkeley: Ramparts Press, 1971.

Bookchin, M. (1982) *The Ecology of Freedom – The Emergence and Dissolution of Hierarchy*. California: Cheshire Books.

Bookchin, M. (1987) 'Social Ecology versus Deep Ecology: A Challenge for the Ecology Movement.' *Green Perspectives*, No. 4-5: http://dwardmac.pitzer.edu/Anarchist_Archives/bookchin/socecovdeepeco.html

Bookchin, M. (1990) *Remaking Society: Paths to a Green Future*. Boston: South End Press.

Boston, T. (1999) 'Exploring Anti-Environmentalism in the Context of Sustainability.' *The Green Electronic Journal* Issue 11, December: http://egj.lib.uidaho.edu/egj11/boston1.html

Bradford, G. (1989) *How Deep is Deep Ecology?* Ojai, CA: Times Change Press.

Brown, L. (2003) *Outgrowing the Earth*. London: W.W.Norton.

Bruner, M. and M. Oelschlaeger. (1994). 'Rhetoric, Environmentalism, and Environmental Ethics.' *Environmental Ethic* 16(4): 377–396.

Bunker, S. (1996) 'Raw materials and the global economy: oversights and distortions in industrial ecology.' *Society and Natural Resources* 9: 419–442.

Castree, N. (2002) 'False Anti-thesis?: Marxism, Nature and Actor Networks.' *Antipode* 34(1): 111–146.

Chapin, M. (2004) 'A Challenge to Conservationists.' *World Watch Magazine*, Nov/Dec., pp. 17–31.

Cole, H., C. Freeman, M. Jahoda and K. Pavitt (1973) *Models of Doom*. New York: Universe Books.

Commoner, B. (1971) *The Closing Circle: Nature, Man, and Technology*. New York: Knopf.

Costanza, R. (1997) *An Introduction to Ecological Economics*. Boston: CRC Press.

Deal, C. (1993) *The Guide to Anti-environmental Organisations*. Berkeley, CA: Odonian Press.

Dryzek, J. (1997) *The Politics of the Earth*. Oxford: Oxford University Press.

Easterbrook, G. (1995) *A Moment on the Earth: The Coming Age of Environmental Optimism*. New York: Random House.

Echeverria, J. and R. B. Eby (eds) (1994) *Let the People Judge: Wise Use and the Private Property Rights Movement*. Washington, DC: Island Press.

Ehrlich, P. R (1968) *The Population Bomb*. New York: Ballantine Books.

Ehrlich, P. and Ehrlich A. (1988) *Betrayal of Science and Reason: How Anti-Environmental Rhetoric Threatens Our Future*. Boston: Island Press.

Elliot, R. (1997) *Faking Nature: The Ethics of Environmental Restoration*. London: Routledge.

Enzenberger, H. M. (1974) A Critique of Political Ecology. *New Left Review* 1/84, March–April.

Foster, J. B. (1999) 'Marx's Theory of Metabolic Rift: Classical Foundations for Environmental Sociology.' *American Journal of Sociology* 105(2):366–405.

Frey, R. S. (2001)'The hazardous waste system in the world system'. *The Environment and Society Reader*. Boston: Allyn and Bacon.

Furedi, F. (1996) *Culture of Fear*. London: Continuum.

Furedi, F. (2005) *Politics of Fear*. London: Continuum.

Gorz, A. (1975) *Ecology as Politics*. London: Pluto Press.

Gottlieb, R. (1993) *Forcing the Spring: The Transformation of the American Environmental Movement*. Washington, DC: Island Press.

Gould, S. (1991) *Bully for Brontosaurus*. London: W.W. Norton.

Grumbine, E. (1994) "Wildness, Wise Use, and Sustainable Development." *Environmental Ethics* 16(3):227–249.

Hager, M. (1993). 'Enter the Contrarians.' *Tomorrow* 3(4):

Haimson, L. and B. Goodman (eds) (1996) *A Moment of Truth. Correcting Scientific Errors in Gregg Easterbrook's A Moment on the Earth*. Environmental Defense Fund. www.environmentaldefense.org/documents/2246_AMomenofTruth_PartOne.pdf.

Hajer, M. A. (1995) *The Politics of Environmental Discourse: Ecological Modernization and the Policy Process*. Oxford: Clarendon.

Haraway, D. J. (1985) "A Manifesto for Cyborgs: Science, Technology, and Socialist Feminism in the 1980s." *Socialist Review* 80:65–107.

Haraway, D. J. (1988) "Situated Knowledges: The Science Question in Feminism and the Privilege of Partial Perspective." *Feminist Studies* 14(3):575–599.

Haraway, D. J. (1992) "The Promises of Monsters: A Regenerative Politics for Inappropriated Others." In *Cultural Studies*, ed. L. Grossberg, C. Nelson and P. A. Treichler. New York: Routledge, pp. 295–337.

Haraway, D. J. (1995) 'Universal Donors in a Vampire Culture, It's All in the Family: Biological Kinship Categories in the Twentieth-Century United States.' In *Uncommon Ground: Toward Reinventing Nature*, ed. W. Cronon. New York: W.W. Norton, pp. 321–366.

Harvey, D. (1974) "Population, Resources and the Ideology of Science." *Economic Geography* 50:256–277.

Harvey, D. (1996) *Justice, Nature, and the Geography of Difference*. Cambridge, MA: Blackwell.

Hawkins, P. A. Lovins and H. Lovins (1999) *Natural Capitalism*. London: Earthscan.

Hays, S. P. (1959) *Conservation and the Gospel of Efficiency: The Progressive Conservation Movement, 1890–1920*. Cambridge, MA: Harvard University Press.

Helvarg, D. (1994) *The War Against the Greens*. San Franscisco: Sierra Club Books.

Hollander, J. (2003) *The Real Environmental Crisis–Why Poverty, not Affluence, is the Environment's Number One Enemy*. California: California University Press.

Jamison, A. (2004) "Learning from Lomborg, or Where Do Anti-environmentalists Come From?" *Science as Culture* 13(2):173–195.

Kaplinsky, J. (2002) "Slick arguments." Spiked Online: http://www.spiked-online.com/Articles/00000006DB4E.htm

Katz, C. (1998) "Whose Nature, Whose Culture? Private Productions of Space and the 'Preservation of Nature'." In *Remaking Reality*, ed. B. Braun and N. Castree. London: Routledge, pp. 46–63.

Lappé, F. M. and J. Collins (1977) *Food First: Beyond the Myth of Scarcity*. Boston: Houghton Mifflin.

Lappé, F. M. and R. Schurman (1990) *Taking Population Seriously*. San Francisco: Institute for Food and Development Policy.

Lappé, F. M., J. Collins and P. Rossett (1998) *World Hunger: Twelve Myths*. Berkeley, CA: Food First.

Latour, B. (1993) *We Have Never Been Modem*. Cambridge, MA: Harvard University Press.

Latour, B. (2002) *War of the Worlds: What about Peace?* Chicago: Prickly Paradigm Press.

Latour, B. (2004) *Politics of Nature: How to Bring the Sciences into Democracy*. Cambridge, MA: Harvard University Press.

Leiss, W. (1972) *The Domination of Nature*. Boston, MA: Littlefield.

Lewis, T. (1995) "Cloaked in a Wise Disguise." In *Let the People Judge: Wise Use and the Property Rights Movement*, ed. J. Echeverria and R. Booth-Eby. Washington, DC: Island Press, pp. 13–20.

Lomborg, B. (2001) *The Skeptical Environmentalist*. Cambridge: Cambridge University Press.

Lovins, A. B. (1977) *Soft Energy Paths: Toward a Durable Peace*. London: Penguin Books.

Luke, T. W. (1997) *Ecocritique: Contesting the Politics of Nature, Economy, and Culture*. Minneapolis: University of Minnesota Press.

Luke, T. W. (1999) *Capitalism, Democracy and Ecology: Departing from Marx*. Urbana: University of Illionois Press.

Mayer, M. and J. Ely (eds) (1998) *The German Greens: Paradox Between Movement and Party*. Trans. M. Schatzschneider. Philadelphia: Temple University Press.

McCarthy, J. (1998) "Environmentalism, Wise Use and the Nature of Accumulation in the Rural West." In *Remaking Reality*, ed. B. Braun and N. Castree. London: Routledge, pp. 126–149.

McCarthy, J. (2002) "First World Political Ecology: Lessons from the Wise Use Movement." *Environment and Planning A* 34(7): 1281–1302.

McCright, A. and R. Dunlap (2000a) "Defeating Kyoto: The Conservative Movements Impact on US Climate Change Policy." *Social Problems* 50(3):348–373.

McCright, A. and R. Dunlap (2000b) "Challenging Global Warming as a Social Problem: An Analysis of the Conservative Movements Counter-Claims." *Social Problems* 47:499–522.

Meadows, D.H. *et al.* (1972) *The Limits to Growth*. New York: University Books.

Milani, B. (2000) *Designing the Green Economy*. Md: Rowman and Littlefield.

Mol, A. P. J. (1995) *The Refinement of Production: Ecological Modernization Theory and the Chemical Industry*. Utrecht: Van Arkel.

Mol, A. P. J. (2003) *Globalization and Environmental Reform: The Ecological Modernization of The Global Economy*. Boston, MA: Institute of Technology Press.

Monbiot, G. (2006) *Heat: How to Stop the Planet Burning*. London: Allen Lane.

O'Connor, J. (1998) *Natural Causes: Essays in Ecological Marxism*. New York: Guilford Press.

Ordway, S. (1956) "Possible Limits of Raw Material Consumption." In *Man's Role in Changing the Face of the Earth*, ed. W. L. Thomas. Chicago: Chicago University Press, pp. 987–999.

Osborn, F. (1954) *The Limits of the Earth*. London: Faber & Faber.

Ray, D. L. (1993) *Environmental Overkill: Whatever Happened to Common Sense?* Washington, DC: Regnery Gateway.

Ross, E. (1998) *The Malthus Factor: Poverty, Politics and Population in Capitalist Development*. London: Zed Books.

Salleh, A. (1984) "Deeper than Deep Ecology: The Eco-Feminist Connection." *Environmental Ethics* 6:339–345.

Salleh, A. (1993) "Class, Race, and Gender Discourse in the Ecofeminism/Deep Ecology Debate." *Environmental Ethics* 15:225–244.

Sandbach, F. (1978) "The Rise and Fall of the Limits to Growth Debate." *Social Studies of Science* 8:495–520.

Schnaiberg, A. (1980) *The Environment: From Surplus to Scarcity*. Oxford: Oxford University Press.

Simon, J. (1981) *The Ultimate Resource*. Princeton: Princeton University Press.

Stern, N. (2007) The Economics of Climate Change – The Stern Review. Cambridge: Cambridge University Press.

Tokar, B. (1995). "The 'Wise Use' Backlash: Responding to Militant Anti-Environmentalism." *The Ecologist* 25(4): 150–157.

Vogt William (1948) *The Road To Survival.* New York: William Sloane Associates.

White, D. F. (2002) 'A Green Industrial Revolution? Sustainable Techonological Innovation in a Global Age.' *Environmental Politics* 11(2):1–26.

York, R., E. A. Rosa and T. Dietz (2003a) "Footprints on the Earth." *American Sociological Review* 68:279–300.

York, R., E. A. Rosa and T. Dietz (2003b) 'STIRPAT, IPAT and ImPACT.' *Ecological Economics* 46:351–365.

Valuing the Environment

9

Fundamental Economic Questions for Choosing Environmental Management Instruments

Thomas D. Crocker

INTRODUCTION: WHY MANAGE COLLECTIVELY?

The human consequences of natural events were considered beyond control until about three centuries ago. People's behaviors were thought to leave all but their immediate environs undisturbed. People lived in subsistence, self-contained communities granting them few options but to adapt to the whims of local natures. Many people were ill-fated and had to endure famine, flood, and pestilence. They did little about what nature did to them. The intellectual turmoil of the 18th century offered a more expansive, less fatalistic vision. As was always acknowledged, nature could affect the social order for good or for ill. The contrasting Enlightenment notion that the social order could affect nature for good or for ill took about two centuries to spread over the intellectual landscape. Nature and the social order are now frequently viewed as jointly determined (e.g. Daly, 1968; Crocker and Tschirhart, 1992; Norgaard, 1994), implying that people can manipulate nature as an asset to do less ill and more good for them.[1]

Though managing nature requires the expenditure of time and effort which could be devoted to other purposes, the expenditure can pay off. Some of nature must be left to support life: people must have air to breathe and water to drink. Nature when disturbed provides the raw materials for finished material goods: wood and stone is used to build houses and plants and animals feed us. Pristine nature promotes peace of mind and health of being: natural scenes often meet human criteria for beauty and they offer venues for exercising mind and body. But differences among people about the relative values of these natural services cause social tensions. One person's sense of beauty or his favorite site for renewal of mind and body may be hurt by another person's desire to build a house or to put meat on his table.

Tensions among people about alternative uses of nature are inevitable because of scarcity, the curse of Adam and Eve. The services of nature are scarce because the finite asset cannot provide everything everybody wants from it. The services must therefore be shared. In addition, some kind of management is called for or the intellectually and physically powerful monopolize nature's services. The worst things in life are often free. When access to and use of scarce natural services is free, predatory behaviors and uncompensated burdens dominate social interactions. But, for reasons to be reviewed below, management systems for nature too often do not self-organize (Coase, 1960) in the sense that individuals will not voluntarily choose to fulfill their duties toward nature or to coordinate their chosen actions in a fashion maximizing their collective well-being. This chapter describes some of the fundamental decisions that must be made when choosing among alternative environmental schemes to fulfill these duties or to overcome these coordination failures. With few exceptions, no attempt is made here to answer these questions. Raising the questions is nevertheless worthwhile because they frequently get answered only by default. These haphazard answers influence the manner in which environmental management problems are structured, thus the solutions derived, the most efficacious ways to implement these solutions, and the natural states and the social orders realized. The chapter presumes it is usually difficult to tell whether a single environmental outcome is due to a management answer or to an exogenous random event. But the likelihood that the process by which the answer was given would lead to the outcome can be ascertained. The structure of this process is the focus of this chapter.

A DUTY OR A CHOICE?

Whether done explicitly or implicitly, environmental management must decide between two quite distinctive motivating visions, the deontological and the individualistic. Each vision highlights different management functions and creates a different dialogue between ecosystem states and human values. The deontological advocates people's duty to preserve nature. One variant, often characterized as 'deep ecology' (e.g., Leopold, 1949; Bateson, 1991; Taylor, 1986), holds that nature or creation has intrinsic moral worth independent of what it does for human well-being.[2] A humanist deontology variant (e.g. Anderson, 1993) brooks no tradeoffs of the 'unity of nature' with broader human purposes because this unity is necessary for the 'common good' and 'civic virtue.' With both variants, management's function is to promote and to assure this imperative (Kelman, 1981).

That paragon of tradeoff institutions, the market and the economic analysis built upon its consequences, is thought to corrode the social order's moral foundations, including this imperative (Marcuse, 1964; Hirsh, 1976; Bowles, 1998). Preference satisfaction via manipulation of nature is seen as moral loss. Some actions are simply not 'right' whatever the consequences for those non-environmental dimensions of interest to human beings. The duties represent constraints binding human choices. No tradeoffs are allowed. Management's proper purpose is to satisfy the constraint by controlling access to nature and by inspiring morally appropriate preferences about its use. This requires making people accountable for any of their choices which cause the constraint to be violated – which violate their duties.

In a real sense the deontological vision is an economics vision. As with any economic problem, it has an objective – the common good, however defined – and it includes a constraint upon choice which, in this case, is to be imposed on ethical grounds by some anointed authority. It is with this ethical posture that it differs from the individualism of standard economic analysis.

Standard analysis adopts individual sovereignty as an ethical posture. The purpose of managing nature is to maximize collective well-being taken as a sum of individuals' well-beings under personal wealth, and biological, technological, and institutional constraints. The individual's behaviors trace what does and does not contribute to his well-being. This well-being can involve a wide variety of preferences, including altruism, community concerns, goodwill, and other collectively oriented motivations involving responsibilities and obligations to others. It can also involve the choice to grant individual discretion in particular problems to those more knowledgeable about them. But where the deontological vision sees duty, the standard approach sees a tradeoff. The ethical posture of the standard approach forces it to do so, given that individuals are sometimes willing to sacrifice nature's unity for mundane commodities such as ease of travel via automobile. The proper purpose of managing nature according to the standard approach is then to maximize collective well-being in accordance with an individualistic stance. This requires making people accountable for any choices they make involving nature which reduce collective well-being. Since collective well-being is a sum of individuals' well-beings, it follows that a person must treat those of his actions which might affect others as if they were to affect himself.

The difference in objectives between deontological and individualistic ethical postures obviously leads to distinct, possibly incompatible solutions about states of nature and the social order. Also, the attitudes the two postures express about implementation strategies differ considerably because of discrepancies in beliefs about whether most individuals are rational in the sense of really being able to perceive and to act on their real interests. If people do not know their real interests, it is a short step from telling people what to do toward nature to telling them how to do it (Russell and Powell, 1999).

ARE PEOPLE RATIONAL?

Not only do adherents of the deontological stance fret about some of the uses to which people put nature; they are also skeptical about whether people can do anything to correct these uses even if they want to. They question whether people are substantively rational and whether they are procedurally rational. The deontologists advance two not mutually exclusive arguments involving human cognitive, communicative, and computational (CCC) limitations for an anointed authority telling people what to do and how to do it. People do not know enough to further their real interests. The first argument focuses on nature alone. Norton (1994) and Vatn and Bromley (1994) are representative examples. With respect to objectives, people are said to be myopic in that they fail to account adequately for the intergenerational and the sustainability consequences of their choices. As for how to meet duties toward nature, environmental risks are thought to be too subtle for untrained people to detect and to heed. Misperceptions are made even more costly by the fact that realizations of these risks can be irreversible. Untrained individuals simply cannot know enough about ecological systems to allow them to decide by their chosen behaviors the fates of their peers and future generations. Experts and those authorities charged with the continuity of society must define and assure appropriate ends for nature and for their implementation.

A more expansive version of CCC limitations questions individuals' rationality in everyday domains, natural and otherwise. In the stripped-down elemental form of rationality that outside commentators nearly always presume to be synonymous with economic analysis, the individual is said to have likes and dislikes with completeness, continuity, and transitivity properties formed as if by immaculate conception. A person's preferences then have a structure with which his reason can work (Sugden, 1991). His means of support and the technologies and the institutions he can access constitute his resources. They are external constraints which bind his choices. He suffers no internal constraints since he possesses logical omniscience – if he knows something, he also knows all its logical implications. He also has negative introspection – if he does not know something, he knows he does not know it. His desires and his actions are always consistent and these actions always work as he wants. In effect, he is hermetically sealed from having to learn, be flexible, or care about the sources of his personal and his institutional legacies. They do not matter to him because they do not influence what he becomes. He is already whole, a general-purpose logic device. This preference processor accurately formulates and then solves optimization problems in the blink of an eye. As he interacts with his similarly rational peers, and this rationality is common knowledge, the consequence is an instantaneous, unique equilibrium characterized by a set of mutually consistent prices, identical for all agents, which define agents' opportunity sets and which exhaust all gains from trade (Arrow and Debreu, 1954) – an economically efficient outcome. The prices alone are sufficient to convey all relevant information about relative scarcities and to provide incentives to cause agents to allocate resources to whom, when, and where they are most highly valued in terms of individuals' preference satisfaction. Clearly, the commentators say, people do not behave like this. It then follows, they say, that skepticism about identifying individuals' versions of personal well-being with collective well-being is justified.

But individual rationality has been shown as unnecessary for collective rationality, where the latter refers to maximization of a sum of individuals' well-being as they themselves perceive well-being. No tasty fruit is left hanging even if people are less than fully rational. The individual's CCC limitations can be recognized by turning the elemental rationality story on its head. Rather than starting from the individual and working up to an exchange institution such as the market, the alternative initially invokes exchange processes and works toward rationality of the individual. Markets and other exchange institutions help individuals construct rationality as opposed to requiring agents to find it on their own. Rationality is an institutional, not individual phenomenon.

By turning the elemental rationality story on its head, individual rationality becomes a proposition induced by postulates of filtering and imitation of rational behavior rather than a set of axioms (Hirsheifer, 1978). Collective rationality asks how rationality is diffused through individuals in a population via accidental or purposeful adaptation to internal and external constraints. More fundamentally, in a Darwinian fashion, collective rationality asks which intrinsic attitudes and behavioral traits of individuals favor imitation and selection for rational behaviors. Thus, the tautology that everything individuals do is rational because they do it is dismissed. An individual's purposefulness and engagement in active learning happen, but he makes technical and allocation blunders because his stand-alone cognition and knowledge are inadequate to match his preferences to his choices. Exchange institutions make a person an entrepreneur of the self better able to align his preferences and his choices. They do so by reducing demands the elemental rationality axiomization makes on a person's CCC skills. Repetition and social interaction pound rationality into one's head.

The prices, whether pecuniary or material, an exchange institution generates are far less complicated for a person to deal with than having to recognize and to figure out, in the absence of prices, all of the natural, psychological, and institutional factors determining them (Hayek, 1945). It thus does not directly follow that the more complicated and removed from his everyday experiences is a person's environment the more an anointed authority's decisions must substitute for that of the person. A person's access to and knowledge of an institutionalized exchange process may compensate for his limited CCC skills regarding the natural world. Numerous controlled experiments in economics fail to reject this proposition.[3] Also, a few recently reported controlled experiments (e.g. Cherry *et al.*, 2003) involving people's interactions with nature reject that justification for imposing the deontological will which holds that people's preferences or objectives mutate with changes in individual circumstances and institutional contexts. Instead, people appear to become more rational because exchange institutions relax their internal constraints rather than mutating their preferences. All these results suggest the deontological vision must seek defense somewhere other than the CCC limitations of individual human beings. The individual's possibly limited elemental rationality need not lead to collective irrationality when that individual has knowledge of and access to exchange institutions.

Recognition that exchange institutions can guide people to their real interests opens an array of management possibilities much broader than universal imposition by anointed authority of the deontological will. At least in North America, the early stages of environmental management acted as if they were exercising this will almost exclusively. Users of nature were told not only what to do but how to do it. Times have changed somewhat though what to do mostly remains left to authority – throughout the world anointed authorities have recently become substantially more friendly toward and sophisticated about environmental management schemes which encourage people to discover and to choose how to attain collective rationality rather than supposing the authority can both root it out for them and directly guide them to it.[4] Mandating the how as well as the what ultimately becomes too expensive, even putting any mandated what beyond reach. These schemes, which focus on how to achieve a mandated goal for nature, nest within one or another configuration of individuals' property rights in nature and its services.

WHICH PROPERTY RIGHT CONFIGURATION?

Environmental management is a search for collective procedural rationality, whether the objective is defined in terms of duties toward or tradeoffs of natural assets. The search involves two questions: (1) how do the CCC burdens for the authority and for natural asset users vary with alternative property right configurations, management instruments, and environmental problems; and (2) how are burdens to be distributed between a central authority and individual users of nature? Or alternatively

stated, what degree of self-organization and authoritatively directed organization for a natural asset will approach collective procedural rationality, given that CCC burdens are always present?

Neither economic theory nor controlled economic experiments suggest a vacuum can produce a self-organization which achieves collective rationality. Even when the conditions to achieve this with natural assets are right, self-organization requires that a legitimized authority have granted individuals rights to, claims over, or permits to be included in uses and to exclude others from uses of the assets. This privatization must include all those dimensions of the asset anyone values. Otherwise, people will treat scarce but unincluded dimensions as if they are free. If the conditions are right and an authority creates and enforces private rights, theory suggests and experiments do not deny that the relative market prices induced by the production, exchange, and consumption of the assets to which the rights are attached will ensure that production involves only those who can generate gains for the collective as well as for themselves, and that consumption is undertaken only by those who most highly value the asset.

Market prices alone will generate all information necessary and sufficient for the coordination of everyone's production, exchange, and consumption plans; and provide incentives for everyone to behave in a manner consistent with collective rationality, whether substantive or procedural. Different people will then value an additional unit of the asset identically and they will treat each other's burdens as their own. An authority need do no more than create, enforce, and be an arbiter of property rights that apply to everyone. But this collective rationality is conditioned by who owns what as determined by the initial distribution of rights to the natural asset and to other assets. Just as do market prices, the initial distribution of rights conveys information and affects incentives about production, exchange, and consumption possibilities. It creates expectations about who will get what from collective rationality. That is, what constitutes collective rationality is conditional upon the structure and, with nonzero CCC costs, the assignment of entitlements to the natural asset. The outcome realized for nature is the result of both price and the organization or means of coordination in production, exchange, and consumption institutions. Prices determine who produces and gets the services of nature – those producers willing to bear the costs and those buyers willing to pay the price. The organization specifies approved contractual relationships between buyer and sellers of natural asset services. An environmental management institution is thus a compound of price rules and approved contractual relations which determine the information and the incentive people have to achieve collective rationality.

Via risk-pooling, the legitimized contractual relations organizations embody may reduce uncertainties about the consequences of random natural events. They also introduce regularity and coherence and thus make reliable predictors of the behaviors of all claimants to an asset. By defining the activities claimants may or may not undertake with respect to an asset, contractual relations specify how claimants must behave with respect to each other through the asset. They specify the rules of the game. Environmental management issues arise when claimants expect contractual relations or the lack of such to misdirect incentives, fail to produce information about the behaviors of the natural asset and of fellow claimants, or produce misleading information or information unevenly distributed among claimants. Expectations like these open the door to predatory behaviors by some individuals and uncompensated burdens for others – externalities in 'econo-speak.'

Natural assets are especially prone to those misdirected incentives and informational shortfalls and asymmetries which inhibit individuals from self-organizing their management so as to achieve collective rationality. The inhibition has four sources. The presence of any one of the four raises the distinct possibility though not the guarantee that a greater degree of centralized collective direction of what to do rather than self-organization via privatization will enhance collective rationality. First, and most obvious, rights for the asset in question may not be defined or are illegitimate in the eyes of those to whom they are meant to apply. Second, the rights may not be defensible such that anyone who holds them cannot at low effort exclude whomever he wishes. Third, assets and the rights attached to them may not be rival such that no two persons can use an asset as cheaply as one can use it. Finally, the asset rights may not be transparent implying that interested parties cannot predict what use, exclusion conditions, and alienation mean. Included in this transparency condition are low CCC costs which ensure that any attempt by an asset right holder to shift the collective's gains from his use of the asset to himself will be successfully challenged by someone willing to take less of the gains.

Consider, for example, the incentive and informational obstacles to self-organization in species, habitat, and ecosystem protection. Landowners may have clear title to commercial uses of the units of an organism present on their sites but title to the genetic reservoir the units represent may be muddy or nonexistent. Even if title to the genetics is not muddy, the organism units and things that can harm them may be mobile, making it costly to defend one's claims to the genetic reservoir. An organism present on one owner's parcel may damage the noncommercial but socially valued, nonprivatized dimensions of habitats on another owner's parcel. Species population dynamics and ecosystem behaviors are incompletely understood, making the habitat consequences of the use of private claims less than transparent. In short, thorough privatization of what to do with natural assets will often not be collectively rational. But substantial room for privatization may well exist for the question of how to achieve this rationality.

WHICH CONTROL INSTRUMENTS TO USE?

Property rights to a natural asset define the limits within which asset users may operate without fear of retribution while permitting them to exercise greater control over future events. The rights are acceptability conditions, which winnow the range of possible outcomes. In effect, the obligations to others the property conditions embody specify the universal features a satisfactory outcome will or will not have. Subsets of individuals and authorities can establish any kind of specialized obligations within the framework of these universal obligations. The universal property rights of all farmers in a given area may, for example, allow any one of them to grow whatever crop he desires. But the authority could have instruments nested in the set of universal rights that would induce farmers to grow less polluting crops voluntarily. Included would be positive (charges) and negative (subsidies) pricing of particular inputs, quantity rationing of inputs, making markets by specifying liability standards and tradeable property rights in dimensions of the natural asset for which none before existed, supplying information to the farmers about the consequences for them and for others of particular production choices, and technical assistance. Numerous variants on these general categories exist. A case in point would be a lump sum subsidy for using at least a given amount of a less polluting input or a subsidy for each unit used of this input.

If the universal property rights are not unduly restrictive for the authority and if the authority is omniscient and clairvoyant such that it is fully informed about the actual and the contingent behaviors of the natural asset and its users, its choice of which instrument to use borders on the irrelevant. Whatever the fully deliberate and aware authority's objectives and its criterion for judging performance, it can usually get there from wherever here is by using any one of or combination of many devices to steer people toward its objective. For example, an *ex ante* and *ex post* deliberate and aware authority for whom the assurance of user compliance is effortless can either specify a pollution quantity for each user which achieves collective rationality or a pollution price that will induce each user to produce that quantity (Pezzey, 1992). The environmental outcome will be the same whatever the instrument used to attain the outcome. The hang-ups appear when the control authority is not fully deliberate about and aware of what is and what will be happening.

The authority and users of the asset then become strategically interdependent. In particular, the authority has to worry about how to organize the monitoring of user behaviors and natural asset states and about encouraging user compliance with control instrument terms. Monitoring requirements and the ease of assuring compliance can differ drastically among instruments and across environmental problems and their settings. The process by which the outcome is sought becomes a prominent question. Learning about behaviors and assuring compliance is not effortless. Though it risks banality to say that a less than fully informed authority must weigh the monitoring and compliance demands an instrument will make upon it, it is only in the last decade that the issue has attracted other than anecdotal attention. An authority is no more likely to be elementally rational than are the individual natural asset users whose behaviors it hopes to control. And the evidence is surely ambiguous that political processes are sufficient to make the control authority an entrepreneur of the self who constructs rationality (e.g. Helland, 1998).[5] More likely, the control authority who will not delegate decisions about how to reach its environmental objective must find procedural rationality more or less on its own.

If the control authority is less-than-fully rational procedurally, the natural asset users whose behaviors are the object of the authority's concern participate in a lottery (Crocker and Shogren, 2001). From the user's perspective for any instrument the authority employs, there are four states worthy of note involving combinations of the user's decision to comply or not to comply with the terms of the instrument and whether or not his compliance behavior is observed by the authority. In many settings the user can influence the odds his behaviors will be observed and thus the likelihood of any state being realized by his choice of output types and of production technology (Crocker, 1984). In making his compliance decision he will weigh the cost of compliance in terms of, say, foregone saleable output against its benefits in terms of penalties avoided. As for the control authority, its systematic attention to users' behavior is a scarce and thus costly resource which it must somehow ration.

It must make three decisions: the control instrument to employ and, conditional on that decision, the degree of effort to devote to monitoring and to compliance and to the allocation of this effort across individual natural asset users. The authority's ex ante choices for these three decisions will influence the ex post rewards users expect from compliance. Consequently, the decisions of the authority and the asset users become strategically interdependent. The cause is the authority's limited procedural rationality. The authority must condition its decisions on what it wants users to do and how it wants them to do it on its guesses about user responses to these decisions. And users will condition their response decisions on their guesses about the authority's decisions. User incentives may now be misdirected toward fooling an authority who will be faced with the additional CCC burden of understanding this misdirection.

The authority's CCC burden becomes especially onerous when it must try to control a problem like nonpoint source pollution, where there is great ambiguity about who did what to the natural asset and substantial spatial variation in the performance of alternative control instruments. On the other hand, a point source of generic dust emissions in an otherwise pristine homogeneous area has no other pollutant with which to mix and its effects on, say, local visibility are readily observed, implying the control authority would have little difficulty in assuring compliance with whatever instrument it chooses to employ.

In the final analysis, there is some point at which the control authority's endeavors to achieve procedural rationality become more trouble than they are worth to it. Its CCC costs exceed the collective benefits of environmental management. Given the property rights configuration and thus the set of control instruments not forbidden to the authority, the lie of this point will vary across control instruments. As in a kind of infinite regress, ascertaining this variation would seem to require detailed data involving the natural asset, its users, and their setting as well as specific explanations regarding those user and asset behaviors and interactions which generate the data. However, the authority might be able to stress its internal constraints less and moderate the strategic interdependence issue by adopting a more general level of analysis. The rationality that exchange institutions induce in the behaviors of users provides a link between the highly specific and the more general. In particular, the link implies general operational guidelines might be developed for a variety of specific control problems.

The guidelines would suggest where and when individual asset users might be trusted to discover and to choose how to achieve procedural rationality. The next section suggests the search for guidelines about this trust issue can even be extended to biodiversity conservation on private lands, an environmental problem notorious for its data and explanatory ambiguities regarding asset and user behaviors and interactions (Innes et al., 1998;

Boyd *et al.*, 1999). The section does so by sketching the operational features and the procedural rationality properties of an instrument which can be very effective at relieving the authority of some of its weightiest CCC burdens. In settings where these burdens are much less extreme, consideration of how asset users might be trusted to find the way to collective rationality becomes less relevant.

CAN ASSET OWNERS BE TRUSTED?

Consideration of so-called economic incentive instruments such as effluent charges, tradeable permits, and liability standards is now *de rigueur* among policymakers. This is evident in the wide variety of environmental problems to which tradeable permit instruments have been applied since the late 1980s. These instruments shift from the control authority to natural asset users many of the most onerous CCC burdens of discovering how to reach an environmental objective in a collectively rational fashion. At the same time, by directing users' incentives toward the environmental objective they can reduce these burdens and increase attainable collective rationality.

The procedures and consequences of the USA's Endangered Species Act (ESA) as applied to private landowners present a vivid contrast to those of a tradeable permit instrument. Because the Act restricts how the owners of land parcels the ESA authority selects can use their land, the owners cannot trade in the restricted land uses. Owner time horizons are thus shortened, owner incentives for investment in the restricted parcels are reduced, and owner ability to collateralize tenure is squeezed. Owners for whom the restrictions are especially harsh are then unable to soften their private losses by reassigning these duties to other owners better able and more willing to bear them. Tradeable permits regarding uses such as development moderate these undesirable owner behaviors by unbundling the claims to land so that owners can trade separate elements of the bundle. Given an authority-defined conservation objective (what to do), owners for whom the cost of contributing to the conservation goal is excessive (how to do it) can shed their costs to those for whom the cost of contributing to the objective is low. Given there usually exists multiple land-use configurations capable of meeting the conservation goal (Ando, 1998), the goal can be met at lower cost than when the authority defines exactly which owners are to contribute and how they are to do it.

A tradeable permit instrument intended to protect biodiversity habitat requires a clear, unambiguous definition of the habitat units that can be traded and a definition of the baseline distribution of habitat units. No one engages in trade unless they think they know what they have and what they are getting

and giving up. The unit to be traded must be transparent and thus accessible to those not expert in ecological systems. It must also be systematic and susceptible to routine calculation. This keeps owner CCC costs low. Any owner is therefore readily informed as to what any habitat configuration means for him. In contrast, the bulk of authority dealings with private landowners under the auspices of the ESA are one-on-one consultations and negotiations, making it difficult for any one owner to compare his treatment by the authority to the treatment his neighbor receives.

The lack of clarity for other owners of the criteria used for and the results of the authority's dealings with any one owner opens the door to favored treatment of especially influential owners. This lack of clarity can also hinder conservation objectives. Other owners can less readily ascertain when another owner is adhering to a conservation agreement if the terms of the agreement are unique to the individual owner. Owners' observations of the behaviors of their fellows can provide a valuable service to a control authority with limited monitoring and enforcement capabilities. The tradeable permit instrument with clearly defined habitat units enhances the ability of owners to recognize the meaning for the authority's conservation objective of what they see other owners doing. If by doing less one owner makes other owners do more, the other owners have an incentive as well to recognize and report this meaning.

Most of the authority's costs in making a market for permits will be up front and fixed, whereas the costs of one-on-one negotiations appear variable both to the authority and to owners. These one-on-one costs increase with the number and the complexity of the negotiations undertaken. The authority's fixed costs for defining a tradeable habit unit and for making and maintaining a market in habitat units likely decline with the volume of permit transactions and with the authority's accumulated experience in making exchange institutions. Moreover, a precisely defined, commonly understood framework for a tradeable permit instrument reduces owner risks, therefore increasing the number of market participants and increasing the scope of the market. This scope can also be broadened by allowing non-owners of land to purchase permits for use they might then set aside to support the authority's conservation objective. The more market participants the better because additional participants offer everyone more trading opportunities, thus causing market prices to reflect owner opportunity costs better (Economides and Siow, 1988). This contrasts with instruments which tell unlucky selected owners how to achieve a conservation objective. In these circumstances, outsiders who desire more habitat protection find it less costly or even free to have the authority carry forward their desires at owner expense.

In sum, well-designed exchange institutions cause asset users to have lower CCC costs for discovering how to reach an objective than does a control authority. It makes sense to assign these burdens to those for whom the burdens are least oppressive. Tradeable permits encourage this assignment. They make user strategies of cooperation with rather than opposition to the authority more rewarding for the user and for the authority. However, the plausible gains in collective procedural rationality caused by a shift and consequent reduction of CCC burdens from the control authority to asset users could be offset by exchange institution imperfections. Stavins (2003) reviews a member of such imperfections including monopolistic advantages in permit or in saleable output markets, nonprofit maximizing behaviors such as staff maximization in public bureaucracies, and the creation of environmental hot spots where asset users spatially or temporally concentrate their activities and environmental damage functions are nonlinear.

In the four decades since Crocker (1966) and Dales (1968) advanced the idea of tradeable environmental permits, two versions have dominated analyses and policy applications. The two are 'cap and trade' and 'credit trading.' Without exception they have been discussed separately and applied separately. With cap and trade the authority specifies an ambient upper limit on environmental loadings (e.g. in air or water) or extraction (e.g. fish), distributes permits in number consistent with the limit, and then passively allows or actively encourages user trading in permits. Rather than setting an ambient limit for the aggregate of users, credit trading sets an upper limit of contributions for each user. Users who do not reach this upper limit receive credits for the amount by which they fall short. These credits can be traded to other users for whom the cost of meeting their personal limits is greater than the market price of the credits.

Neither the cap and trade or the credit trade versions of tradeable permits alone justify trusting asset users to discover and to choose how to achieve full collective rationality. Whatever the reduction in CCC burdens and enhancements in individual user rationality each brings about, neither alone provides users all the incentives needed to attain collective procedural rationality. Furthering individual rationality by simply constructing an arbitrary exchange institution need not mean that collective rationality concerns can be totally dismissed. Greater individual rationality does not guarantee collective rationality. Consider the following intuitive argument for a specific two-part instrument to achieve procedural rationality in biodiversity conservation.

Presume a control authority has specified a conservation objective for habitats on the private lands within a given locale. Let there exist numerous owners, use configurations, and parcel connectivities which could meet this objective, and let the control authority issue use permits in an amount guaranteeing that its objective would be met. Permit distribution and the land-use configuration the initial distribution implies could but need not be founded on biotic and abiotic criteria alone. Permit trade would encourage owners to seek out a configuration less costly to them, if one exists. But neither the initial nor the settled permit trade configuration is likely to be the least-cost, collectively rational configuration. This is because the impure public good nature (Cornes and Sandler, 1986) of land set-asides for conservation makes interdependent the optimal habitat set-aside decisions among the private owners.

An owner who sets aside habitat imposes a private opportunity cost upon himself while providing a positive benefit to his fellow owners. Land set aside by one owner to contribute to the authority-ordained conservation objective means that fellow owners can then set aside less land. Collective procedural rationality requires that this benefit to his fellow owners weigh in each individual owner's land-use decisions. Crocker (2004) and Parkhurst and Crocker (2006) formally show that a bonus (tax) for each of the originally issued permits (not) returned to the authority after the completion of trade in them will induce owners to adopt a collectively rational land-use configuration, whatever the conservation objective. Owners' total bonus (tax) is thus contingent on the extent to which they cooperate via the permit market to find that land-use configuration which is least costly to them in the sense that the sum of the opportunity costs they bear to meet the conservation objective is minimized. The intuition is simply that the bonus (tax) rewards (penalizes) owners for (not) finding via permit trade a landscape configuration which minimizes their opportunity costs while meeting the conservation objective. The fewer the permits the owners use to meet the conservation objective, the greater the bonuses they will receive for permits not used or the less the taxes they will pay for permits used.

CONCLUSIONS

Collective rational solutions to nearly all environmental problems call for having a legitimate public authority tell asset users what to do. Self-organization by natural asset users only occasionally resolves environmental problems.[6] But doing the right thing is a different problem than doing the thing right. Control instruments differ in their CCC burdens for the authority. And, because of the strategic interdependence between authority and asset user these burdens introduce, the instruments differ in the incentives they provide users to be procedurally rational. If imperfections such as monopolistic advantages are not manifest, the construction of exchange institutions in environmental management can both reduce these burdens for everyone – authority and users – and enhance users' incentive to discover how to reach the authority-designated outcome in a collectively rational fashion, thereby increasing the net value of the conservation goals.

These institutions do so because the prices they generate relax the individual user's CCC constraints. Three basic properties describe all collectively rational institutions: (1) agents' promises are credible; (2) supporting market or shadow prices are transparent to all participants; and (3) ready arbitrage opportunities exist such that agents in extramarginal regions quickly displace intramarginal agents who are not efficient. Institutions with these properties provide the gravity to induce procedural rationality. They "...induce that 'wonderful concentration of the mind' akin to the one Samuel Johnson attributed to the prospect of being hanged" (Hirshman, 1970, p. 21).

But CCC burdens make an authority's life difficult. The promises environmental management authorities make about what will be an acceptable way to do something may not always be credible. When these authorities try to command how asset users are to strive to contribute to an environmental objective, it is unusual that the price inferences (benefit-cost analyses) made to inform their commands will be transparent to asset users. And, by definition, commands from on high privilege no arbitrage. Exchange institutions the authority creates and maintains for the environmental dimension of interest stimulate behaviors more likely to satisfy all three criteria than are authoritys' attempts to simulate these behaviors.

The prices exchange generates integrated information about users' private preferences and circumstances. This integral is not otherwise possessed by any individual user and is unlikely to be available in full to any real-world authority. But, as the examples in the preceding section regarding biodiversity conservation illustrate, just any old exchange institution will not do. To maximize the extent to which these three criteria are met, careful up-front attention in design of the institution must be devoted to the nuances of the natural asset, its uses and users, and their interactions. Only when there is an environmental problem where the authority is unable to supply an institutional design to meet any or all of the three criteria is a retreat to having the authority speak to how to resolve the problem as well as what to do for a collectively rational outcome appropriate.[7]

NOTES

1 Academic disciplines if not people's everyday behaviors regularly resist embedding their research activities in a joint determination framework. In the biological sciences, for example, much research extends the perfectly reasonable idea that nature can exist autonomously to the doctrine that nature is in fact autonomous. Vitousek *et al.* (1997) are an exception — they summarize the natural science evidence for a non-autonomous nature. However, they treat the economy as autonomous. The social order affects nature but feedback from nature to the social order is disregarded.

2 William Wordsworth (1770–1850) efficiently captures this posture in his *Ecclesiastical Sonnet*, Number XLIII:

"Tax not the Royal Saint with vain expense ...
Give all thou canst; high heaven rejects the lore
Of nicely calculated less or more ...."

3 See Kagel and Roth (1995) for a detailed survey of these experiments in the 20th century.

4 For a quite thorough recounting of experiments with environmental management schemes in the USA and the increasing acceptance of 'economic incentives' versions, see Anderson and Lohof (1997) and US Environmental Protection Agency (2001). A similar recounting for international experiences can be found in Anderson (2004).

5 Eggertsson (1995) points out that if those for whom the property rights are to be structured and assigned can exert influence upon this structure and assignment, they will dissipate the rents created by the formation of the rights.

6 See Libecap (1989) and Ostrom (1990) for discussions of the exceptions.

7 The retreat may be partial rather than whole. For example, effluent changes are readily observed by asset users but they grant no arbitrage opportunities. If the charges lack credibility because effluent discharges are difficult for the authority to monitor, the retreat may have to go farther, say, to the authority mandating readily observed magnitudes and types of inputs asset users can employ. Nevertheless, each retreat step increases the CCC burden the authority must bear if collectively rational outcomes are sought.

REFERENCES

Anderson, E. (1993) *Values in Ethics and Economics.* Cambridge, MA: Harvard University Press.

Anderson, R.C. (2004) *International Experiences with Economic Incentives for Protecting the Environment.* Report EPA-236-R-04-001, Washington, DC: US Environmental Protection Agency.

Anderson, R.C. and A. Lohof (1997) *The United States Experience with Economic Incentives in Pollution Control.* Washington, DC: Environmental Law Institute (available from USEPA Website as Report EE-0216A).

Ando, A. (1998) 'Species Distributions, Land Values, and Efficient Conservation,' *Science* 279: 2126–2127.

Arrow, K.J., and G. Debreu (1954) 'Existence of an Equilibrium for a Competitive Economy,' *Econometrica* 22: 265–290.

Bateson, G., (1991) *Sacred Unity: Further Steps to an Ecology of Mind*. R.E. Donaldson (ed.). New York: Harper Collins.

Bowles, S. (1998) 'Endogenous Preferences: The Cultural Consequences of Markets and Other Institutions,' *Journal of Economic Literature* 36: 75–111.

Boyd, J., V. Caballero and R. Simpson (1999) "The Law and Economics of Habitat Conservation: Lessons from an Analysis of Easement Acquisitions," *Stanford Environmental Law Journal* 19: 209–255.

Cherry, T., T.D. Crocker and J.F. Shogren (2003) 'Rationality Spillovers,' *Journal of Environmental Economics and Management*. 45: 63–84.

Coase, R.H. (1960) 'The Problem of Social Cost,' *Journal of Law and Economics* 3: 1–44.

Cornes, R. and T. Sandler (1986) 'The Theory of Externalities,' *Public Goods and Club Goods*. Cambridge, UK: Cambridge University Press.

Crocker, T.D. (1966) 'The Structuring of Atmospheric Pollution Control Systems,' In H. Wolozin (ed.) *The Economics of Air Pollution*. New York: W.W. Norton, pp. 61–86.

Crocker, T.D. (2004) 'Markets for Conserving Biodiversity Habitat: Principles and Practices.' In J.F. Shogren (ed.) *Species at Risk: Economic Incentives to Protect Endangered Species on Private Property*. Austin, TX: University of Texas Press, pp. 191–215.

Crocker, T.D. (1984) 'Scientific Truths and Policy Truths in Acid Deposition Research,' In T.D. Crocker (ed), *Economic Perspectives on Acid Deposition Control*. Boston, MA: Butterworth Publishers, pp. 65–80.

Crocker, T.D., and J.F. Shogren (2001) 'Ecosystems as Lotteries,' In H. Folmer, H.L. Gabel, S. Gerking and A. Rose (eds.), *Frontiers of Environmental Economics*. Cheltenham, UK: Edward Elgar, pp. 250–271.

Crocker, T.D. and J. Tschirhart (1992) 'Ecosystems, Externalities, and Economies,' *Environmental and Resource Economics* 2: 551–567.

Dales, J. (1968) *Pollution, Property and Prices*. Toronto, ON: University of Toronto Press.

Daly, H. (1968) 'On Economics as a Life Science,' *Journal of Political Economy* 76: 392–406.

Economides, N. and A. Siow (1988) 'The Division of Markets is Limited by the Extent of Liquidity,' *The American Economic Review* 98: 108–121.

Eggestsson, T.A. (1995) *Economic Behaviors and Institutions*. Cambridge, UK: Cambridge University Press.

Hayek, F. (1945) 'The Use of Knowledge in Society,' *The American Economics Review* 35: 519–530.

Helland, E. (1998) "The Revealed Preferences of State EPA's: Stringency, Enforcement, and Substitution," Journal of *Environmental Economics and Management* 35: 242–261.

Hirsh, F. (1976) *Social Limits to Growth*. Cambridge, MA: Harvard University Press.

Hirshleifer, J. (1978) 'Natural Economy Versus Political Economy,' *Journal of Social and Biological Structures* 1: 319–337.

Hirshman, A.O. (1970) *Exit, Voice, and Loyalty: Responses to Decline in Firms, Organizations and States*. Cambridge, MA: Harvard University Press.

Innes, R., S. Polasky and J. Tschirhart (1998) 'Takings, Compensation and Endangered Species Protection on Private Land,' *Journal of Economic Perspectives* 12: 35–52.

Kagel, J.H. and A.E. Roth (1995) *The Handbook of Experimental Economics*. Princeton, NJ: Princeton University Press.

Kelman, S. (1981) *What Price Incentives? Economists and the Environment*. Boston, MA: Auburn House.

Leopold, A. (1949) *A Sand Sounty Almanac*. New York: Oxford University Press.

Libecap, G.D. (1989) *Contracting for Property Rights*. Cambridge, UK: Cambridge University Press.

Marcuse, H. (1964) *One-Dimensional Man: Studies in the Ideology of Advanced Industrial Society*. Boston, MA: Beacon Press.

Norgaard, R.B. (1994) *Development Betrayed: The End of Progress and a Coevolutionary Revisioning of the Future*. New York, NY: Routledge.

Norton, B.G. (1994) "Economists' Preferences and the Preferences of Economists," *Environmental Values* 3: 311–332.

Ostrom, E. (1990) *Governing the Commons: The Evolution of Institutions for Collective Action*. Cambridge, UK: Cambridge University Press.

Parkhurst, G.M. and T.D. Crocker (2006) *Incentive Design for Private Conservation of Least-Cost Biodiversity Habitat*. Working Paper, Department of Economics and Finance, University of Wyoming, Laramie.

Pezzey, J. (1992) 'The Symmetry Between Controlling Pollution by Price and Controlling It by Quantity,' *Canadian Journal of Economics* 26: 983–999.

Russell, C.S. and P.T. Powell (1999) 'Practical Consideration and Comparisons of Instruments of Environmental Policy,' In J.C.J.M. van der Bergh (ed.) *Handbook of Environmental and Resource Economics*. Cheltenham, UK: Edward Elgar, pp. 307–328.

Stavins, R.N. (2003) 'Experience with Market-Based Environmental Policy Instruments.' In K.-G. Maler and J. Vincent (eds.) *The Handbook of Environmental Economics*. Amsterdam: North Holland / Elsevier Science, Chap. 21.

Sugden, R. (1991) 'Rational Choice: A Survey of Contributions from Economics and Philosophy,' *Economic Journal* 101: 751–785.

Taylor, P.W. (1986) *Respect for Nature: A Theory of Environmental Ethics*. Princeton, NJ: Princeton University Press.

U.S. Environmental Protection Agency (2001) *The United States Experience with Economic Incentives for Managing the Environment*. Report EE-0216 B, Washington, DC: USEPA.

Vatn, A. and D.W. Bromley (1994) 'Choices without Prices without Apologies,' *Journal of Environmental Economics and Management* 26: 129–148.

Vitousek, P. (ed.) (1994) 'Human Determination of Earth's Ecosystems,' *Science* 277: 494–525.

10

Valuing Preferences Regarding Environmental Change

Ian J. Bateman

INTRODUCTION

To include a chapter on what many would term the 'valuation of the environment' may seem an anathema to some readers. Indeed, as Tim O'Riordan kindly pointed out in an editorial introduction to a previous essay on this topic (Bateman, 1995), some critics 'believe most sincerely that monetizing the environment is merely a further step in global degradation of the human spirit, let alone the natural world'. In mitigation, let me submit from the outset that I make no claims that the monetary evaluation methods discussed here are either flawless or a panacea for incorporation of environmental impacts within decision-making. Indeed, as suggested by Diamond and Hausman (1994) in their negative answer to the question 'is some number better than no number?', when applied poorly or in inappropriate circumstances, the information which monetary evaluation techniques provide may be downright misleading. However, where certain conditions prevail, the application of these methods can provide a useful input to project appraisals.

This chapter is structured as follows. In the first section, I compare and contrast environmental impact analysis (EIA) with the project appraisal framework within which monetary values for environmental impacts are most commonly assessed; namely cost–benefit analysis (CBA). Such a comparison is relatively rare yet provides a useful basis for us to consider the fundamental issues which have to be addressed before monetary values and physical unit assessments may be usefully combined within a single project appraisal framework. This discussion also provides an introduction to the third section in which a more detailed discussion of both fundamental and current issues within the field of valuation is presented. Here, the basic economic theory underpinning the concept of value is reviewed, and set against a brief overview of certain pertinent and contemporary research problems. The fourth section provides a review of a wide variety of evaluation methods illustrating each with recent examples of their application. The last section summarises and provides conclusions regarding future research needs and directions. The reader will note that this chapter treats the state of the literature through to the late 1990s; subsequent contributions are not referenced in this review.

CONTRASTING CBA WITH EIA

A comparison of the economic analysis approach underpinning CBA with that of EIA reveals some important contrasts which will assist our understanding of evaluation methods. The theoretical and methodological basis of CBA are discussed in depth elsewhere (for an excellent introduction see Pearce, 1984, or Hanley and Spash, 1993); however, in essence the approach involves the monetisation of all the costs and benefits of a proposed project and its alternatives, and the assessment of the resultant net benefits over a given time horizon. In contrast to EIA, CBA appears more quantitative, assesses impacts using a single unit (money), has a more detailed and developed theoretical basis (particularly with respect to human behaviour and preferences) and, as EIA experts freely acknowledge (Clark *et al.*, 1981), avoids the inherent subjectivity of EIA judgements by implementing a well-defined and highly empirical methodology. On paper then, CBA seems to have numerous advantages over EIA and indeed in the 'perfect' world of full information and well-formed preferences which underpins neo-classical economics then CBA might[1] well be the superior method for project appraisal. Perhaps fortunately we live on Thomas Hardy's blighted star rather than Aldous Huxley's brave new world and, therefore, with the assumption of rational economic man only partially holding, we cannot establish a clear *a priori* superiority between CBA and EIA and have to turn to a comparison of their constituent elements.

The first and most fundamental contrast between the techniques is that they are grounded in differing philosophies. While EIA adopts a highly objective physical science approach to assessment, CBA has grown out of the subjective social science of economics. So as EIA directly assesses environmental impacts, CBA focuses on their socio-economic interpretation as values. As such CBA is overtly anthropocentric and produces assessments which reflect a human rather than physical view of a project. This implies that the two methods must always produce divergent, if not necessarily incompatible, appraisals of a project as they are literally two different ways of looking at the same issue.

These differences are often thrown into sharp relief by their practical implementation. As noted, in comparison with EIA, CBA often appears more quantitative, employs a single as opposed to multiple units of measurement but, partly as a result of this, often covers a narrower scope of assessment. The use of a single money numeraire has both positive and negative consequences. It facilitates an ease of comparison across projects which tends to be absent from physical appraisals and allows ready interpretation of results by decision-makers and various interested parties. Also, in conjunction with the quantitative emphasis of CBA, use of the single monetary numeraire *can* enhance the transparency of assessments by revealing underlying calculations (although obfuscation is also readily attainable, particularly through the use of implicit and unjustified underpinning assumptions). Furthermore, monetary units permit the translation from physical impacts into the corresponding values which are the language of many decision-makers.

Valuation allows the concise assessment of multiple, diverse impacts and as such represents the major theoretical advantage of CBA. Numerous schemes for the weighting and ranking of physical impacts within EIA have been proposed (see e.g., Bisset, 1978; Clark *et al.*, 1981; Wathern, 1988; Rabinowitz *et al.*, 1986; Glasson *et al.*, 1994), but these lack the theoretical rigour underpinning the process of valuation. However, it is not in terms of theory but in the actuality of practice that we find the principal disadvantages of economic valuation and CBA itself. Most noticeably the range of impacts for which values can be readily obtained is markedly smaller than that conventionally considered within EIA. Thus, valuation directly contributes to the relatively restricted scope of CBA as mentioned above. A simple illustration of this drawback is provided by the Department of Transport (DoT) CBA framework for assessing trunk road proposals (COBA) (Cost Benefit Analysis Programme). Table 10.1 details those items which are given monetary values within COBA. While this includes a number of important costs and benefits, the list of items which are not monetised is both more extensive and includes the bulk of environmental impacts. It is interesting to note that these latter impacts are incorporated within the DoT's project appraisal framework via standard EIA techniques.

The diversity of measurement units resulting from such a strategy undermines the single-unit-comparability advantage of CBA. However, such a confluence of monetary and physical assessment units may well reflect the limitations and necessities of project appraisal in practice and we shall return to this theme at the conclusion of this chapter.

We have already mentioned that EIA employs a variety of weighting and ranking techniques in order to compare impacts in a process somewhat analogous to valuation. The weighting of costs and benefits also occasionally occurs in CBA. However, here the technique is employed to a different end; namely to allow both for the effects of CBA outcomes on income distribution and vice versa. Clearly the outcomes of CBA appraisals, if sanctioned by decision-makers, is likely to result in costs to some people and benefits to others with the so-called Hicks–Kaldor potential compensation rule (Hicks, 1939; Kaldor, 1939) being that the sum of benefits should exceed the sum of costs (the latter including the opportunity costs of foregone alternative investments of the resources involved) such that gainers could hypothetically compensate losers and still be better off than if the project had not been sanctioned. However, typically CBA takes the current distribution of income as exogenous to the assessment and is therefore ignorant of the dispersion of benefits and costs amongst the rich and poor of society. Similarly, in assessing the value of these benefits and costs, no account is taken of the ability of individuals to pay and thereby, for example, express value for a given benefit. This latter factor may result in those items which the rich favour being given disproportionately higher values than items preferred by the poor.

Table 10.1 Items monetised and non-monetised in COBA

Item	Present analysis in COBA
Items which are monetised in COBA	
Construction costs	Market prices
Land costs	Market prices (debate as to appropriateness)
Demolition costs	Market prices (debate as to appropriateness)
Compensation costs	Market prices (debate as to appropriateness)
Maintenance costs	Market prices (debate as to appropriateness)
Vehicles' operating costs	Priced from market prices of fuel, operating costs
Time savings	Monetised: debate re value of time
Accident reductions	Monetised: debate re value of life, value of health
Traffic noise	Currently being monetised
Items which are not monetised in COBA	
Recreation/amenity loss	Quantified (area, land, quality)
Visual obstruction	Quantified (expert analysis)
Visual intrusion	Descriptive
Air pollution	Quantified (or unassessed)
Built environment/heritage	Descriptive
Severance	Descriptive/qualitative
Ecological sites	Descriptive/qualitative
Pedestrian/cyclists	Descriptive
Disruption during construction	Descriptive/qualitative

Source: Adapted from Bateman et al. (1993)

There have been attempts to allow for the impacts of income distribution within CBA via weighting and other means (Lichfield *et al.*, 1975; Squire and van der Tak, 1975; Pearce and Nash, 1981; Pearce, 1984; Lichfield, 1988). However, these are the exceptions rather than the rule and such attempts are now very rare. This rejection of such approaches has been on both practical and theoretical grounds. In practice the choice of weighting or allowance scheme is not obvious and can be used to manipulate analysis outcomes. More fundamentally it can be argued both that, given that a CBA result and the project decision are not and should not be synonymous, an assessment based on the current income distribution provides useful information and that the objectives of project appraisal and addressing income inequalities are separate and distinct. Analysis of decision-making across a range of policy spheres has shown that separate goals require separate tools and that joint implementation, despite its attractions, is liable to be inefficient and ill-conceived (Moyer and Josling, 1990; Coleman and Nixson, 1994).

The final contrast between EIA and CBA which we shall consider here is the extent to which they address the temporal effect and sustainability of projects. Both approaches explicitly consider the long-term consequences of projects. In the case of EIA this is achieved via a series of impact prediction techniques. A similar approach is adopted in CBA; however, here future impacts are translated into expected values. Here, the similarity ends as these values are 'discounted', a process by which future values are reduced, the degree of reduction increasing as the time horizon is extended. The justification for such treatment is that individuals exhibit positive time preference; that is, benefits received sooner are valued more than delayed benefits (see Pearce, 1984). The 'discount rate' defines the speed at which future benefits are reduced with a more positive time preference leading to a higher discount rate and a faster reduction of delayed benefits. The advantage of such an approach is that the stream of discounted values can be summed across the entire time horizon to yield a net present value for the project in its entirety. The disadvantage is that the choice of discount rate has long been and continues to be the subject of intense debate,[2] yet its selection can dominate the results of a CBA and can frequently determine whether or not a given project is adjudged to be socially beneficial or not.

The discounting issue has in recent years become inextricably linked to the ongoing debate concerning sustainable development and specifically questions regarding the extent to which CBA can, or cannot, be modified to meet the requirements of sustainability. One of the major complications of this debate is the diversity of views regarding the definition of these terms. Turner and Pearce (1993) show that this diversity can be traced to differences in viewpoints regarding the extent to which economic development can be harmonious with environmental integrity. Borrowing the terminology of O'Riordan and Turner (1983), Turner and Pearce identify two extreme world views: that of the 'technocentrists', for whom sustainability can be assured simply by ensuring that the value of man-made capital created by the use of environmental assets (natural capital) equals or exceeds the value of the latter; and, at the other extreme, the 'ecocentrists' who argue that such capital substitutions are irreversible and unsustainable in that they both reduce the set of opportunities available to future generations and violate the rights of non-human species and of the environment itself. Turner and Pearce formulate these intermediate positions within a continuum of sustainability rules ranging from the very weak to the very strong. Subsequent debate has examined the extent to which these rules may be applied to project appraisal. One example of such an approach argues that strong sustainability may be satisfied within CBA studies if the Hicks–Kaldor potential compensation rule is partially abandoned in favour of actual compensation for the use of natural capital, that is, a move from hypothetical monetary compensation to a rule enforcing actual physical compensation. This could be achieved through the implementation of offsetting 'shadow projects' which might reconstruct, transplant or restore environmental assets used in a development proposal (Buckley, 1989; Bateman, 1991; Turner et al., 1994).

While this debate is well advanced, the explicit inclusion of sustainability offsets or constraints is still very much the exception rather than the rule in CBA practice. In contrast, concerns regarding the sustainability of projects have long been a central feature of EIA studies as evidenced in Ratcliffe's (1977) criteria for evaluating sites, and the derived literature on ecological sustainability within EIAs (Spellerberg, 1992; Newman, 1993). It is clear that, as presently and commonly implemented, the emphasis on physical assessment inherent in EIA makes it the preferred, risk-averse option for ensuring the sustainability of projects. However, this issue is being actively addressed within the literature surrounding CBA, environmental and ecological economics (Costanza, 1991).

In summary we have seen that EIA and CBA are fundamentally different approaches to the issue of project appraisal. However, within this they contain a number of common and comparable elements which provide useful insights into the use of monetary valuations within project appraisal. We now focus down from the general framework in which project appraisals occur towards the specific issues surrounding valuation itself.

MONETARY EVALUATION: THEORY AND DEBATE

The first and most important disclaimer we need to make concerning the monetary evaluation methods discussed subsequently in this chapter is that they are *not* capable of 'valuing the environment'. The environment is a multi-attribute entity which yields a complexity of values only certain of which fall within the remit of human valuation (see subsequent discussion). Instead the most which we can claim is that some of these methods may be able to value humans' preferences for some aspects of certain environmental goods and services.

A second fundamental issue requiring clarification arises from the common and persistent confusion between prices and values. Even the earliest economic commentators recognised that the price of a good need not correspond to its value. Adam Smith, writing more than 200 years ago, noted the extreme disparity between the (very low) price of water and its (very high) value. Yet even today the use of environmental goods and services is commonly dictated by their price rather than their value. A contemporary example of such use is the dumping of UK sewage sludge in the North Sea. In so doing, the sea provides highly valuable but virtually zero-priced waste-assimilation services. These are only now becoming apparent because the phasing out of such dumping (MTPW, 1990) has forced the companies involved to find costly incineration and landfill alternatives.

Returning to the example of water, we can now establish a clear difference between price and value and thereby illustrate the economic meaning of the latter. For simplicity let us consider the consumption of half-litre bottles of mineral water rather than tap water (for which the price is often difficult for the individual to assess). An individual's decision to purchase and consume a bottle of mineral water indicates a surplus of value over price (otherwise they would not purchase the bottle). This 'consumer surplus' can then be added to the price paid to yield the economic value of that bottle of water. In economic parlance we say that value is thereby defined as the maximum amount which the individual would have been willing to pay (WTP) for the bottle of water. We can now formalise this definition of value as per Equation (1):[3]

$$\text{Value} = \text{Willingness to pay (WTP)} = \text{Price paid} + \text{Consumer surplus} \quad (1)$$

This approach to value has some important consequences, not all of which may be immediately apparent. In particular, the equation of value with WTP takes the current ability to pay as exogenous. We have already rehearsed the implications of such an assumption in our discussion of the wider CBA framework and these apply here. But in essence such an approach means that the use of economic values within project appraisal gives us information about the *efficient* allocation of resources (i.e. which projects would yield the highest net benefits), but tells us little about the *equity* implications of a project decision. As discussed above, such considerations *can* be incorporated within the wider CBA framework (via income distribution weights, etc.); however, there is a compelling argument that project appraisal should be restricted to issues of efficiency so leaving equity concerns to the realm of policy and decision-making.

Continuing with our example of the value of mineral water, let us now consider the total value of all the bottles consumed by an individual in some time period, say one hot summer day. Suppose that our individual decides to purchase a second bottle of water. This tells us that again the value of that bottle either equals or exceeds its price. Again using Equation (1) to determine this value, however, in comparing values for bottles 1 and 2 a new phenomena is observed, namely that the value (WTP) of the former exceeds that of the latter. This occurs because the thirst which our individual had before consuming bottle 1 was greater than that before consuming bottle 2. In more general terms the value of any subsequent unit of consumption will, in all but a few rare instances, be lower than that of the preceding unit.[4] This trend will extend over all additional units, so that for our individual the value of a third bottle may be just equal to its price (i.e. consumer surplus is reduced to zero) while for a fourth bottle value falls below price. In this example, therefore, our individual would purchase the first three bottles of water but no more.

We can now determine the total value of mineral water consumption for our individual by summing up the value (WTP) for each of the three bottles bought. Notice that this exceeds the price paid by the sum of consumer surplus. A common way to describe visually the relationships set out above is to graph the value of each unit of consumption. This yields the 'demand curve' shown in Figure 10.1. Comparing this with the horizontal price line we can see that the area under the demand curve and above the price line describes the products' consumer surplus. Adding this to the price paid area under the price line gives the total value of goods consumed. The graph also shows that the quantity consumed is determined by the intersection of the demand curve and price line. Beyond that point additional units have values which are lower than their price and are therefore not consumed.

Our mineral water example refers to a 'private' good, that is, one which has private property rights through which a seller can extract purchase price from the consumer, that price being generally determined by the interaction of demand and supply in a market. However, the theory of economic valuation set out above can, in principle, also be extended to non-market, unpriced, 'public' goods for which only common property rights exist.[5] Examples of such include clean air, sea views or walking in public parks. These goods are often referred to as open-access because they have no market price associated with their use (although in practice as a result of location such goods are more readily available to some rather than others). Again we can see that such public goods are

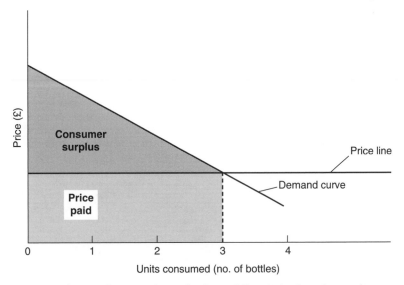

Figure 10.1   Demand curve for a market priced good (bottled mineral water)
From: Bateman, I.J. (1999) Environmental impact assessment, cost-benefit analysis and the valuation of environmental impacts, in Petts, J. (ed.) *Handbook of Environmental Impact Assessment*, Vol.1, Blackwell Science, Oxford, pp. 93–120. Reprinted with permission.

valued by individuals; in economic terms they have a WTP for consumption of such goods. However, here as the market price paid is zero (there may be a non-zero cost of getting to such goods, of which more later), value (WTP) is entirely composed of consumer surplus.

Figure 10.2 describes the demand curve for such a good; walks in a public woodland. Here, the entrance price is zero but our individual derives a significant value from her first walk in the wood during a given period, say one week. As in the case of our private good, the value of subsequent walks declines relative to previous visits. The demand curve mapped out need not be a straight line, indeed such a relationship is unlikely in practice. In Figure 10.2 the first visit is very highly valued after which WTP initially declines rapidly but then at a slower pace. Such a demand curve might be representative of someone who takes their dog to the wood, the initial walk being valued both because it provides exercise for the dog (which the individual values) and for its aesthetic qualities. After this, additional visits are only of use because they provide exercise for the dog. Note that increasing visits above a certain level would again drive the value of additional visits to zero.

Notice that in the above example our public good was in fact providing the individual with two types of value: aesthetic beauty and a place to exercise a dog. In fact a complex good such as a wood provides a variety of values.

The basic division of values within TEV (total economic value) is into either those which are, or are not, derived from use of goods and services by the valuing individual. The first of these, known

generically as 'use values' can be further subdivided into:

1 those which involve direct, present-day use (note here that for our example a woodland can generate both market-priced private goods such as timber and unpriced public goods such as recreation[6]);
2 the option value of future direct use (Weisbrod, 1964);
3 the bequest value of providing use and/or non-use values for present and/or future others.

Pure non-use values are most commonly identified with the notion of valuing the continued existence of entities such as certain species of flora or fauna or even whole ecosystems (Young, 1992). In theory then, in valuing complex goods and such as those provided by the environment we should aim to assess their TEV rather than simply their more obvious direct use or just market-priced attributes (for examples, see Barde and Pearce, 1991).

Wider definitions of value have been argued for. An important issue concerns the extent of the 'moral reference class' (Turner et al., 1994) for decision-making. One question here arises from the treatment of other (both present elsewhere and future) humans while another concerns whether animal, plant and ecosystem interests should be placed on an equal footing with human preferences. The modern origins of such a view can be traced to Goodpaster (1978) and Watson (1979) who take the Kantian notion of universal laws of respect for other persons and extend this to apply to non-human others. Watson feels that those higher animals such as chimpanzees (which he argues are capable of reciprocal behaviour) should

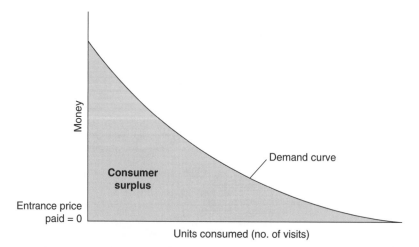

Figure 10.2   Demand curve for an unpriced public good (walks in a wood)
From: Bateman, I.J. (1999) Environmental impact assessment, cost-benefit analysis and the valuation of environmental impacts, in Petts, J. (ed.) *Handbook of Environmental Impact Assessment*, Vol.1, Blackwell Science, Oxford, pp. 93–120. Reprinted with permission.

be accorded equal rights with humans. Hunt (in Perman *et al.*, 1996) and Rollston (1988) build on the land ethic of Leopold (1949) to extend this definition of moral reference even further to include all extant entities, an approach which Singer (1993) defines as the 'deep ecology' ethic. Such a paradigm argues that these entities possess an 'intrinsic' value separate from anthropocentric existence values. A further departure from conventional utilitarianism is proposed by Turner (1992) who argues that all the elements of TEV can be seen as secondary to a primary environmental quality value which is a necessary prerequisite for the generation of all subsequent values. Sidestepping the theoretical case for such philosophical extensions, a practical problem with these non-TEV values is that they are essentially beyond the scope of conventional, anthropocentric, preference-based economic valuation.

Given this, we separate these non-human values from our definition of TEV in Figure 10.3. Such values may be of moral significance (and from our discussions in the second section it may be that the physical assessment emphasis of EIA makes more allowance for these values than does CBA) but by definition they are beyond economics and not considered further here.

The complex array of values encapsulated within the TEV concept need not always be in harmony. In our woodland example, a logging project may increase the private good, use values of timber received by a woodland owner but at the cost of reducing the public good recreation values enjoyed by that same individual. This issue highlights not only the complex trade-offs within a project but, more fundamentally, the conflict of values within the individual herself. Some commentators have

argued that the individual should be seen as encapsulating two parts: the consumer and the citizen.[7] The preferences which each part holds may be very different and lead to distinct values. For example, as a citizen I may value a well-funded public transport system and traffic calming measures, while as a consumer I may wish to drive my private car unimpeded into the central shopping district. The embodiment of these conflicting desires within the individual gives rise to a continuum of preferences and respective values. Where on that continuum an individual will choose to be is, critics argue, likely to be context specific. Furthermore this context is multi-dimensional, encompassing the policy, institutional and economic aspects of the assessment, and the social, psychological and cultural attributes pertaining to the individual.

Neo-classical economic theory implicitly challenges such a view by arguing that the individual will weigh the various costs and benefits of a given project and thereby yield a non-contextual value. Any individual exhibiting variations in values for a given good across contexts can therefore be seen as simply reacting to differences in the information provided or in changes to the good itself (e.g. in changes regarding how the project will affect other people, a factor which will often be of importance to the valuing individual) which occur when the context changes. Critics of this view argue that material changes in context not only instil these objective shifts but can also result in a movement along the consumer–citizen preference continuum such that resulting values are non-commensurate with the individual (as pure consumer) based values found in a CBA. In particular, commentators such as Sagoff (1988) and Jacobs (1994, 1997) argue that the move from assessing man-made capital

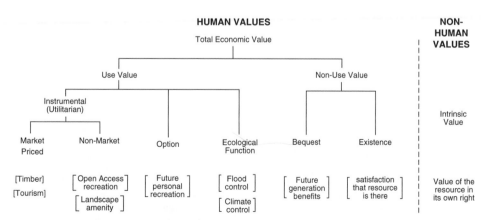

Figure 10.3 Total economic value (TEV)
From: Bateman, I.J. (1999) Environmental impact assessment, cost-benefit analysis and the valuation of environmental impacts, in Petts, J. (ed.) *Handbook of Environmental Impact Assessment*, Vol.1, Blackwell Science, Oxford, pp. 93–120. Reprinted with permission.

goods to environmental goods entails a shift from citizen to consumer values. In such a case reliance on economic valuation techniques which place the individual solely within the role of consumer will yield inappropriate measures when applied to the evaluation of environmental goods.

Sagoff and Jacobs argue that the use of one-to-one economic surveys (such as the contingent valuation method described below) for valuing environmental goods entails a high risk of such category mistakes. They maintain that such goods fall within the natural domain of citizen preferences and that these will be inhibited by such a methodology. Instead such values should be explicitly activated through a social valuation process reliant on agreement and consensus rather than isolated individual opinion. Expanding on this Johnson (1996) proposes the use of citizens' juries and focus groups as an alternative way of providing information to decision-makers regarding the use of environmental goods.

A practical drawback of the citizen jury approach is that such information is incompatible with CBA appraisals although it may more readily be included within EIAs. A more fundamental question concerns the current lack of empirical support for the often stated contention that public goods need public decisions. Few studies have examined the assertion that individually based values are less appropriate for environmental goods than are group-based assessments or that the former activate consumer preferences while the latter yield citizen preferences or that these are significantly different. Furthermore, what evidence is available is somewhat conflicting in its content and conclusions (Mitchell and Carson, 1989; Blamey, 1995; Bateman and Langford, 1997; Brouwer et al., 1997; Burgess et al., 1998).

Nevertheless, this is an important and ongoing debate which has been well heeded. Given this and the problem that these two approaches are mutually exclusive, it seems prudent to suggest that the use of citizens' juries as an additional safety check on economic valuation alone is a sensible and risk averse strategy providing a useful sensitivity analysis across methodologies.

Up to this point we have only measured value in terms of WTP. However, economic theory shows that value can also be assessed by examining how much an individual is willing to accept (WTA) in compensation for a change in the provision of a good. Furthermore, as we can assess both gains and losses in provision, we can define in total four equally valid measures of value: WTP to secure a gain; WTP to avoid a loss; WTA to suffer a loss; WTA to forgo a gain. Following economic theory it can be shown that for most common private goods there should be little difference in the values each measure estimates for a given unit of such a good (Willig, 1973, 1976). However, in an important extension to this work, Hanemann (1991) shows that for goods such as those provided by the environment where individuals have high income elasticity (i.e. they consume relatively more environmental goods as they become richer) and/or low elasticity of substitution (i.e. they are relatively unwilling to trade environmental goods for substitute man-made or other goods) then this WTA/WTP divergence can become very significant.

Hanemann's work provides a justification in economic theory for a commonly observed large empirical divergence between measures such as WTP for a gain and WTA for a loss. However, such a difference has important implications for CBA or the use of monetary values in EIA. In particular it means that, say, a measure of an individual's WTP for a project delivering a specified gain cannot necessarily be taken as indicative of that same individual's WTA compensation for what is in physical terms an identical loss. This problem is far more likely to occur with respect to environmental public goods which exhibit the Hanemann conditions than for common private goods which do not.

While this of itself is a major problem for the practice of economic appraisal, a more fundamental challenge comes from an alternative theoretical explanation of the above phenomena. In their 'Prospect Theory', Kahneman and Tversky (1979) argue that in making decisions and determining values individuals exhibit loss aversion. Such a theory is based on the psychological rather than economic reasoning that in such situations losses loom larger than corresponding gains. Tversky and Kahneman (1991) extend this argument into a theory of reference-dependent preferences where the value of changes in the provision of a good is dependent not only on the magnitude of that change and standard economic factors, but also on the initial allocation or reference point of that good. Here, the combined impact of loss aversion and reference point effects mean that all four of the valuation measures defined above could yield differing estimates of value for what was essentially the same unit of provision.

In a test of these competing theories the author and colleagues found that reference-dependent theory provided a better explanation of individuals' responses when all four measures were applied to real money valuations of various private goods (Bateman et al., 1997a). Given that the Hanemann explanation may apply in conjunction (i.e. elasticity effects may exacerbate valuation differences) then we should not be surprised that surveys yield large differences between alternative value measures when applied to environmental public goods.

Before drawing any conclusions we shall consider one further issue which is current in the literature. This is the problem of the part–whole effect. This occurs when the values given by individuals for the constituent parts of a good exceed the value given to the good as a whole. Such a result is reported by Kahneman and Knetsch (1992) in respect of survey of individuals' WTP to preserve fish stocks, either in the small proportion of Ontario lakes around Muskoka, or for all the lakes in the province. Here, the median WTP for all lakes was found to be only slightly higher than that for the small proportion of lakes and Kahneman and Knetsch argue that this result arose because subjects saw both parts and wholes as interchangeable vehicles for charitable giving, the inference being that respondents were purchasing 'moral satisfaction' rather than the good nominally under valuation. While the conduct of this particular survey was the subject of fierce criticism, subsequent work has suggested that the part–whole issue may be a significant problem, not just for evaluations of environmental goods but for economic appraisals as a whole.[8]

In summary, this section demonstrates that economics has developed a theory of value which applies to both market-priced private goods and the unpriced public goods provided by the environment. Furthermore, this theory has been extended to encompass the complexity of services provided by the environment. However, our review of current issues shows that a vigorous debate is ongoing which challenges the simple notions of individual-based valuation inherent in much traditional economic theory. Complex problems surrounding the issues of WTP–WTA asymmetry, citizen versus consumer preferences, loss aversion, reference-point effects and part–whole problems are all active research issues and it is to be hoped that, rather than succumbing, economic theory will embrace these challenges and

extend itself. It is the author's belief that to do so economists must be prepared to learn from other disciplines so as to integrate rather than ignore evidence. In the meantime those engaging in the process of valuation need to be aware of these issues and assess the extent to which they impinge on specific valuation tasks and contexts. In particular, the use of sensitivity analysis across valuation scenarios, assumptions and even methodologies (including non-valuation approaches) is to be advised.

We now turn to consider the various valuation tools and methods available to the analyst.

METHODS FOR THE MONETARY EVALUATION OF ENVIRONMENTAL PREFERENCES

Research into methods for placing monetary values on preferences for environmental public goods can be traced back to at least the 1940s (Bateman, 1993); however, burgeoning interest in the field over the past two decades has resulted in the development of a wide variety of assessment methods. These can be classified in a number of different ways, none of which is entirely satisfying (Peace and Markandya, 1989). For convenience we adopt a somewhat pragmatic initial division between those methods which are usually applied in order to estimate theoretically consistent values, and a range of generally simpler techniques which are usually used to produce estimates which in theoretical terms are analogous to prices in that they inform about the cost of attaining a given good but not about its value[9]. Figure 10.4 illustrates this

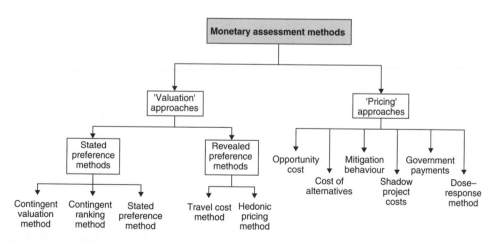

Figure 10.4 Methods for the monetary assessment of non-market and environmental goods. For an overview of the link between stated preference and travel cost methods, see Adamowicz *et al.* [1999]

From: Bateman, I.J. (1999) Environmental impact assessment, cost-benefit analysis and the valuation of environmental impacts, in Petts, J. (ed.) *Handbook of Environmental Impact Assessment*, Vol.1, Blackwell Science, Oxford, pp. 93–120. Reprinted with permission.

division and provides an overview of the methods discussed below of which we consider the simpler pricing techniques first.

'PRICING' METHODS

Opportunity costs

One approach is to examine what value would have to be foregone in order to, say, create or enhance a particular environmental asset. A contemporary UK example is the ongoing creation by the Countryside Commission of a 'New National Forest' between Leicester and Burton-on-Trent in the English Midlands (Countryside Commission, 1990). The 150-square-mile site includes a large area of high-quality agricultural land, the loss of which represents an 'opportunity cost'. This can be assessed as the net value of the crop and livestock products of the land taken by the new forest taking into account any subsidy saving (i.e. subsidies previously paid for lost agriculture). The existence of subsidies means that the market price of lost output is likely to overvalue that crop. However, it can be further argued that such prices are already distorted and reflect the structure of markets and the political weight of major interest groups, rather than competitive relationships. Given the complexity of highly intervened markets such as agriculture the link between market prices and underlying values can be difficult to assess and considerable care needs to be exercised if such information is to be used as a guide to project appraisal and decision-making.

Costs of alternatives

Where an environmental resource is being used (or its use is planned) as part of some development or production project, then one strategy for evaluation is to calculate the cost of using some alternative resource. One example is the sewage treatment alternatives to North Sea dumping discussed above. A second example arises from the extension of the M3 motorway at Twyford Down, a recognised area of outstanding natural beauty in mid-Hampshire, UK. Here a scheme for running the motorway through a tunnel under the Down was costed at just under £70 million (Medley, 1992), a cost which the DoT clearly felt outweighed the environmental benefits of preserving the area as the extension was subsequently achieved by digging a large cutting through the middle of the Down, thereby destroying the integral ecological characteristics of the site. This example underlines a general problem with such an approach in that reliance on pricing rather than valuation techniques places environmental goods at risk of undervaluation.

Mitigation behaviour

Here, the prices which individuals pay in order to mitigate environmental impacts are taken as simple monetary assessments of those impacts. A commonly applied example is the use of double-glazing costs as a proxy for the disamenity value of noise intrusion. Such costs have been used in the assessment of road projects. However, this method is at best partial as it only takes account of noise intrusion within a house rather than also around its environs. It also ignores the welfare loss due to an imposed change in peoples habits, for example, their inability to leave windows open without suffering from heightened traffic noise.

Shadow project costs

Another 'pricing' option is to look at the costs of providing an equal alternative environmental good elsewhere, that is the shadow projects discussed earlier in this chapter. One such study (Buckley, 1989) examined such possibilities for a development project which threatened an existing wildlife habitat. Here, three options were highlighted; asset reconstruction (providing an alternative habitat site); asset transplantation (moving the existing habitat to a new site); and asset restoration (enhancing an existing degraded habitat). The costs of the preferred option can then be entered into the project appraisal as the 'price' of the threatened habitat. However, for such an approach to provide suitable compensation the ecological adequacy of shadow projects needs to be discussed.

Government payments

The Government, as arbiters of public preferences, occasionally directly value environmental goods and services by fixing subsidies paid directly to producers (particularly farmers) for adopting environmentally benign production methods. Such values have been used as part of project appraisals, one case being that for the Aldeburgh sea wall in Suffolk where the costs of wall renovation were assessed against various items one of which was the environmental value of protected land as proxied by ESA (Environment Sensitive Area) subsidies to local farmers (Turner *et al.*, 1992).

The dose–response method

Statistical techniques can be used to relate differing levels of pollution (the 'dose') to differing levels of damage (the 'response'). In one such study the environmental costs of a coal-fired power station were assessed by examining various dose–response models relating acidic and photo-oxidant emissions to their impacts on forest, crops, fisheries, etc. (Holland and Eyre, 1992). Once the physical impact of these emissions on, say, crops has been calculated in terms of tonnes per annum lost then this can be given a monetary evaluation by multiplying this damage by the market price per tonne (although our caveats regarding the distortion of prices in heavily intervened markets such as agriculture again apply here).

Summary: 'Pricing' methods

Non-demand curve 'pricing' approaches can be useful in providing rough monetary assessments of environmental goods and services that might otherwise be treated as free. However, they are not without flaws and limitations. The 'opportunity cost', 'cost of alternatives', 'mitigation behaviour' and 'shadow project costs' approaches provide monetary benchmarks against which the value of the environmental good in question can only be subjectively judged. These are not true valuations as the assessment only considers whether the environmental good is of greater value than the opportunity cost. This criticism also applies to the 'government payments' approach but here we have the additional problem that the benchmark value used is set not by the market but by government. The dose–response approach does have potential for wide application (Schultz, 1986; Barde and Pearce, 1991), although some doubts have been raised concerning the complexity of some dose–response relationships and consequent problems of statistical estimation (Turner and Bateman, 1990).

'VALUATION' APPROACHES

The following methods all centre on the estimation of a demand curve for the environmental good in question. As such they provide true valuations rather than simple 'pricing'. Figure 10.4 indicates that these methods may be categorised into two basic types:

1   *Stated preference methods:* These directly ask people about their valuation of environmental goods. This is generally achieved through one-to-one interviewing (although other variants, including group approaches, are becoming more common) and so many of the issues discussed in the third section are of particular relevance here.
2   *Revealed preference methods:* These ascertain individuals' valuations of environmental assets by observing their purchases of market-priced goods which are necessary to enjoy the environmental good in question (e.g. purchasing petrol so that a day trip to the country can be enjoyed).

Stated preference methods

The contingent valuation (CV) method

Here, environmental evaluations are obtained by using surveys to ask people directly what they are WTP or WTA for a given gain or loss of a specified good. The method can therefore yield estimates of all four of the valuation methods discussed previously. Given its flexible one-to-one or group survey methodology, CV is also an interesting tool with which to investigate issues such as consumer versus citizen preferences, WTP–WTA asymmetry, loss aversion, reference points and part–whole effects as discussed in the preceding section, and indeed the vast CV literature which now exists addresses these and further issues. For example, a recent study involving the author (Bateman and Langford, 1997) suggested that while some respondents exhibited 'citizenship' preferences across differing scenarios (notably members of environmental groups; these respondents were consistently more willing to pay for goods which provided benefits across society), others consistently exhibited 'consumer' preferences (particularly members of private associations such as health clubs; these respondents had higher WTP for exclusive goods which endowed only those who paid with benefits). Such findings are helpful in understanding the complex motivations cross-sectioning society. They suggest that the simple consumer–citizen dichotomy propounded by Sagoff and Jacobs may lack a vital individual level component but in turn provide support for the psychological and cultural theories propounded by authors such as Sarah Lichtenstein and Paul Slovic (2006) and Mary Douglas (1992), suggesting that economists might do well to consider incorporating such views within their models.

Contingent ranking (CR) and choice experiment methods (CE)

Whilst methodologically distinct, CR and CE (or conjoint analysis) methods are sufficiently similar to be considered together. Both involve one-to-one interviews in which respondents make choices between goods, each of which are described via a number of attributes of which one will usually be some monetary or proxy monetary measure. In a CR experiment subjects are asked to give a preference rank order across a number of goods. For example, in a recent study, Foster and Mourato (1997a,b) asked respondents to rank their preferences for various types of bread, one of which was produced via conventional intensive farming methods while others were produced via 'green' agriculture with lower levels of pesticide use. Each loaf was described in terms of various attributes including the price of a loaf and measures of the human health and environmental impacts of associated pesticide use. By varying the price of these products and seeing how this affects respondents' rank ordering the authors were able to infer a valuation for reductions in pesticide use and associated improvements in human health.

The CE method has most frequently been applied to the valuation of recreational assets or sites. Here, respondents, who may be interviewed on or off site, are presented with descriptions of two or more sites, each being characterised via a number of attributes of which one will be some monetary analogue (e.g. a travel cost or entrance fee). Respondents are then asked to choose which site they would prefer to visit. Analysts can then see how respondents' choices change as the site attributes and monetary amounts are varied and from this information infer the value placed on each attribute.

CE models appear to offer three potential benefits over CV studies. First, they more readily provide useful information regarding which attributes a respondent rates most highly in assessing a site (a factor which may aid in planning issues). Second, they appear to provide statistically stronger models of respondents' preferences. Third, they are highly compatible with the revealed preference, travel cost model to which we now turn.

Revealed preference methods

The travel cost (TC) method

The TC method uses the costs incurred by individuals travelling to reach a site as a proxy for its recreation value, that is, values are revealed from individuals' purchases of marketed goods. In effect, the travel costs incurred by an individual in visiting a site represent the vertical 'price' axis on a conventional demand curve diagram such as that shown in Figure 10.2. These travel costs are composed of two elements: the travel expenditure (petrol, fares, etc.) and the value of travel time. These travel costs in part determine the number of visits (horizontal axis, Figure 10.2). In effect, by surveying visitors to a site and asking them for information concerning their travel costs, the frequency of visits over a given period and other determining factors, we can map out the demand curve for the site. As with any good, the value of the site will be equal to the area under this demand curve.

The demand curve is derived from a 'trip-generating function' which simply explains the number of visits (V) as a function of a constant (α), the travel cost (TC) and any other relevant explanatory variables (X) as shown in Equation (2).

$$V = a + b_1 \text{ TC} + b_2 X \qquad (2)$$

Here, the coefficients b_1 and b_2 show the nature and strength of the relationship between the explanatory variables (TC and X, respectively) and the dependent variable (V).

The hedonic pricing (HP) method

Another revealed preference method, in practice the HP method, has often been applied to the valuation of environmental goods such as landscape amenity, noise and air quality as reflected in local house prices. HP models require a two-stage analysis. In stage one the price paid for the environmental good is estimated by statistical analysis of variations in house price in relation to changes in the amount of the environmental good in question. This provides observations along the vertical axis of a demand curve graph. For example, suppose that we were interested in estimating the amenity value of broadleaved trees. Our first-stage analysis would be to see how house price (Hp) varied according to the numbers of broadleaved trees (B) near houses as well as other relevant explanatory variables such as structural characteristics (S; e.g. house size), neighbourhood characteristics (N; e.g. accessibility to workplaces) and other relevant explanatory variables (X). To do this we need to estimate the statistical function shown in Equation (3).

$$\text{Hp} = \alpha + b_1 B + b_2 S + b_3 N + b_4 X \qquad (3)$$

where

α = constant
b_1 = coefficient on number of broadleaves (B)
b_2 = coefficient on structural characteristics (S)
b_3 = coefficient on neighbourhood characteristics (N)
b_4 = coefficient on other explanatory variables (X)

We are interested in b_1 the coefficient on the 'focus' variable B. Analysis of this equation indicates the part of the house price which was paid in respect of the presence (or absence) of broadleaved trees near the house, that is the 'price' of broadleaves (P_B).

Using this information, stage two of the HP analysis maps out the demand curve for the amenity value that broadleaved trees create. To do this we relate the number of broadleaves (B) around houses to the price paid for them (P_B) and any other relevant variables (which we can denote as Z). This gives our demand curve Equation (4):

$$B = \alpha + \gamma_1 P_B + \gamma_2 Z \qquad (4)$$

where

α = constant
γ_1 = coefficient on price paid for broadleaves (P_B)
γ_2 = coefficient on other explanatory variables (Z)

Analysis of Equation (4) maps out the demand curve as the relationship between the price of broadleaves (P_B) and their quantity (B) as shown by the coefficient γ_1. We would expect that $\gamma_1 < 0$, revealing that the characteristic negative relationship between price and quantity demanded is true for amenity views as for other goods.

Garrod and Willis (1992) undertook just such a study of the amenity value of broadleaved trees in Britain. They found a significantly higher price being paid for houses with views of broadleaved trees [i.e. $b_1 > 0$ in Equation (3)] and estimated a characteristic downward sloping demand curve for this amenity [i.e. $\gamma_1 < 0$ in Equation (4)]. Summing the area under this demand curve they duly estimated the value for a variety of scenarios such as planting broadleaved trees around houses or replacing conifer with broadleaved trees.

SUMMARY AND CONCLUSIONS

This chapter has in essence presented two reviews: the first comparing and contrasting the CBA and EIA approaches to project appraisal, while the second has reviewed the various methods available for either pricing or valuing the environmental benefits and costs considered in such appraisals. In this latter review we have seen that, while empirically the 'valuation' methods are sometimes more problematic to apply than the simpler 'pricing' approaches discussed, they do have both wider applicability and directly address the problem of the potentially large divergence between values and prices. Extending CBA by reference to such approaches allows us a greater theoretical ability to incorporate, via a common money unit, those environmental goods and services which otherwise would go unvalued and often by default zero-valued.

In conclusion we return to the mixture of monetary values and physical unit assessments inherent in many practical appraisals and illustrated with respect to the DoT's COBA framework as detailed in Tables 10.1. Should this mixing of CBA and EIA appraisal techniques be avoided? The answer depends on how the resulting information is handled by the decision-maker. In the case of COBA, critics have claimed that assessments made in monetary values may be given greater weight in project appraisals than those measured in physical units (Bateman et al., 1993). However, for many of the items listed in Table 10.1 monetary values are the logical unit of assessment, and physical appraisal would be artificial and unhelpful. Given that the COBA example may well be typical of a large range of real-life appraisals (i.e. where full evaluation of all impacts via either monetary or physical units is impractical, unhelpful or infeasible), investigation of rules for the evenhanded and considered appraisal of the varied information within a CBA/EIA hybrid may well be the most realistic route for future research into the creation and practical implementation of a comprehensive approach to project appraisal.

NOTES

1 Note the use of 'might' rather than 'would'. CBA would only definitely be superior if we were solely interested in maximising the welfare of present-generation humans.

2 See discussions in Baumol (1968); Pearce and Nash (1981); Goodin (1982); Sen (1982); Lind (1982a,b); Sagoff (1988); Pearce and Turner (1990); Price (1993); Henderson and Bateman (1995); Pearce and Ulph (1995). Note that some critics question the basic motivational behaviour and morality underpinning the assumption of a positive time preference (Price, 1993).

3 Equation (1) is in fact a simplification as it can be shown that consumer surplus is only an approximation of the welfare gap between price paid and WTP. In brief, while WTP and related welfare measures (see below) compensate for the effect of changing consumption upon available real income, consumer surplus does not. The magnitude of error is discussed further in Just et al. (1982).

4 This is known as the rule of diminishing marginal utility; that as consumption of a good increases so the utility (roughly speaking the pleasure) which an additional (marginal) unit provides is less than that of the preceding unit. It generally applies not just to goods but also to income and to trades between differing goods.

5 The characteristic differences between private and public goods are more complex than simply whether or not they possess a market price (see Pearce and Turner, 1990, for further discussion).

6 Note also that by changing property rights and restricting entry, unpriced, open-access public goods like recreation can be transformed into market-priced private goods.

7 I am grateful to Roy Brouwer (IVM, Amsterdam) for assistance regarding this issue.

8 The argument runs through Harrison (1992), Smith (1992), Carson and Mitchell (1993), Hausman (1993), Boyle et al. (1994), Diamond and Hausman (1994), Hanemann (1994), and Bateman et al. (1997b).

9 In theory the 'price' information provided by such techniques could be used as the basis of an investigation of underlying demand curves and hence values. However, this would require repeated observations and is unlikely to outperform explicit 'valuation' methods.

REFERENCES AND BIBLIOGRAPHY

Adamowicz, W.L., Boxall, P.C., Louviene, J.J., Swait, J., and Williams, M. (1999) Stated preference methods for evaluating environmental amenities, in Bateman, I.J. and Willis, K.G. (eds) *Valuing environmental Preferences: Theory and Practice of the Contingent valuation Method in the US, EU, and Developing Countries*, Oxford University Press pp. 460–482.

Barde, J.-P. and Pearce, D.W. (eds) (1991) *Valuing the Environment*, Earthscan, London.

Bateman, I.J. (1991) Social discounting, monetary evaluation and practical sustainability, *Town and Country Planning*, 60: 176–178.

Bateman, I.J. (1992) The economic evaluation of environmental goods and services, *Integrated Environmental Management*, 14: 11–14.

Bateman, I.J. (1993) Valuation of the environment, methods and techniques: revealed preference methods, in Turner, R.K. (ed.) *Sustainable Environmental Economics and Management: Principles and Practice*, Belhaven Press, London, pp. 192–265.

Bateman, I.J. (1995) Environmental and economic appraisal, in O'Riordan, T. (ed.) *Environmental Science for Environmental Management*, Longman, New York.

Bateman, I.J. (1996) An economic comparison of forest recreation, timber and carbon fixing values with agriculture in Wales: a geographical information systems approach, *Ph.D. Thesis*, Department of Economics, University of Nottingham.

Bateman, I.J. (1999) Environmental impact assessment. Cost-benefit analysis and the valuation of environmental impacts, in Reffs, J. (ed.) *Handbook of Environmental Impact Assessment*, Vol. 1 Blackwell Science, Oxford, pp. 93–120.

Bateman, I.J. and Bryan, F. (1994) Recent advances in the monetary evaluation of environmental preferences, in proceedings of the conference *Environmental Economics, Sustainable Management and the Countryside*, Countryside Recreation Network, Cardiff.

Bateman, I.J., Garrod, G.D., Brainard, J.S. and Lovett, A.A. (1996) Measurement, valuation and estimation issues in the travel cost method: a geographical information systems approach, *Journal of Agricultural Economics*, 47(2): 191–205.

Bateman, I.J. and Langford, I.H. (1997) Budget constraint, temporal and ordering effects in contingent valuation studies, *Environment and Planning A*, July 1997: 1215–1228.

Bateman, I.J., Turner, R.K. and Bateman, S.D. (1993) Extending cost benefit analysis of UK highway proposals: environmental evaluation and equity, *Project Appraisal*, 8(4): 213–224.

Bateman, I.J., Munro, A., Rhodes, B., Starmer, C. and Sugden, R. (1997a) A test of the theory of reference-dependent preferences, *Quarterly Journal of Economics*, 112(2): 479–505.

Bateman, I.J., Munro, A., Starmer, C. and Sugden, R. (1997b) Does part–whole bias exist? An experimental investigation, *Economic Journal*, 107(441): 322–332.

Baumol, W.J. (1968) On the social rate of discount, *American Economic Review*, 58: 788–802.

Bisset, R. (1978) Quantification, decision-making and environmental impact assessment in the United Kingdom, *Journal of Environmental Management*, 7: 43–58.

Blamey, R.K. (1995) Citizens, consumers and contingent valuation: an investigation into respondent behaviour, *Ph.D. Thesis*, Australian National University, Canberra.

Boyle, K., Desvourges, W.H., Johnson, F.R., Dunford, R.W. and S.P. Hudson (1994) An investigation of part–whole biases in contingent valuation studies, *Journal of Environmental Economics and Management*, 27, 64–38.

Brouwer, R., Powe, N., Langford, I.H., Georgiou, S., Bateman, I.J. and Turner, R.K. (1997) Combining qualitative and quantitative social research in environmental valuation, paper presented at the *Symposium on Environmental Valuation*, Centre d'Economie et d'Ethique pour l'Environnement et le Développement, Domaine des Vaux de Cernay, France, 4–7 October 1997, Centre for Social and Economic Research on the Global Environment, University of East Anglia and University College London.

Buckley, G.P. (ed.) (1989) *Biological Habitat Reconstruction*, Belhaven Press, London.

Burgess, J., Clark, J. and Harrison, C.M. (1998) Respondents' evaluations of a CV survey: a case study based on an economic valuation of the wildlife enhancement scheme, Pevensey Levels in East Sussex. *Area*, 30(1): 19–27.

Carson, R.T. and Mitchell, R.C. (1993) The issue of scope in contingent valuation studies, *American Journal of Agricultural Economics*, December, 1265–1267.

Carson, R.T. and Mitchell, R.C. (1995) Sequencing and nesting in contingent valuation surveys, *Journal of Environmental Economics and Management*, 28: 155–173.

Cicchetti, C.J. and Freeman, A.M. III (1971) Optimal demand and consumer's surplus: further comment, *Quarterly Journal of Economics*, 85: 528–539.

Clark, B.D., Chapman, K., Bisset, R., Wathern, P. and Barrett, M. (1981) *A Manual for the Assessment of Major Development Proposals*, Report prepared for the Scottish Development Department, the Department of the Environment and the Welsh Office, Project Appraisal for Development Control Research Team, Department of Geography, Aberdeen University, HMSO, London.

Coleman, D. and Nixson, F. (1994) *Economics of Change in Less Developed Countries*, Harvester Wheatsheaf, London.

Costanza, R. (ed.) (1991) *Ecological Economics*, Columbia University Press, New York.

Countryside Commission (1990) *A Countryside for the 21st Century*, Countryside Commission, Cheltenham.

Diamond, P.A. and Hausman, J.A. (1994) Contingent valuation: is some number better than no number?, *Journal of Economic Perspectives*, 8(4): 45–64.

Douglas, M. (1992) *Risk and Blame: Essays in Cultural Theory*, Routledge, Oxford.

European Community (EC) (1992) *CAP Monitor: 24.6.92*, EC, Brussels.

Foster, V. and Mourato, S. (1997a) Behavioural consistency, statistical specification and validity in the contingent ranking method: evidence from a survey of the impacts of pesticide use in the UK, *CSERGE Global Environmental Change Working Paper GEC 97–09*, Centre for Social and Economic Research on the Global Environment, University College London and University of East Anglia.

Foster, V. and Mourato, S. (1997b) Are consumers rational? Evidence from a contingent ranking experiment, paper presented at the *Eighth Annual Conference of the European Association of Environmental and Resource Economists (EAERE)*, Center for Economic Research, Tilburg University, the Netherlands, 26–28 June 1997.

Garrod, G.D. and Willis, K.G. (1992) The environmental economic impact of woodland as two stage hedonic price model of the amenity value of forestry in Britain, *Applied Economics*, 24: 715–728.

Glasson, J., Therivel, R. and Chadwick, A. (1994) *Introduction to Environmental Impact Assessment: Principles and Procedures, Process, Practice and Prospects*, UCL Press, London.

Goodin, R.E. (1982) Discounting discounting, *Journal of Public Policy*, 2: 53–72.

Goodpaster, K.E. (1978) On being morally considerable, *The Journal of Philosophy*, 75: 308–325.

Hanemann, W.M. (1991) Willingness to pay and willingness to accept: how much can they differ?, *American Economic Review*, 81(3): 635–647.

Hanemann, W.M. (1994) Valuing the environment through contingent valuation, *Journal of Economic Perspectives*, 8(4): 19–43.

Hanley, N.D. and Spash, C. (1993) *Cost–Benefit Analysis and the Environment*, Edward Elgar, Aldershot.

Hanley, N., MacMillan, D., Wright, R.E., Bullock, C., Simpson, I., Parsisson, D. and Crabtree, R. (1997) Contingent valuation versus choice experiments: estimating the benefits of Environmentally Sensitive Areas in Scotland, paper presented at the *Annual Conference of the Agricultural Economics Society*, Edinburgh.

Harrison, G.W. (1992) Valuing public goods with the contingent valuation method: a critique of Kahneman and Knetch, *Journal of Environmental Economics and Management*, 23: 248–257.

Hausman, J. (ed.) (1993) *Contingent Valuation: A Critical Assessment*, North Holland, Amsterdam.

Henderson, N. and Bateman, I.J. (1995) Empirical and public choice evidence for hyperbolic social discount rates and the implications for intergenerational discounting, *Environmental and Resource Economics*, 5: 413–423.

Hicks, J.R. (1939) The foundations of welfare economics, *Economic Journal*, 49: 696–712.

Hill, B. (1990) *An Introduction to Economics for Students of Agriculture*, 2nd edn., Pergamon Press, Oxford.

Holland, M.R. and Eyre, N. (1992) Evaluation of the external costs of a UK coal fired power station on agricultural crops, *Proceedings of the 2nd International Conference on the External Costs of Electrical Power*, September 1992, Racine, WI.

Howorth, L.S. (1991) Highway construction and wetland loss: mitigation banking programs in the Southeastern United States, *The Environmental Professional*, 13: 139–144.

Hutchinson, W.G., Chilton, S.M. and Davis, J. (1995) Measuring non-use value of environmental goods using the contingent valuation method: problems of information and cognition and the application of cognitive questionnaire design methods, *Journal of Agricultural Economics*, 46(1): 97–112.

Jacobs, M. (1994) The limits of neo-classicism: towards an institutional environmental economics, in Redclift, M. and Benton, T. (eds), *Social theory and the global environment*, Routledge, London.

Jacobs, M. (1997) Environmental valuation, deliberative democracy and public decision-making institutions, in Foster, J. (ed) *Valuing Nature?*, Routledge, London.

Johansson, P-O. (1993) *Cost–Benefit Analysis of Environmental Change*, Cambridge University Press, Cambridge.

Johnson, A. (1996). 'It's good to talk': the focus group and the sociological imagination, *The Sociological Review*, 517–538.

Just, R.E., Hueth, D.L. and Schmitz, A. (1982) *Applied Welfare Economics and Public Policy*, Prentice Hall, Englewood Cliffs, NJ.

Kahneman D. and Knetsch, J.L. (1992) Valuing public goods: the purchase of moral satisfaction, *Journal of Environmental Economics and Management*, 22(1): 57–70.

Kahneman, D. and Tversky, A. (1979) Prospect theory: an analysis of decisions under risk, *Econometrica*, 47(2): 263–291.

Kaldor, N. (1939) Welfare propositions of economics and interpersonal comparisons of utility, *Economic Journal*, 49: 549–552.

Land Use Consultants and the Environmental Appraisal Group (1994) *Good Practice on the Evaluation of Environmental Information for Planning Projects: Research Report*, Department of the Environment Planning Research Programme, HMSO, London.

Leopold, A. (1949) *A Sand County Almanac and Sketches Here and There*, Oxford University Press, New York.

Lichfield, N. (1988) *Economics in Urban Conservation*, Cambridge University Press, Cambridge.

Lichfield, N., Kettle, P. and Whitbread, M. (1975) *Evaluation in the Planning Process*, Pergamon, Oxford.

Lichtenstein, S, and Slovic, P. (eds) (2006) The construction of preference, Cambridge University Press, Cambridge.

Lind, R.C. (1982a) A primer on the major issues relating to the discount rate for evaluating national energy options, in Lind, R.C. (ed.) *Discounting for Time and Risk in Energy Policy*, Johns Hopkins University Press, Baltimore.

Lind, R.C. (1982b) Introduction, in Lind, R.C. (ed.) *Discounting for Time and Risk in Energy Policy*, Johns Hopkins University Press, Baltimore.

McClelland, G., Schulze, W., Lazo, J.K., Waldman, D., Doyle, J.K., Elliot, S.R. and Irwin, J. (1992) *Methods for Measuring Non-Use Values: A Contingent Valuation Study of Groundwater Cleanup*, Center for Economic Analysis, University of Colorado, Boulder.

Medley, G. (1992) Nature, the environment and the future, paper presented at *The University of the Third Age, 1992 International Symposium: The Challenges of the Future*, 13–18 September 1992, Kings College, Cambridge.

Mitchell, R.C. and R.T. Carson (1989) *Using Surveys to Value Public Goods: The Contingent Valuation Method*. Resources For the Future, Washington, D.C.

Morris, P. and Therivel, R. (eds) (1995) *Methods of Environmental Impact Assessment*, UCL Press, London.

Moyer, H.W. and Josling, T.E. (1990) *Agricultural Policy Reform: Politics and Process in the EC and US*, Harvester Wheatsheaf, Hemel Hempstead.

MTPW (Ministry of Transport and Public Works) (1990) *Formal declaration of the Third International Conference for the Protection of the North Sea: The Hague 1990*, MTPW, The Netherlands.

Newman, E.I. (1993) *Applied Ecology*, Blackwell Scientific, Oxford.

Nix, J. (1993) *Farm Management Pocketbook,* 23rd edn, Wye College, University of London.

OECD (Organisation for Economic Cooperation and Development) (1992) *Tables of Producer Subsidy Equivalents and Consumer Subsidy Equivalents 1978–1991*, OECD, Paris.

O'Riordan, T. and Turner, R.K. (eds) (1983) *An Annotated Reader in Environmental Planning and Management*, Pergamon Press, Oxford.

Pearce, D.W. (1984) *Cost Benefit Analysis*, 2nd edn, Macmillan, London.

Pearce, D.W. and Markandya, A. (1989) *The Benefits of Environmental Policy*, Organisation of Economic Cooperation and Development, Paris.

Pearce, D.W. and Nash, C.A. (1981) *The Social Appraisal of Projects: A Text in Cost–Benefit Analysis*, Macmillan, Basingstoke.

Pearce, D.W. and Turner, R.K. (1990) *The Economics of Natural Resources and the Environment*, Harvester Wheatsheaf, Hemel Hempstead.

Pearce, D.W. and Ulph, D. (1995) A social discount rate for the United Kingdom, *CSERGE GEC Working Paper 95–01*, Centre for Social and Economic Research on the Global Environment, University of East Anglia and University College London.

Pearce, D.W., Markandya, A. and Barbier, E.B. (1989) *Blueprint for a Green Economy*, Earthscan, London.

Perman, R., Ma, Y. and McGilvray, J. (1996) *Natural Resource and Environmental Economics*, Longman, Harlow.

Price, C. (1993) *Time, Discounting and Value*, Blackwell, Oxford.

Rabinowitz, D., Cairns, S. and Dillon, T. (1986) Seven forms of rarity and their frequency in the flora of the British Isles, in Soulé, M.E. (ed.) *Conservation Biology: The Science of Scarcity and Diversity*, Sinauer, Sunderland, MA, pp. 182–204.

Randall, A. and Stoll, J.R. (1980) Consumer's surplus in commodity space, *American Economic Review*, 70(3): 449–455.

Ratcliffe, D.A. (ed.) (1977) *A Nature Conservation Review:* Vols 1 & 2, Cambridge University Press, Cambridge.

Rawls, J. (1972) *A Theory of Justice*, Oxford University Press, Oxford.

Rollston, H. (1988) *Environmental Ethics*, Temple University Press, Philadelphia.

SACTRA (Standing Advisory Committee on Trunk Road Assessment) (1992) *Assessing the Environmental Impact of Road Schemes*, HMSO, London.

Sagoff, M. (1988) *The Economy of the Earth*, Cambridge University Press, Cambridge.

Schkade, D.A. and Payne, J.W. (1994) How people respond to contingent valuation questions: a verbal protocol analysis of willingness to pay for an environmental regulation, *Journal of Environmental Economics and Management*, 26: 88–109.

Schultz, W. (1986) A survey on the status of research concerning the evaluation of benefits of environmental policy in the Federal Republic of Germany, paper presented at the *OECD Workshop on the Benefits of Environmental Policy and Decision-making*, Avignon.

Sen, A. (1982) Approaches to the choice of discount rates for social benefit–cost analysis, in Lind, R.C. (ed.) *Discounting for Time and Risk in Energy Policy*, Johns Hopkins University Press, Baltimore.

Singer, P. (1993) *Practical Ethics*, 2nd ed., Cambridge University Press, Cambridge.

Smith, V.K. (1992) Comment: arbitrary values, good causes, and premature verdicts, *Journal of Environmental Economics and Management*, 22: 71–79.

Spash, C.L. and Hanley, N. (1995) Preferences, information and biodiversity preservation, *Ecological Economics*, 12: 191–208.

Spellerberg, I.F. (1992) *Evaluation and Assessment for Conservation*, Chapman & Hall, London.

Squire, L. and van der Tak, H. (1975) *Economic Analysis of Projects*, Johns Hopkins University Press, Baltimore.

Turner, R.K. (1992) Speculations on weak and strong sustainability, *CSERGE Global Environmental Change Working Paper 92–96*, Centre for Social and Economic Research on the Global Environment, University of East Anglia and University College London.

Turner, R.K. (1999) The place of economic values in environmental valuation, in Bateman, I.J. and Willis, K.G. (eds) *Valuing Environmental Preferences: Theory and Practice of the Contingent Valuation Method in the US, EU, and Developing Countries*, Oxford University Press, Oxford.

Turner, R.K. and Bateman, I.J. (1990) *A Critical Review of Monetary Assessment Methods and Techniques*, Report to the Transport and Road Research Laboratory, Environmental Appraisal Group, University of East Anglia.

Turner, R.K. and Pearce, D.W. (1993) Sustainable economic development: economic and ethical principles, in Barbier, E.B. (ed.) *Economics and Ecology: New Frontiers and Sustainable Development*, Chapman & Hall, London.

Turner, R.K., Bateman, I.J. and Brooke, J.S. (1992) Valuing the benefits of coastal defence: a case study of the Aldeburgh sea defence scheme, in Coker, A. and Richards, C. (eds) *Valuing the Environment: Economic Approaches to Environmental Evaluation*, Belhaven Press, London.

Turner, R.K., Pearce, D.W. and Bateman, I.J. (1994) *Environmental Economics: An Elementary Introduction*, Harvester Wheatsheaf, Hemel Hempstead, and the Johns Hopkins University Press, Baltimore.

Tversky, A. and Kahneman, D. (1991) Loss aversion in riskless choice: a reference-dependent model, *Quarterly Journal of Economics*, 106: 1039–1061.

Vatn, A. and Bromley, D.W. (1995) Choices without prices without apologies, in Bromley, D.W. (ed.) *The Handbook of Environmental Economics*, Blackwell Handbooks in Economics, Cambridge, MA.

Wathern, P. (ed.) (1988) *Environmental Impact Assessment: Theory and Practice*, Unwin Hyman, London.

Watson, R.A. (1979) Self-consciousness and the rights of non-human animals, *Environmental Ethics*, 1(2): 99.

Weisbrod, B.A. (1964) Collective-consumption services of individual consumption goods, *Quarterly Journal of Economics*, 78: 471–477.

Willig, R.D. (1973) Consumers surplus: a rigorous cookbook, *Technical Report No. 98*, Institute for Mathematical Studies in the Social Sciences, Stanford University, CA.

Willig, R.D. (1976) Consumer's surplus without apology. *American Economic Review*, 66(4): 587–597.

Young, M.D. (1992) *Sustainable Investment and Resource Use*, UNESCO/Parthenon, Carnforth.

11

Economic Valuation of Ecosystem Services

Randall A. Kramer

INTRODUCTION

When the well is dry, we know the worth of water. (Benjamin Franklin, *Poor Richard's Almanac*, 1746).

On an ever more crowded planet, the challenges of maintaining healthy ecosystems seem ever more complex. More and more natural ecosystems have been dramatically altered by the activities of humans through their development, manufacturing, and consumption activities. While there is cause for concern about the declining health of ecosystems, there is also an emerging recognition that natural ecosystems make significant contributions to economic and human well-being (Heal, 2000). Or, in a term increasingly used by economists as well as environmental scientists, nature provides important "ecosystem services."

In *Nature's Services: Societal Dependence on Natural Ecosystems*, Gretchen Daily (1997) defines ecosystem services as "the conditions and processes through which natural ecosystems, and the species that make them up, sustain and fulfill human life" (p. 3). For instance, wetlands and meandering streams act as a sort of filter, removing or neutralizing pollutants that might otherwise end up in oceans and lakes, affecting the seafood we eat and the water we drink. Protecting tropical forests can conserve unique biodiversity that may be the source of future pharmaceutical or agricultural products and can store carbon to offset increasing carbon emissions of our manufacturing and transportation activities. The importance of nature's services lends credence to the need to do a better job of managing the systems.

The United Nations' sponsored Millennium Ecosystem Assessment (2005) recently reviewed the current status and trends of the world's ecosystems. It found that ecosystem services are under unprecedented threats from economic activities, and the loss of services derived from ecosystems constitutes a significant barrier to achieving the UN Millennium Development Goals to reduce disease, hunger, and poverty. The assessment concluded that "We must learn to recognize the true value of nature – both in an economic sense and in the richness it provides to our lives in ways much more difficult to put numbers on" (Millennium Ecosystem Assessment, p. 5).

This chapter begins by describing the different components that make up the total value of ecosystems and examines the tools that economists use to estimate these values.[1] The chapter then focuses on presenting several case studies that show these tools in use. The case studies focus on valuation of water services, but the cases illustrate how economic valuation might apply to other ecosystem goods and services such as maintenance of biodiversity. Finally, a concluding section discusses implications of this literature for the improved management of ecosystems.

The world's human population depends on healthy ecosystems for a variety of important services. Our decisions to protect, restore, and manage ecosystems should be based in part on a better understanding of how humans benefit from ecosystems and how human behavior can be modified through regulation, economic incentives, and other policy initiatives. Those who advise policymakers on matters of ecosystem management can learn much from the field of environmental economics on the valuation of ecosystem services. With a better understanding of the often neglected and undervalued ecological benefits of improved management of our natural resources, we will be able to design more effective resource management policies.

ECOSYSTEM VALUATION

Water has economic value in all its competing uses and should be recognized as an economic good (The Dublin Principles, 1992).[2]

Total economic value of ecosystems

Government agencies and environmental organizations increasingly recognize that it is useful to measure the benefits and costs of different policy actions that may improve, protect, or degrade our natural resources. Successful long-term economic development depends on wise use of natural resources, and on avoiding, as much as possible, the detrimental impacts of development activities. These impacts can be avoided with more careful planning and design of transportation, urban development, and other infrastructure projects, and by more careful attention to impacts during implementation of the projects (Dixon *et al.*, 1986). Failure to recognize fully the value of natural resources can lead to policies that discourage wise resource use. For example, deforestation may be accelerated and the opportunities for residual forests may be impaired if the environmental services provided by forests are not valued by governments because those services are not traded in the market place (Kramer *et al.*, 1992).

Economic analysis of the environmental impacts of projects and policies has its roots in a body of theory developed by the economists Arthur Pigou (1920) and John Hicks (1939). They held that policies and projects should be based on the resulting changes in social welfare, where social welfare is the sum of individual welfare. Individual welfare is measured by each person's willingness to pay (WTP) for the changes brought about by a policy or project. The intuition behind monetary measurement of project benefits is rather straightforward: people show their preferences for those things they desire by their willingness to spend money to purchase them. The challenge to those who wish to determine the economic benefits of ecosystem services is to develop measurements of the individual willingness to pay and then to aggregate that over the relevant population of beneficiaries.

The total economic value of an environmental resource can be calculated as a sum of four main components: use value, indirect use value, option

value, and nonuse value (Randall, 1991; Pagiola *et al.*, 2004):

- *Use value* refers to the benefit people receive from direct use of the environment: withdrawing water from a river for drinking or irrigation, collecting medicinal plants in a forest, for example. Use value can also include nonconsumptive uses like camping and hiking in a protected ecosystem area. Use value can be diminished by overharvesting, pollution, or certain types of development.
- *Indirect use value* arises from services that users get indirectly and often some distance away from where they originate. Examples include carbon storage and climate regulation provided by intact forests.
- *Option value* refers to users' WTP to preserve the *possibility* of using a resource in the future. An example is the value of protecting a reservoir from nearby development because it might be needed as a future source of drinking water for a municipality.
- *Nonuse value* reflects what people are willing to pay to protect resources they will never use. Some people may derive satisfaction from endangered species being protected out of a sense of environmental stewardship that is unrelated to direct or indirect use, current or in the future.

These various components of resource value constitute the total economic value:

Total economic value = Use value + Indirect use value + Option value + Nonuse value

For monetary valuation, some of these components are easier to measure than others (Pagiola *et al.*, 2004). Use values are generally the most straightforward to measure because there are observable quantities of products consumed as well as market prices that can be used to determine economic value. Recreational use can also be measured by observing the number of visits and the characteristics of visitors and sites. Indirect use values are more difficult to measure for two reasons. First, quantities are often a challenge to measure, for example, determining the flood control provided by a particular wetland. Second, the indirect uses are not usually traded in marketplaces and therefore have no associated prices. Hence, "shadow values" must be estimated in order to "price" the produced services. Option values and nonuse values are the most difficult to measure because these are not reflected in observable behavior. These values are estimated by using surveys that ask people a series of questions about their willingness to pay for ecosystem services they do not use.

For those environmental services provided by ecosystems that are not priced and traded in a marketplace, environmental economists have developed a set of methods to estimate their economic value. Two major categories of methods are used: stated preference and revealed preference methods. Stated preference methods use surveys to elicit directly from individuals the economic value they assign to nonmarket ecosystem services. Revealed preference methods rely on observations of the choices that people make to infer values of resources they are using. No single method is appropriate for every valuation situation.

Despite a growing interest on the part of the academic, governmental, and nongovernmental sectors, environmental valuation remains controversial. Even those who accept the rationale for environmental valuation have ongoing debates about methodological issues, including choice of method, survey design, and selection of econometric models (McMahon and Postle, 2000). The biggest controversy, however, arises from ethical concerns about placing monetary values on environmental services (Foster, 1997). Mark Sagoff (1988), an environmental philosopher, argues that people hold altruistic "citizen preferences" about environmental resources and hence cannot engage in meaningful monetary valuations of the resources. He maintains that using aggregated WTP estimates

in a benefit–cost analysis is an inappropriate way to inform environmental policymaking. Other critics argue that an emphasis on benefit–cost analysis may skew the political process by giving too much influence to the analysts or to the "questionable" information they provide (Shabman and Stephenson, 2000). David Pearce (1999) responds that using resources to pursue a social objective will always impose opportunity costs. So spending more to clean up mercury in the environment means less funding will be available to preserve critical habitat. Therefore, it is appropriate and useful for policymakers to be able to compare the monetary value of different policy options (McMahon and Postle, 2000).

Environmental valuation methods

Environmental economists have developed a number of valuation methods to value services provided by ecosystems. This section will briefly review five of the more widely used methods and provide examples in the context of services provided by freshwater ecosystems. Table 11.1 summarizes these five methods, showing which water-related services are appropriate for the method, and outlining data requirements and method limitations.

Table 11.1 Economic valuation methods for ecosystem services

Method	Approach	Water service appropriate for method	Data needs	Limitations
Contingent valuation method	Ask people directly their willingness to pay (WTP)	All use values and nonuse values (e.g. drinking water, fishing, protecting species)	Survey with scenario description and questions about WTP for specific services	Potential biases due to hypothetical nature of scenarios
Travel cost method	Estimate demand curve from data on travel expenditures and choices	Recreation: boating, fishing, swimming	Survey on expenditures of time and money to travel to specific sites	Only captures recreational benefits; difficult to apply for multiple destination trips
Hedonic property value method	Identify contribution of environmental quality to land values	Water quality, wetland services	Property values and characteristics including environmental quality	Requires extensive information about ecosystem services at hundreds of specific sites
Change in productivity method	Assess impact of change in water service on produced goods	Commercial fisheries, agricultural uses	Impact of change in water service on production; net value of produced goods	Information on biological impacts of changes in ecosystem services often unavailable
Benefit transfer method	Uses value estimates from past studies to approximate value at a new policy site	Any services valued at comparable sites in previous studies	Valuation estimates for same services at similar sites	Limited accuracy due to wide variety of factors that vary across sites

Adapted from Pagiola *et al.* (2004)

The most widely used approach to measuring the economic benefits of environmental conservation is the *contingent valuation method* (CVM). This is a "stated preference method" that allows a sample of people who benefit from a particular resource to tell researchers directly, through surveys, what they are willing to pay for some improvement in environmental quality. Demand for nonmarket goods such as a particular ecosystem service is established by first describing to individuals the characteristics of a simulated market and then asking what they would be willing to pay to ensure that the good is available. One of the strengths of this method is that it can capture both use value (e.g. drinking water use) and nonuse value (e.g. protection of threatened aquatic species) (Mitchell and Carson, 1993). Because of this versatility, it is the most widely used method, although it is controversial because, critics say, people are reporting *hypothetically* on their WTP rather than observed actually spending the money, possibly biasing the resulting valuation estimates (Hanemann, 1995). These concerns can be addressed with careful survey design and implementation (Carson *et al.*, 2001).

How ecosystem values changed a water policy debate in Southern California

One of the best examples of ecosystem values having an impact on public decision-making is the case of allowing tributary waters to flow into Mono Lake in California versus diverting the flows for municipal and industrial water users in Los Angeles. In 1983, the California Supreme Court ordered a reevaluation of Los Angeles' water rights and a balancing of public trust water uses. A contingent valuation study by John Loomis (1987) showed that people were willing to pay for the protection of birds and fish in Mono Lake and that these benefits far exceeded the replacement cost of water from other sources. As a result of this initial study, California's Water Resources Board required that the state's Environmental Impact Report include non-use ecosystem values in its analysis of water reallocation alternatives. In the analysis, non-use ecosystem values were compared dollar for dollar to the hydropower and water supply benefits. Eventually, the state required that tributary flows to Mono Lake be increased significantly, and Los Angeles' water rights were cut almost in half. Although the driving concerns were air and water quality, the economic analysis showing that the new allocation generated important nonuse economic benefits likely influenced this major policy shift (Loomis, 2000).

Another widely used approach to valuing water ecosystem services is the *travel cost method*, a "revealed preference" approach that is based on how people make recreational choices (Smith and Desvousges, 1986). The underlying principle is that people spend time and money to travel to and use a site for recreation. By measuring the expenditure of time and money, we can determine what the recreational service is worth. There are two main versions of this method (Freeman, 1993). The first version estimates a statistical relationship between the number of visits at a site and the level of travel expenditures by visitors, and uses that relationship to estimate the total value of recreation services provided by the site to all users. The second version uses statistical analysis to examine how specific site characteristics influence decisions to recreate at different sites and then to infer the economic value of those characteristics. The travel cost method can be used to estimate the economic values associated with various types of protected ecosystems that afford recreational opportunities.

Although many environmental goods are not traded in markets, their presence may have an affect on property values. The *hedonic property value method* takes advantage of this connection (Smith, 1993; Taylor, 2003). Land prices are usually higher for land parcels close to lakes or estuaries because of the views and boating or fishing opportunities. By statistical analysis, the part of land values due to these environmental services can be separated out. The method controls for other variables influencing land prices so that any remaining price differential is a measure of the WTP for the unpriced environmental good. As an example, the hedonic method could be used to value the availability of better quality water. Suppose that by protecting a forest in its natural state rather than using it for timber production, downstream water quality is enhanced due to reduced land disturbance. If downstream farmers have higher incomes because of the better quality water, those farms should have higher property values. A hedonic study of farms in both high and low water quality areas could control for other differences between the farms and estimate the marginal contribution of water quality to land values.

The *change in productivity method* recognizes that when changes in environmental quality affect the production of marketed goods, these effects can be captured by observing what happens in a related market (Freeman, 1993). So if water pollution reduces fish catches or acid rain reduces timber harvest, we can value those productivity impacts based on the price of the resource, for example, fish or timber. The basic premise is that changes in ecosystem services can impact the quantity and quality of products being marketed. Once those biological impacts are identified, the productivity impacts can be valued through standard economic analysis based on the prices of related goods that are sold in the marketplace (Kramer *et al.*, 1997). Consider the example of wetlands that provide breeding areas and increased food supply for various nearby fisheries. If these fisheries are commercially exploited, then the value of a wetland can be measured in part by the dollar value of the increase in fish catches resulting from the wetland. This method requires an interdisciplinary approach involving biologists and economists.

Finally, there are many policy situations where it is not feasible to conduct an original ecosystem valuation study, either because of time or resource constraints. In such situations, analysts may want to take advantage of previous valuation studies of the same ecosystem services evaluated at other sites (Desvousges *et al.*, 1992). *Benefit transfer* refers to taking existing knowledge and transferring it into a new context. Data are taken from one or more study sites and applied to a policy site where a decision is being made that would benefit from valuation information. How accurate are valuation estimates derived from other sites? Studies have shown that benefit transfer may have limited accuracy because of the large number of factors that can vary across sites, but when the study and policy sites are in the same geographic area, errors are reduced (Rosenberger and Loomis, 2003).

Examples of valuation studies

In this section, several examples are presented of environmental valuation methods applied to aquatic ecosystems. These studies illustrate a range of applications and provide some details on how the methods are applied. Readers interested in additional details should consult the original publications.

Travel cost example: How does atrazine affect water recreation?

Dietrich Earnhart and Val Smith (2003) examined the effects of the pesticide atrazine on water-based recreation at Lake Clinton, Kansas. Atrazine may enhance recreational enjoyment by inhibiting the growth of nuisance algae and thus encourage greater recreation; but the presence of atrazine in reservoirs may be detrimental to fish populations and hence, reduce recreational use. To quantify and compare these countervailing effects, the authors applied the travel cost method in combination with contingent behavior questions.

The authors conducted a survey of 245 residents of Lawrence, Kansas, about their recreational use of Lake Clinton, collecting data on visitation patterns and socioeconomic characteristics. They calculated respondents' travel costs to the lake as

the sum of transportation cost (at 31.5 cents per mile), their time costs (wage rate times the two-way driving time), and access fees. In addition, the researchers asked respondents how their chosen destinations would change with various changes in water quality. Some changes were described as a decrease in algae, some were described as a decrease in fish, and some were described as a combination of the two effects of atrazine.

They found that the average respondent had a $22 trip cost to the lake and made about three trips in the previous year. An improvement in algae-related water quality would lead to an average increase of 2.7 visits, while a decline in fish-related quality would trigger an average decrease of 0.5 visits. The combination of quality changes would lead to an average decrease of 0.6 visits.

The authors then conducted an in-depth statistical analysis of likely travel behavioral responses to water quality change in the light of countervailing effects of atrazine on algae quality and fish quality. They examined the tradeoffs for recreators between these two quality dimensions and concluded that for each 1 per cent decline in fish-related quality, respondents required a 4.7 per cent increase in algal-related quality so as to maintain their same level of recreational enjoyment. While they did not monetize the overall impact of atrazine, they concluded that "knowing the effective rate of exchange between fish- and algae-related water quality in Clinton Reservoir will allow reservoir managers to estimate recreators' responses to future changes in the watershed" (p. 1089).

Hedonic property value example: effects of water quality on residential land prices

Compared to a large number of studies of air quality, the hedonic property value method has been used only a handful of times to value changes in water-related ecosystem services. An excellent example of the potential usefulness of this approach is illustrated by Christopher Leggett and Nancy Bockstael's investigation (2000) of whether water quality affects residential property values along the Chesapeake Bay. They were able to take advantage of a favorable geographic situation: "a highly irregular estuarine coastline that supports a lively market for waterfront homes and that exhibits considerable variation in water quality within a small area" (p. 122).

The authors used data from waterfront property sales from 1993 to 1997 in Anne Arundel County, Maryland. One of the most challenging aspects of using the hedonic method is measuring environmental quality for each property site. They used fecal coliform data from samples collected at 104 sites along the county's coastline and constructed a water quality measure based on the distance of each property from the nearest monitoring station.

Box 1   Summary of Water Quality Management Plan Presented to Catawba Basin Survey Respondents

This management plan addresses the main water-pollution problems in the basin: sediment and nutrients. It also continues to manage related problems such as pollution by toxic substances and bacteria and viruses. While this specific management plan has not been proposed by state governmental agencies, it is drawn from their best available information. This includes information on the condition of the basin and how to best manage the problems.

This potential management plan includes the following components:

1   Construction and use of best management practises (BMPs) within the basin. These include buffer strips and holding ponds for farms, construction sites, and residential areas.
2   Development of a basinwide land-use plan. This would encourage land uses in the basin that are consistent with the goals for water quality in the basin. Government agencies could use this land-use plan to make decisions that would affect water quality.
3   Improving and increasing the capacity of sewage treatment plants in cities within the basin.
4   Purchasing and setting aside of tracts of land that have been determined as critical to the protection of water quality.

The results showed that coliform levels had a significant and negative impact on property values. Once the researchers established this significant effect of water quality on property values, they demonstrated how their results could be used to value water quality improvements. They illustrated the usefulness of the hedonic model by focusing on a hypothetical localized improvement in water quality on waterfront property values along the Saltworks Creek Inlet northwest of Annapolis. They found that modest reductions in fecal coliform counts in the middle and upper reaches of the inlet increased property values by 2 per cent. While this may appear to be a small impact, the potential gains across all properties in the county could amount to more than $12 million if water quality was improved by a similar amount elsewhere. The study makes a convincing case that waterfront owners exhibit a strong willingness to pay for reducing concentrations of fecal coliform bacteria.

Contingent valuation example: the economic value of water quality in the Catawba River basin

A Duke University study used the contingent valuation method to estimate the economic value of protecting water quality in the Catawba River basin at its current level (Eisen-Hecht and Kramer, 2002; Kramer and Eisen-Hecht, 2002). Telephone interviews were conducted with 1085 randomly selected households in 16 counties within the Catawba River basin in North and South Carolina. Before the interviews, the survey respondents were mailed a short information booklet that described a water quality management plan (summarized in Box 1). Respondents were then asked if they would support the management plan. The management plan was offered to respondents at one of eight different price levels, ranging

from $5 to $250 per year for five years (Box 2). The contingent valuation scenario was developed through reviews of other studies and refined during focus groups and pretests.

The survey results indicated that, besides showing a high level of concern about water quality, area residents placed a significant monetary value on protecting water quality in the Catawba basin. Two-thirds of the respondents expressed a WTP, through an increase in state income taxes, for the management plan described in the presurvey booklet. The willingness to pay expressed by respondents puts a dollar value on the well-being they receive from the protection of water quality in their region. This well-being translates into an annual economic benefit of $139 per Catawba River basin taxpayer and more than $75 million

Box 2   Contingent Valuation Question for Valuing Water Quality Management Plan

Now, assume a vote is being held today to approve or reject this management plan. Your payment for this plan would be collected through an increase in your usual state income taxes. All residents in counties within the Catawba River basin would make identical payments. This money would only be used for implementing this management plan for the Catawba River basin. If a majority of Catawba basin county residents vote in favor of this management plan, it will go into effect. Before you answer the following question, please consider your current income, as well as your expenses.

Suppose that this management plan would cost you $—(5, 10, 25, 50, 100, 150, 200, 250) each year for the next five years in increased state income taxes. Would you vote in favor of the management plan?

Table 11.2   Willingness to pay to protect Catawba River water quality

Respondent group	Mean willingness to pay ($)
Total sample	139
Comparison across states	
North Carolina residents	135
South Carolina residents	150
Comparison across income levels	
Household income $30,000 and under	116
Household income between $30,001 and $75,000	157
Household income above $75,000	180

for all taxpayers in Catawba basin counties. Table 11.2 shows a distribution of willingness to pay values. South Carolina residents, living near the more polluted downstream portion of the river, were willing to pay more than North Carolina residents were. For residents in both states, willingness to pay rose with household income.

After the contingent valuation question, the survey questionnaire contained various questions designed to elicit additional information from respondents regarding their votes on the management plan. One of these questions sought to uncover the most important reasons why respondents might value the management plan. The highest-rated reason was quality of area drinking water, followed by the knowledge that the waters in the basin were being protected, regardless of respondents' use of them. These results show that their willingness to pay was a function of both use and nonuse values.

The annual benefits from the CVM survey were used as part of a cost–benefit analysis of implementing the water management plan (Table 11.3). Detailed costs were estimated for each component of the management plan. The results showed a net present value of $95 million that would result from implementing the plan, indicating that benefits far outweighed the costs.

IMPLICATIONS FOR IMPROVED ECOSYSTEM MANAGEMENT

It is increasingly recognized that the natural wealth embodied in the Earth's various ecosystems

Table 11.3   Benefit–cost analysis of implementing the Catawba basin management plan

Net present value of benefits over time	$340 million
Net present value of costs over time	$245 million
Benefits minus costs	$95 million

provides critical support of human well-being and economic activity. The economic worth of these ecosystem services often goes unrecognized. Why should we worry about the economic valuation of these services? In part, because there will always be competing demands for the use of both natural resources and limited public financial resources. Benefit–cost analysis has proven to be a useful tool to guide public decision-making in the face of competing interests. Environmental organizations and other public interest groups may find it useful to turn to benefit–cost analysis, including the analysis of nonmarket values, to advocate for a complete accounting of the impacts of public policies. Those concerned about fiscal responsibility of public investments in natural resource related projects may find that the discipline provided by impartial weighing of benefits and costs can contribute to a wiser use of public funds. Because many of the services provided by ecosystems are outside the realm of market transactions, the value of these service flows is best evaluated with nonmarket methods developed by environmental economists. It is important for environmental professionals to be better informed about these methods, when they are called for, and their strengths and weaknesses (Braden 2000). Environmental valuation is a mature and rapidly growing field, with thousands of applications now complete, many of them applied to ecosystem services. Research in this field has documented a large willingness to pay for improvements in environmental quality and protection of ecosystem services.

The enterprise of producing valuation information and feeding it into the public policy process is unlikely to slow down in the future. As environmental economist Kerry Smith (1993) noted in his appraisal of nonmarket valuation methods, "Environmental resources are increasingly recognized as assets providing services that are no longer readily available. Indeed, demands to measure their values and incorporate them into our decisions is precisely what we would expect as their scarcity increases" (p. 1).

Our expanding knowledge about how to quantify ecosystem services provides opportunities for improved policy formulation regarding the management of those resources. Obviously, the valuation studies will not be the only factor when decisions are being made that will affect the management of ecosystems. But in cases where political decisions are relatively "close calls," estimates of nonmarket values as part of a benefit–cost analysis of policy alternatives may be influential and lead to improved policy formulation (Bennett, 2003).

ACKNOWLEDGMENTS

This review was funded in part by a grant from a coalition of organizations including Environmental Defense, Southern Environmental Law Center, Sierra Club, the North Carolina Conservation Network, the Conservation Council of North Carolina, and the North Carolina Public Interest Research Group. The author appreciates advice and assistance provided by David McNaught, Lisa Dellwo, Rebecca Madsen, Joel Sholtes, Greg Harper, and Jon Eisen-Hecht. The opinions expressed are those of the author only.

NOTES

1 See Chapters 9 and 10 for an in-depth discussion of economic surplus measures and other valuation concepts encountered in estimating total economic value.

2 The 1992 International Conference on Water and Environment in Dublin, Ireland, developed a set of four principles to guide management of freshwater that became known as the Dublin Principles. The other three principles are: (1) fresh water is a finite and vulnerable resource, essential to sustain life, development, and the environment; (2) water development and management should be based on a participatory approach, involving users, planners, and policymakers at all levels; (3) women play a central part in the provision, management, and safeguarding of water (World Meteorological Organization).

REFERENCES

Bennett, Jeff. (2003). "Environmental Values and Water Policy." *Australian Geographical Studies* 41(3): 237–250.

Braden, John B. (2000). "Value of Valuation: Introduction." *Journal of Water Resources Planning and Management* 126: 336–338.

Carson, Richard T., Nicholas E. Flores, and Norman F. Meade. (2001). "Contingent Valuation: Controversies and Evidence." *Environmental and Resource Economics* 19: 173–210.

Daily, Gretchen (ed.) (1997). *Nature's Services: Societal Dependence on Natural Ecosystems*. Washington, DC: Island Press.

Desvousges, W., M. Naughton, and G. Parsons. (1992). "Benefits Transfer: Conceptual Problems in Estimating Water Quality Benefits Using Existing Studies." *Water Resources Research* 28: 657–663.

Dixon, John A., Louise F. Scura, Richard A. Carpenter, and Paul B. Sherman. (1986). *Economic Analysis of Environmental Impacts*. London: Earthscan.

Earnhart, D. and V. Smith. (2003). "Countervailing Effects of Atrazine on Water Recreation: How do Recreators Evaluate Them?" *Water Resources Research* 39: WES 2.1.

Eisen-Hecht, Jon I., and Randall A. Kramer. (2002). "A Cost–Benefit Analysis of Water Quality Protection in the Catawba Basin." *Journal of the Water Resources Association* 38: 453–465.

Foster, John. (1997). *Valuing Nature? Economics, Ethics, and Environment*. London: Routledge.

Freeman, A. Myrick, III. (1993). *The Measurement of Environmental and Resource Values: Theory and Methods*. Washington, DC: Resources for the Future.

Hanemann, W.M. (1995). "Contingent Valuation and Economics." In K.G. Willis and J.T. Corkindale (eds) *Environmental Valuation: New Perspectives*. Cheltenham, UK: CAB International.

Heal, G. (2000) *Nature and the Marketplace: Comparing the Value of Ecosystem Services*. Washington, DC: Island Press.

Hicks, John R. (1939). *Value and Capital: An Inquiry into Some Fundamental Principles of Economic Theory*. Oxford: Clarendon Press.

Kramer, Randall A., and Jon I. Eisen-Hecht. (2002). "Estimating the Economic Value of Water Quality in the Catawba River Basin." *Water Resources Research* 38: 1–10.

Kramer, Randall, Robert Healy, and Robert Mendelsohn. (1992). "Forest Valuation." In Narendra Sharma (ed) *Managing the World's Forests*. Arlington, VA: Kendall Hunt.

Kramer, Randall, Daniel Richter, Subhrendu Pattanayak, and Narendra Sharma. (1997). "Ecological and Economic Analysis of Watershed Protection in Eastern Madagascar." *Journal of Environmental Management* 49: 277–295.

Leggett, C.G. and N.E. Bockstael. (2000). "Evidence of the Effects of Water Quality on Residential Land Prices." *Journal of Environmental Economics and Management* 39: 121–144.

Loomis, John B. (1987). "Balancing Public Trust Resources of Mono Lake and Los Angeles' Water Rights. An Economic Approach." *Water Resources Research* 23(8): 1449–1456.

Loomis, John B. (2000). "Environmental Valuation Techniques in Water Resource Decision Making." *Journal of Water Resources Planning and Management* 126 (6): 339–344.

McMahon, Paul, and Meg Postle. (2000). "Environmental Valuation and Water Resources Planning in England and Wales." *Water Policy* 2: 397–421.

Millennium Ecosystem Assessment. (2005). *Living Beyond Our Means: Natural Assets and Human Well-Being*. New York: United Nations Environmental Program.

Mitchell, Robert C. and Richard T. Carson. (1993). "The Value of Clean Water: The Public's Willingness to Pay for Boatable, Fishable, and Swimmable Quality Water." *Water Resources Research* 29(7): 2445–2454.

Pagiola, Stefano, Konrad von Ritter, and Joshua Bishop. (2004). *Assessing the Value of Ecosystem Conservation* (World Bank Environmental Department Paper, no. 101), The World Bank Environmental Department, in Collaboration with the Nature Conservancy and IUCN – The World Conservation Union.

Pearce, David W. (1999). "Valuing the Environment." In David W. Pearce (ed) *Economics and the Environment: Essays in Ecological Economics and Development*. Cheltenham, UK: Edward Elgar.

Pigou, Arthur C. (1920). *The Economics of Welfare* (1952, 4th edn). London: Macmillan.

Randall, Alan. (1991). "Total and NonUse Values." In J.B. Braden and C.D. Kolstad (eds) *Measuring the Demand for Environmental Quality*. Amsterdam: North-Holland.

Rosenberger, Randall S., and John B. Loomis, (2003). "Benefit Transfer." In P.A Champ, K.J. Boyle and T.C. Brown (eds) *A Primer on Nonmarket Valuation*. Boston: Kluwer Academic.

Sagoff, Mark. (1988). *The Economy of the Earth*. Cambridge: Cambridge University Press.

Shabman, Leonard and Kurt Stephenson. (2000). "Environmental Valuation and Its Economic Critics." *Journal of Water Resources Planning and Management* 126: 382–388.

Smith, V. Kerry. (1993). "Nonmarket Valuation of Environmental Resources: An Interpretive Appraisal." *Land Economics* 69(1): 1–26.

Smith, V. Kerry and William H. Desvousges. (1986). *Measuring Water Quality Benefits*. Norwell, MA: Kluwer Academic.

Taylor, Laura O. (2003). "The Hedonic Method." In P.A Champ, K.J. Boyle and T.C. Brown (eds) *A Primer on Nonmarket Valuation*. Boston: Kluwer Academic.

World Meteorological Organization. (1992). "The Dublin Statement on Water and Sustainable Development." Accessed June 2004 at http://www.wmo.ch/web/homs/documents/english/icwedece.html

12

Assessing Environment–Development Tradeoffs: A Developing Country Perspective

David R. Lee

INTRODUCTION

Many of the most globe's most vexing environmental problems occur in developing countries. These are highly familiar to most observers: deforestation of tropical forests, soil degradation, contamination of surface water and groundwater sources, uncontrolled urban sprawl, air pollution, and so forth. Yet, it is in these same countries that sustainable solutions appear the most challenging. This is for several reasons. First, of course, financial resources are, *ipso facto*, heavily constrained in most poor countries, limiting the ability of governments to respond to environmental problems as well as a multiplicity of other public policy challenges. But reinforcing the lack of resources is a variety of other limitations: weak institutions, poor infrastructure, highly imperfect markets, and educational and health systems which are underdeveloped and underfunded, to mention just a few. Solving environmental policy and management problems must, then, be accomplished recognizing these constraints and within a broader context of simultaneously addressing many other critical social goals: raising food production and enhancing food security, reducing chronic poverty and malnutrition, achieving higher rates of economic growth and improving household livelihoods, all in addition to enhancing environmental outcomes.

The policy question is how best to achieve these goals simultaneously in an environment of limited resources.

Much of the thinking from the international community regarding how to do so has been reflected in a series of commission reports, white papers, planning documents and other high-visibility efforts from the 1970s onward, including, perhaps most notably, the *Report of the World Commission on Environment and Development* (the "Brundtland Report") in 1987. This report focused global attention on sustainable development – development that "meets the needs of the present without compromising the ability of future generations to meet their own needs" – as the measure by which future development strategies should be assessed (WCED, 1987). The Brundtland Report deserves credit for raising the profile of sustainability concerns in the environment–development debate over the past two decades. However, like similar efforts both before and after, the focus on how to achieve sustainable development and "win–win" environment–development outcomes has often morphed into assertions that the existence of synergistic outcomes among multiple objectives is self-evident, that solutions are easily achievable, and that tradeoffs are inconsequential.

This chapter explicitly addresses the question: How can environmental objectives be addressed in the heavily resource-constrained situations facing most developing countries in which environmental problems are only one of many challenges requiring resolution? This is of course a very broad issue, assuming different forms in different countries. To illustrate the broader policy and management challenges, we focus specifically on rural areas of developing countries where issues of poverty and malnutrition are often the greatest and where the most serious environmental challenges include deforestation, soil degradation, water supply and quality, and sustainable food production for burgeoning populations. The chapter first addresses the issues presented by the existence of tradeoffs at the macro or economy-wide level, both conceptually and in the light of empirical evidence. The chapter then turns to assessing tradeoffs at the micro or household level. It ends with a discussion of the factors that are most likely to contribute to successfully achieving environmental sustainability objectives given the multiplicity of other social goals.

ASSESSING ENVIRONMENT AND DEVELOPMENT TRADEOFFS AT THE MACRO LEVEL

At the national or regional level, the issue of tradeoffs among competing social goals is seemingly more straightforward than when considered at the level of the individual rural household. This is in large part due to spatial heterogeneity and the ability of countries to address multiple social goals by focusing on the achievement of one goal in one geographical area or economic sector, and simultaneously focusing on achieving other goals elsewhere. For example, consider the often competing objectives of food production and biodiversity conservation. Increasing food production – whether achieved through more intensive use of purchased inputs (inorganic fertilizers, pesticides, irrigation, etc.) or, in particular, through the clearing of previously undisturbed or minimally disturbed landscapes – typically reduces biodiversity due to the conversion of these landscapes to crop monocultures or polycultures (Swift *et al.*, 2004). However, many countries, at the same time as they are promoting food production in some areas, choose to foster biodiversity conservation through creating national parks, wildlife refuges and other protected areas that serve to enhance the conservation of biological resources. Thus, at a broad geographical scale, these two societal goals are not mutually exclusive (except by degree) and simultaneously achieving them can be accomplished through "win–win" strategies that promote both objectives.

One specific illustration of the environment–development tradeoff is the debate over the "land-sparing" hypothesis in the context of the environmental consequences of the "Green Revolution." The sharp increases in agricultural productivity and enhanced food security experienced in many developing countries since the 1960s, commonly known as the Green Revolution, were due to the adoption of modern high-yielding seed varieties and accompanying packages of purchased inputs, notably pesticides and fertilizers, and associated investment in irrigation systems. The type of intensive agriculture characteristic of Green Revolution systems has been criticized by many observers for the localized problems it has created, including soil salinization and waterlogging, subsoil compaction, changes in soil nutrient balances, groundwater depletion, pest build-up, and an overdependence on purchased inputs which may be unaffordable to the farmer (see review in Pingali and Rosegrant, 1994). Yet as Borlaug and Dowswell (1994), Waggoner *et al.* (1996) and others have pointed out, the vastly increased yields on existing cropland permitted by the Green Revolution have freed up tens of millions of hectares of land elsewhere in developing (and industrialized) countries for other land uses, including conservation and land protection. Thus, taking a global view of land use patterns is critical to accurately understanding the overall environmental consequences of the Green Revolution.

At the same time, achieving multiple goals at the aggregate level is, in at least two senses, equally perplexing as it may be at the household level. First, there is the distributional question of winners versus losers. In the foregoing example, promoting increased food production in one area of a country may inevitably destroy biological resources *in that area*, even if a park or protected area is created elsewhere. So a key corollary question for policy is whether *unique* biodiversity is destroyed in the conversion of land to agriculture and other uses and the extent to which offsetting gains may be created in the decision to locate a protected area elsewhere. Similarly, if creating a national park or protected area involves the forced dislocation of households that formerly lived on those lands, then they are likely to be clear losers (without the availability of offsetting compensatory mechanisms). So even at the aggregate level, spatial heterogeneity will not necessarily create "win–win" outcomes. Second, even if these outcomes can be generated in principle, processes of social choice and the vagaries of the political process will not assure that they are necessarily accomplished. Social choice processes are often complicated and messy and there is no guarantee, even in a democracy, that outcomes desired by the majority will be achieved.

The environmental Kuznets curve

Perhaps the best known approach over the past decade or more to evaluating macro-level tradeoffs involving environmental outcomes is the so-called "Environmental Kuznets Curve" (EKC) approach. The "inverted U" relationship was initially proposed by the economist Simon Kuznets (1955) as a means of explaining the relationship between per capita income and measures of income inequality. In the early 1990s, this "inverted U" relationship was extended by Grossman and Krueger (1991) in their innovative attempt to explain measures of urban air pollution (sulfur dioxide, particulate matter) and average income levels. Shafik and Bandyopadhyay (1992) examined other measures of environmental degradation in their background contribution to the 1992 *World Development Report* of the World Bank, whose publication further raised the profile of this approach. The basic EKC hypothesis is that an "inverted U" exists between some measures of environmental degradation (*Y* axis in Figure 12.1) and per capita income (*X* axis). The underlying argument is that beginning at very low per capita income levels, environmental degradation is low, but as national incomes rise, environmental degradation increases until, at a certain level of income, environmental degradation begins to decline for further increases in incomes. This turning point – point "A" in Figure 12.1 – is commonly called the "environmental transition," the point at which increasing income begins to be associated with improving environmental conditions rather than worsening conditions.

There are many factors that have been hypothesized to account for the EKC relationship, which are only briefly summarized here. The primary underlying argument is a "stages of growth" argument: that the engine of national economic growth shifts over time from an initial reliance on natural resource extraction and management in poor subsistence economies, where pollutants are "clean" and environmental degradation is very low, to a greater reliance on manufacturing, including heavy manufacturing with an associated scaling up of production, and accordingly, of waste, water and air pollution, and toxic pollutants. Eventually, over time, national per capita income rises and economic growth comes to increasingly depend on "post-industrial" sectors such as services, light manufacturing, technology and the information sector, most of which are much less harsh on the environment and less polluting than is a manufacturing-based economy (Panayotou, 1997). The environmental transition wherein development becomes less, rather than more, polluting, is caused by a number of factors, including changes in the mix of outputs, the mix of inputs, greater production efficiency and decreased emission levels per unit of output (Stern, 2003).

Concomitant with this central argument are several related claims. Simultaneous with industrialization, population growth may increase rapidly as mortality rates decline prior to the reduction in birth rates, putting pressure on the existing natural resource base. Eventually, as countries become richer, population growth rates decline. Additionally, environmental quality is claimed to be a luxury good at higher levels of income, suggesting that only rich countries can "afford" environmental quality; thus, it is at these relatively high levels of income that environmental degradation will decline. Income elasticities of demand are low for basic goods like food products (produced in a poor but relatively "clean"

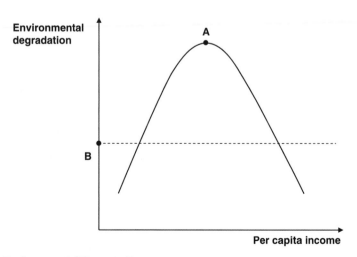

Figure 12.1   Environmental Kuznets Curve

agricultural economy), are higher for most manufactured products, and are highest for services like environmental quality; thus, patterns in demand elasticities contribute to and reinforce the shape of the EKC. Finally, there is the political economic argument that only at a certain level of income can countries afford to possess and financially support institutions that give serious attention to monitoring, regulating and enforcing environmental quality. It is in these countries that those institutions are likely to be the most mature and their mandates consistent with the demands of the populace. All these factors, and others, contribute to the claimed "inverted U" relationship.

These arguments notwithstanding, the all-important question is what does the empirical demonstrate with regard to the hypothesized "inverted U" relationship? The first empirical tests of the EKC were those of Grossman and Krueger applied to patterns of sulfur dioxide and particulate matter emissions in air pollution. Over the past decade and a half, a vast record of empirical research has arisen attempting to replicate and estimate EKC-type relationships for specific indicators measuring a host of environmental problems: air pollution, water contamination, tropical deforestation, public health, public sanitation, solid waste production, and so forth. The attractiveness of the EKC hypothesis is clear: if demonstrated to be true, the existence of the EKC confirms that countries can effectively "grow their way" out of environmental degradation problems. In the short run, economic growth achieved through industrialization may result in worsening environmental indicators. But in the long run, the existence of the declining portion of the EKC suggests that improvements in environmental quality are synergistic with, and not antithetical to, improved economic growth. The policy message is a clear and attractive one.

Unfortunately, the predominance of research on the EKC reveals that empirical support for its existence is, at best, mixed (Stern, 2003; Yandle et al., 2004). Some indicators of environmental degradation – the absence of public hygiene facilities, for example – are monotically decreasing with per capita income. Others such as solid-waste production are mostly increasing with income. Interestingly, the types of income–environment relationships for which the evidence for EKC-type relationships has often been estimated to be strongest are those for which it was first hypothesized by Grossman and Krueger: urban air contamination. This is for several reasons: the geographic areas are often limited and defined; the measures of environmental contamination are technically simple and widely used; and there appear to be fewer confounding socio-economic and biophysical factors likely to complicate estimated EKC relationships.

But for most other indicators, the evidence does not show clear EKC relationships. Tropical deforestation is both one of the most commonly analyzed and one of the most complicated. Some analysts have estimated the existence of an "inverted U" (Cropper and Griffiths, 1994; Antle and Heidebrink, 1995). More recent studies using larger, cross-country data sets have found few discernible relationships (Koop and Tole, 1999; Bhattarai and Hammig, 2001). The application of the EKC hypothesis to deforestation is particularly interesting not only from a policy standpoint, given the prominence of tropical deforestation in international policy debates, but as an empirical counter-example to the conditions typically characterizing urban air pollution. With the case of deforestation, the spatial dimensions are often diffuse and changing over time, the measures are multiple and flawed, and there are many confounding factors that complicate empirical estimation, both over space and time. It is scarcely surprising that the resulting statistical estimates of EKC relationships for deforestation and income, as with other macro-level environmental indicators, are so highly inconsistent.

ASSESSING MICRO-LEVEL ENVIRONMENT–DEVELOPMENT TRADEOFFS

At the micro or household level, assessing tradeoffs among environmental, production and livelihood goals is often more straightforward than at the broader societal level. At the household level, spatial heterogeneity is less, household livelihood strategies are more circumscribed, and the opportunity costs of achieving one goal vis-à-vis another are typically clearer. For example, for poor households with limited land resources, expanding food production through cropland expansion directly competes with setting aside that land to promote long-term forestry or other uses. Similarly, using limited household labor for what may be more highly remunerative off-farm uses will necessarily mean less family labor available on the farm for increased own production (hired labor may still be employed, however). Key tradeoffs in the use of inputs and the outputs achievable from those inputs are often self-evident (though nonetheless difficult to quantify). Additionally, the choice set for allocating resources at the household level is much more limited than at the aggregate level, so the impacts of private resource management decisions on the local landscape are more direct. The challenge – to researchers and environmental and development NGOs, as well as farmers and landowners – is how to promote sustainable solutions to the overriding problems of poverty and malnutrition while simultaneously improving (or not irretrievably worsening) environmental outcomes.

All this is not to suggest, however, that assessing micro-level tradeoffs yields *aggregative* results that are necessarily clear-cut when viewed from the perspective of the broader society. Many individual household units may choose to pursue specific objectives – improving the nutritional status of family members or increasing family income, for example – while other households may pursue alternative objectives. From the onset of the Industrial Revolution, pursuing these household objectives has typically meant a narrowing and specialization of individual functions and the household activities and livelihood strategies built upon that specialization. Adam Smith's celebrated discussion of the specialized functions of the "tailor and the shoemaker" and the improved livelihoods of *both* that were permitted by this specialization might today be reformulated in terms of the computer programmer and the physician. But the implications are the same. When one considers the scope of geographical and sectoral differentiation that is possible at the regional or national level, a society may, in the aggregate, successfully pursue a multiplicity of social goals that individual households may be incapable of replicating at the micro-level (and vice versa, of course). "Win–win" solutions at the household level accordingly may look very different from those promoted at the national or regional level.

An important reason for identifying this distinction between macro- and micro-levels as it relates to environmental management is that it provides a context for evaluating much of the applied household- and village-level research that has accrued in recent years regarding the analysis of tradeoffs and complementarities among environment and development objectives in the rural areas of developing countries. Much of this literature has emerged from the work of applied researchers – and often practitioners – focused on documenting technologies, systems and strategies for jointly achieving environmental and livelihood goals, or, in short, focused on achieving "sustainable development" in rural areas. Typically, these strategies and systems approaches attempt to address multiple goals simultaneously within one farm, household, watershed or ecosystem. Due to the prominence of these approaches in the literature, several different empirical approaches are briefly reviewed here. These are illustrative of a range of approaches, from research-based methodologies to the applied approaches of practitioners, but in no way reflect the full breadth of those experiences.

Micro-level approaches to environment and development tradeoffs

Much of the empirical research on environment–development tradeoffs is reported in one-off, stand-alone studies, using unique, often nonreplicated indicators, and lacking the resources or institutional mechanisms to pursue the type of long-term analysis essential to genuinely analyzing the sustainability of proposed technological, management or policy solutions. A notable exception is the Alternatives to Slash and Burn Program[1] (ASB) of the Consultative Group on International Agricultural Research (CGIAR). This research program has engaged in a decade and a half of applied research on environment–poverty interactions in twelve benchmark sites in tropical forest margin areas in the Amazon basin, the Congo basin, Sumatra (Indonesia), Mindanao (Philippines), and northern Thailand. Much of the ASB's best documented research has focused on long-term research at sites in Brazil, Cameroon, and Sumatra. The ASB analytical framework is notable for addressing all three of the above-mentioned limitations, specifically through employing consistent measures of environmental indicators (biodiversity and carbon sequestration), agronomic and production indicators, and economic indicators such as profitability (returns to land and returns to labor) and employment. By working across multiple sites, the project has employed a consistent framework in its research, a major limitation of most research projects. And through engaging in long-term research in these multiple sites, the project has yielded a much more complete understanding of sustainability dimensions.

The ASB research approach has additionally focused on explicitly analyzing the tradeoffs arising among the above-mentioned indicators at specific project sites. One example is given in Figure 12.2, which examines the results for two indicators – plant biodiversity and returns to labor – estimated across several different systems at two sites (in Acre and Rondônia) in the western Amazon (Vosti *et al.*, 2001). It is evident that, among the systems examined, smallholder forest-based and annual/fallow systems contain the highest measures of biodiversity indicators but generate the lowest returns to labor, a key measure of economic profitability from the standpoint of the smallholder. However, two other systems, the coffee/*bandarra*[2] system and the coffee/rubber system generate the highest returns to labor but are characterized by the lowest biodiversity indicators. These two systems plus the traditional pasture system are the only systems which generate returns to labor above its opportunity cost levels, as measured by the typical wage in local hired-labor markets (vertical line). Overall, the negative relationship between the two indicators included in Figure 12.2 shows that there appears to be a distinct tradeoff between achieving environmental and economic objectives for rural households in western Brazil. Using a different environmental indicator to measure time-averaged above-ground carbon sequestration (not shown) demonstrates a

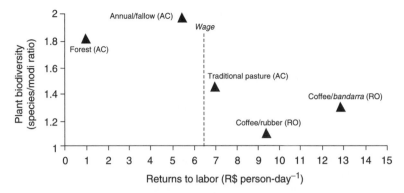

Figure 12.2   Plant biodiversity versus returns to labor, western Brazil
From: Vosti S.A., J. Witcover, C.L. Carpentier, S.J. Magalhaes de Oliveira, and J.C. dos Santos (2001)
"Intensifying Small-scale Agricultura in the Western Brazilian Amazon: Sigues, Implications, and
Implementation" in D.R. Lee and C.B. Barrett (eds) Tradeoffs or Synergies: Agricultural Intensification,
Economic Development and the Environment. Wallingford, CABI: 245–266. Reprinted with permission.

much less distinct empirical relationship (Vosti *et al.*, 2001). Thus, even among two alternative empirical specifications of the same general indicator – environmental quality – fundamentally different conclusions emerge regarding the existence of tradeoffs between environmental and economic indicators. This in itself is an important result.

A second empirical approach is the "Tradeoffs" methodology developed under the auspices of the Ecoregional Research Program of the International Potato Center of the CGIAR, Montana State University (USA) and other program collaborators, and applied at research sites in Ecuador, Peru, Kenya and several other countries (Tradeoffs Analysis Project, 2006). Like the ASB framework, the Tradeoffs methodology attempts empirically to identify and model specific environment–economic

tradeoffs arising in smallholder systems in selected countries. Figure 12.3 demonstrates an example from a research site in the highlands of Ecuador, where household livelihoods are dependent on a potatoes-and-pasture system and in which pesticides (carbofuran) are heavily used in potato production. The figure plots carbofuran leaching into groundwater (the environmental indicator) on the *Y* axis against the value of production (the economic indicator) on the *X* axis, for two different production technologies: the standard production system employing high levels of pesticide applications, and an alternative system using integrated pest management (IPM) and lower levels of pesticides. The figure demonstrates, at each production value, the lower estimated carbofuran leaching under the IPM technology. This type of tradeoff

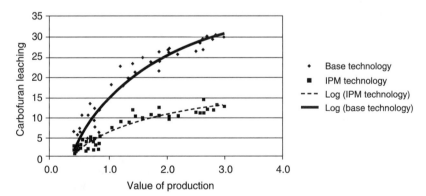

Figure 12.3   Carbofuran leaching versus production value, Highland Ecuador
From: Crissman, C. (2001) "Tradeoffs in Agriculture, the Environment and Human Health: Decision Support for Policy and Technology Mangers" in D.R. Lee and C.B. Barrett (eds), Tradeoffs or Synergies: Agricultural Intensification, Economic Development and the Environment. Wallingford, CABI: 130–150. Reprinted with permission.

curve could be used by policymakers in deciding whether, for example, the environmental gains resulting from lower levels of groundwater contamination are worth the additional investment in research, training and outreach activities to promote IPM among farmers and thus decrease pesticide use.

A third approach is the "ecoagriculture" framework proposed by McNeely and Scherr (2003) and supported by the Ecoagriculture Partners group of nongovernmental, research, and international development organizations. This approach seeks to promote a joint environmental conservation and rural development strategy at the landscape level which focuses on achieving three goals: enhancing biodiversity and environmental services, improving rural livelihoods, and developing more productive and sustainable farming systems to address the food needs of rural smallholders. The focus in this broad-based applied research and development framework is to identify, document and promote applied research and farm household adoption of specific agroecosystems, technologies and policy changes that contribute to all three goals, with a particular emphasis on systems that enhance biodiversity outcomes. A recent assessment of the scientific foundations of the ecoagriculture approach finds considerable scientific merit for the individual elements of this approach; however, there is abundant evidence of the existence of tradeoffs among the multiple goals in specific systems. There is thus far only limited direct research evidence for the simultaneous achievement of these objectives in individual systems which can be broadly replicated (Buck *et al.*, 2004).

The fourth and last approach mentioned here is the Integrated Conservation and Development Project (ICDP) approach to conservation and development promoted by many conservation and development nongovernmental organizations and international donors over the past two decades. This approach is based on the assumption that achieving conservation objectives – notably, biodiversity conservation and improved management of parks and protected areas – can only be accomplished by simultaneously addressing the economic and livelihood needs of those communities and households who live in or near these areas and who are dependent on the natural resources contained therein. Growing out of UNESCO's Man and the Biosphere program of the 1970s, the ICDP paradigm typically emphasized the protection of "core" conservation areas, often containing unique animal and plant biodiversity, by identifying "buffer zones" circumscribing these core areas where some land and resource restrictions exist but where some low-intensity economic activities are permitted: hunting, fishing, collection of firewood, low-level grazing of livestock, etc. (Wells and Brandon, 1992). Other key parts of the ICDP approach have involved community development activities in and around buffer zones as a means to compensate households for losses suffered from environmental protection and to generate sustainable sources of income. These efforts have included: improvements in local resource management practices; ecotourism; local infrastructure development; exploitation of nontimber forest products; promoting agroforestry and improved soil and water management; environmental education programs; and many other programs and projects having the goal of relieving pressure on key biodiversity and other natural resources contained in the core protected areas. The ICDP approach has had some notable successes but has also been subject to many substantive criticisms as a generalizable tool for jointly achieving environmental and economic objectives (Brandon, 1997).

There are many other examples of analytical frameworks, institutional research programs, specific research projects, and applied development efforts that have addressed and continue to address similar sustainable development goals, including environmental management and policy objectives. Space does not permit a full discussion or extensive review of all of these. What many of these approaches have in common is the attempt to address environmental goals in the context of broader economic development objectives. The existence of the tradeoffs which arise is sometimes addressed directly, as mentioned above, but often is a secondary consideration or not addressed at all. Yet, when the existence of tradeoffs is explicitly addressed, they are typically shown to exist, as would be expected. Environment and development strategies which do not recognize this fact are likely to overstate the results of those strategies.

There are other limitations that characterize many of these micro-level approaches in both research and development spheres. The site specificity of agroecological and environmental conditions makes "scaling up" from field-based studies inherently problematic. Further challenges to scaling up occur at the level of broad-based institutions and markets – institutions that function well at the local level may function imperfectly at regional and national levels, while the production strategies of households may encounter price-depressing effects at the aggregate market level. Additionally, the existence of multiple indicators associated with multiple objectives means that different indicators may reveal different and contradictory outcomes (such as the Brazilian example cited above). It can be difficult to reach a consensus among researchers and practitioners regarding which indicators are key; this is particularly true for environmental indicators where there are dozens of possible indicators and there is often less of a consensus (compared to economic or agronomic indicators) regarding which are particularly indicative of environmental quality. Policy distortions, market failures, and high transactions costs further serve to reduce the efficacy of sustainability strategies.

FACTORS CONDITIONING ENVIRONMENT–DEVELOPMENT OUTCOMES[3]

Discussions of environment–development trade-offs often result in conclusions that are unsatisfying to academicians and policymakers alike because of the lack of generalizable results and conclusions. Even at a theoretical level, it is difficult to assert unambiguous complementarities between environmental and development goals (Angelsen and Kaimowitz, 2001; Pagiola and Holden, 2001). But are there empirical regularities in the debate over environment–development tradeoffs that are correctly generalizable? Are there clear-cut "win-win" solutions that apply broadly across the board?

The literature at both levels suggests that any such solutions are few and far between. The macro-level literature on the empirical evidence on the EKC hypothesis is, as reviewed above, mostly inconclusive. Regarding micro-level tradeoffs, many of the "win–win" solutions that have been identified by applied researchers and practitioners that simultaneously address environmental (mostly biodiversity conservation), food security, and livelihood generation goals focus on complex farming systems oriented toward high-value crops and/or agroforestry products (Lee *et al.*, 2001). These tend to be diversified systems incorporating perennial crops, rather than crop monocultures, and are characterized by flexible labor requirements, reduced exposure to risk (stemming from economic, agronomic and environmental sources), high levels of on-farm biodiversity, and possessing substantial agronomic and environmental benefits (reduced sediment transport, improved soil quality, etc.). Even for these systems, questions can be raised about long-term economic sustainability and/or their generalizability to other agroecosystems (Buck *et al.*, 2004).

But all is not lost. One useful way of characterizing the literature on environment and development tradeoffs has focused on the identification of "conditioning factors" and the roles these factors play in determining which specific environment–development outcomes apply under specific circumstances (Vosti and Reardon, 1997; Lee, *et al.*, 2001). Among the conditioning factors that

have been identified as key to determining these outcomes are:

- *Population pressure:* specific agroecosystems which are sustainable under low population pressure may not be so under higher population concentrations, and the type of induced intensification hypothesized by Boserup (1965) may not be possible when irreversible thresholds in environmental degradation have been reached.
- *Agroecological conditions* of rainfall, temperature, soil slope and quality, solar radiation, water and nutrient supply substantially determine agronomic outcomes for given levels of crop prices and input availabilities, and thus indirectly affect alternative land uses.
- *Infrastructure:* both physical insfrastructure (roads, bridges, etc.) and market infrastructure substantially influence market access, household risk, and the profitability of alternative investments.
- *Technological change:* it is widely acknowledged that the *type* of technological change – labor-intensive versus labor-using, for example – can have a major impact on the nature of environment–development outcomes in specific instances.
- *Local labor markets* influence employment opportunities, the opportunity costs of household labor, seasonality in labor demand, and household demand for leisure, and accordingly, household investments in labor-saving technologies and practices.
- At the macro- or economy-wide level, an additional set of factors, including (1) the *macroeconomic* and *sectoral policy environment*, (2) *land-tenure and property rights' regimes*, and (3) *institutional frameworks*, all of which play major roles in helping determine specific environment–production–livelihood outcomes.

Regarding the last set of factors, the environment for price policy determination and institutional development, for example, can lead to vastly different outcomes under given conditions with respect to the other above mentioned variables; hence, the widespread attention in policy circles to "getting prices right" (Timmer, 1986) and "getting institutions right" (Rodrik, 2004).

Given these conditioning factors, perhaps the most productive way to promote synergistic rather than tradeoff relationships among key environmental, production and economic variables is to encourage policies and management strategies that address a wide range of sustainable development goals. There is no unique "one size fits all" technology or policy solution. In the conditions facing many developing countries, the likelihood of achieving sustainable environment–development solutions would appear to be greatest if the following were simultaneously promoted: improved technologies and management practices, tailored to local conditions; infrastructure investments that support smallholders' access to inputs and markets; policies that promote intensification of production in appropriate agroecosystems, mitigating the pressure on fragile ecosystems elsewhere; investments that support diversification into high-value products and alternative employment and income-generating activities; security of land tenure; supportive market and price incentives; and policy, tenure and institutional reforms that create incentives for the informed sustainable management of resources.

Given the abundance of different environmental indicators, in situations where environmental outcomes are emphasized among alternative goals, clear identification of specific priority environmental indicators is important. So too are the development of innovative mechanisms for valuing environmental services, and transfer schemes to enable beneficiaries to compensate those who provide those services, especially where public and quasi-public goods are involved. These mechanisms are explored in other chapters in this volume.

NOTES

1 Recently renamed "ASB – Partnership for the Tropical Forest Margins."

2 *Bandarra* is a fast-growing tree species native to western Brazil.

3 This section draws substantially from Lee *et al.* (2001).

REFERENCES

Angelsen, A. and D. Kaimowitz (2001) "When Does Technological Change in Agriculture Promote Deforestation?" Ch. 6 in D.R. Lee and C.B. Barrett (eds.) *Tradeoffs or Synergies: Agricultural Intensification, Economic Development and the Environment*. Wallingford, UK: CABI Publishing, pp. 89–114.

Antle, J.M. and G. Heidebrink (1995) "Environment and Development: Theory and International Evidence". *Economic Development and Cultural Change* 43(3): 603–625.

Bhattarai, M. and M. Hammig (2001) "Institutions and the Environmental Kuznets Curve for Deforestation: A Cross-country Analysis for Latin America, Africa, and Asia". *World Development* 29: 995–1010.

Borlaug, N.E. and C.R. Dowswell (1994) "Feeding a Human Population that Increasingly Crowds a Fragile Planet". In *Transactions of the 15th World Congress of Soil Science*. International Soil Science Society, Chapingo, Mexico.

Boserup, E. (1965) *The Conditions of Agricultural Growth: the Economics of Agrarian Change Under Population Pressure*. Chicago: Aldine.

Brandon, K. (1997) "Policy and Practical Considerations inLand-use Strategies for Biodiversity Conservation". In R.A. Kramer, C. van Schaik, and J. Johnson (eds.) *Last Stand: Protected Areas and the Defense of Tropical Biodiversity*. New York: Oxford University Press.

Buck, L.E., T.A. Gavin, D.R. Lee, and N.T. Uphoff (2004) *Ecoagriculture: A Review and Assessment of its Scientific Foundations*. Ithaca, NY: Cornell University, Virginia Tech, and Ecoagricultural Partners.

Crissman, C. (2001) "Tradeoffs in Agriculture, the Environment and Human Health: Decision Support for Policy and Technology Managers". Ch. 8 in D.R. Lee and C.B. Barrett (eds.) *Tradeoffs or Synergies: Agricultural Intensification, Economic Development and the Environment*. Wallingford, UK: CABI Publishing, pp. 135–150.

Cropper, M. and C. Griffiths (1994) "The Interaction of Population Growth and Environmental Quality". *American Economic Review Papers and Proceedings* 84: 250–254.

Grossman, G.M. and A.B. Krueger (1991) *Environmental Impacts of a North American Free Trade Agreement*. National Bureau of Economic Research Working Paper 3914, NBER, Cambridge, MA.

Koop, G. and L. Tole (1999) "Is There an Environmental Kuznets Curve for Deforestation?" *Journal of Development Economics* 58: 231–244.

Kuznets, S. (1955) "Economic Growth and Income Inequality". *American Economic Review* 45: 1–28.

Lee, D.R., C.B. Barrett, P. Hazell, and D. Southgate (2001) "Assessing Tradeoffs and Synergies among Agricultural Intensification, Economic Development and Environmental Goals: Conclusions and Implications for Policy". Ch. 24 in D.R. Lee and C.B. Barrett (eds.) *Tradeoffs or Synergies: Agricultural Intensification, Economic Development and the Environment*. Wallingford, UK: CABI Publishing, pp. 451–464.

McNeely, J.A. and S.J. Scherr (2003) *Ecoagriculture: Strategies to Feed the World and Save Wild Biodiversity*. Washington, DC: Island Press.

Pagiola, S. and S. Holden (2001) "Farm Household Intensification Decisions and the Environment". Ch. 5 in D.R. Lee and C.B. Barrett (eds.) *Tradeoffs or Synergies: Agricultural Intensification, Economic Development and the Environment*. Wallingford, UK: CABI Publishing, pp. 73–87.

Panayotou, T. (1997) "Demystifying the Environmental Kuznets Curve: Turning a Black Box into a Policy Tool". *Environment and Development Economics* 2: 465–484.

Pingali, P.L. and M. Rosegrant (1994) *Confronting the Environmental Consequences of the Green Revolution in Asia*. Environment and Production Technology Division Discussion Paper No. 2, International Food Policy Research Institute, Washington, DC.

Rodrik, D. (2004) "Getting Institutions Right". Unpublished manuscript. Harvard University.

Shafik, N. and S. Bandyopadhyay (1992) *Economic Growth and Environmental Quality: Time Series and Cross-Country Evidence*. Background paper for the *World Development Report*. Washington, DC: The World Bank.

Stern, D.I. (2003) "The Environmental Kuznets Curve". International Society for Ecological Economics, Rensselaer Polytechnic Institute, Troy, NY.

Swift, M.J., A.-M.N. Izac, and M. van Noordwijk (2004) "Biodiversity and Ecosystem Services in Agricultural Landscapes – Are We Asking the Right Questions?" *Agriculture, Ecosystems and Environment* 104: 113–134.

Timmer, C.P. (1986). *Getting Prices Right: The Scope and Limits of Agricultural Price Policy.* Ithaca, NY: Cornell University Press.

Tradeoffs Analysis Project (2006) www.tradeoffs.montana. edu/

Vosti, S.A. and T. Reardon (eds.) (1997) *Sustainability, Growth and Poverty Alleviation: a Policy and Agroecological Perspective.* Baltimore, MD: Johns Hopkins University Press.

Vosti, S.A., J. Witcover, C.L. Carpentier, S.J. Magalhaes de Oliveira, and J.C. dos Santos (2001) "Intensifying Small-scale Agricultura in the Western Brazilian Amazon: Sigues, Implications, and Implementation". Ch. 13 in D.R. Lee and C.B. Barrett (eds.) *Tradeoffs or Synergies: Agricultural Intensification, Economic Development and the Environment.* Wallingford, UK: CABI Publishing.

Waggoner, P.E., J.J. Ausubel, and I.K. Wernick (1996) "Lightening the Treat of Population on the Land: American Examples". *Population and Development Review* 22: 531–545.

Wells, M. and K. Brandon (1992) *People and Parks: Linking Protected Area Management with Local Communities.* Washington, DC: World Bank, World Wildlife Fund, and US Agency for International Development.

World Bank (1992) *World Development Report 1992: Development and the Environment.* New York: Oxford University Press.

World Commission on Environment and Development (WCED) (1987) *Our Common Future.* Oxford, UK: Oxford University Press.

Yandle, B., M. Bhattarai, and M. Vijayaraghavan (2004) *Environmental Kuznets Curves: A Review of Findings, Methods, and Policy Implications.* PERC Research Study RS-02-1a. Property and Environment Research Center, Bozeman, MT.

13

Water Policy, Economics and the EU Water Framework Directive*

Joe Morris

INTRODUCTION

Increased demand for water to meet human needs, including its use as a conduit and receptor for waste, has led to deterioration in the state of inland surface water, groundwater and coastal waters in many parts of Europe with consequences for people and the water environment (EEA, 1999, 2001a, b; OECD, 2003). EU legislation on water initially focused on specific environmental problems associated with public health risks, such as drinking and bathing water quality. During the late 1990s, in response to calls for concerted action to improve the ecological quality of surface waters and the state of groundwaters, the European Commission introduced the Water Framework Directive (WFD) (CEC, 2000) in order to promote an integrated and strategic approach to the sustainable management of water resources.

Although the WFD is essentially a regulatory policy which sets targets for ecological water quality, it also uses economic principles and methods to help achieve sustainable water management. This chapter analyses the WFD, and explores in detail the role of economics in policy analysis and implementation. The chapter serves

as a case study in determining the value of water use, cost recovery, incentive pricing and cost-effectiveness analysis. It describes the benefits and costs of the WFD and who can be expected to pay for its implementation. It concludes that, in spite of data management and methodological challenges, the WFD can serve to promote an integrated approach to water resource and related environmental management, with economics playing a useful role in this process.

THE WATER FRAMEWORK DIRECTIVE: AIMS AND METHODS

The WFD 'aims to establish a framework for the protection of inland surface waters, transitional waters, coastal waters and groundwaters' (CEC, 2000). More specifically, its purposes (Article 1) are:

- to prevent deterioration of, and where necessary enhance, the status of aquatic and related ecosystems;
- to promote sustainable water use;
- to aim progressively to reduce, and for priority substances eliminate, pollution from hazardous substances;
- to ensure reduction/prevention of groundwater pollution;
- to contribute to the mitigation of floods and droughts.

*Based on a paper presented to the Second Annual Conference of Applied Environmental Economics, Royal Society, London, 26 March 2004.

In the language of European Directives, terms such as 'prevent' and 'ensure' indicate a strong obligation to achieve a desired outcome whereas 'promote' and 'contribute' imply a weak obligation. The term 'aim' rests somewhere between the two. The Directive commits Member States (MS) to put in place a 'framework' which will achieve 'good' surface water and groundwater status by 2015. Although protecting the ecological quality of the aquatic environment is the main focus of the WFD, the Directive seeks to provide 'a sufficient supply of good quality surface water and groundwater as needed for sustainable, balanced and equitable water use'.

The WFD adopts a 'pressure–state–response' approach to sustainable water resource management, setting ecological standards for a desired 'state' of water, identifying the human induced 'pressures' responsible for failure to meet these standards, and undertaking 'responses' in the form of corrective actions. The 'Framework' prescribes a Common Implementation Strategy and timetable for EU MS (Table 13.1). At the time of writing, most MS have progressed to an understanding of pressures on water resources and environment associated with water use, and have set in place or reinforced previous systems to monitor water quality (e.g. see Defra, 2005).

The WFD is predominantly a 'regulatory', command and control regime, although it does advocate the use of economic instruments to achieve its targets. It uses two main regulatory methods. First, as explained below, it uses environmental quality objectives to set standards to ensure that a particular water characteristic, function or use continues unimpaired. Second, it uses emission limit values to control the discharge of potentially hazardous substances as well as the processes that are associated with such discharges. The latter approach is similar to that contained within the prescriptions for Best Available Techniques under the EU Integrated Pollution Control Regulation (96/61/EC) (HMSO, 2000). This combined approach has the explicit intention (Article 10) of controlling both point source and diffuse pollution of the water environment.

The selection of criteria to define water quality standards, and the target or 'reference' levels for 'good ecological status' that define perceived safe minimum standards, is a critical step in the WFD process. The Directive places entitlement for the definition of standards (and targets to be met) firmly within the natural science community, acting through a responsible implementing agency (the Environment Agency in the case of England and Wales). The involvement of other stakeholders is largely confined to deciding how these standards will be met given that their activities are associated with existing water quality status and the achievement of required improvements. While this is consistent with other European environmental directives that set environmental standards and require measures to prevent or minimise pollution, it is seen by some as an 'ecologist's charter' which could result in costs borne by society which are disproportionately high compared to the benefits obtained. There are, however, provisions in the Directive to guard against this, as discussed below.

The Directive sets out (Annex V) the process for assessing the status of surface and groundwaters, defining the 'normative' characteristics of water associated with high, good and moderate qualities. For river water, for example, it does this in terms of biological (such as the composition and abundance of aquatic flora and fauna), hydromorphological (such as river flows and levels and channel features) and physicochemical (such as oxygen balance, nutrient loads and pollutant concentrations) qualities. Emphasis is placed, however, not only on biodiversity but also on how the biological community interacts with itself and other environmental components to produce self-sustaining ecosystems functions and processes. A similar approach is defined for lakes (Moss *et al.*, 2003), groundwater and transitional waters such as estuary and coastal waters.

Most EU MS had methods in place to define water quality standards prior to the WFD. The approach in England and Wales (E&W), for example, has been to define water quality objectives in terms of 'fitness for purpose', such as for freshwater fisheries or bathing. For rivers, these have been captured in a set of River Quality Objectives (subsequently relabelled as River Ecological Standards 1 as highest to 5 as lowest) which classifies according to biological quality assessments based on oxygen status, related organic pollution and ability to support specified types of environmentally sensitive macro-invertebrates. However, these objectives are not regarded as sufficiently complete for the purpose of the WFD. The latter takes a much broader view of water flora and

Table 13.1 Implementation timetable for the WFD

2003	Identify river basins and districts
2004	Characterise river basins in terms of pressures, impacts and uses
2006	Monitor water status
2009	Identify programme of measures, and publish river basin management plans
2010	Implement water-pricing polices
2012	Implement measures
2015	Completion of implementation

fauna and their part played in ecological communities and processes, of potential pollutants and of environmentally damaging activities including diffuse pollution from farm land.

In E&W, as in other countries, it has been possible to draw on previous water quality monitoring and assessments. Here, 'River Ecology (RE) Standard 2' as presently defined is likely to conform to the WFD standard of 'good' ecological status. According to the Environment Agency for E&W, about 36% of total river length fails to meet the RE2 standard, mostly in the lower reaches of rivers. Identifying the reasons for this shortfall and taking appropriate remedial action is the essence of the WFD. Information on the status of lakes and artificial water bodies, however, is currently not sufficiently complete to determine the extent of the challenge.

The WFD requires that relevant measures of good ecological status are set for waters at the scale of river basin districts. These comprise an area (or a collection of neighbouring areas) that share a common catchment and surface water regime. Following conditions laid down in the Directive, MS are required (Article 5), for each river basin district, to characterise physical and hydrological attributes, and the biological 'reference' conditions for water quality. The intention is to define universal 'reference' conditions which can be applied across all waters with similar characteristics, for example, for all mountain streams or all estuarine waters, allowing of course for specific local context. This task is prescribed to a competent authority; the Environment Agency in the case of E&W, and, as suggested, will reflect predominantly ecological criteria. Anthropogenic uses of water must also be identified together with the pressures on water quantity and quality that arise as a consequence. This analysis necessarily involves the economic analysis of water use as discussed below.

Where existing standards fall short of good status, responsible authorities must develop and implement a programme of measures to put this right. These measures include 'basic' (regulatory) measures that are, for the most part, already in force under various EU Directives and policies, such as: Urban Waste Water Treatment (91/271/EEC), Sewage Sludge (86/278/EEC), Bathing Waters (76/160/EEC) Freshwater Fish (78/659/EEC), Habitats (92/43/EEC), Groundwater (80/68/EEC), IPPC (96/61/EC), Shellfish (79/932/EEC), Abstraction (75/440/EEC), Drinking Water (76/160/EEC) and Nitrates (91/676/EEC). Thus, the WFD is sometimes referred to a 'mother' Directive with a large cohort of 'daughter' regulations.

Where basic measures are perceived to be insufficient, additional 'supplementary' measures may be used to help deliver 'good' ecological status. These include a range of extra regulatory actions such as controls on water abstraction, economic and market instruments such as pollution charges and tradeable permits, and voluntary measures such as the promotion of good practices. The framework may also support other measures such as research, technical assistance, and education and training, as well as projects to address specific pressures such as wetland restoration.

Thus, in some cases the WFD may seek to achieve water quality standards that are above and beyond that required by the existing and substantial raft of water-related directives. It does this by including a long-term ecological perspective that accounts for the inherent dynamics of the natural environment. The implicit assumption here is that the perceived benefits of existing directives exceed their costs, and that this assumption also applies to the incremental benefits and costs of the WFD. This said, it was noted by one UK Government environmental committee (HCEFRAC, 2003a), that it would be helpful if MS agreed to new Directives 'after the practical and economic implications of their implementation have been fully assessed and costed', implying that some aspects of the WFD may not be cost beneficial, at least within the proposed timeframe.

Reflecting a commitment to public participation, once competent authorities have set the standards to be met they are required (Article 14) to 'encourage the active involvement of all interested parties'... 'especially concerning the identification and review of measures to deliver the targets at river basin level'. Stakeholder interests find expression, therefore, in the choice of delivery mechanism rather than in the setting of standards. Stakeholders are, however, likely to influence the setting of water quality standards for artificial, 'heavily modified water bodies' (HMWB) and other cases where the cost of delivering the standards can be shown to result in disproportionate costs or an unacceptable social burden.

The Directive sets down how all this should be done, hence the label 'Framework'. Given the magnitude of the task, the schedule is tight, and there is a feeling that not all MS will achieve 'full and proper' implementation within the timescales set out in the Directive (HCEFRAC, 2003a).

ECONOMICS AND THE WFD

The Directive promotes the concept of water as an economic commodity and the use of economic principles to guide decisions in accordance with the objectives of the WFD. The term 'economics' is used twenty-two times in the Directive, seeking to apply economic principles in four main respects, namely:

- The estimation of the demand for, and the valuation of, water in its alternative uses (Article 5).
- The identification and recovery of costs associated with water services having regard for the polluter pay principle and the efficient use of water (Article 9).
- The use of methods of economic appraisal to guide water resource management decisions (Article 11).
- The use of economic instruments to achieve the objectives of the WFD, including the use of incentive pricing and market mechanisms (Article 11).

Annexe III of the Directive refers to the requirements for economic analysis. Although short on detail, it is long in terms of the challenge of implementation. It requires that 'the economic analysis (of water use) shall contain enough information in sufficient detail':

- To apply the principle of *recovery of costs* of water services (Article 9), taking into account long-term forecasts of *supply and demand for water* in river basin districts, and where necessary:
  (i) estimates of *volumes, prices and costs of water services*;
  (ii) estimates of *relevant investments*.
- To make judgements about the most *cost-effective combinations* of measures (Article 11).

Thus, Annexe III is at the heart of the implementation of the WFD. It recognises that: anthropogenic use of water determines water quality (and quantities); that understanding and influencing water use is a key to sustainability; that decisions on water management must take into account benefits and costs to the environment and society; and that economic instruments can help deliver sustainable water resource management. Recognising the brevity yet significance of the reference to economics in the Directive, the EC produced for guidance a Common Implementation Strategy for the economic analysis which maps out what needs to be done, and how best to do it (WATECO, 2003). The procedures are being tested in a number of selected case study areas (Defra, 2005; Environment Agency, 2007).

The application of economic principles to water management in this way is relatively new in practice, although arguments have been made for some time, for example, in the Dublin Principles (ICWE, 1992). In the UK, for example, the economic value of water and the use of economics to guide water resource decision-making have received recent attention as part of the switch to demand management (Defra, 2001; Environment Agency, 2001a, b). However, until recently prompted by the need to justify environmental spending by privatised water companies (Defra, 2004), the application of economics to the water sector has been somewhat muted, partly because the information, methodologies and resources to apply economic analysis have not been in place.

The WFD goes some way to addressing this deficit by using economic principles to help assess the value of water in use, the costs associated with the degradation of water resources and the water environment and the best ways of delivering pre-defined targets. Thus, economics has the role of shaping the delivery mechanisms rather than defining the standards to be met: that is a focus on cost-effective rather than economically efficient delivery of environmental standards. This perhaps reflects the dominant view, shared by a number of economists, that economics and the subset of environmental economics is insufficiently complete or robust to provide a total basis for a sustainable water management strategy: water is much too valuable to be left to economists. Although the economic perspective may be incomplete, in the E&W case, it is helping to justify considerable spending on environmental protection in the water industry's Asset Management Plans as referred to below (OFWAT, 2003).

ECONOMIC ANALYSIS OF WATER USE: DEMAND, SUPPLY AND PRICING

According to the original timetable, WFD required an economic analysis of water use to be completed for each river basin district by the end of 2004, together with an assessment of the balance of demand and supply, and the pressures and impacts on the water environment. The WFD recognises the relevance of economic principles applied to water, namely those of scarcity, value and, in theory at least, prices which reflect cost of supply and benefit in use.

These principles are illustrated in Figure 13.1 which can either be taken to represent the total market demand for and supply of water in any one sector, such as public water supply or agriculture, or across all sectors taken together.[1] The demand 'curve' shows the relationship between the demand for water and the price of water, other things such as income levels and income distribution, prices of other goods and consumer preferences, remaining constant. It reflects the marginal (extra) benefit or utility obtained from consuming extra units of water and the associated willingness to pay by users for those additional units of water. The latter tends to decline as consumption increases and water is committed to less valuable uses (Rees, 1993; Merret, 1997; Green, 2003). The shape and steepness of the curve reflect the extent to which

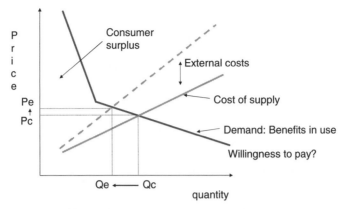

Figure 13.1   Theoretical demand, supply and price relationships for water

demand for water is sensitive to changes in water price only.

It is possible to derive estimates of demand for water in different sectors. With respect to public water supply, for example, broad categories of demand can be distinguished between essential, typically 'in-house', use of water such as for drinking, and non-essential, typically 'out-of-house' uses, such as for car washing. These uses infer different benefits or utility on users and, for any one user, demand for water for essential uses is likely to be relatively less responsive to price changes (i.e. relatively price 'inelastic' with a steep downward sloping demand curve) than for non-essential uses (i.e. relatively price 'elastic' with a relatively flat downward sloping demand curve). These relationships are evident in the way water is consumed. It is unusual to find, for instance, people washing their cars with high-priced bottled water, although they often do so with lower priced water of drinking quality from the household tap.

The demand by an individual for water at a given price is influenced by a range of other factors as mentioned above, especially income and ability to pay. In the case of water for household consumption, for example, poor people may not be able to satisfy adequately essential needs for water and their health suffers as a result. The thirsty pauper has no 'effective' demand for water, even though marginal utility for water is extremely high. Here, willingness to pay is unlikely to provide reliable estimates of the real marginal value of water, especially from a societal viewpoint. The converse may apply in agriculture where farmers, in receipt of government production subsidies for sugar beet (UK) or maize (France), are willing and able to pay higher prices for irrigation water than if there were no subsidies.

Critically, as far as water resource management is concerned, the WFD requires that estimates of demand and user-derived benefits are derived for all major users of water within a river basin. Recognising the important welfare aspects of water management, WFD also requires explicit assessment of affordability and the extent to which changes in water prices could impact on income or welfare.

Figure 13.1 also shows the marginal (extra) cost of supplying additional units of water and the willingness of agents to supply water at given prices. In theory, supplying additional water usually involves rising extra costs per unit supplied as the cheapest sources are usually exploited first. Additional supplies are, however, likely to involve higher unit costs associated with abstraction, for example, from groundwater rather than surface sources, storage, transport and distribution, and administration. There may be scope for economies of scale which can reduce unit costs, potentially giving large-scale operators competitive advantage in water markets. Costs also include water treatment or disposal of dirty water, and those due to impacts on other water users and the environment associated with abstractions and discharges. The extent to which the latter costs are fully borne by water users or alternatively passed on as 'external', uncompensated costs to others, varies considerably. Examples here include the negative impact of discharges of contaminated water on the income of downstream fishermen, on public health or on the amenity value of recreational waters.

WFD clearly advocates that external costs, whether in the form of damage or mitigation measures, should be included in the assessment of water supply costs. The effect of this will be to raise unit costs and, other things remaining constant, increase price (Pc to Pe in Figure 13.1) and reduce consumption (Qc to Qe). Guidance (WATECO, 2003) recommends a detailed assessment of the financial, resource and environmental costs associated with the provision of water services. It is also appropriate to consider how these might vary over time as a result of technology change, changes in demand and the possible strategic responses of supply organisations.

In theory, the market for water would tend towards an equilibrium price for water at which the demand for water is equal to supply. This position theoretically maximises economic efficiency defined in terms of the sum of welfare gain to water consumers and suppliers. Here, in theory, marginal benefits (£ million) equal marginal costs, marginal benefit per unit of water is equal for all uses, and social welfare is maximised. In practice, and particularly for water as a commodity, demand, supply and prices rarely conform to this perfect model. The characteristics of water, as well as features of the water services industry, mean that economic valuation of water is problematic and unregulated market mechanisms are unlikely to prove economically efficient from a welfare perspective. Water does not readily conform to the economist's model of a conventional, traded commodity, because it:

- is essential for life, without close substitute: a 'need' rather than a 'want';
- is a fugitive, reusable resource which can be difficult to control and account for;
- is often a common property, with open access and ill-defined property rights;
- provides public goods, such as the public health benefits of clean water;
- is used in ways which often result in 'external' consequences;
- is subject to uncertain supply associated with climatic variation;
- has significant economies of scale associated with its managed supply;
- is an integral part of the functioning of eco-systems.

Three main points arise here: access and affordability, ecosystems functions and risks. On the first point, access to 'traded' water services depends on ability to pay, and thus prices and consumption may reflect income distribution rather than real benefit (and welfare) to users. There is concern that social welfare may be compromised by reliance on market processes. The costs of delivery to some users may be high, such as those in drier or remote areas, and may exceed their ability to pay. To reduce access could further exacerbate the welfare of vulnerable groups, be unethical and possibly lead to greater costs borne by society as a whole (HCEFRAC, 2003b).

With respect to eco-system functions, the WFD draws attention to the diverse functions of water which include those of consumption (e.g. drinking water), production (e.g. irrigation), regulation (e.g. nutrient cycling), carrier (e.g. transport), habitat (e.g. wetlands and wildlife) and information (e.g. water space amenity) (Turner et al., 2000; de Groot et al., 2002; FAO, 2004; MEA, 2005). These in turn support uses (and indeed some 'non-uses' such as preserved and protected water bodies) which are of value to individuals and society as a whole. This classification of functions in this way can help to provide a framework for the characterisation and valuation of uses within river basins as part of the requirements of Annex III of the Directive. The WFD, by specifying good ecological status as a minimum environmental standard, emphasises the importance of non-consumptive, non-production uses of water. It draws attention to the synergies and tradeoffs amongst the different functions of water. For the most part, and probably wisely, the WFD avoids the need to derive economic values for good ecological status by setting this as a target to be achieved.

With respect to the management of risks, the WFD recognises the vulnerability of eco-systems by setting reference standards and adopting a precautionary approach to the specification of emission levels. Regarding vulnerability of human systems, WFD recognises the need to allow for affordability and the avoidance of disproportionate costs in special cases.

The WFD requires that the amount and value of water used by major user groups is identified for each river basin. By way of example, Table 13.2 summarises water use and the possible basis for valuation of water uses by major sector.

For surface waters, it is useful to distinguish instream (or in-lake) and off-stream categories of use. Off-stream uses involve withdrawals from the water system for consumptive use and delayed return of water to the hydrological cycle. They are commonly associated with relatively intensive use, greater benefits per unit of water and higher unit costs of supply compared with in-stream uses. They also tend to result in greater change in the quality of water during use, and greater impacts when water is eventually returned to source. As referred to earlier, estimates of benefits in use and likely demand are easier to determine for some uses than for others. Furthermore, price elasticities of demand tend to be much lower in the short than the long term, reflecting user ability to adjust consumption habits given time.

Estimates of water values and guidance for methods of evaluation are available but tend to be context specific (Rees, 1993; FWR, 1996; Merret, 1997; Knox et al., 1999; Green, 2003;

Table 13.2 Water uses, benefits and valuation methods

Sector	Function, and use and type of benefit	Basis for estimating the value of water (per unit of water at the margin of supply)
Off-stream uses		
Agriculture	Production function: Irrigation, livestock, on-farm processing. Value added by irrigation water in terms of type, quantity, timing and quality of produce. Very seasonally dependent	Gain/loss of value added, including contracts to supply high-value domestic produce. Costs of alternative sourcing. Winter storage reservoirs. Water-saving practices and technologies. Willingness to pay (WTP), water pricing and trading
Industry: by sub-sector: food processing, chemicals, manufacturing, etc.	Production function: Use in processing and production, cleansing agent, heat and waste transfer. Expenditure on water as % of industry costs	Gain/loss of value added due to curtailed/extra supply. Alternative sourcing cost in short and long term. Cost of water saving and recycling technologies. WTP
Domestic water, water companies	Consumption function: Drinking water, other in-house uses and sanitation, out-of house less-essential uses. Grey water uses. Benefits of treated water returned to water system	Public health impacts, especially on vulnerable groups. Costs of alternative provision, costs of water-saving technologies. Supplements to surface and groundwater Value in use/WTP, loss of consumer surplus
In-stream uses		
Waste assimilation	Regulation function: Discharges to water and land. Water as a conduit for waste. Savings in alternative waste disposal routes	Impacts on water quality (and values) Cost of pre-treatment, costs of alternative disposal methods, costs of flow supplementation
Commercial fisheries	Production/carrier function: Water as medium for production, food source and waste sink. Value added by fisheries production	Gain/loss of value added by operators due to changes in river flow and quality. Alternative sourcing and water-treatment technologies
Navigation	Carrier function: Transport function and related value added	Gain/loss of value added, costs of alternative transport provision
Hydro-power	Production function: Energy production	Value of power generation, alternative sourcing. Water mills as tourist attraction
Recreation amenity, and heritage	Information function: Benefits to users of the water environment (e.g. recreational anglers) and non-users. Property values. Tourist and visitor attractions (wetland sites, water space amenity, water mills)	Value added by operators, WTP (contingent valuation), travel cost based estimates, hedonic (market based) price differentials. Costs of protection and mitigation
Nature conservation (incl. abstraction to off-stream sites)	Habitat/carrier function: Indirect benefits associated with hydrological and ecological processes, biodiversity. Non-user option, existence, altruism and bequest values	Cost of protection and mitigation, replacement cost, WTP
Flood risk management	Regulation function: River and coastal management. Hydrological control, catchment management, impoundment, water resource benefits	Damage avoidance, savings in flood defence costs

Morris *et al.*, 2004). Significant progress has been made on the valuation of environmental aspects as part of the Water Industry Periodic Review (PR04) (Defra, 2003a).

WFD recommends that estimates for water demand are derived according to a 'business as usual' scenario at river basin level. The approach adopted by the Environment Agency for E&W has application here. Estimates of water demand were derived for E&W and its constituent regions (Environment Agency, 2001) for the main sectors of Public Water Supply, Industry and Agriculture,

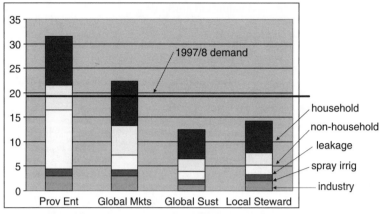

Source: EA, 2001, based on modified UKCIP type scenarios

Figure 13.2   Estimates of demand (109 litres/day) for water for E&W under alternative future scenarios (2025)

modified according to assumptions about the degree of control of leakage in supply networks. These estimates reflect a mix of likely economic, technology, water policy and sector-specific policies (such as the EU Common Agricultural Policy) under alternative future scenarios. Figure 13.2 contains estimates of future demand derived by this method for E&W for the year 2025. The scenarios are based on those used by the Foresight Programme (OST, 2003) to represent possible futures distinguished in terms of social motivation and governance.

The WFD is compatible with the water policy drivers implicit in Global Sustainability and Local Stewardship scenarios which contain a commitment to wise use of natural resources and environmental protection. By comparison, Global Markets and Provincial Enterprise scenarios are more utilitarian and demonstrate high rates of water consumption and less intervention to protect water quality and environment. This type of analysis is being carried out at river basin level, for the Business as Usual case that assumes implementation of current policy (Defra, 2005). In the longer term, beyond 2050, climate change will also make an impact on water demand and supply, especially if measures are not taken to control it.

WFD: OPTIMUM LEVELS OF POLLUTION AND COST-EFFECTIVE MEASURES

The WFD avoids the issue of economic efficiency and a so-called 'optimum' level of water pollution by predefining the environmental standards to be met. An assessment of benefits and costs of alternative standards of environmental protection is

required only where there appears to be a case for 'derogation', that is where meeting the standards will result in disproportionate costs (or rather disproportionate benefits foregone).

The concept of an economic optimum level of water quality is illustrated in Figure 13.3. The figure shows the tradeoff between marginal private benefits (MPB) (net of private costs) enjoyed by users of a resource and the marginal external costs (MEC), including those to the environment, associated with that use but borne by third parties without compensation (Hanley *et al.*, 2001; Teitenberg, 2003). Unrestricted water use (including abstractions and discharges) results in consumption at Qpw, where MPB < MEC. This could, for example, represent 'poor' water quality status with consequences for the environment and society at large. An economic optimum, whereby some external costs are 'internalised' to water users to the point where MPB = MEC, results in consumption at Qmw and 'moderate' water quality status. 'Pm' indicates the marginal value of water to users and a price for water which would, if charged, balance MPB and MEC.

Assuming that the WFD seeks to achieve, everything else remaining the same, 'good' water quality status by limiting water use to Qgw – this results in low environmental costs but high opportunity costs of potential user benefits foregone, shown by Pg as the value of water at the margin of use, and hence an implicit price. Here, MPB > MEC. It is here that some stakeholders may express concern that the WFD imposes intolerable costs on them, especially in the short term. This may also be the situation for Heavily Modified Water Bodies (HMWB) where 'good' status is an expensive enhancement rather than a cost-effective

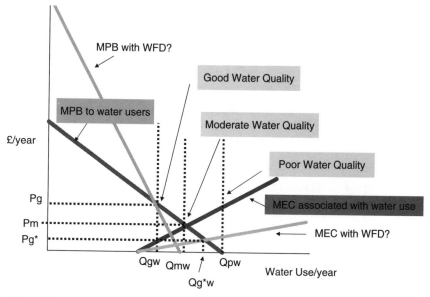

Figure 13.3 WFD and economically optimum standards of water quality

reinstatement. Thus, the challenge of the WFD directive is to deliver water quality objectives at minimum economic cost.

Figure 13.3 also provides a framework for the design and appraisal of intervention measures. Although the basic measures of WFD are regulatory (whereby Qgw is predefined) supplementary measures may use economic instruments such as polluter taxes which could shift the MPB line towards the origin and hence reduce the margin of water use.

Over time, however, the WFD seeks to increase the slope of the MPB line by achieving greater efficiency thereby greater benefits per unit of water use. Measures to achieve this might include precision irrigation which gives 'more crop per drop', water recovery and re-use, and adoption of water-saving practices.

The Directive also seeks to reduce the slope of the MEC line through decoupling of water use and environmental damage through, for example, use of cleaner technologies which generate less pollution at source and avoid the use of water as a conduit for and receptor of potential pollutants. This might include improved sewage treatment or measures taken by farmers to reduce nutrient transfer to rivers. By appropriate measures, the purpose is to achieve a water use (Qg*w) and related benefits currently associated with moderate to poor water quality, but with the low environmental damage costs associated with 'good' water quality (Pg*). The challenge is to identify measures which reduce environmental damage without

unduly compromising the 'profits' of water users: increasing the slope of the MPB curve while reducing that of the MEC curve, and reducing the extent to which they overlap.

This is the essence of cost-effectiveness analysis: identifying options which will deliver good water quality and sustainable water use at minimum total cost. Guidance from WATECO (2003) suggests that the approach to cost-effectiveness analysis involves identifying:

- The pressures responsible for gaps in water quality (such as abstraction or nutrient loads).
- Measures to address these (such as restrictions on abstraction, limits on fertiliser use).
- The extent to which these measures will be successful (such as maintained river and groundwater levels, reduced nutrient load), including consideration of time taken, probability of success, and flexibility.
- Expected costs of the measure (in terms of financial, resource and environmental costs).
- A derived measure of cost effectiveness (such as £/unit increase in river level, £/unit reduction in nutrient load) including allowance for risk and uncertainty.

This is a similar approach to that recommended in the IPC regulations for the selection of Best Available Techniques (HMSO, 2000). Figure 13.4 shows how this might apply. The procedure is complicated by the fact that in any one location, gaps in water quality are usually due to a variety

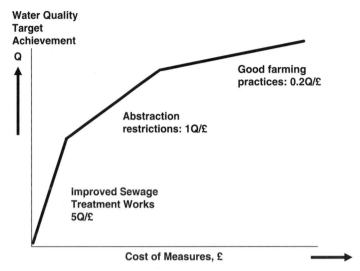

Figure 13.4   Selection of cost-effective measures for achieving water quality objectives

of different pressures which will respond differently to different measures, with different costs and degrees of certainty. Furthermore, the potentially different impact of alternative measures on different water users and interests means that some assessment of benefits (or rather loss of benefits) may also be required, with implications for the use of cost–benefit analysis.

MEASURING ENVIRONMENTAL COSTS AND THE POTENTIAL BENEFITS OF THE WFD

Overall, given that the EU WFD is an output of a negotiated political process (Chave, 2001; Kallis and Butler, 2001), the outcomes of the Directive must by implication exceed the costs of implementation. The overlying assumption of positive net benefit justifies the use of cost-effectiveness analysis rather than cost–benefit analysis. However, in so much as Europe is a large diverse area, the Directive must be capable of local adaptation. Cost–benefit analysis is only required where it is seems that the achievement of target water status will result in disproportionate costs: that is where MPB is persistently much greater than MEC in Figure 13.3.

Recent attempts have been made to place values on the economic functions of the water environment in order to justify environmental protection and enhancement of the kind advocated by the Directive. Specifically with respect to the water sector, high-level estimates from the Water Industry's Periodic Review made by the Environment Agency

for England and Wales, identify annual environmental damages affecting the use of water bodies of between £393 million and £423 million per year (Fisher *et al.*, 2003), of which damage to fishing and bathing are the largest items. Additional non-use loss associated with biodiversity impacts brings the overall estimate to between £1.2 billion and £2.6 billion per year, equivalent to a net present value over 25 years of between £19 billion and £43 billion at a 3.5% discount rate. About half of these damages are considered attributable to the activities of water supply companies. Environmental measures proposed under the water industry's five-year Asset Management Plan are predicted to deliver present value benefits of between £4 billion and £12 billion at 3.5% discount rate. This leaves a gap of about £15 billion to be picked up later by the WFD, to be shared by water companies and by agriculture, which is now perceived to be 'the number one polluter of water in the country' (PCFFF, 2002).

Regarding water pollution by agriculture, a number of studies have used secondary sources of data to estimate the external costs of farming (Hartridge and Pearce, 2001; Pretty *et al.*, 2000; Environment Agency, 2002; Eftec, 2004). These studies, which vary in the way they define and value external costs, estimate average annual damage costs to water which range between about £300 million and £500 million in 2005 prices. In the case of farm land, damage to water by diffuse pollution is closely associated with damage to soils and biodiversity (English Nature, 2002), particularly through surface water and soil–water movement (Evans, 1996), such that particular environment effects and their mitigation should not be seen in isolation. Assuming an average of £400 million per year water damage for E&W implies a present capital sum of about £6.4 billion over 25 years at 3.5% discount rate. The justification for measures to reduce environmental damage from farming (RPA, 2003) is all the greater given the annual £3 billion subsidy to UK agriculture currently provided by the Common Agricultural Policy.

COSTS OF WFD IMPLEMENTATION, COST RECOVERY AND WHO PAYS?

A preliminary assessment of the costs of implementing WFD for the UK was made by WRc (1999) on the basis of knowledge at the time of the likely gap in water quality status, and a mix of measures to control municipal and industrial point-source pollution and diffuse pollution from agriculture and to protect habitats. Estimated incremental present value costs at 6% discount rate (in 1999 prices) ranged between £3.2 billion and £11.2 billion comprising about £0.2 billion for administration, planning and monitoring, and intervention measures accounting for between £3 billion and £11 billion. Of these latter costs, about 15% related to point-source reductions for industry, 45% to water supply companies, about 30% for the control of diffuse pollution from agriculture, and about 10% for habitat protection and low-flow alleviation. The large range in the estimates reflected the high degree of uncertainty with respect to the scale of the challenge. The estimates also tend to reflect end-of pipe rather than in-process solutions that have the potential to be more cost effective in the long term. Reliable estimates of the costs of the WFD can only be obtained once the characterisation of river basins has been completed, confirming the challenge of acquiring robust pre-implementation appraisals of environmental policy (Defra, 2003b).

The WFD requires that charges for water supply and treatment services should adopt the principles of full-cost recovery and polluter pay. This means that users should also pay for costs associated with restoring and protecting environmental quality. Common methods to estimate these costs are currently being devised (WATECO, 2003; Defra, 2005). It is necessary to distinguish the economic costs used to determine cost-effective measures and the financial costs incurred by water supply organisations, inclusive of costs to protect the environment. The two are not necessarily the same. Economic costs (net of taxes and subsidies and inclusive of non-water resource costs such as energy) and environmental costs (such as carbon emissions) are used to select the most appropriate measures, such as improved sewage treatment works. Cost recovery will then be based on the actual financial cost of delivering these most economically cost-effective services.

Member States must 'ensure' that water-pricing policies should be designed to give 'adequate incentives for users to use water efficiently'. Furthermore, there must be 'adequate contribution of the different uses, disaggregated into at least industry, households and agriculture, to the recovery of the cost of water services, based on the economic analysis of uses and the taking account of the polluter pay principle' (Article 9). However, a MS can have regard to the 'social, environmental and economic effects' of recovery regimes and it 'will not be in breach of the Directive if it decides in accordance with established practices not to apply the(se) provisions', as long as they do not compromise the purpose of the Directive. So Article 9 is a recommendation which can be rejected if a case can be made.

Recovery of full capital and operating costs for water services, with or without environmental costs, will be quite a challenge in a number of MS. There are considerable subsidies to water in the domestic and agricultural sectors in some countries, funded mainly by the tax payer. In Spain, Greece and Portugal, for example, according to OECD (1999), domestic users have typically paid between 18 and 25% of total financial costs of water services, whereas in Italy, France, Germany, UK and Netherlands, this ranges between 80 and 100%. On the whole, industry tends to pay full costs, but agricultural irrigation (which accounts for over 70% of abstraction in some Mediterranean MS) enjoys high levels of subsidies, with limited recovery of the capital costs of publicly funded supply systems. The move towards greater privatisation in some parts of the European Water sector will, however, further promote greater recovery of costs as part of the process of self-financing.

The WFD requires that pricing mechanisms should provide incentives to use water efficiently, encouraging water to move to its most rewarding use. In theory, this implies 'benefit pricing' based on willingness to pay. Given the characteristics of water, however, this might not achieve maximum economic welfare. For example, profit-seeking private 'monopolistic' suppliers could limit water supply to push up prices to their advantage. Similarly, public sector water supply companies could do the same to raise revenue for government exchequers. While increased water prices can encourage wise use, where the demand for water is strong (and price inelasticity low), doing so may impact on the disposable income of water users (and make them poorer) before it changes their behaviour.

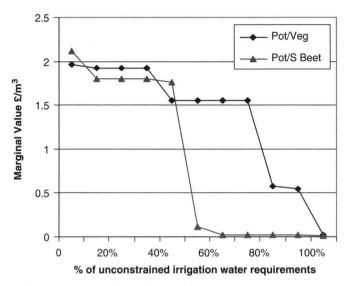

Figure 13.5   The marginal value of irrigation water in eastern England

The strong demand for irrigation water for vegetable production by farmers is a case in point. Modelling representative irrigation farming systems in eastern England (Morris *et al.*, 2004), Figure 13.5 shows that the marginal value of irrigation water (the extra benefit less extra cost) varies considerably between crop types and cropping systems. On high-value potatoes and vegetables it is typically around £1.5 per cubic metre, whereas on sugar beet it is less than £0.10 per cubic metre. Marginal value indicates a theoretical WTP – a demand curve for irrigation water. The relatively high price elasticity of demand for water on sugar beet means that even a small increase in price for water renders sugar beet irrigation infeasible and demand for water would reduce sharply to about 60% of current use on a typical potato and sugar beet farm. However, the relatively low price elasticity of demand for water on potatoes and vegetables means that a relatively large increase in water prices would be required to have a significant effect on the demand for water. Indeed, in the short term, potato farmers facing higher water prices are likely to pay up and suffer reduced incomes as a consequence. Here, increased abstraction charges may serve to transfer incomes to the supply agency without significantly reducing water consumption. In the longer term, of course, farmers would seek cost-reduction strategies such as improved scheduling, drip irrigation or farm reservoirs to store water during winter when abstraction charges are lower.

Increased water-use efficiency could in theory be achieved by trading between water buyers and sellers. With reference to Figure 13.5, in the face of rising abstraction charges, sugar beet farmers might find it attractive, if possible, to sell their 'surplus' water to other farmers, or to non-agricultural users, including wetland conservation interests. New legislation in E&W under the 2003 Water Act facilitates such trading subject to a number of safeguards (HMSO, 2003). There is a risk in water-deficient areas, however, that actions to improve the efficiency of water use may actually reduce the amount and quality of water that is 'returned' to the environment, unless specific actions are taken to prevent this.

A review of abstraction charges levied by the Environment Agency in E&W (Defra, 2001) considered pricing options which varied according to objectives, namely whether cost recovery, revenue generation, water-use efficiency or internalising environmental costs. The outcome was to remain largely with the existing arrangements which recover the cost of administering the licensed system, strengthening other measures to improve water-use efficiency. This reflects the view that charging users of water 'services' rather than direct abstractors of raw water is probably more effective, efficient and practical, especially given the complications associated with entitlement to abstract.

By way of example, the practice of full cost recovery including environmental costs is currently being rehearsed in the water industry's Periodic Review for England and Wales. Indeed, the privatisation of the water industry in E&W has helped to transfer this funding responsibility from the public purse to water companies under economic regulation by the Office of Water Services (OFWAT). With pressure from the Environment Agency and English Nature, and legally bound to meet the requirements of existing statutory obligations, water companies submitted their draft spending programmes for the period 2005–2010 (PR04) (OFWAT, 2003). These included a substantial capital component totalling £4.6 billion for 'environmental improvement', justified against the avoidance of damage costs (estimated between £4 billion and £12 billion). These and other extra costs were expected to raise prices to households by an average 31% in real terms during this period (OFWAT, 2003). Commenting on this, one UK government committee said that 'customers must not be expected to pay for every improvement to the aquatic environment, but where water companies are responsible for damage to the environment, they and their customers should pay to repair that damage' (HCEFRAC, 2003b). With an eye on agriculture, the committee went on to suggest that 'those responsible for diffuse pollution should pay for it' and also that more should be done to 'manage the demand for water'. All these points ring true with the sentiments of WFD.

Meanwhile, OFWAT, representing the interests of water consumers in E&W, has expressed concern that customers cannot be expected to meet ever-increasing bills to pay for environmental improvement, especially when sources of diffuse pollution remain relatively unchecked. The perception is that households are an easy target for recovering the cost of pollution caused by others, and that this is facilitated by a regulatory regime that uses the privatised water industry to reduce the burden on the public purse. WaterVoice (2003) representing the domestic payer argues that special provision should be made, perhaps through tax and benefit credits, to ensure that water is affordable for the poorest groups. At present, UK domestic payers spend about 1.3% of their disposable income on water services. Water prices could double in some areas in real terms under WFD, with consequences for poorer people who find themselves caught by a potentially regressive environment tax, especially if there is a move away from a flat rate (based on house value) to meter-based charging regimes. These issues of cost burden and affordability will no doubt feature strongly during WFD implementation, especially in areas where geographical and climatic factors result in high water costs per capita.

CONCLUSIONS

The preceding discussion has explored the role and potential of economics in the implementation

of the WFD, one of the most significant pieces of environmental legislation in the European Union. The main applications relate to: the assessment of the demand for water; the values associated with its use; the determination of the costs of water services; provision for full recovery; and incentive pricing. Economics is a key criterion in the determination of programmes of cost-effective measures to implement the Directive, including the use of economic instruments. It has an important role in assessing the case for derogation on grounds of disproportionate costs.

The task for economists in the WFD is sufficiently novel that data and methods are developing as part of the implementation process in order to establish a 'framework' that is itself fit for purpose, makes good sense and is robust and cost effective. Most of the challenges at present are operational: transposing the Directive into practical guidance, getting data, sorting out methods, applying them in selected cases, engaging stakeholders and in the process exposing devils that are in the detail. Progress has been made with respect to characterising water use and the resulting pressures and impacts on the water environment. Sufficient information is not yet available on costs and benefits to help design programmes of cost-effective measures to achieve water quality standards or to assess the validity of 'less stringent environmental objectives' (Defra, 2005).

The economic regulation of the privatised water sector in E&W has undoubtedly sharpened economic swords in terms of linking theory and practice and has helped to prepare the way for the economic assessments for WFD. But as the water industry cleans up its act, more attention is placed on the significant environmental costs due to diffuse pollution from agriculture. The search for cost-effective measures to address these 'land management' issues will become an important aspect of the WFD. Indeed, the WFD may in due course become the Land and Water Framework Directive, and economics can help support such an integrated approach.

ACKNOWLEDGEMENTS

Thanks are expressed for helpful comments from colleagues Drs Peter Howsam and Andrew Gill of the Department of Natural Resources, Cranfield University, Bedford, UK.

NOTE

1 The underlying demand analytic concepts are described in further detail in Chapter 11 of this volume.

REFERENCES

Chave, P. (2001) The EU Water Framework Directive, IWA Publishing, Padstow.

Commission of the European Communities (CEC) (2000) Directive of the European Parliament and of the Council establishing a framework for Community action in the field of water policy. 1997/0067 (COD), C5-0347/00.

Defra (2001) Tuning Water Taking: Government decisions following consultation on the use of economic instruments in relation to water abstraction. Department for Environment, Food and Rural Affairs, London, June.

Defra (2003a) Guidance: Assessment of Benefits for Water Quality and Water Resources Schemes in the PR04 Environment Programme. Department for Environment, Food and Rural Affairs, London, www.defra.gov.uk

Defra (2003b) Policy Impact Assessment, Water Framework Directive. Department for Environment, Food and Rural Affairs, London, www.defra.gov.uk

Defra (2004) Principal Guidance from the Secretary of State to the Director General of Water Services: periodic review of water price limits, Department for Environment, Food and Rural Affairs, London, March.

Defra (2005) Water Framework Directive Economic Analysis; Article 5 Supporting Documents. Department for Environment, Food and Rural Affairs, London, March 2004. http://www.defra.gov.uk/environment/water/wfd/economics/index.htm

Eftec/IEEP (2004) Framework for Environmental Accounts for Agriculture, Economics for Environment Consultancy (Eftec) and Institute for European Environmental Policy (IEEP), Report for Department for Environment, Food and Rural Affairs, London.

English Nature (2002) Policy Mechanisms for the Control of Diffuse Agricultural Pollution, with Particular Reference to Grant Aid. English Nature Research Report No. 455, English Nature. Peterborough.

Environment Agency (2001) Water Resources for the Future – A Strategy for England and Wales. Bristol: The Environment Agency.

Environment Agency (2002) Agriculture and Natural Resources: benefits, costs and potential solutions. Bristol: The Environment Agency.

Environment Agency (2007) The Water Framework Directive: Water for Life and Livelihoods /www.environment-agency.gov.uk/subjects/waterquality

European Environment Agency (EEA) (1999). Sustainable water use in Europe: Part 1 - Sectoral Water Use. Environmental Assessment Report No. 1, EEA, Copenhagen.

European Environment Agency (EEA) (2001a) Sustainable water use in Europe: Part 2 - Demand Management, Environmental Issue Report No. 19, EEA, Copenhagen.

European Environment Agency (EEA) (2001b) Sustainable water use in Europe: Part 3 -Extreme Hydrologic Events: Floods and Droughts, Environmental Assessment Report No. 21, EEA, Copenhagen.

Evans, R. (1996) Soil Erosion and its Impacts in England and Wales. London: Friends of the Earth Trust.

Food and Agriculture Organisation (FAO) (2004) Economic Valuation of Water Resources in Agriculture. FAO Water Reports 27, FAO, United Nations.

Fisher, J., Sunman, H. and Tambe, N. (2003) The Environmental Benefits of the Environmental Programme in the Periodic Review of the Water Industry (PR04) Environment Agency.

Foundation for Water Research (FWR) (1996) Assessing the Benefits of Surface Water Quality Improvements. Manual, FR/CL 0005, Foundation for Water Research, The Listons, Marlow, Buckinghamshire.

Green, C. H. (2003) *Handbook of Water Economics: Principles and Practice*. Chichester: Wiley.

Grout, R. S., de Wilson, M. A., and Boumans, M. J. (2002) A typology for the classification, description and valuation of functions, goods and services. *Ecological Economics* 41: 393–408.

Hanley, N., Shogren, J. and White, B. (2001) *Introduction to Environmental Economics*. Oxford: Oxford University Press.

Hartridge, O. and Pearce, D. (2001) Is UK Agriculture sustainable? Environmentally adjusted economic accounts for UK agriculture. CSERGE-Economics, University College, London.

HCEFRAC (2003a) The Water Framework Directive, House of Commons, Environment, Food and Rural Affairs Committee, Fourth Report of Session 2002–2003, March 2003, The Stationery Office, London.

HCEFRAC (2003b) Water Pricing, House of Commons, Environment, Food and Rural Affairs Committee, First Report of Session 2003–2004, December 2003, The Stationery Office, London.

HMSO (2000) The Pollution Prevention and Control (England and Wales) Regulations 2000. (Statutory Instruments 2000 no. 1973). Her Majesty's Stationery Office, London.

HMSO (2003) Water Act 2003. Her Majesty's Stationery Office, London.

International Conference on Water and the Environement (ICWE) (1992) The Dublin Statement on Water and Sustainable Resources. Hydrology and Water Resources Department, World Meteorological Organization, Geneva.

Kallis, G. and Butler, D. (2001) The EU water framework directive: measures and implications. *Water Policy* 3:125–142.

Knox, J. W., Morris, J., Weatherhead, E. K. and Turner, A. P. (1999) Mapping the financial benefits of spray irrigation and potential financial impact of restrictions on abstraction: a case study in Anglian Region. *Journal of Environmental Management* 58: 45–49.

MEA (2005) Millennium Ecosystem Assessment. http://www.maweb.org/en/index.aspx

Merret, S. (1997) *Introduction to the Economics of Water Resources, an International Perspective*. London: UCL Press.

Morris, J., Weatherhead, E. K., Knox, J.,Vasilieou, K., de Vries, T., Freeman, D., Leiva, F. and Twite, C. (2004) The Case of England and Wales. In Burbel, J. and Martin, C.G. (eds) *The Sustainability of European Irrigation under Water Framework Directive and Agenda 2000*. Brussels: EC Director General for Research, Global Change and Eco Systems.

Moss B *et al.* (2003) The determination of ecological status in shallow lakes – a tested system (ECOFRAME) for implementation of the European Water Framework Directive. *Aquatic Conservation: Marine and Freshwater Ecosystems* 13: 507–549.

OECD (1999) The Price of Water in OECD Countries. Organisation for Economic Cooperation and Development, Paris.

OECD (2003) Improving Water Management: Recent OECD Experience. Organisation for Economic Cooperation and Development, Paris.

OFWAT (2003) Setting Water and Sewage Price Limits for 2005–2010. Overview of Companies Draft Business Plans, Office of Water Services, Birmingham, October.

OST (2003) Foresight Scenarios. Office of Science and Technology. Department of Trade and Industry: www.foresight.gov.uk. See also Progress Through Partnership: 11 Agriculture, Natural Resources & Environment.

Policy Commission on the Future of Farming and Food (PCFFF) (2002) Farming and Food: a Sustainable Future. Report of the Policy Commission on the Future of Farming and Food, January. Cabinet Office.

Pretty, J. N., Brett, D., Gee, D., Hine, R. E., Mason, C. F., Morison, J. I. L., Raven, H., Rayment, M. D., van der Bijl, G. A. Williams, S., Atkins, J.P., Hammond, C. J. and Trotter, S. D.

(2000) An assessment of the total external costs of UK agriculture. *Agricultural Systems* 65: 113–136.

Rees J. (1993) Economics of Water Resource Management, NRA R&D Note 128. NRA, Bristol.

RPA (2003) Water Framework Directive Indicative Costs of Agricultural Measures, Report for Defra. Risk and Policy Analysis, Loddon, Norfolk, July.

Tietenberg, T. (2003) *Environmental and Natural Resource Economics*, 6th edn. London: Pearson Education International, Addison Wesley.

Turner, R. K., van den Bergh, J. C. J. M., Soderqvist, T., Barendregt, A., van der Straaten, J., Maltby, B. and van Ierland, E. K. (2000) Ecological–economic analysis of wetlands: scientific integration for management and policy. *Ecological Economics* 35: 7–23.

WATECO (2003) Economics and the Environment: the Implementation Challenge of the Water Framework Directive. A Guidance Document, WATECO, June 2002.

WaterVoice (2003) Review of Water Companies' Price Limits, WaterVoice, Birmingham.

WRc (1999) Potential Costs and Benefits of Implementing the Proposed Water Framework Directive. Report of the Department of the Environment, Transport and the Regions, London.

Knowledges and Knowing

14

Ecological Design
and Education

David W. Orr

Ask the animals, and they will teach you; the birds of the air, and they will tell you; ask the plants of the earth, and they will teach you; and the fish of the sea will declare to you (Job, 7–9).

When you build a thing you cannot merely build that thing in isolation, but must also repair the world around it, and within it so that the larger world at that one place becomes more coherent and more whole; and the thing which you make takes its place in the web of nature as you make it (Christopher Alexander).

BACKGROUND

Imagine living in a random world without order in which no rules applied, and effects followed no discernible pattern of cause. Such a world would be alien to intelligence, morality, and foresight, governed instead by caprice and whimsy, which is to say that it would be a kind of Hell. Design presumes, on the contrary, that matter is ordered and that order matters. But to the questions of exactly what is ordered and how there is no one answer. The more we know, the more mysterious the world appears to be. Beyond the regularities of changing seasons, birth, and death, the world that we experience is often chaotic and violent, governed as much by fate as by foresight. But even that awareness fuels the effort to discover larger patterns, mastery of which will permit us to establish a safe haven or, for some, heaven on Earth. For the builders of megalithic monuments like Stonehenge, the clues to order lay in the observed regularities of the night sky and the movements

of the Sun, Moon, and stars. The Greeks, believers in the possibility of reason, discovered geometrical proportions and mathematical harmony in the world. Some believed that cultivation of reason might lead to societies in which reasonable men might collaborate to manage public affairs democratically, yet another level of harmony. For the ancient Jews, the basis of order was otherworldly, a moral order evident in the Laws God delivered to Moses. For the builders of the great cathedrals, that belief was extended into architectural form, blending Greek geometry with Judeo-Christian theology in service to the idea that inspired humans could design so artfully as to create sacred spaces that were a portion of heaven on Earth.

The fourth great design revolution, built on modern science, presumed a more remote God who had once created a clockwork universe and had the good sense thereafter not to meddle with it. Isaac Newton deciphered the scientific laws God had presumably used and rendered these into the metaphor of a cosmic machine. Adam Smith took that metaphor to describe our tendency to truck and barter as the working out of an invisible hand administering the laws of supply and demand in a mechanistic world. In the fourth revolution, the economy of Adam Smith is the ultimate machine, godlike in its ability to sift order from the chaos of individual self-interest. We continue to live in that faith, now extended to a further abstraction called the global economy.

Each of the design revolutions in some degree persists like geologic layers. Unlike the scientific revolutions described by Thomas Kuhn in which one paradigm overthrows another, less adequate, our sense of order is a kind of lamination in which earlier thinking persists whether in science, social structures, language, or even commonplace superstitions. Each transformation in our understanding of how to make the human presence on earth surrendered in due course to time, human frailty, and their particular flaws, but did not thereby disappear. The megalithic belief in a larger order evident in the rising and setting of the Sun, lunar cycles, and movements of the stars survives in the belief that patterns of ecology represent a larger ordering applicable to human systems. So too, the belief that human reason might yet bring order from unreason and caprice. The Greek experiment in rationality survived and flourished in the Christian era as part of what Arthur O. Lovejoy once described as "the great chain of being" (1936/1974). If humans had the capacity of reason, might they not also discern the very mind of God. The neo-platonism of the medieval world, in Lovejoy's words, "rested at bottom upon a faith … that the universe is a rational order … a coherent, luminous, intellectually secure and dependable world, in which the mind of man could go about its business of seeking an understanding of things in full confidence" (pp. 327–328). Faith in a rational order and the powers of rationality survives into our time, magnified by the Enlightenment of the 18th century into the creed of inevitable progress. The faith of the medieval churchmen survives not just in the millenarian assumptions of nearly every ideological movement, but in the belief that what we made on earth ought to reflect higher obligations than those of self-interest. That, too, is an echo of the ancient belief in a divine order that would lead to a final triumph of right.

The increasingly homogeneous industrial civilization that now stretches around the earth is the signature accomplishment of the fourth revolution, but its future is troubled for reasons that any moderately well-informed high-school student could recite. Its prospects are clouded, first, because it is inflicting a rising level of ecological damage evident as impaired ecological functions, the loss of biological diversity, mutilated ecosystems, spreading blight, pollution, and climate change. For the scientists who study earth processes and ecology the facts are well known. Due to the loss of habitat and pollution, the number of species on Earth will decline by a quarter to one-third in this century. The carbon content of the atmosphere has increased by more than a third from its pre-industrial level of 280 parts per million and is rising at a rate now over 2 parts per million per year, a harbinger of worse to come. The human population has increased six-fold in the last two centuries and will peak at perhaps nine billion. The number of large predatory fish in the oceans has decreased by ninety per cent. Worldwide soil loss is estimated to be 20 to 25 billion tonnes per year. Forests, roughly the size of Scotland, are disappearing each year. Within a few years or maybe in a decade or two, we will reach the peak of the era of cheap oil where supply and demand diverge and start down the backside of the curve. That transition could be the start of an era of bitter geopolitical conflicts. Harvard biologist Edward O. Wilson refers to the decades ahead as a "bottleneck" an uncertain passage through constraints caused by the loss of species, climatic change, and population growth (Wilson, 2002). The scientific evidence documenting the decline of the vital signs of the Earth is overwhelming, so too the burden of pondering such complicated and dire things which may help to explain the growing popularity of escapism, religious zealotry, hyper-consumption, and other modes of denial.

The industrial experiment is failing, too, because of growing inequities and violence. In spite of nearly a century of economic growth, a majority of people on Earth experience life close to the bone. Over one billion people live at the edge of starvation in absolute poverty. Their daily reality is hunger, insecurity, and hopelessness. At the other end of the spectrum another billion live in affluence and suffer the consequences of having too much. Powered by cheap fossil energy, their world is one of traffic jams, suburban malls, satiation, fashion, fad diets, addiction, boredom, excitement, and commercial entertainment. In spite of high rates of economic growth, the trend is toward greater and greater inequity that is leading to a world dominated by a handful of corporations and a few thousand super-wealthy. These two worlds appear to be diverging, but in fact their destinies are colliding. Security, once a function of distance and military might, has been radically changed by terrorism and the diffusion of heinous weaponry. National borders no longer provide safety. The powerful and wealthy are vulnerable now precisely because their power and wealth makes them targets for terrorists and malcontents. And ethics, once a matter of individual behavior, now includes the conduct of whole societies and entire generations whose choices about energy and resource use cast long shadows across the planet and into the far future.

The inability to solve ecological and social problems points to deeper flaws. Like the proverbial fish unaware of the water in which it swims, we, too, have difficulty perceiving fatal flaws in our ideas, paradigms, and behavior that we take for granted until it is too late. In Jared Diamond's words: "human societies and smaller groups may make disastrous decisions for a whole sequence of reasons: failure to anticipate a problem, failure to perceive it once it has arisen, failure to attempt to solve it after it has been perceived, and failure to succeed in attempts to solve it" (2005, p. 438). In our time the inability to perceive and to solve problems is often related to our faith in technology that leads some to believe that we are masters of nature and smart enough to manage it in perpetuity. That presumption, in turn, rests on an improbably rosy view of human capabilities and the faith, as Robert Sinsheimer once put it that nature sets no traps for unwary species (1978).

Our optimism is, I think, a product of a particular era in human history shaped by the one-time drawdown of cheap fossil fuels, the "age of exuberance" in William Catton's words (1980). Our politics, economics, education, as well as personal expectations were shaped by the assumption that we had at last solved the age-old problem of energy. Ancient sunlight fueled rapid economic growth, vastly increased mobility and agricultural productivity, and a level of affluence that our ancestors could not imagine. But it also weakened social cohesion, encouraged overconsumption, polluted our air and water, and contaminated our politics, while creating a fragile and temporary energetic basis for the most complex human civilization ever.

Unfortunately, complex societies are vulnerable to breakdown for many reasons. Anthropologist Joseph Tainter summarizes these by saying that "as stresses necessarily arise, new organizational and economic solutions must be developed, typically at increasing costs and declining marginal return. The marginal return on investment in complexity accordingly deteriorates slowly at first and then accelerates. At this point, a complex society reaches the phase where it becomes increasingly vulnerable to collapse" (1988, p. 195). In other words, even with foresight we fail to anticipate problems which outrun solutions thereby aggregating into crises, then into a system-wide crisis of crises, the sense of care, always a limited resource, falters, human ingenuity, however considerable, fails, and things come tumbling down (Homer-Dixon, 2000). The story is an old one – lack of vision, the intoxication of power, tragedy, arrogance, stupidity, and angry gods.

TOWARD A DESIGN SCIENCE

The fox, it is said, knows many things but the hedgehog knows one big thing. Ecological designers, like the hedgehog, know one big thing – that everything is hitched to everything else as systems within still larger systems and patterns that connect across species, space, and time. Ecological design begins in the recognition that the whole is more than the sum of its parts, that unpredictable properties emerge at different scales, and as a result that we live in a world of surprise and mystery.

Those who design with nature work in the recognition that the world is one and indivisible, that what goes round comes around, that life is more paradoxical than we can ever know, and that health, healing, wholeness, and holy, too, are inseparable. Ecological design is the careful meshing of human purposes with the patterns and flows of the natural world and the study of those patterns to inform human intentions, leaving a margin for error, malfeasance, and the unknown. Ecological design requires an efficiency revolution in the use of energy and materials, a transition to renewable energy, changes in land use and community design, the transition to economies that preserve natural capital, and a recalibration of political and legal systems with ecological realities.

The origins of ecological design can be traced back into our prehistoric ancestors' interest in natural regularities of seasons, Sun, Moon, and stars, as well as in the Greek conviction that humans, by the application of reason, could discern the laws of nature. Ecological design also rests on the theological conviction that we are obliged, not merely constrained, to respect larger harmonies and patterns. The Latin root word for the word religion – bind together – and the Greek root for ecology – household management – suggest a deeper compatibility and connection to order. Ecological design, further, builds on the science and technology of the industrial age, but for the purpose of establishing a partnership with nature, not domination. The first models of ecological design can be found in vernacular architecture and the practical arts that are as old as recorded history. It is, accordingly, as much a recovery of old and established knowledge as discovery of anything new. The arts of building, agriculture, forestry, health care, and economy were sometimes practiced sustainably in cultures that we otherwise might dismiss as primitive. The art of applied wholeness was implicit in social customs, such as the observance of the Sabbath and Holy days and the Jubilee year, or the practise of potlatch in which debts were forgiven and wealth was recirculated. It is evident still in all of those various ways by which communities and societies gracefully cultivate the arts of generosity, kindness, prudence, love, humility, compassion, gentleness, forgiveness, gratitude, and ecological intelligence.

In its specifically modern form, ecological design has roots in the Romantic rebellion against the more extreme forms of modernism, particularly the belief that humans armed with science and a bit of technology were lords and masters of Creation. Francis Bacon, perhaps the most influential of the architects of modern science, proposed the kind of science that would reveal knowledge by putting nature on the rack and torturing her secrets from her, a view still congenial to some who have learned to say it more correctly. The science that grew from Bacon, Galileo, and Descartes over-threw older forms of knowing based on the view that we are participants in the forming of knowledge and that nature is not dead (Merchant, 1982). The result was a science based on the assumptions that we stand apart from nature, that knowledge was to be judged by its usefulness in extending human mastery over nature, and that nature is best understood by reducing it into its compo-nents pieces. "The natural world," in the words of philosopher E. A. Burtt, "was portrayed as a vast, self-contained mathematical machine, consisting of motions of matter in space and time and man with his purposes, feelings, and secondary quali-ties was shoved apart as an unimportant spectator" (1954, p. 104). Our minds are so completely stamped by that particular kind of science that it is difficult to imagine another way to know in which comparably valid knowledge might be derived from different assumptions and something akin to sympathy and a "feeling for the organism" (Keller, 1983).

Among the dissidents to the directions of modern science, Goethe, best known as the author of *Faust*, stands out among the first theorists and practitioners of the science of wholeness. In contrast to a purely intellectual empiricism, what physicist and philosopher Henri Bortoft (1966) calls the "onlooker consciousness," Goethe stressed the importance of relying on observation beginning with intuition that allowed the object being investigated to speak to the observer. Descartes, in contrast, reportedly began his days in bed by withdrawing his attention from the con-taminating influence of his own body and the cares of the world, to engage in deep thinking. He aimed, thereby, to establish the methodology for a science of quantity established by pure thought. Goethe, on the other hand, practiced an applied science of wholeness in which "the organizing idea in cognition comes from the phenomenon itself, instead of from the self-assertive thinking of the investigating scientist" (1966, p. 240).

Instead of the intellectual inquisition proposed by Bacon and practiced subsequently, Goethe pro-posed something like a dialog with nature by which scientists "offer their thinking to nature so that nature can think in them and the phenomenon disclose itself as idea" (p. 242). Facilitation of that dialogue required "training new cognitive capaci-ties" so that Goethean scientists "far from being onlookers, detached from the phenomenon, or at most manipulating it externally ... are engaged with it in a way which entails their own develop-ment" (p. 244). In Bortoft's words, "the Goethean scientist does not project their thoughts onto nature, but offers their thinking to nature so that nature can think in them and the phenomenon

disclose itself as idea," (p. 242) which requires overcoming a deeply ingrained habit of seeing things as only isolated parts not in their wholeness. The mental leap, as Bortoft notes, is similar to that made by Helen Keller who, blind and deaf, was nonetheless able to wake to what she called the "light of the world" without any preconceptions or prior metaphoric structure whatsoever. Goethe did not propose to dispense with conventional science, but rather to find another, and complementary, doorway to the realm of knowledge in the belief that Truth is plural, not the monopoly of one method, one approach, one time, or one culture.

Implicit in Goethe's mode of science is the old view, still current among some native peoples, that the Earth and its creatures are kin and in some fashion sentient, able to communicate to us, that life comes to us as a gift, and that a spirit of trust, not fear, is essential to knowing anything worth knowing. That message, in Calvin Martin's words "is riveting ... offering a civilization strangled by fear, measuring everything in fear, the chance to love everything" and to rise above "the armored chauvinism" inherent a kind of insane quantifica-tion (1999, pp. 107, 113). It is, I think, what Albert Einstein meant in saying that:

> A human being is part of a whole, called by us the universe, a part limited in time and space. He experiences himself, his thoughts and feelings, as something separated from the rest – a kind of optical delusion of his consciousness. This delusion is a kind of prison for us, restricting us to our per-sonal desires and to affection for a few persons nearest us. Our task must be to free ourselves from this prison by widening our circles of com-passion to embrace all living creatures and the whole of nature in its beauty.

Goethe proposed a kind of jailbreak from the prison of Cartesian anthropocentrism and from beliefs that animals and natural systems were fit objects to be manipulated at will. His intellectual heirs include all of those who believe that the whole is more than the sum of its parts, including systems' thinkers as diverse as mathematician and philosopher Alfred North Whitehead, politician and philosopher Jan Smuts, biologist Ludwig von Bertalanffy, economist Kenneth Boulding, and ecologist Eugene Odum. Goethe's approach con-tinues in the study of nonlinear systems in places like the Santa Fe Institute. Biologist Brian Goodwin, for one, calls for a "science of qualities" that complements and extends existing science (1994, p. 198). Conventional science, in Goodwin's view, is incapable of describing "the rhythms and spatial patterns that emerge during the develop-ment of an organism and result in the morphology and behavior that identify it as a member of a particular species ... or the emergent qualities that are expressed in biological form are directly linked to the nature of organisms as integrated wholes" (198–199). Goodwin, like Goethe, calls for a "new biology ... with a new vision of our relationships with organisms and with nature in general ... [one] that emphasizes the wholeness, health, and quality of life that emerge from a deep respect for other beings and their rights to full expression of their natures" (p. 232). Goodwin, Goethe, and other systems' scientists aim for a more scientific science, predicated on a rigor commensurate with the fullness of life in its lived context.

While Goethe's scientific work focused on the morphology of plants and the physics of light, D'Arcy Thompson, one of the most unusual polymaths of the 20th century and one who "stands as the most influential biologist ever left on the fringes of legitimate science" approached design by studying how and why certain forms appeared in nature (Gleick, 1998, p. 199). Of his *magnum opus On Growth and Form*, Sir Peter Medawar said that it was "beyond comparison the finest work of literature in all the annals of science that have been recorded in the English tongue" (Gleick, 1988, p. 200). Thompson seems to have measured everything he encountered, most notably natural forms and the structural features of plants, and animals. In so doing he discovered the patterns by which form arises from physical forces, not just by evolutionary tinkering as proposed by Darwin. Why, for example, does the honeycomb of the bee consist of hexagonal chambers similar to soap bubbles compressed between two glass plates? The answer Thompson discovered was found in the response of materials to physical forces, applicable as well to "the cornea of the human eye, dry lake beds, and polygons of tundra and ice" (Willis, 1995, p. 72). Thompson challenged the Darwinian idea that heredity determined all by showing the physical and mechanical forces behind life forms at all levels. His work inspired subsequent work in biomechanics, evolutionary biology, architecture, and biomimicry, including that by Paul Grillo, Karl von Frisch, and Steven Vogel.

Frisch, for example, explored the ingenuity of animal architecture evolved by birds, mammals, fishes, and insects. African termite mounds a dozen feet high, for example, maintain a constant temperature of approximately 78°F in tropical climates (Frisch, 1974, pp. 138–149). Nests are ventilated variously by permeable walls that exchange gases and by ventilation shafts opened and closed manually as needed with no other instructions than those given by instinct. Interior ducts move air and gases automatically by convection. The system is so ingeniously designed that chambers

deep underground are fed a constant stream of cool, fresh air that rises as it warms before being ventilated to the outside. Termite nests are constructed of materials cemented together with their own excretions, eliminating the problem of waste disposal. Desert termites, with no engineering degrees as far as is known, bore holes to depths of 40 meters below their nests to find sources of water. Beavers construct large dams of 1000 feet or more in length; their houses are insulated to remain warm in sub-zero temperatures. Other animals, less studied, build with comparable skill (see Tsui, 1999, pp. 86–131). Human ingenuity, considerable as it is, pales before that of many animals that design and build remarkably strong, adaptable, and resilient structures without toxic chemicals, machinery, fossil fuels, and professional engineers.

The idea of that nature is shaped by physical forces as much as by evolution is also evident in the work of Theodor Schwenk who explored the role of water as a shaper of Earth's surfaces and biological systems. Of water, Schwenk wrote that:

In the chemical realm, water lies exactly at the neutral point between acid and alkaline, and is therefore able to serve as the mediator of change in either direction. In fact, water is the instrument of chemical change wherever it occurs in life and nature ... In the light-realm, too, water occupies the middle ground between light and darkness. The rainbow, that primal phenomenon of color, makes its shining appearance in and through the agency of water ... In the realm of gravity, water counters heaviness with levity; thus, objects immersed in water take on buoyancy ... In the heat-realm water takes a middle position between radiation and conduction. It is the greatest heat conveyer in the earth's organism, transporting inconceivable amounts of warmth from hot regions to cooler ones by means of the process known as heat-convection ... In the morphological realm, water favors the spherical; we see this in the drop form. Pitting the round against the radial, it calls forth that primal form of life, the spiral ... In every area, water assumes the role of mediator. Encompassing both life and death, it constantly wrests the former from the latter (1989, p. 24).

Moving water shapes landscapes. As ice it molds entire continents. At the micro-scale, its movement shapes organs and the tiniest organisms. But at any scale it flows, dissolves, purifies, condenses, floats, washes, conducts, and some believe that it even remembers. Our language is brim full of water metaphors and we have streams of thought or dry spells. The brain literally floats on a water cushion. Water in its various metaphors is the heart of our language, religion, and philosophy. We are much given to the poetry of water as mists, rain, flows, springs, light reflected, waterfalls, tides, waves, storms. Some of us have been baptized in it. But all of us stand before the mystery that D. H. Lawrence called "the third thing," by which two atoms of hydrogen and one of oxygen become water and no one knows what it is.

"Form patterns," Schwenk wrote, "such as those appearing in waves with new water constantly flowing through them, picture on the one hand the creation of form and on the other the constant exchange of material in the organic world" (1996, p. 34). Water is a shaper, but the physics of its movement is also the elementary pattern of larger systems "depicting in miniature the great starry universe" (p. 45). Water is the medium by which and through which life is lived. Turbulence in air and water have the same forms and mechanics as vortices whether in the ocean, atmosphere, or in space. Sound waves and waves in water operate similarly. Schwenk's great contribution to ecological design, in short, was to introduce water in its fullness as a geologic, biological, somatic, and spiritual force, a reminder that we are creatures of water, all of us merely eddies in one great watershed.

The profession of design as a practical art probably begins with the great British and European landscapers such as Capability Brown (1716–1783) famous for developing pastoral vistas for the rich and famous of his day. In our own history the early beginnings of design as applied ecology are apparent in the work of the great landscape architect and creator of Central Park in New York, Frederick Law Olmsted, and, later, in that of Jens Jensen, who pioneered the use of native plants in landscape designs in the Midwest. Ian McHarg, a brilliant revolutionary, merged the science of ecology with landscape architecture aiming to create human settlements in which "man and nature are indivisible, and that survival and health are contingent upon an understanding of nature and her processes" (1969, p. 27) His students, including Pliny Fisk, Carol Franklin, and Ann Whiston Spirn, continued that vision armed with sophisticated methodological tools of geographic information systems and ecological modeling applicable to broader problems of human ecology.

While the degree of influence varied, many early efforts toward ecological design were inspired by the arts and crafts movement in Britain, particularly the work of William Morris and John Ruskin. In US architecture, for example, Frank Lloyd Wright's attempt to define an "organic architecture" has clear resonance with the work of Morris and Ruskin as well as the transcendentalism of Ralph Waldo Emerson. Speaking before the Royal Institute of British Architects in 1939, Wright described organic architecture as "architecture of nature, for nature … something more integral and consistent with the laws of nature" (1993, pp. 302, 306). In words Morris and Ruskin would have applauded, Wright argued that a building "should love the ground on which it stands" reflecting the topography, materials, and life of the place (p. 307).

Organic architecture is "human scale in all proportions," but is a blending of nature with human created space so that it would be difficult to "say where the garden ends and where the house begins … for we are by nature ground-loving animals, and insofar as we court the ground, know the ground, and sympathize with what it has to give us" (p. 309). Wright's vision extended beyond architecture to a vision of the larger settlement patterns that he called "Broadacre City," arguing that organic architecture had to be more than an island in a society with other values. In his interest on harmonizing site, form, and function, and using natural materials and solar energy, in Wright is a precursor to the green building movement and the larger endeavor of ecological design. And in his often random musings about an "organic society" he foreshadowed the present dialogue about the sustainability of modern society.

Ecological design, however, is not just about calibrating human activities with natural systems. It is also an inward search to find patterns and order of nature written in our senses, flesh, and human proclivities. There is no line dividing nature outside from inside; we are permeable creatures inseparable from nature and natural processes in which we live, move, and have our being. We are also sensual creatures with five senses that we know and others that we only suspect. At its best, ecological design is a calibration, not just of our sense of proportion that the Greeks understood mathematically, but a finer calibration of the full range of our sensuality with the built environment, landscapes, and natural systems. Our buildings are thoughts, words, theories, and entire philosophies crystallized for a brief time into physical form that reveal what's on our mind and what's not. When done right, they are a form of dialogue with nature and our own deeper, sensual nature. The sights, smells, texture, and sounds, of the built environment evoke memories, initiate streams of thought, engage, sooth, provoke, bind or block, open or close possibilities. When done badly, the result is spiritual emptiness characteristic of a great deal of modern design that reveals, in turn, a poverty of thought, perception, and feeling.

More specifically, we are creatures shaped inordinately by the faculty of sight, but seeing is anything but simple. Oliver Sacks once described a man blind since early childhood whose sight once restored found it to be a terrible and confusing burden, preferring to return to blindness and his own inner world of touch. "When we open our eyes each morning," Sacks writes, "it is upon a world we have spent a lifetime *learning* to see" (Sacks, 1993, p. 64). And we can lose not only the faculty of sight, but the ability to see as well. Even with 20–20 vision, our perception is always selective because our eyes permit us to see only within certain ranges of the light spectrum and because personality, prejudice, interest, and culture further filter what we are able to see. Sacks notes that individual people can choose not to see and I suspect the same is true for cultures as well. The affinity for nature, a kind of sight, is much diminished in modern cultures.

Collective vision cannot be easily restored by more clever thinking, but, as David Abram puts it, only "through a renewed attentiveness to this perceptual dimension that underlies all our logics, through a rejuvenation of our carnal, sensorial empathy with the living land that sustains us" (1996, p. 69). Drawing from the writings of Merleau-Ponty, Abram describes perception as interactive and participatory in which "perceived things are encountered by the perceiving body as animate, living powers that actively draw us into relation … both engender(ing) and support(ing) our more conscious, linguistic reciprocity with others" (p. 90). Further, sight as well as language and thought are experienced bodily as colors, vibrations, sensations, and empathy, not simply as mental abstractions. The ideas that viewer and viewed are in a form of dialogue and that we experience perception bodily runs against the dominant strains of Western philosophy. Plato, for illustration, has Socrates say that "I'm a lover of learning, and trees and open country won't teach me anything whereas men in the town do" (*Phaedrus*, 479). Plato's world of ideal forms existed only in the abstract. Similarly, the Christian heaven exists purely somewhere beyond earthly and bodily realities. Both reflected the shifting balance between the animated sacred, participatory world and the linear, abstract, intellectual world. Commenting on the rise of writing and the priority of the text, Abrams says that "the voices of the forest, and of the river began to fade … language loosen(ed) its ancient association with the invisible breath, the spirit sever(ed) itself from the wind, and psyche dissociate(d) itself from the environing air" (p. 254). As a result, "human awareness folds in upon itself and the senses – once the crucial site of our engagement with the wild and animate earth – become mere adjuncts of an isolate and abstract mind" (p. 267).

Through the designed object we are invited to participate in seeing something else, a larger reality. The creators of Stonehenge, I think, intended worshippers to see not just circles of artfully arranged stone, but the cosmos above and maybe within. The Parthenon is a temple dedicated to the goddess Athena, but also a visible testimony to an ideal existing in mathematical harmonies, proportion, and symmetry discoverable by human reason. The builders of Gothic cathedrals intended not just monumental architecture but a glimpse of heaven and a home for sacred presence. For all of the crass, utilitarian ugliness of the factories, slums, and glittering office towers, the designers and builders of the fourth revolution intended to reveal a world of abundance and human potentials in a world they otherwise deemed uncertain and violent, ruled by the economic laws of the jungle.

Finally, the practise of ecological design is rooted in the emerging science of ecology and the specific natural characteristics of specific places. The fifth revolution is not merely a more efficient recalibration of energy, materials, and economy in accord with ecological realities, but a deeper and more coherent vision of the human place in nature. Ecological design is, in effect, the specific terms of a declaration of peace with nature that begins in the science of ecology and the recognition of our dependence on the web of life (Capra, 1996). In contrast to the belief that nature is little more than a machine and its parts merely resources, for designers of the fifth revolution it is, as Aldo Leopold put it:

A fountain of energy flowing through a circuit of soils, plants, and animals. Food chains are the living channels which conduct energy upward; death and decay return it to the soil. The circuit is not closed; some energy is dissipated in decay, some is added by absorption from the air, some is stored in soils, peats, and long-lived forests; but it is a sustained circuit, like a slowly augmented revolving fund of life. There is always a net loss by downhill wash, but this is normally small and offset by the decay of rocks (1949, p. 216).

Energy flowing through the "biotic stream" moves "in long or short circuits, rapidly or slowly, uniformly or in spurts, in declining or ascending volume," what ecologists call food chains. For designers, the important point is that the internal processes of the biotic community, the ecological books in effect, must balance so that energy used or dissipated by various processes of growth must be replenished (Leopold, 1953, p. 162). Leopold proposed three basic ideas:

1 that land is not merely soil;
2 that the native plants and animals kept the energy circuit open; others may or may not;
3 that man-made changes are of a different order than evolutionary changes, and have effects more comprehensive than is intended or foreseen (p. 218).

Ecological design, as Leopold noted, begins in the recognition that nature is not simply dead material or a simply a resource for the expression of human wants and needs, but rather "a community of soils, waters, plants, and animals, or collectively: the land" of which we are a part (Leopold, 1949, p. 204). But Leopold did not stop at the boundary of science and ethics, he went on to draw out the larger implications. For reasons of both necessity and right, the recognition that we are members in the community of life "changes the role of *Homo sapiens* from conqueror of the land-community to plain member and citizen of it" (p. 204). The "upshot" is Leopold's classic statement that "a thing is right when it tends to preserve the integrity, stability, and beauty of the biotic community. It is wrong when it tends otherwise" (224–25). We will be a long time understanding the full implications of that creed, but Leopold, late in his life, was beginning to ponder the larger social, political, and economic requisites of a fully functioning land ethic.

Like Leopold's land ethic, ecological design represents a practical marriage of ecologically enlightened self-interest with the recognition of the intrinsic values of natural systems. Once consummated, however, the marriage branches out into a family of possibilities. Economics rooted in the realities of ecology, for example, requires the preservation of natural capital of soils, forests, and biological diversity; which is to say economies that operate within the limits of the Earth's carrying capacity (Daly, 1996; Hawken *et al.*, 1999). An ecological politics requires the recalibration of the complexities and time scales of ecosystems with the conduct of the public business. An ecological view of health would begin with the recognition that the body exists within an environment, not as a kind of isolated machine (Kaptchuk, 2000). Religion grounded in the operational realities of ecology would build on the human role as stewards and the obligation to care for the Creation (Tucker, 2003). An ecological view of agriculture would begin with the realities of natural systems, aiming to mimic the way natural systems function (W. Jackson, 1980). An ecological view of business/industry would aim to create solar powered industrial and commercial ecologies so that every waste product cycles as an input in some other system (McDonough and Braungart, 2002).

In whatever manifestation, the goal of ecological design is to go "from conqueror of the land community to plain member and citizen of it" (Leopold). But there is no larger theory of ecological design, nor is there a textbook formula that works for practitioners across many different fields and at varying scales. And neither should we presume agreement on what it means for humankind to become a "plain member and citizen of the biotic community." In other words, we have a compass but no map. Architect Samuel Mockbee, founder of the Rural Studio, enjoined his students working with the poor in Hale County, Alabama, only to make their work "warm, dry, and noble." Warm and dry are easy for the most part because they are felt somatically, noble is hard because it requires us to make judgments about what we ought to do relative to some standard higher than creature comfort. But in the best sense of the word it is synonymous with decent, worthy, generous, magnificent, proud, and resilient. And it ought to be synonymous with ecological design as well.

Ecology, the "subversive science," begins with the recognition of our practical connections to the physical world, but it does not stop there. The awareness of the many ways by which we are connected to the web of life would lead intelligent and scientifically literate people to protect nature and the conditions necessary to it for reasons of self-interest. But our knowledge, always incomplete and often wrong, is mostly inadequate to the task of knowing what's in our interest, whether we wish to define that as "highest" or "lowest," let alone discerning exactly what parts of nature we must accordingly protect and how to do it. Science notwithstanding, often we do not know what we are doing and why. More subversive still are questions concerning the interests and rights of lives and life across the boundaries of species and time. Since they cannot speak for themselves, their only advocates will be those willing to speak on their behalf.

There are many clever arguments used to explain why we should or should not be concerned about those whose lives and circumstances would be effected by our action or inaction. Like so many tin soldiers, arrayed across the battlefield of abstract intellectual combat, they assault frontally or by flank, retreat only to regroup, and charge again, each battle giving rise to yet another. But in the end, I think, such questions will not be decided by intellectual combat and argumentation, however smart, but rather more simply and profoundly by affection – all of those human emotions that we try to capture in words like compassion, sympathy, and love. Love, in other words, neither requires nor hinges on intellectual argument. It is a claim that we recognize as valid but for reasons we could never describe satisfactorily. In the end it is a nameless feeling that we accept as both a limitation on what we do and a gift we offer. Pascal's observation that the heart has reasons that reason does not know sums the matter. Love is a gift but the giver expects no return on the investment and that defies logic, reason, and even arguments about selfish genes.

So, after all of the intellectualization and clever arguments, whether we choose to design with nature or not will come down to a profoundly simple matter of whether we love deeply enough, artfully enough, and carefully enough to preserve the web on which all life now and in the future depends. Ecological design is simply an informed love applied to the dialogue between humankind and natural systems. The origins of the practice of ecological design can be traced far back in time, but there are deeper origins found in the recesses of the human heart.

TOWARD DESIGN EDUCATION

The basic principles of ecological design are these (Van der Ryn and Cowan, 1996; McDonough and Braungart, 2002):

- use sunshine and wind;
- preserve diversity;
- account for all costs;
- eliminate waste;
- solve for pattern;
- protect human dignity;
- leave wide margins for error, malfeasance, and ignorance.

The basic principles of modern education appear to be these:

- The purpose of education is to extend human mastery of nature.
- Learning is intellectual, not emotional.
- Curriculum is organized by disciplines and divisions.
- Analytical reasoning (reductionism) and quantification are superior to other modes of knowing.
- Schools are best organized like factories to maximize efficiency.
- Success is measured first by tests, later by careers in the industrial world.
- Academic architecture is a function of cost and efficiency.

The recalibration of education with ecology, and specifically one aimed to inform our role as designers, has large implications for the substance and process of education and the expectations that we bring to it. But what follows is perhaps best regarded as notes for a seminar on design education, a scouting expedition, toward that end rather than a set of firm conclusions or a blueprint.

First, in contrast to assumptions of human mastery of nature, the starting point for ecological design education is a more humble and serious consideration of the 3.8 billion years of evolutionary history. Nature, for ecological designers, is not something just to be mastered, but a tutor and mentor for human actions. Janine Benyus, author of *Biomimicry*, points out, for example, that spiders make biodegradable materials stronger than steel and tougher than Kevlar without fossil fuels or toxic chemicals and the product is biodegradable (Benyus, 1998). From nothing more than substances in seawater, mollusks make ceramic-like materials that are stronger and more durable than anything we know how to make. These and thousands of other examples are models for manufacturing, the design of technologies, farming, machines, and architecture that are orders of magnitude more efficient, elegant, and durable than our best industrial capabilities. But the foundational pedagogy begins with nature as tutor and mentor.

Ecological design, further, is not simply a mimicking of nature toward a smarter kind of industrialization, but rather a deeper revolution in the place of humans in nature. In Wendell Berry's words, design begins with questions "What's here? What will nature permit us to do here? What will nature help us do here?" The capacity to question presumes the humility to ask, the good sense to ask the right questions, and the wisdom to follow the answers to their logical conclusions. Ecological design is not a monologue of humans talking to nature, but a dialogue that requires the capacity to listen, discern, and learn from nature. When we get it right, the results in John Todd's words are "elegant solutions predicated on the uniqueness of place." The industrial standard, in contrast, is based on the idea that nature can be tortured into revealing her secrets, as Francis Bacon so revealingly put it. Brute force and human cleverness, not coevolution and cooperation are at the heart of the modern worldview. So too, standardization and a one size fits all strategy making industrial design look the same and operate by the same narrow logic everywhere. But this is no great victory for humankind because the mastery of nature, in truth, represents the mastery of some men over other men using nature as the medium, as C. S. Lewis once put it (1947).

Second, pedagogy informed by an ecological perspective does not begin with the assumption that humans are infinitely plastic. On the contrary, our sense of order and affinity for design are bounded by our long evolutionary history and our dawning sensations of life. The first safe haven we sense is our mother's womb. Our first awareness of regularity is the rhythm of our mother's heartbeat. Our first passage way is her birth canal. Our first sign of benevolence is at her breast. Our first awareness of self and other comes from sounds made and reciprocated. Our first feelings of ecstasy come from bodily release. The first window through which we see is the eye. The first tool we master is our own hand. The world is first revealed to us through the senses of touch and taste. Our first world view is formed within small places of childhood.

Our ancestors' first inkling that they were not alone was the empathetic encounter with animals. The first music they heard were sounds made by birds, animals, wind, and water. Their first source of wonder, perhaps, was the undimmed night sky. Their first models of shelter were those created by birds and animals. The first materials humans used for building were mud, grass, stone, wood, and animal skins. Their first metaphors were likely formed from daily experiences of nature. The first models for worship were found in what early humans perceived as cosmic harmony were those of the dwelling.

We are creatures shaped by the interplay between our senses and the world around us. We know of five senses and have reason to believe that there are others. For example, some evidence suggests that we have a rudimentary awareness of being watched and there are other possibilities. Aboriginal peoples can walk with unerring accuracy across trackless landscapes in the dark of night. Across all cultures and times, good design is a close calibration of our sensuality with inspiration, creativity, place, form, and materials. Good design feels right and is a pleasure to behold and experience for reasons that we understand at an intuitive level, but have difficulty explaining.

Third, all design involves decisions about how society provides food, energy, shelter, materials, water, and waste cycling, and distributes risks, costs, and benefits and is thereby unavoidably political. Design education, by the same logic, is political, having to do with decisions about energy, forests, land, water, biological diversity, resources, and the distribution of wealth, risks, and benefits. Often cast as "liberal" or "conservative," such decisions in our time are, in fact, often about how the present generation orients itself to the interests of its children and grandchildren. One can arrive at a decent regard for their prospects as either a conservative or as a liberal. These are not opposing positions so much as they are different sides of a single coin. The point is that harmonizing social and economic life with ecological realities will require choices about energy technologies, agriculture, land use, settlement patterns, materials, the handling of wastes, and water that are inescapably political and will distribute risks and benefits in one way or another.

Further, as the Greeks understood, design entails choices that enhance or retard civic life and the prospects for citizenship. But in our time "We are witnessing the destruction of the very idea of the inclusive city" and with it the arts of civility, citizenship, and civilization (Rogers, 1997, p. 10). By including or excluding possibilities to engage each other in convivial dialogue the creators of urban spaces enhance or diminish civility, urbanity, and the civic prospect. It is no accident,

I think, that crime, loneliness, and low participation became epidemic as spaces such as town squares, street markets, front porches, corner pubs, and parks were sacrificed to the automobile, parking lots, and urban sprawl. Better design alone cannot cure these problems, but they can help to engage people with their places as thoughtful and engaged citizens.

Fourth, ecological design implies a better and more robust economics. In an age much devoted to the theology of the market, disciples of the conventional wisdom believe it imprudent to design ecologically if the costs are even marginally more than conventional design. Based on incomplete and highly selective accounting, that view is almost always wrong because it overlooks the fact that we – or someone – sooner or later will pay the full costs of bad design, one way or another. In other words, society pays for ecological design whether it gets the benefits of it or not. Honest accounting, accordingly, requires that we keep the boundaries of consideration as wide as possible over the long term and have the wit to deduct the collateral benefits that come from doing things right. For example, ignoring the costs of wars fought for "cheap" oil, or those of climate change, air pollution, and the health effects of urban sprawl, an SUV (sport utility vehicle) is cheap enough. But price and cost should not be confused. It is the height of folly to believe that we can eliminate forests, pollute, squander resources, erode soils, destroy biological diversity, remodel the biogeochemical cycles of the Earth, and create ugliness, human and ecological, without consequence. The truth is that sooner or later, the full costs will be paid one way or another. The problem, however, is that the costs of environmental dereliction are diffuse and often can be deferred to some other persons and to some later time, but they do not thereby disappear. The upshot is that much of our apparent prosperity is phony and so too the intellectual and ideological justifications for it.

The application of short-term economics to architecture, in particular, has been little short of disastrous. "The rich complexity of human motivation that generated architecture," in Richard Rogers' words, "is being stripped bare. Building is pursued almost exclusively for profit" (1997, p. 67). By such logic we cannot afford to design well and build for the distant future. The results have been evident for a long time. In the mid-19th century, John Ruskin noted that "Ours has the look of a lazy compliance with low conditions" (1880/1989, p. 21). But even Ruskin could not have foreseen the blight of suburban sprawl, strip development, and urban decay, driven by our near terminal love affair with the automobile and inability to plan sensibly. The true costs, however, are passed on to others as "externalities" thereby privatizing the gains while socializing the costs. The truth is, as it has always been, that a phony prosperity is no good economy at all. False economic reckoning has caused us to lay waste to our countryside, abandon our inner cities and the poor, and build auto-dependent communities that are contributing mightily to climatic change and rendering us dependent on politically unstable regions for oil.

An economy judged by the narrow industrial standards of efficiency will destroy values that it cannot comprehend. Measured as the output for a given level of input, maximizing efficiency creates disorder, that is to say, inefficiency at higher levels. The reasons are complex but have a great deal to do with our tendency to confuse means with ends. As a result efficiency often becomes an end in itself while the original purposes (prosperity, security, benevolence, reputation, etc.) are forgotten. The assembly line was efficient for the manufacturing firm, but its larger effects on workers, communities, and ecologies were often destructive and the problems for which mass production was a solution have been compounded many times over. Neighborliness is certainly an inefficient use of time on any given day, but not when considered as a design principle for communities assessed over months and years. For engineers, freeways are efficient at moving people up to a point, but they destroy communities, promote pollution, cause congestion, create dependence on foreign oil, and eliminate better alternatives including design for access that precludes the need for transportation. WalMart, similarly, is an efficient marketing enterprise, but eliminates its competitors and many things that make for good communities, including jobs that pay decent wages. Success on such terms will eventually destroy WalMart and a great deal more. And, of course, nuclear weapons are wonderfully efficient and quick devices as well. Ecological design, in contrast, implies a different standard of efficiency oriented toward ends, not means, the whole, not parts, and the long term not the short term.

Fifth, design education must be grounded in an honest assessment of human capabilities. Ecological design, like all human affairs, has to be carried out in the full recognition of human limitations, including the discomfiting possibility that we are incurably ignorant. T. S. Eliot put it this way:

Human kind
Cannot bear very much reality (1971, p. 119).

In other words, we are inescapably ignorant and the reasons are many. We are ignorant because reality is infinite relative to our intellectual and perceptual capacities. We are ignorant because we individually and collectively forget things that we once knew. We are ignorant because every human action changes the very system we aim to understand. We are ignorant because of our own limited intelligence and because we cannot know in advance the unintended effects of our actions on complex systems. We are ignorant even about the proper ends to which knowledge might be put. Not the least, we are ignorant, as Eliot noted, because, sometimes, we choose to.

Alas, many seem to prefer it that way. From the publication of the Global 2000 Report in 1980 to the present there is a veritable mountain of scientific evidence about human impacts on ecosystems and the biosphere and ways to minimize or eliminate them. But our collective sleepwalk toward the edge of avoidable tragedy continues, suggesting that we are not so much rational creatures as we are adept and creative rationalizers.

Similarly, designers must reckon with the uncomfortable probability that the amount of credulity in human societies remains constant. This is readily apparent by looking backward through the rearview mirror of history to see the foibles, fantasies, and follies of people in previous ages (Tuchman, 1984). For all our pretensions to rationality, others at some later time will see us similarly. The fact is that humans, in all ages and times, are inclined to be as unskeptical and sometimes as gullible as those living in any other – only the sources of our befuddlement change. People of previous ages read chicken entrails, relied on shaman, consulted oracles. We, far more sophisticated but similarly limited, use computer models, believe experts, and exhibit a touching faith in technology to fix virtually everything. But who among us really understands how computers or computer models work, or are aware of the many limits of expertise, or the ironic ways in which technology "bites back"? Has gullibility declined as science has grown more powerful? No, if anything it is growing because science and technology are increasingly esoteric and specialized, hence removed from daily experience. Understanding less and less of either, we will believe almost anything. Gullibility feeds on mental laziness and is enforced by social factors of ostracism, social pressures for conformity, and the pathologies of groupthink that penalize deviance.

This line of thought raises the related and equally unflattering possibility that stupidity may be randomly distributed up and down the socioeconomic–educational ladder. As anecdotal evidence for the latter, I offer the observation that I have likely known as many brilliant people without much formal learning as those certified by a Ph.D. And there are likely as many thoroughgoing, fully degreed fools as there are undegreed fools. I am a professional "educator" and an admission of such gravity leads me to think that the gift of intelligence and intellectual clarity can be focused and sharpened a bit, but can neither be taught nor conjured. The numerous examples of the underdegreed or academic failures include Albert Einstein, Winston Churchill, Frank Lloyd Wright, and Amory Lovins. One should not conclude, however, that formal schooling is useless, but that its effectiveness, for all of the puffery that adorns college catalogues and educational magazines, is considerably less than advertised. And there are those, as lawyer John Berry once noted, who have been "educated beyond their comprehension," people made more errant by the belief that their ignorance has been erased by the possession of facts, theories, and the adornment of weighty learnedness.

Nor does the outlook for intelligence necessarily brighten when we consider the limitations of large organizations. These too, are infected with our debilities. Most of us live out our professional lives in organizations or work for them as clients and often discover to our dismay that the collective intelligence of organizations and bureaucracies is often considerably less than that of any one of its individual employees. We are baffled by the discrepancy between smart people within organizations exhibiting a collective IQ less than, say, kitty litter. We understand human stupidity and dysfunction because we encounter it at a scale commensurate with our own. But confronted with large-scale organizations, whether corporations, governments, or colleges and universities we tend to equate scale, prestige, and power with perspicacity and infallibility. Nothing could be farther from the truth. The intelligence of a large-scale organization (if that is not altogether oxymoronic) is limited by the obligation to earn a profit, enlarge its domain, preserve entitlements, or maintain a suitable stockpile of prestige.

Our human frailties infect the design professions as well. Buildings and bridges sometimes fall down (Levy and Salvadori, 1992). Clever designs can induce an astonishing level of illness and destruction. Beyond some scale and limits design becomes guesswork. British engineer A. R. Dykes puts it this way: "Engineering is the art of modeling materials we do not wholly understand, into shapes we cannot precisely analyze so as to withstand forces we cannot properly assess, in such a way that the public has no reason to suspect the extent of our ignorance" (www.ukcivilengineering. co.uk/quotes.html). In various ways the same is true in other design professions and virtually every other field of human endeavor.

The point is simply to say that human limitations will be evident in every design, project, and system, however otherwise clever. From this there are, I think, two conclusions to be drawn. The first is simply that design, whether of bridges, buildings, communities, factories, or farms and food systems, ought to maximize the capacity of a system to withstand disturbance without impairment, which is to say resilience. Ecological design does not assume the improbable: human infallibility, technological perfection, or some *deus ex machina* that magically rescues us from folly. Rather, ecological designers aim to work at a manageable scale, achieving flexibility, redundancy, and multiple checks and balances while avoiding the thresholds of the irreversible and irrevocable (Lovins and Lovins, 1982, ch. 13; Lovins, 2002).

Forewarned about human limitations, we might further conclude that a principal goal of designers ought to be the improvement of our collective intelligence by promoting mindfulness, transparency, and ecological competence. Compared to people of any other time, the public is less aware of how it is provisioned with food, energy, water, materials, security, and shelter, and how its wastes are handled. Industrial design cloaked the ecological fine print of what are often little better than Faustian bargains providing luxury and convenience now, while postponing ruin to some later time. Ecological design, on the contrary, ought to demystify the world, making us mindful of energy, food, materials, water, and waste flows, which is to say the ecological fine print by which we live, move, and have our being.

Design is always a powerful form of education. Only the terminally pedantic believe that learning happens just in schools and classrooms. The built environment in which we spend over 90% of our lives is at least as powerful in shaping our ideas and views of the world as anything learned in a classroom. Suburbs, shopping malls, freeways, parking lots, and derelict urban spaces have considerable impacts on how we think, what we think about, and what we can think about. The practise of design as a form of public instruction is a jail break of sorts, liberating the ecological imagination from the tyranny of imposed forms and relationships characteristic of the fossil fuel powered industrial age. Architecture, landscape architecture, and planning carried out as a form of public education aims to instruct about energy, materials, history, rhythms of time and seasons, and the ecology of the places in which we live. It would help us become mindful of ecological relationships and engage our places creatively.

Six, awareness of human limitations might cause us, perhaps, to look more favorably on past societies and vernacular design skills created by people at the periphery of power, money, and influence. The truth is that practical adaptation to the ecologies of particular places over long periods of time has often resulted in spectacularly successful models of ecological design (Rudofsky, 1964).

It may well be that the ecological design revolution will be driven, at least in part, by experience accumulated from the periphery not from the center, and led by people skilled at solving the practical problems of living artfully by their wits and good sense in particular places. The success of vernacular design across all cultures and times underscores the possibility that design intelligence may be more accurately measured at the level of the community or culture, rather than at the individual level.

Seven, in modern pedagogy, a great deal of art and philosophy has been cut off from the world of nature. The esthetic standard for ecological design, on the contrary, reconnects to the natural world so artfully as to cause no ugliness, human or ecological, somewhere else or at some later time. The standard, in other words, requires a more robust sense of esthetics that rises above the belief that beauty is wholly synonymous with form alone. Every great designer from Vitruvius through Frank Lloyd Wright demonstrated that beauty in the large sense had to do with the effects of buildings on the human spirit and our sense of humanity, and their impacts on specific places. But the standards for beauty must be measured on a global scale and longer time horizon so that beauty includes the upstream physical effects at wells, mines, and forests where materials originate as well as the downstream effects on climate, human health, and ecological resilience. The things judged truly beautiful will in time be regarded as those that raised the human spirit without compromising human dignity or ecological functions elsewhere. Architecture and landscape architecture, in other words, are a means to higher ends.

Eight, the education of professional designers requires substantial changes. As much art as science, the design professions are not simply technical disciplines, having to do with the intersection of form, materials, technology, and real estate. The design professions such as architecture, landscape architecture, and urban planning are first and foremost practical liberal arts with technical aspects. Writing in the first century B.C., Vitruvius proposed that architects "be educated, skilful with the pencil, instructed in geometry, know much history, have followed the philosophers with attention, understand music, have some knowledge of medicine, know the opinions of the jurists, and be acquainted with astronomy and the theory of the heavens" (1960, pp. 5–6). That is a start of a liberal and liberating education. Design education, therefore, ought to be a part of a broad conversation that includes all of the liberal arts. In George Steiner's words:

> Architecture takes us to the border. It has perennially busied the philosophic imagination, from Plato to Valery and Heidegger. More insistently than any other realization of form, architecture modifies the human environment, edifying alternative and counter-worlds in relationships at once concordant with and opposed to nature (2001, pp. 251–252).

In countless ways all design, even the best, damages the natural world. Extraction and processing of materials depletes landscapes and pollutes. Building construction, operation, and demolition creates large amounts of debris. Agriculture inevitably simplifies ecosystems. A new breed of ecological designers, accordingly, must be even more intellectually agile and broader, capable of orchestrating a wide array of talents and fields of knowledge necessary to design outcomes that can be sustained within the ecological carrying capacity of particular places.

Nine, education is aimed in some fashion to heal us by the systematic cultivation of reason, thoughtfulness, and memory from the more egregious problems inherent in the human condition. As a further step in this progression, the design professions ought to be regarded as healing arts, an ideal rooted in Vitruvius' advice that architects ought to pay close attention to sunlight, the purity of water, air movements, and the effects of the building site on human health. The word "healing" has a close affinity with other words such as holy and wholeness. A larger sense of the profession that architect Thomas Fisher deems a "calling" would aim for the kind of wholeness that creates not just buildings but integral homes and communities (2001).

Compare, for example, the idea that "architecture applies only to buildings designed with a view to aesthetic appeal" (Pevsner, 1990, p. 15) with architecture defined as "the art of place-making" and creation of "healing places" (Day, 2002, pp. 5, 10). In the former, design changes with trends in fashionable forms and materials. It is often indifferent to place, people, and time. The goal is to make monumental, novel, and photogenic buildings and landscapes that express mostly the ego and power of the designer and owner. In contrast, the making of healing places signals a larger allegiance to place that means, in turn, a commitment to the health of other places. Place making is an art and science disciplined by locality, culture, and ecology requiring detailed knowledge of local materials, weather, topography, and the nature of particular places and a creative dialogue between past, present, and future possibilities. It is slow work in the same sense that caring and careful have a different clock speed than carelessness. Place-making uses local resources thereby buffering local communities from the ups and downs of the global economy, unemployment, and resource shortages (Sutton, 2001, p. 200).

Practised as a healing art, architects, for example, would design buildings and communities that do not compromise the health of people and places, drawing on the accumulated wisdom of placed cultures and vernacular skills. They would aim to design buildings that heal what ails us at deeper levels. At larger scales the challenge is to extend healing to urban ecologies. Half of humankind now lives in urban areas, a number that will rise in coming decades to perhaps 80%. Cities built in the industrial model and to accommodate the automobile are widely recognized as human, ecological, and, increasingly, economic disasters. Given a choice, people leave such places in droves. But we have good examples of cities as diverse as Copenhagen, Chattanooga, and Curitiba that have taken charge of their futures to create livable, vital, and prosperous urban places – what Peter Hall and Colin Ward have called "sociable cities" (1998). In order to do that, however, designers must see their work as fitting into a larger human and ecological tapestry.

As a healing art, ecological design aims toward harmony which is the proper relation of parts to the whole. As health professionals is there a design equivalent to the Hippocratic Oath in medicine that has informed medical ethics for two millennia? Are there things that designers should not design? What would it mean for designers to "do no harm"?

Looking ahead, the challenge to the design professions is to join ecology and design in order to create buildings, communities, cities, landscapes, farms, industries, and entire economies that accrue natural capital and are powered by current sunlight – perhaps, one day, having no net ecological footprint. The standard is that of the healthy, regenerative ecosystem. In the years ahead we will discover a great deal that is new and rediscover the value of vernacular traditions such as front porches, village squares, urban parks, corner pubs, bicycles, pedestrian-scaled communities, small and winding streets, local stores, riparian corridors, urban farms and wild areas, and well-used landscapes.

Design practised as a healing art is not a panacea for the egregious sins of the industrial age. However well designed, a world of seven to ten billion human beings with unlimited material aspirations will sooner than later overwhelm the carrying capacity of natural systems as well as our own management abilities. There is already considerable evidence that humans now exceed the carrying capacity of Earth. Further, ecological design is not synonymous with building; often the best design choices will require adaptive reuse or more intense and creative uses of existing infrastructure. Sometimes it will mean doing nothing

at all, a choice that requires a clearer and wiser distinction between our needs and wants.

What ecological designers can do, and all they can do, is to help reduce our ecological impacts and buy us time to reckon with the deeper sources of our problems that have to do with age-old questions about how we relate to each other across the boundaries and sometimes chasms of gender, ethnicity, nationality, culture, and time and how we fit into the larger community of life. Ecological design, as a healing art, is only a necessary, but insufficient part of a larger strategy of healing, health, and wholeness which brings us to spirit.

Finally, design education is not purely secular. For designers it is no small thing that humans are inescapably spiritual beings, but only intermittently religious. Philosopher Erazim Kohak once noted, that "Humans can bear an incredible degree of meaningful deprivation but only very little meaningless affluence" (Kohak, 1984, p. 170). In the former condition most of us tend to grow, harden, and mature while being undone in the latter. This is not a case for deliberately incurring misery which tends to multiply on its own with little assistance, but rather to underscore our inevitable spiritual nature that is like water bubbling upward from an artesian spring. Our only choice is not whether we are spiritual or not but whether that energy is directed to authentic purposes or not.

Much of the modern world, however, has been assembled as if people were machines, lacking any deeper needs for order, pattern, and roots. Modern designers filled the world with buildings, artifacts, and systems divorced from their context and living nature, and telling no story of their origins or place in the larger order of things. They seem to exist as if parachuted down from some alien realm disconnected from ecology, history, culture, and place. Ecological design, on the other hand, is a process by which we become more and more rooted in a particular place and citizens in the community of life in that place. It is a form of story-telling by which buildings and landscapes and the uses we make of them artfully reveal the larger story of which they are a part. Modern design seldom tells any story worth hearing and hence fails to connect us to the nature, history, and evolution of the places in which we live and work. Designers as storytellers aim to speak to the human spirit and help ground our lives and labors in the celebration and awareness of specific places and in the larger story of the human journey (Berry, 1988).

REFERENCES

Abram, David (1966) *The Spell of the Sensuous*. New York: Pantheon.

Benyus, Janine (1998) *Biomimicry*. New York: HarperCollins.

Berry, Thomas (1988) *The Dream of the Earth*. San Francisco: Sierra Club Books.

Bortoft, Henri (1966) *The Wholeness of Nature*. Hudson, NY: The Lindisfarne Press.

Burtt, E. A. (1954) *The Metaphysical Foundations of Modern Science*. New York: Doubleday.

Capra, Fritjof (1996) *The Web of Life*. New York: Anchor.

Catton, William (1980) *Overshoot*. Urbana: University of Illinois Press.

Daly, Herman (1996) *Beyond Growth*. Boston: Beacon Press.

Day, Christopher (2002) *Spirit & Place*. Oxford: Architectural Press.

Diamond, Jared (2005) *Collapse: How Societies Choose to Fail or Succeed*. New York: Viking.

Eliot, T. S. (1971) *The Complete Poems and Plays*. New York: Harcourt, Brace, and World.

Fisher, Thomas (2001) "Revisiting the Discipline of Architecture," in Piotrowski, Andrzej and Robinson, Julia (eds) *The Discipline of Architecture*. Minneapolis: University of Minnesota Press.

Frisch, Karl Von (1974) *Animal Architecture*. New York: Harcourt Brace Jovanovich.

Gleick, James (1988) *Chaos: Making a New Science*. New York: Viking Penguin.

Goethe, Johann Wolfgang von (1952) *Goethe's Botanical Writings* (Transl. Bertha Mueller). Honolulu: University of Hawaii Press.

Global 2000 Report (1980) Washington, DC: Government Printing Office.

Goodwin, Brian (1994) *How the Leopard Changed Its Spots: The Evolution of Complexity*. New York: Simon & Schuster.

Grillo, Paul (1975) *Form, Function, and Design*. New York: Dover.

Hall, Peter, and Ward, Colin (1998) *Sociable Cities: The Legacy of Ebenezer Howard*. New York: John Wiley.

Hawken, Paul, Lovins, A., and Lovins., H. (1999) *Natural Capitalism*. Boston: Little Brown.

Homer-Dixon, Thomas (2000) *The Ingenuity Gap*. New York: Knopf.

Jackson, Wes (1980) *New Roots for Agriculture*. Lincoln: University of Nebraska Press.

Kaptchuk, Ted (2000) *The Web that Has No Weaver*. New York: McGraw-Hill.

Keller, Elizabeth Fox (1983) *A Feeling for the Organism*. New York: W. H. Freeman.

Kohak, Erazim (1984) *The Embers and the Stars*. Chicago: University of Chicago Press.

Leopold, Aldo (1949/1987) *A Sand County Almanac*. New York: Oxford University Press.

Levy, Matthys, and Salvadori, Mario (1992) *Why Buildings Fall Down*. New York: Norton.

Lewis, C. S. (1947) *The Abolition of Man*. New York: MacMillan.

Lovins, Amory (2002) *Small is Profitable*. Snowmass: Rocky Mountain Institute.

Lovejoy, Arthur O. (1974) *The Great Chain of Being*. Cambridge, MA: Harvard University Press.

Lovins, Amory and Lovins, Hunter (1982) *Brittle Power*. Andover: Brick House.

Martin, Calvin (1999) *The Way of the Human Being*. New Haven: Yale University Press.

McDonough, William and Braungart, Michael (2002) *Cradle to Cradle*. Washington, DC: North Point Press.

McHarg, Ian (1969) *Design with Nature*. Garden City: Doubleday.

McHarg, Ian (1996) *A Quest for Life*. New York: Wiley.

Merchant, Carolyn (1982) *The Death of Nature*. New York: Harper & Row.

Pevsner, Nikolaus (1943/1990). *An Outline of European Architecture*. London: Penguin.

Rogers, Richard (1997). *Cities for a Small Planet*. London: Faber & Faber.

Rogers, Richard and Power, Anne (2000) *Cities for a Small Country*. London: Faber & Faber.

Rudofsky, Bernard (1964) *Architecture Without Architects*. Albuquerque: University of New Mexico Press.

Ruskin, John (1880/1989) *The Seven Lamps of Architecture*. New York: Dover.

Sacks, Oliver (1993) "To see and not see," *The New Yorker* (May 10) pp. 59–73.

Schwenk, Theodor (1996). *Sensitive Chaos*. London: Rudolf Steiner Press.

Schwenk, Theodor (1989). *Water: The Element of Life*. Hudson, NY: Anthroposophic Press.

Sinsheimer, Robert "The presumptions of science," *Daedalus* 107(2): 23–36.

Spirn, Anne Whiston (1998) *The Language of Landscape*. New Haven: Yale University Press.

Steiner, George (2001) *Grammars of Creation*. London: Faber & Faber.

Tainter, Joseph (1988) *The Collapse of Complex Societies*. Cambridge: Cambridge University Press.

Thompson, D'Arcy (1942/1992) *On Growth and Form*. New York: Dover.

Tsui, Eugene (1999) *Evolutionary Architecture: Nature as a Basis for Design*. New York: John Wiley.

Tuchman, Barbara (1984) *The March of Folly*. New York: Knopf.

Van der Ryn, Sim and Cowan, Stuart (1996) *Ecological Design*. Washington, DC: Island Press.

Vitruvius (1960) *The Ten Books of Architecture*. New York: Dover.

Vogel, Steven (1988) *Life's Devices*. Princeton: Princeton University Press.

Willis, Delta (1995) *The Sand Dollar and the Slide* Rule. Reading, MA: Addison-Wesley.

Wright, Frank Lloyd (1993) "An organic architecture," in Bruce Brooks Pfeiffer (ed) *Frank Lloyd Wright Collected Writings*, Vol. 3. New York: Rizzoli International Publications.

15

Knowing Systems and the Environment

Richard Bawden

INTRODUCTION

The logic behind the 'environmental concern' is seemingly impeccable, and evidence in its support increasingly irrefutable: Through our activities as an ever-expanding and ubiquitous population of human beings, we are despoiling and degrading the environment of the world about us, while depleting it of its finite resources at an ever-increasing rate. This is not only threatening our own well-being and that of future generations of our own species, but also the stability and sustainability of particular biotic communities across the planet, as well as the integrity of the biosphere as a whole. The organization, functions and hierarchical organization of the long-evolved 'natural eco-systems' in which we are embedded as but one component species are all now at very considerable risk. Such could be the level of the malignancy here that it could lead to a loss of hierarchical control across the entire 'systems-of-complex systems' by which 'nature' is seemingly organized, with the prospect of 'sudden and catastrophic failure' (Pattee, 1973) on an unimaginable scale. Ironically and tragically, much of this circumstance represents the unintended consequences of the developmental processes of modernization that we had come to claim as our greatest achievement as a species. This should dictate the need for a critical reappraisal of our whole techno-scientific, neo-liberal approach to the idea of progress and 'betterment' and for us to be much more reflexive about how we should be living our lives in the context of the impacts that we are having on our environment (Beck, 1992).

THE CONUNDRUM

In the face of such a comprehensive and urgent challenge, it is difficult to reconcile our continuing acts of environmental despoliation with our competencies and dispositions as both rational beings and as the 'moral animals' that we claim to be (Wright, 1994). If we know that what we are doing is unsustainable in the long run, and that we ought not to be doing what we are doing, then why do we continue doing it? If, in all conscience, we appreciate that it is somehow ethically 'wrong' to behave in a manner that threatens the 'integrity, stability and beauty of the biotic community', as Leopold (1949) would assert, then why do we fail to heed our conscience and behave otherwise? And perhaps most significantly and irrationally of all, why, as knowledgeable and concerned citizens of the world, are we continuing to ignore the sensible call to 'learn our way out' of the mess that we ourselves are creating (Milbraith, 1989)?

The focus on learning here – on coming to know – is salutary: rather than planning our way forward under circumstances where, as convention has had it, we were confident of the end-state that we sought, we can only learn our way towards the adaptive management of the resilient systems that are key to sustainable futures (Holling and Gunderson, 2002). The very notion of what it is that we consider most worthy of being sustained in our lives, is, and will probably forever remain after all, quintessentially contestable (Davison, 2001). In essence, our quest to come to terms with sustainability and to design modes of development that lead to environmental sustainability must start with learning what each other means when we use those terms. As it is we, the citizenry, who are responsible for the mess that we now perceive we are making of the world about us, so it is up to us collectively to make meaning through our learning as the basis for our collective judgments about what we now need to do. In other words, it is to social forms of learning that we need to now turn to inform the way by which, acting together, we should manage our environments within the context of sustainable futures (Keen *et al.*, 2005). Social learning is central to the processes of the adaptive management (Holling, 1978) which we need to employ in order to reduce both uncertainties regarding matters of fact and disagreements about goals, objectives and values that can all affect management decisions with respect to the search for sustainability (Norton, 2005). A key complication here is that *what* each individual comes to know, through learning, is very much a function of *how* he or she comes to know, and this makes the search for communal meaning and thus consensual judgment singularly difficult (Maturana and Varela, 1987). And this is then further compounded by the communication limitations imposed by language (Norton, 2005) and even more fundamentally by the apparent 'taboo' of Western culture that 'tells us it is forbidden to know about knowing' (Maturana and Varela, 1987).

While it is our reluctance and/or our lack of capability to engage in social learning or even to recognize its significance that arguably pose the greatest impediments to our current dilemma, other factors also clearly contribute.

There is, for instance, the issue of the priorities that we set for ourselves, to say nothing of our denial of the circumstances on the one hand, and of our addiction to our current ways of life and the resources that we need to support these, on the other (Griffiths, 2003). It is indeed as difficult to concentrate on the global greenhouse gas emissions when one is trying to keep warm in a North American winter or has to drive to the shops in the mall or to the school in the suburbs or to church on the Sabbath, as it is to be concerned with the integrity of the proverbial swamp under the duress of fighting off the equally proverbial alligators. As the human population moves inexorably towards the ten billion mark, and upwards, it is sufficiently difficult to see how the quest for ever-increasing production of food can be sustained for instance, or how the growing demand for energy resources can be met in a rapidly 'globalizing' and modernizing world, without also having to worry about sustaining the environment writ large into an indeterminate future.

We are also victims of the priorities that are set for us by others who have their own motives and motivations for so doing, be that short-term economic gain or political expedience or the machinations of the cultural politics in which environmental discourse is now conducted (Hajer, 1996).

Then there is the matter of the abdication, by us as a civil society, of our responsibilities for public judgment through our deferment, in decisions regarding the public good (Giddens, 1979), to the expertise of the scientists, economists, and the policymakers, and to the officers and institutions of governance. After all, a perfectly rational defense for this can be raised with respect to what Fuller (1991) refers to as 'cognitive authoritarianism' where the rationality of thinking for oneself 'diminishes as the knowledge-gathering activities of society expand to the point of requiring the division of cognitive labor into autonomous expertises.' The clear downsides of this phenomenon, however, relate both to the eventual loss of the capacity of humans to participate in any discussions and decisions about the ways by which we should live our lives through sheer lack of practise (Yankelovich, 1991) and to the essential hegemony that an instrumental technical rationality has come to assume over other rationalities, especially in 'developed' capitalist societies where questions of 'how' have come to supersede those of 'why' (Habermas, 1979). These two latter positions echo the equally somber view of Toffler (1984) that the 'political technology' that has emerged along with modernization has not been particularly adaptive, leading instead to a 'a mismatch between our decisional technology and the decisional environment' that has been characterized by 'a cacophonous confusion, countless self-canceling decisions, noise, fury, and gross ineptitude.' Under these circumstances it is difficult to disagree with Dietz and his colleagues that devising ways to 'sustain the earth's ability to support diverse life, including a reasonable quality of life for humans, involves making tough decisions under uncertainty, complexity, and substantial biophysical constraints as well as conflicting human values and interests' (Dietz *et al.*, 2003).

It is this claim that provides a clue to perhaps the most significant answer to questions about our seeming lack of commitment to 'learning our way out' of the enduring, self-inflicted crisis of our relationships with the world about us. While environmental scientists, systems ecologists, social ecologists, economists, sociologists, and other 'experts' have indeed come to know a very significant amount about 'natural' and 'social systems' 'out there', as a civil society we still know very little about how we can collectively come to make knowledgeable decisions and judgments about what we need to do to change the way we live our everyday lives. In focusing so attentively on knowing *systems* 'out there' we scientists have come to ignore the nature, significance, and development of the *knowing* systems by which they can become known to us all and can be collectively and sustainably managed. We have also managed to alienate a significant proportion of our fellow citizens through their perceptions of the epistemic limitations of our own learning systems (though they would probably not phrase it exactly in those terms) and our unwillingness either to appreciate or to accommodate other systems of knowing (Leach *et al.*, 2005).

The introduction of the idea of 'system' here is deliberate and timely, for the systems idea will be the cognitive principle around which the arguments that follow will be organized.

THE SYSTEMS IDEA

It is difficult these days to browse any article or book about the science, management or politics of the environment that is free of any explicit reference to 'systems' or 'systemics' or, at least, implicit embrace of 'holism' as an essential perspective. These range in their scale of reference from expositions of the panarchy theory of *ecosystem* organization (Gunderson and Holling, 2002) through the significance of *knowledge systems* for sustainable development (Cash *et al.*, 2003) to the characteristics of emerging sciences themselves as with *systems* ecology and, most recently perhaps, in *systems biology* (Kitano, 2002). The ideas that unite these disparate endeavors include wholeness, comprehensiveness, interconnectedness, embeddedness, and emergence.

In this manner, environmental science well exemplifies the claim of Ackoff (1974) that 'the systems age' has come to replace 'the machine age' of the industrial revolution, with holism coming to challenge the previously dominant epistemologies of 'reductionism' and 'positivism' of pre-systemic science. As Jackson (2000) has argued, this new-age dawning is a response to the 'complexity and turbulence' that we are all experiencing in our everyday lives, as well as by the additional confusion brought by the 'multiplicity of viewpoints about the direction we should be taking' and by the multitude of concerns about how we should be handling the difficulties that we face. The adoption of systems (= systemic) perspectives allows a cognitive coherence to be brought to bear on our considerations of these circumstances.

All acts of cognition start with a distinction between a thing (or a being or an entity) and its background or its environment, and each time we explicitly refer to anything therefore we specify 'a criterion of distinction' which indicates what we are talking about (Maturana and Varela, 1987). Cognition thus depends on our abilities to distinguish between an 'it' and the 'other'. In its simplest

formulation then, the systems idea makes the distinction between the system as a coherent bounded entity (the 'it'), and the environment (the 'other') from which it can be distinguished, but with which it is structurally coupled through recurrent interactions between the two (Maturana and Varela, 1987). Therefore, the unit of interest of the systemist – where systemists are to systems as economists are to economies – is both the 'it' and the 'other' as well as the relationships between the two. In fact, systemists typically think in three 'it/other' dimensions – the part (subsystem), the whole (the system of interest), and the higher order whole (the environmental supra-system of the system) of which the system is itself a subsystem – plus all the sets of interactions within and between these three levels. The significance of this 'tri-hierarchical' or 'holarchical' conceptual organization of systems-of-systems lies with the belief that at each 'level', surprisingly novel properties, that are unique to that level, emerge as a function of the interrelationships between its component subsystems and the environment that its higher order system presents to them. The fundamental assumption of 'holism', therefore, is that no system can be known, nor its total characteristics nor properties predicted, through a study of any of its component subsystems in isolation from each other or from the system itself: an assumption that is in direct opposition to the basic premise of reductionism.

The systems idea is far from new within the natural sciences of course. Indeed, while an appreciation of the significance of what might be termed an 'essence of wholeness' seems to pervade a wide range of 'indigenous cultures' and can be associated with an intellectual heritage from ancient Greece, most especially through the insights of Aristotle (Russell, 1961), some of the earliest formalizations of thinking in 'systems terms' can be found among scientists writing about biological phenomena and the 'nature of nature'. Smuts (1926), for instance, was the first to write formally about 'holism' – which he defined as 'the tendency in nature to produce wholes' – and of the significance of that to 'the internal organization' of organisms and to the evolution of their species. In the same era, Woodger (1929) further extended these ideas through his emphasis on the hierarchical nature of organization within organisms, while both Canon (1932) and Henderson (1941) reflected on the significance of the capacities of organisms as 'living systems' to adapt in order to maintain their own integrity in the face of challenges from the turbulent environments in which they had to exist.

Perhaps the most influential biologist with respect to the application of the systems idea to biology, however, was von Bertalanffy, who took some of the key concepts of these early pioneers and further extended them into the formulation of what he eventually referred to as a General Systems Theory (GST) (von Bertalanffy, 1968). Earlier, he had made the vital distinction between 'closed' and 'open' systems (von Bertalanffy, 1950) with regard to their respective relationships with their environments. Central features of the 'open systems' of von Bertalanffy included the need for 'cybernetic regulative processes' that were essential for maintaining their 'steady state equilibria.' Tellingly, the concept of the 'ecosystem,' as it was introduced by Tansley (1935), was a much looser notion of a 'living system' than that developed by von Bertalanffy and his 'systems' predecessors, referring, as it did, to a relatively unbounded set of structural and functional relationships between a biotic community and some circumscribed abiotic features. The integrity of such a system was held to be maintained in a state of equilibrium through the flows of energy and matter between its parts. This key focus on the structural and functional nature of ecosystems in equilibrium was later replaced by the more cybernetic notions of dynamic equilibria and steady states (Patten, 1959) in which a much clearer distinction was made between the ecosystem and its environment. Other vital concepts that this view of functional organizational integrity permitted was the possibility of the adaptation of the whole ecosystem with respect to changes in its environment as well as its evolution as a whole entity as a function of changes in the diversity, both of its parts and of the pattern of inter-relationships between them. Thus, while it became possible to talk sensibly of both stability and resilience of ecosystems as prerequisite properties for adaptation in turbulent environments (Holling, 1973), it also made sense to talk of their capacity to evolve as whole, integrated entities, even in the absence of any clear empirical evidence that this indeed did occur in actuality.

While having adaptive capabilities of this type, such self-organizing systems are generally regarded as being very complex in their organization and in the spectrum of their inter-relationships with the environments in which they are embedded and with which they interact. They can be so complex and dynamic in fact, that they can 'move' or be forced to positions that are far from equilibrium (Gleik, 1987), in which state they are regarded as being on the 'edge of chaos' where a small change in one component of the system can result in greatly amplified reactions elsewhere in the system. This can lead either to the demise of such systems or to the emergence of totally different systems (Stacey, 1996). Of even greater significance in this context is the possibility, promoted by hierarchy and panarchy theorists, that entire hierarchies of systems

can be affected under conditions where the complexity of any system within them becomes so great that it overwhelms existing controls across the entire hierarchy (Levins, 1973) with potentially disastrous results. Importantly it has been claimed that these 'natural' phenomena have potentially as much significance to hierarchies of human organizations and 'social systems' as they do for the ecosystems and hierarchies of nature (Wheatley, 1992). Indeed it was an economist who was to add considerable rigor to GST with his conceptualization of a nine-level hierarchical typology of complexity (Boulding, 1956) that ranged from static structures and frameworks that can be studied, through to transcendental systems which are the realm of 'inescapable unknowables.' Boulding's other considerable contribution was to argue that GST could be used for ordering different fields of study through a focus on the 'individual unit of behavior' in addition to its aim of developing a theory of very general principles, which, as Jackson (2000) observes, was the primary concern of von Bertalanffy.

The key feature of a GST which allows this submission is the principle of isomorphism from which many have been able to conclude that what is so for 'natural systems' must be equally so for 'social systems.' And thus all that has been written above with regard to the nature and behavioral characteristics of living systems has been transposed in one form or another over the years to apply to the nature and dynamics of human organizations and societies: these are, as the logic goes, composed of human beings which are, in turn, 'living systems' and much effort has been put into the application of systems approaches to human affairs over many decades past, in the search for knowledge about the nature of 'human systems.' This has a particular relevance to this present context of an environmental challenge that dictates the need to come to know and understand what is claimed to be 'the fundamental character of interactions between nature and society' (Kates et al., 2001). This is a challenge that accepts that the governance of ecosystems as complex adaptive systems requires flexibility and cognitive capacities for knowing and learning in response to 'environmental feedback' (Levin, 1998).

For all of these efforts, however, and for all of the apparent success of the system's focus in ecology, in environmental science, and as a feature of the emerging sustainability sciences (Kates et al., 2001) and systems biology (Kitano, 2002), the systems idea and the styles of systems thinking that it promotes has actually found declining support within the social sciences. As one of its most astute observers posits, systems thinking was in a less secure position within the social sciences at the beginning of the 21st century than it had occupied several decades earlier (Jackson, 2000) at the time that the 'dawn of the systems age' had been acclaimed by Ackoff (1974). This represented a significant change from the situation where systems perspectives on, and systems approaches to the analysis of, social groupings had really been the dominant paradigm within sociology at least, and most especially among those concerned with organizational management and development.

While a host of reasons can be cited for this situation – and Jackson (2000) indeed articulates most of the major influences from the novelty of changing paradigmatic perspectives that range from functionalist through interpretivist and emancipatory to post-modernist – a central distinction between 'natural systems' and their 'social system' analogs is the inherent reflexivity of human beings (Westley *et al.*, 2002) and our critical capacities for knowing, and for changing our minds and the views that we hold of the world about us.

As it happens, Ackoff would have been more accurate to have claimed in the mid-1970s that this was the dawning of a 'new systems age' or the emergence of a 'second wave of systems thinking', as Midgely (2000) describes it, which was being characterized by a very significant change of mind about 'systems' themselves and systemics, among at least some systemists. And this observation provides a useful segue into the next section of this chapter where the focus changes from a consideration of 'systems in the world' to the 'systems of cognition' through which we come to know that world. In essence, in now changing the emphasis from knowing *systems* as it were to *knowing* systems, we reflect what Checkland (1981) referred to with his introduction of the soft systems methodology as 'the shift in systemicity from the world to ways of inquiry into that world' which indeed represented the character of the dawning of the 'new systems age' to which he himself has been such a dominant contributor.

KNOWING SYSTEMS

A useful way of exploring the notion, nature, development, and significance of 'knowing systems' is to reprise the earlier concept of three-dimensional thinking, specifically through the introduction and exploration of the three-level model of cognitive processing developed by Kitchener (1983) in which she suggested that different 'levels of processing' allows individuals to monitor one level of cognitive tasks at another level. Kitchener distinguishes between cognition (or level one), meta-cognition (level two) and epistemic cognition (level three) as a hierarchical sequence through which individuals can monitor the way by which they conduct their own basic cognitive tasks through a meta-cognitive

process, which is itself monitored by a process of epistemic cognition. Applying a system's logic to this model, and expressing it in terms of knowing, it is thus possible to claim that a knowing or learning system can be seen intrinsically to consist of: (i) a cognitive (sub)system, which deals with the process of coming to know about the matter to hand; (ii) a meta-cognitive system that deals with the process by which the matter to hand is, and can be known; and (iii) the epistemic–cognitive (supra)system that deals with the epistemological limits to what can be known about the other two levels as circumscribed by the very nature of knowledge itself. Conceived in this manner, each level in this intrinsic 'system-of-systems' is profoundly interconnected with the others, with each essentially providing the contextual environment for the others. The epistemic suprasystem, furthermore, can be expanded to embrace not just epistemological aspects of knowing and knowledge, but also of ontological assumptions about reality and the nature of being, and about axiological assumptions and beliefs that are expressed as values. In this manner, the essential focus of the epistemic (supra)system are the cognitive frameworks or 'meaning perspectives' or *Weltanschauungen* that represent 'those usually taken-for-granted and often idiosyncratic values, norms, and beliefs that constitute our individual and socialized views of the world' (Plas, 1986). As Kuhn (1962) presented them, such worldviews become paradigms when, as 'entire constellations of beliefs, values, techniques and so on' they are shared and put into practise by given communities. And this emphasis on 'practise' adds the fourth element of 'methodology' to the episteme where it represents the way through which the other three are interpreted into action. Thus, where method, as the process of knowing, is the focus of meta-cognitive inquiry and evaluation, then methodology, as the expression of epistemological, ontological, and axiological assumptions in practice, is the focus for epistemic–cognitive inquiry.

From this perspective then, it is entirely appropriate to claim that the emergence of the 'second wave of systems thinking' and the introduction of profound distinctions between the previous 'hard' systems thinkers and the new 'soft' systems types was indeed equivalent to the introduction of a new paradigm of systemics. In this light, it is perfectly understandable, as Kuhn himself would have predicted, that the shift would be associated with very significant controversy equivalent to an 'open intellectual warfare,' which not only regrettably continues to this day, but has been further compounded by the subsequent introduction of a third 'critical' wave (Midgley, 2000) which focuses attention essentially on to judgments to do with social (and environmental) conditions, with the placement of boundaries, and with reflexivity in knowing. These three 'waves' differ very profoundly from each other in their epistemic assumptions about the nature of nature, the nature of knowledge and the nature of human nature. Ideas from all three waves or schools of systems thinking will be significant in the conceptualization of the critical knowing/learning systems with which this chapter concludes.

The epistemic position adopted by the General Systems Theorists and by a generation of systems practitioners who were significantly influenced by them (including many systems ecologists, simulation modelers, ecosystem and systems biologists, and environmental scientists) was of the 'hard school.' From this dualistic perspective, nature 'really' is organized in the form of coherent 'systems' of integrated parts that, in turn, 'really' are organized as nested system hierarchies. In this regard it is also interesting to note the definition of a human community offered by Flora *et al.* (1992) as 'a place and *a human system*,' the claim by Daley and Netting (1994) that 'communities are living entities ... (that) ... like people, go through a normal life cycle,' and their call for 'understanding complex social systems' as a vital reason for introducing 'systems thinking' into community development (Daley and Netting, 1994).

Such ontological realism leads, almost inevitably, to the adoption of positivist and objectivist epistemological positions: observed systems are considered to be independent of those doing the observations while the knowledge that is gained through this positivist process is considered to be objective in the sense that there must be some 'permanent or ahistoric matrix or framework' to which appeal may be ultimately made 'in determining the nature of rationality, knowledge, truth, reality, goodness or rightness' (Bernstein, 1983). Moreover, because the hard systems approach was, by definition, held to be value free, there was little that could sensibly be known or said about what ought to be done under circumstances where the indications were that something needed to be done (which, ironically enough, was itself a normative judgment, of course). The issue of judgment became a key criterion of distinction for Vickers (1983) in his contention that human systems were so different that it was just not possible to study them using the logic and methods of the natural sciences, precisely because of the significance of judgment to human beings. His introduction of the notion of 'appreciative systems' – those sets of largely tacit standards of judgment by which we both order and value our experiences – as unique to 'human systems' dictated that they depended on shared understandings and shared cultural mores if they were to be effective and stable. A key feature of 'second wave' systemic thinking relates to the significance of human judgment to the placement, as it were, of systems boundaries (Midgley, 2000). This is in direct contrast to the 'boundaries as given' notion that prevailed within the 'hard school,' and that persisted even when the concept was extended beyond the obvious – such as with the extension of the systems idea beyond obviously bounded individual cells and organs and organisms to include the much less evidently bounded 'ecosystems' in the form of the higher order 'biotic communities,' as introduced by Tansley (1935).

These issues are far from trivial for they get to the very heart of the conundrum of why it is that the citizenry seems so reluctant to commit itself to learning its way out of the environmental mess that, paradoxically, many willingly acknowledge needs to happen. They also highlight some of the critical deficiencies of the environmental and ecological 'knowing systems' that are contributing to this paradox. Witness, for instance, the long-standing ambivalence of ecologists with regard to the ontological status of human beings with respect to 'natural ecosystems,' which does little to inspire the confidence of the citizenry. As Berkes (1999) has stated with such eloquence, within ecology, human beings are so frequently regarded either as somehow 'un-natural' components of 'natural eco-systems' or are placed into such mythological categories as the 'Ecologically Noble Savage,' the 'Intruding Wastral,' or the 'Fallen Angel.' To this, as has been already indicated, must be added the issue of the ontological status and organization of 'nature itself' and whether the notion of ecosystems as cybernetically regulated, stability-seeking entities that can evolve in all of their wholeness, can ever be empirically validated beyond mathematical representation. And what a sloppy concept 'nature' turns out to be under such circumstances, and 'society' too for that matter, and yet vocal is the claim that the new field of sustainability science for instance, 'seeks to understand the fundamental character of interactions between nature and society' (Kates *et al.*, 2001).

The environmental sciences meanwhile have an even more difficult epistemic issue with which to contend, for in identifying the environment as their issue of concern, they are in fact nominating the 'other' as their 'it.' This creates a further source of epistemic confusion, for if the 'other' becomes the 'it' in any act of cognition, then the question arises of what now is the 'other' and what is the significance of the inter-relationships between 'it' and whatever the 'other' is deemed to be? As well as being a matter of some epistemic significance to environmental scientists, this matter is also central to the contributions of environmental philosophers in their attempts to bring synthesis to what Belshaw (2001) has highlighted as 'reason, nature and human concern.' This focus well captures the claim that the quest for environmental sustainability must not only embrace what it is that could persist over time, but also, and essentially, what it is that should be allowed to persist (Thompson, 2004). It also illustrates a basic contention of Norton (2005) that what is needed if we are to discuss intelligently and learn about our environmental goals and how to achieve them, we will need a discourse which is rich enough to express and disagree about values – which perforce must include aesthetic notions of beauty as well as ethics – while also incorporating knowledge gained through scientific understanding. And all of this, as Norton particularly insists, demands communicative and cooperative behavior of us within our communities in ways that allows clarification of epistemological and ethical assumptions, as well as accommodating processes by which these can be safely challenged and, when appropriate, changed. As Grove-White (1996) has asserted, modern environmentalism has evolved 'not simply in response to damaging impacts of specific industrial and social practices, but also, more fundamentally as a social expression of cultural tensions surrounding the underlying ontologies and epistemologies which have led to such trajectories in modern societies.'

Such is the nature and focus of epistemic cognition; its importance returns our gaze to the concept of the 'knowing system' and a further application to it of the systemic principle of 'three dimensionality' with the suggestion that the *knowing system* be regarded as an integral subsystem of every system under consideration along with the environmental suprasystem in which that, in turn, is embedded. This fresh 'three dimensional' view is consistent with the emphasis in the hierarchy theory as being propounded by Ahl and Allen (1996) that the observers (knowers) need to be 'reunited' with the observed, for it is they who are indeed responsible for recognizing boundaries around entities as well as proposing the criteria for making those distinctions. It is these 'knowers cum decision makers' who must make the judgments about who is to be included within the system of concern, what sort of knowledge will be needed and who can be relied upon to generate it, who should be the main beneficiaries of any indicated change, and who will speak for those unable to be present (Ulrich, 1983). In this view, systems are indeed in the eye of the beholder, as Checkland (1981) has asserted: they are abstract constructions of a concrete reality that cannot be directly accessed through the senses, or of a coherent set of processes for collective learning within a community of interest about what might be done to improve situations that they experience and appreciate as problematic to them in one form or another (and perhaps also to others).

These perspectives, it is submitted here, are of profound significance to environmental concerns, and represent crucial ways by which these concerns can become known and can provide intelligible, trustworthy, and collectively generated knowledge about both what *could* be done in the search for environmental sustainability and what *ought* to be done. In essence, these demand different ways of knowing and reasoning and thus different rationalities, as Habermas (1984) has long insisted. Thus, while his 'instrumental rationality' is entirely appropriate for exploring the 'external natural world,' it needs to be replaced by 'communicative action' whenever and wherever the purpose is 'mutual understanding to realize common goals and values' (Yankelovich, 1991). These will demand access to epistemic cognition for their clarification.

With its own three-dimensional capacities for cognitive processing, the 'knowing system' as now envisaged, brings a learning and critically reflexive capacity as a subsystem to any system or systemically appreciated situation in which it embeds itself. As mentioned earlier, it draws on elements of all three waves of systems of thinking: it can consist of a tangible group of people (a 'hard' systems perspective) who commit themselves to behaving as if they were a coherent 'knowing system.' They will have a clear idea of the matters to hand that they are addressing, an unambiguous understanding of the systemic processes of knowing that they employ (a 'soft' systems perspective), and an inherent appreciation of the need to bring critical reflections (a 'critical' systems perspective) to all of these matters as well as to the epistemic aspects of all that, as a knowing system, it is trying to accomplish. Through their interconnectedness, each individual within the 'knowing (sub)system,' will contribute to the collective process of knowing, to the knowledge that comes to be known, and to the democratic deliberations that are an essential property of that system. Each person learns with and through all of the others.

To behave effectively in this manner, knowing (sub)systems must have capabilities at all three levels of cognition, and must be prepared to allow for their own evolution as a knowing system as evidenced by collective intellectual and moral development. This last issue is of signal importance, for as Salner (1986) has emphasized, there is a strong correlation between the capability to think in any systemic way – and thus effectively use any systems methodology – and an advanced state of epistemic development. Drawing especially on the work of both Perry (1968) and Kitchener (1983), Salner argues that it is essentially not until one has learned the characteristics of an epistemological/ethical stance of what Perry referred to as 'contextual relativism' or 'contextualism' and has developed on from 'dualism,' that one is able to develop effective systemic competencies. Addressing complex issues with any success demands the development of complex meaning perspectives or worldviews (West, 2004) and that, in turn, demands critical attention. It is from a similar position that Bawden (2000, 2005) has made the claim that sustainable acts of development in the material and social worlds are functions of the intellectual and moral development of all of those who ought to be involved in those acts. This therefore brings a fresh, critical epistemic perspective to the calls for social learning as the foundations for the adaptive management of 'the environment,' made by Keen *et al.* (2005), and to support the quest for shared goals for sustainability and policies for greater environmental protection through public discourse which is 'holistic' in both its focus and its nature (Norton, 2005). As Norton readily concedes, the adoption of 'holistic adaptive management' as a social learning approach to environmental sustainability, presents a host of philosophical as well as practical challenges to all concerned, with the need for ethical, aesthetic, ontological, and epistemological considerations to be taken seriously if we are ever to come to really know what to do better with respect to our relationships with the environment about us (Norton, 2003), which represents 'the matter to hand,' in the language of 'knowing systems.'

This perspective on social learning for adaptive management indicates the need for what is referred to as 'transformational learning' (Mezirow, 1991) that involves epistemological challenge and change, in contrast to 'informational learning' which is merely 'a change in behavioral repertoire or an increase in the quantity or fund of knowledge' (Kegan, 2000). A key distinction between the outcomes of these two forms of learning lies with their differential impact on the 'frames of reference or minds' that we use in structuring our knowing. Thus, as Kegan sees it, both kinds of learning are expansive and valuable, 'one within a preexisting frame of mind and the other reconstructing that very frame' or worldview: from the perspective of the three-dimensional knowing (sub)system being promoted here, this focus on 'changing frames' or – 'meaning perspectives' as Mezirow (1991) calls them – is directly analogous to the concept of 'epistemic development.'

It is perhaps most useful to conclude this piece with a brief reference to the meta-cognitive competencies – the learning how to learn and knowing how to know – that remain to be addressed here. In patent contrast to those who see learning as the acquisition of knowledge, Kolb (1984) presents it as 'the process by which knowledge is created through the *transformation* of experience.' From this experiential perspective, knowledge is continuously being created, recreated, and 'used' by individuals as they seek to make conceptual sense of what they are sensing through their own experiences of the ever-changing 'concrete' world about them, as the essential prelude to taking sensible actions to adapt or to adapt to that perceived reality.

As Kolb sees it, it this adaptation that is the essential motivation for coming to know and learning: indeed, as he argues, 'learning is *the* major process of human adaptation' (Kolb, 1984, p. 32). Echoing one of the central themes of this chapter, Kolb insists that experiential learning is a 'holistic process' that involves constant transactions 'between the person and the environment' in a manner that engages 'the integrated function of the total organism – thinking, feeling, perceiving and behaving.'

But with that insistence, of course, Kolb reveals an epistemic position, as well as a systemic orientation and logic, that is far removed from the dualism and reductionism that continues to prevail, ironically enough, within our formal Institutions of Learning and in our conventional knowing institutions which 'extract' the perceptual/sensual from the conceptual, and action from reflection, the subjective from the objective, and so on. This Kolbian view of learning is as a quintessentially participative and transformative process in which the transformative power lies with both 'the whole' knowing system and 'its parts.'

Knowers develop a deep sense of the world that they are experiencing from the perspective of being 'embedded' within it and participating as part of it, even as they are trying to make sense out of it. Thus, participation is an 'implicit aspect of wholeness,' as Skowlimowski (1985) claims, which, in this learning/knowing sense includes, as Bohm (1987) emphasized, 'thoughts,' 'felts' and 'feelings' as well the 'state of the body.' Importantly, as Bohm also argued, it is a social process with 'thought passing back and forth between people in a process by which thought has evolved from ancient times.' This continuous 'unfolding' and 'enfolding' of meanings, thoughts, 'felts' and even intentions and 'urges to do things,' cannot be anything other than a dynamic, systemic process of individuals and social groupings alike – and indeed of the mind itself. And all of this is very reminiscent of Goethe's participatory approach to a science through which he strove 'to enliven and deepen our understanding of nature' (Barnes, 2000). The Goethean scientist, claimed Bortoft (1996), 'does not lose himself or herself in nature, but finds nature within himself/herself in fully conscious experience.' Such conscious participation is seen as a synergistic condition in which humanity and nature work together in such a way that 'each becomes more fully itself through the other'; a mutual enhancement.

A similar claim has been made for the synergy between the 'experiential knowing' processes of Kolb (as a subsystem) and what has been called 'inspirational knowing' by Bawden (1998): where the former refers to the transformation of experience into knowledge, inspirational knowing accesses 'innate insights' as its focus for transformation. It can be postulated that it is through inspirational knowing that we come to know our positions on 'rights' and 'virtues' and 'aesthetics' which are then synergistic with our experientially derived contexts and instrumental knowledge of the world, which we bring to bear in our communicative actions with others, to change our ways with it!

And all of this occurs within a learning (sub)system which has an internal ambience of emotions and dispositions which are embraced as essential to the transformative functions of that system.

CONCLUSION

As the above has indicated, a fourth, vital category can, and ought to be added to Berkes' (1999) typology: to the Ecologically 'Noble Savage,' the 'Intruding Wastral,' and the 'Fallen Angel,' can/ought now be included the intrinsically three-dimensional 'Knowing Being.' Through their cognitive competencies, humans are capable of coming to know about matters to hand that concern them, coming to know how they come to know that, and coming to know the epistemic contexts in which these two 'lower order' processes operate (Kitchener, 1983). Such a knowing system can refer both to individuals and to social collectives of individuals. In functional terms, the 'triarchical' organization of this intrinsic human knowing system allows cognitive (level one) processing, meta-cognitive (level two) processing, and epistemic–cognitive (level three) processing to proceed in a synergistically interconnected manner. The epistemic dimension embraces all three of the essential 'elements' of human worldviews and paradigms – epistemology, ontology, and axiology.

A key conceptual implication of the knowing system is the adaptation of the lower-order systems to changes in the epistemic suprasystem, which, it is suggested, tends to evolve (or be deliberately developed) from the relative simplicity of 'dualism' to the much more complex 'contextualism.' The paradox here, or at least the enigma, is that until and unless the knowing system evolves or is developed to this position, it cannot appreciate its own systemic nature (Salner, 1986).

The systems image can be further extended to present this intrinsically three-dimensional 'knowing

system' as the key subsystem within an extrinsic, three-dimensional system-of-systems. A knowing (sub)system attempts to make sense out of what it senses in both 'the system' which it construes or 'brings forth' (Maturana and Varela, 1987) and of which it sees itself as an essential component part, and the environmental suprasystem in which that system is construed to operate, and with which it is 'structurally coupled' (Maturana and Varela, 1987). In this manner it is possible to understand and present the 'environmental concern' as either a concern of state of the immediate system, or of the environmental suprasystem at large – but, and most significantly, only when that knowing (sub)system reaches an epistemic state that supports such a construction!

Therein lies the challenge of engagement with the quest for environmental sustainability and for the design of systems that are appropriately stable, resilient, and influential to that end. Therein also lies the foundations for the provocative claim that all systemic acts of sustainable development in the material and social worlds are quintessentially functions of the epistemic development of those actors who need to engage critically with the issue. And the source of the provocative claim that all acts of development in the context of systemic sustainability will depend on the systemic appreciation (and thus epistemic development) of all of those who need to act in those circumstances.

REFERENCES

Ackoff, R.L. (1974) *Redesigning the Future*. Wiley, New York.

Ahl, V. and Allen, T.F.H. (1996) *Hierarchy Theory: A Vision, Vocabulary, and Epistemology*. Columbia University Press, New York.

Barnes, J. (2000) Participatory science as the basis for a healing culture. In R. Steiner (ed.) *Nature's Open Secret: Introductions to Goethe's Scientific Writings* (trans. John Barnes and Mado Spiegler). Anthroposophic Press, Barrington, MA.

Bawden, R.J. (1998) The community challenge: the learning response. *New Horizons* 99: 40–59.

Bawden, R.J. (2000) Valuing the epistemic in the search for betterment. *Cybernetics and Human Knowing* 7: 5–25.

Bawden, R.J. (2005) Systemic development at Hawkesbury: Some personal lessons from experience. *Systems Research and Behavioural Science* 22: 151–164.

Beck, U. (1992) *Risk Society: Towards a New Society*. Sage, London.

Belshaw, C. (2001) *Environmental Philosophy: Reason, Nature and Human Concern*. McGill-Queen's Press, Montreal.

Berkes, F. (1999) *Sacred Ecology: Traditional Ecological Knowledge and Resource Management*. Taylor & Francis, Philadelphia.

Bernstein, R.J. (1983) *Beyond Objectivism and Relativism: Science, Hermeneutics and Praxis*. University of Pennsylvania Press, Philadelphia.

Bohm (1987) *Unfolding Meaning: A Weekend Dialogue with David Bohm.* Ark Paperbacks, London.

Bortoft, H. (1996) *The Wholeness of Nature: Goethe's Way Toward a Science of Conscious Participation in Nature.* Lindisfarne Books, New York.

Boulding, K.E. (1956) General systems theory – the skeleton of science. *Management Science* 2: 197–203.

Canon, W.B. (1932) *The Wisdom of the Body.* Kegan Paul, Trench, Trubner and Co., London.

Cash, D.W., Clark, W.C., Alcock, F., Dickson, N.M., Eckley, N., Guston, D.H., Jäger, J. and Mitchell, R.B. (2003) Knowledge Systems for Sustainable Development. www.pnas.org/cgi/doi/10.1073/pnas/1231332100 (last accessed on 13 March 2006).

Checkland, P.B. (1981) *Systems Thinking Systems Practices.* John Wiley, Chichester.

Daley, J.M. and Netting, F.E. (1994) Mental maps for effective community development. *Journal of the Community Development Society* 25: 62–79.

Davison, A. (2001) *Technology and the Contested Meanings of Sustainability.* State University of New York Press, New York.

Dietz, T., Ostrom, E. and Stern, P.C. (2003) The struggle to govern the commons. *Science* 302: 1907–1912.

Flora, C.B., Flora, J.L., Spears, J.B., Swanson, L.E., Lapping, M.P. and Weinberg, M.L. (1992) *Rural Communities: Legacy and Change.* Westview Press, Boulder, CO.

Fuller, S. (1991) *Social Epistemology.* Indiana University Press, Bloomington, IN.

Giddens, A. (1979) *Central Problems in Social Theory.* Macmillan, London.

Gleik, J. (1987) *Chaos and the Making of a New Science.* Abacus, London.

Griffiths, T. (2003) The Humanities and an Environmentally Sustainable Australia. www.humanities.org.au/Policy/NRP/expandingRPpapers/GriffithsRP.pdf (last accessed on 14 March 2006).

Grove-White, R. (1996) Environmental knowledge and public policy needs: on humanising the research agenda. In S.Lash, B.Szerszynski, and B.Wynne (eds) *Risk, Environment and Modernity: Towards a New Ecology.* Sage Publications, London, pp. 269–286.

Gunderson, L.H. and Holling, C.S. (eds) (2002) *Panarchy: Understanding Transformations in Human and Natural Systems.* Island Press, Washington, DC.

Habermas, J. (1979) *Communication and the Evolution of Society* (trans. Thomas McCarthy). Beacon Press, Boston.

Habermas, J. (1984) *The Theory of Communicative Action* (trans. Thomas McCarthy) (vol. 1, *Reason and Rationalization;* vol. 2, *Lifeworld and Systems: A Critique of Functionalist Reason.* Beacon Press, Boston.

Hajer, M. (1996) Ecological modernisation as cultural politics. In S.Lash, B.Szerszynski, and B.Wynne (eds) *Risk, Environment and Modernity: Towards a New Ecology.* Sage Publications, London, pp. 246–268.

Henderson, L.J. (1941) Sociology lectures. In B. Barber (ed) *L.J. Henderson on the Social System. Selected Writings.* The University of Chicago, Chicago.

Holling, C.S. (1973) Resilience and stability of ecological systems. *Annual Review of Ecology and Systematics* 4: 1–23.

Holling, C.S. (1978) *Adaptive Environmental Assessment and Management.* John Wiley, New York.

Holling, C.S. and Gunderson, L.H. (2002) Resilience and adaptive cycles. In L.H. Gunderson and C.S. Holling (eds) *Panarchy: Understanding Transformations in Human and Natural Systems.* Island Press, Washington, DC.

Jackson, M.C. (2000) *Systems Approaches to Management.* Kluwer Academic/ Plenum Publishers, New York.

Kates, R.W., Clark, W.C., Corell, R., Hall, J.M., Jaeger, C.C., Lowe, I., Deve, J.J., Schellnhuber, H.J., Bolin, B., Dickson, N.M., Faucheux, S., Gallopin, G.C., Grübler, A., Huntley, B., Jäger, J., Jodha, N.S., Kasperson, R.E., Mabogunje, A., Matson, P., Moopney, H., Moore 111, B., O'Riordan, T. and Svedin, U. (2001) Environment and development. *Science* 292: 641–642.

Keen, M., Brown, V.A. and Dyball, R. (2005). Social learning: a new approach to environmental management. In M. Keen, V.A. Brown and R. Dyball (eds) *Social Learning in Environmental Management: Towards a Sustainable Future.* Earthscan, London.

Kegan, R. (2000) What "Form" transforms? A constructive-developmental approach to transformative learning. In J. Mezirow and Associates (eds) *Learning as Transformation.* Jossey Bass, New York.

Kitano, H. (2002) Systems biology: a brief overview. *Science* 295: 1662–1664.

Kitchener, K.S. (1983) Cognition, metacognition, and epistemic cognition: a three level model of cognitive processing. *Human Development* 26: 222–232.

Kolb, D.A. (1984) *Experiential Learning: Experience as the Source of Learning and Development.* Prentice Hall, Upper Saddle River, NS.

Kuhn, T. (1962) *Structure of Scientific Revolutions.* University of Chicago Press, Chicago.

Leach, M., Scoones, I. and Wynne, B. (2005) *Introduction: science, citizenship and globalisation.* In M. Leach, I. Scoones and B. Wynne (eds) *Science and Citizens.* Zed Books, London.

Leopold, A. (1949) *A Sand County Almanac and Sketches Here and There.* Oxford University Press, Oxford.

Levin, S.A. (1998) Ecosystems and the biosphere as complex adaptive systems. *Ecosystems* 1: 431–436.

Levins, R. (1973) The limits of complexity. In H.H. Pattee (ed.) *Hierarchy Theory: The Challenge of Complex Systems.* George Braziller, New York.

Maturana, H.R. and Varela, F. J. (1987) *The Tree of Knowledge: The Biological Roots of Human Understanding.* New Science Library, Boston.

Mezirow, J. (1991) *Transformative Dimensions of Adult Learning.* Jossey Bass, San Francisco.

Midgley, G. (2000) *Systemic Intervention: Philosophy, Methodology and Practice.* Kluwer Academic/Plenum Publishers, New York.

Milbraith, L.W. (1989) *Envisioning a Sustainable Society: Learning Our Way Out.* State University of New York Press, New York.

Norton, B.G. (2003) *Searching for Sustainability: Interdisciplinary Essays in the Philosophy of Conservation Biology.* Cambridge University Press, Cambridge.

Norton, B.G. (2005) *Sustainability: A Philosophy of Adaptive Ecosystem Management*. The University of Chicago Press, Chicago.

Pattee, H.H. (1973) The physical basis and origin of hierarchical control. In H.H. Pattee (ed.) *Hierarchy Theory: The Challenge of Complex Systems*. George Braziller, New York.

Patten, B.C. (1959) An introduction to the cybernetics of the ecosystem: the trophic–dynamic aspect. *Ecology* 40(2): 221–231.

Perry, W.G. (1968) *Forms of Intellectual and Ethical Development in the College Years*. Holt, Rinehart and Winston, New York.

Plas, J.M. (1986) *Systems Psychology in the Schools*. Pergamon Press, New York.

Russell, B. (1961) *History of Western Philosophy*. George Allen and Unwin, London.

Salner, M. (1986) Adult cognitive and epistemological development in systems education. *Systems Research* 3: 225–232.

Skowlimowski, H. (1985) The co-operative mind as a partner of the creative evolution. *Proceedings of the First International Conference on the Mind–Matter Interaction*. Universidad Estadual de Campinas, Brazil.

Smuts, J.C. (1926) *Holism and Evolution*. Macmillan, London.

Stacey, R.D. (1996) *Complexity and Creativity in Organizations*. Berret-Kohler, San Francisco.

Tansley, A.G. (1935) The use and abuse of vegetational concepts and terms. *Ecology* 16: 284–307.

Thompson, P.B. (2004) Sustainable agriculture: philosophical framework, In R. M. Goodman (ed.) *Encyclopedia of Plant and Crop Science*. Marcel Dekker, New York.

Toffler, A. (1984) Introduction on future-conscious politics. In C. Bezold (ed.) *Anticipatory Democracy: People in the Politics of the Future*. Vintage Books, New York.

Ulrich, W. (1983) *Critical Heuristics of Social Planning: A New Approach to Practical Philosophy*. Haupt, Bern, Switzerland.

Vickers, G. (1983) *Human Systems are Different*. Harper and Row, London.

von Bertalanffy, L. (1950) The theory of open systems in physics and biology. In. F.E. Emery (ed.) *Systems Thinking*. Penguin, Hardmonsworth.

von Bertalanffy, L. (1968) *General Systems Theory*. Penguin, Hardmonsworth.

West, E.J. (2004) Perry's legacy: models of epistemological development. *Journal of Adult Development* 11: 61–70.

Westley, F., Carpenter, S.R., Brock, W.A., Holling, C.S. and Gunderson, L.H. (2002) Why systems of people and nature are not just social and ecological systems. In L.H. Gunderson and C.S. Holling (eds) *Panarchy: Understanding Transformations in Human and Natural Systems*. Island Press, Washington, DC.

Wheatley, M.J. (1992) *Leadership and the New Science: Learning About Organizations from an Orderly Universe*. Berret-Kohler, San Francisco.

Woodger, J.H. (1929) *Biological Principles*. Keegan Paul Trench and Trubner, London.

Wright, R. (1994) *The Moral Animal. Why We Are the Way We Are: The New Science of Evolutionary Psychology*. Vintage Books, New York.

Yankelovich, D. (1991) *Coming to Public Judgment: Making Democracy Work in a Complex World*. Syracuse University Press, Syracuse, NY.

16

Volunteer Environmental Monitoring, Knowledge Creation and Citizen–Scientist Interaction

Max J. Pfeffer and Linda P. Wagenet

INTRODUCTION

Volunteer environmental monitoring offers the possibility to involve citizens more directly in environmental decision-making, especially in economically advanced countries with highly rationalized systems of environmental management. Within the context of management systems based on expert control, the public expresses high levels of concern about the environment, but has relatively few avenues for making meaningful inputs into environmental management decisions. Thus, an important consideration in this chapter is the potential contribution of volunteer monitoring to the democratization of environmental decision-making.

Volunteer environmental monitoring offers the potential to help bridge the chasm between science- and technology-based environmental management and public involvement in environmental decision-making. Volunteer monitoring may also reinforce public confidence in science-based decision-making by allowing members of the public to be directly involved in the generation of data used in environmental management. As a result, environmental decision-making may also become more democratic. Finally, volunteer monitoring may offer one means for increasing more direct human/environment interactions in societies where such interactions are dominated by science- and technology-based rationality.

Volunteer environmental monitoring is likely to be especially important in countries with extensive environmental regulations and clear compliance standards, and where concerned citizens have the time and resources to participate. Indeed, we claim that volunteer environmental monitoring is most likely to appear in economically advantaged countries with higher standards of living and an age and/or class structure that assures an adequate supply of potential volunteers. Volunteers might be retired individuals concerned with the quality of the environment, wishing to be civically engaged, and having the time and freedom to participate in environmental monitoring. Regardless of the age or economic status of volunteers they are likely to have relatively high levels of education and a familiarity with environmental science and its importance in the conservation and protection of natural resources. Indeed, the environmental discourse is largely a scientific discourse, and the democratizing potential of volunteer monitoring is only fully realized when it enables citizens to be meaningfully involved in environmental decision-making. Even challenges to official environmental policies are most effectively mounted when cast in the language of science.

The characterization we offer above is not an empirically based statement of fact, but is intended to serve as a heuristic device that allows us to examine the potential and limitations associated with volunteer environmental monitoring as one means of encouraging more meaningful human/environment interactions. In examining the current literature, we found no comprehensive review of volunteer environmental monitoring worldwide.

For this reason, limiting our focus to the USA helps us begin the task of evaluating the social and environmental significance of volunteer monitoring. For at least the past fifteen years in many parts of the world, there has been increasing interest in more participatory environmental management approaches, and this tendency has been especially important in American water resources management. Because volunteer water quality monitoring in the USA is fairly widespread, we focus on it for more detailed consideration.

We begin by outlining how the American social context is conducive to the growth of volunteer monitoring. We go on to briefly review the range of volunteer environmental monitoring in the USA. Then we outline a set of questions that helps us to evaluate the potential significance of volunteer environmental monitoring as an instrument to further more democratic environmental decision-making and consider these questions with a focus on volunteer water quality monitoring in the USA.

THE SOCIAL CONTEXT AND VOLUNTEER ENVIRONMENTAL MONITORING

As suggested above, the social context within which this activity is likely to thrive is fairly limited, and for this reason we focus our attention on the USA. Interest in volunteer activities is more likely to appear in wealthier countries where people have more leisure time. They are often interested in using this time to enhance their own and the community's quality of life (Applebaum, 1992; Stebbins, 2004). In this context, Beck (2000) predicts that an increasing amount of society's work will be completed by volunteers. More than 50% of Americans are engaged in some sort of volunteer activity (Silverberg, 2004). The Independent Sector (2005), a coalition advocating voluntary citizen action, estimates each hour of volunteer work in the USA to be worth $17.55 on average, and the total annual value of volunteer service to be $272 billion. This level of volunteer activity reflects the favorable social and economic conditions for it in the USA.

Volunteer monitoring can be thought of as a form of stakeholder participation. In the USA, there is local, state and national interest in strengthening stakeholder participation in environmental management, and this interest builds on a long history of civic engagement. Commentators have long noted that American society is distinguished by the prominent role civic engagement plays in the economic, political, social and cultural spheres. Over time some forms of this engagement have grown and declined (Putnam, 1995, 1996, 2000), but new forms of stakeholder involvement have also arisen, as in the environmental domain.

As the influence of the environmental movement grew in the 1960s, it stimulated the enactment of a variety of landmark environmental legislation. The expectation of public inputs into environmental decision-making was established in the 1969 National Environmental Policy Act (NEPA) and was included in virtually all the environmental legislation enacted subsequently. In this way, civic engagement in environmental decision-making was formalized. However, the insertion of citizen or stakeholder engagement into public policy formation and implementation has been difficult, as more than thirty years of experience attest (Bryner, 2001; Petts, 2003).

Volunteer monitoring is a form of stakeholder engagement in American environmental policy and management. It is a form of civic environmentalism that is rooted in grassroots concerns about environmental quality (Weber, 2000; Nerbonne and Nelson, 2004; Silverberg, 2004). Environmental monitoring efforts involving citizens range from simple (e.g. visual inspection of stream conditions) to complex (e.g. use of aerial photography and interdisciplinary teams of hydrologists, biologists and soil scientists) (Linkenhoker, 2001, personal communication). The US Environmental Protection Agency (EPA) has encouraged volunteer environmental monitoring as a way to complete basic monitoring tasks, to promote more active citizen participation in environmental protection, and to create greater awareness and knowledge about environmental processes. The promotion of volunteer monitoring is consistent with EPA's commitment to community-based decision-making through a variety of programs, especially its Community-Based Environmental Planning process (EPA, 2003). This planning process is a vision of a two-pronged system: "one with strong centralized standard-setting and oversight alongside pragmatic, bottom-up decision-making" (John and Mlay, 1999, p. 354; Weber, 2000; Rubin, 2002; Nerbonne and Nelson, 2004).

The involvement of citizens in environmental management has evolved considerably over the years. Although the past thirty years have seen stakeholder participation codified in environmental planning (e.g. NEPA) and developed into a technical and ethical expectation, the early part of the 20th century saw federal agencies acting unilaterally, focusing public relations efforts on gaining citizen approval of specific projects rather than incorporating citizens into the decision-making process. The Administrative Procedures Act of 1946 was an attempt to change the regulatory atmosphere by equalizing the government and citizens in a legal sense, but public engagement that existed until the 1960s tended to be formal and highly structured with more attention to information dissemination than to developing an engaged citizenry (Rosenbaum, 1991; NRC, 1996, 2004; Shapiro, 1996). In recent decades there has been increasing public demand for engagement in environmental protection. This demand, rooted in environmental concern, has been an important aspect of American public opinion at least since the 1970s. Indeed, the institutional commitment to increased environmental protection was in large measure a result of continued public concern about environmental quality (Kraft, 1999). While this concern ebbs and flows, it has remained high for decades and support for environmental protection spans conventional political ideologies (Dunlap and Mertig, 1992).

This environmental concern creates a strong demand for public inputs into environmental decision-making. It is widely acknowledged that lack of meaningful public input may lead to a variety of barriers to successful environmental management, including public opposition. Increases in citizen participation in environmental policy and management have often been a response to dissatisfaction with the level and type of engagement possible. The basic challenge has been to provide effective channels for citizen inputs and agency response to those inputs in the formation and implementation of environmental policies. Such local involvement, it is claimed, is especially effective because there can be an ongoing process of community involvement and collective learning (John, 1994; NRC, 1999). Environmental monitoring and data collection by volunteers and the use of those data by governmental agencies in making management decisions is a concrete response to the challenge of providing more meaningful opportunities for citizen inputs. In recent years there has been a heavy emphasis on local-level stakeholder engagement.

Finally, another factor that needs to be considered in the growth of volunteer monitoring in the USA is the increasing emphasis on devolving a variety of government responsibilities from the federal to the local levels. This devolution which became common in the 1980s, encouraged development of local initiatives for environmental protection. With limited budgets, some localities began to find volunteer inputs very attractive in helping mount environmental initiatives. The inclusion of volunteers in local watershed and other initiatives began to be more actively encouraged by federal agencies in the 1990s (Weber, 2000; Nerbonne and Nelson, 2004; Silverberg, 2004).

For the reasons mentioned above, volunteer monitoring might be expected to flourish in the USA, and its examination in this context sheds light on its potential and limitations for changing and strengthening human/environment interactions. Of course, the concern with strengthening

these interactions is grounded in the broader historical and socio-economic context. American society is highly scientifically and technologically rationalized (Beck, 1992; Dryzek, 1997; Fischer, 2000). By this we mean that decisions about the management of natural resources and the environment are heavily informed by scientifically based information and technical criteria. The decisions are typically made by experts whose approach to environmental problems is grounded in scientific training. With growing regulation, environmental decision-making becomes increasingly monopolized by these experts, expanding the distance between the average citizen and management of the environment. The extent of this chasm is measured by differences in the type and degree of scientific knowledge held.

The monopolization of environmental decision-making by experts contradicts a commitment, mentioned above, to democratic decision-making codified in American environmental legislation. Furthermore, the inability of experts to resolve certain environmental problems, or to determine whether environmental pollutants were the cause of certain human health problems, has undermined the credibility of the experts (Freudenberg, 1993; Irwin, 2001). The lack of confidence in expert knowledge led to calls for greater public inputs and scrutiny in environmental decision-making (Dryzek, 1997; Smith, 2003). Yet, in a science- and technology-based environmental management system, limited scientific literacy among citizens has hampered the possibilities for meaningful public involvement in environmental decision-making.

The scientific literacy of the lay public becomes even more critical as environmental management evolves from being dominated by technical decisions made by bureaucrats to involving the public in environmental problem solving. Technical decisions based in professional resource management bureaucracies are expert driven and have an emphasis on regulation and the rationalization of environmental planning by means of observation, measurement and analysis. On the other hand, the more democratic approach involves elements of public consultation, alternative dispute resolution, policy dialogue, public engagement and an emphasis on local decision-making and control (Dryzek, 1997; Fischer, 2000; Weber, 2000). Although our discussion focuses on the USA, this transition from the expert driven to the more democratic management approach has been underway in varying degrees around the world (Frank,1997; Ewing, 1999). The important point here is that the democratization of environmental management requires that citizens adopt the language of science. With the rationalization of human/nature interactions all options and even points of disagreement are cast in scientific terms (Frank and Gabler, 2006).

Many authors see this movement toward broad-based input in environmental management decisions as a positive occurrence for both ecosystem and human well-being (Zazueta, 1995; Weber, 2000; Ravindra, 2001; Williams, 2002). Other authors are more critical in their appraisal of the potential for linking scientists and citizens. For example, Ward (1996) sees a conflict between what the public wants to know and the various models used by scientists. A number of observers are dissatisfied with public involvement as a means of generating better citizen/scientist inter-actions in environmental management. Baker (1998) views this connection in a negative light by claiming that "the inability of the general public to understand science is only matched by the inability of most practicing physical scientists to under-stand the social world," and Firth (1999, p. 489) cites VanderVink's blunt characterization of the disconnect as "the scientifically illiterate versus the politically clueless."

With these issues in mind, volunteer monitoring might be thought of as community science, which is an interaction between professionals and lay people (Kellert *et al.*, 2000; Carr, 2003). Volunteer monitoring is seen as one element of community science, which strongly emphasizes place. Carr (2003, p. 2) notes that there needs to be a link between community science and traditional sci-ence: "Only through buying into that well-estab-lished positivist mindset can community science expect to carry any weight or influence outside a specific community." As noted by Heiman (1997, p. 297): "The main value of community-initiated science ... is through the basic education provided, enabling residents and workers to test and defend – rather than just trust – their own common sense ... Scientific knowledge acquired through actual participation becomes a part of a people's culture, no longer an alien product to be accepted as an article of faith."

Thus, environmental management is challenged to identify how best to link the inputs of citizens and scientists in a meaningful way. A National Research Council (1999) report claims that watershed management that provides explicitly defined roles and responsibilities for the public and scientists is likely to be the most robust over time. Volunteer monitoring initiatives could provide an explicit role for citizens and specific mechanisms for inter-actions with environmental scientists. Specific inputs from volunteers and scientists could provide a mix of local, and biological, physical and scien-tific knowledge (Born and Genskow, 2000; Fleming and Henkel, 2001). But perhaps most important in terms of creating meaningful interactions is for these efforts to have an identifiable result.

Research has shown that the experiences of stake-holders are most meaningful or valuable when they become involved in activities that have tangi-ble environmental outcomes that they can relate to their own efforts (Byron and Curtis, 2002; Lubell, 2002; Santos and Chess, 2003). Before turning to a closer look at volunteer water quality monitoring, we review several types of volunteer monitoring found in the USA.

THE VARIETIES AND CHARACTERISTICS OF ENVIRONMENTAL MONITORING IN THE USA

To place volunteer water quality monitoring in perspective, it is useful to get a sense of the history and the range of environmental media and issues involved. Volunteer environmental monitoring in the USA appears to have begun in 1890 with citi-zens tracking climate data and reporting it to the National Weather Service (Lee, 1994; Heiman, 1997; Dates, 1999; Hanahan and Cottrill, 2004). This Cooperative Observer Program currently has more than eleven thousand volunteers at five hundred stations and over one hundred years of continuous data (Dates, 1999; National Weather Service, 2004). By the early 1900s, the National Audubon Society sponsored the Christmas Bird Count which continues today (Lee, 1994; Dates, 1999; Hanahan and Cottrill, 2004; National Audubon Society, 2005a). Related to that program, the US Fish and Wildlife Service's Bird Banding Program started in the 1920s with an original intent of setting limits on hunting. Now, data from that program are stored in a national database (Lee, 1994). In the early 1950s, the National Marine Fisheries Service sponsored game-fish tagging (Dates, 1999).

Water quality monitoring, which is now the most common form of volunteer monitoring, did not take hold until the late 1960s and early 1970s, starting as a grassroots effort from various lake associations and groups interested in stream restoration and preservation (Lee, 1994; Dates, 1999). Implementation of the Clean Water Act in 1972 and the Izaak Walton League's program, Save Our Streams, are credited with an increase in citizen water quality monitoring (Lee, 1994; Hanahan and Cottrill, 2004). At that time, water quality monitoring was restricted to measure-ments for water clarity and was used to raise public awareness of environmental resources rather than to engage citizens directly in the moni-toring activity. Scott (cited in Lee, 1994, p. 30) noted that "most people in the water quality busi-ness were chemists and engineers who believed testing needed to be done by professionals. It was like being a heretic to suggest volunteers could collect data." Nevertheless, the EPA has been a strong supporter of volunteer water monitoring programs and provides a place on its website for group reports, where there are hundreds of programs listed, and each state is represented (EPA, 2005). In 1996, the EPA reported that states with citizen monitoring programs more than doubled from fourteen in 1988 to thirty-two in 1992. By 1994, EPA reported that there were five hundred seventeen programs described on its website in 2005 the number had risen to eight hundred and seventy-two (EPA, 2005).

Recently, the results of volunteer monitoring are sometimes being used in the wider public policy arena. For example, the National Weather Service's Cooperative Observer Program described above is helping "to define the climate of the United States and to help measure long-term climate changes" as well as "to support forecast, warning and other public service programs of the National Weather Service" (National Weather Service, 2005). Volunteer participants in the program are not required to take any test, and they gather data from equipment supplied by the National Weather Service.

The National Audubon Society's Christmas Bird Count, mentioned above, started as a holiday tradition hunt with ninety participants and has evolved into more than fifty thousand volunteers tracking early-winter bird populations (National Audubon Society, 2005a). This "citizen science in action" has resulted in the longest running database in ornithology, representing over a century of unbroken data. In the case of the Christmas Bird Count, the focus is less on education or public awareness and more on "the status and distribution of bird populations across the Western Hemisphere" (National Audubon Society, 2005a). Conservationists use the data from the bird count to examine habitat issues or environmental threats. An offshoot of this program is the Great Backyard Bird Count, which is in its eighth year. For this program, individuals are encouraged to count the number of bird species in their "backyard" that can include private residences, national wildlife refuges, national forests or parks, and federal wildlands (National Audubon Society, 2005b). The data are collected over the internet and used by researchers to better understand bird populations and distribution. In 2004, more than four million birds of five hundred fifty-four species were reported from over forty-two thousand checklists (Peet, 2005).

The Digital Monarch Watch (Pathfinder Science, 2004) is a collaborative effort among students, teachers, volunteers and scientific researchers to study the Monarch butterfly. Citizens report Monarch sightings, and the data are kept on a comprehensive website database that is used to track Monarch habitat and migration patterns. Since Monarch butterflies are sensitive to human intervention and environmental perturbations, changes in their migration patterns can signal larger environmental change. Although the data in the Digital Monarch Watch have some scientific utility, the overall goal of the program is to increase appreciation of and concern about this insect.

The examples of volunteer monitoring given here have been national or even global in scope. Volunteer monitoring can be especially significant locally, however, where it might influence policy more readily. For example, odor problems stemming from wastewater treatment plants in North Carolina were investigated by researchers who incorporated neighborhood residents in an advisory as well as a sampling role (Aitken and Okun, 1991). Collaboration among the researchers, neighborhood residents and the wastewater treatment facilities' managers was crucial to this study. Not only did they comprise an Advisory Board, but there were public meetings, analysis of wind and odor data, and examination of plant records. Volunteers were asked to rate the odor level at certain times during the day for a period of six months. In addition, wastewater treatment plant personnel reported odor levels from the facility several times per day over the entire six month time span. Once the source of the odors was well-documented and understood, emission reduction efforts could be implemented.

Another example of local influence in environmental monitoring occurred in an oil refinery area in California (Kuhn, 1999). A group known as Communities for a Better Environment worked with county government officials to enable citizens to collect air samples from a five-gallon bucket. These samples are "reliable, scientifically and legally valid 'grab samples' of air quality" (Kuhn, 1999, p. 657). A Good Neighbor Ordinance was enacted to hold government agency and private economic entities accountable for the air quality of the area.

This brief review of volunteer monitoring's history and examples of national and local monitoring programs shows that this activity is targeted at an array of natural resources. The involvement of citizens in environmental monitoring generally has multiple purposes, and it can be evaluated for its efficacy in achieving desired outcomes. Education and public awareness are the most common objectives associated with volunteer monitoring (Dates, 1999; Bliss et al., 2001; Hubbell, 2002; Gouveia et al., 2004; Hanahan and Cottrill, 2004; Nerbonne and Nelson, 2004). There is also value in monitoring as a way to bring community groups together when contentious issues have been divisive or when trust has been lost between citizens and agencies. In this way, volunteer monitoring, like other forms of civic engagement, can build social capital within the community (Kaufmann, 1999). However, evaluation of the effectiveness of volunteer environmental monitoring, as well as public involvement in environmental decision-making more generally is lacking, and there are a number of questions about its potential as a democratizing force in environmental policy and management (Sewell and Phillips, 1979; Chess, 2000; Palerm, 2000; Fore et al., 2001; Nerbonne and Nelson, 2004; O'Leary et al., 2004; Silverberg, 2004).

Although citizen monitoring has become a more common practise associated with environmental management, skepticism exists about the quality of the data collected as well as its usefulness in environmental policy (Nerbonne and Nelson, 2004). Yet, the manifest purpose of volunteer monitoring is to collect data that can be used in environmental decision-making. These observations lead us to pose the following questions:

1 Does the evidence support the use of volunteers to generate data that meet the scientific standards of official environmental managers?
2 If acceptable data are collected by volunteers, are they used by agencies charged with environmental management?
3 In general, does this activity reduce the chasm between environmental science and the lay public?

We now address these questions with a focus on volunteer water quality monitoring.

VOLUNTEER WATER QUALITY MONITORING

Official recognition and use of water quality data collected by volunteers

As indicated above, organized volunteer water quality monitoring is a relatively recent development that comes on the heels of a fairly extensive history of attempts to make water resources management more responsive to the needs of the lay public. There has long been sensitivity about the implications of water resources management practices for democracy. Feldman (1991, p. 9) posits that using army engineers in the late 19th and early 20th centuries to develop water resources "was linked to the desire to avoid creating an aloof and highly educated military establishment that might become contemptuous of democratic values." This strategy made sense in light of a civilian engineering bureaucracy that was inadequately developed.

However, dependence on a technically proficient body of narrowly trained specialists retarded a fuller appreciation of the social, economic, and environmental impacts of water resources development. It also minimized public participation, excluded consideration of non-engineering solutions to water problems, and ignored the relation between water and other natural resources (Feldman, 1991, p. 9).

There were more aggressive moves toward greater stakeholder involvement in water quality protection in the 1960s, primarily due to citizen concerns about environmental degradation as well as the lack of civic involvement in command and control regulatory activities (Feldman, 1991; Sexton et al., 1999; Wagenet and Pfeffer, 2003). The Water Resources Council, created in 1961, had authority to seek justification for new projects and required "the participation of a variety of stakeholders in policy making" (Feldman, 1991, p. 13).

While appeals for public participation in water resource management are long standing and pervasive, adequate attention is seldom devoted to the details involved in the integration of scientific and public inputs. For example, the EPA (2005) describes its watershed approach as founded on sound science and broad participation of stakeholders, but there is no discussion of the technical requirements for effectively bringing the two together. In the light of such shortcomings, some observers have suggested that scientists or technicians must work with local volunteers for quality control (Heiman, 1997).

To encourage such collaboration, the EPA developed volunteer citizen monitoring manuals and protocols to increase quality assurance and control. Programs focusing on data collection must emphasize credibility, collect temporally consistent data, use standardized techniques and be comparable to other assessment programs. "These projects must adopt protocols that are straightforward enough for volunteers to master and yet sophisticated enough to generate data of value to resource managers" (EPA, 2001, p. 3). For example, the Chesapeake Bay Citizens Monitoring Program initiated a Quality Assurance Project Plan in the mid-1980s. This program was a major step in increasing the credibility of volunteer-collected data among state and local agencies (Hanahan and Cottrill, 2004). The fact that this program was supported by the EPA also increased recognition of the value of citizen monitoring (Lee, 1994; Hanahan and Cottrill, 2004). However, there are few proven models for citizen monitoring activities that effectively link scientists and the public to produce reliable data.

A number of nongovernment organizations, academic institutions, and government agencies have published manuals on citizen monitoring. While many of these manuals can serve as useful resources on data collection methods, most are formulaic distillations of professional monitoring techniques. There are a few manuals that provide guidance on how to organize citizen monitoring programs. The EPA (1997), for example, developed a series of manuals on implementing and maintaining volunteer monitoring programs. Other examples are the Pacific Streamkeepers Federation monitoring handbook and training modules (Pacific Streamkeepers Federation, 1997), the Adopt-A-Stream Foundation's Streamkeeper's Field Guide (Murdoch and Cheo, 1996), and Maryland's Stream Waders Manual (Maryland Department of Natural Resources, 2005). While these manuals discuss data-gathering methods and address data quality concerns, they do not describe how to plan monitoring efforts in ways that link scientists with volunteers, nor do they assess the appropriate level of sophistication for volunteer data collection and analysis.

Typically, volunteer monitoring efforts are focused on overcoming gaps in regulatory agency data collection (Gouveia *et al.*, 2004). For example, nonpoint source pollution in rural and agricultural areas requires comprehensive monitoring to identify its extent and sources, but resources to conduct such extensive monitoring are often lacking. Carefully organized citizen monitoring that collects water quality data using biological, physical, and chemical indicators can provide needed inputs. This volunteer monitoring can complement the work of government agencies by providing them with a means of targeting more sophisticated water quality monitoring efforts, resulting in a more efficient allocation of resources. Table 16.1 shows the number of volunteer water quality monitoring programs in each state in 2005. While there were 872 programs nationwide, relatively few had quality assurance and control measures. Fewer than half the programs (391) had written quality assurance programs; 238 had state approval and 163 were EPA approved. Only 142 programs (16%) had both state and EPA approval for their quality assurance and control (EPA, 2005).

Table 16.1 reveals interesting variation across states in quality assurance measures for volunteer monitoring programs. Although some states, such as California with 52, have many programs registered in the EPA database, only three of California's programs include state- and EPA-approved quality assurance control plans. Similarly, Massachusetts has 60 programs, but only four are state and EPA approved. On the other hand, Alabama has only 23 programs listed but nearly half include state and EPA approval. Likewise, Texas, with 35 programs has 23 that include state- and EPA-approved quality assurance

and control. This variation highlights the important local role in establishing quality assurance standards.

More important than its proliferation, volunteer water quality monitoring is growing in acceptance as a data source for agency decision-making, at least at the state level. There is recognition, however, that such acceptance necessitates greater training and increased emphasis on quality control and assurance (Hanahan and Cottrill, 2004; Nerbonne and Nelson, 2004). Heiman (1997, p. 301) notes the importance of volunteer monitors and agencies working together, resulting in public confidence "in decisions made with – and not just by – agency staff." Citizens bring contextual knowledge to the monitoring process while agencies and scientists contribute more technical knowledge. Evaluations of the effects of stakeholder involvement show that it generally resulted in acceptable environmental decisions (Beierle and Cayford, 2002). These findings offer some basis to expect that volunteer monitoring data more particularly might also play a positive role in environmental decision-making.

Certain types of water quality monitoring are especially well suited to volunteer efforts. For example, biological monitoring of indicator species and basic chemical parameters provide preliminary information about water quality. These water quality criteria are well suited to public involvement because of the relative ease with which they can be measured. Abundance and diversity of biological indicator species such as fecal coliforms and benthic macroinvertebrates can be readily observed and quantified by citizen monitors and are of interest to aquatic scientists for their usefulness as long-range indicators of water quality (Roux *et al.*, 1993; Kerans and Karr, 1994; Heiman, 1997; Nerbonne and Nelson, 2004). Benthic macroinvertebrate communities are widely used indicators for assessment of water quality in streams (Cairns, 1974), and a number of indices based on them have been developed (Cairns, 1974; Hilsenhoff, 1977; Bode *et al.*, 1991). Some of these indices are especially appropriate for use by groups of laypersons (Plafkin *et al.*, 1989; Bode *et al.*, 1991). Several studies have found that data collection and stream assessments by properly trained citizens were largely comparable to professional results (Ely, 2001; Fore *et al.*, 2001; O'Leary *et al.*, 2004).

A study to assess volunteer monitoring in streams revealed that trained volunteers using professional techniques can produce data that are comparable to those of scientists. This study compared volunteers to professionals at two stages in the data-collection process – field sampling and laboratory processing. However, Fore *et al.* (2001, p. 120) also noted that volunteer programs need

Table 16.1 Number of self-reported citizen monitoring programs by type of quality assurance program and state, 2005*

State	Type of quality assurance program				Total
	Written	State approved	EPA approved	Both state and EPA approved	
Alabama	19	14	14	12	23
Alaska	6	3	3	3	8
Arizona	1	0	0	0	2
Arkansas	2	2	2	2	3
California	15	7	3	3	52
Colorado	3	2	2	2	3
Connecticut	6	2	3	2	14
Delaware	2	2	1	1	3
District of Columbia	0	0	0	0	3
Florida	14	10	5	5	19
Georgia	21	19	12	12	23
Hawaii	1	1	0	0	4
Idaho	4	4	3	3	9
Illinois	10	9	2	2	17
Indiana	7	6	0	0	16
Iowa	7	4	2	2	16
Kansas	4	3	3	2	10
Kentucky	5	5	2	2	13
Louisiana	1	1	1	1	2
Maine	19	12	8	5	39
Maryland	9	3	4	2	29
Massachusetts	19	5	9	4	60
Michigan	10	5	4	3	36
Minnesota	16	9	4	4	33
Mississippi	2	2	1	1	3
Missouri	7	6	3	2	16
Montana	4	1	1	1	15
Nebraska	0	0	0	0	4
Nevada	0	0	0	0	1
New Hampshire	9	6	6	5	19
New Jersey	12	5	3	2	24
New Mexico	2	1	1	1	4
New York	11	3	3	2	30
North Carolina	6	1	0	0	13
North Dakota	0	0	0	0	1
Ohio	12	9	5	5	45
Oklahoma	3	3	3	3	3
Oregon	6	2	0	0	16
Pennsylvania	17	7	3	3	44
Rhode Island	2	0	1	0	10
South Carolina	1	0	0	0	5
South Dakota	1	1	1	1	1
Tennessee	0	0	0	0	10
Texas	31	30	23	23	35
Utah	1	0	0	0	2
Vermont	3	2	1	1	6
Virginia	19	11	9	8	30
Washington	27	9	5	5	61
West Virginia	6	6	4	4	10
Wisconsin	8	5	3	3	26
Wyoming	0	0	0	0	1
USA	391	238	163	142	872

*Fifty states plus District of Columbia.

Source: EPA website (http://yosemite.epa.gov/water/volmon.nsf/Home?openform) 2005.

to be guided by experienced researchers and that "laboratory analysis by volunteers will never equal that of professional taxonomists."

O'Leary *et al.* (2004) compared water quality monitoring data collected by volunteers and professional biologists. Volunteers were first trained by the scientists to identify benthic macroinvertebrates to the family level. Participants in the study (volunteers and scientists) collected samples from a local stream over a period of several weeks. Each group then independently analyzed its own samples, identifying the various macroinvertebrates to determine the overall quality of the stream. Scientists subsequently analyzed all the samples collected by all the groups to the genus level, comparing their findings to those of the volunteers. The scientists' classifications at the genus level differed from the findings of the volunteers at the family level. O'Leary *et al.* (2004) note that volunteer and scientist stream quality ratings based on bacteriological and chemical constituent tests were very well correlated, but that metrics of water quality based on family-level analysis did not always correlate well with those based on genus-level analysis. On the other hand, they observed that ratings of overall water quality were either identical or differed little between the family and genus levels of analyses, indicating that the classification errors made did not affect the water quality assessments.

Another study of volunteer macroinvertebrate monitoring was undertaken by Nerbonne and Nelson (2004). The authors note that there is little research that examines the organization of groups monitoring macroinvertebrates and the role of centralized support in enhancing the impact of monitoring efforts. Nerbonne and Nelson (2004) found that state government leaders see the usefulness of volunteer macroinvertebrate monitoring as increasing water quality awareness, and many also see the value of the data collected for enhancing state and local databases. However, the impacts of volunteer water quality monitoring were limited by lack of commitment to state quality assurance goals and procedures. Nerbonne and Nelson (2004) note many quality assurance procedures for volunteer monitoring do not meet professional standards.

Nerbonne and Nelson's work is quite revealing in terms of macroinvertebrate sampling, but there are few other evaluations of volunteer water quality monitoring and its effectiveness in generating data acceptable by scientific standards. Still, the evidence that is available suggests that volunteer monitors can generate data that are useful and reliable. Unfortunately, there is no systematic review of environmental agencies and their confidence in and use of volunteer-generated data. But there are examples of the use of volunteer monitoring by government agencies, and we turn to some of these now.

In the mid-1990s, Virginia created a statewide citizens water quality monitoring coordinator position (Brown *et al.*, 2000). This was the institutional foundation for volunteer monitoring networks within the state, quality assurance and control efforts, and support for the use of volunteer monitoring data by state agencies. Most importantly, Virginia Save Our Streams and government agencies developed a data use matrix and identified the appropriate agency response to pollution events discovered in the analysis of citizen monitoring data. Brown *et al.* (2000, pp. 53–55) conclude that "the agencies' management needed to buy into the value of citizen collected data," and they highlight four ways agencies can use citizen-generated information: (1) background information; (2) assessment information; (3) red flag for pollution events; and (4) special studies.

In Pennsylvania there is an extensive network of volunteer monitors and an emphasis on including the elderly population through the Environmental Alliance for Senior Involvement, the Pennsylvania Department of Aging and the Pennsylvania Senior Environment Corps (Wilson, 2000). Although the Pennsylvania Department of Environmental Protection (DEP) has a widespread monitoring network that encompasses one hundred fifty sampling sites, they utilize volunteer monitoring data to supplement this program at more than three thousand sampling sites (Wilson, 2000). As in Virginia, many of these data are used as red flags for potential pollution problems, and quality assurance and control are overseen by the state. Volunteer-collected data are also used to support grant proposals, to locate obstructions in streams that might cause flooding problems, to assess watersheds that are undergoing site remediation due to abandoned mine drainage, and to determine the success of nonpoint source pollution abatement projects (Wilson, 2000).

Volunteer monitoring in Maine resulted from local concerns about shellfish harvesting and harmful algal blooms. The citizen groups in this case forged a strong connection with the University of Maine Cooperative Extension system and received technical support from state agencies (Stancioff, 2000). The programs are sufficiently effective so that the Maine Department of Marine Resources "is the only state agency which is a member of the Interstate Shellfish Sanitation Conference that uses volunteers in the collection of data to classify shellfish growing areas" (Stancioff, 2000, p. 60).

Similar to Maine, in Oregon volunteer monitoring of water resources stemmed from a 1997 initiative for restoration and protection of fish. There is a strong tie between the volunteer monitoring program and the Oregon State Testing Laboratory. Most importantly, the laboratory oversees the volunteer coordinator. A primary role of the volunteer monitors is to "carry out public education and encourage local participation in watershed issues" (Williams, 2000, p. 62). Still, data collected by volunteers are integrated into state agency monitoring programs. This is supported by the state environmental agency's purchase of high-quality equipment for volunteer groups to use in their sample collection. There is also a broad training program in order to assure quality control. Williams (2000, p. 65) notes that this training helps "overcome skepticism about volunteer data quality." The biggest challenge to the Oregon volunteer monitoring program is the actual management of the collected data. The sole volunteer monitoring coordinator bears this responsibility, which includes hundreds of sampling sites.

Each of the four cases referred to here has a strong link to the state environmental agency for funding and oversight. Dissemination and management are often listed as major challenges to the widespread use of the collected data. Gouveia *et al.* (2004) note a variety of techniques, primarily internet based, to assist in these outreach and communication efforts. The technical sophistication of volunteer monitoring groups has increased dramatically over the years. Geographic information systems, global positioning systems and remote sensing systems are utilized by volunteers throughout the USA to collect and manage data. However, formal evaluation of volunteer use of these technologies is scarce.

The quality of volunteer monitoring data is typically difficult to assess, especially since metadata (information about the data collection itself) is not always included, data across various groups and landscapes are not always comparable, the data are not necessarily complete in terms of sample frequency or longevity of the sampling record, and unspecified logistical issues are ever present in terms of training, supervision, and organization of the volunteers.

The variety and characteristics of volunteer water quality monitors and programs

There is little reporting of the results of formal evaluations of volunteer water quality monitoring in the USA. According to Gouveia *et al.* (2004, p. 137), "Information on the less formalized initiatives is dispersed and non-organized making the analysis of their contribution more difficult." For this reason Gouveia *et al.* (2004) reviewed the literature dating back to 1986 to identify benefits and disadvantages of including volunteers in environmental monitoring. Based on the review,

they concluded that citizen monitoring efforts result in a more aware and educated public; increased collaboration among stakeholders; financial benefits associated with the cost effectiveness of utilizing volunteers for data collection and management; and the development of an "early warning" system for environmental degradation.

Nerbonne and Nelson (2004) found that volunteer macroinvertebrate monitoring groups ranged from two to seven thousand members, averaging between two hundred and three hundred. However, the number of actual stream monitoring volunteers averaged around thirty per group working approximately four years. Groups in states with centralized support tended to have been monitoring for longer periods and had more stream monitors than other states. The main goals of the groups were public education, awareness and gathering baseline data whereas direct influence on policy decisions was less important. Typically, budgets were around five thousand dollars, and two-thirds of the groups sent data to a larger organization for publication and dissemination. Volunteer macroinvertebrate monitoring groups produced less than one official report each year, presented data to the public less than twice a year and had their data summarized in the newspaper only every other year. Just as important, the data were used in land use planning decisions only once every four years (Nerbonne and Nelson, 2004, p. 836).

Because there are so few published evaluations of volunteer monitoring, we turned to the so-called "gray" or non-academic literature. Information on the cases we now describe is less formally and widely disseminated, yet provides useful information about volunteer monitoring. Because of limited peer reviewed publications evaluating the results of volunteer water quality monitoring, we conducted a search on the web (www.google.com) to identify the fugitive literature on the topic. Our keywords were effectiveness, volunteer monitoring, citizen participation, and watershed management. This procedure revealed approximately one hundred webpages. Another search that excluded watershed management resulted in approximately one hundred seventy hits, and there was substantial overlap with the first search. In fact, the search that did not use watershed management as a keyword was still predominantly related to water issues, demonstrating, again, that this is the most common type of volunteer monitoring. Also, several sites included "how to" manuals, which we did not include in our database. We restricted our search to programs within the USA.

The search revealed many annual reports, educational materials, newsletters, and website homepages that describe volunteer monitoring, its effectiveness and some of the challenges. Altogether we discovered approximately fifty homepages that provided useful information about volunteer water quality monitoring programs or projects. We will not describe these programs in detail, but we will provide an overview of some efforts to profile characteristics of volunteer monitors and evaluate whether such programs lead to increased environmental awareness and knowledge. Finally, we will discuss the influence of volunteer monitoring on bridging the chasm between science- and technology-based environmental management and public involvement in environmental decision-making.

The Tennessee Water Resources Research Center conducted an analysis of water monitoring programs in the southeastern USA (Georgia, Kentucky, and Alabama) in order to draw lessons for its own programming. The authors of the study note that "developing a program that can be responsive to shifting political and social landscapes and technological advancements may be one of the most important lessons learned from these programs' 30 plus years of volunteer monitoring experience" (Hanahan and Cottrill, 2004, p. ii). To complete this study, Hanahan and Cottrill reviewed volunteer monitoring literature, conducted personal interviews with program managers in the three states mentioned above and reviewed the program sites on the internet as well as relevant documents. They then interviewed past and current volunteer monitors in Georgia, Kentucky, and Alabama, subsequently interviewing individuals in Tennessee with an interest in volunteer monitoring.

Findings from Hanahan and Cottrill (2004) revealed a variety of mechanisms for structuring a volunteer monitoring program where the approach chosen matched the desired outcomes. In the three states studied, a "typical" volunteer is between the ages of thirty-five and sixty-four, has two years of involvement in the program and has a college degree. There was no predominant occupation among the volunteers (Hanahan and Cottrill, 2004). In addition, volunteers felt that the benefits of participation in the monitoring program surpassed any limitations and that the major benefit was the opportunity to educate fellow citizens. Those interviewed felt that insufficient funding and community support were the most significant drawbacks. Typically, data from these programs were used for red-flagging potential water quality problems.

The evaluations of programming in Georgia, Kentucky, and Alabama informed the developing volunteer monitoring program in Tennessee. The Tennessee program was distinguished from the others in that most participants were highly educated scientists or individuals who work in an environmental field, and viewed quality assurance/control as the primary concern rather than lack of community support, as had been true in the other states.

An evaluation of the Citizen Science Partnership in Minnesota (Minnesota Department of Natural Resources Access Year, 2005) involved more than one hundred fifty individuals using questionnaires, personal interviews, and focus groups. Three state-wide conferences also provided information for the evaluation. In this study, a typical watershed partnership covered approximately sixty-four thousand acres, had six members in the group and had been in existence for five to ten years. This evaluation found that lack of data on the watersheds in Minnesota is not an issue – "[r]ather citizens are frustrated by the lack of data integration and accessibility in readily understood formats" (Minnesota Department of Natural Resources Access Year, 2005, p. 6).

In Minnesota, citizens become involved in watershed monitoring and management because of a sense of stewardship or a focus on watershed well-being. Typically, the conservation values that citizens bring to the associations have developed from childhood experiences and are carried into adulthood. "Other motivators include: (a) perceiving direct threats to valued resources; (b) finding an alternative to government intervention through information, incentives, and education; and (c) fostering a local sense of place, community, and stewardship" (Minnesota Department of Natural Resources Access Year, 2005, p. 6). The danger of burn-out by the volunteers is high, and low morale seems to be an issue for watershed groups due to lack of funding, a feeling that agencies view the citizen role with some skepticism, limited scientific background among the volunteers, and few opportunities for training.

Findings from this case point to challenges that volunteer groups might encounter. Significant among them is a lack of a shared language. Reviewers felt that it is important for scientists to reach out to citizens making sure that the messages conveyed are understood and interpreted correctly. Not only must understanding be achieved, but the information must also be compelling and motivating so that citizens can feel truly engaged in the process. Sustainability of the monitoring effort is achieved when there is a sense of history and place developed among the stakeholders, but individuals have often lost this because of migration into cities and increased isolation. The Minnesota evaluation found that watershed group volunteers are more likely than others to feel strongly attached to surrounding natural resources.

In New York State, an analysis to assess a potential volunteer monitoring program in the

New York City watershed focused on determining what forest resources citizens could effectively monitor; who the potential volunteers were; where data collection could occur; and who could manage the data. Interviews with individuals as well as groups included landowners, agency staff, employees of an environmental education center, scientific researchers, and leaders of existing volunteer monitoring groups (deBoer *et al.*, 2000). Results from this study determined that citizens can effectively undertake a forest inventory, monitor the prevalence of animals such as salamanders, white tail deer and bird populations and track invasive species. In terms of water quality monitoring, the study showed that citizens can monitor the chemical and biological characteristics of water in addition to physical parameters such as flow rate and stream depth.

The study found that potential participant groups range from public school children to landowners to members of watershed organizations. deBoer and her colleagues determined that private forest landowners were a likely target group for monitoring activities. Their interest in monitoring water quality is about equal to their interest in undertaking tree surveys and tracking deer populations (deBoer *et al.*, 2000), and the most ideal location for monitoring is the volunteer's own property. The project resulted in a number of specific recommendations for volunteer monitoring of forest resources in order to maintain water quality. Educating and informing citizens, providing an outlet for citizens to become environmental stewards and making contributions to science and understanding of the natural environment were primary program goals (deBoer *et al.*, 2000). They also found that targeting landowners in the area maximizes participation and increases the consistency of the data-collection efforts long term.

The Volunteer Water Quality Monitoring National Facilitation Project is a nationwide effort sponsored by the USDA's Cooperative State Research Education and Extension Service (CSREES), and focuses on building a support system for Cooperative Extension volunteer monitoring efforts nationwide. A nationwide assessment of Extension projects that undertake volunteer water quality monitoring indicated that most programs are community driven and result from community groups or individuals responding to a local water quality issue (CSREES, 2005). As volunteers explore the issue in question, they uncover data gaps in monitoring programs. The assessment determined that Extension plays a key leadership role in many of these programs, and there is often a cooperative relationship with state environmental agencies.

Although the National Facilitation Project is the most ambitious program to coordinate and facilitate volunteer water quality monitoring projects across the USA, it excludes monitoring programs that are not directly affiliated with CSREES. Thus, there remains the need for a comprehensive and systematic review and evaluation to determine the effectiveness of volunteer monitoring in the USA.

CONCLUSION

Volunteer environmental monitoring is a fairly widespread and growing phenomenon in the USA. Given the relatively high standard of living, a population with the time and resources to volunteer, and a system of regulations that establishes clear criteria for the measurement of environmental quality, the USA represents a context within which volunteer environmental monitoring can flourish. Volunteerism is a form of grassroots environmental activism that to some extent furthers the ideals of democratic participation and establishes a basis for direct interactions between citizens and scientists. Programs of volunteer monitoring typically aspire to make participants more environmentally aware and knowledgeable, thereby facilitating more meaningful interactions with scientists. Earlier in this chapter we asked: does volunteer monitoring reduce the chasm between environmental science and the lay public? In the ways we just indicated, we conclude that volunteer monitoring is a useful practise for reducing the chasm between citizens and scientists and is one useful way to link humans and their environment.

Volunteer monitoring achieves these connections as a form of civic environmentalism which unites grassroots activity with government regulation. These volunteers are rooted in place, and are concerned about local environmental conditions. Governmental science-based regulations and standards provide them with a framework within which to organize monitoring activities. Given the extensive environmental monitoring needed to protect the environment effectively, volunteers offer a low-cost means of providing the data needed to meet effectively the intentions of environmental policies.

While there is great promise for this form of civic environmentalism, often it does not meet up to its full potential. Earlier in this chapter we asked: does the evidence support the use of volunteers to generate data that meet the scientific standards of official environmental managers? All the available evidence we reviewed supports an affirmative response to this question. We also asked: if acceptable data are collected by volunteers, are they used by agencies charged with environmental management? The answer to this question is less certain. There are certainly examples of the effective use of the data by government agencies, but there is also evidence of skepticism. One of the most important limitations to the use of volunteer-generated data is doubts scientists and regulators harbor about the quality of the data generated by volunteers.

Such doubts reduced the use of volunteer-generated data in environmental decision-making. The main source of this skepticism appears to be a lack of adequate quality assurance and control. The adoption of quality assurance and control measures acceptable to government agencies appears to be a relatively simple and effective organizational innovation that increases agency use of volunteer monitoring data. Unfortunately, the proportion of such officially recognized programs in the USA is relatively low nationally and varies by state. Closer ties between government regulatory agencies and volunteer monitoring groups that establish quality standards for volunteer environmental data are needed. Our review of environmental monitoring programs clearly indicates the importance of organizational linkages between governmental agencies and local volunteer programs. These not only assure that volunteer-generated data are used in environmental decision-making, but that they also buoy the morale of participants thereby contributing to the longevity of such efforts.

The social dimensions of volunteer monitoring are also an important aspect in the success of such programs. The character of programs is shaped by the characteristics of volunteers. While people from a wide variety of backgrounds volunteer, it seems that volunteers in many programs are slightly older and more educated than nonparticipants. Still the composition of programs is varied in terms of environmental background and knowledge, making environmental education an important objective alongside data collection.

Volunteer environmental monitoring holds promise as one means of generating stronger links between humans and the environment. However, this phenomenon is also an artifact of its social and historical context. As indicated above, its success depends on systematic observation, officially recognized by government agencies charged with regulating particular environmental media. Thus, volunteer monitoring links people to their local environment, but it is likely to succeed as a sustained effort when it is conducted under the terms and conditions established by the state. However, volunteer monitoring also runs the risk of being co-opted by special interests. Individuals and groups with a stake in presenting environmental conditions as either better or worse than others

would think, might attempt to manipulate the official terms and conditions of environmental monitoring. The most important safeguard against the co-optation of environmental monitoring by special interests is transparency and an open process with opportunities for public inputs during the process of establishing official standards.

As indicated at the outset of this chapter, we believe that volunteer monitoring is most likely to flourish and impact environmental decision-making only under limited circumstances. That said, this activity may support existing regulations of activities impacting the environment, or it may challenge political and economic elites whose decisions impact the natural environment. Whether supporting the status quo or challenging it volunteer monitoring is most likely to succeed when it communicates in the dominant language of science. In addition, for such efforts to be sustained and for them to have a lasting impact, they need resources and an organizational structure that supports systematic collection, processing, and analysis of environmental data. As we suggested above such efforts are most likely to appear in special circumstances that include the following conditions: abundant resources, an educated citizenry able to engage in scientific discourse and some sort of legitimacy (legal, ethical or otherwise) in the eyes of some agents of the state.

REFERENCES

Aitken, M.D. and M.A. Okun (1991) "Investigation of Odor Problems Associated with Wastewater Treatment Facilities in North Carolina." North Carolina State University/Water Resources Research Institute.

Applebaum, H. (1992) *The Concept of Work: Ancient, Medieval, and Modern.* Albany, NY: State University of New York Press.

Baker, V.R. (1998) "Hydrological Understanding and Societal Action." *JAWRA* 34:819.

Beck, U. (1992) *Risk Society: Towards a New Modernity*, ed. by M. Featherstone. London: Sage.

Beck, U. (2000) *The Brave New World of Work.* New York: Polity Press.

Beierle, T.C. and J. Cayford (2002) *Democracy in Practice: Public Participation in Environmental Decisions.* Washington, DC: Resources for the Future.

Bliss, J., G. Aplet, C. Hartzell, P. Harwood, P. Jahnige, D. Kittredge, S. Lewandowski and M.L. Soscia (2001) Community-based ecosystem monitoring. *Journal of Sustainable Forestry* 12:143–167.

Bode, R.W., M.A. Novak and L.E. Abele (1991) "Methods for Rapid Biological Assessment of Streams." ed. by New York State Stream Biomonitoring Unit Water: New York State Department of Environmental Conservation.

Born, S.M. and K.D. Genskow (2000) "Toward Understanding New Watershed Initiatives: A Report from the Madison

Watershed Workshop." In *Madison Watershed Workshop.* Madison, WI: University of Wisconsin.

Brown, S.T., J. Gilliam and J. Johns-Cason (2000) "Cooperation and Partnerships: Virginia's Citizen Monitoring Program, Getting Data to Use." In *Sixth National Volunteer Monitoring Conference.* Austin, TX, pp. 52–56.

Bryner, G. (2001) "Cooperative instruments and policy making: Assessing public participation in US environmental regulation." *European Environment* 11:49.

Byron, I. and A. Curtis (2002) "Maintaining volunteer commitment to local watershed initiatives." *Environmental Management* 30:59–67.

Cairns, J. Jr. (1974) "Indicator species vs. the concept of community structure as an index of pollution." *Water Resources Bulletin* 10:338–347.

Cairns, J. Jr. (1998) "Consilience or consequences: Alternative scenarios for societal acceptance of sustainability initiatives." *Renewable Resources Journal* 16:6–12.

Carr, A. (2003) "A social scientist's perspective on community science." *Volunteer Monitor* 15(2):6.

Chess, C. (2000) "Evaluating environmental public participation: Methodological questions." *Journal of Environmental Planning and Management* 43:769.

CSREES (2005) "Volunteer Water Quality Monitoring National Facilitation Project," Retrieved 2005 (http://www.usawaterquality.org/volunteer/).

Dates, G. (1999) "Watershed Monitoring," Retrieved February, 2005 (water.usugs.gov/wicp/may99mins/att27_Riverwatch.html).

deBoer, Y.E., R.H. Germain and V.A. Luzadis (2000) "Citizen Volunteer Monitoring of Forest Resources in the New York City Watershed." In *Sixth National Volunteer Monitoring Conference.* Austin, TX.

Dryzek, J.S. (1997) *The Politics of the Earth.* Oxford, UK: Oxford University Press.

Dunlap, R. and S. Mertig (1992) "The evolution of the U.S. environmental movement from 1970 to 1990: An overview." In *American Environmentalism: The U.S. Environmental Movement 1970–1990*, ed. by R. Dunlap and A. Mentigo. Philadelphia: Taylor & Francis.

Ely, E. (2001) "Macroinvertebrate data." *Wild Earth* 24–27.

EPA (1997) "Top 10 watershed lessons learned." US Environmental Protection Agency, Washington, DC.

EPA (2001) "Monitoring and Assessing Water Quality" (http://www.epa.gov/owow/monitoring/volunteer).

EPA (2003) "EPA's Framework for Community-based Environmental Protection," Retrieved (http://www.epa.gov/CBEP).

EPA (2005) "National Directory of Volunteer Environmental Monitoring Programs."

Ewing, S. (1999) "Landcare and community-led watershed management in Victoria, Australia." *JAWRA* 35:663.

Feldman, D.L. (1991) *Water Resources Management: In Search of an Environmental Ethic.* Baltimore: Johns Hopkins University Press.

Firth, P. (1999) "The importance of water resources education for the next century." *JAWRA* 35:487.

Fischer, F. (2000) *Citizens, Experts, and the Environment: The Politics of Local Knowledge.* Durham, NC: Duke University Press.

Fleming, B. and D. Henkel (2001) "Community-based ecological monitoring: A rapid appraisal approach." *Journal of the American Planning Association* 67(4):456–465.

Fore, L.S., K. Paulsen and K. O'Laughlin (2001) "Assessing the performance of volunteers in monitoring streams." *Freshwater Biology* 46:109–123.

Frank, D.J. (1997) "Science, nature, and the globalizations of the environment, 1870–1990." *Social Forces* 76: 409–435.

Frank, D.J. and J. Gabler (2006) *Reconstructing the University: Worldwide Shifts in Academia in the 20th Century.* Stanford, CA: Stanford University Press.

Freudenburg, W.R. (1993) "Risk and recreancy: Weber, the division of labor, and the rationality of risk perceptions." *Social Forces* 71:909–932.

Gouveia, C., A. Fonseca, A. Camara and F. Ferreira (2004) "Promoting the use of environmental data collected by concerned citizens through information and communication technologies." *Journal of Environmental Management* 71:135–154.

Hanahan, R.A. and C. Cottrill (2004) "A Comparative Analysis of Water Quality Monitoring Programs in the Southeast: Lessons for Tennessee," Retrieved 02/07/05,(eerc.ra.utk.edu/waht-new/mainbook.pdf).

Heiman, M.K. (1997) "Science by the people: Grassroots environmental monitoring and the debate over scientific expertise." *Journal of Planning Education and Research* Summer:291–299.

Hilsenhoff, W.L. (1977) "Use of Arthropods to Evaluate Water Quality of Streams." University of Wisconsin, Department of Natural Resources, Madison, WI.

Hubbell, S. (2002) "A (slightly) dissenting view." *Volunteer Monitor* 14(2):3.

Independent Sector (2005) "Independent Sector Announces New Estimate for Value of Volunteer Time," Retrieved 05/31/05 (http://www.independentsector.org).

Irwin, A. (2001) "Constructing the scientific citizen: Science and democracy in the biosciences." *Public Understanding of Science* 10:1–18.

John, D. and M. Mlay (1999) "Community-based Environmental Protection: Encouraging Civic Environmentalism." In *Better Environmental Decisions*, ed. by K. Sexton, A.A. Marcus, K.W. Easter, and T.D. Burkhardt. Washington, DC: Island Press.

John, D. (1994) *Civic Environmentalism: Alternatives to Regulation in States and Communities.* Washington, DC: Congressional Quarterly Press.

Kaufmann, J. (1999) "Three view of associationalism in 19th century America: An empirical investigation." *American Journal of Sociology* 104:1296–1345.

Kellert, S.R., J.N. Mehta, S.A. Ebbin and L.L. Lichtenfeld (2000) "Community natural resource management: Promise, rhetoric, and reality." *Society and Natural Resources* 13:705–715.

Kerans, B.L. and J.R. Karr (1994) "A benthic index of biotic integrity for rivers of the Tennessee Valley." *Ecological Applications* 4:768–785.

Kraft, M. (1999) "Making decisions about environmental policy." In *Better Environmental Decisions*, ed. by K. Sexton, A.A. Marcus, K.W. Easter, and T.D. Burkhardt. Washington, DC: Island Press.

Kuhn, S. (1999) "Expanding public participation is essential to environmental justice and the democratic decisionmaking process." *Ecology Law Quarterly 25*.

Lee, V. (1994) "Volunteer monitoring: a brief history." In *The Volunteer Monitor*, vol. 2005: USEPA.

Lubell, M. (2002) "Environmental activism as collective action." *Environment and Behavior* 34:431–454.

Maryland Department of Natural Resources (2005), "Volunteer Water Monitoring," Retrieved 2005 (http://www.dnr.state.md.us/streams/volunteer/vol_index.html).

Minnesota Department of Natural Resources (2005) "Citizen, Science, Watershed Partnerships, and Sustainability: The Case in Minnesota," Retrieved January 14 (http://www.sustain.org/collaborative/CITSCI.pdf).

Murdoch, T.B. and M. Cheo (1996) *Streamkeeper's Field Guide: Watershed Inventory and Stream Monitoring Methods*. Everett, WA: Adopt-A-Stream Foundation.

National Audubon Society (2005a) "The 105th CBC, December 14, 2004–January 5, 2005," Retrieved February 2005 (http:www.audubon.org).

National Audubon Society (2005b) "8th Annual Backyard Bird Count," Retrieved February, 2005 (http://www.audubon.org).

National Weather Service (2005) "National Weather Service Training Page," Retrieved February, 2005.

Nerbonne, J.F. and K.C. Nelson (2004) "Volunteer macroinvertebrate monitoring in the United States: Resource mobilization and comparative state structures." *Society and Natural Resources* 17:817–839.

NRC (1996) *Understanding Risk: Informing Decisions in a Democratic Society*, ed. by N.R. Council. Washington, DC: National Academy Press.

NRC (1999) *New Strategies for America's Watersheds*, ed. by N.R. Council. Washington, DC: National Academy Press.

NRC (2004) *Analytical Methods and Approaches for Water Resources Project Planning*, ed. by N.R. Council. Washington, DC: The National Academy Press.

O'Leary, N., A.T. Vawter, L.P. Wagenet and M.J. Pfeffer (2004) "Assessing water quality using two taxonomic levels of benthic macroinvertebrate analysis: Implications for volunteer monitors." *Journal of Freshwater Ecology* 19:581–586.

Pacific Streamkeepers Federation (1997) "The Streamkeepers Handbook and Modules," Retrieved 2005 (http://www.pskf.org).

Palerm, J.R. (2000) "An empirical–theoretical analysis framework for public participation in environmental impact assessment." *Journal of Environmental Planning and Management* 43:581.

Pathfinder Science (2004) "Digital Monarch Watch," Retrieved 2005 (http://pathfinderscience.net/monarch/index.cfm).

Peet, M. (2005) "Nationwide bird-counting event begins." In *Ithaca Journal* 17 February, p. 18.

Petts, J. (2003) "Barriers to deliberative participation in EIA: Learning from waste policies, plans and projects." *Journal of Environmental Assessment Policy and Management* 5:269.

Plafkin, J.L., M.T. Barbour, K.D. Porter, S.K. Gross and R.M. Hughes (1989) "Rapid bioassessment protocols for use in streams and rivers: Benthic macroinvertebrates and fish," ed. by EPA: US Government Printing Office.

Putnam, R.D. (1995) "Bowling alone: America's declining social capital." *Journal of Democracy* 6:65–78.

Putnam, R.D. (1996) "The strange disappearance of civic America." *The American Prospect* 24:34–48.

Putnam, R.D. (2000) *Bowling Alone: The Collapse and Revival of American Community*. New York: Simon & Schuster.

Ravindra (cited in S.S. Negi) (2001) *Participatory Natural Resource Management*. New Delhi: Industrial Publishing Company.

Rosenbaum, W. A. (1991) *Environmental Politics and Policy*. Washington, DC: Congressional Quarterly Press.

Roux, D.J., H.R. van Vliet and M. van Veelen (1993) Towards integrated water quality monitoring: Assessment of ecosystem health. *Water SA* 19(4):275–280.

Rubin, C.T. (2002) "Civic Environmentalism." In *Democracy and the Claims of Nature*, ed. by B.S. Minteer and B.P. Taylor. Lanham, MD: Rowman & Littlefield, pp. 335–351.

Santos, S.L. and C. Chess (2003) "Evaluating citizen advisory boards: The importance of theory and participant-based criteria and practical implications." *Risk Analysis* 23:269–280.

Sewell, W.R.D. and S.D. Phillips (1979) "Models for the evaluation of public participation programmes." *Natural Resources Journal* 19:337–358.

Sexton, K., A. Marcus, K. Easter and T. Burkhardt (1999) "Conclusion: Strategies for Integrated Decision-making." In *Better Environmental Decisions*, ed. by K. Sexton, A. Marcus, K. Easter, and T. Burkhardt. Washington, DC: Island Press.

Shapiro, M. (1996) "A Golden Anniversary? The Administrative Procedures Act of 1946," Retrieved 2004 (http://www.cato.org/pubs/regulation/reg19n3i.html).

Silverberg, K.E. (2004) "Understanding American Parks and Recreation Volunteers Utilizing a Functionalist Perspective." In *Volunteering as Leisure/Leisure as Volunteering: An International Assessment*, ed. by R.A. Stebbins and M. Graham. Willingsford, Oxfordshire, UK: CAB International.

Smith, W. (2003) "Science, technical expertise and the human environment." *Progress in Planning* 60:321.

Stancioff, E. (2000) The Maine Shore Stewards Program Use of Data. In *Sixth National Volunteer Monitoring Conference*. Austin, TX, pp. 60–61.

Stebbins, R.A. (2004) "Introduction." In *Volunteering as Leisure/Leisure as Volunteering: An International Assessment*, ed. by R.A. Stebbins and M. Graham. Willingsford, Oxfordshire, UK: CABI International.

Wagenet, L.P. and Pfeffer, M.J. (2003) "Stakeholder participation in watershed management I & II." *Impact* 5(6):8–14.

Ward, R.C. (1996) "Water quality monitoring: Where's the beef?" *Water Resources Bulletin* 32(4):673–680.

Weber, E.P. (2000) "A new vanguard for the environment: Grass-roots ecosystem management as a new environmental movement." *Society and Natural Resources* 13:237–259.

Williams, K.F. (2000) "Oregon's Volunteeer Monitoring Program." In *Sixth National Volunteer Monitoring Conference: Moving into the Mainstream*. Austin, TX, pp. 62–66.

Williams, W.L. (2002) *Determining Our Environments: The Role of the Department of Energy Citizen Advisory Boards*. Westport, CN: Praeger.

Wilson, D. (2000) "Pennsylvania's Citizens' Volunteer Monitoring Program." In *Sixth National Volunteer Monitoring Conference*. Austin, TX, pp. 57–59.

Zazueta, A.E. (1995) "Policy Hits the Ground: Participation and Equity in Environmental Policy-Making." Washington, DC: World Resources Institute.

17

Environmental Ethics

Val Plumwood

ENVIRONMENTAL ETHICS: CAN ONLY HUMANS MATTER?

Relentlessly, it seems, humans are driving other species from the planet, in labours that not merely interrupt but reverse the great diversity work of the Earth through evolutionary time. Western culture seems locked into an ecologically destructive form of rationality which is human-centred, or "anthropocentric", treating non-human nature as a sphere of inferior and replaceable Others. In this context, environmental ethics raises concerns about whether we should see humans as the supreme species, and the only possible value of the non-human as the service of human ends, as Passmore (1974), and Norton (1991) assert, or whether non-humans, including animals, plants and elements, have some claims of their own, perhaps even some sort of equality. Can the lives of these non-human others count only indirectly, as instruments or resources, as means to human ends, or can these lives have meaning and worth in themselves? These are among the questions addressed in the subject area of environmental ethics.

Human existence is confronted every day by such questions – not only directly in where we live but in how we work, consume or otherwise impact on our shared world. As I sit writing this, my senses are alert for signs of a small lizard I saw a while ago on the floor. It strayed into my house, a hostile and potentially fatal place for it, and hid before I could catch it. Should I try to find it and carry it to safety outside among its fellows before it dies of cold, starvation, or being walked on? Should I help it if I can? As I would a human, or in some other way? Does its life matter at all, or do only human lives and harms count?

These questions test us every day. Directly or indirectly, they are questions about the needs of the lives, beings and elements around us, and how we can balance them with those of our own species. Sharing worldspace and place with non-human others that surround us is a delineating issue for environmental ethics, along with the urgent questions of human sustainability. Yet until recently (and in most places still) such questions were foreclosed in a reflective world long dominated by monotheist traditions authorising a singularised, anthropoform[1] god who anoints humans the chosen species, privileged to pursue their own development at the expense of inferior elements in the domain of nature. This licence removed the need for much philosophical or ethical consideration of the natural world, since only humans ultimately mattered or could receive the final rewards of paradise. We certainly need a major rethink of these traditions, and environmental ethics and philosophy can offer a useful disciplinary place to begin.

INSTRUMENTALISM AND ENVIRONMENTAL ETHICS

Under the heading instrumentalism cluster a set of questions and problems that are among the most important foundational issues for ethics and environmental ethics. Here, the way we answer can make a big difference. The normal instrumentalist answer is that the plight of my lizard could safely be ignored unless it suited me to practice an unusually high and unnecessary (or, for some, perhaps irrational) degree of compassion. For many anti-instrumentalists, on the other hand, not only does this particular lizard need to be considered, and if possible, helped, but the lives and safety of lizards and other species in general need to be considered – in many spheres including house safety design and road design, for example.

The issue of instrumentalism is primary but preliminary for environmental ethics because if we answer its major questions in one way – the way of conventional wisdom on which many of our institutions are premised – no proper environmental ethics is possible or needed.[2] This way is to affirm that the world is emphatically divided into a realm of ends and a contrasting realm of means available to serve to those ends – or roughly, people and things. Items in the realm of ends have intrinsic value, and everything outside that realm has only instrumental value – they can legitimately be used as commodities and traded, for example, by members of the first group.

The instrumental assumption Western modernity has made, and one of the key ones environmental thought has sought to bring into question, is that all humans, and humans alone, are ends and have intrinsic value, the non-human world being merely a means to human ends. To put it another way, humans need only consult their own species' interest or convenience in deciding what courses of action to follow. Humans are part of a separate, superior order apart from the rest of nature, and only their concerns can be reasons. Nature has no direct claims of its own that could interfere with or constrain human projects.

Instrumentalism towards non-humans (which some also call the "utilitarian" mode)[3] is the commonsense (or default setting) of the dominant culture of modernity. For the instrumentalist, animals and other non-humans are freely available for experimentation and other deleterious uses, a paradigm example of instrumental treatment, in a way humans are not considered to be. This modern "humanist" instrumentalism was naturalised with the onset in modernity, when it replaced older classical default frameworks authorising an intrahuman as well as an interspecies hierarchy (Plumwood, 1993). The older default recognised, along with lower non-human beings, lower categories of humans who were instrumentalisable, for example, as domestic servants or domestic slaves. For Cartesian modernity, humans as a group were set apart as rational and mindful agents against mechanistically conceived nature elements of "nature", reduced to mere matter. Intrahuman hierarchies lost legitimacy, just as humans became individual and equal bearers of their supposed identifying feature, consciousness and choice. Meaning and mind inhabited the human sphere, rather than the chaotic or deterministic zone of matter. This replacement was one of several shifts in Enlightenment concepts that naturalised and deepened Western traditions of

human/nature dualism hyper-separating humans from the rest of nature. The new default set humans as conscious beings even more emphatically apart from the natural world than the monotheist traditions. Even as it supported humanist calls for human equality and freedom (Plumwood, 1993, 2002), its rallying project was development, "the empire of man over mere things".

This definitional context reveals how and why the question of instrumentalism is a crucial one for environmental ethics, but also why it is only the beginning of the subject. For if something can never be more than a means, and is completely instrumentalisable, no ethical constraints on its treatment are admitted. No space for an environmental ethics in relation to such mere "things" can then exist.[4] On the other hand if we answer these questions in the opposite way, rejecting the instrumental treatment of the non-human world, we have merely opened a door. The main task of imagining and negotiating a viable and realistic environmental ethic that will address the crisis in our relations with the non-human world still lies ahead of us. This is why the issues of intrinsic value and moral consideration are preliminary, and should never be seen as exhausting environmental ethics or come to monopolise its attention.

Many have taken consideration of ethical questions posed by interspecies relationships to be *the* key task of environmental ethics (Hay, 2002). But if an environmental ethic, at a minimum, considers the claims to ethical attention of non-human elements as co-constitutors of the world, such an ethic must also be open to challenge conventional concepts of human identity and the associated default, the fundamental Cartesian–humanist division of the world into conscious and unconscious substances, humans (people) and mere things. So the other side of the conventional concern of ecological ethics with the non-human is the project of what might be called "deep sustainability", developing a more ecologically grounded concept of human identity. This has wide implications for many areas of philosophy such as moral epistemology, as well as for human ecological behaviour and ethical reflection.

INSTRUMENTALISM AND UTILITY

The first obvious question up for discussion is: what do we really mean by instrumentalism and why is it a problem? The issues need careful analysis because they are often confused through people adopting conflicting weaker and stronger meanings for the leading terms, instrumentalism and its contrast term intrinsic value. Partly because of this complex tangle of meanings, much debate in philosophy never gets far past this initial issue, so it is the first question we should consider, but we will certainly not be stopping there.

A common tendency has been to equate instrumental value with utility value, with making use of something or simply using it as a means to some end, and so to go on to declare instrumentalism to be inevitable and innocent.[5] We are free to define terms in any way we like, or so it seems until the consequences of novel definitions emerge: if we stipulate definitions significantly at variance with usual or established meanings, we may find ourselves misunderstood, and help to obscure rather than elucidate the issues. To define instrumental value as equivalent to utility value is to adopt a very weak sense of the term which misrepresents what is at stake and obscures alternatives to the dominant framework. Since no one could ever survive without making some use of the natural world, the identification of instrumentalism with mere use (utility with instrumental value) quickly leads to the conclusion that instrumentalism is inevitable and harmless.

Shared meanings for terms are usually derived from larger common narratives in which they appear. The larger philosophical narrative that has problematised instrumentalism was inspired in the first instance not by the instrumental treatment of nature but by that of human beings. Many philosophers have wanted to reject instrumental treatment, at least for the limited class they thought worthy of respect in their own right, the class of humans. The philosopher Kant, for example, exhorts us to treat others (i.e. other humans) always as ends, never as means. The instrumental treatment of human beings, their use in damaging experiments for example, is generally seen as violating this precept. This could hardly be the case, however, if instrumentalism is defined weakly, to mean simply utility, making use of something or deriving a benefit or advantage from this relationship with them. In fact we constantly make use of other humans without anyone seeing anything morally objectionable in it. We might make use of a waiter to access food, or of a parent as a means of support. A child may drink from its mother's breast to get nourishment, for example, or sit on its father's shoulders to get a better view of the carnival. In this sense, it is hard to see what Kant could possibly be concerned about in the use of others as means. This weak sense is a redefinition, and is best marked by a different term such as utility, since it does not connect in an appropriate way with the philosophical narrative problematising instrumentalism.

If mutual use and exchange between living beings is a condition of ecological embodiment, the relevant ethical focus for an ethics of use should not be on a choice between use and no use, but on discerning patterns of use and their ethical meaning.

The pattern the term "instrumentalism" problematises is that of treating others as *no more than* a means to our ends, refusing to acknowledge excess, the respects in which the other outruns our knowledge and purposes.[6] This is also commonly a pattern of reducing the other that is not mutual or reciprocal, that sees others as available for our use, but does not like to see us as available for their use. So what is problematic for Kant in treating others as means seems to be this pattern of treating them only or purely as means, *reducing* them to instruments, as nothing more than the transparent enabler of our projects.[7] Once the instrumentalised other becomes nothing more than a means to meet our needs, those characteristics or potentialities of usefulness to us come to dominate our construction of them, to the exclusion of other ethically relevant features and facts, such as the extent of their own harm or suffering. It is this kind of pattern of reduction to use that is held to be objectionable, not the use itself or the fact that we can make use of others and get some benefit from this use.

Turning to examples, it has often been held that instrumentalism poisons human relationships like friendship, mutualistic relationships from which several parties derive (ideally equal) practical and psychological benefit or utility. Utility is always involved here, but instrumentalism enters the picture in an objectionable way particularly when we *reduce* our friends to a means to our own use or benefit. For example, when we cultivate someone we don't really care for in order to advance our career, or drop a friend when they cease to be of use to us socially. Instrumentalism enters where utility is *the only motive*, or, in graduated form, appears when it becomes the main motive, or when excess over and above usefulness is not recognised or valued. Recognising excess here indicates the contrasting mode to the instrumental – the intrinsic mode: liking and valuing friends for their own sake, for who they are, not just for the use they have for us.

It is therefore essential to distinguish instrumentalism from mere use or utility.[8] Use is inevitable and potentially respectful, compatible with recognising excess, that the used person or item is far more than their usefulness for us. Instrumentalism prizes and prioritises the useful aspects and qualities of the instrumentalised, often to the point of denying, discounting or ignoring other relevant and important features. In the human sphere, the institution of slavery perhaps most clearly exemplifies the reduction of a person to mere use; the best slave is a transparent facilitator of the master's projects. For the Roman philosopher Cato, who advises masters of slaves too old or weak to work to turn them out to die, slaves are to be conceived and managed as nothing over and above a work capacity. (Yes, the intelligence that produces the battery hen was also active 2000 years ago.) Most of us now claim to find such blatantly reductive and instrumental treatment for humans revolting, but that does not mean more disguised or milder forms of instrumentalism towards humans are not still normalised in contemporary societies. In disguised forms an instrumental orientation is often incorporated into criteria for being a good worker, a good servant or a good wife. When people speak about some individual, system or motivation as being "utilitarian", they are often pointing to some area where an instrumental disposition holds sway.[9]

What is instrumentalism for the non-human sphere? The instrumental worldview (or instrumental rationality) sees nothing over and above resources, and casts the universe into reductive categories of use, implicitly reducing the other to little more than product or items for consumption. We will say more about this below, but an especially clear set of examples come from factory farming: the battery hen is reduced to a means of producing eggs, and the reductive intent is revealed by the minimising of resources devoted to her comfort and development over and above what is required for profitable performance of the egg-producing function.

It is not only animals but all aspects of the Earth and of life that can be injured by instrumental rationality. Instrumentalism drains the life, meaning and wonder from the world as expertly as Dracula drains his victims, producing alienation, cynicism and anomie. An instrumental study of the Earth, for example, would focus primarily on where profitable minerals were found and how they could be commodified, rather than on unravelling the wondrous narratives of Earth history. The instrumental outlook is normalised for nature as part of the process of commodification. Instrumentalism towards animals and the natural world is the default setting or commodity culture, admitting some partial and not very consistent exceptions (e.g. for pets). Without social intervention, a competitive, cost-minimising market tends to operate in the instrumental mode, and commodified items ("resources") are treated reductively as means to the end of profit. The forest destruction happening now in the Amazon Basin brings home the truth that political and economic systems that aim to maximally expand the reach of commodification ("resources") give permission to do great damage to nature. Because its associated worldview denies and disappears excess, instrumentalism obscures important alternatives for respectful use, for example, treating farm animals generously and gratefully, use that recognises and respects the more, the excess over and above "productive" function, the whole being. Respectful use

proscribes waste, and aims to qualify and, in most cases, minimise use, thus supporting major reductions in meat eating and forest felling.

Many philosophers defend instrumentalism with the argument that instrumental thinking confining direct moral concern to humans is not completely incompatible with certain types of conservation ethic; it does not automatically mean that non-humans can be entirely ignored or eliminated, since creatures like bats may be seen as having some sort of secondary value for humans, for example, as pollinators or pest controllers, or if rare, as items of scientific or aesthetic interest (Passmore, 1974; Norton, 1991). Instrumentalists usually assume that (and occasionally try to show that)[10] non-humans can get by on this secondary, lower grade of consideration. In an age of mass extinction, when we humans are driving so many other species from the planet, there is good reason to doubt this, and aiming for a higher grade of concern does not seem overly ambitious. This instrumental defence also neglects the ties of instrumentalism to the larger political framework of annexation and commodification, its deadening impact on recognition and conceptual frameworks, and the spiritual impoverishment arising from its fundamentally monological engagement with the world. Nevertheless, the "enlightened self-interest" instrumentalist approach remains the default setting for most contemporary environmental discussion, especially in the sciences.

Critics point out that the "secondary consideration" argument amounts to a statement of a "humans first" claim in the slightly variant terminology of primary and secondary consideration, that it supports human supremacism and bases ethics on a human species version of selfishness (in the qualified form often dubbed "enlightened self-interest") (Plumwood, 1993, 2002). Alternative ethical frameworks and forms of rationality that do not thus prioritise the self and deprioritise the ethical other are available and need to be considered, especially in our dire ecological situation.

INTRINSIC VALUE

There are several complementary arguments that instrumentalism is inevitable that trade on similar confusions and ambiguities, often about the meaning of the contrast term *intrinsic value*. Although there is some variation in terminology, similarly employing weaker and stronger senses of what is normally meant by "intrinsic",[11] the meaning of "intrinsic" adopted here is simply that the item is of value for itself, or is a final end. In terms of value theory, it is valued for its own sake, rather than as a means or route to something else, for example, that it has some sort of value that is not to be reduced to use to promote human ends – to instrumental value. Again, this does not have to mean that the item is unavailable for use, but that its value is not exhausted by its (human) use. In this weak sense of intrinsic value, it implies a rejection of instrumentalism and its corollary: the contraction of reasons (for or against an undertaking) to human reasons.

Some critics of the concept of intrinsic value have identified the concept with objective value, or with value that is somehow detached from all human perspectives and exists objectively in the world.[12] They argue that objective or "detached" concepts of value are incoherent. All value must be human, since generated by human valuers, and only humans or the human like can therefore be the direct source of value. If it suits humans to count the natural world valuable, this can only be indirectly, because it serves their interests. What environmental advocates have to show, on this view, is that conservation of the non-human world is in the human interest, that it will benefit humans more than the alternative. Only reasons of human benefit, short or long term, can count, and no other reasons or arguments are rational. For an intrinsic value position, on the other hand, a much wider range of conditions and a much larger range of beings can provide reasons and count morally.

There is a whole family of similar arguments, which really are variations on the main argument for egoism, or for what might be seen as a sort of "human egoism". They are all based on questionable assumptions. First, we should not grant the premise that it is only human preferences and perspectives that generate value, since animals at least can surely be said to value their lives – that is in many ways the whole point of being an animal and being able to move away from danger. The main point though is that even if valuers are limited to humans (which is highly questionable) this does not mean that humans can only value themselves, or that only humans can make or be the locus of value. Similar arguments would demonstrate that individual valuers can only value their own self-interest, that males can only place value on other males, Europeans value only their own people or culture, etc. – in short, that egoism, sexism and racism are inescapable.

Philosophers have deployed a number of ingenious arguments to reject the idea of "intrinsic value" for nature. These have taken up a lot of discussion space, and have usually involved redefining the concept of "intrinsic" in a very strong sense. The assumption that intrinsic implies objectivity, for example, is taken by Thompson to include the demand, incompatible with recognising value relationality, that any informed person would make the same valuation. The problems of these strong, more problematic senses of "intrinsic" have been the main focus of discussion. Largely ignored have been the normal, weak senses of "intrinsic" where what is involved is simply the distinction, familiar from classical philosophy, between things considered to be of value in themselves (directly) and those valued indirectly, for the sake of or as a means to something else taken to be valuable. That is, intrinsic value is simply what is allowed to be a stopping point in a chain of means–ends reasoning, relative to some public who may accept or contest this point of reference.

The major flaw with all arguments for instrumentalism focusing on the difficulties of strong senses of intrinsic value, after we have cleared up all these conceptual tangles, ambiguities, misunderstandings and repositionings, emerges clearly if we consider this normal, weak sense of intrinsic value as the final end or stopping point for an instrumental chain. It is that ultimately every value-positing position must admit intrinsic values in this weak sense, and that the weak sense is sufficient to carry the argument against instrumentalism. All indirect or instrumental values ultimately require a direct or intrinsic value, some final end or reason to which they are the means, a stopping point to the question "to what end?", "useful to who?" There is, plainly, an infinite regress involved in assuming that all values can be instrumental. This regress is usually broken by the assumption, too general, deep and protected for conscious contestation, that humans are the natural, "intrinsic" stopping point, the ones who matter. But if intrinsic values as stopping points in ethical reasoning must be recognised at some point by every consistent value theory, there cannot be a fundamental difficulty for interspecies ethics with the concept of intrinsic value itself. If there is such a difficulty (which has not been clearly demonstrated) it is not peculiar to an environmental ethics and is not evaded by "normal" intra-human ethics.

Interspecies and environmental ethics that decline to limit ethical consideration to the human sphere do not require some completely new or outlandish concept of intrinsic value intra-human ethics somehow avoids. Intra-human ethics needs the same basic concept of instrumental and intrinsic value as interspecies ethics. What is new in the idea of intrinsic value for non-humans is not the concept of intrinsic value or stopping point itself, *but the challenge to the assumption that humans and only humans can act as stopping points*, the deeply rooted background assumption that only humans matter, and that everything else is available for their use, at their disposal. And despite the heavy focus on various strong senses of the terms instrumental and intrinsic (such as "inherent") in much of the published argument, all that is really needed to establish the ethical case against instrumentalism is the weak sense of reducing the ethical other to usefulness to the human self or species: as we will see in the section "Anthropocentrism and Sustainability", this sense is quite sufficient for showing that instrumentalism is unjust and unethical, as well as impoverishing and limiting for both instrumentaliser and instrumentalised.

After instrumental frameworks have been rejected, the way is cleared to try to develop an environmental ethic as if non-human nature mattered. The next thing on the agenda should be some self-examination, to understand why and how we have been seduced by the instrumental worldview, and how it has come about that modernity has identified instrumentalism with rationality. An important component of ethics is to try to understand our biases, how and why we have reduced and misrepresented the more than human world, especially since the problem we have identified seems to lie at least in part in reductionism and the ethics of recognition. The conception of the human is a key issue here, since humanness is treated in the dominant culture as the crucial divide between those it is and is not permissible to instrumentalise. On examination, we may find that the supposed "obviousness' or "intuitiveness" of limiting ethics to the human or the sentient simply reflects the powerful historical cultural influence of traditions of human centredness and human/nature dualism.

INSTRUMENTALISM AND HUMAN APARTNESS

Some philosophers, ecofeminists especially, argue that the intrinsic/instrumental debates, which have become stereotypical in environmental philosophy, are framed in too narrow terms, and try to address the issues in a larger framework (Warren, 2000; Plumwood, 2002). For them, abstract value and the status of the non-human are too limited as a problem focus; value is only one part of ethics, one only too easily co-optable by economic theory. These critics of contemporary environmental philosophy argue that human instrumentalism is just one strand in a larger web of ethical assumptions concerned with human identity and the human place in the Earth community, assumptions like human–nature dualism and human-centredness. Value questions in interspecies ethics are therefore best addressed in the context of analysing the frameworks of human and male domination, which have consequences for non-humans as well as for humans themselves. Ecofeminists see enlightening parallels between interspecies (human domination of other species) and intra-human domination, especially racial/ethnic and gender domination. For example, one of the most problematic aspects of the instrumentalist default is its simplistic, polarised division of the world into human means and non-human ends. These forms of division have strong parallels in the oppressive forms of division we create between a variety of "us" and "them" groups within human society, including men and women.

Human/nature dualism typically operates with a highly polarised "us"/"them" understanding in which the human and non-human spheres correspond to two quite different substances or orders of being in the world. In the mechanistic model, these orders are thought of as minds and machines. Mechanistic worldviews especially deny nature any form of agency of its own. This supports instrumentalism, since if the non-human sphere is empty of its own purpose, it is appropriate that the human coloniser impose his or her own purposes. Since nature itself is thought to be outside the ethical sphere and to impose no moral limits on human action, we can deal with nature as an instrumental sphere, provided that we do not injure other humans in doing so.

Human/nature dualism is a key, linking part of the network of culture/nature, spirit/matter, mind/body and reason/nature dualisms that have shaped Western culture, and is an active force in contemporary life. As a Western-based cultural formation going back thousands of years, it sees the essentially human as part of a radically separate order of reason, mind or consciousness, set apart from the lower order that comprises the body, the woman, the animal and the pre-human.[13] Inferior orders of humanity, such as women, slaves and ethnic Others ("barbarians"), partake of this lower sphere to a greater degree, through their supposedly lesser participation in reason and greater participation in lower "animal" elements such as embodiment and emotionality. Human/nature dualism conceives the human as not only superior to but as different in kind from the non-human, which is conceived as a lower non-conscious and non-communicative purely physical sphere that exists merely to aid the higher human one. The human essence is not the ecologically embodied "animal" side of self, which is best neglected, but the higher disembodied element of mind, reason, culture and soul or spirit. The polarities of dualism we encounter here are mutual or double sided, and are maintained by the logical structure of hyper-separation, an emphatic form of separation feminists have identified which denies kinship and continuity, often overemphasising differences and distorting both sides. The denial of non-human minds is matched by the denial of human embodiment. Human/nature dualism is a system of thought where the human and non-human are seen in highly separated (indeed hyper-separated) terms as part of different, hierarchically related, categories or orders of being. Often, overestimation of the virtues and abilities of the human side is matched by a reductionist stance that systematically underestimates the non-human side, resulting in systems of beliefs and methodologies about the world that elevate humans to godlike status. A reductionist stance minimises ethical constraints and thus maximises what is available for commodification and unconstrained use as "resources".

Thus, non-humans are taken to be naturally inferior, and seen as lacking the qualities that are supposed to confer ethical mattering on humans – mind, rationality, individuality, etc. The attribution of mindlike characteristics to non-humans is ridiculed and condemned as "anthropomorphism", although the assumption that such qualities are confined to the human lacks any good empirical or logical basis. It is one thing to claim that non-humans should not be presented in inappropriately humanised form (as if they were exactly like humans, penguins as "little people in black and white suits") and another to conclude that this excludes mentalistic properties, for this conclusion requires the assumption that only humans have such properties. But this is exactly what is at issue. The charge of anthropomorphism, when deployed to restrict vocabularies of recognition to the human, is usually[14] entirely circular, being little more than a covert and scientised way to enforce mechanistic and reductionist stances towards non-humans.

Reductive concepts of non-humans, like those of slaves, naturalise their supposed inferiority – they make it seem natural that the other is treated as inferior. Via surrounding narratives of mind and body, instinct and choice, science, rationality, sustainability, human and non-human, user and used that restrict even the vocabulary of mindfulness and moral sensibility to humans, human supremacy is naturalised. The opportunity to correct dominant mythologies by direct experience is rarely available for most people, because, as with the slave, the limited contexts in which modernity encounters non-human others are structured by colonising and belittling relationships with them. An insulated urban majority only rarely encounters situations where experience can disrupt the dominant myths of nature's muteness, blindness and inferior sensibility, or shake the Cartesian reduction to machinery. Reductionism, as Weston (1996) notes, is self-validating, because closed minds cannot experience nature in its fullness and because nature is often reduced to fit impoverished conceptions of it.

Animals, the methodology of reductionism assumes, are mere bodies. Similarly, the slave was assumed to be naturally inferior, and often cast as mere body, while true initiative, creativity and prestige was reserved for the higher level of mind exemplified in the master. Because the slave lacks appropriate qualities of mind (according to Greek apologists for slavery such as Aristotle), and is also seen as lacking agency and individuality, slavery is necessary; the slave requires a master who has the rationality and agency the slave lacks. The slave, like the animal, has no rights of his/her own, but is a transparent mediator of the master's projects, maximally available for production, like the battery hen.

Where did this curious narrative of higher souls or spheres come from? Ancient Greek society was androcentric, anthropocentric and built on slavery. Its democracy was confined to the privileged 30% of the population allowed to vote – the elite males who saw themselves as representing the higher sphere of spirit, mind and reason. The remainder, women and slaves, were identified with material, bodily labour and led highly separated, confined and devalued lives. Correspondingly, in ancient Greek philosophy, the earthly world of materiality and embodiment is seen as not only inferior but also corrupting, and those who leave it behind on death pass to a higher and purer realm of immateriality beyond the earth. Plato's philosophy (a precursor of Christianity) is dematerialising, treating reason – lodged in a pure realm of immaterial, timeless ideas – as opposed to or threatened by the corrupted material world of "coming to be and passing away", the biological world of nature and the body. In the ascent to a better world of spirit beyond earthly, embodied life, matter, death and decay would ultimately be conquered by the opposing elements of spirit and reason in a process of dematerialisation.

This ideology of dualism and human apartness can be traced down through Western culture through Christianity and modern science. With modernity, reason as modern science began to rival and replace religion as the dominant belief system. Western science replaced but also built on this earlier religious foundation, transforming the idea of conquering nature as death by subordinating nature to the realm of scientific law and technology. Modern science, now with religious status, has tended to inherit and update rather than supersede these oppositional and supremacist ideals of rationality and humanity. In the scientific fantasy of mastery, the new human task becomes that of remoulding nature to conform to the dictates of reason to achieve salvation – here on Earth rather than in Heaven – as freedom from death and bodily limitation. This project of controlling and rationalising nature, "the empire of man over mere things" as Descartes termed it, has involved the technological–industrial conquest of nature made possible by reductionist science and the geographical conquests of empire.

The idea of human apartness from the sphere of animality emphasised in culture, religion and science was, of course, shockingly challenged by Charles Darwin in his argument that humans evolved from non-human species. But the insights of continuity and kinship with other life forms that exploded the apartness dogma were the real scandal of Darwin's thought. They remain only superficially absorbed in the dominant culture, even by scientists. The modern scientific project of technological control is justified by continuing to think of humans as a special superior species, set apart and entitled to manipulate and commodify the earth and other species for their own exclusive benefit. Ignoring the evidence that other animals such as birds are just as evolved as humans, human-centred culture assumes that humans are the apex of creation, more intelligent, more communicative, more important and much more evolved that other species (Midgley, 1983; Rogers, 1997).

This traditional sense of human apartness is challenged to some degree by the new science of ecology that stresses the importance of bio-sphere services and ecological processes, and the dependence of humans, like other animals, on a healthy biosphere and a thriving more-than-human, material world. But the influence of dualistic traditions continues in a false consciousness and mode of life which fails to situate human identity, human life and human places in material and ecological terms and which treats consciousness, rather than ecological embodiment, as the basis of human identity and society.[15] The ideology of human apartness has been functional for Western culture in enabling it to colonise the non-human world and so-called "primitive" cultures with little constraint, but it also creates dangerous illusions in denying embeddedness in and dependency on nature, which are implicated in our denial of human inclusion in the food web and in our poor response to the ecological crisis.

ANTHROPOCENTRISM AND SUSTAINABILITY

Our assumption that non-humans are here for our benefit (a strong form of instrumentalism) and that only humans matter is part of a dominant, default worldview which unduly devalues the non-human in areas such as justice, ethics and value. This worldview is human-centred (or anthropocentric) and its misunderstandings of human identity as, for example, above and 'outside' nature, pose risks to human as well as non-human survival. Human supremacism in its strongest forms refuses ethical recognition to non-humans, treating nature as just a resource we can make use of however we wish. It sees humans, and only humans, as ethically significant in the universe, and derives those limited ethical constraints it admits on the way we can use nature and animals entirely indirectly, from harms to other humans. More subtle kinds of human-centredness can appear even among the environmentally concerned, for example, the way dominant versions of environmental philosophy (Singer, 1974, 1998) limit ethical consideration to higher animals and exclude other kinds of beings, giving higher value to those that are *most like* humans (Plumwood, 1998; Warren, 2000; Weston, 2004). Anthropocentric worldviews rely on the historical tradition of human/nature dualism to deny kinship and empathy, endorsing the treatment of nature as radically other, and humans as emphatically separated from nature and from animals. Nature is a hyper-separate lower order lacking continuity with the human, and those features which make humans different from nature and animals, rather than those they share, are stressed as constitutive of a truly human identity. Anthropocentric culture thus leads to a view of the human as outside of and apart from a plastic, passive and "dead" nature which is conceived in reductionist and mechanical terms as completely lacking in qualities such as mind and agency that are seen as exclusive to the human. A strong ethical discontinuity is created around the human species boundary. An anthropocentric culture will tend to adopt concepts of what makes a good human being which reinforce this discontinuity by devaluing those qualities of human selves and human cultures it associates with nature and animality in the human self, and often also to associate with nature inferiorised social groups and their characteristic activities, real or supposed. Thus, women are historically linked to "nature" as reproductive bodies and through their supposedly greater emotionality, and the colonisers' indigenous people are seen as a primitive, "earlier stage" of humanity. At the same time, dominant groups associate themselves with the overcoming or mastery of nature, both internal and external, and the management of colonised groups. For all those classed as nature, as Other, identification and sympathy are blocked by these structures of Othering.

Not only concepts of kinship but concepts of difference are also distorted by the human-centred model. Nature and animals are defined as lack in relation to the human, stereotyped as lacking individuality. They are amorphous and alike in their lack of reason, mind and consciousness, qualities withheld from the non-human sphere through the enforcement of a hyper-separated "rational" vocabulary that pictures them as machines or automata. Recognition of the respects in which non-humans are superior to humans is rare. Assimilative and instrumental models promote insensitivity to the marvellous diversity of nature, attending to differences in nature only if they are likely to contribute in some obvious way to human interests, conceived as hyper-separate from those of natural species and systems. Epistemic and ethical frameworks of commodification lead to a serious underestimation of the complexity and irreplaceability of nature. Thus, scientists assume their own genetically engineered replacements for natural species and varieties are always superior, although they may only be superior in terms of a narrow range of qualities relevant to human use, and have not been tested by survival over a large range of conditions nearly as rigorously as naturally evolved varieties.

Understanding anthropocentrism helps to explain why, for example, nature has been reduced to commodity functions, taken for granted and starved of resources for its own maintenance. Human-centred culture represents nature as inessential and invisible, as the unconsidered background to technological society, its needs systematically omitted from consideration. When dependency on nature is denied, systematically, nature's order, resistance and survival requirements are not perceived as imposing a limit on human goals or enterprises. Correspondingly, hegemonic accounts of human agency that background nature as a collaborative co-agency feed hyperbolised concepts of human autonomy and independence of nature. The popular saying "With every human mouth, a pair of hands", for example, assumes that nature is completely elastic, imposing no ultimate constraints on human expansion. Thus, crucial biospheric and other services provided by nature and the limits they might impose on human projects are neglected in accounting and decision-making.

Denial is often accomplished via a perceptual politics of what is worth noticing, of what can be acknowledged, foregrounded and rewarded as "achievement" and what is relegated to the background. Women's traditional tasks in house labour and childraising provide an example, being treated as inessential, as the background services that make "real" work and achievement possible, rather than as achievement or as work themselves. With nature, backgrounding goes to a further stage of disappearance – we "lose track of it", in a world of growing remoteness, often in culture-wide ways, even when it is performing essential services for us. This process of backgrounding provides an important part of the answer to the why questions with which we started this section, why and how have we come to do something so

dangerous as to take our ecological support systems for granted and to starve them of the resources they need.

Sustainability is a project aimed at countering the rapid exhaustion of our planet's resources for life at the expense of the future. Sustainable development is often defined as development that "meets the needs of the present without compromising the ability of future generations to meet their own needs" (Brundtland, 1989). This refers to the needs of *future humans*. This would make sustainable development a matter of ecological rationality and *environmental justice* – distributive justice for future people. So environmental justice is the fundamental underlying concept for sustainable development,[16] where the concept of environmental justice has been employed especially to raise questions about the distribution of environmental risks, harms and benefits among human populations (Bullard, 1994).[17] We will discuss this concept of justice further in the last section of this chapter. Most philosophers now concede that we do have obligations to future people, and that questions about our impacts on the environments of the future must be considered.

Our obligations towards the future are often assumed to arise only in relation to distributive justice, but other concepts of justice also have application in relation to the generations of the future. The main distributive issues here are whether the consumption and life patterns of the present generation should be allowed to inflict serious environmental risks and costs on the people of the future, either by depriving them of resources they would benefit from that previous generations have enjoyed, or by leaving them a legacy of pollution, impoverishment or other environmental damage. This issue is particularly serious where losses are irreversible, as in the case of species extinction, land degradation, loss of biodiversity and nuclear-waste production. The concept of environmental justice for future generations thus raises questions that overlap with questions of sustainability.

One suggested answer is that the ethical position of people who are removed from us in time has much in common with that of people who are removed from us in space (Routley and Plumwood, 1978). If justice is to be done, the impact of environmental policies on future generations must be considered and given due weight, but the main problem that arises in the case of both temporal and spatial removal is that of uncertainty. In some cases, there is uncertainty about what the needs of future people will be, whether they will be the same as or very different from our own. But even if future people and their needs are very different, this does not excuse neglect of future impacts, since however much the future may change socially and technically, the basic needs of future generations *for a healthy biosphere* are unlikely to be substantially different from our own.

The main problem with the narrow concept of sustainability outlined in Brundtland (1989) is that it is too human-centred and involves an overly exclusive focus on future humans, ignoring the needs of non-humans, present and future. Environmental justice concerns are not confined to the human and have an interspecies distributive aspect. We need to share the Earth not only with future humans but with other species – including difficult and inconvenient ones such as snakes, wolves, bears and crocodiles. Interspecies distributive justice asks us to provide adequate habitat for species life and reproduction, objecting to the use of so much of the Earth for exclusively human purposes that non-humans cannot survive or reproduce their kind. We must recognise not just human but also non-human needs as part of the concept of sustainability, indeed central to it. Indeed, we should consider sustainability as being primarily about sustaining the systems and processes, especially the ecological ones, that sustain all life, and will sustain future human and non-human ones. Thus, its most basic meaning involves sustaining human societies by sustaining the fundamental non-human ecosystems and services they rely on.

Obviously sustainability in this sense will be difficult to achieve if these sustaining systems are invisible to us, if we see these fundamental enabling systems as replaceable, tradeable "resources", or if we have embraced an economic system that operates by assuming the continued supporting role of systems that provide materials, clean up wastes and regulate climate but minimises the resources allocated to maintain them. Clearly, too, genuinely sustainable relationships with service providers cannot be systems that allocate merely minimum resources for providers' well-being or survival. This rules out instrumental, servant or slave-like relations as well as competitive market relations, to name a few of those that define rationality so as to encourage cost cutting at the provider's expense. An ecological rationality must be one where ecological providers are, at a minimum, reliably sustained and strengthened, and not subject to the forms of minimisation, denial and forgetting of creativity, agency and contributions characteristic of hegemonic relationships and monological rationality. Monological relationships are thus ecologically irrational, because they lead to distorting, hegemonic forms of recognition of agency that eventually weaken the provider. Requirements of sustainability rule out monological slavelike relationships and select for dialogical relationships of mutual adaptation and dialogue between mutually recognising and supporting agents.[18]

It is not unreasonable, therefore, to conclude that instrumentalism, dualism and anthropocentrism are limiting and desensitising bases for both ethical and ecological relationships. Instrumental outlooks distort our sensitivity to and knowledge of nature, blocking humility, wonder and openness in approaching the more-than-human, and producing narrow types of understanding and classification that reduce nature to raw materials for human projects. The epistemic and moral dualisms associated with human-centredness are harmful and limiting, even in their subtler and weaker forms. People under their influence, such as those from the Western cultural traditions in which anthropocentrism is deeply rooted, develop conceptions of themselves as belonging to a superior sphere apart, a sphere of "human" ethics, technology and culture dissociated from nature and ecology. This self-enclosed outlook has helped us to lose touch with ourselves as beings who are not only cultural but also as natural, embedded in the earth and just as dependent on a healthy biosphere as other forms of life. Thus, we come to take the functioning of the "lower" sphere, the ecological systems which support us, entirely for granted, needing some grudging support and attention only when they fail to perform as expected. (The dualist assumption that ethics and value are exclusively concerned with and derived from the human sphere is an ethical expression of this same project of human supremacy and self-enclosure.[19]) Although these ethical exclusions have in the modern age helped Western culture achieve its position of dominance, by maximising the class of other beings that are available as "resources" for exploitation without constraint, they are now, in the age of ecological limits, a danger to the survival of human and planetary life.

DEEP OR SHALLOW? THE HUMAN PRICE OF HUMAN-CENTREDNESS

Given that they tend to encourage the ecological equivalents of "delusions of grandeur", human-centred worldviews are in conflict with the requirements of human sustainability and thus with general and long-term human welfare. Clearly, it is a major argument against human-centred worldviews that they distort our conception of ourselves in such potentially dangerous ways. In advancing this standpoint or recognition argument, however, we strike a difficulty; this type of argument is ruled out in the classifications that have come to dominate environmental thinking. Many of the radical critiques of human

centredness in environmental ethics have historically classified environmental positions as deep or
shallow (Naess, 1973) or in a slightly more complex classification, as light green (shallow), mid-
green and deep green, depending on whether their
concern is directed to humans or non-humans.
(Sylvan and Bennett, 1994; Curry, 2006) Thus it is
often claimed that a concern is anthropocentric or
"shallow" if it is concerned with human welfare or
human self-interest (Curry, 2006). A "deep" position in contrast would be one making no reference
to considerations of human welfare, but considering only non-humans, concentrating on defending
"the existential interests of other life forms"
(Hay, 2002).[20]

This way of classifying theories is confusing
and problematic for practical environmental
activism because it assumes a major disconnec-
tion or polarisation between human and non-
human issues. If only non-human concern counts
as deep, this means that truly "deep" green arguments cannot be concerned with human consequences such as sustainability. But for practical
activism this is disastrous. Many, perhaps most,
environmental issues involve both humans and
non-humans, often in connected ways that are
hard to disentangle. Typical environmental struggles are concerned *both* with situating human lives
or living places ecologically and also with winning more security for non-human life and places.
Global issues like climate change and overfishing
have major implications for both human and non-
human species. A local activist aiming at defending her local place may work on improving human
energy, sewerage and garbage systems in that
place to reduce pollution, on conservation of
shoreline habitat for endangered shorebirds, and
on forest protection for the benefit of both the
human environment (e.g. water supplies) and that
of the forest itself (including but exceeding forest-
dwelling animals). When they are so entangled in
the defence of place, it seems unreasonable to
condemn one set of concerns as "shallow" and
laud the other as "deep". This polarised way of
drawing the distinction drives a wedge between
human and non-human issues and marginalises
connections between them. It obliges us to condemn as "shallow" struggles concerned with situating human life in ecological terms, while issues
of wilderness and the defence of non-humans are
treated as "deep", and are set apart from human
environmental justice and sustainability issues.
The environmental problematic becomes identified with just one side of it, compassion for other
life forms.[21]

There can be a useful point in the deep/shallow
distinction, depending on interpretation, but
this polarised way of understanding it creates
many difficulties, because instead of offering an

alternative integrating the human and non-human,
this form of classification tends to promote the
*reverse polarity to anthropocentrism, a reduction
to non-human issues.*[22] It implies that we cannot
avoid "shallowness" or human-centredness unless
we avoid concern with the welfare of humans,
including future humans. This is like saying that
one is egocentric unless one completely ignores
or sacrifices one's own interests and acts only
from pure concern for the other. Such a criterion
equates prudence with selfishness, and makes it
much too hard to avoid being counted as egocentric. Similarly, any requirement that human interests be ignored brings out a sensible and prudent
concern with human fate as anthropocentric, making almost any reasonable position human-
centred. Conversely, theorists inclined to justify
anthropocentrism often make the same assumption that prudence means a selfish and exclusive
concern with human interests, concluding that
only a human-centred ethic is fully rational (or, as
we have seen, possible) and that concern with the
fate of whales or with nature for its own sake is
irrational.[23] But there is a major false dichotomy
here in the implicit choice between human and
non-human interests and needs, between prudence
and altruism, care for self and care for the other.
It is prudent for humans, as ecological beings, to
be concerned with both human and non-human
fates and with their interrelationship.

Ecofeminist perspectives that draw on feminist
theory can emphasise the double-sided character
of the problem created by anthropocentrism and
other hegemonic centrisms. Systems of domination of race and gender distort the identity and fate
of both sides, dominator and dominated, often to
the detriment of both. Thus, traditions of male-
centredness that form the identity of both men and
women create limitations for both genders,
although, of course, the weight of the disadvantages falls to women. Thus, men can be limited
by the demand that they leave emotionality to
women, just as women can be limited by their
supposed exclusion from rationality and extra burdens of emotionality. Human-centredness similarly is double sided, since the advantage it gives
humanity in exploiting the non-human world is
matched by the potentially fatal disadvantage of
misunderstanding human ecological identity and
dependency. A conceptual reduction or devaluation of the Other that licences "purely instrumental" relationships can distort our perceptions and
enframings, impoverish our relations and make us
insensitive to limits, dependencies and interconnections. These conceptual frameworks are a
direct hazard to the Other, but are in turn often an
indirect prudential hazard to self. The treatment of
concern with human issues as "shallow" prevents
a double-sided focus on anthropocentrism as a

problem, not only for non-humans but for humans too, as a factor which prevents us situating ourselves as ecological beings and makes us insensitive to ecological dependencies and interconnections. Thus, some ways of drawing the deep/shallow distinction would brand as "shallow" any attempt to reflect on the dangers an anthropocentric course poses to humans – which is one of the deepest and most useful things we can do.

Given a fuller account of human-centredness (see preceding section), we can discern a degree of integration and radical challenge in the environment movement as a whole that both deep and shallow ecology and environmental ethics miss. Because *both* human and nature sides of the dualism are affected by hyper-separation, resolving the dualism gives rise to two distinct but intertwined projects, the project of situating human life ecologically and the project of situating non-human life ethically and culturally. It is the first concern with situating human life itself in ecological terms that motivates the important and familiar range of environmental struggles for liveability and sustainability some environmentalists neglect and disparage as "shallow". An analysis which notes the way the agency of the non-human world has been rendered invisible can also help to explain our current mode of denial, which simple frameworks of identification cannot do. Because they challenge different aspects of a deeply entrenched conceptual structure of denial, both human and non-human struggles can be subversive and provoke resistance. If Aldo Leopold (1949), who formulated an ethic of sharing the land with non-humans, is "deep", why isn't Rachel Carson, who also reminded us of our denied inclusion in ecosystems as animals?

The first of the two tasks thrown up by human/nature dualism, that of situating human life ecologically, has often been taken to be the more urgent and self-evident, the one of prudence or sustainability, while the other task of situating non-humans ethically is presented as optional, as supererogation, the inessential sphere of altruism. But this is an error; the two tasks are interconnected and cannot be addressed properly in isolation from each other. Human and non-human struggles and ethics are thoroughly interlinked because when we hyper-separate ourselves from nature and reduce it conceptually (in order to justify domination), we not only lose the ability to empathise and to see the non-human sphere in ethical terms, but also get a false sense of our own character and location that includes an illusory sense of agency and autonomy – which are thus in turn a prudential hazard.

We can resolve some of these difficulties of polarisation by being more careful about the meaning of deep and shallow, as well as of anthropocentrism.

The "deep/shallow" distinction could still have some use in a reformulated problematic, following through the common meaning of "deep" as radically or thoroughly challenging conventional dogma – as, for example, giving a "deep" challenge to human-centredness. But thus interpreted, the distinction cannot mark a polarised division between human and non-human concerns, or automatic privilege one type of concern over the other. Some forms of non-human concern can be decidedly "shallow" or non-radical, for example, those that are guided by human apartness and human supremacist assumptions.[24] A focus on situating human lives and settlements ecologically might be very challenging to accepted ways of thinking about both the human and the non-human. On our new definition, *a position that refused to take ethics beyond the human would be considered shallow*, but a concern with human welfare would not necessarily itself be shallow unless it refused to go beyond the human. Both human and non-human issues can be approached in ways that challenge, or, alternatively, support human-centredness. We do not have to place human and non-human allegiances in competition or ignore their connections, and we can understand the environmental problematic as a problem for both the human and non-human spheres. Similarly, we do not have to choose between valuing individuals and valuing ecological wholes, or between valuing animals and valuing trees.

We can avoid deep/shallow polarisation by reinterpreting the depth of a position in terms of the degree of challenge it offers to the structures of anthropocentrism outlined in the preceding section. Since human-centredness distorts the conception of both sides of what it splits apart, human and non-human, "deep" challenges to human-centredness can occur on both sides of the human/non-human divide. The attempt to situate human life and settlements ecologically can offer a "deep" challenge to the historical dualism that denies the ecological aspects of human life, and understands "nature" as exclusive of the human home of "culture". The feminist-inspired historical narrative sketched in the section "Instrumentalism and Human Apartness" links the human and non-human sides of the problem and gives us a different perspective on what it is about environmentalism that is radical or challenging to the weight of cultural tradition. This approach, based on giving a careful analysis of anthropocentrism, enables us to see what current divisions of the field have obscured, that *the environmental problematic is double sided*, with denial of our own embodiment, animality and inclusion in the natural order being the other side of our distancing from and devaluation of that order. Human hyper-separation from nature establishes a discontinuity based on denying both the human-like aspects of nature and the nature-like aspects of the human, as the denial of the sphere of "nature" within the human matches the devaluation and denial of nature without. The key insight here, as Rachel Carson, the great pesticide campaigner, understood in the 1960s, and the work of Mary Midgley and ecofeminist Rosemary Ruether suggested in the 1970s, is that the resulting conception of ourselves as ecologically invulnerable, beyond animality and "outside nature" (as a separate and pure sphere which exists "somewhere else") leads to the failure to understand human ecological identities and dependency on nature, a failure that lies behind so many environmental catastrophes, both human and non-human.

Under the old criterion of depth, in which consideration of human costs is inevitably "shallow", it is not possible to raise consistently the question of how far human-centredness is a disadvantage to humans themselves. Such a question about the human price of human shallowness would inevitably be classed as itself shallow. But given our new criterion of depth, that of challenging human-centredness in its full sense, we can consistently consider this important set of concerns without incurring the charge of shallowness, or more precisely, of endorsing or supporting anthropocentrism. Contrary to the assumption that human-centredness is the only prudent or rational course, more detailed consideration of its nature and impacts supports what those still stuck in a polarised framework would see as the paradoxical conclusion that human-centredness is not in the interests of humans or non-humans, that it is even dangerous and irrational. It is not shallow then to conclude that human-centredness can have severe costs for humans as well as non-humans.

COUNTER-HEGEMONIC INTERSPECIES ETHICS

Leaving behind the human-centred default, new thinking in environmental ethics has been exploring a variety of ethical approaches to interspecies relationships that can be broadly characterised as counter-hegemonic – that is, aimed at countering the human-centred structure we have discussed in the fifth and sixth sections. The very heavy, and often exclusive, emphasis academic philosophy gives abstract and formal questions of value is an impoverished approach for issues and contexts that call for a wider and richer range of specific ethical approaches such as virtue ethics, care ethics, solidarity and friendship ethics, ecological and food web ethics of reciprocity. Working from a broader conception of ethics not centred on rational principle, we can focus instead on concepts such as care, communication and negotiation with the non-human other, for whom *ecological justice* demands a better share of the world (Walker, 1995, 1998; Warren, 2000; Plumwood, 2002; Weston, 2004).

Primary concepts of justice as giving others their due, and as distributional and proportional justice, are not confined to intra-human relationships, and have an application to the non-human sphere and interspecies relationships. An important concept of injustice as "prejudice" is concerned with the impediments to justice presented by prior reductive or oppressive conceptions of the other, as in colonialism, racism and sexism, and this concept of justice has, I shall argue, a clear application to the non-human sphere. The denial of concepts of justice to the non-human sphere, which is thus treated ethically as "Other", is itself a form of injustice. Thus, the reductive and Othering modes of conceiving the non-human that are characteristic of human-centredness are both unjust in themselves, and herald other forms of injustice, such as distributive injustice. They stop us conceiving non-human others in ethical terms, distort our distributive relationships with them, and legitimate insensitive commodity and instrumental approaches.

We must take much more seriously concepts of distributive justice for non-humans, as not inferior or lower in priority to human justice issues. Interspecies distributive justice principles should stress the need to share the earth with other species (including difficult and inconvenient ones like snakes, crocodiles, bears – animals that are predators of humans or animals under human protection) and provide adequate habitat for species' life and reproduction. Distributive injustices to non-humans fostered by the Othering framework include the use of so much of the Earth for exclusively human purposes that nonhumans cannot survive or reproduce their kind. They include also rationalistic farming systems that reduce the share of resources allowed to Earth others to the minimum required for productive survival, such as veal crate or battery egg systems. Just remedies for these oppressive and unjust distributions involve measures for a fairer sharing of the Earth with other species. Assigning more land to Earth others, whether in the form of areas exclusively for their use (as in some wilderness areas and national parks), prioritising their welfare in many multiple-use areas and requiring human behaviour to adjust, and encouraging more non-human use of exclusively human areas like cities and suburbs are examples. These are all matters of interspecies justice that are jeopardised by human-centredness.

In the context of strong centric traditions, counter-hegemonic strategies can be seen as corrective. These are ethical stances which resist distorting centric constructions, helping to counter the influence of the oppressive ideologies of domination and self-imposition that have formed our conceptions of both the other and of ourselves. A positive programme concerned to counter human-centredness would suggest some practical virtues and practices to combat specific aspects of its structure and its politics of instrumentalism, exclusion and denial. Instrumentalism, for example, can be countered by respectful practices of acknowledging and giving full value to the excess over and above usefulness to us. Human apartness (hyper-separation) can be countered by openness to kinship and continuity with the non-human. Since how we treat others is very much a function of epistemic relations, of how we see them, both at the conceptual and empirical level, we need to enter the field of moral epistemology. We counter reductionism, denial and backgrounding with recognition, and its corollaries in virtue ethics, appreciation, gratitude and generosity. Approaches in virtue ethics stress the importance of the more modest and particularistic virtues and values, such as attentiveness, surprise, generosity and respect, which already have a foundation in good ethical practice toward nature (see also Plumwood, 1993).

Feminist philosophers have identified the "universalist/impersonalist tradition" as part of their problem. The excessive concern with abstraction and universal principle to the exclusion of more relationship-based forms of ethics is, not coincidentally, a major reason for the inadequacy of moral philosophy for feminist concerns (Held, 1995). Areas of ethics undervalued or suppressed in andro/anthropocentric treatments of ethics that stress the rational abstract and universal include the ethics of special relationships, "contextual and narrative" aspects of ethics (Gilligan, 1987, p. 141), virtue ethics and moral epistemology, for example, the ethical requirements in certain contexts for attention and openness (Walker, 1995) and for patient and just discernment. An alternative feminist paradigm emphasises an ethics which is a "lattice of similar themes – personal relationships, nurturance and caring, maternal experience, emotional responsiveness, attunement to particular persons and contexts, sensitivity to open-ended responsibilities" (Walker, 1995, p. 140). "This view does not imagine our moral understandings [as] congealed into a compact theoretical instrument of impersonal decision, [such as rights or value] but as deployed in shared processes of discovery, expression, interpretation, and adjustment between persons" (Walker, 1995, p. 140).

Important resources for such adjustment and for recognising and sustaining relationships are the communicative virtues and rationalities

(Dryzek, 1990). The virtues, writes Anthony Weston, are "those traits that sustain and deepen relationship", while Carol Gilligan speaks of morality as arising "from a recognition of relationship", "a perception of the need for response".[25] These counter-hegemonic virtues include philosophical stances and methodologies that maximise our ethical sensitivities to other members of our ecological communities and openness to their agency; they are antagonistic to stances of reduction, superiority and scepticism that minimise the kinds of beings to whose agency and communicative potential we are open. Communicative ethics suggests a further way to interpret the concept of respect: as being able and willing to hear the other, to encounter them dialogically as presences, as positively-other-than, as subject rather than object (and not just in terms of economically cooptable concepts such as value) and to consider their welfare, as both individuals and communities, in all we do.

Counter-hegemonic stances include especially communicative virtues:

- recognising continuity with the non-human to counter dualistic construction of human/nature difference as radical discontinuity;
- reconstructing human identity in ways that acknowledge our animality, decentre rationality and abandon exclusionary concepts of rationality;
- acknowledging difference, nonhumans as "other nations", as "positively-other-than", including a non-hierarchical conception of more-than-human difference;
- decentring the human/nature contrast to allow a more inclusive, interspecies ethics;
- dehomogenisation of both "nature" and "human" categories;
- openness to the non-human other as potentially an intentional and communicative being (the intentional recognition stance);
- listening to the other (attentiveness stance);
- active invitation to communicative interaction;
- redistribution, generosity stance;
- ethical consideration without closure directed towards an excluded class;
- non-ranking stance minimising interspecies ranking and ranking contexts;[26]
- "studying up" in problem contexts (self-critical stance);
- negotiation, a two-way, mutual adjustment stance;
- attention to the other's complexity, outrunning of our knowledge.

Perhaps the most important task for human beings is not to search the stars to converse with cosmic beings but to learn to communicate with the other species that share this planet with us. The possibility of interspecies communication is of course just as contested as that of interspecies ethics. Attempts at serious communication between humans and other species are almost completely precluded by the arrogance and human-centredness of a culture that is convinced that other species are simpler and lesser, and only grudgingly to be admitted as communicative beings. Methodology based on these assumptions more or less guarantees that communication will not take place. Or alternatively, that when it does occur, it takes place on exclusively human terms such that the non-human species is required to learn a human language but not vice versa. This arrangement severely disadvantages the non-human party and allows us to confirm our delusion that other species are inferior. Studies of animal communication usually focus on teaching relatives such as primates human language. The real communication challenge at this level of interspecies communication is for we humans to learn to communicate with other species on their terms, in their own languages, or in common terms, if there are any.

The politics of communicative concepts is one of conflict between tendencies to try to shrink and opposing tendencies to expand the extension of the concept.[27] Shrinkage strategies suit those motivated by the desire to maximise the category of exploitable resources, to maximise what is available for unrestrained and reductive forms of use. Reducing or minimising the category of potential communicants licences forms of use that are unconstrained by considerations of the other's well-being, that are unreflective (because as object or resource the other does not need to be given an account of) and reductive, because the less the other is perceived to be, the less the perceived injustice in their treatment as reducible to mere commodities.

Most people have had some experience of communication with animals even if they do not call it by that name. Nevertheless, communicative acts, models, projects and virtues and concepts of communicability have application and virtue in relation to a much wider group of communicative beings than animals. The category of potential communicants and the concept of communication and communicability can spread out very widely beyond the human to take in not only living inhabitants of the Earth and of space, but also places, experiences, processes, encounters, projects, virtues, situations, methodologies and forms of life. In a dialogical methodology, the other is always encountered as a potentially communicative other. To treat the other as a potentially intentional and communicative being and narrative subject is part of moving from monological modes of encounter (such as those of anthropocentrism) to dialogical modes of encounter. Communicative models of relationships with nature and animals can improve our receptivity and responsiveness, which clearly need much improvement. They seem likely to offer us a better chance of survival in the difficult times ahead than dominant mechanistic models which promote insensitivity to the others' agency and denial of our dependency on them. This clash of models is critical for our times.

NOTES

1 We use the term "anthropoform" here as a synonym for "humanoid" and to replace the seriously confusing and problematic term "anthropomorphism", which has become the chief legitimiser of the reductionist agenda.

2 This has of course been the argument of a number of would be refuters of environmental ethics – including Passmore (1974), Grey (1993) and Thompson (1990).

3 Not to be confused with "utilitarianism", although this doctrine does assume a strong separation of conscious beings from non-conscious beings, the latter allocated to the sphere of ends, often on the grounds that "it makes no difference". The outcome is usually an ethical extensionist and semi-reductionist approach that includes a few "higher" non-humans ethically on the basis of their similarity to humans, for example, their capacity to suffer (Singer, 1998).

4 This is why defences of instrumentalism, such as those of Passmore (1974) and Thompson (1990), focus on the question of whether any environmental ethics is possible (or necessary).

5 See, for example, Grey (1993) and Hayward (1995, 1998).

6 On the concept of excess see Smith (2001).

7 So one form of Kant's Categorical Imperative reads: "Act always so as to treat humanity, whether in yourself or in another, as an end or never merely as a means" (Weston, 2001, p. 90).

8 Nevertheless, many philosophers who defend instrumentalism have not (see, e.g. Hayward, 1998).

9 Although in philosophy there is another important sense of "utilitarianism" as the philosophy of Bentham and Mill advocating "the greatest happiness for the greatest number".

10 For example, Norton (1991) claims to prove that an instrumental ethic will have the same consequences as a non-instrumental one, so it will all come out the same.

11 Thus for example Callicott (1984) introduces the term "inherent" for "intrinsic", thus reinforcing the impression that an objective value is posited.

12 Most notably Thompson (1990).

13 See especially Plumwood (1993), Spelman (1988) and Lloyd (1984).

14 There are some cases where it raises a useful point about respecting non-human difference, but

this is not its normal use. See the discussion in Plumwood (2002, pp. 56–61).

15 Thus contemporary, globalising culture understands "our place" in accord with mind/body dualism, as our conscious homeplace, nation or identity group, rather than in terms of the many global places we draw upon in the global market for the material support for our lives, especially the places of our ecological footprint.

16 On the case against identifying questions of justice with questions of distribution see especially Young (1990) and Warren (2000). These objections, although in the main well taken, do not seem to me to disturb the claim that distributional issues are important and in certain contexts seriously neglected.

17 As one of its major theorists, Robert Bullard, puts it, environmental justice raises questions of "differential exposure and unequal protection … the ethical and political questions of 'who gets what, when, why, and how much'".

18 For more details on this argument see Plumwood (2002).

19 The concept of human self-enclosure is discussed in Abram (1996). Self-enclosure is an important aspect of human-centredness.

20 "Deep" positions are often said also to be ecocentric (or concerned with ecological wholes rather than with the fates of individuals, whether human or non-human, although Curry (2006) allows them to be concerned with individuals as well).

21 Thus, in his first chapter introducing the "ecological impulse" and motivating the movement, "deep" environmental historian Peter Hay proceeds immediately to set up the paradigm of environmental activism as non-human and wilderness defence. "The cornerstone of the environment movement", he writes (p. 25), "may well be the impulse to defend …the existential interests of other life-forms". The ethical and ecological failures involved in other kinds of environmental struggles emerge as peripheral, and are assumed not to challenge major traditions or norms (to be "shallow" or "not radical"). Unsituated in any larger historical and social context, these struggles are treated as semi-technical problems of sustainability that can be solved in terms of better political and economic organisation.

22 We certainly see this reverse reduction in Leopold's criterion (1949) of the ethical as what promotes the flourishing of ecosystems.

23 See, for example, Flannery (2003). When you get to the discussion of the "real issues" (sustainability) it is something of an anti-climax to discover that after all these have not been neglected because our attention has been distracted by whales and koalas but because the market values that have been allocated to ecosystem maintenance by economic rationalists ignore environmental services and environmental flows.

24 Examples might be positions that invariably privilege human pets like cats or dogs over other animals, or which treat human pets as the paradigm of animality.

25 Weston (2001, p. 9), Gilligan (1987, p. 120).

26 Non-ranking is the way I would interpret biospheric egalitarianism.

27 A low redefinition eases criteria for class membership so that too many items fall under it, whereas a high redefinition tightens them so that too few items qualify. On high and low redefinitions see Wittgenstein's *Investigations* (1953).

REFERENCES

Abram, David (1996) *The Spell of the Sensuous*. Pantheon, New York.

Brundtland, Gro Harlem (1989) *Our Common Future*. World Commission on Environment and Development (also known as the Brundtland Report).

Bullard, Robert (1994) *Unequal Protection: Environmental Justice and Communities of Color*. Sierra Club Books, San Francisco.

Callicott, J. Baird (1984) "Non-anthropocentric Value Theory and Environmental Ethics". *American Philosophical Quarterly* 21: 299–309.

Carson, Rachel (1962) *Silent Spring*. Houghton Mifflin, New York; Penguin Books, Harmondsworth (1999).

Curry, Patrick (2006) *Ecological Ethics: an Introduction*. Polity Press, Cambridge.

Dryzek, John (1990) "Green Reason: Communicative Ethics for the Biosphere". *Environmental Ethics* 12: 195–210.

Flannery, Tim (2003) *Beautiful Lies: Population and Environment in Australia*. Quarterly Essay Issue 9, pp. 1–73.

Gilligan, C. (1987) "Moral Orientation and Moral Development" in Feder Kittay, E. and Meyers, D. (eds) *Women and Moral Theory*. Rowman and Littlefield, Totowa, NJ.

Grey, W. (1993) "Anthropocentrism and Deep Ecology" *Australasian Journal of Philosophy* 71(4): 463–475.

Hay, Peter, (2002) *Main Currents of Environmental Thought*. UNSW University Press, Sydney.

Hayward, Tim (1995) *Ecological Thought: An Introduction*. Polity Press, Cambridge.

Hayward, Tim (1998) *Political Theory and Ecological Values* Polity Press, Cambridge.

Held, Virginia (1993) *Feminist Morality*. University of Chicago Press, Chicago.

Held, Virginia (1995) *Justice and Care: Essential Readings in Feminist Ethics*. Westview Press, Boulder, CO.

Leopold, Aldo (1949) *A Sand County Almanac*. Oxford University Press, Oxford.

Lloyd, Genevieve (1984) *The Man of Reason*. Methuen, London.

Midgley, Mary (1980) *Beast and Man*. Methuen, London.

Midgley, Mary (1983) *Animals and Why They Matter*. University of Georgia Press, Athens, GA.

Naess, Arne (1973) "The Shallow and the Deep, Long-range Ecology Movement: A Summary". *Inquiry* 16(1): 95–100.

Norton, Brian (1991) *Towards Unity Among Environmentalists*. Oxford University Press, New York.

Passmore, John (1974) *Man's Responsibility for Nature*. Duckworth, London.

Plumwood, Val (1993) *Feminism and the Mastery of Nature*. Routledge, London.

Plumwood, Val (2002) *Environmental Culture: the Ecological Crisis of Reason*. Routledge, London.

Rogers, Lesley J. (1997) *Minds of Their Own: Thinking and Awareness in Animals*. Allen and Unwin, Sydney.

Routley, R. and Plumwood, Val (1978) "Nuclear Energy and Obligations to the Future" *Inquiry* 21: 133–79.

Ruether, Rosemary R. (1975) *New Woman, New Earth*. Seabury Press, Minneapolis.

Singer, Peter (1974) *Animal Liberation*. Basic Books, New York.

Singer, Peter (1998) "Ethics Across the Species Boundary" in Nicholas Low (ed), *The Global Ethics of Environmental Justice*. Routledge, London, pp.146–157.

Smith, Mick (2001) *An Ethics of Place: Radical Ecology, Postmodernity, and Social Theory*. State University of New York Press, Albany, NY.

Spelman, Elizabeth (1988) *The Inessential Woman*. Beacon, Boston, MA.

Sylvan, Richard and Bennett, David (1994) *The Greening of Ethics*. White Horse Press, Cambridge.

Thompson, Janna (1990) "A Refutation of Environmental Ethics". *Environmental Ethics* 12: 147–160.

Walker, Margaret U. (1995) "Moral Understandings: Alternative 'Epistemology' for a Feminist Ethics" in Held, V. (ed), *Justice and Care: Essential Readings in Feminist Ethic*. Westview Press, Boulder, CO, pp.139–152.

Walker, Margaret U. (1998) *Moral Understandings: A Feminist Study in Ethics*. Routledge, New York.

Warren, Karen J. (2000) *Ecofeminist Philosophy: A Western Perspective on What it is and Why it Matters*. Rowman and Littlefield, New York.

Weston, Anthony (1996) "Self-Validating Reduction: Toward a Theory of Environmental Devaluation". *Environmental Ethics* 18: 115–132.

Weston, Anthony (1998) "Universal Consideration as an Originary Practice". *Environmental Ethics* 20: 279–289.

Weston, Anthony (2001) *A 21st Century Ethical Toolbox*. Oxford University Press, Oxford.

Weston, Anthony (2004) "Mult-centrism: A Manifesto". *Environmental Ethics* 26(1): 25–40.

Wittgenstein, Ludwig (1953) *Philosophical Investigations*. Macmillan, New York.

Young, Iris (1990) *Justice and the Politics of Difference*. Princeton University Press, Princeton, NJ.

18

Biocultural Diversity and Sustainability

Luisa Maffi

INTRODUCTION

Conventional approaches to environmental conservation have tended to consider the role of humans only or mostly in terms of the threats that the intensification of human extractive and transformative activities poses for the environment. From this perspective, finding solutions to environmental problems largely means seeking to put a halt to those activities by "taking human hands off" what is seen as the last remaining pristine environments on the planet (Terborgh, 1999). Underlying this perspective is a philosophical view that depicts humans as external to, and separate from, nature, and interacting with it mostly in an effort to establish dominion over it (Eldredge, 1995). Complementarily, nature is seen as separate from humans and as existing in a primordial, "virgin" state unless and until they are encroached upon by humans.

That the exponential increase in the pace and scale of human activities has come to constitute the prime threat to the environment is undeniable – both through the direct effects of extraction and transformation of natural resources, and through the indirect effects of these activities (such as global climate change). It is now widely recognized that we have entered an era in which massive species extinctions, habitat deterioration, and loss of ecosystem functions are all due principally to human intervention (Millennium Ecosystem Assessment, 2005). From the 1980s onwards, however, several paradigms have challenged the philosophical perspective described above, presenting a different view of human relationships

with the environment and thus of the relationships between the state of the environment, the threats or pressures on it, and the response options to counter or alleviate the threats. Three will be mentioned here in particular.

In the natural sciences, the field of ecosystem health (Rapport, 1998, 2007) embraces a "humans-in-environment" approach which, while acknowledging that the global commons are severely imperiled by human action, takes as its main goal to address Aldo Leopold's challenging question: how can we humanly occupy the Earth without rendering it dysfunctional? In the social sciences, the field of biocultural diversity (Maffi, 2001a, 2005) – drawing from anthropological, ethnobiological, and ethnoecological insights about the relationships of human language, knowledge, and practices with the environment – takes as its fundamental assumption the existence of an "inextricable link" between biological and cultural diversity. And in the realm of policy, the sustainable development paradigm that emerged in the 1980s proposes that the key to sustainability resides in balancing three "pillars:" environment, society, and economy (Bruntland, 1987). The documents spawned by the 1992 Rio Summit on Environment and Development (Rio Declaration, Agenda 21, Convention on Biological Diversity) also recognize the relevance of traditional environmental knowledge for the conservation of biodiversity.

This chapter reviews the field of biocultural diversity, its history and main contributions thus far, as well as the gaps and needs for future research and application. It then explores this field's relationships to the idea of sustainability and points to its current and future opportunities to contribute to achieving a sustainable world.

SOME HISTORY AND DEFINITIONS

The idea of an "inextricable link" between biological and cultural diversity was perhaps first expressed in those terms in the 1988 Declaration of Belém of the International Society of Ethnobiology (http://ise.arts.ubc.ca/declareBelem.html). Several decades of ethnobiological and ethnoecological work had accumulated evidence about the depth and detail of indigenous and local knowledge about plants and animals, habitats, and ecological functions and relations, as well as about the low environmental impact, and indeed sustainability – historically and at present – of many traditional forms of natural resource use. The evidence also pointed to a variety of ways in which humans have maintained, enhanced, and even created biodiversity through culturally diverse practices of management of "wild" resources and the raising of domesticated species (such as the use of fire, protection and dissemination of culturally important "wild" species, agroforestry, horticulture, animal

husbandry, etc.). This countered the image of "pristine" environments, unaffected by humans, that could be brought back to their "original" state by fencing them off to protect them from human activity. (See, e.g. Heckenberger *et al.*, 2003, for recent evidence about the anthropogenic nature of even parts of so-called "virgin" tropical rainforests.)

These findings suggested the following conclusion: the sum total and cumulative effect of the variety of local interlinkages and interdependencies between humans and the environment worldwide means that at the global level biodiversity and cultural diversity are also interlinked and interdependent, with significant implications for the conservation of both diversities. Pioneering global cross-mappings of the distributions of biodiversity and linguistic diversity (taken as a proxy for cultural diversity as a whole) provided independent support for this conclusion, revealing significant geographic overlaps between the two diversities, especially in the tropics, and a strong coincidence between biologically and linguistically megadiverse countries (Harmon, 1996). Map 1 shows some of these correlations, with a focus on endemism in both languages and higher vertebrate species (languages and species only found within the borders of an individual country).

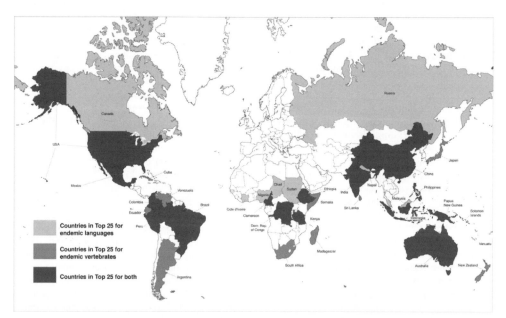

Map 1   Endemism in language and higher vertebrates: comparison of the top 25 countries
Source: Harmon (1996), based on data from Groombridge (1992) (pp. 139–141, for species) and Grimes (1992) (for languages). Figures for Ethiopia include Eritrea. Higher vertebrates include mammals, birds, reptiles, and amphibians; reptiles not included for USA, China, and Papua New Guinea because the numbers were not reported in the source table.

Interest in this topic grew through the 1990s, drawing from a variety of sources in the natural, social, and behavioral sciences, humanities, applied sciences, policy, and human rights. Out of these converging interests, a new field of research and applied work has emerged that has been labeled "biocultural diversity" (Posey, 1999; Maffi, 2001a, 2005; Harmon, 2002; Stepp *et al.*, 2002; Carlson and Maffi, 2004). This label is actually a short form for "biological, cultural, *and linguistic* diversity." Proponents of this field argue that the diversity of life is comprised not only of the variety of species and cultures that have evolved on earth, but also of the variety of languages that humans have developed over time. This approach also highlights the role of language as a vehicle for communicating and transmitting cultural values, traditional knowledge and practices, and thus for mediating human–environment interactions and mutual adaptations. (On the specific aspect of linguistic diversity, see Maffi, 1998, 2001b; Maffi *et al.*, 1999; Harmon, 2002.)

Although the theoretical and methodological bases of the field of biocultural diversity are still being refined, and an explicit, agreed-upon conceptual framework has not been fully worked out yet, it is possible to glean some definitions and key elements based on how the concept has been generally used by its proponents. Biocultural diversity may be defined as follows:

Biocultural diversity comprises the diversity of life in all of its manifestations: biological, cultural, and linguistic, which are interrelated (and possibly coevolved) within a complex socio-ecological adaptive system.

The above definition comprises the following key elements:

1 The diversity of life is made up not only of the diversity of plants and animal species, habitats, and ecosystems found on the planet, but also of the diversity of human cultures and languages.
2 These diversities do not exist in separate and parallel realms, but rather they interact with and affect one another in complex ways.
3 The links among these diversities have developed over time through mutual adaptation between humans and the environment at the local level, possibly of a coevolutionary nature.

A possible representation of these complex relationships at different scales is suggested by the diagram in Figure 18.1.

Taken together, the above assumptions raise important issues of history, pattern, and causality: how have the links among diversities developed and changed over time, how are the relationships

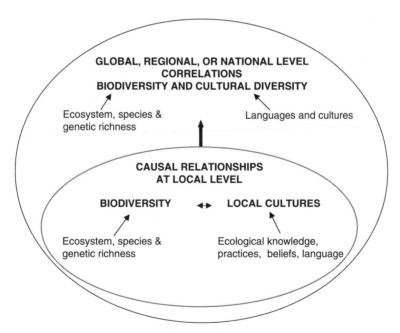

Figure 18.1   Relationship between national/regional/global correlations of cultural and biological diversity and causal relationships between cultures and biodiversity at the local level
Source: original figure by Ellen Woodley, 2005.

manifested today, and how does one form of diversity affect the others? Is local biodiversity, at least to some extent, culture specific? These assumptions also raise significant issues of scale and levels of analysis: how do these diversities and the relationships among them manifest themselves and play out at different degrees of resolution, from the local to the global, and how are patterns and processes connected across scales? This stresses the need for, on the one hand, in-depth studies of the global distributions of biological, cultural, and linguistic diversity, both currently and over time; on the other, detailed case studies of the links between the environment and language, cultural beliefs, knowledge, and practices at regional and local levels.

A corollary of the definition of biocultural diversity is that the trends in biological, cultural, and linguistic diversity are also interrelated, potentially with mutual beneficial or detrimental effects. This corollary points to the importance of having tools, or indicators, to measure and compare the state and trends in these diversities globally and regionally, in order to assess whether the current and historical status of one is mirrored by the current and historical status of the others. Work carried out from a biocultural perspective during the past decade has sought to tackle some of the questions and needs mentioned above. A review of the key literature follows, along with a discussion of some of the gaps that call for further study.

ADVANCES AND GAPS IN BIOCULTURAL DIVERSITY RESEARCH

The main lines of work that have so far been developed in this field can be grouped under three headings: global and regional analyses of the correlations between linguistic and biological diversity; tools for measuring and assessing the state of biocultural diversity; and studies about the persistence and loss of biocultural diversity.

Harmon's initial work on global biodiversity–linguistic diversity correlations (Harmon, 1996) pointed to several large-scale biogeographic factors that might account for these correlations, in that they might comparably affect the development of both biological and linguistic diversity (such as extensive land masses with a variety of terrains, climates, and ecosystems; island territories, especially with internal geophysical barriers; tropical climates, fostering higher numbers and densities of species). In addition, Harmon hypothesized a process of coevolution of small human populations with their local ecosystems. Such a process would have developed over time, as humans interacted closely with the environment,

modifying it as they adapted to it and developing specialized knowledge of it, as well as specialized ways of talking about it. Thus, the local languages, through which this knowledge was encoded and transmitted, would in turn have become molded by and specifically adapted to their socioecological environments.

On the other hand, Mühlhäusler (1996) called attention to the fact that linguistic and cultural distinctiveness can develop also in the absence of mutual isolation: for example, among human groups who belong to the same broadly defined cultural area (i.e. groups sharing many cultural traits), or whose languages are considered to be historically related or to have undergone extensive mutual contact, and who occupy the same or contiguous ecological niches. Such circumstances – high concentrations of linguistically distinct communities coexisting side by side in the same areas and communicating through complex networks of multilingualism – appear to have occurred frequently throughout human history (Hill, 1997), and still exist today in many parts of the world (the Pacific being a prime example). This points to the role of sociocultural factors, along with biogeographic factors, in the development of linguistic diversity.

Numerous other researchers, using databases of the world's languages or the world's cultures, have sought to correlate the global or regional distribution of linguistic or cultural diversity with both environmental and social factors (Nichols, 1990, 1992; Chapin, 1992 [2003]; Mace and Pagel, 1995; Wilcox and Duin, 1995; Nettle, 1996, 1998, 1999; Oviedo *et al.,* 2000; Lizarralde, 2001; Smith, 2001; Collard and Foley, 2002; Moore *et al.,* 2002; Manne, 2003; Skutnabb-Kangas *et al.,* 2003; Sutherland, 2003; Stepp *et al.,* 2004, 2005). Some of the same geographic and climatic factors, such as low latitude, higher rainfall, higher temperatures, coastlines, and mountains, have been repeatedly identified as positively correlated with both high linguistic diversity and high biological diversity. Higher latitudes, plains, and drier climates tend to correlate with lower diversity in both realms.

One of the social factors that have been invoked to account for these patterns is the difference in modes of subsistence (more localized versus ranging over larger territories) influenced by how geography and climate affect the carrying capacity of a given area and access to resources for human use. Ease of access to abundant resources found in place seems to favor localized boundary formation and diversification of larger numbers of small human societies (and languages). Where resources are scarce, the necessity to have access to a larger territory to meet subsistence needs favors smaller numbers of widely distributed

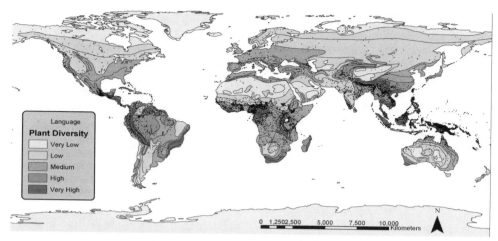

Map 2 Plant diversity and language distribution
Source: from Stepp *et al.* (2004), based in part on data from Barthlott *et al.* (1999).

populations (and languages). The development of complex societies and large-scale economies, which tend to spread and expand beyond their borders, has also been found to correlate with a lowering of both linguistic and biological diversity. There is a significant overlap between the location of threatened ecosystems and threatened languages (Skutnabb-Kangas *et al.*, 2003). On the other hand, low population density, at least in tropical areas, seems to correlate positively with high biocultural diversity. Map 2 shows the overlap of global language distribution and plant biodiversity zones.

It is important to note that, while the correlations in the distribution of biological and linguistic/cultural diversity show clear patterns at the global level, analysis at smaller scales reveals significant variation from region to region and sometimes presents a mixed picture in terms of the patterning of these diversities. While Central and South America, West and Central Africa, South and Southeast Asia, and the Pacific consistently stand out as "hotspots" of biocultural diversity, even in these regions these correlations may occasionally weaken or even disappear when "zooming in" at higher degrees of resolution.

Researchers involved in these studies point to the need for finer-grained analyses that will be more sensitive to the role of local biogeographic and sociocultural factors in producing deviations from global patterns of diversity. More detailed studies on a regional scale will improve our ability to identify correlations and mutual influences and perhaps to discern causal factors that link cultural diversity with biodiversity and affect the development, maintenance, and loss of biocultural diversity. Especially needed is a historical

perspective both on processes of environmental change and on human population movements and expansions and other social, economic, and political factors that may have affected the location and numbers of human populations and their relationships with and effects on the environment. Other critical issues highlighted by this research are the need for a better understanding of how environmental factors may similarly or differentially affect cultural groups and species, and the role of scale and degree of resolution in the analysis of biodiversity–cultural/linguistic diversity correlations. Advances in the use of GIS as a research tool promise to propel this agenda forward in new and insightful ways (Stepp *et al.*, 2004, 2005).

Issues of scale and level of analysis also arise in another realm of the field of biocultural diversity, that related to the development of indicators for the joint measurement and assessment of global conditions and trends of biodiversity and cultural diversity. The earliest efforts to develop such tools were carried out by David Harmon in the early 1990s (Harmon, 1992). Indicators of biodiversity were by then commonly used to monitor the state of the natural world. Harmon set out to identify indicators that might allow for gauging the state of cultural diversity in relation to the state of biodiversity, and thus for determining whether cultural diversity is indeed diminishing – as various reports were suggesting (e.g. Krauss, 1992; Miller, 1993) – and whether it is diminishing in tandem with biodiversity. He proposed a number of aspects of culture for which indicators might be developed: from language, ethnicity, and religion to diet, crops, land management practices, medical practices, social organization, and forms of artistic expression.

The ability to develop indicators depends, of course, on the availability of reliable data sets on the entity to be measured, at whatever scale is appropriate. In order to match biodiversity data on a global scale, therefore, the choice for cultural indicators had to fall on those aspects of culture for which global data exist: languages (Grimes, 2000; now see Gordon, 2005), and ethnicities and religions (Barrett *et al.*, 2001). In a collaborative effort, Harmon and Loh (Harmon and Loh, 2004; Loh and Harmon, 2005) developed the blueprint for an Index of Biocultural Diversity (IBCD), whose purpose is to measure the condition and trends in biocultural diversity on a country-to-country basis (the level at which the available data sets are organized). This is accomplished by aggregating data on the three cultural indicators with data on the diversity of bird/mammal species and plant species as indicators for biodiversity (also selected on the basis of global data availability). The IBCD features three components: a "biocultural diversity richness" component (IBCD-RICH), which is the sheer aggregated measure of a country's richness in cultural and biological diversity; an "areal" component (IBCD-AREA), which adjusts the indicators for a country's land area and thus measures biocultural diversity relative to the country's physical extent; and a "population" component (IBCD-POP), which adjusts the indicators

for a country's human population and thus measures biocultural diversity in relation to a country's population size. For each country, the overall IBCD then aggregates the figures for these three components, yielding a global picture of the state of biocultural diversity in which three areas emerge as "core regions" of exceptionally high biocultural diversity: the Amazon Basin, Central Africa, and Indomalaysia/Melanesia (see Map 3). This largely confirms the geographical correlations found in other work reviewed above, in which only either languages or ethnicities were used as proxies for cultural diversity.

Harmon and Loh point to a number of limitations of the IBCD and caveats concerning its use. They make it clear that this index, like any index, should only be used to measure general conditions and trends and should not be expected to provide an in-depth analysis of the phenomenon at hand, particularly as concerns within-country variation in biocultural diversity. They also point out that, in its current version, the IBCD only provides a "snapshot" of the state of biocultural diversity at the beginning of the 21st century, while data on trends are as yet missing and are the object of future research (see below). They conclude that these latter data, used in conjunction with careful qualitative analyses, will ultimately provide a more adequate and accurate picture of the global state of biocultural diversity. At the same time,

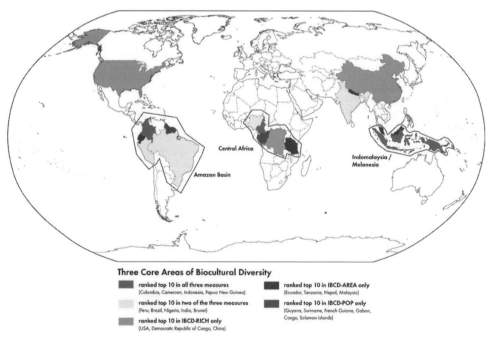

Three Core Areas of Biocultural Diversity

■ ranked top 10 in all three measures
(Colombia, Cameroon, Indonesia, Papua New Guinea)

□ ranked top 10 in two of the three measures
(Peru, Brazil, Nigeria, India, Brunei)

■ ranked top 10 in IBCD-RICH only
(USA, Democratic Republic of Congo, China)

■ ranked top 10 in IBCD-AREA only
(Ecuador, Tanzania, Nepal, Malaysia)

■ ranked top 10 in IBCD-POP only
(Guyana, Suriname, French Guiana, Gabon, Congo, Solomon Islands)

Map 3   The "core areas" of global biocultural diversity
Source: Loh and Harmon (2005).

they acknowledge that the main value of such an index will be largely practical and political, such as to raise awareness about biocultural diversity among decision-makers, opinion-makers, and the general public, and promote needed action for its protection and restoration.

It is in fact notable that the Convention on Biodiversity (CBD) has recently set a goal to develop indicators to monitor progress toward its "2010 Target" of significantly reducing the loss of biodiversity by the year 2010 (Balmford *et al.*, 2005), and among these indicators has included one of the status and trends of linguistic diversity. This is because the CBD has within its mandate the protection and promotion of indigenous and local knowledge relevant to the conservation of biodiversity. Global data on the status and trends of traditional knowledge do not currently exist. Because of the close link between language and knowledge, the status and trends of languages is to be used in this context as a proxy for the status and trends of traditional knowledge. Efforts are now underway to develop time-series data on linguistic diversity, as well as the methodology for a locally appropriate, globally applicable indicator directly focused on trends of retention or loss of traditional environmental knowledge (TEK) over time (Terralingua, 2006). Both of these indicators might contribute to the needs of the CBD and other stakeholders, and will illuminate one of the basic questions in the field of biocultural diversity: is the world's cultural diversity indeed in decline, and, if so, how fast? Correlated with time-series data on biodiversity, these new indicators will also show whether trends in cultural diversity and biodiversity mirror each other.

Cultural indicators that provide valuable information at global, regional, or country levels need to be complemented by others that will be useful in elucidating the relationships among language, culture, and the environment at local scales. Indicators that will be relevant here include ones for measuring persistence and resilience of institutions for knowledge transmission and language vitality, livelihoods and subsistence practices, resource use and management practices, land and resource tenure, social organization and decision-making capacity, and so forth. The development of such indicators is still in its infancy, but there is a growing interest in this endeavor in both academic and international circles.

Very relevant in this connection is some of the recent quantitative work carried out by ethnobiologists to measure the retention and loss of TEK. Researchers such as Zent (1999, 2001), Lizarralde (2001), Ross (2002), Zarger and Stepp (2004), Zent and López-Zent (2004), and others are contributing to the development of quantitative methods for the investigation of the acquisition and transmission of ethnobotanical and ethnoecological knowledge and for the identification of factors (such as age, formal education, bilingual ability, length of residency, change in subsistence practice, and so forth) that may affect the maintenance or loss of TEK. Also, an expert group on language endangerment and language maintenance gathered by UNESCO has put forth a set of recommendations for the assessment of linguistic vitality (UNESCO, 2003) that should provide useful guidance also for the development of linguistic diversity indicators. They point out that sheer trends in "language richness" (number of different languages) are not a fully adequate indicator of the state of languages. Better data on numbers of speakers over time and other sociolinguistic "vital statistics," particularly on intergenerational language transmission, contexts of use, availability of mother tongue education, and so forth, will be needed for this purpose. A methodology has recently been developed for testing linguistic vitality at the local level and identifying the factors (such as age, gender, special roles, etc.) that affect linguistic ability (Florey, 2006).

Last but not least, explaining the links between language, knowledge, beliefs, practices, and the environment at the local level also requires delving into indigenous and other local societies' understandings of human–environment relationships. A view of humans as part of, rather than separate from, the natural world is pervasive in indigenous societies, and so is the perception of a link between language, cultural identity, and land (rather than an abstract notion such as "nature;" see e.g. Blythe and Brown, 2004). It is no surprise, then, that many of the most explicit efforts to jointly maintain and revitalize cultural resilience, linguistic vitality, and biological diversity are grassroots efforts, whether entirely endogenous or promoted and assisted by national and international organizations. Learning about these worldviews and efforts and making the lessons as widely available as possible is one of the goals of ongoing work in biocultural diversity (Maffi and Woodley, in preparation).

Such studies of the factors of persistence or loss of local languages and traditional knowledge and practices, and of how these dynamics relate to the maintenance or erosion of biodiversity locally, are increasingly needed to ground the conceptual framework of the biocultural diversity field. They will significantly help address the many open questions of causality, history, scale and levels of analysis, as well as means of representation, measurement, and assessment, that confront this field. Undoubtedly, they will also contribute to a new generation of biocultural studies at the global level, injecting bottom–up data into what so far has been largely, due to the nature of available data, a top–down approach. This should in turn promote a better understanding of the nature, state, and trends of biocultural diversity, and thus foster the development of better policies in support of biocultural diversity nationally and internationally.

Ultimately, the most fundamental impetus for the protection and maintenance of biocultural diversity can only come not from top–down efforts, but from the ground–up action of indigenous and other societies worldwide whose languages, cultural identities, and lands are being threatened by national and global forces. Wilhelm Meya, Director of the Lakota Language Consortium (Meya, 2006), puts it eloquently, speaking of the situation in his country, the USA – but his words have universal applicability:

> In the same way that a healthy planet requires biological diversity, a healthy cultural world requires linguistic diversity. Yet, language is also an elaborate phenomenon tied to real people and cultures. Language loss threatens a fundamental human right – that of expression of the life and life ways of a people.

Each language relates ideas that can be expressed in that language and no other. Thus, when an indigenous community is no longer allowed to pray, sing, or tell stories in its language, it is denied a fundamental human right. Unfortunately, linguistic rights have been seriously abused for hundreds of years by banning specific languages and indirectly by assaulting language-support structures such as land, economies, and religions.

> ... Languages today are the next frontier in setting the country [indeed the world] into moral and environmental symmetry (Meya, 2006).

In this connection, the field of biocultural diversity has not adopted the conventional academic "neutrality." From its inception, it has embraced a strong ethics and human rights component, and has promoted a vision in which the protection of human rights (both individual and collective) is intimately connected to the affirmation of human responsibilities toward and stewardship over humanity's heritage in nature and culture (Skutnabb-Kangas, 2000; Maffi, 2001c; Posey, 2001; Harmon, 2002). In this view, the biocultural diversity of life has intrinsic value, as diversity is the expression of life's evolutionary potential, and it ought to be protected and maintained. Any damage to it ought to be remedied, and any further damage ought to be prevented. This requires a complex but necessary, and ultimately winning, balance between nature conservation and human development, and between the rights of nature and

the rights of humans. The ethical stance of the field of biocultural diversity posits that, when doubts arise about potential damage to the web of biocultural diversity, a precautionary approach ought to be taken.

BIOCULTURAL DIVERSITY FOR SUSTAINABILITY

This ethical component aligns the biocultural diversity field with other paradigms such as ecosystem health and sustainable development in terms of a shared perception of the relevance of seeking directly to influence policy and public opinion. This approach gives these paradigms a characteristic mixture of theory and practice, research and advocacy, knowledge production and knowledge dissemination. The common goal, of course, is to help achieve (or recover) a sustainable world for the sake of future generations.

From the perspective of biocultural diversity, a sustainable world means a world in which not only biological diversity, but also cultural and linguistic diversity thrive, as critical components of the web of life and contributing factors in the vitality, organization, and resilience of the ecosystems that sustain life. Harmon (2002) points to the interwoven (and possibly coevolved) diversity in nature and culture as the "preeminent fact of existence," the basic condition of life on earth. The continued decrease of biocultural diversity, he warns, would "staunch the historical flow of being itself, the evolutionary processes through which the vitality of *all* life has come down to us through the ages" (Harmon, 2002, p. xiii).

Others have similarly stressed the evolutionary significance of diversity not only in nature but also in culture and language, as a way of "keeping options alive" for the future of humanity and the Earth (Maffi, 1998, 2001a). Bernard (1992, p. 82) has suggested that "[l]inguistic diversity ... is at least the correlate of (though not the cause of) diversity of adaptational ideas" and that therefore "any reduction of language diversity diminishes the adaptational strength of our species because it lowers the pool of knowledge from which we can draw." Mühlhäusler (1995, p. 160) has argued that convergence toward majority cultural models increases the likelihood that more and more people will encounter the same "cultural blind spots" – undetected instances in which the prevailing cultural model fails to provide adequate solutions to societal problems. Instead, he proposes, "[i]t is by pooling the resources of many understandings that more reliable knowledge can arise;" and "access to these perspectives is best gained through a diversity of languages" (p. 160). Along similar lines, Krauss (1996) has proposed that global linguistic diversity constitutes an intellectual web of life, or "logosphere," that envelops the planet and is as essential to human survival as the biosphere – a concept of course reminiscent of Teilhard de Chardin's "noosphere" and of the classic notion of the Logos. Over the past decade, international organizations concerned with the conservation of biodiversity and cultural heritage have begun to listen. In particular, UNESCO, the United Nations Environment Program (UNEP), the CBD, and IUCN – The World Conservation Union, have variously included in their priorities and action plans the improved understanding of the links and synergies between biological diversity and cultural diversity, as well as the role of culture and traditional knowledge and local languages in the conservation of biodiversity. We are perhaps approaching a stage at which it will be recognized that the traditional "three-legged stool" of sustainable development – environment, society, and economy – should be turned into a four-legged one by the addition of a fourth "pillar" culture. These are significant achievements for a field like that of biocultural diversity, that is barely more than a decade old. Yet, there is no singing victory. The momentum may be building in some quarters, but the global political will to act to protect and restore biocultural diversity has yet to materialize. Even within conservation and other international organizations that have adopted the topic, the idea that the conservation of biodiversity should go hand in hand with support for the maintenance and revitalization of local cultures and languages remains sometimes controversial. And many countries still balk at the very idea of acknowledging the cultural and linguistic diversity within their borders, or are only prepared to celebrate it as a treasure from the past, disconnected from present realities and irrelevant to issues of environmental protection and sustainability (if such issues are on the agenda at all, in a world that continues to be dominated by materialism and the pursuit of unbridled economic growth). The general *Zeitgeist* also causes public opinion – the key to political will – to continue to be largely unaware or even oblivious of the growing deterioration of our biocultural world.

If there is hope that the efforts of grassroots communities and of those researchers, practitioners, and activists who embrace a biocultural perspective will be more broadly supported, this hope resides in capacity building and education about conservation in both nature and culture. "[The concept of] conservation reminds us of our duties to nature and to the future, without which the pace of economic growth will merely be a measure of the speed at which we approach the abyss" (Sacks, 2002, p. 174). If people "in the biocultural trenches" can have access to the conceptual and political tools they need, and if a new generation of people can be raised with a firm understanding that, as Harmon (2002) puts it, "diversity in nature and culture makes us human," then humanity will have a chance to pull back from the brink of the abyss and go on to chart a new path toward ecological and cultural sustainability.

ACKNOWLEDGMENTS

This paper is based in part on materials prepared by the author for the workshop "Gaps and Needs in Biocultural Diversity Research?," co-organized by Terralingua and the University of Florida and held in Gainesville, Florida, USA, on April 21–22, 2005, with support from The Christensen Fund (TCF). It benefited from discussions with and contributions by the participants. These inputs and TCF's support are gratefully acknowledged.

REFERENCES

Balmford, A. *et al.* (2005) The convention on biological diversity's 2010 target. *Science* 307: 212–213.

Barrett, D.B., G.T. Kurian, and T.N. Johnson (2001) *World Christian Encyclopedia: A Comparative Survey of Churches and Religions in the Modern World*, 2nd edn, 2 vols. Oxford: Oxford University Press.

Barthlott, W., N. Biedinger, G. Braun, F. Feig, G. Kier, and J. Mutke (1999) Terminological and methodological aspects of the mapping and analysis of global biodiversity. *Acta Botanica Finnica* 162: 103–110.

Bernard, R. (1992) Preserving language diversity. *Human Organization* 51: 82–89.

Blythe, J., and R. McKenna Brown (eds) (2004) *Maintaining the Links. Language, Identity and the Land. Proceedings of the 7th Conference of the Foundation for Endangered Languages, Broome, Australia*. Bath, UK: Foundation for Endangered Languages.

Borrini-Feyerabend, G., K. Macdonald, and L. Maffi (2004) History, culture and conservation. *Policy Matters* 13 (special issue).

Bruntland, G. (ed.) (1987) *Our Common Future. The World Commission on Environment and Development.* Oxford: Oxford University Press.

Carlson, T.J.S. and L. Maffi (eds) (2004) *Ethnobotany and Conservation of Biocultural Diversity*, Advances in Economic Botany Series, Vol. 15. Bronx, NY: New York Botanical Garden Press.

Chapin, M. (1992) The co-existence of indigenous peoples and environments in Central America. *Research and Exploration* 8(2), inset map. [Revised as: Chapin, M. (2003) Indigenous peoples and natural ecosystems in Central America and Southern Mexico, Map insert. *National Geographic Magazine*, February 2003.]

Collard, I.F. and R.A. Foley (2002) Latitudinal patterns and environmental determinants of recent human cultural diversity: Do humans follow biogeographical rules? *Evolutionary Ecology Research* 4: 371–383.

Eldredge, N. (1995) *Dominion.* New York: Henry Holt and Co.

Florey, M. (2006) Assessing linguistic vitality in Central Maluku. Paper presented at the *Tenth International Conference on Austronesian Linguistics,* Palawan, Philippines, 17–20 January 2006.

Gordon, R.G. (ed.) (2005) *Ethnologue: Languages of the World,* 15th edn. Dallas: SIL International.

Grimes, B.F. (1992) *Ethnologue: Languages of the World,* 12th edn. Dallas: SIL International.

Grimes, B.F. (2000) *Ethnologue: Languages of the World,* 14th edn. Dallas: SIL International.

Groombridge, B. (ed.) (1992) *Global Biodiversity: Status of the Earth's Living Resources.* London: Chapman & Hall.

Harmon, D. (1992) Indicators of the World's Cultural Diversity. Presented at the *4th World Congress on National Parks and Protected Areas.* Caracas, Venezuela.

Harmon, D. (1996) Losing species, losing languages: connections between biological and linguistic diversity. *Southwest Journal of Linguistics* 15: 89–108.

Harmon, D. (2002) *In Light of Our Differences: How Diversity in Nature and Culture Makes Us Human.* Washington, DC: Smithsonian Institution Press.

Harmon, D. and J. Loh (2004) The IBCD: A measure of the world's biocultural diversity. *Policy Matters* 13: 271–280.

Heckenberger, M., A. Kuikuro, U.T. Kuikuro, J.C. Russell, M. Schmidt, *et al.* (2003) Amazonia 1492: Pristine forest or cultural parkland? *Science* 301: 1710–1714.

Hill, J.H. (1997) The meaning of linguistic diversity: knowable or unknowable? *Anthropology Newsletter* 38(1): 9–10.

Krauss, M. (1992) The world's languages in crisis. *Language* 68(1): 4–10.

Krauss, M. (1996) Linguistics and biology: threatened linguistic and biological diversity compared. In *CLS 32, Papers from the Parasession on Theory and Data in Linguistics.* Chicago: Chicago Linguistic Society, pp. 69–75.

Lizarralde, M. (2001) Biodiversity and loss of indigenous languages and knowledge in South America. In Maffi, L. (ed.) *On Biocultural Diversity: Linking Language, Knowledge, and the Environment.* Washington, DC: Smithsonian Institution Press, pp. 265–81.

Loh, J. and D. Harmon (2005) A global index of biocultural diversity. *Ecological Indicators* 5: 231–241.

Mace, R. and M. Pagel (1995) A latitudinal gradient in the density of human languages in North America. *Proceedings of the Royal Society of London B: Biological Sciences* 261: 117–121.

Maffi, L. (1998) Language: A resource for nature. *Nature and Resources: The UNESCO Journal on the Environment and Natural Resources Research* 34(4): 12–21.

Maffi, L. (2001a) *On Biocultural Diversity: Linking Language, Knowledge, and the Environment.* Washington, DC: Smithsonian Institution Press.

Maffi, L. (2001b) Linking language and environment: A co-evolutionary perspective. In Crumley, C.L. (ed.) *New Directions in Anthropology and Environment: Intersections.* Walnut Creek, CA: AltaMira Press.

Maffi, L. (2001c) Language, knowledge, and indigenous heritage rights. In *On Biocultural Diversity: Linking Language, Knowledge, and the Environment.* Washington, DC: Smithsonian Institution Press, pp. 412–432.

Maffi, L. (2005) Linguistic, cultural, and biological diversity. *Annual Review of Anthropology* 34: 599–617.

Maffi, L. and E. Woodley (in preparation) *Global Source Book on Biocultural Diversity.* Progress report available at http://www.terralingua.org/GSBBCD.htm

Maffi, L., T. Skutnabb-Kangas, and J. Andrianarivo (1999) Linguistic Diversity. In Posey, D.A. (ed.) *Cultural and Spiritual Values of Biodiversity.* London and Nairobi: Intermediate Technology Publications and UNEP, pp. 21–57.

Manne, L.L. (2003) Nothing has yet lasted forever: current and threatened levels of biological diversity. *Evolutionary Ecology Research* 5: 517–527.

Meya, W. (2006) Letter to *The Financial Times,* London, March 11, 2006.

Millennium Ecosystem Assessment (2005) *Ecosystems and Human Well-Being: Synthesis Report.* Washington, DC: Island Press.

Miller, M.S. (ed.) (1993) *State of the Peoples. A Global Human Rights Report on Societies in Danger.* Boston: Beacon Press.

Moore, J.L., L. Manne, T. Brooks, N.D. Burgess, R. Davies, *et al.* (2002) The distribution of cultural and biological diversity in Africa. *Proceedings of the Royal Society of London B: Biological Sciences* 269: 1645–1653.

Mühlhäusler, P. (1995) The interdependence of linguistic and biological diversity. In Myers, D. (ed.) *The Politics of Multiculturalism in the Asia Pacific.* Darwin: Northern Territory University Press, pp. 154–161.

Mühlhäusler, P. (1996) *Linguistic Ecology: Language Change and Linguistic Imperialism in the Pacific Rim.* London: Routledge.

Nettle, D. (1996) Language diversity in West Africa: An ecological approach. *Journal of Anthropological Archaeology* 15: 403–438.

Nettle, D. (1998) Explaining global patterns of linguistic diversity. *Journal of Anthropological Archaeology* 17: 354–374.

Nettle, D. (1999) *Linguistic Diversity.* Oxford: Oxford University Press.

Nichols, J. (1990) Linguistic diversity and the first settlement of the New World. *Language* 66: 475–521.

Nichols, J. (1992) *Linguistic Diversity in Space and Time.* Chicago/London: University of Chicago Press.

Oviedo, G., L. Maffi, and P.B. Larsen (2000) *Indigenous and Traditional Peoples of the World and Ecoregion Conservation: An Integrated Approach to Conserving the World's Biological and Cultural Diversity.* Gland, Switzerland: WWF-International and Terralingua.

Posey, D.A. (ed.) (1999) *Cultural and Spiritual Values of Biodiversity.* London and Nairobi: Intermediate Technology Publications and UNEP.

Posey, D.A. (2001) Biological and cultural diversity: The inextricable, linked by language and politics. In Maffi, L. (ed.) *On Biocultural Diversity: Linking Language, Knowledge, and the Environment.* Washington, DC: Smithsonian Institution Press, pp. 379–396.

Rapport, D.J. (2007) Healthy ecosystems: An evolving paradigm. In Pretty, J. *et al.* (eds) *The Sage Handbook of Society and Environment*. London: Sage.

Rapport, D.J., R. Costanza, and A. McMichael (1998) Assessing ecosystem health: Challenges at the interface of social, natural, and health sciences. *Trends in Ecology and Evolution* 13(10): 397–402.

Ross, N. (2002) Cognitive aspects of intergenerational change: Mental models, cultural change, and environmental behavior among the Lacandon-Maya of southern Mexico. *Human Organization* 61: 125–138.

Sacks, J. (2002) *The Dignity of Difference: How to Avoid the Clash of Civilizations*. London and New York: Continuum.

Skutnabb-Kangas, T. (2000) *Linguistic Genocide in Education – Or Worldwide Diversity and Human Rights?* Mahwah, NJ: Lawrence Erlbaum.

Skutnabb-Kangas, T., L. Maffi, and D. Harmon (2003) *Sharing a World of Difference: The Earth's Linguistic, Cultural, and Biological Diversity* and map *The World's Biocultural Diversity: People, Languages, and Ecosystems*. Paris: UNESCO.

Smith, E.A. (2001) On the coevolution of cultural, linguistic, and biological diversity. In Maffi, L. (ed.) *On Biocultural Diversity: Linking Language, Knowledge, and the Environment*. Washington, DC: Smithsonian Institution Press, pp. 95–117.

Stepp, J.R., F.S. Wyndham, and R. Zarger (eds) (2002) *Ethnobiology and Biocultural Diversity*. Athens, GA: University of Georgia Press.

Stepp, J.R., S. Cervone, H. Castaneda, A. Lasseter, G. Stocks, and Y. Gichon (2004) Development of a GIS for global biocultural diversity. *Policy Matters* 13 (special issue): 267–270.

Stepp, J.R., H. Castaneda, and S. Cervone (2005) Mountains and biocultural diversity. *Mountain Research and Development* 25(3): 223–227.

Sutherland, W.J. (2003) Parallel extinction risk and global distribution of languages and species. *Nature* 423: 276–279.

Terborgh, J. (1999) *Requiem for Nature*. Washington, DC: Island Press.

Terralingua (2006) Global Indicators of the Status and Trends of Linguistic Diversity and Traditional Knowledge. Project funded by The Christensen Fund (2006–2008).

UNESCO (2003) Language Vitality and Endangerment. International Expert Meeting of the UNESCO Intangible Cultural Heritage Unit's Ad Hoc Expert Group on Endangered Languages, UNESCO, Paris-Fontenoy, 10–12 March 2003. The recommendations approved by the participants are available at http://www.unesco.org/culture/heritage/intanglible/meetings/paris_march2003.shtml. The document "Language Vitality and Endangerment" is available at http://portal.unesco.org/culture/en/ev.php-URL_ID=135848&URL_DO=DO_TOPIC&URL_SECTION=201.html.

Wilcox, B.A. and K.N. Duin (1995) Indigenous cultural and biological diversity: Overlapping values of Latin American ecoregions. *Cultural Survival Quarterly* 18(4): 49–53.

Zarger, R.K. and J.R. Stepp (2004) Persistence of botanical knowledge among Tzeltal Maya children. *Current Anthropology* 45(3): 413–418.

Zent, S. (1999) The quandary of conserving ethnoecological knowledge: A Piaroa example. In Gragson, T.L. and Blount, B.G. (eds) *Ethnoecology: Knowledge, Resources, and Rights*. Athens, GA: University of Georgia Press, pp. 90–124.

Zent, S. (2001) Acculturation and ethnobotanical knowledge loss among the Piaroa of Venezuela: Demonstration of a quantitative method for the empirical study of TEK change. In Maffi, L. (ed.) *On Biocultural Diversity: Linking Language, Knowledge, and the Environment*. Washington, DC: Smithsonian Institution Press, pp. 190–211.

Zent, S. and E. López-Zent (2004) Ethnobotanical convergence, divergence, and change among the Hoti of the Venezuelan Guayana. In Carlson, T.J.S. and Maffi, L. (eds) *Ethnobotany and Conservation of Biocultural Diversity*, Advances in Economic Botany Series, Vol. 15. Bronx, NY: New York Botanical Garden Press, pp. 37–78.

Political Economy of Environmental Change

19

Representative Democracy and Environmental Problem Solution

Ron Johnston

INTRODUCTION

Several characteristics of contemporary environmental problems – especially the more intractable – mean that they can only be tackled through collective action. Such action may be voluntary, with those agreeing to act then meeting their contractual obligations. More generally, however, action is neither likely to be put in place nor successfully followed through unless either there are strong and enforceable sanctions for non-compliance or the incentives to comply are such that signatories to an agreement ensure that the goals are met.

Enforceable and enforced collective action, at almost all scales above the most local and/or involving relatively small numbers of actors, invariably involves states. Indeed, one of the main metaphors used to illustrate the necessity of collective action organized and enforced by a state – the tragedy of the commons – is based on the need for such action to ensure that environmental resource users act in their own and the environment's long-term interests. Thus, understanding how environmental problems can be approached requires an appreciation of the nature of the state and the constraints on state action.

For many contemporary environmental problems, such appreciation extends beyond an understanding of how individual states act to a consideration of inter-state relationships – involving either groups of states defined through common interests wherever they are located (as in whaling) or those occupying one part of the Earth's surface. This raises complex issues regarding the exercise of state power.

This chapter discusses the role of the state/ states in the attempted resolution of environmental issues. The first section outlines the characteristics of those issues which require the state as an institution to address their resolution and the spatial constitution of states which has important impacts on their ability to undertake such tasks. An increasing proportion of contemporary states are liberal democracies; the nature of those institutions, especially the temporal and spatial constraints to the allocation of power through electoral processes – and how those are changing with recent transformations in the nature of capitalism – is the focus of the next sections. These provide the framework for the final sections, which deal with environmental policy in a democratic context – both within individual countries and internationally.

ON ENVIRONMENTAL ISSUES

Many environmental problems share three common characteristics. First, although they are produced through individual actions (with the individuals including corporate actors, and actions intentional or not), the problems' intensity may be more than the sum of those individual contributions. There is very often an exponential increase as more add to the production, especially if the producers are concentrated in relatively confined areas.

Second, most problems affect others in addition to the producers; there are spatial overspills which are frequently unavoidable. As a consequence, solution – or resolution, as the two are not the same – of a problem requires that all, or at least a great majority, of those involved must reduce (if not end) their contributions. This leads to a third common characteristic: individual contributors can gain an advantage over others by declining to participate in a programme designed to solve the problem, but with which most others conform.

The nature of these three defining characteristics is frequently demonstrated through the metaphor of the tragedy of the commons (Hardin, 1968). A piece of common land is shared by a number of graziers, who regularly increase their herd sizes for so long as the marginal returns are positive. This may continue even though the average return per animal is declining and the resource is being depleted. Indeed, individual graziers will feel impelled to increase their herds for so long as the others do, since if they do not their overall returns will decline because of the falling average rate of return. Efficient use and long-term conservation of the common pool resource requires the number of animals to be regulated but individual graziers will not limit their herd size unless all others not only agree to do so but also then stick to their bargain. To act otherwise would not even be altruistic, because in the end the resource will be depleted. Even when a bargain is struck, with all graziers agreeing to limit herd sizes, it is in the interests of each of them individually to then break that agreement and 'free-ride' on the others, reaping the benefit of extra animals in the short term even though in the long term the resource is depleted. If one can benefit from free-riding, however, so can others so what is required is an agreement which all graziers both sign up to and then respect.

A contract is of little value unless it is adhered to – which perhaps requires that it is enforced. If individual graziers are unsure how others will act, their best interests are served by increasing herd size – and what proves the best for one is the best for all, at least in the short term. The graziers know that in the long term all will lose, because the resource is being eroded, but cannot act individually for its long-term preservation if they cannot trust everybody else to do so as well. This paradoxical situation is formally illustrated and analysed through the prisoner's dilemma game (Taylor, 1987; Poundstone, 1992; Laver, 1997; Barrett, 2003) and illustrated by a number of resource depletion issues, such as fishing (see Clover, 2004).

In small-scale situations – within families and households, for example, or in communities comprising only a few actors – generating that trust and enforcing agreements voluntarily entered into may be feasible, as Ostrom (1990, 1995) has illustrated. In others, an alternative mechanism to such self-regulating collective agreements is required. Laver (1984, 1986; see also Herrera, 2002) identified two:[1]

1 privatizing the commons;
2 external regulation by a body with the power to ensure compliance.

The former may be a viable short-term solution. It does not, however, guarantee that, if the resource depletion process is slow, individual owners will not prioritize short-term gains over longer-term benefits and thus eventually create the environmental problem that privatization (as with enclosure in the UK) was designed to avoid.

This leaves external regulation by a body with the power to ensure compliance, of which there is only one in the contemporary world – the state. This is the institution which (since the 1648 Treaty of Westphalia) has exercised internationally recognized sovereign power over a designated territory, within which it arrogates to itself the ultimate source of all regulation and a monopoly over the use of violence. Those controlling the state apparatus ('governments') operate through the 'rule of law', deploying the power to regulate what individuals do within the sovereign territory and ensuring compliance with their regulations, through predetermined sanctions.[2] If individuals know that sanctions will be enforced, they are likely to comply with the regulations – especially if the long-term benefits from doing so have been explained as the rationale for their implementation.

This role of the state as regulator of environmental use has become increasingly crucial in recent decades. Whereas individual citizens are readily made aware of the negative impacts of some of their actions, because they are short term, local and visible – such as depositing pollutants into a water body used for potable water and/or fishing, for example – this is not necessarily the case. Many contemporary issues may be both longer term in their impact, so that the consequences of actions are not immediately apparent – indeed may not be apparent during an individual's life time, as with greenhouse gas warming and long-term climate change – and 'invisible', as with destruction of the ozone layer and declining biodiversity (although their 'impacts', such as increased rates of skin cancer because of ozone-layer depletion, may be).

In such situations the state's role is much more than regulating to tackle a problem: it also has to ensure appreciation of emerging problems with potential deleterious effects. Many contemporary problems are not immediately apprehendable by the majority of citizens, and their production may not be readily appreciated because of the complex science needed to understand their nature. As Hajer (1995) has argued, contemporary ecological crises are discursive constructions: they are problems constructed in such a way – by scientists and others – so that they are perceived as challenges to existing institutions.

Trudgill (1990) identified a number of barriers to be overcome when tackling environmental problems, although the sequence need not take that form in all situations:

1 identification of an issue;
2 recognition that there is a problem associated with it;
3 acceptance of the problem's postulated cause;
4 acceptance that the problem can be tackled and remedied;
5 identification of resources necessary for tackling the problem;
6 winning acceptance for the proposed solution;
7 gaining political commitment to the solution;
8 implementation of the solution;
9 evaluation of the implementation.

The first involves an issue being raised as of potential significance. Unless a problem's existence is recognized, no action may be taken: governments have to be convinced that there is a (perhaps latent) problem. Furthermore, no problem can be tackled satisfactorily unless its causes are understood: governments will be reluctant to act unless they know what they are dealing with. Such knowledge has to come from disinterested sources – which with most environmental issues means scientists with the specialist skills to appreciate a problem's existence and genesis.[3] Even so, such knowledge is insufficient: problems will only be tackled if a viable solution is available, which again probably requires scientific inputs. Governments are major investors in science and engineering – especially that conducted in universities and public sector laboratories where practitioners are not committed to a particular interest group which may be either advantaged or disadvantaged by an environmental regulation – and have to ensure sufficient investment in scientific capacity to ensure that they get the relevant advice.[4]

Surmounting the first four barriers involves debate over scientific arguments: is there a problem, what causes it, and can it be remedied? The next four are political. Tackling a problem involves the commitment of resources, which are invariably in short supply so that there is competition among a wide range of state policies. Resources have to be obtained from citizens through taxation (which may impact all taxpayers or just be directed at those involved in the problem-creation mechanisms, through 'user pays'). For a solution to be implemented there has to be both political and public support, therefore. Governments may decide that neither can be obtained and so do not address the problem.

Recent debates over global warming – which focused on Tony Blair's attempts to get agreement at the July 2005 G8 Summit in Scotland on a way forward to tackle the issue – starkly illustrate the political problems involved in getting agreement over policy direction. The US government, in particular, was reluctant to agree to any communiqué which indicated clear recognition that global warming is a serious immediate problem and that human activity is a clear contributor to the observed trends; it thus made for problems at the first three stages of the above sequence. Further, although the US President was prepared to accept that there may be some evidence of climate change, he challenged the proposed solution that appeared to be accepted by the governments of the other major countries involved. Whereas the latter acknowledged that 'developed' world countries had to date made a greater contribution to global warming than those in the 'developing' world, and that they should therefore bear the initial brunt of programmes to reduce greenhouse gas emissions, the US government insisted that the 'developing' world should be fully involved from the outset. Finally, whereas most countries' governments appeared to accept that a reduction of emissions through a combination of targets and their trading was a feasible way forward, the Americans insisted that it was not and that the focus should be on the development of new and cleaner technologies – clearly indicating that they saw less of an immediate problem. With such debates – over the science, the technology, the politics and the economics – the potential for resolving what many see as the pressing problem for long-term environmental sustainability was remote.

On boundaries and environments

The discussion so far has been aspatial, assuming that environmental problems and states have no geographies. All environmental problems have geographies, however, of both generation and existence – which may not be the same: some of the depletion of the ozone layer over parts of the southern hemisphere has been generated by pollutants dispersed into the northern hemisphere atmosphere. Further, the generating points may be both punctiform and concentrated, as with the impact of a major industrial plant, or they may be widely scattered, as with the ozone-layer-threatening aerosols and refrigerators which emit CFCs into the atmosphere.

States also have geographies; indeed they are defined spatially, with international law allowing them virtually unfettered power within their bounded space. Those bounded spaces rarely conform to the geography of the environmental problems, however. The global environment is a single integrated system and although parts that are close to each other are more likely to be intensively integrated, so that what happens in one substantially impacts on the other, everywhere is connected to everywhere else. A problem generated somewhere may eventually be felt a great distance away; therefore, the environment respects no state boundaries. Some problems may be contained and tackled within an individual state's territory, but many – including most of the contemporary world's major and least tractable environmental problems – are not.

Tackling many environmental problems thus requires international state action. This is more complicated than just winning cooperation among states to counteract problems which (potentially) threaten them all. There is a tragedy of the commons situation in which the actors are states and a prisoner's dilemma game in which states are the players. Furthermore, those interactions concern commons that are outside states' bounded territories. Much of the Earth, especially the deep oceans and the atmosphere, comprises several global commons. International law divides these (and others, such as the moon, the planets and outer space) into three zones (Johnston, 1992; Vogler, 2000):

1 *res nullius* – resources which are not currently claimed by any state, but which are susceptible to such appropriation in principle.
2 *res extra commercium (REM)* – resources which cannot be claimed by any state as its own, over which its laws and regulations apply, but which are treated as a commons on which any state (or citizen of any state) can 'graze'.[5]
3 *res communis humanitatis (RCH)* – resources which are not only beyond national state claims but are recognized as the common property of all and therefore cannot be grazed for individual benefit.

The parameters of these three are not fixed but subject to continued inter-state conflict, as illustrated by the Law of the Sea (Steinberg, 2001). Initially, claims for *res nullius* were restricted to the Earth's land surface plus a narrow zone of three miles around most state's coastlines (being the range of the most powerful guns and hence regulatable by the claimant state). Beyond this, an increasingly wide zone was treated as *RCM* and subject to (often disputed) claims by adjacent states wishing to exploit the resources of what

were known as the contiguous zone and the continental shelf: eventually it was agreed to allow claims to an exclusive economic zone extending up to 200 miles from a state's coast. This left the remainder as *RCH*. This has been the focus of considerable international conflict between, on the one hand, states that wish to see any profits from exploitation of its resources widely distributed and, on the other, those wanting the benefits to remain with the states (and their citizens) who invest in and conduct the exploitation (i.e. they want to treat them as *RCM* not *RCH*).

There are three categories of environmental problem, therefore:

1 those confined within individual states;
2 those involving two or more states and (largely) confined to their sovereign territories;
3 those involving *RCM* and *RCH*, which involve interactions among (potentially) all states.

Tackling them all involves a state acting either within its own territory or in combination with others. Despite its power, however, a state's actions are constrained in a variety of ways. One relates to the way in which those controlling the state apparatus attain and retain that position – through democratic processes. These are the subject of the next major section, followed by a final section which explores the influence of democratic constraints on approaches to environmental issues.

ON STATES AND DEMOCRACY

Francis Fukuyama (1989, 1992) welcomed the triumph of democracy at the end of the 1980s – notably through the fall of the Iron Curtain and the rapid spread of democratic forms of government through much of Africa, Asia and Latin America – as the end of history: 'liberal democracy remains the only coherent political aspiration that spans different regions and cultures around the globe' (Fukuyama, 1992, p. xiii). Not all countries (involving a large proportion of the world's population) had adopted democratic forms of governance by then, whereas in others democracy was not firmly entrenched in daily practices and there were many examples where its basic tenets were recognized more in rhetoric than in practice. Nevertheless, Fukuyama believed there would be no substantial retreat from democratic governance – as there had been in previous decades (in Latin America, for example). Democracy would remain the predominant form of government across the world, and thus the context within which environmental problems would have to be tackled. Thus, the end of history was hailed because liberal democracy appeared to constitute the 'end point of mankind's ideological evolution' and 'the final form of human government' (Fukuyama, 1989, p. 2).

The role of the democratic state

Democracy has become the dominant state form in contemporary capitalism which is characterized by globalization processes and a growing neo-liberal consensus regarding the state's basic tasks. The core state roles remain as summarized by O'Connor (1972):

1 *Securing a consensus from all groups within society* around the mode of production and its particular local formation. This ensures order and stability within the territory governed by and through the relevant state apparatus.
2 *Sustaining and enhancing the conditions for capitalism's successful reproduction*, ensuring continued profit-making and generating support for the state and its actions by the 'capitalist fraction' within society.
3 *Guaranteeing social integration and the welfare of all* by, for example, ensuring that all fractions of society enjoy the fruits of capitalist wealth-production and are protected from its vicissitudes.

Failure to ensure the second of these can stimulate a *rationality crisis* (Habermas, 1976; see also O'Connor, 1987); a lack of confidence in state actions stimulates reductions in capital investment, with consequences for employment and well-being. Failure to ensure the third can generate *legitimation crises*, whereby disadvantaged groups challenge the state apparatus' legitimacy. Failure at both the second and third can lead to an overall *crisis of accumulation*, whereby the state retains support from neither fraction of society and hence suffers a failure in the first role. The likely outcome of such an eventuality is, at worst, collapse of the state apparatus and an anarchic condition until a new order is installed.

The state – or a comparable institution – is necessary to capitalism. As Marx observed and subsequent commentators have accepted, capitalism contains within itself the seeds of its own destruction, and a body apparently independent of society's various interest groups is needed to regulate capital's operations and prevent the self-destructive tendencies prevailing. The state has thus become the institution which prevents individuals and groups from taking actions which may not be in their long-term best interests with regard to the wider 'public good'; however, they may appear in the immediate term. (On the state and trust, see Marquand, 2004.) Its ability to achieve that goal, however, is constrained by its democratic nature.

The second and third of the three state roles – ensuring the conditions for capitalist reproduction and the social integration of all fractions of society – are mutually complementary and yet also to a considerable extent in opposition: overemphasis on one can generate potential crisis in the other.[6] For example, a government operating a high-tax regime to redistribute wealth away from the capitalist fraction and benefit the relatively disadvantaged can create a milieu perceived as unfavourable to investment. This may stimulate a rationality crisis, with reduced levels of investment generating a slow-down in economic activity and job creation, creating the potential for a legitimation crisis among those supposed to benefit from the original policy. Alternatively, a government which over-emphasizes creating a favourable milieu for investment and profit-taking can alienate those who benefit relatively little from such policies, by reducing the size and scope of the welfare state which sustains those who prosper least and have fewest resources with which to compete in labour and other markets. Such a nascent legitimation crisis can, in turn, stimulate unease among investors, who will be wary of making long-term commitments in places where there is a threat to socio-economic and political stability.

Elected governments must thus be continually concerned with the balance of their policies, trying to sustain both favourable conditions for economic growth and prosperity and the needed social integration and stability. If the balance is tilted too far towards one of those two roles, the potential for alienating a major group within society is accentuated. This would be particularly problematic if a considerable imbalance remains for some time without the state deploying alternative modes of ensuring stability and order, for example, through coercive powers not readily available in a representative democracy.

States are intrinsically geographical institutions operating within prescribed territories. That spatial construction is among their key defining characteristics and deployment of territoriality strategies is central to a government's ability to rule within its bounded space (Mann, 1984; Sack, 1986). This geography of the state is key to appreciating important aspects of its role in the regulation of neo-liberalism, as recognized in recent works seeking to 'bring the state back in' to studies of globalization (e.g. Held and McGrew, 2002, 2003; Brenner *et al.*, 2003; Brenner, 2004; Jones and Jones, 2004; Pratt, 2004), but the role of representative democracy in this is largely ignored. Weiss (1999, 2003), for example, argues that how states act within the constraints of globalizing trajectories will reflect local conditions and circumstances, but does not link those degrees of freedom to the operation of democratic systems. Similarly, King and Kendall (2003) and Jessop (2002) do not explore the democratic constraints to state action in the economic sphere – let alone the environmental.

On liberal representative democracy

Democracy has a considerable number of meanings. One commonly deployed portrays control of the state apparatus as defined by Lincoln: 'government of the people, by the people, and for the people'. Governments are elected to manage the state apparatus in the general good and, most importantly, are accountable to their electorate, which can replace them if they believe the general good can be better pursued by other politicians. Democratically elected governments are thus constrained in their use of state power by their need to pursue re-election.[7]

The nature of representative democratic institutions varies greatly across states, with, for example, a wide range of electoral systems which may be reflected in how governments are constructed (Norris, 2004). There is a foundation of common constitutional principles, however, which should guarantee: (1) freedom in the formulation of preferences, through joining organizations, expression, distributing information, voting and competing for votes; (2) freedom in the signification of preferences, through free and fair elections and the right to stand for public office; and (3) an equal weighting of preferences so that, as expressed in Article 21 of the 1948 *United Nations Declaration of Human Rights*, 'the will of the people shall be the basis of the authority of government; this will shall be expressed in periodic and genuine elections which shall be by universal and equal suffrage' (Dahl, 1978).

Whereas democracy's heartland is in classical Europe, the USA has one of the longest traditions of democratic practices associated with these foundational ideas. This 200-year tradition in which the transfer of power between governments as reflecting the will of the people has been unchallenged and is at the core of American capitalist rhetoric.[8] The USA has prospered and, with the (albeit major) exception of its Civil War, been largely conflict free. The will of the people, and their desire for economic success, plus 'the pursuit of happiness' that this enables, has underpinned a stable and prosperous society. American governments have pressed others to adopt the same framework.

That pressure was often violated (or, at least, overlooked) through much of the 20th century, however. The USA and other democratic states (notably in Western Europe) were prepared to tolerate non-democratic regimes elsewhere if their existence was seen to be in the Americans' best interests – and this even extended to regimes which practiced virtually no democracy at all. They were also prepared – covertly if not openly – to assist in undermining popularly elected governments perceived to be operating against their interests (see Johnston, 1999). Nevertheless, the number of countries with democratic forms of government has increased substantially in recent decades, a trend for which the USA has claimed much credit (O'Loughlin *et al.*, 1998; O'Loughlin, 2004).

As representative democracy evolved, particularly with the extension of the franchise to all adults, political parties became fundamental to its operations. These mobilize sections of the electorate around key sets of beliefs and provide governments with relatively secure support when seeking to implement their policy goals. Parties provide order, stability and continuity to the political scene, within and outside the legislature. Without them, each election would be an ad hoc exercise: electors would choose their representatives but have little say on how majorities were assembled to sustain legislative programmes. Without them, too, all policy proposals would be subject to bargaining and 'hidden deals', and governments could fall with considerable regularity. As the business of government extended into more spheres of activity and became crucial to the success of the burgeoning capitalist enterprise, continuity of policy became more important, in which party government was crucial.

The emergence of party government in Europe was associated with sets of ideologies that emphasized one of the two main roles of the capitalist state over the other. On the one hand were parties focused on wealth creation, promoting capitalist class interests; on the other were parties promoting the redistribution ideology of the social integration task, with policies oriented towards the non-capitalist fraction from which they drew much of their electoral support (Harvey, 2005). These were known as right- and left-wing, or conservative and socialist, parties, respectively. As the franchise was widened to incorporate an increasing proportion of the 'working class' fraction, the potential for intra-society conflict was enhanced. In some countries, attempts to mitigate this were taken by the capitalist fraction and its supporters (the 'middle class') through the introduction of (relatively) proportional electoral systems which it was hoped would restrict the freedoms of working-class-based governments to create the potential conditions for a rationality crisis. This was not done elsewhere, however – notably in the UK. As working-class support was mobilized behind socialist-based parties there, the potential for the election of governments with ideologies and programmes that could severely constrain, if not entirely challenge, capitalist operations, was increased. By the 1970s this was identified as a severe threat to Britain's continued economic prosperity (Finer, 1974). The potential for major switches in government policy generated insecurity for investors, to which they responded by transferring their capital and/or expertise elsewhere (either geographically or, as Harvey, 1982, argued, to other, less-productive spheres of activity).

This potential see-sawing of policy directions and its likely impact on well-being has been reflected in the democratic experience in several parts of the world – notably Latin America and some African and Asian countries until relatively recently, but now in the newly democratized states of the former Soviet Union and its 'empire' also. Taylor argued that failure of elected governments to fulfil the expectations they generated when mobilizing electoral support produced a cyclical pattern of shifts in and out of democracy (see Taylor and Flint, 2000; see also Dix, 1984; Osei-Kwame and Taylor, 1984; Werz, 1987). A government may be elected on promises of prosperity for all, which means that it emphasizes the third state role. Success may enable it to maintain popular support, but at an increasing cost: its policies are seen as 'too expensive' by the capitalist fraction and threatening to profit creation. A nascent rationality crisis is stimulated; as support and income generation (via taxation) from the capitalist fraction decreases, the state is unable to fulfil its promises to its main electoral supporters in the working class.

A legitimation crisis follows, and the scene is set for an accumulation crisis as the government loses support across the class spectrum. In the mid-20th century, a common response to this situation – especially in Latin America – was a military coup, forming a government lacking (explicit) popular support which followed policies designed to create a favourable environment for capitalist investment. Its rhetoric was that non-democratic resolution of conflict was necessary for the national good, but as prosperity was regained it would be possible to reinstate a larger welfare state and, eventually, return to popular democratic government. In some cases, the latter was yielded without extensive popular demands: elsewhere, pressure on the military-backed government (perhaps with outside assistance) was needed to gain the concession of renewed government by the people.

Military intervention has not been the only response to the cycle caused by governments which tilted the balance too far towards one of the two roles. In India, for example, when the dominant political party (Congress for much of the first four post-independence decades) failed to meet its supporters' aspirations generated at one election, it mobilized a new support base at the next (Taylor, 1986). In many European countries of the former Soviet bloc since the fall of communism, there have been major changes in the party system between elections. A party may win power but then so fail to meet supporters' aspirations that support disintegrates – perhaps almost to nothing. New parties emerge to fill the gap, and they in turn fail. Democracy is maintained – usually because the armed forces are too weak to intervene – but is unable to stimulate major improvements in economic performance and general well-being.[9] Finally, in a number of states – most of them relatively poor – democracy has only been sustained in a weak form through the predominance of a single party, with a contest between parties allied to the two main tasks (neo-liberal and social democrat, respectively) proving irrelevant to the country's needs.

Western European and a small number of other countries have not experienced these cycles to anything like the extent currently occurring in eastern Europe let alone, with a few exceptions, involving non-democratic seizures of power. Although occasionally governments have been elected which have tilted the balance too far towards one of the main class fractions, the overall system has been robust enough to sustain economic growth and, eventually, a new balance has been created as the potential for a crisis becomes more apparent.

The USA has been almost immune from such experiences throughout its history, in part because its party system has never been constructed on predominantly class lines – as throughout most of western Europe and the former British Commonwealth (Lipset and Rokkan, 1967). Instead, American politics have been largely non-ideological, based to a considerable extent on (spatial) sectional claims to partisan support, especially since the Civil War.[10] The composition of the sections has varied, with series of normal elections (especially Presidential elections), during which each party has continuity of support from certain parts of the country, punctuated by realignments, as one or both of the two dominant parties sought to extend their electoral bases. (The theory was developed by V. O. Key in 1955 and substantially adapted for geographical study by Archer and Taylor, 1981: see also Mayhew, 2002; Gimpel and Schuknecht, 2003.) Although this geographical basis to party support has resulted in certain parts of the country being favoured over others in the promotion of economic prosperity and social well-being (as shown in studies of the pork barrel: Johnston, 1983), this has never threatened either a rationality or a legitimation crisis.

Both main parties have, with some differences of emphasis, been committed to the promotion of capitalism – hence, perhaps, both the general American commitment to democratic rhetoric and at least a partial lack of appreciation of the constraints to democracy operating in Europe and elsewhere, where class-based politics prevailed for most of the 20th century.

A new politics?

That century's last two decades saw major changes in the nature of both capitalism and democracy in Europe, however. Capitalism was responding to the failure of the Fordist mode of regulation to ensure continued growth, replacing it with a new form, which Harvey (1989) termed flexible accumulation. Representative democracy changed with it, as economists and others pressed the case for new forms of regulation – or governance – more favourable to the post-Fordist mode and its globalizing, neo-liberal agenda. Their impact is often traced to the policies developed by right-wing governments in the UK ('Thatcherism') and the USA ('Reaganism') and a (relatively) 'left-wing' government in New Zealand ('Rogernomics'),[11] but the arguments and related policies spread fairly rapidly – with a readier reception in some countries than others.

The core of this new ideology involved a considerable transcending of the existing political scene. In most cases, the party system has remained the same but individual parties – such as Labour in both New Zealand and the UK – have substantially restructured their agenda. The class conflict that underpinned electoral politics for some decades has been considerably diluted as a response to electoral realities: with the decline of traditional manufacturing industries, the working class base that sustained left-wing parties has been substantially eroded, and party leaders have realized the necessity of broadening their appeal to large segments of the middle classes for future political success (Crewe, 1986). They have justified this to their traditional constituencies by stressing the nature of the changing world and contending that the greater good of all is best served by promoting wealth creation. A residual commitment to ensuring that the fruits of economic growth are available to all – by improving the quality of public services, for example – has also been used to sustain support, while at the same time the rhetoric of rights (other than 'civil and human rights') has been replaced by one of obligations and duties – as in the workfare state policies developed in parts of the USA and exported to the UK and elsewhere (Peck, 2001).

Class-based voting has been substantially replaced by issue voting, whereby electors decide whether to return the incumbent government or replace it by an alternative on the basis of evaluations of the country's economic performance and the government's competence in managing the economy, both retrospective and prospective.[12] Increasingly, the answer to questions on what determines how people vote is 'It's the economy, stupid!' The dominant issue is no longer which section (class?) of the population gains and which loses when a particular party is in power, but whether the country as a whole benefits or suffers. Democracy's threats to capitalism have been very substantially reduced by this shift, and the balance between the second and third state roles has become a less important component of the electoral agenda. All governments seek to encompass both, though their stress is on the promotion of capitalism as the *sine qua non*: without it, the third role cannot be pursued.

TIME, SPACE AND THE CONSTRAINTS TO DEMOCRACY

Given these shifts, with many more countries approaching the American norm of 'non-ideological' party systems, the way would seem open for a governments to focus more on global, including environmental, issues. Although one constraint has been removed by the restructuring of democracy, however, others remain. These are inherent to how liberal representative democracy is currently practised, and some are intrinsically geographical in their nature and scope.

Elections are contests between political parties seeking to prove that they are best able to manage capitalism (Schumpeter, 1943). The successful are accountable to the electorate. If they are perceived to have failed by a substantial proportion, relative to what others claim they could do, they may be replaced. Elected administrations seek to avoid that indignity, heeding three constraints to their degrees of freedom to act – one temporal and the other two spatial, although all three interact.

Time and elections

Most governments have relatively short-time horizons. They have to return to the electorate to be judged on their performance a few years (rarely more than five) after they have gained, or regained, office.[13] For some legislators, the horizon is very short – two years for the US House of Representatives, for example.[14] It is also fixed in most cases, although in some countries the incumbent government has the power to determine when it seeks re-election, within a maximum term.

Within these timetables, short-term considerations dominate government actions. Longer-term goals may be important to their rhetoric but, if plausible alternative governments are available, perceived recent performance is most likely to be vital to re-election prospects. Whatever a government's medium- and longer-term aspirations, if short-term performance has disappointed a substantial proportion of the electorate, it may not be given the chance to implement policies that could lead to those ultimate goals.[15] A government may therefore manipulate the political economic cycle in order to ensure favourable economic indicators in the months leading up to an election (see Lewis-Beck *et al.*, 2004). Where the timing of an election is in the incumbent government's control, it may decide to call an election when the potential for re-election seems high because of good economic performance.

The need to convince the electorate of their economic and other successes means that governments can only pursue longer-term goals to the extent that either they do not significantly impact on short-term outcomes or their policies can be 'sold' as so desirable that 'pain now is worth the gain later'. This situation is increasingly exacerbated. Whereas until relatively recently governments, once elected, were largely immune to public pressure until re-election time (including pressure through the media), they are increasingly continuously monitoring and responding to such pressure. Frequent opinion polls, focus groups and other sources of appreciation of public views are being used not only to monitor responses to policies but also to fine-tune and, if necessary, alter those policies in order to maintain positive ratings in the polls. Further, given the important role of the mass media in the presentation of government policies and developing the electorate's understanding of, and reactions to, those policies, governments are increasingly taking account of the agenda of those who control the media, and manipulating their actions to win media approval.

Of space, place and democratic practice

Interacting with this temporal constraint on government actions are two spatial constraints. One concerns the inherently spatial construction of the contemporary state, the other its internal coherence.

Most states have immense power to act as they see fit within their own territories – subject to the constraints of their own legal systems and any international obligations they have entered into and are prepared to meet. In some situations, those controlling the state apparatus may perceive no alternative but to enact a neo-liberal agenda whatever its likely impacts on their electorate – as happened

in New Zealand from 1984 on. Some may be virtually obliged to do so by pressure from more powerful outsiders – including the policies of bodies such as the IMF and the World Bank (Peet, 2003). But in many states for some of the time, and in some for all of the time, the dominant goal is to promote local as against wider interests.

The crucial issue in all of this is the balance between the second and third state roles. Governments wishing to be re-elected are well aware that they must not substantially alienate either or both of society's two main fractions. If the capitalist fraction and its (middle-class) supporters feel that the state is not serving it well, and that the processes of wealth production are failing to deliver against expectations, they may withdraw support – as may the non-capitalist fraction and its (working class) supporters if they feel ill-served by state policies. Most political parties, especially in long-established democracies, have a bedrock of support they can call on, but this is rarely enough to guarantee success at a sequence of elections unless a significant component of both fractions is sufficiently positive and optimistic to vote for the incumbent party(ies) rather than an opposition. To survive, governments have to deliver policies which keep a substantial majority of their electorates contented – which may well mean acting in ways that constrain the overall, global neo-liberal goal. Local, short-term inefficiency may be a necessary price for electoral support.

The first of the state's roles has received relatively little attention in this discussion, although it is overall the most important (without success at it, accomplishment of the other two is very unlikely) – ensuring support for the state apparatus and its actions across the national territory. In part, this is pursued through ideological means – many associated with nationalism – but unless rhetorical claims for unity are accompanied by economic success, long-term cohesion will be threatened.

The implication is that governments must not stimulate – or fail to correct – conditions which lead to the creation of a relatively permanent disadvantaged 'class', whose response to their situation threatens the state's legitimacy. There are circumstances where such classes exist but their radical potential has not been realized (Galbraith, 1979): most governments accept that ignoring the interests of a substantial proportion of their electorate can be dangerous, however.[16]

Threats to the state's cohesion are particularly troublesome when the disadvantaged are spatially concentrated. Uneven development can become a major political cause – especially where it can be associated with other, cultural, movements, such as nationalism. A spatially divided country faces potential political problems which if unresolved can have major economic impacts: this situation may be exacerbated if the country's elected government is drawn overwhelmingly from certain parts of its territory only. This may do no more than stimulate political unrest – as with the UK's north–south divide which was very substantially exacerbated in the 1980s (Hudson and Williams, 1989; Johnston, 1991). But where tensions between regions threaten a country's overall stability, rationality crises may well be the outcome, with legitimation and, possibly, accumulation crises not far behind.

DEMOCRACY AND ENVIRONMENTAL PROBLEMS

The previous section highlighted the constraints on state action created by the role of elections in democracy's operations. Although electoral power gives governments considerable freedom – within constitutional and other constraints – to act as they wish, knowledge that the electorate has to be faced again in the relatively near future provides a considerable restraint on exercise of that freedom. This is clearly illustrated with regard to environmental policy. It is not, of course, an absolute constraint: governments do enact environmental policies and tackle environmental problems, but perhaps not as wholeheartedly and rapidly as some might wish, and as might be in the country's best long-term interests.

Environmental policy at home

Many environmental policies enacted by governments are domestic. Their impacts – as with the problems the policies tackle – may have international ramifications, especially for neighbouring countries, but the issues are very largely internalized to the state's own territory. In such situations, when evaluating a policy proposal, the sixth, seventh and eight barriers listed above are of particular importance.

In determining whether a proposed environmental policy will win acceptance, one consideration facing a government is whether it will become an electoral issue of sufficient import that enough people will decide whether to vote for or against its return to office if the policy is enacted. Few environmental issues reach the top of the political agenda and become determining influences. When asked the main issues which will affect how they vote, most people identify the economy and/or jobs, taxes, pensions, public services – especially education and health. That being the case, the easiest path for parties is to include environmental policies in their rhetoric – they lose nothing by being committed to such

'good causes' – but to do very little when in office. If all of the main parties do the same, the environment becomes a non-issue. If all are committed to the same cause (though each may seek to convince voters that it is 'greener' than all of the others) then it is not a distinguishing characteristic of their manifestos, and unlikely to be salient to the electorate (on which see Brotherton, 1986, and Owens, 1986).

One consequence of this may be the establishment of a ('green') party which puts environmental issues at the top of its agenda, arguing that its opponents are, in effect, complacent about environmental issues and allowing environmental deterioration. (On green parties, see Rudig and Lowe, 1986; Hay and Hayward, 1988; Porritt and Winner, 1988: on green political thought more widely, see Dobson 2000, 2003. On the possible 'environmental transformation' of the state, see Dryzek et al., 2002.) Because of the low salience that environmental issues have for most people, however, such 'green' parties are unlikely to attract wide support unless they have credible policy positions on the dominant issues. Furthermore, under many electoral systems any support that they gain is unlikely to be translated into representation in the legislature: single-member constituency systems such as the UK's discriminate very unfavourably against small parties, who generally win no parliamentary seats (Norris, 2004). Such 'minority, predominantly single-issue' parties usually produce their best performances at second-order elections of low salience to most voters. At these contests, they are able to pick up protest votes against the major parties and, by turning out their supporters in a context of relative apathy, perform well when there is a high abstention rate – as at the 1989 European Parliament elections in the UK.

In proportional representation electoral systems, however, 'green' parties may win both a legislative presence and, where coalition governments are being formed, considerable influence over policy. In Germany, *die Grünen* have won seats in the Bundestag and places in a left-leaning coalition government with the Social Democrats; in New Zealand, the Green Party has cooperated with the minority Labour government in return for concessions on policymaking in certain areas. Bargaining over coalition membership can place environmental issues on the government's agenda. As a counter to this, however, 'greens' have to recognize the necessity of limiting their claims so that the government can successfully negotiate the balancing act across the main state roles. This can create internal tensions within a 'green party'. Should it remain true to its foundational beliefs and refuse to participate either in the implementation of policies that may have harmful environmental effects or in rejecting policies that will ameliorate if not remove environmental harm, in order to sustain the government's overall programme? Or should it compromise on some (many?) issues, in order to win on others?

Governments can largely ignore environmental issues if they have little salience for most electors. On some occasions, however, a large section of the electorate may be mobilized around a certain issue – as happened with a proposal to dam Lake Manapouri in New Zealand to create power-generation capacity for a proposed aluminium smelter. The government can then react, even though it may believe that the issue is unlikely to be a win-or-lose issue at the next election. In any case, many such issues are localized, and an incumbent government can evaluate whether to act or not will be crucial for enough voters in enough places in order to have an impact on its electoral performance. Such calculations were undoubtedly employed by the UK's Labour governments of 1997–2005 over a number of 'countryside' issues, such as reaction to a foot-and-mouth disease outbreak (Ward, 2002; Woods, 2002), protests about the impact of fuel tax increases on rural life, and the banning of hunting with dogs (portrayed by those in favour of the pursuit as the most humane way of controlling a pest).

Once in power, governments are open to a range of influences. Interest groups invest much time, energy and money lobbying governments, seeking to influence legislation and action. Much lobbying favours 'business' interests and, given that most governments, as discussed above, accept that in a neo-liberal, globalizing world sustaining economic growth is key to overall prosperity – and thus to their re-election prospects – these are often the most important influences. (The main opposition to business lobbies are the trade unions, whose influence has been considerably emasculated in recent decades.) Environmental policies are often costly for businesses and threaten profits, so they are wary of any which appear to have such an impact, especially if this will put them at a disadvantage relative to competitors – particularly competitors located in other jurisdictions where less stringent policies apply.

This suggests that most governments will be relatively inactive on the environmental front. The electorate is rarely concerned enough about the issues to make them salient when they next vote, and businesses, whose continued investment is crucial to economic and thence social prosperity, are generally opposed to potentially expensive policies. In the balancing act between the second and third state roles, governments can favour the capitalist fraction considerably.

It is far from the case that governments rarely act with regard to environmental issues, however. Many of the problems raised are localized and their resolution either relatively inexpensive or seen as necessary by a majority of the people affected. With larger issues, once the problem and potential solutions are identified, support for action being taken can be mobilized, especially if most sections of society gain some benefits – there is something in it for everybody and it is not too expensive. Such was the case, for example, with the *Clean Air Act* 1956 which involved all sections of the country accepting the need for a solution to the problem of 'smog', even though the legislation was only introduced to Parliament through a Private Member's Bill (Brimblecombe, 1987). For businesses, costs related to their employees' ill-health were substantially reduced by the requirement that only smoke-free fuels be used, whereas individuals had the costs of replacing their polluting fireplaces met through government grants. More recently, the negative impacts of emissions from internal combustion engines powered by leaded petrol were realized and the UK government, by manipulating tax rates and making the newly available unleaded petrol cheaper, was able to ensure an almost complete switch to the environmentally less harmful fuel over a relatively short period. This had clear benefits for public health and no substantial negative impact on the oil companies, which were able to meet the new regulations.

International environmental policy

Some of the most pressing and least tractable contemporary environmental problems are global in their scope, and many more are clearly international. Action by one country alone is unlikely to have a substantial impact and so cooperation is crucial to problem resolution. The metaphor of the prisoners' dilemma game is even more apt in that arena: it highlights the absence of an institution that can play the state's roles.

The message of the tragedy of the commons metaphor is that prevention of free-riding by all contributors to the generation of an environmental problem requires an institution which has the power to regulate individual behaviour through both persuasion and sanctions. There is no comparable body to the state in the international arena, however, to which individual states are prepared to yield substantial elements of their sovereignty.[17] This could potentially (and substantially) impede their ability to promote their own residents' interests and thence, in a democracy, the likelihood that those in charge of the state apparatus will be re-elected to that position.

There are many international bodies at a variety of scales, however, from institutions involving a few countries only created to tackle local issues, through regional bodies – such as the European Union – to those with global reach, notably the United Nations and its associated agencies.[18] All have only restricted powers which the member states are prepared to grant them and have few sanctions they can impose if one or more members decide not to participate in joint action. This has been made very clear in recent debates over environmental issues – as with the refusal of one US President (George Bush Sr.) to sign the *Biodiversity Convention* developed at the 1992 Rio United Nations Conference on Environment and Development because it threatened 'American jobs and American families ... we cannot permit the extreme in the environmental movement to shut down the United States' (Pepper, 1993, p. 219: George Bush Sr. was running for re-election at the time) and of another (George W. Bush) to sign the *Kyoto Protocol* on reducing carbon emissions for similar reasons (including the perceived relatively favourable treatment given to 'developing economies', which might then be able to compete 'unfairly' with American producers in international markets).[19] Bush's intransigence, clearly defending American interests, later acted as a block to firm international agreement on how to tackle global warming – indeed, even if it was accepted as a problem that needed to be tackled – at the 2005 G8 summit in the UK.

These and many other examples illustrate the problems of gaining international agreement to tackle perceived environmental problems. Even where agreement can be obtained, individual countries find ways of evading their obligations – or getting the agreement changed to their benefit: as when land-locked countries are encouraged ('bribed') by others to subscribe to the International Whaling Convention and then vote to promote those countries' interests, and – since sanctions under that Convention can only be imposed by the 'flag state' (i.e. the country with which a whaling vessel is registered) – encouraging vessels to register with states that are not signatories to the Convention. The pursuit of environmental goals is substantially impeded by such tactics whereby individual countries put their own short-term interests (and their governments' concerns to be re-elected) first.

Nevertheless, there are many international environmental agreements, some with global scope and considerable success: Barrett (2003) lists 225 agreements currently in force and a further 72 which had been agreed and could come into force if ratified by sufficient parties. Whereas many of those multilateral agreements have less than ten signatories – which in most cases means that they are restricted in their scope to defined areas of the Earth's surface (such as the *Regional Convention for the Conservation of the Red Sea and Gulf of Aden Environment*) – others have many more: Barrett lists 9 agreements in force which have more than 150 signatory states, and a further 10 with between 100 and 149 signatories. Some are outline agreements only, with little if any regulatory force, and others lack the signatures of the most important states and so are largely impotent. (Neither the USA nor the former USSR signed the *Moon Treaty*, for example. Only they have sent spacecraft to the Moon.)

Why do states sign up to some environmental agreements and treaties, and then accept the imposed regulation? In a comprehensive analysis of the game-theoretic underpinnings of such actions, Barrett (2003) argues that this is because it is possible – at least in some circumstances – to create agreements in which all of the parties gain. He terms such agreements *self-enforcing*, with three central conditions:

1 They must be *individually rational* – no signatory can gain from withdrawing, no non-signatory can gain from later accession (i.e. all parties should sign up at the outset) and no signatory can gain from failing to comply with the agreed regulations.
2 They must be *collectively rational* – the promises and threatened sanctions must be credible so that states want to sign up to them and gain the benefits that should accrue.
3 They must be *fair*, so that all parties believe that they are reasonably treated.

Creating such agreements is a difficult task – especially if they involve large numbers of participants. Many of the successful ones involve only a small number of countries, because of their limited spatial scope and number of interested parties – Barrett uses the example of *The North Pacific Fur Seal Treaty*, a 1911 agreement to ban pelagic seal-hunting involving Canada, Japan, Russia and the USA, to illustrate a treaty from which none would gain by withdrawing and which is seen as fair.

Barrett also cites treaties with many signatories which meet his three conditions, among which the most notable is the *Montreal Protocol on Substances that Deplete the Ozone Layer*, and various follow-up agreements. This protocol was established within the framework of the earlier *Vienna Convention for the Protection of the Ozone Layer*, which emerged from discussions in the wake of theoretical data which predicted the likely consequences of continued production and consumption of CFCs (no hole had then been discovered in the ozone layer). The negotiations (fully described in Benedick, 1991, 1998: see also Litkin, 1994; Susskind, 1994) took place in a context which apparently 'sounded the death knell for an important part of the international chemical industry, with implications for billions of dollars in investments and hundreds of thousands of jobs in related sectors' (Benedick, 1991, p. 1). Because of this, vested interests – such as the UK chemical giant ICI – were initially strongly opposed to any curbs.[20] But the USA supported them, and its giant chemical firm du Pont encouraged that position by stating that it was phasing out CFC production, largely because replacement alternative, apparently non-toxic, chemicals were by then available. The negotiators were thus able to evolve a protocol which was eventually acceptable to all parties: benefit–cost tables convinced many producer countries that the costs of replacing CFCs would be met by savings in health care and other costs (which would be equitably distributed) and developing countries – which were consumers rather than producers – were provided with aid to assist the transfer to alternative technologies, notably for refrigerators, although the USA insisted on a virtual veto over the financing rules (see Johnston, 1996, p. 228).

The lesson drawn by Barrett from the *Montreal Protocol* is that it illustrates how international environmental agreements can be reached and implemented if they meet certain important criteria. The offered incentives must be attractive to all countries from the outset, so that they want to participate. The nascent treaty must deter non-participation, so that all states want to be involved in its construction, and participation levels should be set to encourage this. If a treaty comes into force with a relatively small number of participants, there is a greater incentive to stay out and free-ride. The incentives must contain clear strategies that will induce behavioural changes, setting realistic and credible targets that can be readily measured and monitored so that compliance can be verified. In addition, if necessary, the incentive structures should encourage countries to participate by, for example, offering side-payments to those which would find compliance most difficult (as with the *Montreal Protocol*) and linking treaties so that compliance with one brings benefits from another. Such agreements are very difficult to negotiate, especially with a large number of parties involved. Barrett's key argument is that they should be based on incentives rather than threats. Full participation and cooperation is most likely if all parties feel that they will benefit, not least because that will allow them to 'sell' participation to their electorates.

The difficulties inherent in such a task are exemplified by Barrett's discussion of the *Kyoto Protocol*, which required ratification from 55 countries before coming into force, including 38 listed as together contributing at least 55 per cent of global carbon dioxide emissions in 1990. The USA and Australia have declined to ratify this, so that although 94 countries had ratified it by the end of 2002, insufficient of the larger carbon dioxide producing countries had done so for it to come into force until Russia did so in late 2004.[21] For Barrett, the biggest problem with the *Protocol* as originally drafted was that it failed to address the enforcement issue; when this was tackled at later meetings, the agreed mechanism was defective. The negotiators, he argued, should have built the *Protocol's* architecture around an agreed practicable enforcement procedure if they were to gain widespread support internationally.

Would Kyoto be effective? Barrett (2003, 380) reports an economic study for the Clinton administration suggesting that the overall benefits would match the costs, *if* the 'emission limits are met in a globally-effective manner'. The proposed trading mechanism for emissions would not facilitate this, however, and the compliance targets agreed in Bonn in 2001 had five major flaws according to Barnett, not least being that it was not legally binding – and to make it so would undoubtedly induce some countries to decline to ratify it. Nevertheless, at the beginning of 2005 the EU implemented a mandatory greenhouse gas emissions trading scheme which is linked to member countries' Kyoto obligations.

The problems over Kyoto illustrate a key difference between international and intranational action to tackle environmental problems: as Barrett puts it:

> the US government can enforce a *domestic* law implementing Kyoto. But it cannot make other countries comply let alone participate (2003, p. 398).

State power is bounded by territorial limits, and the nature of sovereignty precludes one state requiring another to act in a particular way. That goal can only be achieved in a sustainable way when states freely agree to participate, which they are only likely to do if they perceive that benefits will flow so that the action can be justified to their electorate.

CONCLUSIONS

Democracy, in Churchill's much cited words, is 'the worst form of government except all those other forms that have been tried from time to time'. Its application may – as George Bernard Shaw and others contended – be only an experiment, but in the contemporary world it is an experiment that an increasing number of countries is prepared to undertake. We may not have reached Fukuyama's 'end of history', but as the world faces up to crucial decisions regarding the state of its environment democracy provides the framework within which they will be taken. As such, appreciating the nature of democracy as currently practised – that is, government by the consent of the people – is a necessary element in the preparation of environmental policies for which collective action, especially international collective action, is accepted as the only way forward.

Government by consent of the people could take a variety of forms. That currently promoted – representative democracy – involves regular elections whereby those controlling the state apparatus seek re-endorsement by the electorate whereas others present alternative scenarios in their quest to replace those currently in power. Elections play a pivotal role in state operations, therefore: indeed, an increasing part of political life focuses on them. In particular, political parties in power structure their policies and actions in order to ensure electoral victory and their opponents present their alternatives for the same purpose. This agenda setting involves continued interactions between political parties and the electorate, with the former seeking both to identify the salient issues for the latter and to manipulate their priorities so as to coincide, as far as possible, with the issues on which they are most comfortable. In a predominantly capitalist world, those issues inevitably focus on economic concerns: people vote for the parties that are most likely to bring prosperity – to their country and region as well as to themselves.

Environmental issues rarely appear near the top of political agenda in that context. As a consequence, there is little pressure on governments to act, even though it is generally accepted that tackling environmental problems requires collective action of the sort that can only be mobilized and regulated by the state. This is because of the barriers to such action identified here: not only has a government to be convinced of a problem's existence, an appreciation of its causation, and the existence of a fully costed viable solution, it also has to believe that it can win acceptance for that solution. Its evaluation on the last point may rest on whether it thinks it will win or lose a forthcoming election if it pursues the intended policy. More likely to be important, however, is whether it feels that implementing a policy will alienate key supporters and/or have an impact on living standards. In contemporary capitalism, a government's prime role is to sustain economic growth from which most people benefit. If it feels that its popularity may be significantly dented because of the pursuit of particular environmental policies, especially

those whose impact is focused on key constituencies (which may be spatially defined in many electoral systems), then it may be reluctant to act.

This reluctance to act because of the electoral constraints does not mean that substantial environmental policies are never pursued. There are examples, as with clean-air legislation in many countries, where governments have acted, without their electoral prospects being significantly affected. There are three reasons for this. First, the problem being tackled is widely appreciated and the potential benefits of action understood – perhaps because of educational policies promoted by the state. Second, the policy is constructed so that as many actors (individual and corporate) as possible receive benefits and the allocation of costs is perceived to be fair. And third, because of the state's sovereignty, people realize that compliance will be ensured, and that there are no free-riders.

The difficulties that governments face in surmounting the barriers to environmental action over issues that are confined within their own territories are small relative to those encountered when tackling issues on an international scale – especially if they are global in their scope and require multilateral agreement across a large number of countries. They are the same three as those faced with internal policies but, as debates on global climate change have shown, gaining appreciation of the nature (even existence) of a problem may not be straightforward, let alone then winning acceptance that there is a viable solution. Even when that is the case, however, construction of a policy perceived to be fair to all parties and thus attracting not only their participation but also their compliance with the regulations is intensely fraught. The final difficulty is even more so, because no institution can ensure compliance by, if necessary, overriding an individual state's sovereignty. Indeed, it is because of the absence of such an institution, and the unlikelihood of one being established with the needed powers in the current anarchic international regime of sovereign states, that overcoming the second of the difficulties is so important. As the *Montreal Protocol* showed, this is feasible with careful negotiation but, even so, the likelihood of that agreement being reached and sustained would probably have been unlikely if the most affected actors – chemical industries – had not found replacements for CFCs. If that had not been the case, the probability of governments being convinced that to comply with the *Protocol* would have damaged companies and jobs would probably have led to a failure to surmount the barrier of political will.

The resolution of environmental problems of the scale that are now impacting the Earth calls for political will and skill. Identifying viable solutions scientifically is a difficult task, but is only the first stage in a process which involves governments 'selling' those solutions to their electorates (in several senses of that term). They need to be convinced not only that enacting the policy is necessary but also that the costs will not outweigh the benefits (short as well as long term) and that all will contribute in a fair way. If they cannot be convinced – or, more likely in many cases, if governments believe that they cannot convince them – then the policies will be ignored and selfish local gains put above collective benefits. Governments may well recognize the existence of the issues in their rhetoric, and tackle them where the political costs are slight, but their actions may not be commensurate to their words. Democracy, as practised in an increasing number of countries around the world, is presented as a major bulwark of individual freedom – including, it seems, the freedom to harm the environment irrespective of the long-term damage that may cause. There are examples of successfully negotiating effective environmental policies within that freedom, but perhaps too little evidence that enough is being done to promote that cause over a wide enough front, in sufficient time.

NOTES

1 Dryzek (1987) identifies nine, some of which are variants on Laver's basic set.

2 Those in control of the state apparatus may prefer encouragement to conform to punishment for not conforming – carrots rather than sticks – but without the latter, the former may be ineffective.

3 The scientists need not be disinterested once they are convinced by their research findings: they may then become advocates arguing that governments should acknowledge the existence of a problem (and then convince citizens in general accordingly), and then for a preferred solution or resolution.

4 Of course, university scientists are increasingly encouraged to obtain research contracts from such interested parties, in which case their policy advice will be partial – albeit based on credible scientific knowledge. For a more detailed discussion of these issues see Johnston and Plummer (2005).

5 Such resources – such as fish – may be mobile and move in and out of a state's territorial claims.

6 Harvey (2005) goes further and sets the two up as a binary pair of ideal types: the neo-liberal state is associated with the second task and the social democratic state with the third.

7 The nature of that search for re-election varies according to the exact form of the representative democracy. In US Presidential elections, for example, the incumbent may not be seeking re-election (perhaps because he cannot after two terms in office)

and his deputy may not be seeking to replace him: nevertheless, their party will want to retain power and – in most cases – build on the incumbents' achievements.

8 The 2000 Presidential election was a partial exception to this, with the outcome in Florida being subject to several court challenges and many people believing that the wrong candidate was declared elected there.

9 In some countries – as in the central Asian former Soviet republics and several African states – there has been a return to autocratic rule (or, at least, a diminution in the strength of the democratic process).

10 There are some arguments that ideological factors became more important in the 2004 Presidential election, though it is unclear whether this has percolated to other levels of government.

11 On the impact of 'Rogernomics' see a special issue of *GeoJournal* (2003).

12 There are considerable debates regarding the 'death' of class voting and its replacement by issue-based, responsive voting: see Evans (1999).

13 The upper houses of many legislatures have longer periods of office – for life in some cases, such as the UK House of Lords – but few have more than residual power, and those that have more tend to be the ones subject to relatively frequent election, such as the US Senate.

14 There are also issues in some states, where the legislature is not sovereign, regarding the separation of powers between various arms of the state apparatus – basically, the executive, legislature and judiciary.

15 They may – as occurred in New Zealand in 1987 – seek re-election on the grounds that their programme is as yet unfinished (Johnston and Honey, 1988).

16 Of course, a potential danger can be marginalized by excluding a latent radical element from the electorate – which is what occurred with blacks in the southern USA for over a century after the Civil War, despite the passing of the 14th Amendment to the Constitution.

17 The European Union (EU) is a clear exception to this, although there is much debate over the yielding of sovereignty and its implications for national interests.

18 On the negotiation of environmental policy in the EU see, e.g. Kellow and Zito (2002).

19 The UK Government's Department for Environment, Food and Rural Affairs defended its 2004 decision to allow greenhouse gas emissions to increase by 7.5 per cent over the next three years by stating that 'We are balancing the need to protect the competitive position of UK industry while moving us beyond our Kyoto Protocol commitment towards our tougher national goal' (quoted in a letter to *The Times*, 4.11.2004, p.18).

20 As Benedick (1991, p. 64) expressed it, 'some governments allowed self-interest to influence their

scientific positions and use the scientific uncertainty as an excuse for delaying difficult decisions. Many political leaders were long prepared to accept future environmental risks rather than to impose the certain short-term costs entailed in limiting use of products seen as necessary to modern standards of living'.

21 President Clinton signed the Protocol in 1998 in the knowledge that the Senate (which has to approve such a treaty by a two-thirds majority) would certainly refuse to ratify it. President George W. Bush has overridden Clinton's signature by rejecting the Protocol and deciding not to submit it to the Senate for ratification.

REFERENCES

Archer, J. C. and Taylor, P. J. (1981) *Section and Party*. Chichester: John Wiley.

Barrett, S. (2003) *Environment and Statecraft: the Strategy of Environmental Treaty-making*. Oxford: Oxford University Press.

Benedick, R. E. (1991) *Ozone Diplomacy*. Boston: Unwin Hyman.

Benedick, R. E. (1998) *Ozone Diplomacy*, 2nd ed. Cambridge, MA: Harvard University Press.

Brenner, N., Jessop, B., Jones, M. and McLeod, G. (eds) (2003) *State/Space: a Reader*. Oxford: Blackwell.

Brenner, N. (2004) *New State Spaces: Urban Governance and the Rescaling of Statehood*. Oxford: Oxford University Press.

Brimblecombe, P. (1987) *The Big Smoke: a History of Air Pollution in London Since Medieval Times*. London: Methuen.

Brotherton, D. I. (1986) Party political approaches to rural conservation in Britain. *Environment and Planning A* 18: 151–160.

Clover, C. (2004) *The End of the Line: How Overfishing is Changing the World and What We Eat*. London: Ebury Press.

Crewe, I. (1986) On the death and resurrection of class voting: some comments on *How Britain Votes*. *Political Studies* 35: 620–638.

Dahl, R. A. (1978) Democracy as polyarchy. In R. D. Gastil (ed.) *Freedom in the World: Political Rights and Civil Liberties*. Boston: G. K. Hall, pp. 134–146.

Dix, R. H. (1984) Incumbency and electoral turnover in Latin America. *Journal of Interamerican Studies* 26: 435–448.

Dobson, A. (2000) *Green Political Thought*, 3rd ed. London: Routledge.

Dobson, A. (2003) *Citizenship and the Environment*. Oxford: Oxford University Press.

Dryzek, J. (1987) *Rational Ecology*. Oxford: Basil Blackwell.

Dryzek, J. S., Hunold, C., Schlosberg, D., Downes, D. and Hernes, H.-K. (2002) Environmental transformation of the state: the USA, Norway, Germany and the UK. *Political Studies* 50: 659–682.

Evans, G. (ed.) (1999) *The End of Class Politics? Class Voting in Comparative Context*. Oxford: Oxford University Press.

Finer, S. E. (ed.) (1974) *Adversary Politics and Electoral Reform*. London: Anthony Wigram.

Fukuyama, F. (1989) The end of history. *The National Interest* Summer: 2–18.

Fukuyama, F. (1992) *The End of History and the Last Man*. New York: The Free Press.

Galbraith, J. K. (1979) *The Affluent Society*, 3rd ed. London: Penguin Books.

GeoJournal (2004) Special issue on 're-inventing government: emerging geographies in the aftermath of the 1984 reforms in New Zealand'. *GeoJournal* 59: 91–166.

Gimpel, J. G. and Schuknecht, J. E. (2003) *Patchwork Nation: Sectionalism and Political Change in American Politics*. Ann Arbor: University of Michigan Press.

Habermas, J. (1976) *Legitimation Crisis*. London: Heinemann.

Hajer, M. (1995) *The Politics of Environmental Discourse: Ecological Modernization and the Policy Process*. Oxford: The Clarendon Press.

Hardin, G. (1968) The tragedy of the commons: the population problem has no technical solution; it requires a fundamental extension in morality. *Science* 162: 1243–1248.

Harvey, D. (1982) *The Limits to Capital*. Oxford: Basil Blackwell.

Harvey, D. (1989) *The Condition of Postmodernity*. Oxford: Basil Blackwell.

Harvey, D. (2005) *A Brief History of Neo-liberalism*. Oxford: Oxford University Press.

Hay, P. R. and Hayward, M. G. (1988) Comparative green politics: beyond the European context? *Political Studies* 36: 433–448.

Held, D. and McGrew, A. (ed.) (2002) *Governing Globalization: Power, Authority and Global Governance*. Cambridge: Polity Press.

Held, D. and McGrew, A. (ed.) (2003) *The Global Transformations Reader: an Introduction to the Globalization Debate*. Cambridge: Polity Press.

Herrera, G. (2002) The politics of bandwidth: international political implications of a global digital information network. *Review of International Studies* 28: 93–122.

Hudson, R. and Williams, A. (1989) *Divided Britain*. London: Belhaven.

Jessop, B. (2002) *The Future of the Capitalist State*. Cambridge: Polity Press.

Johnston, R. J. (1983) *The Geography of Federal Spending in the United States of America*. Chichester: John Wiley.

Johnston, R. J. (1991) *A Question of Place: Exploring the Practice of Human Geography*. Oxford: Basil Blackwell.

Johnston, R. J. (1992) Laws, states and super-states: international law and the environment. *Applied Geography* 12: 211–228.

Johnston, R. J. (1996) *Nature, State and Economy: a Political Economy of the Environment*. Chichester: John Wiley.

Johnston, R. J. (1999) The United States, the 'Triumph of Democracy' and the 'End of History'. In D. Slater and

P. J. Taylor (ed.) *The American Century: Consequences and Coercion in the Projection of American Power*. Oxford: Blackwell Publishers, pp. 149–165. (ed.) *Worldminds: Geographical Perspectives on 100 Problems*. Boston: Kluwer, pp. 3–8.

Johnston, R. J. and Honey, R. (1988) Political geography of contemporary events: the 1987 general election in New Zealand: the demise of electoral cleavages? *Political Geography Quarterly* 7: 363–368.

Johnston, R. J. and Plummer, P. (2005) What is policy-oriented research? *Environment and Planning A* 37: 1521–1526.

Jones, M. and Jones, R. (2004) Nation states, ideological power and globalisation: can geographers catch the boat? *Geoforum* 35: 409–424.

Kellow, A. and Zito, A. R. (2002) Steering through complexity: EU environmental regulation in the international context. *Political Studies* 50: 43–60.

Key, V. O. Jr. (1955) A theory of critical elections. *Journal of Politics* 17: 3–18.

King, R. and Kendall, G. (2003) *The State, Democracy and Globalization*. London: Macmillan.

Laver, M. (1984) The politics of inner space: tragedies of three commons. *European Journal of Political Research* 12: 59–71.

Laver, M. (1986) Public, private and common in outer space. *Political Studies* 34: 359–373.

Laver, M. (1997) *Private Desires, Political Action: an Invitation to the Politics of Rational Choice*. London: Sage Publications.

Lewis-Beck, M. S., Nadeau, R. and Bélanger, R. E. (2004) General election forecasts in the United Kingdom: a political economy model. *Electoral Studies* 23: 279–290.

Lipset, S. M. and Rokkan, S. E. (1967) Cleavage structures, party systems and voter alignments: an introduction. In S. M. Lipset and S. E. Rokkan (ed.) *Party Systems and Voter Alignments*. New York: The Free Press, pp. 3–64.

Litkin, K. (1994) *Ozone Discourses: Science and Politics in Global Environmental Cooperation*. New York: Columbia University Press.

Mann, M. (1984) The autonomous power of the state. *European Journal of Sociology* 25:185–213.

Marquand, D. (2004) *Decline of the Public: the Hollowing-out of Citizenship*. Cambridge: Polity Press.

Mayhew, D. R. (2002) *Electoral Realignments: a Critique of an American Genre*. New Haven, CT: Yale University Press.

Norris, P. (2004) *Electoral Engineering: Voting Rules and Political Behavior*. Cambridge: Cambridge University Press.

O'Connor, J. (1972) *The Fiscal Crisis of the State*. New York: St Martin's Press

O'Connor, J. (1987) *The Meaning of Crisis: a Theoretical Introduction*. Oxford: Basil Blackwell.

O'Loughlin, J. (2004) Global democratization: measuring and explaining the diffusion of democracy. In C. Barnett and M. Low (eds) *Spaces of Democracy: Geographical Perspectives on Citizenship, Participation and Representation*. London: Sage Publications, pp. 23–44.

O'Loughlin, J., Ward, M. D., Lofdahl, C. L., Cohen, J. S., Brown, D. S., Reilly, D., Gleditsch, K. S. and Shin, M. T. (1998) The diffusion of democracy, 1946–1964. *Annals of the Association of American Geographers* 88: 545–574.

Osei-Kwame, P. and Taylor, P. J. (1984) A politics of failure: the political geography of Ghanaian elections. *Annals of the Association of American Geographers* 74: 574–589.

Ostrom, E. (1990) *Governing the Commons: the Evolution of Institutions for Collective Action*. Cambridge: Cambridge University Press.

Ostrom, E. (1995) Constituting social capital and collective action. In R. O. Keohane and E. Ostrom (ed.) *Local Commons and Global Interdependence: Heterogeneity and Cooperation in Two Domains*. London: Sage, pp. 125–160.

Owens, S. (1986) Environmental politics in Britain: new paradigm or placebo? *Area* 18: 195–201.

Peck, J. A. (2001) *Workfare States*. New York: Guilford Press.

Peet, R. (2003) *Unholy Trinity: the IMF, World Bank and WTO*. London: Zed Books.

Pepper, D. (1993) *Eco-socialism: from Deep Ecology to Social Justice*. London: Routledge.

Porritt, J. and Winner, D. (1988) *The Coming of the Greens*. London: Collins Fontana.

Poundstone, W. (1992) *Prisoner's Dilemma*. New York: Doubleday.

Pratt, N. (2004) Bringing politics back in: examining the link between globalization and democratisation. *Review of International Political Economy* 11: 311–336.

Rudig, W. and Lowe, P. (1986) The "withered" greening of British politics: a study of the Ecology Party. *Political Studies* 34: 262–284.

Sack, R. D. (1986) *Human Territoriality: Its Theory and History*. Cambridge: Cambridge University Press.

Schumpeter, J. A. (1943) *Capitalism, Socialism and Democracy*. London: George Allen & Unwin.

Steinberg, P. E. (2001) *The Social Construction of the Ocean*. Cambridge: Cambridge University Press.

Susskind, L. E. (1994) *Environmental Diplomacy: Negotiating More Effective Global Agreements*. Oxford: Oxford University Press.

Taylor, M. (1987) *The Possibility of Cooperation*. Cambridge: Cambridge University Press.

Taylor, P. J. (1986) An exploration into world-systems analysis of political parties. *Political Geography Quarterly* 5: S5–S20.

Taylor, P. J. and Flint, C. (2000) *Political Geography: World-economy, Nation-state and Locality*, 4th ed. London: Pearson.

Trudgill, S. T. (1990) *Barriers to a Better Environment: What Stops Us Solving Environmental Problems*. London: Belhaven.

Vogler, J. (2000) *The Global Commons: Environmental and Technological Governance*. Chichester: John Wiley.

Ward, N. (2002) Representing rurality? New Labour and the electoral geography of rural Britain. *Area* 34: 171–181.

Weiss, L. (1999) Globalization and national governance: antinomy or interdependence? *Review of International Studies* 25: 59–88.

Weiss, L. (ed.) (2003) *States in the Global Economy: Bringing Domestic Institutions back in*. Cambridge: Cambridge University Press.

Werz, N. (1987) Parties and party systems in Latin America. In M. J. Holler (ed.) *The Logic of Multi-party Systems*. Munich: Springer-Verlag.

Woods, M. (2002) Was there a rural rebellion? Labour and the countryside vote in the 2001 general election. *British Elections and Parties Review* 12: 206–228.

20

Political Ecology from Landscapes to Genomes: Science and Interests

Ronald J. Herring

INTRODUCTION

The genomics revolution in biology would seem to create novel analytical and policy questions for political ecology. Beyond the academic world, the capacity of genetic engineering to create wholly new organisms that cannot be produced in nature has generated movements, global networks and international soft law in anticipation of environmental effects from transgenic organisms. Complex state–market boundaries are being created, centering on bio-safety regimes in response to the global spread – both official and unmonitored – of transgenic organisms. The genomics revolution likewise enables new forms of property in nature, at smaller and smaller scale, invigorating a global debate on bioproperty, bioprospecting, and biopiracy. Does understanding these novel developments require new analytical tools? How might these technological developments sharpen our understanding of frameworks for understanding the political economy of nature in development?

This chapter seeks to draw and analyze continuities with venerable problems of intersecting politics and markets, property and livelihoods, in landscapes. First, there is the irreducibly normative character of distinctions between public goods and bads in nature. Second, transgenic politics reinforces the centrality of science to all political ecology. Science continually presents new challenges to the way interests in nature are understood by citizens and political classes that control states. "Interests" are thus not so easy to deduce from

structure as much political economy assumes. In turn, science itself is enmeshed in the politics of interested parties: states, corporations, universities, NGOs, and scientists themselves. In these contests over science, some knowledge claims win politically while others lose. Of special importance to public policy, ideologies incorporating "junk science" seem to gain political advantage, at least in the short term. The core characteristics of science-as-method – skeptical agnosticism, tentative conclusions, replicability, validation in epistemic communities – is politically weaker than reductionist explanations that evoke anxiety in spheres of low information and cognitive complexity.

POLITICAL ECOLOGY, "DEVELOPMENT" AND INTERESTS

Political ecology has been conceptualized in different ways.[1] The general project is to provide some coherent framework for the political economy of nature in the context of human transformations. These transformations are typically captured by the concept of "development."

Political economy of all varieties is fundamentally a study of who gets what and how, or the dynamics of interests within structures. Agentless structures and structureless agents are equally inadequate as explanatory frameworks (Elster, 1985). The concept of "interest," however, turns out to be more problematic than much social theory suggests;

interests in nature are especially problematic. Individual and collective interests in nature are multidimensional; these are frequently unrecognized prior to degradation that reveals interests. Cultural and aesthetic interests overlay and often contradict material and instrumental interests, with valences that vary over time and landscape.[2] Interests are indeterminate because the alternative ways in which nature is conceptualized, valued, and understood – framed, in short – differentiate interests; interests in nature therefore inevitably introduce a cultural and cognitive problematic. This relationship is widely accepted, though easier to recognize than theorize.[3]

Political economy privileges a smallish number of primary and irreducible concepts: structure, interest, power, collective action. Parsimony necessitates a restricted field of explanatory variables. In the hardest of political economy, ideas are epiphenomenal: this is the original meaning of ideology. Interests are typically derivative of structure. One of the most common conclusions is that the powerful and wealthy prefer market outcomes, the poor and weak prefer state regulation – the rigging of a sphere in which they would otherwise have no power.[4]

Ideas occupy a fragile space in materialist epistemology and method; ideologies are held to reflect interests.[5] The strongest empirical argument against a purely materialist political economy of nature is that its constituents sometimes win politically. Not all wins are without material explanations – the Montreal Protocol's market-rigging effects solidified a global niche for firms capable of the research and production costs of filling regulated markets for ODS (Ozone-depleting substances) substitutes.[6] Nevertheless, in capitalist society, market power usually wins, which means that preservation depends on ideas influencing definitions of interests: will people contribute to NGOs protecting wild places or not? In some cases, preservation wins over degradation embodying powerful interests. Though "wilderness" has enormous opportunity costs, its protection engages civil society, states, international organizations and treaties.[7]

Much indeterminacy of interests in nature is knowledge based; radically different levels of ecological knowledge characterize mass publics, political actors, and administrative managers over time. The sea change in redefinitions of interests – of individuals, firms and states – introduced by the atmospheric science of ozone holes and climate change is archetypal. Interests emerged to the cognitive level, and eventually to the policy level, only when the science became in rational discourse undeniable, even though transparently interested actors continued in denial.

"Development" complicates these relationships. Etymologically, development means "unfolding," as in the development of an embryo. Because societies differ on what the end state can or should be, development itself is a profoundly normative concept. Nothing illustrates this second rethinking of development so well as critiques of GNP measures on grounds of sustainability and environmental quality of life. Developmental transformation of landscapes frequently involves environmental degradation (Blaikie and Brookfield, 1987) and often displacement of the weakest citizens (Kothari, 1995a,b). Yet, even with this knowledge clearly available, the association of development with economic growth has become so naturalized that we often forget critical caveats, both in rigorous analytics and in common language. "Developing countries" became the common designation for a vast range of societies with low incomes: some doing better over time, some doing worse, some collapsing.

Growth-centric conceptualizations confronted a fundamental critique with the rise of ecological science in mainstream discourse. Environmental externalities are unequivocally market failures; market failures in theory call forth public policy. Moreover, environmental resources – and perhaps more critically but less understood, environmental services – are demonstrably necessary for economic activities: growth prospects depend on natural systems, however one values human health or biodiversity (Daily, 1997).

Ecological science has irrevocably complicated the comforting notion of "sustainable development."[8] Developmentalists confront the challenge of interdependent systems in which there may be threshold effects, variable fragility and resilience, cascading causation, indeterminacy – and continuous change. Science is the only conceivable method for beginning to specify these variables, but our collective knowledge is thin, and contested.

Nevertheless, contemporary states typically, often grudgingly, accept that whatever development means, there are public goods in nature; sustainability means paying some attention to them. But claims to public-goods provision in nature presuppose authority – the basis for governance.[9] This is true about both village commons and international treaty regimes. Public authority derives in part from authoritative science and is legitimated by market failure: the inability of self-seeking individuals to provide the level of public goods they individually desire.

Market failure becomes especially prominent as nature itself is commoditized. Karl Polanyi, wrote in 1944:

> Production is interaction of man and nature; if this process is to be organized through a self-regulating

mechanism of barter and exchange, then man and nature must be brought into its orbit; they must be subject to supply and demand, this is, be dealt with as commodities, as goods produced for sale (1957, p. 130).

Commodities are used according to the logic of exchange value; developmental processes typically transform nature to produce high-value commodities. Wilderness was valueless, and dubbed *res nullius* – not only belonging to no one, but not really worth the transaction costs of owning. Commoditization of all components of natural systems altered this situation. Colonial rule often put Polanyi's *Great Transformation* on a forced march.[10] William Cronon's pioneering treatment (1983) of ecologies of New England under British rule is entitled *Changes in the Land;* the final chapter is tellingly titled "That Wilderness Should Turn a Mart." The title emphasizes both the contemporary astonishment that such a transformation occurred and our own retrospective surprise that it happened so fast.

Individuated, market-valued interests in nature may well produce the "tragedy of the commons" if no system of authority evolves or can be imposed. Otherwise, there is no mechanism for preventing individual maximization of interests from collectively undermining the interests of all in preventing degradation of natural systems. Conservationist states are intermittently the solution to this macro collective-action problem. But even conservationist states are no guarantee of environmental protection. First, states have interests of their own. These derive from the logic of state-ness itself – reproduction of systems of power, altering vulnerabilities of position in a global hierarchy of states, generating acceptable material well-being. Moreover, authority to govern nature encounters resistance from those whose livelihood routines are criminalized by conservationist policy. There is suspicion in the villages that states' claims of special expertise and disinterested concern for public goods ring false.[11] Internationally, the plea of rich nations for protection of a "global commons" encounters the same skepticism in the periphery: "sacred groves for the rich."[12]

Conservationist policy is also undermined by incommensurate boundaries of authority and ecology. Administrative and political divisions seldom match up with the scale of ecological dynamics (Herring, 2001). Most evidently, planetary interests in values such as biodiversity or climate change have ramifications in fragmented polities, unconnected institutionally or intersubjectively. There exist no authoritative mechanisms to address common crises, even if cognitive agreement is reached.[13] Yet all environmental governance depends on behavior further down the scale of authority, from global to national to local. Finally, political support for preservation of nature *per se* is vulnerable in every system of authoritative power when markets dictate otherwise. This weakness – and the subsequent vulnerability of natural systems – is produced by interactive effects of ideas about and interests in nature. It is important to recognize this contingency as we consider what is new in the politics of nature under conditions of technical change introduced by genomics.

Biopolitics: The Nature of the Unnatural, Public Authority and Science

The previous section has argued that evocations of the natural demonstrably have some political clout, however limited in contest with material interests. Evocation of the value of nature *per se* draws money and administrative energies, citizen action and removal of some terrain from the market through decommodification and protection. The importance of the natural as ideational interest constitutes one of the major political forces against transgenic organisms.

It is truly fascinating, then, that the global conflict over transgenic organisms has focused on their ecological threats in the sphere of agriculture – an endeavor quite destructive of nature. Transgenic pharmaceuticals seem to be immune to protest. But agricultural biotechnology is still agriculture, and the historic trajectory of agriculture has been to reduce biodiversity in favor of controlled environments for elite lines of useful plants. There are proposals for, proclamations of, "GMO-free zones," but not "insulin-free zones," though human insulin has been made by a transgenic organism since 1978. Cornflakes rather than insulin draw the Franken-prefix in dramaturgy of resistance to "GMOs."[14] Remarkably, the political strategy assumes that mass publics will find consumption of a breakfast cereal more risky than injection of alien biological agents.

There are political reasons for this selectivity: miracle drugs save lives and would make ineffective targets for opposition.[15] Threats to the global poor and their fragile environments are strategically more effective. Indeed, more generally, political ecology demonstrates rural bias. Victor Magagna writes:

It is ironic that the late twentieth century has seen a renaissance of rural history. The march of industrial society continues to change the institutional fabric of every region on the globe; yet, intellectual interest in rural life has perhaps never been more pronounced (1991, p. 1)

Interest in "peasant society" among elites was a cold-war phenomenon. As fears of rural breeding grounds of communist insurrection subsided, worries over degradation of landscapes increased, driven by the core understanding of ecology – interconnectedness within complex systems – theorized largely in cities with grist from reports about people living in remote places.[16] Biotechnology has been defined as a rural problem through an inversion of Michael Lipton's (1977, 1988) "urban bias." People in cities do not want to give up life-saving pharmaceuticals, whether transgenic or not.[17] But talk of threats from transgenic crops to rural environments raises contentious politics that are among the most disputed of our time. The dispute centers on what constitutes a natural means of breeding plants and what constitutes an unnatural means; whether exchange of small amounts of DNA across species is more or less unnatural than exchange of large amounts within species (Herring, 2007b).

Political economy deals with nature in terms of market failure and externalities. If nature provides public goods, preservation becomes a plausible policy objective. If externalities of development threaten those same public goods, means of redress become plausible policy objectives. The notion of public goods in nature is thus central to the analytics of political ecology. Were there no public goods – in nature or elsewhere – there would be no need for governments. Imposition of collective authority soaks up resources that have high opportunity costs; individual liberties are inevitably constrained, no matter how constitutionally limited is public authority. Yet it should be clear by now that public goods and bads in nature are not self-evident nor consensually perceptible, but rather are embedded in a normative logic that is culturally anchored – though demonstrably fluid. Swamps were once unhealthy, and for draining; wetlands purify water and are for preserving (Herring, 1990). The great normative transformation was to convert draining of swamps from a public good to a public bad. This transformation would not have been possible without the notion of ecology. Transgenic organisms represent a further elaboration of this dynamic, with much more contentious politics and much more disputed science (Hilgartner, 2002).

One deep continuity across historical struggles for definition and control of nature and contemporary social protests around transgenic organisms is then precisely the nature of the natural. Assessment of what is natural and what is not is not a question for science, but for ontology and normative theory.

This continuity does not mean that nothing has changed. The potential for creation of property at ever smaller scales, and for breeding new organisms

across the lines of species – even kingdoms – and therefore the possibilities of inducing environmental change in novel and unanticipated ways at unknown probabilities has come with the genomics revolution. Here the ground does shift to contested science. We now turn to what is new in the biopolitics of nature.

New Value in Nature

The genomics revolution created potential, but contested, economic value in biodiversity *per se*. The common lowly soil bacterium *Bacillus thuringiensis* lends its name in the form of *Bt* to large political struggles as well as deployment in ever-expanding acres of cropping systems. For its supporters, *Bt* represents environmental progress: one means of reducing the poisoning of ecosytems by pesticides. For opponents, it represents at least uncertainty, perhaps risk, perhaps disaster. But without recombinant DNA technology, *Bt* would have nothing to contribute to corn or cotton, even though its toxin in foliar applications has been part of organic agriculture for decades. With insertion of the *Bt* gene into particular plants through genetic engineering, a kind of poetic justice enters the fields: only those insects that attack the plant receive a dose of the plant-made toxin, which stays within the insects and kills them. Other fauna in the fields, and humans downstream, are not affected, in sharp contrast to alternatives. Whether this contribution of *Bt* represents progress or the first step toward genetic Armageddon is one strand of the transgenic debate.[18] But there is undeniably new value in nature created by genetic engineering, hence the raging debates over patents and bio-piracy.

So long as biodiversity is valued only in normative terms, as a desirable thing, in and of itself, its political base is fragile, everywhere in the world. This interest-deficit fragility belies global happy-talk about conservation of biodiversity. The greatest collective material interest in biodiversity is probably eco-system integrity, providing services that are public goods – clean air and water, for example. But ecosystem services are notoriously difficult to measure (Daily, 1997). Even measured in ways that meet some threshold level of agreement socially, a means to pay for environmental services is difficult to conjure. Here, political economy of public goods suggests formidable obstacles; the confirmation is the brute fact that societies keep destroying eco-system services: it is not just neo-liberalism or the *Zeitgeist* of privatization – though neither helps.

Genetic engineering has enabled commodification of nature in new ways: value is no longer simply notional, but vigorously contested by interested players. Valuable genetic information may depend on actually existing biodiversity.[19] In the contentious politics surrounding that normative spectrum from "bio-partnerships," to "bio-prospecting" to "bio-piracy," new but variable and unpredictable relationships between value and new forms of property emerge (King *et al.*, 1996; Shiva, 1997). Political conflicts around creation of property from nature at the landscape level were common. Polanyi's "great transformation," via commoditization of nature, greatly intensified these dynamics.[20] In parallel, the genomics revolution creates possibilities for conversion of nature to property on a scale unimaginable a generation ago. The property regime enabled by TRIPS (trade-related intellectual property rights) provisions of the WTO is novel in some ways, but disputed claims to property in nature and its knowledge are not.[21]

Whether or not genomics will ultimately undermine the notional market value of *in situ* nature is unclear, but in the first instance the genomics revolution created new value in natural landscapes. Yesterday's pest or obscure invertebrate could harbor tomorrow's miracle gene. Who is opposed to a cure for cancer or Alzheimer's disease? Via this knowledge-based revaluation of nature, a certain monetary incentive was thus introduced into the political struggle to prevent wholesale destruction of endangered ecosystems. For a while, the Environment Minister had something to say to the Commerce Minister when the question of affordability and opportunity costs of conservation surfaced.[22]

The most important consequence of this new commodification of nature and its knowledge – at a very small scale – is new interests in scientific inquiry. There are potentially high stakes in small findings — hence the restrictions on international fieldwork for fear of "bio-piracy" – a concept long understood by colonial officials (Grove, 1995) but given new urgency by the genomics revolution.[23] The assumption of objectivity in research and findings at the frontier of knowledge then becomes more problematic: science is more easily conceptualized as captured by monetary interests and politicized. This new value in nature has become the subject of new forms of local and global controversy, new regulatory restrictions and new claim-jumping tactics of transnational firms. Most important for consideration of politicized science, the value of these new commodifications themselves are dependent on social acceptability of the enabling science that certifies the safety of products and procedures, judges their uniqueness and utility, and their non-obvious nature in the patent arena (Hilgartner, 2002).

CONTINGENCY OF PROPERTY RIGHTS

Property rights – of states, individuals, and communities – in landscape-level nature are empirically fluid and contingent, and therefore more usefully viewed as claims than as rights. Property rights exist on a spectrum, from one's automobile to state ownership of forests to Monsanto's patent on a genetic event. Property rights in natural systems and products, whether of states or firms, are in practice less a matter of fee-simple ownership than outcomes of dynamic negotiations over time. In practice, regulatory states claiming property rights in nature evoke resistance on the ground for their paramilitary occupation of local turf, criminalization of subsistence routines and ineffectiveness (Herring, 2002). Nancy Peluso criticizes the "custodial paramilitary" model of state resource control which produces "secret wars and silent insurgency" inimical to conservation (1992, pp. 235–236). In political–economy terms, ineffectiveness of state property in nature results from very high information and enforcement costs and low state capacity. In cultural terms, claims of extra-local states and firms are not accepted locally, where both use rights and subsistence needs have priority (Kothari and Parajuli, 1993) and propinquity counts more than distant law: "our rule in our villages." [24] The precise meaning of property in nature is as contextual as structural.

State property in nature is then the vector sum of forces in contention, not shading on maps or lines in gazettes. States negotiate property claims on the ground in episodic clashes with people using nature. Over time, the social ecology model of incorporating human claims and routines into "nature" has taken on more legitimacy, just as the preservationist model of "deep ecology" became less and less tenable politically. What states once called trespassers and poachers are increasingly called stakeholders; the criminalization of subsistence routines introduced by colonial law in much of the world is being reversed by developmentalism. Rather than breaking the law, indigenous people are held to be asserting traditional property rights, usually based on propinquity (our village) or subsistence rights and typically the conjuncture of the two. The fluid and contingent nature of state property regimes in nature is instrumentally misrecognized in official discourses of national parks and international treaties.[25]

Claims to rights in genetic information and products parallel the dynamics of property relations in landscapes. Both normative ambiguity and stealth tactics condition property claims on the ground. Perhaps the most telling example is the global phenomenon of what I have called "stealth seeds" – transgenic seeds that evade claims of both firms and states by spreading underground, farmer to farmer. The underground spread of transgenic seeds by farmers in both India and Brazil forced states to legitimize the farmers' praxis *de facto* – in effect looking the other way – despite lingering doubts about the science (Herring, 2007a). Cotton farmers on the ground in Gujarat have "naturalized" locally made illegal transgenic seeds, incorporating them into their conventional approaches to risk simply as new hybrids (Roy *et al.*, 2007). Misrecognition via reification of property undermines elite attempts to oppose transgenics in the name of farmers through constructions such as "suicide seeds," "terminator technology" and "Monsanto monopoly" (Herring, 2005a).

International social movements use patents and TRIPS as points of contestation, and take appropriation of nature by multinational corporations as embodied in reified patents as evidence of a new dominance of nature. There is a curious "urban bias" to the notion that software and pharmaceuticals are globally appropriated without payment of fees and royalties but farmers are somehow too simple to circumvent firms and states in the same ways. Stealth movements of transgenic seeds suggest a different property structure, more akin to that of users of landscapes and copiers of pirated software. Property in the biological content of field crops can mean only what enforcement techniques can make it mean, which is often not much. Rather, accommodations are struck among claimants, illustrating the fundamentally contingent nature of claims to property in nature. Intellectual property claims that seem hard facts in TRIPS discussions turn soft on the ground; seeds, like forest plants, are hard to police. This is true for the same reason that protected conservation zones often exist as little more than lines on maps – assuming away the problem of biopolitics allows presentation of the local in terms useful to elites of both states and civil society. Property is fundamentally relational: the strength of contesting claims determines the meaning of the property on the ground. Strength in turn is highly situational: Monsanto cannot now enforce property claims in transgenic soy or cotton seeds but may one day be able to through genetic use restriction technology, aka *The Terminator*, if the technology actually works in practice or is anywhere approved for commercialization.

The ability of transgenic seeds to go underground via farmer stealth strategies undermines the surveillance of states and firms assumed in much of two divergent discourses about GMOs internationally. Opponents of transgenic crops fear monopolization of property, with consequent crushing of poor farmers, poor nations and biodiversity alike (Shiva, 1997; Shiva *et al.*, 1999). Proponents of biotechnology reify biosafety

regimes and assure societies that agreements such as the Cartegena Protocol can control nature. Neither discourse is proving robust on the ground. This is in part because both discourses are instrumental, and misrecognize the interests of real actors in real settings. Surveillance of nature is no mean task, either macro or nano.[26] In both landscapes and genomes, property claims are difficult to enforce because:

1 strong interests militate against the formal-legal regime – stakes are high;
2 normative claims of end users defy the property claims of states and firms – legitimacy is contested;
3 nature is difficult to access and assess in technical terms, whether numbers of tigers or altered DNA in plants;[27]
4 monitoring in practice requires intense surveillance of great socio-political difficulty – seed police and forest guards are embedded in local society, costly and difficult to control.

Genomics has enabled claims of property in nature at smaller and smaller scale, but with high levels of indeterminacy. Because of public goods in biosafety, property and regulatory regimes converge: the only legal property is that which passes biosafety tests. Ironically, India's biosafety regime proved unable to detect underground transgenics but did for a time give Monsanto's partner quasi-property rights in *Bt* cotton; only the official varieties had passed biosafety testing and the competition underground seeds were technically illegal. It was the biosafety regime that privileged Monsanto's Bollgard over the indigenous Navbharat transgenic and its offspring, much to the disappointment of farmers (Roy *et al.*, 2007). Given public perceptions that transgenics require special regimes of testing, deep pockets contribute to restricting the market, and thus the distribution of intellectual property, even when, as in the case of *Bt* cotton in India, there was no patent protection of plants.

Property rights in micro-nature depend on close monitoring mediated by high technology. Whatever else the genomics revolution may bring to society, it will certainly bring higher levels of surveillance and regulation, as did scientific forestry before it.

Structural Indeterminacy of Interests

Political economy analyzes interests within structures. Structures define interests. Anyone who doubts structuration of interests should discuss with a sharecropper interests in agrarian class structure, or an adjunct lecturer interests in academic class structure. Natural systems cooperate to some extent: the lumberjack, fish worker, herbalist, poacher – all have interests given by their place in a complex and dynamic biological structure. Yet scaling up interests in nature produces a great deal of indeterminacy. Ecosystem health, global climate change, biodiversity, potential for gene flow from transgenic plants – all are sufficiently removed from ordinary conceptual and practical knowledge of most people that reliance on experts, or congealed knowledge, or persuasive framing, or "common sense" becomes inevitable. Some combination of these common tools reduces both information costs and uncertainty – both of which are aggravated by widespread scientific illiteracy.

Interest indeterminacy certainly characterizes politics outside nature. The effect of free trade, devaluation, or macro-economic policy on the interests of ordinary citizens is disputed and mediated by imprecise and contested economic science only dimly perceived in mass publics and legislatures alike.[28] Still, calculating interests in nature is arguably of a different character. Effects take a long time to become apparent; human lives are short in relation to potentially irreversible ecological change (Narindar Singh, 1976). Any normatively acceptable characterization of interest must therefore consider intergenerational effects. Moreover, causality in complex over-determined chains of dynamics is difficult to parse and disputed among experts. If everything is really connected to everything else – a core tenet of ecology — figuring the effects of particular changes is very difficult, except at the extremes – no one doubts that pollution destroys aquatic ecologies, for example, but it is hard to know how much pollution a system can take, or how resilient its components will be, or which components are vital to systemic functioning.

These characteristics of interests in nature raise doubts about the valorization of local knowledge in conflicts in and about nature (Agrawal, 1995; Herring, 2001). Inability to anticipate ecological consequences of short-term interests and poor information is common and predictable. Moreover, science presents a moving target for interests.[29] Counter-intuitive links between refrigerator gases and skin cancer could not have been imagined by political actors even a generation ago, in a pre-Montreal Protocol world (Herring, 1999b). Virtually no one knew they had an interest in an ozone regime, but everyone did.

At the other end of the scale, genetic engineering has introduced large uncertainties and politically mobilized anxieties about horizontal gene flow through agro-ecological systems (Winston, 2002, pp. 93–99). It would be very hard for most people to determine rationally their specifically ecological interest in transgenic crops, with the exception of organic farmers.[30] At the same time, ecological damage from pesticides is a known threat, potentially reduced by at least the *Bt* varieties of transgenics. Comparative weighing of uncertainty against known risks is difficult even with full knowledge and good theory; counterintuitive complexities characterize transgenic science, where studies are thin (Thies and Devare, 2007).

For social theorists, interest indeterminacy introduces epistemological and methodological caution: interests in nature cannot be read off or deduced from structural position in any straightforward way. Often our most basic interests are dependent on both known unknowns and unknown unknowns. We as a species do know that the timing and distribution of changes from global warming is unknown, but we do not know what else we do not know. Yet there was a time when climate change was on no one's horizon, a time when there was consensual scientific uncertainty and then a time of relative certainty among climatologists – but still denial for transparently political reasons in some circles. Interests are knowledge dependent, but the knowledge as a body is constantly in flux at the margins. Worse still, knowledge is mediated by science which is seldom authoritative, but rather held to be political.

Political Science

In both landscape and molecular politics, conflicts among interests make science a field of legitimation and conflict rather than an agnostic method. There is a political–economic reason for the politicization of science: science is expensive, and has material consequences. There is thus continuous suspicion that who pays the piper calls the tune. With growing privatization of scientific research (Pray and Naseem, 2007), the sphere of disinterested or public-interested science shrinks. With scientific literacy stagnant or declining in the face of increasingly complex theory, mass publics are *per force* drawn to take sides through framings of interested parties. Moreover, states have both interests in science and the capacity to sequester or distort both process and findings.[31]

There is an interactive effect of expensive knowledge, material interests and epistemological complexity. Because of constraints of time and money, ecological dynamics of large systems remain unknown, perhaps unknowable; much of the argument about use, resilience, recovery, and collapse rests on an uncertain empirical base. Reified *science* is called in as authoritative arbiter for public-policy disputes, but quickly becomes more arena than judge. The imperative for political actors is to claim certainty, not ambivalence. Authoritative action requires authoritative knowledge. It is hard for states to require sacrifice hydrocarbon taxes, off-limit protected areas – without authority but authoritative knowledge in nature is hard to come by. More problematically, science has no answers at all to normative questions. Is it acceptable to lose species? What is the time scale? Europe lost a lot of species, and now is regaining some – is that an acceptable price for an industrialization that permitted world domination for a time? Should we expect certainty on issues of gene flow, or is some risk acceptable given that agricultural plants often share genes with wild relatives? Is there any incremental risk from transgenics in comparison with other forms of moving DNA around plants as conventionally done? How precautionary is precautionary enough?

Even normative claims are mediated by science. Assuming a baseline of normative agreement (biodiversity is good, agree a lot of people), the logics from that position diverge with the science. Though preposterous to critics of biotechnology, transgenic organisms may make significant contributions to preservation of biodiversity (Horsch and Fraley, 1998) or crop diversity (Zilberman, *et al.*, 2007). To return to *Bt* crops, the alternatives to toxins targeted to *Lepidopterans* and internal to the plants' own biological machinery typically require indiscriminate destruction of fauna in agro-ecological systems. Pesticide loads in India have reached intolerable levels, with severe externalities, typically for those least able to protect themselves and for environmental integrity. At a more macro-scale, the biological question in India is often: how compatible is ecosystem integrity with human use? Would many small protected areas be as ecologically valuable as several very large ones on the American model? If so, a conservation solution might be politically feasible; the alternative is not. Ecologists in India are more likely to say yes than are ecologists in the USA. In landscapes, public science proclaims as necessary state-proclaimed restrictions on the market in nature in the name of ecosystem health. But the measures of ecosystem health, and measures of its resilience, depend on models that are nowhere fully established. The need for states to act almost always outruns the science, or requires answers to questions science does not presume to ask.

Nevertheless, the science of ecology frustrates policy. Unexpected and often imperceptible interconnections among parts of systems, across vast scalar differences, from microbial to atmospheric, keep being discovered. Moreover, the core *desiderata* of real science – methodological and epistemological commitment to hypothetico-deductive empirical investigation that is replicable and exhibits high standards of validation – frequently

combine to produce political impotence. The history of global warming represents a telling case in point. Real science remained sufficiently uncertain for the most powerful political actor in the international system – the USA – to sustain a plausible claim that "more research is needed before action is justified." The problem with real science is that more research is always needed; the need of human beings to act raises a frequently unbridgeable gap between uncertainty and risk, and hence subjectivity.

Where ecological science is incomplete, junk science has ready answers. In burning test trials of transgenic crops meant to find out if there is an environmental threat, opponents of transgenics presume to know the answers before collecting the evidence. NGOs in India demanded field-trial data on *Bt* cotton tests for biosafety while simultaneously proclaiming that no one would believe the data if they were released: the science was that of a state already committed to biotechnology. Likewise, despite sharply increasing sales of Mahyco/Monsanto Bollgard *Bt* seeds, and the scramble for both licensed and underground competing *Bt* seeds, "the failure of *Bt* cotton" has continued to fill press accounts. This conclusion became established fact in the international network opposing biotechnology. NGOs continue to declare that "*Bt* cotton has failed" and that all arguments to the contrary are biased by corporate sponsorship of the research.[32] Indian farmers were not listening; their livelihood depends on real empiricism, not ideology. Dr Shanthu Shantaram published the simple question: "If Bt-Cotton is Failing in India, Why is it So Popular Among Farmers?" P. V. Sateesh of the Deccan Development Society wrote: "Bravo Dr Shantharam, you have done a yeoman service to your masters but on the day of judgement in a future not so far away, scientists like you will be remembered as 'Enemies of the People.'"[33]

Despite the triumphalism of the pro-biotech interests globally, based on increasing farmer acceptance in both rich and poor countries, there remain unresolved questions of incomplete science and unknown risk.[34] But it is not in the interest of *Bt* farmers to investigate; they understand the dead-end nature of the pesticide treadmill and seem largely unwilling to take the large risks of organic alternatives. Uncertainties are externalized to society as a whole. For all the romanticization of local knowledge and the *Volk*, it is clear that sons of the soil do not always know best. Farmers adopted pesticides to solve their insect problems before insecticides became biologically and economically unviable. They did so not after considering science and social externalities, but rather from perceived necessity of protecting their crops.

Uncertainty is the most powerful political weapon of the anti-transgenic movement, and honest science is always incomplete at the frontier. The elision of and escalation from uncertainty to anxiety meets little cognitive resistance. Fearing the unknown is not only the first response, but to some extent the rational response. Second, it is logically impossible to prove an empirical negative.[35] One cannot, for example, imagine the evidence that would convincingly prove that horizontal transgene flow will not cause major ecological damage somewhere, sometime. What is unknown is the incremental risk, compared to alternatives, and that is the relevant question. That there is any uncertainty at all buttresses the rationality of consumers in Europe and Japan who rightly see little gain in transgenics, but seek to restrict the technology even in areas where there may be significant benefits – both to poor farmers and to environments. It is only through this framing dynamic that the same grains Americans consume daily with little thought can be rejected as "poison" in Africa in the midst of famine.[36]

Because science is suspected of being political, its authoritative power declines even without the partisan devaluation frequently observed and widely understood.[37] The increasingly dominant standard narrative of genetic engineering and development is precautionary: there is much uncertainty, perhaps risk, but alternatives are likewise risky and costly (Pinstrup-Andersen and Schiøler, 2000; Conway, 1997). In comparing uncertainties and risks, the question of *whose* science becomes intensely relevant. In some anti-globalization mobilization, "Western science" is first conjured, then attacked as "imperialist" or "reductionist." The countervailing attack from proponents of biotechnology targets "junk science" – of which there is a great deal.[38] Whether or not transgenic foods or crops are safe or good depends fundamentally on which science one endorses and consumes.[39] The global landscape of social movements and developmental options hinges on science, but the authority of science has been altered by the political processes into which it has been thrust.

Convergence of Landscapes and Genomes: Protected Areas and Biosafety Regimes

Genetic engineering and landscape preservation converged in bio-prospecting. Commodification of micro-nature in the form of bio-partnerships and bio-prospecting promised revaluation of nature. The connection has not proved robust, for both economic and political reasons. The reciprocal connection is a perceived threat to landscape ecological integrity from horizontal gene flow.

Genetic engineering depends on access to useful genes. Reduction in landscape-level biodiversity restricts options. Bio-prospecting as a new institutional arrangement promised that new science could lay the base for a pro-poor development strategy that benefits local people and validates local knowledge (Weiss and Eisner, 1998). In the widely discussed Merck-INBio deal in Costa Rica, upfront money was paid for bio-prospecting rights; the contract specified sharing of benefits from any commercialized product from biota so studied. "Para-taxonomists" would use local knowledge and new skills to characterize biodiversity. Opponents of bio-prospecting characterize the global flow of biota as "biopiracy."[40] Vandana Shiva's (1997) book on the subject is informatively entitled *Biopiracy: The Plunder of Nature and Knowledge*. The intersection of new biological possibilities with pressures for harmonization of intellectual property rights through the WTO, WIPO (World Intellectual Property Organization) and bilateral negotiation has produced polar political positions on the possibility that biotechnology might offer aid to conservation.

How much environmental threat comes from transgenic crops is unknown. There is in nature a great deal of gene flow among plants. To be of concern, the movement of a single gene would have to increase the fitness of some wild relative significantly. This may happen with traditional agriculture: the super-weed "Johnson grass" appeared long before rDNA technology; as with many domesticated crops, there is natural exchange between the agricultural plant [Sorghum in this case] and weedy relatives (Winston, 2002, pp. 94-95). More critically, whole genomes are continually inserted into agricultural and other ecologies via invasive plants (Pimentel *et al.*, 1998); cases of enormous economic losses are documented and continuing. The Center for Biological Informatics of the US Geological Survey estimates the costs of invasive species at $138 billion per year in the USA (http://invasivespecies.nbii.gov/). An invasive species brings a whole genome to an ecological system. Nevertheless, the political construction of recombinant DNA technology as a threat has dominated in biopolitics the spread of demonstrably destructive invasive species. If one had a marginal dollar to spend on regulation, would it be better spent in monitoring invasive species or transgenic crops?

The phrase "paper parks" has long been recognized as the triumph of symbolic politics and administrative hubris over biological reality. Paper parks exist only on paper; inside the protected areas are guerillas, drug smugglers, poachers of endangered species, slash-and-burn agriculturalists, hunting/gathering peoples, and ordinary bandits hiding from the law.[41] GMO-Free Zones are being declared all over the world, sometimes by states (as in Brazil),[42] sometimes via local legislation (the USA), often by social movements. The great irony is of course that these zones cannot and will not exist until there really is some gene-use restriction technology – dubbed "The Terminator" by global civil society's leading-edge opposition to genetic engineering.[43] Though the standard narrative posits a seemingly sensible, and unobjectionable "risk–benefit analysis," buttressed by comforting assurances of a "bio-safety regime," the reality is that risks are unknown and seeds seem virtually impossible to police, as evidence from Gujarat to Rio Grande do Sul indicates (Herring, 2007b). In the optimistic narrative, state technicians will be able to make for society the cost–benefit analysis necessary to decide on which transgenics when and where. Insititutions will enforce a bio-safety regime, serving the same functions as paper parks: assuring mass publics of state competence and concern for their welfare.

CONCLUSIONS

Explaining environmental outcomes begins with attention to market and state. There is great variance in how and to what extent states decommodify nature and rig markets for conservation or preservation. Particular pieces of nature are granted reservations and restrictions, bans on commodification and regulatory protection, both within nation states and internationally through treaties [e.g. CITES (Convention on International Trade in Endangered Species), ITTA (International Tropical Timber Agreement), WHC (World Heritage Convention)].[44] How much states intend these paper policies to alter environments depends on the vector sum of operative interests weighted by their power. Because dominant interests are expressed in terms of market-measured value – whether of labor or capital – nature seldom wins.[45] Profits and jobs are difficult to answer with ideas alone. Moreover, whatever states claim they seek to do, what states can do depends on elusive, but real, differences in capacity:[46] are forest guards outgunned by poachers? Are they more responsive to local society than to the central state? What clout does the ministry of environment have compared to the ministry of commerce?

A political economy of nature then helps explain the long-term trend toward simplification and degradation of natural systems by commodification, market exchange and ineffective or symbolic state action. To move from a political economy of nature to robust political ecology requires attention to what counts as an interest. Science continually presents new challenges to the way interests in nature are understood by

citizens and political classes that control states; the sea change in redefinitions of interests – of both individuals and states — introduced by the atmospheric science of ozone holes and climate change is archetypal. Transgenic organisms represent a particularly compelling illustration of these propositions: threats to ecological integrity are largely unknown, but constitute condensation points for strongly held political positions. Much then depends on the science, but interpretations of authoritative science vary with interests: who pays for the studies? How much is suppressed or altered? What gets studied, what remains unexamined? As science itself faces postmodern attacks, what begins as a mode of inquiry becomes enmeshed in partisan projects and suspicions. As science is inevitably a work in progress, and therefore incomplete, uncertain if honest – misrecognition of interests *via* obfuscating ideologies renders real science politically vulnerable.

A robust political ecology will of necessity parse these complicated intersections of cultural logics and structural positions in defining and attaining human interests in nature. The great ideational divide in conservation/preservation of landscapes revolves around the degree of acceptable commodification: to what extent should "nature become a mart" in Cronon's (1983) phrase? The comparable axis for genomes is risk versus uncertainty. In neither case does science pretend to have an answer. But in both cases, public goods and bads in nature – the legitimation of state intervention — are of necessity embedded in normative logics that are culturally anchored but demonstrably fluid. The very notion of Frankenfoods depends on cultural understanding of monsters and the hubris of science going where it should not tread; Pandora married into the family of Prometheus and brought her box that was not to be opened.

How do societies deal with uncertainties of change that promise public goods when there are known risks in continuing with the *status quo?* The prior question is fundamental to development studies: at whose risk? To whose benefit? With regard to both landscape-level ecologies and micro-level transgenics, the normative answer depends on how we conceptualize public goods and bads in nature, how we couch alternatives, how we construct differential uncertainties and risks, and how we project from a body of science that will never be as complete or certain as we would like (Herring, 2007).

NOTES

1 See Paul Robbins (2004, pp. 6–7) for a schematic discussion.

2 For illustration, including state interests that vary with time, see Schama (1996): especially

Introduction, Prologue, Chap 3 (The Liberties of the Greenwood); Gold (1997).

3 This position on the ideational is pervasive in what are otherwise quite different approaches; see, e.g. Dryzek (1997); Peluso and Watts (2001); Peet and Watts (1996); Robbins (2004).

4 Beginning with Karl Polanyi (1944/57).

5 Outside political economy, many analysts simply assume that ideas have power. As assumption, this take is obviously problematic: ideas may well be epiphenomenal, instrumental, ephemeral, reactive, fleeting. Yet the need for reintegrating ideas into political–economic analysis is increasingly evident: even our seemingly "hard" data-built facades of reality often depend in the last instance on the tenuous relationship between real interest and representation; e.g. Herring, (2003).

6 For one of many treatments of this phenomenon, see Herring (1999b).

7 To illustrate: the second administration of George W. Bush in the USA has made a frontal appeal to interests – in cheap fuel, in national independence from the "Middle East" to "open up" the Alaskan wilderness for oil drilling. This position deflects attention from interests that are probably much more responsible: corporate oil from which the administration draws heavily in personnel and financial support.

8 A foundational document is *Our Common Future* from the World Commission on Environment and Development (1987), but the normative intrusion can be traced at least as far as Karl Polanyi (1957, pp. 34, 41), who subtitled a section of Chap 3 of *The Great Transformation:* "Habitation Versus Improvement." The great transformation was in part the substitution of "habitation," which we now call "subsistence" as *telos* with "improvement," or in contemporary language "development."

9 For expansion, see Herring "Politics of Nature," Harvard University Center for Population and Development Studies, Working Paper Series Number 7 (Cambridge, October 1991).

10 David Arnold (1996): the view is acknowledged, but complicated by the introduction of ideas of conservation via colonial science in Richard H. Gove (1995). An illustrative case is explained in Ramchandra Guha (1985, 1990). Contrary to nationalist readings, colonialism is certainly not a necessary condition for degradation; see Richard M. Eaton (l990), and D. Raghunandan (l987).

11 See Herring (2002) Kothari and Parajuli (1993) Nancy Peluso (1992) and Bandyopadhyay and Shiva (1988). On violent local opposition to state preservation of natural areas in the USA, see David Helvaarg (1994).

12 I attribute the concept to Mahhav Gadgil, who challenged my worldview on this point, pers. com. (1991). See Subir Sinha and Ronald Herring (1993); Subir Sinha *et al.* (1997); Ramachandra Guha (1989b).

13 At the village scale, see Gold and Gujar 2002. It may be that recognition of common interests will be more spurred by a recognition of common interest in defensive reaction to planetary threats: the globalization of pathogens (e.g. Pimentel *et al.,* 1998). This prospect seems more plausible than global revaluation of nature in landscapes when economics presses in other directions.

14 "GMO" ("genetically modified organisms") is a political label, utilized mainly by opponents of genetic engineering: see Herring (2005b).

15 See Lezaun (2004). Neither diabetic patients nor their physicians typically resist synthetic human insulin. There are biological reasons as well, as laboratories are assumed to contain transgenes better than open fields.

16 Baviskar (1995) was widely and I think wrongly criticized for making this point in India. For an illuminating treatment of surprises around imagined villagers for the north–south alliance against globalization and biotechnology, see Madsen (2001).

17 Sharad Joshi, leader of the farmer association *Shetkari Sanghatana,* in protesting against Delhi's effort to burn the unauthorized transgenic Bt cotton fields in Gujarat posed the question as one of farmers' freedom and, implicitly, urban bias: "Development should not be locked up in the cities. The marvel of technology should reach the villages." Joshi quoted in Sajid Shaik (2001), "Farmers Decide to Defend their Bt Gene Cotton Crops," *The Times of India.*

18 Shiva (1997); Shiva *et al.,* (1999); McHughen (2000); Pinstrup-Anderson and Schiøler (2000); Paarlberg (2001); Winston (2002); Herring (2007b).

19 Tanksley and McCouch (1997) Weiss and Eisner (1998).

20 On pre-market struggles over valuation, see Schama (1996, especially chap 3); on intensification, Polanyi 1957 (especially pp. 1–30; 43–55; 68–76; 130–219; 223–258); also Cronon (1983, 1991); on British India, Guha (1990).

21 Suneetha Subramanian found in her dissertation research that indigenous healers in Kerala would either restrict specialized knowledge to verbal communications or write in the sand, to protect their intellectual property (pers. comm., 2003).

22 See e.g., A. Gupta (1998); Weiss and Eisner (1998); Varley and Scott (1998). After the famous Merck-INBio deal's novelty wore off, however, this prospect has decidedly dimmed. For economic reasons, see Simpson *et al.* (1996).

23 The Convention on Biological Diversity, which entered into force in 1994, diagnosed one cause of loss of biodiversity as poorly defined rights of ownership in nature. Before 1992, biota were regarded as the "common heritage of mandkind." Appropriation from this common pool was common, with no consultation with indigenous peoples connected to these resources. The CBD granted nations sovereign

rights over biological resources and mandated sharing returns with indigenous peoples.

24 For a summary, Sinha and Herring, 1993; Kothari et al., 1996; Menon, 1996; Herring, 2001; Madhav Gadgil et al. (1993).

25 For example, see the "bank" metaphor and comparable analogies in Ministry of Environment and Forests, Government of India (1994; 1996).

26 The parallel to James Scott's views (1998) of the astigmatism of high modernism is clear; Scott's state needs visibility, but finds it hard to attain.

27 Technology may improve surveillance capacity. Simple, inexpensive kits that test for the Bt protein have been developed in India, but seem to be unreliable and not much in use.

28 Likewise, reservation wages do have a cultural logic and the effects of higher wages on demand for labor over time in a dynamic system are unknown with any degree of certainty to individual workers. Yet workers have long-standing repertories of political claims, predicted and explained by their position in a class structure, and consistently pursue some public policies. See Elster (1985).

29 The New York Times of June 27, 2004, carried a report on a new danger to the Endangered Species Act: what constitutes a species is under challenge, in part from new tools of genomics. Since the ESA is used to protect habitat, and is under constant assault from the political right, decertifying the unique identity of small-population species could have major consequences at the landscape level.

30 Movement of pollen through plant communities will put that lucrative market niche at risk, so long as there is global market segmentation of organic and non-organic foods depending on transgenic content of end product. Organic farmers in the USA have defined organic as "GMO free," but this conjunction is by no means necessary: organic connotes cultivation techniques, not seed germplasm. Roy (2005) presents evidence for dispute among Indian farmers on whether or not a plant that makes its own pesticide – Bt – is organic or not; some organic farmers think so, others not.

31 Revkin (2006) reports on the Bush administration's efforts to rein in and control climate scientist James E. Hansen of the National Aeronautics and Space Administration on global warming. The Bush administration has by most accounts interfered more in science than is typical of American presidencies. Mooney 2005.

32 Sivramiah Shantharam has posted two useful commentaries, "If Bt-Cotton is Failing in India, Why is it So Popular Among Farmers?" and "The Brouhaha about Bt-Cotton in India" on AgBioView. A withering critique was posted by P.V. Satheesh on GM WATCH daily http://www.gmwatch.org. Shantaram's position is confirmed by Jayaram 2001; 2004.

33 See previous note and http://www. agbioworld.org/newsletter_wm/index.php?caseid= archive&newsid=2330, 7 March 2005.

34 To take the Bt case again, recognizing that there are many Bt events with different possible effects, see Thies and Devare (2007) on soil structure and root exudates.

35 On this point with regard to transgenics and the environment, see Pinstrup-Andersen and Schiøler (2000); Winston (2002).

36 Zambia and Zimbabwe in 2002 rejected United Nations food aid containing some transgenic maize kernels in the midst of famine, terming the shipment "poison." Scientific American, August 2004, Vol. 291, Issue 2, p. 8.

37 For example, the association of "Vedic Science" with the Bharatiya Janata Party in India (Nanda 2003) or "Creation Science" with fundamentalist Christian politics in the USA; see Mooney (2005).

38 The website www.junkscience.com is, however, ideologically identifiable as an anti-regulation operation. What it considers junk science is what much public policy considers the only valid science, illustrating the point in the text. Likewise, Björn Lomborg's The Skeptical Envirnomentalist (2001) has generated a global reassessment of how much we really know about ecological dynamics.

39 Contrast the positions and examples of Nandy (1988) and Nanda (2003).

40 Svarstad and Dhillion (2000). Ironically, despite the vigorous condemnation in India of biopiracy, the BBC characterized Navbharat's appropriation of Monsanto's Bt gene in Gujarat as reverse biopiracy. The assumption that flow can move only upstream, from South to North, is clearly problematic. Navbharat's chief is widely celebrated in India as "Robin Hood" for enabling underground seeds with Bt technology. Herring (2005a).

41 Roy (1994) presents a chilling portrait; see also Menon (1997). Peluso and Watts (2001) find it possible to entitle an entire book Violent Environments.

42 Neto (2003); Paarlberg (2001, pp. 67–72, passim).

43 These campaigns (falsely) declared the existence of terminator technology on which Monsanto (again falsely) had a patent that threatened nature universally. This disinformation campaign spread from a Canadian website (RAFI) to achieve extraordinary effectiveness on a global scale. Its story has yet to be written authoritatively.

44 Decommodification means removal of some tracts of nature from the market; rigging markets means limiting what markets can and cannot do – e.g. what species are endangered and therefore protected from legal international trade. See Herring (2002) and Herring (1999a).

45 Madhav Gadgil and Ramachandra Guha (1995) argue that there are different ways to classify classes and their interests, in which environmental degradation is driven by "omnivores" at the expense of "eco-system people."

46 This is a slippery concept, easily operationalized *post factum,* but at that point the variable becomes tautological. Nevertheless, there are ways to make comparisons. See Edith Brown Weiss and Harold Jacobson (1998). See Menon (1997b) as well.

REFERENCES AND BIBLIOGRAPHY

Agrawal, A. (l995) "Dismantling the Divide Between Indigenous and Scientific Knowledge," *Development and Change* 26: 413–439.

Arnold, David (1996) *The Problem of Nature: Environment, Culture and European Expansion.* Oxford: Blackwell.

Bandyopadhyay, J. and Shiva, V. (1988) "Political Economy of Ecology Movements," *Economic and Political Weekly* June 11, 1223–1332.

Baviskar, A. (1995) *In the Belly of the River: Tribal Conflicts over Development in the Narmada Valley.* Delhi: Oxford University Press.

Blaikie, P. and Brookfield, H. (l987) *Land Degradation and Society.* London: Methuen.

Bromley, Daniel W. and Chapagain, Devendra P. (1984) "The Village Against the Center: Resource Depletion in South Asia," *American Journal of Agricultural Economics* December, 868–873.

Conway, Gordon (1997) *The Doubly Green Revolution: Food for All in the 21st Century.* New York: Penguin Books.

Cronon, William (l983) *Changes in the Land: Indians, Colonists and the Ecology of New England.* New York: Hill and Wang.

Cronon, William (1991) *Nature's Metropolis.* New York: W.W. Norton & Company.

Daily, Gretchen (ed) (1997) *Nature's Services: Societal Dependence on Natural Ecosystems.* Washington, DC: Island Press.

Dryzek, John, (1997) *The Politics of the Earth.* New York: Oxford University Press.

Eaton, Richard M. (l990) "Human Settlement and Colonization of the Sundarbans, 1200-1750," *Agriculture and Human Values* VII(2). 6–16.

Elster, Jon (1985), *Making Sense of Marx.* Cambridge: Cambridge University Press.

Ferguson, James (1994) *The Anti-Politics Machine: Development, Depoliticization, and Bureaucratic Power in Lesotho.* Minneapolis: University of Minnesota Press.

Gadgil, M. and Guha, R. (l995) *Ecology and Equity: The Use and Abuse of Nature in Contemporary India.* New Delhi: Penguin.

Gadgil, M, N.V. Joshi, and Suresh Patil, (l993) "Power to the People: Living Close to Nature," *The Hindu:* Survey of the Environment, pp. 58–62.

Gold, A. G. (l997) "Wild Pigs and Kings: The Past of Nature and the Nature of the Past in a Rajasthani Princedom," *American Anthropologist* 99: 1.70–84.

Gold, Ann Grodzins and Bhoju Ram Gujar, (2002) *In the Time of Trees and Sorrows: Nature, Power,and Memory in Rajasthan.* Durham: Duke University Press.

Gove, Richard, H. (1995) *Green Imperialism.* Cambridge: Cambridge University Press.

Guha, Ramachandra (1985) "Forestry and Social Protest in British Kumaun, c. 1893–1921," *Subaltern Studies* IV: 54–100.

Guha, Ramachandra (l989a) *The Unquiet Woods: Ecological Change and Peasant Resistance in the Himalaya.* Delhi: Oxford University Press.

Guha, Ramachandra (l989b) "Radical American Environmentalism and Wilderness Protection: A Third World Critique," *Environmental Ethics* ll:l.

Guha, Ramachandra (l990) "An Early Environmental Debate: The Making of the l878 Forest Act," *The Indian Economic and Social History Review* 27:l.

Gupta, Anil (1998) "Rewarding Local Communities for Conserving Biodiversity: The Case of the Honey Bee." *"Protection of Global Biodiversity: Converging Strategies.* Lakshman D. Guruswamy and Jeffery A. McNeely (eds). Durham: Duke University Press, pp.180–189.

Guruswamy, Lakshman, D. and Jeffrey A. McNeely (eds) (1998) *Protection of Global Biodiversity: Converging Strategies.* Durham: Duke University Press.

Helvaarg, David (l994) *The War Against the Greens.* San Francisco: Sierra Club Books.

Herring, Ronald J. (1990) "Resurrecting the Commons: Collective Action and Ecology," *Items* 44:4.

Herring, Ronald (1991) "Politics of Nature" Harvard University Center for Population and Development Studies, Working Paper Series No. 7,October.

Herring, Ronald J. (1999a) "Nature, State and Market: Implementing International Regimes in India." In Stig Toft Madsen (ed.) *South Asian Environments in a State-Society Perspective.* Copenhagen and London: Nordic Institute of Asian Studies and Curzon Press; Delhi: Manohar (2003).

Herring, Ronald J. (l999b) "Market-Structuring Regulation and the Ozone Regime: Politics of the Montreal Protocol." In Timothy Mount and Mohammed Dore (eds) *Global Environmental Economics: Equity and the Limits to Markets.* Oxford: Blackwell.

Herring, Ronald J. (2001) "Authority and Scale in Political Ecology: Some Cautions on Localism." In Louise E. Buck, Charles C. Geisler, John Schelhas, and Eva Wollenberg (eds) *Biological Diversity: Balancing Interests through Adaptive Collaborative Management.* Boca Raton, FL: CRC Press.

Herring, Ronald J. (2002) "State Property Rights in Nature (With Special Reference to India)." In John F. Richards (ed.) *Land, Property and the Environment.* Oakland, CA: Institute for Contemporary Studies.

Herring, Ronald J. (2003) "Data As Social Product." In Ravi Kanbur, (ed) *Q-Squared: Combining Qualitative and Quantitative Methods in Poverty Appraisal.* New Delhi: Permanent Black.

Herring, Ronald J. (2005a) "Miracle Seeds, Suicide Seeds and the Poor: GMOs, NGOs, Farmers and the State." *In Social Movements in India: Poverty, Power and Politics.* R. Ray and M.F. Katzenstein (eds). Lanhams, MD: Rowman & Littlefield; New Delhi: Oxford University Press.

Herring, Ronald J. (2005b) "The Political Biology of Labels." *Agnet,* March 8, http://archives.foodsafetynetwork.ca/agnet/2005/3-2005/agnet_march_8.htm.

Herring, Ronald J. (2007a) "Stealth Seeds: Bioproperty, Biosafety, Biopolitics", *Journal of Development Studies* 43(1).

Herring, Ronald J. (2007b) "The Genomics Revolution and Development Studies: Science, Politics and Poverty," *Journal of Development Studies* 43(1).

Herring, Ronald J. and Erach Bharucha, 1998, "Embedded Capacities: India's Compliance with International Environmental Accords," In Edith Brown Weiss and Harold Jacobson, (eds) *Engaging Countries: Strengthening Compliance with International Environmental Accords.* Cambridge: MIT Press.

Hilgartner, Stephen (2002) "Acceptable Intellectual Property." *Journal of Molecular Biology* 319: 943–946.

Horsch, Robert B. and Robert T. Fraley, (1988) "Biotechnology Can Help Reduce the Loss of Biodiversity". *Protection of Global Biodiversity: Converging Strategies.* Lakshman D. Guruswamy and Jeffery A. McNeely (eds). Durham: Duke University Press.

Jayaraman, K. S. (2001) "Illegal Bt Cotton in India Haunts Regulators," *Nature Biotechnology,* 19: 1090.

Jayaraman, K.S. (2004) "India Produces Homegrown GM Cotton," *Nature Biotechnology* 22(3): 255–256.

King, Steven R., Carlson, T.J. and Moran, K. (1996) "Biological Diversity, Indigenous Knowledge, Drug Discovery and Intellectual Property Rights: Creating Reciprocity and Maintaining Relationships," *Journal of Ethnopharmacology* 51: 45–57.

Kothari, Ashish (1997) *Understanding Biodiversity: Life Sustainability and Equity.* New Delhi: Orient Longman Limited.

Kothari, Ashis, Neena Singh, Neena and Saloni Suri (1996) *People & Protected Areas.* New Delhi: Sage Publications.

Kothari, Ashis, Suri, S. and Singh, H. (eds) (1997) *Building Bridges for Conservation: Toward Joint Management of Protected Areas in India.* New Delhi: Indian Institute of Public Administration.

Kothari, Smitu (1995a) 'Whose Nation is It? The Displaced as Victims of Development'. *Lokayan Bulletin* 11(5): 1–8.

Kothari, Smitu (1995b) 'Developmental Displacement and Official Policies: A Critical Review'. *Lokayan Bulletin* 11(5): 9–28.

Kothari, Smitu and Pramod Parajuli (1993) 'No Nature without Social Justice: A Plea for Cultural and Ecological Pluralism in India'. In Wolfgang Sachs (ed.) *Global Ecology: A New Arena of Political Conflict.* London: Zed.

Lomborg, Bjorn (2001) The Skeptical Environmentalist: Measuring the Real State of the World. London: Cambridge University Press.

Lipton, Michael (1977, 1988) *Why Poor People Stay Poor: Urban Bias and World Development.* London: Temple Smith; Cambridge, MA: Harvard University Press.

Madsen, Stig Toft (2001) "The View from Vevey," *Economic and Political Weekly.* September 29, pp. 3733–3742.

Magagna, Victor V. (1991) *Communities of Grain.* Ithaca, NY: Cornell University Press.

McHughen, A. (2000) *Pandora's Picnic Basket: The Potential and Hazards of Genetically Modified Foods.* Oxford, New York: Oxford University Press.

Menon, Vivek (1996) *Under Seige: Poaching and Protection of the Greater One-Horned Rhinoceroses in India.* Delhi: TRAFFIC-India, WWF-I.

Ministry of Environment and Forests (Government of India), (1994), Report of the Committee on Prevention of Illegal Trade in Wildlife and its Products (New Delhi).

Ministry of Environment and Forests (Government of India), (1996), Recommendations of the Committee Appointed by the Honorable High Court of Delhi on Wildlife Preservation, Protection and Laws (New Delhi, February 1996).

Mooney, Chris (2005) *The Republican War on Science.* New York: Basic Books.

Nadkarni, M. V. (1987) "Agricultural Development and Ecology: An Economist's View," *Indian Journal of Agricultural Economics* 42(3): 359–375.

Nandy, A. (1988) *Science, Hegemony and Violence.* Delhi: Oxford University Press.

Neto, R. B. (2003) "GM confusion in Brazil," *Nature Biotechnology* 21(11): 1257–1258.

North, D. C. (1990) *Institutions, Institutional Change, and Economic Performance.* Cambridge and New York: Cambridge University Press.

Omvedt, G. (1987) "India's Green Movements," *Race and Class* XXVIII(4): 29–38.

Paarlberg, R. L. (2001) *The Politics of Precaution: Genetically Modified Crops in Developing Countries.* Baltimore: Johns Hopkins University Press.

Peet Richard, and Watts, Michael (1996) *Liberation Ecologies: Environment, Development and Social Movements.* London: Routledge.

Peluso N, (1992), *Rich Forests and Poor People: Resource Control and Resistance in Java.* Berkeley: University of California Press.

Peluso Nancy and Watts, Michael (eds) (2001) *Violent Environments.* Ithaca, NY: Cornell University Press.

Pimentel, David *et al.* (1998) "Ecology of Increasing Disease," *BioScience* 48(10): 817–826.

Pinstrup-Anderson, Per and Ebbe Schiøler (2000) Seeds of Contention: World Hunger and the Global Controversy over GM Crops. Baltimore: Johns Hopkins University Press.

Polanyi, K. (1957) *The Great Transformation.* Boston: Beacon Press.

Pray, Carl and Naseem, Anwar (2007) "Supplying Crop Biotechnology to the Poor: Opportunities and Constraints," *Journal of Development Studies* 43(1).

Raghunandan, D. (1987) "Ecology and Consciousness," *Economic and Political Weekly* XXII(28): 545–549.

Revkin, Andrew C. "Climate Expert Says NASA Tried to Silence Him," *New York Times,* January 29, p. 1.

Robbins, Paul (2004) Political Ecology. Oxford: Blackwell.

Roy, Devparna (2005) "To Bt or Not To Bt," Indian Cotton: Biology and Utilities, Histories and Meanings. Workshop of Syracuse and Cornell Universities, April 29–30. http://www.einaudi.cornell.edu/SouthAsia/conference/cotton/

Roy, Devparna. Herring, Ronald J. and Geisler, Charles C. (2007) "Naturalizing Transgenics: Loose Seeds, Official Seeds, and Risk in the Decision Matrix of Gujarati Cotton Farmers," *Journal of Development Studies* 43(1).

Roy, S. Deb (l994) Manas National Park: A Status Report (New Delhi: unpublished ms).

Schama, S. l996, *Landscape and Memory*. New York: Vantage.

Scott, James C. (l998) *Seeing Like a State: How Certain Schemes to Improve the Human Condition Have Failed*. New Haven: Yale University Press.

Shaik, Sajid (2001) "Farmers Decide to Defend their Bt Gene Cotton Crops," *The Times of India*, Oct. 31.

Shiva, Vandana (1997) *Biopiracy: The Plunder of Nature and Knowledge*. Boston, MA, USA: South End Press.

Shiva, Vandana, Ashok Emani and Afsar H. Jafri (1999) "Globalization and Threat to Seed Security: Case of Transgenic Cotton Trials in India" *Economic and Political Weekly*, March 6–12 and 13–19.

Simpson, R. David, Roger A. Sedjo and John W. Reid (1996) "Valuing Biodiversity for Use in Pharmaceutical Research," *Journal of Political Economy* 104(1): 163–185.

Singh, Narindar (1976) *Economics and the Crisis of Ecology*. Delhi: Oxford University Press.

Sinha, Subir and Herring, Ronald (1993) "Common Property, Collective Action and Ecology," *Economic and Political Weekly*.

Sinha, Subir, Shubhra Gururani, and Brian Greenberg (1997) "The 'New Traditionalist' Discourse of Indian Environmentalist," *The Journal of Peasant Studies*, Vol. 24, No. 3, April 1997, pp. 65-99.

Sivaramakrishnan, K., 1995, "Colonialism and Forestry in India: Imagining the Past in Present Politics". *Comparative Studies in Society and History* 37(1):3–40.

Svarstad, Shiva Hanne and Dhillion, Shivcharn S. (eds) (2000) *Responding to Bioprospecting*. Oslo: Spartacus Forlag AS.

Tanksley, S.D. and McCouch, S.R. (1997) Seed Banks and Molecular Maps: Unlocking Genetic Potential from the Wild. *Science*, 277(5329): 1063.

Thies, Janice E. and Devare, Medha (2007) "An Ecological Assessment of Transgenic crops," *Journal of Development Studies* 43(1).

Varley, John D. and Preston T. Scott (1998) "Conservation of Microbal Diversity a Yellowstone Priority," *American Society of Microbiology News* 64(3), pp. 147–151.

Weiss, Charles and Eisner, T. (1998) "Partnerships for Value-Added Through Bioprospecting," *Technology In Society* 20: 481–498.

Weiss, E. B. and Jacobson, H. (1998) *Engaging Countries: Strengthening Compliance with International Environmental Accords*. Cambridge, MA: MIT Press.

Winston, Mark (2002) *Travels in Genetically Modified Zone*. Cambridge, MA: Harvard University Press.

World Commission on Environment and Development, 1987, *Our Common Future*. New Delhi: Oxford University Press.

Zilberman, David Holly Ameden and Matin Qaim (2007) "The Impact of Agricultural Biotechnology on Yields, Risks and Biodiversity in Developing Countries," *Journal of Development Studies* 43(1).

21

Protest Movements, Environmental Activism and Environmentalism in the United Kingdom

Steven Griggs and David Howarth

INTRODUCTION

Let us begin with the thought that the demands of 'environmental politics' – broadly construed here, as a range of challenges to the logic of industrialization[1] – are *popular*. It is unusual, though not of course unprecedented, for citizens or politicians to declare themselves against cleaner air and unpolluted drinking water, to reject publicly rationales for fuel-efficient transport systems, protected wilderness areas or 'sustainable growth'. There are not many who are in favour of less pollution controls, more greenhouse gases, unregulated transport systems, greater species depletion, or economic growth regardless of environmental cost. In short, the quintessentially public goods struggled for by numerous environmental organizations and movements are goods desired by a large majority of the people living in particular societies and communities.

However, despite the obvious popular appeal of many environmental demands, the translation of such popularity into a populist form of politics that builds broader political projects and alliances has been far from unproblematic for the environmental movement in the United Kingdom (UK). Collective action problems, the prevalence of 'Nimbyism', the difficulties of articulating environmental demands into coherent political programmes, diverse conceptions of the problems and solutions to environmental issues, and so on, have conspired to impede the construction of political projects, coalitions and alliances that can sustain demands for a better environment, and which can influence the framing, passing and implementation of more environmentally friendly public policies. Popularity has thus not translated necessarily and unproblematically into populism. Whilst environmental concerns and demands are very often *popular* – at the local, regional, national and even global levels – they are very rarely *populist*.

In order to investigate this apparent paradox, this chapter examines the difficulties of – and limits to – constructing a populist form of politics to advance environmental demands and interests. We begin with a few remarks on the concept of populism as a distinctive form of politics, after which we outline three waves of environmental protest in the UK. These three waves are then characterized in terms of a populist/non-populist spectrum of possibilities, as we seek to position different manifestations of protest and activism along this continuum. In particular, we focus our examination on a series of new forms of environmental activism – the recent struggles against road building and airport construction in the UK, which brought together local residents and radical environmentalists – asking why and how this particular logic of campaigning was pursued, while also seeking to evaluate these new phenomena in the light of our discussion of populism. We offer in conclusion reflections on the implications of our analysis for the practice and critical evaluation of environmental politics and campaigns.

A NOTE ON POPULISM

We often speak of something being 'populist' when it is perceived to be popular by a large majority of a country's population. A politician's 'populist gesture' – a call to stop fox hunting, for example, or a policy designed to 'get tough on criminals' – is usually taken to be a pejorative description of a politician's speech or decision that appeals to the more baser instincts of the mass of a country's *populus*. However, while there is often a good deal of slippage between ethnographic and theoretical usages of the term, it is important not to confuse our ordinary common-sense meanings of the word with a precise theoretical concept. Here, and to avoid such confusion, we draw upon Ernesto Laclau's efforts (2005) to construct a rigorous notion of populism, which centres on the construction of equivalential linkages between dispersed social and political demands. The latter logic involves the provision of specific means of representation – what Laclau calls 'empty signifiers' – which can serve as points of subjective identification to hold together a diverse set of agencies in a precarious and contingent unity. In this conception, a populist form of politics is thus not to be confused with a specific ideological *content* – a rhetorical appeal to 'the people' for instance – nor with certain *types* of movements or organizations, such as the People's Party in the USA for instance, but is understood as a *dimension* of social relations that is peculiarly political. If, in general, the political refers to the production, reproduction and transformation of social relations, that is, a simultaneous and ongoing process of instituting and contesting the social, then populism names that dimension which contests social relations by dividing society – or at least various sites of it – into opposed camps in the endless struggle for hegemony. A populist politics involves the construction of a collective agency – 'the people' for instance – that can establish a political frontier by the creation of antagonistic relations between subjects.

Our picture of populism as a form of politics thus consists of three basic features. In the first instance, populist discourses appeal to a collective subject such as 'the people' or 'the nation' as the privileged subject of interpellation. In Michael Freeden's terms, populist discourses and movements aim to decontest the meaning of 'the people' or its functional equivalent, using such appeals as the principal means to constitute political identities and recruit subjects (1996, p. 76). Second, populist discourses are grounded on the construction of an underdog/establishment frontier, which, if successful, opposes 'the people' to an enemy or adversary, say the elite or the power holders. Third, in order to constitute this political frontier dividing the people from the establishment – the production of equivalential effects amongst the particularities that make up the people – populist discourses are necessarily predicated on a certain passage through the universal. There is, in other words, an appeal to *all* the people within a delimited sphere or domain, and the elaboration of ideologies and symbols designed to realize these goals (Howarth, 2005).

Alongside these three features, we introduce a further theoretical assumption and four implications. First, we assume that the political orientation and character of a populist movement is dependent on the kinds of hegemonic articulation available and practised within a given historical context. Here, we build on Laclau's claim (Laclau, 1977; Laclau and Mouffe, 1985) that the content of a political discourse is not determined by an *a priori* analysis of its ideological elements, but from the way in which their meanings are conferred by the hegemonic projects that articulate them. In other words, if a political force wishes to articulate and pursue its demands democratically, it must inscribe them into a broader discourse, and help to build organizational forms and levels of accountability, which can ensure that its interests are represented and pursued in political ideologies and public policies. How it does so, and indeed whether it is able to do so, depends not only on agency, but also on the particular articulations and discourses available within a given time and political space.

The first implication that needs to be added to our short theoretical account is that not all 'populist appeals' constitute a populist politics. While a discourse may indeed consists of populist appeals and demands, this by itself does not constitute a populist politics. For example, John Major's 'back to basics' campaign in the UK during the mid-1990s – when he urged people in 1993 'to get back to basics: to self-discipline and respect for the law, to consideration for others, to accepting responsibility for yourself and your family, and not shuffling it off on the state' (www.johnmajor.co.uk, accessed 12 March 2007) – might be said to comprise a set of populist appeals and gestures, but it did not constitute a populist politics that was able to divide the social by instituting a new political frontier. Instead, Major's rhetoric represented the attempt by a weak and failing government to shore up

Thatcherite hegemony with a new, more moderate and appealing discourse. As against our conception of populist politics, Major's discourse represented no more than a form of transformism designed to negate, domesticate and displace challenges and demands to maintain the status quo. A similar thing could be said about William Hague's 'commonsense revolution', which had a populist ideological content, but did not partake of a populist form of politics, as well as Blair's Third Way politics (though the latter case is much more contentious).

A second implication is that populist forms of politics do not necessarily have to have recourse to the name of 'the people' to be a populist movement or discourse, as a wide range of subjects can be appealed to in order to constitute a populist struggle. Furthermore, the populist form of politics is not restricted to struggles and demands at the level of the 'nation-state', as populist struggles may be carried out at the local, regional and even global levels. Indeed, even at these different levels of analysis, populist struggles may occur in different spaces or sites of struggle. The fourth and final implication of our approach is to sharpen the distinction between populist and non-populist forms of politics, in which the former is characterized by the *degree* of division and contestation brought about by the populist mobilization. In other words, to put it in quantitative terms, the greater the number of demands articulated into an equivalential chain, which in turn serves to split different social spaces and sites into opposed camps, the greater the degree of populist politics. On the other hand, the failure to articulate different demands, or the struggle to disarticulate equivalential demands, is a characteristic of non-populist forms of politics. Having briefly introduced what we mean by populism, and set out the contours of our approach, we turn now to a brief genealogy of environmental protest in the UK.

THREE WAVES OF ENVIRONMENTAL PROTEST

It is possible, we believe, to characterize the historical development of environmental movements and protest in terms of our approach to populism. More precisely, three broad phases of environmental politics can be discerned. These are what we shall call respectively 'conservation environmentalism', 'ecological environmentalism' and 'radical environmentalism', respectively. We explore each in turn, concentrating our remarks on the last phase of development, that of 'radical environmentalism'.

The first phase of the development of environmental movements, beginning in the latter part of the nineteenth century, was marked by what might be termed 'conservation environmentalism'. It witnessed the emergence of associations such as the *Royal Society for the Protection of Birds* (RSPB) (1889), the *National Trust* (1895) and the *Council for the Protection of Rural England* (CPRE) (1926), groupings that modelled themselves to some extent on earlier entities such as the *Royal Society for the Prevention of Cruelty to Animals* (RSPCA), which was founded in 1824 (Byrne, 1997, pp. 129–130). This early conservation movement lost momentum after 1900 as legislation arguably responded to the demands voiced by conservation groups, and the end of the First World War shifted public attention towards the challenges of reconstruction. Indeed, the CPRE was the only significant conservation organization established during the interwar period in the UK (Dalton, 1994, p. 33).

These early conservation groups concentrated their efforts principally on the protection and preservation of flora, fauna and habitats that were being jeopardized by the developing logic of industrialization. While seizing on popular issues animating a range of citizens, their politics was decidedly anti-populist, choosing an approach that may be deemed reformist and anthropocentric, stressing the value of the natural environment *for* human well being (Byrne, 1997, p. 130; Richardson, 1995). In more theoretical terms, these first-wave organizations are best conceptualized in terms of various types of interest group theory, that is, as reformist organizations that work within existing institutions advance a particular sectional interest or cause, but without the bigger aspiration of forming a government or being part of a government (see Grant, 2000).

This first phase of environmental politics contrasts rather sharply with the phase of what might be called 'ecological environmentalism', which constituted a more radical and broader way of defining and contesting environmental issues at the end of the 1960s and 1970s. As against the more anthropocentric orientation of its predecessors, it became associated with a 'biocentric' philosophy that located human beings within a larger natural system – the planet – and advocated the notion that human beings have to learn to coexist with the diverse lifeforms that constitute its shared universe (Byrne, 1997, p. 130). Organizations such as *Friends of the Earth* (FoE) and *Greenpeace* inscribed new environmental problems onto the political agenda, challenging the use of nuclear power, the adverse consequences and long-term implications of industrial pollution, as well as quality of life issues. They were also able to internationalize environment struggles, privileging the adoption of citizen engagement and forms of direct-action protests.

There are many interpretations for the emergence of 'ecological environmentalism' in the UK and elsewhere, which cannot be explored in detail here (see, e.g. Rootes, 2002a, b). Suffice to say, first, that this new form of protest was grounded on a greater theoretical and practical awareness of the dangers posed to the environment (some of which had been exposed, for instance, in 1972 with the publication of *The Limits to Growth*), especially the dawning realization that the Earth comprised a system of finite resources that could not support the infinite production and consumption required by advanced industrial orders (Meadows, 1974). Second, the movement shared an elective affinity with the other radical movements of the time, challenging the perceived materialism, indeed decadence, of capitalist social relations, as well as the more statist and bureaucratic modes of social control that were alleged to exist at that time. Indeed, emerging alongside a series of other 'new social movements' during the period – the peace movement, the women's movement, various struggles for sexual liberation and the student movement – ecological politics partook of a 'second wave' of social movements that rocked the apparent social democratic consensus prevalent in many advanced capitalist societies. 'Ecology' became an exemplary component of the so-called 'counter-culture', and its demands for a return to nature and for a more harmonious coexistence with the natural world. Of course, this emergence of 'ecological environmentalism' did not as much usurp first-wave conservation groups as develop in parallel to them. During this second wave of environmentalism, we also witnessed the rapid growth of the membership of conservation groups such as the *National Trust* and the RSPB (Dalton, 1994, p. 41).

However, understood in these broad terms of new social movements in the 1960s and 1970s, ecological environmentalism served to problematize the orthodox 'Left–Right' distinctions of the time, which were generally predicated on conflicts over the productivist distribution of resources, raising issues of greater democratic participation, the dissemination of alternative values as core aspects of political activity, and the construction of new identities and subjectivities. As such, this wave of protest can in certain respects be understood in terms of the social movement theory that arose in the wake of these new types of phenomena, which is exemplified by theorists such as Jürgen Habermas (1981), Alain Touraine (1985) and Alberto Melucci (1989). Unlike their interest group conservation predecessors, the emerging ecological social movement was alleged to rest like other social movements of the time upon the reproduction of a common identity, the threat of mass mobilization and the adoption of a radical programme of change and contestation of the existing order (see Scott, 1990, p. 6).

Indeed, for Touraine, such new social movements organized against the logics of programmed societies, in which the stakes of conflict comprise an all encompassing and technocratic state versus the potentiality of a democratic and libertarian civil society. In the field of ecological politics, such movements can be seen to challenge the prioritization of economic development and growth over the protection of the environment and human beings' quality of life. Thus, for theorists such as Habermas (1981, 1987), the ecological movement was a means of defence against the ongoing 'colonization of the lifeworld' wrought by the unchecked hegemony of instrumental reason and capitalist relations of production.

As such, it is evident that this so-called second wave of environmental activism contained an important populist dimension. Standing for the most part outside the existing institutional structures of governance and decision-making, the ecological movement sought to transform the values and norms of civil society, and eschewed their possible co-option into positions of power and compromise. However, while it attempted to translate its apparent popularity into a form of populist politics that could challenge the existing system by constructing a popular pole, there were limits to this type of politics. One important obstacle was the difficulty of connecting together the disparate demands and orientations of different social movements so as to mount a general challenge to the existing power structure. Given the uneven nature of the constituencies of those broadly supportive of ecological values – the socio-economic differences, the different class positions, the different political values that supporters exhibited with respect to other issues, and so forth – the ecological movement was unable to weld together and then organize a coherent bloc of support that could effectively challenge the existing power structures at different levels of the system. As commentators such as Jonathan Porritt (2005) have recently acknowledged, from its early days the ecological movement found it difficult to move from a position of opposition and critique to articulating a positive vision for the future. Speaking to *The Observer*, he argued that 'Environmental organisations for many years [were] saying "no" and protecting and stopping, because in a way that became part of the culture of the movement. There's still a lot of criticising and blame-laying and not enough saying what solutions are available' (6 October, 2005).

More precisely, it was difficult to articulate a common set of values and an overall ideological stance that could form the basis of what Gramsci (1971) would call a counter-hegemonic project. On the one hand, such an operation carried the intrinsic dilemma that the particularity of one set of demands – demands, for instance, to deal with environmental issues such as deforestation or nuclear power – would be dissolved and weakened in a more universal, and by definition more general, set of demands. This logic of linking the more general sentiments and values of ecologism to the more concrete demands and struggles put forward by specific groupings required the linking of general values to particular struggles in ways which would permit groups to understand their interests as part of a wider struggle, and then to act for particular goals, and for the wider struggle to be rearticulated in the name of particular struggles. However, on the other hand, the issues themselves were often incorporated into the policies and programmes of existing organizations, whether they be political parties or pressure groups. Thus, the capacity of the British Labour Party to incorporate the demands of feminists and peace activists, as well as the absence of a more radical left alternative, drew some environmental activists to invest in party politics and influence through the Labour Party. The absence of ties between the environmental movement and the alternative Left, and the absence of radicalization through protest, meant there was no forging of opposition linkages between environmental activists and feminist and peace movements (Rootes, 1992; Doherty, 1999, p. 279).

The final weakness in mounting an effective challenge to power structure was, and remains, the powerful logic of institutionalization and organization that blunts the more radical aspirations and activities of members of oppositional movements. As theorists such as Piven and Cloward (1977) have insisted, employing arguments that share important affinities with an elite theorist such as Robert Michels (1959), a constant threat to radical organizations such as 'poor people's movements' is a logic of organization in which resources get directed more at the reproduction of the organization and its leadership than in the campaigns which advance the interests of those supposedly represented by the organizations. Similar arguments are advanced by rational choice theorists who posit 'cycles of protest' in which enthusiastic activists and supporters become disillusioned with the slow pace of institutionalized protest, or the delivery of small victories and compensations, and decide either to opt out of the organization, and political activity more generally, or look to other forms of protest such as direct action (Hirschman, 1982; Chong, 1991).

One way to conceptualize these tendencies is to distinguish between social movements and 'social movement organizations' (SMOs), where the latter consist of 'organizations within organizations which mobilize resources for the wider movement' (Byrne, 1997, p. 45). The difficulty arises when SMOs come to supplant the activity of the wider social movements of which they are a part, a weakness that is evident in the history of environmental activism in the UK. Take for instance the emergence and development of FoE. It was established in 1969 in San Francisco by David Brower, who coined the phrase 'think globally, act locally', with the aim of marrying conventional lobbying politics alongside more unconventional forms of politics such as direct action. A UK wing was founded in 1970 and its membership, like that of *Greenpeace*, grew rapidly, witnessing a seven-fold increase between 1985 and 1993 (Doherty, 1999, pp. 277–278). By 1990 it boasted a membership of 190,000 and an annual income of £4.5 million. By 2004 the FoE as a whole claimed to unite a million environmental activists in 70 countries (FoEI, 2005, p. 14). In so doing, FoE enlarged the focus of 'ecological environmentalism' to include a wide range of issues – pollution, deforestation, nuclear power, unsustainable development, and so on – linking their causes to the wider context of the capitalist system, allegedly controlled by corporate interests and 'big government', and employed a wide range of tactics, ranging from traditional lobbying to more direct and participatory forms, while encouraging new types of organization and decision-making. Over time, however, as the organization became more sedimented and institutionalized, so it was perceived to be distancing itself from the more radical demands and ideas of the ecological movement. For Jordan and Maloney, FoE had from its origins in environmental activism transformed itself into a 'mail order protest business' with a relatively autonomous leadership divorced from its grassroots memberships and local activists (1997, pp. 106–122). It was 'neither very democratic nor participatory in practice and [...] offered no opportunity for radicals to express alternative ideologies through protest' (Doherty, 1999, p. 278). In short, whilst FoE can claim to have strengthened the infrastructure of the environmental movement, it is alleged to function more as a social movement organization and as a professionalized centralized organization (Doherty, 1999) than as an organization rooted in the wider ecological movement, and its role as an expert adviser to government has been called into question (Rootes, 1992). Indeed, with other organizations following similar trajectories, by the turn of the 1990s, the environmental movement in Britain could be characterized as 'well-organized and well-supported but politically moderate' (Doherty, 1999, p. 275).

As a consequence of this process of institutionalization, more radical groups and movements have been encouraged to supplement the supposed deficiencies of the likes of FoE with different forms of political protest. One such development has been the emergence of *Earth First!*, which was formed in the USA in early 1980s, with a UK branch established in 1991 (Wall, 1999). *Earth First!* defined itself against FoE and other environmental organizations such as *Greenpeace* with its top-down, almost business-like or military style of organization and campaigning, arguing that their institutionalization had blunted their radicalism and turned them into 'insider groups' with reformist ideologies and agendas. Instead, *Earth First!* is principally a loose alliance of relatively autonomous radical activists and groups, whose stance is resolutely anti-organizational and anti-institutional, eschewing all forms of conventional lobbying, as well as efforts to become insider groups dispensing expert advice to government or business. Their ideology, such that it is consciously elaborated, is largely anti-capitalist, sometimes anarchistic, and oriented to concrete projects involving overt political action. Indeed, Earth Firsters (as they sometimes refer to themselves) are most visible in their direct-action protests against the building of new roads, airports and housing developments, where they are prepared to use techniques of bodily risk to prevent physically the implementation of the projects and infrastructures they oppose. Their networks of organization, their repertoires of protest and their counter-cultural ideas has led Doherty to argue that these protest groups are 'the first significant example of environmental protest by a new social movement in Britain' (1999, p. 276).

Struggles and demands since the late 1980s and early 1990s have thus brought a further twist to the story of populism in the environmental field. As organizations rooted in the populist mobilizations of the second wave of protest were increasingly institutionalized in the 1980s, so new forms of populism in the form of direct-action struggles began to emerge in the early 1990s. This resulted in the construction of new alliances between what we might term conventional forms of 'conservation environmentalism', usually in the form of NIMBY groups opposed to local developments, and the more radical environmentalists associated with organizations like *Earth First!*, *Road Alert!* and *Reclaim the Streets* (Wall, 1999, 2005). This new kind of protest caught the older ecological organizations off balance, and they were initially ambivalent, sometimes even opposed, to the new logics of activism (Rootes, 2002a, pp. 5–7). This meant that while groups like FoE were not entirely absent from such struggles, the organizing and directive role that might have been expected from them was no longer as prominent. Instead, struggles against new road-building projects and airports were spearheaded by self-styled 'eco-warriors',

radicalized community activists and local residents groups.

Such new forms of environmental protest that linked together eco-protesters and local residents have been characterized as a 'third wave' of grassroots environmentalism, which though critical of previous institutionalized patterns of protest are prepared to work alongside and in conjunction with mainstream environmental organizations and conservation groups (Bosso, 1997; see also Carter, 2001, pp. 141–146). The emergence of such radical environmental protest in the 1990s challenges the 'conventional wisdom' which holds that the political culture or institutional structures of 'bureaucratic accommodation' in Britain have 'constrained environmentalists to moderate action and the pursuit of broad political alliances' (Rootes, 2002a, p. 9). Instead, as Rootes suggests, it concentrates our understanding of environmental activism on the contingent dynamics of political competition, and the group identities, strategies and tactics of actors engaged in grassroots protests (2002a, p. 9; 2002b, p. 52).

He thus explains the adoption of new forms of environmental protest in the 1990s as a response to the changing political conjuncture of neo-liberal Thatcherism, its campaign of road building, changing public opinion, and the example offered to activists by the campaign against the poll tax, concluding that 'nowhere else in the European Union was a government so determinedly committed to a controversial, large-scale programme of road-building, so resistant to hostile public opinion, or so imaginatively confronted by so heterogeneous an environmental movement' (Rootes, 2002b, p. 52). Such a focus on changing political conjunctures pushes to the fore the importance of time and the articulation of political spaces in accounts of environmental activism. However, whilst novel identities and discourses may become available in a changing political conjuncture, their successful articulation still depends in part not only on how actors seek to forge political frontiers and construct a collective agency, but also on their capacity to do so. In order to examine such processes and evaluate these new forms of radical environmentalism, we shall compare struggles against the development of roads and airports in the UK since the 1990s. It is to the struggles against road building that we turn first.

THE STRUGGLES AGAINST ROAD BUILDING

In May 1989, the Conservative Party of Mrs Thatcher published the Transport White Paper *Roads for Prosperity*, promising 'the biggest road-building programme since the Romans'. Conforming to the model of 'predict and provide' the programme envisaged the construction of more than 2700 miles of new roads and more than 150 by-passes to relieve congested towns and cities; road building was to be the centrepiece of Britain's transport policy and was deemed a necessary component of sustained economic growth (DoT, 1989). Few would have predicted that by 1996 the proposals to resolve the UK's growing transport problems would lay in tatters following a rolling series of protests that drew direct-action protesters, new-age travellers, students and radical environmentalists alongside local residents and dignitaries in what has been dubbed an alliance of 'Vegans and Volvos' (*The Times*, 19 May, 1997).

The protests had an inauspicious beginning. In the middle of 1992 a small band of protesters took up residence on the top of Twyford Down near Winchester in Hampshire, which was to be the site of the first of the proposed road schemes. The aim was to extend the M3 motorway from Winchester to the coast, thus speeding up car journeys from London to Southampton, a task which required the literal destruction of the down with its ancient Iron Age trackways, grasslands, butterfly breeding grounds and sites of special scientific interest, not to mention its stunning scenic value (Kingsnorth, 2004). And while local residents had for many years opposed the scheme through the conventional channels of public inquiries, legal submissions, petitions, letter-writing campaigns in local newspapers and public lobbying, the cause appeared lost (Bryant, 1995). However, the numbers of those prepared to use their bodies to save the down grew to such an extent that in December 1992 security guards forcibly ejected protesters from the site, causing outrage amongst like-minded opponents of the scheme and, importantly, galvanizing local residents and environmental SMOs such as FoE to take a stronger interest in challenging the construction. In fact, this symbolic event triggered a 'bandwagon effect' that brought together local residents, students at Winchester college and local politicians with the radical environmentalists who were digging themselves in on the site. The construction of the new road, which began the following year, was met with concerted resistance that was to last over a year and a half, as large numbers of protesters harried bulldozers, occupied offices and buildings, bound themselves to trees, immobilized machinery and protested in the streets (Kingsnorth, 2004).

This spiral of protest attracted much local and national media attention, turned eco-protesters such as 'Swampy' into national heroes, and resulted in the arrest of many hundreds of protesters (Paterson, 2000). It was, however, eventually defeated, and the road was finally built. Nevertheless, it initiated what might be termed a new logic of protest that was imitated in a series of protests across the country during the mid-1990s. Protests against the Cradlewell by-pass outside Newcastle, where a road was to be built through a forested conservation area, involved the construction of treehouses. Other tactics included the building of tunnels, the use of lock-ons, and the development of increasingly sophisticated systems of lookouts and alarm systems that warned protesters about the dangers of police or security guards (Doherty, 1999). The camp at Solsbury Hill, outside Bath, which was created in late 1993 to prevent the Batheaston by-pass, was almost home-like, where people lived, ate and slept in reasonable safety and security (Kingsnorth, 2004). In short, as the tactics of bodily risk became increasingly more sophisticated and daring, and as the numbers of protesters of all political and social hues grew, so did the economic and political costs of road building.

Resorting to these repertoires of bodily risk, protesters also began to forge novel alliances with local residents. Consider for instance the campaign by local residents against the Newbury by-pass in 1995 and 1996, which was named the 'Third Battle of Newbury'. This campaign had the effect of not only generating and then maintaining a relatively high degree of organization amongst residents and radical protesters, but it also instigated a process of individual and collective radical policy learning (Dudley, forthcoming). In the run-up to the building of the by-pass, the residents formed a strong and successful alliance with both radical activists and national environmental groups such as FoE, adopting a novel organizational strategy, which reflected more the open and non-hierarchical model of new social movements than the organizational order of sectional interest groups. During the height of the direct action, in the early months of 1996, the local residents were integrally involved in the organization and execution of the protest campaign. Even after the by-pass was actually built, a core of local residents continued active campaigning on such issues as the environmental effects of the road and future development in Newbury. Members of the 'Third Battle', who had initially been sceptics or had not been involved in environmental campaigns prior to the campaign against the by-pass, went on to become chairwomen of the local branch of the *Council for the Protection of Rural England* (CPRE), coordinator of Newbury FoE and founder of *Local Voices*, which campaigned on the controversial question of development between the by-pass and the town of Newbury. Indeed, further evidence of continued local activism was indicated by the determination of a group of local business people to follow up the environmental effects of the by-pass. They raised £50,000 to pay for a team of academics to monitor atmospheric and water pollution levels (Dudley, forthcoming).

The sum total of these processes and logics was the beginnings of a populist form of politics that linked together different demands and identities in an equivalential chain that was pitted against the government and its transport policies. Not only had thousands of protesters endeavoured to stop the new schemes, but they had also managed with local residents to stimulate a national public debate about the future of transport, and about sustainable development as such. By 1996, with John Major's Conservative Party headed for electoral defeat and its transport policy in tatters, the proposals for the massive road-building campaign were effectively abandoned by the incoming New Labour administration.

THE STRUGGLES AGAINST AIRPORT EXPANSION

Paradoxically, perhaps, the victory of residents and radical environmentalists served to quell the nascent populist politics in the environmental field in the UK at the end of the 1990s. (We use the word 'perhaps' to draw attention to the fact that the disappearance of a political enemy – in this case the road-building programme of the Conservative Party – which functions as the target of protest, can in many instances serve to weaken the identities and commitments of those in opposition.) However, while the anti-roads protests were important in leading the Conservative Government to abandon its *Roads for Prosperity* programme, similar logics were at work in Britain's airport policy. Indeed, in the struggles against the expansion of airports since the middle of the 1990s, which have been greatly accelerated by the 2003 Government White paper to expand massively Britain's airport capacity, it is possible to discern at least three concrete logics of protest as models with which to organize campaigns and define their objectives, strategies and tactics.

First, in the mid-1990s local residents in Knutsford and Mobberley had opposed the building of a second runway at Manchester Airport. At least two distinctive forms of protest manifested themselves in this campaign. Initially, the main emphasis was to mount a narrowly conceived technical and legal challenge against the proposed new runway. This was focused on the lengthy Public Inquiry, which was convened by the government to decide whether or not to go ahead with a new runway. The protest action consisted primarily of a logic of rational argumentation by which the residents, via their team of legal representatives, sought to concentrate principally on planning issues and disputes over the commercial viability of expansion at the airport to persuade the government inspector of their anti-expansion case (Griggs and Howarth, 2002). In short, the local residents saw themselves as members of 'Middle England', who as responsible citizens chose to invest in the process of the Public Inquiry.

This investment chimed well with their strategy of 'civic virtue' and depended crucially on the support of key elite individuals in the villages, as well as activities such as fund-raising, village meetings and the signing of petitions. Such elite forms of campaigning were consonant with the social and political identities of residents who felt at this stage that the Inquiry was a legitimate and favourable terrain to contest the injustice they clearly felt. However, following their failure to convince the Public Inquiry's Inspector of the reasonable merits of their case through rational deliberation and argumentation, local residents joined forces with more radical environmentalists, who were prepared to employ direct-action tactics so as to prevent physically the construction of the new runway. This ushered in a second logic of protest that centred on alliance building, direct-action campaigning, where protesters were prepared to employ strategies of bodily risk, and a high-publicity media campaign.

These two logics of protest, corresponding to the different phases of the Manchester campaign, each revealed significant difficulties for those opposed to airport expansion. If the focus on the Public Inquiry disclosed a too narrowly conceived approach, which relied heavily on the use of considerable financial and other resources to bring in an expensive London-based legal team to fight the expansion, then the so-called 'Vegans and Volvos' alliance never amounted to more than a short-term and precarious coalition of forces that quickly dissipated after activists were removed from the Bollin Valley surrounding the airport. As activists involved in the campaign were quick to point out, the failure to engage fully in a high-profile lobbying and media campaign, directed at Westminster and the national press, quite probably weakened the overall thrust of the protest movement and weakened the campaign.

In contrast to the different logics of protest at Manchester Airport, a third logic is evident amongst those campaigning against the growth of Heathrow Airport, which is the UK's largest airport. This protest campaign was spearheaded by *HACAN ClearSkies*, which, as the name suggests, brought together the *Heathrow Association for the Control of Aircraft Noise* (HACAN) – a series of local residents' associations struggling to ameliorate the inexorable expansion of Heathrow, and *ClearSkies*, a community initiative in south-east London opposed to rising noise from circling planes waiting to land at Heathrow airport. Over time, this group had evolved from a 'classical residents' group focused on the single issue of containing the Heathrow juggernaut to a more broadly based movement that sought to widen its demands to include proposals to manage the ever-growing and subsidized demand for air travel at the national and regional levels.

In so doing, they sought with others to build a broad coalition of forces, *AirportWatch*, which opposed airport expansion at a number of actual and potential sites of airport development, while simultaneously linking-up with other environmental groups and activists (such as *FoE, Woodland Trust, CPRE*, and so on) to widen the overall appeal of the movement. The latter logic involved a broadening of its ideological repertoire to include issues of social justice, equality, and demands for an integrated transport network alongside its more narrow demands to stop the unprecedented growth of Heathrow (Griggs and Howarth, 2004).

Returning to our previous discussion of populist politics, these three logics can be plotted along a populist/non-populist continuum. As will be recalled, the *degree* of populism in any particular context can be determined by the number of equivalential links and demands that can be constructed across a number of contiguous sites of struggle. With respect to the three logics of airport protest, the first phase of the local residents campaign to overturn the desire to build a second runway at Manchester Airport represents a clear non-populist form of mobilization and protest. The residents and their campaign groups *MAJAG* (Manchester Airport Joint Action Group) and *KAMJAG* (Knutsford and Mobberley Joint Action Group) stuck resolutely *within* the existing institutional rules of the game, and sought to win the argument through rational and reasonable arguments. Little was done to link their opposition to other demands and to other struggles, even though there existed attempts to link different groups and demands together in other contexts (e.g. the road protests at Twyford Down, Newbury). Initially, local residents actively shunned the direct-action protesters, viewing their presence as detrimental to their image and overall aims.

Given the failure of this mode of protest, however, there were attempts to forge alliances with direct-action groups, and here we see a nascent form of populist politics, as *MAJAG* and *KAMJAG* extended their demands to include more direct environmental concerns. Widening the scope of their campaign means that we can plot the second phase of the Manchester protest as closer to the populist pole of the spectrum. However, given the precarious character of the equivalential link, and the still limited set of demands, it constituted a very restricted form of populism. Turning, finally, to *HACAN ClearSkies* we see a much more developed form of populist politics. Not only was this movement instrumental in linking different groups together under the banner of *HACAN ClearSkies*, but it was also integral in the formation of *AirportWatch*, which brought together various anti-airport groups into a single campaign.

Even more so, in seeking to widen the scope of the airport struggle by connecting it with broader environmental issues (such as air pollution and the need for a more efficient and integrated national transport network), social justice, demands for equality and better quality of life for all, the logic of protest pursued by those involved in *HACAN ClearSkies* represents a much more developed form of populist politics that sought to transcend the more narrow confines of aviation policy and politics. [Departing from the UK context for a moment, this most developed form of populist politics is evident in the struggles around the building of Narita Airport in the Sanrizuka area of Japan during the 1970s. Here, the airport struggle condensed a wide range of demands and struggles in the Japanese context, responding to numerous contradictions in Japanese society (Apter and Sawa, 1984).]

CONCLUSIONS

This chapter has explored the different trajectories of environmental politics and protest in the UK. It has employed a distinction between 'being popular' and populist politics to characterize and analyse different manifestations of environmental protest and activism in the UK. Inevitably, perhaps, in this light the picture may appear to be bleak, if not entirely forlorn: environmental issues have not yet constituted a populist form of politics that can radically challenge the juggernaut of modern capitalist industrialism. This is not, however, to decry the various attempts to articulate radical demands and to construct equivalential linkages between different sorts of demands and identities, both amongst those who share interests around an issue, and between environmental demands and wider issues, such as social justice and economic equality, in society as a whole. Environmental activism that has been institutionalized and environmental activism that seeks to challenge the system have secured important victories and change, and the dialectic of populism will no doubt deliver further victories in the future.

Nonetheless, implicit in our argument is the normative stance that the political success of environmental concerns and demands depends not just on their popularity, but also on their articulation as popular demands embedded in equivalential chains which link environmental demands to other social and economic demands. The capacity to engage in such populist forms of politics will, as we have argued throughout, depend in part on agency, that is, in terms of the capacity and willingness of leaderships to engage in populist politics, but also on the historical context of the campaign in terms of time and the construction of political space. At any point in time, the political orientation and character of a populist movement is dependent on the kinds of hegemonic articulation available and the construction by competing hegemonic projects of the discursive terrain in which decisions take place. Articulating this movement from being popular to a populist form of politics is thus far from unproblematic in the domain of environmental movements, but the wider success of the environmental movement will arguably rest upon such transformative political strategies.

NOTES

1 This broad definition follows the work of Torgerson who defines environmentalism as 'an overarching category to capture a range of responses arising in the latter part of the twentieth century to challenge an exuberant industrialism' (1999, p. 3).

REFERENCES

Apter, D.E. and Sawa, N. (1984) *Against the State: Politics and Protest in Japan*, Cambridge, Mass.: Harvard University Press.

Bosso, C. (1997) 'Seizing Back the Day: The Challenge to Environmental Activism in the 1990s', in N. Vig and M. Kraft (eds) *Environmental Policy in the 1990s*, Washington: CQ Press, 3rd edn.

Bryant, B. (1995) *Twyford Down*, London: Spon Press.

Byrne, P. (1997) *Social Movements in Britain*, London: Routledge.

Carter, N. (2001) *The Politics of the Environment. Ideas, Activism, Policy*, Cambridge: Cambridge University Press.

Chong, D. (1991) *Collective Action and the Civil Rights Movement*, Chicago: Chicago University Press.

Dalton, R.J. (1994) *The Green Rainbow. Environmental Groups in Western Europe*, New Haven: Yale University Press.

Department of Transport (1989) *Roads for Prosperity*, London: HMSO.

Doherty, B. (1999) 'Paving the Way: the Rise of Direct Action against Road-Building and the Changing Character of British Environmentalism', *Political Studies* 47(2): 275–291.

Dudley, G. (forthcoming) 'Individuals and the Dynamics of Policy Learning: The Case of the Third Battle of Newbury', *Public Administration*.

FoEI, (2005) *Friends of the Earth International Annual Report, 2004*, Amsterdam: Friends of the Earth.

Freeden, M. (1996) *Ideologies and Political Theory: A Conceptual Approach*, Oxford: Clarendon Press.

Gamsci, A. (1971) *Prison Notebooks*, London: Lawrence and Wishant.

Grant, W. (2000) *Pressure Groups and British Politics*, Basingstoke: Macmillan.

Griggs, S. and Howarth, D. (2002) 'An Alliance of Interest and Identity? Explaining the Campaign against Manchester Airport's Second Runway', *Mobilization* 7(1): 43–58.

Griggs, S. and Howarth, D. (2004) 'A Transformative Political Campaign? The New Rhetoric of Protest Against Airport Expansion in the UK', *Journal of Political Ideologies* 9(2): 167–187.

Habermas, J. (1981) 'New Social Movements', *Telos* 49: 33–37.

Habermas, J. (1987) *The Theory of Communicative Action*, Cambridge: Polity Press.

Hirschman, A.O. (1982) *Shifting Involvements*, Princeton: Princeton University Press.

Howarth, D. (2005) 'Populism or Popular Democracy? The UDF, Workerism and the Struggle for Radical Democracy in South Africa', in F. Panizza (ed.) *Rethinking Populism*, London: Verso.

Jordan, G. and Maloney, W. (1997) *The Protest Business? Mobilizing Campaign Groups*, Manchester: Manchester University Press.

Kingsnorth, P. (2004) 'Roadrage'. *The Ecologist*, 1st March, available at www.theecologist.org/archive.asp accessed 12 March 2007.

Laclau, E. (1977) *Politics and Ideology in Marxist Theory*, London: New Left Books.

Laclau, E. (2005) *On Populist Reason*, London: Verso.

Laclau, E. and Mouffe, C. (1985) *Hegemony and Socialist Strategy*, London: Verso.

Meadows, D. (1974) *The Limits to Growth: A Report for the Club of Rome's Project on the Predicament of Mankind*, New York: University Books.

Melucci, A. (1989) *The Nomads of the Present*, London: Hutchinson.

Michels, R. (1959) *Political Parties: A Sociological Study of the Oligarchial Tendencies of Modern Democracy*, New York: Dover.

Paterson, M. (2000) 'Swampy Fever', in B. Seel, M. Paterson and D. Doherty (eds) *Direct Action in British Environmentalism*, London: Routledge.

Piven, F.F. and Cloward, R.A. (1977) *Poor People's Movements: How They Succeed, How They Fail*, New York: Pantheon Books.

Porritt, J. (2005) *Capitalism: As if the World Matters*, London: Earthscan.

Richardson, D. (1995) 'The Green Challenge', in D. Richardson and C. Rootes (eds) *The Green Challenge*, London: Routledge.

Rootes, C. (1992) 'The New Politics and the New Social Movements: accounting for British exceptionalism', *European Journal of Political Research* 22(2): 171–191.

Rootes, C. (2002a) 'The Transformation of Environmental Activism: An Introduction', in C. Rootes (ed.) *Environmental Protest in Western Europe*, Oxford: Oxford University Press.

Rootes, C. (2002b) 'Britain', in C. Rootes (ed.) *Environmental Protest in Western Europe*, Oxford: Oxford University Press.

Scott, A. (1990) *Ideology and the New Social Movements*, London: Unwin Hyman.

Torgerson. D. (1999) *The Promise of Green Politics. Environmentalism and the Public Sphere*, Durham: Duke University Press.

Touraine, A. (1985) 'An Introduction to the Study of Social Movements', *Social Research* 52(4): 747–787.

Wall, D. (1999) *Earth First! and the Anti-Roads Movement: Radical Environmentalism and Comparative Social Movements*, London: Routledge.

Wall, D. (2005) *Babylon and Beyond*, London: Pluto Press.

22

Faces of the Sustainability Transition

Tim O'Riordan

INTRODUCTION

Sustainable development has become a universal phrase. It means everything, and is in danger of meaning nothing. Engineering, architecture, tourism, fashion – all carry the tag "sustainable". Governments use it to mean "long lasting" whatever the merits of the policy they are proposing. Before long, sustainable development will split into a view promoted by "diehards" who like the rigour of its intellectual and scientific foundations, and those "modernists" who transform it into catchy and more amenable phrases such as "quality of life", "well-being", "social justice" and "localism". In essence, it will narrow and blossom in the same age. My guess is that the "modernists" will win.

This chapter does not offer a comprehensive perspective on the evolution of the sustainable development idea. Plenty of others have done this. The reader is encouraged to look at Bill Adams revised text on *Green Development* (2003), my own collection on *Globalism, Localism and Identity* (2000), Jonathon Porritt's recent analysis *Capitalism: as if the world matters* (2005), an excellent anthology in *Scientific American* (September, 2005), Lester Brown's recent analysis *Plan B 2.0* (2005) and the UN Global Biodiversity Outlook 2 (UNCB, 2006). All of these have extensive and helpful bibliographies and web links to major research institutions such as the World Resources Institute, International Institute for Sustainable Development and International Institute for Environment and Development.

All of these texts offer a variety of statistics and studies of how the planet is changing, how and why humans are altering fundamental life-support systems, and how the scope for change is possible financially and in practical delivery terms, if there is sufficient international will and common desire. Sadly, despite the efforts by the United Nations (much maligned by a number of rich countries and not always helping its cause in its own mismanagement and poor middle-level leadership) and by the group of eight rich nations (also ideologically divided and essentially unwilling to share wealth on any noticeable scale), there is no serious and sustained global direction in favour of truly sustainable development. Thus, despite the attention to exciting local initiatives in the text that follows, the absence of any global recognition and support of these commendable local actions, and the failure to provide a financial and legal framework to enable them to flourish means they will remain well-intentioned, evidence of what can be done, but isolated, possibly damaging to others nearby, and unsupported. The wider "scaling up" issues linked to these local initiatives will be given more attention at the end of the chapter.

According to Bob Kates and his colleagues (Kates *et al.*, 2005, p. 9), if you hit "Google" you will find 8.72 million linked web pages to the phrase "sustainable development". So at one level, sustainable development has overwhelmed us. But Kates and his colleagues offer a more incisive point. Since 1945, they comment (2005, p. 10), four global movements have emerged – peace, democracy, development and environment.

Sustainable development binds them all. Without peace, social justice, human and environmental care and a sense of the well-being of all generations of life to come, there can be no human activity that is sustainable. While we struggle for human rights, for genuine democracy, for freedom from terror and oppression, and for a decent living in health-creating surroundings, we are really struggling for sustainable development. In a world that is widening its poverty (UNDP, 2005), destroying its natural resources and the life-supporting ecological interconnections which we still do not fully understand (UNEP, 2002) we are moving further from sustainable development by the day. Maybe we are making some inroads into greater democracy and human rights. But with insecurity almost ubiquitous, where there is poverty and injustice, and where desperate people remove natural resources just to continue living, sustainable development has become a necessity for the good of all humanity.

Sustainable development originated in the mid-20th century as a basis for managing wildlife in Africa for the continuation of survival of very prominent species whose symbolism of existence provided a metaphor for the human condition. Lying deep within this early concept was a belief in human interference with natural resources so that both "nature" could be secured and economic development spread across poor livelihoods. At the heart of the very early thinking in sustainable development was a powerful belief that well-managed nature, sensitive to natural systems, functioning and aware of the limits and tolerances of ecosystem functioning, would result in robust and spreading economic gain and social advance.

Bill Adams (2003) carried these ideas through the subsequent 20 years to the emergence of the United Nations involvement on the theme in its First Global Conference on the Human Environment held in Stockholm in 1972. This meeting began a global and multilateral perspective on sustainable development that has led to the UN Conference on Environment and Development in Rio de Janeiro in 1992, and the World Strategy on Sustainable Development in Johannesburg in 2002. For all of their earnest diplomacy, these huge conferences did begin to focus minds and forge global environmental agreements, on global scientific assessments and on global mobilisation of a whole host of new governmental organisations. It is these "foot soldiers" of the sustainability transition that remains the most exciting outcome of this transformation, though we must also pay recognition to the major agreements and science assessments.

Let us be quite clear from the outset. The planet itself is not in peril. Humans are. Lenton *et al.* (2004, pp. 30–35) summarise the latest state of scientific analysis around the "Gaia Theory". This is the proposition that life on Earth, linked to physical processes involving transformations of materials and energy, acts in such a manner as to maintain itself even when faced with potentially catastrophic change. Lenton *et al.* recognise that any abiotic system contains self-regulating properties. What is interesting about the planet Earth is that both biotic and abiotic (life forms and energy–material forms) seem to interact to produce global resilience to external perturbation.

Whether this is a matter of chance ("lucky" Gaia) or design ("probable" Gaia) will likely never be determined. And in many ways, it does not matter which is the interpretation. For 3.8 billion years, the Earth has evolved to be truly sustainable in self-regulating, wholly recycling, evolving new forms of life, maintaining stable and dynamic conditions out of equilibrium with the gases and high-temperature circumstances found in its neighbouring planets.

The Gaia idea continues to evolve and to provoke the scientific community. Its originator, Jim Lovelock (2006) has just produced a more controversial analysis where he regards Gaia almost as a judgemental entity, prepared to extract revenge for erring humans. This is a step out of his earlier propositions where he invested in Gaia as a scientific theory, not a source of moral censure. But he seems to be a little carried away by Greek mythology in that Themis, the daughter of Gaia, was also the Goddess of Justice. Hence, the apparent scope for humanity to alter the life-support processes of the planet beyond the point where these processes can sustain a growing and continuously developing humanity, may result in a "Gaia 3" lieing beyond Gaia 1 (no oxygen) and Gaia 2 (oxygen and humans) where humanity becomes forcibly less influential in Gaian processes. This is a point for further reflection at the end of the chapter.

Humans have begun to reach the point where our collective influence is so large as to be a significant factor in destabilising these resilient self-regulatory systems. John Schellnhuber *et al.* (2004) summarise this state as the "anthropocene era". The phrase is attributable to the Nobel Laureate, Paul Crutzen (2002). It refers to a new geological epoch in which humankind has emerged as a globally transformative species with the intelligence to understand what it is doing and how it could stop it. Whether humanity can turn around and become "Gaian" in its outlook and behaviour is quite another matter. Jared Diamond (2004) produces evidence that in closed societies, human civilisations can actually knowingly destroy themselves, even when "get out" options are available. Diamond's analysis applies to relatively isolated societies in microcosms of the planet.

It is a moot point whether humanity as a whole can actually recognise and act on a series of sustainability transitions that will result in a peaceable, free, innovative and just collection of communities that can adjust to the rhythms and opportunities of natural systems.

Diamond himself is not quite ready to apply his "doom" doctrine to humanity as a whole. Yet if one needs his analysis of why civilisations fail, the conditions could apply to the current global economy and politics. Diamond suggests five conditions where a group may obliterate itself, even when aware of its predicament:

• hostile neighbours which reduce the scope for migration;
• no friendly support mechanisms within the community;
• external threats such as climate change;
• lack of technological inventiveness to overcome the loss of resources;
• inability to manage common pool resources.

We may speculate whether such conditions apply to the globe as a whole. Certainly there is no escape from the planet. Social tensions within the human family are increasing. Climate change is becoming a serious threat, as is the loss of species and habitats. Our technological inventiveness may simply not be successful enough to put us on the path to sustainability. And we are still very poor at devising predictive and sensitive measures for managing ecosystems, the web of life-support functions that sustain a habitable planet.

Though the science of a changing future is more alarming than ever before in its predictions, there is still huge uncertainty as to whether there are global "tipping points" that would seriously make the Earth uninhabitable for the most vulnerable. So conditions, namely evidence of apocalypse, are not yet fully accepted, despite the signs on the planetary map. Then the management of the common property, as helpfully summarised by Tom Deitz and his colleagues (Deitz et al., 2003), requires conditions of legal accommodation, collective will, decent information, incentives and disincentives for straying into non-sustainability that frankly are not being met as part of the global commons. One reason is that no one knows the "true" ecosystem functioning boundaries in any given "commons", so no one can be sure what a non-sustainable initiating act is before it is initiated. More relevant, however, is the case that the vulnerable are being systematically excluded from the commons, nor enabled to care for them.

At this point, a debate is emerging that we simply cannot respond to comprehensively. Maybe humanity will splinter into many fractions of relative survival, based largely on defensiveness, aggressive self-interest and constant struggle to live over and above nature yet succumb eventually to its limits in a myriad of ways. Such a possibility implies no global "government", no global "security" and no global "justice". Sustainable development would atrophy and disappear, but a variant of non-sustainable, very diverse, humanity would remain.

This is the start and end point of this chapter. The "tipping point" thesis of Schellnhuber and his colleagues is gaining ground. The level of damage to ecosystem services as summarised by the Millennium Ecosystem Assessment (2005) is now such that over 60 per cent of natural systems are damaged or destroyed. Climate change impacts as summarised in a recent UK Government Report (Schellnhuber et al., 2006) are now such that impacts or existing natural processes are beginning to be judged as genuinely alarming.

Nevertheless this chapter has to take an optimistic view. It looks at how localism can work to create sustainability. It accepts that global institutions of development, security and trade will not work to promote the outcome for sustainability. But at the very local level, people can form communities where safety, security and sustainability can flourish and form livelihoods that offer hope and recommendation for all involved.

Yet it must also be noted, here and later, that localism is not a panacea. It offers scope for hope and provides the evidence of uplift. But set outside a supportive political, regulatory and economic framework, it is by no means a solution. In addition, local activity, especially where promoted by well-to-do and earnest people, may add to the poverty and vulnerability of others less able to recognise the dangers those well-intentioned efforts may bring. Local organic food may sound cosy, but if magnified across the landscape, the excellent organic food produced by poor communities elsewhere may simply not be marketable on the scale required.

So this chapter is not a primer on sustainable development. The Kates et al. analysis (2005) is as good as any in assessing the history and prognoses for sustainable development. It is worthy of note that the major global summit at Rio in 1992 and Johannesburg aftermath in 2002 produced worthy statements but tragically little real improvement in the lot of the poor and the marginalised, nor for nature as a whole. Report after report since (notably the two UN documents published by the Environment Programme in 2002 and the Development Programme in 2005) and by the science community, summarised by Schellnhuber et al. (2004), all point to the same outcomes:

1   We do not know anything like enough about the changing state of planetary life support. So we are altering huge assemblages of natural systems with no clue as to the long-term consequences.

2 Uncertainty in scientific prediction means that we cannot reliably forecast where any given initiating state (e.g. additions of greenhouse "atmosphere warming" gases into the whole of the global atmosphere) might lead. In addition, changes of natural "phases" of apparent stability (ocean currents, salinity, removal of plankton and the oceanic surface due to increasing acidity of the oceans caused by excessive carbon dioxide absorption) may result in new natural phases (emergent conditions) that cannot be predicated either with regard to timing or consequence.

3 Combine uncertainty with unpredictability with expectations for changing human behaviour or contentious policy (e.g. pricing carbon) in political systems involving many competing nations and argumentative citizens and lobbies, and no political leader can make headway. Where the pay off is not only highly unclear, and where it may be two generations away, then politicians prevaricate. Yet the delay in action may result in unavoidable distress for huge swathes of people and national systems in generations to come. Admittedly we did something about curbing the damage to stratospheric zone and to acid rain, but these were targetable processes with focal policy on particular emissions, and where business was broadly compliant. The global totality of damage is nothing like as targetable and the science is hotly contested. So early and sustained political action is all but impossible.

4 We are also not good at delivering well-being for people and nature. Economies are designed to give people livelihoods and consumption, not well-being. That elusive phrase applies to security, esteem, sociability, trust, companionship, spirituality and peace of mind, to family pleasures, to aesthetic wonder and to meaningful employment in socially friendly settings. I develop this point later in the chapter.

5 Sustainability is holistic and integrated. It cuts across all manner of human-devised decision, financing and regulatory policy structures. The advantages of interconnectedness and internal "pay-off" are lost in the style of most governmental and private sector analysis and delivery. A health benefit for more natural areas surrounding hospitals or within the poorer parts of dense urban areas may be "lost" if the group advancing the health gain comes from nature conservation or the promotion of biodiversity, and the health gains are distant and not readily targeted to a vulnerable low health group.

Felix Dodds and Tom Pippard (2005) have taken these points further to examine a global agenda for "all-round" security. The authors look at health, AIDS/HIV, natural hazards, loss of natural resources (water, soil, fish, agriculture), terrorism, oppression and violence at all levels. They find examples of insecurity everywhere, even amongst the wealthy. As is the case throughout this analysis, the life-saving Millennium Development goals need to be pursued. Sadly, all the signs are that they are not being so.

The UK Government has recently published one of the best sustainable development strategies of any modern government (HM Government, 2005). This lays out five principles for promoting sustainable development:

1 Living with nature's limits and using natural processes for the betterment of people and nature. This requirement places nature first when threatening human survival, and demands an appropriate costing of natural functioning.

2 Creating a fair and just and secure society that is confident in itself. This means that sound justice extends to all generations and that improving the lot of the disadvantaged along sustainable lines is a critical element of sustainability

3 Establishing an economy that delivers both of these objectives. This suggests that the economy is subservient to the wider goals of a resilient nature and a just society. It also means that non-sustainability should pay its way.

4 Creating a form of governance that is open, learning, participatory and sincere. This is the toughest nut to crack. But at least it is a core principle.

5 Using science to ensure that decisions and outcomes are based on the soundest information available, using the precautionary principle where necessary to be safe and to provide room for error.

The UK strategy provides all the right signals for promoting the cause of sustainable development. It even advocates that each minority will establish sustainable development (SD) action plans so that they can be monitored for effective performance. So far so good. But there is, as yet, no capacity in the UK Government for the change in direction. There are no champions in ministries, agencies or ministerial cabinets that can push these ideas forward. So the strategy may look glossy, but its delivery will be hampered by institutional inertia and a chronic lack of leadership and willingness to innovate. Good messages do not add up to good delivery.

It is still early days. Cultures of bureaucracy do not change rapidly. All governments at any scale, may well depend on non-sustainable outcomes to maintain electoral support. Indeed, under the pressures of inevitable short-termism, it may not be possible for national governments to deliver a long-term vision of ecological existence that will be electorally popular. Even the green parties have to compromise their basic values when they get into power (see Dobson, 2003). We may not be able to produce either a national or international version of an"ecological democracy" that places the well-being of posterity into its reckoning. And, if it ever does, it may be too late to reverse the interim destructive trends. All the signs currently are that carbon-dependent societies cannot get off the carbon "fix". The longer they depend on the carbon input, the more difficult will be the carbon withdrawal process.

Sustainable development is an illusion. We cannot agree on what it is, or should lead to. We cannot get our way there for the reasons listed above. We do not have the science, the communications or the mindset about "satisfaction" to take us there. We cannot grasp what we are doing to undermine the resilience of planetary life-support systems. So we need to build from below, from communities and from the myriad of experiences where people buck the trend. This is the experience outlined next.

"SAFE SPACE" IN SOUTH AFRICA[1]

Joubert Park, in the heart of Johannesburg, is a beguilingly pleasant open space. Surrounded by skyscrapers and revealing some 30 acres of mown grass and flower beds, it looks the picture of urban respectability. Lovers come to share lunch, children play improvised football, elderly couples snooze in the sun. Yet five years ago Joubert Park was a "no-go" area. Violent crime, rape, drug trafficking and overwhelming danger were commonplace. Behind the apparently reassuring thin line of skyscrapers lie ranks of poor housing, disjointed communities, and disease. Left alone, families within 200 metres of the Park would have no sense of safety, let alone sustainability.

In the heart of Joubert Park are two remarkably special spaces. One is The Greenhouse: The People's Environmental Centre. The other is Lapeng: Child and Family Resource Service.[2] The Greenhouse is championing non-carbon energy technology for local small businesses, community housing schemes and local schools. This non-profit group grows organic and permaculture food plants for nearby residents and helps them to learn how to plant and to tend in micro communal gardens. It has teamed up with a local photovoltaic cell manufacturer to establish a number of chain companies all creating jobs and entrepreneurial skills for young inner city blacks, where unemployment stands at 40 per cent.[3] It is even supplying the city of Johannesburg with some electricity. Every week twenty to thirty representatives of community organisations from around this sprawling city visit the Greenhouse. Here, they gain access to the web, to do-it-yourself information on sustainable

energy and food production, and how to start up micro-businesses in sustainable product design.

Lapeng is the neighbour of Greenhouse. It provides a safe and nurturing space for three hundred hugely disadvantaged children from nearby, often violent, communities. Almost all children have been exposed to trauma, to disease and to fear. Lapeng brings parents and children together to share art, group learning, opportunities to create new learning skills in science, maths and literacy. Above all, it is a space for play, for free expression and for security.

Writing a foreword to a remarkable set of essays commemorating ten years of post-Apartheid transformation (Pieterse and Meintjies, 2004), Albie Sachs, one of the most respected of South Africa's freedom lawyers, and a Justice of the Constitutional Court, wrote:

> there is a difference between an empty society and an open one. The first might be full of goods and refer to itself as a winning nation, yet be void of true human concern. The second might have relatively few material goods but be rich in expressive human interaction (2004, p. ix).

Both Lapeng and The Greenhouse are typical of the particular transition to sustainability followed by South Africa. Both rely hugely on charitable contributions from business, international development agencies, private foundations and residents' associations. Both are formally supported by national, regional and city government policies, finances and the overriding drive towards poverty alleviation and black economic empowerment. Both depend on volunteers who earn income by benefiting from the healthy food, the power and the small payments for their services.

Above all, both have created *safe space*. In South Africa, security from violence, HIV/AIDS, rape and personal loss of self-esteem is the bedrock of the sustainability transition (O'Riordan *et al.*, 2000). One important reason why Joubert Park is now so safe is because these two projects have generated a whole new concept of "space" in the heart of an otherwise alienating metropolis. Here, people care; here people hope; here people watch out for strangers and ensure they are kept at bay.[4] Here, above all, sustainability embraces family togetherness, group learning, skills development – food and non-carbon energy management, and a sense of reliable continuity towards a better and more peaceful life.

Writing of their experiences in creating a similar ecologically sound child–adult school near Stellenbosch (close to Cape Town) Eve Annecke and Marke Swilling captured the mission of this remarkable element of the South African transition to sustainability:

> The Lynedoch Eco Village must create a safe space where South Africans from all backgrounds can live in peace with each other and in harmony with nature. It must also be a place where people from all over the world can come and share in the life of the community while they learn, think and create works of art and knowledge that will contribute to the making of a better world. It must, above all else, be a place where all life is celebrated and beauty in all of its forms treasured for this and future generations (2004, p. 295).

WIDER LESSONS FROM THESE FACES OF THE SUSTAINABILITY TRANSITION

Bob Kates and his colleagues (2005, p. 12) distinguish between what has to be sustained and what has to be developed. The first involves values such as nature in the round, its ecosystem support services and the broader concept of well-being for people as individuals and as social groups. The South African examples exhibit the intensely local combination of the two. Providing safety, community, esteem and capacity for making a worthwhile livelihood leads to sharing of responsibility, mentoring of the less fortunate, and building a setting where good can repulse evil. The South African experience reveals how a vision of a hopeful future, that links parent to child, official to resident, business to volunteer and health to art, really begins a process of enormous sustainability potential. What is vital is that the sunlight of such schemes, mere pinpoint on the communal map of the nation, are allowed to shine forth and be supported by those who can give some of their non-sustainable earnings to illuminate further such transformations.

The South Africa experience also highlights the importance of local solutions, within broad frameworks of support and guidance, that enable people to create such sustainable space. This is happening against a current of economic development in Southern Africa, supported by the South African President Mbeki and his advisors, which is widely criticised as running counter to such initiatives. These points will be raised at the concluding part of this paper. Such initiatives, driven by vision and exuberance, provide the essence of sustainable identity and well-being. This combination is possibly the most important measurement concept in the current transition.

Yet a purely localist and populist interpretation of such initiatives is misleading. There is a strong case in both academic and policy analysis for

looking and assessing such projects to see why they work and what is the role of community leaders and the willingness to "buck the trend". Yet if nothing is learned by national governments and local authorities cannot nurture them or safeguard them, if there are no funds or people resources for sustainable housing community use and clean technologies, even when both are possible and accessible for local people, the outcome can be more frustration, not more sustainability. The main advantage of local case studies such as the ones outlined here is that they show what is possible and how it is concernable to establish some form of improvement step by step, beyond the locality. This requires NGO champions, good media coverage and a strong SD framework.

SUSTAINABLE NORTH WEST: A KINDER AMERICAN FACE

Oregon is one of an influential group of US states which is championing the sustainability transition. It has created a Sustainability Board, run from the Governor's office, and backed by former Governor Klitzhopper's vision of "leap to sustainability in one generation".[5] This Board, comprising leading government officers, businesses and civil organisations, has met five tests of a successful transition:

- Leadership from the top of government, with a clear vision of progressive environmental improvement, social justice and preserving viable opportunities for future generations.
- A commitment by all state agencies to prepare a sustainability action plan monitored by the Oregon Progress Board (the official audit body) and subject to measures that encompass well-being as well as profit.
- Attracting major state businesses on board so that they connect through business links to a wider corporate community.
- Raising the profile of sustainability audits across the spectrum of all the affairs of the state, so that the culture of the public and private service is slowly changing.
- Promoting an image of the state as a champion of renewable technologies, healthy eating and efficient use of resources.
- Providing an incentive rather than an introductory regulative framework for progressive transition. This is achieved in part by a Cascadia (Sustainability) Scorecard (Northwest Environment Watch, 2004). The Scorecard is an index of trends in seven arenas – health, economy, population, energy, sprawl, forest and pollution. It is an early document with much sophistication to go. But it provides the basis for a more universal audit that is beginning to have bite for the Oregon Progress Board.

Sustainability is not universally welcomed in bossy Oregon. In the more conservative rural areas, it is seen as a guise for more environmental interference by government. Hence the need to present the concept through incentives and a more profound change in the culture of business and public service. Here is where the unique composition of the Oregon Sustainability Board may be made or broken. If it succeeds in shifting the basis of good public and corporate practice, then it may survive. The adoption of a sustainable procurement policy by the Portland suite of city, county and metro governments is an example of how this approach is beginning to gain broader support in the public service.

Of critical significance here is the change in the culture of "best practices". In Montnomah County, around Portland, all middle managers receive regular training to build a common outlook in sustainable procurement based in part on Natural Step Training (Waage, 2003, pp. 40–61). These training programmes result in the formation of progress improvement teams who channel their energies in energy management, waste water recycling, waste reuse and minimisation, low-energy buildings and "green" transport schemes. All of this spreads out to local schools and hospitals so that the "sustainability culture" is steadily taking hold. Of critical importance is the continuing supportive framework of the Oregon Sustainability Board. The name of the game is a strategic framework of accepted outlooks and practices and local-level implementation diffused through the ethos of employees and citizens.

Sustainability Northwest,[6] headed by Martin Goebel, has championed three initiatives for rural Oregon. These are Healthy Forests, Healthy Communities;[7] Oregon County Brief;[8] and Stahlbush Island farms.[9] Each of these captures the spirit of the new Oregon Sustainability drive. Each also captures the spirit of "sustainability space" that also characterises the South Africa examples. Goebel puts the case well:

> Local entrepreneurs choose to reinvent traditional natural resource based industries rather than abandon them. The value they place on sustaining rural landscapes and livelihoods resonates with an urban market increasingly interested in environmental and social responsibility. By posting resources devoted to marketing and/or by combining products for larger orders, they were collectively able to reach urban markets that their individual members could not have accessed (in Danks et al., 2004, p. 123).

"Sustainability space" may be an important new notion. It combines the traditional notions of landscape and community identity[10] with a fresh approach to defining well-being, to adding ecological and esteem value to products and services, and to improving the fundamental health and security of all citizens in a common bond of creative sharing. These are the faces of the sustainability transition that may hold most promise.

The essence of well-being is the linking of the bases for total satisfaction (happiness, contentment, pleasure) to resilience (adaptiveness, learning, coping, renewing) through the processes of personal esteem and inner growth for better capability and improved social care and connectedness for collective effectiveness. It is critical to see the interrelationships between the larger millennial goals and the more intimate well-being criteria. Right now, there are few measures to offer over a comprehensive approach to well-being. In the UK the New Economics Foundation and the UK Sustainable Development Commission (Porritt, 2005) are championing such moves.

At the heart this is Albie Sachs' notion of "true human concern". This is epitomised by what is happening in Joubert Park, in parts of Oregon and in a myriad of other sustainable spaces. The New Economics Foundation argues for a set of well-being indicators that encompass the following:

- Creates a sense of engagement with people, neighbourhoods, nature and social care; volunteering, community involvement, business networks, training days all add to a sense of sharing and learning.
- Recognise the quality of work in terms of its meaningfulness and social value, not just personal income. A reformed smoker or drug addict may be the best bet for helping would-be smokers or drug users to enjoy healthier and more fulfilling lives. Maintaining and reinforcing social connectedness may well be increasingly recognised as very important activities, built into every school experience.
- Reclaims slow time. The adjective "slow" is beginning to be coined to represent quality, health, connectedness and environmental empathy. Slow food[11] is a term to encompass the rediscovery of regional flavour and savours, of developing a taste for good, rather than banishing it, and for joining food production to locality to consumption. So it is with time. Speed removes reflection, caring and concern for others: speed disconnects action from consequence, and global to local. Reclaiming slow time means offering more opportunity for family, friends, caring needs and contemplation. Much of this may be achieved by more flexible working conditions, by home-working, by job sharing and by various forms of teleconferencing to reduce avoidable travel. There will be implications for pensions and overall family income.
- Establishes schools for sustainability. There is a golden opportunity to create sustainability spaces for schools. The Department of Education and Skills[12] and the Higher Education Funding Council

for England[13] have both issued new guidance on introducing active sustainability into schools and universities. Over the next five years the performance indicators that guide professional development and the management of the educational estate will begin to bear the integrated approaches to sustainability living. In a speech delivered on 14 October 2004 the UK Prime Minister, Tony Blair,[14] called for every school to be a living laboratory for sustainability.

- Extends patient-supported health care. There is a steady realisation that caring for people through their social relationships and not from their individual illnesses is beginning a process of holistic patient centredness. One study, quoted by the New Economics Foundation (2004) noted that when doctors discussed empathetically with their

patients over how to proceed with health care, the patients showed signs of being much more proactive, engaged in selecting diets, exercise regimes and medication, and sought fewer visits.

There is a huge amount to be done on the well-being front. What is becoming clearer is the connection between personal shared responsibility and social entitlements. These public service agencies need to be held to account for effective and efficient delivery. In return, they should expect individual citizens and social groups to display greater personal responsibility over such matters as their health, their social behaviour, their learning and their safety.

Figure 22.1 summarises how Jonathon Porritt presents well-being based on the New Economics

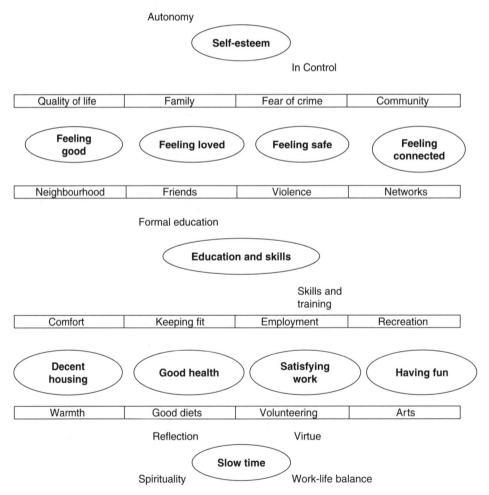

Figure 22.1   The components of well-being
Source: Porritt (2005, p. 221).

Foundation and emerging research elsewhere. Basically we are at a threshold where a "better balance" in life, neighbourliness and environment is beginning to be seen by many as a basis for future prosperity.

In this context the UK Sustainable Development Commission (Porritt, 2005) has begun to champion the notion of invest-to-save. This means applying whole-life costing to the value of procurement and building design, for energy efficiency and for avoiding ill-health or social disruption through pooled preventative programmes. Slowly the procurement professionals are taking a lead in regarding the act of buying goods and services as a process for creating sustainable space. Rutgers University[15] for example, runs a $700 million business on the basis of sustainable procurement. It creates local jobs, it stimulates innovative technology especially in recycling and refurbishment of goods, and it generates new business opportunities.

All this text suggests is that there is immense scope for reanalysing the notion of locality into sustainable space. The path to sustainability is far more likely to come through local and regional initiatives; indeed it is already doing so throughout the world, then at the global, intergovernmental level. Yet it is at this level that ultimately critical political and economic steps are to be taken. Tragically, for the moment, there is only a glimmer of light on this particular horizon.

PROSPECTS AND DIAGNOSES

This volume covers a myriad of concepts and experiences that have a bearing on sustainable development. Indeed, this is the essence of the whole text. We appear to have a philosophy that embraces sustainability. But we fall short on the tools. The processes of valuing natural systems are immensely restricted by our fundamental lack of understanding of resilience, trigger points and catastrophic transformations in natural systems. So the good will of economists is stretched when the scientific tools of prognosis and beneficial or malign interaction are simply unknowable. If we were to remove all tropical forests over the coming 50 years, who can predict now just what would be the consequences for global systems and human livelihoods all over the planet? In the absence of any political or resource ownership arrangement that puts care into a tropical forest or a marine margin, then global life-support significance is of apparently no value until the damage occurs.

Sue Owens (2003) complains that the notion of sustainable development contains fundamental contradictions between a new-liberal opportunism

and a naturalistic values of living within nature. She is not sanguine that the two constructions can be resolved. Nor am I.

Similarly, we may have ways of knowing about sustainable futures, but still lack the "knowledge" upon which policy will be designed to act. The 2002 Millennium Development Goals argue for a 50 per cent reduction in global poverty in a decade. Yet the evidence is illusionary: poverty is deepening if well-being becomes the yardstick rather than income or health. Just because we have "knowledge" about poverty alleviation tells us all too little about how to gain agreement on the necessary actions. Global institutions are similarly ill-designed for effective sustainability. They profess all the right values, but the delivery is moving economies further and further away from designing for ecosystem resilience. The emergence of the Chinese economy places a fresh burden on illegal natural resources trading and global carbon dioxide emissions. It is not possible to see how the planet can fare sustainably in the face of the particular nature of Chinese economic expansion, when human and natures' rights get short shift.

Sustainability should encompass all of the values and ideas enshrined in this book. It is not the fault of the authors or editors that it does not. It is precisely the impression, the unfamiliarity, the unsuitability of the sustainability "message" with inadequate politics and international action that cause it to remain marginalised. So we return to the case of this chapter. Sustainability can flourish at the local level. As it expands due to sheer dedication and commitment, so the larger political, economic and social structures will lead its call for a "better way".

By way of offering some hope in an otherwise gloomy assessment, here are five zones of sustainability engagement that deserve more attention:

- Businesses are beginning to realise that gradual resource depletion and human suffering are neither good for business nor are morally acceptable to business practice. Coalition of public, private and civil interests are beginning to coalesce around key resources under stress such as fish, water and critical ecosystems. The mobilisation is likely to grow and coalesce around sustainable development frameworks at the regional level.
- Consensus are noticeably shifting their purchasing towards environmentally and socially friendly products, dragging the producers and retailers along with them. This appears to be a progressive trend.
- Tipping points around climate change impacts are beginning to get noticed. This may begin a move towards "civic virtue" that can be supported in the schools.

- Well-being is getting on to the political agenda, and it may well become an important focus of activity and behaviour in schools.
- Using ecosystem processes as a basis for economic analysis in resources management is gaining ground. More and more promising case studies are being cited.

There is a huge amount that is sustainable at all manner of local levels. The key is to identify these, to nurture these and to ensure that the bigger "systems" of transformation do not swamp them. I personally am not sanguine that we will succeed but I will put much faith and effort into championing local sustainability initiatives.

NOTES

1 For information on The Greenhouse, see www.greenhouse.org.za

2 For more information in Lapeng e-mail: lapeng@cordial.co.za

3 See City of Johannesburg Integrated Development Plan (2003/4, 2004, p. 18).

4 A core of 12 volunteers patrol the part with video cameras looking out for anyone who is not familiar who is acting suspiciously. The role of the "park guard" for unemployed blacks and Afrikaaners is common place in South Africa. In Joubert Park these guards are local people, which is less common elsewhere.

5 See www.sustainableoregon.net

6 See www.sustainabilitynorthwest.og

7 See www.hfhe.org

8 See www.oregoncountrybeef.com

9 See www.stahlbush.com

10 See, e.g, Manuel Castells (1997), Doreen Massey (1994), Andrew Weigert (1997), and Anthony Giddens (1997).

11 The Ecologist "Slow food: a movement to save the world", *The Ecologist Magazine,* April 2004, 40–53.

12 Higher Education Funding Council for England. A Sustainable Development Action Plan for the Higher Education Sector. Committee Paper. www.hefce.ac.uk/publications (2005).

13 Department of Education and Skills *Sustainable Action Plan* London DfES (2004) www.dfes.gov.uk

14 Tony Blair "Climate change and sustainable development" www.nd10.gov.uk/speeches

15 See www.rutgers.edu/procurement

REFERENCES

Adams, W. (2003) *Green Development.* Routledge, London.

Annecke, E. and Swilling, M. (2004) An experiment in living and learning in the Boland. In E. Pieterse and F. Meintgies (eds) *Voices of the Transition: the Politics, Poetics and Practices of Social Change in South Africa.* Heineman, London, pp. 293–303.

Brown, L.C. (2005) *Plan 2.0 Rescuing a Planet Under Stress and a Civilisation in Trouble.* Norton, New York.

Castells, M. (1997) *The Power of Identity.* Blackwell, Oxford. City of Johannesburg Integrated Development Plan (2003/4, 2004) p. 18.

Crutzen, P.J. (2002) The anthropcene: geology of mankind. *Nature* 415: 23.

Danks, C., Geobel, M. and Steer, K. (2004) Recreating natural resource-based businesses: sustaining the land and communities in the North West. *In Daage* (2004) 109–125.

Department of Education and Skills (2004) *Sustainable Action Plan.* London DFES. www.dfes.gov.uk

Diamond, J. (2005) *Collapse: How Societies Choose to Fall or Survive.* Penguin/Allen Lane, Harmondsworth, Middlesex.

Dietz, T., Osborn, E. and Stern, P.C. (2003). The struggle to manage the commons. *Science* 302 (12 December): 1907–1912.

Dobson, A. (2003) *Citizenship and the Environment.* Oxford University Press, Oxford.

Dodds, F. and Pippard, T. (eds) (2005) *Human and Environmental Security: an Agenda for Change.* Earthscan, London.

Giddens, A. (1997) *Modernity and Self Identity: Self and Society in Late Modern Age.* Polity Press, Cambridge.

HM Government (2005) *Securing the Future: UK Sustainable Development Strategy.* The Stationery Office, London.

Kates, R.W., Parris, T. and Leiserowitz (2005) What is sustainable development? Goals, indicators, values. *Environment* 47(2): 9–21.

Lenton, T., Caldeira, K.G. and Szathmary, E. (2004) What does history teach us about the major transitions and role of disturbances in the evolution of life and the earth system? In H.J. Schellnhuber *et al. Earth Systems Analysis for Sustainability.* MIT Press, Cambridge, MA, 29–52.

Lovelock, J. (2006) *The Revenge of Gaia.* Penguin Books, London.

Massey, D. (1994) *Space, Place, Place and Gender.* Routledge, New York.

Millgnium Ecosystem Assessment (2005) *Ecosystems and Human Well-Being.* Island Press, Washington, DC.

New Economics Foundation (2004) *A Well-being Manifesto for a Flourishing Society.* London.

Northwest Environment Watch (2004) *Cascadia Scorecard,* Seven key trends, *Shaping the Northwest.* Seattle.

O'Riordan, T. (ed.) (2000) *Globalism, Localism and Identity.* Earthscan, London.

O'Riordan, T., Preston-Whyte, R. Hamann, R. and Manquele, M. (2000) The transition to sustainability: a South Africa Perspective. *The South Africa Geographical Journal* 82(2): 1–10.

Owens, S.E. (2002) Is there a meaningful definition of sustainability? *Plant Genetic Resources* 1(1): 5–9.

Pieterse, E. and Meintjies, F. (eds) (2004) *Voices of the Transition: the Politics, Poetics and Practices of Social Change in South Africa.* Heinemann, London.

Porritt, J. (2004) *Redefining Prosperity: Resource Productivity, Economic Growth and Sustainable Development*. UK Sustainable Development Commission, London.

Porritt, J. (2005) *Capitalism: As If the World Matters*. Earthscan, London.

Sachs, A. (2004) Forward. In E. Pieterse and F. Meintjies (eds) *Voices of the Transition: the Politics, Poetices and Practics of Social Change in South Africa*. Heinemann, London, p. ix.

Schellnhuber, H.J., Crutzen, P.J., Clark, W.C., Claussen, M. and Held, H. (2004) *Earth Systems Analysis for Sustainability*. MIT Press, Cambridge, MA.

Schellnhuber, H.J., Cramer, W., Wigley, T. and Yohe, G. (2006) *Avoiding Dangerous Climate Change*. Cambridge University Press, Cambridge.

Scientific American (2005) Crossroads for Planet Earth. September, Vol. 293(3).

The Ecologist (2004) Slow Food: a movement to save the world. *The Ecologist Magazine,* April, 40–53.

UN Convention on Biodiversity (UNCB) (2006) *Global Biodiversity Outlook 2*. Cambridge University Press, Cambridge.

UN Development Programme (UNDP) (2005) *Human Development Report*. Oxford University Press, New York and Geneva.

UN Environment Programme (UNEP) (2002) *Global Environmental Outlook*. Norton, New York.

Waage, S. (ed) (2003) *Arts, Galileo and Ghandi: Designing the Future of Business Through Nature, Genius and Enterprise*. Greenleaf Publishing, Sheffield.

Weigent, A. (1997) *Self, Interaction and Natural Environment: Refocusing our Eyesight*. New York University Press, New York.

23

The Greening of Business: Opportunity or Contradiction?

Christina Page and Amory Lovins

THE OPPORTUNITY: BEYOND "COMMAND AND CONTROL" ENVIRONMENTALISM

Can businesses be green and make a profit? Can corporations grow and prosper by the traditional standards of the marketplace *and* behave as active and effective stewards of the planet?

The historical record of environmental policy and corporate behavior might suggest that these two goals are mutually exclusive. In the period between 1970 and 1990, protection of the environment in the developed world fell largely to government. In the USA, an "alphabet soup" of regulations – RCRA, CAA, CWA, ESA, and TOSCA[1], to name a few – served to protect the public and the natural world from environmental threats. By and large, government created rules of behavior and imposed them upon business, with varying degrees of effectiveness (see Box 1). In the early 1990s the spotted owl controversy in the USA, in which the ESA clashed with the interests of the timber industry, reinforced a stereotype through which some still view sustainability:[2] it's either the jobs or the owls.

However, in the 1990s, an important shift began to gain momentum: ever more people in industry (and government, and the public in general) began to reject the assumption that the health of the economy and the health of the environment are inevitably at odds with one another. Business, and society as a whole, has slowly begun the move beyond "command and control" environmentalism towards a mindset where the most forward-thinking companies see a competitive advantage in pursuing sustainability.

We are a long way from rendering environmental regulations obsolete. Within a regulatory environment, there are ways of achieving compliance that will cost companies money (including large public-relations and legal fees should they choose to lobby and litigate), and there are ways of doing so that can actually improve a company's competitive advantage. Government regulation, if crafted to encourage the holistic approach to ecosystem protection that William Reilly advocates, can enable creative companies both to do well financially and do right by the planet. And, since private business and companies are generally the source of most innovation and invention, they have the ability to create solutions more quickly – solutions that are more flexible, efficient, and holistic – than if they were merely obeying the letter of the law.

For some companies, this was not a new idea in the 1990s. When the threat of chlorofluorocarbons (CFCs) to the ozone layer became clear in the 1980s, DuPont saw an opportunity to move from its lackluster position in a heavily commoditized market into a lucrative leadership role in developing and selling patentable, cutting-edge CFC substitutes. To this end, it aggressively supported the international Montréal Protocol against CFCs (Benedick, 1998; Reinhardt, 2000). Carrier introduced a non-ozone-depleting alternative into its high-end, energy-efficient air conditioners eight years ahead of the Montréal Protocol's schedule, providing a unique marketing angle. The units cost less to maintain and will remain serviceable after the final phase-out of ozone-depleting chemicals under the Protocol in 2010. Designing proactively protects Carrier's air conditioners against

Box 1   Space invaders and the US Environmental Protection Agency

In a 1990 speech to the National Press Club, EPA chief William K. Reilly reflected on the evolution of environmental policy in the USA:

… Perhaps some of you have played a somewhat primitive, pre-Nintendo video game called "Space Invaders." In that game, whenever you see an enemy ship on the screen, you blast at it with both barrels – typically missing the target at least as often as you hit it. The last two decades of environmental policy in this country have been similar in some ways to that video game: every time we saw a blip on the radar screen, we unleashed an arsenal of control measures to eliminate it. In the late 1960s we saw that we had an air pollution problem, so we enacted ambitious clean air laws. At about the same time, we became aware of serious water pollution and passed the equally ambitious Clean Water Act … And so it went through the 1970s and 1980s: drinking water, radiation, pesticides, hazardous waste, medical wastes – each problem dealt with essentially in isolation, without reference to all the others. As I noted, many of those efforts have been successful – up to a point. But the upshot of this piecemeal approach to pollution control has been that we have set our pollutant- and medium-specific goals over the last twenty years without adequately addressing our overall environmental quality objectives. Rarely did we evaluate the relative importance of individual chemicals or individual environmental media. We didn't assess the combined effects on ecosystems and human health from the total loadings of pollutants deposited through different media, through separate routes of exposure, and at various locations. We have never been directed by law to seek out the best opportunities to reduce environmental risks *in toto;* nor to employ the most efficient, cost-effective ways of proceeding (Reilly, 1990).

obsolescence and makes them more attractive to consumers (Parrot and Ottman, 2003).

A panoply of recent trends has continued to support the idea that sustainable business practices constitute a competitive advantage to forward-thinking corporations. Socially and environmentally responsible investing is a growing force in the business world. Assets involved in socially responsible investing (SRI) through the screening of retail and institutional funds, shareholder advocacy, and community investing has grown 40% faster than all other professionally managed investment assets in the USA from 1995 to 2003. Over these years, investment portfolios involved in SRI grew by more than 240%, non-SRI portfolios by 174% (Social Investment Forum, 2003). Global sales of organic food in 2006 were $39 billion, up from $36 billion in 2005 (Reuters, 2007) and in places from Dominica to Wales, Austria to Sweden, organic

Box 2   Emerging markets: the winds of change

Wind power is the fastest growing source of electricity on the planet. Wind power accounted for just 0.3% of global electricity supply in 2002, but this share is expected to be ten times higher in 2030 (International Energy Agency, 2004). Thanks to advances in technology, now being driven by such giant firms as General Electric, windpower is rapidly becoming cheaper and more efficient; in the USA, current windpower costs per generated kilowatt-hour are competitive with gas-fired plants (American Wind Energy Association, 2004).

Wind energy is not only interesting to small, alternative green enterprises. Increasingly, it is part of mainstream corporate business strategies. This is particularly true in Japan. With its severe shortage of space and robust offshore winds, Japan is a natural candidate for windpower. As a small island nation, it has considerable investment in slowing the effects of global warming and rising tides, and so the argument for windpower use in Japan may extend beyond mere short-term financial benefits. At the end of 2003, Japan had increased its cumulative capacity for wind power generation to 686 MW (American Wind Energy Association, 2004). Offshore windpower turbines are becoming increasingly attractive – they tap into stronger and more consistent wind patterns and don't take up valuable real estate. Hitachi Zosen Corporation hopes to be selling wholesale energy from a 3-MW floating wind-energy platform in the near future. Ishikawwajima-Harmina Heavy Industries has plans for a similar facility with 10 MW of capacity. Hokkaido and Tohoku Electric Power Companies are energy generators in Japan, which already possess enormous working wind generators.

The leaders in windpower are countries like Germany, Denmark, and Spain, with India and China coming up rapidly. Land-based wind turbines have long been a common sight in places like California, produce a fifth of Denmark's total electricity, and are very widespread in northern Europe. Though still a small fraction of global energy supply, windpower's rapid installability, falling cost, and freedom from fuel-price volatility promise to make it an important part of future energy markets. As concern for the problems of global warming increase, windpower is becoming both the cutting edge of alternative energy technology and a source of profitable carbon credits.

food is being adopted or considered as a universal practice and a national branding strategy. Windpower is the fastest-growing source of electricity globally (Box 2). The global windpower industry grew from 4800 megawatts of capacity in 1995 to over 74,000 megawatts in 2006. In 2006 alone, over 15,000 megawatts were added worldwide (Global Wind Energy Council, 2007) – almost five times the 3200 megawatts that nuclear power added annually, on average, through the 1990s (International Energy Annual, 2002).

Green marketing is booming; LOHAS (Lifestyles of Health and Sustainability) represent a growing demographic of considerable interest to marketing experts. The Toyota *Prius* hybrid vehicle was 2004's surprise success story; sales of hybrid vehicles in the United States increased by 570% between 2000 and 2004 (*Canadian Driver*, 2004), and many customers still sign up for year-long waiting lists to get one.

Additionally, the challenge of mitigating the problem of climate change is drawing increasing attention from the financial community. In 2001, a group of thirty-five large institutional investors (representing more than US$4.5 trillion) created the Carbon Disclosure Project (CDP) in an effort "*to assess and provide hard data on a company's exposure to climate change risks through impacts of both extreme weather events and regulation of greenhouse gas emissions.*" In 2002, the London-based CDP surveyed the chairs of the world's 500 largest companies and found that 80 per cent of respondents were willing to acknowledge that climate change posed an important financial risk.

"*The financial impact of climate change extends well beyond the obvious, emissions-intensive sectors such as oil and gas and electric utilities,*" stated a CDP project press release, predicting that the financial services, transportation, semiconductor, telecommunications and electronic equipment sectors, among others, would also be significantly affected. The report noted that firms that are quick to reduce greenhouse gas emissions "stand to gain competitive advantage, in terms of both cost and market risk management" (Burns, 2003).

In a 2004 meeting between the UN Environment Programme and twelve leading investment firms, participants agreed that problems related to environmental, social, and corporate governance affect long-term shareholder value – and that ignoring such problems could harm their stock prices (Reuters, 2004). Recent analysis suggests that anticipated regulations designed to curtail global warming and encourage fuel efficiency will hit some automakers harder than others – General Motors and Ford more than Honda, for example, based on conventional technology (*New York Times*, 25 July, 2004; WRI, 2003). But firms like Honda that invested in fuel efficiency early could

also lose that advantage if laggards choose to leapfrog to more advanced technologies, such as ultralight autobody materials.

What, then, are strategies for companies that wish to achieve sustainability and retain or increase their economic security and prosperity?

GETTING THERE: THE NATURAL CAPITALISM FRAMEWORK

The natural capitalism framework (Hawken *et al.*, 1999) provides some guidelines for how businesses can begin to capture the opportunity *to turn sustainability into a strategic advantage.* Industrial capitalism, as currently practiced, typically does not value, but rather is liquidating, very important forms of capital. It productively uses and reinvests in financial capital (money and goods), but inadequately values human capital (individuals, cultures, and communities) and largely ignores natural capital (the intact ecosystems that provide free services and abundant resources). Largely or wholly omitting the most valuable of the four forms of capital, and accounting the resources only as an extraction cost, is a material omission. Firms that play with a full deck – productively using and reinvesting in all four kinds of capital, not just two – not only make more money and have more fun, they also help to protect and enhance the ecosystem services that underpin life itself.

Ecosystem services include things like atmospheric and ecological stability, water and nutrient cycling, pollination and biodiversity, topsoil and biological productivity, and the ability to assimilate and detoxify natural and human wastes. As Senator Gaylord Nelson once remarked: "The economy is a wholly owned subsidiary of the environment, not the other way round." (Nelson, 2002). A 1997 *Nature* article estimated the total value of the Earth's ecosystem services at US$33 trillion annually, close to the total gross world product (Costanza *et al.*, 1997; Balmford *et al.*, 2002). Most ecosystem services have no known substitute, and we have no idea how to produce them ourselves.

The consequences of neglecting the importance of ecosystem services and natural capital are significant. For example, during the 1980s, average losses to property insurers for weather-related disasters were $2 billion per year. In the 1990s, they were $12 billion per year. In 1996, the President of the Reinsurance Association of America indicated that climate change, left unchecked, could bankrupt the insurance industry (Nutter, 1996). Swiss Reinsurance, a $29 billion company, has started querying its clients regarding their corporate strategies for addressing climate change regulation (Linden, 2006).

Box 3 "Environmental injury is deficit spending"

Robert F. Kennedy Jr., speaking at a Canadian Bar Association meeting in 2003, scoffed at the "inevitable" excuse that the time has come where it is necessary to choose between environmental protection and a strong economy:

> That is a false choice ... [We] treat the planet as if it was a business in liquidation, to convert our natural resources to cash as quickly as possible, to have a few years of pollution-based prosperity, we can generate an instantaneous cash flow and the illusion of a prosperous economy. But our children are going to pay for our joy ride. And they're going to pay for it with denuded landscapes and poor health and huge cleanup costs that are going to amplify over time ... Environmental injury is deficit spending. It's the way of loading the costs of our generation's spending on to the backs of our children ... In a true free market economy, you can't make yourself rich without making your neighbors rich and without enriching your community. What polluters do is make themselves rich by making everybody else poor (Wyatt, 2003).

The conventional prescription of "environmental economists" is to internalize nature's value so we all make better decisions. But although the right value clearly isn't zero, it's not known or agreed upon, and reaching agreement could take decades. How, therefore, can companies behave *as if* natural capital were properly valued, but without needing to know or signal that value? Here are four operational principles – the basis of "natural capitalism" – that offer a practical path to become sustainable at a handsome profit.

First, radically increase the productivity of resources. In other words, do more, better, with far less, for longer. Radically increased resource productivity is the cornerstone of natural capitalism. Using resources more effectively has three significant benefits: slower resource depletion, less pollution, and a way to create good, lasting, and meaningful jobs. Through fundamental changes in production design and technology, leading companies are making natural resources stretch enormously further. The book *Natural capitalism* describes how to win expanding, not diminishing, returns to investments in advanced resource efficiency, so that very large savings can cost *less* than small ones.

Second, practice biomimicry. Design industrial systems to produce the way nature does, with closed loops, no waste, and no toxicity. *Biomimicry* – the title of Janine Benyus' book, subtitled "Innovation Inspired by Nature" – seeks not merely to reduce waste but to eliminate the entire *concept* of waste. Closed-loop production systems, modeled on nature's designs, either return every output harmlessly to the ecosystem as a nutrient or create valuable inputs for other manufacturing processes. Industrial processes that emulate nature's benign chemistry allow constant reuse of materials in continuous closed cycles, eliminate toxicity, and promote more efficient production (Benyus, 1997).

Third, shift from a product-based economy to a solutions-based economy. The business model of traditional manufacturing rests on the sporadic sale of goods. The natural capitalism model delivers value as a continuous flow of services – leasing an illumination service, for example, instead of selling light bulbs, much as Xerox leases "document management services" instead of simply selling a copier. This shift rewards both the provider and the consumer for delivering the desired service in cheaper, more efficient, and more durable ways. Ultimately, it will lead to an economy where wealth is measured not by the acquisition of goods but by the continuous receipt of a high-quality, high-value, continuously evolving service. Moving to a "service and flow" economy reduces waste from unwanted input and output, tempers inventory and revenue fluctuations, and lowers the risk of disposal liabilities; it also encourages greater resource efficiency and closed loops as described in the first two principles.

Fourth, reinvest in natural capital. Capitalism reinvests earnings back into productive capital. Businesses are ultimately dependent on the planet's ecosystem services if they are to survive and prosper. It is therefore logical and advantageous, when designing out waste, to reinvest some of the resulting profits into the kind of capital that is now scarcest – natural capital. This reverses worldwide ecological destruction by sustaining, restoring, and expanding stocks of natural capital, so that they can keep on producing the ecosystem services and resource flows that we cannot live without.

Below are some real-life, market-based examples of these ideas, and a more in-depth exploration of the principles behind them.

PRINCIPLE 1: RADICALLY INCREASING RESOURCE PRODUCTIVITY

Historically, the first Industrial Revolution arose when a relative scarcity of laborers was limiting progress in exploiting seemingly boundless nature. The mechanized textile mills introduced during 1770–1812, which enabled one Lancashire spinner to do the work previously done by 200 weavers, were only one of many technologies that increased the productivity of workers by roughly one-hundredfold.

Anyone who announced in 1750 that it would be possible to increase labor productivity by a hundredfold would have been laughed at; indeed, that whole concept was not known then. Yet that is just what the Industrial Revolution did. Profit-maximizing capitalists economized on their scarcest factor of production – namely, skilled people – by radically increasing the productivity of the limited numbers of workers and the efficiency of labor. The logic of economizing on one's scarcest resource, as economics teaches, remains perennially valid. Yet today the pattern of scarcity has reversed. Now we have abundant people and scarce nature, not the other way around; so now it is no longer people but nature that needs to be used far more productively. Future profits are likely to be limited by scarcities of forests, not chainsaws; or fish, not nets and boats; of fertile soil, not plows; of fresh water, not pumps.

Using the constraints and challenges of our current global condition, we have to change our strategy: use more people and more brains to wring four, ten, or even one hundred times as much benefit from each unit of energy, water, and material. Numerous companies are incorporating these ideas into their operations. The carpet company Interface Incorporated achieved a cumulative savings of $231 million between 1995 and 2003 through aggressive elimination of waste (Interface Sustainability Report, 2004). 2004 savings from waste reductions accounted for approximately 28% of the firm's operating profits (Anderson, 2005). It has reduced its use of non-renewable energy by 18% since 1996, and reduced its water consumption per unit of product by 26% (Interface Sustainability Report, 2004). The aluminum producer Alcoa, as of 2003, had reduced its greenhouse gas emissions by more than 23.5% from 1990 levels while expanding production. DuPont offset initial costs for its own investment in reductions by selling both carbon credits and its technical expertise in greenhouse gas reduction to other companies (Feder, 2003). In 2004, DuPont Vice President Paul Tebo announced that the company had reduced energy consumption by 9% while boosting production 30%, and he predicted that DuPont would save $3 million to $4 million in the coming year (Wisby, 2004). Dow Chemical Company's Louisiana Division launched its Waste Reduction Always Pays (WRAP) program in 1981 to encourage employees to provide in-house projects for reducing waste. Over the twelve years of the program's life, Dow implemented over 900 worker-suggested energy-saving projects. The measures achieved average annual returns on investment of over 200% – with both returns and savings rising in the latter years, even after the annual savings had surpassed $110 million. BP surpassed even its own goals when it began an aggressive program to reduce greenhouse gas emissions in 2000 (Box 4). In all, firms like IBM, DuPont, and STMicroelectronics have lately been cutting their energy use per unit of production by 6% *per year*, with typical paybacks of two to three years from improving their existing plants.

Box 4   Radically increased resource productivity at BP Amoco

On 1 January 2000, BP Amoco initiated the first corporate greenhouse gas (GHG) trading system in the world with the stated goal of reducing internal greenhouse emissions by 10% below 1990 levels by 2010 through an innovative internal trading program.

BP has 127 individual business units (BUs) throughout the world, ranging from exploration to retail sales. Prior to the launch of BP's tradable permits' program, each BU was allocated a certain emission limit and given a preliminary indication of expected annual allocations through 2005. The sum of all the emission limits, called a Group Cap, equaled BP's total GHG footprint in 1998. The Group Cap is lowered annually, based on expectations for the coming year and necessary progress toward the target, and each BU's limit is reallocated accordingly. The key to this system is that BUs' trade emissions – or, more precisely, the ability to emit – with one another. Emission allowances are bought and sold through an internal website-based market that tracks the account of each BU. This system allows for the most economically efficient allocation of emissions as those BUs with low or negative abatement costs trade against those with high abatement costs. As opportunities arise for reduction, each BU has the incentive to pursue emission-reducing measures that cost it less than the current price for emission credits offered by other BUs. Once a BU has achieved those reductions, it is free to sell to the highest-bidding BU emissions credits equal to the difference between its allocated limit and its actual emissions. On 11 March 2002, Chairman Lord John Browne announced that BP had already surpassed its 10% reduction target at no net cost, and was focusing on maintaining that emission level through 2012 while increasing oil and gas production by 5.5% per year (Browne, 2002). In 2004, he noted that the initiatives had

Box 4   Radically increased resource productivity at BP Amoco—cont'd

actually increased shareholder value by approximately $650 million by using less energy to produce its products (Browne, 2004).

The rapid achievement of BP's goal is the result of hundreds of initiatives taken throughout the company rather than one universal solution. Each BU is allowed to create its own mixture of technology and efficiency to meet emission targets. This allowed grassroots ideas to flourish and individuals to initiate programs based on their local knowledge and hands-on experience with the systems they were seeking to improve. Initiatives were broad ranging and highly lucrative. A new, combined-cycle cogeneration heat and power plant in Hull, UK, saved 150,000 tons of carbon dioxide equivalent *per year* and reduced the site's energy costs. Installation of eight air/fuel ratio controllers on large compressor engines in BP's Canadian natural gas systems reduced emissions by 50,000 tons per year.

These successes illustrate the gratifying gains in efficiency and environmental performance that are possible with the right combination of effort and technology. BP has created a framework that specifically allows employees to improve directly the company and rewards units that acted on opportunities as they arise. But the big surprise – not yet widely practiced by BP or most other firms – is that optimizing whole-system design to achieve multiple benefits from single expenditures can often "tunnel through the cost barrier," as described in *Natural Capitalism*, often making efficiency's capital costs *negative* in new facilities.

Challenges to sustainability through resource efficiency

Resource efficiency initiatives, by their nature, are attractive. They are easy to understand, often require little capital expenditure, and have the capacity for fast and dramatic cost savings. Thus, it is easy to overlook some of their drawbacks.

However, studies in systems thinking (Jones *et al.*, 2002) suggest that *just* using natural resources more efficiently may have unintended and undesirable consequences for sustainability. Rather than reduce the overall pressure on limited resources, efficiency may merely increase throughput if used in isolation. For example, if a lumber mill finds a way to extract twice as many board-feet from a single tree, the mill may experience a temporary increase in revenue. But the resulting increased supply of wood on an open market will decrease the price the mill is able to command per board-foot, causing it to increase production (and the number of trees milled) just to maintain existing revenue. If we assume that increasing our throughput *alone* will allow us to avoid resource depletion, we may instead be achieving mere efficiency by ultimately sacrificing those resources – and, with them, the prospect of sustainability, true wealth, and global security. Thus, it is important to use resource efficiency as a starting point in sustainable and competitive business practices, in conjunction with the other three principles of natural capitalism – not by itself.

PRINCIPLE 2: BIOMIMICRY

Resource efficiency is a cornerstone of natural capitalism and of sustainable business practices, but it is only one component. The concept of biomimicry (Benyus, 1997) goes beyond merely reducing waste; it means eliminating the entire *concept* of waste by adopting biological patterns, processes, and often materials. Biomimicry imitates nature at any or all of three levels: what things do, how they're made, and how they fit. It's a design revolution only just beginning, but already showing enormous promise for applying the design genius of 3.8 billion years of evolution, in which the roughly 99% of designs that didn't work already got recalled by the Manufacturer, leaving only highly successful designs from which a great deal can be learned.

The common industrial approach to creating things has been a combination of age-old processes that can be described as "take, make, and waste" and "heat, beat, and treat." Raw materials are extracted from the ground, either by mining or chemically intensive monocultural farming; a small fraction of the material is turned into a product (99% of which either never gets into a product or is discarded within six months (Hawken *et al.*, 1999). The remaining material is landfilled, incinerated, or otherwise disposed of as non-usable product. In order to turn raw extracted material into the desired end product, one heats it to hundreds or thousands of degrees, beats it into the desired shape, and then treats it with (frequently toxic) substances to ensure a certain property, such as durable color, suppleness, or strength.

Instead of this traditional, energy-intensive, highly wasteful method, biomimicry challenges us to ask: how would nature solve this problem? For example, the surface of the lotus flower is microscopically bumpy and sheds dirt, allowing it to stay luminously clean in the middle of a swamp; the German company IPSO has applied this "lotus effect" to its Lotusan paint. Since the bumpy surface is self-cleaning when it rains, Lotusan comes with a five-year guarantee of cleanliness without the aid of sandblasting or detergents. The company is exploring other applications, including auto paint and roof shingles (Robbins, 2001).

Engineers designed a building for hot, dry Harare, Zimbabwe, based on the architecture of a tropical termite mound. Termites dig special vents around the base of their mounds to allow cool air in; as this air warms, it is exhausted out the top of the mound. The system is so effective it is able to keep the interior of the mound at a steady 31°C during 35° nights and 105° days. By copying the termites' heating and cooling system, architect Mick Pearce and his Arup partners were able to design a nine-story, 26,000-square-meter building that during its first nine months of operation used a tenth of the cooling energy a conventional building its size would have used.

Another firm, Pax Scientific, was inspired by the naturally occurring Fibonacci Series (the mathematics of spirals found in many sea creatures and terrestrial organisms) to rewrite the book on pumps, fans, turbines, propellers, and anything else that moves or is moved by fluids. By mimicking the elegance of nature, Pax has developed new pump and fan designs that can dramatically reduce energy requirements, noise, and other problems, potentially saving a substantial chunk of the world's electricity (Lovins, 2004).

The biomimicry framework also encourages us to look at how nature would fit processes into an existing ecosystem. The field of industrial ecology studies examples and opportunities for transforming the waste stream of one industry into the feedstock of another, successfully converting the "take-make-waste" model into a closed-loop system (Box 5).

PRINCIPLE 3: SHIFT TO A SOLUTIONS ECONOMY: FROM PRODUCT TO SERVICE

All good marketing textbooks teach us to ask: What is the problem the customer is trying to solve? Principle 3 of the natural capitalism framework recommends focusing on the solution and offering "less stuff and more service." If you look at end goals, you see that people don't really want heating oil and refrigerant; they want hot showers and cold beer. They want the cheapest

---

Box 5   Industrial symbiosis and the legacy of Kalundborg

"Industrial symbiosis," also known as "industrial ecology" and "by-product synergy," is an innovative form of industrial collaboration that redefines waste and by-products as inputs to other industrial operations. The term "symbiosis" is borrowed from biology and refers to the close association of different species for mutual benefit. With this approach, waste is no longer a risk or operational liability that requires treatment and disposal; rather, it presents an opportunity for reducing costs, gaining access to cheaper materials and energy, and generating income from production residues.

Perhaps the world's best known and successful example of industrial symbiosis is in Kalundborg, an industrial zone located on the coast of Denmark. The Kalundborg zone example comprises six non-competitive core partners: Denmark's largest coal-fired power station, an oil refinery, a plasterboard manufacturer, a pharmaceutical company, the municipality of Kalundborg, and a soil remediation company.

These six companies are involved in over fifteen symbiotic business relationships that have developed organically over several decades. These include the power station delivering process steam to the oil refinery and drywall (plasterboard) manufacturer; the municipality receiving excess heat from the power station to heat district homes; the power station delivering gypsum derived from waste sulfur from its stack scrubbers to the drywall manufacturer; and the pharmaceutical company delivering sludge to local farmers for fertilizer use. As reported by the companies, waste exchange has resulted in a shared financial gain of over US$16.8 million per year on an overall investment of US$84.1 million. The many annual savings include a 60% water use reduction for the power plant and an estimated 130,000 tons of carbon dioxide avoided thanks to energy sharing (Ehrenfeld and Chertow, 2001).

Industrial symbiosis has many benefits for the economy, environment, and community including job and business creation, improved competitiveness through reduction in operating costs in existing businesses, stimulating innovative product creation, accelerated use of declining industrial sites, attraction of industries that fit material flows, reduced risk from waste treatment and disposal, reduced dependence on municipal infrastructure such as landfills and wastewater treatment plants, and a positive image.

For more information on industrial symbiosis, see www.symbiosis.dk (the official Kalundborg website) and www.sustainable.doe.gov/business/indeco.shtml. Similar examples are emerging in other countries.

end-use solution. If done correctly, a service- and solutions-based business approach can reduce or eliminate waste and cost while improving customer satisfaction.

Focusing on the end result requires a shift in the business model. If I run a paint shop and you come in wanting to paint your house, my likely goal is to sell you as much paint as I can without blatantly cheating you. Meanwhile, you are trying to buy the least amount of paint necessary to get the job done. Odds are good that there will be excess paint, or an extra trip to the paint store if the housepainter runs short.

If instead you pay me a set amount to keep your house the color you want, with minimal peeling and fading, for the next ten years, I will be motivated to use as little paint as I can to get the job done and keep you satisfied, because it is part of my cost. I might even look for alternative ways of keeping your house the color that you want, if I can find a solution that is cheaper, more efficient, and less wasteful, starting with more durable paint more carefully applied.

Thinking in terms of solutions rather than products can also change traditional notions of ownership and liability as well. If there are regulatory restrictions around certain kinds of paint – such as those that many homeowners now experience around lead-based paint on old houses – I will increasingly retain responsibility for the impact of the paint for the duration of its life, rather than transferring responsibility to the customer once the cans are sold. As legislation around extended producer responsibility emerges – for example, Europe's end-of-life vehicle agreement under which auto companies assume the cost of disposal of certain heavy metals at the end of a vehicle's useful life, just as packaging and appliance makers own their products forever – companies will be forced to look at models of increased responsibility for their products. This is an opportunity not just for better design, but also for shifting to a solutions-based business model that aligns producer with customer incentives – a good idea in any enterprise.

Focusing on solutions rather than products can help rearrange our mental furniture and inspire innovations based on other principles. If we apply principles of biomimicry and green chemistry to the problem of providing satisfactory house pigmentation services I might look for permanent, non-toxic alternatives to paint, such as refracting light to create color, as peacocks do with the structure of their feathers.

In a solutions economy, the customer and the provider have the same incentives instead of opposing ones. Walter Stahel of the Product-Life Institute in Geneva, Switzerland, points out the shift in incentives with something as simple as a

disposable camera. Although not a strict leasing arrangement, the consumer has an incentive to return the camera to the developer because it is the only way to retrieve what they value – namely, its images (Stahel, 2002). The camera company, in turn, has an incentive to make the parts easy to disassemble and reuse. Stahel points out some of the other differences between sale of a product and sale of a service in Table 23.1.

There are many examples of solutions-based business models. Typically, companies that use chemicals for heavy industrial cleaning and processing spend ten dollars to manage and dispose of the chemicals for every dollar they spend to buy them in the first place. Chemical management services companies (including traditional chemical companies such as Castrol and PPG Chemical) provide the service of the chemicals while retaining ownership and liability, and are thereby able to achieve the same goal with less waste, more efficiency, and generally a cost reduction of 5 to 20% annually. Such services are common in the electronics, automotive, and aerospace industries and are starting to catch on in other sectors (Westervelt, 2003). Between 2000 and 2004, the chemical management services market has grown 50% to approximately $1.22 billion (Chemical Strategies Partnership, 2004).

Energy services companies (ESCOs) provide a performance-based energy efficiency consulting model. The ESCO determines a plan for increasing energy efficiency, for a client, then installs and maintains any equipment involved. Most ESCOs assume financial and liability risks for the projects. In return, they are paid a portion of the cost savings the customer realizes. The size of the market is hard to estimate, but between 1992 and 1995, total installed costs of energy efficiency projects by ESCOs went from $500 million to $2.3 billion in the USA (World Energy Efficiency Association, 1999).

Leasing is hardly a new idea. Xerox began leasing its copiers and charging by the page once it realized that customers wanted copies, not copiers; today, 50% of its revenues are from equipment leasing. The company's takeback business model influenced the way Xerox designed its product. Instead of designing for obsolescence, so a customer had to reinvest in a new model every few years, Xerox designed its product for easy value recapture when the customer returned the unit. The company increased ease of disassembly, compatibility of parts for remanufacturing, and overall durability. When the company realized how much residual value was in the products its customers were returning, Xerox started an asset recovery program. The firm created an infrastructure for equipment "take-back" that provided a steady and predictable flow of used machines,

Table 23.1 Selling performance versus selling products

Sale of a product *(Industrial economy)*	Sale of a service *(Service economy)*
Object of the sale is a product	Object of the sale is performance, customer satisfaction is the result
Liability of the seller for the manufacturing quality (defects)	Liability of the seller for the quality of the performance (usefulness)
Payment is due for and at the transfer of the property rights ("as is where is" principle)	Payment is due pro rata if and when the performance is delivered ("no fun, no money" principle)
Work can be produced centrally/globally (production), products can be stored resold, exchanged	Work has to be produced *in situ* (service), around the clock, no storage or exchange possible
Property rights and liability are transferred to the buyer	Property rights and liability remain with the fleet manager
Advantages for the buyer: • right to a possible increase in value • status value as when buying performance	**Advantages for the user:** • high flexibility in utilization • little own knowledge necessary • cost guarantee per unit performance • zero risk (if firm stays in business or result is insured by a creditworthy third party) • status symbol as when buying product
Disadvantages for buyer: • zero flexibility in utilization • own knowledge necessary • no cost guarantee • full risk for operation and disposal (e.g., driver's license)	**Disadvantages for user:** no right to a possible increase in value
Marketing strategy: publicity, sponsoring	**Marketing strategy:** customer service
Central notion of value: high short-term exchange value at the point of sale	**Central notion of value:** constant utilization value over long-term utilization period

Source: Adapted from Stahel (2001)

saving money through the reuse of valuable components and reducing warehousing and disposal costs. These efforts resulted in the Document Center 265 copier – 97% recyclable and 80% remanufacturable – and an innovative internal program with the slogan "zero to landfill – for our children" (Parrot and Ottman, 2003). Because it is cost effective for Xerox to reuse the parts, the company offers trade-in credits to encourage customers to return Xerox equipment that they had purchased. As of 2000, 25% of Xerox's production came from remanufacturing (Fishbein *et al.*, 2000). In 2003 alone, Xerox diverted 144 million pounds of waste from landfills and saved approximately $200 million through product remanufacturing (Box 6).

The solutions economy also includes the increasingly popular trend of car sharing. ZipCar is a for-profit US company based in Cambridge, Massachusetts, with growing offices in New York City, Washington DC, and Denver. The company provides mobility service without the hassles of

car ownership (upkeep, fueling, etc.), or, as its ads boast, "wheels when you want them." In the Zipcar business model, members pay a monthly fee, and can reserve a car online at short notice. An electronic key card gives access to the car, parked in a convenient, reserved location nearby; the electronic tracking system records time and distance traveled and automatically bills for the trip, saving parking space, money, and hassle. While car sharing is not flawless (Box 7), it provides a flexible alternative to single-car ownership.[3]

Challenges of product to service

Does shifting to a solutions-based, service-oriented business model guarantee increased sustainability? Probably not. A study by INFORM suggests that several conditions are necessary. First, the lessor must be the manufacturer of the product or a captive leasing company, so that it has an incentive to recapture the product at the end

Box 6 Xerox and the solutions economy

"Everything that Xerox delivers to its customers is designed to be returned – whether it's a machine, a cartridge, a spare, or packaging. All of these items, once returned, are processed for reuse or recycling. The only thing we want to leave with our customers is – *THE DOCUMENT*" (Xerox Corporation, 1997).

Box 7   How big is your ecological wheelprint?

Car sharing has lately become another environmental "answer" to the multitude of problems created by the automobile. Proponents of car sharing promote it as a solution that yields benefits over private car ownership for both the customer and the environment. While the financial benefits are clear and quantifiable for many car-share participants, it remains to be seen whether this business model inevitably reduces environmental impacts.

Proponents of car sharing assert that it reduces total vehicle miles traveled (VMT). In Europe, studies show significant VMT reductions. A survey for the Swiss Department of Energy showed that an average automobile "co-operative" program member drives approximately a quarter of the average distance driven by a family-car owner in Switzerland.

However, a central problem with the argument that car sharing reduces VMTs lies is the assumption that the majority of people signing up for car-share memberships *already drive* private cars. Furthermore, the assumption that car-share members will *give up* their private cars in favor of car sharing, or at least substitute car-sharing trips for private car use, may be flawed. In San Francisco, a majority of people joining City CarShare have no car and did not own one before joining. Rather, many members are substituting car-share trips for trips that they formerly took by other means, such as biking or walking – thereby actually *increasing* VMTs among car-share members. Behavior may differ in Europe, which accounts for 75% of current car-share memberships, and where culture and infrastructure may support sustainable use of car-sharing programs. In Switzerland, half of all Swiss train stations provide car-sharing lots, promoting the kind of intermodal transportation that helps to make car sharing efficient and environmentally beneficial.

Another oft-cited benefit is a reduction in cars on the road. Various studies suggest that there are five to nine parking places for every car in the USA. Under many car-sharing schemes, a single car can meet the needs of 10 to 15 car-share members. In 1995, two Swiss car-share programs, ShareCom and ATG, jointly owned 330 cars that served 6000 members. The small fleet of 330 replaced about 1500 privately owned cars which members had before they joined ShareCom and ATG. Such a shift thus reduces the overall amount of energy and materials involved in the manufacture of cars.

When considering the environmental impact of a car-sharing program, it is worth thinking about the cars themselves. Car-sharing programs may encourage the adoption of new, "green" technologies. For instance, Zipcar has electric hybrid vehicles in its Boston fleet, allowing members to test-drive them.

Certainly, car sharing is not the answer to all transportation problems, nor to the environmental problems caused by our current system, but, appropriately applied, it may offer some very sensible short-term solutions to problems that otherwise remain unaddressed. Like other environmental solutions, it is important to test it against the reality of human behavior.

---

of its life. Second, the lessor must have a realistic and cost-effective means of collecting the product at the end of its life. In handling the product after recapture, the company needs to prioritize refurbishing and remanufacturing ahead of recycling or sending the product to the landfill. The company must be aggressively committed to product design for durability, reusability, and recapture. Lacking these criteria, a company could easily achieve a highly successful leasing model that cranks out disposable products and contributes little to improving sustainability in their industry (Fishbein *et al.*, 2000). Like everything else, it is important to keep the big picture – and the end goal – in mind when considering a shift to a solutions-based business model.

PRINCIPLE 4: REINVESTING IN NATURAL CAPITAL

If our ultimate goal is truly sustainable businesses practices, merely reducing toxicity and waste is insufficient; rather, it is necessary to design business models and practices that are actively restorative to natural ecosystems and natural capital. In some business sectors, the connection between sustainability and industry is obvious. The National Ski Areas Association has launched a nation-wide "Keep Winter Cool" campaign in collaboration with the Natural Resources Defense Council. The program includes carbon-reduction initiatives and clean-energy projects at member resorts and public education programs around the effects of global warming.

Another example of the business sense in reinvesting in natural capital is the story of Perrier Vittel. Bottled water represents the fastest-growing segment of the beverage industry, with double-digit growth in Europe in 2003 alone. Nestlé, Perrier's parent company, was the world leader in bottled water in 2004. Perrier water was threatened by agricultural pollution from pesticides and fertilizer. Instead of setting up a mechanical filtering system or moving the plant to a new location, the company launched an aggressive watershed management effort. It purchased 1500 hectares of agricultural land for $9 million and entered into long-term contracts with forty local farms to improve farming practices, placing an emphasis on pasture-based dairy farming, improved animal waste management, elimination of corn cultivation and agrochemicals, and composting manure. The program cost $24.5 million over the first seven years.

The program has reportedly had a greater success in reducing non-point source pollution than government programs. Farmers willingly adopted new practices, partly because of favorable market conditions for organic foods. An external study determined that plan was more financially justifiable than filtration. In addition, there may be a brand benefit to staying "natural" for Perrier, instead of resorting to water treatment or reverse osmosis techniques for filtering its water (Perrot-Maître *et al.*, 2001).

The connection between the health of natural systems and economic prosperity has surfaced in some rather unexpected quarters. Pork farmers, after years of mass factory farming, are losing an estimated $90 million per year to bad meat. The stress of overbreeding and overcrowding has been scientifically linked to drastic declines in the quality of livestock meat. Farmers are therefore looking at traditional, lower-stress, free-range farming alternatives to protect the health of their revenue stream (Casau, 2003).

The multinational corporation Unilever recognized the connection between the long-term health of their company and the health of ecosystem services. Currently, Unilever is the largest purveyor of fish sticks and frozen peas in the UK. The company has launched a Sustainable Pea Initiative in collaboration with the NGO Forum for the Future, to collaborate with its farmer suppliers in pursuit of more sustainable farming practices. In partnership with the Worldwide Fund for Nature, Unilever launched the Marine Stewardship Council, a nonprofit, autonomous institution that helps to create and certify sustainable fishing grounds and fisheries globally. As Vice President of Marketing Christopher Pomfret remarked, "... Sustainability is about the long-term security of our supply chain ... [so] if our business is to continue, then we need to sustain our sources of supply – and the only way to do that is to make them sustainable" (Pomfret, 2002).

THE PATH FORWARD: CHALLENGES AND OPPORTUNITIES

Thinking in whole systems

Companies wishing to integrate sustainability into business practices with real staying power cannot afford a piecemeal approach. Individual firms and industries should embrace a wide variety of strategies to increase sustainability. This can involve examining the processes behind product design, the buildings in which business is conducted, purchasing policies, accounting practices, recruitment, reward systems, and corporate culture. Such broad-based approaches have been pioneered in a variety of industries, from process efficiency in breweries to better electronic design (see Boxes 8 and 9). This can be daunting: when the carpet company Interface began pushing beyond resource efficiency and waste reduction to the possibility of making carpets from corn-based polymers, it quickly realized that true whole-systems sustainability could mean tackling issues of organic farming, genetically modified corn, and rural communities.

For businesses to measure the full benefits of sustainable practices, they need to look at the big picture; shifting the mindset around accounting practices is a good place to start. Monetary benefits are often found in unexpected places. A study in the journal *Energy* looked at 77 case studies of energy efficiency programs. Of the 52 monetized studies, the average energy payback on the program investment was 4.2 years when program officials measured only energy savings. Yet when other, non-energy benefits to the efficiency programs were taken into account – such as increased worker productivity from better lighting conditions, reduced maintenance, or better safety – the average payback dropped by more than half, to just 1.9 years (Worrell *et al.*, 2003).

One potential technique for overcoming such barriers is *appropriate cost accounting*. Environmentalists have been saying for years that to account properly for the damage being done to the environment we must measure external costs and internalize them into traditional accounting systems. Likewise, green building experts, energy efficiency initiatives, and other ideas with big long-term payoffs sometimes fail to meet short-sighted hurdle rates for investment, even though these investments are far less risky than normal investments in production or marketing to gain or keep market share (so they deserve a more favorable discount rate) and almost always cost less than the firm's marginal cost of capital. For a firm with a 12% cost of capital to pass up a 14% return on a virtually riskless investment simply misallocates capital and shortchanges shareholders.

Merely using different accounting terms may change a decision-maker's perception of a proposed sustainability innovation. For example, savings can be expressed in terms of simple payback (SPB) or internal rate of return (IRR) (Hawken *et al.*, 1999). While a three-year payback would be rejected by many engineers (who often evaluate the adoption or rejection of a technology

Box 8 Breweries hopping with efficiency

Though beer has fueled many late-night discussions, the brewing industry's large impact on the environment is rarely considered. Fortunately, breweries are proactively identifying profitable strategies for reducing their effects on the planet.

The environmental impact of the beer industry is significant by virtue of the sheer volume of the beverage consumed worldwide. World beer consumption is expected to top 180 billion liters by 2010. In China, beer production increased by 600% in the past decade and in recent years has rivaled US consumption in total volume. Every bottled or canned liter of beer requires between 5 and 20 liters of water to produce. Japan's Asahi brewery consumes 4.9 billion gallons annually. Brewery wastewater has a high biological oxygen demand, sometimes straining municipal wastewater treatment plants; in 2000, a 2500-barrel spill from a Coors Brewing Company plant in Colorado killed 54,000 fish in a nearby creek.

However, breweries are making progress. The industry's average energy and water consumption has dropped drastically in the past decade. A European brewery optimized its hot water system, reducing annual heating oil consumption by 340 tonnes and water by 40,000 cubic meters. An Asian brewery implemented a closed-loop cooling device, saving 50,000 cubic meters of water per year.

To reduce water consumption, some breweries use an air-rinse technology in their canning lines that reduces a can line's rinsewater needs by 79% (Hawken *et al.*, 1999). Ionized air is injected into the can, neutralizing the static charge that causes dirt and debris to stick to the interior. The unstuck material is promptly vacuumed off. The technology cleans better and eliminates moisture that can foster microbial contamination. On the wastewater end, Anheuser-Busch has installed an anaerobic digester in its Houston brewery's wastewater system. Methane produced by the digester heats the brewery's boilers, saving $57,000 per week.

Other breweries target their supply chain. Neumarkter Lammsbraeu brewery in Germany invokes a 400-year-old Bavarian "purity" law requiring beer to be made from only the purest materials to support organic farming by its malt and grain suppliers. This not only contributes to the quality of raw materials, but also stabilizes local labor markets. In Canada, the trade organization Brewers of Ontario, which provides 2 billion servings of beer annually, has achieved a 99% bottle take-back rate. With a 10-cent bottle deposit, the system reduced disposal costs by 89% in 1997.

Bottle reuse generated manufacturer savings of CAN$150 million (US$100 million) and avoided CAN$30–35 million in municipal waste management costs. New Zealand-based DB Breweries, owned by Heineken, achieved a 19% reduction in energy consumption in 18 months at its Waitemata brewery, thanks in part to a system that bases annual employee bonuses on energy management performance. Energy metering, tuning boilers, and better attention to energy consumption during peak demand times (when energy is most expensive) helped the brewery surpass its goals (McEwan *et al.*, 2004).

These activities help brewers produce the same bottle of beer with less water, energy, and material – helping to conserve the planet's resources and often improving their own bottom line. And that should make beer from these companies taste even better.

in SPB terms), a 33% before-tax IRR would generally be considered quite a reasonable investment (Hawken *et al.*, 1999; Swisher, 2001).

In the arena of energy-efficient building design, there have been efforts to overcome the barriers of split incentives and incompatible payback time periods via various strategies. One of these is performance-based leasing, in which the builder or owner is given a fee based on proven reductions in their buildings' energy costs. This effectively "unsplits" the incentives for efficiency measures up-front. Similarly, experiments facilitated by the Rocky Mountain Institute have shown that paying architects and engineers for what they save, not what they spend, has the potential for significant benefits to building design (Eley Associates, 1997).

While the opportunities are vast, implementing sustainability in a manner that is effective, enduring, and beneficial to the bottom line has its share of challenges. Some of the best, most elegant ideas fall prey to short-term thinking, fragmented management practices, and draconian payback thresholds. For example, in spite of the tremendous success of the Dow WRAP program, the initiative ended and monitoring stopped after its internal champion retired. DuPont successfully created CFC substitutes and backed the Montréal Protocol – but the returns fell short of the company's hopes as prospective clients found other substitutes for CFCs (e.g. cleaning circuit boards with carbon dioxide instead of fluorocarbons) (Benedick, 1998; Reinhardt, 2000). Shortcuts, such as reducing costs through resource efficiency and then using the savings to produce additional unsustainable products, can lead to merely breaking even, or creating even more damage to the ecosystem. Thus, keeping perspective on the entire system is crucial to identifying the best, most breakthrough opportunities – as well as crafting initiatives that will bear up under organizational change and financial challenges.

THE FUTURE

In spite of the increase in voluntary, proactive sustainability initiatives from corporations in the past decade, external pressure still plays a significant role in affecting behavior in the business sector. Shareholder advocacy around social and environmental issues increased by 15% between 2001 and 2003 alone (SRI, 2004). There is a growing movement supporting the restriction and revocation of corporate charters in specific situations (Cray, 2002). If others choose to take up this model, corporations in the future may need to pay closer attention to justifying their existence. Already, their "franchise to operate" or "license to grow" depends on the goodwill of millions of email-empowered activists.

Of the largest one hundred economies in the world, over half are corporations rather than nation-states. They are responsible for many of

Box 9   Whole systems thinking and product design

All the biggest design mistakes are made on the first day. And those design mistakes in the computer and electronics industry are causing millions of pounds of toxic waste from electronic gadgets to end up in developing Asian countries. They also represent an enormous missed opportunity for business. Discarded computers and other electronic devices from the USA, UK, Japan, Australia, and Singapore are finding their way to India, Pakistan, and China. A report issued in 2002 found that 50 to 80% of waste from electronic devices collected for recycling in the USA is being exported to developing countries (Puckett *et al.*, 2002).

A typical case of the fallout from this exportation is the Chinese town of Guiyu, in Guangdong province. Electronic waste is burned in open pits at night to salvage the valuable metals inside. Handling and breathing these carcinogenic substances without proper protection causes significant health problems. Guiyu's groundwater has become so polluted that drinking water has to be trucked in from 18 miles away. The Chinese government officially bans the import of electronic waste. Meanwhile, computer companies are encountering pressure from activists and international legislation to take responsibility for end-of-life disposal of their products. With such trends gaining momentum, it is becoming increasingly difficult for computer and electronics companies to treat the fate of their products as someone else's problem.

Creating new end-of-life strategies for their products offer these industries a perfect opportunity to do the right thing, reduce waste, and realize additional profits at the same time. If we apply the principles of natural capitalism to this problem, reducing electronic waste changes from a nuisance to a vast business opportunity. Here are two ways to do it:

1. *Build it right the first time.* Heavy metals such as mercury and cadmium are integral to traditional computers' electronic systems, while most computer cases are made of mold-formed plastic that is hard to disassemble at the end of the computer's life. These substances create both a toxic manufacturing environment and a disposal problem – and since increasing amounts of electronic goods are manufactured in Asia *and* dumped there, designing out the hazardous material could improve company performance at both ends. In 2001, Sony had to replace peripheral cables for 1.3 million Playstation consoles containing small amounts of cadmium when the Dutch government blocked their sale due to environmental concerns (Reuters, 2001).

Some companies are already starting to address such problems. Apple computers have latches and snap-on components that make them easy to disassemble. European and Japanese companies have made progress in phasing out persistent, bioaccumulative, and toxic chemicals (PBTs) in manufacturing. However, there's still a long way to go. Flat-panel LCD display screens perform better, are easier to read, are freer of potential electromagnetic bioeffects, and use less space, energy, and materials (including far less lead) than the older cathode-ray-tube (CRT) monitors, which are one of the main components of exported electronic waste. But the flat panels contain mercury in the fluorescent bulbs that provide screen backlighting, so they aren't the ultimate solution – yet.

2. *Sell information management – not hardware.* Nobody wants plastic, heavy metals, and silicon chips; they want fast, reliable, upgradeable information. Plenty of companies already have creative and lucrative leasing arrangements for office computers. Leasing encourages design for reuse and anticipates the cost of disposal – something that regulation and consumer pressure increasingly require anyway. The company that figures out the optimal way to deliver solutions rather than products and the best way of eliminating unwanted side-effects will gain significant competitive advantage.

our current environmental ills. They can also, over the next century, play a decisive role in halting and repairing some of the environmental degradation that is the legacy of the first Industrial Revolution. This may be because they anticipate the liability risks if they do not embrace sustainability, or because they see vast business opportunities. It is doubtful and perhaps naive to think that business can or will do this single-handedly; it will take the combined effort of government, business and civil society, and probably more than a little pressure in the form of shareholder activism, innovative regulation, and simply the fear of losing out to the competition. But the fastest working and most rewarding of these are seizing the business opportunity before competitors do.

The best environmental strategy is not (as many business schools teach only in elective courses) to manipulate regulation to disadvantage competitors; rather, as core-strategy courses *should* be teaching, it is making regulation relevant only to competitors, not to oneself. This means designing new industrial systems and business models that (as Ray C. Anderson, Chairman of Interface, Inc. states his ultimate goal) "take nothing, waste nothing, do no harm, and do very well by doing good." Tomorrow's business winners will be the firms that take their designs from nature, their values from their customers, and their discipline from the marketplace. Traditional environmental regulation may still be necessary in special cases, but it will become increasingly anachronistic, since the firms that need it will already be out of business, having spent too much money and time making things that nobody wants. Things that in the 20th century were called "wastes" and "emissions," should instead be known as "unsaleable production," because that term focuses our minds on asking: "Why are we producing things that nobody wants? Let's stop making them. Let's design them out." That fundamental redesign is the wellspring of world-class innovation.

NOTES

1 Respectively, the Resource Conservation and Recovery Act, Clean Air Act, Clean Water Act, Endangered Species Act, and Toxic Substances Control Act.

2 The Brundtland Report, produced in 1987, described sustainable development as "development that meets the needs of the present without compromising the ability of future generations to meet their own needs."

3 For further information on the product to service business model, see White *et al.* (1999).

REFERENCES

American Wind Energy Association (2004). Global Market Reports 2004. http://www.awea.org/pubs/documents/globalmarket2004.pdf

Anderson, Ray (2005). Personal Communication with Amory Lovins.

Balmford, Andrew *et al.* (2002): "Economic Reasons for Conserving Wild Nature," *Science* magazine, August 9.

Benedick, Richard Elliot (1998). *Ozone Diplomacy: New Directions in Safeguarding the Planet.* Cambridge: Harvard University Press.

Benyus, Janine (1997). *Biomimicry.* New York: Random House.

Browne, J. (2002). Speech to Stanford University Graduate School of Business. "Beyond Petroleum: Business and the Environment in the 21st Century," March.

Browne, J. (2004). "Beyond Kyoto," *Foreign Affairs* July/ August.

Burns, Cameron (2003). "Carbon Disclosure Project Looks at Businesses and Climate Change," *Energy Pulse*, July 29. http://www.energypulse.net

Canadian Driver (2004). "U.S. Hybrid Vehicle Sales Increase 25 percent in 2003," April 23. (see www.canadiandriver.com/news/040423-5.htm)

Casau, Armelle (2003). "When Pigs Stress Out," *New York Times,* 7 October.

Chemical Strategies Partnership (2004). "Chemical Management Services Industry Report 2004."

Costanza, R., d'Arge, R., de Groot, R. *et al.* (1977). "The Value of the World's Ecosystem Services and Natural Capital," *Nature* 387:253–260.

Cray, Charlie (2002). "Chartering a New Course: Revoking a Corporation's Right to Exist," *Multinational Monitor*, 23 (10, 11) http://www.multinationalmonitor.org/mm2002/02oct-nov/oct-nov02corp1.html

Ehrenfeld, J. and Chertow, M. 2001. "Industrial Symbiosis: The Legacy of Kalundborg". In *Handbook of Industrial Ecology* (eds) Ayres, R. and Ayres, L. Eheltenham, UK: Edward Elgar.

Eley Associates (1997). "Energy Performance Contracting for New Buildings." Prepared by Eley Associates for the Energy Foundation.

Energy Information Administration (2002). International Energy Annual, 2002. World Nuclear Electricity Installed Capacity, January 1, 1980–January 1, 2002. www.eia.doe.gov/iea/elec.html

European Wind Energy Association (2004). "Global Wind Power Growth Continues to Strengthen." Press Release, March 10.

Feder, Barnaby J. (2003), "Some Businesses Take Initiative to Voluntarily Reduce Emissions," *New York Times*, December 1.

Fishbein, B. K. McGarry, L. S., and Dillon P. S. (2000). "Leasing: A Step Towards Extended Producer Responsibility." Inform, Inc.

Frey, Darcy (2002). "How Green is BP?" *New York Times*, December 8.

Global Market reports, 2004. American Wind Energy Association. At http://www.awea.org/pubs/documents/globalmarket2004.pdf

Global Wind Energy council (2007). "Global Wind Energy Markets Continue to Boom – 2006 Another Record Year," Press release, February 2.

Hawken, P., Lovins, L., and Lovins, H.L. (1999). *National Capitalism: Creating the Next Industrial Revolution*. Boston, New York, London: Little Brown.

Interface Sustainability Report (2004). See www.interfacesustainability.com

International Energy Agency (2004). World Energy Outlook 2004. http://www.worldenergyoutlook.org/

Jones, A, Seville, D., and Meadows, D.L. (2002). "Resource Sustainability in Commodity Systems: The Sawmill Industry in the Northern Forest," *System Dynamics Review* 18(2), 171–204.

Leatherhead Food International (1998). "The European Organic Foods Market," September.

Linden, Eugene (2006). "Cloudy with a Chance of Chaos: Climate Change May Bring Violent Weather Swing – and Sooner than Experts Had Thought," *Fortune* Magazine, 17 January.

Lovins, A.B., Datta, E., Bustnes, O. *et al.* (2004). *Winning the Oil End Game: Innovation for Profits, Jobs, and Security*. Snowmass, CO: Rocky Mountain Institute.

McEwan, D., Coup, T., and McHugo, K. (2004). "Energy Efficiency Pays off for DB Staff," *EnergyWise News*, October, pp.17–22.

Nelson, G., Campbell, S., and Woznaik, P. (2002). Beyond *Earth Day: Fulfilling the Promise*. Madison, WI: University of Wisconsin Press.

New York Times (2004). "Catching Up on the Cost of Global Warming," July 25.

Nutter, F. (1996). The Insurers. In *Financing Change: The Financial Community, Eco-Efficiency, and Sustainable Development*. Schmidheiny, Stephan, and Zorraquin, Federico J.L. (eds). Cambridge, MA: The MIT Press.

Parrot, Kate and Ottman Jacquelyn (2003). "Sustainability and the Marketplace: Products, Markets and New Marketing Approaches Working Group." Society for Organizational Learning Sustainability Consortium. April 15 (see www.solonline.org).

Perrot-Maître, D. and Davis, P. (2001). "Case Studies of Markets and Innovative Financial Mechanisms for Water Services from Forests." From Forest Trends and the Katoomba Group (www.forest-trends.org), May.

Pomfret, Christopher (2002). "Can Sustainability Sell?" March 20, 2002 speech to Institute of Practitioners in Advertising. See http://www.rmi.org/sitepages/pid830.php

Puckett, J., Byster, L. Westervelt. S. *et al.* (2002). "Exporting Harm: The High-Tech Trashing of Asia." Basel Action Network and Silicon Valley Toxics Coalition.

Reilly, William K (1990). "Aiming Before We Shoot: The Quiet Revolution in Environmental policy." Address to the National Press Club, 26 September.

Reinhardt, F. (2000). *Down to Earth: Applying Business Principles to Environmental Management*. Cambridge, MA: Harvard Business School Press.

Reuters (2001). "Sony Swaps Playstation One Cables," Reuters Ltd, December 5.

Reuters (2004). "Companies Ignoring Social Woes Risk Stock Slide: UN." June 26.

Reuters Ltd. (2007). "Organic Food Sales Blossoming, Industry Says," Reuters, February 16.

Robbins, Jim (2001). "Engineers ask Nature for Design Advice," *New York Times*, December 11.

Social Invesment Forum (2003). "Report on Socially Responsible Investing Trends in the United States," SIF Industry Research Program, December, Washington, DC: SIF.

Stahel, Walter (2001). "From Design For Environment to Designing Sustainable Solutions," *Our Fragile World: Challenges and Opportunities for Sustainable Development*. EOLSS Publishers, p.1563.

Stahel, Walter (2002). Presentation, Collaborative Innovation for Sustainability Conference. Aspen, CO.

Swisher, J. (2001). Rocky Mountain Institute, personal communication. March 31.

Westervelt, R. (2003). "Chemical Management Services Gains Traction: New Industries, Offerings Expand Initiative." *Chemicalweek* (June 4 issue).

White, A., Stoughton, M. and Feng, L. (1999). "Servicizing: The Quiet Transition to Extended Producer Responsibility." Boston, MA: Tellus Institute for U.S. EPA.

Wisby, Gary (2004). "Environmentalism no Threat to Growth: Panelists. Chicago *Sun Times*, January 30.

World Resources Institute (WRI) and Sustainable Asset Management (2003). "Changing Drivers: the Impact of Climate Change on Competitiveness and Value Creation in the Automotive Industry."

World Energy Efficiency Association (1999). "Briefing Paper on Energy Services Companies." http://www.naesco.org/meminfo.htm – What is an ESCO.

Worrell, E., Laitner, J.A., Ruth, M., and Finman, H. (2003). "Productivity Benefits of Industrial Energy Efficiency Measures." 28 *Energy* 1081, September.

Wyatt, Nelson (2003). "Economy and Environment are Linked: Kennedy," CNews, Sunday, August 17.

Xerox Corporation (1997). "The environmental Call: What on Earth Are We Doing For Customers?" Xerox Environment, Health and Safety.

Xerox Environmental Health and Safety Report (2004). http://www.xerox.com/downloads/usa/en/e/ehs_2004_progress_report.pdf

Environmental Technologies

24

The Human Dimensions of Global Environmental Change

Thomas J. Wilbanks and
Patricia Romero-Lankao

INTRODUCTION

Global environmental change is a term that covers a wide range of processes and phenomena: from geological subsidence to industrial waste disposal, from climate change to land use change. In some cases, such as earthquakes and volcanoes, the causation seems to have little connection with human action, while in other cases the changes are direct results of human action, such as urbanization. But there are very few cases where global environmental change is of very broad interest unless it has implications for human societies and decisions.

In the contemporary world, with notable exceptions such as major natural disasters, most of the significant global environmental changes for human societies and economies reflect the hand of humanity as it transforms the "natural" world (Thomas, 1956; Turner *et al.*, 1990): extracting resources, emitting wastes, reshaping the earth's surface as we seek to control environmental processes that undermine our comfort, convenience, and productivity. As a result, the term "natural environment" refers to conditions that no longer exist on this Earth; for instance, most people in industrialized countries live their lives in very largely controlled "environments," where contact with "nature" is an aspect of recreation and tourism rather than a part of daily life, and even "nature preserves" have been affected by such human interventions as air pollution and recreational land use.

Reflecting this reality, human dimensions of global environmental change include three major categories: human *driving forces* that lead to environmental change, human *impacts* of environmental change, and human *responses* to environmental change. More recently, a fourth cross-cutting category has been added as well in some circles: *human decision support*, which links information about driving forces and impacts with decisions either to moderate driving forces or to reduce impacts.

Perhaps the most significant development of the past half-century has been a new recognition that human societies cannot take nature for granted. An egregious lack of sensitivity to environmental processes invariably leads sooner or later to consequences that are bad for society, either in terms of resource availability or human health (e.g. Carson, 1962; Hardin, 1968).

The sections that follow will briefly describe these four categories and their most salient subdimensions, in the light of new social perspectives that emphasize the nature–society balance rather than human dominance. These dimensions will then be illustrated by three cases: human settlements; economic growth and development; and governance, public policy, and society. Finally, human dimensions of global environmental change will be related to three fundamental challenges for the 21st century: sustainability, equity, and peace.

THE PRINCIPAL HUMAN DIMENSIONS OF GLOBAL ENVIRONMENTAL CHANGE

Human systems as driving forces for global environmental change

In the history of the Earth, human systems are relative newcomers as driving forces for large-scale environmental change. Early examples might include human roles in the disappearance of Quaternary mega-fauna and, much later, impacts of European colonization on biota (Crosby, 1986; Steffen et al., 2003). But human systems have emerged as a significant force in the dynamics of the global environment with the population and industrial growth of the past three centuries.

One key driver has been industrialization, supported by the development and diffusion of technologies and by widespread organizational innovations (Mumford, 1963; Grübler, 1994). Industrialization has threatened local, regional, and global environments through the emission of wastes, the overexploitation of natural resources, an introduction of hosts of new toxic substances, and profound modifications in biogeochemical cycles. On the other hand, two processes intrinsic to industrialization, technological change and affluence, have created capacities to solve near-term problems and design longer-term alternatives.

Demographic dynamics are another driver of environmental change. The world's population grew from 1.6 billion at the beginning of the 20th century to about 6 billion at the beginning of the 21st century (Demeny, 1990). Such an increase in human needs is linked to the number of people demanding food, housing, and other basics, whose satisfaction requires the use of natural resources and results in the emission of pollutants.

These factors operate differently in regions and local areas according to such other drivers as affluence, cultural context, and institutional settings. For instance, individuals in industrialized countries consume three times as much meat and eleven times as much gasoline as people living in developing countries (Steffen et al., 2003). Consumption in affluent countries spreads footprints across the world in its demands (and competition) for resources, its impacts on economic systems, and the cumulative effects of its associated waste production; and the resulting economic globalization relates to global environmental change in complex ways. For example, along with increasing resource extraction and in some regions reducing local control over environmental management decisions, in some areas multinational corporations are more sensitive to environmental and health issues because of the risk of liabilities in the event of disasters such as Bhopal.

An aspect of these changes is urbanization, globally and regionally, which accounts for less than two per cent of the Earth's surface (Lambin et al., 2001) but which is associated with a very large and growing share of basic human needs, resource consumption (drawn very largely from other areas), and waste production (often exported to other areas). Perhaps at least as strong a driver in some regions is that people and societies low on the development ladder often see environmental preservation as a lower priority than job and income creation (MA, 2005). Environmental responsibility can be seen as a luxury, unless it has direct connections with human health or other direct issues such as the sustainability of current livelihoods.

Technological change has encouraged consumption by providing attractive options, by increasing capacities to extract and transform resources, and by increasing the global reach of economic markets and information systems. Combined with a global shift toward a greater reliance on market forces, this process stimulates competition in which environmental management is often a secondary consideration. In some cases, it reduces interest in investing in environmental protection in order to keep production costs as low as possible. In others, regulatory interventions can make economic competition a powerful driver for environmental management, as in the case of requirements by certain European countries that imported products come from production systems that meet international ISO standards.

Finally, institutional change is a driving force in global environmental change, associated with changes in how decisions are made, options are developed and evaluated, and relationships are formalized in organizational structures. For instance, the scale of decision-making in government and business is in flux, affected by such developments as the information technology revolution. Institutions affecting environmental management include those related to economic growth and trade, although environmental goals are seldom addressed coherently (Wade, 1997; Gibbs, 2000; Varady et al., 2001). A second set of institutions is focused on particular nature–society issues with environmental management implications, such as the international CGIAR (Consultative Group on International Agricultural Research) centers which promoted a "Green Revolution" in global agriculture, where economic gains were in some cases accompanied by social and environmental costs (e.g. erosion and decreased fertility of soils, increased ineffectiveness of pesticides against pests) (Wright, 1986; Simonian, 1988). A third set explicitly addresses environmental issues, such as stratospheric ozone depletion, dumping of wastes in the North Sea, and dumping of radioactive wastes, where results have been influential, and issues related to the world's natural and cultural heritage, tropical timber, and many fisheries, where the results have generally been judged as less effective (Brown *et al.*, 1998; Victor *et al.*, 1998; Miles *et al.*, 2001).

Impacts of global environmental change on human systems

Because human well-being is related to environmental conditions, changes in the global environment affect human systems. Some of the effects are likely to be positive for certain areas, such as shifts in regional advantages for agricultural production or tourism. But, for understandable reasons, most of the literature on impacts concerns vulnerabilities to possible adverse consequences of environmental changes, especially those caused by human activities. Arguably, global environmental changes are affecting the quantity and quality of food, air and soil quality, and water resources in many areas, threatening human health and in fact jeopardizing the stability of the Earth system as a whole (MA, 2005).

Vulnerability to environmental change is related to *exposure* to changes such as in climate extremes or storms, *sensitivity* to such changes, and the *capacity to cope* with such changes (Clark *et al.*, 2000). These variables depend in part on socioeconomic and institutional context: for example, prior stressors, the distribution of resources within a population, and the institutions which mediate both exposures and coping with the stressors in question. In many cases, impacts of global-change processes such as climate change *"will fall disproportionately upon developing countries and the poor persons in all countries"* (IPCC, 2001, p. 12).

The most fundamental types of impacts are focused on basic human needs: food, water, shelter, health, security, and knowledge/skill development; and the most clearly defined examples are associated with extreme weather events and other natural disasters such as storms, fire, and drought. Food and water supply are salient examples. Food supply needs are growing with global population increases, affluence, and demands to assure nourishment for all of the world's people, including in such chronic problem areas as Africa. A number of analyses suggest that changes in the world's climate and ecosystems threaten sustained food production in many parts of the world (Parry *et al.*, 1999; IPCC, 2001; MA, 2005). Besides land-based systems, fisheries are impacted by human activities in coastal zones and in the oceans and by

possible changes in ocean temperature and circulation. Both the quantity and quality of water resources for meeting human needs can be affected by global environmental change, from land-use changes to pollutant discharges and river basin engineering. Projections of climate change indicate that the amount and intensity of precipitation will change in many regions over the next century, and severe storms and other weather events will change in their locations and intensities.

Another example of impacts is relationships between global environmental change and health conditions, from epidemics to environmental pollution associated with waste disposal. For instance, ultraviolet radiation related to the "ozone hole" has potential health impacts, and changes in temperatures and humidity levels in many areas due to climate change (especially in cities) are likely to have implications for the spatial and temporal distribution of pests and disease vectors (Epstein, 1999).

More generally, global environmental change can reshape regional comparative advantage for many kinds of economic activities, and therefore patterns of trade. Results can include impacts on an area's sense of place as its activity patterns change, and impacts on population distributions due to rural–urban and interregional patterns of migration.

Other patterns of migration may be more directly related to changes in environmental conditions, although it is often difficult to determine how environmental conditions interact with other forces such as economic and demographic dynamics or political instability. The famine in 1975–1985 in Ethiopia, for instance, was the result of diverse factors (e.g. insufficient distribution systems, governmental failure to take action, international trade factors, and insufficient response by donor countries). There is evidence, however, pointing to an important role in migratory flows of environmental problems such as severe reductions in food production, flooding, or droughts (Döös, 1997).

Political stability and instability may itself, in fact, relate to environmental conditions. The classic example in world history is the relationship between stable governance and settlement and control over water resources. Where water is scarce and water rights are contested, any changes in the global environment that affect water supply can be expected to provoke political disputes and conflicts.

Possible impacts of global environmental change are often studied with such tools as scenario-driven impact assessments, as well as with impact and vulnerability assessments. Scenarios, for example, have been applied to project implications of climate change in such sectors as agriculture, forestry, and water for ecological and human systems. Impact and vulnerability assessments generally focus on a particular geographic, social, or ecological entity, explore possible outcomes of projected changes, and consider factors that may shape the ability to respond to these possible outcomes.

Responses of human systems to global environmental change

Human systems respond to concerns about environmental changes in a variety of ways. Some responses arise from relatively localized bottom-up judgments about directions of change, while other responses reflect reactions to top-down signals, whether policy changes or market conditions.

In general, responses can either be directed at avoiding changes or at reducing the impacts of changes (sometimes called "mitigation" and "adaptation," respectively). In reducing the impacts of changes, responses can be aimed at either reducing the sensitivity of human systems to the changes (e.g. relocating activities from particularly vulnerable areas) or increasing the preparedness to deal with impacts if they are experienced.

In most cases, the literature on responses of human systems to global environmental change is more concerned with negative impacts of change than with positive impacts, although positive impacts are not only possible but also quite common: for instance, "carbon fertilization" of crops and forests associated with higher concentrations of carbon dioxide in the Earth's atmosphere. In many cases, one location's loss is another's gain; for example, shifts in the distribution of rainfall result in drier conditions in some areas but wetter conditions in others. Since impacts of environmental changes can mean both winners and losers, responses to observed or projected impacts can be diverse, and in some cases they can provoke controversy.

Where impacts are negative, Burton (1996) has suggested eight categories of adaptive responses: bearing loss, sharing loss, modifying the events, preventing the effects, changing land or resource use, changing location, doing research to increase options, and using education to encourage behavioral change. Choices of strategies from among this menu depend on such variables as institutional and financial capacities, who exercises control, and relationships with other kinds of social and economic processes under way.

In many connections, especially related to responses to global climate change, "adaptation," or coping with change, is distinguished from "mitigation," or avoiding change (Wilbanks et al., 2003). Mitigation, however, is essentially a form of adaptation: responding to concerns about impacts by seeking to reduce those impacts and their causes.

In general, responses are differentiated in several ways:

1 Voluntary (or "autonomous") or driven by policy interventions. Some responses occuring as individual or institutional decisions, observing environmental change or considering projections of change in a context of risk avoidance, act on their own. Other responses are encouraged or required by policy initiatives that might range from tax incentives to regulations (e.g. related to land use).
2 Individual or collective. Some responses are based on the decision-making of single entities, from individuals and families to organizations, while others depend on collaborative agreements among a number of entities.
3 Large scale or small scale. Some responses are large in geographical, temporal, or financial extent (such as global carbon emission trading); others are quite localized in time, space, and cost (such as a decision to move from a vulnerable location to a less vulnerable one).
4 Top down or bottom up. Responses may arise from actions by large institutions, affecting many embedded components, or they may arise from local agency, with a potential for larger cumulative effects.

In these senses, responding to concerns about climate change through the Kyoto Protocol of the United Nations Framework for Climate Change is an example of a policy intervention undertaken top down at a global scale through collective agreement among many of the world's more industrialized countries. By contrast, responding to concerns about climate change through self-motivated actions by individual cities or business firms is an example of voluntary, individual action, generally but not always small scale, undertaken from the bottom up.

Supporting decisions by human systems about global environmental change

A subcategory of human dimensions, linking information about driving forces and impacts with responses, is infrastructures and information systems for informing decisions about responses. Often referred to as "decision support" and related to the literature on decision-making as well as environmental change, this dimension is an innovative field for combining practice and knowledge. It includes systems for observing change, developing structures for understanding change and representing it in scenarios of possible futures, and communicating with decision-makers and other interested parties (often termed "stakeholders," meaning that they have a vested interest in decisions that might be reached). Ideally, it also includes feedbacks from users to decision support system developers (GEOSS, 2004), although in many cases these feedbacks are poorly developed in practice.

Issues in this field of human dimensions include processes for effective interactions between "experts" and stakeholders, human resource needs for useful linkages, and implications of emerging information technologies. In particular, where decision support systems are intended to provide social benefits, it is important to consider categories of benefits and metrics for documenting them in designing observational systems, which suggests a critical need for bottom-up information about user decision needs and priorities.

THREE ILLUSTRATIVE CASES

The human dimensions of global environmental change can be seen in a great many processes rooted in nature–society interactions. We consider three cases: human settlements, carbon, and climate change; economic growth and development; and governance and society.

Human settlements, carbon, and climate change

A human settlement is a complex dynamic system bringing together within its territory a host of linkages among socioeconomic, geopolitical, and environmental processes operating at diverse scales (including changes in market conditions, transformations in the carbon cycle, and climate change). Three issues illustrate how human dimensions work in these regards: bidirectionality, increasing urbanization, and multiple stresses operating at diverse scales in space and through time.

Interactions of human settlements with global environmental change are bidirectional. On the one hand, a very large proportion of human impacts on global environmental transformations such as climate change originates in human settlements, especially in urban areas. Greenhouse gas (GHG) emissions are driven by factors such as income, lifestyles, consumption patterns, and institutional settings, which depend in part on whether the settlement of reference is urban or rural, developed or poor (Romero-Lankao, 2004). For instance, cities (which are home to almost half the global population) consume about three quarters of the world's natural resources (including fossil fuels). The 744 largest cities, hosting 20% of the world's population, emit 32% of global CO_2 emissions. Again, the level of GHG emissions differs with the level of development of urban areas.

On the other hand, human settlements can be affected by the impacts of global environmental change in a wide variety of ways. Over history, human settlements have been resilient to variability in environmental conditions (drought, flooding) that are part of their normal experience. But when environmental changes are more extreme or persistent than that experience (e.g. hurricane Katrina), or when changes are not foreseen or adaptive capacity is limited, human settlements become vulnerable. In such cases, human settlements can be affected by climate variability and change through impacts on certain physical infrastructures, on the health of the population, or on the market demands for the products and services they offer. The vulnerability of human settlements tends to be a function of factors such as: (a) location (human settlements situated in coastal or riverine areas for instance are at most risk; (b) economy (settlements dependent on agriculture or tourism are more prone to risks); (c) level of development; and (d) institutional capacity (IPCC, 2001).

As mentioned, *urban* settlements are especially significant in a discussion of human dimensions of the carbon cycle as one instance of environmental change. Cities are centers of key activities (e.g. transportation) inducing transformations of the carbon cycle and the climate system. By demanding food, energy, and cement, and by producing wastes that must go somewhere, they create an ecological footprint extending to distant and remote places. Furthermore, they are targets of impacts of climate change (e.g. health effects of heat waves, or damages to urban infrastructure caused by flooding). At the same time, cities are centers of cultural opportunities and sometimes centers for explorations of changing lifestyles capable of inducing transformations in development that could contribute both to decarbonization of societies and economies and to enhancing adaptive capacities.

As in other instances of global environmental change (e.g. desertification), climate change is one of a set of possible stressors (e.g. continued urban growth, rural to urban migration, institutional limitations, and political instabilities) that may confront settlements, especially in less-developed areas. Industrialized countries and developing countries often differ in the mix of stresses due to characteristic differences in economic and human resources available to respond to the most salient stresses. As illustrated by the 2003 heatwave in Europe, industrialized countries' ability to adapt can be affected by interconnections among inadequate infrastructure (air conditioning), inappropriate or poorly prepared health services, social relationships (elderly were left alone while their families were on vacation), and the magnitude of climate variability, which exceeded experience with climate variability. Settlements in less developed countries and regions are constrained by another mix of factors: lack of income and assets, lack of access to public services, and inadequate institutional capacities to deal with significant impacts of global environmental change, particularly when changes are abrupt.

Economic growth and development

Connections between global environmental change and regional economic growth and development are at the heart of the concept of "sustainable development." The fundamental issue is that in most cases social and economic well-being depends in part on environmental conditions, from healthy conditions for life to the availability of environmental resources for economic products and services. If the environment deteriorates beyond some point, then economic development and human well-being suffer. In that kind of case, a lack of attention to environmental management eventually presents threats to development (Wilbanks, 1994).

Nature–society linkages are especially important in this regard in two ways. First, economic growth depends heavily on access to physical and chemical building-blocks drawn from the environment, such as energy sources, building materials, and nutrients in food. Because one of the most familiar types of global environmental change in this era is the depletion of many types of Earth resources, from oil and gas to copper, such changes present a challenge for sustained development.

Second, economic growth returns waste substances to the environment. In fact, one measure of the level of economic development is the volume of wastes generated by economic production and consumption. Perhaps the pre-eminent environmental policy issue for industrialized countries is how to handle this reality. One set of questions is, if the wastes should not be emitted to the air or water, where should they go? Another set of questions is about who is responsible, who watches, and who pays? Yet another has to do with potentials for technological change either to reduce the volume of waste flows or to develop ways to convert wastes into useful forms of energy and other resources.

Governance and society

Any discussion of human dimensions of global environmental change must be placed within the context of global trends toward democratic decision-making in place of top-down "command and control" decision-making. What this trend implies for environmental change is unclear. On the one hand, it can mean increased attention to environmental quality, especially as that relates to human health (in the past quarter-century, command and control political economies in Europe and Asia have been among the least sensitive to environmental quality). On the other hand, relatively localized democratic decision-making can give a higher priority to job and income creation than to environmental preservation, especially in lower-income areas. For instance, the Response Component of the Millennium Ecosystem Assessment (MA, 2005) reports considerable disagreement about relationships between ecosystem protection and human well-being: one group of experts confident that the relationship is quite positive, the other (mainly from developing areas) quite confident that the relationship is often negative.

Other foci for attention to human dimensions

Other cases are numerous and diverse. For instance, human driving forces, impacts, and responses to environmental change extend through systems for agriculture and food production, waste disposal, water resource management and use, health threats and management approaches, energy and materials supply and use, recreation and tourism, and financial and risk management in a number of connections. In many cases, a naturally varying environment is taken as a given and managed as needed for human well-being, based on extensive and often culturally rich historical experience.

When global environmental change results in departures from this experience at local and regional scales, familiar practices in linking society and nature face pressures to adapt. Where the needed adaptations go beyond gradual, incremental change to more substantial changes – reshaping institutions and their relationships, modifying patterns of cost and benefit, and provoking controversy and conflict – then human dimension issues become unusually, sometimes painfully salient.

CHALLENGES FOR THE 21st CENTURY

Putting these connections into focus, human dimensions of global environmental change have the potential to be profoundly important for such fundamental challenges to our time as sustainability, equity, and peace.

Sustainability

It is widely recognized that the sustainability of global and more localized economic growth depends on a balanced relationship with the environment, and moving toward more sustainable patterns of life on Earth without sacrificing comfort and convenience, while reducing economic gaps between the world's rich and poor, is one of the pre-eminent challenges of this century (NAS, 1999). For instance, analyses prepared for the 1992 United Nations Conference on Environment and Development in Rio de Janeiro suggested that energy services need to be increased by a factor of seven or eight if economic gaps between the "North and South" are to be reduced, meaning not only efficiency improvements but also massive increases in energy resource development, energy production, and energy consumption. Meeting this kind of need while keeping environmental impacts from the world's energy system no greater than at present is one of the great challenges of this century, and materials requirements are similar in magnitude and implications.

Other great challenges in achieving a sustainable nature–society balance include accelerating current trends in fertility reduction, coping with rapid urbanization in developing countries, reversing declining trends in agricultural production in Africa while sustaining historic trends elsewhere, restoring degraded ecosystems while conserving diversity elsewhere – all without sacrificing aspirations for economic development or turning away from democratic decision-making (NAS, 1999; MA, 2005), and reshaping resource-consumptive trends in consumption.

Many of the transitions that need to be well underway in the next half-century seem difficult to imagine under current conditions for decision-making and action. Most likely, progress will require a combination of technological development to produce substitutes and reduce waste generation and of increased reliance on relatively local adaptation and problem solving (e.g. Speth, 2002). A particular challenge to sustainability is abrupt changes in environmental or human system conditions, from abrupt climate change to epidemics to terrorism.

Equity

Equity is a crucial challenge related to global environmental change. World income inequalities are increasing. For instance, measured as a percentage of the core's GNP per capita, the GNP per capita of Sub-Saharan Africa fell from 5 in 1960 to 2 in 1999 (Wade, 2005, p. 21 and Table 1). This has two very direct environmental implications: the richest countries and sectors consume and emit a higher share of natural resources and pollutants, respectively; and the poorest regions and sectors are ill-equipped to deal with hurricanes, droughts, and other impacts of global

environmental change. More broadly, it poses other challenges as well: developing strategies to reshape patterns of consumption and production, and promoting policies aimed at enhancing more sustainable development pathways by the poor, and thereby reducing economic and social gaps.

Addressing equity concerns generally calls for promoting economic, social, and sustainable development actions in and for relatively disadvantaged countries, regions, and populations; diversifying economic production and upgrading available technologies for environmentally sustainable development; building social capital to enable effective actions; building institutional capacity; and combining empowerment with enhanced human and resource availability. The challenge is that such initiatives are immersed in processes such as economic globalization, political and institutional processes, and power-based relationships that can undermine attention to equity issues. In some cases, prospects for reducing gaps between the rich and the poor appear to depend on identifying strategies that do not imply major sacrifices on the part of those with political and economic power: that is, avoiding a need to make painful choices between rich and poor, environment and jobs, and participation and decisiveness.

Despite the challenges, progress is being made in many areas, often through parties other than large-scale government. For example, NGOs and certain international organizations are heavily involved in projects to enhance social participation in development projects, often working with local communities, and some successes are being seen in improving economic opportunities, health, education, and the quality of life of disadvantaged groups and areas without significantly undermining environmental management.

Peace

No social objective is more fundamental than peace: security against warfare, terrorism, and other threats to the survival of individuals, families, and societies. Historically, "environmental capital" – in the sense of environmental services or attributes that have value for achieving human objectives – has often been a focus of conflict, whether it takes the form of locational attributes such as points of control over transportation routes or of environmental resources such as water or coal. In other cases, environmental characteristics combined with human system changes – such as population growth – can induce motivations for expansion or migration.

Environmental conditions, therefore, have the potential to be a factor in sparking conflict when they are related to control over scarce environmental resources, when they contribute to social and political stresses that have multiple causes (e.g. in a search for whom to blame for health crises), or when they are a factor in encouraging human migration (rural–urban or interregional). The challenge for this century is to assure that environmental conditions are not roots of conflict and, conversely, that conflict does not seriously endanger the environment (e.g. use of nuclear devices).

SUMMARY

Global environmental change includes a wide variety of processes and phenomena, but these subjects of interest are increasingly interconnected with human society and its activities. Human agency is a powerful driving force for environmental change, especially as the world's population grows and its appetite for environmental goods and services grows even more rapidly. Human societies, economies, and institutions are impacted by environmental change, in some cases potentially very significantly, and responses to these impacts – or concerns about risks of them – in turn shape further changes. In these ways, the issue is not so much "human dimensions of global environmental change," which implies that environmental changes are the primary focus and human interactions are secondary, but complex relationships between human aspirations and decisions on the one hand and the environmental setting for "Spaceship Earth" on the other. Both are a part of a single living reality, and improving the understanding of dynamic connections is a high priority in assuring a sustainable future for both parts of the nature–society calculus.

REFERENCES

Brown Weiss, E., and H.K. Jacobson (eds) (1998). *Engaging Countries: Strengthening Compliance with International Environmental Accords.* Cambridge, MA: MIT Press.

Burton, I. (1996). "The Growth of Adaptation Capacity: Practice and Policy," in J. Smith *et al.* (eds) *Adapting to Climate Change: An International Perspective.* New York: Springer-Verlag, pp. 55–67.

Carson, R. (1962). *Silent Spring.* Boston: Houghton Mifflin.

Clark, W., J. Jaeger, R. Corell, R. Kasperson, J. McCarthy, D. Cash, S. Cohen, P. Desanker, N. Dickson, P. Epstein, D. Guston, J.M. Hall, C. Jaeger, A. Janetos, N. Leary, M. Levy, A. Luers, M. MacCracken, J. Melillo, R. Moss, J. Nigg, M. Parry, E. Parson, J. Ribot, H-J. Schellnhuber, D. Schrag, G. Seielstad, E. Shea, C. Vogel, and T. Wilbanks (2000). "Assessing Vulnerability to Global Environmental Risks," Discussion Paper 200–12, Environment and Natural Resources Program, Kennedy School of Government. Cambridge, MA: Harvard University Press.

Crosby, A. (1986). *Ecological Imperialism. The Biological Expansion of Europe.* Cambridge, UK: Cambridge University Press.

Demeny, P. (1990). "Population," in B. Turner II *et al.* (eds) *The Earth as Transformed by Human Action.* Cambridge: Cambridge University Press, pp. 41–54.

Döös, B. R. (1997). "Can Large-scale Environmental Migration Be Predicted?" *Global Environmental Change* 7 (1): 41–61.

Epstein, P. (1999). "Climate and Health," *Science* 285: 347–348.

GEOSS (Global Earth Observation System of Systems) (2004). *IEOS/GEOSS Implementation Issues.* Washington, DC: American Meterological Society, December.

Gibbs, D. (2000). "Ecological Modernization, Regional Economic Development, and Regional Development Agencies," *Geoforum* 31: 9–19.

Grübler, A. (1994). "Industrialization as a Historical Phenomenon," in R. Socolow, C. Andrews, F. Berkhout, and V. Thomas (eds) *Industrial Ecology and Global Change.* Cambridge: Cambridge University Press.

Hardin, G. (1968). "The Tragedy of the Commons," *Science* 162: 1243–1248.

IPCC (International Panel on Climate Change) (2001). *Climate Change 2001: Impacts, Adaptation, and Vulnerability.* Cambridge: Cambridge University Press.

Lambin. E., B. Turner, H. Geist, S. Agbola, A. Angelsen, J. Bruce, O. Coomes, R. Dirzo, G. Fischer, and C. Folke (2001). "The Causes of Land-use and Land-cover Change: Moving Beyond the Myths," *Global Environmental Change* 11: 261–269.

MA (2005). *Synthesis Report,* Millennium Ecosystem Assessment. Washington, DC: Island Press.

Miles, E.L., A. Underdal, S. Andresen, J. Wettestad, J.B. Skjærseth, and E.M. Carlin (eds) (2001). *Environmental Regime Effectiveness: Confronting Theory with Evidence.* Cambridge, MA: MIT Press.

Mumford, L. (1963). *Technics and Civilization.* New York: Harcourt Brace.

NAS (U.S. National Academy of Sciences) (1999). *Our Common Journey: A Transition Toward Sustainability.* Washington, DC: National Academy Press.

Parry, M., C. Rosenzweig, A. Iglesias, G. Fischer, and M. Livermore (1999). "Climate Change and World Food Security: A New Assessment," *Global Environmental Change* 9 (Suppl.1): S51–S67.

Romero-Lankao, P. (2004). "Pathways of Regional Development and the Carbon Cycle," in C. Field, and M. Raupach (eds) *Toward CO₂ Stabilization: Issues, Strategies, and Consequences.* Island Press Washington DC.

Simonian, L. (1988). "Pesticide Use in Mexico: Decades of Abuse," *The Ecologist* 18: 82–87.

Speth, J. (2002). "A New Green Regime: Attacking the Root Causes of Global Environmental Deterioration," *Environment* 44: 16–25.

Steffen W., A. Sanderson, P. Tyson, J. Jaeger, P. Matson, B. Moore III, F. Oldfield, K. Richardson, H-J. Schellnhuber, B. Turner II, and R. Wassson (eds) (2003). *Global Change and the Earth System: A Planet Under Pressure.* Berlin: IGBP Springer.

Thomas, W. L. (ed) (1956). *Man's Role in Changing the Face of the Earth.* Chicago: University of Chicago Press.

Turner, B. II, W. Clark, R. Kates, J. Richards, J. Mathews, and W. Meyer (eds) (1990). *The Earth as Transformed by Human Action.* Cambridge: Cambridge University Press.

Varady, R., P. Romero-Lankao, and K. Hankins (2001). "Managing Hazardous Materials Along the U.S.–Mexico Border," *Environment* 43: 22–37.

Victor, D.G., K. Raustiala, and E.B. Skolnikoff (eds) (1998). The Implementation and Effectiveness of International Environmental Commitments. Cambridge, MA: MIT Press.

Wade, R. (1997). "Greening the World Bank: the Struggle Over the Environment, 1970–1985," D. Kapur, J.P. Lewis, and R. Webb (eds) *The World Bank: Its First Half-Century.* Washington, DC: Brookings Institutions Press, pp. 611–734.

Wade, R. (2005). "Failing States and Cumulative Causation in the World System," *International Political Science Review* 26(1): 17–36.

Wilbanks, T. (1994). "'Sustainable Development' in Geographic Context," *Annals of Association of American Geographers* 84: 541–57.

Wilbanks, T., S. Kane, P. Leiby, R. Perlack, C. Settle, J. Shogren, and J. Smith (2003). "Possible Responses to Global Climate Change: Integrating Mitigation and Adaptation," *Environment* 45(5): 28–38.

Wright, A. (1986). "Rethinking the circle of poison: the politics of pesticide poisoning among Mexican Farmers," *Latin American Perspectives* 3: 26–59.

25

Healthy Environments

Howard Frumkin

WHAT IS ENVIRONMENTAL HEALTH?

The chapters of this *Handbook* define "environment" in many ways. Nonscientific sources such as *Webster's Dictionary* offer a useful general definition: "the circumstances, objects, or conditions by which one is surrounded." But *Webster's* offers a second, more intriguing definition of "environment:" "the complex of physical, chemical, and biotic factors (as climate, soil, and living things) that act upon an organism or an ecological community and ultimately determine its form and survival." Both definitions emphasize that the environment exists in reference to something else – presumably the creatures that occupy it. Key among those creatures – and central to most people's concern for the environment – is *Homo sapiens*.

In fact, members of the public care deeply about the impact of the environment on human health. A 1999 survey of 1234 randomly selected respondents across the USA asked about a range of public health attitudes and beliefs. When asked about environmental factors such as pollution and their relation to public health, 85% said that they considered environmental factors to be important determinants of health problems (and 38% considered them very important). Respondents identified numerous health problems as importantly affected by environmental factors, including sinus and allergy problems (54%), childhood asthma (54%), childhood cancer (39%), colds and influenza (35%), and birth defects (36%). When queried about specific environmental exposures, a high proportion of respondents opined that they had a "great deal" of impact on health; these exposures included contaminated drinking water (58%), toxic waste (56%), air pollution (53%), foods contaminated with bacteria

(53%), and pesticides in foods (47%) (Hearne *et al.*, 2000).

The human impact of environmental exposures must be considered broadly, in at least three respects. First, the environment affects people along many dimensions, including their medical status, psychological well-being, and even spirituality. This corresponds to the broad definition of health in the 1948 constitution of the World Health Organization: "A state of complete physical, mental, and social well-being and not merely the absence of disease or infirmity." Second, while environmental exposures can be toxic, the focus of much current scientific and popular attention, other environmental exposures can also be health promoting. Third, environmental exposures affect health on a range of spatial scales. Some environmental factors that affect health operate very locally, and the Environmental Health professionals who address them work on a local level; think of the restaurant and septic tank inspectors who work for the local health department, or the health and safety officer at a manufacturing facility. Other environmental factors affect health at a regional level, and the professionals who address these problems work at a larger spatial scale; think of the state and provincial officials responsible for air pollution or water pollution enforcement. At the global level, such problems as climate change require professional responses on a national and international scale. These are crafted by professionals in organizations such as the Intergovernmental Panel on Climate Change.

Environmental Health is defined in many ways (see Box 1). Some definitions make reference to the relationship between people and the environment, evoking an ecosystem concept, while others focus more narrowly on addressing particular environmental conditions. Some focus on abating

Box 1   Definitions of environmental health

World Health Organization: "Comprises those aspects of human health, including quality of life, that are determined by physical, chemical, biological, social and psychosocial factors in the environment. It also refers to the theory and practice of assessing, correcting, controlling, and preventing those factors in the environment that can potentially affect adversely the health of present and future generations."

Agency for Toxic Substances and Disease Registry: "Environmental Health is the branch of public health that protects against the effects of environmental hazards that can adversely affect health or the ecological balances essential to human health and environmental quality."

European Charter on Environment and Health: "Environmental Health comprises those aspects of human health and disease that are determined by factors in the environment. It also refers to the theory and practice of assessing and controlling factors in the environment that can potentially affect health. It includes both the direct pathological effects of chemicals, radiation and some biological agents, and the effects (often indirect) on health and well-being of the broad physical, psychological, social and aesthetic environment, which includes housing, urban developmental land use and transport."

National Center for Environmental Health: "Environmental Health is the discipline that focuses on the interrelationships between people and their environment, promotes human health and well-being, and fosters a safe and healthful environment."

Sources:
- For ATSDR and NCEH definitions: US DHHS, Environmental Health Policy Committee, Risk Communication and Education Subcommittee. An Ensemble of Definitions of Environmental Health. November, 1998. Available: http://web.health.gov/ enviornment/DefinitionsofEnvHealth/ehdef2.htm
- For the WHO definition: WHO, http://www.who.int/phe/en/
- For the European Charter definition: Environment and Health: The European Charter and Commentary. WHO Regional Office for Europe, Copenhagen, 1990. WHO Regional Publications European Series No. 35.

hazards, while others focus on promoting health-enhancing environments. Some focus on physical and chemical hazards, while others extend more broadly to aspects of the social and built environments. In the aggregate, the definitions in Box 1 make clear that Environmental Health is many things: an interdisciplinary academic field, an area of research, and an arena of applied public health practice.

THE EVOLUTION OF ENVIRONMENTAL HEALTH

Ancient origins

The notion that the environment can have an impact on comfort and well-being – the core proposition of Environmental Health – must have emerged in the early days of human existence. The elements can be harsh, and our ancestors sought shelter in caves, or under trees, or in crude shelters they built. The elements can still be harsh, both on a daily basis and during extraordinary events, as the European heatwave of 2003, the Asian tsunamis of 2004, and the American hurricanes of 2005 reminded us.

Our ancestors confronted other challenges that we would now identify with Environmental Health. One was food safety; there must have been procedures for preserving food, and people must have fallen ill and died from eating spoiled food. Dietary restrictions in ancient Jewish and Islamic law, such as bans on eating pork, presumably evolved from the recognition that certain foods could cause disease. Another challenge was clean water; we can assume that early peoples learned not to defecate near or otherwise soil their water sources. In the ruins of ancient civilizations from India to Rome, from Greece to Egypt to South America, archeologists have found the remains of water pipes, toilets, and sewage lines, some dating back more than 4000 years (Rosen, 1958). Still another environmental hazard was polluted air; the sinus cavities of ancient cave dwellers show evidence of high levels of smoke in the caves (Brimblecombe, 1988), foreshadowing modern indoor air concerns in homes that burn biomass fuels or coal.

An intriguing passage in the Biblical book of Leviticus (14:43–46) may refer to an environmental health problem well recognized today: mold in buildings. When a house has a "leprous disease" (as it is usually translated),

…then he who owns the house shall come and tell the priest, "There seems to me to be some sort of disease in my house." Then the priest shall command that they empty the house before the priest goes to examine the disease, lest all that is in the house be declared unclean; and afterward the priest shall go in to see the house. And he shall examine the disease; and if the disease is in the walls of the house with greenish or reddish spots, and if it appears to be deeper than the surface, then the priest shall go out of the house to the door of the house, and shut up the house seven days. And the priest shall come again on the seventh day, and look; and if the disease has spread in the walls of the house, then the priest shall command that they take out the stones in which is the disease and throw them into an unclean place outside the city; and he shall cause the inside of the house to be scraped round about, and the plaster that they scrape off they shall pour into an unclean place outside the city; then they shall take other stones and put them in the place of those stones, and he shall take other plaster and plaster the house. If the disease breaks out again in the house, after he has taken out the stones and scraped the house and plastered it, then the priest shall go and look; and if the disease has spread in the house, it is a malignant leprosy in the house; it is unclean. And he shall break down the house, its stones and timber and all the plaster of the house; and he shall carry them forth out of the city to an unclean place.

This passage raises the interesting possibility that ancient dwellings suffered mold overgrowth. It also implies that the "unclean place outside the city" was an early hazardous waste site. Who hauled waste materials there, and what did that work do to their health?

Still another ancient environmental health challenge, especially in cities, was rodents. European history was changed forever when infestations of rats in 14th century cities led to the Black Death (Herlihy and Cohn, 1997; Cantor, 2001; Kelly, 2005). Modern cities continue to struggle periodically with infestations of rats and other pests, whose control depends in large part on environmental modifications.

Industrial awakenings

Modern Environmental Health further took form during the age of industrialization. With the rapid growth of cities in the 17th and 18th centuries, "sanitarian" issues rose in importance. *"The urban environment,"* wrote one historian,

> fostered the spread of diseases with crowded, dark, unventilated housing; unpaved streets mired in horse manure and littered with refuse; inadequate or nonexisting water supplies; privy vaults unemptied from one year to the next; stagnant pools of water; ill-functioning open sewers; stench beyond the twentieth-century imagination; and noises from clacking horse hooves, wooden wagon wheels, street railways, and unmuffled industrial machinery (Leavitt, 1982, p. 22).

The provision of clean water became an ever more pressing need, as greater concentrations of people increased both the probability of water contamination and the impact of disease outbreaks. Regular outbreaks of cholera and yellow fever in the 18th and 19th centuries highlighted the need for water systems, including clean source water, treatment including filtration, and distribution through pipes. Similarly, sewage management became a pressing need, especially after the provision of piped water and the use of toilets created large volumes of contaminated liquid waste (Duffy, 1990).

The industrial workplace – a place of danger and even horror – gave additional impetus to early Environmental Health. Technology advanced rapidly during the late 18th and 19th centuries, new and often dangerous machines were deployed in industry after industry, and mass production became common. The conditions in the mills and factories were often abominable.

Charles Turner Thackrah (1795–1833), a Yorkshire physician, developed an interest in the diseases he observed among the poor in the city of Leeds. He described many work-related hazards in a short 1831 book with a long title: *The Effects of the Principal Arts, Trades and Professions, and of Civic States and Habits of Living, on Health and Longevity, with Suggestions for the Removal of many of the Agents which Produce Disease and Shorten the Duration of Life.* In it he proposed guidelines for the prevention of certain diseases, such as the elimination of lead as a glaze in the pottery industry and the use of ventilation and respiratory protection to protect knife grinders. Public outcry, and the efforts of early Victorian reformers such as Thackrah, led to passage of the Factory Act in 1833 and the Mines Act in 1842. Occupational health did not blossom in the USA until the early 20th century, pioneered by the remarkable Dr. Alice Hamilton (1869–1970). A keen first-hand observer of industrial conditions, she documented links between toxic exposures and illness among miners, tradesmen, and factory workers, first in Illinois (where she directed that state's Occupational Disease Commission from 1910 to 1919) and later from an academic position at Harvard. Her books, including *Industrial Poisons in the United States* (1925) and *Industrial Toxicology* (1934) helped establish that workplaces could be microenvironments that threatened worker health.

A key development in the 17th to 19th centuries was the quantitative observation of population health – the beginnings of epidemiology. With the tools of epidemiology, observers could systematically attribute certain diseases to certain environmental exposures. John Graunt (1620–1674), an English merchant and haberdasher, analyzed

London's weekly death records – the "Bills of Mortality" – and published his findings in 1662 as *Natural and Political Observations Upon the Bills of Mortality*. Graunt's work was one of the first formal analyses of this data source, and a pioneering example of demography. Almost two centuries later, when the English Parliament created the Registrar-General's Office (now the Office of Population Censuses and Surveys) and William Farr (1807–1883) became its "Compiler of Abstracts," the link between vital statistics and Environmental Health was forged. Farr made observations about fertility and mortality patterns, identifying rural–urban differences, variations between acute and chronic illnesses, and seasonal trends, and implicating certain environmental conditions in illness and death. Farr's 1843 analysis of mortality in Liverpool led Parliament to pass the Liverpool Sanitary Act of 1846, which created a sanitary code for Liverpool and a public health infrastructure to enforce it.

If Farr was a pioneer in applying demography to public health, his contemporary Edwin Chadwick (1800–1890) was a pioneer in combining social epidemiology with Environmental Health. At the age of 32 Chadwick was appointed to the newly formed Royal Commission of Enquiry on the Poor Laws, and helped reform Britain's Poor Laws. Five years later, following epidemics of typhoid fever and influenza, he was asked by the British government to investigate sanitation. His classic report, *Sanitary Conditions of the Labouring Population* (1842), drew a clear link between living conditions – in particular overcrowded, filthy homes, open cesspools and privies, impure water, and miasmas – and health, and made a strong case for public health reform. The resulting Public Health Act of 1848 created a Central Board of Health, with power to empanel local boards that would oversee street cleaning, trash collection, and water and sewer systems. As Sanitation Commissioner, Chadwick advocated such innovations as urban water systems, toilets in every house, and transfer of sewage to outlying farms where it could be used as fertilizer (Hamlin, 1998). Chadwick's work helped establish the role of public works – essentially applications of sanitary engineering – to protecting public health. As eloquently pointed out by Thomas McKeown more than a century later, these interventions were to do far more than medical care to improve public health and well-being during the industrial era (McKeown, 1979).

Doctor John Snow (1813–1858) was, like William Farr, a founding member of the London Epidemiological Society. Snow gained immortality in the history of public health for what was essentially an environmental epidemiology study. During an 1854 outbreak of cholera in London, he observed a far higher incidence of disease among people who lived near or drank from the Broad Street pump than among people with other sources of water. He persuaded local authorities to remove the pump handle, and the epidemic in that part of the city soon abated. (There is some evidence that it may have been ending anyway, but this does not diminish the soundness of Snow's approach.) Environmental epidemiology was to blossom during the 20th century and provide some of the most important evidence needed to support effective preventive measures.

Finally, the industrial era led to a powerful reaction in the worlds of literature, art, and design. In the first half of the 19th century, Romantic painters, poets, and philosophers celebrated the divine and inspiring forms of nature. In Germany, painters such as Caspar David Friedrich (1774–1840) created meticulous images of the trees, hills, misty valleys, and mercurial light of northern Germany, based on a close observation of nature, and in England, Samuel Palmer (1805–1881) painted landscapes that combined straightforward representation of nature with religious vision. His countryman John Constable (1776–1837) worked in the open air, painting deeply evocative English landscapes. In the USA, Hudson River School painters such as Thomas Cole (1801–1848) took their inspiration from the soaring peaks and crags, stately waterfalls, and primeval forests of the northeast.

At the same time, the New England transcendentalists celebrated the wonders of nature. "Nature never wears a mean appearance," wrote Ralph Waldo Emerson (1803–1882) in his 1836 paean, *Nature*. "Neither does the wisest man extort her secret, and lose his curiosity by finding out all her perfection. Nature never became a toy to a wise spirit. The flowers, the animals, the mountains, reflected the wisdom of his best hour, as much as they had delighted the simplicity of his childhood." Henry David Thoreau (1817–1862), like Emerson a native of Concord, Massachusetts, rambled from Maine to Cape Cod and famously lived in a small cabin at Walden Pond for two years, experiences that cemented his belief in the "tonic of wildness." And America's greatest landscape architect, Frederick Law Olmsted (1822–1903), championed bringing nature into cities. He designed parks that offered pastoral vistas and graceful tree-lined streets and paths, intending to offer tranquility to harried people and to promote feelings of community. These and other strands of cultural life reflected yet another sense of "Environmental Health," forged in response to industrialization: the idea that pristine environments were wholesome, healthful, and restorative to the human spirit.

The modern era of environmental health

The modern field of Environmental Health dates from the mid-20th century, and perhaps no landmark better marks its launch than the 1962 publication of Rachel Carson's *Silent Spring*. *Silent Spring* focused on DDT, an organochlorine pesticide that had seen increasingly wide use since the Second World War. Carson had become alarmed at the ecosystem effects of DDT; she described how it entered the food chain and accumulated in the fatty tissues of animals, how it indiscriminately killed both target species and other creatures, and how its effects persisted for long periods after it was applied. She also made the link to human health, describing how DDT might increase the risk of cancer and birth defects. One of Carson's lasting contributions was to place human health in the context of larger environmental processes. "Man's attitude toward nature," she declared in 1964, "is today critically important simply because we have now acquired a fateful power to alter and destroy nature. But man is a part of nature, and his war against nature is inevitably a war against himself…[We are] challenged as mankind has never been challenged before to prove our maturity and our mastery, not of nature, but of ourselves."

The *recognition of chemical hazards* was perhaps the most direct legacy of *Silent Spring*. Beginning in the 1960s Dr. Irving Selikoff (1915–1992) and colleagues at the Mount Sinai School of Medicine intensively studied insulation workers and other occupational groups, and showed that asbestos could cause a fibrosing lung disease, lung cancer, mesothelioma, and other neoplasms. Outbreaks of cancer in industrial workplaces – lung cancer in a chemical plant near Philadelphia due to bis-chloromethyl ether (Figueroa *et al.*, 1973; Randall, 1977; hepatic hemangiosarcoma in a vinyl chloride polymerization plant in Louisville (Creech and Johnson, 1974); and others – underlined the risk of carcinogenic chemicals. With the enormous expansion of cancer research, and with effective advocacy by such groups as the American Cancer Society (Patterson, 1987), environmental and occupational carcinogens became a focus of public, scientific, and regulatory attention (Epstein, 1982).

But cancer was not the only health effect linked to chemical exposures. Dr. Herbert Needleman (1927–), studying children in Boston, Philadelphia, and Pittsburgh, showed that lead was toxic to the developing nervous system, causing cognitive and behavioral deficits at levels far lower than had been appreciated. When this recognition finally helped achieve the removal of lead from gasoline, population blood lead levels plummeted, an enduring public health victory. Research also suggested that chemical exposures could threaten reproductive function. Wildlife observations such as abnormal genitalia in alligators in Lake Apopka, Florida, following a pesticide spill (Guillette *et al.*, 1994), and human observations such as an apparent decrease in sperm counts (Carlsen *et al.*, 1992; Swan et al., 1997), suggested that certain persistent, bioaccumulative compounds (persistent organic pollutants, or POPs) could affect reproduction, perhaps by interfering with hormonal function. Emerging evidence showed that chemicals could damage the kidneys, the liver, the cardiovascular system, immune function, and organ development.

Some knowledge of chemical toxicity arose from toxicologic research in laboratories, and other insights resulted from long-term epidemiologic studies. But catastrophes – reported first in newspaper headlines and only later in scientific journals – also galvanized public and scientific attention. The discovery of accumulations of hazardous wastes in communities across the USA – Love Canal in Niagara Falls, New York (Gibbs, 1998; Mazur, 1998); Times Beach, Missouri, famous for its unprecedented dioxin levels; Toms River, New Jersey, and Woburn, Massachusetts, where municipal drinking water was contaminated with organic chemicals; "Mount Dioxin," a defunct wood-treatment plant in Pensacola, Florida; and others – raised concerns about many health problems, from nonspecific symptoms to immune dysfunction to cancer to birth defects. And acute disasters such as the isocyanate release that killed hundreds and sickened thousands in Bhopal, India, in 1984, made clear that industrialization posed real threats of large-scale, acute chemical toxicity (Kurzman, 1987; Dhara and Dhara, 2002; Moro and Lapierre, 2002).

Even as the awareness of chemical hazards grew, supported by advances in toxicology and epidemiology, environmental health during the second half of the 20th century was developing in a different direction altogether: *environmental psychology*. This field arose as a subspecialty of psychology, building on advances in perceptual and cognitive psychology. Scholars such as Stephen and Rachel Kaplan (1982a, 1982b, 1989) at the University of Michigan carried out careful studies of human perceptions of and reactions to various environments. An important contribution to environmental psychology was the theory of *biophilia*, first advanced by Harvard biologist E.O. Wilson in 1984 (Wilson, 1984).

Wilson defined biophilia as "the innately emotional affiliation of human beings to other living organisms." He pointed out that for most of human existence, our ancestors had lived in natural settings, interacting daily with plants, trees, and other animals. As a result, Wilson maintained, the affiliation with these organisms had become an innate part of human nature. Other scholars extended Wilson's concept beyond living organisms, postulating a connection with other features of the natural environment – rivers, lakes, and ocean shores, waterfalls, panoramic landscapes, and mountain vistas (Kellert and Wilson, 1993; Kellert, 1997). Environmental psychologists studied not only natural features of human environments, but also such factors as light, noise, and way-finding cues, to assess their impact. They increasingly recognized that people responded to various environments, both natural and built, in predictable ways. Some environments were alienating, disorientating, or even sickening, while others were attractive, restorative, and even salubrious.

A third development in modern Environmental Health is the continued *integration of ecology with human health*. Ancient wisdom in many cultures had recognized the interrelationships of the natural world and human health and well-being. But with the emergence of formal complex systems analysis and modern ecological science, the understanding of ecosystem function advanced greatly. As part of this advance, the role of humans in the context of ecosystems was better and better delineated. On a global scale, for example, the concept of carrying capacity (Rees and Wackernagel, 1995) helped clarify the impact of human activity on ecosystems, and permitted evaluation of how ecosystem changes in turn affected human health and well-being (Aron and Patz, 2001; McMichael, 2001; Millenium Ecosystem Assessment, 2003; Waltner-Toews, 2004; Brown *et al.*, 2005). Ecological analysis was also applied to specific areas relevant to human health. For example, there were advances in medical botany (Lewis and Elvin-Lewis, 2003; van Wyk and Wink, 2004), in medical geology (Skinner and Berger, 2003; Selinus *et al.*, 2005) in the understanding of biodiversity and its value to human health (Grifo and Rosenthal, 1997), and in the application of ecology to clinical medicine (Aguirre *et al.*, 2002; Ausubel and Harpignies, 2004). These developments, together, reflected a progressive synthesis of ecological and human health science, yielding a better understanding of the foundations of Environmental Health.

A fourth feature of modern Environmental Health was the expansion of *clinical services* related to environmental exposures. Occupational medicine and nursing had been specialties within their respective professions since the early 20th century, with a traditional focus on returning injured and ill workers to work and, to some extent, on preventing hazardous workplace exposures. In the last few decades of the 20th century, these professional specialties incorporated a public

health paradigm, drawing on toxicologic and epidemiologic data, utilizing industrial hygiene and other primary prevention approaches, and engaging in worker education. In addition, the occupational health clinical paradigm was broadened to general environmental exposures. Clinicians began focusing on community exposures such as air pollutants, radon, asbestos, and hazardous wastes, emphasizing the importance of taking an environmental history, identifying at-risk groups, and providing both treatment and preventive advice to patients. Professional ethics expanded to recognize the interests of patients (both workers and community members) as well as those of employers, and in come cases even those of unborn generations and of other species. Finally, a wide range of alternative and complementary approaches arose in occupational and environmental health care. For example, an approach known as "clinical ecology" postulated that overloads of environmental exposures could impair immune function, and offered treatments including "detoxification," antifungal medications, and dietary changes purported to prevent or ameliorate the effects of environmental exposures (Randolph, 1976, 1987; Rea, 1992–1998).

Environmental Health policy also emerged rapidly. With the promulgation of environmental laws beginning in the 1960s, national, state, and provincial governments created agencies and assigned them new regulatory responsibilities. These agencies issued rules that aimed to reduce emissions from smokestacks, drainpipes, and tailpipes, control hazardous wastes, and achieve clean air and water. While many of these laws were oriented to environmental preservation, the protection of human health was often an explicit rationale as well. Ironically, the new environmental regulations created a schism in the Environmental Health field. Responsibility for Environmental Health regulation had traditionally belonged to health departments, but these were now transferred to the new environmental departments. In the USA, the Environmental Protection Agency (EPA) assumed some of the traditional responsibilities of the Department of Health, Education, and Welfare (now Health and Human Services), and corresponding changes occurred at the state level. Environmental regulation and health protection became somewhat estranged from each other.

Environmental regulatory agencies increasingly attempted to ground their rules in evidence, using quantitative risk assessment techniques (National Research Council, 1983). This signaled a sea change in regulatory policy. The traditional approach had been simpler; dangerous exposures were simply banned. For example, the 1958 Delaney clause, an amendment to the 1954 Federal Food, Drug and Cosmetic Act, banned carcinogens in food. In contrast, emerging regulations tended to set permissible exposure levels that took into account anticipated health burdens, compliance costs, and technological feasibility. Perhaps in response to the technical and ethical limits of risk assessment, the Precautionary Principle – *Vorsorgeprinzip*, or "forecaring principle"– emerged in European environmental policy in the 1970s. This principle has been expressed in many ways. The Rio Declaration of 1992 declared that lack of "full scientific certainty shall not be used as a reason for postponing cost-effective measures to prevent environmental degradation," and the 1998 Wingspread Statement on the Precautionary Principle made direct reference to human health:

> When an activity raises threats of harm to human health or the environment, precautionary measures should be taken even if some cause and effect relationships are not fully established scientifically. In this context the proponent of an activity, rather than the public, should bear the burden of proof. The process of applying the precautionary principle must be open, informed and democratic and must include potentially affected parties. It must also involve an examination of the full range of alternatives, including no action.

While the Precautionary Principle has not been widely adopted in the USA, it became the basis for European environmental law in the 1992 Treaty on European Union and has been implemented in the national policy of many other countries (Raffensberger and Tickner, 1999; Foster *et al.*, 2000; Harremoes *et al.*, 2002).

At the dawn of the 21st century, then, the Environmental Health field had moved well beyond its traditional sanitarian functions. Awareness of chemical toxicity had advanced rapidly, fueled by discoveries in toxicology and epidemiology. At the same time, the complex relationships inherent in environmental health – the effects of environmental conditions on human psychology, and the links between human health and ecosystem function – were better and better recognized. In practical terms, clinical services in Environmental Health had developed, and regulation had advanced through a combination of political action and scientific evidence.

EMERGING ISSUES IN ENVIRONMENTAL HEALTH

Environmental Health is a dynamic, evolving field. As the 21st century unfolds, traditional sanitarian functions remain keenly important, and chemical hazards will continue to be a focus of scientific and regulatory attention. Looking ahead, we can identify at least five trends that will further shape Environmental Health: environmental justice, a focus on susceptible groups, scientific advances, global change, and moves toward sustainability.

Beginning around 1980, African-American communities in the USA identified exposures to hazardous waste and industrial emissions as matters of racial and economic justice. Researchers documented that these exposures disproportionately affected poor and minority communities, a problem that was aggravated by disparities in the enforcement of environmental regulations. The modern *Environmental Justice* movement was born, a fusion of environmentalism, public health, and the civil rights movement (Bullard, 1994; Cole and Foster, 2000). Historians have observed that Environmental Justice represents a profound shift in the history of environmentalism (Gottlieb, 1993; Shabecoff, 1993; Dowie, 1995). In the USA, this history is commonly divided into "waves." The first wave was the conservation movement of the early 20th century, the second wave was the militant activism that blossomed on Earth Day, 1970, and the third wave was the emergence of large, "inside-the-beltway" environmental organizations such as the Sierra Club, the League of Conservation Voters, and the Natural Resources Defense Council, which gained considerable polity influence by the 1980s.

Environmental Justice, then, represents a fourth wave, one that is distinguished by its decentralized, grass-roots leadership, its demographic diversity, and its emphasis on human rights and justice. Moreover, the principles of Environmental Justice have been applied to global environmental health challenges, from resource extraction (Douglas *et al.*, 2005) to hazardous waste transport (Marbury, 1995) to biodiversity conservation (Wells, 1992), and it is clear that inequities in exposures, access to remedies, and health impacts, vary across ethnic, racial, and economic groups across the world (Kothari, 1996; Bullard *et al.*, 2005). The vision of Environmental Justice – eliminating disparities in economic opportunity, environmental exposures, and health – is one that resonates with public health priorities. It emphasizes that Environmental Health extends well beyond technical solutions to hazardous exposures, to include human rights and equity as well. It is likely that this vision will be an increasingly central part of Environmental Health in coming decades.

Environmental Justice is one example of a broader trend in Environmental Health – a focus on *susceptible groups*. For many reasons, specific groups may be especially vulnerable to the adverse health effects of environmental exposures. In the case of poor and minority populations, these reasons include disproportionate exposures, limited access to legal protection, limited access to health care, and in some cases compromised baseline health status. Children comprise another susceptible population, for several reasons. They eat more food, drink more water, and breathe more air per unit of body weight, and are therefore heavily exposed to any contaminants in these media. Children's behavior – crawling on floors, placing their hands in their mouths, and so on – further increases their risk of exposure. With developing organ systems and immature biological defenses, children are less able than adults to withstand some exposures. And with more years of life ahead of them, children have more time to manifest delayed toxic reactions.

These facts have formed the basis for research and public health action on children's environmental health. Women bear some specific environmental exposures risks, both in the workplace and in the general environment, due both to disproportionate exposures (e.g. in health care jobs) and to unique susceptibilities (e.g. to reproductive hazards). Elderly people also bear some specific risks, and as the population ages, this group will attract further Environmental Health attention. For example, urban environments will need to take into account the limited mobility of some elderly people, and provide ample sidewalks, safe street crossings, and accessible gathering places, to serve the elderly. People with disabilities, too, require specific Environmental Health attention, to minimize the risks they face. In coming decades, Environmental Health will increasingly take account of susceptible groups, as the risks they face and their needs for safe, healthy environments become better recognized.

A third set of emerging issues in Environmental Health is introduced by *scientific advances*. In toxicology, better detection techniques have already enabled us to recognize and quantify low levels of chemical exposure, and have supported major advances in the understanding of chemical effects. Advances in data analytic techniques have supported innovative epidemiologic analyses and the use of large databases. In particular, the use of Geographic Information Systems (GIS) has yielded new insights regarding spatial distribution of environmental exposures and diseases. Perhaps the most promising scientific advances are occurring at the molecular level, in the linked fields of genomics, toxicogenomics, and proteiomics (Schmidt, 2003; Mattes *et al.*, 2004; Pesch *et al.*, 2004; Pognan, 2004; Waters and Fostel, 2004). New genomic tools such as microarrays (or "gene chips") have enabled scientists to characterize the effects of chemical exposures on the expression of thousands of genes. Databases of genetic responses, and resulting protein and metabolic pathways, will yield much information on the effects of

chemicals, and on the variability in responses among different people. Scientific advances in Environmental Health will have profound effects on the field in coming decades.

Moving from the molecular scale to the global scale, a fourth set of emerging issues in Environmental Health relates to *global change*. This broad term has many components, including population growth, climate change, urbanization, and the increasing integration of the world economy. These trends will shape Environmental Health in many ways.

The world population is currently just over 6 billion, and is expected to plateau at something like 8.5 to 9 billion during the 21st century. Most of this population growth will occur in developing nations, and much of it will be in cities. Not only the population growth, but also the increasing per capita demand for resources such as food, energy, and materials, will strain the global environment, in turn affecting health in many ways. For example, environmental stress and resource scarcity may increasingly trigger armed conflict, an ominous example of the links between environment and health (Homer-Dixon, 1999; Klare, 2001). Global climate change, which results in large part from increasing energy use, will threaten health in many ways, from infectious disease risks to heatwaves to severe weather events. As more of the world's population is concentrated in dense urban areas, features of the urban environment – noise, crowding, vehicular and industrial pollution – will come to be important determinants of health (UN Centre on Human Settlements, 2001). And with integration of the global economy – the complex changes known as "globalization" – hazards will cross national boundaries (Ives, 1985), trade agreements and market forces will challenge and possibly undermine national Environmental Health policies (Low, 1992; Sand, 1992; Runge, 1994; Brack, 1998; Victor *et al.*, 1998), and global solutions to Environmental Health challenges will increasingly be needed.

Sustainability has been a part of the vernacular since the 1980s. In 1983, the United Nations formed the World Commission on Environment and Development to propose strategies for sustainable development. The Commission, chaired by Norwegian Prime Minister Gro Harlem Brundtland, issued its report, *Our Common Future*, in 1987. The report included what has become a standard definition of sustainable development: "development that meets the needs of the present without compromising the ability of future generations to meet their own needs." Several years after the publication of *Our Common Future*, the United Nations Conference on Environment and Development (UNCED), commonly known as the "Earth Summit," convened in Rio de Janeiro. This landmark conference produced, among other documents, the Rio Declaration on Environment and Development, a blueprint for sustainable development. The first principle of the Rio Declaration placed Environmental Health at the core of sustainable development: "Human beings are at the centre of concerns for sustainable development. They are entitled to a healthy and productive life in harmony with nature."

Like environmental justice, the concept of sustainable development blends environmental protection with notions of fairness and equity. As explained on the web site of the Johannesburg Summit, held ten years after the Earth Summit,

> The Earth Summit thus made history by bringing global attention to the understanding, new at the time, that the planet's environmental problems were intimately linked to economic conditions and problems of social justice. It showed that social, environmental and economic needs must be met in balance with each other for sustainable outcomes in the long term. It showed that if people are poor, and national economies are weak, the environment suffers; if the environment is abused and resources are over consumed, people suffer and economies decline. The conference also pointed out that the smallest local actions or decisions, good or bad, have potential worldwide repercussions (http://www.johannesburgsummit. org/html/basic_info/unced.html).

The concept of sustainability has emerged as a central theme, and challenge, not only for environmentalism but also for Environmental Health as well. In the short term, sustainable development will permit improvement in the living conditions, and therefore the health, of people across the world, especially in the poor nations. In the long term, sustainable development will protect the health and well-being of future generations. Some of the most compelling thinking in Environmental Health in recent years offers social and technical paths to sustainable development (Hawken et al., 1999; Brown, 2001, 2003; McDonough and Braungart, 2002; Ehrlich and Ehrlich, 2004). These approaches build on the fundamental links among health, environment, technological change, and social justice. Ultimately they will provide the foundation for lasting Environmental Health.

THE FORCES THAT DRIVE ENVIRONMENTAL HEALTH

Public health professionals tell the classic story of a small village perched alongside a fast-flowing river. The people of the village had always lived near the river, they knew and respected its currents, and they were skilled at swimming, boating, and water rescue. One day, they heard desperate cries from the river and noticed a stranger being swept downstream past their village. They sprang into action, grabbed their ropes and gear, and pulled the victim from the water. A few minutes later, as they rested, a second victim appeared, thrashing in the strong current and gasping for breath. The villages once again performed a rescue. Just as they were commenting on the coincidence of two near-drownings in one day, a third victim appeared, and they also rescued him. This went on for hours. Every available villager joined in the effort, and by mid-afternoon all were exhausted. Finally, the flow of victims stopped, and the villagers collapsed huffing and puffing in the town square.

At that moment, one of the villagers strode whistling onto the town square, relaxed and dry. He had not been seen since the first victims were rescued, and had not helped with any of the rescues. "Where were you?" his neighbors challenged him. "We've been pulling people out of the river all day! Why didn't you help us?"

"Ah," he replied. "When I noticed all the people in the river, I thought there must be a problem with that old footbridge upstream. I walked up to it, and sure enough, a gaping hole had broken through it. So I patched the hole, and people stopped falling through."

Upstream thinking has helped identify the root causes of many public health problems, and this is nowhere more true than in Environmental Health. Environmental hazards sometimes originate far from the point of exposure. Imagine that you inhale a hazardous air pollutant. It may come from motor vehicle tailpipes, from power plants, and/or from factories. As for the motor vehicle emissions, the amount of driving people do in your city or town reflects urban growth patterns and available transportation alternatives, and the pollutants generated by their cars and trucks vary with available technology and prevailing regulations. As for the power plants, the amount of energy they produce reflects the demand for energy by households and businesses in the area they serve, and the pollution they emit is a function of how they produce energy (coal? nuclear? wind?), the technology they use, and the regulations that govern their operations. Hence, a full understanding of the air pollutants you breathe must take into account urban growth, transportation, energy, and regulatory policy, among other "upstream" determinants.

These ideas are at the core of a useful model created by the World Health Organization, called DPSEEA (Driving Forces – Pressures – State – Exposure – Effects – Actions) (see Figure 25.1) (WHO, 2004). The DPSEEA model was developed both to help analyze Environmental Health hazards and to help design indicators useful in decision-making. The *driving forces* component

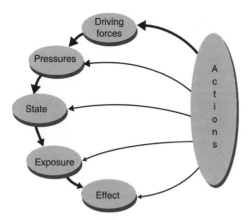

Figure 25.1 The DPSEEA model: Driving forces, pressures, state, exposure, effects, actions

refers to the factors that motivate the environmental processes. In the air pollution example, these might include population growth, consumer preferences for energy-consuming homes, appliances, and vehicles, and urban sprawl that requires driving long distances. The driving forces result in *pressures* on the environment, such as the emission of oxides of nitrogen, hydrocarbons, particulate matter, and other air pollutants. These emissions, in turn, modify the *state* of the environment, accumulating in the air and combining to form additional pollutants such as ozone. However, this deterioration in the state of the environment does not invariably threaten health; human *exposure* must occur. In the case of air pollutants, exposure occurs when people are breathing when and where the air quality is low. (Some people, of course, sustain higher exposures than others; an outdoor worker, an exercising athlete, or a child at play receive relatively higher doses of air pollutants than a person in an air-conditioned office.) Finally, the hazardous exposure may lead to a variety of *health effects*, acute or chronic. In the case of air pollutants, these effects may include coughing and wheezing, asthma attacks, heart attacks, and even early death.

To eliminate or control environmental hazards and protect human health, society may undertake a wide range of *actions*, targeted at any of the upstream steps. For example, protecting the public from the effects of air pollution might include encouraging energy conservation to reduce energy demand, and designing "live–work–play" communities to reduce travel demand (addressing driving forces), providing mass transit or bicycle lanes to reduce driving, requiring emissions controls on power plants, or investing in wind turbines to reduce emissions from coal-fired power plants (addressing pressures), requiring low-sulfur fuel

(to improve the state of the environment), warning people to stay inside when ozone levels are high (to reduce exposures), and providing maintenance asthma medications (to mitigate health effects). The most effective long-term actions, however, are those that are preventive in approach, aimed at eliminating or reducing the forces that drive the system. This theme is universal in public health, applying both to environmental hazards and to other health hazards as well.

Because many upstream forces define Environmental Health, many professionals can be said to practice Environmental Health, albeit indirectly. Certainly, the Environmental Health Director at a local health department, the Director of Environment, Health, and Safety at a manufacturing firm, an environmental epidemiology researcher at a university, or a physician working at an environmental advocacy group, would recognize themselves and be recognized by others as Environmental Health professionals. But many other people work in fields that have an impact on the environment and human health. The engineer who designs power plants helps protect the respiratory health of asthmatic children living downwind if she includes sophisticated emissions controls. The transportation planner who enables people to walk instead of drive also protects public health by helping clean up the air. The park superintendent who maintains urban green spaces may contribute greatly to the well-being of people in his city.

A wide range of environmental factors affects human health and well-being. Understanding environmental change and protecting ecosystem integrity require understanding the role of humans, and ultimately, protecting human health and well-being. These challenges fall to environmental professionals, to members of many other professions, to policymakers, and to the general public.

REFERENCES AND BIBLIOGRAPHY

Aguirre AA, Ostfeld RS, Tabor GM, House C and Pearl MC (eds.) (2002). *Conservation Medicine: Ecological Health in Practice.* New York: Oxford University Press.

Aron JL and Patz JA (2001). *Ecosystem Change and Global Health: A Global Perspective.* Baltimore: Johns Hopkins University Press.

Ausubel K and Harpignies JP (eds.) (2004). *Ecological Medicine: Healing the Earth, Healing Ourselves.* San Francisco: Sierra Club Books.

Brack D (ed.) (1998). *Trade and Environment: Conflict or Compatibility?* London: Earthscan.

Brimblecombe P (1988). *The Big Smoke.* London: Routledge.

Brown LR (2001). *Eco-Economy: Building an Economy for the Earth.* New York: W.W. Norton.

Brown LR (2003). *Plan B: Rescuing a Planet under Stress and a Civilization in Trouble.* New York: W.W. Norton.

Brown VA, Grootjans J, Ritchie J and Townsend M (2005). *Sustainability and Health: Supporting Global Ecological Integrity in Public Health*. London: Earthscan.

Bullard RD (1994). *Dumping in Dixie: Race, Class, and Environmental Quality*. Boulder: Westview Press.

Bullard RD, Johnson GS and Torres AO. Addressing global poverty, pollution, and human rights. In: Bullard RD (ed.) (2005). *The Quest for Environmental Justice: Human Rights and the Politics of Pollution*. San Francisco: Sierra Club Books.

Caldwell M (1988). *The Last Crusade: The War on Consumption 1862–1954*. New York: Atheneum.

Cantor N (2001). *In the Wake of the Plague: The Black Death and the World It Made*. New York: Free Press.

Carlsen E, Giwercman A, Keiding N and Skakkebaek N. Evidence for decreasing quality of semen during past 50 years. *BMJ* 1992;305:609–613.

Chase M (2003). *The Barbary Plague: The Black Death in Victorian San Francisco*. New York: Random House.

Cole LW and Foster SR (2000). *From the Ground Up: Environmental Racism and the Rise of the Environmental Justice Movement*. New York: New York University Press.

Creech JL and Johnson MN. Angiosarcoma of the liver in the manufacture of PVC. *J Occup Med* 1974;16:150–151.

Dhara VR and Dhara R (2002). The Union Carbide disaster in Bhopal: a review of health effects. *Arch Environ Health*, 57(5):391–404.

Douglas O, Kemedi V, Okonta I and Watts M. Alienation and militancy in the Niger Delta: Petroleum, politics, and democracy in Nigeria. In: Bullard RD (ed.) (2005). *The Quest for Environmental Justice: Human Rights and the Politics of Pollution*. San Francisco: Sierra Club Books.

Dowie M (1995). *Losing Ground: American Environmentalism at the Close of the Twentieth Century*. Cambridge: MIT Press.

Duffy J (1990). *The Sanitarians: A History of American Public Health*. Urbana: University of Illinois Press.

Ehrlich P and Ehrlich A (2004). *One with Nineveh: Politics, Consumption, and the Human Future*. Washington, DC: Island Press.

Epstein S (1982). *The Politics of Cancer*. New York: Random House.

Figueroa WG, Raszkowski R and Weiss W (1973). Lung cancer in chloromethyl methyl ether workers. *New Eng J Med*, 288(21):1096–1097.

Foster KR, Vechia P and Repacholi MH (2000). Risk management. Science and the precautionary principle. *Science*, 288(5468):979–981.

Gibbs LM (1998). *Love Canal: The Story Continues*. Gabriola Island, BC: New Society Publishers.

Gore A (1992). *Earth in the Balance*. Boston: Houghton Mifflin.

Gottlieb R (1993). *Forcing the Spring: The Transformation of the American Environmental Movement*. Washington, DC: Island Press.

Grifo F and Rosenthal J (eds.) (1997). *Biodiversity and Human Health*. Washington, DC: Island Press.

Guillette LJ Jr, Gross TS, Masson GR, Matter JM, Percival HF and Woodward AR (1994). Developmental abnormalities of the gonad and abnormal sex hormone concentrations in juvenile alligators from contaminated and control lakes in Florida. *Environ Health Persp*, 102(8):680–688.

Hamilton A (1925). *Industrial Poisons in the United States*. New York: Macmillan.

Hamilton A (1934). *Industrial Toxicology*. New York and London: Harper & Brothers

Hamlin C (1998). *Public Health and Social Justice in the Age of Chadwick Britain 1800–1854*. Cambridge: Cambridge University Press.

Harremoes P, Gee D, Macgarvin M, Stirling A, Keys J, Wynne B and Vaz SG (eds.) (2002). *The Precautionary Principle in the 20th Century: Late Lessons from Early Warnings*. London: Earthscan.

Hawken P, Lovins A and Lovins LH (1999). *Natural Capitalism: Creating the Next Industrial Revolution*. Boston: Little, Brown.

Hearne SA, Locke PA, Mellman M, Loeb P, Dropkin L, Bolger G, Fink N and Byrnes M (2000). Public opinion about public health—United States, 1999. *Morbid Mortal Weekly Report*, 49(12):258–260.

Herlihy D and Cohn SK (1997). *The Black Death and the Transformation of the West*. Cambridge: Harvard University Press.

Homer-Dixon TF (1999). *Environment, Scarcity and Violence*. Princeton: Princeton University Press.

Hurley A (1994). Creating ecological wastelands: Oil pollution in New York City, 1870–1900. *J Urban History* 20:340–364.

Ives JH (ed.) (1985). *The Export of Hazard: Transnational Corporations and Environmental Control Issues*. Boston: Routledge & Kegan Paul.

Kaplan S and Kaplan R (1982a). *Humanscape: Environments for People*. Ann Arbor: Ulrichs.

Kaplan S and Kaplan R (1982b). *Cognition and Environment: Functioning in an Uncertain World*.

Kaplan R and Kaplan S (1989). *The Experience of Nature: A Psychological Perspective*. New York: Cambridge University Press.

Kellert SR and Wilson EO (1993). *The Biophilia Hypothesis*. Washington, DC Island Press.

Kellert SR (1997). *Kinship to Mastery: Biophilia in Human Evolution and Development*. Washington, DC: Island Press.

Kelly J (2005). *The Great Mortality: An Intimate History of the Black Death, The Most Devastating Plague of All Time*. New York: HarperCollins.

Klare MT (2001). *Resource Wars: The New Landscape of Global Conflict*. New York: Henry Holt.

Kothari S. Social movements, ecology and justice. In: Hampson F, Reppy J (eds.) (1996). *Earthly Goods: Environmental Change and Social Justice*. Ithaca: Cornell University Press.

Kurzman D (1987). *A Killing Wind: Inside Union Carbide and the Bhopal Catastrophe*. New York: McGraw-Hill.

Larsen LH (1969). Nineteenth-century street sanitation: A study of filth and frustration. *Wisconsin Magazine of History* LII: 239–247.

Leavitt JW (1982). *The Healthiest City: Milwaukee and the Politics of Health Reform*. Princeton: Princeton University Press.

Lewis WH and Elvin-Lewis MPF (2003). *Medical Botany: Plants Affecting Human Health*, 2nd ed. New York: Wiley.

Low P (ed.) (1992). *International Trade and the Environment*. World Bank Discussion Papers 159. Washington, DC: The World Bank.

Marbury HJ (1995). Hazardous waste exportation: The global manifestation of environmental racism. *Vanderbilt Journal of Transnational Law,* 28:251–294.

Mattes WB, Pettit SD, Sansone SA, Bushel PR and Waters MD (2004). Database development in toxicogenomics: issues and efforts. *Environ Health Persp,* 112(4):495–505.

Mazur A (1998). *A Hazardous Inquiry: The Rashomon Effect at Love Canal.* Cambridge: Harvard University Press.

McDonough W and Braungart M (2002). *Cradle to Cradle: Remaking the Way We Make Things.* New York: North Point Press.

McKeown T (1979). *The Role of Medicine: Dream, Mirage, or Nemesis?* Princeton: Princeton University Press.

McMichael T (2001). *Human Frontiers, Environments and Disease.* New York: Cambridge University Press.

Melosi M (2000). *The Sanitary City: Urban Infrastructure in America from Colonial Times to the Present.* Baltimore: Johns Hopkins University Press.

Millenium Ecosystem Assessment (2003). *Ecosystems and Human Well-Being: A Framework for Assessment.* Washington, DC: Island Press.

Moro J and LaPierre D (2002). *Five Past Midnight in Bhopal: The Epic Story of the World's Deadliest Industrial Disaster.* New York: Warner.

National Research Council (1983). *Risk Assessment in the Federal Government: Managing the Process.* Washington, DC: National Academy Press.

Patterson JT (1987). *The Dread Disease: Cancer and Modern American Culture.* Cambridge: Harvard University Press.

Pesch B, Bruning T, Frentzel-Beyme R, Johnen G, Harth V, Hoffman W, KOY, Ranft U, Traugott UG, Thier R, Taegen D and Bolt HM (2004). Challenges to environmental toxicology and epidemiology: where do we stand and which way do we go? *Toxicol Lett,* 151(1):255–266.

Pognan F (2004). Genomics, proteomics and metabonomics in toxicology: hopefully not 'fashionomics'. *Pharmacogenomics,* 5(7):879–893.

Raffensberger C and Tickner J (1999). *Protecting Public Health and the Environment: Implementing the Precautionary Principle.* Washington, DC: Island Press.

Randall W (1977). *Building 6: The Tragedy at Bridesburg.* Boston: Little, Brown.

Randolph TG (1976). *Human Ecology and Susceptibility to the Chemical Environment.* Springfield, IL: Charles C. Thomas Publisher.

Randolph TG (1987). *Environmental Medicine: Beginnings and Bibliographies of Clinical Ecology.* Fort Collins, CO: Clinical Ecology Publications.

Rea WJ (1992–1998). *Chemical Sensitivity* (4 vols). Boca Raton: Lewis Publishers.

Rees WE and Wackernagel M (1995). *Our Ecological Footprint: Reducing Human Impact on the Earth.* Gabriola Island, BC: New Society Publishers.

Rosen G (1958). *A History of Public Health.* New York: MD Publications (reprinted Baltimore: Johns Hopkins University Press, 1993).

Rosenberg C (1962). *The Cholera Years: The United States in 1832, 1849, and 1866.* Chicago: University of Chicago Press.

Runge CF (1994). *Freer Trade, Protected Environment: Balancing Trade Liberalization and Environmental Interests.* New York: Council on Foreign Relations Press.

Sand PH (ed.) (1992). *The Effectiveness of International Environmental Agreements: A Survey of Existing Legal Instruments.* Cambridge, UK: Grotius.

Schmidt CW (2003). Toxicogenomics: An emerging discipline. *Environ Health Persp,* 110:A750–A755.

Selinus O, Alloway B, Centeno JA, Finkelman RB, Fuge R, Lindh U and Smedley P (2005). *Essentials of Medical Geology: Impacts of the Natural Environment on Public Health.* Burlington, MA: Elsevier Academic.

Shabecoff P (1993). *A Fierce Green Fire: The American Environmental Movement.* New York: Hill & Wang.

Sullivan R (2004). *Rats: Observations on the History and Habitat of the City's Most Unwanted Inhabitants.* New York: Bloomsbury.

Swan SH, Elkin EP and Fenster L (1997). Have sperm densities declined? A reanalysis of global trend data. *Environ Health Persp* 105(11):1228–1232.

Tarr J (2002). Industrial waste disposal in the United States as a historical problem. *Ambix: The Journal of the Society for the History of Alchemy and Chemistry,* 49:4–20.

Tarr JA (1996). *The Search for the Ultimate Sink: Urban Pollution in Historical Perspective.* Akron: University of Akron Press.

UN Centre for Human Settlements (Habitat) (2001). *Cities in a Globalizing World: Global Report on Human Settlements.* London: Earthscan.

van Wyk B-E and Wink M (2004). *Medicinal Plants of the World: An Illustrated Scientific Guide to Important Medicinal Plants and Their Uses.* Portland, OR: Timber Press.

Victor DG, Raustiala K and Skolnikoff EB (eds.) (1998). *The Implementation and Effectiveness of International Environmental Commitments: Theory and Practice.* Cambridge: MIT Press.

Waltner-Toews D (2004). *Ecosystem Sustainability and Health: A Practical Approach.* New York: Cambridge University Press.

Waters MD and Fostel JM (2004). Toxicogenomics and systems toxicology: aims and prospects. *Nature Reviews Genetics,* 5(12):936–948.

Wells M (1992). Biodiversity conservation, affluence, and poverty: Mismatched costs and benefits and efforts to remedy them. *Ambio,* 21:237–243.

Wilson EO (1984). *Biophilia.* Cambridge: Harvard University Press.

World Commission on Environment & Development (1987). *Our Common Future.* New York: Oxford University Press.

World Health Organization, Regional Office for Europe (2004). Environment and health information system. The DPSEEA model of health-environment interlinks. Available: http://www.euro.who.int/EHindicators/Indicators/20030527_2

Zinsser H (1935). *Rats, Lice and History: Being a Study in Biography, Which, After Twelve Preliminary Chapters Indispensable for the Preparation of the Lay Reader, Deals with the Life History of Typhus Fever.* Boston: Little, Brown.

26

Air Pollution: History of Actions and Effectiveness of Change

Ian Colbeck

INTRODUCTION

People are increasingly concerned about the impact that air pollution has on health, and on the urban and rural environment. The health impacts of air pollution are very serious and, currently, in the urban areas of developing countries, second only to the impacts of water and sanitation. As shown in Table 26.1, air pollution imposes a heavy burden on the health of urban populations throughout the developing world. Every year, there are an estimated 0.5-1 million premature deaths by air pollution worldwide. An assessment of health damages from exposure to the high levels of particulates in 126 cities worldwide where the annual mean levels exceed 50 μg m^{-3} reveals that these damages may amount to near 130,000 premature deaths, over 500,000 new cases of chronic bronchitis and many more lesser health effects each year (Lvovsky et al., 2000; Lvovsky, 2001).

In aggregate terms, this is equivalent to 2.8 million disability-adjusted life years (DALY) (Homedes, 1996) lost for this sample of nearly 300 million people or 9 DALYs lost per 1000 exposed residents. Urban air pollution is a serious problem worldwide. The gravity of the urban air pollution problem is largely attributed to the complex and multi-sectoral nature of everyday air polluting activities as well as the earlier inadequate actions of governments. The levels of exposure and the associated health burden in low- and middle-income countries are much higher than in rich

countries despite lower energy consumption. This is because, although development fosters energy use, it brings with it policies, technologies and institutional capacity for combating the adverse environmental effects of fuel consumption on local, national and regional scales. In many countries when economic factors are considered, pollution control takes a backseat to jobs and business development.

The World Health Organisation (WHO) has recently estimated that air pollution from particulate matter claims an average of 8.6 months from the life of every person in the European Union (EU). Current policies to reduce emissions of air pollutants in the EU by 2010 are expected to save 2.3 months of life for the EU population; the equivalent of preventing 80,000 premature deaths and saving over 1 million years of life (WHO, 2005). In the EU, the estimated annual monetary benefit from decreased population mortality attributed to particulate matter is €58–161 billion, and savings on the costs of diseases attributed to particulate matter account for €29 billion.

In the UK a report by the Committee on the Medical Effects of Air Pollutants (COMEAP, 1998) concluded that the deaths of between 12,000 and 24,000 vulnerable people may be brought forward each year and that between 14,000 and 24,000 hospital admissions and re-admissions may also result from poor air quality. These effects are attributed to just three: particulate matter (PM$_{10}$) (which is estimated to bring forward 8100 deaths annually), sulphur dioxide (3500 deaths) and ozone (from 700 to 12,500 deaths).

Table 26.1 Premature mortality and burden of disease due to air pollution, by region (projected annual averages for 2001–2020, World Bank 2000)

Region	Premature deaths (thousand per year)			Burden of disease (million DALYs per year)		
	Indoor	Outdoor (urban)	Total	Indoor	Outdoor (urban)	Total
China	150	590	740	4.5	14.0	18.5
East Asia and Pacific	100	150	250	3.5	3.8	7.3
Established market economies	0	20	20	0.0	0.5	0.5
Former socialist economies	10	200	210	0.2	3.8	4.0
India	490	460	950	17.0	10.1	27.1
Latin America and Caribbean	10	130	140	0.3	3.7	4.0
Middle East crescent	70	90	160	2.4	2.5	4.9
South Asia	220	120	340	7.6	2.6	10.2
Sub-Saharan Africa	530	60	590	18.1	1.2	19.3
World	1570	1810	3480	53.4	42.2	95.6

Air pollution also has other effects on our environment; forests, lakes, crops, wildlife and buildings can all suffer significant damage from high levels of airborne pollutants. Oxides of nitrogen (NO_x), for instance, can be transported over hundreds or even thousands of kilometres before being deposited as acid rain, which can acidify soil and, because of its ability to fertilise the soil, can cause changes in species composition and biodiversity. NO_x also react with volatile organic compounds in the atmosphere in the presence of sunlight to form ground level ozone, a significant component of summertime smog. Ozone is also a long-range pollutant which can cause direct effects on sensitive vegetation. It has been associated with reduced yields in crops and forestry, as well as with changes in species composition and biodiversity in natural and semi-natural ecosystems.

AIR POLLUTION IN PRE-INDUSTRIAL TIMES

We often assume that air pollution is a modern phenomenon, and that it has become worse in recent times. However, since the dawn of history, mankind has been burning biological and fossil fuel to produce heat. The walls of caves, inhabited millennia ago, are covered with layers of soot and many of the lungs of mummified bodies from Palaeolithic times have a black tone (McNeill, 2001).

Air pollution problems in ancient Rome appear in many documents (Hughes, 1993; Makra and Brimblecombe, 2004). As residents of what had become the largest city in the world, ancient Romans were well aware of the problem of air pollution. They called it *gravioris caeli* (heavy heaven) or *infamis aer* (infamous air). 'The smoke, the wealth, the noise of Rome …' held no charms for the Roman poet Horace (65 BC–8 AD) who described the blackening of buildings by smoke. Seneca (4 BC–65 AD), in one of his letters (104) to Lucilius in 61 AD states that

I expect you're keen to hear what effect it had on my health, this decision of mine to leave [Rome]. No sooner had I left behind the oppressive atmosphere of the city and that reek of smoking cookers which pour out, along with a cloud of ashes, all the poisonous fumes they've accumulated in their interiors whenever they're started up, than I noticed the change in my condition at once. You can imagine how much stronger I felt after reaching my vineyards (Costa, 1997).

Indoor air pollution, and in particular particulate matter, was also a significant problem. Animal and vegetable oils were burned to provide artificial light, and wood, vegetal materials and animal dung were used to heat their homes. All these materials produced high quantities of soot and toxic gases. Capasso (2000) examined skeletons buried by the volcanic eruptions of Vesuvius and found evidence of inflammation of the pulmonary tract. Histological assessment of the lungs of ancient human mummies has shown that anthracosis was a regular disorder in many ancient societies, including the Egyptian, Peruvian and Aleutian.

The Roman Senate introduced a law about 2000 years ago stating that 'Polluting air is not allowed' (Makra and Brimblecombe, 2004). In 535 AD the Roman emperor Justinian issued the *Institutes,* intended as a sort of legal textbook for law schools, which under the section Law of Things states that: 'By the law of nature these things are common to mankind – the air, running water, the sea, and consequently the shores of the sea' (see http://www.fordham.edu/halsall/basis/535institutes.html). Mamane (1987) discusses how the Hebrew Mishnah and its interpretation through the Jerusalem and Babylonian Talmud details pollution issues around 200 AD (Mamane, 1987) while Makra and Brimblecombe (2004) describe general environmental awareness in ancient Israel. Typically, smoke and soot dominated air pollution in earlier times although natural sources such as the long-range transport of dust (Chun, 2000) could also have an adverse impact on the environment.

Emissions from mining and metallurgy caused both local and regional air pollution (McNeil, 2001). Analysis of ice cores from Greenland reveals that the atmospheric concentration of lead began to increase between 400 BC and 300 AD with four times as much atmospheric lead compared with earlier times (Boutron et al., 1994; Boutron, 1995). At the height of the Roman Empire some 4000 tonnes of lead a year were emitted into the atmosphere (Hong et al., 1994). Levels fell after the fall of the Roman Empire before rising again during the Medieval and Renaissance periods to double that detected earlier. Levels continued to increase during the industrial revolution with a sharp increase between the 1930s and 1960s as a result of lead additives in petrol. In the 1960s atmospheric lead concentrations were 200 times higher than natural background levels (Boutron, 1995).

With the increasing use of unleaded fuels lead emissions in developed economies have declined steeply since the 1970s. For instance in the USA the reduction between 1975 and 1997 was 98% with a similar figure reported for the UK (Dore et al., 2004). Analysis of Greenland ice also shows that copper emissions peaked twice before the industrial revolution, once during the Roman period as a result of the use of copper alloys for the military and coinage, and later during the Sung dynasty where production peaked at 13,000 tonnes per year (Hong et al., 1996). As the smelting technology was primitive, approximately 15% of the smelted copper was emitted into the atmosphere. From the industrial revolution onwards (Newell, 1997) copper production has continued to grow with present-day production around 9 million tonnes per year. However, technological developments have resulted in significant reductions in atmospheric emissions with a factor of 0.25 today (Hong et al., 1996).

Urban pollution varied with the population and extent of the city and probably more importantly on the nature of industrial activity and fuel used. Initially, industrial activities were sited near forests since transportation of large quantities of fuel would have been difficult. Port cities were the exceptions since ships could transport cheaply wood and charcoal. Maimonides (1135–1204), a philosopher and physician, wrote that in cities 'the air becomes stagnant, turbid, thick, misty and foggy' and that these conditions result in 'dullness of understanding, failure of intelligence and defect of memory' (Turco,1997).

Extensive burning of coal did not begin until the 18th century although port cities short of wood might turn to coal as a fuel. London tried this initially in the 13th century and to a great degree in the 16th century. Air pollution was recognised as a public health problem and in 1285 a commission was set up to remedy the situation. Edward I issued a royal proclamation prohibiting the use of sea coal in kilns: '… after the complaint of his citizens … now burn them [kilns] and construct them of sea-coal instead of brushwood and charcoal, from the use of which sea-coal an intolerable smell diffuses itself throughout the neighboring places and the air is greatly infected, to the annoyance of the magnates, citizens and others there dwelling and to the injury of their bodily health'. To enforce the proclamation two weeks later a royal commission authorised 'to punish offenders by grievous ransoms'. By 1329, the ban had either been lifted or lost its effect (Brimblecombe, 1987).

By the second half of the 16th century Queen Elizabeth was so 'greved and annoyed with the taste and smoke of sea cooles' that in 1578 the Company of Brewers promised only to use wood in their brewing operations (te Brake, 1975). In 1661 John Evelyn produced *Fumifugium, or The Inconvenience of the Air and Smoke or London Dissipated.* Though not the first author to note the contaminating effects of London's unhealthy practices, Evelyn's work is one of the most well known early writings on air pollution. His work described the effects of pollution on London. To reduce the pollution, Evelyn proposed moving such industries as breweries and lime-burners to locations far outside of London to prevent the soot from settling in the city. In addition to relocating polluting industries, Evelyn also encouraged gardens and orchards to be planted on the city's periphery. Common law was used to protect individuals and when Thomas Legg of London complained, in 1691, of the smoke from his neighbour's bakehouse, the baker was ordered to put up a chimney '*soe high as to convey the smoake clear of the topps of the houses*' (Ashby and Anderson, 1981). However, such laws protect people and not the air. For protection of the air the State must intervene.

AIR POLLUTION IN THE NINETEENTH CENTURY

Air quality worsened in the 18th and 19th centuries due to the invention of the steam engine and by 1870 Britain had approximately 100,000 coal-fed steam engines (Clapp, 1994). The early 19th century saw a growing interest in the health of towns because they had become the focus of the population as people moved into cities, stimulated by rapidly increasing employment opportunities. The severity of pollution in the UK was sufficient that the issue was brought to parliament. A commission was appointed to investigate whether the users of steam engines and furnaces should be compelled to erect them in a manner less harmful to public health and comfort. The findings were published in 1819 and 1820 and reflected a new attitude towards public health and indicates changes in the thinking on social and economic issues. The report gave prominence to the question of the impact of smoke on health. Several physicians were convinced that smoke was damaging to health, one observing that sick people recovered more quickly in fresh country air.

Others did not agree and argued that smoke interfered with the spread of airborne diseases and could even act as a cure. The 1821 Smoke Prohibition Act tried to encourage the prosecution of a public nuisance from smoke but added little to the law and simply served to state that parliament was against smoke. Property owners were not willing to vote for legislation, in the interest of health and comfort, whose implementation required regulations and financial sacrifices. By 1843 air pollution had worsened and a select committee was again appointed to report on the situation. In 1845 a third report was issued to supplement the 1843 report. Both reports highlighted the medical and social consequences of pollution and gave information on the feasibility of preventative measures. Between 1844 and 1850 no fewer than six bills were introduced into parliament to compel furnaces to consume their own smoke. All failed to pass into law (Ashby and Anderson, 1976).

A smoke clause was written into the City of London Sewers Bill which was enacted in 1851 whilst, in 1853, the Smoke Nuisance Abatement (Metropolis) Act passed through parliament. This latter act applied to steam boats plying above London Bridge. Enforcement fell to the police and in 1854 secured over 150 convictions (Clapp, 1994). The Act was strengthened three years later and appears to have lead to some reduction in the amount of industrial smoke in London, but enforcement was uneven and many factories remained exempt from regulation. The Alkali Act, passed in 1863, had two main provisions. First, it instructed all alkali manufactures to condense 95% of their hydrochloric acid gas and second it provided for the appointment of alkali inspectors – the first national regulatory environmental body in the world. Latter amendments to the Act resulted in the inclusion of other processes and pollutants including sulphur compounds. However, the inspectorate lacked the authority to regulate sulphur emissions as a result of coal burning. This could lead to the situation in which coal-fired furnaces could emit more sulphuric acid than was allowed to be released from works manufacturing the chemical.

When the Government surveyed the enforcement of local laws against smoke in large towns and cities they found that only Liverpool appeared to be enthusiastic for enforcement (Hawes, 1998). Much smoke abatement legislation occured within UK sanitary legislation. The Sanitary Act of 1866 permitted local authorities throughout England and Wales to prosecute the owners of smoky factories. The wording of this statute, which only applied to chimneys that released 'black smoke in such quantity as to be a nuisance', allowed many defendants to evade fines by arguing that their furnaces produced dark brown smoke rather than black smoke (Ashby and Anderson, 1981). The 1875 Public Health Act gave all local authorities the opportunity to combat smoke if they were so inclined. This Act specified that black smoke was to be prevented only 'as far as practicable', again allowing many factory owners to avoid prosecution. The 1891 Public Health (London) Act took the regulation of smoke away from the police and passed it local parishes. The Act narrowed the range of what constituted illegal smoke and initially caused the smoke problem to increase.

Air pollution across Britain may well have been at its worst towards the end of the 19th century. There are many descriptions from writers but there is a lack of measurements. The early sanitary inspectors were keen to record the number of times smoke was observed and some estimates of soot fall appear to be unreasonably high. Brimblecombe (2004) has shown that there is a strong similarity between fog frequency in London and modelled pollution load. The incidence of London fogs peaked at nearly 70 per year by the 1890s (Brimblecombe, 1982). The literature and press of the time were full of descriptions of the severity and impact of the fog (Brimblecombe, 1987). In particular, the book *London Fogs* by Russell (1880) drew attention to the fact that the fog was responsible for a considerable number of deaths every year. In 1873 he calculated that 500 additional deaths per week could be attributed to fog and for three weeks in

1880, 2000 deaths were the result of the fog. In a lecture in 1889 Russell estimated that the annual cost of the damage due to fogs was of the order of £5.2 million, equivalent to the wages of 100,000 labourers (Ashby and Anderson, 1977a). Smoke from domestic fires was implicated, which was not covered by any laws to abate smoke. In addition to 1872 London experienced a number of deadly fogs in January 1880, February 1882 and December 1891. All had a common thread: a combination of a stagnant fog and smoke. In 1905 Des Voeux used the term smog to describe such conditions. It was popularised in 1911 when Des Voeux presented to the Manchester Conference of the Smoke Abatement League of Great Britain a report on the deaths that occurred in Glasgow and Edinburgh in the autumn of 1909 as a consequence of smoke-laden fogs.

In an attempt to tackle this problem, physicians, engineers, sanitarians and lay reformers, particularly from the middle and upper classes, joined forces eventually leading to the formation of the Smoke Abatement Committee. The anti-smoke lobby failed to get particularly effective laws through parliament, but it did cultivate a new social attitude toward air pollution aided by allies in the press. It performed a valuable service in keeping the issues before the public and lifting social norms for the environment to a level which would eventually, albeit some 75 years later, make it practicable to bring in laws to control domestic as well as industrial smoke (Ashby and Anderson, 1977b). One way this was achieved was through exhibitions which provided an important form of mass communications. The Fog and Smoke Committee, aware of the emissions from domestic smoke, organised a Smoke Abatement Exhibition at the Exhibition Buildings in South Kensington in 1881 and Manchester in 1882 to demonstrate the existence and practicability of smoke control equipment (Ranlett, 1981). However, the traditional British open fireplace was also widely held to be essential to the health of the nation. Contemporary anxieties about harmful air pollutants accumulating within the home, particularly the build up of contaminants emanating from people's bodies, led to an almost obsessive preoccupation with the efficient ventilation of interior spaces. The open fireplace, despite its many defects, was recognised to be the primary ventilating agent in most homes (Mosley, 2003).

Public health issues were not restricted to London, and a growing group of enthusiastic professionals began to emerge in cities throughout the country. One of Britain's first smoke abatement societies to be established was the Manchester Association for the Prevention of Smoke which was founded in 1842 along with similar groups in Leeds and Huddersfield (Mosley, 2001, 2004;

Platt, 2004). This may reflect Manchester's strong stand against air pollution. It modified complex medieval administrative practices to address industrial pollution and created new bodies that considered environmental matters (Bowler and Brimblecombe, 2000). The Nuisance Inspectors, introduced in the 1820s, held an increasingly important position in improving the environment of towns and cities of the 19th century. Reformers had little to show for all their hard work and any incremental improvements in air quality were more than overwhelmed by gains in the consumption of fossil fuels. Smoke pollution made a significant impact on Victorian life. This was not only by its direct physical and medical effects but also on its impact on social perceptions as reflected in the language, literature and art of that period (Brimblecombe, 2004; Mosley, 2004).

It is difficult to assess accurately the pollution concentrations in 19th century cities. One of the first observations was made in London in 1885 and during foggy weather 860 μg m^{-3} of 'organic matter' was reported. In 1897, 420 μg m^{-3} was measured in Leeds (Cohen and Rushton, 1925). By the beginning of the 20th century estimates of soot fall were being published with figures of 220 tons per square mile per annum for Leeds (Cohen and Rushton, 1925) and 259 tons per square mile per annum for London (Des Voeux and Owens, 1912).

Current-day researchers have been using a variety of techniques to estimate air pollution in the 19th century. Harrison and Aplin (2002) have studied the link between the electricity in the air and smoke pollution; particles of soot in the smoke profoundly change the electrical conductivity of the air such that smoke-laden air is much less electrically conductive than clean air. Using this relationship and atmospheric electricity records from Kew Observatory they determined that the mean smoke concentration in 1863 was 170 μg m^{-3}. Further analysis revealed significant differences between 19th and 20th century, diurnal variations in smoke. A common feature was an afternoon minimum, but the morning maximum during the 19th century was much broader and commenced earlier than in the 20th century, probably the result of the lighting of domestic fires. With the advent of motor traffic the morning peak is nowadays more sharply defined. Knowledge of air quality, visual range and smoke levels can be obtained from Monet's 'London Series' of paintings, landscapes of the Westminster area around 1900. Thornes (personal communication) has assessed the visibility in Monet's paintings by estimating distances from bridges, buildings, chimneys and church spires to determine air quality at the time of the paintings.

TWENTIETH-CENTURY CONCERNS

Smoke abatement remained at the forefront of anti-pollution efforts to the 1950s with some minor successes such as the Smoke Abatement Act of 1926. It was recognised that smoke need no longer be black to be a nuisance and was defined as including soot, ash, grit and gritty particles. A common defence against a claim of nuisance was that the best practicable means to prevent emission of smoke had been applied and that the emission was due to lighting up from cold, temporary mechanical failure or a suitable fuel not being available. However, reformers were unable to legislate against domestic coal burning and emissions from domestic fires continued to pollute the air.

An anticyclone covered much of the UK from Friday 5th to Tuesday 9th December 1952. The resulting low temperatures meant that residents were burning large quantities of coal and thus emitting large amounts of smoke and sulphur dioxide through low-level chimneys. The smoke-laden fog that shrouded London over this period brought premature death to thousands and inconvenience to millions. An estimated 4000 people died because of it, and cattle at Smithfield were asphyxiated. Road, rail and air transport were almost brought to a standstill and a performance of La Traviata at the Sadler's Wells Theatre had to be suspended when fog in the auditorium made conditions intolerable for the audience and performers.

On 18th December the Minister of Health announced that there had been 4703 deaths compared with 1852 over the same period the previous year. Men suffered more than women and those over 45 years old suffered the most. Many who died already suffered from chronic respiratory or cardiovascular complaints. Mortality from bronchitis and pneumonia increased more than sevenfold as a result of the fog. A recent reassessment of the data has suggested that the mortality count was nearer 12,000 rather than the 4000 generally reported for the episode (Bell *et al.*, 2004). During the smog peak concentrations of sulphur dioxide and particulate smoke were estimated to be 1.4 ppm and 4460 μg m^{-3}, respectively.

Public outcry and government inquiry followed, leading to the Clean Air Act of 1956, which closely regulated domestic coal smoke. As these residents and operators were necessarily given time to convert to different fuels, fogs continued to be smoky for some time after the Act of 1956 was passed. In 1962, for example, 750 Londoners died as a result of a fog. The Clean Air Act was seen both in the UK and abroad as a seminal development in environmental legislation although there was no mention in the Act of air concentrations, standards or guidelines, nor was there any explicit recognition in policy instruments of any concept of exposure to people or to the environment.

Following the 1956 Act coal consumption in domestic and industrial sources began to decline with domestic consumers moving to gas. Commercial and industrial sources tended to use fuel oil and gas oil which reduced smoke emissions but continued to emit significant amounts of sulphur. The oil shortages of the mid-1970s had a significant impact on fuel usage. Over the past 50 years transport activity has increased significantly and is now the dominant source in terms of exposure of people to air pollution. In 1950 there were 4 million vehicles registered in the UK, 50% of them cars whilst at the turn of the 21st century there are over 28 million vehicles on the roads, 85% of them cars. Changes in overall fuel patterns, other than the shift away from coal, were determined by broader-scale economic conditions and influences rather than by environmental policies (Williams, 2004). In addition to air quality improvements, social, economic and technological changes have also helped to reduce smoke emissions.

The growth in emissions from coal combustion in the 19th century is paralleled by the rise in petroleum combustion emissions in the twentieth century. This shift towards mobile combustion of fossil fuels has moved attention away from primary pollutants such as sulphur dioxide to nitric oxide and carbon monoxide. More importantly it has made us aware of the ability of the atmosphere to generate secondary pollutants – so-called photochemical smog. Such photochemical smog was first observed in Los Angeles in the 1940s but it was not until the early 1950s that the origin of this pollution, basically the formation of ozone from the action of sunlight on volatile organic compounds, was recognised (Haagen-Smitt, 1952; Haagen-Smit *et al.*, 1953). This kind of pollution, and associated health effects, is now found across the globe. The highest ever recorded levels of ozone in the UK were in the hot summer of 1976 when mortality in London was increased by 40–60% (Office for National Statistics, 1997).

AIR POLLUTION REGULATION

The EU has now become the driver of most of the UK's environmental policy and legislation. Since 1997, the EU has had an explicit goal on integrating environmental concerns into areas of policy such as energy, agriculture and transport. Air quality is one of the areas in which Europe has been most active in recent years. The European Commission has aimed to develop an overall strategy. Member States are required to transpose and implement new directives on air quality which set long-term quality objectives. The first directive, in 1980, related to sulphur dioxide and suspended particulate matter in urban areas (80/779/EEC). At this time, Europe was setting the pace in terms of environmental legislation with the UK following behind.

In 1996, the EU adopted Framework Directive 96/62/EC on ambient air quality assessment and management. This Directive covered the revision of previously existing legislation and the introduction of new air quality standards for previously unregulated air pollutants, setting the timetable for the development of daughter directives on a range of pollutants. The list of atmospheric pollutants to be considered included sulphur dioxide, nitrogen dioxide, particulate matter, lead and ozone – pollutants governed by already existing ambient air quality objectives – and benzene, carbon monoxide, polyaromatic hydrocarbons, cadmium, arsenic, nickel and mercury.

The Framework Directive was followed by daughter directives (Table 26.2), which set the numerical limit values, or in the case of ozone, target values for each of the identified pollutants. Besides setting air quality limit and alert thresholds, the objectives of the daughter directives are to harmonise monitoring strategies, measuring methods, calibration and quality assessment methods to arrive at comparable measurements throughout the EU and to provide for good public information. These daughter directives replaced the earlier air quality directives. Although goals and criteria are set out by the EU, it is the individual member countries which have the responsibility to monitor and assess air quality and to implement the directive.

In the 1990s, WHO updated its air quality guidelines for Europe (WHO, 2000) to provide detailed information on the adverse effects of exposure to different air pollutants on human health.

The prime aim of these guidelines was to provide a basis for protecting human health from the effects of air pollution. These guidelines are not standards in themselves but provide starting points for countries to set standards, taking local socioeconomic factors into account. The guidelines do not specify thresholds of concentrations which, if exceeded, will result in adverse effects on health. They actually specify concentrations at which effects are not expected even among sensitive groups in the population

In the UK the 1995 Environment Act created a number of new agencies, such as the Environment Agency, and set new standards for environmental management. It also required the Secretary of State to prepare a National Air Quality Strategy. The first national air quality strategy was published in 1997; following a review in 1998, a revised strategy 'The Air Quality Strategy for England, Scotland, Wales and Northern Ireland' was published in January 2000. This set out the framework for improving air quality and set air quality standards and objectives for eight key air pollutants to be achieved between 2003 and 2008. For two of the pollutants, NO_2 and SO_2, it also set objectives for protecting vegetation and ecosystems. The Strategy also identified what needs to be done at international, national and local level to achieve the objectives. It provided a framework that allows all stakeholders to identify the contributions they can make. Following a review and public consultation in 2001, an addendum to the strategy was introduced in 2003. This set tighter objectives for particles, benzene and carbon monoxide and a new objective for polycyclic aromatic hydrocarbons (Table 26.3). The Addendum includes different targets of different parts of the UK reflecting different local conditions, although it is difficult to see air quality on one side of the Scottish border could meet the objectives and on the other side fail.

Table 26.2   EU air quality framework directive daughter directives

Directive	Details	Implementation date
First Daughter Directive (1999/30/EC)	Limit values for NO_x, sulphur dioxide, lead and PM_{10}. The limit values for NO_x for the protection of vegetation must be met by 2001. The health limit values for sulphur dioxide and PM_{10} must be met by 2005, those for NO_2 and lead must be met by 2010	19 July 2001
Second Daughter Directive (2000/69/EC)	Limit values for benzene and carbon monoxide. The limit value for carbon monoxide must be met by 2005 and that for benzene 2010	13 December 2002
Third Daughter Directive (2002/3/EC)	Target values for ozone to be attained where possible by 2010	9 September 2003
Fourth Daughter Directive (2004/107/EC)	Limit values for arsenic, cadmium, mercury, nickel and polycyclic aromatic hydrocarbons	15 February 2007

Table 26.3 UK air quality standards

Pollutant	Objective Concentration	Objective Measured as	Objective To be achieved by
Benzene	3.25 µg m^{-3} in Scotland and NI	Running annual mean	31 December 2010
	5 µg m^{-3} in England and Wales		
1,3-Butadiene	2.25 µg m^{-3}	Running annual mean	31 December 2003
Carbon monoxide	11.6 mg m^{-3}	Running 8-hour mean	31 December 2003
Lead	0.5 µg m^{-3}	Annual mean	31 December 2004
	0.25 µg m^{-3}		31 December 2008
Nitrogen dioxide	200 µg m^{-3}, not be exceeded more than 18 times a year	1-hour mean	31 December 2005
	40 µg m^{-3}	Annual mean	31 December 2005
Ozone	100 µg m^{-3}, not to be exceeded more than 10 times per year	Daily maximum of running 8-hour means	31 December 2005
Particles (PM$_{10}$)	50 µg m^{-3}, not to be exceeded more than: 7 times per year in UK except London 10 times per year in London	24-hour mean	31 December 2010
	18 µg m^{-3} in Scotland 20 µg m^{-3} in England outside London, Wales and NI 23 µg m^{-3} in London	Annual mean	31 December 2010
Sulphur dioxide	266 µg m^{-3}, not to be exceeded more than 35 times per year	15- minute mean	31 December 2005
	350 µg m^{-3}, not to be exceeded more than 24 times per year	1-hour mean	31 December 2004
	125 µg m^{-3}, not to be exceeded more than 3 times a year	24-hour mean	31 December 2004
Polycyclic aromatic hydrocarbons	0.25 ng m^{-3} benzo[a]pyrene	Annual mean	31 December 2010

During the 1970s evidence began to emerge that a significant element of acid deposition over Scandinavia was the result of long-range, transboundary air pollution. This implied that cooperation at the international level was necessary to solve problems such as acidification. In 1979 the Convention on Long-range Transboundary Air Pollution was signed by thirty-four Governments and the European Community. The Convention was the first international legally binding instrument to deal with problems of air pollution on a broad regional basis. Besides laying down the general principles of international cooperation for air pollution abatement, the Convention set up an institutional framework bringing together research and policy. The Convention on Long-range Transboundary Air Pollution entered into force in 1983. It has been extended by eight specific protocols (Table 26.4). The initial Protocols adopted flat-rate emission reduction policies. The NO$_x$ Protocol required the application of an effects-based approach using the concepts of critical loads approach. This approach, which aims at gradually attaining critical loads, sets long-term targets for reductions in emissions, although it has been recognised that critical loads will not

Table 26.4 EU air quality framework directive daughter directives

Year	Protocol	Entered into force
1999	To abate acidification, eutrophication and ground-level ozone	17 May 2005
1998	Persistent organic pollutants (POPs)	23 October 2003
1998	Heavy metals	29 December 2003
1994	Further reduction of sulphur emissions	5 August 1998
1991	Control of emissions of volatile organic compounds or their transboundary fluxes	29 September 1997
1988	Control of nitrogen oxides or their transboundary fluxes	14 February 1991
1985	Reduction of sulphur emissions or their transboundary fluxes by at least 30%	2 September 1987
1984	Long-term financing of the cooperative programme for monitoring and valuation of the long-range transmission of air pollutants in Europe (EMEP)	28 January 1988

be reached in one single step. The most recent Protocol deals with a range of pollutants and addresses three environmental issues: acidification, eutrophication and ground-level ozone.

The Protocol sets emission ceilings for 2010 for four pollutants: sulphur, NO_x, volatile organic compounds (VOCs) and ammonia. These ceilings were negotiated on the basis of scientific assessments of pollution effects and abatement options. Parties whose emissions have a more severe environmental or health impact and whose emissions are relatively cheap to reduce will have to make the biggest cuts. Once the Protocol is fully implemented, Europe's sulphur emissions should be cut by at least 63% (Figure 26.1), its NO_x emissions by 41%, its VOC emissions by 40% and its ammonia emissions by 17% compared to 1990.

CONCLUSIONS

The response to air pollution has changed over time with, for example, a change from permissive to mandatory legislation together with the setting of specified air quality standards. Increasing economic globalisation has forced the development of international regulatory regimes with a change from addressing single issues to integrated pollution control. However, the localised nature of many of the remaining air pollution problems will require action at local or regional levels in order to secure effective solutions.

Air quality is generally improving. In UK urban areas in 2004, air pollution was recorded as moderate or higher on 22 days on average per site,

compared with 50 days in 2003, 20 days in 2002 and 59 days in 1993 (Defra, 2005). In general, there has been a long-term decline in the number of air pollution days, largely because of a reduction in particles and sulphur dioxide, but fluctuations from one year to the next can occur because of differences in weather conditions (Figure 26.2). In rural areas, the figure for 2004 was 42 days on average per site, compared with 61 in 2003 and 30 in 2002. The number of days has fluctuated between 21 days in 1987 and the 2003 figure of 61 days. The series can be volatile from one year to the next, and there is no clear trend. This reflects the variability in levels of ozone, the main cause of pollution in rural areas. More ozone is produced in hot, sunny weather, as was the case during 2003. Looking towards 2025, mathematical models predict that most pollutant levels will continue to fall, but targets for nitrogen dioxide, PM_{10} and ozone may be breached in some areas. Thus, adverse health effects will continue, especially with long-term exposure. This raises questions over whether and how air quality can be further improved, taking costs and benefits into account. Developing strategies and reduction measures will require consideration of economic and social as well as environmental issues.

Vehicle emissions, in particular, will reduce further as new vehicles and fuels become cleaner, and more polluting older vehicles fall out of the vehicle pool. But it is important to maintain and build on these improvements to air quality to reduce the impact that air pollution has on public health and meet our national objectives and the air quality limit values set under European legislation.

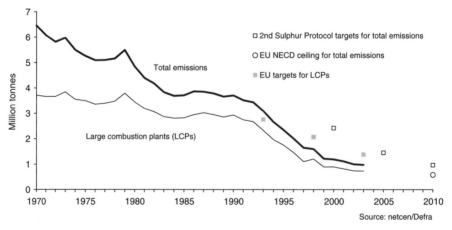

Figure 26.1   Sulphur dioxide emissions and targets: 1970–2010
Source: Defra e-digestenvironmental statistics
Website: *http://www.defra.gov.uk/environment/statistics/*
Crown copyright material is reproduced with permission of the Controller of the HMSO.

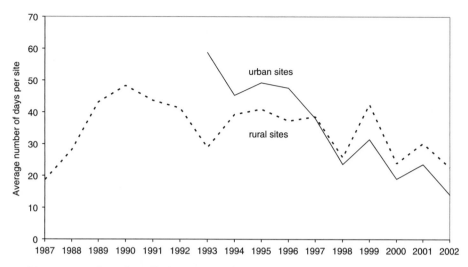

Figure 26.2 Days when air pollution was moderate or higher, 1987–2004

Today's complexity of emission sources and strengths means that the air pollution picture, and resultant management challenge, is much more complicated than in the past. Various actions are being planned to combat the threat of climate change. If these are successfully implemented then this may have a significant impact on pollutant emissions, and hence air quality in the 21st century.

REFERENCES

Ashby, E. and Anderson, M. (1976) Studies in the politics of environmental protection: The historical roots of the British Clean Air Act, 1956: I. The awakening of public opinion over industrial smoke, 1843–1853. *Interdisciplinary Science Reviews* 1: 279–290.

Ashby, E. and Anderson, M. (1977a) Studies in the politics of environmental protection: The historical roots of the British Clean Air Act, 1956: III. The ripening of public opinion, 1898–1952. *Interdisciplinary Science Reviews* 2: 190–206.

Ashby, E. and Anderson, M. (1977b) Studies in the politics of environmental protection: The historical roots of the British Clean Air Act, 1956: II. The appeal to public opinion over domestic smoke, 1880–1892. *Interdisciplinary Science Reviews* 2: 9–26.

Ashby, E. and Anderson, M. (1981) *The Politics of Clean Air.* Clarendon Press, Oxford.

Bell, M.L., Davis, D.L. and Fletcher, T. (2004) A retrospective assessment of mortality from the London smog episode of 1952: the role of influenza and pollution. *Commentary Environmental Health Perspectives* 112: 6–8.

Boutron, C.F. (1995) Historical reconstruction of the Earth's past atmospheric environment from Greenland and Antarctic snow and ice cores. *Environmental Review* 3: 1–28.

Boutron, C.F., Candelone, J.P. and Hong, S. (1994) Past and recent changes in the large scale tropospheric cycles of Pb and other heavy metals as documented in Antarctic and Greenland snow and ice: a review. *Geochimica et Cosmochimica Acta* 58: 3217–3225.

Bowler, C. and Brimblecombe, P. (2000) Control of air pollution in Manchester prior to the Public Health Act, 1875. *Environment and History* 6: 71–98.

Brimblecombe, P. (1982) Long term trends in London fog. *Science of the Total Environment* 22: 19–29.

Brimblecombe, P. (1987) *The Big Smoke. A History of Air Pollution in London Since Medieval Times.* Methuen, London, New York.

Brimblecombe, P. (2004) Perceptions and effects of late Victorian air pollution. In *Smoke and Mirrors*, E.M. Du Puis (ed.). New York University Press, New York.

Capasso, L. (2000) Indoor pollution and respiratory diseases in ancient Rome. *The Lancet* 356: 1774.

Chun, Y. (2000) The yellow-sand phenomenon recorded in the Chosunwangjosilok. *Journal of the Meteorological Society of Korea* 36: 285–292.

Clapp, B.W. (1994) *An Environmental History of Britain.* Longman, London.

Cohen, J.B. and Rushton, A.G. (1925) *Smoke.* Edward Arnold, London.

COMEAP (1998) The Quantification of the Effects of Air Pollution on Health in the United Kingdom. HMSO, London.

Costa, C.D.N. (ed. and trans.) (1997) *Dialogues and Letters: Seneca.* Penguin Books, London.

Defra, (2005) e-Digest of Environmental Statistics, http://www.defra.gov.uk/environment/statistics/airqual/index.htm

Des Voeux, H.A. and Owens, J.S. (1912) The soot fall of London. *Lancet* Jan. 6, 47.

Dore, C.J., Watterson, J.D., Goodwin, J.W.L., Murrells, T.P., Passant, N.R., Hobson, M.M., Baggott S.L., Thistlethwaite, G., Coleman, P.J., King, K.R., Adams, M. and Cumine, P.R. (2004)

UK Emissions of Air Pollutants 1970 to 2002, UK National Atmospheric Emissions Inventory.

Haagen-Smit, A.J. (1952) Chemistry and physiology of Los Angeles smog. *Industrial Engineering Chemistry* 44: 1342–1346.

Haagen-Smit, A.J., Bradley, C.M. and Fox, M.M. (1953) Ozone formation in photochemical oxidation of organic substances. *Industrial Engineering Chemistry* 45: 2080–2089.

Harrison, R.G. and Aplin, K.L. (2002) Mid-nineteenth century smoke concentrations near London. *Atmospheric Environment,* 26: 4037–4043.

Hawes, R. (1998) The municipal regulation of smoke pollution in Liverpool, 1853–1866. *Environment and History* 4: 75–90.

Homedes, N. (1996) The Disability-Adjusted Life Year (DALY) Definition, Measurement and Potential Use. World Bank, HCDWP 68.

Hong, S., Candelone, J.P., Patterson, C.C. and Boutron, C.F. (1994) Greenland ice evidence of hemispheric lead pollution two millennia ago by Greek and Roman civilizations. *Science* 265: 1841–1843.

Hong, S., Candelone, J.P., Patterson, C.C. and Boutron, C.F. (1996) History of ancient copper smelting pollution during Roman and medieval times recorded in Greenland ice. *Science* 272: 246–249.

Hughes, J.D. (1993) Pan's Travail: Environmental Problems of the Ancient Greeks and Romans. Johns Hopkins University Press, Baltimore.

Lvovsky, K. (2001) Health and Environment 2001. Environment Strategy Papers Series, no. 1. World Bank.

Lvovsky, K., Hughes, G., Maddison, D., Ostro, B. and Pearce, D. (2000) Environmental costs of fossil fuels: A rapid assessment method with application to six cities. Environment Department Paper No. 78. Washington, DC: World Bank.

Makra, L and Brimblecombe, P. (2004) Selections from the history of environmental pollution, with special attention to air pollution. Part 1. *International Journal of Environment and Pollution* 22: 641–656.

Mamane, Y. (1987) Air-pollution control in Israel during the 1st and 2nd century. *Atmospheric Environment* 21: 1861–1863.

McNeill, J.R. (2001) *Something New Under the Sun. An Environmental History of the Twentieth Century World.* W.W. Norton, New York, London.

Mosley, S. (2001) *The Chimney of the World.* White Horse Press, Cambridge.

Mosley, S. (2003) Fresh air and foul: the role of the open fireplace in ventilating the British home, 1837–1910. *Planning Perspectives* 18: 1–21.

Mosley, S. (2004) Public perceptions of smoke pollution in Victorian Manchester. In *Smoke and Mirrors*, E.M. Du Puis (ed.), New York University Press, New York.

Newell, E. (1997) Atmospheric pollution and the British copper industry, 1690–1920. *Technology and Culture*, 28, 655–689.

Office for National Statistics (1997). In *The Health of Adult Britain: 1841–1994.* J. Charlton and M. Murphy (eds). The Stationery Office, London.

Plat, H.L. (2004) The 'invisible evil': noxious vapor and public health in Manchester during the age of industry. In *Smoke and Mirrors*, E.M. Du Puis (ed.). New York University Press, New York.

Ranlett, J. (1981) The smoke abatement exhibition of 1881. *History Today*, 31, 10–13.

Russell, F.A.R. (1880) *London Fogs.* Edward Stanford, London.

te Brake W.H. (1975) Air pollution and fuel crises in preindustrial London, 1250–1650. *Technology and Culture* 16, 337–359.

Turco, R.P. (1997) *Earth Under Seige: From Air Pollution to Global Change*, Oxford University Press, Oxford.

WHO (2000) Air Quality Guidelines for Europe, 2nd edn. Copenhagen, WHO Regional Office for Europe, 2000 (WHO Regional Publications, European Series, No. 91).

WHO (2005) Particulate Matter Air Pollution: How It Harms Health. Fact sheet EURO/04/05.

Williams, M. (2004) Air pollution and policy – 1952–2002. *Science of the Total Environment*, 15–20, 334–335.

World Bank (2000). Fuel for Thought: An Environmental Strategy for the Energy Sector. Annex 2, Washington DC.

27

Terrestrial Environments, Soils and Bioremediation

Andrew S. Ball

ENERGY AND MATTER FLOW IN SOIL SYSTEMS

Terrestrial environments remove atmospheric carbon dioxide (CO_2) by plant photosynthesis during the day, which results in the growth of plant roots and shoots and increases in soil microbial biomass. When a plant sheds leaves and roots die, this plant litter decays, but some of it can become protected physically and chemically as dead organic matter (OM) in the form of humus, which can be stable in soils for even thousands of years. The decomposition of soil carbon by soil microbes (decomposers) also releases CO_2 to the atmosphere. Soil represents the largest carbon pool on the Earth's surface (2,157,000–2,293,000 billion kg). Of this, around 70% is organic and the remaining consists of carbonates (Batjes, 1996). The soil organic carbon pool is twice that present in the atmosphere and two or three times larger than that present in all living matter (Prentice *et al.*, 2001). Decomposers or detritivores are heterotrophs that obtain their energy either from dead organisms or from organic compounds dispersed in the environment.

It can be seen that once fixed by plants, organic energy can move within the soil system through the consumption of living or dead OM. Upon decomposition the chemicals that were once organized into organic compounds are returned to their inorganic form and can be taken up by plants once again. But what if the compound present in soil is not naturally present? What if it is a compound not naturally found in the environment? Does the compound get broken down or does it persist? If it is a toxic compound what effect does it have on the biotic community? All the questions must be considered because more and more of the Earth's terrestrial environments are being contaminated with human-made (xenobiotic) compounds. In this chapter the effects of such pollutants on the terrestrial environments are considered along with the opportunities to remove or remediate the compounds in soil.

SOURCES OF NUTRIENTS IN SOIL

The Earth as a whole is a natural recycling system. It has to be, because it is a closed matter system. No new matter is being added to the Earth, so all new biomass must be made from existing matter. Terrestrial ecosystems generally have a high storage capacity for nutrients. This storage capacity exists largely in and on organic material which is slowly decayed. Decomposition of OM may take several months to several years to complete. Within the soil, nutrients are stored on the soil particles, in dead OM or in chemical compounds (Foth and Turk, 1972).

Nutrients are gained and lost by terrestrial communities in a variety of different ways. Nutrient budgets can be produced if gains and losses of nutrients are calculated. In some terrestrial environments these budgets may be more or less in balance, while in others inputs exceed outputs leading to the accumulation in the compartments of living biomass and dead OM. Alternatively, outputs may exceed inputs if the biota is disturbed by an event such as deforestation or crop harvest.

Weathering of parent rock and soil is the dominant source of nutrients such as calcium, phosphorus and potassium, which may then be taken up by plants. Mechanical weathering is caused by processes such as freezing of water and growth of roots. Chemical weathering however, is the most important process, particularly carbonation, in which carbonic acid (HCO_3) reacts with minerals to release ions such as potassium and calcium. Simple dissolution of minerals in water also makes nutrients available from rock and soil.

Atmospheric CO_2 is the source of carbon in terrestrial communities. Similarly, gaseous nitrogen from the atmosphere provides most of the nitrogen content of communities. Several types of bacteria possess the enzyme nitrogenase and convert N_2 to NH_4^+, which can then be taken up by plant roots. Communities containing plants such as legumes and alder trees (*Alnus* spp.), with their root nodules containing symbiotic nitrogen-fixing bacteria, may receive a substantial proportion of their nitrogen in this way ($100–300$ kg ha^{-1} year^{-1}). Other nutrients from the atmosphere become available to communities as wetfall (rain, snow and fog) and dryfall (settling of particles during dry periods). Rain contains a number of chemicals from a variety of sources:

- solutions of trace gases such as sulfur and nitrogen oxides;
- solutions of aerosols containing particles rich in sodium, sulfate, magnesium and chloride;
- dust particles from fires, storms rich in calcium, sulfate and phosphate.

The constituents of rainfall that serve as nuclei for raindrop formation make up the rainout component, whereas other constituents, both particulate and gaseous, are collected from the atmosphere as rain falls (the washout component). Nutrients dissolved in precipitation become available to plants when the water enters the soil and reaches plant roots, although some absorption by leaves is also involved. Dryfall is an important process, particularly in terrestrial habitats with long, dry seasons, and can account for up to 50% of the atmospheric input of sulfate, nitrate, calcium and potassium. In some cases streamflow can provide a significant input of nutrients when material is deposited in flood plains.

The two other main inputs in terrestrial environments are:

- decomposition of dead OM;
- inputs from man of compounds not usually present at the concentration found.

The processes involved in these two inputs are the subject of further discussion in the following sections.

DECOMPOSITION

The breakdown of OM is not a single chemical transformation but a complex process, with many steps, sequential and concurrent. There is chemical alteration of OM, physical fragmentation and finally release of mineral nutrients. Many species of small animals live in the soil and contribute to the mechanical decomposition of OM as well as to its chemical decomposition through digestion. Figure 27.1 shows the cycling of nutrients common to all terrestrial systems. Breakdown starts almost immediately after the organism, or part of it, dies.

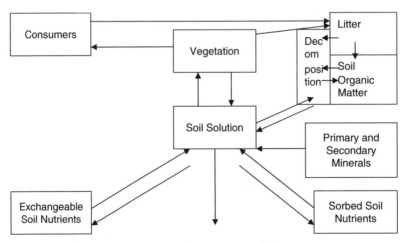

Figure 27.1   Common forms of nutrient cycling in terrestrial systems

The OM is colonized by microorganisms that use enzymes to oxidize the OM to obtain energy and carbon. The surfaces of leaves and roots (and often their interiors) are colonized by microorganisms even before they die. Despite their small volume in soil, microorganisms are key players in the cycling of nitrogen, sulfur, and phosphorus, and in the decomposition of organic residues. Thereby they affect nutrient and carbon cycling on a global scale. The organic residues are in this way converted to biomass or mineralized to CO_2, H_2O, mineral nitrogen, phosphorus and other nutrients (Bloem et al., 1997). Nutrients immobilized in microbial biomass are subsequently released when microbes are grazed by microbivores such as protozoa and nematodes. Microorganisms are further associated with the transformation and degradation of pollutants, waste materials and synthetic organic compounds.

Soil animals such as earthworms assist in the decomposition of OM by incorporating it into the soil where conditions are more favorable for decomposition than on the surface. Earthworms and other larger soil animals such as mites, collembola and ants by fragmenting organic material increase its surface area, enabling more microorganisms to colonize the OM and decompose it more rapidly (Begon et al., 1995).

During decomposition the organic molecules in OM get broken down into simpler organic molecules that require further decomposition or conversion into mineralized nutrients (Ball and Allen, 1991). The compounds in OM vary in the ease with which microorganisms can break them down (McCarthy and Ball, 1991). The first organic compounds to be broken down are the easiest – simple sugars and carbohydrates such as glucose and sucrose. These compounds are also the first products of photosynthesis and are high-quality substrates for decomposition; the molecules are small and their chemical bonds are energy rich (Ball, 1993). This results in the release of much more energy than would be required to create the enzymes necessary to break down the sugar. Further, small soluble molecules can readily be taken inside the microorganism and metabolized internally (Berrocal et al., 1997).

Cellulose represents a polymer of glucose, being composed of β 1–4 linked glucose units. Cellulose is the main component of primary cell walls in plants. Cellulose, the most widely found molecule in terrestrial ecosystems, is resistant to degradation, and a range of extracellular enzymes, for example endoglucanases, are required to cleave the large polymer into smaller chains (Ball and Trigo, 1997). Other cellulolytic enzymes (exoglucanases) cleave smaller oligosaccharide units from the end of the chain. Finally, β-glucosidases

separate individual glucose units from the end of the polymers to complete the disassembly (Ball and McCarthy, 1989).

A number of naturally occurring phenolic compounds containing double (unsaturated) C=C bonds are found in plant litter. These compounds, which have very important roles in plant structure and function, also affect decomposition. Two types of these compounds are recognized. Polyphenols (or tannins) and small phenol polymers made from a number of phenolic acids, including caffeic, chlorogenic, cinnamic, p-coumaric, ferulic and protocatechuic acids. In plants, polyphenols are thought to serve as primarily defense mechanisms against grazing animals and pathogenic microorganisms. Lignins are the second type of phenolic compounds containing C=C bonds and the second most abundant compound present in plant litter (Betts et al., 1991). These large, amorphous and very complex compounds are among the more complex and variable of natural compounds; they have quite variable structure and no precise chemical description. Lignin is one of the slowest of the plant components to decay, and its decomposition results in almost no energy gain to the microorganism since large amounts of energy are required to complete its decomposition. Certain complex enzyme systems that can decompose the cellulose, hemicellulose and lignin fractions of plant litter simultaneously have been identified (Trigo and Ball, 1996). The chemical qualities that provide strength and rigidity to the plant are the cause of a very slow rate of decomposition.

We can therefore see that there exists in a natural soil community the potential for the breakdown and recycling of a whole range of compounds. So far we have concerned ourselves with the breakdown and removal of natural products in the soil. But what about the breakdown of other compounds which can be found to be present in soil as a result of anthropogenic activities? The cleanup of environments contaminated with such compounds is generally termed bioremediation.

BIOREMEDIATION

A general definition of the term bioremediation is "the use of living microorganisms to return an object or area to a condition, which is not harmful to plant or animal life." A good example of bioremediation is the treatment of wastewater such as sewage. This is a long-standing and well-understood example of using living organisms to return material to its original state.

However, over the last two decades there has been a more specific use of the term bioremediation.

This is reflected in the two more commonly used and perhaps more specific definitions:

- The use of living organisms to degrade environmental pollution or to prevent pollution through waste treatment (Atlas, 1995).
- The application of biological treatment to the cleanup of hazardous chemicals (Cookson, 1995).

The advantages of using bioremediation rather than digging up the contaminated soil and placing it elsewhere are that only moderate capital investment is required as is only low energy input. In addition the processes are environmentally safe, do not generate waste, and are self-sustaining. In many cases bioremediation not only offers a permanent solution to the problem but also is cost effective. Cleaning up existing terrestrial environmental contamination in the USA alone can cost as much as $1 trillion. Bioremediation can help reduce the costs of treatment as follows:

- *Treating contamination in place*: Most of the cost associated with traditional cleanup technologies is associated with physically removing and disposing of contaminated soils. Because engineered bioremediation can be carried out in place by delivering nutrients to contaminated soils, it does not incur removal–disposal costs.
- *Harnessing natural processes*: At some sites, natural microbial processes can remove or contain contaminants without human intervention. In these cases where intrinsic bioremediation (natural attenuation) is appropriate, substantial cost savings can be realized.
- *Reducing environmental stress*: Because bioremediation methods minimize site disturbance compared with conventional cleanup technologies, post-cleanup costs can be substantially reduced.

From the above discussion it can be concluded that bioremediation is generally more cost effective than landfilling or incineration, the two other widely used methodologies to remove contaminants from soil. Table 27.1 shows the cost effectiveness of bioremediation alongside alternative treatments.

Specific terms used to describe the activity of the microorganisms and the way these organisms are:

- *Biodegradation* – where the breaking down of a compound or substance is achieved with living organisms such as bacteria or fungi. These could be indigenous to the area, or could be introduced.
- *Biostimulation* – where the natural or introduced population of microbes in an area are enhanced through addition of nutrients, engineering, or other manipulation of an area. This speeds up the natural remediation process.
- *Bioaugmentation* – where specific living organisms are added to a site or material to achieve a desired bioremediation effect.
- *Biorestoration* – restoration to original or near original state using living microbes.

The technology used to treat a polluted soil is dependent on the site and the pollutant. Therefore, each bioremediation treatment is site specific. However, the different treatments rely on a number of basic technologies which are described below:

In situ *bioremediation* (ISB) is the use of microorganisms to degrade contaminants in place with the goal of obtaining harmless chemicals as end products. Most often, *in situ* bioremediation is applied to the degradation of contaminants in saturated soils, although bioremediation in the unsaturated zone can occur. ISB has the potential to provide advantages such as complete destruction of the contaminant(s), lower risk to site workers, and lower equipment/operating costs. ISB can be categorized by metabolism or by the degree of human intervention. At a high level, the two categories of metabolism are aerobic and anaerobic. The target metabolism for an ISB system will depend on the contaminants of concern. Some contaminants (e.g. fuel hydrocarbons) are degraded via an aerobic pathway, some anaerobically (e.g. carbon tetrachloride), and some contaminants can be biodegraded under either aerobic or anaerobic conditions (e.g. trichloroethene).

Accelerated in situ *bioremediation* is where substrate or nutrients are added to an aquifer to stimulate the growth of a target consortium

Table 27.1 A comparison of the costs of bioremediation with the costs of traditional methodologies

Method	Cost of bioremediation per m^3 ($)		
	Year 1	Year 2	Year 3
Incineration	583	None	None
Landfill	737	None	None
Thermal desorption	220	None	None
Bioremediation	193	29	22

of bacteria. Usually the target bacteria are indigenous; however, enriched cultures of bacteria (from other sites) that are highly efficient at degrading a particular contaminant can be introduced into the aquifer (termed bioaugmentation). Accelerated ISB is used where it is desired to increase the rate of contaminant biotransformation, which may be limited by lack of required nutrients, electron donor, or electron acceptor. The type of amendment required depends on the target metabolism for the contaminant of interest. Aerobic ISB may only require the addition of oxygen, while anaerobic ISB often requires the addition of both an electron donor (e.g. lactate, benzoate) as well as an electron acceptor (e.g. nitrate, sulfate). Chlorinated solvents, in particular, often require the addition of a carbon substrate to stimulate reductive dechlorination. The goal of accelerated ISB is to increase the biomass throughout the contaminated volume of aquifer, thereby achieving effective biodegradation of dissolved and sorbed contaminants.

Monitored natural attenuation (intrinsic bioremediation) is the other method of applying *in situ* bioremediation. One component of natural attenuation is the use of indigenous microorganisms to degrade the contaminants of concern without human intervention (such as supplementing the available nutrients). Site characterization and long-term monitoring comprise the activities required to implement natural attenuation. Long-term monitoring is used to assess the fate and transport of the contaminants compared against the predictions. The reactive transport model can then be refined to obtain better predictions. Natural attenuation processes are typically occurring at all sites, but to varying degrees of effectiveness depending on the types and concentrations of contaminants present and the physical, chemical, and biological characteristics of the soil and groundwater. Natural attenuation processes may reduce the potential risk posed by site contaminants in three ways:

- The contaminant may be converted to a less toxic form through destructive processes such as biodegradation or abiotic transformations.
- Potential exposure levels may be reduced by lowering of concentration levels (through destructive processes, or by dilution or dispersion).
- Contaminant mobility and bioavailability may be reduced by sorption to the soil or rock matrix.

APPLICATION OF BIOREMEDIATION

The rapid expansion and increasing sophistication of the chemical industries in the past century and particularly over the last thirty years has

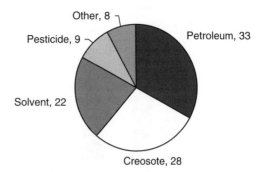

Figure 27.2 The major types of waste chemicals that are applicable to bioremediation (%)

meant that there has been an increasing amount and complexity of toxic waste effluents. At the same time, fortunately, regulatory authorities have been paying more attention to problems of contamination of the environment. Industrial companies are therefore becoming increasingly aware of the political, social, environmental and regulatory pressures to prevent escape of effluents into the environment. The occurrence of major incidents (such as the Union-Carbide (Dow) Bhopal disaster or the release of radioactive material in the Chernobyl accident, etc.) and the subsequent massive publicity due to the resulting environmental problems has highlighted the potential for imminent and long-term disasters in the public's conscience. Even though policies and environmental efforts should continue to be directed towards applying pressure to industry to reduce toxic waste production, bioremediation presents opportunities to detoxify a whole range of industrial effluents (Figure 27.2).

Bacteria can be altered to produce certain enzymes that metabolize industrial waste components that are toxic to other life, and also new pathways can be designed for the biodegradation of various wastes. Bioremediation is therefore capable of dealing with a wide range of waste chemicals.

BIOREMEDIATION AS A BUSINESS

Over the last few years, a number of companies have been established already to develop and commercialize biodegradation technologies. For an example of a bioremediation company, Envirogen (New Jersey) has developed recombinant PCB (polychlorinated biphenyl)-degrading microorganisms with improved stability and survivability in mixed populations of soil organisms. The same

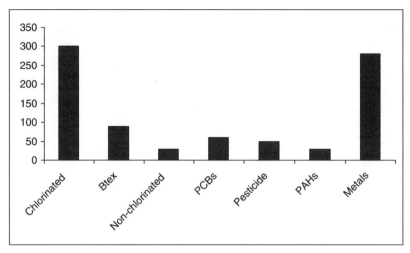

Figure 27.3   The number of sites in the USA that require treatment for pollution (N.B. Btex = benzene, toluene, ethylbenzene, xylene)

company also has developed a naturally occurring bacterium that degrades trichloroethylene (TCE) in the presence of toluene, a toxic organic solvent killing many other microorganisms. Use of microbes for bioremediation is not limited to detoxification of organic compounds. In many cases, selected microbes can also reduce the toxic cations of heavy metals (such as selenium) to the much less toxic and much less soluble elemental form. Figure 27.3 summarizes the potential size of this industry by examining the number of polluted sites that have been identified in the USA that require pollution cleanup.

THE SOCIAL ACCEPTABILITY OF BIOREMEDIATION

As we have seen, bioremediation offers the possibility of technically effective and relatively less expensive remediation. Assuming that the promised bioremediation strategies are realized, why would anyone object to using these natural treatments? A failure to anticipate issues that can derail plans to deploy any technology, including bioremediation can be catastrophic (Axelrod, 1994; Davison et al., 1997). While some issues may revolve around technical aspects of bioremediation, others may derive from non-technical, social concerns. Site-specific bioremediation decision-making can be viewed as a social process that is informed by scientific and technical data, rather than as a physical process that is determined by scientific and technical data. While it is not asserted that bioremediation represents a controversial technology, the use of a simple clean-up

option may become controversial (Priest, 1994). Bioremediation encompasses a suite of potential remediation options whose remediation targets, mechanisms and capabilities differ. Therefore, generic questions about the suitability of bioremediation have limited applicability to the particular situations in which it might be considered for deployment. Yet, neither is every possible permutation of contaminant, site, remediation mechanism and remediation goal likely to produce a unique social response. The approach probably lies somewhere in the middle – an exploration of the generic factors that may influence patterns of social responses to specific bioremediation applications.

To date there have been relatively few systematic studies of social responses to bioremediation. Therefore, despite increasing applications of bioremediation, social issues related to its deployment have not been documented. While bioremediation may prove to be socially acceptable for cleaning up contamination, it may not fully be acceptable either across the suite of approaches it encompasses or across the range of sites at which it is proposed for deployment (Stern and Dietz, 1994). Experience with other technologies, ranging from nuclear facilities to agricultural biotechnology to incineration, reveals the ability of individuals and groups to delay technology deployment. However, the rejection of these technologies is not universal; they are considered seriously as options in some locations and in a smaller subset of cases, deployed. This variation in social responses to technologies makes it difficult to generalize. For example, rather than asking whether bioremediation is socially acceptable or using general opinion surveys to predict site-specific responses to a particular proposed remediation option, it is better to ask under what circumstances might bioremediation be acceptable. Further, the acceptability of this technology should be viewed as multidimensional instead of one-dimensional (e.g. as only as a matter of risk, or risk communication, or education). Acceptability evolves over time through interactions with individuals and organizations, and in response to new technical and non-technical information. Without systematic data, complete analysis of the social dimensions of bioremediation cannot be undertaken. Instead a systematic approach to identifying and analyzing the social determinants of the acceptability of bioremediation can be made. This approach relies on a conceptual framework and draws from published literature to illustrate the attributes of bioremediation and its use.

A FRAMEWORK FOR CONSIDERING THE SOCIAL ASPECTS OF BIOREMEDIATION ACCEPTABILITY

To gain a better understanding of social acceptability issues and to improve our ability to predict outcomes in deliberations over the social acceptability of controversial technologies, Wolfe and Bjornstad (2002) developed a conceptual framework for organizing what was perceived to be the most important issues. The resulting framework, PACT (Public Acceptability of Controversial Technologies) provides a common logic through which to view site-specific decision-making about remediation technologies (Figure 27.4).

PACT is built around dimensions that operate to influence decision-oriented dialogs over controversial remediation technologies in any location. These dimensions, which operate simultaneously (Figure 27.4) are:

- *Dialogue dimension* – the decision-oriented dialogs themselves.
- *Technology dimension* – the technology of concern relative to other technology options.
- *Constituent dimension* – the parties involved in remediation decision-making.
- *Context dimension* – the social, institutional, physical context within which the contaminants are located and remediation decision-making occurs (Wolfe *et al.*, 2002).

Each dimension consists of several factors that are likely to influence the degree of acceptability. The factors relevant to specific decision settings and technologies vary from situation to situation. This PACT-based analysis focuses on an array of attributes that could strongly influence acceptability. In this context acceptability refers to participants' willingness to consider the technology in question as a viable alternative, rather than to whether the technology ultimately is deployed. PACT provides a framework through which to see how participants' position changes over time, from absolute positions of support or opposition at one extreme to completely negotiable positions at the other. Changes in positions may be related to any of PACT's dimensions – from decisions about who should or should not participate in decision-making to the kinds of technologies worth considering.

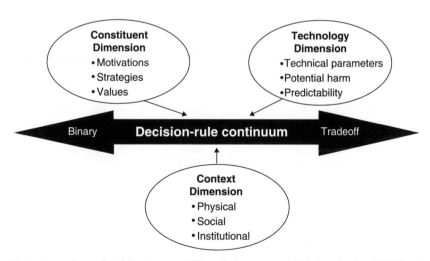

Figure 27.4   Overview of Public Acceptability of Controversial Technologies (PACT), used to assist in site-specific decision-making about remediation technologies. From Wolfe and Bjornstad (2002)

APPLYING THE CONCEPTUAL FRAMEWORK TO BIOREMEDIATION

Clearly, better data generated from the application of bioremediation would enhance the decision-making process. However, in the absence of those data, the application of PACT can assist the decision-making process.

Dialog dimension

Technology, constituent and context dimensions are all important in terms of how they influence participants' positions during the course of decision-making. It is through dialog that the social acceptability of bioremediation is negotiated or determined. Decision-oriented dialogs may be formal or informal, taking any number of forms such as public meetings, open houses, or site tours; private face-to-face meetings, social events; news media reports and editorials, etc. Further, remediation decision-making may last for several years. Participants' decision rules may shift during the course of dialog, as participants respond to interactions with other involved parties and to new information. During the course of decision-making, participants may apply different decision rules to the different issues about which they are concerned. Some issues may be totally negotiable, while others may be absolutely non-negotiable.

Within and among dialog fora, the formal and informal rules of engagement vary, influencing such issues as:

- whether majority views prevail or consensus is sought or required;
- the degree to which dissenting or minority opinions are elicited and considered;
- the power different groups or factions have in promoting or blocking specific actions.

Technology dimension

This dimension consists of the attributes of the technology or technologies in question that affect their acceptability, that is, that affect involved parties' willingness to consider the technology seriously as an option. Attributes include the technology's technical parameters, potential harm to human health and the environment and the predictability of the technology's performance and of any associated harm to human health or well-being, the community, or the environment. Because many bioremediation technologies exist, their lack of predictability may also condition the degree of their acceptability. Among the factors that influence predictability are the specific bioremediation strategies under discussion, similarity of site conditions for proposed application with past application and the reliability of financial cost estimates.

Constituent dimension

Each party involved in bioremediation decision-making brings different values, motivations and strategies to its involvement. These have an enormous impact on the ensuing dialog about using particular bioremediation methods at specific sites. Values affect bioremediation acceptability primarily by influencing and distinguishing those issues that are negotiable from those that are not. Therefore, although environmental values have been the subject of considerable scrutiny and analysis (e.g. Axelrod, 1994; Stern and Dietz, 1994; Eagly and Kulsea, 1997), a core versus trade-off value distinction is probably the most important in the context of the PACT framework. Threats to core values which are deeply held and resistant to change create consternation and, potentially, outrage. For example, the idea of using genetically engineered microbes in bioremediation may become a fundamental issue among those constituents whose core values are affronted (Hoban et al., 1992; Hagedorn and Allender-Hagedorn, 1997). Among the most challenging and, currently, least predictable situations are those in which core values or trade-off values that constituents think are important conflict. These kinds of conflicts need not be framed as either/or choices; more of one item need not mean less of another. Further, layers of conflict may emerge as constituents consider bioremediation in detail.

Context dimension

Physical, social, and institutional context create the settings in which bioremediation strategies may be selected and deployed. A site's physical context – its topography, hydrogeology, climatic regime, the nature and extent of contamination – clearly is important in determining the potential technical effectiveness of bioremediation techniques. A site's physical context also influences the potential social acceptability of bioremediation options. As an example, constituents' potential concerns about bioremediation may be different at sites isolated from human habitation than at sites adjacent to residences, recreation areas, or schoolyards. While ecological issues may take precedence at isolated sites, human health concerns likely would predominate at the latter. Social context refers to a community's demographic and social composition, its internal dynamics and interactions with outsiders (e.g. corporations, state and federal governments, advocacy groups). A community's social context evolves; it is influenced by the past and it responds to internal and external influences. One crucial way in which social context may influence bioremediation acceptability is through trust. If the organizations promoting the use of bioremediation are distrusted by community members (e.g. if those same organizations polluted the ground or water but, for a while, denied the pollution or appeared to avoid taking responsibility for cleaning up the site), statements about bioremediation may have little credibility. Bioremediation may be socially unacceptable because of the historic interactions among involved parties rather than because of its technical attributes (Slovic, 1987, 1997). Institutional context refers to the nature of organizations that influence or are affected by the contaminated site, decisions regarding its remediation and the consequences of those decisions. For example, many bioremediation strategies require continued monitoring and maintenance. These kinds of activities, in turn, require continuing institutional support. If constituents believe that long-term funding or institutional responsibility for monitoring and maintenance are unlikely, then those bioremediation alternatives may not be considered as serious options.

CONCLUSION

In this chapter the flow of nutrients and energy through the terrestrial ecosystem has been examined. The importance of the soil biota in global nutrient cycling has been described. The range of compounds present in soils, varying from simple sugars to complex polyphenolic lignins, suggests that soil microbes are capable of degrading a range of chemicals both natural and human made, enabling suitable conditions to exist within the soil. The use of living microorganisms to return an object or area to a condition which is not harmful to plant or animal life is a general description of the process of bioremediation. Bioremediation is now a successful environmental biotechnology, having a number of advantages (e.g. cost, environmental friendly means of disposal) over any alternative treatment of contaminated land such as landfilling or incineration. There exist large areas of the world where contaminated land can be found, constituting an environmental and health hazard. Bioremediation offers the opportunity to utilize the natural microbial population to treat the contaminated site, returning the elements making up the contaminants to the natural nutrient cycling.

PACT analysis predicts that the social acceptability of bioremediation would be conditional rather than absolute. The technologies that constitute bioremediation are neither acceptable nor unacceptable. Rather, acceptability is derived from the setting in which it might be used. Applying PACT to bioremediation reveals flaws in the typical one-dimensional method often used for gaining technology acceptance (e.g. educating the public about the technology and its benefits or communicating effectively the attributes of the technology in question).

REFERENCES

Atlas RM (1995). Bioremediation. *Chemical and Engineering News* 73(14): 32–42.

Axelrod L (1994). Balancing personal needs with environmental preservation: Identifying the values that guide decisions in ecological dilemmas. *Journal of Social Issues* 50(3): 85–104.

Ball AS (1993). Carbohydrates. In: Chambers A and Rickwood D (eds) *Biochemistry Labfax*. Bios Scientific Publishers, Oxford, pp. 305–315.

Ball AS and Allen M (1991). Solubilisation of wheat straw by actinomycetes. In: Wilson WS (ed.) *Advances in Soil Organic Matter Research: The Impact of Agriculture on the Environment*. The Royal Society of Chemistry, Cambridge, UK, pp. 275–286.

Ball AS and McCarthy AJ (1989). Comparative analysis of enzyme activities involved in straw saccharification by actinomycetes. In: Grassi G, Pirrowitz D and Zibetta H (eds) *Energy from Biomass*, 4th edn. Elsevier Applied Science, London, pp. 271–274.

Ball AS and Trigo C (1997). The role of actinomycetes in plant litter decomposition. *Recent Developments in Soil Biochemistry* 1: 9–13.

Batjes NH (1996). Total carbon and nitrogen in the soils of the world. *European Journal of Soil Science* 47: 151–163.

Begon M, Harper JL and Townsend CR (1995). *Ecology Individuals, Populations and Communities*, 2nd edn. Blackwell Science, Oxford.

Berrocal MM, Rodriguez J, Ball AS *et al.* (1997). Solubilisation and mineralisation of [^{14}C] lignocellulose from wheat straw by *Streptomyces cyaneus var. viridochromogenes* during growth in solid-state fermentation. *Applied Microbiology and Biotechnology* 48: 379–384.

Betts WB, Dart RK, Ball AS *et al.* (1991). Biosynthesis and structure of lignocellulose, In: Betts WB (ed.) *Biodegradation: Natural and Synthetic Materials*. Springer Verlag, London, pp. 139–156.

Bloem J, de Ruiter P and Bouwman LA (1997) Food webs and nutrient cycling in agro-ecosystems. In: van Elsas JD, Trevors JT and Wellington EMH (eds) *Modern Soil Microbiology*. Marcel Dekker, New York, pp. 245–278.

Cookson, JT (1995). *Bioremediation Engineering; Design and Application*. McGraw Hill, New York.

Davison A, Barns I and Schibeci R (1997). Problematic publics: a critical review of surveys of public attitudes to biotechnology. *Science Technology and Human Values* 22(3): 317–348.

Eagly A and Kulsea P (1997). Attitudes, attitude structure and resistance to change. In: *Environment, Ethics, and Behavior: The Psychology of Environmental Valuation and Degradation*. The Lexington Press, San Francisco, pp. 122–153.

Foth HD and Turk LM (1972). *Fundamentals of Soil Science*. John Wiley, New York.

Hagedorn C and Allender-Hagedorn S (1997). Issues in agricultural and environmental biotechnology: identifying and comparing biotechnology issues from public opinion surveys, the popular press and technical/regulatory sources. *Public Understanding of Science* 6: 233–245.

Hoban T, Woodrum E and Czaja R (1992). Public opposition to genetic engineering. *Rural Sociology* 57(4): 476–493.

McCarthy AJ and Ball AS (1991). Actinomycete enzymes and activities involved in straw saccharification. In: Betts WB. (ed.) *Biodegradation: Natural and Synthetic Materials*. Springer Verlag, London, pp. 185–200.

Prentice IC, Farquhar GD, Fasham MJR *et al.* (2001) The carbon cycle and atmospheric carbon dioxide. In: Houghton JT, Ding Y, Griggs DJ *et al.* (eds) *Climate Change: The Scientific Bases*. Cambridge University Press, Cambridge, UK, pp. 183–237.

Priest SH (1994). Structuring public debate on biotechnology: Media frames and public response. *Science Communication* 16(2): 166–179.

Slovic P (1987). Perception of risk. *Science* 236: 280–285.

Slovic P (1997). Public perception of risk. *Journal of Environmental Health* 59: 22–23.

Stern PC and Dietz T (1994). The value basis of environmental concern. *Journal of Social Issues* 50(3): 65–84.

Trigo C and Ball AS (1996). Production and characterisation of humic-type solubilised lignocarbohydrate polymer from the degradation of wheat straw by actinomycetes. In: Clapp CE, Hayes MHB, Senesi N and Griffith SM (eds) *Humic Substances and Organic Matter in Soil and Water Environments*. IHSS Press, Birmingham, UK, pp. 101–106.

Wolfe AK and Bjornstad DJ (2002) Why would anyone object? An exploration of social aspects of phytoremediation acceptability. *Critical Reviews in Plant Sciences* 21(5): 429–438.

Wolfe AK, Bjornstad DJ, Russell M *et al.* (2002). A framework for analyzing dialogs over the acceptability of controversial technologies. *Science Technology and Human Values* 27(1): 134–159.

28

Regenerating Aquaculture – Enhancing Aquatic Resources Management, Livelihoods and Conservation

Stuart W. Bunting

INTRODUCTION

Aquaculture, broadly defined as the farming of aquatic species, has emerged as an important food producing sector. The Food and Agriculture Organisation of the United Nations (FAO, 2004) noted that since 1970 total aquaculture output grew at an average of 8.9% per year until 2002, as compared with 2.8% for terrestrial livestock farming and 1.2% for capture fisheries over the same period. Furthermore, finfish and shellfish farming expanded at rates of 9.1 and 5.8% per year between 1995 and 2000, reaching production levels globally of 23.2 and 12.4 million tonnes, respectively; farmed fish, crustaceans and molluscs represented 27% of global supplies in 2000 (FAO, 2002a, b). Despite rapid population growth, aquaculture production per capita increased from 0.7 kg in 1970 to 6.4 kg in 2002; the number of people economically active in the sector in 2002 was 9.8 million and production (including plants) was worth US$60 billion (FAO, 2004).

However, global figures, dominated by freshwater aquaculture production in China, conceal the fact that farming marine finfish accounts for a small proportion, less than 5% in 2000, of that consumed by humans. Moreover, aquatic foods do not necessarily readily substitute for meat and livestock products and farming marine fish and

shrimp can constitute a net protein consumer (Naylor et al., 2000; Muir, 2005). Aquaculture also appropriates a wide range of environmental goods and services (Berg et al., 1996; Beveridge et al., 1997; Kautsky et al., 1997; Folke et al., 1998; Bunting, 2001a) and where demand exceeds the environmental carrying capacity adverse impacts are observed: physical and chemical environmental parameters may be affected; eutrophication can occur; species abundance and diversity may be changed. Escapees from the culture system may directly compete with and predate upon wild stocks; aquaculture has been cited as causing disease and genetic degradation in native populations. Furthermore, despite broad positive socioeconomic benefits associated with appropriate aquaculture development (ODI, 1999) locally aquaculture development is often contentious, occasionally resulting in conflict, especially where traditional or *de facto* access rights and resource use patterns are disrupted.

Problems associated with poorly planned and inappropriate aquaculture development, discussed further below, have resulted in calls for future development initiatives to focus on low-impact, ecologically sound and sustainable aquaculture (Naylor et al., 1998; Naylor et al., 2000). Notably, Principle 6 of the UNEP 'Guiding principles for post-tsunami rehabilitation and reconstruction'

(UNEP, 2005) recommended 'Ecosystem based management measures' that 'promote sustainable fisheries management in over-fished areas, and encourage low impact aquaculture'. The principles also 'Encourage investment in community-based aquaculture and other livelihoods that bring benefits to local populations and do not degrade coastal ecosystems', specifying that 'Rehabilitated aquaculture must adopt environmentally sound management practices that do not pollute, damage habitats or cause long-term harm, including use of feed that is taken from sustainable sources and seeds that are raised in environmentally sound hatcheries or taken from sustainable fisheries'.

The guiding principles outlined above aim to address unsustainable fisheries and aquaculture activities in conjunction with post-tsunami reconstruction efforts throughout the tsunami-affected area. However, similar problems exist in many other situations, both coastal and freshwater. Consequently, an urgent need has been identified for a review that: provides a clear synopsis of problems associated with existing aquaculture practices; profiles high-potential strategies to realise the goal of low-impact aquaculture; highlights opportunities for low-impact aquaculture using case studies; identifies policy initiatives that would promote approaches to regenerating aquaculture.

SYNOPSIS OF PROBLEMS ASSOCIATED WITH AQUACULTURE

Articles in popular science journals have focused public attention on negative environmental impacts of aquaculture, in particular intensive production of salmon and shrimp (*New Scientist*, 1996, 1998; *Nature,* 1997; Naylor *et al.*, 1998, 2000). However, as Boyd (1999) noted, the debate over impacts associated with aquaculture can become distorted by stakeholder groups with markedly different agendas, and consequently, he advocated impact assessment based on a rational appraisal, using the best available data.

Environmental impacts associated with aquaculture development include: habitat and ecosystem loss associated with site development; appropriation of environmental goods and services; physical and chemical degradation of water resources; nutrient enrichment in receiving waterbodies and subsequent eutrophication and altered ecological status; release of escapees, disease and parasites into the environment. Principal causes and affects associated with these environmental impacts are discussed below and summarised in Table 28.1; Bunting (2001b) provides a more comprehensive review. Consequences of these negative environmental impacts can include self-pollution, restricted amenity, reduced functionality and impacts on option and non-use values.

Reduced functionality

Natural wetland functions support a wide array of environmental goods and services that sustain economic activities and societal systems (Burbridge, 1994). However, aquaculture development can damage the functional integrity of wetlands, disrupting the supply of environmental goods and services. Considering shrimp farm wastewater, Robertson and Phillips (1995) noted that where the assimilative capacity of a mangal ecosystem was exceeded and excessive loads of ammonia and organic matter resulted in anaerobic sediments, mangrove trees might die and various functions may be lost. Reduced biomass production and decreased nutrient assimilation and cycling could potentially lead to increased nutrient export and consequently adverse impacts on the receiving environment. Loss of the mangrove root system could decrease sediment stability, leading to erosion, which could increase saline intrusion and the risk of flooding inland. Loss of mangrove habitat also represents a loss of nursery areas for juvenile fish and shrimp, which would affect fisheries and threaten biodiversity.

Self-pollution

Wastewater from land-based aquaculture is routinely discharged to streams and rivers supplying other aquaculture operations downstream, whilst waste discharged from pen and cage farms may be conveyed to other farms by currents and tides. Moreover, for pen and cage aquaculture and pump-ashore facilities there is a danger that discharged wastes may contaminate water intended to supply the farm. Problems of self-pollution have been associated with shrimp farming in the intertidal area surrounding the Mundel–Dutch Canal lagoon system in the northwestern province of Sri Lanka (Corea *et al.*, 1998). Farms were discharging wastewater containing high concentrations of ammonia, nitrite, nitrate and metals to the same areas used to supply their own or neighbouring culture systems. The problem was compounded as mixing in the shallow lagoon was limited and a sandbar prevented exchange between the lagoon and ocean.

Furthermore, destruction of mangroves and salt marshes to build shrimp farms resulted in a loss of flood-buffering capacity and elevated sediment loads which compounded water quality problems in the lagoon. Self-pollution caused disease outbreaks, encouraged parasite infestations, increased gill fouling with suspended solids, retarded growth and resulted in poor quality

Table 28.1 Negative impacts associated with aquaculture development

Impact	Consequence	Key features	References
Physical	Modified hydrology	Reduced flow rates, modified channel morphology and flow regimes and salinisation of surface waters in coastal areas	Beveridge and Phillips (1993); Phillips et al. (1993); Tran et al. (1999)
	Sedimentation	Reduced interstitial water flow, increased embeddedness and sediment anoxia	
Chemical	Pollutants	Excreted ammonia toxic to invertebrates; waste nutrients lead to hypernutrification; therapeutants and their residues affect non-target organisms leading to antibiotic resistant strains	Nature Conservancy Council (1990); Phillips et al. (1993); Weston (1996); Davies et al. (1997)
	Reduced oxygen concentrations	On-farm respiration reduces oxygen levels leading to exclusion of sensitive species; biological and chemical oxygen demand further depletes oxygen levels	Kelly and Karpinski (1994); Gillibrand et al. (1996)
Nutrient enrichment	Eutrophication	Increased phytoplankton and periphyton production near cages, including possible stimulation of toxic algae blooms; increased epiphyte growth downstream of land-based farms; elevated respiration during decomposition	Nature Conservancy Council (1990); Bonsdorff et al. (1997); Selong and Helfrich (1998); New Scientist (1999)
Shifting trophic status	Modified species assemblages	Elimination of pollution sensitive invertebrates and fish; increased abundance and biomass of tolerant species and eutrophication leading to trophic cascades	Oberdorff and Porcher (1994); Loch et al. (1996); Selong and Helfrich (1998)
Escapees	Predation, competition and ecological impacts	Ecosystem disruption through foraging and consumption of native flora and fauna; escapees breed with resident populations leading to genetic degradation	Welcomme (1988); Arthington and Bluhdorn (1996); Bardach (1997)
Disease and parasites	Loss of native species	Viruses, bacteria and parasites infest native populations, exotic parasites may also devastate non-resistant indigenous populations	Arthington and Bluhdorn (1996); McAllister and Bebak (1997)
Self-pollution	Reduced production and product quality	Upwelling of anoxic water causing fish-kills in cages; reduced water quality leading to disease outbreaks and stimulation of toxic algae blooms	Lumb (1989); Black et al. (1994); Corea et al. (1998)
Restricted amenity	Decline in capture fisheries and water quality	Competition and disease can damage capture fisheries; sedimentation and plant growth restricts water flow in navigation and irrigation canals; reduced water quality affects access of livestock and humans to water causing social unrest	Nature Conservancy Council (1990); Phillips et al. (1993); Primavera (1997); Tran et al. (1999)
Reduced functionality	Loss of ecological functions	Discharged wastewater can degrade ecosystems leading to habitat loss, decreased diversity, restricted storage capacity for nutrients and water and disruption to flows of environmental goods and services	Burbridge (1994); Robertson and Phillips (1995)
Impacts on option and non-use values	Reduced perception of aquatic resources	Degraded aquatic environments and stakeholder conflicts lead to a negative perception of aquaculture, with reduced values attributed to the ecosystem	Turner (1991); Folke et al. (1994); Muir et al. (1999)

Source: Adapted from the review by Bunting (2001b)

shrimp production (Corea *et al.*, 1998). Similarly, wastewater discharged from Indonesian shrimp farms caused water quality problems in adjacent small-scale *tambak* shrimp farms that were unable to regulate the exchange of water between ponds (Muluk and Bailey, 1996).

Increased oxygen consumption by microbial communities in sediments below cage farms may lead to anoxic conditions and the evolution of methane and hydrogen sulphide; release of which has been implicated in causing gill disease on fish farms (Black *et al.*, 1994). Furthermore, respiration by benthic invertebrates and microbial communities in sediments receiving organic inputs from cage farms may depress oxygen concentrations in the water column. Such depressions have been observed beneath cage farms (Lumb, 1989; Gillibrand *et al.*, 1996) whilst upwelling of this anoxic water poses a serious threat to the health of cultured fish and has been implicated in causing fish kills (Beveridge, 1996). Nutrients from aquaculture have also been implicated in stimulating toxic algae blooms, which in certain situations have resulted in restrictions on shellfish fisheries (*New Scientist*, 1999).

Social tension and conflicts

Aquaculture development is promoted to enhance poor livelihoods, contribute to food security, income diversification, greater employment opportunities, reduced food prices and more efficient water and nutrient management within households or communities (ODI, 1999). However, aquaculture development can be contentious, resulting in social tension and conflict, especially where traditional access rights and resource use patterns are disrupted. In Bangladesh, Honduras and Thailand, emergence of shrimp farming was associated with growing inequity (Dewalt *et al.*, 1996; Deb, 1998; Flaherty *et al.*, 1999), whilst in the Bay of Fundy, Canada and southern Chile, expansion in the cage culture of salmon reportedly caused conflict (Ridler, 1997; Marshall, 2001; Barrett *et al.*, 2002).

Aquaculture development often occurs around sheltered coastal bays and lagoons, in marine and freshwater wetlands, along rivers, on floodplains and in reservoirs, which constitute important centres for social and economic activity, and as such is likely to come into conflict with existing users. For example, rapid, uncontrolled expansion of aquaculture in bays can result in access to fishing grounds and navigational routes used by shore-based communities being severely disrupted, which can in turn lead to social tension (Beveridge and Phillips, 1993). In the case of Laguna de Bay, rapid expansion of pen-based aquaculture resulted in severe conflicts between fish-pen operators and fishing communities, the fish-pens also appropriated environmental goods and services affecting the productivity of the natural ecosystem (Yap, 1999). According to this author, in the early 1980s pens occupied 35,000 ha of the 90,000 ha lake area, and interfered directly with navigation and traditional capture fisheries.

Conflicts relating to cage-based aquaculture development in human-made reservoirs and the multiple uses of water resources were observed in west and central Java. Pollnac and Sihombing (1996) noted that overlapping jurisdictions and a lack of environmental monitoring and controls resulted in overcrowding of cages, fish kills and social inequalities. The risk of theft and vandalism is prevalent in situations where aquaculture development has come into conflict with existing or even potential future users of water resources. Pollnac and Sihombing (1996) reported that fish cage nets had been cut at Silalitoruan, and that this had been attributed to local fishermen who wanted to remove competition for space and resources and also release fish that they could subsequently capture. Aquaculture operations in reservoirs can also come into conflict with normal operating procedures when sudden or unexpected drawdown occurs and this either restricts the area accessible for culture or in severe cases damages infrastructure. Enclosures and structures for aquaculture can interfere with other uses, including navigational routes and recreational activities, such as swimming, boating and angling. Physical structures associated with aquaculture, including shore-based trestles, pens, cages and brushparks and reefs can also impede fishing operations, limiting the diversity of useful fishing gear; in certain cases driftnets, longlines and traps may be practical; however, trawling, bottom dragging and seine netting may be impossible.

Conflicts over the positioning of aquaculture operations can also arise through indirect competition; for example, culture units can have various negative impacts on the scenic quality of an environment. Developments of intensive fish farms in remote and unspoilt wilderness areas, areas of open water, famous historic and scenic views and close to monuments and listed structures have all given rise to concern. Practices at the site can cause further anxiety, for example, employing cages with storage huts and other tall structures, vividly coloured buoys and unsightly or untidy buildings and work areas. In the case of Scotland, controls now exist to restrict the development of culture units in scenically sensitive areas and where required an Environmental Impact Assessment should consider the scenic impact of a proposed culture unit, restricting development in sensitive areas.

Minor changes in the planning and management of culture sites can often avoid major intrusions into the landscape; screening of land-based facilities is often possible, planting trees or building earthen banks around a site can often render it almost invisible, especially if close attention is given to footpaths, tracks and roads. Selecting geographically secluded areas, for example, wooded locations, valleys or sheltered foreshore areas can help reduce the visual impact. When considering a cage site, it is possible to incorporate it into the natural environment in several ways: avoiding open water locations and positioning the cages in an irregular manner, giving them a more organic appearance. Improved management practices, using subdued colours, removing tall structures and generally ensuring the site and associated areas and buildings are clean and tidy can be adopted to help reduce the impact.

Wastewater discharged from land-based aquaculture systems can also affect other aquatic resources users. Deteriorating water quality in the Mundel–Dutch Canal lagoon system, attributed to increased shrimp farming, was blamed for declining capture fisheries; this caused resentment in local fishing communities, which manifested itself as poaching (Corea et al., 1998). Furthermore, during a survey in the local community several people complained of skin diseases possibly associated with poor water quality and although no direct link was established, such concerns may strengthen opposition to shrimp farming. Saline wastewater from shrimp farms in Songkhla, Thailand has been implicated in causing the death of livestock drinking from canals (Primavera, 1997). Furthermore, this author reported that following salinisation of surface water by shrimp farm discharges in Nellore district, India, women were forced to spend longer collecting drinking water from distant sources.

Yields on farms cultivating rice have been affected by saline water released from neighbouring shrimp farms; however, this situation arose due to leaching during the dry season and bunds being breached during the growing season (Tran et al., 1999). These authors noted that shrimp farming may increase the sediment load to local canals and rivers, and Phillips et al. (1993) reported that sediment in shrimp farm wastewater in Thailand and Sri Lanka caused irrigation canals to become silted.

Aquaculture wastewater may encourage nuisance growths of macrophytes that interfere with navigation and recreational activities, for example, angling, swimming and boating. Disease transmission from cage farms to native fish stocks was blamed for declining returns to recreational fisheries (Nature Conservancy Council, 1990). Reductions in native crayfish populations owing to diseases introduced with farmed animals resulted in a proliferation of aquatic macrophytes, eliminating habitat for game fish (Thompson, 1990). Declining recreational fisheries and habitat loss can have severe implications for rural economies that receive income from visiting anglers and tourists. People not directly involved with aquaculture development probably ascribe other benefits to the environment where it is proposed, for example, for moorings for recreational and fishing boats, a site for fishing or collecting shellfish and seaweed, for recreational activities or conservation purposes. Consequently, planning and managing aquaculture development must take into account the demands and priorities of different stakeholder groups.

Impacts on option and non-use values

Reduced quality of the aquatic environment can influence the value an individual attributes to preserving the resource to allow the individual, other individuals and future generations the option of using the resource at a later date (Muir et al., 1999). The impact of an activity on this *option value* may be estimated by assessing the willingness-to-pay (WTP) of an individual to preserve the environment. Folke et al. (1994) extrapolated marginal costs of Sek 50–100[1] and Sek 20–30 kg^{-1} for nitrogen and phosphorus removal, respectively, from sewage in Sweden, to represent the WTP of Swedish society to limit nutrient discharges from salmon aquaculture. Based on a comparison of waste production presented as person equivalents, it was estimated in 1994 that the cost to society of eliminating nitrogen and phosphorus discharges originating from salmon aquaculture equated to Sek 4–4.5 kg^{-1} of production. These authors also calculated that internalising this cost increased production costs for salmon to Sek 31–31.5 kg^{-1}, and although the 1994 farm gate price for salmon was not given, such an increase would reduce profits and possibly threaten the viability of salmon farming.

Environments also have non-use values, the intrinsic or *existence value* of environments is unrelated to humans and their present, or potential, direct or indirect use of the resource (Turner, 1991; Muir et al., 1999). People, although unlikely ever to visit a region may attribute value to its existence and feel a sense of loss when the ecosystem is damaged or degraded through inappropriate aquaculture development. Environmental degradation would also reduce the value ascribed to passing the asset on to future generations, termed the *bequest value*. Therefore, although changes in non-use values of environments due to aquaculture development have not been described, they may be expected to be negative.

Summary

Environmental impacts, cases of self-pollution, restricted amenity and functionality and decreased non-use values associated with aquaculture development are reviewed above. Growing recognition that aquaculture development may be responsible for a wide range of environmental costs and stakeholder conflicts has led to increased awareness regarding the need for more rational, low-impact aquaculture development. The following sections detail some of the most promising approaches to regenerating aquaculture, provide pertinent examples and highlight possible limitations.

REGENERATION STRATEGIES AND POLICIES

Considering aquaculture development in terms of poverty alleviation, social development, equitable use of natural resources and environmental protection several high potential strategies for low-impact aquaculture are reviewed below.

Resource efficient production

The poor resource base of small-scale farms in developing countries means that unexploited nutrient sources, for example, crop by-products, terrestrial weeds, aquatic plants and manure, represent important production-enhancing inputs to fishponds (Edwards et al., 1996). Little and Edwards (1999) reviewed the strategies that have evolved to integrate production of livestock with aquaculture; manure from cattle, buffalo, sheep, pigs and poultry has been employed to enhance production in aquaculture systems. Where access to on-farm nutrient resources is limited and the cost of inorganic fertilisers and supplementary feed prohibitive, food processing by-products, brewery waste, offal and human waste have been used to stimulate production in fishponds. Employing unexploited resources derived from within farming systems or from the immediate area to intensify production in fishponds has been termed *integrated aquaculture* (Little and Muir, 1987). By-products of aquaculture, notably nutrient-rich water and sediments, may also be exploited to enhance production, either in agriculture or other aquatic farming systems. Integrated livestock–agriculture–aquaculture production systems that make efficient use of resources, enhance production and spread risks constitute a promising approach to low-impact aquaculture, for example, the *vuon–ao–chuong* (garden–pond–livestock) or

VAC system developed in Vietnam; periphyton-based aquaculture introduced to farmers participating in the CARE-LIFE project in Bangladesh constitutes another.

Trials with periphyton-based aquaculture in freshwater ponds in Benin, West Africa, gave significantly higher annual fish yields, as compared with production from other rural ponds managed for aquaculture (Hem et al., 1995). Consequently, this innovative pond management strategy was proposed as a suitable technique to increase fish production in rural ponds in south Asia, in particular in Bangladesh and India. Studies in Bangladesh demonstrated significantly higher fish production over controls with the addition of various substrates (Azim, 2001; Azim et al., 2002). Trials in freshwater ponds stocked with 6000 rohu (*Labeo rohita*), 4000 catla (*Catla catla*) and 1500 kalbaush (*L. calbasu*) per hectare, and containing substrate with a surface area roughly equal to the pond area, resulted in an extrapolated annual production of 7000 kg ha^{-1}, almost a threefold increase on average pond production in Bangladesh (Azim, 2001). In addition to enhanced production, farmers trialling the approach noted other benefits; they reported seeing fish rub against the branches in their pond to dislodge parasites, and poaching, a serious concern for many, was also believed to have decreased significantly (Bunting et al., 2005).

Within the CARE-LIFE project a range of interventions were introduced to farmers to help them improve fish production, including stocking polycultures, regular feeding and fertiliser application, dike-cropping and introducing branches as a substrate for periphyton. Combined, these interventions resulted in higher production; however, the relative contribution of periphyton to increased production was unclear. Findings from controlled trials represent a valuable benchmark in this respect, demonstrating the principle and potential of the strategy (Azim, 2001). However, what was evident from the CARE-LIFE participants was that they had adapted and refined the approach developed by researchers to their specific situation. This indicates that guidelines prepared to promote new approaches with poor farmers should acknowledge and account for the logistical and resource limitations they face. Due consideration should also be given to prevailing and indigenous knowledge, guidelines should not be too prescriptive and highlight where alternative or more modest inputs could be applied; although potential gains may be reduced so might the associated degree of risk. There may also be unforeseen environmental and public or animal health impacts associated with adoption of modified management regimes. Widespread deployment of brushparks in Benin, West Africa, to enhance periphyton-based fisheries resulted in local forests being overexploited. Concern has also been raised over the possible role of integrated fish–livestock production systems in mediating the spread of avian flu.

Horizontally integrated production

Horizontally integrated production has been defined as 'the use of unexploited resources derived from primary aquaculture activities to facilitate the integration of secondary aquaculture practices' (Bunting, 2001b). Horizontal integration has the potential to perform several important functions, the most valuable being the assimilation of wastes, reducing discharges to the receiving environment, whilst at the same time producing aquatic products that can be marketed. Reducing waste discharges through horizontal integration will contribute to environmental protection and reduce the risk of negative feedback mechanisms, limiting the possibility of self-pollution. Practical experiences with this approach include: culturing green mussels (*Perna viridis*) on bamboo sticks in wastewater draining from a commercial shrimp farm in the Upper Gulf of Thailand (Lin *et al.*, 1993); cultivating macroalgae in shrimp farm effluent (Briggs and Funge-Smith, 1996); marine and freshwater constructed wetlands to treat wastewater and produce food and biomass crops (Bunting, 2001b; Bunting *et al.*, 2003).

Cultivating hairy cockle (*Scapharca inaequivalvis*) and seaweed (*Gracilaria* spp.) in wastewater from a commercial shrimp farm in Kota Bharu, Malaysia, reduced the concentration of ammonium, total nitrogen (TN) and total phosphorus (TP) by 61, 72 and 61%, respectively (Enander and Hasselstrom, 1994). Wastewater was passed through two ponds in series, one with a surface area of 30 m^2 and 0.3 m deep stocked with 60 kg of cockles and one with a surface area of 18 m^2 and 0.7 m deep containing 5 kg of seaweed. Two sets of ponds were operated in parallel and each received 5.5 m^3 day^{-1} of wastewater. Mean TN and TP concentrations in shellfish ponds were reduced by 55 and 67%, respectively, whilst seaweed ponds reduced mean TN and TP concentrations by 36 and 10%, respectively. A mass balance model predicted that during one month, bivalve production (30 kg) would sequester 260 g of nitrogen and 9 g of phosphorus, whilst seaweed production (60 kg) would assimilate 5 g of nitrogen and 7 g of phosphorus. Enander and Hasselstrom (1994) noted that bivalves and seaweed are both potentially useful products for shrimp farmers, as they may either be recycled for shrimp feed or sold for supplementary income; another advantage is that diversifying species may disperse financial risks associated with farming shrimp.

Horizontally integrated systems have mostly been developed at the pilot scale and been assessed based on technical issues such as nutrient retention, wastewater treatment capacity and production; few have attempted to evaluate physical parameters in association with managerial and financial indicators (Bunting, 2001b; Bunting *et al.*, 2003). Management of horizontally integrated systems may represent an increase in workload as compared with traditional practices and where this was justifiable through increased revenue it could constitute an important source of employment. Diversification through horizontal integration may help spread financial, environmental and disease risks associated with intensive production, but demands refined management approaches, planning and marketing. Lack of vision, limited access to venture capital, inappropriate policy, stakeholder pressure and poorly defined benefits for the operator all constrain innovation and diversification in aquatic farming systems.

Sustainable feed supplies

Widespread concern has been expressed regarding the large volumes of fishmeal used to formulate feeds for commercial shrimp and salmon farming (Naylor *et al.*, 1998, 2000). Furthermore, Naylor *et al.* (2000) noted that 'if the growing aquaculture industry is to sustain its contribution to world fish supplies, it must reduce wild fish inputs in feed and adopt more ecologically sound management practices'. Fishmeal is not the only issue, however, rapid expansion of the prawn farming industry in Bangladesh resulted in the overexploitation of freshwater snail populations for feed, thus demonstrating that sustainable feed supplies are a must, even for smaller components of the aquaculture sector. Research is underway to replace fishmeal in commercial feeds with meat by-products, microbial proteins and oilseeds enriched with omega-3 fatty acids.

Striving for organic production managers at Bahía de Caráquez shrimp farm, Ecuador noted that it was necessary to find a substitute for the high proportion of fishmeal used in commercial feeds. Experiments conducted to develop an alternative feed formulation indicated that vegetable protein could be used to replace fishmeal. However, it was necessary to source organically produced vegetable protein; the Encarnacion Organic Farm was established to supply feed ingredients. Seed pods from leguminous tree species (*Prosopis* sp. and *Leucaena* sp.) planted on the embankments of the ponds provided a further source of high-protein feed ingredients; owing to the harsh environment of the Caráquez estuary and saline growing conditions it was necessary to plant a combination of native and exotic tree species. Other species planted on the embankments produce aloe vera, almonds, fruit and flowers, generating additional income, supporting organic honey production and providing habitat for wildlife. Harvesting seed pods from the trees also provides employment for local community members, an important development as disease outbreaks and the abandonment of many conventional shrimp farms resulted in widespread unemployment.

Concern over fishmeal use in aquaculture must be balanced against the benefits it conveys as compared with other major users, namely pig and poultry production. Static fishmeal production globally and rising demand from aquaculture and livestock farming will probably mean an increase in the relative cost of fishmeal. Kristofersson and Anderson (2006) predicted that higher fishmeal prices are likely to result in innovation within the farming sector, including greater substitution with new soyabean meal products, enhanced diet formulation, increased production efficiency, a shift in cultivation to species less reliant on fishmeal, and breeding programmes and genetic engineering for novel plant-based products to substitute for fishmeal. These authors also noted that sustainable management of the pelagic fisheries from which fishmeal is derived would mean that coastal communities could benefit from the rising relative cost of fishmeal, whilst more efficient processing of the catch, with less waste and discarded bycatch, would improve fishmeal supplies without adding to fishing pressure. Fish oil has been used as a substitute for fishmeal in aquaculture feeds; however, as with fishmeal, supplies are finite, whilst complete substitution of vegetable oil for fish oil is impossible as the digestibility, amino acid profile and micronutrient composition differ.

Sustainable seed supplies

Sustainable access to fry and fingerlings can constitute a significant constraint on aquaculture development. Several traditional aquaculture practices evolved based on the collection of gravid females or seed from the wild; however, harvest of wild seed was often unsustainable and unable to support higher production. Collection of seed, in particular, shrimp seed also involved a significant by-catch of larval fish and crustaceans that was discarded, further damaging wild stocks; Larsson et al. (1994) estimated that 872–2300 km^2 of mangrove was required to supply post-larvae to Colombia shrimp farms in 1990, equating to 20–50% of the country's mangrove forest. In response, state-run hatcheries, often supported with external assistance, were established to

supply seed to emerging aquaculture sectors; however, in many cases these hatcheries were poorly managed, producing low numbers of poor quality seed, production cycles were poorly matched to farmers needs and the timely distribution of seed problematic. Considering these problems recent initiatives have aimed to enhance seed supplies, matching supply with demand, and ensuring quality and timely availability. Government support to private hatchery development in Thailand including training, extension and marketing assistance helped ensure fish seed was readily available in northeast Thailand (Ingthamjitr et al., 1997). Seed for prawn farming under the Greater Noakhali Aquaculture Extension Project come from disease-free broodstock maintained in hatcheries and consequently no antibiotics are required (Lecouffe, 2005); post-larvae are sold to farmers with an identity card to permit traceability, opening up the opportunity for environmental certification.

Limited or unreliable supply of seed to rural and upland areas can constrain the development of aquaculture. However, in the Lao PDR promotion and support of fish seed supply networks meant that upland rice farmers in Dong Village, Sepone district, were able to obtain fish seed to stock into the numerous bomb craters on the hillsides and raise fingerlings to sell on locally or to farmers travelling from more remote villages (Lithdamlong et al., 2002). Asian Institute of Technology Aqua Outreach, working in collaboration with the Savannakhet Livestock and Fisheries Section, Department of Livestock and Fisheries, established the 'nursing network' to facilitate the distribution and decentralised nursing of fish seed supplied by the provincial level government. Fry from a central hatchery were supplied to nodal farmers who then distributed it to their neighbours or nursed it in net enclosures or hapas thus increasing its size and value, and making it available for longer to prospective buyers in often remote areas where communications are difficult. The potential impact of this initiative was limited by the capacity of the central hatchery and subsidised nature of fry distribution to nodal farmers. In response a 'spawning network' was established to promote decentralised hatchery production, although this too required significant institutional support, training and the supply of premixed hormones to induce spawning, a step which was only possible owing to the 'cold chain' for animal vaccines developed previously.

Even where fry–trader networks are well established there can be problems; traders have a vested interest in the stocking densities adopted by farmers, and in some cases it has been suggested that farmers have been persuaded to buy excessive numbers of fish (Bunting et al., 2005).

Furthermore, traders without access to certain species may be tempted to encourage farmers to stock an inappropriate combination of fish; practical difficulties in identifying juvenile fish also constrain farmers from achieving preferred stocking rates. The quality of fish available to farmers may also represent a constraint; during a meeting with fish farmers in Kotiadi, Bangladesh, participants mentioned that having to stock small fingerlings resulted in poor growth and production (Bunting *et al.*, 2005). Where failings in the distribution of seed are addressed by the development of hatcheries in remote and inaccessible areas there is a risk, as with poorly managed hatcheries and local programmes elsewhere, that limited stock numbers will result in inbreeding (Muir, 2005).

Community-based management

Community-based management usually centres on common pool resources which, in the context of India, for example, have been defined as 'non-exclusive resources to which the rights of use are distributed among a number of co-owners, generally identified by their membership of some group such as a village or community' (Chopra and Dasgupta, 2002) and thus include: community pastures, grazing lands and forests, wastelands, dumping grounds and threshing areas, watershed drainage schemes, village ponds, rivers and other common pool wetlands. Definition and characterisation are likely to vary between countries and regions, as will associated use rights and management arrangements. Recent experiences with community-based management of aquatic resources include: Integrated Floodplain Management in Bangladesh; Local Resource User Groups (LRUGs) for management of farmer-managed aquatic systems (STREAM, 2006a); Participatory Action Plan Development (PAPD) as a tool to plan the management of wetland and common pool resources.

Considering the establishment of LRUGs in Vietnam and Bangladesh, aquatic animals collected from interconnected rice plots, trenches, canals, ponds and open water bodies often contribute to household incomes and food security in poor rural communities. As such it was recognised that community-based management was required to improve and sustain the exploited aquatic animal populations; the management approach developed was based on forming LRUGs to 'work together in managing a specific area in the community to enhance productivity and yield of aquatic animals' (STREAM, 2006a). LRUGs help members identify interconnections and explore joint management options; facilitate listening and knowledge sharing; enhance participation in decision-making for improved aquatic resource management; and promote accountability within the group and the broader community. Four steps to group formation were identified: discuss the situation and management practices that might be applied with the community; identify the area to be managed during transect walks and discussions with households with land in the area; present management options identified in step one to the group and agree on the most appropriate one to implement; members of the LRUG implement management of the area. Management activities identified in trials included deepening water bodies, habitat creation, maintaining broodstock and nursing juveniles in hapas and household ponds.

Changing patterns of access to natural resources associated with aquaculture development, even where these changes are based on models of community-based management can have negative social impacts. Ellis and Allison (2004) note that current approaches to community-based natural resource management are flawed as they are often based on a particular sector, for example, fisheries, forestry and wildlife, and often act to exclude new or different users from accessing the resource. It should also be acknowledged that it is difficult to ensure effective participation and inclusion of the most marginalised men and women in communities. As Chambers (2005) says 'The challenge is how to give voice to those who are left out and to make their reality count' and that 'The tendency for local elites to capture projects and programmes and use them for their own benefit should indeed be recognised as a fact of life'.

However, dealing with such realities and understanding the roles of leadership, patronage, unions, political parties and frequently coercion and extortion, may present opportunities to achieve more effective implementation and sustainable livelihoods, enhancements for poor people; discussing the probability that local elites will monopolise initiatives, Chambers (2005) comments 'there are benefits as well as costs in this. Leaders are often leaders because they have ability, and projects may be better managed through their participation. Leaders, especially where there is an active political party, may seek support and legitimacy and so have an incentive to spread the benefits of projects to more rather than fewer people'.

Organic and fair-trade production

Intensive production in aquaculture can result in negative environmental and social consequences such as those outlined in Table 28.1. In response, organic and fair-trade production have emerged as widely advocated mechanisms to redress the balance, promoting environmental protection, animal welfare, public health, ethical trading and corporate social responsibility. There are various interpretations of what constitutes organic or fair-trade production and only a limited number of certified producers, but growing markets demonstrate that people are willing to pay a premium for such products, and for some this may represent a pathway out of poverty.

Shrimp farming in Ecuador, once the world's largest producer has been affected by numerous problems, self-pollution, shrimp disease outbreaks, contamination with chemicals and social injustice, as traditional resource users were excluded and land ownership was transferred to a few powerful families. In response to these problems and protests and boycotts in Europe the owners of EcoCamaronera Bahía sought advice to address the situation, and began by planting mangrove trees on the embankments and in ponds; other vegetation in the ponds was also allowed to develop as a wildlife habitat. In the late 1990s the initiative was taken to convert the farm to organic production, multi-cropping of embankments was established, organic feeds were formulated and effluent monitoring initiated. A notable problem in converting to organic production was identifying a suitable replacement source of protein for feed formulation; the solution found was to use the protein rich seed pods of leguminous trees planted on embankments. Following these developments the farm was inspected and certified by Naturland, Germany, as organic; subsequently, two neighbouring farms adopted similar management approaches, and were inspected and certified. Adopting organic practices lead to wider benefits; modified management approaches, such as collecting seed pods, resulted in increased employment for local community members and the diverse bird fauna resident at the farm attracted birdwatchers and tourists that contributed to the local economy. Development of fair-trade standards for prawns, such as those produced as part of the Greater Noakhali Aquaculture Extension Project, Government of Bangladesh, supported by the Danish International Development Agency, also promises to enhance the livelihoods of small producers and favour equitable and environmentally sound farming.

Despite recent demand growth, organic and fair-trade goods only supply niche markets and as such only a relatively small number of farmers stand to benefit (Siriwardena, 2005). With a diverse sector such as aquaculture, the transaction costs of converting to organic production can be significant; the onus rests with the producer to formulate a strategy to convert to organic production, demanding considerable investment in researching organic standards and approaches, sourcing or formulating organic inputs, modifying management practices and commissioning the certification process. Poor small-scale farmers, by definition, will not have the resources to convert to organic production. Fair-trade agreements, such as that advocated by the Fairtrade Foundation require small farmers to organise themselves into groups, and group formation may be one way that producers could share the transaction costs of converting to organic production.

Environmental management

Phillips *et al.* (1993) noted that appropriate site selection, improved management, wastewater treatment and effective planning and monitoring would reduce environmental impacts associated with farming shrimp; broader consideration of such aspects would undoubtedly reduce environmental impacts associated with other forms of aquaculture. Jones (1990) noted that water abstracted for fish farming reduced river flows in stretches between the farm intake and outflow to such an extent that the movement of migratory fish could be hampered. Such situations should be avoided through improved site selection and restricting abstraction to safeguard ecological flows. Water conservation measures such as those outlined by Boyd and Gross (2000) would help limit abstraction from both surface and groundwater sources: seepage control through employing good construction practices; limiting water exchange; providing storage for rain and runoff water.

Supplementary aeration and water reuse also have potential for improving water use efficiency in aquaculture. Measures to protect culture water quality would further reduce the need for water exchange; promoting vegetated pond margins would facilitate nutrient uptake and stabilise embankments, reducing erosion and eliminating scour which can be an important source of suspended solids in pond-based aquaculture (Funge-Smith and Briggs, 1998). In general, the use of chemicals in temperate aquaculture is closely regulated to ensure operator safety, protect the environment and safeguard product quality. In tropical developing countries where aquaculture production has increased dramatically, the rate of expansion in the industry has frequently overwhelmed attempts to regulate and monitor the industry. Consequently, problems associated with chemical use in aquaculture may be more acute in developing countries with limited resources and poorly defined regulatory frameworks. However, as institutional arrangements evolve operators of aquaculture facilities will require appropriate management strategies and treatment technologies to limit or avoid the release of chemicals in wastewater.

Where water quality does decline during the culture process a range of strategies for reducing pollutant concentrations in discharges have been proposed, ranging from ecologically based lagoon and wetland systems to technologically advanced filters and sterilisation units (Nature Conservancy Council, 1990; Cripps and Kelly, 1995; Bunting, 2001b). However, strategies developed by researchers and commercial enterprises with a vested interest in uptake are sometimes constrained by concerns regarding reliability, practical limitations, possible risks and financial demands (Bunting, 2001b). Awareness of available strategies should be enhanced and operators of aquaculture facilities supported in selecting the most appropriate and effective option.

Policy initiatives to promote low-impact aquaculture systems

Building on recommendations presented by Siriwardena (2005) for shrimp farming, policy initiatives to support low-impact aquaculture systems should include: low-impact aquaculture in integrated coastal areas or watershed planning and management, enhancing and securing access to resources for poor farmers and supporting community-based planning and management initiatives, strengthening existing institutions, providing technical support and better management of the aquatic environment and implementing or enhancing monitoring and evaluation, strengthening legislation and law-enforcement and eliminating corruption.

Promotion of resource-efficient production, notably integrated farming systems, will require government assistance, training, sustainable seed supplies and management advice (*New Agriculturalist*, 1998); although it is also noted that 'much has still to be learned about the factors that influence the success or failure of efforts to establish integrated agriculture–aquaculture systems'. Bioeconomic modelling offers a promising approach to exploring the performance of integrated farming systems under different management scenarios (Bunting, 2001b; Bunting *et al.*, 2003); however, as Prein noted 'If they [farmers] are not able to observe the benefits, to understand how their farming system can be realistically enhanced and to participate in discussing how that process should happen, then the successful integration of aquaculture into existing agricultural systems, however beneficial, will not occur' (*New Agriculturalist*, 1998). Invoking the DFID Sustainable Livelihoods Framework (DFID, 2001) in structuring the analysis of periphyton-based aquaculture in Bangladesh enabled environmental, institutional and social constraints to adoption by poor farming households to be realistically identified, and enabled optimal system-wide opportunities for improved management practices to be elicited during discussions with farmers (Bunting *et al.*, 2005).

Conversion to organic production requires financial investment and administration, both of which may be beyond poor small-scale farmers; policies are required that support smaller farmers in adopting organic production. Furthermore, even some of the largest certifiers of organic food, such as the Soil Association, have to date only produced interim standards for organic aquaculture (Soil Association, 2005). Fundamental concerns regarding the use of fishmeal and oil from fisheries that have not been certified as sustainable constitute a significant barrier to the development of organic aquaculture; other contentious issues include the interaction of farmed and wild stocks, nutrient recycling and animal welfare. Naturland, another body that presides over standards for organic aquaculture was criticised for certifying a shrimp farm in Ecuador as organic when the legacy of unsustainable development of the industry was still in evidence, and the farm continued to use land previously covered by mangroves (Swedish Society for Nature Conservation, 2006). Prospects for the organic aquaculture of carnivorous species such as salmon, where production depends largely on fishmeal and oil from uncertified fisheries, appear poor, although those producers that can meet the required standards stand to gain a worthwhile premium. For species such as omnivorous carp and tilapia, shellfish and seaweed, with appropriate institutional support and raised awareness amongst consumers, prospects for more widespread organic production seem promising. However, there is a recognisable need for consistency amongst standards and an internationally recognised certification scheme (*The Economist*, 2003).

Considering the Fairtrade Foundation there are two sets of generic producer standards, one for smallholders organised in cooperatives or other organisations with a democratic, participative structure and one for workers on plantations and in factories. These generic standards specify the minimum requirements which producers must meet to be certified Fairtrade. Principles of the Fairtrade Foundation aim to encourage producer organisations to continuously improve working conditions and product quality, increase the environmental stability of their activities and invest in the development of their organisations and the welfare of their producers or workers. For organised workers, employers should pay decent wages, guarantee the right to join trade unions and provide good housing when relevant. On plantations and in factories, compliance with minimum health and safety and environmental standards is expected and no child or forced labour can occur. Furthermore, traders dealing in Fairtrade goods must: pay a price to producers that covers the costs of sustainable production and living; pay a 'premium' that producers can invest in development; make partial advance payments when requested by producers; sign contracts that allow for long-term planning and sustainable production practices. However, there are significant barriers to the expansion and diversification of the fair-trade movement, notably subsidies to domestic producers and on exports and tariffs on imports (Fairtrade Foundation, 2006). The Trade Justice Movement constitutes a coalition of more than seventy organisations working towards reform of the rules and institutions governing international trade (Trade Justice Movement, 2006) and further support to such movements would enhance prospects for fairly traded goods.

Participatory planning for enhanced natural resources management

Discussion concerning aquaculture development has the potential to become distorted by individuals or groups with different agendas, as Boyd (1999) noted people have strong opinions, but often contradictory perspectives when it comes to environmental issues; only by understanding the reasons for such differences and recognising them as legitimate will it be possible to formulate management approaches that are acceptable and feasible. As Adams *et al.* (2002) noted 'Different stakeholders in a common pool resource bring to their decision-making different assumptions, knowledges and goals for that resource which are not always explicit'. Consequently, engagement with the full range of stakeholders is critical as often the perspectives of poorer groups are missed, and even within these poorer groups the voices of women, children and the powerless often remain unheard (Bunting, 2001; Punch *et al.*, 2002).

The degree of representation or engagement achieved is critical. According to DFID (2001) 'Involving those who stand to win or lose from policy or institutional reform, or who may influence the reform process, helps to make the interests of key stakeholders transparent and to build ownership of the reform process'. If the process is not fully representative it is flawed from the outset; only by ensuring the interactive participation of all groups and ensuring their voices are heard will shared learning and understanding be achievable. Participation in planning and decision-making should be regarded as a right (Pretty, 1995). The importance of having stakeholders fully represented in participatory decision-making is widely acknowledged; however, social context and norms constitute inherent flaws to participation. Consequently, the stakeholder Delphi,

capitalising on the knowledge and opinions of stakeholders, with their quasi-anonymity guaranteed, has been proposed as a promising approach to capturing multiple perspectives, conducting joint analysis and facilitating group decision-making (Bunting, 2001b).

Furthermore, considering culture-based fisheries and stock enhancement programmes, it was noted in the Bangkok Declaration resulting from the Conference on Aquaculture in the Third Millennium, hosted by the Network of Aquaculture Centres in Asia-Pacific (NACA), that the potential of such strategies would only be realised by 'creating conducive institutional arrangements to enable and sustain investment in common pool resources' (NACA/FAO, 2000). In this regard approaches for the decentralised governance and management of natural resources are required; however, as Chopra and Dasgupta (2002) noted regional variations in prevailing institutional arrangements must be considered, and 'centralised' drafting of 'decentralised participatory governance' avoided. These authors also noted that steps to enhance access and facilitate change should be transparent and first take into account pre-existing institutions of natural resources management; policy, institutional structures and legislation should complement interventions and foster and support wider uptake of equitable co-management arrangements.

Development planning and investment

Article 9 of the FAO Code of Conduct for Responsible Fisheries (FAO, 1995) proclaims that 'States should promote active participation of fishfarmers and their communities in the development of responsible aquaculture management practices'. Furthermore, inclusion of all resource users in the development process is necessary to 'ensure that the livelihoods of local communities, and their access to fishing grounds, are not negatively affected by aquaculture development'. Dialogue between key stakeholders and policy formers is necessary for State-level agencies to 'establish, maintain and develop an appropriate legal and administrative framework which facilitates the development of responsible aquaculture'. Research investment to support the development of methods and protocols for the participatory monitoring and evaluation of environmental health and livelihoods impacts is also needed to 'promote responsible development and management of aquaculture, including an advanced evaluation of the effects of aquaculture development on genetic diversity and ecosystem integrity, based on the best available scientific information' and to 'establish effective procedures specific to aquaculture to undertake appropriate environmental assessment and monitoring with the aim of minimising adverse ecological changes and related economic and social consequences'. The Code of Conduct also notes that policy formulation should promote 'responsible aquaculture practices in support of rural communities, producer organisations and fish farmers'.

The Bangkok Declaration (NACA/FAO, 2000) called for collaborative multidisciplinary research; stakeholder participation in research; improved linkages between researchers, extension services and producers; efficient communication networks; regional and inter-regional cooperation; and a continued effort to build the skills of researchers involved in aquaculture development. Furthermore, it was proposed that food security and poverty alleviation could be improved through 'promoting poor-people-centred development focus in aquaculture sector policies'; using participatory approaches to identify and assess the needs of the poor; developing and extending appropriate aquaculture strategies; promoting sustainable small-scale household production in rural areas where it may be the only source of fish for vulnerable groups such as pregnant and lactating women and families with infants and pre-school children.

The Bangkok Declaration also calls for greater investment in people, research and aquaculture development; improved environmental sustainability, notably through integration of aquaculture into coastal area and inland watershed management plans; integration of aquaculture into rural development and poverty alleviation programmes; strengthening of institutional support to implement transparent and enforceable policy and regulatory frameworks; application of innovations in aquaculture, including sustainable stock enhancement; improving culture-based fisheries and enhancements; better management of aquatic animal health; improved nutrition in aquaculture; application of genetics and biotechnology; improved food quality and safety; promotion of market development and trade. The need for enhanced information flows at the national, regional and inter-regional levels to encourage policymaking, planning and the application of rules and procedures was also highlighted; initiatives such as AFGRP Strategy Studies, FMSP Key Programme Lessons and STREAM Better Practice guides constitute important developments in this regard (AFGRP, 2006; FMSP, 2006; STREAM, 2006b).

CONCLUSIONS

Based on the above discussion it may be concluded that a number of promising technical, social and institutional approaches to regenerating aquaculture have been identified and to some degree tested; however, strategies are required that promote and support uptake and where necessary adaptation. Awareness of promising approaches to low-impact and regenerating aquaculture should be promoted amongst target institutions including national and local government authorities, extension agents, development practitioners, educational establishments and communities that stand to benefit. Furthermore, policy, institutions and processes should support regenerating aquaculture and where appropriate, incentives and disincentives implemented to help change perspectives and behaviour.

ACKNOWLEDGEMENTS

An earlier version of this manuscript was prepared for the 'Communication for Aquaculture Development' project funded by the Aquaculture and Fish Genetics Resources Programme (AFGRP) of the UK Department for International Development. This publication is an output from a project funded by the UK Department for International Development (DFID) for the benefit of developing countries. The views expressed are not necessarily those of the DFID. Comments on this earlier draft by two anonymous reviewers were greatly appreciated. Subsequent revision of the manuscript was supported by the EC INCO-DEV MANGROVE project (http://www.streaminitiative.org/Mangrove/index.html) which received research funding from the European Community's Sixth Framework Programme (Contract: INCO-CT-2005–003697). This publication reflects the author's views and the European Community is not liable for any use that may be made of the information contained herein.

NOTE

1 1US$ = approximately 6 Sek (Folke *et al.*, 1994).

REFERENCES

Adams, B., Brockington, D., Dyson, J. and Vira, B. (2002) Analytical framework for dialogue on common pool resource management. Common Pool Resource Paper 1. Department of Geography, University of Cambridge, UK.

AFGRP (2006) Aquaculture and Fish Genetics Research Programme. DFID AFGRP website: http://www.dfid.stir.ac.uk/Afgrp/index.htm (accessed 20 July 2006).

Arthington, A.H. and Bluhdorn, D.R. (1996) The effect of species introductions resulting from aquaculture operations. In: Baird, D.J., Beveridge, M.C.M., Kelly, L.A. and Muir, J.F. (eds) *Aquaculture and Water Resource Management*. Blackwell Science, Oxford, UK, pp. 114–139.

Azim, M.E. (2001) The potential of periphyton-based aquaculture productions systems. PhD thesis. Wageningen University, Wageningen, The Netherlands.

Azim, M.E., Verdegem, M.C.J., Khatoon, H., Wahab, M.A., van Dam, A.A. and Beveridge, M.C.M. (2002) A comparison of fertilization, feeding and three periphyton substrates for increasing fish production in freshwater pond aquaculture in Bangladesh. *Aquaculture* 212: 227–243.

Bardach, J.E. (1997) Aquaculture, pollution and biodiversity. In: Bardach, J.E. (ed.) *Sustainable Aquaculture*. John Wiley & Sons, New York, pp. 87–99.

Barrett, G., Caniggia, M. I. and Read, L. (2002). "There are more vets than doctors in Chiloé": social and community impact of the globalization of aquaculture in Chile. *World Development* 30: 1951–1965.

Berg, H., Michélsen, P., Troell, M., Folke, C. and Kautsky, N. (1996) Managing aquaculture for sustainability in tropical Lake Kariba, Zimbabwe. *Ecological Economics* 18: 141–159.

Beveridge, M.C.M. (1996) Cage Aquaculture (2nd edn). *Fishing News Books*. Blackwell Science, Oxford, 346 pp.

Beveridge, M.C.M. and Phillips, M.J. (1993) Environmental impact of tropical inland aquaculture. In: Pullin, R.S.V., Rosenthal, H. and Maclean, J.L. (eds) *Environment and Aquaculture in Developing Countries*. ICLARM, Conf. Proc. 31, pp. 213–236.

Beveridge, M.C.M., Phillips, M.J. and Macintosh, D.J. (1997) Aquaculture and the environment: the supply and demand for environmental goods and services by Asian aquaculture and the implications for sustainability. *Aquaculture Research* 28: 797–807.

Black, K.D., Ezzi, I.A., Kiemer, M.C.B. and Wallace, A.J. (1994) Preliminary evaluation of the effects of long-term periodic sublethal exposure to hydrogen sulphide on the health of Atlantic salmon (*Salmo salar* L). *Journal of Applied Ichthyology* 10: 362–367.

Bonsdorff, E., Blomqvist, E.M., Mattila, J. and Norkko, A. (1997) Coastal eutrophication: causes, consequences and perspectives in the archipelago areas of the northern Baltic Sea. *Estuarine, Coastal and Shelf Science* 44(Suppl. A): 63–72.

Boyd, C.E. (1999) Aquaculture sustainability and environmental issues. *World Aquaculture* 30(2): 10–13, 71–72.

Boyd, C.E., Gross, A. (2000) Water use and conservation for inland aquaculture ponds. *Fisheries Management and Ecology* 7: 55–63.

Briggs, M.R.P. and Funge-Smith, S.J. (1996) Coastal Aquaculture and Environment: Strategies for Sustainability. Final Technical Report, ODA Research Project R6011, Institute of Aquaculture, University of Stirling, Stirling, Scotland.

Bunting, S.W. (2001a) Appropriation of environmental goods and services by aquaculture: a re-assessment employing the ecological footprint methodology and implications for horizontal integration. *Aquaculture Research* 32: 605–609.

Bunting S.W. (2001b) A design and management approach for horizontally integrated aquaculture systems. PhD thesis. University of Stirling, Scotland, 279 pp.

Bunting, S.W., Bostock, J.C., Lefebvre, S. and Muir, J.F. (2003) Assessing prospects for horizontally integrated aquaculture using bioeconomic modelling. In: Chopin, T. and Reinertsen, H. (eds) *Beyond Monoculture*. European Aquaculture Society Special Publication, vol. 33, European Aquaculture, Society, Oostende, pp. 140–141.

Bunting, S.W., Karim, M., Wahab, M.A. (2005) Periphyton-based aquaculture in Asia: livelihoods and sustainability. In: Azim, M.E., Verdegem, M.C.J., van Dam, A.A., Beveridge, M.C.M. (eds), *Periphyton: Ecology, Exploitation and Management*. CABI Publishing, Wallingford, UK.

Burbridge, P.R. (1994) Integrated planning and management of freshwater habitats, including wetlands. *Hydrobiologia* 285: 311–322.

Chambers, R. (2005) *Ideas for Development*. Earthscan, London.

Chopra, K. and Dasgupta, P. (2002) Common pool resources in India: evidence, significance and new management initiatives. Common Pool Resource Paper 2. Institute of Economic Growth, Delhi.

Corea, A., Johnstone, R., Jayasinghe, J., Ekaratne, S. and Jayawardene, K. (1998) Self-pollution: a major threat to the prawn farming industry in Sri Lanka. *Ambio* 27: 662–668.

Cripps, S.J. and Kelly, L.A. (1996) Reductions in wastes from aquaculture. In: Baird, D.J., Beveridge, M.C.M., Kelly, L.A. Land Muir, J.F. (eds) *Aquaculture and Water Resource Management*. Blackwell Science, Oxford, pp. 166–201.

Davies, I.M., McHenery, J.G. and Rae, G.H. (1997) Environmental risk from dissolved ivermectin to marine organisms. *Aquaculture* 158: 263–275.

Deb, A.K. (1998) Fake blue revolution: environmental and socio-economic impacts of shrimp culture in the coastal areas of Bangladesh. *Ocean & Coastal Management* 41: 63–88.

Dewalt, B.R., Vergne, P. and Hardin, M. (1996) Shrimp aquaculture development and the environment: people, mangroves and fisheries on the Gulf of Fonseca, Honduras. *World Development* 24: 1193–1208.

DFID (2001) Sustainable livelihoods guidance sheets. Department for International Development, London.

Edwards, P., Demaine, H., Innes-Taylor, N. and Turongruang, D. (1996) Sustainable aquaculture for small-scale farmers: need for a balanced model. *Outlook on Agriculture* 25: 19–26.

Ellis, F. and Allison, E.H. (2004) Livelihood Diversification and natural Resource Access. Livelihood Support Programme, LSP Working Paper 9, FAO, Rome. 37 pp.

Enander, M. and Hasselstrom, M. (1994) An experimental wastewater treatment system for a shrimp farm. *Infofish International* 4: 56–61.

Fairtrade Foundation (2006) Fairtrade standards. Fairtrade Foundation website: http:www.fairtrade.org.uk/about_ standards. htm (accessed 11 July 2006).

FAO (1995) Code of Conduct for Responsible Fisheries. FAO, Rome.

FAO (2002a) Aquaculture development and management: status, issues, and prospects. COFI:AQ/1/2002/2. FAO, Rome.

FAO (2002b) The state of world fisheries and aquaculture 2002. FAO, Rome.

FAO (2004) The state of world fisheries and aquaculture 2004. FAO, Rome.

Flaherty, M., Vandergeest, P. and Miller, P. (1999) Rice paddy or shrimp pond: tough decisions in rural Thailand. *World Development* 27: 2045–2060.

FMSP (2006) Key Programme Lessons. Fisheries Management Science Programme website: http://www.fmsp.org.uk/KeyLessons.htm (accessed 11 July 2006).

Folke, C., Kautsky, N. and Troell, M. (1994) The costs of eutrophication from salmon farming: implications for policy. *Journal of Environmental Management* 40: 173–182.

Folke, C., Kautsky, N., Berg, H., Jansson, A. and Troell, M. (1998) The ecological footprint concept for sustainable seafood production: a review. *Ecological Applications* 8: 63–71.

Funge-Smith, S.J. and Briggs, M.R.P. (1998) Nutrient budgets in intensive shrimp ponds: implications for sustainability. *Aquaculture* 164: 117–133.

Gillibrand, P.A., Turrell, W.R., Moore, D.C. and Adams, R.D. (1996) Bottom water stagnation and oxygen depletion in a Scottish sea loch. *Estuaries, Coastal and Shelf Science* 43: 217–235.

Hem, S., Avit, J.B.L.F., Cisse, A. (1995) Acadja as a system for improving fishery production. In: Symoens, J.J. and Micha, J.C. (eds) *The Management of Integrated Freshwater Agropiscicultural Ecosystems in Tropical Areas.* Seminar Proceedings, 16–19 May (1994), Technical Centre for Agricultural and Rural Co-operation (CTA), Royal Academy of Overseas Sciences, Brussels, pp. 423–435.

Ingthamjitr, S., Phromtong, P. and Little, D. (1997) Fish seed production and marketing in northeast Thailand. *Naga* 20(3–4): 24–27.

Jones, J.G. (1990) Pollution from fish farms. *Journal of the Institution of Water and Environmental Management* 4: 14–18.

Kautsky, N., Berg, H., Folke, C., Larsson, J. and Troell, M. (1997) Ecological footprint for assessment of resource use and development limitations in shrimp and tilapia aquaculture. *Aquaculture Research* 28: 753–766.

Kelly, L.A., Karpinski, A.W. (1994) Monitoring BOD outputs from land-based fish farms. *Journal of Applied Ichthyology* 10: 368–372.

Kristofersson, D. and Anderson, J.L. (2006) Is there a relationship between fisheries and farming? Interdependence of fisheries, animal production and aquaculture. *Marine Policy* 30: 721–725.

Larsson, J., Folke, C. and Kautsky, N. (1994) Ecological limitations and appropriation of ecosystem support by shrimp farming in Colombia. *Environmental Management* 18: 663–676.

Lecouffe, C. (2005) Freshwater prawn (*Macrobrachium rosenbergii*) farming in Greater Noakhali Districts. Programme Report. Aquaculture and Fish Genetics Research Programme, Institute of Aquaculture, University of Stirling, Stirling, UK.

Lin, C.K, Ruamthaveesub, P. and Wanuchsoontorn, P. (1993) Integrated culture of the green mussel (*Perna viridis*) in wastewater from an intensive shrimp pond: concept and practice. *World Aquaculture* 24(2): 68–73.

Lithdamlong, D., Meusch, E. and Innes-Taylor, N. (2002) Promoting aquaculture by building the capacity of local institutions: developing fish seed supply networks in the Lao PDR. In: Edwards, P., Little, D.C. and Demaine, H. (eds) *Rural Aquaculture*. CABI Publishing, Wallingford, UK.

Little, D.C. and Edwards, P. (1999) Alternative strategies for livestock–fish integration with emphasis on Asia. *Ambio* 28: 118–124.

Little, D. and Muir, J. (1987) A Guide to Integrated Warm Water Aquaculture. Institute of Aquaculture, University of Stirling, Scotland. 238 pp.

Loch, D.D., West, J.L. and Perlmutter, D.G. (1996) The effect of trout farm effluent on the taxa richness of benthic macro ivertebrates. *Aquaculture* 147: 37–55.

Lumb, C.M. (1989) Self-pollution by Scottish fish-farms? *Marine Pollution Bulletin* 20: 375–379.

Marshall, J. (2001) Landlords, leaseholders & sweat equity: changing property regimes in aquaculture. *Marine Policy* 25: 335–352.

McAllister, P.E. and Bebak, J. (1997) Infectious pancreatic necrosis virus in the environment: relationship to effluents from aquaculture facilities. *Journal of Fish Disease* 20, 201–207.

Muir, J. (2005) Managing to harvest? Perspectives on the potential of aquaculture. *Philosophical Transactions of the Royal Society B* 360: 191–218.

Muir, J.F., Brugere, C., Young, J.A. and Stewart, J.A. (1999) The solution to pollution? The value and limitations of environmental economics in guiding aquaculture development. *Aquaculture Economics and Management* 3: 43–57.

Muluk, C. and Bailey, C. (1996) Social and environmental impacts of coastal aquaculture in Indonesia. In: Bailey, C., Jentoft, S. and Sinclair, P. (eds) *Aquaculture Development: Social Dimensions of an Emerging Industry*. Westview Press, Boulder, Co, pp. 193–209.

NACA/FAO (2000) *Aquaculture Development Beyond 2000: the Bangkok Declaration and Strategy*. Conference on Aquaculture in the Third Millennium, 20–25 February 2000, Bangkok, Thailand. NACA, Bangkok and FAO, Rome, 27 pp.

Nature (1997) Aquaculture: a solution, or source of new problems? *Nature* 386: 109.

Nature Conservancy Council (1990) Fish Farming and the Scottish Freshwater Environment. Nature Conservancy Council, Edinburgh, 285 pp.

Naylor, R.L., Goldburg, R.J., Mooney, H., Beveridge, M., Clay, J., Folke, C., Kautsky, N., Lubchenco, J., Primavera, J. and Williams, M. (1998) Nature's subsidies to shrimp and salmon farming. *Science* 282: 883–884.

Naylor, R.L., Goldburg, R.J., Primavera, J.H., Kautsky, N., Beveridge, M.C.M., Clay, J., Folke, C., Lubchenco, J., Mooney, H. and Troell, M. (2000) Effect of aquaculture on world fish supplies. *Nature* 405: 1017–1024.

New Agriculturalist (1998) Integrating aquaculture – more to fish farming than fish. *New Agriculturalist* 98/5. http://www.new-agri.co.uk/98–5/focuson/focuson6.html (accessed 3 July 2006).

New Scientist (1996) Blue revolutionaries. *New Scientist* 152(2059): 32–36.

New Scientist (1998) Danger, shrimps at work. *New Scientist* 157(2122): 11.

New Scientist (1999) Forget the shellfish. *New Scientist* 163(2197): 5.

Oberdorff, T. and Porcher, J.P. (1994) An index of biotic integrity to assess biological impacts of salmonid farm effluents on receiving waters. *Aquaculture* 119: 219–235.

ODI (1999) *Aquaculture.* Key sheets for Sustainable Livelihoods. Overseas Development Institute, London.

Phillips, M.J., Kwei Lin, C. and Beveridge, M.C.M. (1993) Shrimp culture and the environment: lessons from the world's most rapidly expanding warmwater aquaculture sector. In: Pullin, R.S.V., Rosenthal, H. and Maclean, J.L. (eds) *Environment and Aquaculture in Developing Countries.* ICLARM Conf. Proc. 31, pp. 171–197.

Pollnac, R.B. and Sihombing S. (1996) Cages, controversies and conflict: carp culture in Lake Toba, Indonesia. In: Bailey, C., S. Jentoft and P. Sinclair (eds), *Aquaculture Development: Social Dimensions of an Emerging Industry.* Westview Press, Boulder, CO, pp. 249–261.

Pretty, J.N. (1995) Participatory learning for sustainable agriculture. *World Development* 23: 1247–1263.

Primavera, J.H. (1997) Socio-economic impacts of shrimp culture. *Aquaculture Research* 28: 815–827.

Punch, S., Bunting, S.W. and Kundu, N. (2002). Poor livelihoods in peri-urban Kolkata: focus groups and household interviews. U/T Government's Department for International Development Project R 7872, Working Paper 5, University of Stirling, UK.

Ridler, N.B. (1997) Rural development in the context of conflictual resource usage. *Journal of Rural Studies* 13, 65–73.

Robertson, A.I. and Phillips, M.J. (1995) Mangroves as filters of shrimp pond effluent: predictions and biogeochemical research needs. *Hydrobiologia* 295: 311–321.

Selong, J.H., Helfrich, L.A. (1998) Impacts of trout culture effluent on water quality and biotic communities in Virginia headwater streams. *Progressive Fish-Culturist* 60: 247–262.

Siriwardena, S. (2005) Shrimp farming at the cross roads. id21 Research Highlight. id21 website. http://www.id21.org (accessed 19 June 2005).

Soil Association (2005) Fish farming and organic standards. Soil Association, Bristol, UK.

STREAM (2006a) Local Resource Users' Groups? What are they? STREAM website: http://www.streaminitiative.org/Library/pdf/bpg/BPGSRS_EN.pdf (accessed 4 May 2006).

STREAM (2006b) Better-Practice Guidelines – What are Better-Practice Guidelines? STREAM website: http://www.streaminitiative.org/Library/pdf/bpg/WhatareBPGs.pdf (accessed 11 July 2006).

Swedish Society for Nature Conservation (2006) Eco-labelling of shrimp farming in Ecuador. Swedish Society for Nature Conservation website. http://www.snf.se/pdf/rap-inter-shrimp-ecuador.pdf (accessed 10 July 2006).

The Economist (2003) The promise of a blue revolution. *The Economist* 368(8336): 19–21.

Thompson, A.G. (1990) The danger of exotic species. *World Aquaculture* 21: 25–32.

Tran, T.B., Le, C.D. and Brennan, D. (1999) Environmental costs of shrimp culture in the rice-growing regions of the Mekong Delta. *Aquaculture Economics & Management* 3: 31–42.

Trade Justice Movement (2006) About the Trade Justice Movement. Trade Justice Movement website: http://www.tradejusticemovement.org.uk/about.shtml (accessed 8 July 2006).

Turner, K. (1991) Economics and wetland management. *Ambio* 20, 59–63.

UNEP (2005) Annotated guiding principles for post-tsunami rehabilitation and reconstruction. Global Programme of Action for the Protection of the Marine Environment from Land-based Activities, United Nations Environment Programme.

Welcomme, R.L. (1988) International Introductions of Inland Aquatic Species. FAO Fisheries Technical Paper 294, FAO, Rome, 318 pp.

Weston, D.P. (1996) Environmental considerations in the use of antibacterial drugs in aquaculture. In: Baird, D.J., Beveridge, M.C.M., Kelly, L.A. and Muir, J.F. (eds), *Aquaculture and Water Resource Management.* Blackwell Science, Oxford, pp. 140–165.

Yap, W.G. (1999) Rural aquaculture in the Philippines. RAP Publication. FAO, Bangkok.

29

Shopping for Green Food in Globalizing Supermarkets: Sustainability at the Consumption Junction

Peter Oosterveer, Julia S. Guivant
and Gert Spaargaren

INTRODUCTION

As the market for sustainable (or 'green') food expanded worldwide in the 1990s, supermarkets took up dominant roles as channels for its commercialization. Alternative natural food and grocery stores and farmers' markets were forced to assume a secondary role. Countries where most organic products are sold via supermarket chains tend to be the countries where the organic market shares are the highest as well (Willer and Yussefi, 2004). And although the organic food movement in Europe dates back more than fifty years, it is only since the 1990s that organic foods are achieving mainstream status largely through these supermarket sales (Van der Grijp and den Hond, 1999).

Supermarket retailers thus have become key players and their strategies and goals can be said to be of crucial and even further increasing importance with respect to the future provisioning of green food products worldwide. But, although supermarkets are playing a growing role in our daily lives as consumers, academic research on its social, economic and political implications are still incipient. Environmental and rural sociology, as Marsden *et al.* (2000) observed, have left important issues such as analyzing new trends in food provisioning, including the relationship

between changing consumer demands for ethically and environmentally acceptable products, the responses from companies through new products and new information and marketing approaches, to environmental economists and marketing specialists.

This chapter aims at filling this omission and contributing to the analysis of the roles played by supermarkets from the perspective of sustainability transitions in the food sector. We use the plural to refer to transitions as we consider these not being one essential trajectory, because the possibilities for new developments are open and involve a complex set of issues, especially when levels beyond the individual nation-state or region within the global network society are included. We start by introducing a theoretical and disciplinary outlook for understanding the emerging societal trends in the transitions towards sustainability in food provision and in particular the roles of consumers and retailers therein. We will characterize the sustainable food consumer and add four story lines to show some of the complexities involved in this. A review of concrete provider and consumer strategies is presented that will allow us to start developing an analysis of social practices at the shopping floor of retail outlets. We conclude by presenting a first outline of a research agenda on supermarkets as consumption junctions that are of crucial importance for the future greening of food consumption in global modernity.

CONCEPTUAL TOOLS FOR UNDERSTANDING CONSUMER BEHAVIOR

What exactly signifies the growing role of retailers in the provision of sustainable food? In answering this question we look at the theory of ecological modernization. This theory has been developed in the 1980s to make sense of the processes of environmental change emerging in modern industrial (mostly OECD) societies from the 1970s onwards. Among the core tenets of this theory is the claim that there are – within industrial production and consumption systems – emerging sets of criteria to be used by actors within these systems to assess and judge the environmental performance of products, technologies and processes. The ecological performance becomes part of the game, next to and parallel to economic performance indicators. By taking on board criteria for ecologically rational production and consumption, actors become involved in the further modernization of the organization of production and consumption from an environmental point of view. Because of their central position, economic or market actors such as producers, retailers and consumers are assigned important roles in this modernization process. From the mid-1980s onward, governments and environmental NGOs are pressurizing, facilitating and regulating these key economic actors on the basis of horizontal governance networks, applying policies and (economic, voluntary) instruments which are attuned to their needs and possibilities (Mol and Sonnenfeld, 2000).

While ecological modernization theory has been developed originally to analyze changes in production processes and providers' strategies at the up-stream ends of production–consumption chains, from the mid 1990s onwards the theory has also been applied to consumer behavior at the bottom end of production–consumption chains (Spaargaren, 2003). In its application to the sphere of consumption, the theory had to be complemented and adapted in some specific ways, since the rationalities governing everyday life and consumption are different from the rationalities dominant in the production sphere. The criteria for ecologically rational consumption behavior are to be embedded in the life–world rationalities which shape daily routines. This asks for a 'translation' of many of the technical goals and regulatory schemes used in the expert systems involved in environmental policy making. In order for people

to 'recognize and understand' the kind of behavior involved in sustainability transitions, a series of 'environmental heuristics' needs to be developed at the level of ordinary, everyday life consumption routines like shopping for food, traveling from home to work, going for a weekend holiday, etc. (Spaargaren and Martens, 2005). Such environmental heuristics facilitate two processes at the same time. They provide a definition or indication of sustainability goals to be realized in these specific consumption domains and they present an action frame or action perspective which people themselves can apply in the specific context concerned, contributing to sustainability transitions.

In the case of retail shopping for sustainable food, these heuristics can take different forms, ranging from devices for sustainable packaging of products to reading authorized and controlled labeling schemes attached to sustainable products, or the use of special discount and saving systems bringing together groups of more sustainable products and services. What kind of heuristics (to be understood as short-hand versions of the 'story lines' as they figure in discourse theory) will become the dominant ones in specific situations and societies depends very much on the actors involved in their construction and cannot be analyzed without taking into account the power relations between the central actors in the provision system on the one hand and groups of citizen-consumers with specific lifestyles and CCC demands on the other.[1]

The retail outlet is an appropriate setting to study the (re)construction and change of the sets of heuristics used for the sustainability transitions in the food sector. The retail outlet is the proper unit of analysis since it functions as the 'locale for interaction' between providers and consumers. The retail outlet, in the words of Schwartz-Cowan (1987), is an example of the consumption junction as the meeting point of system- and life–world rationalities. The consumption junction as 'locale' is not just functioning as a physical setting for interaction but is also constitutive for this interaction in the first place (Goffman, 1963; Giddens, 1984). By approaching shopping practices as they occur in the retail outlet as consumption junction, it becomes possible to combine actor-oriented and social-structural analyses in studying the greening of food production and consumption.

During the 1970s and 1980s, many attempts have been made in social theory to confront the separation between micro and macro studies (cf. Bourdieu, 1977; Giddens, 1979). This so-called structurationist approach argued that the relationship between social action and social structure should be studied at the level of social practices. Using a series of new and redefined concepts researchers could study long-term changes in institutions without losing sight of the human-made character of social structures, and investigate interests, motives and lifestyles of individual human beings not in 'isolation' but in their situated 'contextness' of social structures. The notion of 'duality of structure,' as introduced by Giddens, has gained wide acceptance in sociology and other social sciences as a key concept and a vehicle for bridging the gap between micro and macro studies. Although welcomed by many as an elegant conceptual framework, structuration theories are nevertheless criticized for the lack of empirical research showing the fruitfulness of their conceptual apparatus. If applied in research, so it is suggested implicitly or explicitly by many critics, this framework will turn out to be overly actor-centered and voluntaristic and therefore unable to deal adequately with the long-term structural changes taking place in globalizing modernity (Archer, 1982; Stones, 2005).

In the field of consumption research, the division between micro and macro studies has taken its own, specific form in the distribution of tasks between micro-economic and social–psychological models on the one hand and structural approaches in transition studies and sociological studies on 'systems of provision' on the other. Consumer research networks in the UK (Southerton *et al.*, 2003) and in the Nordic countries (Boström *et al.*, 2005) have done path-breaking work to bridge this gap, giving some examples of the heuristical quality of structuration theory for empirical research. Those networks study mundane technologies and behaviors from a contextual perspective, looking at the different ways in which the minutiae of everyday life (using the fridge or the stove, cooking and lighting practices) connect to long-term (technological) changes in the systems of provision. Substantial contributions to consumption research have also been made by putting forward the notion of 'political consumerism' to analyze the new (sub- and trans-national) political frameworks for 'individual' commitment to sustainable consumption patterns in globalizing modernity (Micheletti, 2003).

THE SUSTAINABLE FOOD CONSUMER

There is a trend both in academic and market research to classify people consuming sustainable (and particularly organic) food as one uniform segment of the population. Essentially these views are based on the belief that when people behave similarly this should be explained through a correspondence in their attitude, or that consuming particular products requires the presence of similar socio-economic or cultural traits. Richter (2002)

pointed at the continuous gap between consumers' responses in research and their real life practices. Thus, data projected from those studies can indicate higher demand rates than the ones that would be obtained considering actual consumption practices. Searching for more complex characterizations about who are the organic food consumers, several marketing studies, undertaken by international consultancy companies, do not exhibit this problem. These recent studies are progressively replacing the 'rational information processor consumer model' by new non-positivist perspectives, where the symbolism involved in the act of consuming is taken into account and analyzed through ethnographic and qualitative research (Murcott, 1999).

Spaargaren (2003) agrees with critics on the need to improve the analytical perspective currently prevailing in consumption research. The current social–psychological models use individual attitudes to predict concrete and future behavior employing, for example, several fixed indicators to identify environmental awareness. As an alternative for the individualist approaches, the social practices' model is offered. In this sociological model social structures are not considered as external variables, but are taken as crucial for the analysis of consumption behavior. Instead of taking the individual and his/her attitudes as central to understand a certain aspect of his/her consumption practices, the social practices' model highlights the actual consumption practices, located in the space and time shared by the individual and other social actors. And, instead of focusing on isolated aspects of behavior, the model aims at establishing the way in which a group of social actors relate to the many everyday practices in order to reduce environmental impacts. Whether or not citizen-consumers actually engage with sustainability transitions in the food sector depends on many different factors, some of which are easier to detect and analyze than others.

Individual consumer choices should thus be approached as part of a wider context (Belk, 1995) and changes in consumer behavior should therefore not only be related to psychological (attitudinal) mechanisms, but also to wider changes in society. Macnaghten (2003) identified three dimensions in societal transition processes towards sustainable consumption. The first process relates to transformations in the production sphere and the retailing sector. The second one concerns macro-social developments, such as demographic changes, and the third refers to changes in the form and content of social practices. These processes, according to Macnaghten, must be understood in a framework that interprets consumption as practical, stratified and relational.

Thus, consumption of sustainable food products in supermarkets should not be detached from transformations in these three different dimensions. Our interest is to understand the supermarket orientation to the sustainable food products on offer and the strategies proposed for the retail sector, as part of a complex and dynamic process. This process captures and stimulates transformations in the consumers' food choices, which does not necessarily imply coherent social practices, making it possible to oppose *the sustainable* and *the non-sustainable* food consumer. Lifestyles and social practices are like twin social concepts: 'Each individual's lifestyle is built using a series of blocks corresponding to a set of social practices that individuals evoke in their routine' (Spaargaren, 2003, p. 689). This definition agrees with the one presented by Giddens (1991), to whom lifestyle is a set of social practices assumed by an individual, together with the narrative regarding self-identity which follows it.

Do food consumption practices constitute a particular category in the wider field of consumption behavior? According to Halkier (2001), yes, since food is literally incorporated into the body, or purposefully kept out of it. It is a daily experience that cannot simply be compared with consumption of other goods and is a necessary ingredient of all peoples' everyday life. One characteristic of these consumer practices is that they are negotiated socially, so they are intersubjective, compound and contingent and not close to a rationalist model. As Warde (1997) defines, food practices belong to the unspectacular side of consumption. If we follow this idea, those practices are not easily identified with a search for status or with the communication of meanings to those in a position to witness the products consumed. Food consumption might have to do with decisions that can be related with this 'exhibition,' but remains a private practice for the most part. Individuals do not have to evaluate continuously every minute consumption decision but they are nevertheless increasingly made aware that they are making a 'choice.' This involves a mix of decisions and routines. A tension and ambiguity between them is what Halkier observes for highly industrialized countries, when she states that:

> consumers become concerned by a television show that exposes the poor quality of meat products but this experience is filtered out within a couple of days. They then return to the habit of buying a particular sausage (that contains little meat) because it is one their children like. Consumers would like to have better quality foods but feel at the same time that public information about food risks disturbs their experience of cooking and eating (2001, p. 208).

Halkier takes the relationship between food consumption and ambiguity one step further. She defines it in recuperating Bauman's (1993) concept of ambiguity, which refers to the indeterminate and open processes in social life, especially in modern societies, where it is impossible for individuals to achieve secure and unambiguous order with respect to knowledge about society and themselves.

Ambiguity, or the balance of trust and risk in food systems, evolves over time and can acquire many different shapes depending on many different factors. One obvious conclusion can, however, already be formulated: trust-generating mechanisms used in traditional local settings will not work effectively in the global circuits of food provisioning. Talking to the farmer at the local food market and visiting the farms where our daily food stuffs are produced, can no longer remain the most dominant and relevant trust-generating mechanism in reflexive modernity. People have to rely on abstract systems, scientific expertise and various information systems, to make long-distance assessments on the quality of the products and the reliability of the information flows which come along with them. With the growing significance of global food chains for our everyday food practices, the need for trust-building mechanisms based on abstract systems and expert knowledge forcibly increases. Relevant expert systems include medical professions, health services, state organisms, social care, etc. But trust is not necessarily blind. From the analyses of Beck *et al.* (1994) and Giddens (1990) on risk, it can be derived that, in conditions of reflexive modernity, trust in (abstract) expert systems:

- is fragile (since people are aware that systems considered safe today can be hit by some food crisis tomorrow) and needs constant monitoring/work, commitment;
- is related to the (shop and production) systems and their organizational principles as well as to the people/experts who make these systems work;
- is reproduced/disturbed /re-established especially by processes occurring at the so-called 'access points,' where lay-people meet the experts (or their representatives) of the systems in a regular and more or less organized way.

The awareness of the need to make daily choices in food consumption and of the presence of uncertainties and ambiguities in trusting food products constitute what can be seen as important drivers behind the considerable growth in sustainable food consumption we witness nowadays on a worldwide scale.

DEFINING SUSTAINABLE FOOD

It is important to avoid the use of exclusive definitions of sustainability, for example, when concentrating primarily or exclusively on science-based (life-cycle) assessments of the environmental impact of provisioning particular food products. Some observers claim that a sustainable lifestyle or food consumption pattern can be rather clearly defined in technical terms, but such an essentialist approach is not very helpful. The different story lines with respect to sustainable food production and consumption emerging over the past two or three decades are to a certain extent based on science but always mixed up with broader societal issues. Applying a sociological definition of sustainable food is therefore required, making the definition dependent from the evolving ways in which consumer concerns about food are interpreted in specific societies.

Before elaborating such a definition it deserves paying attention to a perspective on sustainable food provisioning applied within environmental and rural sociology that has attracted broad support, that is, alternative agro-food networks. The growth of green food has interested the social sciences but most studies concentrated on the analysis of the proliferation of alternative agro-food networks (AAFNs) operating at the margins of mainstream industrial food circuits. This bias may be understandable as AAFNs[2] provide many opportunities for the renewed interest in local, determinedly microanalytical and ethnographic elements in the study of sustainable food production and consumption practices. These studies are essentially based on a dichotomy between the food production of the 'industrial world,' with its heavily standardized quality conventions and logic of mass commodity production, on the one hand, and the 'domestic world' on the other, where quality conventions embedded in trust, tradition and place support more differentiated, localized and 'ecological' products and forms of economic organization. The concept of 'quality' evokes the cultural aspects of this model but remains mainly production centered. The analysis of AAFNs makes it possible to express strong normative commitments to the social movements contesting mainstream, corporate industrial agro-food systems and the related hegemonic agricultural techno-scientific complex. In this arena, AAFNs figure as material and symbolic expressions of alternative eco-social imaginaries, and the literature emphasizes its capacity to wrest control from corporate agribusiness and create a domestic, sustainable, and egalitarian food system. It can be regarded as a form of resistance to the disruptive effects of global competition in the food market.

These loaded normative assumptions cause problems in the analysis of large-scale and industrialized organic production. Several observers consider the entrance of agribusiness into the organic market a misconception of what organic principles should be, and therefore, as this trend is not desirable, it should not be studied (Michelsen, 2002). Others focus on the role of the 'conventionalization' of the organic industry as a crucial process in the transformation of the organic sector and are prepared to consider redefinitions of public policies in relation to family farming (Guthman, 2002, 2004; Raynolds, 2004). Goodman (2003) states that new localized economic arrangements are often uncritically seen as precursors of an associative economy by virtue of their embeddedness in interpersonal ties of reciprocity and trust. In this way, local personal relations can, and also tend to be, idealized in the evaluation of rural development strategies based on territorial value added.[3] The AAFNs' perspective is not only used to characterize local markets, but also points at cultural aspects of global and mainstream markets. Culture values are attributed to the local consumer, while the consumer in conventional mainstream markets is depicted as just following a narrow economic rationality.

Opening up such normative definitions requires the recognition that different story lines with respect to sustainable food production and consumption are emerging in different countries and among different groups of consumers. Applying a sociological, historical perspective means conceiving green consumerism as a multidimensional category, covering a number of different 'consumer concerns' about food, all of them including sustainable foods but with different significance. We can identify four dimensions that are not mutually exclusive:

(1) *Naturalness*. Key characteristics are unadulterated food and the use of natural processes during the production process. Examples are organic food consumption which belongs historically among the most clearly defined categories of sustainable food concerns. Also in this category can be included whole foods, considering the ones that support a healthy lifestyle, offer high nutritional value, promote long-term good health, and are free of artificial ingredients and preservatives. The practices involved in producing natural foods can cover a wide range of farming methods, including certified organic production. The orientation to 'natural' foods currently can be related to the search for nutrition, enhancing health and a broad identification of food quality, and it is present globally.

(2) *Food-safety* concerns originated in many food crises and scares like those on BSE ('mad cow' disease) and genetically modified organisms (GMOs),

mainly in Europe (where the retail sector had a very determinant role), on avian flu, or on pesticide contamination. Without having read Ulrich Beck (1992) on the emergence of the risk society, consumers nowadays are aware of contemporary food risks as a new form of risk. This means that these food risks are difficult to assess from a lay perspective and impossible to safeguard completely from a (national, science-based) expert point of view. Today, flows of food are organized and regulated at global levels whereby no one can escape the products (and the risks) of food produced and consumed in the space of flows.

(3) *Animal welfare* constitutes a controversial but rather well-circumscribed dimension of consumer concerns, although very unevenly developed throughout the world, it seems. In response to widespread public concerns about the specific ways of bio-industrialized production of chicken and eggs after World War II, mainstream markets for fresh eggs in Western Europe have taken animal welfare issues into account. Furthermore, the radical tactics of animal welfare activists in Europe (e.g. in the UK) have contributed to the high visibility of these concerns, although giving them in some countries a controversial character as well. In most Asian countries animal welfare issues, however, are considered at best as a secondary priority, only to emerge when issues of survival and poverty are satisfactorily dealt with.

(4) *Environmental* (or *eco-system related*) concerns related to modern industrial food production and consumption, mainly activated and campaigned for by environmental organizations and social movements all over the world. They argue that food production and consumption should be sustainable in the Brundtland report meaning of the word: producers (farmers) should manage ecosystems in such a way that future generations are not deprived of a well-functioning sustenance base to human life and consumers should include these concerns in their consumption practices, including fair trade. In many cases these *eco-system concerns* are interlinked with one or more of the other dimensions, most notably in the case of pesticide use with human health.

Most people share some of the above-mentioned dimensions of food concerns at some moments in their lives. Which of the concerns worry people most varies between different groups of consumers and different countries, while the overall level of consumer concerns differs as well between different parts of the world. Instead of trying to determine and explain the many different possible contents and specific profiles of consumer concerns in different parts of the world as a particular phenomenon in itself,[4] it seems more promising to take a dynamic, process-oriented and contextual perspective to green

consumerism in the global network society. From this perspective, the emphasis is on the interaction between emerging green consumer concerns on the one hand and developing retailer strategies for green food provision on the other. This interaction process is reciprocal but not well balanced in terms of power relations. Retailers are more powerful in many respects than consumers when it comes to shaping green food consumption. On the other hand, consumer power has increased considerably not only as a result of a series of food crises but also because of the emergence of private-interest-based regulation of food quality and food safety (Ponte and Gibbon, 2005). If indeed it is 'up to the consumers to decide,' retailers and food producers implicitly acknowledge that consumer interests have to be taken seriously. Consumer demands for green products are taken into account also because they are articulated and supported by a growing number of organizations and movements which claim to act on behalf of the consumers and for that reason demand access to networks making decisions on the future provision of green food.

CLOSE ENCOUNTERS AT THE SHOPPING FLOOR

Whether or not consumers actually engage with sustainability transitions in the food sector depends on many different factors, some of which are easier to detect and analyze than others. One interesting opportunity to study this phenomenon is the retail outlet where we can approach consumers and their shopping practices while establishing a balance between macro and micro approaches.

First, there is the visual level concerning the more sustainable products and services on offer, that is, the ways in which these products are presented to the consumer as well as the information systems attached to them. Visual indicators are important for analyzing emerging consumers buying sustainable foods, but in order to really gain an in-depth understanding of the consumption practices implied in shopping for sustainable food in retail outlets, these physical devices are only a first step. For information, images, messages, products and services to be really accepted, bought or 'appropriated' by citizen-consumers, they have to be embedded in a vital and active system of trust relations which involves both providers and citizen-consumers. When applied to our object of analysis, the retail outlets, these assumptions help to 'read from the shelves' what kind of social relations and strategies are reflected and mirrored in the specific physical setting of the retail outlet. This notion of trust and power relations 'being mirrored' or 'reflected' in physical characteristics of the setting should not be inter-preted in any mechanistic or static way. To be able to read and decipher these inscribed trust relations one needs a social theory on the ways in which relations behind the product and information flows are organized in the context of reflexive modernity. In the language of structuration theory, these trust and power relations are said to be *instantiated* at the very moment when people enact – with the help of the physical characteristics or technologies included in the shop setting – the social practice of shopping.

When a set of valid indicators for shop-level assessments is available, they can be used not just for assessing environmental policies but also for the evaluation and construction of market-based forms of citizenship involvement in the greening of food chains. Product images and information exchange about production circumstances of cer-tain foods provided by NGOs, public media or Internet and e-mail communication, in combina-tion with supermarkets in-shop policies on infor-mation and communication, may influence consumer shopping practices. Micheletti (2003) refers to particular forms of engagements in terms of 'political consumerism,' and she shows that reliable sets of environmental performance indica-tors on a retail level can be used for many different forms of environmental action and pressure. Power relations equally get specific characteris-tics in the retail outlet. Supplementing the more conventional notions of economic power of food producers and consumers, political power at the

shop floor and information control acquire increasing importance. Viewed from this perspec-tive, shopping practices are directly linked to supermarket decisions on how they organize the provisioning of food in their shops.[5] The main cornerstones of our framework for the analysis of consumer practices when buying sustainable food in retail outlets are summarized in Figure 29.1.

When operationalizing this conceptual model into strategic variables, we distinguish between three basic sets. At the right-hand side of the con-ceptual model, we discern a set of variables and indicators referring to the environmental strategies of the main actors in the provision system. The second set of theoretical variables and items refers to the processes at the shop-floor level. Here, we make a further distinction between variables describing the physical characteristics of the green product and information flows on the one hand, and variables and items referring to the relation-ships of power and trust as they are reproduced in the shop-floor setting on the other. Finally, we use a set of variables describing the lifestyles and con-sumption patterns of the groups of consumers shopping for green food. We try to describe their environmental performance beyond the specific food-shopping practice and we look for ways to relate the revealed preferences for green food to basic characteristics of their lifestyles and overall consumption levels, and to the involvement of global civil society actors such as consumer NGOs.

Both the physical characteristics of the retail outlet and the social relations governing the

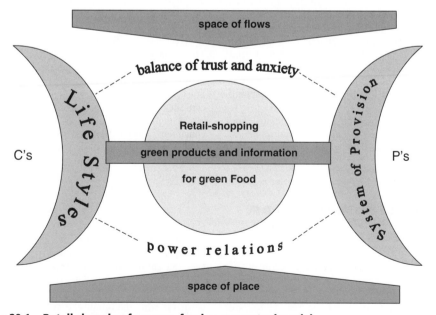

Figure 29.1   Retail shopping for green food: a conceptual model

shopping practices are the result of a specific articulation of local and global forces or dynamics. The distinction introduced by Castells (1996) between the space of place and the space of flows can help to make sense analytically of these dynamics, without incurring a simplistic dualism. So, these concepts are helpful provided that one does not relapse into an interpretation of this scheme which situates the sustainability transitions exclusively or primarily in the 'local dynamics' of the space of place, while regarding the globalizing forces and dynamics of the space of flows as a threat and negative factor for sustainability transitions (Oosterveer, 2005a). It is, for example, an exciting dimension of sustainability transitions in the food sector when one witnesses the globalization and standardization of the modes of production and consumption of organic foods formerly restricted to niche markets. These unorthodoxies can only be understood and properly analyzed when possibilities for environmental change at the level of the space of flows are taken into account as well and analyzed with respect to the many diverging ways of being connected to local factors and dynamics in the space of place.

To be in the condition to face the challenge of globalization for the social sciences and, in our case to be able to capture the complexities in the new role of retail chains, we need a new mapping of space and time, that will not exclude national specificities, but will avoid any type of dualism between the national and the global level, between the 'inside' and the 'outside' (Beck and Willms, 2004).[6] Assuming a cosmopolitan view, the transitions to more sustainable food-consumption practices are related to the regional origin of the products and their concomitant food-miles, the (EU, WTO, national governments) standardized norms for packaging, safety and environmental quality, the environmental strategies of the retail chains and the orientation of the shopping public. While shopping for sustainable food, people can engage with production processes and colleague–consumer groups and environmental NGOs worldwide, thereby performing different forms of what Beck refers to as banal cosmopolitanism (Beck and Willms, 2004).

PROVISION OF SUSTAINABLE FOOD IN THE RETAIL OUTLET: SUPERMARKET STRATEGIES

The dominant position of the retail sector in the processes of change in food provision can basically be explained by two reasons. First, retailers can substantiate their claim to be 'closest' to food consumers in many important respects by pointing to the fact that they meet on a regular, almost daily basis with major segments of mainstream food consumers (Seth and Randall, 2001). Second, the organization of food production and consumption has become a global affair, and consequently supermarket chains operating worldwide have a privileged position. These reasons are further elucidated below.

Retailers 'know best' the concerns of their clients since meeting them in the retail outlet leads to first-hand information about what consumers want and what their concerns are. These regularized and frequent interactions at the shop floor also provide retailers with the possibilities to experiment with new (green/healthy) food products and practices. As a consequence, retailers claim to have the power to make or break the market for sustainable products and services in the food sector. They 'create' and 'control' not just green consumers but also – 'on behalf of the consumer' – the suppliers of green products (Bevan, 2005, p. 7). The increasingly significant presence of supermarkets in the green-food sector is part of transformations induced by the supermarkets themselves in the food consumption sphere, by providing new options and taking initiatives regarding product innovations and food quality. During the 1990s a fundamental shift has taken place in retailing in Western countries from selling highly standardized and packaged brand-name food products to loyal customers, towards increasingly fragmented micro-marketing strategies increasingly selling perishable foods such as fruits, vegetables, dairy and meat (Guptill and Wilkins, 2002). This transition forces retailers to expect more and more from their suppliers in terms of the policing of food delivery as well as the type and specifications of the food produced. This stands to give retailers a market advantage with customers and it demonstrates to governments that they are taking existing food regulations seriously (Flynn et al., 2003). Retailers constructed so-called private-interest or market-based regimes for quality control, offering the consumer individual choice also with respect to food quality 'beyond basic standards.' Although many governments still play an important role in the regulation of food markets, retailers are 'at the apex of this quality construction; being able to absorb and transmit regulatory changes, customer reactions and supply chain quality assurance parameters' (Marsden et al., 2000, p. 8).

The second reason for the growing retailer dominance is that the organization of food production and consumption has increasingly become a global affair. To keep up with the high dynamics of food consumption and production in globalizing markets – with the (quality) regulation of food in the 'space of flows' (Oosterveer, 2005) – the resources of local, independent shop owners or small (organic) farmers' cooperatives are far from sufficient. Through concentration and internationalization over the past decades retailers have gained competitive advantage, resulting for example in many countries in Europe in five major retail chains accounting for considerable shares in the overall food sales. In 1990 no retailers were included in the Fortune 500 list of the largest global companies, but in 2002 more than 50 were. By that time, Wal-Mart had become the largest of all companies, considering the size of sales (Reynolds and Cuthbertson, 2004, pp. 1–22). This process was related to the closure of small shops and independent retailers (Dobson *et al.*, 2003). In 2005, the top 10 global food retailers accounted for combined sales of $840 billion – 24% of the estimated $3.5 trillion global market (up from 18% in 2001). See Table 29.1.

One example of these large retail firms is the French hypermarket chain Carrefour, selling food through its super- and hyper-markets in France, but also in many other countries around the world (see Box 1).

Consequently, food quality and safety issues stretch far beyond the local or national level. While for a long time quality control rested primarily in the hands of public regulators, we witnessed a major change in the 1990s when retailers assumed a more active role in the development of food safety standards and procedures, like HACCP (Hazard Analysis and Critical Control Point).

In order to attract consumer attention, supermarkets refer to different storylines from within

Box 1   Carrefour: global mega-grocer

Carrefour operates over 11,000 stores (430,000 employees) in more than 30 countries in Europe, Latin America and Asia. France accounts for about half of the company's sales. At the beginning of 2005, Carrefour planned to open 70 hypermarkets, including 15 in China, 7 in Brazil, 6 in Colombia, 5 in Indonesia, 4 in Thailand and 3 in Poland.

Source: ETC Group (2005, p. 8)

the general frame of sustainable food consumption. Also, combinations of different storylines are developed, like the combination of the 'naturalness' and the 'safe food' storylines. The growing importance of this specific combination can be illustrated by considering some of the transformations that are taking place in the area of food marketing. In the 1980s, a new perspective on consumer research split the academic marketing field into two coexisting perspectives. The conventional perspective assumes a positivist approach, employing quantitative research and focusing on the purchasing process. The newly emerging non-conventional perspective adopts a non-positivist methodology, employing also ethnographic and qualitative methods in dialog with sociology and anthropology, assuming a cultural perspective where consumers are not considered as rational (economic) actors. This approach emphasizes the cultural signification process as it is intertwined with consuming practices at different levels and within specific social contexts. The constellations of meaning and practices characterizing subcultures of consumption and styles of food consumption are not based on socio-economic circumstances exclusively or primarily, since even members of one subculture can belong to several socioeconomic groups (Thompson and Troester, 2002).

An empirical example of these non-positivist trends in marketing research is provided by the work of the Hartman Group in the USA (Hartman Group, 2000; http://www.hartman-group.com, 2003). In studies of this Group on organic food consumption, it was found that – with an annual growth of 15 to 20 per cent – organic food products are becoming part of mainstream food consumption practices in the USA, while no longer being restricted to just 'market niches.' In this new generation of marketing research, the stages of organic food consumption are explored in a qualitative way, working from the periphery to the center of the market, attempting to establish a comprehensive characterization of organic consumer lifestyles, consumer behaviors, distribution channels and information sources. These reports contextualize the organic food consumer as being

Table 29.1 Top 10 global food retailers

Company	2004 revenues (US$ millions)	Percentage global market share (grocery retail)
1. Wal-Mart* (USA)	287,989	8
2. Carrefour (France)	99,119	3
3. Metro AG (Germany)	76,942	2
4. Ahold (Netherlands)	70,439	2
5. Tesco (UK)	65,175	2
6. Kroger (USA)	56,434	2
7. Costco (USA)	52,935	2
8. ITM Enterprises (France)	51,800	1
9. Albertson (USA)	39,897	1
10. Edeka Zentrale (Germany)	39,100	1

* Wal-Mart does not report grocery sales separate from total revenues. Market research firm, Retail Forward, estimates that Wal-Mart sold $109 billion in groceries in 2004.

Source: ETC Group (2005 p. 6)

part of cultural changes where they are becoming more concerned with quality of life and health, and thereby transforming consumption practices (Barry, 2004). This phenomenon has also been the focus of a study about 'The Wellness Lifestyle Shopper: Mapping the Journeys of Wellness Consumers' (http://www.hartman-group.com, 2000). This study states that consumers, as well as their paths to achieve wellness, are complex social entities because they have to consider an enormous diversity in products appealing to health entering the market. Data showed that the American population spends around $66 billion per year on healthy products, a significantly growing market (http://www.hartman-group.com, April 2003). People's concerns with health and the nutritional quality of food were mentioned by 66 per cent of the organic consumers interviewed in this research. Concerns about pesticide risks was the reason given by 38 per cent, food safety by 30 per cent, while only 26 per cent mentioned environmental reasons, contradicting the belief that organic food consumers are essentially environmentally conscious citizens. Instead of understanding consumers as being informed solely by economic and scientific arguments, these consultancy reports recognize that changes in the cultural sphere impact the way in which people consume products and services and deal with related experiences and information.

The National Grocers Association (NGA) of the USA has recently established the organic market to be the fastest growing sector of food products in the supermarkets. Also, the NGA relates this development to the growing importance consumers attach to well-being and health.[7] At a conference on organic food consumption in the USA in 2003 the NGA concluded:

> As the fastest growing category in the food industry today, and public concern about health and wellness expected to continue, no retailer can safely ignore this increasingly important facet of the industry. The $5 billion organic market is growing at a rate of five times larger than the growth rate of the overall food industry and doubling in size every 3½ years since 1990. By 2003, the market for organic products is estimated to be over $13 billion.[8]

On the basis of a research among 146 representatives of supermarket chains, the NGA underlined the importance of organic or natural foods for supermarkets and offered a number of recommendations for interested supermarkets (see Box 2). Paying close attention to these recommendations, Wal-Mart plans to roll out at the end of 2006 a complete selection of organic foods – food certified by the USDA in its nearly 4000 stores in the USA. Just as significant, the company says it will price all this organic food at an eye-poppingly tiny premium over its already cheap conventional food: the organic Cocoa Puffs and Oreos will cost only 10 per cent more than the conventional kind (*New York Times*, 4, June 2006). The entrance of Wal-Mart into this sector will challenge the argument that organic food is elitist.

By way of comparison, let us now consider the situation in a less industrialized country like Brazil. The rising trend in the consumption of

---

Box 2   Recommendations for supermarkets initiating the sales of organic and natural foods

It may seem simple, but it is not as easy as simply adding a few new product lines to the store shelves or setting up a separate natural foods' section inside the store. To draw this business and meet the needs of these customers, retailers need to become as knowledgeable about natural and organic foods and products as the people they hope to sell to. This segment of consumers knows more, and asks more, and the retailer who will enjoy the long-term benefits of these natural sales will be the one who can answer their questions. Your naturals' section will be most successful, and most profitable, if you follow a few basic guidelines to cater to the natural products customer.

First, when conceptualizing your naturals' section, work with a knowledgeable natural products' distributor who can assist you with product selection, competitive pricing, promotional strategies and merchandizing. Working in partnership with a distributor who understands this industry will help you build a solid foundation for your own natural products' section. Second, realize the importance of product information and education for this new customer group. Natural products' magazines, shelf talkers, product demos and other consumer information are vitally important to the success of your section. Third, assign a natural products' section manager who is knowledgeable about the products, knows how they differ from commercial brands, can answer customers' questions, and is committed to your customers' natural products' education. Most of all, associates and customers must be encouraged to take time to savor what those in the natural products' industry have grown to love: the flavor, the quality, and the absolute uniqueness of natural foods. Knowledge of and enthusiasm for these products, perhaps more than anything else, will be the deciding factors in the success of your store's natural products' section.

Source: Jonathan M. Seltzer. Natural Foods: A Natural Profit Opportunity (*National Grocer Magazine*)[9]

healthy food has been detected in a number of market studies. One segment of this category can be called 'diet/light' and has witnessed annual growth rates of 30 per cent over the last few years, according to ABIAD (Brazilian Dietetic and Special Purpose Foods Association). According to the Brazilian Food Industry Association sales of light and diet products alone already corresponded to about US$ 1.7 billion of the total of US$ 47 billion in food sales in 2002. This entails a growth rate of 952.5 per cent over a period of 11 years. Since 1990, the average number of new products released jumped from 40 to 80 per year. These data for Brazil follow the wider global trend of searching for a better quality of life, directly associated with better nutrition, in combination with food that is tasty and pleasurable (Frutifatos, June 2002). We can see this global trend mirrored and supported by retail practices for the provision of green foods in Brazil (see Box 3).

The Brazilian Supermarket Association confirmed the presence of a trend comparable with the one in the supermarkets in Europe and USA, where interest in the broad category of healthy food (including organic products) is growing fast: 'The (food) sector knows that consumers want good health and longevity. Supermarkets can meet this demand by including certified natural products, organic and functional food, as well as "diet" and "light" products and, what is very important,

by giving consumers information on what they are buying. Ultimately your store can "sell" health!'[11]

Provider strategies in sustainable food are related to systems of provision (i.e. the relations with farmers), marketing strategies (involving one or more of the identified dimensions of sustainable food) and with the positioning of these products within the shop. Surprising is the observed variety in supermarket strategies in this regard. For example, French retail groups that publish a sustainable development report (Carrefour and Auchan) consider quality chain ('filière'), organic and fair trade products as indicators for their company's performance regarding social and environmental responsibilities in the area of food products. Leclerc and Intermarché remain fragmented and display much less information but they consider organic, fair trade (Leclerc) and integrated farming (Intermarché) as sustainable food products. This variability in company strategies deserves further elaboration.

After presenting these different results of empirical research on provider strategies in the field of sustainable food provision, it becomes clear that this interesting perspective demands further elaboration. Based on our conceptual framework and the review of empirical trends in retail policies worldwide, we think the following variables to be of crucial importance for (research into) future supermarket strategies:

Box 3   Green food provisioning in Brazilian supermarkets

In order to supply their retail outlets with sufficient quantities of green food in response to the increasing demand, supermarkets have to organize systematically their provisioning, pushing the growth of organic food production. In 2001, the total sales of formally certified organic food products in Brazil reached US$200 million in 2003 and is estimated to be around US$ 250 to 300 millions for 2004 (Globo Rural, November 2002; Ormond et al., 2002). To supply this demand, the country had more than 800,000 hectares certified as organic in 2003, which grew from 275,576 hectares in 2001. In addition, there is also a large quantity of 'informally certified' or non-certified organic production, especially in the southern states of Rio Grande, Paraná and São Paulo. The estimated number of organic producers is around 14,000 and among them small-sized family farms are responsible for up to 80 per cent of the production (Folha de São Paulo, 10/11/2002). Greater São Paulo represents half of the national consumption of organics and is also the main pole of production. The annual growth of organic production is calculated as being between 30 and 50 per cent. An important part is production for export (soybeans, coffee, juices, sugar, nuts, oils, banana, guaraná, etc.), which is around 70 per cent of the total certified volume (there are 12 national and about 9 international certifying agencies active in the country), generating in 2001 US$ 130 million (Exame, 28/05/2002). But the sector is also growing in the domestic market which belongs, together with Argentina, the most developed in Latin America. Around 45 per cent of the organic food sales in the domestic market are done through supermarkets, 26 per cent through fairs and 16 per cent in specialized stores.[10] Most of the products are fresh vegetables and fruits, but a growing number of companies and small family units is processing tea, coffee, mate tea, jams, oils, breakfast cereals, and dairy products. Fresh vegetables are presented in different forms thereby allowing the producer to increase the value. Provisioning supermarkets with organic food from farmers or farmers' associations takes different forms but two stand out as most important. The first strategy is to enter into direct contracts with organic farmers (e.g. applied by regional supermarkets in the city of Florianópolis, Santa Catarina State) and the second one is to rely on intermediaries (more adopted by national and international chains, e.g. in the cities of Rio de Janeiro and São Paulo). The first strategy allows a more diversified offer of products, higher quality, more space of exhibition and less difference in price between conventional and organic products than the second one does.

- product information strategies (what information is available in the shop, on the shelves and on the product; which sustainable food storyline, or combination thereof, is referred to);
- price settings in relation to other food products;
- physical location within the shop (separate section, separate shelves, separate section within a shelf, mingled among other products);
- linkages with suppliers ((in)formal contracts, certification/labeling, retailer supervision of production process, farmers within the shop);
- company communication (adds, other forms of publicity, which sustainable food storyline, or combination thereof, is referred to);
- company strategy: is sustainable food considered an essential part of the corporate image or only one category of products on sale?

With the use of these variables in the context of cross-national, comparative research, it will be possible to identify and analyze the different retailer strategies with regard to sustainable food provision and consumption in globalizing food chains, also for sustainable food.

BUYING SUSTAINABLE FOOD: CONSUMER STRATEGIES

Making sense of consumers shopping for sustainable food in supermarkets cannot only be done through reference to economic variables and attitude–behavior relations. Attention to other issues is required for comprehending consumer strategies for buying sustainable food, such as concerns about environment, food safety and health, the importance of different lifestyles and the changes in consumer trust in food. In particular, the issues of lifestyle and trust will be further elaborated.

Life in modern-day societies is characterized by an increasing plurality of different lifestyles and this diversity is also reflected in the varying consumer choices and marketing strategies in the food sector (Slater, 1997). Giddens (1991), Warde (1994) and Beck *et al.* (1994) point at the plurality in lifestyles and consider reflexivity related to consumption practices as a key element for understanding this variety. Consumer reflexivity becomes particularly visible in the significance attached to health and bodily well-being in the face of conflicting expert systems.

Searching for health and buying sustainable food can be part of different lifestyles. Even within the clearly identifiable sector of organic food, a marketing study in Germany by Biohandel, March 2006, found a large diversity in consumer behavior. This study points at the presence of different reasons for different categories (combining socio-economic backgrounds and cultural attitudes) of

German consumers in buying organic food. They distinguished three (out of the ten in total) consumer categories that are interested in buying organic food, together representing some 35 per cent of the German population:

- *Post-materialists* constitute the traditional basis for organics. They buy organic products because they care for the environment, but also for pleasure, taste, feeling well and health. Organic labeling gives the certainty of buying the 'right' products.
- *Middle-class consumers* have recognized organics as a trend and buy it on rational grounds and partly also on status. Important for them is pesticide- and chemical-free food and although they are less informed about organics they do trust seals and labels. They generally buy organic food in conventional stores.
- *Modern performers* consider organics as being associated with fitness and energy and prefer to buy these products, of which they do not know much, in conventional supermarkets around the corner. Without attaching too much value to these labels and categorizations, this study points very clearly at the diversity in lifestyles of which organic food consumption can be part. A Dutch study on lifestyle profiles for consumers of biological products (Wertheim, 2005) indicated that for the broader category of biological or sustainable food this variety may even be larger. In modern society diversity does not stop at the front door of the consumers' homes as, even within the context of the household food, habits may be different, exposing different lifestyles between people that are in close social relationship. To explain this cultural fragmentation of our dietary preferences Richardson (2004) points at the increased individualization of culinary preferences (especially among pre-teen children) and the growing presence of dietary restrictions (due to food sensitivities, allergies and forms of vegetarianism). Cultural forces beyond the household are increasingly potent in fragmenting any united dietary patterns existing within it.

As in other countries, in Brazil concerns about health and lifestyle play a role in selecting vegetables next to their cosmetic appearance and flavor. Thus, consuming sustainable food can be occasional and only one among several other health-oriented practices. Guillon and Willequet (2003) identified this trend as the 'ego-trip' way of consuming sustainable food. Since the beginning of the 1990s, this trait seems to be present in individuals' decisions towards self-protection (e.g. the safety and sanitary quality of food) and self-promotion (beauty, healthiness and fitness). Many consumers who could not be identified as being environmentally aware or socially responsible were concerned with buying alternatively produced food. Similarly, data from research undertaken in Europe in 1998 (Antoine, 1998) found that 76 per cent of consumers considered food as the best medicine, though they did not specify any particular food, either organic or conventional. Consumer food choices can hardly be explained by their consciousness of the relation between pesticide use and product appearance. Many consumers were not even very well informed about the qualities of organic food products, but they are still interested, often for health reasons. In the *SuperHiper Magazine* (June, 2002) research, 92.5 per cent of the people interviewed were interested in knowing more about organic food, particularly about the composition and nutritional value and disease prevention potential. Many consumers also wanted to know recipes and ways to consume fruits and vegetables. These results signal the recommendation for producers, suppliers, and supermarkets to make the benefits and advantages of 'green' products more visible and comprehensible and create a better fit between their information strategies and consumer practices and concerns.

Obviously, not all choices are available for all consumers. Nevertheless, it is essential to be aware that buying sustainable food is not limited to the richer echelons of Western societies. In many other countries consumer concerns about food are growing as well, although not necessarily applying the same dimensions of sustainability (or storylines) as in the EU or the USA. Aprilia (2005), for example, showed in her research on Thailand that in this country 'green' food provisioning was initially intended for export but later a domestic demand emerged as well. Currently, organic food has become a niche market already available in supermarkets targeted at high-class consumers but the creation of new markets for the majority of middle-class Thai has not yet taken place. Most Thai consumers select their food primarily on the basis of its freshness and taste, whereas the organic food consumers state that they consume organic food mainly for health and safety reasons.[12] Local Thai certification schemes thus pay as much attention to organic production practices as to hygiene: hygienic certification is food produced with hygienic processes that may include chemical substances provided at an acceptable level that is harmless to human health. Hygienic food provides better opportunities for the average Thai to consume less-polluting, less-contaminated food produced at more affordable prices than organic food.

In general, trust is an essential element of consumer strategies with regard to sustainable food. The presence of risks associated with food (pesticide residues, bird flu, BSE, etc.) that may endanger human health in combination with the necessary dependence on experts to provide information about their occurrence and danger, necessitates some form of trust relation to be developed and sustained.

Trust in the food system as abstract system refers both to the products and technologies applied and the people at work in (global) food chains. Consumers need to trust the health and safety of the food they consume, the information provided about the production process involved, and the people producing and retailing the food. As a personal relationship with the producer of food is not possible in supermarkets, consumer trust in sustainable food bought at retail outlets has some specific characteristics. The relationship between consumers and members of staff is necessarily superficial, so consumers' trust in sustainable food products is generally based on the (environmental) image of the company (and its obligation to uphold its image) and on the active (visible through labels) presence of independent controlling agencies supervising certifications and other indicators of good environmental performance. This trust is nevertheless precarious so in case of acute problems consumers are inclined to (temporarily) look for additional external and independent sources of information to orient themselves in order to make the right choices, for example, the choices for sustainable food.

This short discussion of consumer concerns, trust and lifestyles is based on the preliminary results of an emerging strand of empirical research worldwide in the field of consumer strategies in buying sustainable food. They point at the importance of further elaboration of transnational, comparative research in this domain. For this future research, we have again tried to identify a number of variables and factors we think to be important:

- Dimensions of consumer concerns as contained in specific lifestyles. Different (combinations of) storylines can be connected with different lifestyles' characteristics also outside the consumption domain of food.
- Product information strategies. What information is asked for by consumers and which formats of information provision (in the shop, on the shelves, on the products) fit best to the lifestyles of consumer groups.
- Social relations of anxiety and trust: the preferences of consumers for specific company information strategies targeted at establishing and maintaining a meaningful trust relation.
- (Premium) prices for green products: 'willingness to pay,' although not isolated from concrete shopping and eating practices, nor from cultural/lifestyle characteristics and always in relation to other food products.

- Preferences of consumers for specific formats for the physical location and presentation of green products within the shop (separate section, separate shelves, separate section within a shelf, mingled among other products)
- Preferences of consumers for specific ways of constructing 'hierarchies for green qualities.'

These variables can help to move beyond the isolated, individualist approaches to consumer behavior which tend to dominate research for a long time. By using these kinds of variables, consumer strategies with regard to buying sustainable food are researched in direct relation to and linked with provider strategies.

THE SHOPPING FLOOR AS LOCALE FOR SUSTAINABLE FOOD TRANSACTIONS

The shopping floor constitutes the locale where supermarket chains and consumers interact in the selling and buying of sustainable food, or where provider and consumer strategies actually meet. Although the social practices in the retail outlet cannot be understood without including the different strategies presented above, so far, very little empirical research has been done applying this perspective.

Nevertheless, some empirical findings are available and one interesting starting point is the physical lay-out of the shop. The way different sustainable food products are positioned in the shop setting proves very important in the success of such products. For an example on the shelf space, see Box 4. Items such as the assortment of sustainable food products (quantitative as well as qualitative), the positioning of the green assortment in overall assortment, the spatial structure of green provision, and the dimension of sustainability referred to are characterizing differences on the shopping floor and thereby facilitating or complicating the enactment of particular provider and consumer strategies.

A new strategy introduced by hypermarket chains in France since the mid-1990s included the selling of regional and organic food products, combined with the development of quality systems ensuring product traceability and the rearrangement of the stores in 'consumption universes' (i.e. the thematic regrouping of products not according to the product's nature but to the consumers' use) (Laurenceau, 2005). Shop managers consider regional or quality labeled food products (e.g. Label Rouge, AOC) as equivalent to other products and thus do not need separate treatment, except for organic products (mostly because they are a more recent phenomenon in these shops and formally demand strictly separated 'filières').

Box 4   The importance of supermarket shelf space for sustainable food consumption

Experimental research provided evidence for the observation that not only the price of a product or its characteristics determines consumer interest but also that the context in the supermarket plays an important role as well.

While testing consumer reactions to the presentation of sustainable food in supermarkets, researchers ascertained that consumers are sensible to the way a product is offered. When a product is presented in a supermarket in such a way that it gives the impression of being popular and of good quality, consumers are more interested in buying this product. For example, when sustainable foods are offered more space in the supermarket shelves they seem to be popular and therefore consumers will buy more.

Source: Dagevos et al. (2005)

When several labels are available for one category of products, these can correspond to different levels of quality and therefore with different prices. Retailers use food labels as general signs of quality according to their perceptions of consumers' needs and preferences (e.g. consumers may choose different labels for different kinds of meat). In most cases, labels represent a higher quality and are thus displayed on higher shelves, thus more expensive, but this is not the case in certain stores, where either the display is vertical or where the higher quality has become common and widely purchased. In France, the notion of 'quality' seems to create coherence between the retailers' and consumers' concerns for sustainable food. In the store this 'quality' is materialized into the mix of 'terroir' (regional product) and sustainability labels. Consumers are familiar with some labels, such as organic agriculture but not with others and, evidently, they ask for more information figuring on the product itself as well as in the store and in the catalog.

Supermarkets may use different strategies in their shops when commercializing organic food products (Richter et al., 2001). How prominent and strong is the attempt from the firm to persuade or facilitate the consumer in his or her green choices, preferences and routines? How consumer oriented is the strategy when compared to internal (profits/market) interests of the providers? How cosmopolitan is the sustainable food product in terms of being originated and regulated and standardized from a local or an international/global perspective? Some essential differences characterizing these strategies are the number and diversity of products in the retail outlet, the motivation and qualification of the workers at the selling point, and the presentation and positioning of the products in the store. During interviews in several European supermarkets, Richter *et al.* (2001) observed that the person in charge of the organic food sector usually emphasized that organics were part of a broader environmental and social responsibility strategy from the retailing chain. However, when looking deeper into the data, they found out that these commitments can lead to different practices and are diffuse in their scope.

In the Brazilian case, Guivant *et al.* (2003) concluded that, although the main international and national retail chains, like Carrefour, Pão de Açucar and Wal-Mart, have invested significantly in the organic food supply, they basically assumed only minimum and basic strategies. The growing presence of organic products can not necessarily be considered the result of elaborate concern-wide strategies to replace conventional food. Organic foods are dispersed in the area for products associated with well-being and quality of life. For this reason, organic products are normally placed on shelves where, without appropriate information to consumers, they are mixed with hydroponics (food products, with lower pesticide presence and benefiting from a lower price and a 'clean' image), conventional products packed very similarly to organic ones (with colorful labels identifying producers and strengthening the notion of being commercialized directly from producers, looking like 'natural' products), and conventional vegetables in packages with misleading statements (such as, 'free of conservatives,' 'natural product,' etc.). Only certain regional supermarkets have special, refrigerated stands, with signs, clearly separating organics from hydroponics, conventional, etc. In these supermarkets, the demand for healthy foods is contributing significantly to the growth of organic food consumption and production, contradicting the negative forecasts from a part of the organic agriculture movement about sales through supermarkets. It may be expected that this provision strategy from the retail sector is 'converting' consumers to become more 'green.'

Although again based on scarce and scattered empirical data, these findings nevertheless point at the following variables and indicators that might be useful for studying the shopping floor practices implied in the consumption of sustainable food:

- The availability of products – number and diversity of products and departments in the retail store with a green profile.
- Location and presentation of green products – products' location on the shelf (vertical as well as horizontal).
- Trust enhancing strategies in everyday shopping for sustainable products, both from the side of

providers and from the side of consumers (in situ information strategies; communication devices; potential for mutual feedback and control).
- Motivation, training and qualifications of the workers at the selling point, resulting in passive, defensive or pro-active strategies for confronting the consumer with sustainable products.
- The framing of the price setting: the position of sustainable food in comparison with other food products in terms of relative prices but also in terms of image, etc.
- Shopping connected special actions and devices (eco-saving systems; eco-bonus cards, etc.) for the promotion of sustainable food.

Again the list is indicative instead of exhaustive and is meant to contribute to the development of a future research agenda that takes as its starting point the practice of shopping at the retail outlet as a relevant case of a consumption junction.

CONCLUSION

Viewed from a global perspective, green food consumerism is on the increase, although its development can be considered uneven in different respects. With the help of a theoretical framework and its operationalization into three specific categories of factors and variables, we are able to identify the heterogeneous paths that lead to increased sustainability. With the help of the research outline as suggested, it is possible to investigate different forms of 'fits' and 'misfits' between retailer strategies on the one hand and consumer strategies on the other. Guided by our typology of four dimensions of sustainability, we were able to reconstruct some different ways of framing sustainable food consumption in different settings in different parts of the world. When Brazilians are 'going organic' to improve their lifestyles and well-being through the consumption of food that suits the image of a sportive, healthy, and modern life, they put pressure on retailers to provide a hierarchy of food (quality) choices structured according to our first dimension in particular. Guivant (2003) talks about an 'ego-trip' – instead of an eco-trip – discourse as being specifically relevant for the Brazilian case. In this ego-trip discourse on the greening of food production and consumption, fit and healthy personal bodies are connected in a direct way to fit and healthier food flows worldwide. When, in the aftermath of bird flu, Malaysian and Thai food consumers are 'going organic' or start buying at least 'pesticide-controlled' food, they start looking for a government and retail-controlled provision of 'safe' food along the lines of our second dimension of consumer concerns as well. Most likely, however, their concerns are not framed in terms of an ego-trip discourse of the kind found in Brazil. Nevertheless, adherence to safe and health food standards in both parts of the world could very well turn out to be quite comparable with respect to the ways in which consumers respond to strategies by retailers who use the provisioning of green food to create new markets. What puts the Asian consumers and retailers in a different position, compared to their Brazilian or European counterparts, is the lower level of the overall provisioning of green food in their societies. In all regions, however, at least some food is sold and bought under the heading of green or sustainable food.

International trends situate supermarkets as central stakeholders in the expansion of organic food consumption, notwithstanding the conflicts that may exist with farmers about the conditions of supplying sustainable food, as well as with consumers about the framing of sustainability and about the price. According to what has been observed in recent research, worldwide most supermarkets have approached the organic food sector as part of a wider strategy aimed at appealing to those consumers interested in green consumption, including four types of motivations mentioned in this chapter. This inclusion of organic foods into the conventional market and especially on supermarket shelves is not always welcomed as a desirable development by the organic agriculture movement. Their concerns and criticisms are related to the resulting demand for large-scale production, commercialization in supermarkets, consumption restricted to higher income classes due to high prices, etc. An alternative response would be to hold on to small-scale organic food production and consumption distributed through local markets. However, according to the arguments presented in this chapter, a significant growth of the sustainable food market depends on the inclusion of such products in supermarkets. Among the most important challenges in this respect are the negotiations taking place between producers and the retail sector, and the framing of sustainability in and through shopping practices at the consumption junction of supermarkets. As the quality and quantity of sustainable products on offer increase and consumption is stimulated along with changes in lifestyle patterns, the dynamics of sustainable production may be strengthened, breaking away from the negative projections coming from the organic agriculture movement and opening up more alternatives for green-food production and consumption. Accepting the presence of other strategies in the provisioning of sustainable food, next to and as an alternative to the traditional channels of organic food supply, makes it possible to study the strengths and weaknesses of the different options. This chapter has identified some of the key indicators for organizing such empirical studies, as well as formulated a wider conceptual framework to guide this research.

NOTES

1 CCC refers to citizen/consumer demands for 'convenience, comfort and cleanliness' (Shove, 2003).

2 We follow here Goodman's (2003, 2004) presentation of the main arguments around AAFNs.

3 A reply to these criticisms was presented by Van der Ploeg and Renting (2004).

4 As it is done in the many research projects on attitudes and value-orientations of groups of consumers. See Torjusen et al. (2004).

5 For example, retailers can choose (not) to establish specific contracts with organic farmers in the region, or submit contracts under specific (favorable) conditions. This can be illustrated using the case of, for example, Brazilian supermarkets. While in some main cities, like São Paulo and Rio de Janeiro, contracts are not different for organic farmers associations, in Florianópolis regional retail chains offer specific conditions, including refrigeration for the exposure of fresh products (Guivant, 2003).

6 'Through this perspective we agree on the need to avoid methodological nationalism, which blinds conventional sociology to the multidimensional process of change that has irreversibly transformed the very nature of the social world and the place of states within that world' (Beck and Sznaider, 2006, p. 2).

7 http://www.nationalgrocers.org/MarketCenter.html# ConsumerSolutions, April 2003.

8 Ibid.

9 http://www.nationalgrocers.org/ NGNaturalFoods.html

10 Willer and Yussefi (2004), p. 134.

11 SuperHiper Magazine, June 2002.

12 In particular, bird flu surfacing in Thailand, as in other countries of Southeast Asia in 2002, caused widespread consumer concerns about the presence of food risks. Publicity about the presence of high quantities of pesticides in food in supermarkets and stalls is another cause for concern.

REFERENCES AND BIBLIOGRAPHY

Antoine, J.-M. (1998) Les aliments fonctionnels: La perspective de l'industrie alimentaire. Forum sur les aliments fonctionnels. Conseil de l'Europe. Strassbourg, Editions du Conseil de l'Europe: 170.

Aprilia, A. (2005) Analysis of Sustainable Food in Bangkok, Thailand: Production, Consumption and Communication. Environmental Policy Group. Wageningen, Wageningen University. MSc thesis.

Archer, M. (1982) "Morphogenesis versus structuration: on combining structure and action." *British Journal of Sociology* 33 (4): 455–482.

Barry, M. (2004) "The symbolic power of 'organic'." *Hartbeat Newsletter* April.

Bauman, Z. (1993) *Modernity and Ambivalence*. Cambridge, Polity Press.

Beck, U. (1992) *Risk Society: Towards a New Modernity*. London, Sage Publications.

Beck, U. and N. Sznaider (2006) "Unpacking cosmopolitanism for the social sciences: a research agenda." *British Journal of Sociology* 57 (1): 1–23.

Beck, U., and J. Willms (2004) *Conversations with Ulrich Beck*. Cambridge, Polity Press.

Beck, U., A. Giddens and S. Lash (eds) (1994) *Reflexive Modernization. Politics, Tradition and Aesthetics in the Modern Social Order*. Cambridge, Polity Press.

Belk, R. W. (1995) Studies in the new consumer behaviour. In: D. Miller (ed.) *Acknowledging Consumption. A Review of New Studies*. New York, Routledge, pp. 58–95.

Bevan, J. (2005) *Trolley Wars. The Battle of the Supermarkets*. London, Profile Books.

Boström, M., A. Follesdal, M. Klintman, M. Micheletti and M. P. Sorenson (eds) (2005) *Political Consumerism: its Motivations, Power and Conditions in the Nordic Countries and Elsewhere*. 2nd International Seminar on Political Consumerism. Oslo, TemaNord.

Bourdieu, P. (1977) *Outline of a Theory of Practice*. Cambridge, Cambridge University Press.

Castells, M. (1996) *The Rise of the Network Society. Volume I of The Information Age: Economy, Society and Culture*. Malden (MA) and Oxford, Blackwell Publishers.

Dagevos, H., E. van Herpen and M. Kornelis (2005) *Consumptiesamenleving en Consumeren in de Supermarkt. Duurzam voedselconsumptie in de context van markt en maatschappij*. Wageningen, Wageningen Academic Publishers.

Dobson, P. W., M. Waterson and S. W. Davies (2003) "The patterns and implications of increasing concentration in European food retailing." *Journal of Agricultural Economics* 54 (1): 111–125.

Dries, L., T. Reardon and J. F. M. Swinnen (2004) "The rapid rise of supermarkets in Central and Eastern Europe: implications for the agrifood sector and rural development." *Development Policy Review* 22 (5): 525–556.

ETC Group (2005) Oligopoly, Inc. 2005. *Concentration in Corporate Power*. Communiqué, ETC Group: 18.

Flynn, A., T. Marsden and E. Smith (2003) "Food regulation and retailing in a new institutional context." *The Political Quarterly* 74 (1): 38–46.

Giddens, A. (1979) *Central Problems in Social Theory. Action, Structure and Contradiction in Social Analysis*. Berkeley, University of California Press.

Giddens, A. (1984) *The Constitution of Society. Outline of the Theory of Structuration*. Cambridge, Polity Press.

Giddens, A. (1990) *The Consequences of Modernity*. Stanford, Stanford University Press.

Giddens, A. (1991) *Modernity and Self-Identity: Self and Society in Late Modern Age*. Cambridge, Polity Press.

Goffman, E. (1963) *Behaviour in Public Places*. London, Free Press.

Goodman, D. (2003) "The quality 'turn' and alternative food practices: reflections and agenda." *Journal of Rural Studies* 19 (1): 1–7.

Goodman, D. (2004) "Rural Europe redux? Reflections on alternative agro-food networks and paradigm change." *Sociologia Ruralis* 44 (1): 3–16.

Guillon, F. and F. Willequet (2003) *Les aliments santé: marché porteur ou bulle marketing?* Paris, Armand Colin.

Guivant, J. (2003) "Os supermercados na oferta de alimentos orgânicos: apelando ao estilo de vida ego-trip." *Ambiente e Sociedade* 6 (2): 63–98.

Guivant, J., M. Fernanda de A. C. Fonseca, F. Sampaio, V. Ramos and M. Scheiwezer (2003) *Os supermercados e o consumo de frutas, legumes e verduras org,nicos certificados*. Relatório final de pesquisa, CNPq projeto 520874/01–3.

Guptill, A. and J. L. Wilkins (2002) "Buying into the food system: Trends in food retailing in the US and implications for local foods." *Agriculture and Human Values* 19: 39–51.

Guthman, J. (2002) "Commodified meanings, meaningful commodities: re-thinking production–consumption links through the organic system of provision." *Sociologia Ruralis* 42 (4): 295–311.

Guthman, J. (2004) "The trouble with 'organic lite' in California: a rejoinder to the 'conventionalisation' debate." *Sociologia Ruralis* 44 (3): 301–316.

Halkier, B. (2001) "Consuming ambivalences. Consumer handling of environmentally related risks." *Journal of Consumer Culture* 1 (2): 205–224.

Hartman group (2000) *Organic Lifestyle Shopper: Mapping the Journeys of Organic Consumers*. Bellevue, The Hartman Group.

Laurenceau, M. (2005) *Sustainable Food Consumption and Retailer Strategies in France: A Matter of Quality?* Environmental Policy Group. Wageningen, Wageningen University. MSc thesis.

Macnaghten, P. (2003) "Embodying the environment in everyday life practices." *The Sociological Review* 51 (1): 63–84.

Marsden, T., A. Flynn and M. Harrison (2000) *Consuming Interests. The Social Provision of Foods*. London, UCL.

Micheletti, M. (2003) *Political Virtue and Shopping. Individuals, Consumerism, and Collective Action*. New York, Palgrave MacMillan.

Michelsen, J. (2002) "Recent developments and political acceptance of organic farming in Europe." *Sociologia Ruralis* 41 (1): 3–20.

Mol, A. P. J. and D. A. Sonnenfeld (eds) (2000), *Ecological Modernization Around the World. Perspectives and Critical Debates*. Ilford, UK, Frank Cass.

Murcott, A. (1999) " 'Not Science but PR': GM food and the making of a considered sociology." *Sociological Research Online* 4 (3).

Neven, D. and T. Reardon (2004) "The rise of Kenyan supermarkets and the evolution of their horticulture procurement systems." *Development Policy Review* 22 (6): 669–699.

Oosterveer, P. (2005a) *Global Food Governance*. Wageningen, Wageningen University. PhD thesis.

Oosterveer, P. (2005b) "Global regulation of food and consumer involvement: labelling of sustainable fisheries using the Marine Stewardship Council (MSC)." *TemaNord* 517: 339–363.

Ormond, P. J., S. R. Lima de Paula, P. Faveret Filho and L. Thibau M. da Rocha (2002) *Agricultura Orgânica: Quando o passado é futuro.* Rio de Janeiro, BNDS Setorial.

OTA (2006) *2006 Manufacturer Survey.* Greenfield, OTA (Organic Trade Association).

Ponte, S. and P. Gibbon (2005) "Quality standards, conventions and the governance of global value chains." *Economy and Society* 34 (1): 1–31.

Raynolds, L. T. (2004) "The globalization of organic agro-food networks." *World Development* 32 (5): 725–743.

Reynolds, J. and C. Cuthbertson (ed.) (2004) *Retail Strategy. The View from the Bridge.* Oxford, Elsevier.

Richardson, J. (2004) "What's for dinner? Understanding meal fragmentation as a cultural phenomenon." *Heartbeat: Taking the Pulse of the Marketplace.* http://www.hartman-group.com/products/HB/archives2005.html

Richter, T. (2002) *Conceptual Basics for National Standardized Data Gathering Concerning Organic Consumption and Influencing Factors.* 14th IFOAM Organic World Congress, Victoria, Canada.

Richter, T., O. Schmid, U. Meier, D. Halpin, P. van der Berge and P. Damary (2001) *Marketing Approaches for Organic Products in Supermarkets: Case Studies from Western Europe and the United States of America Conducted in 2000.* Basel, Research Institute of Organic Agriculture.

Schwartz-Cowan, R. (1987) "The consumption junction: a proposal for research strategies in the sociology of technology." In: W. E. Bijker, T. P. Hughes and T. J. Pinch (eds) *The Social Construction of Technological Systems: New Directions in the Sociology and History of Technology.* New York, The Guilford Press.

Seltzer, J. M (2004) "Natural foods: a natural profit opportunity" *National Grocer Magazine.* http://www.nationalgrocers.org/NGNaturalFoods.html (accessed Nov. 2004)

Seth, A., and G. Randall (2001) *The Grocers. The Rise and Rise of Supermarket Chains.* London, Kogan Page.

Shove, E. (2003) *Comfort, Cleanliness and Convenience. The Social Organization of Normality.* Oxford, Berg.

Slater, D. (1997) *Consumer, Culture and Modernity.* London, Polity Press.

Southerton, D., H. Chappels and B. van Vliet, (eds) (2003) *Sustainable Consumption; the Implications of Changing Infrastructures of Provision.* Cheltenham, Edward Elgar Publishing.

Spaargaren, G. (2003) "Sustainable consumption: a theoretical and environmental policy perspective." *Society and Natural Resources* 16: 687–701.

Spaargaren, G., and S. Martens (2005) "Globalisation and the role of citizen-consumers in environmental politics." In: F. Wijen, K. Zoeteman and J. Pieters (eds) *A Handbook of Globalisation and Environmental Policy. National Government Interventions in a Global Arena.* Cheltenham, Edward Elgar Publishing, pp. 211–245.

Stones, R. (2005) *Structuration Theory.* Houndmills, Basingstoke and New York, Palgrave MacMillan.

Thompson, C. J. and M. Troester (2002) "Consumer value systems in the age of postmodern fragmentation: the case of the natural health microculture." *Journal of Consumer Research* 28 (4): 550–571.

Torjusen, H., L. Sangstad, K. O'Doherty Jensen and U. Kjaerness (2004) *European Consumers' Conceptions of Organic Food: A Review of Available Research.* SIFO Professional Report. Oslo, SIFO.

Van der Grijp, N. M and F. den Hond (1999) *Green Supply Chain Initiatives in the European Food and Retailing Sector.* Amsterdam, IVM (Institute for Environmental Studies).

Van der Ploeg, J. D. and Henk Renting (2004) "Behind the 'redux': a rejoinder to David Goodman." *Sociologia Ruralis* 44 (2): 234–242.

Warde, A. (1994) "Consumers, identity and belonging: reflections on some theses of Zygmunt Bauman." In: R. Keat, N. Whiteley and N. Abercrombie (eds) *The Authority of the Consumer.* London, Routledge, pp. 58–74.

Warde, A. (1997) *Consumption, Food and Taste. Culinary Antinomies and the Commodity Culture.* London, Sage Publishing.

Wertheim S. (2005) *Bio-Logisch!? In the Eye of the beHolder.* Research Report LEI/ WUR. Wageningen, WUR.

Willer, H., and M. Yussefi (eds) (2004) *The World of Organic Agriculture. Statistics and Emerging Trends, 2004.* Bonn, International Federation of Organic Agriculture Movements.

Zhang, X., X. Fu and J. Yang (2005) *The Vegetable Supply Chain of Supermarkets in Sichuan, China.* Vegsys Project Report 29. Den Haag, LEI.

Redesigning Natures

30

Healthy Ecosystems:
An Evolving Paradigm

David J. Rapport

INTRODUCTION

The concept of ecosystem health has evolved considerably over the past decade. Its beginnings can be found in obscure and until the mid-1990s all but forgotten writings of the American naturalist, Aldo Leopold. Independently conceived in the late 1970s, it has now blossomed to become a guidepost for many national and international agencies concerned with assessing and rehabilitating the state of the environment at national and international levels. It is, at base, a very simple notion – namely that ecosystems can become unhealthy, if overstressed by anthropogenic activities. But the assessment of ecosystem health has proved far more difficult – partly because ecosystems are themselves highly complex, ever changing entities, and difficult to define and delimit. Further complexities arise when it is fully recognized that humans are part and parcel of these complex systems rather than simply being an 'external' pressure on the system. Human societies and whole civilizations have prospered or collapsed depending on whether they were able to sustain the health of the ecosystems of which they were a part.

Here, I review the evolution of the concept of ecosystem health and its potential to motivate and guide the politics of the environment, so that humankind might better cope with what appears to be an increasingly desperate challenge. In the next section, I set the stage for the discussion by reviewing the character of ecosystem dynamics and differentiating between the dynamics of normal perturbation-dependent ecosystems and those of anthropogenically stressed systems. In the third

section, I discuss the expansion of the concepts of 'health' and 'illness' as they have successively been applied to organisms, populations and now ecosystems and indeed the entire biosphere. I then review diagnostic indicators of ecosystem pathology and health that are commonly used in assessing the state of the environment. These indicators generally pertain to the biophysical parameters, leaving out consideration of a host of social indicators of ecosystem health. In the fifth section, the notion of ecosystem health is treated more holistically so that it encompasses sociocultural aspects as well as the biophysical. Here it is important to consider how the health of ecosystems might be linked to human health, as well as how cultural practices and governance systems impact ecosystem health. The sixth section provides an overview of the politics of the environment and various impediments to the implementation of eco-health practices. The seventh section contains a guide for the design of healthy ecosystems, and the concluding section suggests what is needed if the challenge of restoring health to the world's ecosystems is to be met.

THE DYNAMICS OF HUMAN-DOMINATED ECOSYSTEMS

Human-dominated ecosystems are ecosystems that are heavily influenced by the cumulative impacts of human activities (Vitousek *et al.*, 1997). Such ecosystems behave in a radically different way from ecosystems that have evolved in the absence of such influences. To begin with, the behavior of human-dominated ecosystems generally does not exhibit the classic recovery-and-reorganization cycle (Holling, 1986, 2001) that is characteristic of most perturbation-dependent systems (Vogl, 1980). In perturbation-dependent systems, for example, in fire-controlled ponderosa pine forests, periodic disturbances (fires) are a source of renewal. These episodic events serve to recycle nutrients, free up space and thus provide new and fertile ground for adaptation to changing ecological conditions (Yazvenko and Rapport, 1997).

In contrast, in stressed ecosystems (and here stress may be induced by artificial removal of natural perturbations, such as in some forest-management systems in which fire suppression has been practiced), the ecosystem becomes increasingly vulnerable to catastrophic failure. In such cases, rather than triggering renewal, the transformation propels the ecosystem to a higher degree of disorder – in other words, it entrains an ecosystem breakdown process (Rapport and Regier, 1995). The breakdown of the desert grasslands in the southwestern USA offers a classic example. Before human domination, this ecosystem was well buffered against periodic drought and supported a diversity of native grazers. In the late 19th century, the picture changed dramatically when settlers fenced off the area and introduced cattle at high densities. The introduction of this 'exotic' destabilized the landscape, facilitating erosion of the thin soils (as a result of trampling) and the spread of native creosote bush and mesquite (Schlesinger *et al.*, 1990; Rapport and Whitford, 1999). Once this process becomes entrained, it is enhanced by periodic drought and wind- and rain-driven erosion. In many cases such as this one, even the complete removal of the initial source of anthropogenic stress (cattle) is insufficient to reverse this process. In general, once anthropogenic stresses become chronic, ecosystems progressively slip from slightly to highly degraded states, only to be ultimately relegated, in some circumstances, to urban centers or industrial parks (Rapport and Regier, 1995).

HEALTH AND ILLNESS AT THE POPULATION AND ECOSYSTEM SCALE

The distinction between 'health' and 'illness,' which in the Western tradition can be traced back to Greek and Roman mythology (which had separate goddesses to represent each), continues today, with medicine focused on treating illness, while public health is focused on preventing illness, that is, on maintaining health. The two approaches draw upon very different conceptual underpinnings. That is, recognizing illness and treating it is an entirely different enterprise than characterizing 'health' and maintaining it.

Indeed the concept of 'health' is somewhat enigmatic, insofar as 'health' is most easily recognized in its absence. It is relatively simple to identify biological systems – be they individuals, populations, or ecosystems – with impaired functions, but far more difficult to identify the key attributes that confer 'well-functioning.' It is no wonder that since the establishment of the World Health Organization (WHO) more than half a century ago, there has been a succession of definitions of 'health.' Early on 'health' was defined in terms of physical and mental well-being; later it was recognized that health is a 'resource' for carrying on the processes of life and fulfilling aspirations; today it is understood that 'health' requires not only physical, mental, and psychological well-being, but also social, cultural, and environmental well-being.

At least in Western medicine, the health concept came to be focused primarily on the individual. 'Vital signs' such as pulse, body temperature, and blood pressure were found to be critical in distinguishing not only between 'life' and 'death' (that is their true meaning), but also in recognizing illness, or impaired conditions. There are a myriad of other signs that point to potential or real ills. Many of these signs are rather obvious: appearance of rashes, hives, boils, discharge of bodily fluids, particularly blood, color/smell of urine, general energy levels (or vitality), unusual weight gains or losses, lack of neural functions, impairments in visual or other sensory perceptions, etc. Nowadays sophisticated chemical and microbiological assays probe deeper, pinpointing at the systemic level or organ level various signs and syndromes that allow the diagnostician to 'rule in' or 'rule out' possible causes of illness. Collectively, all of these enable the physician to zero in on the possible causes of illness and the possible treatments to restore health to individuals.

A little more than half a century ago, the notion of 'health' began to be applied to populations. Extending the notion beyond the individual to the population level opened vast new fields in epidemiology as well as veterinary practice. In veterinary practice, particularly in the care of farm animals, no doubt this intuitive notion arose from observations that some flocks or herds 'prospered,' providing robust offspring, while others languished and eventually died off. Similarly, for human populations, longevity and disease burdens have been found to vary considerably among countries or regions. Explanations for these variations are often complex – relating to genetics, nutrition, endemic pathogens, public health services, and many other factors.

Once it was accepted (and in veterinary medicine this took some time) that the concept of 'health' applied to whole populations, the challenge was to identify signs of health at this level. Clearly, the notion of 'vital signs' that pertained to individuals is not meaningful at the population level. Indeed, in extreme cases one can have a population of individuals that are judged perfectly healthy, but as a whole the population may be unhealthy – as for example in a sexually reproducing population, if there are significant imbalances in the sex ratio. On the other hand, it is evident that populations generally comprise a mixture of healthy and unhealthy individuals, and despite this, the population as a whole may prosper. One of the keys to 'health' at the population level is the capacity to reproduce – and thus early measures of population health were connected to reproductive success. Other measures of population health relate to age distribution within the population, genetic and phenotypic diversity, and the relationship of birth to death rates, that is, the so-called 'net' reproductive rate.

The expansion of the notion of 'health' to levels above that of the population, namely to ecosystems, landscapes, and the biosphere was initially strongly rejected, particularly by ecologists. Some argued that 'health' was far too subjective a notion to be considered as useful in the analysis of ecosystem condition; others claimed (although the notion of health had already expanded to the level of populations) that 'health' is intrinsically a property of individuals and has no meaning at higher levels; others still objected that if one grants that ecosystems have the property of 'health,' then it is tantamount to resurrecting the analogy of 'ecosystem as organism,' a comparison long rejected by ecologists. However, this is not a valid inference. To assert that ecosystems can become dysfunctional does not at all imply that ecosystems must behave as organisms. All biological systems must indeed have processes of self-regulation if they are to persist. However, at each level of organization (cellular, organ, organism, community, ecosystem, etc.) the mechanisms differ (Rapport *et al.*, 1985). Regulation in ecosystems (which enables the system to recover from natural perturbations) does not require genetic or physiological controls – rather the control is passive, the result of co-adaptation of the biotic community to changes in the environment (Holling, 2001).

When stresses on ecosystems exceed the ecosystems' inherent capacity to self-regulate, then ecosystems tend to break down – in other words, they tend to become sick. Aldo Leopold (1941), the esteemed American naturalist, observed in his native Wisconsin woodlands numerous signs of what he termed 'land sickness.' Among the major signs Leopold identified were declines in soil fertility, declines in crop yields, invasions of exotic species, loss of biodiversity, loss of native flora and fauna, etc. In a series of essays, published nearly half a century after his death, Leopold also wrote of the 'health of the land,' describing its essential properties (Leopold, 1999).

Leopold's observations in the 1940s followed the dramatic deterioration in the farmlands of the Midwest, brought on by extreme droughts – which resulted in characterizing the region as a giant 'dust bowl.' But the dust bowl was not solely due to 'natural forces.' Its extent and severity were undoubtedly impacted by farming practices, through conversion of large tracts of forest to fields and methods of cultivation that left land fallow for much of the time. This practice, during a drought, created the conditions for massive soil loss and producing the vast 'dust bowl' across the Midwestern agricultural states. By the mid-1970s it was recognized that ecosystem breakdown was a far more universal phenomenon – not only confined to isolated agricultural failures, but also found in many ecosystems in which human activities overwhelmed the systems' capacity to regenerate and retain their full range of functionality.

One of the best documented cases of ecosystem failure is in the Lower Laurentian Great Lakes (Lakes Erie and Ontario) – one of the world's largest freshwater reservoirs. In less than two centuries of human domination, from the time of massive European settlement to the present, the basin has experienced such profound changes as to render the ecosystem almost unrecognizable from its earlier state. Regier and Hartman (1973) provided a rather detailed historic account of the cumulative impact of a century and a half of cultural stress on Lake Erie's fish community. Long gone are many large and once abundant fish stocks such as the lake sturgeon, pike, and other large benthic (bottom-dwelling) fish, and in their stead the lake's food web has become dominated by small pelagic (upper-water) invasive species – in particular the alewife and smelt. These transformations are telltale signs of a fundamental breakdown in ecosystem structure – in which the established species so critical to the organization of the ecosystem are replaced by invasive species that bring about instability. At the same time there has been a wholesale reduction in biodiversity in fish species, particularly at local (bay) scales, but also at the whole basin level. As well, there has been considerable sedimentation (which destroys spawning habitats) owing to land-use changes (particularly clearing of forests for agriculture and urban development). In addition, the loadings of contaminants, particularly pesticides and heavy metals, has led to abnormalities in reproduction of avian and other species and poses considerable risks to human health. Among the many signs of ecosystem distress in the lower Laurentian Great Lakes are the frequency and extent of algal blooms, massive fish die-offs, and for the central basin of Lake Erie, large regions of anoxia (absence of oxygen) in bottom waters during mid-summer, when the waters become 'stratified' (i.e. there is little mixing between surface waters and bottom waters).

Another now well-documented case of ecosystem breakdown on regional scales is that of acidification of forests, rivers and lakes owing to industrial effluents. In the early 1960s, Swedish soil scientist Svanté Odén observed a progressive increase in the acidity of Swedish lakes and their forest soils. He hypothesized that these systems were being acidified, not from local activity, but from industrial emissions in northern Europe, particularly in the UK, carried to Scandinavia via long-range atmospheric transport and precipitated as acid rain (Odén, 1968). This was a bold hypothesis in its day – and it was met with a great deal of skepticism by fellow scientists. But over time, it was shown to be correct. The impacts of 'acid rain' have proven to reduce significantly the vitality of ecological processes, not only in Scandinavia, but also in many regions of North America and other countries. Acidification of waters and soils significantly interferes with physiological processes, (especially reproductive success), and thus over time many areas subjected to acid precipitation have shown vastly reduced diversity, particularly in sensitive taxa, such as fish. In highly acidified waters, few organisms survive in any taxa (Schindler, 1988).

Multiple stresses on the Laurentian Great Lakes and global issues such as acid precipitation are part of a much larger picture in which human activities, through a variety of mechanisms – including the release of waste residuals, overharvesting, land-use change resulting in physical restructuring of ecosystems, and the accidental or purposeful introduction of exotic organisms – have destabilized the Earth's ecosystems. The resulting ecosystem pathology is recognized by the presence of the 'ecosystem distress syndrome' – that is, a group of signs that appear in many types of ecosystems, under a variety of anthropogenic stress conditions (Rapport et al., 1985; Rapport, 1989a).

DIAGNOSTICS FOR PATHOLOGICAL AND HEALTHY ECOSYSTEMS

Evaluating the health of ecosystems often begins with looking for such signs, which differentiate unstressed from stressed ecosystems. Many of these (e.g. loss of biodiversity, loss of soil fertility, declines in primary productivity, increases in invasive species) had already been observed by Leopold in his native Wisconsin landscape (Leopold, 1999), under stress from agricultural practices and land-use change. These and other signs of ecosystem pathology, for example, changes in the size distribution of biota, transformations in community structure to favor 'r-selected' species, increased disease prevalence, bioaccumulation of contaminants, horizontal transport (leaching) of nutrients from terrestrial systems – were documented in a variety of later studies. A review of a number of such studies led to the elucidation of a general 'ecosystem distress syndrome' (EDS) (Rapport et al., 1985).

Further tests for the presence of EDS were carried out on additional ecosystems known to be heavily impacted by human activities. These included the Gulf of Bothnia (Baltic Sea) (Rapport, 1989a) the Lower Great Lakes (Rapport, 1983), Ponderosa pine forests in the southwestern USA (Yazvenko and Rapport, 1997), and desert grasslands (Rapport and Whitford, 1999). In all of these cases, although representing very different ecosystems and types of anthropogenic stress, the signs of EDS were found to be present. These studies as well as a great many others (e.g. Rapport et al., 1995) confirm that EDS provides a good general barometer for the biophysical health of ecosystems that have been severely degraded. An important element in establishing the presence of EDS is an intimate knowledge of the normal dynamics of the ecosystem under examination. Most ecosystems, even those without anthropogenic stress, undergo large fluctuations – and the various parameters that are used to detect EDS are hardly constant. To confirm the presence of EDS, it must be shown that the direction of change and the magnitude of change fall well beyond what might be explained in terms of the normal range of fluctuation over historical time, in both ecological and historical time. While such measures are basic to evaluating the health of ecosystems, they do not go far enough.

As humans are very much part of and not apart from the ecosystem, additional ecosystem health indicators relating to the socio-economic, cultural, and human health dimensions ought to be developed (Rapport, 1995). These indicators would in effect constitute a societal distress syndrome (SDS) that mirrors the biophysical manifestations in EDS. Such indicators should not be included as an 'afterthought' (as is still rather common in *post facto* socio-economic assessments tacked on to conservation or restoration plans) but rather as a central element of a holistic ecosystem health evaluation.

While recognizing EDS is essential to detect or confirm advanced stages of ecosystem pathology, and by inference its absence suggests a state of ecosystem health, there are more targeted measures that focus on 'health' rather than pathology. For ecosystem health there are three generally accepted measures (Rapport *et al.*, 1998):

- *Vitality (or productivity)* – the degree to which an ecosystem is able to carry on its functions of energy conversion and nutrient transport at levels in the normal range for each system (systems vary in these normal ranges over orders of magnitude – for example, primary productivity of a marshland, or grassland, is considerably higher than for a boreal forest).
- *Organization* – the degree to which the ecosystem maintains its biotic diversity and the complexity of interrelationships between the biotic and abiotic components.
- *Resilience* – the degree to which ecosystems can 'buffer' perturbations (such as storms, floods, fires, droughts) and maintain their basic structure and function.

While it is quite possible to have, at least temporarily, a 'healthy' economy, at the expense of a biophysically healthy ecosystem – for example, an economy fueled by aggressive harvesting of natural resources – this obviously is not sustainable in the long term. Neither can one expect to sustain healthy human populations indefinitely within a degrading ecosystem. In both cases, eventually the collapse of one aspect of the system brings about the collapse of the other. Similarly with governance, the absence of a participatory democratic political system – that is a political system that allows freedom of expression, fosters decision-making that respects the beliefs and goals of the community at all scales, and emphasizes 'responsibilities' as well as 'rights' – will generally hamper the necessary social responsiveness needed to adjust human behaviors to achieve healthy ecosystems under constantly changing circumstances.

ECOSYSTEM HEALTH IN A TRANSDISCIPLINARY CONTEXT: THE SOCIO-CULTURAL DIMENSIONS

In the previous section we discussed the biophysical measures of ecosystem pathology and health – that is transformations in the structure and function of ecosystems attributable to cumulative stress from human activities. This has been almost the exclusive focus of most of the early evolution of the notion of ecosystem health (Rapport *et al.*, 1979, 1985, 1995; Schaeffer *et al.*, 1988; Rapport, 1989a,b; Leopold, 1999). Today, the concept of 'healthy ecosystems' has transcended the boundaries of an increasing number of disciplines. It now incorporates integrally, the social, natural, and health science dimensions. Here, we examine a few of the connections between the social and biological aspects.

The human health/ecosystem health interface

In conventional medical practice the focus is on the individual in isolation from the ecosystem of which they are a part. Thus, the diagnosis is directed towards 'fixing the problem' through the use of medicines or other interventions, rather than examining some of the 'upstream' causes – which in many cases turn out to be ecological imbalance (Rapport *et al.*, 1998). An ecosystem approach to human health, in contrast, looks at patients as part of their ecosystem – which in the widest context includes family, socio-economic, cultural, and ecological contexts. As individuals do not live in a vacuum, but rather their capacity to maintain health is very much conditioned by their socio-economic, cultural, and biophysical environment, the long-run solution to an increasing number of human health burdens is not through medicines and other treatments that focus on 'fixing the problem' within the individual, but rather on restoring ecosystem health, so that the human health burdens are, in general, much reduced (Rapport *et al.*, 2003).

It is well known that highly polluted environments increase the likelihood of certain kinds of disease: exposure to toxic substances impact neural functions; exposure to carcinogens may lead to various cancers; exposure to highly polluted (nutrient-enriched) waters is a direct source of many infections from bacteria and other organisms; and increased exposure to UV radiation (owing to the depletion of the ozone layer) results in a detectable increase in cataracts, as well as skin melanomas (McMichael, 1997, 2001). Some of these relationships are relatively easy to quantify and a number of models have related increased human health burdens to one or more types of environmental degradation (McMichael, 2001). In other cases, the connections between human health outcomes and the health of ecosystems are complex and not so easy to quantify, although nonetheless very real. The difficulty in making direct linkages, however, is owing to the complexity of the relationship. We need particularly to take into account the capacity of human societies to buffer adverse health effects (through public health and medical services), inherent lags between environmental changes and organism response, and the fact that health outcomes are the summation of many influences (McMichael and Woodruff, 2005).

Yet it has long been appreciated (ever since the eye-opening work of Rachael Carson; Carson, 1962) that toxic substances, particularly pesticides, bio-accumulate in the food chain, and may accumulate in significant enough amounts to place the top level in the food chain at risk. This is certainly the case nowadays for consumption of a number of fish stocks in the Laurentian Lower Great Lakes and their tributaries.

Less known are the many complex pathways by which human alterations of ecosystems pose public health risks. Among these are changes in the distribution of human pathogens as a result of human-created ecological imbalance. With global warming, for example, there has been in certain regions (in Central and South America and Africa) an increase in the geographic range of malaria, dengue fever, and other vector-borne diseases – from lowlands to highlands as the mosquito vectors find an expanded range. Lyme disease appears to be on the rise in the northeastern USA, partially owing to a significant increase in an intermediate host (deer) of the mites that bear the pathogen. The rise in deer population in turn may be attributed to an ecological imbalance created by wiping out the natural predators (wolves), while at the same time increasing habitat favorable to deer (suburban lawns and abandoned farmlands).

The rise in the prevalence of the extremely debilitating and often fatal Schistosomiasis furnishes another example. This disease is endemic in many regions of Africa and Asia and is spread to humans through contact with intermediate hosts, water-dwelling snails. The creation of large dams and their associated irrigation schemes have led to the spread of the snails, and thus to an increasing prevalence of the disease. Here too there are complex interactions with other human-mediated ecological imbalances, such as the spread of the invasive water hyacinth (*Eichhornia crassipes*). Increased abundance of water hyacinth in Lake Victoria (East Africa) is contributing to an increase in the snails bearing the pathogen causing Schistosomiasis (Plummer, 2005).

The cultural health ecosystem – health interface

Cultural practices, which reflect the beliefs, mores, and social organization of a people, have in recent times been identified as a major source of ecological imbalance, as the drive for material well-being has permeated every aspect of modern society. However, this has not always been the case, and indeed even today is not universally the case. Historically, many cultures evolved practices that maintained (and perhaps enhanced) ecosystem health. These cultures possessed or over time gained the values that placed a high

premium on sustaining their resources. For instance, the Polynesians who colonized Hawaii, as perhaps in the case of all colonizers, at first contributed significantly to the degradation of the ecosystem, clearing lowland forests, with the consequent loss of much biodiversity, particularly in avian species. However, over time there was adaptation, and these early Hawaiian colonizers developed a holistic system of land tenure and management, known as *ahupua'a*. In this system, agricultural practices, forest management, and fisheries were conducted in an integrative fashion – such that, for example, nutrient enrichment from taro crops enhanced the inshore fishery. Alternatively, in situations where cultural practices have failed to respect the health of ecosystems, invariably ecosystem (as well as societal) collapse has ensued. Easter Island provides the classic case, in which total deforestation led to the complete collapse of that once thriving island civilization. The demise of Sumerian civilization in Mesopotamia has also been attributed to deforestation. In this case, the cutting of forests in the upper part of the watershed resulted in catastrophic downstream floods (Wright, 2004).

The healthy governance ecosystem – health interface

Far less attention has been given to this interaction, yet the state of governance systems may lie at the very heart of ecosystem health issues. Where a participatory process prevails, that is, in situations in which people have an opportunity to express their views in the spirit of furthering community interests (as opposed to individual interests), and where a process of community consultation results in a group decision, there is a far greater likelihood that the range of factors determining ecosystem health will be taken into account. Of course there are often considerable barriers in getting people to think in terms of an ecosystem context, rather than focus on individual benefits, or community benefits to the costs of the wider ecosystem. Too often local people or town officials restrict their focus to their local area.

For governance for ecosystem health to be effective, 'environment' cannot be isolated from other societal issues – such as sustainable livelihoods, public health, cultural values, etc. There inevitably are trade-offs, both short and long term, but if the governance system is a genuinely participatory one in which all points of view are heard, these considerations are more likely to come into play. In villages and towns, the scale for true participatory process is manageable. When it comes to the modern 'city states' and 'nation states,' the direct connection between those who govern and what is at stake is often tenuous or lost completely. These are the dominant decision-making structures in place today, and the least favorable for deriving integrative and innovative solutions to complex issues of promoting and maintaining healthy ecosystems. What we often end up with is the rhetoric without the substance. Ultimately it is a matter of power and politics – and here too often the special interests of the few prevail over the needs and interests of the many.

THE POLITICS OF HEALTHY ECOSYSTEMS

The quest for material well-being has been a predominant goal of human societies from very early on. And some early civilizations may well have collapsed owing to the cumulative impacts of their activities on their environments. Certainly this appears to have been the case in the demise of the Polynesians that colonized Easter Island, and no doubt it has also been the fate of other ancient civilizations (Wright, 2004). As dramatic as these transformations must have been for those societies, the environmental impacts were relatively contained – at least by today's standards, although no doubt in the case of the Romans, Mayans and Incas, large territories may have been compromised. Today, humankind stands at the brink of the failure of the Earth's ecosystems at the regional and perhaps even biosphere levels.

The recent Millennium Ecosystem Assessment – the largest undertaking yet to assess the state of the Earth's ecosystems at sub-global and global levels – adds further to the accumulating evidence that ecosystems, worldwide, are in failing health (Table 30.1).

These examples (given in Table 30.1) point to a clear failure nationally and internationally to arrest the continued degradation of the Earth's ecosystems.

Part of the problem is the fragmented approach to the environment reflected in multiple agencies that lack coordination within and among them. While the ecosystem health concept is increasingly being taken 'on board' in mission statements of organizations such as the World Wildlife Fund, International Union for Conservation of Nature (IUCN), and Conservation International (CI), and has been instrumental in mounting federal, state, and provincial programs that specifically monitor the health of rivers, lakes, coastal areas, forests, grasslands, etc., regulatory agencies have by and large focused on single issues, often at cross-purposes, rather than taking a holistic approach that cuts across many areas. For example, one agency may focus on regulations to achieve clean water, while another encourages maximizing agricultural yields by more intensive chemical agriculture (Mooney et al., 2005).

What is required are agencies that focus on the 'ecosystem as patient' – that is, that can take action across a broad spectrum, such that a better harmonization between human activity and nature's capacity to absorb stress without losing its resilience can be achieved. Too often, however, the politics of special interests preclude this. Too often it is a case of 'one step forward, two steps back' – as in the case of the aftermath of the great Midwestern flood of 1993 along 1600 km of the Mississippi and Missouri rivers (Pinter, 2005). Here, despite the Federal Emergency Management Agency's massive buyouts along the floodplain properties, including relocating the floodplain town of Valmeyer, Illinois, developers

Table 30.1 Indicators of regional and global ecosystem pathology

Focus area	Environmental issue	Estimated (2005)
Fresh water	Lack access	1.1 billion people
Fresh water use	Withdrawal by humans	54%, by 2025 70%
Basic sanitation	Lack access	2.6 billion people
Forests	Deforestation	9.4 million ha/year
Forests	Global depletion	50%
Coral reefs	Lost/degraded	40%
Mangrove forests	Disappeared	30%
Earth land surface	Degraded	40–50%
Land cover	Projected reduction	33% over 100 years
World fisheries	Degraded/depleted	60%
Agriculture	2020 demand	Grains by 40% *
Agriculture	2020 demand	Livestock by 60%**

*Projected increase over 1999 levels – for rice, wheat, and maize
**Projected increase over 1999 levels
Source: Based on data from Mooney *et al.* (2005) and Ayensu *et al.* (1999)

have gone ahead with redevelopment on lands that were under water in 1993 to the tune of more than $2.2 billion in the St. Louis area alone. This surge in floodplain development would appear to more than counterbalance the efforts to restore the health of the flood-plain ecosystem.

Of course the dramatic declines in the health of the world's ecosystems that have occurred (including the nearly total demise of the Aral Sea, the Mesopotamian wetlands, and a significant proportion of desert grasslands, coral reefs, etc.) do not go unnoticed. But there remains a huge gap between environmental consciousness and societal response – that is, between knowledge and action. As a consequence, most political responses are relatively weak compared with the forces that continue to erode the Earth's life-support systems. This problem is compounded by the fact that political action in this day and age seems to be more dictated by the very forces that are causing the destruction than by the forces that are conscious of the consequences of that destruction. Consider for example the gigantic seasonal 'dead zone' created in the Gulf of Mexico along the coastal waters of Louisiana. It has long been recognized that the use of chemical fertilizers in the Mississippi basin is the major source of this large-scale ecological degradation – which in some years covers an area as large as 20,000 sq. km. Yet the Gulf's Dead Zone appears to be growing over recent decades as the fertilizer industry continues to turn a blind eye to the science that links fertilizer use with the ecological damage (Kaiser, 2005).

In this and many other cases, political responses to arrest ecosystem degradation have failed to address the root causes – that is, the forces generating threats on ecosystems – largely because powerful economic interests are allied against such interventions. There are few notable exceptions: the Kyoto Protocol (whose acceptance was hard won) is designed to cap greenhouse gas emissions to levels well below what they are at present; the Montreal Protocol prohibits CFC emissions, as these gasses are responsible for breakdown of the protective stratospheric ozone layer.

Most actions on the environment, while billed as 'protection,' do very little in fact to protect ecosystem health. Rather, many are actions that attempt to 'fix' the problem once it arises, and generally with limited success. Thus, billions of dollars have been expended over the past several decades in efforts to 'clean up' the Great Lakes. These funds targeted forty-two 'areas of concern' (bays, harbors, etc.), where contaminants were deemed excessive and thus a threat to the health of the ecosystem. While two of the least impacted areas were subsequently de-listed (Collingwood and Severn), the remaining areas of concern and

the Lower Great Lakes generally remain a highly polluted and toxic body of water, in which one cannot safely consume the remaining otherwise edible fish.

Further, efforts to protect the lakes have neglected a multitude of persisting basin-level sources of stress other than contaminants. In particular, development in the Basin is often at the expense of near-shore habitat (including wetlands that are critical for breeding and foraging for many species). Shipping continues to provide a major source of potentially invasive species, such as the Zebra mussel which have in the past radically transformed and destabilized basin-wide ecology.

In sum, despite decades of political action aimed at protecting this major North American ecosystem, results by and large are meager. Once an ecosystem becomes severely damaged, experience has shown that the problems may well be intractable. The Laurentian Great Lakes ecosystem remains highly contaminated, has lost much of its commercial fishery, no longer yields a safe source of food for local populations, and continues to pose serious health risks to the population living in the basin. Efforts to restore other heavily damaged ecosystems have met with similar limited success (e.g. the Baltic Sea, the Mediterranean Sea, the Kyrönjoki Estuary in Finland).

Thus, while the politics of the environment has in a few domains taken notable steps (mainly in the international protocols and accords), it has proven overall relatively impotent against the short-term economic interests of corporations (multinationals) whose mantra is 'growth' at any cost. It is of course not just corporations – which under the umbrella of their organization may with some notable exceptions act with little regard to their impact on the health of regional ecosystems. It is also consumers who are imbued with short-term thinking, self-interest, and disregard of the responsibilities that comes with their freedoms to consume. In a sustainable world, growth in material well-being would be tempered by the necessity of preserving the vitality of the Earth's ecosystems. It follows directly from the 'precautionary principle' that activities which threaten the resilience (capacity to rebound from disturbance), productivity (vitality), and organization (biodiversity) of ecosystems should be severely constrained or prohibited altogether. It is heartening to see that the United Nations Development Programme, the United Nations Environment Programme, the World Bank, and the World Resources Institute have reconfirmed their commitment to making 'the viability of the world's ecosystems a critical development priority for the 21st century' (United Nations Development Programme et al., 2002).

Unfortunately, such thinking has made little inroads into the practical politics of the environment in today's relentless pursuit of material well-being. Indeed, at the regional scale, new trade agreements – such as the North American Free Trade Agreement (NAFTA) – have given impetus to expanding production with heavy environmental costs, particularly in terms of energy use and air and water contamination, in regions where environmental regulations are the least capable of counteracting these activities. No wonder that NAFTA has stimulated a boom in Industrial Parks in Northern Mexico, just south of the USA/Mexican border, where enforceable environmental legislation is almost non-existent.

Corruption also undoubtedly plays a large role in short-circuiting effective action to maintain ecosystem health. Then too, political action on the environment is often undermined by the lack of opportunity for participation and by public apathy. There appears to be a significant disconnect between, on the one hand, the desirability of maintaining viable ecosystems that are conducive to maintaining a high quality of life, and on the other hand, short-term economic interests. To the extent that short-term economic interests prevail, the disconnect remains one of the key barriers to reversing the worldwide slide towards ecosystem degradation. Where unbridled economic growth remains the sole and dominant societal value, the continued erosion of ecosystem health is inevitably the price to be paid.

DESIGN FOR ACHIEVING HEALTHY ECOSYSTEMS

I recall a conversation I had in the early 1980s with Eugene Odum, one of the great proponents of holistic ecosystem-level theory and practice (Odum, 1997). We were discussing the potential of structuring state of environment reporting on an ecosystem basis rather than on the approach prevailing at that time – which looked at the environment largely in terms of air and water quality. While Odum had the utmost appreciation for the power of the ecosystem approach, he was hesitant about its practicality, asking 'but who speaks for the ecosystem'?

That question lingers on today. True, the ecosystem approach has come to by and large dominate environmental assessments (Mooney et al., 2005), and a number of commissions and communities have adopted an 'ecosystem approach' to their particular region – but giving that approach 'teeth,' that is, providing it with enforceable legislation, requires having jurisdiction (sovereignty) over the region. And political units are no more closely aligned with ecosystems

today than they were two to three decades ago, thus limiting their ability to enforce legislation.

The design for healthy ecosystems thus must become as much about governance and maintenance of cultural practices as it is about restoration of the biophysical environment. With respect to governance, the immediate need is effective realignment of authority to correspond to an ecosystem-based region, rather than a politically or administratively based region. This was made explicit for the Great Lakes, and consequently a host of commissions and international agencies with overlapping basin-wide interests and authority have been established. Yet, none of these commissions or agencies, singly or collectively, has the power accorded to sovereign nations, states, or provinces, and therefore their authority is limited to that of advisory. The root issue here is clearly one of power: where the sovereign authorities have jurisdiction over the ecosystem, they may be able to enact and enforce ecosystem-wide legislation to maintain and restore ecosystem health. As a model, one might look to the practices of the European Union, which has some limited authority to levy significant fines on member states that fail to meet overall environmental standards.

A design for achieving regional eco-cultural health thus becomes a design for reintegrating society with nature through a shift in values, and through more adaptable and holistic governance systems. It becomes a task of designing governance systems which are truly participatory and are rooted in local and regional ecosystems, their peoples, their socio-cultural practices, and their livelihoods. 'Truly participatory' means not only a process that brings community members together to 'have their say,' but rather an educational process whereby community members come to think beyond their individual interests, that is beyond the 'not-in-my-backyard' syndrome, and enter into discussions about what is in the community (local, regional) interest in achieving quality of life in balance with the environment. At every level, the question once posed by Aldo Leopold, 'how can humans occupy the earth without rendering it dysfunctional?' should be of paramount concern.

A design for regional eco-cultural health would address head-on the conflict within the politics of the environment: namely that the drivers of corporate activity cannot be solely the 'bottom line;' corporate accountability must be encouraged to go beyond its shareholders, to the community and to the ecosystem of which a corporation is a part. If these are compromised, the long-term future of the corporate entity is also tenuous.

Finally, a design for regional eco-cultural health must be 'precautionary' rather than reactive. Here there are lessons from the history of ecosystems already compromised. These lessons suggest in unequivocal terms that once systems are severely damaged there is little hope of seeing recovery. The emphasis of eco-cultural health care thus must be prevention, not cure.

ULTIMATE CHALLENGES FOR RECOVERING THE VITALITY OF OUR LIFE-SUPPORT SYSTEMS

If we are to maintain a viable future for humankind, a sea-change in human behavior, values, and attitudes must take place. It is encouraging to see some positive signs in this direction – that is, the growing number of individuals, NGOs, and communities that have recognized their responsibilities toward environmental stewardship and have taken actions accordingly. There are also some exemplary actions taken voluntarily by a small but growing number of corporations, which may serve as a 'bell-weather' of things to come. To maintain or regain ecosystem health, in the face of the economic pressures that are presently destroying it, will take more than these isolated efforts – it will take a major societal transformation such that maintaining healthy ecosystems is seen as the highest goal. In addition, there is a need for a body politic with the capacity for critical thinking in evaluating the various assurances governments give about how they are 'safeguarding' the environment. We must free our collective self from the delusion that environmentally 'everything is under control' because that is what governments and their corporate allies tell us and want us to believe.

Above all there must be the strong will to collectively live differently than in the recent past – to have a lighter 'footprint' on the planet (Rapport, 2003; Palmer *et al.*, 2004; Rapport and Singh, 2006). This requires an ecosystem health ethic that guides our behavior in decisions large and small. Such an ethic, along with a thorough grounding as to what constitutes healthy ecosystems in both the biophysical and socio-cultural realm, should enable people to find innovative ways of integrating their activities within the ecosystem while respecting the maxim 'do no harm.' Formulating such an ethic is as important as any other action if humankind is to succeed in changing course.

ACKNOWLEDGMENTS

I am grateful to Victoria Lee, Luisa Maffi, Bruce Mitchell, and Jorge Nef for their comments on earlier drafts.

REFERENCES

Ayensu, E., Claasen, D., van R., Collins, M. *et al.* (1999). International ecosystem assessment. *Science* 286: 685–686.

Carson, R. (1962). *Silent Spring*. Houghton Mifflin, New York, 368 pp.

Holling, C.S. (1986). The resilience of terrestrial ecosystems: local surprise and global change. In: Clark, W. C. and Munn, R.E. (eds) *Sustainable Development of the Biosphere*. Cambridge University Press, Cambridge, UK, pp. 292–316.

Holling, C.S. (2001). Understanding the complexity of economic, ecological and social systems. *Ecosystems* 4: 390–405.

Kaiser, J. (2005). Gulf's dead zone worse in recent decades. *Science* 308: 195

Leopold, A. (1941). Wilderness as a land laboratory. *Living Wilderness* 6: 3.

Leopold, A. (1999). *For the Health of the Land* (previously unpublished essays and other writings), Callicott, B. and Freyfogle, E.T. (eds). Island Press, Washington, DC, 243 pp.

McMichael, A.J. (1997). Global environmental change and human health: Impact assessment, population vulnerability and research priorities. *Ecosystem Health* 3: 200–210.

McMichael, A.J. (2001). *Human Frontiers, Environments and Disease: Past Patterns, Uncertain Futures*. Cambridge University Press, Cambridge, UK.

McMichael, A.J. and Woodruff, R.E. (2005). Detecting the health effects of environmental change: Scientific and political challenge. *EcoHealth* 2 (1): 1–3.

Mooney, H., Cropper, A. and Reid, W. (2005). Confronting the human dilemma: How can ecosystems provide sustainable services to benefit society? *Nature* 434: 561–562.

Odén, S. (1968). The acidification of air and precipitation and its consequences in the natural environment. Ecology Committee Bulletin, No. 1, Swedish National Science Research Council, Stockholm.

Odum, E.P. (1997). *Ecology: A Bridge Between Science and Society*. Sinauer Associates, Sunderland, MA, USA, 330 pp.

Palmer, M., Bernhardt, E., Chornesky, E. *et al.* (2004) Ecology for a crowded planet. *Science* 304: 1251–1252.

Pinter, N. (2005). One step forward, two steps back on U.S. Floodplains. *Science* 308: 207–208.

Plummer, M.L. (2005). Impact of invasive water hyacinth (*Eichhornia crassipes*) on snail hosts of schistosomiasis in Lake Victoria, East Africa. *EcoHealth* 2: 81–86

Rapport, D.J. (1983). The stress-response environmental statistical system and its applicability to the Laurentian Lower Great Lakes. *Statistical Journal of the United Nations ECE* 1: 377–405.

Rapport, D.J. (1989a). Symptoms of pathology in the Gulf of Bothnia (Baltic Sea): Ecosystem response to stress from human activity. *Biological Journal of the Linnean Society* 37: 33–49.

Rapport, D.J. (1989b). What constitutes ecosystem health? *Perspectives in Biology & Medicine* 33: 120–132.

Rapport, D.J. (1995). Ecosystem health: exploring the territory. *Ecosystem Health* 1: 5–13.

Rapport, D.J. (2003). Regaining healthy ecosystems: the supreme challenge of our age. In: Rapport, D.J, Lasley, W., Rolston, D.E., Nielsen, N.O., Qualset, C.O. and Damania, A.B. (eds) *Managing for Healthy Ecosystems.* CRC Press, Boca Raton, Fl, USA, pp. 5–10.

Rapport, D.J. and Regier, H.A. (1995). Disturbance and stress effects on ecological systems. In: Patten, B.C. and Jorgensen, S.E. (eds) *Complex Ecology: The Part–Whole Relation in Ecosystems.* Prentice Hall PTR, Englewood Cliffs, NJ, pp. 397–414.

Rapport, D.J. and Singh, A. (2006). An EcoHealth-based framework for state of environment reporting. *Ecological Indicators* 6: 409–428.

Rapport, D.J. and Whitford, W. (1999). How ecosystems respond to stress: Common properties of arid and aquatic systems. *BioScience* 49 (3): 193–203.

Rapport, D.J., Thorpe, C. and Regier, H.A. (1979). Ecosystem medicine. *Bulletin of the Ecological Society of America* 60: 180–182.

Rapport, D.J., Regier, H.A. and Hutchinson, T.C. (1985). Ecosystem behavior under stress. *The American Naturalist* 125: 617–640.

Rapport, D.J., Gaudet, C. and Calow, P. (eds) (1995). *Evaluating and Monitoring the Health of Large-scale Ecosystems.* Springer-Verlag, Heidelberg, 454 pp.

Rapport, D.J., Costanza, R. and McMichael, A. (1998). Assessing ecosystem health: Challenges at the interface of social, natural, and health sciences. *Trends in Ecology and Evolution* 13 (10): 397–402.

Rapport, D.J., Howard, J. Lannigan, R. and McCauley, W. (2003). Linking health and ecology in the medical curriculum. *Environment International* 29: 353–358.

Regier, H.A. and Hartman, W.L. (1973). Lake Erie's fish community: 150 years of cultural stresses. *Science* 180: 1248–1255.

Schaeffer, D.J., Henricks, E.E. and Kerster, H.W. (1988). Ecosystem Health: 1. Measuring ecosystem health. *Environmental Management* 5 (12): 445–455.

Schindler, D.W. (1988). Effects of acid rain on freshwater ecosystems. *Science* 239: 149–159.

Schlesinger, W.H., Reynolds, J.R., Cunningham, G.L., Huenneke, L.F., Jarrell, W.M., Virginia, R.A. and Whitford, W.G. (1990). Biological feedbacks in global desertification. *Science* 247: 1043–1048.

United Nations Development Programme, United Nations Environment Programme, World Bank, and World Resources Institute (2002). *People and Ecosystems: The Fraying Web of Life,* p. vii.

Vitousek, P.M., Mooney, H.A., Lubchenco, J. and Melillo, J.M. (1997). Human domination of earth's ecosystems. *Science* 277, 494–499.

Vogl, R.J. (1980). The ecological factors that produce perturbation-dependent ecosystems. In: Cairns, J. Jr. (ed.) *The Recovery Process in Damaged Ecosystems.* Ann Arbor Science Publishers, Ann Arbor, MI, pp. 63–94.

Wright, R. (2004). *A Short History of Progress.* House of Anansi Press, Toronto, 211 pp.

Yazvenko, S.B. and Rapport, D.J. (1997). The history of ponderosa pine pathology: Implications for management. *Journal of Forestry* 97 (12): 16–20.

31

Environment and Human Security

Laura Little and Chris Cocklin

INTRODUCTION

Our conceptions of the relationship between humans and the environment fundamentally shape the way people use, interact with and respond to nature. Since the Enlightenment, the dominant Western conception of this relationship has been one of sharp division between human societies and the natural environment. Human systems, such as economics and politics, have been viewed as separate and distinct from the natural environment, and social constructions of the human/environment relationship have stressed the rightness and capacity of human beings to gain mastery over nature (Hargrove, 1989; Pepper, 1996). O'Riordan (1976) described this as technocentricism.

With the growing awareness of global environmental degradation, there have been criticisms of these dominant ideas and efforts to rethink the human society/environment relationship. Much of this rethinking has involved the many and contested notions of 'sustainable development' (Barnett, 2001; Robinson, 2004). While consensual definitions remain elusive (Cocklin, 1995), discourses of sustainable development have shared, amongst other things, a greater recognition and understanding of the interdependence of human societies and the natural environment. Complementary, but lesser known discourses that have emerged are concerned with the nexus between environment and security.

This chapter explores these latter discourses, providing background to the environment and security debate and outlining the various perspectives that have emerged. Given that concepts of 'sustainable development' remain dominant in much of environmental discourse, we ask the question: What can the viewing of environmental issues through the lens of 'security' contribute to our understanding of the relationship between human societies and the natural environment?

THE CONCEPT OF 'SECURITY'

We frequently use the term 'security' in everyday life, usually presuming it has a clear, unambiguous meaning. On a national level, for example, we typically associate the term with protection against military attack and, more recently, protection from terrorism. At an individual level we may associate 'security' with protection against burglary or with having sufficient income (i.e. *financial* security). Analysing discourses more closely reveals, though, that the concept of security is contested (Lipschutz, 1995, p. 12). Definitions of security vary depending on understandings of: (a) what entities we are trying to make secure – is it individuals, communities, a nation state? and (b) what we define as threats to security. This chapter will explore the various meanings with which the term 'security' has been invested, focusing on how definitions of security incorporate these two elements.

While the meaning of 'security' is ambiguous, we contend that the various definitions of security have basic commonalities. First, the underlying aim of most security thinking, at least in theory, is to ensure the survival and basic welfare (however this is defined) of human beings

(Barnett, 2001, p. 138). Second, discussions of security connote *existential* threats (Wæver, 1995, p. 56). These ideas fundamentally shape the role of security discourse in political discussions. Specifically, connotations of an existential threat lead to the privileging of 'security issues' (Report of the 24th United Nations of the Next Decade Conference, 1989, p. 10, cited in Levy, 1995, p. 44). As Wæver argues in the context of nation state security, labelling a problem as a 'security issue' has the effect of '[raising] a specific challenge to a principled level' and thereby allowing the state to 'claim a special right to use whatever means are necessary to block it [the threat]' (Wæver, 1995, pp. 55–56). Thus, security discourse is a powerful political tool that can be used to channel energy and resources in particular directions. Consequently, definitions of what constitutes 'security' have important implications for both human beings and the natural environment.

DOMINANT CONCEPTIONS OF SECURITY

Although security is a contested concept, there are ideas about what constitutes 'security' that have become dominant in security discourse and these dominant conceptions provide the basic framework for contemporary discussions. Additionally, *re*definitions of security are, in significant part, a response to perceived shortcomings of these dominant views.

The definitions that are dominant in *political* discourse construct security in terms of the nation state as the entity to be secured. During most of the Cold War, the primary threat to security of the nation state was considered to be military attack. More recent definitions of security have expanded the notion of threats to nation-state security to include various non-military threats. Currently, there is no agreement as to which non-military concerns constitute threats to security; for example, threats to a state's energy and food supply and threats to economic prosperity have all been discussed in terms of security (Barnett, 2001, pp. 34–36). Nevertheless, we argue that the tendency in mainstream security discourse is to focus primarily on economic concerns when considering non-military threats to security. Therefore, discussions of dominant conceptions of security in this chapter refer to definitions that construct security in terms of the nation state and that represent threats to security primarily in military and economic terms. These definitions betray certain assumptions about what *should be protected* and what *the relevant risks* are to that object of protection. Specifically, they construct the nation state as the entity that should be protected.

This focus points to an assumption that the nation state is of paramount importance in securing human survival and welfare.

What ideas underlie these assumptions? The focus on the nation state in security discourse stems from the political ideas of Realism, modern liberalism and nationalism, theories that offer differing justifications for the existence of the nation state. In broad terms, modern liberalism sees the importance of the state arising from its role as the vehicle for securing and protecting the collective interest of its citizens (Hirst, 2001, p. 63). Nationalism views the nation state as the embodiment of a natural collective entity and, for that reason, worth securing for its own sake. Realism requires a more detailed explanation.

Classical Realists view the nation state structure as essential for 'security and stability' in human life (Hirst, 2001, pp. 58, 64). These views are based on particular ideas about human nature and human relationships. Classical Realists take a pessimistic view of human nature. They see human beings as driven by a desire for material goods and status, and inherently competitive, self-interested and inclined towards immorality in their quest for these (Donnelly, 2000, pp. 14–15, 24–25). In a classical Realist worldview, the interests and desires of one person will conflict with those of others, and the ability of one person to ensure their interests over others will depend on the margin of power they are able to exercise (Der Derian, 1995, p. 29). Under 'natural' conditions, people are roughly equal and no one person has overwhelming power to impose their will (Der Derian, 1995, pp. 29–30). Consequently, classical Realists see the natural state of human beings as struggle and war, as people continually seek a margin of power over others that will guarantee their own interests (Der Derian, 1995, pp. 29–30; Donnelly, 2000, pp. 14–15, 24–25). The only way to control these tendencies towards conflict and anarchy is through the creation of a higher authority that can exercise control over others (Der Derian, 1995, p. 30; Donnelly, 2000, p. 15). The need for sovereign states – which can impose order through their monopoly on violence – follows almost as a natural resolution to these problems of conflict and anarchy (Der Derian, 1995, p. 30; Burchill *et al.*, 2001, p. 84).[1]

Realist views also inform understandings of dominant threats to security and therefore fundamentally shape security discourse. Post-WWII, the ideas of classical Realists evolved into a loosely defined school of international relations, termed Realism. While this school does not present a single theory of international relations, Realist works tend to be underpinned by a number of common ideas (Donnelly, 2000, pp. 6–9). A key idea of Realism is that it purports to see the world 'as it is' (Butfoy, 1997, p. 4; Burchill, 2001, p. 70) rather than how we would like it to be.

The Realist thinking that emerged shortly after WWII saw relations between states as essentially mirroring classical realists' conception of relationships between people (Donnelly, 2000, p. 10). Nation states, like human beings, were viewed as 'separate, autonomous and formally equal political unit[s]' (Donnelly, 2000, p. 17) driven by the pursuit of their own interests (Burchill, 2001, p. 75). Realists saw national interests as being frequently divergent and therefore assumed that clashes of national interest were inevitable (Burchill, 2001, p. 75).[2] In the international system these clashes played out in an environment of anarchy; there was no supreme international authority that could exercise power over all states and impose order. Consequently, the 'law of the jungle still prevail[ed]' (Donnelly, 2000, p. 10). Later, in the 1960s, Neo-Realists emphasised the structure of the international system – the state of anarchy – as shaping state behaviour rather than inherent human characteristics (Brown *et al.*, 1995, p. x). However, despite this distinction, the lessons conveyed by Realism are the same: states must look to their own interests and 'must count ultimately on [their] own resources to realise [these] interests' as no one else can be 'counted on' to assist them (Waltz, 1979, p. 107; Butfoy, 1997, p. 5).

In Realist thinking, the only way for states to ensure their own interests is by achieving a margin of power over other states (Frédérick, 1999, p. 91). This worldview does not preclude cooperation between states, but means that cooperation occurs within narrowly circumscribed boundaries; states will only cooperate when it 'directly services' their national interests (Frédérick, 1999, p. 92).

How do states achieve 'power' over other states? There is no single definition of what constitutes 'power' in relation to nation states. Commentators have defined power as encompassing, variously, population, economic strength, military power, political systems and so on (Isaković, 2000). Realists, though, have tended to focus on military capacity as the most important measure of power (Waltz, 1959, p. 210; Sullivan, 1990, p. 76; Isaković, 2000, p. 140). While Realist thinking has always viewed non-military concerns such as the economy as important, it has constructed these concerns as important primarily to the extent that they underpin *military power*. Thus, according to Waltz (1959), the use of force presents a reliable means by which states can achieve their aims in an anarchic international system. Consequently, 'the capacity to use force tends to become the index by which the balance of power is measured' (Waltz, 1959, pp. 210, 238). Paradoxically, the capacity of *other* states to use

force in securing their interests means that the use of force is, at the same time, the primary *threat* to the nation state (Barnett, 2001, p. 27).

Changes since the 1970s, particularly emerging economic interdependence – whereby a state's economic interests can be powerfully affected by policies in other states (other than military action) – have challenged these Realist views (Dabelko and Dabelko, 1995, p. 3; Barnett, 2001, pp. 35–36). There has been a growing consensus within mainstream security discourse that military attack is not the sole threat to the interests of the nation state, and that military power may not be sufficient to secure a state's interests (see, e.g. The White House, 1998). Although there is no clear consensus on what non-military threats are relevant, as explained earlier, the tendency at least in mainstream security discourse is to focus primarily on economic issues.

Yet, while the relevant threats to security are being redefined, importantly, the object of security – the nation state – remains largely uncontested in these security discourses. To the extent that the nation state remains the object to be secured, broader definitions of security still operate within the Realist framework (Barnett, 2001, p. 33 and Ch. 4). Thus, the logic of power, national self-interest and protection from external threats persist as central features of contemporary security discourse.

REDEFINING 'SECURITY' IN TERMS OF THE ENVIRONMENT

Whilst developments have been occurring within dominant security discourses, rethinking of 'security' has also taken other directions, outside the mainstream security arena. One of these directions has focused on connections between the environment and security. Commentators first began linking these concepts in the 1980s, with papers by Ullman (1983), Myers (1986), Matthews (1989) and Westing (1989) amongst the first to make the association. Myers (1986, p. 251), for example, argued that: 'The notion of national security can no longer be centred so strongly on simple considerations of military prowess. It increasingly entails key factors of environmental stability that underpin our material welfare'. Shortly after, Westing (1989, p. 129) wrote that 'comprehensive human security has two intertwined components: *political* security on the one hand (with its military, economic, and social/ humanitarian sub-components); and *environmental* security on the other (with its protection-oriented and utilisation-oriented sub-components)'.

Since then, a number of different perspectives on the link between the environment and security

Table 31.1 Definitions of security

Definitions of security		What are the subjects of security discourse?	
		Only nation states	States and non-state entities
What sort of things can threaten the subjects of security discourse	Military threats only	1. The environment is relevant to national security only to the extent that it shapes military threats	(not discussed)
	Military and non-military threats	2. The environment is relevant to national security *per se*	3. The environment is relevant to the security of individual people and communities

Source: Adapted from Page, 2002 (p. 37)

have emerged. Some of these perspectives fall within or intersect with mainstream security debates, while others represent a more radical departure from the dominant security discourse. Based broadly on a classification developed by Page (2002) this section groups these perspectives into three main approaches (Table 31.1). (We acknowledge that these groupings do not represent mutually exclusive categories, and that some perspectives may not fit neatly into one or another approach.)

The first approach (upper left quadrant of Table 31.1) links environmental concerns to the issue of security by focusing on the role of environmental degradation in causing violent conflict. This approach is best represented by the work of Homer-Dixon, who has conducted studies exploring the indirect and direct links between environmental degradation and violent conflict (see, e.g. Homer-Dixon, 1994, p. 94). The value of Homer-Dixon's work lies in its detailed interrogation of the role of the environment in instigating armed conflict, and serves as an important reminder that the environment can indeed be one causal factor of military engagement. This environment and conflict literature does not rethink Realist definitions of security, though, and continues to view environmental concerns within a framework of security defined in terms of the nation state (Finger, 1991; Elliot, 1998, p. 231). While the literature encompasses conflict between non-state entities, this is largely viewed through the prism of national security, with the emphasis being on how this conflict impacts on the integrity and security of the state. Also, while the literature seeks to broaden understandings of what *causes* military conflict – and this is where environmental concerns become relevant – it continues to construct 'security' primarily in terms of military threats. Thus, this approach to security is firmly embedded within the traditional Realist paradigm (Elliot, 1998; Barnett, 2001).

The second approach (lower left quadrant of Table 31.1) integrates environmental concerns into security discourse by arguing that environmental degradation is a threat to national security *per se*. This approach is represented by the work of Myers and Matthews, referred to above. Broadly speaking, the approach is based on the view that national interests are closely linked to the environment. The natural environment shapes the state of the economy – and, in turn, military power, human health and general quality of life. Consequently, environmental degradation can threaten the ability of the state to achieve its interests. Conversely, the impacts of environmental degradation (e.g. weather extremes resulting from climate change) can result in damage to the economy and to human health and, as such, pose a risk to the national interest (Matthews, 1989; Frédérick, 1999, pp. 94–95; Barnett, 2001). So, just as concepts of national security were broadened in the 1970s to encompass economic concerns, authors such as Matthews argue that concepts of national security should be extended 'to include resource, environmental and demographic issues' (Matthews, 1989, p. 162). In contrast with the first, this second approach redefines the concept of security by expanding the notion of what constitutes threats to security. Importantly, though, it does not challenge the centrality of the nation state as the object of security discourse (Barnett, 2001).

The third approach (lower right quadrant of Table 31.1) to security is the most radical. A defining feature of this approach is that it links environmental concerns to notions of security that are not based on protection of the nation state (see, e.g. UNDP, 1994; Lonergan, 1999). There are two major branches of this approach relevant to environmental discourse. The first advocates protection of the global biosphere, an approach known as 'ecological security' (see Barnett, 2001, Ch. 8). Taking the object of security to be the biosphere, environmental degradation is, by definition, a threat to 'security'. The second branch advocates the notion of 'security' as the protection of human beings. There is no clear consensus on what protection of human beings involves – whether it focuses on threats to physical survival or whether it encompasses broader threats to human welfare and development. But there is a general consensus in discussions of human security that the environment affects peoples' survival and welfare. Therefore, environmental degradation can be seen as a threat to 'human security' – either as a threat to human survival or to human welfare and development more generally. The remainder of this chapter will focus on the concept of 'human security' and explore what this perspective can add to understandings of the human/environment relationship.

HUMAN SECURITY

Human security has been defined in a number of ways, reflecting the fact that there is a range of ideas about what is required for the survival and welfare of people and communities. In 1994, the United Nations Development Programme (UNDP) defined human security in terms of protection against chronic threats – such as 'hunger, disease and repression' – and protection against 'sudden and hurtful disruptions in the patterns of daily life' (UNDP, 1994, p. 23). This definition emphasises the importance, to human security, of economic opportunity, a healthy environment, and access to food and natural resources, among other things. The Global Environmental Change and Human Security Project (GECHS) defines human security in similar terms, but focuses more on whether or not people have options *to deal with threats* to their well-being, as determinants of human security. According to the GECHS definition (Lonergan, 1999, p. 29), individuals will enjoy human security if they:

- have the options necessary to end, mitigate or adapt to threats to their human, environmental and social rights;
- have the capacity and freedom to exercise these options;
- actively participate in attaining these options.

These definitions of human security share several important features. First, they focus on basic human needs and the threats to meeting those needs. In relation to the environment, this focus promotes an exploration of how people are dependent on the environment for their basic needs and, in turn, how environmental degradation can impact on peoples' security. Second, the

human security literature has established a focus on social institutions and structures; that is, how social institutions and structures influence whether peoples' basic needs are met, how they facilitate or inhibit the provision of basic needs, and the extent to which they ensure people have options to deal with threats to their well-being. This promotes a consideration of how environmental threats are shaped by social factors. Third, discussions of human security focus strongly on equity concerns, in particular the question of who is more and who is less secure, and why. This leads to analyses of environmental concern that focus on *who* is most exposed to environmental threats and the social and environmental factors influencing peoples' level of exposure.

THE VALUE OF A HUMAN SECURITY PERSPECTIVE

Many of the issues raised by discussions of human security are also considered in other environmental discourses. This raises the question: why seek to discuss environmental issues in terms of *security* (albeit in the broader language of *human* security)? This section argues that there are at least three advantages in considering environmental issues from a human security perspective. In proposing these advantages, we are not arguing that a human security perspective should be preferred over other approaches to environmental issues (e.g. 'sustainable development', 'environmental justice'). Rather, we argue that considering environmental issues from a human security perspective raises important questions that are often overlooked.

The military and security

One benefit of taking a human security perspective is that it facilitates contestation of the ideas and assumptions surrounding concepts of military security (Barnett, 2001, pp. 137–138). This is important because the pursuit of military security – through building military preparedness and engaging in warfare – has significant direct and indirect impacts on the environment. These impacts – and the broader connections between military practices and environmental degradation – are frequently overlooked in environmental discourses.

The direct impacts of war on people and the environment are substantial (e.g. Finger, 1991; Renner, 1991; Leggett, 1992; Elliot, 1998, 2004; UNDP *et al.*, 2003, pp. 25–27). To cite but a few examples: use of chemical agents by US forces during the Vietnam War destroyed 54% of mangrove forests in South Vietnam – an important ecosystem for both environmental and economic reasons (Leggett, 1992, p. 69; Lanier-Graham, 1993, pp. 30–39); landmines and unexploded ordnance kill or injure 15–20,000 people each year (ICBL, 2003, p. 41; Neilsen, 2005, pp. 231–232); and the use of 'scorched earth' policies in countries including Sudan, Ethiopia, Guatemala and East Timor have led to destruction of villages, crops and vegetation (Leggett, 1992, p. 69).

Preparation for war also exacts a significant environmental and human toll (often on the domestic population). For example, practices on military bases – including manoeuvres, bomb-testing and dumping of waste – are variously responsible for contaminating land, soil and local water, destroying vegetation and animal habitats (Renner, 1991, pp. 134–136, 141–150). The area of land affected by these activities is not insignificant. According to *USA Today*, approximately 1 in 10 Americans live within 10 miles of a military site listed as a national priority for hazardous waste clean-up (Eisler, 2004, p. A1).

Perhaps the greatest impact of pursuing military security, though, is the opportunity cost of public and private financial resources (Elliot, 1998, p. 235; 2004, pp. 216–218). In 2003, global military expenditure totalled US$956 billion (SIPRI, 2004, p. 307). When this figure is compared with the US$22.6 billion the World Health Organization estimates is required to provide access to improved water and sanitation services for the global population, it becomes clear that the opportunity costs of military spending are substantial (Hutton and Haller, 2004, p. 2). The opportunity costs are proportionately greatest in low- and middle-income countries, where the percentage of GDP spent on the military tends to be higher than in developed countries (Nielsen, 2005, pp. 221–222).

It is true that, more recently, the role of the military has expanded to include activities that are less damaging or that may even have a positive environmental impact, such as the provision of assistance in natural disasters, and various peacekeeping operations around the world. Nevertheless, the primary role of the military remains the preparation for and conduct of war – activities which have an overwhelmingly negative environmental impact. Furthermore, the military are a measure of 'last resort' in resolving environmental problems, which should, ideally, be ameliorated through attempts to resolve their fundamental causes (Little and Cocklin, *forthcoming*).

In the discourse of military security, these impacts can be justified on the grounds that the risks associated with military attack are high and that therefore military concerns must be prioritised in relation to other issues. Within this mindset, the negative impacts of military practices

on the environment and people are regrettable but necessary. However, an analysis of the risks posed to people by military attack as compared with other risks to human survival and welfare raises questions about these assumptions. For example, in 2001, there were 3547 deaths resulting from terrorism, including those in the September 11 attacks in the US (USSD, 2002, p. 2). Since then, international terrorism has become a major focus within military security discourse and political discourse more generally. Yet, when compared with an estimated 2 million deaths from diarrhoeal diseases in that same year (WHO, 2002, p. 186), it is clear that globally, in 2001 at least, human beings were more at risk from contaminated drinking water than from international terrorism. Indeed, even considering all deaths resulting from armed conflict generally, this figure is still only one-eighth the number of people that died from diarrhoeal diseases in 2001 (WHO, 2002, p. 190).[3]

The concept of human security is useful, then, in that it facilitates a reassessment of how military and non-military concerns are prioritised and how resources are allocated in seeking to achieve security (Stern, 1999, p. 143). The notion of human security decouples the concept of 'security' from military power. In so doing it rejects the assumptions implicit in military security discourse – that military attack is the primary threat to people and that human security is achieved primarily through military power – and instead focuses directly on the underlying aims of security discourse: human survival and welfare (see Barnett, 2001, pp. 124–127). This allows for a reconsideration of the real risks to peoples' survival and welfare, and whether military security practices actually allocate resources in accordance with these risks (Barnett, 2001, pp. 24–25, 43–44). It also exposes the contradictions in military security discourse as it becomes apparent that practices designed to achieve military security can actually lead to *insecurity* by contributing directly to environmental degradation, destroying livelihoods and diverting resources away from investment in the environment and society.

A discourse of human security does not mean that military concerns become irrelevant. Rather, it means that environmental and social concerns assume a more appropriate standing in security discussions, as opposed to the automatic privileging of military interests (Stern, 1999, pp. 143–147). As Stern argues, this levelling of the conceptual playing field makes the tradeoffs between various concerns – military, environmental, social, etc. – explicit and, in so doing, allows for a more open discussion of how to allocate resources (Stern, 1999, p. 143). These discussions have the potential to elevate environmental concerns in the political arena because a focus on the basic underlying aims of security highlights the importance of the natural environment to peoples' security – something which has, in practice, usually been downplayed in political discourse.

The national economic interest and security

As discussed in the second part of this chapter, in recent decades the focus of national security discourse has enlarged beyond military threats to encompass economic concerns. These discourses of national *economic* security often provide the justification for policies that produce significant negative social and environmental impacts, both within the state and in other states, increasing the *in*security of some groups and individuals (see generally, Dalby, 1992, 1994, 1997, 1999). In this context, a human security perspective can offer important benefits by exposing the contradictions between the apparent aims and actual outcomes of national security practices.

The domestic arena

Within nation states, discourses on national economic security are often used to justify policies that do not, in fact, benefit all those within the nation and, indeed, often impact in ways that are highly inequitable. The policies on dam building in India and agricultural adjustment policies in Australia provide examples of this.

India is the third largest builder of dams in the world (Roy, 1999a, p. 3), with 3360 large dams already built (the vast majority since independence) and 695 under construction[4] (Rangachari *et al.*, 2000, p. 19). The building of dams in India is perceived, and frequently discussed, in terms of India's national interest and national economic security. There is a widespread and deeply ingrained view that dams are vital for India's self-sufficiency and modernisation; indeed, for its *national security* (emphasis added) (Kothari, 1995, p. 425). Former Prime Minister, Pandit Nehru, once referred to dams as 'the temples of modern India' (Kothari, 1995, p. 425; Roy, 1999a, p. 3; 1999b, p. 15).

Certainly, dams can deliver important benefits in India, including improved irrigation, the supply of water, flood control and electricity generation, along with the concomitant economic and social benefits (Rangachari *et al.*, 2000, pp. 23–28). However, they can also have significant environmental and social impacts. These impacts include destruction of forests and grazing lands, declining water quality and fish stocks, land degradation and displacement of large numbers of people.[5] Discourses on national economic security suggest that the benefits outweigh the costs and the

negative impacts of dams are justified on these grounds. These discourses analyse costs and benefits at the level of the nation state and focus on whether the aggregate net (economic) benefit to the nation outweighs the costs borne by some groups and individuals (in economics, the notion of Pareto optimality). What this approach overlooks is the issue of equity in the *distribution* of costs and benefits among people within the state. Yet, the example of dams in India demonstrates that questions of equity have fundamental implications for the *security* of many people.

The distribution of costs and benefits of large dams in India is highly inequitable (Kothari, 1995, pp. 434–437; Roy, 1999a, b; Rangachari *et al.*, 2000, pp. 129–145). For example, tribals – a disadvantaged group constituting 8% of the population – make up 47% of the displaced population (Rangachari *et al.*, 2000, p. 131). Tribals are highly vulnerable to the effects of displacement because, while they tend to be rich in resources through access to common property, they typically have low incomes (Rangachari *et al.*, 2000, p. 131).[6] So, displacement tends to leave them both resource poor and income poor (Roy, 1999b, p. 12; Rangachari *et al.*, 2000, p. 131).[7] In contrast, those who benefit from large dams – through irrigation and increased electricity supply – tend to be the more economically secure groups in rural and urban areas.[8] This is not always the case, of course; some dams stimulate economic activity through tourism, they increase fish stocks in some areas, and lead to improvements in sanitation and hygiene – all of which may benefit poor people (Rangachari *et al.*, 2000, pp. 109–110).[9] However, the study on India prepared for the World Commission on Dams confirms that, *overall*, large dams in India tend to promote inequity, with 'tribals and other rural people ... [losing] even the little they have in order to benefit those who already have more than they do' (Rangachari *et al.*, 2000, p. 151).

These inequities have negative implications for the security of people in India. If benefits flow overwhelmingly to those who are relatively secure and the costs are borne by those least secure – then dam building can actually heighten overall levels of insecurity, even though they may provide aggregate economic benefits to the state (although this itself is debatable).

The common response to insecurity produced by national economic strategies is an argument that, in the longer term economic gains will 'trickle down' and improve the security of all people within the state. However, the example of agricultural policy in Australia raises questions about these assumptions. Over the past two decades or so, Australia (along with many Western nations) has enthusiastically pursued what is widely referred to as a neo-liberal economic restructuring agenda. While the neo-liberal strategy defies any simple characterisation, it is based on an ideology that extols the virtues of free enterprise, entrepreneurship, personal responsibility and self-interest (Patterson and Pinch, 1995; Tickell and Peck, 1995; Cerny, 1997; Lewis *et al.*, 2002). In the domain of agricultural policy, implementation of this strategy has involved the pursuit of free trade, the removal of domestic agricultural subsidies, deregulation of prices, the internalisation of risk (e.g. associated with extreme weather events) and market-based allocation of natural resources (such as water).

These policies have created significant burdens in rural areas. While it has been popularly advocated that these burdens would be short term and offset by the 'trickle down' of national economic benefits, evidence suggests that these policies have heightened insecurity in many rural areas. For example, the gap between income and wealth in Australian cities compared with rural areas is now considerable, with the ten highest income regions situated in urban areas while the ten poorest are in rural areas. Pritchard and McManus (2000) attribute this gap to the combined effects of heightened global competition and the rationalisation of services. Additionally, according to Gray and Lawrence (2001), increasingly people living in rural Australia 'are subject to economic and political processes over which they have no control' and which affect not only their incomes but also 'all aspects of their lives, including their family and community, and the biophysical environment in which they live and work' (Gray and Lawrence, 2001, p. 4).

In this context, the concept of human security can be useful because it decouples the notion of national economic security and prosperity from the security of individual people and communities. In so doing it exposes some of the inequities in policies purportedly in the national interest. In relation to the environment this is important because national development policies frequently emphasise the need to trade off environmental concerns for the sake of development of the economy. Yet a focus on the situation of poor people in developing countries reveals that these development policies may in fact heighten the insecurity of these people; many are heavily dependent on natural resources and, therefore, degradation of natural resources often undermines their survival and welfare. Of course, a human security perspective does not imply that human welfare and environmental concerns are always compatible. But it does mean that claims that certain policies are in the national interest – particularly those leading to environmental degradation – *are no longer taken for granted*

(Dalby, 1999; Stern, 1999).[10] Thus, a human security perspective allows for a more rounded discussion of the true costs and benefits of particular development and economic policies.

The international arena

At the international level, discourses on national economic security often provide justification for policies that benefit those within the nation, at the expense of people in other states (Dalby, 1994). Frequently these policies create more insecurity, globally, than they ameliorate.

One example of where pursuit of national economic security has created insecurity in other states is the use of agricultural subsidies. The USA, along with many other countries, provides substantial subsidies to its domestic agricultural producers (IMF and World Bank, 2001; OECD, 2002b; Roberts and Jotzo, 2002).[11] From 1995 to 2003, the USA provided over US$130 billion in subsidies (Environmental Working Group, 2005); the provision of agricultural support is continuing under the 2002 US Farm Act (Allen, 2002).

The US Government argues that provision of agricultural subsidies is in the national economic interest. President Bush stated: 'The success of America's farmers and ranchers is essential to the success of the American economy ... it [the Farm Bill] helps America's farmers, and therefore it helps America' (The White House, 2002a). Certainly, subsidies are in the interests of recipient agricultural producers. Subsidies enable producers to sell at prices below production value and thereby increase the competitiveness of their products in international and domestic markets. There is considerable debate as to whether agricultural subsidies are, however, in the overall national economic interest, with critics variously arguing that subsidies distort the domestic economy, drain money from taxpayers and favour large agricultural producers at the expense of smaller farmers (Allen, 2002; OECD, 2002b; Roberts and Jotzo, 2002; Mutume, 2003; Environmental Working Group, 2005). Nevertheless, the prevailing government view – and the justification for the use of subsidies – is that it they are in the national economic interest.

While subsidies benefit interests within the USA, they can seriously undermine the welfare of groups outside the USA, particularly those in the global South. By enabling producers to sell at below production cost, subsidies encourage excess production and supply and so lead to declines in international agricultural prices (FAO, 2003c). This can have serious impacts in other countries, particularly developing countries, as the economies of these countries are often heavily dependent on international agricultural trade

(OECD, 2002a; FAO, 2003a). For example, many countries in west and central Africa are highly dependent on cotton production; according to the FAO cotton production 'accounts for 5 to 10% of GDP in Benin, Burkino Faso, Chad, Mali and Togo' (FAO, 2003b, Factsheet, 12). These countries have been strongly impacted by the significant drop in international prices for cotton – from a long-term average of 70 cents to a 30-year low of 42 cents per pound in 2001 (ICAC, 2003, p. 1) – caused, in large part, by US subsidies of almost US$4 billion a year[12] (Minot and Daniels, 2002; Oxfam, 2002, p. 6; FAO, 2003b, Factsheet, 12; Mutume, 2003, see also Goreaux, 2004; WTO, 2004). According to Oxfam, Burkino Faso lost 12% of its export earnings due to the drop in prices; Mali lost 8% and Benin lost 9% (Oxfam, 2002, p. 17). Additionally, sub-Saharan producers are estimated to have incurred income losses of $920 million in 2001–2002 and $230 million in 2002–2003 (ICAC, 2003, p. 3). Given the high levels of existing poverty, these economic losses are almost certain to have undermined the welfare of many people in these countries (see, e.g. Minot and Daniels, 2002; Oxfam, 2002).

A second example of the externalisation of insecurity as a result of the pursuit of national economic interests is provided in the context of climate change. Some countries, notably the USA and Australia, have concluded that the Kyoto Protocol is not in their national economic interests. Operating within the dominant security framework these countries have pursued their interests by refusing to ratify the Protocol in its current form.[13] Their refusal has the potential to undermine significantly the security of people in other states by fuelling climate change. Climate change presents a major risk to the security of many people outside the USA, particularly those in small island states, whose very survival is threatened by rising sea levels. Because the USA and Australia have not ratified the Protocol, neither is bound to reduce their emissions. And, while Australia claims it is likely to meet its targets under the Protocol all the same (Australian Greenhouse Office, 2004), the USA – the much larger emitter – is far from doing so (US EPA, 2004).[14]

Discourses on national security suggest that the international impacts of agricultural subsidies and refusal to sign the Kyoto Protocol are irrelevant; it is the interests of people *within* the state that should be considered in a national security analysis. Yet in terms of *global* security, the examples above demonstrate that the international impacts of national security policy are highly relevant. Given the high level of vulnerability in African cotton-producing countries, and comparatively lower levels in the USA, it is arguable that the security gains of US policy are outweighed by the degree of insecurity created elsewhere, leading to *less* security globally. Similarly, given how many people are at risk from climate change on a global scale, USA and Australian economic security policies are likely to lead to more people becoming insecure than the numbers whose (economic) security is maintained. Thus, by ignoring the relevance of people outside the state, discourses on national economic security often promote policies that lead to more *global* insecurity.

One of the benefits of a human security perspective, then, is that it broadens the security analysis beyond the nation state to encompass the security of all human beings (Barnett, 2001). This means that, unlike concepts of national security – where the costs and benefits of a policy are considered only in relation to the citizens of the state – the impacts of policies on people *outside* the state are included in the analysis. By making the security of people outside the state relevant, human security analyses can: (a) highlight current *global inequities* in the distribution of costs and benefits – the fact that national security policies often achieve security of those within the nation at the expense of particular groups outside the nation; and (b) challenge understandings of the costs and benefits associated with particular policies and thereby, perhaps, promote policies that are more in the interests of *global* security.

Providing a fuller account of costs and benefits is particularly important in relation to *environmental* concerns because the environmental impacts of one country's activities are often incurred by other countries; ecological systems do not respect national boundaries. Additionally, with fewer restrictions on international trade and finance, environmentally degrading processes, or wastes – along with their associated impacts – can be 'exported' away from the originating state. Therefore, operating within a national security perspective – where many costs are externalised – states often see the benefits of environmentally degrading practices as outweighing the costs. By internalising these costs, a human security perspective can alter understandings of the costs of environmentally degrading practices. Taking a human security perspective does not mean it will be possible to ensure that people and environments always benefit from national policies. But it does suggest a shift away from the existing, competitive approach to relations with other states – where states achieve security at the expense of people and environments in other states – to an ethic of cooperation and compromise between states in an effort to achieve greater global security (Dalby, 1994; Barnett, 2001).

HUMAN SECURITY AND SUSTAINABLE DEVELOPMENT

A further advantage in taking a human security approach is that it focuses on the links between the environment, and social, political and cultural issues. This is important as the social dimensions of environmental issues have often been over-looked in the dominant environmental discourse on 'sustainable development' (Lehtonen, 2004; McKenzie, 2004). Sustainable development has, historically, focused strongly on the economic aspects of environmental issues (Barnett, 2001). It has tended to emphasise the need for changes to the economy – specifically, the incorporation of environmental concerns into economic models – rather than social, political or cultural change (Barnett, 2001; Lehtonen, 2004). As the concept has evolved, the importance of social, political and cultural factors – for example, poverty, social equity, governance – has increasingly been recognised (Najam *et al.*, 2003; Lehtonen, 2004) to the point that social development is now given appreciable status as one of the three pillars of sustainable development (Najam *et al.*, 2003, p. S13).[15]

However, while social, political and cultural concerns are being integrated increasingly into discussions of sustainable development, importantly, they are still viewed as issues separate from environmental concerns. As Lehtonen explains, the models of sustainable development conceive of social, environmental (and economic) issues as 'independent elements that can be treated, at least analytically, as separate from each other' (Lehtonen, 2004, p. 201). The assumption is that these elements are 'at odds with one another and needing to be carefully balanced' (McKenzie, 2004, p. 11). This approach tends to overlook the many *interconnections* between environmental and social, political and cultural issues (Lehtonen, 2004).

A human security perspective focuses specifically on these interconnections. The concept of human security places human beings, rather than the environment, at the centre of the analysis. Environmental concerns are considered in terms of how they relate to human survival and welfare and are viewed as only one of a matrix of issues – including social, economic, political – affecting human security (Figure 31.1). The figure represents the matrix of environmental, economic, social, political and cultural factors that make up the total environment – both the natural and the human environment – in which human beings live. This focus on human beings leads to an analysis of *how* people and communities are linked to the environment and how these links are mediated by various social, political, economic and other factors. It places environmental concerns into an explicitly social context, with a focus on environmental concerns from the point of view of *how they affect people and communities*. This emphasis on people exposes the fact that people and communities are differentially affected by environmental change – with poor people often disproportionately affected – and promotes an analysis of the social, political, economic, cultural and environmental factors leading to these inequities (see Figure 31.1) (Barnett, 2001, pp. 134–138). For example, a recent analysis of the predicted impacts of climate change on the human security of urban residents suggests that slum residents are likely to be particularly affected by climate change and that this is due to a variety of socio-economic and political factors (Little and Cocklin, *forthcoming*).

By focusing on the interconnections between human and environmental concerns, a human security perspective emphasises the fact that environmental concerns are *human social and political problems* as much as scientific and economic ones (Barnett, 2001; McKenzie, 2004). It emphasises aspects of environmental issues that tend to be different from those that are the focus of sustainable development discourse. A human security perspective can therefore complement efforts to understand environmental problems through concepts of sustainable development (Barnett, 2001).

CRITICISMS OF THE HUMAN SECURITY CONCEPT

While the section above has outlined a number of benefits in considering environmental issues from a security perspective, this approach has not been without its critics. A number of commentators have argued against linking the concepts of environment and security generally (see, e.g. Levy, 1995; Dalby, 1999; Deudney, 1999). Some critics have argued, from a dominant security perspective, that the environmental concerns should not be linked to security because environmental degradation is unlikely to lead to violent conflict (Dalby, 1994; Deudney, 1999). According to this view, environmental concerns do not constitute a security risk within the dominant understandings of security. Also, given that environmental problems have little to do with dominant concepts of national security, linking the terms will create conceptual confusion (Levy, 1995; Deudney, 1999). Thus, expanding the term security to include environmental concerns, it is argued, will simply make the term so broad as to be meaningless and so useless as an analytical tool (Levy, 1995; Deudney, 1999).

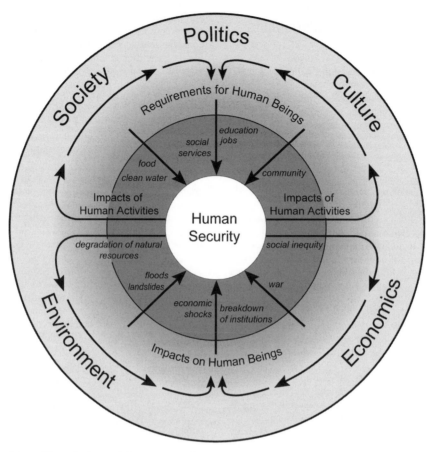

Figure 31.1   The relations of human security

Others have argued that discussing environmental concerns in security terms risks co-option of the environmental agenda into the dominant national security paradigm (Dalby, 1994, 1999; Levy, 1995; Deudney, 1999; Barnett, 2001). This paradigm, they argue, is inappropriate for dealing with environmental issues; the focus on violence and domination, the division of the world into 'us and them' and privileging of state interests over the interests of other groups is seen as contrary to the cooperative, global and peaceful approach required to address environmental problems.

Some of these criticisms are based on arguments that environmental concerns do not fit within the dominant security framework. These arguments are somewhat irrelevant to the concept of *human security* because the very aim of a human security approach is to contest the boundaries of the dominant security framework (Barnett, 2001). Other concerns – about the mismatch in approaches to environment and security – do not necessarily

apply to concepts of human security either. These concerns relate to considering environmental issues within dominant discourses on national security (Dalby, 1994, p. 44). As explained in the section above, it is these very ideas – privileging of the state, the focus on conflict and so on, that concepts of human security aim to contest (Barnett, 2001). Certainly, as Barnett points out, there is a danger in that, what begins as contestation of the dominant security discourse, could in the end result in co-option of the environmental agenda (Barnett, 2001, p. 137). At the same time, however, it is this use of the language of security that makes human security such a powerful tool for contesting the dominant models of security (Dalby, 1994, 1997; Barnett, 2001). It is on this basis that we believe the human security perspective holds value.

Interestingly, perhaps the greatest (and most overlooked) challenge in taking a human security approach to environmental issues may lie not so much in avoiding the concepts of conventional security discourse but rather in dealing with some of the fundamental questions security discourse raises. These questions include (Dalby, 1997): How can human security be promoted? And what level of security is reasonable in a world of inherent uncertainty and change?

CONCLUSION

Security thinking has a pervasive influence in modern life; it shapes state priorities, domestic political discourse and patterns of international relations. As such, security thinking has profound environmental implications. Despite this, dominant environmental discourses have given relatively little attention to the connections between security and the environment. A human security perspective can illuminate these connections. By focusing directly on the needs of individual people and communities, and by broadening the notions of who is relevant in security discourse a human security analysis can reveal the contradictions between the apparent aims of security practices and its negative environmental and social outcomes. From an environmental perspective, this is important in deepening understandings of the interconnections between environmental and human concerns. And, from a security perspective, the concept of human security is important in suggesting alternate visions of domestic and international politics: ones based on understandings of human interdependence rather than solely the human potential for conflict and violence.

NOTES

1 Not all Realists see the 'nation state' as 'the final evolutionary form of political community' (Burchill, 1995, p. 76). For example, both Carr and Morgenthau (key thinkers in the Realist school) conceive of political communities other than the nation state (Burchill, 1995, pp. 76, 81).

2 This view conflicts with those of the liberal internationalist school of international relations, which was dominant between WWI and WWII. Liberal internationalists saw nation states as having common interests and believed international cooperation could form the basis of the international system (see Burchill, 1995).

3 An estimated 230,000 people died as a result of war in 2001, compared with 2 million deaths caused by diarrhoeal diseases (WHO, 2002, p. 190).

4 There are various estimates of the number of large dams in India. The figures cited here were obtained from Rangachari et al. (2000).

5 See Rangachari et al. (2000, pp. 70–154). Forests and grazing lands are destroyed through submergence of land. Land that is not submerged can also be impacted as removal of other land creates more pressure on remaining resources (Rangachari et al., 2000, pp. 74–108). This and other environmental changes resulting from dams can also have significant negative social impacts (Rangachari et al., 2000, pp. 121–129). For example, declines in fish stocks have major impacts on the livelihoods of local fisherman, while land degradation can lead to significant economic and social losses for local people (Rangachari et al., 2000, pp. 121–129). Additionally, the social impacts of displacement are substantial (Kothari, 1995, pp. 427–428; Roy 1999b, pp. 11–12; Rangachari et al., 2000, pp. 110–121). Displaced persons often lose access to common property – resources from which many derive subsistence and income. Displacement is also associated with significant psychological trauma caused by fragmentation of communities, loss of cultural and religious links to the local region and conflict with host communities (Rangachari et al., 2000, pp. 111–115). A report prepared for the World Commission on Dams estimates that tens of millions of people have been displaced by dams in India (Rangachari et al., 2000, p. 116; Roy, 1999a, pp. 20–21).

6 Although, as Kothari notes, not all tribal people are still dependent on the land (Kothari, 1995, pp. 427–428).

7 While appropriate compensation or resettlement programmes may assist tribal peoples in dealing with displacement, the report prepared for the World Commission on Dams suggests that the issues are mostly not addressed (Rangachari et al., 2000, pp. 111–112, 137). Roy confirms this (Roy, 1999b, pp. 10–11).

8 The report on India for the World Commission on Dams suggests that the bulk of domestic electric supply generated from dams goes to well-off families in urban and rural areas (Rangachari et al., 2000, pp. 134, 136). Additionally, the benefits of irrigation tend to flow disproportionately to large, rather than small or marginal landholders (Rangachari et al., 2000, p. 139).

9 Also, in rare instances, where landless and marginal farmers have been allotted good agricultural land as compensation for displacement, dams can impact positively on poor people (Rangachari et al., 2000, p. 137).

10 Dalby argues that 'the assumptions that states really do operate in the interests of their national population needs to be re-examined' (Dalby, 1999, p. 158). Indeed, Stern suggests that it would be difficult to ensure that security policies operate consistently in the interests of all the national population. He argues that 'it is not useful to regard the security concept as inextricably linked to the nation state because problems of security and

security-policymaking cut across the ladder of social aggregation' (Stern, 1999, p. 145).

11 Subsidies are also particularly high in other OECD countries (see IMF and WB, 2001; Supper, 2001; OECD, 2002a). This is despite requirements, under the 1995 WTO Agreement on Agriculture, that countries reduce subsidies on domestic agriculture (Supper, 2001; FAO, 2003b).

12 This figure refers to US subsidy levels in 2001–2002.

13 The Australian Government states, 'The Government has decided not to ratify the Kyoto Protocol at present because it is not in Australia's interest to do so' (Australian Greenhouse Office, 2005). Similarly, the USA has refused to ratify the Protocol, principally on economic grounds (The White House, 2002b).

14 Under the Kyoto Protocol the USA is required to reduce its emissions by 7% below 1990 levels; the US national inventory of greenhouse gas emissions and sinks indicates that US emissions are currently 13% *above* 1990 levels (US EPA, 2004).

15 The three pillars being: the environment, economics and society (see Najam *et al.*, 2003).

REFERENCES

Allen, Mike (2002) 'Bush Signs Bill Providing Big Farm Subsidy Increase', *The Washington Post*, 12 May, p. A01.

Australian Greenhouse Office (2004) *National Greenhouse Gas Inventory 2002*. Australian Greenhouse Office, Canberra.

Australian Greenhouse Office (2005) 'Kyoto Protocol'. Available at http://www.greenhouse.gov.au/international/kyoto/index.html (accessed 31 January 2005).

Barnett, Jon (2001) *The Meaning of Environmental Security: Ecological Politics and Policy in the New Security Era*. Zed Books, London.

Brown, Michael E., Lynn-Jones, Sean M., Miller, Stephen E. (1995) *The Perils of Anarchy: Contemporary Realism and International Security*. MIT Press, Cambridge, MA.

Burchill, Scott (1995) 'Realism and neo-realism' in *Theories of International Relations*, 2nd edn, eds. Scott Burchill, Richard Devetak and Andrew Linklater *et al.*, Palgrave, New York.

Burchill, Scott *et al.* (2001) *Theories of International Relations*, 2nd edn. Palgrave, New York.

Butfoy, Andrew (1997) *Common Security and Strategic Reform: A Critical Analysis*. Macmillan Press, London.

Cerny, P. (1997) 'Paradoxes of the competition state: the dynamics of political globalisation', *Government and Opposition* 32: 251–274.

Cocklin, C. (1995) 'Agriculture, society and environment: discourses on sustainability', *International Journal of Sustainable Development and World Ecology* 2: 240–256.

Dabelko, Geoffrey D. and Dabelko, David M. (1995) 'Environmental security: issues of conflict and redefinition', *Environmental Change and Security Project Report 1*. Woodrow Wilson Center, Washington DC, pp. 3–13.

Dalby, Simon (1992) 'Ecopolitical discourse: "environmental security" and "political geography", *Progress in Human Geography* 16 (4): 503–522.

Dalby, Simon (1994) 'The politics of environmental security' in *Green Security of Militarized Environment*, ed. Jyrki Käkönen. Dartmouth Publishing, Aldershot, UK; Vermont, USA.

Dalby, Simon (1997) 'Contesting an essential concept: reading the dilemmas in contemporary security discourse' in *Critical Security Studies: Concepts and Cases*, eds. Michael C. Williams and Keith Krause. University of Minnesota Press, Minneapolis.

Dalby, Simon (1999) 'Threats from the South: geopolitics, equity, and environmental security' in *Contested Grounds: Security and Conflict in the New Environmental Politics*, eds. Daniel H. Deudney and Richard A. Matthew. State University of New York Press, Albany.

Der Derian, James (1995) 'The value of security: Hobbes, Marx, Nietzsche, and Baudrillard' in *On Security*, ed. Ronnie D. Lipschutz. Columbia University Press, New York.

Deudney, Daniel H. (1999) 'Environmental security, a critique' in *Contested Grounds: Security and Conflict in the New Environmental Politics*, eds. Daniel H. Deudney and Richard A. Matthew. State University of New York Press, Albany.

Donnelly, Jack (2000) *Realism and International Relations*. Cambridge University Press, Cambridge.

Eisler, Peter (2004) 'Pollution cleanups pit Pentagon against regulators', *USA Today,* 14 Oct., p. A1.

Elliot, Lorraine (1998) *The Global Politics of the Environment*. MacMillan Press, London.

Elliot, Lorraine (2004) *The Global Politics of the Environment,* 2nd edn. New York University Press, New York.

Environmental Working Group (2005) 'Environmental working group farm subsidy database'. Available at: http://www.ewg.org/farm/regionsummary.php?fips=00000 (accessed 31 January 2005).

Finger, Matthias (1991) 'The military, the nation state and the environment', *The Ecologist* 25 (5): 220–225.

Food and Agricultural Organization of the United Nations (FAO) (2003a) *Commodity Market Review 2003–2004*. FAO, Rome.

Food and Agricultural Organization of the United Nations (FAO) (2003b) 'FAO fact sheets: input for the WTO ministerial conference in Cancún'. Available at: http://www.fao.org/documents/show_cdr.asp?url_file=/docrep/005/y4852e/y4852e00.htm

Food and Agricultural Organization of the United Nations (FAO) (2003c) 'Subsidies, food imports and tariffs key issues for developing countries', FAO. Available at: http://www.fao.org/english/newsroom/focus/2003/wto2.htm

Frédérick, Michel (1999) 'A Realist's Conceptual Definition of Environmental Security' in *Contested Grounds: Security and Conflict in the New Environmental Politics*, eds. Daniel H. Deudney and Richard A. Matthew. State University of New York Press, Albany.

Goreaux, Louis (2004) 'Prejudice caused by industrialized countries subsidies to cotton sectors in west and central Africa', 2nd edn. Available at http://www.fao.org/es/esc/common/ecg/47647_en_Goreux_Prejudicef.pdf

Gray, I. and Lawrence, G. (2001) *A Future for Regional Australia: Escaping Global Misfortune.* Cambridge University Press, Cambridge.

Hargrove, E. (1989) *Foundations of Environmental Ethics.* Prentice Hall, Englewood Cliffs, NJ.

Hirst, Paul (2001) *War and Power in the 21st Century: The State, Military Conflict and the International System.* Polity Press, Cambridge.

Homer-Dixon, Thomas F. (1994) 'Environmental scarcities and violent conflict: evidence from cases', *International Security* 19 (1): 5–40.

Hutton, Guy and Haller, Laurence (2004) *Evaluation of the Costs and Benefits of Water and Sanitations Improvements and the Global Level.* World Health Organization, Geneva.

International Campaign to Ban Landmines (ICBL) (2003) *Landmine Monitor Report 2003.* Human Rights Watch, USA.

International Cotton Advisory Committee (ICAC) (2003) 'Impacts of government measures on African cotton producers', ICAC, Washington, DC.

International Monetary Fund and World Bank (2001) 'Market access for developing countries' exports' (Prepared by the Staffs of the IMF and the World Bank).

Isaković, Zlatko (2000) *Introduction to a Theory of Political Power in International Relations.* Ashgate, Aldershot.

Kothari, Smitu (1995) 'Damming the Narmada and the politics of development' in *Toward Sustainable Development: Struggling Over India's Narmada River,* ed. William F. Fisher. M.E. Sharpe, New York.

Lanier-Graham, Susan D. (1993) *The Ecology of War: Environmental Impacts of Weaponry and Warfare.* Walker, New York.

Leggett, Jeremy (1992) 'The environmental impact of war: a scientific analysis and Grennpeace's reaction' in *Environmental Protection and the Law of War,* ed. Glen Plant. Belhaven Press, London and New York.

Lehtonen, M. (2004) 'The environmental–social interface of sustainable development: capabilities, social capital, institutions', *Ecological Economics* 49 (2): 199–214.

Levy, M. (1995) 'Is the environment a national security issue?', *International Security* 20 (2): 35–62.

Lewis, N., Moran, W., and Cocklin, C. (2002) 'Restructuring, regulation and sustainability' in *The Sustainability of Rural Systems: Geographical Interpretations,* eds. Bryant C. Bowler and C. Cocklin. Kluwer Academic, Dordrecht.

Lipschutz, Ronnie D. (1995) 'On security' in *On security,* ed. Ronnie D. Lipschutz. Columbia University Press, New York.

Little, L. and Cocklin, C. (2005) 'The vulnerability of urban slum dwellers to global environmental change' in *Global Environmental Change and Human Security,* eds. R. Matthew, M. Brklacich, D. Giannikopoulos and B. McDonald. University of Pittsburgh Press, Pittsburgh (forthcoming).

Lonergan, Steven (1999) *Global Environmental Change and Human Security: GECHS Science Plan,* IHDP Report Series, Report No. 11. Bonn, Germany.

Matthews, Jessica Tuchman (1989) 'Redefining security', *Foreign Affairs* 68: 162–177.

McKenzie, S. (2004) *Social Sustainability: Towards Some Definitions,* Hawke Research Institute Working Paper Series No. 27, University of South Australia.

Minot, Nicholas and Daniels, Lisa (2002) 'Impact of global cotton markets on rural poverty in Benin'. Paper presented at the Northeast Universities Development Consortium Conference (NEUDC) Program, 25–27 October 2002, Williams College, Williamstown, MA.

Mutume, Gumisai (2003) 'Mounting opposition to Northern farm subsidies', *Africa Recovery* 17 (1). Available at http://www.un.org/ecosocdev/geninfo/afrec/

Myers, N. (1986) 'The environmental dimension to security issues', *The Environmentalist* 6 (4): 251–257.

Najam, Adil, Rahman, Atiq A., Huq, Saleemul, and Sokona, Youba (2003) 'Integrating sustainable development into the Fourth Assessment Report of the Intergovernmental Panel on Climate Change', *Climate Policy* 3 (Suppl. 1): S9–S17.

Nielsen, Ron (2005) *The Little Green Handbook: A Guide to Critical Global Trends.* Scribe Publications, Melbourne.

Organization for Economic Co-operation and Development (OECD), Joint Working Party on Agriculture and Trade (2002a), 'Agricultural policies in OECD Countries: a positive reform agenda', COM/AGR/TD/WP(2002)19/FINAL (6 November 2002).

Organization for Economic Co-operation and Development (OECD) (2002b) *Agricultural Policies in OECD Countries: Monitoring and Evaluation.* OECD, France.

O'Riordan, T. (1976) *Environmentalism.* Pion, London.

Oxfam International (2002) 'Cultivating poverty: The impact of US cotton subsidies on Africa' (Briefing Paper).

Page, Edward (2002) 'Human security and the environment' in *Human Security and the Environment: International Comparisons,* eds. Edward A. Page and Michael Redclift. Edward Elgar Publishing, Cheltenham, UK.

Patterson, A. and Pinch, P. (1995) '"Hollowing out" the local state: compulsory competitive tendering and the restructuring of British public sector services', *Environment and Planning A* 27: 1437–1461.

Pepper, D. (1996) *Modern Environmentalism.* Routledge, London.

Pritchard, B. and McManus, P. (2000) *Land of Discontent: The Dynamics of Change in Rural and Regional Australia.* University of New South Wales Press, Sydney.

Rangachari, R., Sengupta, N., Iyer, R. R., Banerji, P., and Singh, S. (2000) *Large Dams: India's Experience.* A WCD case study prepared as an input to the World Commission on dams. Cape Town, www.dams.org

Renner, Michael (1991) 'Assessing the military's war on the environment' in *State of the World (1991),* eds. Lester Brown, Alan Durning and Christopher Flavin *et al.* Allen & Unwin, Sydney.

Roberts, Ivan and Jotzo, Frank (2002) '2002 US Farm Bill: support and agricultural trade'. ABARE Research Report 01.013, Australian Bureau of Agricultural and Resource Economics.

Robinson, John (2004) 'Squaring the circle? Some thoughts on the idea of sustainable development', *Ecological Economics* 48: 369–384.

Roy, Arundhati (1999a) *The Cost of Living*. Flamingo, London.

Roy, Arundhati (1999b) *The Greater Common Good* (The Fourteenth Hugo Wolfsohn Memorial Lecture, 6 October 1999).

Stockholm International Peace Research Institute (SIPRI) (2004) *SIPRI Yearbook (2004): Armaments, Disarmaments and International Security*. Oxford University Press, Oxford.

Stern, Eric K. (1999) *The Case for Comprehensive Security* in *Contested Grounds: Security and Conflict in the New Environmental Politics*, eds. Daniel H. Deudney and Richard A. Matthew. State University of New York Press, Albany.

Sullivan, Michael P. (1990) *Power in Contemporary International Politics*, University of South Carolina Press, Columbia.

Supper, Erich (United Nation Conference on Trade and Development) (2001) 'Policy issues in international trade and commodities' Study Series No. 1, UNCTAD/ITCD/TAB/2, United Nations, New York and Geneva.

Tickell, A and Peck, J. (1995) 'Social regulation after Fordism: regulation theory, neo-liberalism and the global–local nexus', *Economy and Society* 24: 357–386.

Ullman, R.H. (1983) 'Redefining security', *International Security* 8 (1): 129–153.

United Nations Development Programme (UNDP) (1994) *Human Development Report 1994: New Dimensions of Human Security*. Oxford University Press, Oxford.

United Nations Development Programme, United Nations Environment Program, World Bank, and World Resources Institute (2003) *World Resources 2002–2004: Decisions for the Earth, Balance, Voice and Power*. World Resources Institute, Washington, DC.

United States Department of State (USSD) (2002) *Patterns of Global Terrorism*.

United States Environmental Protection Agency (2004) *Inventory of US Greenhouse Gas Emissions and Sinks: 1990–2002*, US Environmental Protection Agency, Washington, DC.

Wæver, Ole (1995) 'Securitization and desecuritization' in *On Security*, ed. Ronnie D. Lipschutz. Columbia University Press, New York.

Waltz, Kenneth N. (1959) *Man, The State and War: A Theoretical Analysis*. Columbia University Press, New York.

Waltz, Kenneth N. (1979) *Theory of International Politics*. Addison-Wesley Publishing, Boston, MA.

Westing, A. (1989) 'The environmental component of comprehensive security', *Bulletin of Peace Proposals* 20 (2): 129–134.

White House, The (1998) *A National Security Strategy for a New Century*.

White House, The (2002a) 'President signs Farm Bill', Remarks by the President upon signing the Farm Bill (Office of the Press Secretary, 13 May 2002).

White House, The (2002b) 'President Announces Clear Skies & Global Climate Change Initiatives' (Office of the Press Secretary, Media Release 14 February 2002).

World Health Organization (WHO) (2002) *World Health Report 2002, Reducing Risks, Promoting Healthy Life*, World Health Organization, France.

World Trade Organization (WTO) (2004) *United States – Subsidies on Upland Cotton – Report of the Panel – Addendum* WT/DS267/R/Add.1 (8 September 2004). Available at: www.wto.org

32

Sustainable Agriculture and Food Systems

Jules Pretty

PERSISTENT AND NEW WORLD FOOD PROBLEMS

Something is wrong with our agricultural and food systems. Despite great progress in increasing productivity in the last century, hundreds of millions of people remain hungry and malnourished. Further, hundreds of millions eat too much, or the wrong sorts of food, and it is making them ill. The health of the environment suffers too, as degradation seems to accompany many of the agricultural systems we have evolved in recent years. Can nothing be done, or is it time for the expansion of another sort of agriculture, founded more on ecological principles, and in harmony with people, their societies and cultures?

In the earliest surviving texts on European farming, agriculture was interpreted as two connected things, *agri* and *cultura*, and food was seen as a vital part of the cultures and communities that produced it. Today, however, our experience with industrial farming dominates, with food now seen simply as a commodity, and farming often organised along factory lines. To what extent can we put the culture back into agriculture without compromising the need to produce enough food? Can we create sustainable systems of farming that are efficient and fair and founded on a detailed understanding of the benefits of agroecology and people's capacity to cooperate?

As we advance into the early years of the 21st century, we have some critical choices (Pretty, 2002). Humans have been farming for some 600 generations, and for most of that time the production and consumption of food has been intimately connected to cultural and social systems.

Yet over just the last two or three generations, we have developed hugely successful agricultural systems based largely on industrial principles. They certainly produce more food per hectare and per worker than ever before, but only look so efficient if we ignore the harmful side-effects – the loss of soils, the damage to biodiversity, the pollution of water, the harm to human health (Conway and Pretty, 1991; Pretty *et al.*, 2000, 2003a).

But why should this idea of putting nature and culture back into agriculture matter? Surely we already know how to increase food production? In developing countries, there have been startling increases in food production since the beginning of the 1960s, a short way into the most recent agricultural revolution in industrialised countries, and just prior to the Green Revolution in developing countries. Since then, total world food production grew by 145% per cent. In Africa, it is up by 140%, in Latin America by almost 200% and in Asia by a remarkable 280%. The greatest increases have been in China – an extraordinary five-fold increase, mostly occurring in the 1980s 1990s. In the industrialised regions, production started from a higher base – yet in the USA, it still doubled over forty years, and in Western Europe grew by 68%.

Over the same period, world population has grown from three to more than six billion. Again, though, per capita agricultural production has outpaced population growth. For each person today, there is an extra 25% of food compared with that for people in 1961. These aggregate figures, though, hide important differences between regions. In Asia and Latin America, per capita food production has stayed ahead, increasing by 76 and 28%, respectively. Africa, though, has fared badly, with food production per person 10% less today than in 1961. China, again, performs best, a trebling of food production per person over the same period. Industrialised countries have seen a 40% increase in food production per person.

Yet these advances in aggregate productivity have only brought limited reductions in incidence of hunger. At the turn of the 21st century, there were nearly 800 million people hungry and lacking adequate access to food, an astonishing 18% of all people in developing countries. Nonetheless, there has been progress to celebrate, as incidence of undernourishment stood at 970 million in 1970, comprising a third of people in developing countries at the time. Since then, average per capita consumption of food has increased by 17% to 2760 kilocalories per day – good as an average, but still hiding a great many people surviving on less: 33 countries, mostly in Sub-Saharan Africa, still have per capita food consumption under 2200 kcal per day. The challenge remains huge (Pinstrup-Anderson *et al.*, 1999).

There is also significant food poverty in industrialised countries. In the USA, the largest producer and exporter of food in the world, 11 million people are food insecure and hungry, and a further 23 million are hovering close to the edge of hunger – their food supply is uncertain but they are not permanently hungry. A further sign that something is wrong is that one in seven people in industrialised countries are now clinically obese, and that five of the ten leading causes of death are diet related – coronary heart disease, some cancers, stroke, diabetes mellitus and arteriosclerosis. Alarmingly, the obese are outnumbering the thin in some developing countries, such as Brazil, Chile, Colombia, Costa Rica, Cuba, Mexico, Peru and Tunisia.

Despite great progress, things will probably get worse for many people before they get better. As total population continues to increase, until at least the mid-21st century, so the absolute demand for food will also increase. Increasing incomes will mean people will have more purchasing power, and this will increase demand for food. But as our diets change, so demand for the types of food will also shift radically. In particular, increasing urbanisation means people are more likely to adopt new diets, particularly consuming more meat and fewer traditional cereals and other foods – what Barry Popkin calls the nutrition transition (Popkin, 1998).

One of the most important changes in the world food system will come from an increase in consumption of livestock products. Meat demand is expected to rise rapidly, and this will change many farming systems. Livestock are important in mixed production systems, using foods and by-products that would not have been consumed by humans. But increasingly farmers are finding it easier to raise animals intensively, and feed them with cheap though energy-inefficient cereals and oils. Currently, per capita annual food demand in industrialised countries is 550 kg of cereal and 78 kg of meat. By contrast, in developing countries it is only 260 kg of cereal and 30 kg of meat. These food consumption disparities between people in industrialised and developing countries are expected to persist (Delgado *et al.*, 1999).

THE DEVELOPMENT OF IDEAS ABOUT AGRICULTURAL SUSTAINABILITY

All commentators agree that food production will have to increase substantially in the coming years. But there are very different views about how best this should be achieved (Avery, 1995; NRC, 2000; Smil, 2000; Pretty, 2002; Tilman *et al.*, 2002; Trewevas, 2002; McNeely and Scherr, 2003; Nuffield Council on Bioethics, 2004):

- Some say agriculture will have to expand into new lands – but this will mean further losses of biodiversity.
- Others say food production growth must come through redoubled efforts to repeat the approaches of the Green Revolution.
- Others still say that agricultural sustainability offers options for farmers to intensify their land use and increase food production.

But solving the persistent hunger problem is not simply a matter of developing new agricultural technologies and practices. Most poor producers cannot afford expensive technologies. They will have to find new types of solutions based on locally available and/or cheap technologies combined with making the best of natural, social and human resources.

Intensification using natural, social and human capital assets, combined with the use of best available technologies and inputs (best genotypes and best ecological management) that minimise or eliminate harm to the environment, can be termed "sustainable intensification" (Pretty *et al.*, 2003). Although farmers throughout history have used a wide range of technologies and practices we would today call sustainable, it is only in recent decades that the concepts associated with sustainability have come into more common use.

Concerns began to develop in the 1960s, and were particularly driven by Rachel Carson's book *Silent Spring* (Carson, 1963). Like other popular and scientific studies at the time, it focused on the environmental harm caused by agriculture. In the 1970s, the Club of Rome identified the economic problems that societies would face when environmental resources were overused, depleted or harmed, and pointed towards the need for different types of policies to generate economic growth.

In the 1980s, the World Commission on Environment and Development, chaired by Gro Harlem Brundtland, published *Our Common Future*, the first serious attempt to link poverty alleviation to natural resource management and the state of the environment. Sustainable development was defined as *"meeting the needs of the present without compromising the ability of future generations to meet their own needs"*. The concept implied both limits to growth and the idea of different patterns of growth (WCED, 1987).

In 1992, the UN Conference on Environment and Development was held in Rio de Janeiro. The main agreement was Agenda 21, a 41-chapter document setting out priorities and practices in all economic and social sectors, and how these should relate to the environment. Chapter 14 addressed Sustainable Agriculture and Rural Development (SARD). The principles of sustainable forms of agriculture that encouraged minimising harm to the environment and human health were agreed. However, progress has not been good, as Agenda 21 was not a binding treaty on national governments, and all are free to choose whether they adopt or ignore such principles (Pretty and Koohafkan, 2002).

The "Rio Summit" was followed by several important actions that came to affect agriculture:

1. The signing of the Convention on Biodiversity in 1995.
2. The establishment of the UN Global IPM (Integrated Pest Management) Facility in 1995, which provides international guidance and technical assistance for integrated pest management.
3. The signing of the Stockholm Convention on Persistent Organic Pollutants in 2001, so addressing some problematic pesticides.
4. The ten years after Rio World Summit on Sustainable Development held in Johannesburg.

The concept of agricultural sustainability has grown from an initial focus on environmental aspects to include first economic and then broader social and political dimensions (Cernea, 1991; DFID, 2002):

• *Ecological* – the core concerns are to reduce negative environmental and health externalities, to enhance and use local ecosystem resources, and preserve biodiversity. More recent concerns include broader recognition for positive environmental externalities from agriculture (including carbon capture in soils and flood protection).
• *Economic* – economic perspectives seek to assign value to ecological assets, and also to include a longer time frame in economic analysis. They also highlight subsidies that promote the depletion of resources or unfair competition with other production systems.
• *Social and political* – there are many concerns about the equity of technological change. At the local level, agricultural sustainability is associated with farmer participation, group action and promotion of local institutions, culture and farming communities. At the higher level, the concern is for enabling policies that target poverty reduction.

WHAT IS AGRICULTURAL SUSTAINABILITY?

What do we understand by agricultural sustainability? Many different terms have come to be used to imply greater sustainability in some agricultural systems over prevailing ones (both pre-industrial and industrialised). These include sustainable, ecoagriculture, permaculture, organic, ecological, low input, biodynamic, environmentally sensitive, community based, wise use, farm fresh and extensive. There is continuing and intense debate about whether agricultural systems using some of these terms qualify as sustainable (Balfour, 1943; Lampkin and Padel, 1994; Altieri, 1995; Pretty, 1995).

Systems high in sustainability are making the best use of nature's goods and services whilst not damaging these assets (Altieri, 1995; Pretty, 1995, 2002, 2005; Conway, 1997; NRC, 2000; Uphoff, 2002). The key principles are to:

1. Integrate natural processes such as nutrient cycling, nitrogen fixation, soil regeneration and natural enemies of pests into food production processes.
2. Minimise the use of non-renewable inputs that damage the environment or harm the health of farmers and consumers.
3. Make productive use of the knowledge and skills of farmers, so improving their self-reliance and substituting human capital for costly inputs.
4. Make productive use of people's capacities to work together to solve common agricultural and natural resource problems, such as for pest, watershed, irrigation, forest and credit management.

The technologies themselves are often incorporated into packages. These include:

1. *Integrated pest management*, which uses ecosystem resilience and diversity for pest, disease and

weed control, and seeks only to use pesticides when no other options are available.

2. *Integrated nutrient management*, which seeks both to balance the need to fix nitrogen within farm systems with the need to import inorganic and organic sources of nutrients, and to reduce nutrient losses through erosion control.
3. *Conservation tillage*, which reduces the amount of tillage, sometime to zero, so that soil can be conserved and available moisture used more efficiently.
4. *Agroforestry*, which incorporates multifunctional trees into agricultural systems, and collective management of nearby forest resources *(joint forest management)*.
5. *Aquaculture*, which incorporates fish, shrimps and other aquatic resources into farm systems, such as into irrigated rice fields and fish ponds, and so leads to increases in protein production.
6. *Water harvesting* in dryland areas, which can mean formerly abandoned and degraded lands can be cultivated, and additional crops grown on small patches of irrigated land owing to better rainwater retention.
7. *Livestock integration* into farming systems, such as dairy cattle and poultry, including using zero grazing.

The idea of agricultural sustainability, therefore, does not mean ruling out any technologies or practices on ideological grounds. If a technology works to improve productivity for farmers, and does not harm the environment, then it is likely to be beneficial on sustainability grounds. Agricultural systems emphasising these principles are also multifunctional within landscapes and economies. They jointly produce food and other goods for farm families and markets, but also contribute to a range of valued public goods, such as clean water, wildlife, carbon sequestration in soils, flood protection, groundwater recharge and landscape amenity value.

As a more sustainable agriculture seeks to make the best use of nature's goods and services, so technologies and practices must be locally adapted and fitted to place. These are most likely to emerge from new configurations of social capital, comprising relations of trust embodied in new social organisations, and new horizontal and vertical partnerships between institutions, and human capital comprising leadership, ingenuity, management skills and capacity to innovate. Agricultural systems with high levels of social and human assets are more able to innovate in the face of uncertainty (Cernea, 1991; Pretty and Ward, 2001; Pretty, 2003).

A common, though erroneous, assumption has been that agricultural sustainability approaches imply a net reduction in input use, and so are essentially extensive (they require more land to produce the same amount of food). All recent empirical evidence shows that successful agricultural sustainability initiatives and projects arise from changes in the factors of agricultural production (e.g. from use of fertilisers to nitrogen-fixing legumes; from pesticides to emphasis on natural enemies). However, these have also required reconfigurations on human capital (knowledge, management skills, labour) and social capital (capacity to work together) (Li Wenhua, 2001; Pretty and Ward, 2001; Uphoff, 2002; Pretty, 2003).

A better concept than extensive, therefore, is to suggest that sustainability implies intensification of resources – making better use of existing resources (e.g. land, water, biodiversity) and technologies. For many, the term intensification has come to imply something bad – leading, for example, in industrialised countries to agricultural systems that impose significant environmental costs (Conway and Pretty, 1991; Tegtmeier and Duffy, 2004). The critical question centres on the "type of intensification". Intensification using natural, social and human capital assets, combined with the use of best available technologies and inputs (best genotypes and best ecological management) that minimise or eliminate harm to the environment, can be termed "sustainable intensification".

THE ENVIRONMENTAL CHALLENGE

Increased food supply is a necessary though only partial condition for eliminating hunger and food poverty. What is important is who produces the food, has access to the technology and knowledge to produce it, and has the purchasing power to acquire it. The conventional wisdom is that, in order to increase food supply, efforts should be redoubled to modernise agriculture. But the success of industrialised agriculture in recent decades has masked significant negative externalities, with environmental and health problems increasingly well documented and costed, including Ecuador, China, Germany, the Philippines, UK and USA (Pingali and Roger, 1995; Crissman et al., 1998; Waibel et al., 1999; Pretty et al., 2000, 2001; Norse et al., 2001; Tegtmeier and Duffy, 2004). These environmental costs change our conclusions about which agricultural systems are the most efficient, and indicate that alternatives that reduce externalities should be sought.

There are surprisingly few data on the environmental and health costs imposed by agriculture on other sectors and interests. Agriculture can negatively affect the environment through overuse of natural resources as inputs or through their use as a sink for pollution. Such effects are called negative externalities because they are usually non-market effects and therefore their costs are not part of market prices. Negative externalities are one of the classic causes of market failure whereby the polluter does not pay the full costs of their actions, and therefore these costs are called external costs (Baumol and Oates, 1988).

Externalities in the agricultural sector have at least four features: (i) their costs are often neglected; (ii) they often occur with a time lag; (iii) they often damage groups whose interests are not well represented in political or decision-making processes; and (iv) the identity of the source of the externality is not always known. For example, farmers generally have few incentives to prevent pesticides escaping to water bodies, to the atmosphere and to nearby nature as they transfer the full cost of cleaning up the environmental consequences to society at large. In the same way, pesticide manufacturers do not pay the full cost of all their products, as they do not suffer from any adverse side effects that may occur.

Partly as a result of lack of information, there is little agreement on the economic costs of externalities in agriculture. Some authors suggest that the current system of economic calculations grossly underestimates the current and future value of natural capital (Costanza *et al.* 1997; Daily, 1997). Such valuation of ecosystem services remains controversial because of methodological and measurement problems (Hanley *et al.*, 1998; Carson, 2000) and because of its role in influencing public opinions and policy decisions. The great success of industrialised agriculture in recent decades has masked significant negative externalities, many of which arise from pesticide overuse and misuse.

There are also growing concerns that such systems may not reduce food poverty. Poor farmers, at least whilst they remain poor, need low-cost and readily available technologies and practices to increase local food production. At the same time, land and water degradation is increasingly posing a threat to food security and the livelihoods of rural people who occupy degradation-prone lands. Some of the most significant environmental and health problems centre on the use of pesticides in agricultural systems (Pretty, 2005).

THE REAL COSTS OF FOOD

When we buy or bake our daily bread, do we ever wonder about how much it really costs? We like it when our food is cheap, and complain when prices rise. Indeed, riots over food prices date back at least to Roman times. Governments have long since intervened to keep food cheap in the shops, and tell us that policies designed to do exactly this are succeeding. In most industrialised countries, the proportion of the average household budget spent on food has been declining in recent decades. Food is getting cheaper relative to other goods, and many believe that this must benefit everyone as we all need to eat food. But we have come to believe a damaging myth. Food is not cheap. It only appears cheap in the shop because we are not encouraged to think of the hidden costs of damage caused to the environment and human health by certain systems of agricultural production. Thus, we actually pay three times for our food. Once at the till in the shop, a second time through taxes that are used to subsidise farmers or support agricultural development, and a third time to clean up the environmental and health side-effects. Food looks cheap because we count these costs elsewhere in society. As economists put it, the real costs are not internalised in prices.

This is not to say that prices in the shop should rise, as this would penalise the poor over the wealthy. Using taxes to raise money to support agricultural development is also potentially progressive, as the rich pay proportionally more in taxes, and the poor, who spend proportionally more of their budget on food, benefit if prices stay low. But this idea of fairness falters when set against the massive distortions brought about by modern agricultural systems that additionally impose large environmental and health costs throughout economies. Other people and institutions pay these costs, and this is both unfair and inefficient. If we were able to add up the real costs of producing food, we would find that modern industrialised systems of production perform poorly in comparison with sustainable systems. This is because we permit cost-shifting – the costs of ill-health, lost biodiversity and water pollution are transferred away from farmers, and so not paid by those producing the food nor included in the price of the products sold. Until recently, though, we have lacked the methods to put a price on these side-effects.

When we conceive of agriculture as more than simply a food factory, indeed as a multifunctional activity with many side-effects, then this idea that farmers do only one thing must change. Of course, it was not always like this. It is modern agriculture that has brought a narrow view of farming, and it has led us to crisis. The rural environment in industrialised countries suffers, the food we eat is as likely to do as much harm as good, and we still think food is cheap. The following words were written more than fifty years ago, just before the advent of modern industrialised farming. *"Why is there so much controversy about Britain's agricultural policy, and why are farmers so disturbed about the future?... After the last war, the people of these islands were anxious to establish food production on a secure basis, yet, in spite of public goodwill, the farming industry has been through a period of insecurity and chaotic conditions".* These are the opening words to a national enquiry that could have been written about a contemporary crisis. Yet they are by Lord Astor, written in 1945 to introduce the Astor and Rowntree review of agriculture. This enquiry was critical of the replacement of mixed methods with standardised farming. They said, *"to farm properly you have got to maintain soil fertility; to maintain soil fertility you need a mixed farming system".* They believed that farming would only succeed if it maintained the health of the whole system, beginning in particular with the maintenance of soil fertility: *"obviously it is not only sound business practice but plain common sense to take steps to maintain the health and fertility of soil"* (Astor and Rowntree, 1945).

But in the enquiry, some witnesses disagreed, and called for a *"specialised and mechanised farming"*, though interestingly, the farming establishment at the time largely supported the idea of mixed farming. But in the end, the desire for public subsidies to encourage increases in food production took precedence, and these were more easily applied to simplified systems rather than mixed ones. The 1947 Agriculture Act was the outcome, a giant leap forward for modern, simplified agriculture, and a large step away from farming that valued nature's assets for farming. Sir George Stapledon, British scientist knighted for his research on grasslands, was another perceptive scientist well ahead of his time (see Conford, 1988). He too was against monocultures and in favour of diversity, arguing in 1941 that *"senseless systems of monoculture designed to produce food and other crops at the cheapest possible cost have rendered waste literally millions of acres of once fertile or potentially fertile country".* In his final years, just a decade after the 1947 Act, he said *"today technology has begun to run riot and amazingly enough perhaps nowhere more so than on the most productive farms ... Man is putting all his money on narrow specialisation and on the newly dawned age of technology has backed a wild horse which given its head is bound to get out of control".*

AGRICULTURE'S UNIQUE MULTIFUNCTIONALITY

We should all now be asking: what is farming for? Clearly, in the first instance, to produce food, and we have become very good at it. A great success, but only if our measures of efficiency

are narrow. Agriculture is unique as an economic sector. It does more than just produce food, fibre, oil and timber. It has a profound impact on many aspects of local, national and global economies and ecosystems. These impacts can be either positive or negative. The negative ones are worrying. Pesticides and nutrients leaching from farms have to be removed from drinking water, and these costs are paid by water consumers, not by the polluters. The polluters, therefore, benefit by not paying to clean up the mess they have created, and have no incentive to change their behaviour. What also makes agriculture unique is that it affects the very assets on which it relies for success. Agricultural systems at all levels rely for their success on the value of services flowing from the total stock of assets that they control, and five types of asset, natural, social, human, physical and financial capital, are now recognised as being important (For further discussions, see Bourdieu, 1986; Coleman, 1988; 1990; Putnam, 1995; Benton, 1998; Carney, 1998; Flora, 1998; Grootaert, 1998; Ostrom, 1998; Pretty, 1998; Scoones, 1998; Uphoff, 1998; Costanza et al., 1999; Pretty and Ward, 2001).

Natural capital produces nature's goods and services, and comprises food, both farmed and harvested or caught from the wild, wood and fibre; water supply and regulation; treatment, assimilation and decomposition of wastes; nutrient cycling and fixation; soil formation; biological control of pests; climate regulation; wildlife habitats; storm protection and flood control; carbon sequestration; pollination; and recreation and leisure. *Social capital* yields a flow of mutually beneficial collective action, contributing to the cohesiveness of people in their societies. The social assets comprising social capital include norms, values and attitudes that predispose people to cooperate; relations of trust, reciprocity and obligations; and common rules and sanctions mutually agreed or handed down. These are connected and structured in networks and groups.

Human capital is the total capability residing in individuals, based on their stock of knowledge skills, health and nutrition. It is enhanced by access to services that provide these, such as schools, medical services and adult training. People's productivity is increased by their capacity to interact with productive technologies and with other people. Leadership and organisational skills are particularly important in making other resources more valuable. *Physical capital* is the store of human-made material resources, and comprises buildings, such as housing and factories, market infrastructure, irrigation works, roads and bridges, tools and tractors, communications, and energy and transportation systems, that make labour more productive. *Financial capital* is more of an accounting concept, as it serves as a facilitating role rather than as a source of productivity in and of itself. It represents accumulated claims on goods and services, built up through financial systems that gather savings and issue credit, such as pensions, remittances, welfare payments, grants and subsidies.

As agricultural systems shape the very assets on which they rely for inputs, a vital feedback loop occurs from outcomes to inputs. Donald Worster's three principles for good farming capture this idea. It is farming that makes people healthier, farming that promotes a more just society, and farming that preserves the Earth and its networks of life. He says *"the need for a new agriculture does not absolve us from the moral duty and common-sense advice to farm in an ecologically-rational way. Good farming protects the land, even when it uses it"* (Worster, 1993). Thus, sustainable agricultural systems tend to have a positive effect on natural, social and human capital, whilst unsustainable ones feed back to deplete these assets, leaving less for future generations. For example, an agricultural system that erodes soil whilst producing food externalises costs that others must bear. But one that sequesters carbon in soils through organic matter accumulation helps to mediate climate change. Similarly, a diverse agricultural system that enhances on-farm wildlife for pest control contributes to wider stocks of biodiversity, whilst simplified modernised systems that eliminate wildlife do not. Agricultural systems that offer labour-absorption opportunities, through resource improvements or value-added activities, can boost economies and help to reverse rural-to-urban migration patterns.

Agriculture is, therefore, fundamentally multifunctional. It jointly produces many unique non-food functions that cannot be produced by other economic sectors so efficiently. Clearly, a key policy challenge, for both industrialised and developing countries, is to find ways to maintain and enhance food production. But the key question is: can this be done whilst seeking both to improve the positive side-effects and to eliminate the negative ones? It will not be easy, as past agricultural development has tended to ignore both the multifunctionality of agriculture and the pervasive external costs.

This leads us to a simple and clear definition for sustainable agriculture. It is farming that makes the best use of nature's goods and services whilst not damaging the environment. It does this by integrating natural processes such as nutrient cycling, nitrogen fixation, soil regeneration and natural enemies of pests into food production processes. It also minimises the use of non-renewable inputs that damage the environment or harm the health of farmers and consumers.

It makes better use of the knowledge and skills of farmers, so improving their self-reliance, and it makes productive use of people's capacities to work together to solve common management problems. Through this, sustainable agriculture also contributes to a range of public goods, such as clean water, wildlife, carbon sequestration in soils, flood protection and landscape quality.

CAN SUSTAINABLE AGRICULTURE WORK IN DEVELOPING COUNTRIES?

Can these ideas about sustainability work in practice? At the University of Essex, we recently completed the largest survey of sustainable agriculture improvements in developing countries. The aim was to audit progress towards agricultural sustainability, and assess the extent to which such initiatives, if spread on a much larger scale, could feed a growing world population that is already substantially food insecure. More than 280 projects in 57 countries were investigated (Pretty *et al.*, 2003b, 2006). We calculated that almost 9 million farmers were using sustainable agriculture practices on about 29 million hectares, more than 98% of which emerged since 1990. These methods are working particularly well for small farmers, as about half of those surveyed are in projects with a mean area per farmer of less than one hectare, and 90% in areas with less than two hectares each.

Improvements in food production were occurring through one or more of four different mechanisms:

1. The first involves the intensification of a single component of a farm system, with little change to the rest of the farm, such as home-garden intensification with vegetables and/or tree crops, vegetables on rice bunds, and introduction of fish ponds or a dairy cow.
2. The second involves the addition of a new productive element to a farm system, such as fish or shrimps in paddy rice, or agroforestry, which provides a boost to total farm food production and/or income, but which does not necessarily affect cereal productivity.
3. The third involves better use of nature to increase total farm production, especially water (by water harvesting and irrigation scheduling) and land (by reclamation of degraded land), so leading to additional new dryland crops and/or increased supply of additional water for irrigated crops, and thus so increasing cropping intensity.
4. The fourth involves improvements in per hectare yields of staples through the introduction of new regenerative elements into farm systems, such as legumes and integrated pest management, and

new and locally appropriate crop varieties and animal breeds.

Thus, a successful sustainable agriculture project may be substantially improving domestic food consumption or increasing local food barters or sales through home gardens or fish in rice fields, or better water management, without necessarily affecting the per hectare yields of cereals. Home-garden intensification occurred in a fifth of projects, but given its small scale accounted for less than 1% of the area. Better use of land and water, giving rise to increased cropping intensity, occurred in a seventh of projects, with a third of farmers and a twelfth of the area. The incorporation of new productive elements into farm systems, mainly fish and shrimps in paddy rice, occurred in 4% of projects, and accounted for the smallest proportion of farmers and area. The most common mechanisms were yield improvements with regenerative technologies or new seeds/breeds, occurring in 60% of the projects, by more than half of the farmers and about 90% of the area.

What is happening to food production? It was found that sustainable agriculture has led to an average 93% increase in per hectare food production (Pretty *et al.*, 2003b, 2006). The relative yield increases are greater at lower yields, indicating greater benefits for poor farmers, and for those missed by the recent decades of modern agricultural development. The increases are quite remarkable, as most agriculturalists would be satisfied with any technology that can increase annual productivity by even one or two per cent. It is worth restating: these projects are seeing a close to doubling of per hectare productivity over several years, and this still underestimates the additional benefits of intensive food production in small patches of home gardens or fish ponds.

These aggregate figures understate the benefits of increased diversity in the diet as well as increased quantity. Most of these agricultural sustainability initiatives have seen increases in farm diversity. In many cases, this translates into increased diversity of food consumed by the household, including fish protein from rice fields or fish ponds, milk and animal products from dairy cows, poultry and pigs kept in the home garden, and vegetables and fruit from home gardens and farm micro-environments. Although these initiatives are reporting significant increases in food production, some as yield improvements, and some as increases in cropping intensity or diversity of produce, few are reporting surpluses of food being sold to local markets. This is because of a significant elasticity of consumption amongst rural households experiencing any degree of food insecurity. As production increases, so also does domestic consumption, with direct

benefit particularly for the health of women and children. In short, rural people are eating more food and a greater diversity of food, and this does not show up in the international statistics.

RESOURCE-CONSERVING TECHNOLOGIES IN INDUSTRIALISED COUNTRIES

There are also an increasing number of proven and promising resource-conserving and regenerative technologies for pest, nutrient, soil, water and energy management in industrialised countries. These have emerged from many sources: from traditional agricultural systems; from both the long history of organic agriculture and from recent experiences with "precision farming"; and from farms in both developing and industrialised countries. Many of these have arisen on farms where steps have already been taken to reduce the adverse environmental effects. Natural processes are favoured over external inputs, and by-products or "wastes" from one component of the farm are emphasised as they can become inputs to another. In this way, farms remain productive as well as reducing negative impacts on the environment (Pretty, 1998).

These technologies do two important things. They conserve existing on-farm resources, such as nutrients, predators, water or soil. And they introduce new elements into the farming system that add to the stocks of these resources, such as nitrogen-fixing crops, water-harvesting structures or new predators, and so substitute for some or all external resources. Most represent low- or lower-external input options than is currently the norm.

Many of the individual technologies are also multifunctional. This implies that their adoption will result in favourable changes in several aspects of the farming system and natural capital at the same time. For example, hedgerows encourage wildlife and predators and act as windbreaks, so reducing soil erosion. Legumes are introduced into rotations to fix nitrogen, and also act as a break crop to prevent carry-over of pests and diseases. Clovers in pastures can reduce fertiliser bills and lift sward digestibility for cattle. Grass contour strips slow surface run-off of water, encourage percolation to groundwater, and are a source of fodder for livestock. Catch crops prevent soil erosion and leaching during critical periods, and can also be ploughed in as a green manure. The incorporation of green manures not only provides a readily available source of nutrients for the growing crop but also increases soil organic matter and hence water retentive capacity, further reducing susceptibility to erosion. Low-lying grasslands managed as water meadows that are good for wildlife also provide an early-season yield of grass for lambs.

The principle of integration, a central theme of sustainable agriculture, implies a focus on increasing the number of technologies and practices, and the positive, reinforcing linkages between them. If a technology, such as a nitrogen-fixing legume, is taken by farmers and adapted to fit their own cropping systems, and this leads to increases in output or reductions in costs, then this is the strongest evidence of success.

What is clear is that resource-conserving and regenerative technologies are spreading. In Denmark, some 150 farms have in-field weather stations to help predict disease outbreaks in potatoes, leading to cuts in fungicide use, with some growers able to postpone first applications for five or more weeks. In the UK, some 150,000 hectares of cereal farms were computer mapped in the early 2000s, enabling inputs to be targeted more precisely and total use of pesticide and fertiliser to be cut. Also in the UK, three-quarters of crops grown in glasshouses use natural predators to control pests rather than pesticides. In France, there are 700 farms in the national network researching and implementing "agriculture durable". In the state of Baden-Württemberg in southern Germany, 100,000 farms are using sustainable practices and technologies, though not all are integrated at the whole farm level. In Australia, one-third of all farmers are members of Landcare groups. The organic revolution also continues, with demand from consumers growing and with the number of farmers converted entirely to organic practices in industrialised countries continuing to grow rapidly.

WHAT IS THE PLACE FOR GENETIC MODIFICATION?

Only a few years after the development of the first genetically modified crops for agriculture, opinions on benefits and risk are sharply divided. Some argue that genetically modified organisms are safe and essential for world progress; others state that they are not needed, and hold too many risks. The first group believes that media manipulation and scare-mongering are limiting useful technologies; the second that scientists, private companies and regulators are understating hazards for the sake of economic returns.

Neither view is entirely correct, for one simple reason. Genetically modified organisms are not a single, simple technology (Pretty, 2001; GM Science Review, 2003; Nuffield Council on Bioethics, 2004). Each product brings different potential benefits for different stakeholders; each poses different environmental and health risks.

It is, therefore, useful to distinguish between different generations of genetically modified technologies. The first-generation technologies came into commercial use in the late 1990s, and have tended not to bring distinct consumer benefits, one reason why there is so much current public opposition. The realisation of promised benefits to farmers and the environment has only been patchy. First-generation technologies include herbicide tolerance, insect resistance, long-life tomatoes, bacteria in containment for production of cheese and washing-powder enzymes, and flowers with amended colour.

The second-generation technologies comprise those already developed and tested, but not yet commercially released, either because of uncertainties over the stability of the technology itself, or over concerns for potential environmental risks. Some of these applications are likely to bring more public and consumer benefits, and include a range of medical applications. These include viral resistance in rice, cassava, papaya, sweet potatoes and pepper; nematode resistance in various cereal and other crops, such as banana and potato; frost tolerance in strawberry and *B.t.* clover; trees with reduced lignin; vitamin A rice; and "pharming" with crops and animals for pharmaceuticals.

The third-generation technologies are those that are still far from market, but generally require better understanding of the whole gene complexes that control such traits as drought or salt tolerance and nitrogen fixation. These are likely to bring more explicit consumer benefits than the first generation. These include stress tolerance in cereals, such as thermo, salt and heavy metal tolerance, drought resistance, physiological modifications of crops and trees to increase efficiency of resource use (nutrients, water, light) or delaying of ageing in leaves, neutraceuticals (crops boosted with vitamins/minerals), vaccine crops (such as banana and potato), designer crops modified to produce oils or plastics, the development of new markers to replace antibiotics, and legumes with increased tannins for bloat control in cattle.

The first-generation technologies have tended only to provide substantial private benefits for the companies producing them and farmers using them. Many of the later generation genetically modified organisms are, by contrast, more multifunctional and public-good oriented, though like all technologies clearly none are without risk. Modifications of crops with low value in rotations, such as legumes and oats, will make them more attractive to farmers because of high protein and energy content. Others will be more efficient in nitrogen use, so reducing nitrate leaching, or modifications of rhizobia could improve the nitrogen-fixing capacity of a wide range of crops. Both options would reduce the need to use nitrogen fertilisers.

Although the pace of change in the development of GM has provoked many debates, there has been relatively little said about the potential benefits for developing countries. Many concerns are about important indirect effects, such as the growing centralisation of world agriculture. These represent structural changes in agriculture in which GM crops are a contributor to change, but not necessarily the driver. These contested positions raise important questions. Will GM crops contribute to the further promotion of technological approaches to agricultural development? Could such technologies bring environmental benefits and so promote sustainability? Are GM technologies essential for feeding a hungry world, or is hunger more a result of poverty, with poor consumers and farmers unable to afford modern, expensive technologies?

Some say emphatically yes, often raising the spectre of famine and excessive population growth as a way to gain greater support for GM as a whole. But GM crops can only help to feed the world if attention is paid to the processes of technology development, to benefit sharing and to low-cost methods of production. Most commentators agree that food production will have to increase, and that this will have to come from existing farmland. But past approaches to modern agricultural development have not been successful in all parts of the world.

In most contexts, people are hungry because they are poor. They simply do not have the money to buy either the food they need or the modern technologies that could increase their yields. What they need are readily-available and cheap means to improve their farm productivity. So a cereal crop engineered to have bacteria on the roots to fix free nitrogen from the air, or another with the apomixis trait, would be a great benefit for poor farmers. But unless such technologies are cheap, they are unlikely to be accessible to the very people who need them most.

Agricultural sustainability is now an increasingly viable option for many farmers in developing and industrialised countries alike. But where there are no alternatives to specific problems, then GM could bring forth novel and effective options. If research is conducted by public-interest bodies, such as universities, non-government organisations and governments themselves, whose concern is to produce public goods, then biotechnology could result in the spread of technologies that have immense benefits.

Two questions need to be asked when considering the potential benefits of GM crops in developing countries:

1. Does the new GM crop replace an existing technology or practice that is more harmful to the environment or human health?

2. Does the new GM crop address a problem that has not been solved by existing research?

B.t. cotton was widely grown in China, India and South Africa in 2002. In China, insecticide use fell by 60–80% on 2001 levels of use, resulting in substantial health and cost benefits for farmers. In South Africa, farmers in KwaZulu-Natal increased gross margins, even though seeds are more expensive. There remain controversies, centred on the possibility of insect resistance and the corporate control of seed markets (one company has made 90% of the patent applications for cotton genes).

Golden rice is a strain of rice that produces enhanced levels of β-carotene, which is converted to vitamin A in the body. Severe vitamin A deficiency affects 100 to 250 million pre-school children worldwide. Many develop childhood blindness, and all are subject to increased morbidity from measles and diarrhoea. If they could access varied and adequate diets, then this would not occur. But many are in families too poor for this. In many countries, cereal products are augmented with micronutrients and vitamins, but rice is consumed as grain, and so this cannot be done. Thus, vitamin A rice represents an opportunity to deal with a significant public health problem. However, there remain uncertainties over the bio-availability of β-carotene, and field trials of golden rice have not yet been scheduled.

Rice yellow mottle virus is a major factor limiting African rice production, often reducing yields by 50–95%. It has not been possible to introduce resistance into local varieties through conventional breeding, but GM has led to the development of novel resistant varieties. These have been tested in five countries, with resulting complete resistance to the virus. Viral-resistant cassava has been developed by the public sector in South Africa (universities) and is now grown by small farmers. It offers protection against viral diseases that conventional breeding had been unable to solve. In Kenya, sweet potato resistant to feathery mottle virus has been developed by the government's research institute, and this is expected to increase farmers' yields by 20%.

CHANGING WHOLE SYSTEMS

What we do not yet know is whether a transition to sustainable agriculture, delivering greater benefits at the scale occurring in these projects, will result in enough food to meet the current food needs in developing countries, let alone the future needs after continued population growth and adoption of more urban and meat-rich diets. But what we are seeing is highly promising.

There is also scope for additional confidence, as evidence indicates that productivity can grow over time if natural, social and human assets are accumulated.

Sustainable agriculture systems appear to become more productive when human capacity increases, particularly in the form of farmers' capacity to innovate and adapt their farm systems for sustainable outcomes. Sustainable agriculture is not a concretely defined set of technologies, nor is it a simple model or package to be widely applied or fixed with time. It needs to be conceived of as a process for social learning. Lack of information on agroecology and necessary skills to manage complex farms is a major barrier to the adoption of sustainable agriculture.

A problem is that we know much less about these resource-conserving technologies than we do about the use of external inputs in modernised systems. So it is clear that the process by which farmers learn about technology alternatives is crucial. If they are enforced or coerced, then they may only adopt for a limited period. But if the process is participatory and enhances farmers' ecological literacy of their farms and resources, then the foundation for redesign and continuous innovation is laid.

But successes are still regrettably in the minority. Yet time is short, and the challenge is simply enormous. This change to agricultural sustainability is clearly benefiting poor people and environments in developing countries. People have more food, are better organised, are able to access external services and power structures, and have more choices in their lives. But change may also provoke secondary problems. For example, building a road near a forest can help farmers reach food markets, but also aid illegal timber extraction. Equally, short-term social conflict may be a necessity for overcoming inequitable land ownership, so as to produce better welfare outcomes for the majority.

Projects may be making considerable progress on reducing soil erosion and increasing water conservation through adoption of zero tillage, but still continue to rely on applications of herbicides. In other cases, improved organic matter levels in soils may lead to increased leaching of nitrate to groundwater. If land has to be closed off to grazing for rehabilitation, then people with no other source of feed may have to sell their livestock; and if cropping intensity increases or new lands are taken into cultivation, then the burden of increased workloads may fall particularly on women. Also, additional incomes arising from sales of produce may go directly to men in households, who are less likely than women to invest in children and the household as a whole.

GETTING THE POLICIES RIGHT

Three things are now clear from evidence on the recent spread of agricultural sustainability:

1. Some technologies and social processes for local-scale adoption of more sustainable agricultural practices are increasingly well tested and established.
2. The social and institutional conditions for spread are less well understood, but have been established in several contexts, leading to more rapid spread in the 1990s and early 2000s.
3. The political conditions for the emergence of supportive policies are least well established, with only a very few examples of real progress.

As indicated earlier, agricultural sustainability can contribute to increased food production, as well as make an impact on rural people's welfare and livelihoods. Clearly much can be done with existing resources. A transition towards a more sustainable agriculture will not, however, happen without some external help and money. There are always transition costs in learning, in developing new or adapting old technologies, in learning to work together, and in breaking free from existing patterns of thought and practice. It also costs time and money to rebuild depleted natural and social capital.

Most agricultural sustainability improvements seen in the 1990s and early 2000s arose despite existing national and institutional policies, rather than because of them. Although almost every country would now say it supports the idea of agricultural sustainability, the evidence points towards only patchy reforms. Nonetheless, recent years have seen some progress towards the recognition of the need for policies to support sustainable agriculture. Yet only three countries have given explicit national support for sustainable agriculture – putting it at the centre of agricultural development policy and integrating policies accordingly.

These are Cuba, Switzerland and Bhutan. Cuba has a national policy for alternative agriculture; Switzerland has three tiers of support to encourage environmental services from agriculture and rural development, and Bhutan has a national environmental policy coordinated across all sectors.

Several countries have given sub-regional support to agricultural sustainability, such as the states of Santa Caterina, Paraná and Rio Grande do Sul in southern Brazil supporting zero tillage, catchment management and rural agribusiness development, and some states in India supporting participatory watershed and irrigation management. A larger number of countries have reformed parts of agricultural policies, such as China's support for integrated ecological demonstration villages, Kenya's catchment approach to soil conservation, Indonesia's ban on pesticides and programme for farmer field schools, Bolivia's regional integration of agricultural and rural policies, Sweden's support for organic agriculture, Burkina Faso's land policy, and Sri Lanka and the Philippines' stipulation that water users' groups be formed to manage irrigation systems.

A good example of a carefully designed and integrated programme comes from China. In March 1994, the government published a White Paper to set out its plan for implementation of Agenda 21, and put forward ecological farming, known as *Shengtai Nongye* or agro-ecological engineering, as the approach to achieve sustainability in agriculture. Pilot projects have been established in 2000 townships and villages spread across 150 counties. Policy for these "eco-counties" is organised through a cross-ministry partnership, which uses a variety of incentives to encourage adoption of diverse production systems to replace monocultures. These include subsidies and loans, technical assistance, tax exemptions and deductions, security of land tenure, marketing services and linkages to research organisations. These eco-counties contain some 12 million hectares of land, about half of which is cropland, and though only covering a relatively small part of China's total agricultural land, do illustrate what is possible when policy is appropriately coordinated.

An even larger number of countries has seen some progress on agricultural sustainability at project and programme level. However, progress on the ground still largely remains despite, rather than because of, explicit policy support. No agriculture minister is likely to say they are against sustainable agriculture, yet good words remain to be translated into comprehensive policy reforms. Agricultural systems can be economically, environmentally and socially sustainable, and contribute positively to local livelihoods. But without appropriate policy support, they are likely to remain at best localised in extent, and at worst simply wither away.

AREAS OF DEBATE AND DISAGREEMENT

What we do not yet know is whether progress towards more sustainable agricultural systems will result in enough food to meet the current food needs in developing countries, let alone the future needs after continued population growth and adoption of more urban and meat-rich diets. But what is occurring should be cause for cautious optimism, particularly as evidence indicates that

productivity can grow over time if natural, social and human assets are accumulated.

A more sustainable agriculture which improves the asset base can lead to rural livelihood improvements. People can be better off, have more food, be better organised, have access to external services and power structures, and have more choices in their lives. But like all major changes, such transitions can also provoke secondary problems. New winners and losers will emerge with the widespread adoption of sustainable agriculture. Producers of current agrochemical products are likely to suffer market losses from a more limited role for their products. The increase in assets that could come from sustainable livelihoods based on sustainable agriculture may simply increase the incentives for more powerful interests to take over. Not all political interests will be content to see poor farmers and families organise into more powerful social networks and alliances.

Many countries also have national policies that now advocate export-led agricultural development. Access to international markets is clearly important for poorer countries, and successful competition for market share can be a very significant source of foreign exchange. However, this approach has some drawbacks:

1. Poor countries are in competition with each other for market share, and so there is likely to be a downward pressure on prices, which reduces returns over time unless productivity continues to increase.
2. Markets for agri-food products are fickle, and can be rapidly undermined by alternative products or threats (e.g. avian bird flu and the collapse of the Thai poultry sector).
3. Distant markets are less sensitive to the potential negative externalities of agricultural production and are rarely pro-poor (with the exception of fair-trade products).
4. Smallholders have many difficulties in accessing international markets and market information.

There is indeed very little clear evidence that export-led poverty alleviation has worked. Even Vietnam, which has earned considerable foreign exchange from agricultural development, has had to do so at very low prices and little value added.

More importantly, an export-led approach can seem to ignore the in-country opportunities for agricultural development focused on local and regional markets. Agricultural policies with both sustainability and poverty-reduction aims should adopt a multi-track approach that emphasises five components: (i) small farmer development linked to local markets; (ii) agri-business development – both small businesses and export-led;

(iii) agro-processing and value-added activities – to ensure that returns are maximised in-country; (iv) urban agriculture – as many urban people rely on small-scale urban food production that rarely appears in national statistics; and (v) livestock development – to meet local increases in demand for meat (predicted to increase as economies become richer).

A differentiated approach for agricultural policies will become increasingly necessary if agricultural systems themselves are to become more productive and sustainable whilst reducing negative impacts on the environment.

REFERENCES AND BIBLIOGRAPHY

Altieri M A (1995). *Agroecology: The Science of Sustainable Agriculture*. Westview Press, Boulder, CO.

Association for Better Husbandry (2000). Project reports and annual reports. [At URL www.ablh.org]

Astor, Vicount and Rowntree (1945). *Mixed Farming and Muddled Thinking. An Analysis of Current Agricultural Policy*. Macdonald and Co., London.

Avery D (1995). *Saving the Planet with Pesticides and Plastic*. The Hudson Institute, Indianapolis.

Balfour E B (1943). *The Living Soil*. Faber and Faber, London.

Baumol W J and Oates W E (1988). *The Theory of Environmental Policy*. Cambridge University Press, Cambridge.

Benbrook C M (2003). Impacts of Genetically Engineered Crops on Pesticide Use in the United States. Northwest Science and Environmental Policy Center, Ames, IA.

Benton T (1998). Sustainable development and the accumulation of capital: reconciling irreconcilable? In Dobson A (ed) *Fairness and Futurity*. Oxford University Press, Oxford.

Bourdieu P (1986). The forms of capital. In Richardson J (ed) *Handbook of Theory and Reserch for the Sociology of Education*. Greenwood Press, Westport, CT.

Carson R (1963). *Silent Spring*. Penguin Books, Harmondsworth.

Carson R T (2000). Contingent valuation: a user's guide. *Environmental Science and Technology* 34: 1413–1418.

Cernea M M (1991). *Putting People First*, 2nd edn. Oxford University Press, Oxford.

Cleaver K M and Schreiber G A (1995). *The Population, Agriculture and Environment nexus in Sub-Saharan Africa*. World Bank, Washington, DC.

Coleman J (1988). Social capital and the creation of human capital. *American Journal of Sociology* 94(suppl.) S95–S120.

Coleman J (1990). *Foundations of Social Theory*. Harvard University Press, Cambridge, MA.

Conford P (ed) (1988). *The Organic Tradition. An Anthology of Writing on Organic Farming*. Green Books, Bideford, Devon.

Conway G R (1997). *The Doubly Green Revolution*. Penguin, London.

Conway G R and Pretty J (1991). *Unwelcome Harvest*. Earthscan, London.

Costanza R, d'Arge R, de Groot R et al. (1997 and 1999). The value of the world's ecosystem services and natural capital. *Nature* 387: 253–260.

Crissman C C, Antle J M and Capalbo S M (eds) (1998). *Economic, Environmental and Health Tradeoffs in Agriculture: Pesticides and the Sustainability of Andean Potato Production*. CIP, Lima, and Kluwer, Boston.

Daily G (ed) (1997). *Nature's Services: Societal Dependence on Natural Ecosystems*. Island Press, Washington, DC.

de Freitas H (1999). Transforming microcatchments in Santa Caterina, Brazil. In Hinchcliffe F, Thompson J, Pretty J, Guijt I and Shah P (eds) *Fertile Ground: The Impacts of Participatory Watershed Development*. IT Publishing, London.

Delgado C, Rosegrant M, Steinfield H et al. (1999). *Livestock to 2020: the Next Food Revolution*. International Food Policy Research Institute, Washington, DC.

DFID (2002). *Sustainable Agriculture*. Resource Management Keysheet 10. Department or International Development, London.

FAO (1999). *The Future of Our Land*. FAO, Rome.

Feder G, Murgai R and Quizon J B (2004). Sending farmers back to school: the impact of Farmer Field Schools in Indonesia. *Review of Agricultural Economics* 26(1): 45–62.

Flora J L (1998). Social capital and communities of place *Rural Sociology* 63(4): 481–506.

GM Science Review (2003). Office of Science and Technology, London. At URL http://www.gmsciencedebate.org.uk/report/default.htm#first

Hanley N, MacMillan D, Wright R E et al. (1998). Contingent valuation versus choice experiments: estimating the benefits of environmentally sensitive areas in Scotland. *Journal of Agricultural Economics* 49(1): 1–15.

Kenmore P E, Carino F O, Perez C A et al. (1984). Population regulation of the brown planthopper within rice fields in the Philippines. *Journal of Plant Protection in the Tropics* 1(1):1937.

Lampkin N and Padel S (eds) (1994). *The Economics of Organic Farming: An International Perspective*. CAB International, Wallingford, Oxfordshire.

Landers J N, De C Barros G S-A, Manfrinato W A et al. (2001). Environmental benefits of zero-tillage in Brazil – a first approximation. In Garcia Torres L, Benites J and MartinezVilela A (eds) *Conservation Agriculture – A Worldwide Challenge, Vol. 1*. XUL, Cordoba, Spain.

Lang T and Heasman D (2004). *Food Wars*. Earthscan, London.

Li Wenhua (2001). *Agro-Ecological Farming Systems in China*. Man and the Biosphere Series, Vol. 26. UNESCO, Paris.

McNeely J A and Scherr S J (2003). *Ecoagriculture*. Island Press, Washington, DC.

Norse D, Li Ji, and Zhang Zheng (2001). *Environmental Costs of Rice Production in China*. Aileen Press, Bethesda.

NRC (2000). *Our Common Journey: TransitionTowards Sustainability*. Board on Sustainable Development, Policy Division, National Research Council. National Academy Press, Washington, DC.

Nuffield Council on Bioethics (2004). *The Use of Genetically Modified Crops in Developing Countries*. Nuffield Council on Bioethics, London.

Ostrom E (1998). Social capital: a fad or fundamental concept? Center for Study of Institutions, Population and Environmental Change. Indiana University, USA.

Petersen P, Tardin J M and Marochi F (2000). Participatory development of non-tillage systems without herbicides for family farming: the experience of the center–south region of Paraná. *Environmental Development and Sustainability* 1: 235–252.

Pingali P L and Roger P A (1995). *Impact of Pesticides on Farmers' Health and the Rice Environment.* Kluwer, Dordrecht.

Pinstrup-Anderson P, Pandya-Lorch R and Rosegrant M (1999). *World Food Prospects: Critical Issues for the Early 21st* Century. IFPRI, Washington, DC.

Popkin B (1998). The nutrition transition and its health implications in lower-income countries. *Public Health Nutrition* 1(1): 521.

Pretty J (1995). *Regenerating Agriculture.* Earthscan, London.

Pretty J (1998). *The Living Land.* Earthscan, London.

Pretty J (2001). The rapid emergence of genetic modification in world agriculture: contested risks and benefits. *Environmental Conservation* 28(3): 248–262.

Pretty J (2002). *Agri-Culture: Reconnecting People, Land and Nature.* Earthscan, London.

Pretty J (2003). Social capital and the collective management of resources. *Science* 302: 1912–1915.

Pretty J (ed) (2005). *The Pesticide Detox. Towards a More Sustainable Agriculture.* Earthscan, London.

Pretty J N and Koohafkan P (2002). Land and Agriculture: From UNCED Rio to WSSD Johannesburg. FAO, Rome.

Pretty J and Waibel H (2005). Paying the price: the full cost of pesticides. In Pretty J (ed) *The Pesticide Detox.* Earthscan, London.

Pretty J and Ward H (2001). Social capital and the environment. *World Development* 29(2): 209–227.

Pretty J N, Brett C, Gee D et al. (2000). An assessment of the total external costs of UK agriculture. *Agricultural Systems* 65(2): 113–136.

Pretty J, Brett C, Gee D et al. (2001). Policy challenges and priorities for internalising the externalities of agriculture. *Journal of Environvironmental Planning and Management* 44(2): 263–283.

Pretty J N, Mason C F, Nedwell D B et al. (2003a). Environmental costs of freshwater eutrophication in England and Wales. *Environmental Science and Technology* 37(2): 201–208.

Pretty J, Morison J I L and Hine R E (2003b). Reducing food poverty by increasing agricultural sustainability in developing countries. *Agriculture, Ecosystems and the Environment* 95(1): 217–234.

Pretty J, Noble A D, Bossio D, Dixon J, Hine R E, Penning de Vries F W T and Morison J I L (2006). Resource-conserving agriculture increases yields in developing countries. *Environmental Science and Technology* 40(4): 1114–1119.

Pretty J, Peacock J, Sellens M et al. (2005). The mental and physical health outcomes of green exercise. *International Journal of Environmental Health Research* 15(5): 319–337.

Putnam R (1995). Bowling alone: America's declining social capital. *Journal of Democracy* 6(1): 65–78.

Reicosky D C, Dugas W A and Torbert H A (1997). Tillage-induced soil carbon dioxide loss from different cropping systems. *Soil and Tillage Research* 41: 105–118.

Reij C (1996). Evolution et impacts des techiques de conservation des eaux et des sols. Centre for Development Cooperation Services, Vrije Univeriseit, Amsterdam.

Sanchez P A (2000). Linking climate change research with food security and poverty reduction in the tropics. *Agriculture, Ecosystems and the Environment* 82: 371–383.

Scoones I (1998). *Sustainable Rural Livelihoods: A Frame work for Analysis,* IDS Discussion Paper, 72, University of Sussex.

Smil V (2000). *Feeding the World: A Challenge for the 21st Century.* MIT Press, Cambridge, MA.

Stephens P A, Pretty J N and Sutherland W J (2003). Agriculture, transport policy and landscape heterogeneity. *Trends in Ecology and Evolution* 18(1): 555–556.

Tegtmeier E M and Duffy M D (2004). External costs of agricultural production in the United States. *International Journal of Agricultural sustainability* 2: 155–175.

Tilman D, Cassman K G, Matson P A et al. (2002). Agricultural sustainability and intensive production practices. *Nature* 418: 671–677.

Trewevas A (2002). Malthus foiled again and again. *Nature* 418: 668–670.

Uphoff N (1998). Understanding social capital: learning from the analysis and experience of participation. In Dasgupta P and Serageldin L (eds) *Social capital: A Multiperspective Approach.* World Bank, Washington, DC.

Upoff N (ed) (2002). *Agroecological Innovations.* Earthscan, London.

Waibel H, Fleischer G and Becker H (1999). The economic benefits of pesticides: A case study from Germany. *Agrarwirtschaft* 48H(6): S219–S230.

Wilson C and Tisdell C (2001). Why farmers continue to use pesticides despite environmental, health and sustainability costs. *Ecological Economics* 39: 449–462.

World Commission on Environment and Development (1987). *Our Common Future.* Oxford University Press, Oxford.

Worster D (1993). *The Wealth of Nature: Environmental History and the Ecological Imagination.* Oxford University Press, New York.

33

Animals and Society

Henry Buller and Carol Morris

INTRODUCTION

... the relationships between animals are bound up with the relationship between man and animal, man and women, man and child, man and the elements, man and the physical and microphysical universe (Deleuze and Guatarri, 1988, p. 259).

Animals and humans are rarely, if ever, wholly apart, even though the spaces they occupy are increasingly differentiated. They share common origins, common biologies and a long history of interaction which, in many cases, engenders forms of interdependence (as well as exploitation). They eat each other and compete for natural resources, creating related geographies of spatial practice (Serpell, 1995, 1996; Quammen, 2005). Moreover, human spaces (including human bodies) are also occupied, defined and constructed, to varying degrees, by non-human animals as is human society, and human social organisation fundamentally contingent upon human–animal relations. The early domestication of herbivores accompanied the formation of sedentary and hierarchicised human society, with herd size becoming an early sign of social power. Domestication has been, above all, a form of commodification and hence the groundwork for the later capitalist 'production' of nature. As much if not more than a source of food, animals are a source of wealth. Paradoxically, animals thus became one of the means by which power in such emergent human societies moved away from its 'animal' basis (physical strength and attributes) to more formal ('humanised') social systems and structures.

Animals have always provided the metaphorical reference points for forms of human differentiation (moral, physical, and so on) as well as servitude. Thomas, for example, asserts that the tendency to see in each species some socially relevant human quality was very ancient for men had always looked to animals to provide categories with which to describe themselves (Thomas, 1983, p. 64). Early on, human values were ascribed to wild and domesticated animals alike, giving rise to a durable and symbolic classification that went well beyond simple utilitarian criteria. Wild beasts and wolves, in particular, came to embody a variety of evils, both social and political including sexual predation, vagrancy, marginality and freedom.

Animals enter political philosophy not only as the concrete dogs, cats, pigs, horses and cattle with whom we share our lives, and the equally concrete wolves, lions and bears that once we feared, but also as symbols of such forms of life as we free, adult, male [sic] members of society may seek to exclude, idealise or tame. It is through our fleshy kinship with the animals that we may grow like wolves, faithless, treacherous and hurtful or like lions, or foxes (from the Discourses of Epitectus, quoted in Clark, 1999, p. 111).

Thomas, again, remarks that 'it is an enduring tendency of human thought to project upon the natural world (and particularly, the animal kingdom) categories and values derived from human society and then to serve them back as a critique or reinforcement of human order' (1983, p. 61). Extolling the virtues of both animalian and human characteristics, Machiavelli advises:

It is necessary for a prince to know how to make use of both natures, and that one without the other is not durable. A prince, therefore, being compelled knowingly to adopt the beast, ought to choose the fox and the lion; because the lion cannot defend himself against snares and the fox cannot defend himself against wolves. Therefore, it is necessary to be a fox to discover the snares and a lion to terrify the wolves. Those who rely simply on the lion do not understand what they are about (1998, Ch. 18).

In a more material vein, De Planhol, in his magnificent treatise on the historical geography of large fauna (2004), distinguishes a whole series of human/animal interactions. These include the destructive (eradications and extinctions), the shared (those animals who have 'chosen' to approach humanity – examples of the quasi-symbiotic relationship of wildlife to certain human practices and spaces from field mice in wheat fields to pigeons in cities), the domesticated (from farm animals to pets), the protected (zoos, restituted species, and so on) and the new 'anthro-pogenic fauna' who owe their existence (or at least their proliferation) to indirect human actions, whether through the elimination of their predators or their competitors, through the impact of human use of other animals or finally as a result of direct 'predation' upon an expanding human population. De Planhol writes:

> If humanity has destroyed much, it has also, with-out doubt also created. Humanity has favoured inumerable species over others. It has substituted new creatures for those that were there before. The supreme predator, humankind, has also been, for many species, as a simple result of its presence, a source of attraction, a 'pole of reference' sought out by some as much as fled from by others (De Planhol, 2004, p. 209).

Though it is common to conceive civilisation and urbanisation as lengthy processes of anthro-pocentric dissassociation from nature and the wild (Tester, 1991), the built environment has in fact acted as a stimulus for many forms of natural (and 'unnatural') colonisation, from rats to cats, from bats to blatts (Blanc, 1999). Indeed, current interest in the new urban ecologies has drawn attention to the contradictions inherent in perceptions of the city as a de-wilded space (Whatmore and Hinchliffe, 2003). The countryside too stands as an archetypically constructed yet shared space in which nature, and wild things, invest a key part (Jones, 2003; Buller, 2004). Although human society has been highly selective both in retaining certain animals as composing its domesticated rural aesthetic and functionality, and in excluding those that threaten its stability, the countryside, maintains Murdoch (2003), after Whatmore (1999), is more than just human.

Animals are, therefore, and always have been, co-constituents (though not necessarily equal co-constituents), both materially and metaphorically, of the societies, practices and spaces that we label as human. Our interest, how-ever, lies in the visibility and changing nature of that co-constitutionism. We may be 'cultures and formations of humans and non-humans' (Franklin, 2003, p. 255) but recognition of this has been long in coming. It is accepted that many of humanity's social and spatial practices are, to a greater or lesser extent, contingent upon our relationships to non-human animals. It is perhaps surprising then that human intellectual endeavour, and the social and natural sciences in particular, has been, for the bulk of its history, rooted in an anthropocentrism and humanism that has largely excluded the non-human from consideration in close association with the human. The study of non-human animals has been confined (and this only relatively recently in the history of science) to specific disciplines with their equally specific natural sci-ence epistemologies. As Lestel (2004, p. 16) writes: 'there is no place in universities where one can study humans and animals together'. In a similar vein, Tovey (2003, p. 196) argues that 'ani-mals remain largely invisible in social science texts'. The key influence here is that of Hiedegger who forwarded a classification that distinguished stones as having no world, plants and animals as being poor in world – the mere recipients of exter-nal influences – and humans who have a world and can therefore make worlds [see also Berger and Luckman's seminal 1967 text which, for Tovey (2003), using this distinction promoted a 'radical discontinuity' between human and non-human animals in contemporary sociology].

Inspired by Lestel's (2004) call for the develop-ment of an authentic 'ethno-ethology' which would permit the study of human and non-human animals together, by Benton's suggested 're-alignment of the human social sciences with the life-sciences' (1991, p. 25) and by Tovey's argument that sociology should rethink society to include animals (2003), we explore in this chapter three different sets of contemporary approaches to understanding and accounting for the relations between human and non-human animal society each based on a distinct ontological tradition. The first draws on the assertion that humans are animals and that much of human social organisation, behaviour and relations to non-human animals can be explained in terms of this human animality. The second taking as its starting point contempo-rary evocations of the 'otherness' of animals and the need for human–non-human animal relations to be based on a less anthropocentric conceptuali-sation of the non-human world. The third looks to post-modern literatures to explore the contem-porary material relationality of human and non-human animals as operated through networks and less essentialist understandings of both 'human' and 'animal'. Before undertaking this, however, we need to appreciate the sense of rupture that has occurred in recent times as new ethical and ontological challenges have emerged to counter the traditional view of the status of non-human animals and their role in human society (Singer, 2004). The following section thereby traces the salient elements of the modernist legacy of society/animal relations with a view to providing a backdrop to our principal endeavour, the exploration of new understandings of the relationship between (human) society and (non-human) animals that emerges from recent and contemporary challenges to modernist ontology.

CHALLENGING THE MODERNIST LEGACY

We don't think of hogs as animals, Bob, not in the same way as cats and dogs and deer and squirrels. We say 'pork units'. What they are Bob, is 'pork units' – a crop like corn or beans (Ribeye Clark in Annie Proulx, 2002, p. 302).

The dichotomous and dialectical separation of humans from non-human animals, and indeed from Nature, justified and sanctioned through a wide variety of dominant cultural, religious, scientific, ideological and philosophical positionings, arguably stands as one of the singularly most fundamental ontological divides of, particularly Western and Christian, humanist thought. From Aristotle to Kant, from Descartes to Marx, animals have been, in one shape or form, 'lesser' beings, 'objects', 'outsiders' and 'others'. Animal interests are considered subservient to human needs, and their independent subjective existence has long been largely ignored or explicitly denied. Critically, animal non-humanness ultimately serves to define humans and 'human nature'. These assumptions have been among the many foundation stones of civilisation and modernity. Ideas of Nature as divine and romantic – as Eden lost – shifted during the Enlightenment to a disenchanted nature as material, orderly, scientific and objective – as a humanised Eden to be stewarded, yet also improved to yield bigger and more plentiful crops and livestock through human endeavour, science and technology, whether it be in the fields of contemporary farming or engineering. At one and the same time, this represented an ontological, sociological and spiritual shift that, first, sought to embed truth and hence emancipation in objective scientific observation of natural events and material exploitation of nature as resource, second, permitted the interpretation of nature in cultural and hence sociological terms,

and third, allowed a more rational interpretation of what was still largely seen as an essentially benevolent God's will.

Animals were always a key component of modernism's reconfiguration of Nature. Descartes' 'animal machines' are an overused but nonetheless powerful evocation of the place farm animals in particular came to occupy in the new scientisation and materialisation of the non-human. For Count Buffon, whose *Histoire Naturelle*, Glacken (1967, p. 139) describes as the most impressive natural history written during the 18th century, modern civilisation was ultimately contingent upon animal domestication. But, while domestication was, after all, a form of human–animal interaction dating back to the Neolithic, modernity brought to it a particular intensity, techno-scientific ingenuity and moral and philosophical justification that de-deified (and dehumanised) animality. It also sought to emancipate and progress the human 'species being' through production and through supremacy over non-humans and thereby transformed animals into the quintessential 'others' of the global biosphere. In doing so, modernism provided the stage for the more explicitly exploitative regimes of accumulation (or indeed redistribution) under which animals were quickly divided, in the most utilitarian manner, into the useful (and edible) and the useless (or threatening). The former were subjected to systems of increasingly aggressive husbandry methods and selectivity (Harrison, 1964), while the latter were widely eradicated from the newly colonised lands (Dunlap, 1999) and confined either to more remoter and less productive spaces or maintained, as observable others, in specific reserves and zoos.

The modernist legacy is, therefore, a strong one and continues to pervade humanity's relations to the non-human through the maintenance of ontological distinctiveness, through ethical postures and through the supremacy of humanism and anthropocentrism. No matter how 'clever' a pet dog or how 'articulate' a trained chimpanzee, non-human animals remain quintessentially and characteristically distinct from modern human society for the latter has constructed itself on this being so. Animals are 'owned' like things. Animals are bought and sold as commodities. Animals are considered less important for they are not humans. Benton (1993) has argued, mirroring Foucault (1965), that 'the situation of animals with respect to human social practices is analogous to that of socially or economically disadvantaged or dependent human groups' (p. 161), and others have equated domesticated animals with the condition of human slavery (notably Spiegel, 1997) or sought to define them simply in Aristotelian terms as 'lesser humans'. Nevertheless, such analogous relationships mask the essence of what has been an unequivocally dominant sense of 'difference' driven, as pointed out by Dupré (2002), by pernicious Cartesianism in society's assumptions about animal being (or lack of it from Heidegger's point of view), self-hood and consciousness.

However, the ontological certainties upon which modernism was founded now appear to be a lot less secure and with them the longstanding separation of humans and non-humans and, at broader level, 'the modernist dichotomy of nature/society' (Goodman, 1999, p. 33). In his introduction to *Social Nature*, Castree (2004, p. 5), considers two things as axiomatic, first 'that nature has never been simply "natural"… Rather, it is intrinsically social'; second, that 'the all-too-common habit of talking of nature "in itself"', as a domain which is by definition nonsocial and unchanging, can lead not only to confusion but also the perpetuation of power and inequality in the wider world'. In philosophy, particularly environmental and ethical philosophy, in sociology and in geography, to name but three of a far wider disciplinary engagement, the relationship of non-human animals to human society is currently being interrogated afresh. A number of drivers can be identified: first, by discourses of environmentalism, ethics, feminism, welfarism and animal liberationism; second, by new advances in biology and genetic science; third, by post-modernism, social constructivism and a gamut of new cross-over contrivances that allow the relationality between humans and non-human animals to be mapped and centred as subject; fourth, by cultural studies and their focus on what Lippit (2000) calls 'animetaphor', and finally, by technoscience itself (Haraway, 1997). Human society and animals are increasingly intertwined into new configurations of similitude, difference, relational materiality, shared space, ethical responsibilities and ultimately, to use Latour's (2001) phrase, a 'common collective'. For the sake of simplicity, we have ordered these recent and less dualistic approaches to human/animal relations into three groups: humans as animals, animals as others and human/animal. It is to the first of these new ontologies that we now turn.

HUMANS AS ANIMALS

Culture is ultimately a biological product (Wilson, 1996, p. 107).

Taking the issue of 'difference' and human exceptionalism away from its traditional philosophical preserve by building notably upon Darwin and the natural sciences, the first of these new approaches to animal/society relations challenges the conventional separation of human and (non-human) animals by reinforcing both the animalian attributes of human society and the 'human-like' attributes of animals (which, though fascinating, is beyond the scope of this chapter). As an essentially scientific and empirically founded debate, this has been championed by Edward Wilson and Richard Dawkins, amongst others, but is also extended by the experiences and accounts of people who have spent much of their lives in the company of animals (from Gerald Durrell to Jane Goodall). It has, in doing so, provoked vigorous controversy on both sides and revealed the limitations of longstanding epistemological divisions. Noske (1997) observes that 'satisfied as they have been with the gulf between human subjects and animal objects, most social scientists have done nothing to fill the gap between humans and animals […] never bothering to ask what human–animal continuity means in their own sphere' (p. 92). Yet, the human sciences are not alone in their limitation. Hence, Noske also notes that 'the animal sciences are simply not equipped to deal with those characteristics in animals which, according to the social sciences, make humans human' (1997, p. 90).

At one level, the most obvious challenge to the dualist orthodoxy separating human society from animals (and, consequently, social and natural science) has come from the natural sciences; from natural history, evolutionary theory and Darwinism. Yet, while all three fundamentally contested the hitherto hegemonic Western theological belief in the defining original difference between animals and humans, they nevertheless stand as accomplishments of modernism and, as such, have accompanied the intensification of anthropocentrism in society's attitude towards and use of animals. Knowing that human animals shared an evolutionary voyage with monkeys and fish, or that human beings are, after all, an animal species, created no new ethical relations. On the contrary, it reinforced humanity's sense of its own unique progress and post-animalisation as evidenced in the works of philosophers and commentators as diverse as Hobbes, Marx, Spinoza and Heidegger. Yes we are animals, but are also so much more than animals.

Recent years have seen a significant return to the discontinuities and continuities of human–animal linkages (Noske, 1997; Gray, 2002). This is inspired by the debates and, in some cases, controversies associated with sociobiology and neo-Darwinism and by a new sense of interrogation not of human–animal differences but of their relativity, something we come back to in the final section of this chapter. 'The most decisive political conflict, which governs all other conflict in our culture, is that between the animality and the humanity of man' writes Agamben (2002, p. 56).

Without wishing to enter the 'what makes us human' debate, the nature of our difference to animals is, in some sense, at one and the same time both clearer and less clear today than it has ever been. For Serres (2000, p. 8) 'we became the men [*sic*] we are by learning – how, we may never know – that we are going to die', a knowledge that separates humanity utterly from the animals that 'have neither death nor objects' (p. 8). Yet, countless modern examples of the social organisation of animal groups lay contest to Hobbes' belief that society emerged only with mankind.[1] Meanwhile, Aristotle's oft-cited insistence upon language as one of the critical distinguishing features of humans is today challenged by our contemporary knowledges of complex forms of animalian communication.

Advances made in genetic science and genetic mapping have drawn attention to the high degrees of similitude between the human genome and that of many, particularly, mammalian species (98% in the case of humans and chimpanzees), leading in some quarters to calls for certain primates to be reclassified as *Homo troglodytes* (Wildman *et al.*, 2003). It is, however, sociobiology that has confounded, in the most dramatic way, the longstanding belief not only in 'human nature' as being something entirely separate from animal nature but also in human social science explanation as being fundamentally ontologically distinct from that of the natural sciences. Sociobiology, as advanced by Wilson (1975), in essence sought as its ultimate project to explain human behaviour (both individually and socially) in biological and evolutionary terms. The contribution of culture, society, environment, nurture, and so on, the traditional domains of the social sciences, were thereby reduced or negated to the sidelines of explanation. Even altruistic behaviour, the apparent selflessness of which would appear to oppose the Darwinian assumption that the fittest survive, has been enrolled into sociobiological explanation (Dawkins, 1979) as an act of sibling gene preservation. Sociobiology partly dehumanises human difference by lumping being human and being animal together in a manner that, for some critics, ends up by concluding that 'man [*sic*] is a machine' (Blackwell, 1980), an unequivocal return to modernist science. To quote Gray (2002, p. 188): 'Descartes described animals as machines. The great cogitator would have been nearer the truth if he had described himself as a machine'.

Sociobiology has undoubtedly taken Darwinism forward. Even its most vitriolic opponents accept the importance of genetic evolution and indeed many of the basic assumptions (though not necessarily the proof) of sociobiology. As in so many things, the battle lines, drawn up in the heat of an increasingly politicised debate

(Segerstråle, 2000), have since blurred. Gould, one of the major critics of Wilson, writes:

> For Linnaeus, Homo Sapiens was both special and not special. Unfortunately … special and not special have come to mean nonbiological and biological, or nature and nurture. These later polarisations are nonsensical. Humans and animals and everything we do lies within out biological potential (Gould, 1997, p. 251, quoted in Dupré, 2001, pp. 38–39).

In this way, evolutionary psychology, the contemporary re-invention and re-scientisation of sociobiology has become the partial possibilism to sociobiology's partial determinism. Hence, for Gray (2002, p. 79) 'the idea that we can rid ourselves of animal illusion is the greatest illusion of all. […] We are far more than the traces that other humans have left in us'.

Where does this get us in terms of our new understanding of human–animal relations? Not very far ultimately for sociobiology offers few genuine insights, for our purposes here, into interspecies relationality. Although Wilson and others believed that sociobiology would eventually replace sociology and anthropology in accounting for human social behaviour and organisation, this has not happened. On the contrary, we might argue that just as Darwinism played into humanity's efforts to demonstrate its superiority over nature, so sociobiology has ironically reinforced emphasis on humanity's cultural exceptionalism giving the 'cultures' of humanity a renewed focus in the social sciences. More significantly for the debates presented here, sociobiology (and its critics) offer few arguments for a more emancipatory stance with respect to human–animal relations. Wilson argues for a 'robust and richly textured anthropocentric ethic […] one based on the hereditary needs of our own species' (1996, p. 176). Once again, Noske sums up the position effectively: 'Sociobiology may be biaised, but so is anti-sociobiology. Together with the sociobiologists, many anti-sociobiologists have an ideological stake in the status quo: the object status of animals' (Noske, 1997, p. 101).

ANIMALS AS OTHERS

> When a cow is just a cow, McDonalds becomes possible (Steeves, 1999, p. 2).

The most well-explored intellectual terrain in recent years for society/animal relations is probably that of animal welfare, animal rights and animal ethics. Although proponents of the second and third may dismiss the former as a moral contrivance that allows concerned consumers to have their lamb and eat it, all three represent what Singer describes as a 'new consciousness' (1990, p. viii), one largely prompted, it should be acknowledged, by his 1975 book *Animal Liberation*. Franklin (1999, p. 175) sees animal rights as the 'apotheosis of change' concerning 20th century human–animal relations. Animal rights have become 'a major cultural phenomenon', one that has 'challenged long-standing, traditional views about non-human animals' (DeGrazia, 2002, p. 2). Significantly, the animal rights issue has penetrated both sides of the onto-logical frontier from philosophers and ethicists to ethologists and biologists (De Grazia, 1999). Thus, the demonstration of animal sentience through scientific ethology and behavioural studies (Dawkins, 1998) has gone hand in hand with the emerging social science engagement in animal well-being, the moral status of non-human animals and their equality in moral rights as sentient beings. Common to both is a de-essentialising of animals as 'objects', defined wholly in terms of their anthropocentric relationship to, and use by, humans, and an increasing sense, and acceptance, of their value status as subjects and individuals in their own right.

To review the literature on animal rights and ethics would be a misplaced endeavour in the con-text of the current chapter. The attribution, and the defence, of animal rights has become a global social movement as well as a wide-ranging moral debate. Consequently, the literature is vast, extending from the popular to the prosaic, the metaphysical to the pragmatic. It encompasses not only the traditional beasts of burden, the farm ani-mals, but also pets and companion species, wild animals, whether in captivity or not, laboratory animals, pests and fish. The movement is unequiv-ocal in its abhorrence of the way humanity treats animals in general, and certain animals, such as livestock, in particular, and generally united in its espousal of some sort of moral engagement *vis á vis* non-human animals. Nevertheless, significant dif-ferences exist within the movement and literature regarding the extent to which the rights argument has repositioned the relationship between humans and non-human animals. This section focuses on the debates within the social sciences over the status of animals as 'others' and the degree to which that 'otherhood' influences our own ethical positionality with respect to them.

An important fault line separates the acknowl-edgement and recognition of animals as recognis-able 'others' from the position that accepts no real distinction between non-human and human ani-mals in terms of access to moral consideration. Such a distinction, as Singer (1975) argues, might be termed 'speciesism'; belonging to one species, in the same way as belonging to one race or gender, does not confer particular rights over and above all other species. Tester (1991), in his essentially historical account of the animal rights issue, labels these two positions, the 'demand for difference' and the 'demand for similitude'. The former emerges from late 18th century bour-geoisie as an 'attempt to enhance the privilege of being human through the projection of social extirpation of animality' (1991, p. 88) that acknowledged and celebrated human (and in par-ticular bourgeoisie human) superiority. The latter represents a more emancipatory and egalitarian romantic bourgeois morality based on a less arro-gant view of humanity's place within nature.

Tester sees animal rights as emerging from the reconciliation and alliance of these two very different sets of socio-cultural attitudes in the form of an 'epistemological break' (1991, p. 149). He does not set out to establish the link between this late 19th century 'break' and the emergence of a genuine animal rights movement in the latter half of the 20th century, an omission that Franklin (1999) specifically seeks to address. Yet is clear that the difference/similitude distinction continues to have a major relevance today with the senti-mentalism of the 'demand for difference' still marking, according to Franklin (1999, p. 32), 'the most central and perhaps the most common attitude' of humans to animals. 'Most human beings' after all, wrote Singer in 1975 (p. 9), 'are speciesists'. In a similar vein, writes Gray (2002, p. 3), 'most people today', think that they belong to a species that can master its destiny'.

Arguably animal rights does not resolve the 'difference'/'similitude' distinction and many would claim that it cannot. Mary Midgley (1984), in an oft-quoted argument, makes the point that all species are far from being equal in their capacity to suffer pain and discomfort, in their claims to sentiency, in their inherent value or, as a result, in their identification as legitimate recipients of rights and due moral consideration. Those that fulfill these criteria most readily are, unsurprisingly, similar to us, sentient, warm blooded, terrestrial and mammalian. They are, dare we say, almost human. Singer writes in defence of the wolf's morality, that this has been shown to be 'a highly social animal, a faithful and affectionate spouse – not just for a season, but for life – a devoted parent, and a loyal member of the pack' (1975, p. 223). The starting point for animal rights appears there-fore fundamentally anthropomorphic and anthro-pocentric, suggesting a 'sliding scale' of independent moral value (DeGrazia, 2002). Those animals that are the most 'similar' to us in sentience, cognition, social organisation and biol-ogy are accorded the higher moral status. In this way, animal ethics becomes a further extension of human ethical engagement that has, throughout human history, spread out on the basis of growing acceptance of similarity. For Tester, this is significant:

> Animal rights only wants to talk of the animals with which people are most familiar and it only talks of animals to the extent that we do something to them. It is not a morality founded on the reality of animals, it is a morality about what it is to be an individual human who lives a social life. The crucial facet of animal rights is precisely that it states claims which we are asked to do something about: animal rights is a social problem (1991, p. 16).

At one level, there is something reassuringly humanist and modernist in this persistent human centrism and in the reaffirmation of the importance of human and non-human similarities. Yet, at another level, it reveals its own constructivist limitations and what Wood (2004, p. 143) refers to as 'species tribalism'. By grounding itself in anthropocentrism, it never gets away from the idea that the value of animals is only determined and scaled in human terms (viz. Singer's faithful, affectionate, devoted and loyal wolves). Even where significant rights are accorded to our more proximal animalian confrères, these are contingent upon our own centrality. Of course, to some degree this is inevitable. As Benton (1993, p. 162) puts it 'ascription of rights to (non-human) animals is always other-ascription'. Dogs do not join the RSPCA and no lion is going to stop killing gazelles out of any concern for the latters' rights.

Yet, there is a growing tension here between this pragmatic constructivist, anthropomorphic and sentimentalist approach to rights and the more post-modern conviction that human–non-human relations need to be reconstructed on the basis of a morality that *is* (unlike Tester's assertion above) founded upon the reality of animals. Such an approach recognises animalian otherness and difference while rejecting the sentimentality (Baker, 2000) and the subject/object dualism of modernism. In short, it calls for an alternative, 'other' animal ethics. For Aaltola (2002) the emergence of such 'other animal ethics' (p. 194), which 'respect difference and abandon exclusions based on similarity' (p. 207), represents, for many (e.g. Birch, 1993; Steeves, 1999; Baker, 2000), the way forward for animal ethics in general. In the remainder of this section, we explore some of the trajectories of these other ethical positions.

We begin, though, with what is a profound difficulty. For natural scientists, the issue of animal subjectivity and its consequences for assessing suffering has long been problematic. Wemelsfelder (1999, p. 38), for example, writes: 'current standards for objective measurement *a priori* deprive an animal of its status as a subject, as a being with

its own point of view'. For the social sciences, the issue is equally fraught. Whilst we acknowledge modernism's blind spot with regard to animal 'otherness', animal histories and the independent reality of animalian individuals, there nonetheless remains not only the fundamental ontological challenge for contemporary engagements with animal/society relations to establish and allow for non-human animalian reality but also the epistemological challenge of understanding it and acting upon it. Nagel's famous essay (1974) *What is it like to be a bat* offers a tantalising glimpse of the possibilities of 'bat-hood' before slamming the door in our face by arguing that cross-species comprehension may never be possible because of the limits of our own nature.[2] Animal minds remain closed to us (Haynes, 2001). How might we construct non-anthropocentric values and attitudes to underpin our ethical relations with animal others? A 'feeling of being bound' (Caputo, 1993, quoted in Aaltola, 2002, p. 199) might, along with 'awe', 'wonder' and 'letting be' (Aaltola, 2002, p. 199), be suitably deconstructivist as attitudes towards non-human animals, but can they provide a viable alternative to more traditional moral concepts and actions ? Aaltola concludes that '"other" animal ethics is in trouble, for it cannot explain what exactly our attitudes to other beings are based on, what is their scope and what they tell us to do' (2002, p. 202). Nevertheless, an ethics of otherhood should not be rejected. The task of 'other animal ethics' is to circumscribe a means of 'knowing' animals that does not depend (entirely) on our pre-set human conceptions, classifications and positionality. Derrida's 'abyssal rupture' (Derrida, 2004) between men and animals depends, argues Wood (2004, p. 137), 'on our mode of mutual engagement'.

There is considerable evidence that new forms of 'knowing' are indeed emerging. In his aforementioned essay, Nagel (1974) offered two possible pathways across interspecies boundaries; these are imagination and 'objective phenomenology'. The first we might lightly skip over, to return to later.[3] The second suggests greater potential though its precise sense is elusive. Throwing aside 'supernatural exercises in identity-shifting', Acampora (1999, p. 118) interprets Nagel's phrase as 'what it means to be with other individuals of different yet related species', proposing that 'existential residency might be a primordial world-relation ubiquitous throughout the biotic (or at least animal) realm' (p. 119).

Recent work on the interaction between farmers and the animals in their charge emphasises new and different ways of interspecies understanding that emanate through shared practices, coexistence and co-enrollment (Holloway, 2001, 2003; Porcher, 2002; Masson, 2003; Wilkie, 2005).

Holloway (2001), for example, in his study of pets and livestock on hobby farms, suggests that 'the expression of alternative subjectivities by animals is evident in, for example, their abilities to confound the expectations and knowledges of the humans involved, and their participation in communication with humans' (p. 305). In a different light, Haraway's *Companion Species Manifesto* (2003) seeks to 'tell stories about relating in significant otherness, through which the partners come to be who we are in flesh and sign' (p. 25). Her objective here is to develop an 'ethics and politics committed to the flourishing of significant otherness' (p. 3). For Haraway, co-constitution and co-evolution are hence the rule rather than the exception. This is something both Jones (2000) and Holloway (2001) strongly endorse with reference to rural spaces where physical and cultural animal presences are formative of these 'naturecultures' (Haraway, 2003), creating a series of situated moral understandings and ethical relations. Finally, Whatmore's (2002) inspiring account of wildlife networks combines both imagination and the detailed mapping of the heterogeneous networks that co-configure 'wild' animals and human animals in an effort to 'glimpse the animal experience of becoming *leopardus* and *Caiman latirostris*' (p. 32). Sharing space, place and co-constituted nature-cultures with animal others is recognition of non-human forms of subjectivity and agency without having to become bats.

Can we go any further? In a radical departure from the static sharing or binary of human/animal, Deleuze and Guattari's (1988) notion of 'becoming' is a mutual process (and resolutely not a metaphor) of alliances and assemblages that not only transcends subject/object but also hierarchies: 'the rat and the man are in no way the same thing, but Being expresses them both in a single meaning in a language that is no longer that of words, in a matter that is no longer that of forms, in an affectability that is no longer that of subjects' (p. 285). In this way, 'rat' and 'man', along with the other classifications are displaced (or deterritorialised) by the ongoing, continual and real process of becoming. What emerges is, for Urpeth (2004), a non-moral, de-anthropomorphised vision of nature; a set of 'interkingdoms' (Deleuze and Guattari, 1988); the relation of human societies and animal packs through becoming is one of alliance (Baker, 2000). How such alliances might yield tangible 'rights' is a complex problem in itself, though Birke and Parisi (1999) offer the following:

> To argue for a politics of affiliation is to see politics as networks and connections that are not rooted in essential kinds or boundaries. [...] With regard to non-human animals [...], we must learn to experience connections and to think in terms of becoming. [...] This is not just about being able to imagine their suffering. It is also about an imagining that can become their knowing (p. 69).

Perhaps, though, we think too much about society/animal relations and thereby fall prey to an over-emphasising of, and an overdependence upon, the cognitive and the intelligible. As a result, we are trapped by the analytical conventions and behavioural intentions that cognition implies. This is certainly the message of non-representational theory (Thrift, 1996, 2000) which privileges 'everyday moments of encounter' so as to 'build an ethics of generosity by stimulating affective energy and by refining the perceptual toolkits necessary to build moral stances' (Thrift, 2004, p. 93). 'What we have here, then, is an Aristotelian ethical coping based upon pre-theoretical everyday convictions, indeed, very often upon mental processes that cannot be brought to consciousness at all' (2004, p. 93). Instead of theory building and acknowledged pre-positionalities, human/animal encounters are affective, performed, improvised, lived practices through relations and associations. It is fitting then, to end this section with Hélène Cixous (1993). Commenting on Cixous's tale *Love of the Wolf* (1998), Baker (2000, p. 188) writes: 'Cixous's story is exemplary, not as a model of how to think about animals but as an instance of not allowing rhetoric or theory to disguise the contradictory and comprised manner in which people live with animals. They do live inexpertly with animals'.

HUMAN/ANIMAL HYBRIDS

> If belief in the stable separation of subjects and objects ... was one of the defining stigmata of modernity; the implosion of subjects and objects in the entities populating the world at the end of the Second Millennium – and the broad recognition of this implosion in both technical and popular cultures – are stigmata of another historical configuration (Haraway, 1997, p. 42).

In his treatise on the future of human nature, Habermas makes the point that the categories of what has *'come to be by nature'* and what is *manufactured*, a fundamental distinction that is constitutive of our self-understanding as a species, are being 'de-differentiated' (2003, p. 46). Biotechnogical innovation is crossing the frontiers between animal and society, blurring the distinction between nature and culture and creating new or hybrid forms. In this third section, we explore an understanding of animals and humans as relational, hybrid or co-constructed *achievements*.

In this view, the familiar notion of 'animal' as recognisably distinct and different in *essence* from 'human' is troubled and problematised as animals are re-envisioned and imagined as being adapted or 'made' through the labour of humans working with, and through, all sorts of non-human beings and objects. Nothing comes to be *by nature* alone, neither humanity nor animality.

The current vexed debates surrounding the genetic engineering of animals illustrate this position very well. For Haraway (1997, p. 100), genetically engineered 'mice and humans in technoscience share too many genes, too many work sites, too much history, too much of the future not to be locked in familial embrace'. Of course, while genetically engineered 'Oncomouse'[4] and its transgenic siblings from elsewhere in the animal kingdom might provide unequivocal evidence of animal/human hybrids, they could also be dismissed as 'monstrous' cases; the exceptional products of the scientific laboratory. However, more familiar creatures also bear the marks of their co-construction, relationality and ultimately hybridity. We only have to think about companion animals, not to mention the typical range of farm animals – pigs, cows, chickens and sheep – to know that humans have long been involved in their domestication and breeding, even before the advent of genetic, and now genomic, knowledge and techniques. Even 'wild animals' no longer represent pure, unadulterated nature and animality, seemingly untouched by human hand. Though they may occupy 'places apart', they nonetheless

> find their place in the world less than secure. The radio collars and tags which adorn the remotest parts of the animal kingdom, no less than their daily exhibition in the wildlife documentaries that occupy TV screens in millions of homes around the world disturb the geometry of distance and proximity (Whatmore, 2000, p. 270).

The provenance of these ideas lies in contemporary science and technology studies (Latour, 1993, 1999) which recognised that science was a thoroughly heterogeneous (or hybrid) cultural practice – involving promiscuous mixtures of 'organic beings, technological devices, discursive codes and people' (Whatmore, 1999). It became meaningless therefore to think in advance about commonly held categories such as nature (animal) or society. Instead, these categories were reconceptualised as the end-point or outcome of scientific practice, and implicated the working together of all sorts of things, both human and non-human. Put another way, this new way of thinking entails the rejection, or, as Haraway describes in the opening quotation, the 'implosion', of familiar binary categories such as subject and object,

nature and society, animal and human A further, highly significant implication of this hybrid, heterogeneous and imploded way of seeing is that actors are no longer just 'human'. As Haraway (1997, p. 68, emphasis added) argues 'implosion does not imply that technoscience is "socially constructed", as if the "social" where ontologically real and separate; "implosion" *is* a claim for heterogeneous and continual construction through historically located practice, *where the actors are not all human*'.

Hybridity has thus been proposed as a way through the impasse between 'realist' and 'constructivist' approaches to the 'question of nature' (Whatmore, 1999; Castree and MacMillan, 2001; Murdoch, 2001). It is through this diffusion of ideas about hybridity within the academy that we now find them applied to the study and understanding of animals, and their changing relationship with humans. While at first sight a hybrid, or relational, account of animals might suggest the disappearance of the creatures we have come to understand as 'animals' (and 'humans'), the intellectual project of hybridity has actually been driven in large part by a desire to recognise and accommodate the active and creative presence of non-humans within the lexicon of social analysis.

There are two significant implications to be derived from such arguments. The first concerns the autonomy or sovereignty of the individual, whether they be human or non-human; the second concerns the interconnectivity of subjects. Bataille (1992) writes that 'the animal, like the plant, has no autonomy in relation to the rest of the world' (quoted in Atterton and Callarco, 2004, p. 34). In seeking to breach what she refers to as 'individualistic ethics' (p. 47), Whatmore (1997) dissects the autonomy of the human self to show how intersubjectivity and the entwining of human self with the non-human world become prerequisites for establishing ethical communities. Whilst networks have emerged as the metaphor *de préférence* for the elucidation and exploration of relationality and hybridity, it is important to note that two slightly different versions of the human/animal are evident within what we are labelling the hybrid or relational approach. First, animals appear notably as 'literal' technoscientific hybrids within Donna Haraway's accounts of technoscience. In other words, genetically engineered creatures such as Oncomouse are understood as the outcome of historically located, situated (scientific) practices. Such creatures are at once undeniably and distinctively mousey – furry and squeaky, etc. – but also human and technological, and reveal how 'species being is technically and literally brought into being by transnational, multibillion dollar, interdisciplinary long term projects' (pp. 58–59). Not only do transgenic animals, according to Haraway, cross the line between nature and artifice they also 'greatly increase the density of all other kinds of traffic on the bridge between what counts as nature and culture ... transgenic creatures, which carry genes from "unrelated" organisms, simultaneously fit into well-established taxonomic and evolutionary discourses and also blast widely understood senses of natural limit. What was distant and unrelated becomes intimate' (p. 56).

Actor Network Theory (ANT), as developed by Latour (2005), Callon (1986), Law and Hassard (1999) and others, offers a second means of re-envisaging animals as human/animal, albeit one that is relatively more conceptual and abstract than Haraway's technoscientific hybrids. Here, the emphasis is on animals as achievements of socio–material–natural *network* building.[5] Networks are built, or engineered, through the enrolment of a whole variety of actants, both human and non-human. 'When a phenomenon [such as a particular type of animal] "definitely" exists, this does not mean that it exists forever, or independently of all practice and discipline, but that it has been entrenched in a network' (Latour, 1999, pp. 155–156, quoted in Murdoch, 2001, p. 119). Applying these abstract ideas to rural animals, Jones describes:

> the cows that produce milk or meat, the hens that produce eggs, the boars and sows that produce meat and other pigs for meat, the horse that learns to jump, the dog that learns to control sheep, or hunt, and so on are *in relation* with humans, technology, information and science, productive actants who contribute vital affordances to achievements of one kind or another emanating from networks. These are important conceptualisations which begin to place animals more visibly and precisely in these achievements (Jones, 2003, p. 292).

Similarly, Whatmore (2002, p. 37) argues for a way of understanding wildlife as a relational achievement 'spun between people and animals, plants and soils, documents and devices in heterogeneous social networks which are performed in and through multiple places and fluid ecologies'.

All of the understandings of animals that we have sketched out in this chapter have been the subject of some form of critique and the hybrid or relational account is no exception. While there appears to be some agreement that ANT has made, or at least has the potential to make, animals more visible within social scientific accounts, its symmetrical (making no *a priori* assumptions about the role of humans and non-humans within analysis) and relational (entities have no inherent qualities or characteristics, or internal agency; rather, these emerge out of relations with other

network entities) assumptions are the focus of some contestation. These features of the theory deny animals, according to Jones (2003), their inherent qualities and characteristics which need to be accounted for in relational processes. He goes on to argue that animals do have forms of intentionality, autonomous agency and 'otherness' (all of which, as we have shown above, are increasingly revealed and demonstrated through environmental philosophy, animal science and ethology) and these characteristics inevitably generate certain effects in relation to their enrolment into networks. In particular, the 'otherness' of animals has meant that very often they have had to be 'harshly enrolled' through a variety of restraining techniques. Jones (and others) observes a lack of concern within ANT for the fate of organic beings (as opposed to the range of technological artefacts and devices that are more often than not the focus of ANT analysis) enrolled into networks, 'it being too studiously neutral and as a result bypasses questions of unequal power'.

Echoing the debate on difference which we highlighted above, Jones advocates a redrawing of the distinction between nature/animal and society that the hybrid approach has attempted to do away with, in order that the unequal treatment of animals within networks can be revealed and addressed. Interestingly, Murdoch (2001, 2003), reviewing a number of ANT critiques, reaches the same conclusion, albeit from the point of view of the social rather than the animal. Social entities, so the ANT critics assert, have stable and immutable characteristics (i.e. that cannot be reduced to network relations) and these have implications for nature/animals. Of particular importance is that awareness of humans (towards other humans and non-humans) 'provides a moral and ethical dimension to human action' (Murdoch, 2001, p. 127). For all human embeddedness 'in the complex relations of "nature – culture", [they] continue to carry responsibility for the fate of nonhumans' (2001, p. 113).

Rather than throw the baby out with the bath water, however, Murdoch aligns himself with Hacking (1999) to argue for an awareness of heterogeneous relationships alongside a recognition of human exemptionalism (and to this we could add in Jones' recognition of animal otherness). This 'middle way', so Murdoch (2001) drawing on Whatmore (1997) proposes, is one that describes a

> 'relational ethics' [which] require attention to the heterogeneous composition of human action, the non-humans that lend themselves to this action and the ecosystem in which it unfolds. It also implies a very human sense of responsibility towards both non-humans and ecosystems.

Having taken the bold step of radically refor-mulating understanding of the nature–culture/animal–human relationship, at least some of the proponents of hybridity appear to be reviewing the situation and re-introducing/accommodating aspects of the 'Animal as Other' that we considered in the second part of this chapter.

CONCLUSION

It has not been our intention in this chapter to identify and elucidate the myriad ways in which human society exploits, uses, enjoys, hunts, observes, kills and eats non-human animals. Rather, our starting point is that such interactions are so unavoidable and so universal that they are and have been constitutive of human society from its very origins. What becomes important is not the distinctions and categories that classify and separate (and from which we nonetheless trace instrumental interaction between us and them, subject and object, inside and outside, nature and society, humans and non-humans), but the net-works that intertwine and confuse such binaries. These turn what are often neglected objects into subjective agents or actors (both physical and non-physical) and give rise to 'hybrids' as the achieve-ment of a co-constitutive and reciprocal relationality within a more heterogeneous society/nature. 'It is the network [...] rather than the binary', argues Watts (2005, p. 167), 'that captures modern nature'. We want to end this chapter by drawing attention to the frameworks for two such capturing networks: the first, post-humanism which arguably permits the enjoining and intermixing of humans and non-humans in practice and in thought; the second, 'crossing the divide', which seeks to link together social and natural, con-structed and 'realist' conceptions of the living world.

Singer ends *Animal Liberation* (1975) with a call for 'genuine altruism' in the way humans act towards animals. He is, in effect, asking us to show greater 'humanity' in recognising the 'moral indefensibility' of our treatment of animals. Yet for an increasing number of environmentalist and academic writers, it is humanism and the 'human' itself that is being revealed as incompatible with a more enlightened relationship between society and nature. They argue that in becoming less distinctively human, we need to become less humanist, to move away from the 'arrogance of humanism' (Ehrenfeld, 1978), to enter an era of 'super' or 'post' humanism (Wolfe, 2003). For if humanism is based on the ontological separation of society from nature, upon 'the emancipation of humans from the determinism of nature' (Murphy, 1997, p. 270) and upon human mastery over

nature (Ferry, 1992), and thereby its objectifica-tion, post-humanism is, for Braun (2004, p. 273), 'another name for *non-anthropocentrism*, for recognising a "vital topology" that extends far beyond us and that is not of our making alone' (emphasis in original). Murphy sees post-human-ism as 'pro-nature humanism' which 'transcends the limited conception of only humans and their constructions of objects of value' (p. 291). Over the last few years, as this chapter has shown, animals have been awarded an increasing subjec-tivity, biography and instrinsic value. With such subjectivity comes a history long denied by the 'framing myth' of their 'ontological status' (Haraway, 1989, p. 146). Of course, post-human also suggests 'post-nature' (Castree, 2005) or 'post-animal', even de-animalisation, regrouping all into a holism defined by relations and spaces of interaction rather than divisions and categories of differentiation. Derrida writes:

> I would like to have the plural of animals heard in the singular. There is no animal in the general sin-gular, separated from man by a single indivisible limit. We have to envisage the existence of 'living creatures' whose plurality cannot be assembled within the single figure of an animality that is simply opposed to humanity (2004, p. 125).

At one level, this is provoking stuff which can 'alert us to a world existing under our noses but which we fail ... to see if we divide it into natural and social things' (Castree, 2005, p. 225) though its translation into a genuinely effective code of 'relational ethics' remains frustrating unclear. Mapping the cartography of our interconnectivity with animals may reveal our communality but does it require us to act any differently towards them? As Lestel (2004, p. 78) points out, animals can only achieve their subjectivity through their relations with humans (at least to the extent that we are aware of it). It is, as ever, up to us to take the initiative.

The second and linked framing of networks we wish to draw attention to in this conclusion crosses an equally rigid divide, that between the social and the natural sciences. As with post-humanism, a reintegration of social and natural sciences, particularly with respect to the study of society/animal relations, is a contemporary debate of some vitality dating from Catton and Dunlap's (1978) call for a paradigmatic shift away from human exemptionalism (Redclift and Benton, 1994; Soper, 1995; Dickens, 2001; Newton, 2003). In challenging the historical elimination of nature and human/animal interdependences from social science discourse (Grundman and Stehr, 2000), while seeking to avoid the twin ghosts of determinism and sociobiology, this debate parallels post-humanism in its eagerness to employ 'explanatory strategies which, instead of centering themselves upon distinctive human attributes … would give a central place to what is shared by humans and (other) animals, thereby situating the distinctively human within the overall context of the natural history of our species' (Benton, 1991, p. 3). Yet, while we can acknowledge the undeniable progress that has been made in 'closing the "great divide"' (Goldman and Schurman, 2000) both in environmental sociology and in the sociology of science, Newton (2003) continues to doubt whether 'we are on the same ontological and epistemological terrain when moving from the natural to the social'. The challenge persists. As a first step at least, Tovey (2003) calls for the integration of domestic animals into sociological accounts as a move towards circumventing the society/nature divide. Whatever the means, these ontological and epistemological interogations point to a fundamental shift not only in the mechanisms and practices of human relations to non-humans but also to the understanding and definition of human and animal selves and the ethical positioning that thereby results.

NOTES

1 Though of course, animal society, as Callon and Latour (1981) point out, has no Leviathan in the Hobbesian sense for it has no materials and instruments.

2 Prompting Laycock (1999) to ask whether Nagel means what is it like for a human to be a bat or for a bat to be a bat?

3 Noting in passing the relatively recent growth of a peculiarly literary form, the animal novel. Paul Auster's *Timbuktoo* (1997), Neil Astley's *End of my Tether* (2002), Roy Hattersley's *Buster* (1992) and Matt Haig's *The Last Family in England* (2004) are but some of the recent breed whose genre is celebrated in France by the annual *Prix littéraire 30 millions d'amis*, often refered to as the *Goncourt des animaux* after the more famous literary prize. Of course, these writings join a long cultural tradition of animals being represented in film, from the overly anthropomorphised Disney and Pixar characters to more recent examples (Baker, 2000; Burt, 2002).

4 Oncomouse is a patented medical model; a mouse that has been genetically engineered to develop cancer.

5 In spite of the appeal of ANT to scholars concerned to make non-humans more visible within social science, the application of these ideas within studies of animal–human relations has been limited to date. Apart from Jones (2003) and Whatmore (2003) notable examples of looking at animals through an ANT lens include: Donaldson *et al.* (2002),

Evans and Yarwood (2000) and Woods (1997, 2000).

REFERENCES

Aaltola, E. (2002) 'Other animal ethics' and the demand for difference. *Environmental Values* 11: 193–209.

Acampora, R. (1999) Bodily being and animal world: toward a somatology of cross-species community. In P. Steeves (ed.), *Animal Other: On Ethics, Ontology and Animal Life*. Albany State University of New York Press, p. 118.

Agamben, G. (2002) *L'Ouvert – De l'homme à l'Animal*. Paris, Rivages, p. 56.

Proulx, Annie (2002) *That Old Ace in the Hole*. New York, Harper, p. 302.

Atterton P. and Callarco, M. (eds) (2004) *Animal Philosophies: Ethics and Identity*. London, Continuum, p. 34.

Baker, S. (2000) *The Postmodern Animal*. London, Reaktion Books.

Bataille, G. (1992) *Theory of Religion*. New York, Zone Books.

Benton T. (1993) *Natural Relations: Ecology, Animal Rights and Social Justice*. London, Verso.

Benton, T. (1991) Biology and social science: why the return of the repressed should be given a (cautious) welcome. *Sociology* 25 (1): 1–29.

Berger, P. and Luckman, T. (1967) *The Social Construction of Reality*. New York, Anchor Press.

Birch, T.H. (1993) Moral considerability and universal consideration. *Environmental Ethics* 15 (4): 313–332.

Birke, L. and Parisi, L. (1999) Animals, becoming. In P. Steeves (ed.) *Animal Other: On Ethics, Ontology and Animal Life*. Albany State University of New York Press, pp. 55–74.

Blackwell, R. (1980) *Sociobiology, the New Religion*. Paper presented at the ITEST Conference on The State of the Art in March, 1980, available at http://itest.slu.edu/ articles/90s/blackwell2.html (accessed 5th March 2005).

Blanc, N. (2000) *L'animal dans la ville*. Paris, Odile Jacob.

Braun, B. (2004) Querying posthumanisms. *Geoforum* 35: 269–273.

Buller, H. (2004) Where the wild things are: the shifting iconography of animals in rural space. *Journal of Rural Studies* 20: 131–141.

Burt, J. (2002) *Animals in Film*, London, Reaktion Books.

Callon, M. (1986) Some elements of a sociology of translation: domestication of the scallops and the fishermen of Saint Brieuc Bay. In J. Law (ed.) *Power, Action and Belief: a New Sociology of Knowledge*? London, Routledge, pp. 196–233.

Callon, M. and Latour, B. (1981) Unscrewing the Big Leviathan: how actors macrostructure reality and how sociologists help them to do so. In K. D. Knorr-Cetina and A. V. Cicourel (eds) *Advances in Social Theory and Methodology: Toward an Integration of Micro- and Macro-Sociologies*. Boston, MA, Routledge, pp. 277–303.

Caputo, J. (1993) *Against Ethics*, Bloomington, Indiana University Press.

Castree, N. (2004) Socializing nature: theory, practice and politics. In N. Castree and B. Braun (eds) *Social Nature*. Oxford, Blackwell, pp.1–21.

Castree, N. (2005) *Nature*. London, Routledge.

Castree, N. and MacMillan, T. (2001) Dissolving dualisms: actor-networks and the reimagination of nature. Ch. 11 in N. Castree and B. Braun (eds) *Social Nature: Theory, Practice and Politics*. London, Blackwell, pp. 208–224.

Catton, W. and Dunlap, R.E. (1978) Environmental sociology: a new paradigm. *American Sociologist* 13: 41–49.

Cixous, H. (1993) *Three Steps on the Ladder of Writing*. New York, Columbia University Press.

Cixous, H. (1998) Shared at dawn (renamed *Love of the Wolf* in 2nd Ed, 2005). In Cixous, H. *Stigmata: Escaping Texts*. London, Routledge.

Clark, S. (1999) *The Political Animal*. London, Routledge.

Dawkins, R. (1979) Defining sociobiology. *Nature* 280: 427–428.

Dawkins, M. (1998) *Through Our Eyes Only: the Search for Animal Consciousness*. Oxford, Oxford University Press.

DeGrazia, D. (1999) Animal ethics around the turn of the twenty-first century. *Journal of Agricultural and Environmental Ethics* 11, pp. 111–129.

DeGrazia, D. (2002) *Animal Rights: a Very Short Introduction*. Oxford, Oxford University Press.

Deleuze, J. and Guatarri, F. (1988) *A Thousand Plateaus*. London, Athlone Press.

De Planhol, X. (2004) *Le Paysage Animal*. Paris, Fayard.

Derrida, J. (2004) The animal that therefore I am. Reprinted in P. Atterton and M. Callarco (eds) *Animal Philosophies: Ethics and Identity*. London, Continuum, pp. 113–127.

Dickens, P. (2001) Is capital modifying human biology in its own image ? *Sociology* 35 (1): 93–110.

Donaldson, A., Lowe, P. and Ward, N. (2002). Virus–crisis–institutional change: the foot and mouth actor network and the governance of rural affairs in the UK. *Sociologia Ruralis* 42 (3): 201–214.

Dunlap, T. (1999) *Nature and the English Diaspora*. Cambridge, Cambridge University Press.

Dupré, J. (2001) *Human Nature and the Limits of Science*. Oxford, Oxford University Press.

Dupré, J. (2002) *Humans and Other Animals*. Oxford, Oxford University Press.

Ehrenfeld, D. (1978) *The Arrogance of Humanism*. Oxford, Oxford University Press.

Evans, N. and Yarwood, Y. 2000. The politicisation of livestock: rare breeds and countryside conservation. *Sociologia Ruralis* 40 (2): 228–248.

Ferry, L. (1992) *Le Nouvel Ordre Ecologique*. Paris, Grasset.

Foucault, M. (1965) *Madness and Civilisation*. London, Tavistock.

Franklin, A. (1999) *Animals and Modern Culture*. London, Sage.

Franklin, A. (2003) *Nature and Social Theory*. London, Sage.

Glacken, C (1967) *Traces on the Rhodian Shore*. Berkeley, University of California Press.

Goldman, M. and Schurman, R. (2000) Closing the 'Great Divide': new social theory on society and nature. *American Sociological Review* 26: 563–584.

Goodman, D. (1999) Agro-food studies in the 'Age of Ecology': nature, corporeality, biopolitics. *Sociologia Ruralis* 39 (1): 17–38.

Gould, S.J. (1997) quoted in Dupré (2001) *Human Nature and the Limits of Science*. Oxford, Oxford University Press.

Gray, J. (2002) *Straw Dogs: Thoughts on Humans and Other Animals*. London, Granta.

Grundman, R. and Stehr, N. (2000) Social science and the absence of nature: uncertainty and the reality of extremes. *Social Science Information* 39 (1): 155–179.

Habermas, J. (2003) *The Future of Human Nature*. Cambridge, Polity Press.

Hacking, I. (1999) *The Social Construction of What?* Cambridge, MA, Harvard University Press.

Haraway, D. (1989) *Primate Visions: Gender, Race and Nature in the World of Modern Science*. New York, Routledge.

Haraway, D. (1997) *Modest_Witness@Second_Millennium. FemaleMan_Meets Oncomouse. Feminism and Technoscience*. New York and London, Routledge.

Haraway D. (2003) *The Companion Species Manifesto*. Chicago, Prickly Paradigm Press.

Harrison, R. (1964) *Animal Machines*. London, Vincent Stuart.

Haynes, R. (2001) Do regulators of animal welfare need to develop a theory of pschological well being? *Journal of Agricultural and Environmental Ethics* 14: 231–240.

Holloway, L.E. (2001) Pets and protein: placing domestic livestock on hobby-farms in England and Wales. *Journal of Rural Studies* 17 (3): 293–307.

Holloway, L. (2003). 'What a thing, then is this cow ...': positioning domestic livestock animals in the texts and practices of small scale 'self sufficiency'. *Society and Animals* 11: 145–165.

Jones, O. (2000) Inhuman geographies: (un)ethical spaces of human and non-human relations. In C. Philo and C. Wibert (eds) *Animal Geographies*. London, Routledge, pp. 268–291.

Jones, O. (2003) The restraint of beasts: rurality, animality, actor network theory and dwelling. Ch. 16 in P. Cloke (ed.) *Country Visions*. Harlow, Pearson Education.

Latour, B. (1993) *We Have Never Been Modern*. Hemel Hempstead, Harvester Press.

Latour, B. (1999) *Pandora's Hope*. Cambridge, MA, Harvard University Press.

Latour, B. (2001) *Politiques de la Nature*. Paris, La Découverte.

Latour, B. (2005) *Reassembling the Social: an Introduction to Actor-Network Theory*. Oxford, Oxford University Press.

Law, J. and Hassard, J. (eds) (1999) *Actor Network Theory and After*. Oxford, Oxford University Press.

Laycock, S. (1999) The animal as animal. In P. Steeves (ed.) *Animal Others: on Ethics, Ontology and Animal Life*. New York, State University, pp. 271–284.

Lestel, D. (2004) *L'Animal Singulier*. Paris, Seuil.

Lippit, A.M. (2000) *Electric Animal: Toward a Rhetoric of Wildlife*. Minneapolis, University of Minnesota Press.

Machiavelli (1998) *The Prince*. Harmondsworth, Penguin Books.

Midgley, M. (1984) *Animals and Why They Matter*, Athens, GA, University of Georgia Press.

Masson, J.F. (2003) *The Pig Who Sang to the Moon : The Emotional World of Farm Animals*. New York, Ballantine Books.

Murdoch, J. (2001) Ecologising sociology: actor-network theory, co-construction and the problem of human exemptionalism. *Sociology* 35 (1): 111–133.

Murdoch, J. (2003) Co-constructing the countryside: hybrid networks and the extensive self. Ch. 15 in P. Cloke (ed.) *Country Visions*. Harlow, Pearson Education.

Murphy, R. (1997) *Sociology and Nature: Social Action in Context*. Scranton, Westview Press.

Nagel, T. (1974) What is it like to be a bat? *Philosophical Review* 83: 435–450.

Newton, T. (2003) Crossing the great divide: time, nature and the social. *Sociology* 37 (3): 433–457.

Noske, B. (1997) *Beyond Boundaries: Humans and Animals*. Montreal, Black Rose Books.

Porcher, J. (2002) *Eleveurs et animaux, réinventer le lien*. Paris, Presses Universitaires de France.

Proulx, A. (2002) *That Old Ace in the Hole*. London, Harper.

Quammen, D. (2005) *Monster of God: The Man Eating Predator in the Jungles of History and the Mind*. London, The Harvill Press, Random House.

Redclift, M. and Benton, T. (eds) (1994) *Social Theory and the Global Environment*. London, Routledge.

Segerstråle, U. (2000) *Out there and In There: Defenders of the Truth: The Sociobiology Debate*. Oxford, Oxford University Press.

Serpell, J. (1995) *The Domestic Dog: Its Evolution, Behaviour and Interactions with People*. Cambridge, Cambridge University Press.

Serpell, J. (1996) *In the Company of Animals: A Study of Human–Animal Relationships*. Cambridge, Cambridge University Press.

Serres, M. (2000) *Retour au Contrat Naturel*. Paris, Bibliothèque nationale de France.

Singer, P. (1975) *Animal Liberation*. Random House, New York.

Singer, P. (2004) Preface. In P. Atterton and M. Callarco (eds) *Animal Philosophies: Ethics and Identity*. London, Continuum, pp. xi-xii.

Soper, K. (1995) *What is Nature? Culture, Politics and the Non-Human*. Oxford, Blackwell.

Spiegel, M. (1997) *The Dreaded Comparison: Human and Animal Slavery*. New York, Mirror Books.

Steeves, P. (ed) (1999) *Animal Others: on Ethics, Ontology and Animal Life*. New York, State University Press.

Tester, K. (1991) *Animals and Society : the Humanity of Animal Rights*. London, Routledge.

Thomas, K. (1983) *Man and the Natural World*. London, Allen Lane.

Thrift, N. (1996) *Spatial Formations*. London, Sage.

Thrift, N. (2000) Steps to an ecology of place. In D. Massey, J. Allen and P. Sarre (eds) *Human Geography Today*. Cambridge, Polity Press, pp. 295–322 .

Thrift, N. (2004) Summoning life. In P. Cloke, M. Crang and M. Goodwin (eds) *Envisioning Human Geography*. London, Arnold, pp. 81–103.

Tovey (2003) Theorising nature and society in sociology: the invisibility of animals. *Sociologia Ruralis* 43 (3): 1–20, 196.

Urpeth J. (2004) Animal becomings. In P. Atterton and M. Callarco (eds) *Animal Philosophies: Ethics and Identity*. London, Continuum, pp. 101–110.

Watts, M. (2005) Nature : culture. In P. Cloke and R. Johnston (eds) *Spaces of Geographical Thought*. London, Sage, pp. 142–174.

Wemelsfelder, F. (1999) The problem of animal subjectivity and its consequences for the scientific measurement of animal suffering. In F. Dolins (ed.) *Attitudes to animals: views in animal welfare*. Cambridge, Cambridge University Press, pp. 37–53.

Whatmore S. and Hinchliffe, S. (2003) Living cities: making space for urban nature. *Soundings* 22: 137–150.

Whatmore, S. (1997) Dissecting the autonomous self : hybrid cartographies for a relational ethics. *Environment and Planning D: Society and Space* 15: 37–53.

Whatmore, S. (1999) Hybrid geographies: rethinking the 'human' in human geography. Ch. 2 in D. Massey, J. Allen, and P. Sarre (eds) *Human Geography Today*. Cambridge, Polity Press, pp. 22–39.

Whatmore, S. (2000) Heterogeneous geographies: reimagining the spaces of N/nature. In I. Cook, D. Crouch, S. Naylor and J. Ryan (eds) *Cultural Turns/Geographical Turns*. Pearson Education, Harlow, Essex, pp. 265–272.

Whatmore, S. (2002) *Hybrid Geographies. Natures, Cultures, Spaces*. London, Sage.

Wildman, D.E., Uddin, M. Liu, G. Grossman, L.I. and Goodman, M. (2003) Implications of natural selection in shaping 99.4% nonsynonymous DNA identity between humans and chimpanzees: Enlarging genus Homo. *Proceedings of the National Academy of Sciences* 100 (12): 7181–7188.

Wilkie, R. (2005) Sentient commodities and productive paradoxes: the ambiguous nature of human–livestock relations in northeast Scotland. *Journal of Rural Studies* 21 (2): 213–230.

Wilson, E.O. (1975) *Sociobiology: The new synthesis*. Cambridge, MA, Belknap Press.

Wilson, E.O. (1996) *In Search of Nature*. Washington, DC, Island Press.

Wolfe, C. (2003) *Animal Rites: American Culture, the Discourse of Species and Posthumanist Theory*. Chicago, University of Chicago Press.

Wood, D. (2004) Thinking with cats. In P. Atterton and M. Callarco (eds) *Animal Philosophies: Ethics and Identity*. London, Continuum, pp. 129–144.

Woods, M. (1997) Researching rural conflicts: hunting, local politics and actor-networks. *Journal of Rural Studies* 14 (3): 321–340.

Woods, M. (2000) Fantastic Mr Fox? Representing animals in the hunting debate. In C. Philo and C. Wilbert (eds) *Animals Spaces, Beastly Places*. London, Routledge, pp. 183–200.

34

Social Change and Conservation

Madhav Gadgil

MAN, A PRUDENT ANIMAL

Man, n. An animal so lost in rapturous contemplation of what he thinks he is as to overlook what he indubitably ought to be. His chief occupation is extermination of other animals and his own species, which, however, multiplies with such insistent rapidity as to infest the whole habitable earth and Canada (Ambrose Bierce: *Devil's Dictionary*).

Yet, paradoxically, man might be the only prudent species of animal on Earth. In the late 1960s, Slobodkin (1968) asked if there were any such animal species. By prudent he meant predators that exhibit restraints on harvests such that long-term yields from the prey populations are enhanced at the cost of some immediate harvests. His answer was in the negative. Animals always seem to behave as optimal foragers, concentrating at any time on prey that maximizes the energy or nutrient returns per unit time, or minimizes the risks incurred. If animals leave some species, or age class, or patches of prey alone, it is only because they have better options. Humans too behave as optimal foragers much of the time (Borgerhoff-Mulder, 1988). In the Torres Strait, fishing may be stopped in localities where fish yields are observed to have declined, or in parts of New Guinea, the hunting of Birds of Paradise may be temporarily abandoned if their population declines (Eaton, 1985; Nietschmann, 1985). These responses may merely indicate that the returns from that prey species or those localities are lower than the returns possible from alternative

species or localities. But in other cases humans seem to refrain from harvesting resources that might provide higher returns than the alternatives exploited. Thus, residents of the village Kokre Bellur near Bangalore in south India have been strictly protecting painted storks and spotbilled pelicans breeding in the midst of their village for hundreds of years, although the same birds may be hunted outside the breeding season. Obviously the nesting birds are far easier prey, which is nevertheless left alone, very likely enhancing the long-term availability of the prey population. Modern resource management practices too include examples of deliberate restraints on resource harvests, whether these are mesh-size regulations, closed seasons, or protection to endangered species (Gadgil and Berkes, 1991).

Man owes these occasional displays of prudence, and more importantly, his dominant position in the biological world to his occupation of the *cognitive niche* (Tooby and DeVore, 1987; Pinker, 1997). People attempt to reach their goals by constructing complex chains of behavior in response to specific situations. Such behaviors are planned on the basis of cognitive models of the causal structure of the world. These models are learnt during the lifetime and communicated through the medium of the language, which allows accumulation of knowledge over generations within groups of people. These capabilities enable humans to undertake deliberate actions, either to conserve or destroy natural resources. Occasional prudence is the outcome of deliberate actions aimed at long-term conservation. Of course, as the "Devil's Dictionary" notes man's chief occupation is extermination of other animals (and plants) often undertaken very deliberately. Thus, large-scale felling of forests in the northeastern USA and the massacre of hundreds of thousands of buffaloes of the prairies by early European colonizers of the Americas were instances of such deliberate destruction.

Modern resource management practices are based on explicitly stated rationale; they are implemented on the basis of written prescriptions, rules, and regulations. It is of course possible that the motivations underlying particular practices may be different from the stated ones, and that the prescriptions and rules may not be adhered to in practice. But such practices are accompanied by explanations cast in an idiom acceptable to modern science-based societies. Traditional resource management practices, on the other hand, are not supported by such explanations. The concerned community members may accept certain restraints on the use of biological resources so as not to offend deities, or because their violation would attract social sanctions. It is unlikely that they would state that all members of the community agree to a given practice because it fulfills some secular purpose, such as provision of an ecosystem service. It has, therefore, been often argued that the conservation consequences of traditional resource use practices are totally unintended, merely incidental consequences (Diamond, 1993). It is of course possible that this is so in certain cases. But it is also plausible that on occasions traditional societies might have arrived at practices that promoted sustainable resource use through a trial and error process based on some simple rules of thumb. If sustainability of resource use conferred an advantage on the concerned community, other communities might also acquire such practices through imitation. Such practices of ecological prudence may then spread as a part of a system of religious beliefs or social conventions without their secular function being explicitly recognized.

After all, man is very capable, not only of spelling out what he intends, but also of deliberate and elaborate cheating, and therefore of pursuing a hidden agenda, quite at variance with some declared agenda (Pinker, 1997). Thus, the burning, along with every living creature of the great Khandva forest, that once stood at the site of present-day Delhi, was declared as an offering to please the fire god, Agni. It was most likely motivated by the need to displace the shifting cultivating Nagas occupying that forest and clear the land for invading pastoral Pandavas and Kauravas (Karve, 1967). Similarly, the British reservation of huge tracts of community-managed forests in the 19th century was justified as necessary for forest conservation. It was most likely motivated by the desire to acquire very cheaply

wood resources to support the expansion of railways. Indeed, as one of the British revenue administration officials put it, this was "*confiscation, not conservation*" (Gadgil and Guha, 1992). As a consequence, human societies display a whole range of deliberate and incidental, conservation and destruction practices, involving to variable degrees different components of the society, with different declared and hidden agendas.

Humans are also accomplished tool users and as bearers of elaborate culture, human societies have exhibited substantial technological progress over time. As a highly social animal, humans engage in extensive exchange of goods and services and display elaborate social differentiation (Lenski and Lenski, 1978). These social relationships have continued to change over time, often driven by technological progress. Changes in conservation and destruction practices, deliberate as well as incidental, involving to variable degrees different components of the society, with diverse declared and hidden agendas, have gone hand in hand with the social changes. It is the intention of this chapter to explore these manifold developments.

OF ECOSYSTEM PEOPLE, BIOSPHERE PEOPLE AND ECOLOGICAL REFUGEES

Dasmann's (1988) insightful analysis of the different ways in which people relate to their natural resource base provides an appropriate framework in which to view these developments. He identifies people at the two extremes of what is undoubtedly a continuum as *ecosystem people* and *biosphere people*. Ecosystem people largely depend on their own muscle power, and that of their livestock to gather, produce, and process most of the resources they consume. The bulk of these therefore come from a limited resource catchment; from an area of some 50 square kilometers around their homesteads. In other words, they are people with rather small ecological footprints. Moreover, ecosystem people have characteristically used their resource catchments over long periods, often several generations. Many tribal, peasant, pastoral, rural artisan communities of India fit this description.

It is appropriate to further distinguish the ecosystem people into two categories, namely (i) autonomous ecosystem people, and (ii) subjugated ecosystem people. In some of the more inaccessible corners of the world, the ecosystem people are still autonomous. However, over most of the Earth they have been subjugated by the biosphere people and have very limited control over their own resource base of natural and semi-natural ecosystems. They gather and produce little that can fetch value in markets and therefore have very limited access to produce of intensively managed and artificial ecosystems. Biosphere people, on the other hand, have extensive access to additional sources of energy such as fossil fuels or hydroelectricity. This enables them to transport and transform large quantities of material resources from all over the world for their own use. Their resource catchments are vast. They are people with huge ecological footprints. They are continually tapping newer and newer resources, for example, nuclear power, from newer localities, for instance, drilling for oil in the deeper seas. They are very much part of an increasingly globalizing market economy. Most citizens of the First World, and the Third World elite, behave as biosphere people.

To the two categories of Dasmann, one may add yet one more, that of *ecological refugees*. Ecological refugees are ecosystem people, deprived of access to their traditional resource base, who are forced to colonize new localities where they continue to depend largely on human and animal muscle power to gather, produce, and process resources. Their resource catchments remain limited, but these are no longer ecosystems with which they have been integrated over generations. They therefore neither have the attachment, nor the knowledge, nor the motivation to use the resources of these new catchments in a prudent fashion. White colonizers of North America were largely such ecological refugees of the resource crunch that afflicted Europe during the Little Ice Age (Crosby, 1986). So must have been the people who gradually colonized whole series of Pacific islands over the last millennium (Diamond, 1991). Examples of present-day ecological refugees include peasants moving into the rain forest of the Amazon in Brazil or Western Ghats in India. Another wave of ecological refugees today creates the shantytowns of Third World cities, or lives as illegal, immigrant labor working on farms in the USA.

CULTURAL EVOLUTION OF CONSERVATION PRACTICES

People, by and large, pursue self-interest. They are unlikely to be motivated to use a resource base in a prudent, sustainable fashion (a) if their resource catchments are vast, so that degradation of any particular part of the catchment affects them very little; or (b) if they have open before them possibilities of substitution as any one resource element is depleted; or (c) if their control over the resource base is tenuous, so that others may at any time deplete a resource they value, even if they themselves used it in a restrained

fashion. Indeed, profligacy is likely when any one of these three conditions obtains. It is only when people perceive their resource catchments as limited, possibilities of substitution of an exhausted resource element as remote, and their own control over the resources as secure, will they be motivated to use the resource base in a prudent fashion.

It is possible to think of historical scenarios satisfying these conditions and thereby favoring cultural evolution of deliberate conservation measures. Many hunter-gatherer societies, as well as shifting cultivator or horticultural societies significantly dependent on foraging, are known to have been highly territorial, with each endogamous group constantly struggling with neighboring groups (Vayda, 1974). In comparison with other mammals of their size, humans take an extraordinarily long time to mature sexually and begin reproduction. Human population sizes cannot therefore increase rapidly in response to increased availability of resources. The well-being of a human group therefore requires the availability of resources, and possibly a wide diversity of resources, at a minimal level over periods of several years.

For territorial groups, this implies the need to sustain resource levels on a long-term basis within their own territory. Any group that failed to achieve this would find itself weakened and subject to the territorial aggression of neighboring endogamous groups, and be culturally, even if not genetically, exterminated. Cultural group selection that may be quite effective under these special conditions, might then favor behavioral traits that would ensure sustainable use of the biological resources of the territory (Boyd and Richerson, 1985). In addition, within-group cooperative behavior promoting prudent resource use is also expected to prevail in endogamous groups where a relatively small number of individuals repeatedly interact with each other over long periods. Such practices may once have been particularly common in groups inhabiting stable, productive habitats where territoriality is likely to have been strong.

Today, few communities of ecosystem people retain control over resources; such control has been or is being usurped by the more powerful biosphere people. But in the pre-industrial world, many communities of ecosystem people are likely to have fulfilled all the three conditions promoting prudence. It must be emphasized that when newly colonizing a locality people with hunting–gathering–fishing–subsistence agriculture technologies are unlikely to have fulfilled these conditions. At the frontier, they are likely to have perceived potential resource catchments as extensive, and faced with abundant resources perceived many possibilities of substitution. Such people would not be ecosystem people rooted in a locality in our terminology; they may be thought of as ecological refugees unlikely to behave prudently. The many examples of extinctions of species on islands such as Madagascar or in Polynesia are probably cases of profligate resource use by early colonizing communities (Diamond, 1991).

All human communities would at some point in their history have been such colonizers. Initially they may have neither the motivation, nor the knowledge of the resource base to arrive at regimes of sustainable use. In this period they may be responsible for, and witness the elimination of many biological populations. If they do become rooted in a locality, and come to control its resource base effectively, they are likely to see themselves as being adversely affected by resource overuse and gradually become motivated to use the resources in a prudent fashion. When so motivated they may develop some simple rules of thumb to promote conservative use through a process of trial and error. Joshi and Gadgil (1991) suggest that such a process of trial and error may be based on comparing levels of harvesting effort and yields in the recent past. This could lead to a decision rule such as the following: (a) enhance harvesting effort, if an enhancement of harvesting effort in the past was accompanied by increased harvests (or in some other way left one better off), or if a reduction in the harvesting effort in the past was accompanied by reduced harvests (or in some other way left one worse off); (b) step down harvesting effort, if an enhancement of harvesting effort in the past had left one worse off, or a reduction in harvesting effort had led to one being better off.

Such a decision rule, Joshi and Gadgil (1991) show, can lead to a sustainable harvest, especially when the reduction in the harvesting effort takes the form of total protection of some resource element. Such protection may be afforded by creating refugia such as sacred groves or sacred ponds from which no harvests are made; it may involve total protection to all individuals of some taxa such as species of monkeys widely protected in India; or total protection to especially vulnerable life-history stages such as colonial birds breeding at a heronry. Such measures evidently tend to promote conservation of biodiversity coupled with its sustainable use. Ecosystems people in many parts of the world continue to exhibit a variety of such cultural traditions of conservation practices, in spite of recent loss of control over the resource base (Gadgil and Berkes, 1991; Gadgil et al., 1993). Thus, in southern India, the large fruit bat *Pteropus giganteus* may be protected at the daytime roost, although it will be hunted at night away from the roost. Other plants and animals may be totally protected at all times, for instance as kin accorded totemic status.

Thus, a farming community called the Bishnois of Rajasthan in northwestern India will not cut a Khejari (*Prosopis cineraria*) tree even if it grows in the midst of a field. They also give total protection to all antelope species and peafowl around their villages. Amongst the most notable of such instances of total protection is that accorded to tree species of the genus *Ficus* in many parts of Africa and Asia. Because of this, *Ficus* trees are often left standing when forests are clear-cut in India, and huge trees of *F. religiosa* dot India's thickly settled rural countryside, and persist even in city centers. *Ficus* is considered a keystone resource of tropical forests since it often fruits in months when none of the other plants are in fruit and therefore promotes the persistence of frugivorous birds and primates (Terborgh, 1986). Today, many forest dwellers of India appear to be aware of this role of *Ficus* trees, and it is quite plausible that the widespread protection to *Ficus* came to be accorded in the interest of sustaining populations of favored prey species of humans such as fruit-eating pigeon.

Such societies promote maintenance of a diversity of habitats by protecting samples of natural communities on sacred sites (e.g. sacred groves, sacred ponds). These sacred sites may be associated with nature spirits resident in trees or pools of water or with more formalized worship as with Buddhist temple groves of Thailand or groves associated with Shinto shrines in Japan. Sacred groves, ponds and lagoons persist to this day in many parts of Asia, Africa and some of the Pacific islands; a sacred cacao grove in Mexico has also been described (Gomez-Pompa et al., 1990). Indeed, even Polynesian islanders seem to have developed a variety of such practices, following the initial spate of exterminations (Ruddle and Johannes, 1985). So also Madagascans came to protect the lemur species as sacred animals after having been earlier responsible for the extinction of giant lemurs (Jolly, 1980; Diamond, 1991).

While in the long-term interests of the group, such restraint is often likely to be against the immediate, short-term interest of individual group members. Small-scale societies of ecosystem people achieve acceptance of such restraint largely through attribution of sacred qualities to the individual plants, animals, or whole biological communities to be protected, coupled with social sanctions against violation of the regulations. Thus, in southern India many sacred groves are dedicated to cobras. People believe that they will be safe from death from snakebite so long as they respect protection of the serpent grove, but will incur the wrath of cobras, or some associated deities if they violate it. However, the protection of the sacred groves need not always be absolute; in the case of a special calamity such as a fire consuming houses in the village, the deity, through the agency of the priest, may permit selective harvesting of the trees (Gadgil and Vartak, 1976). Of course, the sanctions may become entirely social sanctions divorced from religious context as in the case of the revival of sacred groves as safety forests after conversion to Christianity in Manipur and Mizoram (Malhotra, 1990). In all cases, however, regulation is enforced primarily on the basis of fear of undesirable repercussions visited either by supernatural forces or one's own social group.

AGRARIAN SOCIETIES

As small-scale gatherer–horticulturalist societies gave way historically to larger agrarian societies, regimes of regulation of the use of the biodiversity resources underwent a transformation. The agrarian societies are characterized by far greater levels of social stratification, with an elite supported by the surplus generated by the peasantry (Lenski and Lenski, 1978). The elite, now belonging to Dasmann's (1988) category of biosphere people, has access to surplus produced from many different localities. It therefore has little motivation to promote sustainable resource use in any particular locality. While the ecosystem people may hold on to some of their traditions of sustainable resource use and conservation of biodiversity in such a stratified society, many such traditions may be weakened. At the same time, the possibility of bringing resources from distant localities opens up to the biosphere people the option of not utilizing resources close to home.

It has been suggested that humans have an inherent love of certain forms of natural landscapes – topophilia and of a variety of living organisms – biophilia (Wilson, 1984; Kellert and Wilson 1993; Pretty, 2004). Biosphere people then potentially have the choice of, and innate interest in, maintaining biologically diverse communities around themselves by lightly exploiting their immediate surroundings, while transferring the pressure of resource exploitation to more distant tracts. Indeed they have often engaged in such geographic discounting; since the level of any people's concern for the health of the ecosystem appears to fall off, often quite rapidly with distance from their habitation.

Thus, Kautilya's Arthashastra, the 2000-year-old Indian manual of statecraft, prescribes the maintenance of princely hunting preserves just outside the capital (Kangle, 1969). Such hunting preserves played an important role in biodiversity conservation in medieval Europe and Asia. The Mughal emperors of India, for instance, maintained large areas of several hundreds of square kilometers each as hunting preserves in northern India; in fact, these hunting preserves have served as the nuclei of India's modern system of wildlife sanctuaries and national parks (Gadgil, 1991a). Again it was fear of punishment that ensured protection of these hunting preserves; fear of punishment at the hand of the aristocracy and their armed forces, rather than the wrath of deities or sanctions of one's own social group.

These stratified agrarian societies gave rise to several organized religions. Those of the East, such as Buddhism, absorbed some elements of the conservation traditions of the ecosystem people, as witness the temple groves of Thailand. On the other hand, religions of the Middle East, Christianity and Islam, rejected the attribution of any sacred quality to nature (Whyte, 1967). In fact, refugia in the form of sacred groves covered much of the Middle East and Europe before the spread of Christianity and Islam. The famous "Epic of Gilgamesh" describes the destruction of a sacred cedar grove protected by a wild giant to build the king's palace in the city of Uruk in Mesopotamia. The Greek and Roman landscape was dotted like a leopard skin with hundreds of sacred spaces, which usually contained groves of trees and springs of water. The groves varied in size from small plots with a temple and a few trees to those covering several square kilometers. The third sacred war in Greece (355–347 BC) was fought over the issue of illegal cultivation of Apollo's sacred ground at Crisa (Hughes, 1994). As Christianity spread, these groves became an object of religious zeal. The emperor Theodosius II (5th century AD) issued an edict directing that the pagan groves be cut down unless they had already been appropriated for some purpose compatible with Christianity. A few of them became monastery gardens and churchyards. Others were razed to ground (Hughes, 1984). But even today, of the thousands of sacred sites to which Christian pilgrims are attracted to in Western Europe, nearly a sixth are groves and rock formations, a third are sacred springs and the rest high places in the hills (Nolan, 1981).

EUROPEAN EXPANSION

Much of the European expansion beginning in the 15th century carried this spirit of rejection of any attempts at restraints on the harvest of biological resources or transformation of natural habitats. The European society was also caught up at this time in a process of "atomization," involving a rejection of community control over management of resources. This led to movements such as "enclosure of the commons" by the feudal chieftains. It was this rejection of the ethic of conservation, and of community management that fueled a wave of overfishing and overhunting sweeping over North America, along with the weakening of cultures of indigenous people at the time of the colonization of that continent by European populations. Its manifestations included the fur trade by the Hudson Bay Company dramatically increasing the hunting pressure on Canadian wildlife (Crosby, 1986). Berkes (1989) has documented the response of Amerindians drawn into this trade as commercial suppliers of pelt; their involvement led to a breakdown of their traditional hunting practices which rejected killing except for consumption as food or pelt for their own use. They did, however, develop new sets of rules of thumb coupled with the demarcation of territories that promoted more sustainable use even under pressures of commercial harvests.

The European impacts also related to large-scale introduction of exotic plants and animals on the many continents and islands newly colonized by them. Domesticated animals such as cattle and sheep, and rabbits running wild, were responsible for a significant decline of Australia's indigenous marsupial fauna (Wilson, 1990). Interestingly enough, one of the few placental mammals that had reached Australia before the Europeans, the dog dingo, became a major pest for the European ranchers. The dingo then attracted bounty, a reward paid on presentation of its tail as evidence of having killed one. The indigenous people of Australia took to this as a welcome source of income, but while hunting dingos they ensured that they would not kill lactating females or disturb the dens, so that the dingo population would continue to thrive. The Australian aborigines had thus arrived at simple rules of thumb for ensuring sustainable harvests of a biological resource that supported them (Meggitt, 1965).

Notions of conservation arose amongst Europeans only in the second half of the 19th century as deforestation led to serious negative consequences in parts of Europe such as Switzerland, and as frontiers of new lands to be colonized seemed to be closing. Notably, Switzerland was one of the European countries that retained practices of community ownership and management of forest resources. Switzerland successfully revived the protection of forests under community control. Elsewhere, in colonial possessions watershed protection came to be declared as the function for which forests were conserved (Grove, 1992). It is, however, plausible that this was a situation where the hidden agenda was quite different, and was simply supplies of cheap timber for colonial needs without compensating communities for confiscation of resources. That is why in India many of the hill-slope forests supposedly

being protected for their watershed function were clear felled and converted to teak plantations.

The closing of the western frontier in the USA also led to the National Park movement with the setting aside of large scenic areas such as Yellowstone National Park (Koppes, 1988). Notably enough, these areas were thought of as primeval nature untouched by man. In reality, they were humanized landscapes where nature was lightly trodden upon and much of its diversity protected over thousands of years of use by indigenous people. So in the National Park system the emphasis was on regulating human access by deploying bureaucratic machinery. For the first half of the 20th century there was little interest in the conservation of biodiversity for its own sake in Europe, or in Neo-Europes like the USA or Australia, or in the state-sponsored efforts which came to be influenced by the European world view in all countries of the world.

COLONIAL EXPERIENCE AND ITS AFTERMATH

The European expansion led to the small-scale communities with autonomous control over the local resource base being ever more effectively absorbed into the large-scale, stratified societies all over the world, including in India. First the colonial, and then the independent Indian state, has privatized, or more often brought under state control the common property resources earlier under *de facto* community control. The forest and other resources of the privatized or state-controlled lands have then been mobilized to support the larger economy. The local communities, no longer in control over their own resource base, and having to deal with rapidly changing levels of resource stocks influenced much more by outside demands, have tended to lose their motivation for sustainable use, and along with that many systems of conservation of biodiversity, such as protection to sacred groves. The larger commercial interests working in league with the state apparatus have no stake in sustainable use of resource stocks from any particular locality and have therefore promoted a process of sequential exploitation. During such a process resource harvests at any one time are focused on the most easily accessible profitably exploitable elements. When these are exhausted, the focus shifts to the next most readily accessible, most profitably exploitable element. When the resulting resource exhaustion leads to serious resource shortages technological innovations permit tapping of new kinds of resources, or opening up of entirely new localities for resource exploitation (Gadgil, 1991b).

It is instructive to examine these developments over the last century along two axes; one of the extent to which ecosystems have been transformed through human interventions, and the second of the manner in which people relate to the world of nature, to the world of human-made artifacts and to each other. For this purpose, the terrestrial and aquatic ecosystems may be classified into four major categories: (i) natural ecosystems that are subject to very low levels of human demands because of their inaccessibility, (ii) natural and semi-natural ecosystems including low-input agricultural systems subject to higher levels of human demands, (iii) ecosystems managed intensively for biological production, and (iv) largely artificial ecosystems dedicated to industrial production and organized services. The people, too, may be assigned to four major categories: (i) autonomous ecosystem people, (ii) subjugated ecosystem people, (iii) biosphere people, and (iv) ecological refugees. Table 34.1 summarizes the relationship and notes the examples that are discussed further. All of these come from India; however, they do represent broader patterns encountered in other parts of the world.

Autonomous people

There are relatively few examples today of truly autonomous ecosystem people, people fully in control of largely natural ecosystems with very light human demands, and in consequence with high levels of biodiversity. Inhabitants of Sentinelese island (lat. 11° 30′ N long. 92° 15′ E) in the Andaman–Nicobar chain in the Bay of Bengal provide one such example. These relatively inaccessible islands harbor tropical rainforest biota with high levels of endemism, that is, of species restricted to this chain of islands. They were inhabited by a number of hunter–gatherer tribal groups, without knowledge of metal tools, with the exception of the Shompens of the Nicobars who had a more advanced fishing economy. The British colonized the islands in the

Table 34.1 Variety of people–ecosystem contexts, with specific examples discussed in the chapter

	People			
Ecosystems	Autonomous ecosystem people	Subjugated ecosystem people	Biosphere people	Ecological refugees
Inaccessible natural/ semi-natural ecosystems	Poor in modern economic sense, with access to high levels of biodiversity e.g. Sentinelese islanders	—	—	—
Accessible natural/ semi-natural ecosystems	—	Poor in modern economic sense, serve as agents for destruction of biodiversity, e.g. Gangtes of Manipur, sometimes as stewards, e.g. Village Forest Committees, protectors of Saranas	Visitors for recreational/ commercial use	Poor in modern economic sense, serve as agents of destruction of biodiversity, e.g. Maldharis of Gir
Ecosystems managed intensively for biological production	—	—	Well-off in modern economic sense, promote low biodiversity production systems	Poor in modern economic sense, alienated from biodiversity, e.g. Jharkhand tribals
Artificial ecosystems dedicated to industrial production/ organized services	—	—	Well off in modern economic sense, promote high biodiversity systems for recreational purposes	Poor in modern economic sense, alienated from biodiversity, e.g. Jharkhand tribals

mid-19th century, an attempt that was strongly resisted by the indigenous people. By the 1870s the resistance was overcome and as a consequence many of the tribal populations either drastically declined or were exterminated. During British times the islands primarily served as a convict colony, with many of the released convicts settling down to agriculture (Superintendent of Government Printing, 1909).

After independence in 1947 the islands were used to create agriculture-based settlements of many people displaced from the then East Pakistan (now Bangladesh), as well as to supply forest resources for rapidly growing forest-based industries, especially plywood (Saldanha, 1989). As these were progressively overused and exhausted, leading to substantial erosion of biodiversity, pressure has built up to overcome the resistance of the tribal groups still holding out to open up their territories to commercial forest exploitation. Two of the tribal groups, however, do continue to retain hold over their territories; these are the Jarwas and Sentinelese. Jarwas live on the larger South Andaman and Baratung islands (lat. 12° 0′ N–2° 20′ N long. 92° 45′ E–92° 55′ E), about two-thirds of which has been colonized by immigrants and subjected to forest exploitation. Their territory is therefore easily accessible; it is, however, stoutly defended by Jarwas with their bows and arrows. There are ongoing attempts to overcome this with the aid of a major road passing through their territory. This Jarwa territory remains much richer in biodiversity than the rest of the islands; Jarwas are, however, poorer in a modern economic sense with access to very few and simple artifacts. The Sentinelese live on a smaller island, very rich in biodiversity, which remains entirely under their control (Bhattacharyya, 1993; Gadgil, 1998).

Subjugation: political and economic

The vast majority of the ecosystem people of the world are, however, no longer isolated like the Sentinelese. In no case do they seem to have voluntarily sought to integrate with the larger society dominated by the biosphere people. Rather, the biosphere people have sought to integrate them in order to access resources of their territories, as well as to utilize their labor. The process of such integration tends to proceed through several stages involving political subjugation, followed by economic subjugation. Such subjugation is accompanied by a reduction in levels of access of the people to the biodiversity resources of their own localities, though simultaneously they may have enhanced access to other resources through market channels. The process is also generally accompanied by an erosion of

biodiversity driven by overexploitation of natural resources to meet outside demands with the subjugated ecosystem people often serving as agents of such erosion.

The case of Gangtes, one of the Kuki tribes of Churchandpur district (lat. 23° 4′–25° 50′ N long. 93°–94° 47′ E) of the northeastern state of Manipur, illustrates the complete sequence from autonomy through political and then economic subjugation (Hemam, 1997; Gadgil et al., 1997). Like many other Kuki and Naga groups, Gangtes continued head hunting well into 19th century. At that time, there were no roads so that people had to walk for several days to reach markets where they may have exchanged hand-woven cloth or honey for iron tools or rice. These journeys were hazardous as they may have involved passage through territories of alien tribes. Each settlement was in consequence a largely self-sufficient and self-governing entity. Their base of subsistence was shifting cultivation with long fallows of about 15 years. There were no fixed village boundaries and individual social groups presumably shifted around from time to time. Amongst Gangtes hereditary village Chiefs determined through primogeniture made all group-level decisions in consultations with a council of elders. They assigned lands for cultivation and annually received some grain from each family in the group as a tribute. In the course of cultivation the densest, tallest forest tracts were left alone as requiring too much labor for conversion to fields; the more recent fallows were also avoided. The people worshipped many natural elements including mountain peaks, streams, plants and animals. They strictly protected patches of sacred groves called *gamkhal* as also other spirit-possessed lands called *nungens*. Also protected were bamboo groves called *mavuhak* from which bamboo may be extracted only for house construction, but the shoots, much relished as food, were left alone. This preserved luxuriant vegetation which led Captain Pemberton to remark in 1835: "I know no spot in India, in which the products of the Forests are more varied and magnificent but their utility is entirely local" (Pemberton, 1835).

The process of political subjugation of Gangtes began in the late 19th and early 20th centuries with the development of incipient links with mainstream India, then ruled by the British. Initially, the British did not so much want access to resources of Churchandpur compared with the richer Myanmar. Hence, they wanted to ensure safe movements of British troops in the hill tracts of Manipur bordering Myanmar. A people without fixed, well-defined villages and with traditions of killing aliens coming into their territories were an obvious threat. The British therefore concentrated on fixing village boundaries and assigning land ownership. Given the Gangte system of hereditary village Chiefs, they decided to confer on the Chiefs all ownership over land, legally converting others into tenant farmers. However, during the Biritish reign this made little operational difference in most areas, with no acceptance of private property in land by Gangtes, except in the Churchandpur town area beginning 1930s. The late 19th and early 20th centuries also witnessed the gradual spread of Christianity that began to slowly erode the traditional belief system underlying the conservation practices.

In this phase of political subjugation there was little change in the economy. The road network remained extremely limited and the Gangte communities remained almost totally self-sufficient in terms of resource use. The fallow cycles for shifting cultivation were still long, and substantial areas of forest retained protection in forms of sacred groves. However, by introducing the legal concept of private ownership and a religion that questioned attribution of sacred qualities to nature, this phase did set the stage for the radical changes that followed independence; but unlike on the mainland, this political subjugation had limited impact in the absence of an access to markets.

Market forces

During the Second World War, the British strengthened the road network of Manipur, laying the foundation for the rapid development of transport, communication and commodification that began on Independence in the 1950s. This was consistent with the policies of economic development and national integration adopted at the time of independence. These development policies also encouraged forest-based industries by offering them many resources including wood at highly subsidized rates (Gadgil and Guha, 1992, 1995). As the production of these industries grew at a rapid pace the demands outstripped supplies, leading to severe overexploitation. Such overexploitation was facilitated by the fact that there was no social group sufficiently motivated to ensure a more sustainable pattern of resource use with a secure enough control over the resource stocks. Access to markets meant that a local community no longer suffered an inevitable shortage of resources if those from their own vicinity were overused and exhausted. Neither would such a difficulty be experienced by an industry drawing resources from a larger spatial scale.

Market access therefore reduced the motivation of the various parties concerned, local communities, such as those of Gangtes, as well as other consumers such as the plywood industry, for restraints on levels of harvests. At the same time conversion to Christianity eroded the conservation ethos of local communities grounded in attribution of sacred qualities to various natural elements. Linkages to the larger national society also diluted the control of any one agency over any given resource base by bringing a variety of actors into play. Thus, in the autonomous phase where a person might even be killed while passing through alien territory, there was clear-cut control by a local community. In the phase of political subjugation this was legally recognized as the control in terms of ownership by the local chieftains. But in the post-independence phase of economic subjugation the state Forest Department has stepped in to claim control over forest lands and forest resources. How this claim is to be reconciled with the claim over land ownership by the Gangte Chiefs has not been clarified. This uncertainty has affected the security of resource control by both parties. At the same time the traditional authority structure of a Gangte village community headed by a hereditary Chief with a council of elders has also been affected by the institution of a democratically elected village council. There are continuing contradictions in this system as well since the hereditary chief remains at the head of the elected village council. Under these circumstances, forest resources of Gangte villages have been affected by demands greatly exceeding sustainable yield levels. These demands are no longer local, but reflect the markets not only of remaining parts of the state of Manipur, but also the rest of the country, and even beyond national borders.

Land use changes

This process of overexploitation of forest resources has also shaped patterns of land-use. With forest resources bringing in highest levels of net profits closest to the market town of Churchandpur, all tree stocks have been exhausted in this area. In response settled cultivation on terraced fields with very low levels of crop diversity such as monocultures of pineapple has replaced the much more biologically diverse shifting cultivation. In more remote areas too commercial monoculture plantations have been taken up of species like Agor (*Aquilaria aquatocha*), the highly valuable wood that was the first to be exhausted even from areas far from market.

The traditional land-use pattern of Gangtes included leaving aside substantial areas as *gamkhals* or so-called village forest reserves (VFRs), as well as bamboo reserves and other spirit lands. These practices have been completely abolished in villages close to market towns where land and its produce have acquired high commercial value. In some of the more remote villages the practice of protection of a VFR encircling the settlement has been revived following recognition of its value as a firebreak preventing the spread of fire to the settlement during the slash and burn operations (Gadgil *et al.*, 1997). The traditions of protection of such forest patches still continue in other more remote areas.

Links to markets has meant access to many new goods such as soaps and transistor radios for the Gangtes; goods that need cash for purchase. Sale of timber, fuelwood and other forest produce such as cane is their only source of cash. But generation of this cash has led to a reduction in the return of nutrients to the fields in slash and burn cycles with a depletion of the standing tree stocks. This has reduced the levels of productivity of shifting cultivation (Ramakrishnan, 1992). The spurt in population growth, following the introduction of modern health care systems, especially control of malaria, has also increased the pressure on land, and led to a shortening of fallow cycle and a reduction in productivity of shifting cultivation (Hemam, 1997).

Ecological refugees

The next Indian case study is that of the tribal populations of the Jharkhand (lat. 22° 0′ – 24° 0′ N long. 84° 0′ – 87° 40′ E). This region adjoins the old, thickly settled Gangetic plains that were the nucleus of the 2000-year-old Magadha empire extending over much of the Indian subcontinent. Emperor Ashoka of Magadha is famous for his conquest of and subjugation of the tribals in the extensive hilly tracts that border the Gangetic plains (Gadgil and Guha, 1992). This part of the country also fell rather early to British colonialists, who quickly established control over the region during the second half of the 18th century. The forests of Jharkhand are rich in sal (*Shorea robusta*), an important timber for the production of railway sleepers. This was a resource much valued by the British who therefore acted firmly to ban all shifting cultivation by the tribal people in the early 19th century.

The forests were then taken over either as a property of the British Government, or that of private landlords. Where the tribal villages were a part of the Government-controlled forest lands, the tribals were totally deprived of any rights to forest resources. Unlike with Gangtes, the landlords too did not come from amongst the tribals themselves. They came from other castes at a higher level in the hierarchy. These landlords exploited the lower caste or tribal tenants on their lands quite ruthlessly. The tribals were therefore subjected to extreme political as well as economic subjugation as early as the first half of the 19th century. Their only earnings came from work

as poorly paid tenant farmers or as forest laborers along with the sale of some minor forest produce at extremely low rates. They thus have a long history as poor people serving as agents of destruction of biodiversity.

These Jharkhand tribals are today amongst the most striking example of ecological refugees of the country. They have served as the most mobile source of very inexpensive, unskilled labor in many contexts. One of the major economic enterprises under British rule was the development of tea estates replacing the rain forest of the Brahmaputra valley in northeastern India. This involved wholesale destruction of biodiversity. While the plantation owners were British, the laborers responsible for actual destruction of biodiversity were mostly Jharkhand tribals working under conditions that have been described as being close to slavery (Gadgil, 1942).

Elsewhere, these tribals were resettled in so-called forest villages. The primary focus of the plantations they worked on was to replace the natural sal-dominated, fairly diverse humid forest with monocultures of teak (*Tectona grandis*). Teak is an excellent timber resistant to termite and fungal attacks, highly valued for ship building and gun carriages in the 19th century and for furniture and house construction in the 20th. But teak is a hard wood, little used traditionally; it also does not provide any other product of local utility. On the other hand, sal leaves are used to make plates and sal seed has value as an oil seed. Many associates of sal, such as mahua (*Madhuca indica*) and tendu (*Diospyros melanoxylon*) have great local utility. So replacement of natural sal forest by teak plantations deprives local people of access to a diversity of biological resources of value (Gadgil and Guha, 1995).

Costs of conservation

Conservation practices of the ecosystem people with their limited resource catchments are organized on limited spatial scales. Thus, the sacred groves have sizes ranging from a fraction of a hectare, to tens, at best hundreds of hectares. The biosphere people have their own conservation practices. In keeping with their larger resource catchments, they tend to protect much larger areas as nature reserves (Gadgil, 1996). Agrarian states had their tradition of hunting preserves of aristocracy, ranging in size from a few hundreds to thousands of hectares. The modern-day nature reserves are even larger in spatial scale, in keeping with the even larger resource catchments of modern-day societies of biosphere people. Thus, the largest national parks of the day extend over hundreds of thousands of hectares. Many of them, as in India, have their roots in the hunting preserves of the aristocracy. Indeed, the state-sponsored conservation efforts in India, begun in the 1950s, have been led principally by erstwhile princes, with the Maharaja of Mysore being the first president of the Indian Board for Wild Life and the Maharaja of Baroda the first president of the Indian branch of the World Wildlife Fund. The whole nature conservation approach has tended to focus on protection of a few flagship species such as the Indian elephant, rhinoceros, tiger or lion, with the firm belief that such protection can be achieved primarily through elimination of the subsistence demands of India's ecosystem people by the force of guns and guards.

Consider as an example, Gir National Park, dedicated to the conservation of the lion. At the time of the British conquest, the lion was distributed over much of the northern peninsula. It was a prime hunting trophy and was eliminated through most of its range during the 19th century. In 1920 the Nawab of Junagarh had to pretend that it had become extinct in his hunting preserve of Gir in order to resist the pressure of organizing a lion hunt for the Governor General of India. In fact, just 22 lions still survived in Gir at that time, and by the 1950s their population had grown to over 200 (Seshadri, 1969). When Gir was constituted a National Park, an important measure immediately introduced was removal of Maldharis, buffalo keepers who had coexisted with the lion in the Gir forest for centuries. Maldharis traditionally accepted occasional kills of their livestock by lions as inevitable; their animals were an important source of food, especially for the lazier male lions. Being removed from the Gir forest has meant serious hardships and a decline in living standards for Maldharis. They are currently living on the periphery of the National Park, denied access to the rich grazing within the park. But the lions are addicted to buffaloes and come out at night to feed on them. Now when a buffalo is killed, the impoverished Maldharis are unwilling to tolerate the loss and poison the carcass to kill the lion when it comes to feed on it a second time. So the displacement of Maldharis has served little positive purpose from the perspective of the flagship species in whose interest the National Park is being managed (Gadgil, 1991a).

In Keoledev national park (lat. 27° 15′ N long. 77° 35′ E) famous for its water birds, modern conservation attempts have also involved exclusion of grazing in the extensive shallow wetlands. When grazing was thus banned in early 1980s, without any alternatives being made available, there were protests leading to police firing, with some deaths. But the ban was implemented forcing local peasants to substantially reduce their livestock holdings, impoverishing them to a significant extent. But this cessation of grazing has

had an adverse impact on wetlands as a water-bird habitat. This is because the wetlands have now become choked by an excessive growth of a grass, *Paspalum*. This particular conservation measure has thus tended to impoverish people as well as the bird life, which was at the center of the conservation concern (Vijayan, 1987).

Jharkhand tribals too have suffered from being excluded in the interest of nature conservation. One of the major recreation areas for the wealthy from Calcutta city is the Betla National Park (lat. 23° 40′ N long. – 84° 40′ E) in the Daltonganj district of Bihar. The tribals who were earlier settled in many forest villages in what now constitutes this national park are in serious difficulties. With the constitution of the national park the forestry operations employing them have been halted. Further restrictions are imposed on their collection of forest produce. Moreover, the tribals are now being asked to move out of the forest villages altogether, without any provisions for alternative livelihoods. So the tribals, who were earlier instruments of destruction of biodiversity to help meet the economic demands of the wealthy, are now being further impoverished to meet their recreational demands for access to biodiversity (Tiger Task Force, 2005).

There has been little interest in the broader objective of conserving biodiversity on the part of the Indian state. Indeed the National Commission on Agriculture recommended in 1976 the replacement of all mixed species forests of India by more productive monoculture plantations (National Commission on Agriculture, 1976). In consequence, the state forest departments have resorted to clear-cutting sacred groves, considering them, as one Chief Conservator of Forests once put it to me, as merely *"stands of over-mature timber."*

The inevitable consequence has been the ever-growing conflict between the state machinery striving to protect nature reserves by force of arms and the local communities all over the country. Given this hostility, there is little political support for the ongoing conservation effort in India, outside of a narrow circle belonging to part of the urban middle classes. There have therefore been moves towards denotification of some of the wildlife sanctuaries. There have been similar experiences in other parts of the Third World as well, with management focusing on flagship species, often in the interests of ecotourism, and concentrating on excluding local communities, with little participation or benefit sharing on the part of the latter (Gomez-Pompa and Kaus, 1992). Only in a few exceptional cases, such as Papua New Guinea, has there been some recognition of the traditions and rights of the local communities and attempts to involve them in the conservation of biodiversity.

THE AMERICAN EXPERIENCE

The 1960s witnessed the beginnings of an interest in biodiversity issues on the part of the developed countries. This interest began with a realization that pesticides persisting in the environment were reaching excessive levels of concentrations in the bodies of animals high up in the food chain, with serious consequences, such as reproductive failure in the case of the peregrine falcon. The discovery of DDT in the bodies of penguins from Antarctica also highlighted the ubiquity of these substances. The result was the passage of the Endangered Species Act of 1973 in the USA (Primack, 1993). This act committed the US Federal Government to protect critical habitats of endangered species, thereby promoting conservation of a diversity of natural habitats outside the National Park system. The 1970s also saw the coming on the scene of Nature Conservancy, a Washington-based NGO with considerable influence. Nature Conservancy initiated the process of systematizing the assignment of conservation priorities to various elements: individual taxa, or populations of particular taxa in a given region, as well as land and water elements supporting populations of one or more species of high conservation priority. The habitats so identified may be of very limited size, of a few hectares or less.

Nature Conservancy has followed up on the identification of such habitats by either their outright purchase or by organizing agreements with the owners compensating them in some way for giving up the option of developing the property in ways incompatible with the conservation objectives. These developments have brought onto the modern biodiversity management scene the practice of protecting large numbers of patchy, widely dispersed elements of the landscape, analogous to the traditional practices of protection of sacred groves or water bodies (Grove, 1988). The Nature Conservancy approach also shifted attention from the protection of flagship species and spectacular landscapes to habitats rich in taxa of high conservation value. This habitat focus also obtains in programs in other countries, such as protection of *Sites of Special Scientific Interest* in Great Britain. These programs have also brought in the new element of paying individual landowners for behaving in ways conducive to the conservation of biodiversity.

The 1970s saw the beginning of serious attention being paid not only to protecting specific areas, but also to regulating processes that have an impact on biodiversity levels. Such regulation pertains to the production of harmful substances such as pesticides and a variety of industrial effluents that directly affect living organisms, as well

as molecules such as chlorofluorocarbons that have an indirect effect through destruction of ozone and consequent increase in levels of ultraviolet radiation. The regulatory process has also broadened to assessment of environmental impacts of a range of developmental activities, such as the damming of rivers or extraction of timber. There is, however, as yet little explicit consideration of broader biodiversity issues in the process of environmental impact assessment, with attention remaining focused on endangered species. The list of these endangered species does include many flowering plants and smaller vertebrates as in the famous case of the snail darter – but there is still very inadequate attention paid to invertebrates or microbes.

The efforts at management of biodiversity in the developed countries have thus begun to incorporate a number of new, significant elements. These include: (1) interest in a broader range of living organisms, (2) focus on a wide range of habitats, often small in extent and widely dispersed, (3) positive rewards to private landowners for adopting biodiversity friendly practices, and (4) overall regulation of processes affecting biodiversity.

IMPERATIVES OF BIOTECHNOLOGY

By the 1970s the remarkable developments that followed the elucidation of the molecular basis of heredity began to be applied to the manipulation of life forms to human ends. These, in combination with a number of other technological advances, have led to the developments that go by the generic name of biotechnology. The biotechnological industry is already an important player on the world economic scene; it is expected to play a much bigger role in the coming decades. Its bag of tricks includes the capability to move genes from one organism to another; from a mouse to a bacterium, from a virus to a tobacco plant. This implies that organisms once believed to be insignificant could turn out to be of considerable commercial value. Industrial concerns are therefore now greatly interested in access to the world's resources of biological diversity. They would also benefit from information on traditional uses of such biodiversity, as drugs, dyes, cosmetics, and so on. Much of this biodiversity is outside industrially developed countries; much knowledge of the uses of plants and animals is with the ecosystem people of the Third World.

Organizing access to this biodiversity and this information is therefore now a priority of biotechnology industry, and of the industrial nations. These players are also interested in establishing monopolistic control over these resources and this information to the greatest extent possible (Reid *et al.*, 1993). In the meantime, at least until the biotechnological applications are in place and monopolistic control is established over biodiversity resources – perhaps held in *ex situ* collections – as well as information on their uses, the industry and industrial nations are interested in the conservation of natural biodiversity and knowledge of its uses. It is likely that this interest will be evanescent; to be given up once monopoly over biodiversity resources brought into *ex situ* collection is established (Gadgil, 1993). Nevertheless, for the present, there is considerable worldwide concern with managing biodiversity, concern that has led to the signing of the International Convention on Biological Diversity by over one hundred and seventy countries. This convention has constructed a radically different framework for the management of biodiversity than what has prevailed so far (Reid *et al.* 1993).

ECODEVELOPMENT

There are other significant developments as well. As democracy has struck roots in India, and other parts of the world, the paradigm of state-controlled conservation treating local communities as enemies is being questioned. Thus, the difficulties experienced by Jharkhand tribals have triggered a variety of political protests, some non-violent, others violent, resulting in the creation of a separate Jharkhand state. One of the slogans of this Jharkhand movement has an ecological content: "Teak is Bihar, sal is Jharkhand." Their demand is for the retention of natural sal-dominated forest.

A notable development accompanying such awakening is the program of participatory management of forest resources. Under this program villagers are encouraged to organize village forest committees (VFCs) to promote natural regeneration of forests. The VFCs are given certain authority to control forests; an authority that was earlier monopolized by the Forest Department, and are expected to assume responsibility for forest protection. In return they are given full rights over non-timber forest produce such as sal leaves and oil seeds, as well as a share of any timber that may be harvested. This program initiated in 1973 in the tribal villages of Midnapore district (lat. 22° 20' N long. −87° 20' E) of West Bengal has now spread over many other parts of this tribal belt, as well as elsewhere in the country. The program has by and large been a success. It is an important instance of the ecosystem people getting organized to promote ecological restoration, and in the process enhancing their own standard of living (Poffenberger and McGean, 1996; Ravindranth *et al.*, 2000).

The Jharkhand political struggle also incorporates an issue with significant implications for biodiversity conservation. The tribal religious practices throughout the country included protection of sacred plants, animals, ponds, and forests. The system of sacred groves, called "saranas" has been an important ingredient of the traditional tribal religion of Jharkhand. These saranas are variable in size, mostly small, about 0.1 ha, but very numerous, and serve to protect samples of original natural plant life. These were treated with contempt by Christian missionaries who had a significant influence in this tribal region, as also by proponents of high-caste Hinduism. Now protection of these saranas has emerged as an element of the move to restore tribal self-respect and to empower them (Gokhale *et al.*, 1998). Indirectly then, conservation of indigenous biodiversity is beginning to emerge as a part of the agenda of the poor to pull themselves out of the poverty trap.

UNESCO's Biosphere Reserve program is another endeavor to combine efforts at conservation of biodiversity with alleviation of poverty of ecosystem people. However, it has met with limited success largely because of inadequate understanding of local ecology and people's needs. Thus, Nandadevi (lat. 30° 30'–30° 40' N, long. 79° 44'–79° 58' E) in the western Himalayas is one of India's Biosphere Reserves. In this region, the traditional livelihoods depended on summer grazing in these pastures supplemented by collection of medicinal herbs. This grazing has been banned in the core zone of the biosphere reserve, along with mountaineering expeditions, another important source of cash incomes. As a result, local people feel significantly impoverished. The few ecodevelopment programs such as fuelwood plantations that have been brought in as a part of the Biosphere Reserve activities have failed to create sufficient levels of additional incomes. At the same time local people believe that in the absence of grazing the diversity of medicinal herbs in the pastures are now being replaced by extensive growth of a few species such as *Rumex* (Negi, 1999).

The philosophy behind biosphere reserves, that of combining conservation with development efforts, also motivates a number of development programs that go by the name of "Integrated Conservation and Development Programs" (ICDPs), and the many projects funded by the Global Environment Facility (GEF). However, there has been little genuine progress so far in actually realizing these goals. As a review of a GEF program in New Guinea documents, the programs remain ineffective as they continue to be designed by outsiders committed to conservation, but with little understanding of what would motivate local people, especially the poor, to participate in conservation efforts (Global Environment Facility, 1998).

There are also attempts to bring to local people financial benefits, either through ecotourism, or through sustainable harvesting of biodiversity resources. There are limited success stories of combining the two in parts of Zimbabwe where substantial incomes are generated though high-priced hunting licenses for large mammalian game animals, with levels of such harvests kept well within sustainable limits (McNeely, 1995). These incomes are shared with local, relatively poor people. Other attempts at combining poverty alleviation with biodiversity conservation have focused on promoting enterprises such as processing of medicinal plants or wild fruit. Programs in India involving processing of fruit of *Phyllanthus emblica*, a moist deciduous forest tree species, by local Solliga tribals in the B.R.T. (Biligiri Rangan Temple) hills of Karnataka or processing and marketing of wild mango by local women's groups in the Kangra district of Himachal Pradesh have met with varying degrees of success (Chopra, 1998). Unfortunately, this program in the B.R.T. hills has run into difficulties with the insistence of conservationists on total exclusion of any harvests of non-timber forest produce (Tiger Task Force, 2005). Providing alternative means of livelihood to communities depending on a threatened biological resource has also been attempted with success in the Indian ocean islands of the Seychelles where turtle shell artisans were provided financial and technical support to shift to occupations like coconut souvenir carving (Global Environment Facility, 1997).

REFERENCES

Berkes, F. (ed) (1989) Cooperation from the perspective of human ecology. *Common Property Resources: Ecology and Community Based Sustainable Development*. Belhaven Press, London, pp. 70–88.

Bhattacharyya, S. (1993) Ecological Organization of Indian Rural Populations. Ph.D thesis, Indian Institute of Science, Bangalore.

Borgerhoff-Mulder, M. (1988) Behavioural ecology in traditional societies. *Trends in Ecology and Evolution* 3(10): 260–266.

Boyd, R. and Richerson, P.J. (1985) *Culture and the Evolutionary Process*. Chicago University Press, Chicago.

Chopra, K. (1998) Economic aspects of biodiversity conservation. *Economic and Political Weekly* XXXIII (52): 3336–3340.

Crosby, A.W. (1986) Ecological imperialism. *The Biological Expansion of Europe, 900–1900*. Cambridge University Press, Cambridge.

Dasmann, R.F., (1988) Towards a biosphere consciousness. In: D.Worster (ed.) *The Ends of the Earth: Perspective on Modern Environmental History*. Cambridge University Press, Cambridge, pp. 177–188.

Diamond, J. (1991) *The Rise and Fall of the Third Chimpanzee*. Vintage, London, pp. 360.

Diamond, J. (1993) New Guineans and their natural world. In: S. Kellert, and E.O. Wilson, (eds) *The Biophilia Hypothesis*. Island Press, Washington, DC.

Eaton, P. (1985) Customary land tenure and conservation in Papua New Guinea. In: J.A. McNeely, and D. Pitt (eds) *Culture and Conservation: The Human Dimension in Environmental Planning*. Croom Helm, Dublin.

Gadgil, D.R. (1942) *Industrial Evolution of India*. Oxford University Press, New Delhi.

Gadgil, M. (1991a) Conserving India's biodiversity: the societal context. *Evolutionary Trends in Plants* 5(1): 3–8.

Gadgil, M. (1991b) India's deforestation: patterns and processes. *Society and Natural Resources* 3(2): 131–143.

Gadgil, M. (1993) Tropical Forestry and Conservation of Biodiversity, *Norway/UNEP Experts Conference on Biodiveristy*. Trondheim, 24 May.

Gadgil, M. (1996) Managing biodiversity. In: K. J. Gaston (ed.) *Biodiversity: A Biology of Numbers and Differences*. Blackwell, Oxford.

Gadgil, M. (1998) Conservation: Where are the people? In: *Annual Survey of the Environment, '98*, The Hindu, pp. 102–137.

Gadgil, M. and Berkes, F. (1991) Traditional resource management systems. *Resource Management and Optimization* 18(3–4): 127–141.

Gadgil, M. and Guha, R. (1992) *This Fissured Land: An Ecological History of India*. Oxford University Press, New Delhi, and University of California Press, Berkeley.

Gadgil, M. and Guha, R. (1995) *Ecology and Equity: Use and Abuse of Nature in Contemporary India*. Routledge, London.

Gadgil, M. and Vartak, V.D. (1976) Sacred groves of Western Ghats of India. *Economic Botany* 30: 152–160.

Gadgil, M. Berkes, F. and Folke, C. (1993) Indigenous knowledge for biodiversity conservation. *Ambio* XXII (2–3): 151–156.

Gadgil, M., Hemam, N.S., and Reddy, B.M. (1997) People, refugia and resilience. In: C. Folke and F. Berkes (eds) *Linking Social and Ecological System*. Cambridge University Press, Cambridge, pp. 30–47.

Global Environment Facility (1997) *GEF Project Implementation Review*. GEF, Washington DC.

Global Environment Facility (1998) *Study of GEF Project Lessons*. GEF, Washington, DC.

Gokhale, Y., Velankar, R., Chandran, M.D.S. and Gadgil, M. (1998) Sacred woods, grasslands and water bodies as self-organized systems of conservation. In: P.S. Ramakrishnan, K.G. Saxena and U.M. Chandrashekara (eds) *Conserving the Sacred For Biodiversity Management*. Oxford and IBH Publishing, New Delhi, pp. 366–396.

Gomez-Pompa, A. and Kaus, A. (1992) Taming the wilderness myth. *BioScience* 42(4): 271–279.

Gomez-Pompa, A., Flores, J.S. and Fernandez, M.A. (1990) The sacred cacao groves of the Maya. *Latin American Antiquity* 1(3): 247–257.

Grove, N. (1988) Quietly conserving nature. *National Geographic* 174 (January): 818–844.

Grove, R.H. (1992) Origins of western environmentalism. *Scientific American* July, 22–27.

Hemam, N.S. (1997) The Changing Patterns of Resource Use and its Bio-social Implications: An Ecological Study Amongst the Gangtes of Manipur. Ph.D thesis. Calcutta University, Calcutta.

Hughes, J.D. (1984) Sacred groves: the gods, forest protection and sustained yield in the ancient world. In: H.K. Steen (ed.) *History of Sustained Yield Forestry*. Forest History Society, Durham.

Hughes, J.D. (1994) *Pan's Travail: Environmental Problems of the Ancient Greeks and Romans*. Johns Hopkins University Press, Baltimore.

Jolly, A. (1980) *A World Like Our Own. Man and Nature in Madagascar*. Yale University Press, New Haven and London. pp. 272.

Joshi, N.V. and Gadgil, M. (1991) On the role of refugia in promoting prudent use of biological resources. *Theoretical Population Biology* 40 (2): 211–229.

Kangle, R.P. 1969. *Arthashastra*. University of Bombay, Bombay.

Karve, I. 1967. *Yugantha*. Deshmukh, Pune, pp. 287.

Kellert, S. and Wilson, E.O. (ed.) (1993) *The Biophilia Hypothesis*. Island Press, Washington, DC.

Koppes, C.R. (1988) Efficiency, equity, esthetics: shifting themes in American conservation. In: D. Worster (ed.) *The Ends of the Earth*. Cambridge University Press, Cambridge, UK, pp. 230–251.

Lenski, G. and Lenski, J. (1978) *Human Societies: An Introduction to Macrosociology*. McGraw-Hill, New York.

Malhotra, K.C. 1990. Village supply and safety forest in Mizoram: a traditional practice of protecting ecosystems. In: *Abstracts of V International Congress of Ecology, Yokohama, Japan*, p. 439.

McNeely, J.A. (1995) Economic incentives for conserving biodiversity: Lessons from Africa. In: L.A. Bennum, R.A. Aman and S.A. Crafter (eds) *Conservation of Biodiversity in Africa. Local Initiatives and Institutional Roles*. National Museums of Kenya, Nairobi, pp.199–215.

Meggitt, M.J. (1965) The association between Australian aborigines and dingoes. In: A. Leeds and P. Vayda (eds) *Man, Culture and Animals*. American Association for the Advancement of Science, Washington, DC, pp. 7–26.

National Commission on Agriculture (1976) Report of the NCA – Part IX – Forestry. Ministry of Agriculture, Government of India, New Delhi.

Negi, H.S. (1999) Co-variation in Diversity and Conservation Value Across Taxa: A Case Study from Garhwal Himalaya. Ph.D thesis, Indian Institute of Science, Bangalore.

Nietschmann, B. (1985) Torres strait islander sea resource management and sea rights. In: K. Ruddle and R.E. Johannes (eds) *The Traditional Knowledge and Management of Coastal Systems in Asia and the Pacific*. UNESCO, Ja Ranta Pusat, Indonesia.

Nolan, M.L. 1981. Types of contemporary Western European pilgrimage places. In: *Conference on Pilgrimages: The Human Quest*. Pittsburgh.

Pemberton, R.B. 1835. *Report on the Eastern Frontier of India*. Calcutta.

Pinker, S. (1997) *How the Mind Works* (1999 reprint by Penguins, London), pp. 660.

Poffenberger, M. and McGean, B. (eds) (1996) *Village Voices, Forest Choices: Joint Forest Management in India*. Oxford University Press, Delhi.

Pretty J. (2004) How nature contributes to mental and physical health. *Spirituality and Health International* 5(2): 68–78.

Primack, R.B. (1993) *Essentials of Conservation Biology*. Sinauer Associates, Sunderland, MA, p. 564.

Ramakrishnan, P.S. (1992) Shifting agriculture and sustainable development: An interdisciplinary study from North-Eastern India. In: J.N.R. Jeffers, *Man and Biosphere Series*, Vol. 10. The Parthenon Publishing Group, Lancaster, JK.

Ravindranth, N.H., Murali, K.S. and Malhotra, K.C. (eds) (2000) *Joint Forest Management and Community Forestry in India*. Oxford and IBH Publishing Co., New Delhi.

Reid, W.V., Laird, S.A., Meyer, C.A., Gamez, R., Sittenfeld, A., Janzen, D., Gollin, M.A. and Juma, C. 1993. *Biodiversity Prospecting: Using Genetic Resources for Sustainable Development*. World Resources Institute, Washington, DC, p. 341.

Ruddle, K. and Johannes, R.E. (eds) (1985) *The Traditional Knowledge and Management of Coastal Systems in Asia and the Pacific*. United Nations Educational Scientific and Cultural Organization. Jakarta Pusat, Indonesia, pp. 313.

Saldanha, C.J. (1989) *Andaman, Nicobar and Lakshadweep, An Environmental Impact Assessment*. Oxford and IBH Publishing Co., New Delhi.

Seshadri, B. (1969) *The Twilight of India's Wildlife*. John Baker, London, pp. 212.

Slobodkin, L.B. (1968) How to be a predator. *American Zoologist* 8: 43–51.

Superintendent of Government Printing, Calcutta (1909) *Andaman and Nicobar Islands*. Imperial Gazetteer of India, Provincial Series.

Terborgh, J. (1986) Keystone plant resources in the tropical forest, In: M.E. Soule (ed.) *Conservation Biology*. Sinauer Associates, Sunderland, MA.

Tiger Task Force (2005) *Joining the Dots*. New Delhi.

Tooby, J. and DeVore, I. (1987) The reconstruction of hominid evolution through strategic modeling. In: W.G.Kinzey (ed.) *The Evolution of Human Behavior: Primate Models*. SUNY Press, Albany, NY.

Vayda, AP (1974). Warfare in ecological perspective. *Annual Review of Ecology and Systematics* 5: 183–193.

Vijayan, V.S. (1987) Keoladeo national park ecology study. *Bombay Natural History Society, Annual Report, 1987*. Bombay.

Whyte, L. (1967) The historical roots of our ecological crisis. *Science* 155: 1203–1207.

Wilson, A.D. (1990) The effect of grazing on Australian ecosystems. In: D.A. Saunders, A.J.M. Hopkins and R.A. How (eds) *Australian Ecosystems: 200 Years of Utilization, Degradation and Reconstruction*. Proc. Ecol. Soc. Aust. 1990–16. Surrey Beatty and Sons, Chipping Norton, NSW.

Wilson, E.O. (1984) *Biophilia*. Harvard University Press, Cambridge, MA.

35

Coral Reefs and People

David Smith, Sarah Pilgrim and Leanne Cullen

CORAL REEF CONSERVATION

The term coral reef often produces images of warm climates, crystal clear turquoise waters, golden beaches and a huge array of colourful fish and other species. They are well renowned for their beauty, biological diversity and high productivity and it is this latter characteristic that makes them critical to the survival of tropical marine ecosystems and to the welfare of millions of local peoples (Berg *et al.*, 1998; Hoegh-Guldberg, 1999).

Many different species make up a coral reef but the dominant feature is reef-building corals. These *hermatypic* corals are colonial and made of many millions of small anemone-like individuals interconnected to form the characteristic colonies that form the backbone of a coral reef. Hermatypic corals secrete calcium carbonate skeletons that are laid down on top of dead coral. The success of hermatypic corals is largely due to their symbiotic relationship with microscopic algae (zooxanthellae) that live as part of a mutualistic relationship within the surface layers of their coral host. The zooxanthellae are photosynthetic and provide corals with up to 98% of their energetic requirements. Therefore, corals are primarily light dependent, a trait that restricts corals to the tropical and sub-tropical belts where temperatures are constantly warm and light intensity constantly high. Corals are also, in part, heterotrophic, and are dependent on their anemone-like tentacles to feed on zooplanktons mostly during the night.

The secretion of calcium carbonate by corals, as well as a myriad of other algae and animal species, produces a highly complex three-dimensional biogenic structure that enables many different species seemingly to coexist but utilise their environment slightly differently, thereby reducing competitive stress. Corals are therefore key to the diversity of reef-based systems. Coral reefs represent one of the biggest natural structures on the planet and are home to more species than any other marine system. They are vitally important for the socio-economic welfare of hundreds of millions of people and provide a variety of services. For example, coral reefs act as natural seawall defence and buffer the coastline from the erosive forces of large storms and tidal surges. Never has this fact been more evident than when examining the impacts of the 2004 Asian Tsunami and the recent spate of Caribbean hurricane activity.

Their biological diversity alone makes coral reefs exceptionally important, but this, coupled with the other functions they provide such as provision of food, bioactive compounds, other economic benefits, coastal protection and geochemical cycling, results in coral reefs being one of the most important ecosystems on the planet. However, the majority of coral reefs around the world are overexploited and threatened, with up to 60% showing severe signs of decline (Wilkinson, 2002). These disturbing facts are exacerbated by the fact that pressures are likely to double over the next 50 years as populations expand, coastal zones become more developed and other anthropogenic-induced stresses and natural phenomena continue to degrade reefs around the world.

Realisation of the threats that face coral reefs is the starting point for effective strategic conservation management, which needs to be both regional as well as global in perspective. Realisation of the full economic potential of reef systems can also be a very positive outcome of trans-disciplinary conservation research and a powerful persuasive argument for the need for, and benefits of, active conservation and sustainable managed practices. Conservation actions and strategies need to be inclusive in their approach and consider both local and global issues. Direct management action is urgently required if reef resources are to be available, at least in their current state, for future generations. We face a critical time in the world of coral reef conservation with coastal populations rising, climatic conditions changing and the demand for reef-based resources increasing. Predictions suggest that up to 70% of coral reefs could be lost by 2050 and some conservation scientists have suggested that coral-dominated reefs could completely disappear over the next 50–100 years.

The loss of coral reef systems will have a major impact on over half a billion people and will be devastating for many maritime states and countries. Urgent collective action is required but first we need detailed knowledge of reef systems, their biological function and their realised economic and intrinsic value as well as their response to past, present and future conservation management strategies. Numerous studies have demonstrated that the threats to reefs are similar the world around, but the methods required to protect reefs need to consider the traditions and requirements of local user groups as well as take into consideration the global perspective.

Each dependent community has different requirements, ranging from daily subsistence needs to those related to a multi-billion dollar tourist industry. It is essential that we take into consideration these stakeholders, and it is most likely that different approaches, albeit with similar project aims, need to be considered and implemented if we are to be successful in conserving reef-based systems. A truly cross-disciplinary approach is needed to protect reefs as is an understanding of local user group requirements and attitudes towards conservation. Without such approaches it is highly unlikely that restrictive management strategies will be successful.

THE IMPORTANCE OF CORAL REEF-BASED SYSTEMS

Reefs contain more species than any other marine system. For example, approximately 25% of all marine species identified are reef associated despite coral reefs making up less than 1% of the Earth's surface. Their topographic diversity, intermediate levels of disturbance, tight resource partitioning, specialisation and high levels of competition are just some of the reasons why reefs

are so diverse containing representatives from over 95% of all animal phyla as compared with only 25 to 30% characteristic of tropical rainforests. This biodiversity has both intrinsic and extrinsic values but the true biodiversity of reefs is unknown and it has been suggested that currently we only know about 10% of the true species richness, with estimations ranging from 600,000 to 9 million reef-associated species worldwide. With the world facing accelerated species loss (estimated as being up to 1000 times the background rate), factors that significantly contribute to the degradation of reef systems will have global impacts with respect to the Earth's total species and gene pool.

We know that coral reefs are among the most diverse ecosystems on Earth, but the level of diversity held within reef systems is geographically dependent. The greatest coral reef diversity is found at the Indo-Pacific interface, and diversity decreases as you radiate away from this biodiversity hot spot. For example, around Raja Ampat, Indonesia, there are over 600 of the world's corals (out of a total of 800), 400 species are found in the Philippines, about 350 in the northern Great Barrier Reef, about 250 in Fiji and less than 50 in Hawaii. A second centre of biodiversity exists around the western shelves of the Atlantic, and in particular the Caribbean but here only around 60 species of coral can be found. On a more regional scale, patterns of diversity are affected by localised environmental conditions, for example, exposure levels as well as stochastic environmental features (Dornelas *et al.*, 2006) and human practices.

THREATS TO CORAL REEFS

Numerous factors result in the loss of species from a coral reef and the reduction of its physical as well as biological integrity (Graham *et al.*, 2006). Generally speaking we can classify the threats facing reef-based systems into *natural phenomena* and *anthropogenic stressors*. The effects of these phenomena and stressors range from negligible to catastrophic. In part, natural phenomena impacting reefs help maintain biological diversity by reducing dominance and producing temporal and spatial patchiness. Furthermore, coral reefs display a good ability to adapt to short-term natural catastrophic events through an array of physiological and behavioural plastic responses. However, coral reefs are not well adapted to survive exposure to long-term stress and thresholds exist beyond which communities can collapse, resulting in a long-term loss of diversity and productivity (and therefore natural capital). Such shifts in community structure are often termed "phase shifts" as community changes can be dramatic, for example, from a diverse and productive coral-based system to one that is overgrown and dominated by algae and therefore characterised by having reduced physical as well as biological complexity. One of the problems of course is in identifying these thresholds and introducing appropriate measures to prevent or reverse the trend.

Anthropogenic stressors, examples of which include agricultural and industrial runoff, increased sedimentation from land clearing, human sewage as well as direct physical damage, tip the balance in favour of species loss rather than maintenance and have had devastating effects around the world. It is also becoming increasingly difficult to distinguish between the natural and anthropogenic factors as the effects of natural factors could be accelerated by anthropogenic stressors. For example, coral bleaching is in part a natural response to environmental stress, in particular high seawater temperatures and high light intensity. Coral bleaching events are therefore related to prevailing climatic conditions and will be greatly affected by the predicted human-accelerated global climate change. In recent decades bleaching episodes have increased in their frequency and intensity with more than 16% of the world's coral reefs being destroyed in the last big bleaching event during the 1998 El Niño (Wilkinson, 2002).

It seems most likely that future El Niño events will have an even greater effect with predictions suggesting that up to a quarter of all coral reefs could be killed during the next one. Although we are now starting to understand the fundamental mechanisms that cause bleaching (Smith *et al.*, 2005), we are far from suggesting possible remediation actions and our best strategy at present is to ensure that other factors that stress reef systems, which can be more easily managed, are done so, giving reefs the best possible chance to respond favourably to changing climate and environmental conditions.

Other "natural" factors that result in reef degradation include large storms and hurricanes. It has been shown that the growth and recruitment rate of corals are negatively impacted by storm damage (Crabbe and Smith, 2003; Crabbe *et al.*, 2004). Intermediate levels of disturbance help maintain reef diversity by decreasing competitive exclusion and enhancing patchiness. However, increased frequency and intensity of storms may once again tip the balance beyond a species maintenance function to one that greatly reduces the physical and biological integrity of reef systems.

Finally, with respect to "natural phenomena", coral diseases have recently increased in their frequency of occurrence. "Disease" is a phenomenon common to all biological populations and communities and once again probably helps maintain biological diversity in the long term. However, it seems that the frequency of disease occurrence is increasing, which has been attributed to increases in organic pollution (i.e. sewage) and land pollutants washing on to reef systems. So once again we see that human impacts are adding to the weight of natural phenomena that are impacting reefs the world around. It is very difficult if not impossible to start to tackle the natural phenomena; the best alternative available to managers is to identify and alleviate the human-induced pressures thereby giving reefs a better chance of withstanding natural, degrading phenomena.

Many studies have examined anthropogenic stressors and their impacts on reef systems. These include: overexploitation of both fin and shell fisheries, the use of non-sustainable and damaging fishing practices, exploitation of coral rock and physical destruction of reef systems, coastal development leading to loss of feeder habitats including mangrove and seagrass beds, and in terms of coastal terrestrial environments, deforestation and land redevelopment. The relative importance of each of these factors varies dramatically around the world and is highly dependent on the characteristics of local people as well as the geographic location. For example, the majority of coral reefs within southeast Asia are overexploited in terms of fisheries and the use of destructive techniques such as blast fishing and poison fishing represents one of the major problems. In parts of Central America reefs such as the Honduras Bay Island reefs, part of the Meso-American Barrier Reef System, have been negatively impacted by the establishment of large plantations on the mainland, whereas off the coast of Sinai within the Red Sea, reefs are threatened by dive-based tourism.

In the majority of regions of the world coral reef habitats are overfished and/or overexploited for one reason or another. Coral colonies and brightly coloured charismatic reef fishes and invertebrates are collected for the growing aquarium and jewellery trade. Currently, over 1000 species are taken from coral reefs for the aquarium trade and, although tightly regulated in places, it is not in others, and a black market trade in species export exists. Also, quotas as well as collection techniques vary and can have major consequences for the integrity of reef systems. For example, cyanide poisoning is often used in parts of southeast Asia to stun fish either for the pet trade or the live food trade. It is extremely difficult to detect which species are caught legally as compared to species caught in an unsustainable and unregulated manner.

Probably the major threat facing reef systems is overextraction in general. As fishing techniques advance (e.g. become more efficient) and access to these techniques increases, overexploitation of an already exploited system is likely to increase.

This coupled with increases in local population sizes is likely to result in a doubling of current exploitation rates over the next 50 years or so. Fisheries pressure and the resulting impacts vary greatly with the techniques used. Some techniques such as blast fishing are completely unsustainable but, although being highly illegal, they are commonly used in some parts of the world.

Overfishing in general also represents a huge problem for coral reefs. Reefs represent a limited resource in terms of the fish they recruit. Non-targeted fisheries techniques and increased fisheries effort due to technique or the actual number of fishermen could result in the loss of a single species and the collapse of a coral reef system. For example, removal of grazing fish species and hence the reduction in grazing pressure can result in fast-growing algae species overgrowing a reef. Also, non-size selective fisheries techniques, such as the many forms of fish fences common around Indonesia, can dramatically reduce recruitment of new fish biomass into the system and once again result in community as well as population collapse.

The term pollution is of course very general, and in terms of coral reefs the biggest threats really relate to those factors reducing water quality. Pollutants can be in the form of fertilisers or sewage which can elevate otherwise limiting nutrients. Normally nutrient levels are very low and this characteristic gives corals the advantage over the normally fast grown algal species. An increased nutrients status can therefore lead to algal overgrowth (Goreau, 1992). This will lead to reduced light availability to reef-building corals and overall a reduction in physical complexity. Reduced light is also common in areas where there is large run off from the land. Terrestrial soils can limit light availability to reefs and increased input of sediment is common when coastal lands become developed through deforestation or manipulation of the coastal fringe. The removal of mangrove and seagrass beds is common practice and both these habitats are connected with reef systems. They constitute important nursery habitats for juvenile coral reef species whilst also entrapping sediments (marine and land based) which may otherwise impact adjacent reefs.

Greater sediment load coupled with increased deposition rates can greatly alter the dynamics of a coral reef. The various species of coral respond differently to sediment load and deposition. At the individual level, colony corals demonstrate decreased growth rates (Crabbe and Smith, 2002, 2005) and changes in their colony structure (Crabbe and Smith, 2002, 2006). A reduction in species diversity is common and eventually, when natural rates of erosion are greater than rates of calcification, coral reef systems can completely collapse.

Any factor that increases sediment input on to reefs needs to be carefully managed if there is not to be a decrease in coral diversity. Ironically some of these processes have increased due to the aesthetic value of reefs, for example tourism development, which might inadvertently be impacting their most important commodity.

Of course, the level of impact and the potential for recovery depends largely on the method and scale of the exploitation (or pollution). Some exploitative techniques are highly damaging and examples are the aforementioned blast fishing and coral mining. Trends in blast fishing have changed and now it is common practice in some parts of the world to use multiple bombs on, or "carpet" bombing of, reef systems. Many centuries of coral growth can be destroyed and although recovery rates do vary depending on local conditions, impacted reefs can take between 50 and 100 years to recover. Similarly, the use of live coral in the mining trade also reduces the carrying capacity of reefs and is similarly unsustainable.

Coral reefs represent a huge resource of natural goods and services. They are and can be exploited; however, overexploitation and non-sustainable resource utilisation could have dramatic effects on the socio-economic welfare of many millions of people. Protection and active management are required and strategies should revolve around a multidisciplinary approach. In the majority of cases, in one way or another, many of the anthropogenic factors that negatively impact a coral reef stem from economic incentives. How are we to protect reefs for future generations, if we do not provide alternatives to local communities for loss of income derived from the required restrictive conservation policies? It is imperative that we take into consideration the economic characteristics of local stakeholder communities when producing management strategies that are aimed at being (a) biologically successful and (b) socially acceptable (both in terms of agreement and compliance).

Consequently, it is extremely important that we gain a full understanding of the economic value of coral reefs. It is important that we understand not only the realised value of reef systems but also their potential value. Only then can we start to study and suggest alternatives rather than additional income streams that are equivalent in economic benefit but also serve to reduce the pressures on reef systems. Such knowledge is essential if we are to ensure future sustainable coral reef exploitation and also if we are to produce workable management strategies that are agreeable to local as well as global stakeholders. Of course this requires a very much localised approach but lessons learnt from other areas can be cross-transferred to other dependent communities.

CORAL REEF ECONOMICS

Coral reefs represent an important economic resource with benefits accruing to local and global economies (Cesar, 2000). Locally, reef fisheries are a vital source of protein for millions of people, reef-related tourism is a major foreign currency earner, and reefs provide natural coastal protection from wave action and potential storm damage. On the global scale, reefs are valued for their role in the carbon and calcium cycles, their inherent existence, the consumer surplus enjoyed by Scuba divers, and for their bioprospecting potential (Spurgeon, 1992; Pendleton, 1995; Cesar, 2000).

Constanza et al. (1997) estimated the world's marine and terrestrial ecosystem services to be worth US$20,949 and 12,319 billion per year, respectively. Within the marine sector, it was estimated that coral reefs alone are worth US$6075 ha^{-1} per year, which equates to US$375 billion per year on a global scale. More recently it has been estimated that coral reefs provide annually around US$30 billion in net benefits in goods and services to world economies; these goods and services include fisheries, coastal protection, tourism and recreation, and biodiversity (Cesar et al., 2003).

Despite the numerous estimates on the value of reef-based systems the full economic potential of reefs is largely unrealised (Spurgeon, 1992). It is often the case that many of the world's ecosystem services and natural capital are given too little weight in policy decisions because they are not fully captured in commercial markets or adequately quantified (Costanza et al., 1997). Additionally, many of the world's richest ecosystems and poorest people are found together in the tropics so it is not surprising that human aspiration for improved living conditions often clashes with conservation and biodiversity objectives (Randall, 1991). This apparent conflict of interest between economic development and the environment has created worldwide problems. In 1983 the United Nations appointed an international commission to propose strategies for "sustainable development" defined as "development which meets the needs of human well-being in the short term without threatening the local and global environment in the long term." To achieve this aim, some form of accounting for national and international natural resources is required (Spurgeon, 1992).

Additionally, the realisation at the local scale that sustainable management of coral reefs can serve to increase their economic worth whilst also conserving biodiversity would go a long way towards the possibility of successfully implementing sustainable practices. Furthermore, the identification of viable (i.e. economically comparable) and sustainable alternative (not additional) income streams for reef-dependent communities could significantly help decrease current pressures on reefs. An understanding of the realised as well as potential economic value of reef-based systems in terms of the local user groups could help drive forward conflict resolution and help address the balance between exploitation and environmental degradation.

The role of economics in ecological protection

The disciplines of ecology and economics both have important contributions to make to the identification and solving of the global problem of overexploitation and degradation of natural capital (Perrings et al., 1992). Ecology can be used to establish the amount and availability of natural resources and habitats, and identify and monitor population dynamics and system changes. Economics has a major role to play in explaining resource use and degradation, measuring its impact, and designing policies to combat degradation (Pearce and Mäler, 1991; Dixon, 1997). The explanatory role of environmental or ecological economics is that we need to understand why resource degradation occurs and how economic mismanagement through wrong pricing, ill-defined property rights and incentive structures, contributes to environmental loss. Environmental economics can also be used within a policy development role to devise the best practicable solutions for sustainable development (Pearce and Mäler, 1991).

It is agreed by many environmental economists and ecologists that natural resources should be valued in economic terms when decisions involving the loss or preservation of them are made (Dixon, 1986). The main argument for this is that without true evaluation, the existence of these resources will be given zero value suggesting no net loss if they decline (Green and Tunstall, 1991). Those who oppose the economic evaluation of natural resources often misrepresent economics as being about money, which it is not. Money is simply the means of measurement to compare the relative values of different goods and services (Pearce and Turner, 1990). In economics all values given to goods are subjective and the environmental economist is seeking to derive a method of measuring the values different individuals place on different goods so that they can be compared (Green and Tunstall, 1991). Money is used as the basis for comparison because an individual's subjective values are unobservable and often incomparable;

for instance, it is difficult to measure one person's enjoyment of a natural resource as compared to another person's (Robbins, 1935).

The environment is often undervalued. Environmental benefits include many non-marketed goods and services which have no readily available monetary values. Additionally, natural habitats have off-site benefits which occur a distance away from the habitat. For correct valuation to be made, non-marketed goods and services and off-site benefits must be accounted for and recent advances in valuation techniques mean that more of these can be quantified in monetary terms providing more comparable and influential information (Spurgeon, 1992).

The role of economics in reef management

Approaches to natural resource management have been varied and continue to be adapted (Figure 35.1). In the past, natural resource and protected area management focused on understanding and managing ecosystems from a biological viewpoint with very limited user group consultation. Currently, however, management strategies have started to incorporate the wider social and economic factors, specific user groups are now actively encouraged to participate in various management actions, and cost–benefit analyses, environmental impact assessments and bio-economic models are used to support decision-making. It is thought that in the future, financial, business, legal and ethical factors will play an increasingly important role in natural resource management (Spurgeon, 2001). User groups will be more involved in the decision-making and action process, with the hope that this will allow empowerment and perceived local ownership

of resources which in turn leads to a sense of pride and increased local protection of resources. With this changing and dynamic management approach, understanding the existing and potential values of coral reefs is essential to a successful outcome that maintains ecological wealth and develops sustainable utilisation.

There is widespread interest in coral reefs and their conservation, but at the same time there is also a lack of resources to provide even minimal levels of management. Part of the reason for this is the difference between economic values and monetary prices, and that people and governments have been responding to monetary price signals. Additionally, local user groups are largely unaware of the potentially increased monetary benefits of a healthy reef in the long term. Economics can be used to explain this with total economic value (TEV) calculation being one of the most useful tools. This economic value of an ecosystem is often defined as the total value of the goods and services the ecosystem provides (Cesar, 2000).

Assessing the total economic value of coral reefs

Many of the benefits associated with coral reefs are not exchanged in markets, are hard to value and therefore have often been ignored or grossly underestimated. Complete and detailed quantitative data on coral reef ecosystems are rarely available so decisions are often made without these or the involvement of local user groups (Fernandes *et al.*, 1999). One problem with environmental valuation is identifying the various components of value and attaching monetary prices. For this the calculation of a TEV can be a very useful tool. It is calculated as a measure of the current

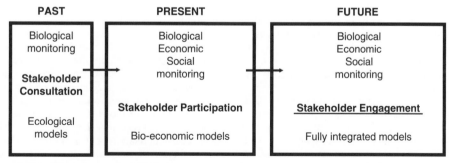

Adapted from Spurgeon (2001)

Figure 35.1   Changing approaches to natural resources management. Understanding the full current and potential values of coral reefs is critical to a successful outcome in a changing management approach

Table 35.1 Total economic value (TEV) of coral reefs

Use values			Non-use values		
Direct (extracted)	*Direct (in situ)*	*Indirect*	*Option*	*Existence and bequest*	*Intrinsic*
Fisheries	Tourism	Biological support	Maintenance	Knowledge of system	Biodiversity
Aquarium trade	Research	Coastal protection	of system for	existence and	Species richness
Curio trade	Education	Global life support	future use	continued existence	Existence with no
Pharmaceuticals	Recreation			for enjoyment by	human use
Construction	Culture			future generations	
materials	Religion				

TEV = use values + non-use values

economic worth of any resource and encompasses all direct, indirect and non-use values of that resource (Table 35.1). Direct use values relate to actual use or the goods (extractive and non-extractive) provided by the resource in question (e.g. fisheries or recreation); indirect use values are the functional uses, or services, provided by a resource (e.g. coastal protection); and non-use values represent the existence, bequest and option values, as well as the intrinsic value of natural resources and the biodiversity they support (Bateman *et al.*, 2002). The economic information contained in a TEV calculation can be very powerful in making the case, to local user groups and decision-makers in particular, for the benefits of protecting and managing coral reefs sustainably (Dixon, 1997). However, it is important to note that, because some benefits of coral reefs cannot be quantified, TEV provides only a minimum estimate for their "true" value (Spurgeon, 1992).

Direct use valuation

Direct use values include both *extractive* (e.g. fisheries, aquarium and curio trade, pharmaceuticals, construction materials) and *non-extractive* (e.g. tourism, research, education, social value) uses. The direct extractive goods are, in most cases, marketable commodities, hence their economic value is relatively simple to calculate utilising existing market prices (Spurgeon, 1992).

The potential economic value of direct harvesting (*extractive uses*) can be calculated using cost–benefit analysis (CBA) where all costs and financial benefits associated with the harvesting and sale of products are taken into account. Due to natural fluctuations in "catch", variable market prices and financial outgoings, CBA is best calculated over long periods (Spurgeon, 1992) which results in the production of economic productivity values. Changes in productivity will be reflected in the monetary values calculated and may be indicative of environmental/ecological changes. It should be noted, however, that the associated costs and benefits of harvesting any

product are difficult to determine and, in some cases, market prices may not reflect the true worth of a product.

Those goods collected purely for subsistence (self-consumption) have no direct financial value attached; therefore, other methods of valuation must be incorporated such as the replacement cost method, which utilises the market value of alternative potential replacement products, or contingent valuation (CV) which essentially values consumer surplus. CV uses hypothetical situations to put monetary values on non-marketed goods and services. To do this, people's hypothetical willingness to pay, or to accept compensation for, the use of a good or service is used (Spurgeon, 1992). For example, people might be asked how much they would be willing to pay for a certain reef product if they could not obtain it elsewhere themselves, or how much they would be willing to accept in compensation to discontinue use of that product. Where money is not perceived in the same way as in the Western world, the costless choice method can be used when the hypothetical bidding uses commonly exchanged goods (Dixon and Sherman, 1990).

Of the direct uses of coral reefs which are *non-extractive*, in most cases, tourism yields the greatest financial benefit, and many small island nations depend on reef-related tourism for their continued economic development. Revenues generated directly by reef-related tourism range from Scuba diving and marine park entrance fees to accommodation, food and travel costs (Spurgeon, 1992). The current value of tourism can be defined using the financial revenue (FR) approach, the contingent valuation method (CVM) or the travel cost method (TCM). The FR approach calculates the direct financial profits provided by reef-related tourism (Berg *et al.*, 1998). In addition to the financial benefits of tourism, there is often a large tourist consumer surplus value, which constitutes the additional satisfaction gained by tourists in excess of what they paid for their trip. In many cases tourists visit specific reef sites for free or pay less for admission and equipment hire than they would be willing to pay (Spurgeon, 1992). To determine the extent of this additional value the CVM can be used. TCM could be employed, which assumes that the number of people travelling to a site is inversely related to the distance travelled to get there. If the number of people visiting the site and their travel costs are known, regression analysis can be used to estimate the value of that site to visitors (Spurgeon, 1992). As CVM includes social values, this approach could yield a higher value than that of FR or TCM (Berg et al., 1998).

Local communities gain additional benefits from reefs in a way similar to tourist consumer surplus; this social value may include cultural and heritage values representing the benefit to communities of traditions and customs that have evolved based on reef associations, or spiritual and aesthetic benefits. For instance, nomadic Bajo tribes of southeast Asia base their cultural traditions, worldviews and spiritual beliefs on the sea and its components. No quantifications currently exist for the extent of social value, but estimations could be made using an adapted CVM, surveying locals on their willingness to pay to maintain the reefs in their current condition (Spurgeon, 1992).

Indirect use valuation

The indirect values of coral reefs are the associated functional benefits, or services, provided by reefs which include coastal protection, bioprospecting potential, and biological and global life support. For example, the physical structure of coral reefs provides people with indirect economic benefits without requiring direct resource extraction. Reefs that fringe the shore provide a natural wave break and protect economically as well as environmentally valuable coastal habitats from storm damage (Berg et al., 1998). Reefs also provide large amounts of beach material essential for the preservation of tropical sandy shores. The coastal protection function of coral reefs can be described through the preventative expenditure approach (or replacement cost method), defined as the cost of replacing the coral reef with protective constructions, for example, groynes or underwater offshore wave breakers (Spurgeon, 1992); or by looking at the loss of property value, defined as the cost of land loss (price of lost land, buildings, roads, etc.) as a result of coastal erosion. This method also includes the loss of income resulting from lost land-use opportunities (e.g. agriculture) (Berg et al., 1998).

Bioprospecting can also be considered under the Indirect Use Valuation banner. There has recently been a sharp increase of interest in bioprospecting, that is, the search for naturally occurring bioactive compounds that may have commercial benefits and applications, for example, agricultural, chemical or pharmaceutical (Simpson et al., 1996). The hit rates, that is, the probability of finding a bioactive compound in a species, is much higher on coral reefs than in comparable (in terms of diversity) terrestrial systems, and as the true biodiversity of reefs is unknown, regulated bioprospecting has high economic potential.

A high degree of connectivity exists between reefs and other tropical marine (and coastal) systems, and this characteristic has economic implications. Connected habitats and systems need to be considered during economic valuation exercises. Accurate valuation of connected systems is difficult and highly localised but estimations of a rough value may be possible using a change in productivity approach, which is essentially the difference in value of a reef-supported economic activity with and without the reef. Alternatively, the biological support value is effectively the value of the supported activity multiplied by an estimated percentage dependence of that activity on the reef's presence, referred to as the "percentage dependence technique" (Spurgeon, 1992).

Non-use valuation

The non-use values of coral reefs refer to the perceived benefits of reefs outside of the value of any goods or services they provide us with. Non-use values include the existence type values placed on reefs by humans, which is measurable, and the intrinsic value of coral reef and associated organism biodiversity.

Existence, Bequest and Option
Existence and bequest are simply the values that people place in the knowledge that a natural resource or individual organism exists and that it will continue to do so for future generations to enjoy. "The proof that these values exist is apparent in the fact that people will pay money to charities such as 'Save the Whale' even though they know they are unlikely to ever experience a whale first hand" (Spurgeon, 1992). Existence and bequest values have not been determined for coral reefs but have been measured for individual species (Pearce and Turner, 1990) and for other ecosystems. The only method of valuation is the CVM using people's stated willingness to pay for an area or species to be preserved. Measurement of the existence and bequest value of coral reefs would require an extensive CVM survey that included local, national and international population representations (Spurgeon, 1992). The greater the quality and the uniqueness of the reef on a national and global scale, the greater its existence

value will be. On the local scale, population size, level of income, education and environmental perception will also greatly influence the overall value (Spurgeon, 1992).

Biodiversity

The valuation of biodiversity in its own right, that is, as removed from any association with human welfare, is complicated, controversial and incredibly difficult to achieve, but methods for its valuation do exist. It is controversial because many people feel that it is wrong for humans to place a value on biodiversity as it is of infinite value, both to human welfare and in its own right. However, making public or private decisions that affect biodiversity implicitly means attaching a value to it. Hence, monetary valuation can be used as a democratic approach to make decisions about public issues, including biodiversity ones (Nunes and van den Bergh, 2001).

Biodiversity also has an instrumental value to human society and the maintenance of biodiversity has become a popular argument for ecosystem protection for this reason (Dixon and Sherman, 1991). There are potentially enormous welfare implications related to biodiversity loss, individual organisms have direct value in terms of consumption or production, and the combination of organisms, and their role in sustaining biophysical cycles, within a framework of ecosystems, makes them of indirect value in satisfying human needs for the services of those ecosystems (Perrings et al., 1992).

To value biodiversity, CVM is the most commonly used technique as it is able to identify and measure passive or non-use values. Existing monetary value estimates seem to give explicit support for the belief that biodiversity has a significant positive social value, but most studies lack a uniform clear perspective on biodiversity as a distinct concept separate from biological resources, that is, the instrumental value of biodiversity. Monetisation of benefits is possible but will always lead to an underestimate of the "real" value, since the primary (intrinsic) value of biodiversity cannot be translated into monetary terms (Nunes and van den Bergh, 2001). Therefore, any calculated value may be used to justify protection measures, but will constitute only a small portion of the total value of biodiversity (Gowdy, 1997).

Empirical work at Montego Bay, Jamaica, was carried out by Ruitenbeek and Cartier (1999) in order to estimate the net present value (NPV) of Montego Bay reefs. Estimated value for direct uses included tourism and recreation (NPV of US$315 million), fisheries (NPV US$1.31 million) and coastal protection (NPV US$65 million). Therefore, the value of the reefs for direct local uses was calculated at US$381 million or

US$8.93 million per hectare of reef. However, these values fail to include the non-use benefits reaped by both local residents and visitors to the area. When estimating the value of non-use benefits, Ruitenbeek and Cartier (1999) assessed the willingness to pay of local residents and tourists. They found that for typical population characteristics, and using typical visitor profiles, the non-use benefits of Montego Bay biodiversity has a NPV of US$13.6 million to tourists and US$6 million to Jamaica residents. The above values amount to a NPV of approximately US$400 million for Montego Bay reefs, or marginal benefits of US$10 million per 1% of coral abundance improvement. However, this is likely to be a lower-bound estimate as no institutional arrangement currently exists for capturing biological prospecting values, although estimated values stand at around US$7775 per species (Ruitenbeek and Cartier, 1999). Compared with reefs of central and southeast Asia, Jamaican reefs have relatively low biodiversity and low resource value as alternate food and medicine sources are available to local peoples. This emphasises the vast importance of even the less diverse reefs to human populations worldwide and the immense cost to society of replacing their goods and services if lost.

Much additional research is needed into natural resource and ecosystem valuation (Costanza et al., 1997). Economic assessments can be used to examine the extent of the benefits directly and indirectly associated with natural resource use (Bunce et al., 1999), but a key problem for policymakers is the lack of quantitative models and procedures to facilitate a comprehensive economic and ecological analysis, including identification, measurement and prediction of the effects of economic activity on the environment. Specifically, the degradation of coral reefs has not been extensively analysed in a framework amenable to economic policy analysis (Ruitenbeek et al., 1999). Hence, there is need for adequate valuation and dissemination of this information to policymakers, and if appropriate economic methods can be developed; the values generated could be used by policymakers to implement long-term economically and ecologically sound management practices.

One of the major reasons why conservation management initiatives fail to reach their goals is lack of compliance of local communities to management rules and regulations. There are three possible reasons for reduced compliance; (a) lack of awareness of rules and regulations; (b) disagreement between stakeholders; (c) stakeholders will be significantly and negatively impacted by new rules and regulations. The actual, perceived, or even just expected, economic losses to communities utilising reefs for livelihoods and subsistence is a major concern. So a key role of management should be to improve, or at the very least maintain, the economic status of local people, which means that management strategies must consider the impact on local people. Monitoring of economic status should be included (in addition to biological monitoring) to ensure no losses occur due to economically inappropriate management schemes. To do this, a simple series of economic performance criteria, that is, testable parameters in which changes could be used to measure the success of management, would be invaluable. Alternative income streams must also be made available where incomes begin to, or are expected to be, negatively impacted. Additionally, economics could, and should, be used as a tool to encourage conservation and sustainable utilisation efforts, as the economic benefits of a healthy reef system will clearly far outweigh those of an impacted and biologically limited system.

CORAL REEF MANAGEMENT

Highly diverse landscapes, the primary focus of conservation efforts, have remained occupied by human populations throughout history, and will continue to be into the future. Conservationists in the field have to make concise decisions on the realistic options available for the management of resources within these systems, aside from the speculation of management options widely discussed in theory (Alcorn, 1993). Despite long-term dependence on reef resources by coastal communities in particular, it has become evident that efforts to govern and sustain reef fisheries have frequently failed, especially in tropical fisheries where exploitation is intense and equipment diverse, and often even destructive. Transferable quotas are difficult to implement in artisanal fisheries whereby data are incomplete, reef ownership ill-defined and landings go unrecorded (Rudd *et al.*, 2003). Thus, in a deviation from previous trains of thought, policymakers, conservationists and academics alike have recently come to consider the importance of community in resource management, as a consequence of the poor outcome of government efforts.

A community is perceived as a spatially defined unit with a well-defined social structure (Agrawal and Gibson, 1999). In the past, intrusive strategies and externally planned regulations excluding community stakeholders have overruled traditional systems of management often at great costs to local communities with limited conservation benefits yielded. Shortcomings have included government instability, monitoring and enforcement

limitations, and inaccessible, unavailable or outdated science. Hence, there has been a recent surge of resources into community-management schemes from international agencies, including the World Bank, Worldwide Fund for Nature and The Nature Conservancy (Agrawal and Gibson, 1999).

Coral reef goods and services fall into the pool of common resources, with high subtractability by its users and difficulty in excludability of external non-authorised exploiters (Rudd *et al.*, 2003). Management of coral reef systems, like many other global resource pools, has predominantly been in the hands of those harbouring its goods and services historically. For ninth-tenths of human existence on Earth, hunting and gathering pressures exerted on any ecosystem have had to be constrained for the survival of the community and wildlife population through to the present day. Local resource users, particularly on isolated oceanic islands, are often keenly aware of their own impact on local ecosystems, observing the differences between natural and anthropogenically derived fluctuations (Drew, 2005). Thus, self-management practices, originated and legitimated locally (Feit, 1988), have evolved. These community–ecosystem reciprocal partnerships evolve from the equitable sharing of environmental benefits and the reinforced exclusion of locals.

Traditional management practices

Self-management practices derive from a sense of ownership and responsibility for the resource combined with an in-depth knowledge of the system, its components and their interactions, reinforced by a set of socially accepted informal rules. These rules comprise the backbone of the community with respect to worldview and are termed social norms. Some have even evolved more recently in a local response of resistance to government-enforced management efforts (Feit, 1988). Losing a portion of their individuality to fight for common interests, through reinforcing internalised behaviour norms, can act to influence the direction of self-management (Agrawal and Gibson, 1999). For instance, where established norms act to limit fishing to certain areas of the reef at certain times and protect areas of regeneration such as aggregation and spawning sites, community norms ensure management is self-sustaining, but where accepted norms promote excessive exploitation, for instance, in Indonesia where certain marginal tribes believe mined coral used in house foundations makes the building stronger, norms may be to the detriment of the environment. Rural Fijians, for instance, believe that reef is a supernatural occurrence, thus the introduction of conservation-oriented practices would prove futile in such a culture (Rudd *et al.*, 2003). Strategies for the intrinsic alteration of long-existing community norms are unknown.

Social norms held by the islanders of Ahus Island, Papau New Guinea, include prohibiting spear and net fishing, and strictly limiting invertebrate harvests, within six demarcated areas, *tambu*, of the reef lagoon upon which they rely. Tambu fishing is only permitted for significant ceremonial occasions that occur up to three times annually. At these times, tambu areas are opened very briefly (two to three hours) for intensive exploitation to provide food for consumption at the ceremony. As a result, target species extracted from the restricted areas are over 20% larger and fish biomass caught, 60% higher. Such management practices are embedded in tradition rather than being conservation oriented, but succeed in yielding the same results. Despite intense resource dependence for both sustenance and economic incentives, compliance is successful through the perceived legitimacy of behavioural norms in benefiting the whole of society, combined with moral influence from peers. Also, exclusive ownership rights over reef resources in the direct vicinity induce a sense of community responsibility reinforcing community compliance and conservation incentives (Cinner *et al.*, 2005). A similar situation was unveiled in New Island province of Papau New Guinea by Wright (1985). This region practises a tradition of restricting fishing access to a reef area during the mourning period for an influential community member (lasting up to several years). This acts to replenish local fish stocks for a ceremonial feast concluding mourning and sustains community harvests into the future.

The treatment of the reef, according to these norms, devolves into a local pattern of customary, and thus expected, behaviour by which everyone conforms as thought to be in the best interests of the community. No government enforcement is required, but instead local taboos exist, reducing cost expenditure by government bodies on enforcement sanctions and the monitoring and punishment of rule breakers that arises with other forms of management (Rudd *et al.*, 2003). This evolution of traditional practices that act to sustain the reef system in its complexity forms a self-management system that is adaptive to phase shifts in the environment as they begin to emerge. Costs of making collective decisions are reduced between individuals sharing a spatial unit and communal norms in frequent contact with one another, and although social stratification is inevitable in any human coercion, even in terms of gender relations, power differences are likely to be far less pronounced and far better understood within an established community (Agrawal and Gibson, 1999).

Many self-management systems have proved successful at maintaining a resource over a long historical period (Feit, 1988). Over a lengthy period, any system of management would have to adapt to the dynamic nature of an environmental system. For a coral reef this may include altered species composition, sedentarisation, nutrient influx and shifts in water temperature. Traditional self-management systems contain a delicate feedback system that allows for subtle changes in the environment to be detected and accounted for in the relevant adjustment of practices. This is the key component by which a management system is deemed successful over time, and requires an intricate understanding of the ecology of the system.

Hence, life by the sea requires a dynamic fluidity of practice attuned to adapting to seasonal and longer-term environmental transitions; although not every community member may hold this intricate knowledge, within the community exists a way of being within the world that allows for respectful coexistence and information transfer (Tyrrell, 2005). In successful self-management systems, ecological knowledge of the coral reef has co-evolved with established practices and changing environmental conditions. This local knowledge is the intellectual antecedent of such practices (Drew, 2005). By observing the delicate balance between natural and anthropogenic pressures impacting upon the reef, resource users can enhance their knowledge *of* the sea whilst *at* sea (Tyrrell, 2005).

The *I-Kiribati* tribe, indigenous to Kiribati Island situated in the tropical Pacific, have a long history of self-management over their reef resources through the sustainment of customary practices. Traditional systems of management diminished during British colonisation, and development projects evolved that interfered with bonefish spawning grounds, a key local food resource, long known to the I-Kiribati. As a result, six of the seven spawning runs were depleted. The elders' knowledge acted to advise researchers on phase shifts that had occurred and subsequent areas of the water that needed restoration. Consequently, the ecosystem was replenished as were its resources. This example exemplifies the importance of local knowledge – ignoring this resource is only to the detriment of the ecosystem (Drew, 2005).

Through traditional biomonitoring techniques that involve observing shifts in catch species size and abundance, artisanal fishers can improvise upon pre-existing practices to accommodate such changes, often by self-imposed restrictions or altered efforts, and thus successfully manage the resources of the reef system for future generations. In addition to biotic changes, traditional management systems have had to adapt to anthropogenic changes as other peoples have come into the area, often fishing local waters with altered incentives. External threats to resources from outsiders have become more frequent in recent decades, but throughout history resource users have been forced to extend management practices into new areas to adapt to these impositions, and will continue to regulate external access into the future (Feit, 1988; Agrawal and Gibson, 1999). Hence, self-management of reef systems does not have a history of continual success in the light of environmental change, but rather a history of disruption, adaptation and self-renewal (Feit, 1988).

Despite an assemblage of successful self-management systems, with survival of the community and reef acting as testimony of not only local knowledge but also its effective application in a particular time and place despite relative isolation in the past (Feit, 1988), sustainability is not an automated response to self-management. Thus, it may be possible to re-establish equitable partnerships that meet similar conservation goals today. As already outlined, for a management system to be successful and self-enforcing it requires four components: intricate knowledge and understanding of the ecosystem, ownership responsibility, adaptive capacity and the social institutions capable of enforcement. The latter was demonstrated by Hoffmann (2002) when reef management in the Rarotonga region of the Cook Islands greatly improved, in terms of both reef system health and species diversity, upon the re-implementation of the traditional marine social institution, responsible for governing social rules and relationships, known as *Ra'ui*. Where one of these components is lacking, self-management is likely to collapse. Thus, self-management of a system is not automatically indicative of sustainability, but rather the capacity of a community to learn from mistakes and feel a continued sense of responsibility in the light of sometimes rapid change, and where communities lose this capacity, management breakdown follows.

The inshore fisheries of Fiji continue to provide a key source of protein to the locals under traditional self-management practices. Here, the government lacks the financial and institutional resources to impose resource restrictions, leading to local fisheries' networks being left to their own management devices. Well-established social channels act to facilitate knowledge devolution, community decision-making and self-enforcement, with non-conformists threatened with harsh treatment. Thus, social capital enforces successful self-management where government capacity is weak and governance is successfully decentralised to local decision-making bodies at the grassroots level (Rudd *et al.*, 2003).

Self-management in itself refers to the competency of the resource users in sustainably exploiting the resources in their local area. It requires no external intervention and is often tied up within the culture of the society. Local devolution of power is thus devoid of regional costs of decision-making and resolution of disputes by government bodies (Agrawal and Gibson, 1999). The long-awaited intrusion from state and market pressures has the capacity to generate despoiling communities from the image of ecological primitives in harmonious balance with their ecosystems (Agrawal and Gibson, 1999). Although this image may be somewhat thwarted by idealism, the shift as a result of external forces is certainly not.

Increased consumption pressures from population expansion, combined with technological innovations and institutional alterations, have the capacity to deplete traditional practices. Additionally, marketisation acts to place cash values on common property systems. This has the potential to increase volumes of resource extraction that may in turn generate environmental degradation. These pressures, combined with weak non-specific property rights, act to renegotiate resource extraction incentives (Agrawal and Yadama, 1997; Agrawal and Gibson, 1999). For instance, the sustainability of giant clam (Tridachnidae) stocks is highly dependent on local knowledge with sedentary populations developing site-specific requirements, and yet community management of these stocks more often than not fails. Rudd *et al.* (2003) assert this exception to the high market prices of giant clam, overwhelming traditional social norms, controlling opportunism, with economic incentives.

Centralised conservation management

Sense of ownership is key to self-management capacity as it evokes a sense of responsibility over resources by which access and user rights remain in the hands of locals (Feit, 1988). Ownership is key to exercising authority over the resource via the construction of regulations, subsequent implementation, and resolutions of disputes that arise, without which the benefits of such efforts are nullified (Agrawal and Yadama, 1997). When ownership changes hands, through privatisation for instance, especially against the will of the resource users, any responsibility felt towards the resource is lost or abandoned, often in spite of the new management body. When privatisation occurs, subtractability of the resource subsequently increases as does the excludability of local users (Figure 35.2). A shift in control implies an incapability of traditional practices and an inferiority of their knowledge compared with

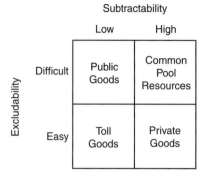

Figure 35.2   A classification of resources based on public ownership in relation to the ease of exclusion of non-authorised users and the degree of subtractability (adapted from Ostrom et al., 1994)

that of outsiders. Although new institutions have in mind a sustainable management concept, by disregarding local knowledge, environmentally and culturally, enforcement is jeopardised.

This situation arose in the Turks and Caicos Islands where marine resources had been sustainably managed for generations under the responsibility, knowledge and cooperation of coastal fisher families. In complete disregard for this system, the central government imposed its authority by allocating marine reserves excluding community involvement without any local consultation. Consequently, locals hold government-imposed restrictions in low regard to the extent that the start of the lobster season is now termed The Big Grab for its free-for-all nature, with everyone maximising their catch illegally. Many lobsters taken now fall 95% under size as a result (Rudd et al., 2003). Thus, despite the mutual desire of all stakeholders for resource sustainment, government-imposed top-down management can evoke a form of rebellion to the detriment of government management bodies, local communities and local ecosystems, although the largest costs will be felt by local resource dependants. This introduces state management as an alternative to self-management.

State management is a form of management deriving from the legal authority of the nation state. When national interests lie in conserving a local resource, management of that resource is often taken out of the hands of local users and transferred to the hands of professional policy-makers and scientists (Feit, 1988). State management exerts authority from a distance and overrides the benefits of territoriality and personal interests held by settlements local to the resource.

Instead of being locally derived, the knowledge system state management relies upon usually evolves further a field, for instance, in Western scientific institutions. Thus, it relies on a different, some might say disconnected, set of components to local systems of community management.

Management practices imposed in state systems are often based on a conceptual theory of sustainability, rather than generations of observations learned *of* the sea combined with personal experience gained *at* sea. In Indonesia, as in many other developing regions, development experts given control of inshore reef fisheries are those with a background of formal education, often with minimal understanding of reef systems. Nevertheless, they view locals, for their lack of literacy skills, as their students rather than their teachers of the local waters. Formal education is assumed justification for the accuracy of their decisions, whilst the absence of the written word from local folk knowledge acts as justification for its own inadequacy (Dove, 1988). This logic assumes that a teenager who has gone through formal schooling is more adept to the changes and needs of the local reef fishery than a community elder who has spent his life fishing those waters. In truth, no outsider can hope to acquire a fraction of the botanical, climatological and ecological knowledge and understanding of the community elder, for in the government's attempt to identify an expert for management advice, it overlooks the greatest one. For as long as governments continue to view traditional folklore as an obstacle to resource management in the light of conservation, state management systems will be doomed to failure (Dove, 1988).

The dynamics of a reef system can never be fully understood by those external to it (Tyrrell, 2005), and since effective sampling of all of the world's ecosystems is impossible, state management theory is based on results generated from periodic surveys. These tend to be isolated in a specific time and place, incorporating few of the environmental parameters at work in an ecosystem. Interpretation of these data outside of their environmental context generally acts as the foundation of imposed state regulations (Feit, 1988). Hence, government agencies lack the ecosystem experience of local resource users and focus on quantitative predictions of potential outcomes (Lundquist and Granek, 2005). Thus, the knowledge available to these two different forms of management is inherently different and culturally bound if government bodies stay on this isolated route (Drew, 2005).

The synthesis of local knowledge on site permits anticipation and response to environmental alteration, unlike state theory whereby changes are only detected when they reach large-scale phase shifts. State management responds to such changes by returning to desktop modelling based on data from similar ecosystems and relying on assumptions to fill in gaps in the data, cutting itself off from local ecological understandings and cultural norms in a situation where it can ill-afford to do so (Feit, 1988). Thus, self-management has the capacity to provide more rapid responses to sudden and often complex environmental changes.

State management combines Western science with government institutions of enforcement. The values derived from a local sense of ownership are replaced by a government's need to assert authority, and little adaptive capacity exists within this imposed externally enforced set of regulations, especially when those in a primary position to monitor resource changes are those with the least power. Where the fish production of a reef is predictable over time and space, users have the ability to predict the levels of resource withdrawal that can be permitted without depleting the system, thus making them effective managers. But where local knowledge and understanding of a reef system has been depleted, the scientific theoretical basis of state management may be the only alternative. As predictability of a system decreases with loss of local knowledge systems, uncertainty as to the levels of withdrawal increases, and thus management tends more towards theoretical insurance functions to prevent environmental degradation. Hence, management capacity shifts from the realms of local control to state control and its basis of scientific theory (Rudd et al., 2003).

State management originally devolved from an illogical misconception that conservation is dependent on exclusive protection, and this is not possible where resources are being utilised within livelihoods, as human communities lack the capacity to exploit with due constraint. And even if this capacity was once a reality, the past is long gone. Protected area designation of Marine National Parks for instance is the strongest example of this Westernised way of thinking (Agrawal and Gibson, 1999). Recent decades have seen a trend in widespread protected-area designation under state management, many employing exclusion theory. This entails excluding all local resource users from future access to reef systems, undermining their ability to exploit at sustainable levels as they have been doing for generations. Thus, local costs of such designations have been considerable, whilst conservation benefits have yet to be seen.

Hence, paper parks abound in which constraints are neither rewarded nor violation punished (Rudd et al., 2003). Paper park designation are routes by which governments are being seen to designate conservation areas that undermine

local management capacity and fulfil policy targets, but fail to manage them effectively to the benefit of global biodiversity (Alcorn, 1993). What these paper designations have shown us though is that the success of forced coercion into culturally external, disciplined resource use strategies is limited. The imposition of state management erases the incentive of community benefits upon good stewardship of the sea. This deficiency is precisely the reason for reconsideration of community involvement and its benefits (Agrawal and Gibson, 1999).

Self- and state-imposed management depict the two ends to the scale that is resource control. A grey area exists between these two extremes under the title of co-management and has the potential to combine the assets of the two systems to differing degrees. For instance, decentralised state control may play a larger role in reef management where local knowledge is required but community capacity is low. Thus, a mutually respectful working relationship between government agencies and resource users is forged to form multicultural action plans (Drew, 2005). This route appreciates the role of stakeholder involvement at every stage of planning, especially if stakeholders are to contribute support and commit to monitoring and enforcement. This is particularly important where governments lack stability or sufficient resources leading to management breakdown. Post-implementation, a common management failure is regulation enforcement due to inaccessibility, funding limitations, poor assignment of responsibilities and lack of public support. Community engagement buffers these effects by continuing management practices and enforcement in the light of government declining support (Lundquist and Granek, 2005).

Successful co-management effectively combines the traditional partnerships between locals and their resources with supporting policy to legally enforce this relationship, although steps have to be taken to establish the institutions, power relations and communications necessary for this balance. Action must be taken to avoid government agencies creating a false sense of power in local communities but in fact withholding decision-making and resolution rights for themselves (Alcorn, 1993). Where this happens successfully, local decision-making combined with government implementation techniques may prove a winning combination (Agrawal and Gibson, 1999).

The Samoan Fisheries Act (1988) succeeds in achieving this balance. By passing an Act of Government legally enforcing local reef-fishing practices dictated by societal norms, and thus securing local ownership rights and resource security, the government has enhanced local incentives to conserve and enforce regulations (Zann, 1999). Trust between stakeholders is vital in creating this relationship essential to co-management. Social norms reinforced through relations of trust ensured successful implementation of a marine reserve by coastal communities of Apo Island, in the Philippines. Where trust fails and each fisher maximises his catch disregarding local authority regulations, management breaks down, as happened on the nearby Sumilon Island (Russ and Alcala, 1999). Thus, both microlevel "community capacity" exhibited through social networks and norms, and macrolevel "institutional capacity" exhibited through designated property and resource rights are necessary for co-management. Lacking one of these, management implementation will fail on community or government level, or both (Rudd *et al.*, 2003).

Changes in technology and demography are not likely to make systems of self-control devoid, but instead are likely to illicit management adaptations as have been made repeatedly in the past in the face of both anthropogenic and natural pressures. More subtle changes instead are likely to draw on co-management options. These include a shift from traditional institutions and social systems responsible for the gathering and transfer of knowledge of local reef dynamics to systems of formal education, locally detached with a view towards commercialism (Feit, 1988). Thus, co-management of coral reef systems is likely to become a more prevalent option in the future as local knowledge systems shift in the light of development. Local input remains essential to fill in state gaps in knowledge, to determine realistic culturally-viable objectives and to build stakeholder ties. Thus, upon recognising their autonomy, authority and in-depth environmental understanding, legal force should be applied to pre-existing community practices for their self-enforcing, adaptive and sustainable success throughout history (Feit, 1988).

SUMMARY

For a long time coral reefs have been recognised for their beauty, biological diversity and high productivity; they are only now being recognised as an important economic resource bringing benefits to both local and global economies. Unfortunately, despite their importance, reefs around the world are being degraded or destroyed by human activities and more damage is expected as anthropogenic impacts, particularly as a result of population growth and economic activities, continue to increase. Realisation of the full economic potential of reefs could provide a strong argument for conservation and sustainable

utilisation of reef resources. Hence, management strategies are required that take into account both ecological and economic impacts. Local understanding of the full economic potential of reefs when sustainably utilised, and provision of socially and economically acceptable alternatives where resource use must be limited, are both important to gaining and strengthening local support. Additionally, successful management requires compliance from local people. Therefore, active engagement of local communities by promoting self-management and continuation of traditional practices is essential along with government support where necessary, in the form of legal backing through policy provision and logistical support with implementation strategies. Self-management strategies require local knowledge, local resource ownership, capacity to adapt to a changing environment and provision of social institutions capable of rule enforcement. These self-management objectives must be instated now whilst generations of ecological knowledge remain prevalent, as future management prospects without local knowledge threaten to rely on theory and state alone, endangering reefs and the diversity of life they support.

REFERENCES

Agrawal, A. and C. C. Gibson (1999). Enchantment and disenchantment: The role of community in natural resource conservation. *World Development* 27: 629–649.

Agrawal, A. and G. N. Yadama (1997). How do local institutions mediate market and population pressures on resources? Forest panchayats in Kumaon, India. *Development and Change* 28: 435–465.

Alcorn, J. B. (1993). Indigenous peoples and conservation. *Conservation Biology* 7: 424–426.

Bateman, I. J., R. T. Carson, B. Day, M. Hanemann, N. Hanley, T. Hett, M. Jones-Lee, G. Loomes, S. Mourato, E. Özdemiroglu, D. Pearce, R. Sugden and J. Swanson (2002). *Economic Valuation with Stated Preference Techniques: A Manual.* Edward Elgar Cheltenham, UK.

Berg, H., M. C. Öhman, S. Troîng and O. Lindén (1998). Environmental economics of coral reef destruction in Sri Lanka. *Ambio* 27: 627–634.

Bunce, L., K. Gustavson, J. Williams and M. Miller (1999). The human side of reef management: a case study analysis of the socioeconomic framework of Montego Bay Marine Park. *Coral Reefs* 18: 369–380.

Cesar, H. S. J. (2000). Coral reefs: their functions, threats and economic value. In: Cesar, H. S. J. (ed.) *Collected Essays on the Economics of Coral Reefs.* CORDIO, Department of Biology and Environmental Sciences, Kalmar University, Kalmar, Sweden. pp. 14–39.

Cesar, H. S. J., L. Burke and L. Pet-Soede (2003). The economics of worldwide coral reef degradation. *Cesar Environmental Economics Consulting* 23.

Cinner, J. E., M. J. Marnane and T. R. McClanahan (2005). Conservation and community benefits from traditional coral reef management at Ahus Island, Papau New Guinea. *Conservation Biology* 19: 1714–1723.

Costanza, R., R. d'Arge, R. de Groot, S. Farber, M. Grasso, B. Hannon, K. Limburg, S. Naeem, R. V. O'Neil, J. Paruelo, R. G. Raskin, P. Sutton and M. van den Belt (1997) The value of the world's ecosystem services and natural capital. *Nature* 387: 253–260.

Crabbe, J. and D. J. Smith (2002). Real-time monitoring of reef health and coral recruitment using digital videophotography and computer image analysis and modelling in the Wakatobi Marine National Park, S.E. Sulawesi, Indonesia. *Coral Reefs* 21 (3): 242–244.

Crabbe, J. and D. J. Smith (2003). Computer modelling and estimation of recruitment patterns of non-branching coral colonies at three sites in the Wakatobi Marine Park, S.E. Sulawesi, Indonesia; implications for coral reef conservation. *Computation Biology and Chemistry* 27 (1): 17–27

Crabbe, M. J. C. and D. J. Smith (2005). Sediment impacts on growth rates of Acropora and Porites corals from fringing reefs of Sulawesi, Indonesia. *Coral Reefs* 24: 437–441.

Crabbe, M. J. C. and D. J. Smith (2006). Modelling variations in corallite morphology of *Galaxea fascicularis* coral colonies with depth and light on coastal fringing reefs in the Wakatobi Marine National Park (S.E. Sulawesi, Indonesia). *Computational Biology and Chemistry* 30: 155–159.

Crabbe, M. J. C., S. Karaviotis and D. J. Smith (2004). Preliminary comparison of three coral reef sites in the Wakatobi Marine National Park (S.E. Sulawesi, Indonesia): Estimated recruitment dates compared with Discovery Bay, Jamaica. *Bulletin of Marine Science* 74: 469–476

Dixon, J. A. (1986). The role of economics in valuing environmental effects of development projects. In: Dixon, J. A. and M. M. Hufschmidt (eds.) *Economic Valuation Techniques for the Environment: A Case Study Workbook.* Johns Hopkins University Press, Baltimore, pp. 3–10.

Dixon, J. A. (1997). Economic values of coral reefs: what are the issues? In: Hatziolos, M., A. J. Hooten, and M. Fodor (eds.) *Coral Reefs: Challenges and Opportunities for Sustainable Management.* World Bank, Washington, DC.

Dixon, J. A. and P. B. Sherman (1990). *Economics of Protected Areas: A New Look at Benefits and Costs.* Island Press, Washington, DC.

Dixon, J. A. and P. B. Sherman (1991). Economics of protected areas. *Ambio* 20: 68–74.

Dornelas, M., S. R. Connolly and T. P. Hughes (2006). Coral reef diversity refutes the neutral theory of biodiversity. *Nature* 440: 80–82.

Dove, M. R. (1988). *The Real and Imagined Role of Culture in Development: Case Studies from Indonesia.* University of Hawaii Press, Honolulu.

Drew, J. A. (2005). Use of traditional ecological knowledge in marine conservation. *Conservation Biology* 19: 1286–1293.

Feit, H. A. (1988). Self-management and state-management: Forms of knowing and managing northern wildlife. In: Freeman, M. M. R. and L. N. Carbyn (eds.) *Traditional Knowledge and Renewable Resource Management in Northern Regions.* Boreal Institute for Northern Studies and International Union for the Conservation of Nature and Natural Resources, Edmonton, pp. 72–91.

Fernandes, L., M. A. Ridgley and T. van't Hof (1999). Multiple criteria analysis integrates economic, ecological and social objectives for coral reef managers. *Coral Reefs* 18: 393–402.

Goreau, T. J. (1992). Bleaching and reef community change in Jamaica: Symposium on long term dynamics of coral reefs. *American Zoologist* 32: 683–695.

Gowdy, J. M. (1997). The value of biodiversity: Markets, society and ecosystems. *Land Economics* 73: 25–41.

Graham, N. A. J., S. K. Wilson, S. Jennings, N. V. C. Polunin, J. P. Bijoux and J. Robinson (2006). Dynamic fragility of oceanic coral reef ecosystems. *Proc. Natl Acad. Sci. USA* (published online).

Green, C. H. and S. M. Tunstall (1991). Is the economic evaluation of environmental resources possible? *Journal of Environmental Management* 33: 123–141.

Hoegh-Guldberg, O. (1999). Climate change, coral bleaching and the future of the world's coral reefs. *Marine and Freshwater Research* 50: 839–866.

Hoffmann, T. C. (2002). The reimplementation of the Ra'ui: Coral reef management in Rarotonga, Cook Islands. *Coastal Management* 30: 401–418.

Lundquist, C. J. and E. F. Granek (2005). Strategies for successful marine conservation: Integrating socioeconomic, political, and scientific factors. *Conservation Biology* 19: 1771–1778.

Nunes, P. A. L. D. and J. C. J. M. van den Bergh (2001). Economic valuation of biodiversity: sense or nonsense? *Ecological Economics* 39: 203–222.

Ostrom, E., R. Gardner and J. Walker (1994). *Rules, Games, and Common-Pool Resources.* University of Michigan Press, Ann Arbor.

Pearce, D. and K. G. Mäler (1991). Environmental economics and the developing world. *Ambio* 20: 52–54.

Pearce, D. and K. Turner (1990). *Economics of natural resources and the environment.* Harvester Wheatsheaf, Hemel Hempstead, UK.

Pendleton, L. H. (1995). Valuing coral reef protection. *Ocean and Coastal Management* 26: 119–131.

Perrings, C., C. Folke and K. G. Mäler (1992). The ecology and economics of biodiversity loss: the research agenda. *Ambio* 21: 201–211.

Randall, A. (1991). The value of biodiversity. *Ambio* 20: 64–67.

Robbins, L. C. (1935). *An Essay on the Nature and Significance of Economic Science.* Macmillan, London.

Rudd, M. A., M. H. Tupper, H. Folmer and G. C. van Kooten (2003). Policy analysis for tropical marine reserves: Challenges and directions. *Fish and Fisheries* 4: 65–85.

Ruitenbeek J. and C. Cartier (1999). Marine system valuation: An application to coral reef systems in the developing tropics. *Issues in Applied Coral Reef Biodiversity Valuation: Results for Montego Bay, Jamaica.* (Project RPO 682-22) Project Task Team Leader: R. Huber, World Bank; Washington, DC, p.149.

Ruitenbeek, J., M. A. Ridgley, S. Dollar and R. Huber (1999). Optimisation of economic policies and investment projects using a fuzzy logic based cost-effectiveness model of coral reef quality: Empirical results from Montego Bay, Jamaica. *Coral Reefs* 18: 381–392.

Russ, G., R. Alcala and A. C. Alcala (1999). Management histories of Sumilon and Apo Marine Reserves, Philippines, and their influence on national marine resource policy. *Coral Reefs* 18: 307–319.

Simpson, R., R. A. Sedjo and J. W. Reid (1996). Valuing biodiversity for use in pharmaceutical research. *Journal of Political Economy* 101: 163–185.

Smith, D. J., D. J. Suggett and N. R. Baker (2005). Is photoinhibition of zooxanthellae photosynthesis the primary cause of thermal bleaching in corals? *Global Change Biology* 11: 1–11.

Spurgeon, J. P. G. (1992). The economic valuation of coral reefs. *Marine Pollution Bulletin* 24: 529–536.

Spurgeon, J. P. G. (2001). Valuation of Coral Reefs: The Next 10 Years "Economic Valuation and Policy Priorities for Sustainable Management of Coral Reefs" an International Consultative Workshop, ICLARM, Penang, Malaysia.

Tyrrell, M. (2005). The social reproduction and growth through practice of marine knowledge and skill. Unpublished.

Wilkinson, C. (2002). *Status of Coral Reefs of the World: 2002.* Australian Institute of Marine Science, Townsville, Queensland.

Wright, A. (1985). Marine resource use in Papua New Guinea: Can traditional concepts and contemporary development be integrated? In: Ruddle, K. and R. Johannes (eds.) *The Traditional Knowledge and Management of Coastal Systems in Asia and the Pacific.* UNESCO, Jakarta Pusat, Indonesia, pp. 79–100.

Zann, L. P. (1999). A new (old) approach to inshore resources management in Samoa. *Ocean and Coastal Management* 42: 569–590.

Institutions and Policies for Influencing the Environment

36

The Role of Science and Scientists in Environmental Policy

Jonathan Hastie

INTRODUCTION

From waste disposal to energy policy, fisheries science and climate change, modern environmental problems pose increasingly complex challenges to the social, economic and political apparatus of modern societies. Scientific research is widely held to be an invaluable tool for policymakers seeking to address these intricate problems in the most productive and efficient manner. However, this seemingly straightforward functional relationship between science, on the one hand, and politics, on the other, belies the complex interdependencies that exist between the two. Scientists, and science in general, can have multiple, sometimes conflicting roles within the political arena, depending on factors such as the ideals of the individual scientist, the nature of the issue, the configuration of power and interests among actors, and established socio-political discourse. This chapter explores these different functions that experts may perform, and the different perceptions of how science is used within policy, and goes on to consider what implications this has for how scientific assessments for policymakers are designed.

THE ROLES OF SCIENCE AND SCIENTISTS

Studies in the sociology of science have explored in great detail the role played by experts in society. Broadly speaking, theories of expertise can be conceptualised as existing on a spectrum between pure, positivist views of science, that is, linear, objective and separated from politics, and more constructivist approaches that view science as a social process. Figure 36.1 shows the key theories and roles of expertise emphasised in this chapter, showing their position on this spectrum.

Science as the objective pursuit of knowledge

The conventional, positivist view of science, as the objective pursuit of knowledge, has its roots in the philosophies that emerged during the Enlightenment period in Western Europe. Philosophers such as Francis Bacon established the notion of science as a linear, progressive force, yielding objective truths about nature (Pepper, 1990). The early concepts of science were dominated by the concept of logical positivism, which sees knowledge as being produced by creating and testing hypotheses to verify their accuracy (Ayer, 1966; Hume, 1999). Karl Popper challenged this verificationist stance, suggesting instead that knowledge was approached gradually by the rejection of alternative hypotheses, resulting in ever bolder conjectures that had survived critical refutation and thus assume better representations of reality – that is, falsificationism (King *et al.*, 1994; Gieryn, 1995; Harrison, 2003). The Popperian notion achieved dominance over the verificationist stance, and now is a key component of what might be seen as the conventional view of science (Wynne, 1995). While verificationist and falsificationist theories differ, both approaches agree on the use of empirical methods to test

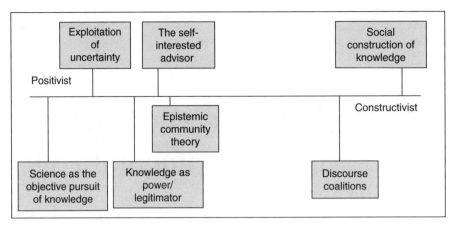

Figure 36.1   Role of experts in society

conflicting hypotheses and to generate objective truths, contributing to the linear progression of science (Harrison, 2003):

> It is confessed, that the utmost effort of human reason is, to reduce the principles, productive of natural phenomena, to a greater simplicity, and to resolve the many particular effects into a few general causes, by means of reasonings from analogy, experience, and observation (Hume, 1999, p. 15).

The idea of differences among scientists (expressed through conflicting hypotheses) is integral to the conventional notion of science. Positivists consider that science is capable of resolving these differences, revealing the nature of reality in an objective fashion and creating robust facts whose validity can be agreed on by all observers.

The conventional idea of the scientific enterprise sees it as inherently different from, and superior to, other systems of knowledge. In Merton's (1968) survey of scientists, he specifies four general "institutional norms" considered to be characteristic of science. The first is *organised scepticism*, whereby scientists suspend judgement until sufficient evidence is available, accepting or rejecting nothing without proof. The second is *universalism*, whereby knowledge claims must be tested with pre-established, universal criteria. The third is *disinterestedness*, whereby scientists support ideas on the basis of scientific merit, not self-interested motives. The fourth norm is *communism*, whereby scientists must share findings, in order for scientific knowledge to progress. Merton identified these four norms as ideals which scientists aspire to in their research (Merton, 1968).

The concept of science as a process of generating objective truths (based on ideals of disinterested scientists and universal criteria) is integral to the ideology of technocracy. Technocracy refers to a system whereby society is governed according to scientific principles. Proponents of technocracy argue that science can (and should) be used to provide objective knowledge about a problem and indicate policy solutions in a way that is free from the influence of political interests (Jasanoff, 1990). The use of technology and expert knowledge, it is argued, has the potential to generate effective policies to solve major social problems (Technocracy Inc., 1960). Science, according to this view, derives its legitimacy independently of the political sphere, and can provide a universal standard which is widely accepted. The conventional view of science suggests a clear line exists between science and policy, which should be upheld (Beneviste, 1973; Jasanoff, 1990; Litfin, 1994; Woodhouse and Niesma, 1997). For positivists, the relationship between these two different fields is linear, with science generating objective knowledge about a problem which is subsequently applied by policymakers. This perspective is typical of early work in the field of science and technology studies (STS) (Jacob, 2005; Layzer, 2006).

The technocratic ideal of increasing the role of science in policy has been highly influential since the early 20th century. During World War II, considerable resources were invested in building scientific networks to allow scientists to contribute to the war effort. Walsh (2004) documents this process at the Scripps marine biology and oceanographic research institute during both world wars, where the military heavily influenced the research agenda. Since this period, scientific expertise has been increasingly used for policy, to the extent that policy is often perceived as illegitimate without it (Greenberger *et al.*, 1976; Andresen and Ostreng, 1989; Litfin, 1994; Hasenclever *et al.*, 1997; Clark *et al.*, 2002). A major increase in the use of scientific expertise has occurred in the field of environmental policy (for a wide range of issues including resource management, acid rain and climate change), both to address existing problems and to identify new issues (Boehmer-Christiansen and Skea, 1991; Yearley, 1995). Clark *et al.* (2002) suggest that the use of scientific experts has been self-reinforcing, as scientists themselves promote the increased use of their expertise, for example, in frequent calls for further study.

While the involvement of scientists in policy continues to increase, the optimistic, technocratic view of a single notion of "science" as an unmitigated benefit for society has been subject to ongoing challenge over the last forty years. Woodhouse and Nieusma (1997) point to problems caused by the increased use of science and technology in the 1960s as instrumental in challenging this view – disasters such as the Three-Mile Island nuclear accident, the chemical explosion in Bhopal, and controversies such as the environmental effects of industrial chemicals like PCBs. Beck (1992) suggests that these increasing risks stemming from the use of science have further increased the demand for science-based policy, as scientific expertise is called upon to address or quantify these risks.

The political use of science

The traditional, optimistic and positivist notion of science is further questioned by the use of scientific evidence to advance political agendas. One way this can happen is through the manipulation of uncertainty and differences of opinion among scientists. As we have seen, the idea that scientists hold differing opinions and hypotheses is inherent to the scientific enterprise (Miller and Edwards, 2001). Where these differences have not been settled, different social groups are likely to favour one or other side of the dispute. Strict positivists might suggest that "science" occurs before politics, with groups forming preferences based on the knowledge that they find most persuasive. However, studies of science in society have found that where scientists disagree, policymakers and interest groups may take advantage of these differences, using only those scientific findings that legitimate their *pre-existing* preferences and interests (Andresen and Ostreng, 1989; Boehmer-Christiansen and Skea, 1991; Barker and Guy-Peters, 1992; Haas, 1992a; Litfin, 1994; Healey, 1997; Rifkin and Martin, 1997; Skodvin, 1999; Wilson, 1999; Alverson, 2002; Harrison, 2003;

Russo and Denious, 2005). This has been evident in recent controversies regarding scientific issues, which have undermined the public perception that science is used objectively, independent of political interests (Martin and Richards, 1995). One example in the UK has been the selective use of scientific advice by the government regarding bovine spongiform encephalopathy (BSE) and its transmission to humans (Barker and Guy-Peters, 1992; Beck *et al.*, 2005; Cummings, 2005).

As noted above, the increased use of science in society means that policy that does not use scientific knowledge is perceived as illegitimate, and thus groups are compelled actively to seek scientific evidence to support their policy preferences. This strategic use of science can be understood as an attempt to mobilise the "cognitive authority" of science in order to bolster certain policy positions (Jasanoff, in Gieryn, 1995; Raman, 2005). A common image used to depict the strategic use of science by different groups is that of a weapon – for example, "artillery" or "ammunition", used in a political war between group interests (Andresen and Ostreng, 1989; Boehmer-Christiansen and Skea, 1991; Miller and Edwards, 2001). Where scientific findings appear to oppose a group's chosen policy, they often seek to undermine the research, by questioning its methods and the assumptions on which it is based (Jasanoff, 1990; Harris, 1999; Harrison, 2003). In her study on the BSE debate in the 1990s, Cummings (2005) explores how industry groups engaged in fallacious reasoning when communicating supportive scientific evidence to the public, by emphasising the credentials of this research while wilfully ignoring the comparative credentials of opposing research. To undermine research, groups also often accuse studies of being politicised, appealing to the common perception that science should be separate from politics (Healey, 1997; Hutchings *et al.*, 1997; Harrison, 2003). While disputes over scientific advice may reflect genuine differences regarding appropriate scientific methods, it may also be primarily a political strategy, the important point to note being that it is difficult and entirely subjective to distinguish between scientific and political disputes.

In some cases, where assessments leave room for interpretation, different interest groups draw different conclusions from the same scientific assessment to support their desired policy (Litfin, 1994). This is frequently the case where the level of uncertainty is so high that assessment authors themselves do not completely agree.[1] In this case, groups in favour of maintaining the status quo may use these differences as a delaying tactic, calling for more conclusive evidence before action can be taken (Boehmer-Christiansen and Skea, 1991; Litfin, 1994; Jung, 1999; Harrison, 2003).

Regarding climate change, for example, groups opposed to reducing greenhouse gas (GHGs) emissions have highlighted disagreements between climate modellers to discredit their validity, and emphasise the need for further understanding before costly action is taken (Miller and Edwards, 2001). In contrast, for other groups, a high degree of uncertainty may be used to justify taking pre-cautionary action (Boehmer-Christiansen and Skea, 1991).

In her study of fisheries management policy in the USA, Layzer (2006) suggests that different interest groups seek to frame scientific research in ways that create a credible definition of a policy problem and potential solutions that match their pre-existing interests. Groups then seek to promote their own "problem definition frameworks" above competing interpretations, through a combination of public, political and legal pressure. As we shall see, such "framing" of scientific knowledge is an important way in which not only interest groups, but also individual scientific advisors seek to advance their own policy agendas.

Knowledge as power

It could be argued that the notion of science being used to legitimate pre-existing policy preferences does not seriously challenge the positivist viewpoint. Science being used in this way does not undermine the concept that scientists are disinterested, merely suggesting that while science is generated independently of politics, the subsequent *products* of science are exploited by politicians. However, this distinction is not so easy to make where political interests get involved *during* the assessment process, for example, commissioning new research or pressuring scientists to deliver evidence to support their policy position (Beneviste, 1973; Harrison, 2003).

The mobilisation of new research or expert advice to legitimate preferences often occurs in institutions where experts are "on tap", readily available to generate supporting knowledge or advice (Barker and Guy-Peters, 1992). This is particularly common in cases where there is some interstate rivalry, where states generate research at the national level to support their negotiating positions in international regimes (Litfin, 1994; Harrison, 2003). Another example of this phenomenon is where paid researchers are used to generate knowledge that supports industry positions (Andresen and Ostreng, 1989).

Where strong interests are involved in funding research, the separation between science and politics becomes blurred, with scientists no longer conforming to the disinterested ideal. In this case, even where research sponsors seek to remove themselves from the scientific process, there may be pressure on the scientists to deliver findings that comply with the interests of the sponsors (Boehmer-Christiansen and Skea, 1991; Alcock, 2004). In some cases this pressure may be explicit, with political interests actively pressuring the scientists to deliver certain results; in others it may simply be implicit or perceived. Where such explicit pressure is identified the validity of research is often widely criticised. One example where alleged tampering with evidence has occurred is in the fish stock assessments undertaken by the Department of Fisheries and Oceans (DFO) in Canada in the late 1980s[2] (Doubleday et al., 1997; Hutchings et al., 1997). A more recent example regards research on reproductive health in the USA, where Russo and Denious (2005) claim that a number of appointments of experts, charged with providing scientific advice or making decisions on the funding of research projects, have been made on the basis of their commitment to a socially conservative, pro-life agenda rather than their scientific credentials.

Where science is used to support pre-existing preferences (whether created for this purpose or not), this could suggest that the role of knowledge *vis-à-vis* political interests in influencing preferences, and thus behaviour, is minimal. According to this perspective, scientific advice has no independent influence on behaviour, acting as a mere "fig leaf" for political conflicts (Andresen and Ostreng, 1989). As a result, it is argued, where an increasing level of consensus among scientists emerges, political conflicts may actually become more explicit, since the potential for the strategic manipulation of differences to disguise political struggles becomes limited (Jasanoff, 1990; Litfin, 1994). Litfin (1994) explores how this occurred in negotiations within the ozone-depletion regime. An opposing, more positivist, perspective might interpret this phenomenon (i.e. increasing consensus resulting in increased political conflicts) as the resolution of genuine scientific, non-political disputes, with politics only becoming important after this occurs.

Implications for positivism

Woodhouse and Nieusma (1997) argue that the increased awareness of science-based risks and the selective use of experts has caused a shift in public perceptions of science from the strongly positivist technocratic ideal to the other extreme, adopting a highly cynical view of science as capable of being employed to further any agenda, and unable to generate any objective knowledge (see also Cozzens and Woodhouse, 1995; Cummings, 2005). However, while in some cases this is true, at present there remains clear support among many scientists, policymakers and the public for the view that science can generate objective "scientific" knowledge of political problems, providing it remains independent of politics (Bolin, 1994; Cozzens and Woodhouse, 1995; Edge, 1995; Gieryn, 1995; Lewenstein, 1995; Sallenave and Cowley, 2006). While not as optimistic as the technocratic ideal regarding the potential benefits of science, this conventional view, embodying the normative concept of a clear division between science and politics, remains rooted in positivism.

In what may be seen as an attempt to resolve the political use of science and the positivist notion of objective knowledge, many point to the unique characteristics of "science for policy" to distinguish it from normal or pure science (Skodvin, 1999; Miller and Edwards, 2001; Harrison, 2003). Similar distinctions have been made between "core" and "frontier" knowledge, "mandated science", "transcience", "postnormal science" and "science-in-progress" (Jasanoff, 1990; Skodvin, 1999; Ebbin, 2004; Jacob, 2005; Sallenave and Cowley, 2006). According to this argument, science for policy is characterised by three key elements – uncertainty, time constraints and high stakes. These are sketched below.

The term "frontier knowledge" conveys the notion that science for policy is conducted at the "margins of knowledge", that is, where the scientific knowledge is incomplete or rests on uncertain assumptions (Skodvin, 1999). In science for policy, scientists are required to generate new knowledge about a policy problem, or to apply existing knowledge to a new problem; knowledge which by its very nature is incomplete. In the literature on environmental policy, uncertainty is discussed in relation to a wide range of issues, including climate change, bovine growth hormones, nuclear radiation, fish stock assessments, acid rain and ozone depletion (Andresen and Ostreng, 1989; Edge 1995; Boehmer-Christiansen and Skia, 1991; Barker and Guy-Peters, 1992; Healey 1997; Lane and Stephensen, 1999; Miles *et al.*, 2001; Alverson, 2002; Botsford *et al.*, 1997; Dimitrov, 2003; Harrison, 2003). While for some issues uncertainty has been reduced following substantial research, even here major areas of uncertainty remain, indicating that it is a persistent feature of science for policy (Harrison, 2003). Where there is uncertainty, scientific advisors are often required to summarise the "best available science", which may be incomplete or insufficiently substantiated (Beneviste, 1973; Jasanoff, 1990; Lane and Stephenson, 1999; Kinzig and Starrett, 2003). Policymakers, however, typically desire a high degree of certainty from scientists in order to lend credence to their decisions (McKitrick, 2005).

A common characteristic of policy science is the time constraints that scientists are forced to work under (Beneviste, 1973; Greenberger *et al.*, 1976; Jasanoff, 1990). This problem stems from the contrasting social environments inhabited by scientists and policymakers. Where scientists proceed slowly, requiring high confidence margins before making strong conclusions, policymakers have limited time horizons, with their position assured only until the next election, and often require quick answers from scientists regarding pressing policy problems (Litfin, 1994; Kinzig and Starrett, 2003; McKitrick, 2005; Sallenave and Cowley, 2006). Scientific advisory processes are often given funding and a political mandate for a limited duration, meaning scientists are rarely given the time to conduct the research required to answer a policy question with a margin of error that they consider acceptable (Miller and Edwards, 2001; PCE, 2004).

For issues of policy where scientific advice is required, the stakes involved are often high, and this can have an impact on the content or presentation of scientific advice (Harrison, 2003; PCE, 2004). For example, potential environmental, economic and social costs associated with climate change, due to both its occurrence and the policy prescriptions to tackle carbon dioxide emissions, are enormous, meaning that significant interests have been mobilised on either side of the debate (Skodvin, 1999). For fisheries, scientific advice can have a direct impact on the livelihoods of fishermen where it forms the basis for regulations; when scientists advise quota restrictions, fishermen may lose their primary source of income. Where the stakes of science-based policy are high, it can be expected that science will be subject to intense political pressures and is more likely to be used competitively[3] (Greenberger *et al.*, 1976; Andresen and Ostreng, 1989).

It is argued that it is the combination of these three characteristics of science for policy – high uncertainty, time constraints and high stakes – that means scientific knowledge can be exploited, influenced or even ignored by strong political interests (Harrison, 2003). The distinction between science for policy and research science can be understood as an exercise in "boundary work", a concept that will be explored in detail later. Essentially, the distinction can be seen as a device through which the threat to the objective image of science is defused. Thus, the selective use of expertise can be seen as limited to policy/mandated/frontier science, meaning the potential of "normal science" to produce objective knowledge remains unchallenged (Gieryn, 1995; Jacob, 2005). While it is widely accepted that a number of constraints are evident in science for policy, as we shall see the idea that these characteristics are unique to policy science is challenged by constructivists.

The value of consensus

Given what is seen as the widespread "manipulation" of science in the face of uncertainty, a common argument to emerge is that science can only influence policy where it is consensual, that is, shared by all but the most peripheral scientists (Andresen and Ostreng, 1989; Haas, 1992a; Hasenclever *et al.*, 1997). Where such consensus is evident, there is less scope for the strategic manipulation of scientific disputes, and scientists are able to draw on the shared credibility of multiple sources. Consensus is a source of authority for scientists and policymakers, as it allows the cognitive authority of scientific knowledge to be mobilised behind a single set of policy recommendations (Hasenclever *et al.*, 1997). Scientific assessments for policy are typically designed to produce a report that reflects a consensus among authors, often embodied in a specific summary for policy makers.[4] In most cases, scientific assessments are tasked with summarising the "state of the art", outlining points of agreement among all scientists (not just assessment authors) on a specific issue, and areas where agreement remains to be reached (Jung, 1999).

As we have seen, the empiricist notion of science sees disagreement as integral to the process of generating knowledge, with more robust knowledge emerging from the testing of competing hypotheses. Attempts to foster compromise could be seen as inhibiting this competition and hence attempts to generate specific answers (i.e. "getting to the truth"). As such, experts are often uneasy about operating in an arena where consensus building is the dominant strategy (Litfin, 1994).

Scientists exercising power/ influencing policy

In the previous model of science in society, the unique aspects of science for policy leaves it open to political manipulation. Scientific advice, and scientists themselves, while they may be disinterested in the conduct of their research, are "exploited", "manipulated" and "pressured" by strong political interests seeking to use the cognitive authority of science to pursue their own interests. Scientists that identify with the norm of disinterestedness are often unaware of the political element of their work, or where they are aware, they may be hostile to the involvement of policymakers and the public in science, seeking to avoid "dangerous" politicisation of research (Beneviste, 1973; Andresen and Ostreng, 1989; Jung, 1999; Skodvin, 1999). In the models to follow, this idea is challenged by looking at how *scientists* actively involve themselves in political struggles, seeking to promote specific policies, either as individuals or groups.

The influence of individual scientific advisors over policymakers is well documented in the field of environmental policy. Senior scientists in the UK such as David Fisk, a scientist at the Department of the Environment in the 1990s, have been identified as influential in persuading the UK government to take an aggressive stance on climate change (Harrison, 2003). It is also suggested that individual scientists in the USA played a key role in influencing the position of EPA administrator Lee Thomas on ozone depletion (Haas, 1992b). Individual scientists in key positions within the government policymaking structure can often be seen promoting specific policy responses to scientific advice. This is true of the UK's current chief scientific advisor, Sir David King, who has published a number of articles on climate change policy, in which he advocates a range of policies including taxation of air travel (King, 2005). However, the influence of such recommendations cannot be simply derived from their publication.

To understand the political role that scientists can actively engage with, it is important to discuss the power held by individual scientists and scientific advisors in the political sphere. This can again be explained by reference to the uncertainty and time constraints acting on science for policy. To provide policy advice, "knowledge brokers" are required, in order to translate existing knowledge into specific recommendations, particularly when facing uncertainty (Litfin, 1994; Skodvin, 1999; Harrison, 2003). Knowledge brokers, who typically come from a scientific background, define and frame problems based on existing knowledge, choosing the facts that are presented and highlighting certain issues above others (Beneviste, 1973; Barker and Guy-Peters, 1992). Through framing and interpretation experts may present data in a way that favours a particular policy choice, or may simply determine the range of policy options and issues that are considered – what Layzer refers to as creating a "problem definition framework" (Woodhouse and Nieusma, 1997; Layzer, 2006). Where knowledge brokers are the dominant source of knowledge for policymakers, they can therefore have a significant impact on the policy preferences identified by policymakers (Litfin, 1994; Dowding, 1996).

The power exerted by scientific advisors, it is argued, is a distinct type of power. While power is often seen as physical and manipulative, knowledge-based power is productive and discursive (Litfin, 1994). Sebenius' (1992) account of epistemic communities within bargaining theory sees power as the ability to influence a zone of potential agreement, meaning that information is as potent a bargaining tool as more forceful forms of action. The power of scientists comes from their scientific credentials, as well as the persuasiveness of their knowledge claims, particularly the perceived credibility of the knowledge used to support their position (Litfin, 1994). Knowledge brokers are therefore constrained to some extent by the available data, which, while open to interpretation, are not completely malleable (Litfin, 1994; Dimitrov, 2003).

The notion of scientists as promoting certain policy choices does not necessarily challenge the positivist notion of science, and is in fact integral to the firmly positivist technocratic ideology of science being used by governments to generate efficient solutions to policy problems. This concept rests on the idea that the interpretation of scientific data is based on an objective, rigorous methodology – that is, scientists uncover the truth, or the most plausible hypotheses based on objective observations, then advise policymakers on the best solutions. However, in areas of high uncertainty, it was noted, existing knowledge may be open to a number of different interpretations. Where this dispute cannot be resolved objectively,[5] policy recommendations become a subjective choice, which may vary between individual scientists. The specific recommendations made may depend on "non-scientific" factors such as political values, culture and other interests (Boehmer-Christiansen and Skia, 1991; Haas, 1992a; Cozzens and Woodhouse, 1995; Woodhouse and Nieusma, 1997; McKitrick, 2005). This point diverges considerably from the conventional view of science. As discussed above, political values and beliefs can produce pressure on scientists to provide knowledge that supports certain political ends. However, it is argued that the subjective factors determining policy preferences can operate as a purely unconscious bias that predisposes the individual scientist to accept a certain explanation or hypothesis over another, with scientists believing their recommendations to be derived purely from objective observations of data (Boehmer-Christiansen and Skia, 1991; Clark *et al.*, 2002).

Epistemic community theory

Epistemic communities theory accepts the notion that scientists are far from disinterested (whether consciously or not) and examines how they actively engage in building a consensus to gain authority, allowing them to advance their policy preferences (Haas 1992a, Sebenius, 1992). Epistemic communities can be defined as:

> A network of professionals with recognised expertise and competence in a particular domain and an authoritative claim to policy-relevant knowledge within that domain or issue-area ... [this community has] a shared set of normative

and principled beliefs ... shared causal beliefs ... shared notions of validity and a common policy enterprise (Haas, 1992a, p. 3).

The theory looks at how such a network is built up, identifying the importance of transnational links between scientists, developed through conferences, journals, research collaboration and informal communications and contacts between experts (Haas, 1992a). Where individual members of the community gain key advisory positions within their respective national policymaking arenas, these transnational links allow them to coordinate their advice with experts in other countries, allowing them to influence the international agenda and states' behaviour (Haas, 1992a).

It is argued that where epistemic communities come to hold an authoritative claim to policy-relevant knowledge, they gain significant influence over policy. States are seen as "uncertainty reducers", who turn to epistemic communities to provide knowledge and advice about a problem (Haas, 1992a). As with individual advisors, this position allows epistemic communities to clarify state interests and behaviour, through defining the problem and delimiting the range of policy options that are given consideration (Haas, 1992a, b).

Hasenclever *et al.* (1997) note that more than one epistemic community may exist in any issue area, and the question of most significance is how any single network comes to have an "authoritative claim to policy-relevant knowledge". Epistemic community theory is thus preoccupied with power and interests – that is, the respective power of competing groups, and how experts exert power to gain influence over policy and pursue their own agenda (Haas, 1992a; Sebenius, 1992; Jung, 1999; Dimitrov, 2003). As indicated above, an expert's power is based on the persuasiveness and perceived credibility of their interpretation of the available knowledge. Thus, where groups of experts are in conflict, power comes from being seen as more credible and/or authoritative than competing groups. This can come from the persuasiveness of their interpretation (either their ideas or the persuasive ability of individual members of the community), the reputations of group members or the size and resources of the community (i.e. reflecting a larger consensus) (Cozzens and Woodhouse, 1995; Haas, 1992b). Epistemic communities could be seen as exerting both types of power identified by Litfin above (1994) – knowledge-based power and more physical, manipulative forms of power, with minimal distinction between the two (Sebenius, 1992).

The theory of competing epistemic communities meshes well with the bureaucratic model of politics (Jung, 1999). This model highlights the conflicts between bureaucratic agencies at the subnational level, driven by efforts by these groups to expand their power and resources within the political system (Litfin, 1994). Expert networks both use and are used by these competing bureaucratic groups. For example, Haas documents how members of the scientific community pushing for regulation of CFCs were positioned as advisors within agencies such as the EPA in the USA, where they were able to recruit new members to the epistemic community (Haas, 1992b). The attempt to expand the role of the EPA in the process can be seen as being driven by the political agenda of the epistemic community. From the perspective of bureaucratic politics, agencies such as the EPA could be seen as the driving force, supporting epistemic communities where their framing of the problem proposes to give these agencies a central role in coordinating policy response (Litfin, 1994). In this example, both explanations are likely to be relevant, and the respective roles of agencies and the epistemic community may be interwoven and difficult to separate.

Discourse coalitions

Litfin (1994) recognises the way in which scientists and policymakers form coalitions of support around particular discourses, in much the same way that epistemic communities mobilise around a set of shared ideas and principles. Litfin defines a discourse as: "a set of linguistic practices and rhetorical strategies embedded in a network of social relations" (Litfin, 1994). Despite some obvious similarities, the theory of discourse coalitions differs from epistemic community theory in two key ways.

First, while epistemic community theory recognises that shared causal beliefs and principles may have a persuasive power independent of material resources and interests, as noted above its main focus is on the action taken by the epistemic community to advance their policy preferences. Litfin (1994) on the other hand places more emphasis on the first element, emphasising the power of ideas and beliefs embedded in discourses. Second, on a related point, the theory of discourse coalitions challenges to a greater extent the positivist notion of a linear relationship between science and politics, which suggests science is created independently of politics, with science only becoming infused with politics when it is translated into policy proposals and framed in a certain way (Litfin, 1994; Miller and Edwards, 2001). While it is important to note that Haas' original elucidation of epistemic community theory does recognise the importance of value-based judgements in the way epistemic communities frame knowledge, the primary focus of the theory is on how science influences politics, not vice versa

(Haas, 1992a, p. 12; Sebenius, 1992). In contrast to this linear focus, Litfin's analysis of discourse coalitions places greater emphasis on the interaction and inseparability of science and politics.

Similarly to epistemic community theory, then, Litfin emphasises the way discourses are actively advanced by coalitions of scientists, policymakers and other stakeholders to support political ends in the same way that epistemic communities advance shared beliefs. Litfin points to the discourses surrounding ozone depletion, for example, the idea that ozone depletion was just a skin-cancer issue, which was used to support a political choice to focus on adaptive policies to depletion such as minimising sun exposure (Litfin, 1994). However, discourses also have their own persuasive power which influences the subsequent development of knowledge. Discourses may be institutionalised, remaining dominant long after the epistemic community has dissolved (Walsh, 2004; Young, 2004). Discourses determine what scientific knowledge is used, and privilege certain models over others – in fisheries management, for example, the discourse of single-species quotas (total allowable catches, or TACs) calls for research into stock abundance, while the discourse of ecosystem management calls for studies into ecosystem dynamics (Walsh, 2004). Discourses may persist even where they begin to lose support, which is evident in the case of the TAC discourse, which persists despite increasing criticism from both fisheries managers and scientists (Holm and Nielsen, 2004).[6]

Theories of epistemic communities and discourse coalitions share a number of common elements, and the original account of epistemic communities can accommodate many of the points made by discourse coalition theory. However, the main differences regard the features of the theory they emphasise. Epistemic communities, with its emphasis on how scientists wield power and influence to advance their agenda, may seem more threatening to the positivist notion of science than the more subtle interactions of discourse coalitions. Yet the focus on discourses problematises the *content* of scientific knowledge, suggesting it to be heavily influenced by the discourse dominant at the time. In essence, science and politics are linked at every stage in the process, with no uniform, uncontentious way to separate the two. This is a major challenge to the positivist ideal, and such a challenge is taken further in the final model of science in society that follows.

Science as a social construct

The constructivist model of science in society sees scientific knowledge as constructed within a social process (Latour, 1987; Jasanoff, 1990; Cozzens and Woodhouse, 1995; Rifkin and Martin, 1997). Knowledge emerges as a product of ongoing negotiations between social groups (primarily scientists) as to what constitutes a scientific fact, and is thus considered to be valid only where it is accepted and subsequently used by others (Bourdieu, 1975; Latour, 1987; Harrison, 2003). The idea of validity as social acceptance can be seen to underpin the concept of peer review, which is integral to scientific knowledge (Miller and Edwards, 2001; McKitrick, 2005; Sallenave and Cowley, 2006). For constructivists, this model is not constrained to "science for policy", but rather can be applied to all forms of science, even what positivists might consider "pure" or research science (Latour, 1987). While part of what determines the social acceptance of knowledge claims is the accuracy of their representation of reality (i.e. how well it performs as either a *post-hoc* explanation or *pre-hoc* predictor of behaviour), constructivists argue that many other factors beyond the persuasiveness of an idea or theory can decide whether scientific knowledge is accepted by peers. Persuasiveness is not the same as the positivist notion of "objective truth", as for constructivists, no objective criteria exist for determining the truth of a knowledge claim, or even whether an experiment has been adequately replicated (Gieryn, 1995).

One of the most notable expressions of the social nature of science comes from Kuhn's (1970) work on scientific paradigms. Kuhn challenged the positivist, linear notion of science by suggesting that science was constructed within paradigms, which determine how phenomena are understood and studied. Paradigms allow communication between scientists, and allow for the accumulation of knowledge, by setting out an accepted base of knowledge and standard methodology (Harrison, 2003). Knowledge that is not embedded in the dominant paradigm is seen by those within the paradigm as invalid or non-scientific (Gieryn, 1995). For Kuhn, scientific progress occurs not in a linear fashion, but through scientific revolutions, whereby paradigms face discrepancies which cannot be resolved by experiment or theory within the existing paradigm, resulting in the emergence and dominance of a new paradigm (Kuhn, 1970; Barker and Guy-Peters, 1992). In the struggle for dominance between paradigms, the persuasiveness of a paradigm (e.g. the extent to which it appears to reflect "reality" or "logic") is only one element used to secure victory (Gieryn, 1995).

Both paradigms and discourses are used to interpret and frame new knowledge (Litfin, 1994; Harrison, 2003). Some theorists go further in suggesting that the social interactions occur not only at the level of paradigms or discourses, but also in the competition of individual knowledge claims (e.g. Latour, 1987). While for positivists, conflicting hypotheses among scientists are resolved by empirical testing, for constructivists these differences are resolved by negotiation to agree what constitutes a fact (Latour, 1987). Latour (1987) suggests that knowledge claims are constantly transformed by scientists, engineers and policymakers who push them either in the direction of fact or fiction. He suggests that knowledge claims are like genes, which must pass themselves on in order to survive as a fact, and sees scientists as strategic actors using a variety of tools to gain recognition (Latour, 1987).

Constructivists emphasise that the "non-scientific" elements that inform scientific judgements (see the section "Scientists Excercising Power ...") play a primary role in these negotiations, whether in science-for-policy or otherwise (Bourdieu, 1975; Latour, 1987; Miller and Edwards, 2001; Harrison, 2003). Thus, scientists have different institutional cultures and structures, as well as different beliefs and values that influence the way in which they react to knowledge claims and what they consider to be scientifically derived knowledge (Cozzens and Woodhouse, 1995; Bäckstrand, 2004; McKitrick, 2005). Studies of peer-review mechanisms have demonstrated that judgements of knowledge claims are often inconsistent, and driven by factors such as the extent to which they reflect the reviewer's methodological leanings, the personal credibility of the author, or place of origin of research (Bourdieu, 1975; Jasanoff, 1990; Miller and Edwards, 2001; McKitrick, 2005). The role of non-scientific, subjective factors in the generation of facts challenges the norms of universalism and organised scepticism inherent in the conventional notion of science. As noted earlier, Merton (1968) himself recognised that these norms were ideals that are not always reflected in reality.

Latour (1987) suggests that once a fact has gained acceptance, it is placed in a "black box" that is used by subsequent scientists without question, allowing scientists to build on the work of their predecessors. Creating black boxes, however, obscures the negotiation process that occurred to establish the fact on which the black box is based, and thus the social content of knowledge (Latour, 1987). Black boxes, it is argued, are "stacked" behind a scientific claim as a defensive move, so that it is only through unpacking these black boxes and identifying the social processes behind the facts, that a knowledge claim can be effectively studied (Latour, 1987). The more black boxes that are stacked, the further back a reader must go and the harder they must work to challenge the knowledge claim (Latour, 1987).

The notion that scientific knowledge is the product of negotiations based on social, non-scientific factors suggests that all knowledge claims are subjective and therefore open to political contestation. As such, the positivist notion that there are objectively observable characteristics of science that make it distinct from and superior to other forms of knowledge is disputed by constructivists (Jasanoff, 1990; Gieryn, 1995). Indeed, Cozzens and Woodhouse (1995) suggest that any boundary that is drawn between science and non-science, or "good" and "bad" science, is artificial, temporary and convenient to the person or group drawing the line. By dismissing a certain discipline or method as "bad" science, actors seek to "denude" it of the cognitive authority that is held by mainstream science (Jasanoff, in Gieryn, 1995). Gieryn (1995) suggests that attempts to define science by Merton and Popper have been driven by the need to discredit different scientific practices.[7]

While negotiations of facts may be driven by non-scientific factors, the accuracy with which scientific explanations represent reality (e.g. the extent to which they can predict phenomena) cannot be dismissed as irrelevant to these negotiations. In many cases the extent to which a theory represents reality is itself a subject of negotiation, but in some cases this can be observed by all. In the field of medicine, for example, it cannot be disputed that scientific knowledge has generated key technologies and treatments capable of fighting off disease, through the consistent application of methods considered "scientific". In this respect, medicine has proven itself superior to knowledge systems such as herbalism (although it is not necessary to discount the value of herbalism altogether). While constructivism does not necessarily reject the idea that science can yield *insights* (not truth) into reality, it does indicate that social factors are integral to knowledge construction, and scientific boundaries must be treated with scepticism. This assertion is vital in fields such as fisheries management where the traditional, ecological knowledge of fishers can yield important insights, yet is typically cast outside the boundary of science and often neglected in policymaking circles.

Constructivism offers a completely different perspective to positivism as to the role of experts in society, since it undermines the notion that there is a clear boundary between science and politics on which everyone can agree. This concept has gained particular support in discussions of science for policy, even if the transferability of the model to knowledge in general (i.e. "pure" science) is more controversial. For constructivists, science and politics are highly interrelated at all levels within society, and develop together in a complex relationship (Bourdieu, 1975; Barker and Guy-Peters, 1992; Harrison, 2003). The boundary

between science and policy is simply another manifestation of the boundary between "science" and "non-science", and as such is temporary and contextual, subject to ongoing negotiation between social actors (Yearley, 1995; Cash and Clark, 2001; Clark *et al.*, 2002; Durant, 2006). The science–policy boundary is shaped and reshaped by these social actors in a process known as "boundary work", a process that is heavily influenced by institutional cultures and structures, as well as normative beliefs and interests of the individuals involved (Cozzens and Woodhouse, 1995; Miller and Edwards, 2001).

By shaping the concepts of "science" and "policy", boundary work seeks to establish processes by which actors on either side of the boundary interact with each other. Negotiations over the science-policy boundary revolve around which issues are scientific and which political, as well as who should be given expert status (Gieryn, 1995; Rifkin and Martin, 1997; Cash and Clark, 2001; Miller and Edwards, 2001; Durant, 2006). The outcome of these negotiations can determine both the content of scientific advice and the influence experts exert, as well as the way in which the resulting scientific advice and policy choices are perceived by other actors. Recent studies of boundary work have begun to examine how these negotiations are carried out within "boundary organisations" that straddle the science–policy divide, such as the Intergovernmental Panel on Climate Change (Guston, 2001; Jacob, 2005; Raman, 2005; Durant, 2006).

Gieryn (1995) suggests that there are limits to how the science–policy boundary can be drawn, based on common, implicit understandings of what may be termed science – thus, witchcraft could never be realistically included in a conception of science (see also Zehr, 2005). Nevertheless, a considerable degree of latitude remains regarding the nature of the science–policy boundary, which may show significant variation across issue areas. An example of this regards how decisions on risk are taken. In fisheries management in the EU, for example, scientists decide on the level of risk that their advice should incorporate, choosing to implement a precautionary approach (Wilson and Degnbol, 2002). In fields such as nuclear radiation or public health, it is argued that decisions on the level of acceptable risk should not be left entirely up to scientists (Barker and Guy-Peters, 1992). For both issues, the question of who should decide on appropriate levels of risk remains a point of debate.

Does science really matter?

All accounts of the role of science in society make the assumption that scientific knowledge plays at least a limited role in policy. However, some argue that there are issues where despite it being available, scientific knowledge will simply have no role to play. For example, where strong political and economic interests are at stake, policymakers may simply ignore scientific advice that challenges their interests, without even mobilising opposing advice (Harrison, 2003; Jasanoff, 1990). This often occurs in fisheries management, where policymakers point to the need to consider "socioeconomic" consequences to justify setting quotas above the level recommended by scientists (Andresen and Ostreng, 1989, p. 240). Even where scientific advice is used, it is argued that it is unlikely to be more important than economic and political factors in explaining policy change (Andresen and Ostreng, 1989; Dimitrov, 2003; Harrison, 2003). In issues such as ozone depletion, for instance, it was argued that the successful negotiation of a treaty can be explained more by the commercial interests of DuPont, one of the key producers of ozone-depleting substances, rather than by growing scientific awareness (Sebenius, 1992).

While science may not provide the primary explanation for behaviour, it is nevertheless argued to retain some influence over policy. This may consist of defining the problem, determining the range of policy options available and clarifying state interests in ways that favour particular policy choices. While scientific knowledge may not have been sufficient for policy changes, in most cases without scientific awareness of the problem there would have been no perceived need for change in the first place.

THE INFLUENCE OF SCIENTIFIC ASSESSMENTS

The above discussion of the role of science in society has attempted to flesh out the complex relationship between knowledge and politics, by providing a broad descriptive summary of the available literature. This chapter has sought to position different models of expertise on the positivist–constructivist spectrum, depending on the extent to which they adhere to key positivist assumptions of science. This has provided a useful structure based on the most salient distinctions between theories; however, in some cases this has required comparing violations of different assumptions, which makes the positioning of theories a somewhat subjective choice. Nevertheless, the order in which I have placed these theories means that as one moves further down the constructivist end of the spectrum, theories typically become wider in the scope of their criticism of the positivist assumptions of science. For example,

the theory of epistemic communities accepts the basic premise that scientific knowledge can be used to support different policy preferences, but goes further by suggesting that scientists themselves also advance their own agendas. Similarly, the notion of the social construction of science accepts this premise, but goes further still in suggesting that social and political interests play an important role not only in the use of knowledge, but also in the creation of that knowledge.

The different concepts of the role played by experts, and the scientific enterprise in general, within society suggest that it is inappropriate to assume that the application of the scientific method will necessarily address social and environmental problems in the most effective way. Where science is used to legitimate pre-existing interests, for example, it may be inhibited from exploring alternative options that may yield greater environmental benefits.[8] However, in other cases, it can be expected that the use of scientific research as a policy guide may provide new methods that are more able to predict and explain environmental phenomena – although the extent to which this reflects an objective notion of "truth" is highly questionable. This idea that science can generate progress in the sense of greater prediction and control of nature, is clearly supported within a range of fields, and unquestionably in the field of medicine. Given this possible value of science to policy, an important question arises as to how scientific advice may be presented so that useful research findings are adopted by policymakers.

In the environmental sphere, one of the most important mechanisms through which science can provide input into policy is through environmental assessments. Scientific assessments and advisory processes can vary considerably, for example, in terms of membership, scope, institutional structure and targeted audience (Clark et al., 2002). Another key area of variance is in the influence they have over policy (Clark et al., 2002). According to these authors, assessments can only be valuable where they are influential in changing behaviour.[9] However, influential assessments are considered to be the exception, and such influence is typically indirect, influencing the framing, goals and options available, and visibility of an issue (Clark et al., 2002).

The influence of environmental assessments: issue characteristics

The above discussion indicates that one of the most important factors determining the influence of science on policy is the configuration of interests among different actors, at both the national and international level (Andresen and Ostreng, 1989). Thus, scientific advice can be expected to be less influential on policy where intense political conflicts are prevalent and powerful interest groups oppose assessment findings. In contrast, where assessments are supported by strong interests (such as powerful industry groups, environmental groups, or the public where issues are highly visible), their influence over policy is likely to be higher, although the extent to which the policy change would have happened without scientific advice must be considered (Goldstein, 1989). However, where environmental groups are guided by strong normative ethics, or where public attention is at the level of outrage, the conclusions of scientific assessments may be considered too conservative or based on unethical assumptions, and hence ignored (Andresen and Ostreng, 1989). This has been the case in the International Whaling Convention, where scientific advice suggesting that whales can be sustainably harvested has not been acted on because of the hostility of strong environmental interests to any form of whaling (Andresen and Ostreng, 1989).

Theories of expertise that highlight the way in which politically active experts operate to advance their preferred policies suggest that important factors to be considered regarding assessment influence are the power and resources that expert groups mobilise behind scientific assessments. Epistemic community theory, for example, implies that assessments are more influential where they are actively supported by a strong epistemic community (Harrison, 2003). In Haas' study on ozone depletion, he emphasises the importance of epistemic communities in promoting the findings of assessments that called for the use of strong regulatory measures (Haas, 1992b). Countries where the epistemic community was strongest were the first actively to encourage CFC controls (Haas, 1992b).

In addition to power and interests, the influence of an assessment may be dependent on external shocks and crises beyond the direct control of individual scientists or policymakers. One such example was the discovery of the "ozone hole" above Antarctica, which Litfin (1994) argues was important in bringing about a new discourse, and hence increasing the acceptance of the scientific hypothesis of ozone depletion. In the literature on the role of ideas in international regimes, a similar trend is suggested, that ideas such as neo-liberalism have tended to be more readily accepted in periods of crisis, when existing structures are unable to meet the demands placed on them (Goldstein, 1989).

The influence of environmental assessments: characteristics of the advice itself

The issue of assessment influence has been the subject of considerable research in the last decade,

which has considered how it is affected by the content and presentation of scientific advice, as well as the institutional setting in which it is generated (Global Environmental Assessment Project, 1997; Jung, 1999; Cash and Clark, 2001; Miller and Edwards, 2001; Clark et al., 2002; Cash et al., 2003; Alcock, 2004). This research starts from the basic premise of science as socially constructed, treating scientific assessments as a process of social communication between actors such as scientists, policymakers, NGOs and the media (Cash and Clark, 2001). Within this process, actors negotiate over issues regarding assessment content (e.g. the questions it asks or methods used) and structure (who conducts the assessment, how close the advisory and decision-making bodies get) (Cash and Clark, 2001).

Advice and assessments are argued to be most likely to be used where they are perceived by their audience as demonstrating salience, credibility and legitimacy (Cash and Clark, 2001; Clark et al., 2002; Cash et al., 2003; Alcock, 2004). Salience refers to the extent to which the assessment is relevant, addresses key problems and provides recommendations that can be effectively implemented within the current policy environment. Credibility refers to the scientific credentials of the assessment, particularly based on the perception of whether producers of knowledge have expertise, trust that producers' interests or biases do not drive knowledge creation, and the idea that the scientific assessment represents a consensual agreement among scientists.[10] Legitimacy refers to the extent to which the procedures of the assessment for creating knowledge and agreement are perceived as fair, and whether the concerns of all parties are represented and considered (Cash and Clark, 2001; Clark et al., 2002; Cash et al., 2003).

It is argued that different audiences (i.e. scientists, decision-makers and stakeholders) perceive these concepts in different ways, and so an assessment must balance the needs of the different groups in order to be influential (Cash et al., 2003; Alcock, 2004). As we have seen in the constructivist theory of science, subjective, "non-science" factors may play a significant role in judgements of knowledge claims, meaning there is no single, uniform way to determine the *credibility* of an assessment – there is no judge who is not a party to the dispute (Bourdieu, 1975; Barker and Guy-Peters, 1992; Rifkin and Martin, 1997; Miller and Edwards, 2001). Salience and legitimacy, while they may be easier to judge (e.g. by considering who participated in the assessment or how recommendations fit into legal requirements), are similarly subjective judgements based on beliefs as to what is appropriate (Cash et al., 2003; Alcock, 2004). While different groups evaluate these characteristics in different ways, each group requires salience, credibility and legitimacy. While scientists typically emphasise credibility, for example, they also require their work to be seen as relevant in order to legitimate the use of their research and secure continued funding (Gieryn, 1995; Yearley, 1995; Jung, 1999).

The need for both salience and credibility in Cash and Clark's sense is well established in the literature on environmental assessments (Andresen and Ostreng, 1989; Bolin, 1994; Cash and Clark, 2001; Sissenwine and Mace, 2001; Harrison, 2003). Awareness of the need for the legitimacy of environmental policy has also increased, often embodied in concepts of social capital or stakeholder participation (Edge, 1995; Pretty and Ward, 2001; Sissenwine and Mace, 2001; Bäckstrand, 2004). These three goals are often somewhat contradictory, and must therefore be kept in balance.

In order to achieve salience, policymakers need to ensure that assessments tackle the issues of most importance to them, and provide best available estimates in the face of uncertainty, even where these may be less robust than scientists would like (Kinzig and Starrett, 2003). Failure to provide closure on these issues encourages the strategic use of scientific advice to delay or dispute policy change, and prevents assessments being perceived as salient (Jung, 1999). To ensure these needs are met, policymakers may seek to control the agenda of the scientists and the methods used to generate results. For example, they may favour quantitative approaches in order to provide numbers policymakers can work with (Harrison, 2003; Wilson and Hegland, 2005). At the minimum policymakers seek a degree of communication with scientists as to assessment goals.

To achieve legitimacy, assessments need to be seen to represent the interests of the groups involved in an issue. Assessments may fail to be perceived as legitimate where groups feel their concerns have been ignored (Cash et al., 2003). For example, the Intergovernmental Panel on Climate Change (IPCC) is perceived by developing countries as lacking legitimacy, because it fails to address issues of most importance to them, such as technology transfer and responsibility (Miller and Edwards, 2001). Thus, recommendations for reductions in emissions are considered to carry an automatic bias in the assumption that all countries should contribute equally. This supports the idea that science is infused with the values and beliefs of the scientists involved, and thus likely to be biased against excluded groups. To secure legitimacy, attempts are often made to provide for the participation of stakeholders, or to achieve geopolitical representation among scientists (e.g. including scientists from developing countries) to ensure that groups are not excluded (Clark et al., 2002). Assessments that provide for such international

representation are generally seen as more legitimate across a wider range of actors (Beneviste, 1973; Andresen and Ostreng, 1989).

However, while necessary for salience and legitimacy, participation of policymakers and stakeholders in the assessment process risks undermining the actual or perceived credibility of scientific advice (Gieryn, 1995; Skodvin, 1999). A similar problem may also emerge where scientists are required to make predictions with a lower degree of certainty than they are accustomed to using, or where they are required to use methods that are not well respected among the scientific community. In some cases it has been seen that where political requirements overtly dominate how scientific knowledge is generated, it can limit the accuracy of scientific advice.[11] In other cases, political involvement is simply perceived as threatening credibility, based on the normative idea that science and politics are distinct, and should always be kept separate in order to protect scientists from political manipulation (Jasanoff, 1990; Skodvin, 1999; Sissenwine and Mace, 2001; Clark et al., 2002). The need to preserve the image of credibility can mean that scientists are reluctant to use methods desired by policymakers, or provide a "best guess" in the face of uncertainty, and may seek to limit the involvement of political interests.[12] The potential for political distortion of scientific results does not necessarily lead to the conclusion that all political involvement in science is bad. Limiting involvement to scientists and using only the most scientifically defensible techniques may result in a loss of salience, that is, a failure to provide useful input into policy (Sissenwine and Mace, 2001; Clark et al., 2002).

The challenge for designing scientific assessments for policy, therefore, is to provide the optimal degree of involvement between scientists and policymakers while preserving both actual and perceived scientific credibility (Jasanoff, 1990; Jung, 1999; Clark et al., 2002). A common argument is that science and policy need to be "close but not too close", that is separated at some level (Gieryn, 1995; Fritz, 1998; Jung, 1999; Skodvin, 1999; Sissenwine and Mace, 2001). Several approaches may be used to achieve this, such as the use of peer review or other quality assurance mechanisms, limiting political involvement to certain stages of the process or by selecting scientists that have good scientific credentials and a broad geopolitical representation (Clark et al., 2002). Particularly important, according to Clark et al. (2002) is that provision must be made for ongoing review, providing an arena for the negotiations that occur between scientists and policymakers. Skodvin (1999) provides a detailed discussion of the strategies used by the IPCC to manage the involvement of scientists and policymakers.

However, even where such strategies are used, the potential for science to be perceived as politicised remains high, as has been evident in the controversy over the wording of the IPCC's Second Assessment Report, where authors were criticised for removing key paragraphs without the explicit agreement of all participants (Miller and Edwards, 2001).

To some extent, the level of salience, credibility and legitimacy commanded by assessments can be seen as a function of the way in which the science–policy boundary is organised (Cash and Clark, 2001; Jung, 1999). Actors attribute these characteristics to assessments through observations of the boundary and how it compares to their expectations (e.g. a belief that science should be separate from politics). Questions of risk and participation outlined above explicitly address how this boundary is drawn. Another vital component of the science–policy boundary is the institutional structure of the assessment process (Clark *et al.*, 2002).

The influence of environmental assessments: institutional setting

Alcock (2004) suggests that actors in an issue area will attribute to an assessment those qualities they attribute to the organisation producing it, thus the institutional setting is an important factor in perceptions of salience, credibility and legitimacy and therefore assessment influence. Aspects such as the reputation or institutional identity and characteristics of an organisation are of particular importance to these perceptions. Institutions may also perform certain functions that shape and contextualise trade-offs between salience, credibility and legitimacy – for example, providing for communication between scientists and policymakers, dispute settlement procedures or by setting research priorities (Clark *et al.*, 2002). While Alcock (2004) recognises the importance of institutional setting on assessment influence, he nevertheless argues that actors may in some cases be able to counteract structural constraints.

One key distinction that can be made regarding institutional setting is between autonomous and embedded advisory mechanisms (Alcock, 2004). Autonomous organisations are those which have an institutional identity separate from the political institution that they provide advice for.[13] Embedded organisations are specifically designed to serve the policy-making organisation, based within its institutional structure.[14] This distinction builds on the contrast between autonomy and involvement introduced by Andresen *et al.* (2000). However, Alcock's use of "autonomous" and "embedded" appears to be related only to the organisational layout – that is, whether scientific organisations are located within policymaking organisations, while for Andresen *et al.*, autonomy relates to a range of

factors such as control over research agendas and a separate source of funding (Skodvin, 1999; Andresen *et al.*, 2000; Alcock, 2004).

Organisations with an independent institutional identity (Alcock's notion of autonomy) are considered to have higher autonomy in the sense that funding is provided and appointments are made by scientific organisations on the basis of merit, and scientists have control over the research agenda (Skodvin, 1999; Alcock, 2004). This provides them with considerable credibility in the eyes of an assessment audience. However, they show a lower level of involvement between scientists and policymakers in the policymaking process – scientists are not involved in formulating policy advice, the functions of research and research coordination are separated, and there is limited communication between scientists and policymakers (Clark *et al.*, 2002). This can negatively impact on the salience of their advice (Alcock, 2004). Embedded organisations, on the other hand, typically show a higher level of involvement and lower autonomy, with policymakers having greater control over the research agenda and scientists communicating more with policymakers in generating advice. Alcock suggests that embedded organisations are perceived as having high levels of salience, credibility and legitimacy within the organisation in which they are embedded, but are perceived as lacking all three among actors outside the organisation (Clark *et al.*, 2002; Alcock, 2004).

Another important element of institutional structure regards whether an assessment is ad hoc, set up to provide a one-off assessment, or more permanent, established to provide ongoing advice to policymakers. Permanent assessments can build up relationships of trust between experts and decision-makers, and learn from experience, to provide assessments that achieve a better balance between salience, credibility and legitimacy (Bolin, 1994; Clark *et al.*, 2002). But such assessments also incur potential "threats to learning", such as a reluctance to risk established reputation by admitting error, or becoming mired in outdated methods and procedures (Clark *et al.*, 2002). Clark *et al.* (2002) suggest that permanent assessment institutions get better at "doing the job right", but over time become less likely to be "doing the right job". Ad hoc advisory processes, on the other hand, can use the most up-to-date methods and provide a fresh insight into the problem, but are rarely influential as they lack experience (Clark *et al.*, 2002).

CONCLUDING REMARKS

In conclusion, several different perspectives exist for viewing the roles played by experts and scientific knowledge within the policy arena. If we are to understand environmental problems and political behaviour, it is vital that we recognise these varied roles played by scientists, and the different conceptions of science that social actors identify with.

In more recent discussions within this issue area, the insights provided by sociologists of science regarding the interrelatedness of science and politics forms the starting point for discussions, such as how to maximise the input scientific assessments have in policy. While this is a useful area worthy of continuing research, care must be taken to avoid the simple conclusion that more scientific input is all that is required for better environmental management. Rather, given these tight linkages between science, on the one hand, and policy, on the other, it becomes clear that in observing environmental policy we must study science, policy and the blurred, shifting boundary between the two with equal intensity.

NOTES

1 Beck et al. (2005) explore how policymakers in the Ministry for Agriculture, Fisheries and Food, when presented with a number of potentially equally credible scientific explanations of BSE and the question of species transferability, chose the explanation that best served the interests of the agricultural policy community, by minimising potential health risks and emphasising the safety of British beef. This scientific position was maintained by MAFF, which subsequently restricted risk-relevant information from the public and the wider scientific community, even in the face of mounting evidence that suggested their position was no longer credible.

2 In Canada, it was argued that the Department of Fisheries and Oceans interfered with the conduct and publication of research, putting pressure on scientists to deliver favourable results (Hutchings et al., 1997).

3 In certain areas, this pressure may be highly visible, for example, in fishing communities where fishermen have engaged in active protests, often with the threat of violence. In one account of a meeting of the New England Fisheries Management Council in the USA involving fisheries scientists and managers, the participants faced a room full of angry fishermen, and required a security presence to maintain order (Dobbs, 2000).

4 According to Skodvin (1999) this occurs in the Intergovernmental Panel on Climate Change.

5 Strong constructivists, we shall see, would make the point that disputes are never resolved objectively.

6 Discourses embody Layzer's notion of a problem-definition framework, but are more wide reaching in

that they often describe a type of worldview that may apply to a whole range of different problems.

7 Gieryn (1995, p. 394) suggests that Merton was trying to discredit Aryan science, while Popper wished to dismiss psychoanalysis and Marxism as unscientific.

8 This is a potential pitfall of any scientific theory, and is not an inevitable consequence of science being used to support pre-existing interests.

9 This is a necessary, but not sufficient condition for assessments to be environmentally valuable; an assessment may be influential, but give poor advice that leads to environmental harm.

10 While the accuracy of the extent to which science adheres to these ideals is questioned by constructivists, they remain important characteristics for many observers against which scientific assessments are measured.

11 This was claimed to be the case in Canada in the 1980s, where it is argued that bureaucratic interference in scientific stock assessments led to the health of the cod stock being overestimated, which played a key role in the subsequent fishery collapse (Botsford et al., 1997).

12 This conflict is common in fisheries stock assessments, where managers require an estimate of sustainable catch levels each year in order to set regulations, yet scientists may be constrained by the quality of the data (Wilson and Hegland, 2005).

13 The International Council for the Exploration of the Sea (ICES), is an example of an autonomous organisation, with a history of independence dating back to 1902. ICES provides advice on fisheries management to the EU and other regional fisheries management organisations.

14 An example of an embedded organisation is the Scientific, Technical and Economic Committee for Fisheries, a committee in the EU tasked with reviewing ICES advice and transmitting recommendations to the EU Commission.

REFERENCES

Alcock, F. (2004). "The Institutional Dimensions of Fisheries Stock Assessments", *International Environmental Agreements*, 4: 129–141.

Alverson, D. (2002). "Factors Influencing the Scope and Quality of Science and Management Decisions (the good, the bad and the ugly)", *Fish and Fisheries*, 3: 3–19.

Andresen, S. and Ostreng, W. (eds) (1989). *International Resource Management: The Role of Science and Politics.* New York: Belhaven Press.

Andresen, S., Skodvin, T., Underdal, A. and Wettestad, J. (2000). *Science and Politics in International Environmental Regimes: Between Integrity and Involvement.* Manchester: Manchester University Press.

Ayer, A. J. (1966). *Logical Positivism.* Glencoel IL: The Free Press.

Bäckstrand, K. (2004). "Civic Science for Sustainability: Reframing the Role of Scientific Experts, Policy-makers and Citizens in Environmental Governance", in Biermann, F., Campe, S. and Jacob, K. (eds) *Proceedings of the 2002 Berlin Conference on the Human Dimensions of Global Environmental Change: Knowledge for the Sustainability Transition. The Challenge for Social Science*, Global Governance Project: Amsterdam, Berlin, Potsdam and Oldenburg, pp. 165–174.

Barker, A. and Guy-Peters, B. (1992). *Politics of Expert Advice: Creating, Using and Manipulating Scientific Knowledge for Public Policy*. Edinburgh: Edinburgh University Press.

Beck, M., Asenova, D. and Dickson, G. (2005). "Public Administration, Science, and Risk Assessment: A Case Study of the U.K. Bovine Spongiform Encephalopathy Crisis", in *Public Administration Review*, 65(4): 396–408.

Beck, U. (1992). *Risk Society: Towards a New Modernity (electronic resource)*. London: Sage Publications.

Beneviste, G. (1973). *The Politics of Expertise*. London: Croom Helm.

Boehmer-Christiansen, S. and Skea, J. (1991). *Acid Politics: Environmental and Energy Policies in Britain and Germany*. London: Belhaven Press.

Bolin, B. (1994). "Science and Policy Making", *Ambio*, 23(1): 25–29.

Botsford, L., Castilla, J. and Petersen, C. (1997). "The Management of Fisheries and Marine Ecosystems", *Science*, 277: 509–515.

Bourdieu, P. (1975). "The Specificity of the Scientific Field and the Social Condition of the Progress of Reason", *Social Science Information*, 14: 19–47.

Cash, D. and Clark, W. (2001). *From Science to Policy: Assessing the Assessment Process*. Faculty Research Working Paper Series, John F. Kennedy School of Government, Harvard University.

Cash, D., Clark, W., Alcock, F., Dickson, N., Eckley, N., Guston, D., Jaëger, J. and Mitchell, R. (2003). "Knowledge Systems for Sustainable Development", in *Proceedings of the National Academy of Sciences (PNAS)*, 100(14): 8086–8091.

Clark, W., Mitchell, R., Cash, D. and Alcock, F. (2002). *Information as Influence: How Institutions Mediate the Impact of Scientific Assessments on Global Environmental Affairs*. KSG Faculty Research Working Paper Series, Harvard University.

Cozzens, S. and Woodhouse, E. (1995). "Science, Government and the Politics of Knowledge", in Jasanoff, S., Markle, G., Petersen, J. and Pinch, T. (eds) *Handbook of Science and Technology Studies*. Thousand Oaks, CA: Sage Publications.

Cummings, L. (2005). "Giving Science a Bad Name: Politically and Commercially Motivated Fallacies in BSE Inquiry", in *Argumentation*, 19:123–143.

Dimitrov, R. (2003). "Knowledge, Power, and Interests in Environmental Regime Formation", *International Studies Quarterly*, 47: 123–150.

Dobbs, D. (2000). *The Great Gulf: Fishermen, Scientists and the Struggle to Revive the World's Greatest Fishery*. Washington, DC: Island Press.

Doubleday, W., Atkinson, D. and Baird, J. (1997). "Comment: Scientific inquiry and fish stock assessment in the Canadian Department of Fisheries and Oceans", in *Canadian Journal of Fisheries and Aquatic Science*, 54: 1422–1426.

Dowding, K. (1996). *Power*. Buckingham: Open University Press.

Durant, D. (2006). "Managing Expertise: Performers, Principals, and Problems in Canadian Nuclear Waste Management", *Science and Public Policy*, 33(3):191–204.

Ebbin, S. (2004). "Black Box Production of Paper Fish: An Examination of Knowledge Construction and Validation in Fisheries Management Institutions", *International Environmental Agreements*, 4: 143–159.

Edge, D. (1995). "Reinventing the Wheel", in Jasanoff, S., Markle, G., Petersen, J. and Pinch, T. (eds) *Handbook of Science and Technology Studies*. Thousand Oaks, CA: Sage Publications.

Fritz, J.S. (1998). *Report on International Scientific Advisory Processes on the Environment and Sustainable Development*. Nairobi: UN System-Wide Earthwatch Coordination, United Nations Environment Programme.

Gieryn, T. (1995). "Boundaries of Science", in Jasanoff, S., Markle, G., Petersen, J. and Pinch, T. (eds) *Handbook of Science and Technology Studies*. Thousand Oaks, CA: Sage Publications.

Global Environmental Assessment Project (1997). *A Critical Evaluation of Global Environmental Assessments: The Climate Experience*. Calventon, MD: CARE.

Goldstein, J. (1989). "The Impact of Ideas on Trade Policy: The Origins of US Agricultural and Manufacturing Policies", *International Organisation*, 43(1): 31–71.

Greenberger, M., Crenson, M. and Crissey, B. (1976). *Models in the Policy Process: Public Decision Making in the Computer Era*. New York: Russell Sage Foundation.

Guston, D. (2001). "Boundary organizations in Environmental Policy and Science: an introduction". *Science, Technology and Human Values*, 26 (4): 399–408.

Haas, P. (1992a). "Introduction: Epistemic Communities and International Policy Co-ordination", *International Organisation*, 46(1): 1–35.

Haas, P. (1992b). "Banning chlorofluorocarbons: Epistemic community efforts to protect stratospheric ozone", *International Organisation*, 46(1): 187–224.

Harris, M. (1999). *Lament for an Ocean* (revised edition). Toronto: McClelland and Stewart.

Harrison, N. (2003). *Science and Politics in the International Environment*. Lanham, MD: Rowman & Littlefield Publishers.

Hasenclever, A., Mayer, P. and Rittberger, V. (1997). *Theories of International Regimes*. Cambridge, UK: Cambridge University Press.

Healey, M. (1997). "Comment: The interplay of policy, politics and science", *Canadian Journal of Fisheries and Aquatic Science*, 54: 1427–1429.

Holm, P. and Nielsen, K. (2004). "The TAC machine (Appendix B, Working Document 1)", in ICES, 2004. *Annual Report of the Working Group for Fisheries Systems (WGFS)*. Copenhagen.

Hume, D. (1999). In Beauchamp. T. (ed.) *An Enquiry Concerning Human Understanding*, Oxford: Oxford Philosophical Texts.

Hutchings, J., Walters, C. and Haedrich, R. (1997). "Is scientific inquiry incompatible with government information control?", *Canadian Journal of Fisheries and Aquatic Science*, 54: 1198–1210.

Jacob, M. (2005). "Boundary work in contemporary science policy: A review", *Prometheus*, 23(2): 195–207.

Jasanoff, S. (1990). *The Fifth Branch: Science Advisers as Policymakers*. Cambridge, MA: Harvard University Press.

Jung, W. (1999). *Expert Advice in Global Environmental Decision Making: How Close Should Science and Policy Get?* ENRP Discussion Paper E-99-14, John F. Kennedy School of Government, Harvard University.

King, G., Keohane, R. and Verba, S. (1994). *Designing Social Inquiry: Scientific Inference in Qualitative Research*. Princeton, NJ: Princeton University Press.

King, D. (2005). "Thirteenth BES lecture: Climate change: The science and the policy", in *Journal of Applied Ecology*, 42: 779–783.

Kinzig, A. and Starrett, D. (2003). "Coping with uncertainty: A call for a new science-policy forum", *Ambio*, 32 (5): 330–335. Royal Swedish Academy of Sciences.

Kuhn, T. (1970) *The Structure of Scientific Revolutions*. Chicago: University of Chicago Press.

Lane, D. E. and Stephenson, R. L. (1999). "Fisheries-management Science: A framework for the implementation of fisheries-management systems", *ICES Journal of Marine Science*, 56: 1059–1066.

Latour, B. (1987). *Science in Action: How to Follow Scientists and Engineers Through Society*. Cambridge, MA: Harvard University Press.

Layzer, J. (2006). "Fish stories: Science, advocacy, and policy change in New England fishery management", *The Policy Studies Journal*, 34(1): 59–80.

Lewenstein, B. (1995). "Science and the Media", in Jasanoff, S., Markle, G., Petersen, J. and Pinch, T. (eds.). *Handbook of Science and Technology Studies*. Thousand Oaks, CA: Sage Publications.

Litfin, K. (1994). *Ozone Discourses*. New York: Columbia University Press.

Martin, B. and Richards, E. (1995). "Scientific Knowledge, Controversy and Public Decision Making", in Jasanoff, S., Markle, G., Petersen, J. and Pinch, T. (eds) *Handbook of Science and Technology Studies*. Thousand Oaks, CA: Sage Publications.

McKitrick, R. (2005). "Science and environmental policy-making: Bias-proofing the assessment process", in *Canadian Journal of Agricultural Economics*, 53: 275–290.

Merton, R.K. (ed.) (1968). "Science and the Social Order", *Social Theory and Social Structure*. New York: The Free Press: 591–695.

Miles, E., Underdal, A., Andresen, S., Wettestad, J. and Skjaerseth, J. (2001). *Environmental Regime Effectiveness: Confronting Theory with Evidence*. The MIT Press.

Miller, C. and Edwards, P. (eds) (2001). *Changing the Atmosphere: Expert Knowledge and Environmental Governance*. The MIT Press.

PCE (2004). *Missing Links; Connecting Science with Environmental Policy*. Wellington: Parliamentary Commission for the Environment.

Pepper, D. (1990). *The Roots of Modern Environmentalism*. London: Routledge

Pretty, J. and Ward, H. (2001). "Social capital and the environment", *World Development*, 29(2): 209–227.

Raman, S. (2005). "Institutional perspectives on science-policy boundaries", *Science and Public Policy*, 32(6): 418–422.

Rifkin, W. and Martin, B. (1997). "Negotiating expert status: Who gets taken seriously", *IEEE Technology and Society Magazine* (Spring): 30–39.

Russo, N. and Denious, J. (2005). "Controlling birth: Science, politics, and public policy", *Journal of Social Issues*, 61(1): 181–191.

Sallenave, R. and Cowley, D. (2006). "Science and effective policy for managing aquatic resources", *Reviews in Fisheries Science*, 14: 203–210.

Sebenius, J. (1992). "Challenging conventional explanations of International Co-operation: Negotiation analysis and the case of epistemic communities", *International Organisation*, 46(1): 323–365.

Sissenwine, M.P. and Mace, P.M. (2001). "Governance for Responsible Fisheries: An Ecosystem Approach". Paper presented to the *Reykjavik Conference on Responsible Fisheries in the Marine Ecosystem*. Reykjavik, Iceland, 1–4 October 2001.

Skodvin, T. (1999). *Science–Policy Interaction in the Global Greenhouse*. Center for International Climate and Environmental Research (CICERO), Working Paper 1999: 3.

Technocracy, Inc. (1960). *An Analysis of Technocracy*. Information Brief, no. 50. http://www.technocracy.org/?p=/documents/briefs/b50

Walsh, V. (2004). *Global Institutions and Social Knowledge: Generating Research at the Scripps Institution and the Inter-American Tropical Tuna Commission, 1900s–1990s*. The MIT Press.

Wilson, D. (1999). "Fisheries Science Collaborations: The Critical Role of the Community". Keynote presentation at the *Conference on Holistic Management and the role of Fisheries and Mariculture in the Coastal Community*, 11–12 Nov, 1999, Tjärnö Marine Biological Laboratory, Sweden

Wilson, D.C. and Degnbol, P. (2002). "The effects of legal mandates on fisheries science deliberations: The case of atlantic bluefish in the United States", *Fisheries Research*, 58: 1–14.

Wilson, D.C. and Hegland, T. (2005). *An Analysis of Some Institutional Aspects of Science in Support of the Common Fisheries Policy*. Project Report for Policy and Knowledge in Fisheries Management. CEC 5th Framework Programme No. Q5RS–2001-01782. Institute for Fisheries Management, Publication No. 126.

Woodhouse, E.J. and Nieusma, D. (1997). "When expert advice works, and when it does not", *IEEE Technology and Society Magazine*, 16(1):23–29.

Wynne, B. (1995). "Public Understanding of Science", in Jasanoff, S., Markle, G., Petersen, J. and Pinch, T. (eds.) *Handbook of Science and Technology Studies*. Thousand Oaks, CA: Sage Publications.

Yearley, S. (1995). "The Environmental Challenge to Science Studies", in Jasanoff, S., Markle, G., Petersen, J. and Pinch, T. (eds) *Handbook of Science and Technology Studies*. Thousand Oaks, CA: Sage Publications.

Young, O. (2004). "Institutions and the Growth of Knowledge: Evidence from International Environmental Regimes", *International Environmental Agreements*, 4: 215–228.

Zehr, S. (2005). "Comparative boundary work: US acid rain and global climate change policy deliberations", *Science and Public Policy*, 32(6): 445–456.

37

Interdependent Social–Ecological Systems and Adaptive Governance for Ecosystem Services

Carl Folke, Johan Colding,
Per Olsson and Thomas Hahn

INTRODUCTION

The pre-analytic vision of this chapter is that human societies and globally interconnected economies are parts of the dynamics of the biosphere, embedded in its processes and ultimately dependent on the capacity of the environment to sustain societal development with essential ecosystem services and support (Odum, 1989; Millennium Ecosystem Assessment, 2005). Throughout history humans have shaped nature and nature has shaped the development of human society (Turner *et al.*, 1990; Redman, 1999). The human dimension has expanded and intensified and become globally interconnected, through technology, capital markets and systems of governance with decisions in one place influencing people and ecosystems elsewhere (Holling, 1994). Reduced temporal variability of renewable resource flows in some parts of the world has resulted in increased spatial dependence on other areas on Earth, reflected in, for example, widespread ecosystem support to urban areas (Folke *et al.*, 1997). Humanity has become a major force in structuring ecosystem dynamics from local scales to the biosphere as a whole (Steffen *et al.*, 2004).

In this context it becomes clear that patterns of production, consumption and well-being develop not only from economic and social relations within and between regions, but to be sustained also depend on the capacity of ecosystems throughout the world to support societal development (Arrow *et al.*, 1995, 2004). Social conditions, health, culture, democracy, and matters of security, survival and the environment are interwoven in a grand panorama of regional and worldwide dependency.

Sometimes change is gradual or incremental. During these periods of steady progress, things move forward in roughly continuous and predictable ways. At other times, change is abrupt, disorganizing or turbulent. During such periods, experience is often insufficient for understanding, consequences of actions are ambiguous, and the future of system dynamics often uncertain (Gunderson and Holling, 2002). Evidence points to a situation where periods of abrupt change are likely to increase in frequency and magnitude (Steffen *et al.*, 2004), which poses new fundamental challenges for science, management, policy and governance.

Theories, models and policies for resource and environmental management has to a large extent been developed for gradual or incremental change situations focusing on a unique state with assumptions of linear dynamics and generally disregarding interactions across scales. Recent research has revealed that the implementation of such theory and policy tends to invest in controlling a few selected ecosystem processes, at the expense of key ecological functions, in the urge to fulfil economic or social goals (e.g. Gunderson *et al.*, 1995; Holling and Meffe, 1996; Allison and Hobbs, 2004). This behavioural pattern causes loss of resilience (capacity to buffer change and continue to develop) of desired states (Holling, 1973; Folke, 2006a). Loss of resilience results in vulnerable systems. Historical loss of resilience has put whole regions and cultures into vulnerable states with constrained options for development (Kasperson *et al.*, 1995; Redman, 1999; Schröter *et al.*, 2005). Vulnerable systems subject to change may easily shift from one state (stability domain, basin of attraction) into another (Walker and Meyers, 2004). When such shifts occur the common focus tends to be on the event that revealed the shift and not on the variables, processes and trajectories that caused loss of resilience prior to the event.

To what extent are human societies adapting their capacity for learning and foresight to deal with this new global and challenging situation? We agree with the findings of the Millennium Ecosystem Assessment that the societal capacities to manage the Earth's ecosystems are evolving more slowly than humanity's (over)use of the same systems. Conventional resource models, based on single resources and linear dynamics, are of limited use for the purpose of navigating society towards sustainability.

The perspective presented here emphasizes the following features:

1 society and nature represent truly *interdependent social–ecological systems;*
2 social–ecological systems are *complex adaptive systems;*
3 *cross-scale and dynamic interactions* represent new challenges for governance and management in relation to interdependent social–ecological systems and ecosystem services.

Research for sustainability increasingly addresses the intricate feedbacks of social–ecological systems, their complex dynamics, and how they play out across spatial and temporal scales. A deeper understanding of coupled systems undergoing change is essential in this context. The implications for current management and policy are challenging.

Here, we provide a brief overview of the three features and than turn into a discussion of systems of adaptive governance (Dietz *et al.*, 2003; Folke *et al.*, 2005) that allow for responding to, adapting to and shaping environmental change. We highlight some features of such governance systems by focusing on management practices of human groups involved with ecosystems and social mechanisms behind such management, including social taboos, social networks, bridging organizations, leadership and actor groups. Societies face the challenge of dealing with unpredictability, uncertainty and change, and how to build resilience to periods of abrupt change and allow for transformations into more desirable social–ecological pathways.

THE THREE FEATURES

Interdependent social–ecological systems

In our view, there are neither natural or pristine systems without people nor social systems without nature (Folke, 2006b). Social and ecological systems are not just linked but truly interconnected and co-evolving across spatial and temporal scales (Berkes and Folke, 1992; Norgaard, 1994). We refer to them as *social–ecological systems* (Berkes and Folke, 1998) emphasizing the humans-in-the-environment perspective (Berkes *et al.*, 2003). It is close to impossible truly to understand ecosystem dynamics and their ability to generate services without accounting for the human dimension. Focusing on the ecological side only, as a basis for decision-making for sustainability, simplifies reality so much that the result is distortions and leads to incomplete and narrow conclusions. For example, an observed shift in a lake from a desired to a degraded state may indicate that the lake has lost resilience, but if there is capacity in the social system to respond to change and restore the lake the social–ecological system is still resilient (Carpenter and Brock, 2004; Bodin and Norberg, 2005).

The same is true for social sustainability. Despite a vast literature on the social dimension of resource and environmental management, studies have predominantly focused on investigating processes within the social domain only, treating the ecosystem largely as a given, an external "black box", assuming that if the social system performs adaptively or is well organized institutionally it will also manage resources and ecosystems in a sustainable fashion. A human society may show great ability to cope with change and adapt if analysed only through the social dimension lens (Huitric, 2005). But such an adaptation may be at the expense of changes in the capacity of ecosystems to sustain the adaptation, and may generate traps and breakpoints in social–ecological systems (Allison and Hobbs, 2004).

There has been substantial progress in understanding the social dimension of ecosystem management, including organizational and institutional flexibility for dealing with uncertainty and change (e.g. Lee, 1993; Grumbine, 1994; Westley, 1995; Berkes and Folke, 1998; Danter et al., 2000; Gunderson and Holling, 2002; Berkes et al., 2003; Dietz et al., 2003; Anderies et al., 2004; Armitage, 2005; Ostrom, 2005) and social capital and conflict (e.g. Adger, 2003; Ostrom and Ahn, 2003; Pretty, 2003; Galaz, 2005). Challenges for the social sciences have been raised in this context (e.g. Scoones, 1999; Abel and Stepp, 2003). Social sources of resilience such as social capital (including trust and social networks) and social memory (including experience for dealing with change) (Olick and Robbins, 1998; McIntosh, 2000) are essential for the capacity of social–ecological systems to adapt to and shape change (Folke et al., 2003).

Complex adaptive systems

In our view, social–ecological systems are complex adaptive systems characterized by historical (path) dependency, non-linear (non-convex) dynamics, regime shifts, multiple basins of attraction and limited predictability (Costanza et al., 1993). Theories of complex systems portray systems not as deterministic, predictable and mechanistic, but as process dependent, organic and self-organizing with feedbacks between multiple scales (e.g. Kauffman, 1993; Holland, 1995; Arthur, 1999; Levin, 1999). The ecosystem-based approach recognizes the role of the human dimension in shaping ecosystem processes and dynamics (Dale et al., 2000) and that human actions have pushed ecological systems into less productive or otherwise less desirable states with negative consequences for human livelihood and security. The existence of "regime shifts" in ecosystems is an area of intense research (Scheffer et al., 2001) with examples from forests, lakes, wetlands, coastal areas, fisheries, coral reefs (Folke et al., 2004), grazing lands (Scholes and Walker, 1993), agriculture (Rockström, 2003) and marine systems (Troell et al., 2005; Grebmeier et al., 2006). In some cases, these shifts may be irreversible or too costly to reverse (Mäler, 2000).

The human dimension reflects properties of complex adaptive systems such as a diverse set of institutions and behaviours, local interactions between actors and selective processes that shape future social structures and dynamics (Holland et al., 1986; Arthur, 1999; Janssen and Jager, 2001; Lansing, 2003). Complexity makes it hard to forecast the future. Not only are forecasts uncertain, but also the usual statistical approaches will likely underestimate the uncertainties since key drivers like climate and technological change are unpredictable and may change in non-linear

fashions (Kinzig et al., 2003; Peterson et al., 2003). Gunderson (2001) nicely illustrates the need for learning and flexibility in the social system when confronted with alternative and uncertain explanations of ecosystem change.

The complex adaptive systems approach shifts the perspective on governance from aiming at controlling change in resource and ecosystems that are assumed to be stable, to enhancing the capacity of social–ecological systems to learn to live with and shape change and even find ways to transform into more desirable directions following change (van der Leeuw, 2000; Berkes et al., 2003; Norberg and Cumming, 2007).

It is in this context that the *resilience* perspective becomes central. Resilience is the capacity to absorb change, reorganize and continue to develop. The concept of resilience was invented to address the paradox of how change and persistence work together (Holling, 1973). Resilience research addresses how systems assimilate disturbance and make use of change for innovation and development, while simultaneously maintaining characteristic structures and processes (Folke, 2006a). It is argued that managing for resilience enhances the likelihood of sustaining and developing desirable pathways for societal development in changing environments where the future is unpredictable and surprise is likely (Gunderson and Holling, 2002; Adger et al., 2005).

Cross-scale and dynamic interactions

Social–ecological systems are linked across temporal and spatial scales and levels of organization. Human capacities for abstraction and reflexivity, forward-looking action and technology development are strikingly different from ecological systems (Westley et al., 2002) and enable human systems to transcend constraints of ecological scale. Local groups and communities are subject to decision from regional levels and connected to global markets and vice versa (Berkes et al., 2006). A social–ecological system can avoid vulnerability at one time scale through the technology it has adopted. Similarly, resilience at one spatial extent can be subsidized from a broader scale, a common pattern in human cultural evolution (Redman, 1999; van der Leeuw, 2000) and exacerbated by technology, capital markets and financial transfers that mask environmental feedback.

Such feedbacks and their cross-scale interactions in relation to resilience are in focus of truly integrated social–ecological systems' modelling of agents and ecosystems with multiple stable states (e.g. Carpenter et al., 1999; Janssen and Carpenter, 1999; Janssen et al., 2000; Carpenter and Brock, 2004; Bodin and Norberg, 2005).

Recent work suggests that complex systems "stutter" or exhibit increased variance at multiple scales in advance of a regime shift (Carpenter and Brock, 2006). Such increases in variance help characterize regime shifts and may even allow early warning indicators of some regime shifts. Furthermore, multiple thresholds and regime shifts at different scales and in different and interacting ecological, economic and social domains are proposed to exist within regional social–ecological systems (Kinzig et al., 2006).

New insights are emerging on cross-scale interactions in social–ecological systems (Gunderson and Holling, 2002; Young, 2002; Cash et al., 2006) including the dynamics of social and economic drivers of land use change (Lambin et al., 2003) and on governance systems that allow for learning and responding to environmental feedback and change (Dietz et al., 2003). Good ecosystem management requires governance and management approaches that can deal with the change and uncertainty inherent in social–ecological systems and match social and ecological structures and processes operating at different spatial and temporal scales (Folke et al., 1998b; Brown, 2003).

ADAPTIVE GOVERNANCE FOR ECOSYSTEM SERVICES

The capacity to adapt and shape change is an important component of resilience in social–ecological system (Berkes et al., 2003). In a social–ecological system with high adaptability the actors have the capacity to reorganize the system within desired states in response to changing conditions and disturbance events (Walker et al., 2004). This includes social sources of resilience for dealing with uncertainty and change and a focus on adaptive capacity (Folke et al., 2003), learning and innovation in social–ecological systems and even the capacity to transform into improved pathways or trajectories (Folke et al., 2005).

Because of cross-scale interplay, positive feedbacks causing non-linear dynamics and possible shifts between alternate states in interdependent social–ecological systems, new approaches to governance will be required for guiding management and policy of ecosystem services towards sustainability. Based on several case studies Folke et al. (2003) identify four critical factors for social–ecological systems that interact across temporal and spatial scales that seem to be required for dealing with ecosystems' dynamics during periods of rapid change and reorganization:

1 learning to live with change and uncertainty;
2 combining different types of knowledge for learning;
3 creating opportunity for self-organization toward social-ecological resilience;
4 nurturing sources of resilience for renewal and reorganization.

Governance and management systems have to be designed to incorporate these factors. The emerging perspective of adaptive governance (Dietz et al., 2003; Folke et al., 2005) represents one such approach. Adaptive governance conveys the difficulty of control, the need to proceed in the face of substantial uncertainty, and the importance of dealing with diversity and reconciling conflict among people and groups who differ in values, interests, perspectives, power and the kinds of information they bring to situations (Dietz et al., 2001). Such governance fosters social coordination that enables adaptive co-management of ecosystems. Adaptive co-management combines the dynamic learning characteristic of adaptive management (Gunderson et al., 1995; Carpenter and Gunderson, 2001) with the linkage characteristic of collaborative management (Gadgil et al., 2000; Wollenberg et al., 2000; Ruitenbeek and Cartier, 2001; Folke et al., 2003, Borrini-Feyerabend, 2004). For such governance to be effective it requires an understanding of both ecosystems and social–ecological interactions.

Adaptive governance relies on multi-level arrangements, including local, regional, national, transnational and global levels, where authority has been reallocated upward, downward and sideways away from central states. It refers to a type of governance that is dispersed across multiple centres of authority (Hooghe and Marks, 2003), "pluricentric" rather than "unicentric" (Kersbergen and van Waarden, 2004), and is characterized by non-hierarchical methods of control (Ostrom, 1998; Stoker, 1998). The common property resource research refers to such nested, quasi-autonomous decision-making units operating at multiple scales as polycentric institutions (Ostrom, 1998; McGinnis, 2000; Dietz et al., 2003).

We have previously proposed that there are, at least, four interacting aspects to be concerned about in adaptive governance of complex social–ecological systems with cross-scale dynamics (Folke et al., 2005):

- *Building knowledge and understanding of resource and ecosystem dynamics*: detecting and responding to environmental feedback in a fashion that sustains the capacity of the environment to provide ecosystem services requires ecological knowledge and understanding of ecosystem processes and functions (Berkes and Folke, 1998). All sources of understanding need to be mobilized and management of complex adaptive systems may benefit from the combination of different knowledge systems.

- *Feeding ecological knowledge into adaptive management practices*: successful management is characterized by continuous testing, monitoring and re-evaluation to enhance adaptive responses acknowledging the inherent uncertainty in complex systems (Carpenter and Gunderson, 2001). It is increasingly proposed that knowledge generation of ecosystem dynamics should be explicitly integrated with adaptive management practices rather than striving for optimization based on past records. This aspect emphasizes a learning environment and knowledge generation with associated institutions (e.g. Brown, 2003). Forming a learning environment that accepts continuous testing and adaptation requires leadership within management organizations (e.g. Danter *et al.*, 2000) and collaboration within social networks (Janssen *et al.*, 2006).
- *Supporting flexible institutions and multilevel governance systems*: the adaptive governance framework is operationalized through adaptive co-management where the dynamic learning characteristic of adaptive management is combined with the multi-level linkage characteristic of co-management (Olsson *et al.*, 2004a). The sharing of management power and responsibility may involve multiple often polycentric institutional and organizational linkages among user groups or communities, government agencies and non-governmental organizations, including support from legal, political and financial sources to ecosystem management initiatives.
- *Dealing with external perturbations, uncertainty and surprise*: it is not sufficient for a well functioning multi-level governance system to be in tune with the dynamics of the ecosystems under management. It also needs to develop adaptive capacity for dealing with change in, e.g. climate, disease outbreaks, hurricanes, global market demands, subsidies and governmental policies. The challenge for the social–ecological system is to enhance the adaptive capacity to deal with disturbance, to face uncertainty and be prepared for change and surprise. A resilient social–ecological system may even make use of disturbances as opportunities to transform into more desired states. Non-resilient social–ecological systems are vulnerable to external drivers and change.

Management practices for dealing with ecosystem change

Holling (1978) proposed "adaptive management" – a constantly changing management system, not only to meet the continuously changing and unpredictable ecosystem, but also to learn from it. Adaptive ecosystem management is an ongoing process, an organized way to deal with uncertainty and learn from management actions (Gunderson

et al., 1995). Basically, such adaptive management can be divided into:

- A conceptual system model, sometimes expressed as a computer simulation model that represents available knowledge and understanding of the system processes, structure and elements.
- A set of strategies that represents management policies or actions.
- A set of criteria for judging the success of the implementation of management actions and policies.
- A process that continuously evaluates and responds to the effects of management actions on the system and incorporates lessons learned in a new set of strategies to improve management.

Walters (1997) in his review of adaptive management argues that a reason for failure lies in management stakeholders showing deplorable self-interest, seeing adaptive-policy development as a threat to existing research programmes and management regimes, rather than as an opportunity for improvement. This is why it becomes important to address the social dimension and contexts for adaptive governance in relation to ecosystem management such as processes of participation, collective action and learning.

Policy increasingly emphasizes involvement of local users and landowners in adaptive ecosystem management. Involving local resource users can improve incentives for ecosystem management (Agrawal and Gibson, 1999; Fabricius and Koch, 2004). In addition, traditional and local knowledge about resources and ecosystem dynamics in communities can provide unique information about local conditions and complement scientific knowledge in ecosystem management efforts (e.g. Berkes *et al.*, 2000; Olsson and Folke, 2001; Becker and Ghimire, 2003; Aswani and Hamilton, 2004; Sheil and Lawrence, 2004).

But still few ecological inventories or stakeholder analyses (that tend to focus on conflicting interests) capture human resources in the landscape or the social structures and processes underlying incentives and values for ecosystem management.

Social–ecological inventories and local stewards of ecosystem services

Social–ecological inventories have been suggested to improve ecosystem management (Schultz *et al.*, in press). Such inventories identify people with ecosystem knowledge that practice ecosystem management. Social–ecological inventories help visualize ecosystem management on the ground in relation to ecosystem services, focusing on local steward groups acting outside official management plans. In a social–ecological inventory, conducted in a river basin of southern Sweden, local steward groups, their ecosystem management activities, motives and social networks were identified. Methods included interviews, participatory observations and review of documents and other written material. The inventory revealed a rich diversity of steward groups that manage and monitor a range of ecosystem services at different spatial scales. Contributions of local stewards include on-site ecosystem management, long-term and detailed monitoring of species and ecosystem dynamics, responses to environmental change, generation of local ecological knowledge employed in management practices, public support for ecosystem management and specialized networks (Schultz *et al.*, in press).

Such local stewards of ecosystem services in dynamic landscapes are to be found also in strongly human-dominated environments like urban areas. For example, Colding *et al.* (2006) demonstrate that green space, like golf courses, allotment areas and domestic gardens, cover more than twice the area of protected lands and are managed by local steward groups below the level of municipalities. These areas play a significant role in urban ecosystem services' generation, but they and their stewards are seldom recognized in this capacity in urban planning schemes.

Practices and ecosystem dynamics

Berkes and Folke (1998, 2002) identified management practices of local groups that make ecological sense and where people have developed practices to deal with ecosystem dynamics including abrupt periods of change (see Table 37.1). Traditional practices for ecosystem management include multiple-species management, resource rotation, ecological monitoring, succession management, landscape patchiness management, practices of responding to and managing pulses and ecological surprises. There exist practices that seem to reduce social–ecological crises in the event of large-scale natural disturbance such as creating small-scale ecosystem renewal cycles, spreading risks and nurturing sources of ecosystem reorganization and renewal (Colding *et al.*, 2003; Folke *et al.*, 2003).

Ecological knowledge and monitoring among local groups appears to be a key element in the development of many of the practices. The practices are linked to social mechanisms such as flexible user rights and land tenure; adaptations for the generation, accumulation and transmission of ecological knowledge; dynamics of institutions; mechanisms for cultural internalization of traditional practices; and associated worldviews and cultural values (Berkes *et al.*, 2000).

There exist numerous resource management practices among local people that have abandoned the steady-state and linear worldview.

Table 37.1 Social–ecological practices and mechanisms of local communities and traditional societies in the case studies of Berkes and Folke (1998) linking social and ecological systems volume

1. **Management practices based on ecological knowledge**
(A) *Practices found in conventional resource management and in local and traditional societies*
Monitoring resource abundance and change in ecosystems
Total protection of certain species
Protection of vulnerable life-history stages
Protection of specific habitats
Temporal restrictions of harvest
(B) *Practices mainly found in local and traditional societies*
Multiple species management
Maintaining ecosystem structure and function
Resource rotation
Succession management
(C) *Practices related to the dynamics of complex systems*
Management of landscape patchiness
Watershed-based management
Managing ecological processes at multiple scales
Responding to and managing pulses and surprises
Nurturing sources of ecosystem renewal
2. **Social mechanisms behind management practices**
(A) *Generation, accumulation and transmission of local ecological knowledge*
Reinterpreting signals for learning
Revival of local knowledge
Folklore and knowledge carriers
Integration of knowledge
Intergenerational transmission of knowledge
Geographical diffusion of knowledge
(B) *Structure and dynamics of institutions*
Role of stewards/wise people
Cross-scale institutions
Community assessments
Taboos and regulations
Social and religious sanctions
(C) *Mechanisms for cultural internalization*
Rituals, ceremonies and other traditions
Cultural frameworks for resource management
(D) *World view and cultural values*
A world view that provides appropriate environmental ethics
Cultural values of respect, sharing, reciprocity, humility and other

Adapted from Folke *et al.* (1998).

For example, there are those that *evoke small-scale disturbances* in ecosystems recognizing that change and also abrupt change is part of development. Such practices trigger small-scale release and create smaller renewal cycles in the local ecosystem and may reduce the impact of large-scale natural disturbances (Holling *et al.*, 1998). Examples include shifting cultivation and fire management for habitat improvement. These practices provide for the regeneration of important resources by creating habitat heterogeneity. Pulse fishing, employed by the James Bay Cree, and pulse grazing, employed by some African pastoralists, represents examples of such disturbance practices (Berkes *et al.*, 2000).

Furthermore, there exist local resource management practices that may be important for *dealing with abrupt change* and disturbance events. The polyculture of Samoa represents an example of such a practice. A minor food crop, yams, became the most important food for an extended period following a large-scale cyclone. Polyculture and "multiple-disturbance tolerant" species among the char-dwellers in Bangladesh serve the same function by reducing the potential impacts of flooding or droughts (Colding *et al.*, 2003). Diversification of livestock species among many pastoral groups in the African Sahel may reduce the effects of various disturbance regimes such as disease outbreaks and droughts.

Locally protected habitats, such as sacred groves, buffer zone areas and range reserves, may be important for the *reorganization* of ecosystems following disturbance events. Such areas may provide dispersal and migration of animals and plants into disturbed ecosystems. Even taboos imposed on populations of common species may have critical functions in the reorganization phase – especially those imposed on mobile link species (Elmqvist *et al.*, 2001; Lundberg and Moberg, 2003; Bodin *et al.*, 2006).

These are examples of practices common in traditional societies and local communities and that help insure the communities to uncertainty in resource flows and make people adaptive to change (Folke *et al.*, 1998a, 2006).

Social taboos and ecosystem services

Successful resource management systems require flexible social mechanisms for continual adjustments to environmental dynamics. Thus, institutional structures (rules and norms in use) are needed to take environmental variability and ecological feedbacks into account and provide capacity for management to respond to such dynamics. We have analysed social taboos in this context, defining a taboo as a prohibition imposed by social custom or as a protective measure. Such institutions are based on cultural norms that are not governed by government for either promulgation or enforcement. In Colding and Folke (2001) social taboos were grouped into six major categories in relation to their resource and

Table 37.2 Resource and habitat taboos (RHTs) and their nature conservation and resource management functions

Category	Function
Segment taboos	Regulate resource withdrawal
Temporal taboos	Regulate access to resources in time
Method taboos	Regulate methods of withdrawal
Life-history taboos	Regulate withdrawal of vulnerable life history stages of species
Specific-species taboos	Total protection to species in time and space
Habitat taboos	Restrict access and use of resources in time and space

Source: Colding and Folke (2001)

ecosystem management functions (Table 37.2). The last two categories of Table 37.2 can be referred to as *non-use taboos*, because they do not allow for human use of biological resources. The other four categories may be referred to as *use taboos* since the taboos permit restrictive use of resources (Colding and Folke, 2001).

Segment taboos apply when a cultural group bans the utilization of particular species for specific time periods for human individuals of a particular age, sex or social status. Thus, certain segments of a human population may be temporarily proscribed from the gathering and/or consumption of a species. This group of taboos exists in a number of traditional societies in, for example, Africa and South America.

Temporal taboos may be imposed *sporadically*, *daily* or on a weekly to seasonal basis. Cases recorded in the literature derive, for example. from Oceania and India. Such taboos are imposed on both aquatic and terrestrial resources. In an ecological context, they function to reduce harvesting pressure on particular subsistence resources and are closely related to the dynamic change of resource stocks.

Method taboos are imposed on certain gear types and extraction methods that may easily reduce or deplete the stock of a resource. Method taboos are common in Southeast Asia and are often fishing related.

Life-history taboos apply when a cultural group bans the use of certain vulnerable stages of a species' life history based on its age, size, sex or reproductive status. Such taboos may be imposed on reproducing and nesting species, and species particularly susceptible to overharvesting, such as slow moving, or sessile, marine species. Examples of such taboos derive mainly from India and Oceania.

Specific-species taboos prohibit any use of particular species and their populations. The reasons for the existence of specific-species taboos vary, ranging from beliefs in species being toxic, serving as religious symbols, representing reincarnated humans, and species being avoided due to their behavioural and physical appearance.

Habitat taboos are often imposed on terrestrial habitats, river stretches, ponds and coastal reefs. Examples of such "socially fenced" ecosystem types (Colding *et al.*, 2003) include "sacred groves" of India and Africa, "spirit sanctuaries" of South America, *waahi tapu* and *ahupua'a* in the South Pacific, and *hima* of Saudi Arabia. Habitat taboos provide for the protection of a number of ecological services on which a local community may depend. These services include the maintenance of biodiversity, regulation of local hydrological cycles, prevention of soil erosion, pollination of crops, preservation of locally adapted crop varieties, habitat for threatened species and predators on noxious insect and pest species of crops, and serving as wind and firebreaks.

An example from southern Madagascar (Bodin *et al.*, 2006) illustrates the significance of recognizing culturally protected and managed areas (Nabhan, 1997) in the generation of ecosystem services, like allotment areas or golf courses in the urban context (Colding *et al.*, 2006). In Madagascar the landscape is heavily fragmented, except for small forest patches that hold an abundance of rare species serving as refuges. Analyses of movements of animals illustrate that the landscape due to the small patches is fairly well connected despite the fragmentation that the forest patches support, for example, pollination of staple crops in local livelihoods (Bodin *et al.*, 2006). A national government or an international conservation NGO may conclude that to conserve biodiversity (the rare species) and ecosystem services of the landscapes these forest patches need to be urgently protected from human use and abuse and transformed into no-take areas through legal protection or governmental intervention. But a social–ecological inventory would reveal that they are in fact already protected by a social taboo system of sacred forests (Schultz *et al.*, in press). As a matter of fact, it is such systems that have sustained the biota and ecosystems services of the landscape. Implementing top-down policies may disrupt such socially and culturally enforced systems that sustain ecosystem services.

Hence, lack of policy recognition of the social and cultural dimension at local scales of ecosystem management may degrade landscapes further. However, with the information in mind of the significance of the sacred forests, the social taboos, the NGO and the national government could help secure such management institutions, for example, through what Ostrom and Schlager (1996) refer to as umbrella organizations or

Alcorn and Toledo (1998) as tenural shells. This becomes increasingly significant in the face of large-scale economic drivers of change or incorporations into global cultures and value systems.

When people comply with self-enforced norms, economic transaction costs may be low relative to formal enforcement measures. During such conditions, institutions, like social taboos, may provide for (1) low monitoring costs, (2) low enforcement costs, and in many cases (3) low sanctioning costs (Colding and Folke, 2001). Incentives should be created that strengthen social networks of steward groups for ecosystem management in multi-level governance systems (Folke *et al.*, 2005). It is time to move conservation beyond confrontation to multi-level collaboration (Wondolleck and Yaffee, 2000; Folke, 2006a) and recognize that ecosystem management is to a large extent people's management (Berkes, 2004).

Social networks, ecosystem management and bridging organizations

Ecosystem management is an information-intensive endeavour and requires knowledge of complex social–ecological interactions in order to monitor, interpret and respond to ecosystem feedback at multiple scales (Imperial, 1999a; Folke *et al.*, 2003). We have earlier argued that it is difficult if not impossible for one or a few people to possess the range of knowledge needed for ecosystem management (Olsson *et al.*, 2004a). Instead, knowledge for dealing with social–ecological systems' dynamics is dispersed among individuals and organizations in society and requires social networks that span multiple levels in order to draw on dispersed sources of information (Olsson *et al.*, 2006).

In this sense, knowledge of ecosystem dynamics resides in networks. A challenge is to identify mechanisms for organizing relations between relatively autonomous, but interdependent actors (Kersbergen and van Waarden, 2004) and avoid fragmented and sectoral approaches to the ecosystem management. Several studies have looked at the role of social networks in inter-organizational collaboration and collective action in relation to natural resource management (e.g. Agranoff and McGuire, 1999, 2001; Mandell, 1999; Carlsson, 2000; Mandell and Steelman, 2003; Imperial, 2005). A challenge is to identify social mechanisms and enabling institutional arrangements that can mobilize knowledge at critical times.

There is a need to increase the understanding of the role of networks in adaptive governance of social–ecological systems and mechanisms for facilitating cross-scale interactions, dealing with uncertainty and change, and enhancing ecosystem management (Scheffer *et al.*, 2002; Bodin and Norberg, 2005; Janssen *et al.*, 2006). Westley (2002) argues that the capacity to deal with the interactive dynamics of social and ecological systems requires networks of interacting individuals and organizations at different levels to create the right links, at the right time, around the right issues.

The Ecomuseum of the Kristianstads Vattenrike is an example of an organization that bridge local actors and communities with other organizational levels (Olsson *et al.*, 2004b). *Bridging organizations* increase the potential to redirect external forces into opportunities, serve as catalysts and facilitators between different levels of governance and bring in resources, knowledge and other incentives for ecosystem management (Folke *et al.*, 2005). A bridging organization like the Ecomuseum provides an arena for trust building, sense making, learning, vertical and horizontal collaboration, and conflict resolution (Hahn *et al.*, 2006). It uses networks of local steward groups to mobilize knowledge and social memory, which in turn help deal with uncertainty and shape change (Folke *et al.*, 2003, 2006). The different networks and the numerous linkages that can be activated when needed contribute to the robustness of the social–ecological system and therefore are sources of social–ecological resilience. They constitute the social memory (in the sense of Macintosh, 2000) that can be mobilized at critical times and increase response options to deal with uncertainty and change.

The adaptive co-management and the "adhocracy" in Kristianstads Vattenrike rely on sleeping links that can be activated when there is a conflict or crisis and the Ecomuseum helps mobilize experience and social memory for dealing with change (Hahn *et al.*, 2006). Thus, bridging organizations can play a crucial role in the dynamic relationship between key individuals, social memory and resilience. Such structures of social capital need to be recognized and nurtured in conservation and ecosystem management efforts.

Leadership and actor groups

A key mechanism behind adaptive co-management is leadership which can come in different forms. For example, key individuals can provide visions of ecosystem management and sustainable development that frame self-organizing processes (Agranoff and McGuire, 2001; Westley, 2002). Key stewards are important in establishing functional links within and between organizational levels and therefore facilitating the flow of information and knowledge from multiple sources to be applied in the local context of ecosystem management. Social networks often emerge as self-organizing processes (i.e. not implemented by external pressure) involving key persons who share some common interests although they represent different stakeholder groups (McCay, 2002). Leadership has been shown to be of great significance for public network management. Network leadership and guidance is very different from the command and control of hierarchical management (Agranoff and McGuire, 2001). It requires steering for holding the network together (Bardach, 1998) and balancing social forces and interests that enables self-organization (Kooiman, 1993).

However, social–ecological systems that rely on one or a few key stewards might be vulnerable to change. This is exemplified by Peterson (2002) who describes the management of the long-leaf pine forest in Florida and how the desirable state or the stability domain of the forest is maintained by fire as a main structuring variable. Fire frequency has decreased in the area and long-leaf pine forest ecosystems therefore risk entering into other less desirable ecosystem states. The forest is within a military base and an air-force general has been a key steward for maintaining the forest through active burning. When the general left his position, a new general that did not share the interests and convictions of his predecessor replaced him. However, some of the personnel who had taken an active part in ecosystem management had developed knowledge and affection for the long-leaf pine forests. They also used a scientist's model of forest dynamics to successfully convince the new general of the importance of fire management for maintaining the desirable stability domain of a long-leaf pine forest ecosystem. This example shows how structures and processes like social networks can provide a social memory of ecosystem management that sustains adaptive capacity in times of change.

The strength of networks depends on the ability of the key persons to exchange information with other stakeholders, identify common interests and gather support for such interests (e.g. ecosystem management) within their own organization or stakeholder group. Bardach (1998) describes how leaders play different roles in systems of strategic interaction which include eliciting common goals, creates an atmosphere of trust, broker organizational and individual contributions, and deploys energies in accord with some strategic plan. Organizations that do not appear to have much in common may develop crucial links, thanks to these key persons who form the nodes of different, loosely connected, networks. In his seminal paper, Granovetter (1973) argued that weak ties, that is, the bridges between different stakeholder groups, may be the most valuable for generating new knowledge and identifying new opportunities and hence create a macro effect: "those to whom we are weakly tied are more likely to move in circles different from our own and will thus have access to information different from that which we receive" (p. 1371). Applied to ecosystem management, we argue that a loosely connected network involving a diversity of stakeholders is important for gathering different types of ecological knowledge, build moral and political support (legitimacy) from "non-environmental" sectors, and attain legal and financial support from various institutions and organizations. Hence, if polycentric cross-level institutions provide the structure for adaptive co-management, multiple-overlapping networks of key persons provide the processes.

As an example, Bebbington (1997) identifies brokers as key stewards in sustainable agriculture intensification in the Andes, including their role in coordinating social networks in the management process. In all the cases of sustainable intensification, outsiders have played a key role in bringing in new ideas, but more importantly they have brought in networks of contacts. These brokers had different backgrounds, including a priest, university professor, European volunteers and funding agencies. The connections they brought with them helped the members of the local communities gain access to non-local institutions and resources, including access to NGOs with technical assistance and financial resources, sources of technology, donors and alternative trading networks. These networks spread across national and international boundaries in ways that would have been hard for the locals to do on their own.

In addition to leaders, we have previously identified other essential actors and actor groups that serve social mechanisms in adaptive co-management networks: knowledge carriers, knowledge generators, stewards and sense-makers. Folke *et al.* (2003), based on several case studies, identified the following actor groups; knowledge retainers, interpreters, facilitators, visionaries, inspirers, innovators, experimenters, followers and reinforcers. In coastal communities of eastern Africa, actor groups like beach recorders of fish catches and middlemen that link fishers to markets are of major significance in shaping exploitation patterns of coastal and marine ecosystems and thereby influencing the capacity of these social–ecological systems to generate and sustain ecosystem services (Crona, 2006; de la Torre-Castro, 2006). Holling and Chambers (1973), in their analyses of social roles in resource management workshops, stressed the importance of also including individuals with opposite views that oppose and criticize. These roles of actor groups are all important components of social networks

and essential for creating the conditions that we argue are necessary for ecosystem management.

TRANSFORMING SOCIAL–ECOLOGICAL SYSTEMS

Crisis, perceived or real, seems to trigger learning and knowledge generation (Westley, 1995) and opens up space for combinations of different social memories and new management trajectories of resources and ecosystems (Gunderson, 2003). Olsson and Folke (2001) described how threats of acidification, overfishing and disease successively initiated learning and generated knowledge and institutions for landscape management among local groups in the Lake Racken catchment in western Sweden. Based on empirical work Olsson *et al.* (2004a) observed the following sequence of local self-organizing toward adaptive co-management of ecosystems.

- A sequence of social responses to environmental events widens the scope of local management from a particular issue or resource to a broad set of issues related to ecosystem processes across scales.
- Management expands from individual actors, to group of actors to multiple-actor processes.
- Organizational and institutional structures evolve as a response to deal with the broader set of environmental issues.
- Knowledge of ecosystem dynamics develops as a collaborative effort and becomes part of the organizational and institutional structures.
- Social networks develop that connect institutions and organizations across levels and scales and facilitate information flows, identify knowledge gaps, and create nodes of expertise of significance for ecosystem management.
- Knowledge for ecosystem management is mobilized through social networks and complements and refines local practice for ecosystem management.
- In the time series of events the ability to deal with uncertainty and surprise is improved which increases the adaptive capacity to deal with future change.

The crises that trigger such self-organization may be caused by external markets and tourism pressure, floods and flood management, shifts in property rights, threats of acidification, resource failures, rigid paradigms of resource management, new legislation or governmental policies that do not take into account local contexts (Berkes *et al.*, 2003). A social–ecological system with low levels of social capital and social memory is vulnerable to such changes and may as a consequence shift into undesired pathways (Gunderson and Holling, 2002).

In contrast, crisis may trigger social capital and social memory to be mobilized and combined into new forms of governance systems with the ability to manage dynamic ecosystems and landscapes. This has been referred to as building social capacity for resilience in social–ecological systems (Folke *et al.*, 2003) and it requires evoking change in social structures (Westley, 1995). Key individuals with strong leadership may catalyze opinion shifts (Gladwell, 2000; Scheffer *et al.*, 2003) and creative teams and actor groups that emerge into a large connected community of practitioners can prepare a social–ecological system for rapid change (Blann *et al.*, 2003; Guimera *et al.*, 2005) and transform it into a new pathway of development.

Transformability means creating and defining a new attractor that directs the development of the social–ecological system by introducing new components and ways of making a living, thereby changing the state variables, and often the scales of key cycles, that define the system (Walker *et al.*, 2004).

Transformations toward alternative forms of governance has been addressed by Kettl (2000), Kuks and Bressers (2004) and Agrawal (2005). Olsson *et al.* (2004b) analysed the emergence of a governance system for adaptive co-management of the wetland landscape of Kristianstad in southern Sweden, a process where unconnected management by several actors in the landscape was mobilized, renewed and moved into a new configuration of ecosystem management within about a decade. The self-organizing process was triggered by the perceived threats to the area's cultural and ecological values among people of various local steward associations and local government. A key individual provided visionary leadership in directing change and transforming governance. The transformation involved four phases:

1 preparing the system for change;
2 the opening of an opportunity;
3 navigating the transition;
4 charting a new direction for building resilience of the new governance regime.

Trust-building dialogues, mobilization of social networks with actors and teams across scales, coordination of ongoing activities, sense-making, collaborative learning and creating public awareness were part of the process. A comprehensive framework with a shared vision and goals that presented ecosystem management as development and turned problems into opportunities was developed and contributed to a shift in values and meaning of the broader agricultural–urban–wetland landscape among key actors. When a window of opportunity at the political level opened, it was possible to tip and transform the governance system into a trajectory of adaptive co-management with extensive social networks of practitioners engaged in multilevel governance. The transformation took place within the existing legal and formal institutional framework (Hahn *et al.*, 2006). Currently, adaptive capacity is built to make the new social–ecological configuration resilient to change. Strategies for adaptive capacity are presented in Table 37.3.

Understanding the sources of resilience that allow for mobilization of social capital and memory to generate novelty and innovation for transformation of social–ecological systems into improved pathways of development is a central issue for sustainability research.

CONCLUSIONS

We have only scratched the surface of an immense research challenge that promises to provide a much richer understanding of not just human–environment interactions but of how the world we live in actually works and the implications it has for current policies and governance. The chapter emphasizes that the social landscape should be approached as carefully as the ecological in order to clarify features that contribute to the resilience of social–ecological systems. In this context, Pretty and Ward (2001) find that relations of trust, reciprocity, common rules, norms and sanctions and connectedness in institutions are critical. We have similar findings that include vision, leadership and trust; enabling legislation that creates social space for ecosystem management; funds for responding to environmental change and for remedial action; capacity for monitoring and responding to environmental feedback; information flow through social networks; the combination of various sources of information and knowledge; sense-making and arenas of collaborative learning for ecosystem management. Our work illustrates that the interplay between individuals (e.g. leadership, teams, actor groups), the emergence of nested organizational structures, institutional dynamics and power relations tied together in dynamic social networks, are examples of features that seem critical in adaptive governance that allows for ecosystem management and for responding to environmental feedback across scales.

An important lesson from the research is that it is not enough to create arenas for dialogue and collaboration, nor is it enough to develop networks to deal with issues at a landscape level. Further investigation of the interplay between key individuals, actor groups, social networks, organizations and institutions in multi-level

Table 37.3 Processes and strategies in Kristianstads Vattenrike that increase capacity for dealing with uncertainty and change

Developing motivation and values for ecosystem management
  Envisioning the future together with actors
  Developing, communicating and building support for the mission
  Identifying and clarifying objectives
  Developing personal ties
  Establishing a close relationship and trust with key individuals
  Fostering dialogue with actors
  Providing arenas for trust building among actors
  Building trust in times of stability to facilitate conflict resolution
  Developing norms to avoid loss of trust among actors
  Continuously communicating success and progress of projects
Directing the local context through adaptive co-management
  Encouraging and supporting actors to perform monitoring, including inventories
  Encouraging and supporting actors to manage ecosystem processes for biodiversity and ecosystem services
  Initiating and sustaining social networks of key individuals
  Mobilizing individuals of social networks in problem-driven projects
  Making sense of and guiding the management process
  Synthesizing and mobilizing knowledge for ecosystem management
  Providing coordination of project and arenas for collaboration
  Encouraging and inspiring actors to voluntary participation
  Initiating projects and selecting problems that can be turned into possibilities
  Creating public opinion and involving local media
Navigating the larger environment
  Influencing decision-makers at higher levels to maintain governance structures that allow for adaptive co-management
    of the area
  Mobilizing new funding when needed
  Mobilizing external knowledge when needed
  Exchanging information and collaboration with local steward associations in Sweden and internationally
  Collaborating with national and international scientists
  Collaborating with national and international non-governmental organizations
  Participating in international institutional frameworks
  Supporting diffusion of the values of KV through social networks
  Providing a buffer for external drivers
  Communicating with national media

Adapted from Olsson *et al.* (2004b).

social–ecological systems in relation to adaptive capacity, cross-scale interactions and enhancement of resilience is needed. We have to understand, support and perhaps even learn how to navigate actively the underlying social structures and processes in the face of change. There will be inevitable and possibly large-scale environmental changes, and preparedness has to be built to enhance the social–ecological capacity to respond, adapt to and shape our common future and make use of creative capacity to find ways to transform into pathways of improved development. We conclude that the existence of transformative capacity is essential in order to create social–ecological systems with the capability to manage ecosystems sustainably for human well-being. Adaptive capacity will be needed to strengthen and sustain such systems in the face of external drivers and events.

REFERENCES

Abel, T. and Stepp, J.R. (2003). A new ecosystems ecology for anthropology. *Conservation Ecology* 7(3): 12. URL: http://www.consecol.org/vol7/iss3/art12/

Adger, W.N. (2003). Social capital, collective action and adaptation to climate change. *Economic Geography* 79: 387–404.

Adger, W.N., Hughes, T., Folke, C., Carpenter, S.R. and Rockström, J. (2005). Social–ecological resilience to coastal disasters. *Science* 309: 1036–1039.

Agranoff, R.I. and McGuire, M. (1999). Managing in network settings. *Policy Studies Review* 16: 18–41.

Agranoff, R.I. and McGuire, M. (2001). Big questions in public network management research. *Journal of Public Administration Research and Theory* 11: 295–326.

Agrawal, A. (2005). *Environmentality: Technologies of Government and the Making of Subjects.* Duke University Press, Durham, NC.

Agrawal, A. and Gibson, A. (1999). Enchantment and disenchantment: the role of community in natural resource conservation. *World Development* 27: 629–649.

Alcorn, J.B. and Toledo, V.M. (1998). Resilient resource management in Mexico's forest ecosystems: the contribution of property rights. In F. Berkes and C. Folke (eds) *Linking Social and Ecological Systems: Management Practices and Social Mechanisms for Building Resilience.* Cambridge University Press, Cambridge, UK, pp. 216–249.

Allison, H.E. and Hobbs, R.J. (2004). Resilience, adaptive capacity, and the "Lock-in Trap" of the Western Australian agricultural region. *Ecology and Society* 9(1), 3. URL: http://www.ecologyandsociety.org/vol9/iss1/art3/

Anderies, J.M., Janssen, M.A. and Ostrom, E. (2004). A framework to analyze the robustness of social–ecological systems from an institutional perspective. *Ecology and Society* 9(1): 18. URL: http://www.ecologyandsociety.org/vol9/iss1/art18/

Armitage, D. (2005). Adaptive capacity and community-based natural resource management. *Environmental Management* 35: 703–715.

Arrow, K., Bolin, B., Costanza, R., Dasgupta, P., Folke, C., Holling, C.S. *et al.* (1995). Economic growth, carrying capacity, and the environment. *Science* 268: 520–521.

Arrow, K., Dasgupta, P., Goulder, L., Daily, G., Ehrlich, P., Heal, G. et al. (2004). Are we consuming too much? *Journal of Economic Perspectives* 18: 147–172.

Arthur, B.W. (1999). Complexity and the economy. *Science* 284: 107–109.

Aswani, S. and Hamilton, R. (2004). Integrating indigenous ecological knowledge and customary sea tenure with marine and social science for conservation of bumphead parrotfish (*Bolbometopon muricatum*) in the Roviana Lagoon, Solomon Islands. *Environmental Conservation* 31: 69–83.

Bardach, E. (1998). *Managerial Craftmanship: Getting Agencies to Work Together.* Brookings, Washington, DC.

Bebbington, A. (1997). Social capital and rural intensification: local organizations and islands of sustainability in the rural Andes. *The Geographical Journal* 163: 189–197.

Becker, C.D. and Ghimire, K. (2003). Synergy between traditional ecological knowledge and conservation science supports forest preservation in Ecuador. *Conservation Ecology* 8(1): 1. URL: http://www.consecol.org/vol8/iss1/art1/

Berkes, F. (2004). Rethinking community-based conservation. *Conservation Biology* 18: 621–630.

Berkes, F. and Folke, C. (1992). A systems perspective on the interrelations between natural, human-made and cultural capital. *Ecological Economics* 5: 1–8.

Berkes, F. and Folke, C. (eds) (1998). *Linking Social and Ecological Systems: Management Practices and Social Mechanisms for Building Resilience.* Cambridge University Press, Cambridge, UK.

Berkes, F. and Folke, C. (2002). Back to the future: ecosystem dynamics and local knowledge. In L.H. Gunderson and C.S. Holling (eds) *Panarchy: Understanding Transformations in Systems of Humans and Nature.* Island Press, Washington, DC, pp. 121–146.

Berkes, F., Colding, J. and Folke, C. (2000). Rediscovery of traditional ecological knowledge as adaptive management. *Ecological Applications* 10: 1251–1262.

Berkes, F., Colding, J. and Folke, C. (eds) (2003). *Navigating Social–Ecological Systems: Building Resilience for Complexity and Change.* Cambridge University Press, Cambridge, UK.

Berkes, F., Hughes, T.P., Steneck, R.S., Wilson, J.A., Bellwood, D.R., Crona, B. *et al.* (2006). Globalization, roving bandits, and marine resources. *Science* 311: 1557–1558.

Blann, K.S., Light, S. and Musumeci, J.A. (2003). Facing the adaptive challenge: practitioners' insights from negotiating resource crisis in Minnesota. In F. Berkes, J. Colding and C. Folke (eds) *Navigating Social–Ecological Systems: Building Resilience for Complexity and Change.* Cambridge University Press, Cambridge, UK, pp. 210–240.

Bodin, Ö. and Norberg, J. (2005). Information network topologies for enhanced local adaptive management. *Environmental Management* 35: 175–193.

Bodin, Ö., Tengö, M., Norman, A., Lundberg, J. and Elmqvist, T. (2006). The value of small size: loss of forest patches and ecological thresholds in Southern Madagascar. *Ecological Applications* 16: 440–451.

Borrini-Feyerabend, G., Pimbert, M., Farvar, M.T., Kothari, A. and Renard, Y. (2004). *Sharing Power: Learning by Doing in Co-management of Natural Resources Throughout the World.* Int. Inst. Environ. Dev./World Conserv. Union/Comm. Environ. Econ. Policy/Collab. Manag. Work. Group/Cent. Sust. Dev., Tehran.

Brown, K. (2003). Integrating conservation and development: a case of institutional misfit. *Frontiers in Ecology and the Environment* 1: 479–487.

Carlsson, L. (2000). Policy networks as collective action. *Policy Studies Journal* 28: 502–520.

Carpenter, S.R. and Brock, W.A. (2004). Spatial complexity, resilience and policy diversity: fishing on lake-rich landscapes. *Ecology and Society* 9(1), 8. URL: http://www.ecologyandsociety.org/vol9/iss1/art8

Carpenter, S.R. and Brock, W.A. (2006). Rising variance: a leading indicator of ecological transition. *Ecology Letters* 9: 311–318.

Carpenter, S.R. and Gunderson, L.H. (2001). Coping with collapse: ecological and social dynamics in ecosystem management. *BioScience* 6: 451–457.

Carpenter, S.R., Brock, W.A. and Hanson, P.C. (1999). Ecological and social dynamics in simple models of ecosystem management. *Conservation Ecology* 3(2): 4. URL: http://www.consecol.org/vol3/iss2/art4

Cash, D.W., Adger, W., Berkes, F., Garden, P., Lebel, L., Olsson, P. *et al.* (2006). Scale and cross-scale dynamics: governance and information in a multilevel world. *Ecology and Society* 11(2): 8. URL: http://www.ecologyandsociety.org/vol11/iss2/art8.

Colding, J. and Folke, C. (2001). Social taboos: 'invisible' systems of local resource management and biological conservation. *Ecological Applications* 11: 584–600.

Colding, J., Elmqvist, T. and Olsson, P. (2003). Living with disturbance: building resilience in social–ecological systems.

In F. Berkes, J. Colding, and C. Folke (eds) *Navigating Social–Ecological Systems: Building Resilience for Complexity and Change*. Cambridge University Press, Cambridge, UK, pp.163–185.

Colding, J., Lundberg, J. and Folke, C. (2006). Incorporating green-area user groups in urban ecosystem management. *Ambio* 35: 237–244.

Costanza, R., Waigner, L., Folke, C. and Mäler, K.-G. (1993). Modeling complex ecological economic systems: towards an evolutionary dynamic understanding of people and nature. *BioScience* 43: 545–555.

Crona, B. (2006). *Of Mangroves and Middlemen: A Study of Social and Ecological Linkages in a Coastal Community*. PhD Thesis, Department of Systems Ecology, Stockholm University, Stockholm.

Dale, V.H., Brown, S., Haeuber, R.A., Hobbs, N.T., Huntly, N., Naiman, R.J. *et al.* (2000). Ecological principles and guidelines for managing the use of land. *Ecological Applications* 10: 639–660.

Danter, K.J., Griest, D.L., Mullins, G.W. and Norland, E. (2000). Organizational change as a component of ecosystem management. *Society and Natural Resources* 13: 537–547.

de la Torre-Castro, M. (2006). Beyond regulations in fisheries management: the dilemmas of the 'beach recorders' *Bwana Dikos* in Zanzibar, Tanzania. *Ecology and Society* 11(2): 35. URL: http://www.ecologyandsociety.org/vol11/iss2/art35

Dietz, T., Ostrom, E. and Stern, P. (2003). The struggle to govern the commons. *Science* 302: 1907–1912.

Elmqvist, T., Wall, M., Berggren, A.L., Blix, L., Fritioff, S. and Rinman, U. (2001). Tropical forest reorganization after cyclone and fire disturbance in Samoa: remnant trees as biological legacies. *Conservation Ecology* 5(2): 10. URL: http://www.consecol.org/vol5/iss2/art10

Fabricius, C. and Koch, E. (2004). *Rights, Resources and Rural Development: Community-based Natural Resource Management in Southern Africa*. Earthscan, London.

Folke, C. (2006a). Resilience: the emergence of a perspective for social–ecological systems analyses. *Global Environmental Change* 16: 253–267.

Folke, C. (2006b). Conservation against development versus conservation for development. *Conservation Biology* 20: 686–688.

Folke, C., Jansson, Å., Larsson, J. and Costanza, R. (1997). Ecosystem appropriation by cities. *Ambio* 26: 167–172.

Folke, C., Berkes, F. and Colding, J. (1998a). Ecological practices and social mechanisms for building resilience and sustainability. In F. Berkes and C. Folke (eds) *Linking Social and Ecological Systems: Management Practices and Social Mechanisms for Building Resilience*. Cambridge University Press, Cambridge, UK, pp. 414–436.

Folke, C., Pritchard, L., Berkes, F., Colding, J. and Svedin, U. (1998b). The problem of fit between ecosystems and institutions. *IHDP Working Paper* No. 2. International Human Dimensions Programme, Bonn URL: http//:www.uni-bonn.de/IHDP/public.htm

Folke, C., Colding, J. and Berkes, F. (2003). Synthesis: building resilience and adaptive capacity in social–ecological systems. In F. Berkes, J. Colding and C. Folke (eds) *Navigating Social–Ecological Systems: Building Resilience for Complexity and Change*. Cambridge University Press, Cambridge, UK, pp. 352–387.

Folke, C., Carpenter, S.R., Walker, B.H., Scheffer, M., Elmqvist, T., Gunderson, L.H. *et al.* (2004). Regime shifts, resilience and biodiversity in ecosystem management. *Annual Review in Ecology, Evolution and Systematics* 35: 557–581.

Folke, C., Hahn, T., Olsson, P. and Norberg, J. (2005). Adaptive governance of social–ecological systems. *Annual Review of Environment and Resources* 30: 441–473.

Folke, C., Fabricius, C., Cundill, G., Schultz, L., Queiroz, C., Gokhale, Y. *et al.* (2006). Communities, ecosystems, and livelihoods. In *Ecosystems and Human Well-being: Multiscale assessments: Findings of the Sub-global Assessments Working Group*. Millennium Ecosystem Assessment, Island Press, Washington, DC, pp. 261–277.

Gadgil, M., Seshagiri Rao, P.R., Utkarsh, G., Pramod, P. and Chatre, A. (2000). New meanings for old knowledge: the people's biodiversity registers programme. *Ecological Application* 10: 1307–1317.

Galaz, V. (2005). Social–ecological resilience and social conflict: institutions and strategic adaptation in Swedish water management. *Ambio* 34: 567–572.

Gladwell, M. (2000). *The Tipping Point: How Little Things Can Make a Big Difference*. Little, Brown, Boston, MA.

Granovetter, M. (1973). The strength of weak ties. *American Journal of Sociology* 78: 1360–1380.

Grebmeier, J.M., Overland, J.E., Moore, S.E., Farley, E.V., Carmack, E.C., Cooper, L.W. *et al.* (2006). A major ecosystem shift in the northern Bering Sea. *Science* 311: 1461–1464.

Grumbine, R.E. (1994). What is ecosystem management? *Conservation Biology* 8: 27–38.

Guimera, R., Uzzi, B., Spiro, J. and Nunes Amaral, L.A. (2005). Team assembly mechanisms determine collaboration network structure and team performance. *Science* 308: 697–702.

Gunderson, L.H. (2001). Managing surprising ecosystems in southern Florida. *Ecological Economics* 37: 371–378.

Gunderson, L.H. (2003). Adaptive dancing: interactions between social resilience and ecological crises. In F. Berkes, J. Colding and C. Folke (eds) *Navigating Social–Ecological Systems: Building Resilience for Complexity and Change*. Cambridge University Press, Cambridge, UK, pp. 35–52.

Gunderson, L.H. and Holling, C.S. (eds) (2002). *Panarchy: Understanding Transformations in Human and Natural Systems*. Island Press, Washington, DC.

Gunderson, L.H., Holling, C.S. and Light, S. (eds) (1995). *Barriers and Bridges to the Renewal of Ecosystems and Institutions*. Columbia University Press, New York.

Hahn, T., Olsson, P., Folke, C. and Johansson, K. (2006). Trust building, knowledge generation and organizational innovations: the role of a bridging organization for adaptive comanagement of a wetland landscape around Kristianstad, Sweden. *Human Ecology* 34: 573–592.

Holland, J. (1995). *Hidden Order: How Adaptation Builds Complexity*. Addison-Wesley, Reading, MA.

Holland, J.H., Holyoak, K.J., Nisbett, R.E. and Thagard, P.R. (1986). *Induction: Processes of Inference, Learning, and Discovery*. MIT Press, Cambridge, MA.

Holling, C.S. (1973). Resilience and stability of ecological systems. *Annual Review of Ecology and Systematics* 4: 1–23.

Holling, C.S. (ed) (1978). *Adaptive Environmental Assessment and Management*. John Wiley, New York.

Holling, C.S. (1994). An ecologists view of the Malthusian conflict. In K. Lindahl-Kiessling and H. Landberg (eds) *Population, Economic Development, and the Environment*. Oxford University Press, Oxford, UK, pp. 79–103.

Holling, C.S. and Chambers, A.D. (1973). Resource science: the nurture of an infant. *BioScience* 23: 13–20.

Holling, C.S. and Meffe, G.K. (1996). Command and control and the pathology of natural resource management. *Conservation Biology* 10: 328–337.

Holling, C.S. Berkes, F. and Folke, C. (1998). Science, sustainability, and resource management. In F. Berkes and C. Folke (eds) *Linking Social and Ecological Systems: Management Practices and Social Mechanisms for Building Resilience*. Cambridge University Press, Cambridge, UK, pp. 342–362.

Hooghe, L. and Marks, G. (2003). Unraveling the central state, but how? Types of multi-level governance. *American Political Science Review* 97: 233–243.

Huitric, M. (2005) Lobster and conch Fisheries in Belize: a history of sequential exploitation. *Ecology and Society* 107(1): 21.

Imperial, M.T. (1999a). Institutional analysis and ecosystem-based management: the institutional analysis and development framework. *Environmental Management* 24: 449–465.

Imperial, M.T. (1999b). Analyzing institutional arrangements for ecosystem-based management: lessons from the Rhode Island Salt Ponds SAM Plan. *Coastal Management* 27: 31–56.

Imperial, M.T. (2005). Using collaboration as a governance strategy: Lessons from six watershed management programs. *Administration & Society* 37: 281–320.

Janssen, M.A. and Carpenter, S.R. (1999). Managing the resilience of lakes: a multi-agent modeling approach. *Conservation Ecology* 3(2), 15. URL: http://www.consecol.org/vol3/iss2/art15/

Janssen, M.A. and Jager, W. (2001). Fashions, habits and changing preferences: simulation of psychological factors affecting market dynamics. *Journal of Economic Psychology* 22: 745–772.

Janssen, M.A., Walker, B.H., Langridge, J. and Abel, N. (2000). An adaptive agent model for analysing co-evolution of management and policies in a complex rangeland system. *Ecological Modelling* 131: 249–268.

Janssen, M.A., Bodin, Ö., Anderies, J.M., Elmqvist, T., Ernstson, H., McAllister, R.R.J., Olsson, P. and Ryan, P. (2006). A network perspective on the resilience of social-ecological systems. *Ecology and Society* 11(1): 15. URL: http://www.ecologyandsociety.org/vol11/iss1/art15/

Kasperson, J.X., Kasperson, R.E. and Turner, B.L. II (eds) (1995). *Regions at Risk: Comparisons of Threatened Environments*. United Nations University Press, New York.

Kauffman, S. (1993). *The Origins of Order*. Oxford University Press, New York.

Kersbergen van K. and van Waarden, F. (2004). Governance as a bridge between disciplines: cross-disciplinary inspiration regarding shifts in governance and problems of governability, accountability and legitimacy. *European Journal of Political Research* 43: 143–171.

Kettl, D.F. (2000). The transformation of governance: globalization, devolution, and the role of government. *Public Administration Review* 60: 488–497.

Kinzig, A.P., Starrett, D., Arrow, K., Bolin, B., Dasgupta, P., Ehrlich, P.R. et al. (2003). Coping with uncertainty: a call for a new science-policy forum. *Ambio* 32: 330–335.

Kinzig, A.P., Ryan, P., Etienne, M., Allison, H., Elmqvist, T. and Walker, B.H. (2006). Resilience and regime shifts: assessing cascading effects. *Ecology and Society* 11(1): 20. URL: http://www.ecologyandsociety.org/vol11/iss1/art20/

Kooiman, J. (ed) (1993). *Modern Governance*. Sage, London.

Kuks, S. and Bressers, H. (ed) (2004). *Integrated Governance and Water Basin Management: Conditions for Regime Change and Sustainability*. Kluwer Academic, Dordrecht.

Lambin, E.F., Geist, H.J. and Lepers, E. (2003). Dynamics of land-use and land-cover change in tropical regions. *Annual Review of Environment and Resources* 28: 205–241.

Lansing, J.S. (2003). Complex adaptive systems. *Annual Review of Anthropology* 32: 183–204.

Lee, K.N. (1993). *Compass and Gyroscope: Integrating Science and Politics for the Environment*. Island Press, Washington, DC.

Levin, S.A. (1999). *Fragile Dominion: Complexity and the Commons*. Perseus Books, Reading, MA.

Lundberg, J. and Moberg, F. (2003). Mobile link organisms and ecosystem functioning: implications for ecosystem resilience and management. *Ecosystems* 6: 87–98.

Mäler, K.-G. (2000). Development, ecological resources and their management: a study of complex dynamic systems. *European Economic Review* 44: 645–665.

Mandell, M.P. (1999). Community collaborations: working through network structures. *Policy Studies Review* 16: 42–64.

Mandell, M.P. and Steelman, T.A. (2003). Understanding what can be accomplished through interorganizational innovations. *Public Management Review* 5: 197–224.

McCay, B.J. (2002). Emergence of institutions for the commons: contexts, situations, and events. In E. Ostrom, T. Dietz, N. Dolsak, P. Stern, S. Stonich and E.U. Weber (eds) *The Drama of the Commons*. National Academies Press, Washington, DC, pp. 361–402.

McGinnis, M. (2000). *Polycentric Governance and Development*. University of Michigan Press, Ann Arbor.

McIntosh, R.J. (2000). Social memory in Mande. In R.J. McIntosh, J.A. Tainter and S.K. McIntosh (eds) *The Way the Wind Blows: Climate, History, and Human Action*. Columbia University Press, New York, pp. 141–180.

Millennium Ecosystem Assessment, (2005). *Synthesis*. Island Press, Washington, DC. URL: http://www.MAweb.org

Nabhan, G.P. (1997). *Cultures of Habitat: On Nature, Culture, and Story*. Counterpoint, Washington, DC.

Norberg, J. and Cumming, G.S. (2007). *Complexity Theory for a Sustainable Future*. Columbia University Press, New York.

Norgaard, R.B. (1994). *Development Betrayed: The End of Progress and a Coevolutionary Revisioning of the Future.* Routledge, New York.

Odum, E.P. (1989). *Ecology and our Endangered Life-Support System.* Sinauer, Sunderland, MA.

Olick, J.K. and Robbins, J. (1998). Social memory studies: from "collective memeory" to historical sociology of mnemonic practices. *Annual Review of Sociology* 24: 105–140.

Olsson, P. and Folke, C. (2001). Local ecological knowledge and institutional dynamics for ecosystem management: a study of Lake Racken watershed, Sweden. *Ecosystems* 4: 85–104.

Olsson, P., Folke, C. and Berkes, F. (2004a). Adaptive co-management for building social–ecological resilience. *Environmental Management* 34: 75–90.

Olsson, P., Hahn, T. and Folke, C. (2004b). Social–ecological transformation for ecosystem management: the development of adaptive co-management of a wetland landscape in southern Sweden. *Ecology and Society* 9(4): 2. URL: http://www.ecologyandsociety.org/vol9/iss4/art2

Olsson, P., Gunderson, L.H., Carpenter, S.R., Ryan, P., Lebel, L., Folke, C. and Holling, C.S. (2006). Shooting the rapids: navigating transitions to adaptive governance of social–ecological systems. *Ecology and Society* 11(1): 18. URL: http://www.ecologyandsociety.org/vol11/iss1/art18

Ostrom, E. (1998). Scales, polycentricity, and incentives: designing complexity to govern complexity. In L.D. Guruswamy and J.A. McNeely (eds) *Protection of Global Biodiversity: Converging Strategies.* Duke University Press, Durham, NC, pp. 149–167.

Ostrom, E. (2005). *Understanding Institutional Diversity.* Princeton University Press, Princeton, NJ.

Ostrom, E. and Ahn, T.K. (2003). *Foundations of Social Capital.* Edward Elgar, Cheltenham, UK.

Ostrom, E. and Schlager, E. (1996). The formation of property rights. In S. Hanna, C. Folke and K.-G. Mäler (eds) *Rights to Nature.* Island Press, Washington, DC, pp. 127–156.

Peterson, G.D. (2002). Forest dynamics in the southeastern United States: managing multiple stable states. In L.H. Gunderson and L. Pritchard Jr. (eds) *Resilience and the Behavior of Large Scale Ecosystems.* Island Press, Washington, DC, pp. 227–246.

Peterson, G.D., Carpenter, S.R. and Brock, W.A. (2003). Uncertainty and management of multi-state ecosystems: an apparently rational route to collapse. *Ecology* 84: 1403–1411.

Pretty, J. (2003). Social capital and the collective management of resources. *Science* 302: 1912–1914.

Pretty, J. and Ward, H. (2001). Social capital and the environment. *World Development* 29: 209–227.

Redman, C.L. (1999). *Human Impact on Ancient Environments.* The University of Arizona Press, Tucson.

Rockström, J. (2003). Water for food and nature in drought-prone tropics: vapour shift in rain-fed agriculture. *Philosophical Transactions of the Royal Society London, Biological Sciences* 358: 1997–2009.

Ruitenbeek, J. and Cartier, C. (2001). The invisible wand: adaptive co-management as an emergent strategy in complex bio-economic systems. *Occasional Paper 34, Center for International Forestry Research.* Bogor, Indonesia.

Scheffer, M., Brock, W.A. and Westley, F. (2000). Mechanisms preventing optimum use of ecosystem services: an interdisciplinary theoretical analysis. *Ecosystems* 3: 451–471.

Scheffer, M., Carpenter, S., Foley, J., Folke, C. and Walker, B. (2001). Catastrophic shifts in ecosystems. *Nature* 413: 591–696.

Scheffer, M., Westley, F. and Brock, W.A. (2003). Slow response of societies to new problems: causes and costs. *Ecosystems* 6: 493–502.

Scholes, R.J. and Walker, B.H. (1993). *Nylsuley: The Study of an African Savanna.* Cambridge University Press, Cambridge, UK.

Schröter, D., Cramer, W., Leemans, R., Prentice, I.C., Araújo, M.B., Arnell, N.W. *et al.* (2005). Ecosystem service supply and vulnerability to global change in Europe. *Science* 310: 1333–1337.

Schultz, L., Folke, C. and Olsson, P. (in press). Enhancing ecosystem management through social–ecological inventories: lessons from Kristianstads Vattenrike, Sweden. *Environmental Conservation.*

Scoones, I. (1999). New ecology and the social sciences: what prospects for a fruitful engagement? *Annual Review of Anthropology* 28: 479–507.

Sheil, D. and Lawrence, A. (2004). Tropical biologists, local people, and conservation: new opportunities for collaboration. *Trends in Ecology and Evolution* 19: 634–638.

Steffen, W., Sanderson, A., Jäger, J., Tyson, P.D., Moore III, B., Matson, P.A. *et al.* (2004). *Global Change and the Earth System: A Planet under Pressure.* Springer Verlag, Heidelberg.

Stoker, G. (1998). Governance as theory: five propositions. *International Social Science Journal* 50: 17–28.

Troell, M., Pihl, L., Rönnbäck, P., Wennhage, H., Söderqvist, T. and Kautsky, N. (2005). Regime shifts and ecosystem service generation in Swedish coastal soft bottom habitats: when resilience is undesirable. *Ecology and Society* 10(1): 30. URL: http://www.ecologyandsociety.org/vol10/iss1/art30/

Turner II, B.L., Clark, W.C., Kates, R.W., Richards, J.F., Mathews, J.T. and Meyer, W.B. (eds) (1990). *The Earth as Transformed by Human Action: Global and Regional Changes in the Biosphere over the Past 300 Years.* Cambridge University Press, Cambridge, UK.

van der Leeuw, S.E. (2000). Land degradation as a socionatural process. In R.J. McIntosh, J.A. Tainter and S.K. McIntosh (eds) *The Way the Wind Blows: Climate, History and Human Action.* Columbia University Press, New York, pp. 190–210.

Walker, B.H. and Meyers, J.A. (2004). Thresholds in ecological and social–ecological systems: a developing database. *Ecology and Society* 9(2), 3. URL: http://www.ecologyandsociety.org/vol9/iss2/art3/

Walker, B.H., Holling, C.S., Carpenter, S.R. and Kinzig, A.P. (2004). Resilience, adaptability and transformability in social–ecological systems. *Ecology and Society* 9(2): 5. URL: http://www.ecologyandsociety.org/vol9/iss2/art5/

Walters, C. (1997). Challenges in adaptive management of riparian and coastal ecosystems. *Conservation Ecology* 1(2): 1. URL: http://www.consecol.org/vol1/iss2/art1/

Westley, F. (1995). Governing design: the management of social systems and ecosystems management. in

L.H. Gunderson, C.S. Holling and S. Light (eds) *Barriers and Bridges to the Renewal of Ecosystems and Institutions.* Columbia University Press, New York, pp. 391–427.

Westley, F. (2002). The devil in the dynamics: adaptive management on the front lines. In L.H. Gunderson and C.S. Holling (eds) *Panarchy: Understanding Transformations in Human and Natural Systems.* Island Press, Washington, DC, pp. 333–360.

Westley, F., Carpenter, S.R., Brock, W.A., Holling, C.S. and Gunderson, L.H. (2002). Why systems of people and nature are not just social and ecological systems. In L.H. Gunderson and C.S. Holling (eds) *Panarchy: Understanding*

Transformations in Human and Natural Systems. Island Press, Washington, DC, pp.103–119.

Wollenberg, E., Edmunds, D. and Buck, L. (2000). Using scenarios to make decisions about the future: anticipatory learning for the adaptive co-management of community forests. *Landscape and Urban Planning* 47: 65–77.

Wondolleck, J.M. and. Yaffee, S.L. (2000). *Making Collaboration Work: Lessons from Innovation in Natural Resource Management.* Island Press, Washington, DC.

Young, O. (2002). *The Institutional Dimensions of Environmental Change: Fit, Interplay and Scale.* Cambridge University Press, Cambridge, UK.

38

Contested Ground in Nature Protection: Current Challenges and Opportunities in Community-Based Natural Resources and Protected Areas Management

Steven R. Brechin, Grant Murray
and Charles Benjamin

INTRODUCTION

The literatures on the sociopolitical dimensions of protected areas and on community-based natural resource management are broad, rapidly growing, and rife with controversies and opportunities. Given the monumental task of attempting to organize these vast readings into a few pages, we lack the space to sufficiently discuss every issue or deserved reading. Instead, we attempt to mark for the reader general trends and key references.

This chapter links together four bodies of literature. Section A concentrates on the social and political issues related to demarcated, land-based biodiversity conservation efforts with particular focus on management issues involving local peoples and communities. Section B takes on a similar discussion but around the rapidly growing interest in marine-based protected areas. Section C focuses on the even broader topic of community-based natural resources management, largely under the frame of common property resources with special attention to state-centered devolution of responsibilities that are redefining community-based efforts. Finally, Section D

reviews the social promises and pitfalls of ecotourism. Ecotourism typically has been viewed as a panacea for improving the economic base of local communities near protected areas but the on-the-ground realities paint a more complex picture. These four areas of scholarship and practice are linked in part by a movement among conservation and natural resource management organizations and agencies to pursue conservation beyond established protected areas to a wider ecosystem or landscape scale. Hence, to an ever-increasing extent the discussion of local-scale dynamics, within or near protected areas, must be placed within the context of large-scale dynamics and conditions. Also, the growing need for conservationists to protect nature outside established protected areas provides distinct opportunities to integrate the findings of the literature on common property resources to this cause. The chapter concludes by arguing that the evidence clearly points to the future of biodiversity conservation, resting on finding more effective and seamless ways of integrating local people and communities into the conservation process and not in their greater separation.

TERRESTRIAL PROTECTED AREAS

Government-owned parks and other forms of protected areas have long been a central element of strategies to protect terrestrial resources. Because of the potential impacts of protected areas on the livelihoods of people living in and around them, the relations between park and resident populations and the links between biodiversity conservation and rural livelihoods have been important themes of protected area management over past decades. The latest opportunities and challenges include: (1) the re-emergence of fortress conservation; (2) questions surrounding the development of corporate conservation; (3) the troubled nature of defining conservation success; (4) the rise of private protected areas; (5) the development of large-scaled approaches to conservation planning; and (6) the emergence of community-conserved areas.

The re-emergence of fortress conservation–revisiting questions of social justice?

A current debate within the international biodiversity conservation community today is whether or not a "Resurgent Protection Paradigm," as Wilshusen et al. (2002) have called it, truly is having a major effect on both conservation policy and practice, especially in regards to impacts on local communities. In recent years there has been renewed rhetoric from a number of well-established scientists and conservation organizations advocating both (1) an abandonment of the social agenda related to conservation efforts, such as Integrated Conservation and Development Projects (ICDPs) that were popular during the 1980s and 1990s (see McShane and Wells, 2004; Hutton et al., 2005; Bray and Anderson, 2005), and (2) a greater emphasis on, or return to strict preservationist practices (e.g. Oats, 1999; Terborgh, 1999; see also Kramer et al., 1997; Brandon et al., 1998). To varying degrees a number of conservation scientists argued that the conservation community has wasted precious time and resources without positive effect while pursing complicated and demanding projects that mixed core conservation efforts with socioeconomic development projects. A fundamental question in this debate is how many conservation scientists and organizations working on the ground truly prescribe to this view? Is it a majority or simply a vocal minority? Hutton et al. (2005), however, indicate that "resurgent protectionism" rhetoric has had an important effect at least in Southern Africa.

The logic behind the people-sensitive approach to conservation, in addition to obvious moral concerns, has been in large part to enlist the support of local people and communities in achieving conservation goals. It rests on the idea of engaging local people in the conservation mission in a host of strategies, ranging from environmental education programs to co-management of the protected areas themselves (Borrini-Feyerabend, 1996). The ICDPs, on the other hand, were a more narrowed attempt to create development activities near these communities with the hope that these efforts would draw community members' attention and energy away from developing resources within the core conservation areas. While the logic was generally sound, the implementation of this approach to date has been largely very poor for a long list of reasons (see Wells and Brandon, 1992; Wells et al., 1999; Wilshusen et al., 2002; McShane and Wells, 2004; Bray and Anderson, 2005).

The rhetoric from the "resurgent protectionist" conservationists forcibly states the need to return to the core mission of biodiversity conservation through protecting species through strict enforcement of protected area boundaries, and to do so without burdening the effort by dealing with weighty and intractable social problems and agendas. What is ironic about the re-emergence of this fortress conservation approach, however, is that the people-sensitive approach to conservation emerged out of a prior time when a strict preservationist approach felt the sting of criticism for both a lack of success and cries of social injustices from disaffected local people and communities who resented greatly the efforts of conservationists. This was largely the result of the conservation community's general insensitivity to local needs and perspectives during the early years (see West and Brechin, 1991). As a consequence local people frequently attempted to undermine conservation efforts. Hence, given this history, one must immediately ask: Are the resurgent protectionists re-inventing a "square wheel"?

If a strict preservationist's approach reinvokes failed strategies from the past,[1] and ICDPs were likewise unsuccessful, then how should one go forward in protecting nature while providing for communities? In an answer to that question, a number of scholars and practitioners argue that conservation community must pursue biodiversity conservation with social justice (West and Brechin, 1991a; Western et al., 1994; Stevens, 1997; Zerner, 2000; Brechin et al., 2002, 2003c; Wilshusen et al., 2002; Borrini-Feyerabend et al., 2004a). A number of publications have outlined the elements to this approach. Brechin et al. (2002) list six key elements: human dignity, legitimacy, governance, accountability, adaptation and learning, understanding of non-local forces. There is not sufficient space to go into each of these in detail here. However, this approach to conservation is built around a core notion of viewing conservation as continuous social and political planning and implementation processes. Instead of seeing the creation of protected areas as the end, a social justice approach views that event as one of many endless steps. Fundamentally this approach is collaborative and cooperative. The goals are to create a shared vision of the conservation mission and how to achieve it with complete transparency, for all stakeholders to share both authority and responsibility and to negotiate continually on both tactical and strategic efforts, as well as the resolution of conflict.

Key publications related to these efforts, in addition to those above, can be found in exploring the many products of the World Conservation Union (IUCN), in particular its Commission of Environmental, Economic and Social Policy (CEESP) but also its World Commission on Protected Areas (WCPA). This approach, however, is not easy. It is time consuming, frustrating and requires both organizational and institutional capacity at many levels. With this capacity lacking in many communities and nation-states, there is a firm role for civil society and indeed the world community to provide the necessary conditions for such an approach to flourish.[2] In spite of the decades of discussion on the conflicted relationships between protected areas and communities, researchers have never systematically documented the social impacts on conservation efforts on local communities. However, this may be changing as there have been recent discussions at international meetings about getting such an effort underway.[3]

In spite of the backlash rhetoric heard today, the conservation practice on the ground, at least so far, appears to be staying largely with a more people-sensitive approach, although there are reports of new problems (see Chapin, 2004; Hutton et al., 2005). While ICDPs have been essentially abandoned as a funding staple, many on the ground efforts seem to continue to attempt to involve local people and communities in both planning and managing protected areas. However, what one means by "participation" or "collaboration" continues to be an important source of difference and controversy. This concern over sincere versus insincere participation and partnering was a fundamental point in Mac Chapin's controversial December 2004 article in World-Watch, in which he argues that the large conservation organizations prefer now to go it alone. In spite of the uncertainty over participation, new strategies for engaging the local community seem to have evolved recently as well, with former strategies receiving new life. The two most discussed today seem to be: (1) market-oriented projects and

(2) conservation concessions. Market-oriented projects tend to focus on income-generating schemes such as shade-tree organic coffee, eco-tourism, and community-based forestry, that is, sustainable logging of community lands (Perfecto et al., 1996; Bray et al., 2002, 2003; Bray and Anderson, 2005). Conservation concessions, especially direct cash compensation, illustrate an innovative approach for protecting biodiversity (e.g. Ferraro and Kramer, 1997; Ferraro and Kiss, 2002). Under this approach, conservation organizations and/or donors provide direct payments to communities for the opportunity costs in loss access to areas now under strict protection. However, such an approach would appear to have the greatest potential under institutional settings of clear property ownership, rights and enforcement. While finding economic substitutes for lost access to resources remain an important step forward in achieving conservation with social justice it likely will not replace the loss of political voice. People and communities typically want to maintain a strong say in protecting and promoting their collective futures. Hence, legitimate, active and full participation in the political practice of conservation as well as negotiated resource access, as allowable, will likely need to remain a key element in promoting biodiversity with social justice into the future. (Brechin et al., 2002, 2003b).

Corporate conservation–controversy over BINGOs and indigenous groups

One of the most interesting developments within the biodiversity community of late and potentially one of the most problematic has to do with the rapid rise of what Chapin (2004) calls "BINGOs" or Big International Non-government Organizations.[4] In this specific context Chapin refers particularly to large and recently very wealthy conservation organizations such as World Wildlife Fund, The Nature Conservancy, and Conservation International. Chapin, an anthropologist, then with the Center for Native Lands and who has worked on conservation issues with indigenous and traditional peoples for many years, reported on the growing discontent among philanthropic foundations and indigenous/local groups regarding the recent behavior of the BINGOs. Chapin presents two main explanations for the BINGOs' recent behavior: (1) increased corporate support for BINGOs leading to greater conflicts of interests (or appearances of conflicts of interests) with regards to conservation activities on the ground; and (2) neglected partnerships or collaborative approaches with indigenous and traditional peoples' organizations. In sum, Chapin's argument is that the BINGOs have become rich and powerful enough to both define and implement their own visions of conservation, which largely fall within the "Resurgent Protection Paradigm" discussed above. Consequently, the BINGOs appear less concerned about or supportive of full partnering or developing shared visions of conservation with other stakeholders. This possible growing hegemonic tendency of BINGOs appears to have been enhanced with a nearly 50% global drop in biodiversity funding while their own resources have grown substantially (Bray and Anderson, 2005).

BINGO officials have dismissed publicly the charges by citing a number of efforts that involve local communities (see letters to the editors published in *World-Watch*, January 2005; Bray and Anderson, 2005). However, these charges were some time in coming. During the 2003 IUCN World Parks Congress in Durban South Africa, the CEOs of the BINGOs were scolded publicly by indigenous groups for their lack of sensitivity toward their needs (Brechin, personal observation; Brosius, 2004). Also in 2003, a number of philanthropic foundations with funding interest in biodiversity conservation met quietly to discuss this very issue (Chapin, 2004).

The discussion on the role of indigenous and traditional peoples in conservation efforts remains essential, however. Janis Alcorn claims that 85% of the world's protected areas, IUCN categories I–VI, are inhabited by local/indigenous peoples (Alcorn, 2000; Bray and Anderson, 2005). Her message here clearly rests on the reality that efforts to protect biodiversity must fully engage these groups in the effort. However, there have been recent writing by prominent biodiversity scientists such as John Terborgh calling indigenous peoples "the danger within" (Terborgh, 1999, p. 40; see also Bray and Anderson, 2005, p. 5). There are reported plans for continued discussions between BINGOs, foundations, and indigenous groups and their supporters. Regardless, the controversy surrounding the BINGOs simply highlights another important issue related to conservation that needs further exploration, that is, the role of organizations more generally. Much has been written about biodiversity conservation and its many complexities. However, very little research has been directed to understanding organizations themselves or the arrangements that deliver conservation efforts.[5] As findings in organizational studies continually show, the capabilities and behavior of organizations matter greatly in creating outcomes.

The troubled nature of defining conservation success

In spite of the recent practical efforts by conservation organizations to measure the impacts of their efforts,[6] one of the most daunting questions before them today is the elusive notion of conservation success (see Brechin *et al.*, in press; Murray, 2005). At first blush this appears to be a straightforward question. While criteria can always be constructed to define success, the harder question is determining what the most appropriate measures should be. The more explicitly sociological question is who determines what those measures are. Disagreement can easily develop among different interest groups and in our analysis lies at the heart of many current conservation controversies. If conservation practice is seen as a multi-stakeholder process then there will likely be multiple notions of success as well. This could include cultural, economic, and political notions of success in addition to ecological ones (Brechin *et al.*, in press). Maximizing more than one goal at a time tends to be extraordinarily difficult. Of course, the conservation community itself, typically led by conservation biologists and ecologists, has tended to view success mostly in biological or ecological terms, such as protecting species, biodiversity or habitat. Other stakeholders may have other criteria such as local communities who might think in terms of political control over natural resources, tourism dollars, or the creation of livelihoods.[7]

Even when focusing on a single ecological measure, biodiversity (or number of species), there appear to be several interesting challenges. Bruner *et al.* (2001) analyzed the impacts of anthropogenic threats to biodiversity in 93 protected areas in 22 tropical countries, and compared the levels of impacts within protected areas against a 10-kilometer belt surrounding the protected areas. Their results showed that, compared to land areas outside of them, most protected areas have been generally successful in halting land clearing and have been effective to a lesser extent in slowing logging, hunting, fire, and grazing. Similarly, Nepstad *et al.* (2006) have found similar results among the rainforest parks and indigenous territories in the Amazon. However, both Bruner *et al.* and Nepstad *et al.* showed that protected areas are effective in protecting nature in *relative terms*, only when compared to lands that were not protected. They also say nothing about the actual state of biodiversity within the protected areas. This observation leads to a set of challenging questions: Should one then define conservation success in *absolute* or in *relative* terms? If in absolute terms, is that realistic? At what scale does one reasonably activate such a measure? How should one consider the forces outside of the protected area compared to those inside of it? These questions are not easily answered and suggest that if success is defined in absolute terms, it would rarely, if ever, be achieved; every conservation effort would likely fail by definition. If one then accepts a "relative measurement" one is left with attempting to determine the near impossible in how much species loss is acceptable.

In critiquing Bruner *et al.*'s findings discussed above, Stern (2001) remarked that in Bruner *et al.*'s effort to evaluate conservation success, Bruner and his colleagues had failed to integrate social factors into their assessment. Bruner *et al.* responded that Stern had "confused ends with means" (Bruner *et al.*, 2001). This response only illustrates the difficult political road ahead in defining conservation success especially among individuals and groups that represent differing perspectives to those of conservation scientists. Should not social issues such as legitimacy, social justice, and meaningful collaboration be important ends as well? How can biodiversity conservation be achieved, even in its narrowest interpretation, if it ignores the local people and communities in and around protected areas? To conclude, one must wonder if the conservation community will ever be able to develop a universally accepted notion of conservation success or even if it should. It is likely that any realistic measures will need to be relative measures. From this logic, it will be up to each independent multi-stakeholder group to form its own notion of success through negotiation and reflection. This particular take on conservation success suggests it remains very much a "local process," which must be worked out on a site-by-site and group-by-group basis. Hence, conservation success would be unique to each site, lacking particular standards beyond basic guidelines, making defining conservation success endlessly controversial and difficult to compare across sites.

Rise of private parks and nature reserves

One of the most recent developments in international conservation has been the rise of private parks–protected areas owned not by the state but by private landowners. Their exact numbers are unknown, but estimates have been made in the thousands, with more being established each year (Langholz and Lassoise, 2001). Jeffery Langholz has been one of the most active observers of this phenomenon (Langholz, 1996, 2002, 2003; Langholz *et al.*, 2000). Private parks have been lauded by some groups as an effective new tool in the effort to protect biodiversity. In an era of privatization, turning to the private sector to solve public problems has become an encouraged approach. Many private parks, however, seem to be associated with eco-tourism enterprises or sponsored research areas. Interestingly, their forms can be as diverse as the nature they attempt to protect (Langholz, 2003). However, as a new tool in the conservationists' toolkit there have been little empirical analyses of how they are faring. The potential implications for local people and communities are noteworthy. In poor tropical countries, private reserves tied to eco-tourism can quickly become isolated playgrounds for the wealthy while locals lack basic resources for their livelihoods. Transnational pharmaceutical firms may purchase tracks of forests or rights to areas for bioprospecting, generating a conflict of interests among these corporations and local communities which traditionally depended on those resources (Dorsey, 2003). Reports from South Africa indicate that wealthy white farmers have turned their ranches into private game parks to avoid land claims by former black farmhands allowed under new post-Apartheid laws (Brinkate, 1996). Hence, the conservation mission itself was not the top priority and the contributions likely limited to actual biodiversity conservation need. Similarly, privately held parks, especially where landowners are foreign individuals or organizations, can create political tensions, especially in countries with troubled colonial pasts. From this particular vantage point, private parks might be seen as a new type of colonialism. Even where political overtones are less prominent, there has been concern that private conservation areas simply make their managers less willing to engage outside communities in working out differences, or more willing to implement stricter enforcement regimes that might sour relationships with neighboring communities, affecting conservation efforts more broadly.

Large-scale/landscape approaches innovation in conservation planning

One of the most important biodiversity conservation innovations of the last decade or so has been the development of landscape-scale approaches (Dinerstein *et al.*, 1995). The names may vary – biological "hot spots", "eco-regions" etc. – but the idea behind them remains largely the same (see Redford, 2003, for an excellent discussion of varied approaches). Instead of focusing on protecting nature at one protected area at a time, a number of conservation organizations are now attempting to look and plan for conservation of larger biologically rich regions. Proponents present this as a "science-based" approach, as it attempts to prioritize critical areas for protection through scientifically identifying biologically rich and coherent regions. Here, the task is to find ways to protect the ecological integrity of a region though establishing a connected system of protected areas and/or creating conservation corridors connecting existing areas or even through efforts to conserve resources outside of protected areas.

While the approach makes planning sense, one of the burning questions today within the conservation community concerns is what effects this will have on local communities (See Gezon, 2003; Bray and Anderson, 2005). Some proponents of this approach see it as a way to distance conservation efforts from community-based approaches, especially from the ICDP model (see Kramer *et al.*, 1997). Some have claimed, however, that framing this approach as "scientific" essentially absolves the conservation organizations from the "politics" of conservation (see Chapin, 2004). However, proponents of people-centered conservation question the ability of conservation organizations to avoid successfully either the politics or needs of local people and communities regardless of the scale (Gezon, 2003; Goodale *et al.*, 2003). It can easily be argued that increasing the scale of conservation requires greater coordination and involvement of more local people and communities, not less. Bray and Anderson (2005) report on some efforts by conservation organizations to do just that.[8] Still, documenting the unknown impacts and interrelationships between local peoples and landscape-level planning remains one of the most important research questions in conservation today.

Community conserved areas

A current focus of international attention is the effort to institutionalize the significant nature conservation efforts of indigenous and traditional peoples into the international classification system of parks and protected areas. Community Conserved Areas (CCAs) attempt to recognize and account for the many conservation efforts by local peoples, indigenous as well as non-indigenous communities, with long stable histories of resource management. These areas are voluntarily conserved largely through customary laws and practices by local community groups or peoples. Examples of CCAs are numerous and diverse (see Borrini-Feyerabend *et al.*, 2004b). They include the preservation of natural areas such as sacred groves for religious or cultural reasons, areas largely untouched by humans, and sites heavily managed for sustainable use, such as marine areas or pasturelands. CCAs are believed to number well into the thousands. It is important to recognize that CCAs are NOT the addition of an "IUCN VII" protected area category. Rather the designation is part of a more innovative redesign of the entire classification system while keeping the established IUCN categories I–VI (see Borrini-Feyerabend *et al.*, 2004, p. 25; Brosius, 2004).

In addition to the existing "Protected Area Categories" I–VI, a new heading of "Governance Types" has been added, reflecting the different ways in which protected areas are governed. These types include: government-managed protected areas, co-managed protected areas, private protected areas, and community conserved areas. IUCN's recognition of governance types is an important innovation in how one should think about and classify protected areas.

Recognizing CCAs as an accepted and recognized form of conservation has been one of the more active efforts of TILCEPA (Theme on Indigenous and Local Communities, Equity, and Protected Areas), a joint working group of two commissions from the World Conservation Union (IUCN), the World Commission on Protected Areas (WCPA) and the Commission on Environmental and Economic and Social Policy (CEESP). At the Vth World Parks Congress in Durban, South Africa, in 2003, CCAs were formally recognized as legitimate conservation efforts. Some nation states have led the way by incorporating CCAs into their own national protected area systems. Still, given the wide variety of CCA types and the many challenges faced by today's traditional communities, CCAs face an uncertain future. Examples of such challenges include rapid changes due to sharp increases in population numbers and configuration, modernization programs that break down traditional ways of relating, or even changes in existing policy and rule enforcement largely from the greater presence of national government systems (see Section C of this chapter).

We now turn to a related discussion related to rapidly growing interest in marine protected areas.

NEW OPPORTUNITIES AND CHALLENGES IN MARINE PROTECTED AREA MANAGEMENT

While the conservation community has focused more attention on terrestrial areas than on marine or coastal areas, worldwide concern over marine resources has increased dramatically over the last few years, as the limits of many resources have been reached (Allison *et al.*, 1998; Roberts *et al.*, 2003a). The coastal zone has become ever more crowded and overused, and anthropogenic impacts – including overfishing – have become increasingly apparent (Botsford *et al.*, 1997; Wilkinson, 2000; Lubchenco *et al.*, 2003; Pauly and Maclean, 2003).

In the wake of continuing fisheries collapses around the world, some authors have placed an increased emphasis on the explicit management of fisheries, claiming that this approach can help to reduce uncertainty and improve fisheries yields

(Botsford *et al.*, 1997; Allison *et al.*, 1998; Roberts *et al.*, 2001, 2003a; Pauly *et al.*, 2002). Marine protected areas (MPAs) have more recently gained importance as a conservation tool, heightening the sense that they often have the "dual objectives" of conservation and fisheries management (NRC, 2000; Christie *et al.*, 2002; Roberts *et al.*, 2003a; Ward and Hegerl, 2003) though a range of "socio-economic" benefits have also been described, including supporting tourism (Gubbay, 1995; Agardy, 1997). In response, over the last several decades the popularity of MPAs as a management tool has dramatically increased (Kelleher, 1995; Allison *et al.*, 1998; NRC, 2000; Christie *et al.*, 2003a; Lubchenco *et al.*, 2003). Virtually every coastal country has now implemented some kind of MPA (Agardy *et al.*, 2003).

A variety of international fora has called for increasing MPA coverage. For example, the World Summit on Sustainable Development in Johannesburg called for the establishment of a global network of MPAs by 2012. Likewise, one of the key recommendations emerging from the 2003 Durban World Parks Congress (in which marine areas was a cross-cutting theme) was to increase dramatically the number and coverage of MPAs.[9] It was further recommended that these networks should be extensive and include strictly protected areas that amount to at least 20 to 30% of each habitat, and should contribute to a global target for healthy and productive oceans (see also Roberts *et al.*, 2003b).

A review of the literature centered on MPAs suggests several key opportunities and challenges including (1) a mixed record of community-based protected-area performance over the last two decades concomitant with a growing acceptance of the efficacy of "no-take" reserves; (2) a need for both more social and ecological information and a move towards ecosystem-based management;

and (3) a move towards marine conservation/protection at larger scales, including networks of MPAs and a need for integrated management beyond the boundaries of protected areas.

Mixed record of CBM success and the proven success of "no-take" marine reserves

The recent growth in the perceived need for MPAs came with the questioning of top-down, centralized approaches to protected area management. The somewhat longer history of stakeholder involvement in fisheries management provided a starting point. The 1980s and 1990s in particular saw the development of a number of CBM approaches to MPA management. Many new MPAs emphasized local communities and participatory management approaches, promising a long list of potential benefits. These are briefly listed in Table 38.1 (see van't Hof, 1994; Agardy, 1997; Pollnac *et al.*, 2001).

Some have argued that greater fisher participation and control of the management process works best in small-scale, artisanal, coastal fisheries that have not been over-capitalized. The calls for increased participation – and the increased sense of legitimacy that is assumed to come with it – have been particularly strong in coral reef fisheries, for example. These areas tend to be nearer to shore, more easily enforced, and more bounded geographically and ecologically. They are also commonly found in countries with weak regulatory and/or enforcement capabilities. The Philippines alone have seen the establishment of over 400 MPAs (Pollnac *et al.*, 2001). However, the performance of many of these reserves can perhaps best be described as "mixed."

Some studies have pointed to the "success stories" in improving fish yields, including Apo

Table 38.1 Perceived benefits of collaborative management in marine protected areas

Benefit	Literature
An increased sense of legitimacy, understanding, and support for resultant policy	Kaza, 1988; White *et al.*, 1994; McCay, 1995; Sen and Nielsen, 1996; and Christie *et al.*, 1994 – case study
Participants can gain a greater sense of control over their own lives	Pinkerton, 1989
Community development	Pinkerton, 1989; White *et al.*, 1995
Maintenance of cultural integrity due to control over important life decisions, and the pace of change	Brechin *et al.*, 1991; Goodland, 1991; Colchester, 1994
Incorporation of locally specific (often complex) knowledge, both ecological and social	West and Brechin, 1991a; Wilson *et al.*, 1994; Gubbay, 1995
Pre-existing management regimes can serve as foundations for future management	Gubbay, 1995; White *et al.*, 1995
Facilitation of monitoring and enforcement of protected area regulations	Christie *et al.*, 1994; Gubbay, 1995

and San Salvador Islands in the Philippines (e.g. Russ and Alcala, 1996; Christie *et al.*, 2003b) or in Florida and the Caribbean (e.g. Roberts *et al.*, 2001). Others have pointed to positive socio-economic benefits (White *et al.*, 1994; Ferrer *et al.*, 1996). Still others have suggested that MPA performance can be enhanced by the devolution of authority for MPA development and management and that collaborative MPA management arrangements can lead to increased effectiveness (Mascia, 2003).

On the other hand, some authors have suggested that as few as 20 to 25% of Philippine MPAs can be considered "successful," leading to some concern about the rejection of the CBM model (Pollnac *et al.*, 2001). Likewise, Christie *et al.* (2002), while continuing to advocate for the CBM approach, have noted mixed success in other aspects, including declines of fish abundances in adjacent areas, as well as declines in target species abundance within MPAs. They have suggested that it is not always realistic for scattered, small no-take areas to improve fisheries yields, particularly in situations where fishing effort increases in areas immediately adjacent to those areas. In some cases, the "successful" model employed in places like Apo Island, Philippines (see Russ and Alcala, 1996) have been replicated in other situations without regard to local context or institutional differences (Christie *et al.*, 2003a, who cite White *et al.*, 2002). In any case, without effective conservation outside of reserve boundaries (see below), community-based MPAs seem unlikely to address underlying environmental issues in tropical countries (Agardy *et al.*, 2003; Christie *et al.*, 2003a).

At the same time, there has been a growing emphasis on the efficacy of "no-take" marine reserves. While the "theory underlying the design of marine reserves ... is still in its infancy" and empirical data are lacking (Botsford *et al.*, 2003, p. S25), there has been an emerging consensus that a particular class of MPAs – often referred to as no-take marine reserves – are most effective at achieving the dual objectives of conservation and fisheries management. For example, a recent Scientific Consensus Statement on Marine Reserves and Marine Protected Areas stated that "reserves result in long-lasting and often rapid increases in abundance, diversity, and productivity of marine organisms." Critically, this same statement goes on to state that "full protection (which usually requires adequate enforcement and public involvement) is critical to achieve this full range of benefits. Marine protected areas do not provide the same benefits as marine reserves" (as quoted in Lubchenco *et al.*, 2003, p. S5; see also Allison *et al.*, 1998; Halpern, 2003).

In sum, despite a great deal of uncertainty, there has been a growing opinion among many that

no-take reserves are superior to less restricted MPAs in achieving fisheries management and conservation goals. This has come at the same time that the track record of the CBM approach has been mixed. In part due to these trends, a debate has erupted around MPAs, where even basic definitions and objectives of MPAs have been discussed (Agardy *et al.*, 2003; Christie *et al.*, 2003). Christie *et al.* (2003a: p. 23) have noted that "MPAs tend to be two sorts: the 'park' model, whereby a government agency declares an area out-of-bounds for some or all activities, and the 'community-based' model, whereby coastal communities assume many of the responsibilities for implementing, monitoring, and enforcing rules for the protection of marine areas." Likewise, Agardy *et al.* (2003) note that basic questions over the definitions and objectives of MPAs remain unanswered, but there is a danger that views on MPAs have become "polarized", with a rift developing between those who advocate that only no-take reserves can confer conservation benefits, and those that stress that other types of MPAs can confer a broader range of benefits. This rift, they fear, may impede the use of MPAs to conserve marine biodiversity.

There is a risk that only restrictive, no-take reserves based on the "park" model will become the dominant mode of dealing with conservation and sustainable issues (Agardy *et al.*, 2003). Indeed, Christie *et al.* (2003a) argue that "there has been an emerging interest (largely based on biological arguments) in abandoning the community-based approach, which typically involves smaller areas, for the park model, which generally involves larger MPAs" (Christie *et al.*, 2003a, p. 24). Critically, they suggest that this has been done without a clear understanding of potential responses by impacted stakeholders or even the implications for long-term conservation objectives. There is a concern, in other words, that MPA models will become driven by overly general "rules of thumb," where no-take reserves are focused on as the only marine and coastal conservation and resource management tool (Agardy *et al.*, 2003; Christie *et al.*, 2003a).

However, it should be noted that even strident advocates of "park"-style MPAs recognize the need for stakeholder participation. A group of influential ecologists (see Roberts *et al.*, 2003b) recently contended that there has been little guidance for prioritizing among the various biological, economic, and social criteria that have been offered for selecting the location of marine reserves. They argue that biological criteria be met before socio-economic ones. Yet they also add that

> we have said little ... about the role of stakeholders in selecting reserves. By placing the emphasis on biology first, socioeconomics later, our scheme

might appear to exclude stakeholders, or at least defer their involvement until later in the process. However, taking this approach would be disastrous. Numerous studies have convincingly demonstrated that efforts to create reserves without close stakeholder involvement will fail (Kelleher and Recchia, 1998; NRC, 2000). It is critical that stakeholders are involved from the very beginning, including during the evaluation of sites according to ecological attributes (Roberts et al., 2003b).

Need for more ecological and social information

Mascia (2003, p. 630) reviewed recent social research on the "human dimensions" of coral reef marine protected areas and has claimed that "... social factors, not biological or physical variables, are the primary determinants of MPA success or failure." Yet, in a 1999 mail survey of MPAs in the wider Caribbean, Mascia (1999) found that MPAs were usually established without comprehensive information about either socio-economic or biological conditions, but that biological assessments had occurred twice as frequently as socio-economic ones. Furthermore, when social science research is conducted, it is often done too late in the design process to have an impact on policies (Christie et al., 2003a; Mascia et al., 2003). Christie et al. (2003a) have suggested that MPA social science should seek to illuminate (a) the characteristics and behaviors of constituencies, (b) what communities want, and (c) what communities know. Clearly, there is also a need for research into effective governance regimes for protected area management (van't Hof, 1994; Mascia, 1999), and for systematic identification of the factors that contribute to protected area effectiveness (Pollnac et al., 2001).

The ecological science behind MPAs is also far newer than for terrestrial areas (Allison et al., 1998; Gerber et al., 2003). There are several differences to consider between marine and terrestrial protected areas that have important implications from an ecological perspective on protected area design (Carr et al., 2003). Marine areas are fluid, three-dimensional, often larger-scale, and far more "open" in terms of the flow of nutrients, organisms, and water masses. The major focus of terrestrial protected areas has been on addressing habitat loss, while the major focus in marine and coastal areas has been on over-exploitation (Allison et al., 1998; Carr et al., 2003).[10] Many basic questions have only begun to be explored, including the question of boundaries and the movement of animals in and out of these areas, questions about recruitment and larval dispersal into and out of reserves, issues of

"adequate" size and connectivity, and how "catastrophic" phenomena affect reserve design.[11] Furthermore, reconciling the goals of conservation and fisheries management can be challenging (Hastings and Botsford, 2003), a challenge that is further complicated when other goals (e.g. tourism) are added to the mix. Additionally, while some authors have suggested ways to standardize approaches to reserve design (e.g. Roberts et al., 2003a), others have stressed that MPA designs will need to be tailored to the particular ecological requirements specific to the area in question (Grantham et al., 2003). The needs of different organisms are different, suggesting a heightened complexity as managers seek to manage multiple stocks or take an ecosystem approach.

A second major part of Recommendation 22 from the Durban WPC was to "implement an ecosystem-based approach to sustainable fisheries management and marine biodiversity conservation." Because they protect geographic areas that include resident species and their biophysical environment, MPAs – particularly strictly protected ones – are seen as inherently offering an "eco-system based" approach to conservation or fisheries management (Mascia, 2003; Ward and Hegerl, 2003). Ecosystem-based approaches are distinct from previous efforts that focused on single-species management or conservation efforts (though the underlying theory often still problematically rests on tenets of single-species management) (Lubchenco et al., 2003). Calls for an ecosystem-based approach have also come from within the fisheries and integrated coastal zone management literature, though development of ecosystem-based approaches have lagged behind developments in terrestrial areas (Botsford et al., 1997; Cicin-Sain and Knecht, 1998; FAO, 2002).

It is also important to note that several scholars have called for an increase in the diversity of sources of knowledge that are drawn upon in the management process. For example, some authors have highlighted the complexity and depth of local knowledge systems, and have called for greater inclusion of local, traditional, or experience-based ecological knowledge in the management of natural resources (Berkes, 1993, 1999).[12] This is true with respect to both terrestrial and MPAs, though calls for the inclusion of information derived from other knowledge systems has been particularly strong in the fisheries literature (e.g. see Neis et al., 1999; Neis and Felt, 2000; Murray et al., 2005). Given the opacity, fluidity, stochasticity, and sheer size of marine ecosystems, accurate and adequate information can be particularly hard to come by and user groups may be uniquely situated to provide or improve the knowledge foundation upon which to base management decisions. Proponents generally claim two types of benefits: (1) that the overall quality of information upon which to base management decisions will be increased; and (2) that "democratizing" the production of knowledge provides a mechanism for participation at critical stages in the management process (St. Martin *et al.*, 2007).

Need for marine resource management at larger scales

Finally, there has been a move towards the conservation/protection of marine resources beyond the scale of single, relatively small areas. This includes a shift to networks, a move towards ecosystem management (described above), and a recognition of the need for management beyond protected area boundaries. While individual protected areas can address many objectives, no one area is likely to address all of them, particularly given the large-scale, "open" nature of many marine ecological processes. The use of networks is seen as a means of maintaining essential large-scale ecosystem processes (Roberts *et al.*, 2003a, b) without increasing the overall size of individual MPAs to unfeasible proportions. Furthermore, for fisheries management purposes, it has been suggested that while large reserves are effective at achieving conservation goals, smaller reserves are more effective at achieving export of fishable stocks, and networks are seen as a way of increasing overall coverage while simultaneously serving to enhance fisheries (NRC, 2000; Roberts *et al.*, 2003b). MPAs are unlikely to work in isolation, and should be connected in networks (Allison *et al.*, 1998; Lubchenco *et al.*, 2003), with careful attention to issues of the span of the network, the size and shape of individual areas, the number of units, and the placement of the units.

Likewise, many argue that MPAs, including no-take reserves and/or community-based MPAs, are fundamentally inadequate without complementary conservation measures that protect species and ecosystems outside of reserve boundaries (Allison *et al.*, 1998; Christie *et al.*, 2002, 2003a; Lubchenco *et al.*, 2003). There are threats, for example, beyond the scope of any MPA, such as widely dispersed pollutants, diseases, or the spread of exotic species (Allison *et al.*, 1998). Nor can they control increasing fishing effort outside of reserve boundaries (Christie *et al.*, 2002, 2003b). Networks of protected areas will likely be most effective when considered as one important tool embedded in an integrated coastal-zone management strategy (Roberts *et al.*, 2003b). Indeed, there is an entire literature that centers on the need for integration across sectors, government departments and levels, and across stakeholders.[13] Spatial management, including the use of MPAs,

is seen as one means of achieving this integration (Ward and Hegerl, 2003).

However, we argue that there is a need to adapt the approach to the context, to tailor the type of MPA to the context, and to expand the tool-kit to include multiple use areas and conservation outside of protected area boundaries. Practitioners commonly propose a no-take rate of 20 to 30%. This has profound implications for other goals. Moreover, there are areas where government control is ineffective, and large, government-ordered no-take MPAs are unrealistic (White *et al.*, 1994; Christie *et al.*, 2002). A greater range of protected area types, including the increasingly popular zoning approach, will be appropriate to accommodate multiple management objectives.

As in terrestrial areas, involving a greater range of stakeholders will not necessarily be an easy task. Even identifying who should be at the table is difficult, as issues of access and ownership are often more complicated in marine systems. This raises complex questions about "who should decide" and "who should pay" (Carr *et al.*, 2003). Furthermore, as Christie *et al.* (2003a, p. 22) argue, "... biological and social goals may be contradictory or unequally appealing to different constituency groups, resulting in controversy and conflict. These dynamics contribute to the high rate of MPA failure ...."

It is also important to recognize that, while linked, resource use and participation in resource management are two separate issues. It may indeed be appropriate for certain areas to be no-take areas, but the selection, design and management of these sites needs to be done in a transparent, participatory, and socially just manner – not by scientists in isolation following strictly biological criteria (Agardy *et al.*, 2003). Pollnac *et al.* (2001) found that high levels of participation were one of the key ingredients leading to MPA effectiveness. Indeed, there seems to be general acceptance that stakeholder participation is critical to the success of any project (Pollnac *et al.*, 2001; Roberts *et al.*, 2003a). Several recent studies stress the need for sharing authority and fostering participatory decision-making (Mascia, 2003). This sentiment was reaffirmed at the Durban WPC, which also emphasized that there should be an increased role for local groups. This is no less true in marine and coastal areas than it is for terrestrial areas.

COMMUNITY-BASED NATURAL RESOURCE MANAGEMENT, COMMON PROPERTY RESOURCES, AND BIODIVERSITY CONSERVATION

The limited geographic extent of protected areas has led to increasing concern with the conservation of biodiversity in rural, working landscapes. While recent efforts to create and reinforce protected areas have targeted areas of high biodiversity – biodiversity "hotspots" – only 12% of the Earth's land surface is contained within protected areas (Chape *et al.*, 2005). Consequently, there has been an increased awareness that a great deal of biodiversity can and must be maintained through appropriate forms of natural resource management outside of protected areas (e.g. Vandermeer and Perfecto, 1995). Community-Based Natural Resources Management (CBNRM), largely based in local experience managing Common Pool Resources (CPRs), encompasses a range of approaches that emphasize local rights, responsibilities and benefits as a vehicle for simultaneously enhancing rural livelihoods and protecting biodiversity.

Growing experience with CBNRM and CPR has highlighted strengths and weaknesses of these approaches. Many local communities have long histories of managing natural resources effectively and sustainably, though not without difficulties. Historically, community protection of biodiversity was more likely a latent function of management efforts around selected species of important economic value. Yet traditional institutions in local communities may provide a foundation for protecting biodiversity conservation today. Researchers and practitioners have started to investigate the role these institutions might play in managing biodiversity conservation, especially as efforts move beyond the boundaries of established protected areas. Such an approach focuses largely on biodiversity conservation in working landscapes rather than wilderness areas, although some communities maintain special CCAs (Community Conserved Areas) noted above.

Community-based natural resource management (CBNRM)

Experience with conservation over the past several decades has underscored the linkages between rural poverty, environmental degradation and governance. It has reinforced the importance of reconciling the needs and aspirations of local people with the objectives of biodiversity conservation (USAID, 2002b; Brosius *et al.*, 2005; WRI, 2005). Community-based approaches encourage the participation of local resource users as a means of harnessing their knowledge, capacities, and incentives and of integrating sustainable use and nature protection. Since the mid-1980s, conservation and development organizations have experimented with a wide range of strategies, reflecting the specific social and environmental circumstances and the priorities of the actors involved. These include biosphere reserves,

multiple-use conservation areas, buffer zones, extractive reserves, social forestry and other approaches (Wells, 1994; Barrow and Murphree, 2001).

Community-based natural resource management (CBNRM) involves the transfer of some managerial rights and responsibilities to local communities of resource users from the central government. It creates economic incentives for conservation through such activities as sustainable agriculture, agroforestry, social forestry, non-timber forest products, community-social services, ecotourism, and biological prospecting (Wells, 1994; Emerton, 2001). The rationale is based on the belief that local populations are both better situated and more highly motivated than outside agencies to manage resources in an ecologically and economically sustainable manner. Community-based approaches have been used around the world to promote sustainable management of a variety of terrestrial and aquatic resources. A few notable examples include numerous iterations of dry-forest management in the Sahel (Thomson and Coulibaly, 1995; Winter 1997, 2000; Heermans and Otto, 1999; Kerkhof, 2000; USAID, 2002); pioneering wildlife management initiatives in southern Africa, such as Zimbabwe's CAMPFIRE and Namibia's LIFE program (Child *et al.*, 1997; Hasler, 1999; Jones, 1999; Fortmann *et al.*, 2001; Jones, and Murphree, 2001; Long, 2004; Murphree, 2005); and forest co-management in South Asia (Poffenberger and McGean, 1996; Agrawal, 2000; Khare *et al.*, 2000; Agrawal and Ostrom, 2001).

Many CBNRM programs are based on some degree of power-sharing, or co-management, between communities and governments (Borrini-Feyerabend *et al.*, 2004a). These initiatives frequently introduce new management structures, but a common theme in the literature is building on local skills and traditional institutional capital (i.e. experience in self-organization). However, the effectiveness of these structures depends on a number of factors, including local capacity, policy environment, and social process.

First, the ability to achieve conservation goals depends on a community's capacity to manage resources sustainably and to function effectively as a group. Communities must be able to mitigate a wide range of possible threats to the natural environment and by extension to their livelihoods (Salafsky and Margoulis, 1999; Salafsky *et al.*, 2002). Threats may be internal or external to the communities. They may be anthropogenic or environmental in nature. Their effects are further complicated through interaction via multiple pathways. Some analysts have questioned whether community-based institutions have the capacity to manage resources and conserve biodiversity in a rapidly changing social, economic, and political setting (Enters and Anderson, 1999; Ruttan and Borgerhoff Mulder, 1999; Kellert *et al.*, 2000). But this is the case for any organization or set of institutional arrangements. Here, communities might need external assistance to help them cope with rapid changes.

Local capacity and motivation (motivation is a critical point) to confront threats to the environment must not be assumed, particularly where this involves biological diversity (Sognorwa, 1999). In their wide-ranging review of conservation among small-scale societies, Smith and Wishnie conclude that, "voluntary conservation is rare" (2000, p. 493). Most examples of conscious conservation behavior by local communities is "habitat or resource specific" and therefore "not necessarily an indicator of the overall human impact on ecosystem processes and biodiversity" (Smith and Wishnie, 2000, p. 515).[14] Thus, while community-based conservation may be appropriate for sustainable species-based management, it may be less effective at protecting ecosystems and the biodiversity they contain. Much debate over the conservation behavior of local communities has focused on how much of their ability to survive in a natural environment without significantly degrading is due to circumstance (such as low population densities, low market penetration, and simple technology) and how much is due to conscious design. The distinction between deliberate and unintentional sustainability of resource exploitation is not merely academic; it permits assessment of local practices and institutions to "support environmental conservation when the demographic parameters and economic incentives change" (Smith and Wishnie, 2000).

Second, local rights over natural resources and local management systems must be understood as they relate to the broader institutional environment, including the laws and policies of the state. These rights are defined by policies that define administrative structures and processes and by specific technical policies governing the natural resource sector. Secure rights can facilitate sustainable resource management; insecure local rights (e.g. where local rights are ambiguous or are not recognized by the state) frequently create disincentives for participation. Research on local management regimes has repeatedly stressed the importance of an enabling institutional environment (Ostrom, 1990, 1999a, 1999c; Gibson *et al.*, 2000a). Conversely, there are numerous examples where national policy has effectively undermined local systems of natural resource management (Lindsay, 1999; Arnold, 2001).

Finally, local social and political processes, rooted in social differentiation within communities, have a profound impact on the distribution of costs and benefits of participation in natural

resource projects. In particular, the notion of "community" has come under increasing scrutiny, as empirical studies have demonstrated that community-level dynamics impact the depth of local participation and the distribution of conservation benefits. Differentiation based on social identity (gender, age, ethnicity) influences access to resources and decision-making structures as well as patterns of natural resource use and dependence (Li, 1996; Belsky, 1999; Leach *et al.*, 1999; Agrawal and Gibson, 2001a). Without appropriate balance of powers, adequate checks and balances, or effective accountability mechanisms, "elite capture" can undermine popular participation by inhibiting fair and equitable outcomes of local governance (Ribot, 1996, 1999, 2002, 2004; Béridogo, 1997; Dème, 1998; Litvack *et al.*, 1998; Enters and Anderson, 1999; Agrawal and Gibson, 2001b). CBNRM initiatives have met with mixed results because they fail to attend systematically to social dimensions of natural resource management (Kerkhof, 2000; Wilshusen *et al.*, 2002, 2003; Brechin *et al.*, 2003c).

Common-pool resources

CBNRM is rooted conceptually in common-pool resource (CPR) theory (Berkes, 1989; Ostrom, 1990; Berkes and Folke, 1998; Dietz *et al.*, 2003; Fabricius, 2004). Historically, biodiversity conservation has likely resulted as an unanticipated outcome of more deliberate action related to sustainable local livelihood generation. As biodiversity conservation has become an issue of global concern, it has become clear that many areas of high biodiversity also fall under common-property regimes. This overlap raises the question of whether the capacity and motivation of local people seeking sustainable livelihoods can be harnessed to conserve biological diversity. It is not entirely clear that this capacity and motivation can be extrapolated from CPRs forming the basis of rural production systems to biodiversity per se, as valued by the conservation community. Yet understanding the strengths and weaknesses related to CPR management practice might lead us to insights for conserving biodiversity at the local level as well.

CPRs share attributes that complicate their governance and management (Ostrom, 1999c). All resources can be distinguished along two axes: the nature of exclusion and the nature of consumption (Thomson, 1992; Ostrom *et al.*, 1994; Ostrom, 2002). CPRs, which include forests, rangelands, and fisheries, are those for which exclusion is difficult or costly and for which consumption is "rivalrous" (Thomson, 1992; Ostrom *et al.*, 1994). That which is consumed by one person is no longer available to another person.

They are therefore "very difficult to protect and very easy to deplete" (McKean, 2000, p. 29). Renewable CPRs consist of stocks and flows of resource units (Ostrom, 1990). The stock produces a benefit stream, which can be appropriated by users. If the rate of appropriation exceeds the flow, then resource users consume the capital and can jeopardize the future productive capacity of the resource. For use to be sustained, off-take rates must be maintained below the production rate in order to preserve the stock; this would be the same in protecting biodiversity.

Common-pool resources, by nature, pose a special set of management problems known as collective action dilemmas. In the interest of maintaining sustainable levels of exploitation, resource users are expected to engage in individually costly behavior, by forgoing immediate gains, in order to produce a collective good. There is therefore a temptation for users to free ride, letting others "bear the costs of providing joint benefits ... resulting in underprovision or degradation of the common resource" (Poteete and Ostrom, in press). The ability of a community of resource users to overcome collective action problems depends on a number of factors, including the adequacy of information, the distribution of interests and the nature of the resource itself. Additionally, "actors must overcome coordination problems, distributional struggles and incentive problems" (Poteete and Ostrom, in press).

Recent research into the management of common-pool resources has highlighted the importance of institutions in facilitating collective action by resource user groups (Ostrom, 1990, 1999b; Thomson, 1992; NEF, 1996; Thomson and Schoonmaker Freudenberger, 1997; Mehta *et al.*, 1999; Varughese, 2000; Gibson *et al.*, 2000a, b; Agrawal, 2001a; Young 2002; Poteete and Ostrom, in press). Institutions have shared formal and informal rules, norms and strategies that shape human incentives and behavior in recurring social situations, such as those encountered as people share and manage resources as members of communities. In natural resource settings, institutions structure how resources may be used, managed and owned and how conflicts are resolved (Agrawal, 2001b; Agrawal and Ostrom, 2001). They influence the distribution of costs and benefits associated with particular decisions and are therefore important tools for shaping the incentives of individuals and for overcoming collective action dilemmas.

CPR theory emphasizes that institutional effectiveness depends on getting the right "fit" between rule systems and their social and environmental contexts – the characteristics of communities of resource users and the characteristics of the natural resources as economic goods (Ostrom,

1990; Thomson, 1992; Thomson and Schoonmaker Freudenberger, 1997; Young, 2002). It also stresses the need to approach institutional arrangements as "complex adaptive systems" which are continuously adjusted by resource managers in response to contextual changes (Ostrom, 1999a). The implications for natural resource policy and management are tremendous – successful CBNRM can be fostered by appropriate institutional arrangements under different conditions. Institutional flexibility can permit decision-makers to adjust the conditions of resource use as necessary.

This premise has led to a large number of studies seeking to identify the appropriate conditions for collective action and conversely the effects of different parameters on collective action. These are most clearly expressed as "design principles" (Ostrom, 1990) or "critical enabling conditions" (Agrawal, 2001a). At the broadest level of generalization, empirical evidence supports the contention that successful self-governance of forest resources requires a group of appropriate size, clear and secure tenure, and local autonomy to develop and implement rules (Gibson *et al.*, 2000a). Based on a growing body of empirical evidence, Ostrom suggests that collective action in forest management is likely to take place: (1) when resources are not too far deteriorated to manage sustainably, (2) when reliable and valid information can be obtained at a reasonable cost, (3) when the availability of resource units is predicable, and (4) when the spatial extent is small enough that users can gain accurate information about the boundaries and internal resource conditions given their own resource limitations (Ostrom, 1999c). Furthermore, self-organization appears more likely where the community of resource users exhibit several key attributes: (1) the resource has salience to local livelihoods, (2) the users share an understanding of the resource, (3) the discount rate is low, (4) the distribution of interests is relatively uniform, (5) the users can self-organize with a degree of autonomy, and (6) the user group has prior experience with self-organization (Ostrom, 1999c).

However, other theorists argue that a "design principles" approach is of limited utility because it cannot be applied universally, because institutions vary widely by functional scope, spatial domain, degree of formalization, stage of development, and interactions with other institutions (Young, 2002). Young (2002) maintains that the ability both to understand existing institutional arrangement and to design new institutional arrangements must account for whether the institutional arrangements are configured in an optimal way to achieve results (fit), whether the institutional arrangements are on an appropriate scale to address problems (scale) and how institutions interact with institutions on the same and difference scales (interplay).

Institutions are embedded in specific social, political, and economic settings. Agrawal and Gibson (2001b) propose that policy-relevant studies of CBNRM must give due attention to the micro-political interactions among local actors with multiple, often divergent, interests and to the institutions that structure these interactions. A focus on institutions overcomes some of the conceptual problems of community structure and dynamics by focusing on the "ability of communities to create and to enforce rules" (Agrawal and Gibson, 1999, p. 628). Institutions are valuable because (1) they "define some of the power relations that define the interactions among actors who created" them, and (2) "help to structure the interactions that take place around resources" (Agrawal and Gibson, 1999, p. 637). Thus, while outcomes of interactions are largely the product of social and political processes based on unequal access to power and resources, institutions provide a basic framework for understanding these interactions. Moreover, institutions remain the primary mechanism for shaping the actions and outcomes of human interactions, particularly in CBNRM, which is largely a process of local institutional development (Agrawal and Gibson, 1999). This applies equally to protection of biodiversity (Barrett et al., 2001).

Decentralization

Successful collective action is most likely where national policy encourages participation and autonomy and facilitates self-organization (Ostrom, 1990, 1999a, c; Gibson et al., 2000a). The trend toward increased participation in conservation and natural resource management has been paralleled by a striking global movement toward political liberalization and democracy throughout the developing world over the past quarter century (Huntington, 1991; Carothers, 2002; Gibson, 2002; Plattner, 2002). Since the early 1990s, this movement has been accompanied by decentralization reforms in a large number of developing countries as a solution to failed centralized regimes (Manor, 1999; UNCDF, 2000).

Decentralization is a process by which powers are transferred from the central government to lower levels in a "territorial hierarchy" (Crook and Manor, 1998, p. 6). The policy literature generally distinguishes between two broad forms of decentralization: *deconcentration*, whereby powers are shifted to lower levels within an administrative hierarchy, and *devolution* (or democratic decentralization), whereby powers are formally ceded by the central government to democratic local government units (Crook and Manor, 1998;

Manor, 1999; Ribot, 1999, 2004). In practice, decentralization encompasses diverse forms of government, distinguished by the specific distribution of political, administrative and fiscal powers among different levels and units of government (Crook and Manor, 1998, 2000; Manor, 1999). Subsidiarity is a central tenet of decentralization. It maintains that problems should not be addressed at any higher level than absolutely necessary, in order to avoid the increased transaction costs of coordination and communication associated with higher levels of organization. By this logic, the state should not intervene unless lower levels of social organization are unable to resolve a problem (Thomson, 2000).

Decentralization has created opportunities for resolving some vulnerabilities of CBNRM by increasing and institutionalizing local control over natural resources, empowering a broader range of actors and creating an enabling institutional framework for self-governance (Thomson, 1994; Brinkerhoff, 1995; Pierce Colfer and Capistrano, 2005). However, the specific configuration of decentralized powers appears to be more important than the mere fact of having decentralized in determining outcomes (Crook and Manor, 1998; Litvack et al., 1998; Azfar et al., 1999; Burki et al., 1999; Ribot, 2002). Ribot (2002) argues that two features in particular are important for positive outcomes: downward accountability of local government to their constituencies to facilitate greater efficiency and equity of public service, and the secure transfer of discretionary powers to local leaders so that they can make meaningful decisions for their constituencies.

In order to understand the impact of decentralization on local-level natural resources management and its potential in protecting biodiversity, it is also necessary to examine not only the distribution of power between central and local governments, but also the relationship between decentralized local government, in which powers are generally vested, and village government, where local knowledge, capacity and institutions are situated. This distinction is implicit in Larson's study of natural resource management by local government in Nicaragua (Larson, 2002). She points out that the "decentralization literature is primarily concerned with the pivotal role of municipal authorities," although it rarely considers the natural resources sector, and that "the NRM literature often gives little attention to local government, placing much greater emphasis on community-based conceptions of decentralization" (Larson, 2002, p. 17). It should not be assumed that the qualities of communities that favor them in participatory approaches to conservation are reproduced in "local" government. The role of local government units in natural resources and the impact of decentralization on community-level natural resource management depend on what powers are transferred, to whom they are transferred and by what means they are transferred (Agrawal and Ribot, 1999; Agrawal and Ostrom, 2001; Ribot, 1999, 2002). The role of local communities themselves depends on the specific powers that are transferred directly to them and the different channels they have to influence decisions of local government.

While local-level institutional arrangements may be able to respond adequately to an array of collective action problems at multiple organizational scales, the rigid structure and procedures of local government may compromise its capacity to manage effectively under complex, site-specific conditions. Thomson (2000, p. 3) explains that in francophone West African countries, for example, "policies of uniformity limit the extent of devolution to district-level, general-purpose governments," thus denying "authority to their populations to organize at the level at which specific problems occur." In this tradition, Sahelian decentralization imposes "inflexible governance frameworks," allowing only general-purpose government, where the smallest unit – the commune – is larger than that needed to address the majority of natural resource and collective action problems, contravening the principle of subsidiarity (2000, p. 3). Additionally, decentralization legislation in this region restricts the issues that general local government units can address, the authorities that they can exercise and the resources they can mobilize. In short, he concludes that the limitations placed on self-governance perpetuate central control, virtually assuring the failure of local natural resources management.

Lessons

The effectiveness of community-based approaches to natural resource management depends on the incentives, rights, and capacities of local resource managers. For resource-dependent populations, incentives for sustainable management are generally rooted in their livelihood systems (i.e. people tend to protect what is important to their survival). Incentives are shaped at least partially by institutional arrangements. Experience has shown that incentives can be created to promote biodiversity protection by communities outside of protected areas (e.g. Namibia's LIFE CBNRM Project, Zimbabwe's CAMPFIRE Project). Experience with CBNRM and decentralized natural resource management offers several very broad lessons for achieving local control and sustainable management of natural resources and possibly for protecting biodiversity.

First, legislation must permit self-organization on a scale that is appropriate to the social and ecological scale of natural resource use. The principle of subsidiarity posits that problems should be managed at the lowest possible level in order to minimize transaction costs. There are frequently economies of scale and scope to be gained by supra-village organizations such as rural communes or municipalities, but the contours of ecological and livelihood problems rarely follow administrative boundaries. Higher levels of "local" government provide structures through which transboundary resource problems can be coordinated, but they also involve higher transaction costs. In the aggregate, local institutions governing resource use are frequently polycentric – they encompass multiple governing authorities at different levels, some of which are specialized and others general purpose (Ostrom, 1999a). Finding the appropriate scale for natural resource institutions is a question of matching social and administrative boundaries – including scopes of power – with those of ecological and livelihood systems.

Second, specific social, environmental, and economic conditions under which natural resources are managed are highly site specific. Institutional theorists have forcefully argued the importance of crafting institutions that "fit" the ecological and social circumstances of natural resource problems (Ostrom, 1990; Young, 2002). Institutions that fit highly variable contexts must themselves be dynamic and flexible. Communities of resource users must have the power to alter resource use in response to fluctuating conditions. Common changes in the biophysical environment include decreased rainfall resulting in resource scarcity, patchiness of resource productivity, subtle changes in water level, or location of migratory (fugitive) resources. Changes in the social and economic conditions include increasing commercial pressure, migration, and restrictions placed on alternative sources. These changes can be quite localized, requiring site-specific adjustments in the conditions of resource use. Institutional flexibility requires that resource users have access to collective choice mechanisms (i.e. the right to make and change rules) based on culturally relevant processes with low transaction costs.

Third, crafting natural resource institutions is not a one-shot activity. It requires ongoing community participation both to assure that management principles are relevant to human behavior and resource ecology and to enhance their legitimacy among resource users. There are no clear procedures for untangling the complexity of social relations and competing demands in local natural resource systems. Effective solutions must be worked out through constructive dialogue and compromise. Involving a range of stakeholders in negotiation and consensus building promotes greater buy-in and enhances the legitimacy of rules and management systems. Through this process, diverse actors can build credible mutual commitments that are more likely to be enforced and upheld when contested (Ostrom, 1990; Putnam, 1993; Ostrom, et al., 1994; Pretty, 2003; Benjamin, 2004).

Fourth, participation and inclusion can foster legitimate and enforceable commitments among local stakeholders and between local stakeholders and government, but conflict is endemic even when rule systems are clear and well adapted to local situations. Turner (1999) stresses that conflict must not be understood to be implicitly degenerative but rather as a driver of change. Negotiation of access and management systems is an ongoing process, part of the dynamic quality of local natural resource systems. Mechanisms for mediating disputes in a positive manner must engage local capacities and foster local participation in order to prevent degenerative conflict and to contribute to the perpetuation of legitimate, credible agreements among all stakeholders.

Finally, changing rules is necessary but insufficient to bring about social change, even under a broad policy of decentralization. Policy and institutional reform must be accompanied by parallel efforts to harness local capacity, knowledge and incentives while increasing the accountability and responsiveness of local government. Where local officials are granted discretionary powers, communities must build the capacity to engage them and advocate for their interests. A strong civil society (community-based organizations, NGOs and NGO coalitions) is important to this process, by acting as a counterbalance to the power of elected officials and by representing diverse interests that may not be effectively expressed through electoral processes.

THE POLITICAL ECOLOGY OF ECOTOURISM

Ecotourism has been a cornerstone of efforts both to improve people–parks relations and to promote CBNRM. It is one of the fastest-growing tourism sectors. It also has been widely promoted as a way of simultaneously promoting nature conservation, local development, and fostering a sense of stewardship among local residents in and around protected areas. The conservation benefits of eco-tourism, however, remain unclear (Boo, 1990; Brandon and Margoulis, 1996; Honey, 1999). Moreover, while there clearly are potential economic benefits associated with tourism (including ecotourism), many of the proponents of

ecotourism overlook forces of political economy, taking a somewhat naïve view of the promise for equitable distribution of economic benefits, and the envisioned support for conservation that would follow. In fact, an emerging literature suggests that the equitable delivery of benefits to local people can be blocked due to forces that operate at both the macro (regional to international) and micro (community) scales.

At the macro-scale, these forces can be broadly lumped under the rubric of political economy, where political economy "... refers to the struggle over power and influence to monopolize benefits from eco-tourism" (West et al., 2003, p. 105). Monopolization in this usage is a structural process created by power and influence relations within the tourism sector, and that is facilitated by competitive advantages derived from such things as economies of scale, financial stability and access to capital (or credit), access to tourism markets and commonality of interests among powerful actors (West et al., 2003). The literature on the political economy of eco-tourism is only just emerging, though there is a wider literature on the political economy of tourism (see Britton, 1982; Machlis and Burch, 1983, for some earlier works). The particular mechanisms by which benefits are concentrated vary, but case studies from the Galapagos (Bailey, 1991), South Korea (Woo, 1991), Benin (West et al., 2003), and Mexico (Murray, 2003) suggest that the equitable distribution of ecotourism benefits to local communities is not guaranteed (see also West and Brechin, 1991b; Campbell, 1999; Honey, 1999; Young, 1999).

Exception theories

West et al. (2003) describe several exception theories to this tendency towards the concentration of benefits in the hands of the few. In theory, these exceptions represent mechanisms whereby local communities can capture some of the economic benefits from ecotourism development. These include the formation of a "cottage industry," involving small enterprises that pre-date the arrival of extra local forces (see Kutay, 1984). It should be noted, however, that several authors (e.g. Machlis and Burch, 1983; Jain, 1999; West et al., 2003) have pointed out that this foothold is tenuous in the face of intruding large-scale capital. Another mechanism offered by West et al., (2003) is comprised of related "niche" theories where small-scale operators can monopolize certain *skills* or *resources* that allow them to compete against larger-scale operators. This mechanism is highlighted in work by Hough (1991) with the Havasupai in the Grand Canyon (a case of resource control), and by Weber (1991) with the

Sherpa in Sagamartha National Park, Nepal (a case of skill monopolization). A corollary (or "symbiotic") niche theory suggests an exception where smaller-scale enterprises coexist as complements to large-scale interests who see it in their own self-interest to allow these enterprises to exist (West and Brechin, 1991b). The authors cite examples of such things as large package tour companies engaging their tourists with small handicraft operators, dance troupes, or local food vendors as part of a package deal offering local flavor, culture, or atmosphere. The third principal exception theory is the use of coalitions of different stakeholders that can be used to "secure favorable policies" (West *et al.*, 2003). This idea is fundamentally premised on the ability of coalitions to level the playing field, and to equalize the balance of power. Acting alone, local communities often have little in the way of real power. Often, these various exception theories can work best synergistically (Jain, 1999).

Micro-level political economy

It is important to recognize that there are more powerful actors *within* a community and less powerful ones, and these actors may be just as likely to use their power, influence, and resources to capture economic benefits as do powerful actors at the macro-level. Indeed, the "community base" for community enterprises is rarely, if ever, all-encompassing. The term "community" itself has been used somewhat generically and simplistically, and does not take things like local histories, heterogeneity, and political stakes into account (Li, 1996; Brosius *et al.*, 1998; Agrawal and Gibson, 1999; Belsky, 1999, 2003).

Those community members with some initial disadvantages, such as those with a lack of capital, insufficient land, or inadequate guiding skills might, for example, be excluded in ecotourism enterprise development. Furthermore, depending on how the enterprise is designed and developed, they may be excluded from sharing in the benefits of ecotourism altogether. Generally speaking, there will be several relevant questions that should be considered, including: the nature and size of the benefits, the particular form (what kind of ecotourism it is?) by which those benefits are captured, the perspective from within the community, the nature of the community and the division/stratification within it, and the fate of the benefits (e.g. do they stay in local systems or "leak" out). In Belizean "bed and breakfast" based ecotourism enterprises, for example, class, gender, and patronage differentiate communities and substantially shape divergent intracommunity practices *vis-á-vis* ecotourism, and tend to concentrate benefits in the hands of a few (Belsky, 1999, 2003).

Likewise, Jain (1999) found that although local communities in Nepal were able to capture economic benefits derived from trekking ecotourism, many, if not most, of the benefiting lodges were organized by ex-Army members that were part of the majority ethnic group, had access to capital and were able to provide the facilities and services that tourists demand, and also had significantly more cross-cultural and market experience.

CONCLUSION

The future of biodiversity conservation and local people, their communities and institutions, whether marine based or terrestrial, are strongly intertwined. If the percentage of land area under protection status is unlikely to change substantially in the coming decades, and 85 percent of these areas *already contain* people and communities (Alcorn, 2000), then the future protection of biodiversity conservation must rest largely on working together with the people and communities inside of those areas. This conclusion only takes on added weight when we consider the future directions of biodiversity conservation efforts. As key conservation organizations adopt landscape or eco-regional approaches that focus beyond existing land and marine protected areas, these efforts likewise by their very nature require conservationists to fully engage local people and communities into the conservation effort as well. So, whether attempting to protect nature inside or outside of protected areas, involving local people and communities in the conservation process remains absolutely essential. While our discussion here does not include reflections on national and transnational policies and other forces that shape local realities, we appreciate their importance. Under such large regimes and limited resources from conservation organizations and agencies, implementing and enforcing conservation objectives will need to be carried out largely by the local communities and organizations themselves. This will not happen unless the local people can identify with and support the larger conservation effort. From our personal experiences, it is our contention that most local people support the basic idea of biodiversity conservation itself. Their biggest complaint comes largely from how it is actually pursued.

Hence, a central question today that will take on even greater urgency in the future is – can the conservation community seamlessly integrate effective efforts to protect nature while allowing for the welfare and interests of an ever growing number of families and communities, especially the poor? While protecting wilderness has been the ideal in protecting biodiversity, these areas with very limited human impact will likely remain in limited numbers; demographic realities state that the world can expect anywhere from two to four billion more humans to sustain over the next thirty years, with most of the growth found in the poorer countries of the tropics.

The debates on whether or not local people and communities add to the conservation process or distract from them must end. Instead, our attention requires us to focus on a much more realistic but difficult challenge: how to integrate successfully the needs of nature with those of local people and engage fully these communities in the conservation mission itself.

NOTES

1 Dan Brockington (2002, 2004) claims that conservation efforts can succeed in spite of social injustices caused by conservation practice itself. He cites his research in the Mkomazi Game Reserve, Tanzania, as an example. We believe this is a much trickier claim than one might think. Immediately one must ask then – what is the definition of success? (credit to Crystal Fortwangler). One could easily argue that conservation at the point of a bayonet is by definition not very successful. However, even if one adopts a very limited notion of success (i.e. very strict boundary control), at what temporal scale does one use? One might keep local people under strict control for some time but then lose it all in an explosion of anger as a case in Togo illustrates (Lowry and Donahue, 1994). While there are likely examples of narrowly defined conservation success on the backs of the local poor, as Brockington has seemly found, is this the rule or the exception? Also, how should one interpret Brockington's point? Should it be viewed as a basis for policy decisions? To be clear, Brockington remains in favor of social justice, he argues it needs to be framed in moral not pragmatic terms. While there is a point here, one must wonder whether it is still more pragmatic as general policy to have local people working with conservationists then against them? Imposing wilderness, as Roderick P. Neumann, (1998), wonderfully shows, employing James Scott, does not come without a moral economy of the villager.

2 See Brechin et al., 2003c (chapter 10) and Wilshusen and Murguia (2003) for more detailed discussion related to social capital formation and scaling up.

3 Contact Dan Brockington at the University of Manchester and Joe Igoe at University of Colorado-Denver. These two researchers are attempting to organize such a study.

4 Chapin's article, "A Challenge to Conservationists," was published by the Worldwatch

Institute in the November/December 2004 issue of *World-Watch*.

5 See Brechin et al. *(2003)* for an introduction on this topic.

6 For Example, see The Nature Conservancy's Conservation by Design and their more recent Conservation Audits (the others are doing something similar); and Nick Salasky and Richard Margoluis' most recent effort to push the success measurement practice with their new Conservation Measurement Partnership program (http://www.conservationmeasures.org/CMP/).

7 To highlight the difficulties in defining success even among conservation biologists we offer the following. IUCN officials at the 2003 World Parks Congress, Durban, South Africa were gleeful in announcing that the world community had surpassed its long term objective of placing 10% of the world land surfaces under protection status. However, in spite of this important accomplishment, the world's problems with biodiversity loss have hardly been solved. The reasons are many. They include fundamental questions such as: How much biodiversity needs to be saved? And how much area needs to be saved to protect that amount of biodiversity? These are extremely difficult scientific questions to answer.

8 See too work completed by the World Wildlife Fund (Oviedo and Maffi, 2000).

9 Recommendation 22, echoing Recommendation 11 from the Caracas conference.

10 We acknowledge that habitat loss or degradation does occur in marine systems (Botsford et al., 1997; Watling and Norse, 1998; Roberts et al., 2003a); we suggest, however that historically this has not been the primary motivation for creating marine protected areas.

11 A February, 2003, Special Issue of *Ecological Applications* entitled "The Science of Marine Reserves" provides an overview of many of these issues.

12 See also an online special issue of *Ecology and Society* **9**(3): 7. URL: http://www.ecologyandsociety.org/vol9/iss3/

13 This literature is too extensive to list here. For a good starting point, see Cicin-Sain and Knecht, (1998). For a specific discussion of community involvement in integrated coastal management, see the 1997 special issue of *Ocean & Coastal Management* (Vol 36, nos 1–3).

14 Smith and Wishnie argue that, "to qualify as conservation, any action or practice must not only prevent or mitigate resource overharvesting or environmental damage, it must be designed to do so. The conditions under which conservation will be adaptive are stringent, involving temporal discounting, economic demand, information feedback, and collective action" (Smith and Wishnie, 2000, p. 493). They discount low population density, low harvest intensity, and exclusion of outsiders as intentional conservation measures.

REFERENCES AND BIBLIOGRAPHY

Agardy, T.S. (1997). *Marine Protected Areas and Ocean Conservation.* Austin, TX: R.G. Landes and Academic Press.

Agardy, T., Bridgewater, P., Crosby, M.P., Day, J., Dayton, P.K., Kenchington, R., Laffoley, D., McConney, P., Murray, P.A., Parks, J.E. and Peau, L. (2003). Dangerous Targets? Unresolved issues and ideological clashes around marine protected areas. *Aquatic Conservation: Marine and Freshwater Ecosystems* 13:1–15.

Agrawal, A. (2000). Small is beautiful, but is larger better? Forest-management institutions in the Kumaon Himalaya, India. In *People and Forests: Communities, Institutions, and Governance,* edited by C.C. Gibson, M.A. McKean and E. Ostrom. Cambridge, MA: MIT Press, pp. 57–85.

—— (2001a). Common property institutions and sustainable governance of resources. *World Development* 29: 1649–1672.

—— (2001b). *The Decentralizing State: Nature and Origin of Changing Environmental Policies in Africa and Latin America, 1980–2000.* Paper prepared for the 97th Annual meeting of the American political science association, San Francisco, August 30 to September 2.

Agrawal, A., and Gibson, C.C. (1999). Enchantment and disenchantment: the role of community in natural resource conservation. *World Development* 27:629–649.

—— (ed). (2001a). *Communities and the Environment: Ethnicity, Gender, and the State in Community-Based Conservation.* New Brunswick, NJ: Rutgers University Press.

—— (2001b). The role of community in natural resource conservation, In *Communities and the Environment.* Edited by A. Agrawal and C.C. Gibson. New Brunswick, NJ: Rutgers University Press.

Agrawal, A., and Ostrom, E. (2001). Collective action, property rights, and decentralization in resource use in India and Nepal. *Politics and Society* 29:485–514.

Agrawal, A. and Ribot, J. C. (1999). Accountability in decentralization: a framework with South Asian and West African cases. *Journal of Developing Areas* 33:473–502.

Alcorn, J. (2000). Preface. In *Indigenous Peoples and Conservation Organizations: Experiences in Collaboration,* edited by R. Weber, J. Butler and P. Larson. Washington, DC: World Wildlife Fund.

Allison, G.W., Lubchenco, J. and Carr, M.H. (1998). Marine reserves are necessary but not sufficient for marine conservation. *Ecological Applications* 8(1): S79–S92.

Arnold, J.E.M. (2001). *Forests and People: 25 Years of Community Forestry.* Rome (Italy): FAO.

Azfar, O., Kähkönen, S., Lanyi, A., Meagher, P. and Rutherford, D. (1999). *Decentralization, Governance and Public Services: The Impact of Institutional Arrangements: A Review of the Literature.* College Park, MD: IRIS Center, University of Maryland.

Bailey, C. (1991). Conservation and Development in the Galapagos Islands. In *Resident Peoples and National Parks: Social Dilemmas and Strategies in International Conservation,* edited by P.C. West and S.R. Brechin. Tucson, AZ: University of Arizona Press.

Barrett, C.B., Brandon, K., Gibson, C. and Gjertsen, H. (2001). Conserving biodiversity amid weak institutions. *Bioscience* 51(6):497–502.

Barrow, E. and Murphree, M. (2001). Community conservation: from concept to practice, edited by D. Hulme and M. Murphree *African Wildlife and Livelihoods.* Oxford: James Currey, pp. 24–37.

Belsky, J.M. (1999). Misrepresenting communities: the politics of community-based rural ecotourism in Gales Point, Belize. *Rural Sociology* 64(4):641–666.

Belsky, J.M. (2003). Unmasking the 'local': gender, community, and the politics of community-based rural ecotourism in Belize. In *Contested Nature: Promoting International Biodiversity Conservation with Social Justice in the 21st Century,* edited by S.R. Brechin, P.R. Wilshusen, C.L. Fortwangler and P.C. West. Albany: State University of New York Press.

Benjamin, C.E. (2004). *Livelihoods and Institutional Development in the Malian Sahel: A Political Economy of Decentralized Natural Resource Management.* Doctoral Dissertation, School of Natural Resources & Environment. University of Michigan, Ann Arbor.

Berdidogo, B. (1997). Processus de décentralisation au Mali et couches sociales marginalisÈes. *Bulletin de l'APAD* 14:21–33.

Berkes, F., ed. (1989). *Common Property Resources: Ecology and Community-Based Sustainable Development.* London: Belhaven Press.

Berkes, F. (1999). *Sacred Ecology: Traditional Ecological Knowledge and Resource Management.* Philadelphia: Taylor & Francis.

Berkes, F. (1993). Traditional ecological knowledge in perspective. In *Traditional Ecological Knowledge: Concepts and Cases,* edited by T. Julian, J. Inglis. Ottawa: Canadian Museum of Nature, International Program on Traditional Ecological Knowledge.

Berkes, F. and Folke, C. (1998). *Linking Social and Ecological Systems: Management Practices and Social Mechanisms for Building Resilience.* Cambridge: Cambridge University Press.

Boo, E. (1990). *Ecotourism: The Potentials and Pitfalls,* Volumes 1 and 2. Washington, DC: World Wildlife Fund.

Borrini-Feyerabend, G. (1996). *Collaborative Management of Protected Areas: Tailoring the Approach to the Context.* Gland, Switzerland: IUCN.

Borrini-Feyerabend, G., Pimbert, M., Farvar, M.T., Kothari, A. and Renard, Y. (2004a). *Sharing Power: Learning by Doing in Co-Management of Natural Resources throughout the World.* London: IIED, Gland: IUCN/CEESP/CMWG; Tehran: Cenesta.

Borrini-Feyerabend, G., Kothari, A. and Oviedo, G. (2004b). *Indigenous and Local Communities and Protected Areas: Towards Equity and Enhanced Conservation.* Gland, Switzerland and Cambridge, UK: IUCN.

Botsford, L.W., Castilla, J.C. and Peterson, C.H. (1997). The management of fisheries and marine ecosystems. *Science* 277: 509–515.

Botsford, L.W., Micheli, F. and Hastings, A. (2003). Principles for the design of marine reserves. *Ecological Applications* 13(1) (Suppl.):S25–S31.

Brandon, K. and Margoulis, R. (1996). The bottom line: getting biodiversity conservation back into ecotourism. In *The Ecotourism Equation: Measuring the Impacts*. Yale Bulletin Series, no.99, edited by J.A. Miller and E. Maek-Zadeh. New Haven: Yale University Press, pp. 28–39.

Brandon, K., Redford, K.H. and Sanderson, S.E. (1998). *Parks in Peril: People, Politics, and Protected Areas*. Washington, DC: Island Press.

Bray, D.B. and Anderson, A.B. (2005). *Global Conservation Non-Government Organizations & Local Communities: Perspectives on Programs and Project Implementation in Latin America*. Working Paper 1. Conservation and Development Series. Institute for Sustainability Science in Latin America and the Caribbean. Latin American and Caribbean Center, Florida International University.

Bray, D.B., Plaza-Sanchez, J.L. and Murphy, E.C. (2002). Social dimensions of organic coffee production in Mexico: Lessons for ecolabeling initiatives. *Society and Natural Resources* 15:429–446.

Bray, D.B., Merino-Perez, L., Negreros-Castillo, P., Segura-Warnholz, G., Torres-Rojo, J.M. and Vester, H.F.M. (2003). Mexico's community-managed forests as a global model for sustainable landscapes. *Conservation Biology* 17 (3):672–677.

Brechin, S.R., West, P.C., Harmon, D. and Kutay, K. (1991). Resident peoples and protected areas: A framework for inquiry. In Resident Peoples and National Parks: Social Dilemmas and Statergies for International Conservation, P.C. West and S.R. Brechin (eds). Tucson, AZ: The University of Arizona Press, pp. 5–28.

Brechin, S.R., Murray, G. and Mogelgaard, K. Conceptual and practical issues in defining conservation success: the political, social, and ecological in an organized world. *Journal of Sustainable Forestry*.

Brechin, S.R., Wilshusen, P.R., Fortwangler, C.L. and West, P.C. (2002). Beyond the square wheel: toward a more comprehensive understanding as biodiversity conservation as social and political process. *Society and Natural Resources* 15:41–64.

Brechin, S.R., Wilshusen, P.R., Fortwangler, C.L. and West, P.C. (eds). (2003a). *Contested Nature: Promoting International Biodiversity Conservation with Social Justice in the Twenty-first Century*. Albany, NY: SUNY Press.

—— (2003b). The road less traveled: toward nature protection with social justice. In *Contested Nature: Promoting International Biodiversity Conservation with Social Justice in the Twenty-first Century*, edited by. S.R. Brechin, P.R. Wilshusen, C.L. Fortwangler, P.C. West, 251–270. Albany, NY: SUNY Press.

Brechin, S.R., Wilshusen, P.R. and Benjamin, C.E. (2003). Crafting Conservation Globally and Locally. In *Contested Nature: Promoting International Biodiversity Conservation with Social Justice in the Twenty-first Century*, S.R. Brechin, P.R.Wilshusen, C.L. Fortwangler and P.C.West (eds.), 159-182. Albany, NY: SUNY Press.

Brinkate, T. (1996). People and parks: Implications for sustainable development in the Thukela Biosphere Reserve. Paper presented at the Sixth International Symposium on Society and Natural Resource Management. Pennsylvania State University, State College.

Brinkerhoff, D. (1995). African state–society linkages in transition: the case of forestry policy in Mali. *Canadian Journal of Development Studies* 16:201–228.

Britton, S.G. (1982). The political economy of tourism in the Third World. *Annals of Tourism Research* 9:331–358.

Brockington, D. (2002). *Fortress Conservation: The Preservation of the Mkomazi Game Reserve, Tanzania*. Bloomington: Indiana University Press.

—— (2004). Community conservation, inequity and injustice: myth of power in protected area management. *Conservation & Society* 2:411–432.

Brosius, J.P. (2004). Indigenous peoples and protected areas at the world parks congress. *Conservation Biology* 18(3): 609–612.

Brosius, J.P., Tsing, A.L. and Zerner, C. (1998). Representing communities: histories and politics of community-based natural resource management. *Society & Natural Resources* 11:157–168.

—— eds. (2005). *Communities and Conservation: Histories and Politics of Community-based Natural Resource Management*. Walnut Creek: Alta Mira.

Bruner, Aaron G. *et al.* (2001). Effectiveness of parks in protecting tropical biodiversity. *Science* 291(5):125–128.

Burki, S.J., Perry, G.E. and Dillinger, W.R. (1999). *Beyond the Center: Decentralizing the State*. Washington, DC: The World Bank.

Campbell, J. (1999). Ecotourism in developing communities. *Annals of Tourism Research* 26(3):534–553.

Carothers, T. (2002). The end of the transition paradigm. *Journal of Democracy* 13:5–21.

Carr, M.H., Neigel, J.E., Estes, J.A., Andelman, S., Warner, R.R. and Largier, J.L. (2003). Comparing marine and terrestrial ecosystems: implications for the design of coastal marine reserves. *Ecological Applications* 13(1) (Suppl).:S90–S107.

Chape, S., Harrison, J., Spalding, M. and Lysenko, I. (2005). Measuring the extent and effectiveness of protected areas as an indicator for meeting global biodiversity targets. *Philosophical Transaction of the Royal Society B* 360: 443–455.

Chapin, M. (2004). "A challenge to conservationists". World Watch: *Vision for a Sustainable World*. Washington, DC: Worldwatch Institute. November/ December.

Child, B., Ward, S. and Tavengwa T. (1997). *Zimbabwe's CAMP-FIRE Programme: Natural Resources Management by the People*. Harare (Zimbabwe): IUCN-ROSA.

Christie, P., White, A.T. and Buhat, D. (1994) Community-based coral reef management on San Salvador Island, Philippines. *Society and Natural Resources* 7:103–117.

Christie, P., White, A. and Deguit, E. (2002). Starting point or solution? Community-based marine protected areas in the Philippines. *Journal of Environmental Management* 66:441–454.

Christie, P., McCay, B., Miller, M.L., Lowe, C., White, A.T., Stoffle, R., Fluharty, D.L., McManus, L.T., Chuenpagdee, R., Pomeroy, C., Suman, D.O., Blount, B.G., Huppert, D., Villahermosa Eisma, R., Oracion, E., Lowry, K. and Pollnac, R. (2003a). Toward developing a complete understanding: a social science research agenda for marine protected areas. *Fisheries* 28(12):22–26.

Christie, P., Buhat, D., Garces, L.R. and White, A.T. (2003b). The challenges and rewards of community-based coastal resources management: San Salvador Island, Philippines, edited by S. Brechin, P. R. Wilshusen, C.L. Fortwangler, and P. West. In *Contested Nature: Promoting International Biodiversity with Social Justice in the Twenty-first Century.* Albany: State University of New York Press.

Cicin-Sain, B. and Knecht, R. (1998). *Integrated Coastal and Ocean Management: Concepts and Practices.* Island Press: Washington, DC.

Colchester, M. (1994. Salvaging Nature: Indigenous peoples, protected areas, and biodiversity conservation, United Nations Research Institute for Social Development Discussion Papers.

Crook, R. and Manor, J. (1998). *Democracy and Decentralisation in South Asia and West Africa.* Cambridge (UK): Cambridge University Press.

—— (2000). *Democratic Decentralization.* Washington, DC: The World Bank.

Dème, Y. (1998). *Natural Resource Management by Local Associations in the Kelka Region of Mali.* London: IIED.

Dietz, T., Ostrom, E. and Stern, P. (2003). The struggle to govern the commons. *Science* 302(12):1907–1912.

Emerton, L. (2001). The nature of benefits & the benefits of nature: why wildlife conservation has not economically benefited communities in Africa. In *African Wildlife and Livelihoods.* edited by D. Hulme and M. Murphree. Oxford: James Currey, pp. 208–226.

Enters, T. and Anderson, J. (1999). Rethinking the decentralization and devolution of biodiversity conservation. *Unasylva* 50:6–11.

FAO (Food and Agricultural Organization) (2002). *FAO Guidelines on the Ecosystem Approach to Fisheries (Final Draft).* Rome: FAO.

Fabricius, C. (2004). The fundamentals of community-based natural resource management. In *Rights, Resources and Rural Development: Community-Based Natural Resource Management in Southern Africa*, edited by C. Fabricius, E. Koch, H. Magome, and S. Turner. London: Earthscan, pp. 3–43.

Ferraro, P.J. and Kramer, R.A. (1997). Compensation and economic incentives: reducing pressure on protected areas. In *Last Stand: Protected Areas and the Defense of Tropical Biodiversity*, edited by R. Kramer, C. van Schaik, and J. Johnson. New York: Oxford University Press.

Ferraro, P.J. and Kiss, A. (2002). Direct payments to conserve bioidiversity. *Science* 298:1718–1719.

Ferrer, E.M., Polotan-Dela Cruz, L. and Agoncillo-Domingo, M. (eds.) (1996). *Seeds of Hope: A Collection of Case Studies on Community-Based Coastal Resources Management in the Philippines.* College of Social Work and Community Development (CSWCD), University of the Philippines, Quezon City.

Fortmann, L. (1997). Voices from communities managing wildlife in southern Africa. *Society and Natural Resources* 10:403.

Fortmann, L., Roe, E. and van Eeten, M. (2001). At the threshold between governance and management: community-based natural resource management in southern Africa. *Public Administration and Development* 21(2):171–185.

Gerber, L., Botsford, L.W., Hastings, A., Possingham, H.P., Gaines, S.D., Palumbi, S.R. and Andelman, S. (2003). Population models for marine reserve design: a retrospective and prospective synthesis. *Ecological Applications* 13(1) (Suppl):S47–S64.

Gezon, L.L. (2003). The regional approach in northern Madagascar: moving beyond integrated conservation and Development. In *Contested Nature: Promoting International Biodiversity Conservation with Social Justice in the Twenty-first Century*, edited by S.R. Brechin, P.R. Wilshusen, C.L. Fortwangler and P.C. West Albany, NY: SUNY Press, pp. 183–194.

Gibson, C.C. (2002). Of waves and ripples: democracy and polical change in Africa in the 1990s. *Annual Review of Political Science* 5:201-221.

Gibson, C.C., McKean, M.A. and Ostrom, E. (2000a). Explaining deforestation: the role of local institutions. In *People and Forests.* edited by C.C. Gibson, M.A. McKean and E. Ostrom. Cambridge, MA: MIT Press, pp. 1–26.

—— (eds.) (2000b). *People and Forests: Communities, Institutions, and Governance.* Cambridge, MA: MIT Press.

Goodale, U.M., Stern, M.J., Margoluis, C., Lanfer, A.G. and Fladeland, M. (eds.) (2003). *Transboundary Protected Areas: The Viability of Regional Conservation Strategies.* New York. Food Product Press.

Goodland, R. (1991). Prerequisites for ethnic identity and survival. In *Resident Peoples and National Parks: Social Dilemmas and Statergeis for International Conservation*, P.C. West and S.R. Brechin (eds). Tucson: University of Arizona Press.

Grantham, B.A., Eckert, G.L. and Shanks, A.L. (2003). Dispersal potential of marine invertebrates in diverse habitats. *Ecological Applications* 13(1) (Suppl):S108–S116.

Gubbay, S. (ed) (1995). *Marine Protected Areas: Principles and Techniques for Management.* New York: Chapman and Hall.

Halpern, B.S. (2003). The impact of marine reserves: do reserves work and does size matter? *Ecological Applications* 13(1) (Suppl):S117–S137.

Hasler, R. (1999). *An Overview of the Social, Ecological and Economic Achievements and Challenges of Zimbabwe's CAMPFIRE Programme.* London: IIED.

Hastings, A. and Botsford, L.W. (2003). Comparing designs of marine reserves for fisheries and for biodiversity. *Ecological Applications* 13(1) (Suppl):S65–S70.

Heermans, J. and Otto, J. (1999). *Whose Woods These Are: Community Based Forest Management in Africa.* Washington, DC: USAID/Africa Bureau—Sustainable Development Office.

Honey, M. (1999). *Ecotourism and Sustainable Development: Who Owns Paradise?* Washington, DC.: Island Press.

Hough, J. (1991). The Grand Canyon National Park and the Havasupai people: cooperation and conflict. In *Resident Peoples and National Parks: Social Dilemmas and Strategies in International Conservation*, edited by P.C. West and S.R. Brechin. Tucson: The University of Arizona Press.

Hulme, D. and Murphree, M. (2001). *African Wildlife and Livelihoods: The Promise and Performance of Community-based Conservation.* Oxford: James Curry.

Huntington, S.P. (1991). *The Third Wave: Democratization in the Late Twentieth Century.* Norman, OK: University of Oklahoma Press.

Hutton, J., Adams, W.M. and Murombedzi, J.C. (2005). Back to the barriers?: changing narratives in biodiversity conservation. *Forum for Development Studies* 2:341–370.

Jain, N. (1999). *Trekking Tourism and Protected Area Management in the Annapurna Conservation Area, Nepal.* Doctoral Dissertation, School of Natural Resources and Environment, University of Michigan, Ann Arbor.

Jones, B.T.B. (1999). *Community-Based Natural Resource Management in Botswana and Namibia: An Inventory and Preliminary Analysis of Progress.* London: IIED.

Kaza, S. (1988). Community involvement in marine protected areas. *Oceanus* 31(1):75–81.

Kelleher, G. and Recchia, C. (1998). Editorial – lessons from marine protected areas around the world. *Parks* 8(2):1–4.

Kelleher, G., Bleakley, C., and Wells, S. (1995). *A Global Representative System of Marine Protected Areas*, Vol I-IV. Washington, DC: GBRMPA/The World Bank/IUCN.

Kellert, S.R., Mehta, J.N., Ebbin, S.A. and Lichtenfeld, L.L. (2000). Community natural resource management: promise, rhetoric, and reality. *Society and Natural Resources* 13:705–715.

Kerkhof, P. (2000). *Local Forest Management in the Sahel.* London: SOS Sahel.

Khare, A., Sarin, M., Saxena, N.C., Palit, S., Bathla, S., Vanya, F. and Satyanarayana, M. (2000). *Joint Forest Management: Policy, Practice and Prospects.* London: IIED.

Kramer, R.A., and van Schaik, C.P. and Johnson, J. (eds.) (1997). *The Last Stand: Protected Areas and the Defense of Tropical Biodiversity.* New York: Oxford University Press.

Kutay, K. (1984). *Cahuita National Park, Costa Rica: A Case Study in Living Cultures and National Park Management.* Doctoral Dissertation, School of Natural Resources and Environment, University of Michigan, Ann Arbor.

Langholz, J. (1996). Economics, objectives, and success of private reserves in sub-saharan and Latin America. *Conservation Biology* 10:271–280.

—— 2002. Privately owned parks. In *Making Parks Work: Strategies for Preserving Tropical Forests*, edited by J. Terborgh, C. van Schaik, L. Davenport and M. Rao. Washington, DC: Island Press, pp. 172–188.

—— 2003. Privatizing conservation. In *Contested Nature: Promoting International Biodiversity Conservation with Social Justice in the Twenty-first Century*, edited by S.R. Brechin, P.R. Wilshusen, C.L. Fortwangler and P.C. West. Albany, NY: SUNY Press, pp. 159–182.

Langholz, J. and Lassoie, J. (2001). Perils and promises of privately owned protected areas. *BioScience* 51(12): 1079–1085.

Langholz J, Lassoie, J. and Schelhas J. (2000). Incentives for biodiversity conservation: Lessons from Costa Rica's private wildlife refuse program. *Conservation Biology* 14(6): 1735–1743.

Larson, A.M. (2002). Natural resources and decentralization in Nicaragua: are local governments up to the job? *World Development* 30:17–31.

Leach, M., Mearns, R. and Scoones, I. (1999). Environmental entitlements: dynamics and institutions in community-based natural resource management. *World Development* 27:225–247.

Li, T.M. (1996). Images of community: discourse and strategy in property relations. *Development and Change* 27: 501–527.

Lindsay, J.M. (1999). Creating a legal framework for community-based management: principles and dilemmas. *Unasylva* 50:28–34.

Litvack, J., Ahmed, J. and Bird, R. (1998). *Rethinking Decentralization in Developing Countries.* Washington, DC: The World Bank.

Long, S.A. (ed.) (2004). *Livelihoods and CBNRM in Namibia: The Findings of the WILD Project.* Windhoek (Namibia): DfID and MET.

Lowry, A. and Donahue, T.P. (1994). Parks, politics, and pluralism: the demise of national parks in Togo. *Society & Natural Resources* 7(4):321–329.

Lubchenco, J., Palumbi, S.R., Gaines, S.D. and Andelman, S. (2003). Plugging a hole in the ocean: the emerging science of marine reserves. *Ecological Applications* 13(1):(Suppl) S3–S7.

Machlis, G.E. and Burch, W.R.J. (1983). Relations between strangers: cycles of structure and meaning in tourist systems. *Sociological Review* 31(4):665–692.

Manor, J. (1999). *The Political Economy of Democratic Decentralization.* Washington, DC: World Bank.

McKean, M.A. (2000). Common property: what is it, what is it good for, and what makes it work? In *People and Forests*, edited by C.C. Gibson, M.A. McKean and E. Ostrom. Cambridge, MA: MIT Press, pp. 27–56.

Mascia, M. (1999). Governance of marine protected areas in the wider Caribbean: preliminary results of an international mail survey. *Coastal Management* 27:391–402.

Mascia, M. (2003). The human dimension of coral reef marine protected areas: recent social science research and its policy implications. *Conservation Biology* 17(2):630–632.

Mascia, M.B., Brosius, J.P., Dobson, T.A., Forbes, B.C., Horowitz, L., McKean, A. and Turner, N.J. (2003). Conservation and the social sciences. *Conservation Biology* 17:649–650.

McCay, B. (1995). Social and ecological implications of ITQ's: an overview. *Ocean and Coastal Management* 28(1–3):3–22.

McShane, T. and Wells, M.P. (2004). *Getting Biodiversity Projects to Work: Toward More Effective Conservation and Development.* New York: Columbia University Press.

Mehta, L., Leach, M., Newell, P., Scoones, I., Sivaramakrishnan, K. and Way, S.-A. (1999). *Exploring Understandings of Institutions and Uncertainty: New Directions in Natural Resources Management.* Institute for Development Studies, University of Sussex.

Murphree, M.W. (2005). Congruent objectives, competing interests, and statergic compromise: concept and process in the evolution of Zimbabwe's CAMPFIRE, 1984–1996, in J.P. Brosius, A.L. Tsing and C. Zerner (eds) Communities and Conservation; Histories and politics of Community-Based Natural Resource Management. AltaMira Press, pp. 105–147.

Murray, G.D. (2003). *Contextual Influences on Protected Area Form and Function in Quintana Roo, Mexico.* Doctoral Dissertation, The University of Michigan, Ann Arbor.

Murray, G.D. (2005). Multifaceted measures of success in two mexican marine protected areas. *Society & Natural Resources* 18 (10):884–905.

Murray, G.D., Neis, B. and Bavington, D. (2005). Local ecological knowledge, science, and fisheries management in Newfoundland and Labrador: a complex, contested and changing relationship. In *Participation in Fisheries Governance*, edited by T.S. Gray. Dordrecht, The Netherlands: Springer-Verlag.

NEF (Near East Foundation) (1996). *Inventaire institutionnel des associations locales de gestion des ressources naturelles du Kelka*. Douentza, Mali: Near East Foundation (PAGRN).

Neis, B. and Felt, L. (eds) (2000). *Finding our Sea Legs: Linking Fishery People and Their Knowledge with Science and Management*, St. John's, Newfoundland,: ISER Books.

Neis, B., Schneider, D.C., Felt, L.F., Haedrich, R.L., Hutchings, J.A. and Fischer, J. (1999) Northern cod stock assessment: what can be learned from interviewing resource users? *Canadian Journal of Fisheries and Aquatic Sciences* 56:1944–1963.

Nepstad, D., Schwartzman, S., Bamberger, B., Santilli, M., Ray, D., Schlesinger, P., Lefebvre, P., Alencar, A., Prinz, E., Fiske, G. and Rolla, A. (2006). Inhibition of Amazon deforestation and fire by parks and indigenous lands. *Conservation Biology* 20(1):65–73.

NRC (National Research Council) (2000). Marine Protected Areas: Tool for Sustaining Ocean Ecosystems. Washington, DC: National Academy Press.

Oats, J.F. (1999). *Myth and Reality in the Rain Forest: How Conservation Strategies are Failing in West Africa*. Berkeley, CA: University of California Press.

Ostrom, E. (1990). *Governing the Commons: The Evolution of Institutions for Collective Action*. Cambridge: Cambridge University Press.

—— (1999a). Coping with the tragedy of the commons. *Annual Review of Political Science* 2:493–535.

—— (1999b). Institutional rational choice: an assessment of the institutional analysis and development framework. In *Theories of the Political Process*, edited by P.A. Sabatier. Boulder, CO: Westview Press. pp. 35–71.

—— (1999c). *Self-Governance and Forest Resources*. CIFOR. http://www.eldis.ids.ac.uti/static/DOC7033.htm

—— 2002. Understanding the complex linkage between attributes of goods and the effectiveness of property-rights regimes. In *Common Goods: Reinventing European and International Governance*, edited by A. Héritier. Lanham, MD: Rowman & Littelfield.

Ostrom, E., Gardner, R. and Walker, J. (1994). *Rules, Games, and Common-Pool Resources*. Ann Arbor: The University of Michigan Press.

Pauly, D. and Maclean, J. (2003). *In a Perfect Ocean: the State of Fisheries and Ecosystems in the North Atlantic Ocean*. Washington, DC: Island Press.

Pauly D., Christensen, V., Guénette, S., Pitcher, T., Sumaila, R., Walters, C., Watson, R. and Zeller, D. (2002). Towards sustainability in world fisheries. *Nature* 418:689–695.

Perfecto, I.R., Rice, A., Greenberg, R. and Van der Voort M.E. (1996). Shade Coffee: A Disappearing Refuge for Biodiversity. *Bioscience* 46 (8):598–608.

Pierce Colfer, C. and Capistrano, D. (eds.) (2005). *The Politics of Decentralization*. London: Earthscan.

Pinkerton, E.W. (1989) Introduction: Attaining better fisheries management through comanagement – prospects, problems and propositions. *Cooperative Management of Local Fisheries: New directions for improved management and community developmented*. E.W. Pinkerton. Vancouver, University of British Columbia Press.

Plattner, M.F. (2002). Globalization and self-governance. *Journal of Democracy* 13:54–67.

Poffenberger, M. and McGean, B. (eds.) (1996). *Village Voices, Forest Choices: Joint Forest Management in India*. Delhi: Oxford University Press.

Poteete, A. and Ostrom, E. (in press). An institutional approach to the study of forest resource. In *Human Impacts on Tropical Forest Biodiversity and Genetic Resources*, edited by J. Poulson. New York: CABI Publishing.

Pollnac, R.B., Crawford, B.R. and Gorospe, M.L.G. (2001). Discovering factors that influence the success of community-based marine protected areas in the Visayas, Philippines. *Ocean & Coastal Management* 44:683–710.

Pomeroy, R.S. (ed.) (1994) *Community Management and Common Property of Coastal Fisheries in Asia and the Pacific: Concepts, Methods and Experiences*. Manila: International Center for Living Aquatic Resources Management.

Pretty, J. (2003). Social capital and the collective management of resources. *Science* 302:1912–1914.

Putnam, R.D. (1993). *Making Democracy Work: Civic Traditions in Modern Italy*. Princeton, NJ: Princeton University Press.

Ribot, J.C. (1996). Participation without representation: chiefs, councils and forestry law in the West African Sahel. *Cultural Survival Quarterly* 20:40–44.

—— (1999). Decentralization, participation and accountability in Sahelian forestry: legal instruments of political–administrative control. *Africa* 69:23–65.

—— (2002). *Democratic Decentralization of Natural Resources: Institutionalizing Popular Participation*. Washington, DC: World Resources Institute.

—— (2004). *Waiting for Democracy: The Politics of Choice in Natural Resource Decentralization*. Washington: World Resources Institute.

Roberts, C.M., Andelman, S., Branch, G., Bustamante, R.H., Castilla, J.C., Dugan, J., Halpern, B.S., Lafferty, K.D., Leslie, H., Lubchenco, J., McArdle, D., Possingham, H.P., Ruckelshaus, M. and Warner, R.R. (2003a). Ecological criteria for evaluating candidate sites for marine reserves. *Ecological Applications* 13(1) (Suppl.): S199–S214.

Roberts, C.M., Branch, G., Bustamante, R.H., Castilla, J.C., Dugan, J., Halpern, B.S., Lafferty, K.D., Leslie, H., Lubchenco, J., McArdle, D., Ruckelshaus, M. and Warner, R.R. (2003b). Application of ecological criteria in selecting marine reserves and developing reserve networks. *Ecological Applications* 13(1) (Suppl.): S215–S228.

Roberts, C.M., Bohnsack, J.A., Gell, F., Hawkings, J.P. and Goodridge, R. (2001). Effects of marine reserves on adjacent fisheries. *Science* 294:1920–1923.

Russ, G.R. and Alcala, A.C. (1996). Do marine reserves export adult fish biomass? Evidence from Apo Island, central Philippines. *Marine Ecology Progress Series* 132:1–9.

Ruttan, L. and Borgerhoff Mulder, M. (1999). Are East African pastoralists truly conservationists? *Current Anthropology* 40(5):621–651.

Salafsky, N. and Margoulis, R. (1999). Threat reduction assessment: a practical and cost-effective approach to evaluating conservation and development projects. *Conservation Biology* 13:830–841.

Salafsky, N., Margoulis, R., Redford, K.H. and Robinson, J.G. (2002). Improving the practice: a conceptual framework and research agenda for conservation science. *Conservation Biology* 16:1469–1479.

Sen, S. and Nielsen, J.R. (1996). Fisheries co-management: a comparative analysis. *Marine Policy* 20 (5):405–418.

Smith, E.A. and Wishnie, M. (2000). Conservation and subsistence in small-scale societies. *Annual Review of Anthropology* 2000:493–524.

Sognorwa, A.R. (1999), Community-based wildlife management (CWM) in Tanzania: are the communities interested? *World Development* 27(12):2061–2079.

Stern, M. (2001). Parks and factors in their success. *Science* 293(5532):1045–1047 [Reply10 August].

Stevens, S. (1997). *Conservation Through Cultural Survival: Indigenous Peoples and Protected Areas.* Washington, DC: Island Press.

St. Martin, K., McCay, B., Murray, G.D., Johnson, T. and Oles, B. (2007). Communities, knowledge, and fisheries of the future in *The Journal of Global Environmental Issues.*

Terborgh, J. (1999). *Requiem for Nature.* Washington, DC: Island Press/Shearwater Books.

Thomson, J.T. (1992). *A Framework for Analyzing Institutional Incentives in Community Forestry. FAO Community Forestry Note.* Rome: FAO.

—— (1994). *Legal Recognition of Community Capacity for Self-Governance: A Key to Improving Renewable Resource Management in the Sahel. Sahel Decentralization Policy Report,* Vol. III. USAID.

—— (2000). Special districts: an institutional tool for improved common pool resource management. *Paper presented at the Eighth Conference of the International Association for the Study of Common Property, Bloomington, Indiana (May 31–June 4, 2000).*

Thomson, J.T. and Coulibaly, C. (1995). Common property forest management systems in Mali: resistance and vitality under pressure. *Unasylva* 46:16–22.

Thomson, J.T. and Schoonmaker Freudenberger, K. (1997). *Crafting Institutional Arrangements for Community Forestry. FAO Community Forestry Field Manual.* Rome: FAO.

Turner, M.D. (1999). Conflict, environmental change, and social institutions in dryland Africa: Limitations of the community resource management approach. *Society and Natural Resources* 12:643–657.

UNCDF (2000). Local development and decentralized management of natural resources. *Local Development and Decentralized Management of Natural Resources, 10–16 December 2000, Cotonou, Benin.*

USAID (2002a). *Investing in Tomorrow's Forests: Toward an Action Agenda for Revitalizing Forestry in West Africa.* Washington, DC:USAID/AFR/SD.

USAID (2002b). *Nature, Wealth, and Power: Emerging Best Practice for Revitalizing Rural Africa.* Washington,DC: USAID/AFR/SD.

Vandermeer, J. and Perfecto, I. (1995). *Breakfast of Biodiversity: The Truth About Rain Forest Destruction.* Oakland, CA: Food First Books.

Van't Hof, T. (1994). *Resolving Common Issues and Problems of Marine Protected Areas in the Caribbean.* Bridgetown, Barbados, Caribbean Conservation Association.

Varughese, G. (2000). Population and forest dynamics in the hills of Nepal: institutional remedies by rural communities. In *People and Forests.* Edited by C.C. Gibson, M.A. McKean, and E. Ostrom. Cambridge, MA: MIT Press, pp. 193–226.

Ward, T. and Hegerl, E.(2003). *Marine Protected Areas in Ecosystem-based Management of Fisheries: a Report to the Department of the Environment and Heritage.* Accessed at http://www.biodiversity.org/wcpa/ev.php?URL_ID=4624& URL_DO=DO_TOPIC&URL_SECTION=201&reload=11066 79476&PHPSESSID=ed5426bb48af24d955e170303a 422d53.

Watling, L. and Norse, E.A. (1998). Disturbance of the seabed by mobile fishing gear: a comparison to forest clearing. *Conservation Biology* 12:1180–1997.

Weber, W. (1991). Enduring peaks and changing cultures: the Sherpas and Sagamartha (Everest) National Park. In *Resident Peoples and National Parks: Social Dilemmas and Strategies in International Conservation,* edited by P. C. West and S.R. Brechin. Tucson, AZ: University of Arizona Press.

Wells, M. (1994). Biodiversity conservation and local peoples' development aspirations: new prospects for the 1990s. In *Biodiversity Conservation: Problems and Policies,* edited by C. Perrings, K.-G. Mäler, C. Folke, C.S. Holling and B.O. Jansson. Dordrecht (The Netherlands): Kluwer Academic Publishers.

Wells, M. and Brandon, K. (1992). *People and Parks: Linking Protected Area Management with Local Communities.* Washington, DC: World Bank.

Wells, M., Guggenheim, S., Khan, A., Wardojo, W. and Jepson, P. (1999). *Investing in Biodiversity: A Review of Indonesia's Integrated Conservation and Development Projects.* Washington, DC: World Bank.

West, P.C. and Brechin, S.R. (eds.) (1991a). *Resident People and National Parks: Social Dilemmas and Strategies in International Conservation.* Tucson, AZ: University of Arizona Press.

—— 1991b. National parks, protected areas, and resident peoples: a comparative assessment and integration. Resident Peoples and National Parks: Social Dilemmas and Strategies in International Conservation, edited by P.C. West and S.R. Brechin., AZ: Tucson, University of Arizona Press.

West, P.C., Fortwangler, C.L., Agbo, V., Simsik, M. and Sokpon, N. (2003). The political economy of ecotourism: Pendjari National Park and Ecotourism concentration in Northern Benin. In *Contested Nature: Promoting International Biodiversity with Social Justice in the Twenty-first Century,* edited by S.R. Brechin, P.R. Wilshusen, C.L Fortwangler and P.C. West. Albany: State University of New York Press.

Western, D., Wright, R.M. and Strum, S.C. (eds.) (1994). Natural Connections: Perspectives in community-based Conservation. Washington, DC: Island Press.

White, A.T., Hale, L.Z., Renard, Y. and Cortesi, L. (eds.) (1994). *Collaborative and Community-Based Management of Coral Reefs. Lessons from Experience*. West Hartford: Kumarian Press.

White, A.T., Salamanca, A. and Courtney, C.A. (2002). Experience with marine protected area planning and management in the Philippines. *Coastal Management* 30:1–26.

Wilkinson, C. (ed.) (2000). *Status of the Coral Reefs of the World: 2000*. Townsville, Queensland: Australian Institute of Marine Science.

Wilshusen, P.R., Brechin, S.R., Fortwangler, C.L. and West, P.C. (2002). Reinventing the square wheel: critique of a resurgent "protect paradigm" in international biodiversity conservation. *Society and Natural Resources* 15:17–40.

—— (2003). Contested nature: conservation and development at the turn of the twenty-first century. In *Contested Nature*, edited by S.R. Brechin, P.R. Wilshusen, C.L. Fortwangler and P.C. West. Albany, NY: SUNY Press.

Wilshusen, P.R. and Murguia, R.E. (2003). Scaling up from grassroots NGO: networks and the challenges of organizational maintenance in Mexico's Yucatan peninsula. In *Contested Nature*, edited by S.R. Brechin, P.R. Wilshusen, C.L. Fortwangler, and P.C. West. Albany, NY: SUNY Press.

Winter, M. (1997). *La gestion décentralisée des ressources naturelles dans trois pays du Sahel: Sénégal, Mali, et Burkina Faso*. Ouagadougou (Burkina Faso): CILSS.

——(2000). *Natural Resource Management Policy in Mali: The Process of Design and the Options for the GDRN5 Network*. Mopti (Mali): GDRN5 Network, SOS Sahel GB, NEF, IIED.

Woo, H.T. (1991). An assessment of tourism development in the national parks of South Korea. In *Resident Peoples and National Parks: Social Dilemmas and Strategies in International Conservation*, edited by P.C. West and S.R. Brechin. Tucson, AZ: The University of Arizona Press.

WRI (2005). *World Resources 2005 – The Wealth of the Poor: Managing Ecosystems to Fight Poverty*. Washington, DC: World Resources Institute.

Wunsch, J. and Olowu, D. (eds.) (1995). *Failure of the Centralized State: Institutions and Self-Governance in Africa*. San Francisco, CA: Institute for Contemporary Studies.

Young, E.H. (1999). Balancing conservation with development in small-scale fisheries: is ecotourism an empty promise? *Human Ecology* 27(4):581–620.

Young, O.R. (2002). *The Institutional Dimensions of Environmental Change: Fit, Interplay, and Scale*. Cambridge, MA: MIT Press.

Zerner, C. (2000). *People, Plants and Justice: The Politics of Nature Conservation*. New York: Columbia University Press.

39

Institutions, Collective Action and Effective Forest Management: Learning from Studies in Nepal

Harini Nagendra and Elinor Ostrom

DRIVERS OF TROPICAL DEFORESTATION

Loss of forest cover represents one of the most serious environmental challenges facing the world today. Catastrophic consequences are predicted for the world's environment and human well-being. The impact has been particularly severe in the tropics, where rates of forest clearing have been among the highest in recent times. Although tropical forests cover less than 10% of Earth's land cover, they harbor between 50 and 90% of the world's animal and plant diversity (World Resources Institute, 1992). Less than two-thirds of the area originally under tropical forest cover remains available to us, and even this is being degraded at alarming rates each year (Geist and Lambin, 2003).

What are the driving forces that contribute to tropical deforestation? Much of the blame has been laid on shifting cultivation and commercial logging as major proximate drivers (Myers, 1993; Rudel, 2002). Population growth is also frequently identified as a primary underlying cause that drives deforestation in the developing tropics (Allen and Barnes, 1985; Mather and Needle, 2000). The IPAT framework, where I (*environmental impact*) is the product of P (*population*), A (*affluence*), and T (*technology*), has long been considered as a comprehensive equation that identifies all major drivers of environmental change (Ehrlich and Ehrlich, 1990; Meyer and Turner, 1992). All these explanations completely ignore institutions and the powerful capacities of people to organize themselves into collective groups to combat problems.

In a recent review of 152 subnational case studies of tropical deforestation taken from publications in scientific journals, Geist and Lambin (2001) found that PAT variables are significant causal factors in less than half (42%) of all cases. Even in these cases, the PAT variables do not act alone. Except in 3% of the cases, the PAT variables were found associated with institutional and/or cultural factors that were crucial co-explanatory drivers of change. As they state, "Our findings reveal that prior studies have given too much emphasis to population growth and shifting cultivation as primary causes of deforestation" (2001, pp. 143–144). Indeed, their meta-analysis clearly demonstrates that institutional innovations can act to set negative feedback loops in motion, decreasing and even reversing forest degradation and clearing (Geist and Lambin, 2003). This has been supported by findings from recent studies in Nepal and Mexico (Schweik *et al.*, 2003; Bray *et al.*, 2004).

Similarly, Kaimowitz and Angelsen (1998) reviewed some 150 quantitative studies (reports, articles, and books) examining diverse factors that affect deforestation and addressing the relative importance of economic, demographic, and biophysical causes of deforestation. General agreement exists in the studies they reviewed that deforestation was more likely to occur in dryer, flatter regions with high-fertility soil where expansion of cropped areas and pasture lands was a key motivator (1998, pp. 89–90). Regarding population pressure – frequently presumed to be a major cause of deforestation – many national studies did find that population density and the percentage of forested land in a country are negatively correlated. The correlation tends to disappear, however, when additional variables are incorporated into a model. Kaimowitz and Angelsen interpret this frequent finding to imply that some third set of variables is likely to be simultaneously affecting both presumed cause and effect. And the subnational studies reported by Geist and Lambin (2001) also lead to a serious questioning of this earlier presumed cause of deforestation.

In the Kaimowitz and Angelsen (1998) review, some forty studies focused on national-level data and generated a regression model that tested various factors posited to affect forested areas. Kaimowitz and Angelsen point to a number of problems with these studies, which include the poor quality of available data, the limited degrees of freedom from problems of cross-national studies where the number of "independent" variables must be kept relatively small, the confusion between correlation and causality, and other statistical problems. In their conclusion, they express "strong doubts about the value of producing more global regression models" (1998, p. 104). They also point out that institutions are frequently mentioned as being important factors that shape the incentives of various agents who engage in deforestation. Yet they point out that "there are few modeling exercises that explicitly take institutions in account" (1998, p. 104).

In Ostrom and Nagendra (2006), we focus on the gap Kaimowitz and Angelsen identified, evaluating the impact of institutions on forests. Major debates have taken place over what types of policy "interventions" best protect forests, with choices of property and land tenure systems being central issues. We provide an overview of findings from a long-term interdisciplinary, multiscale, international research program that analyzes the institutional factors affecting forests managed under a variety of tenure arrangements. Evidence from multiple research methods challenges the presumption that a single governance arrangement will control overharvesting in all settings. When users are genuinely engaged in decisions about rules affecting their use, we find that the likelihood of their following the rules and monitoring others is much greater than when an authority simply imposes rules. We move the debate beyond the boundaries of protected areas into larger landscapes where government, community, and co-managed protected areas are embedded. This approach helps us understand when and why deforestation and regrowth occur in specific regions within larger landscapes. Our findings support a new research frontier on the most effective institutional and tenure arrangements for protecting forests.

In this chapter, we focus in further on prior work on collective-action theory as related to common-pool resources, including forests. Until recently, the dominant theory predicted that individual users of a common-pool resource were trapped into a bad equilibrium of overusing if not destroying the resources they used unless the resources were owned either by the government or privately. We rapidly review the evidence that has been mounting regarding the possibility – not the necessity – that resource users may engage in the costly effort to create rules limiting entry and use patterns. We then address how "blueprint thinking" and easy access to external funds may constitute a serious threat to successful collective action and turn to an examination of these issues in the context of Nepal using methods developed as part of the International Forestry Resources and Institutions research program. Our findings provide support for:

- the need for flexible adaptation rather than blind adoption of blueprint thinking;
- recognizing the importance of contextual variables affecting the relationship between population and resource use (or overuse);
- understanding how financial benefits can serve as incentives for effective management, but may also stimulate conflict and dissension;
- acknowledging that heterogeneity can be positively or negatively associated with successful collective action.

In general, we point to the importance of the institutions used by resource users in affecting the cumulative impact of factors such as population size, external financial incentives, or heterogeneity on resource conditions.

COLLECTIVE ACTION AND COMMON-POOL RESOURCES: A POTENTIAL FORCE FOR SUSTAINABLE LANDSCAPES

The study of common-pool resources, including forests, fisheries, irrigation systems, the atmosphere, and the Internet, has been a vigorous research program since the important early studies of open-access fisheries by Gordon (1954) and Scott (1955) and the dramatic metaphor created by Hardin (1968). Common-pool resources yield finite flows of benefits (e.g. firewood and fish) where it is hard to exclude potential users. Each person's use of a resource system subtracts resource units from the quantity of units available for harvesting. The initial theoretical studies of common-pool resources tended to analyze simple systems using relatively similar assumptions (Baumol and Oates, 1988; Brown, 2000). It is frequently assumed that the resource generates a highly predictable, finite supply of one type of resource unit (e.g. one species) in each relevant period. Users are assumed to be short-term, profit-maximizing actors who have complete information and are homogeneous in terms of their assets, skills, discount rates, and cultural views. In this theory, *anyone* can enter a resource and take resource units.

Users are also assumed to be trapped in a situation where they can use any amount of the resource at any time. Organizing so as to create rules that specify rights and duties of participants creates a public good for those involved. All users benefit from this public good, whether they contribute or not. Thus, getting "out of the trap" is itself a second-level dilemma. Since much of the initial problem exists because the individuals are in a dilemma setting where they generate negative externalities on one another, it is not consistent with the conventional theory that individuals can solve a second-level dilemma when they are already predicted not to be able to solve the initial social dilemma. Thus, extensive free-riding is predicted on efforts to self-organize and govern a resource as a community of users (Olson, 1965).

Contrary to the conventional theory, however, a large number of studies have demonstrated that not only have those facing multiple social dilemmas crafted institutions to govern their own resources, they have in many instances sustained these regimes for very long periods (NRC, 1986, 2002; McCay and Acheson, 1987; Ostrom *et al.*, 1988; Wade, 1988; Ostrom, 1990, 2005). The possibility that the appropriators would find ways to organize themselves was not mentioned in basic economic textbooks on environmental problems until recently (compare Clark, 1976, with Hackett, 1998). The design principles that characterize robust, long-lasting, institutional arrangements for the governance of common-pool resources have been identified (Ostrom, 1990) and supported by further testing (Guillet, 1992; Morrow and Hull, 1996; Abernathy and Sally, 2000; Weinstein, 2000; Moor *et al.*, 2002).

Still other in-depth case analyses have documented the accelerated overharvesting of forests that occurred after national governments declared themselves to be the owners of forested land (Arnold and Campbell, 1986; Ascher, 1995). Several well-crafted empirical studies also have begun to identify variables that are associated with a higher probability of successful organization or failure (Hayami, 1998; Bardhan, 1999; Dayton-Johnson, 2000; Ostrom and Nagendra, 2006). A recent National Research Council report (2002) provides an excellent overview of the substantial research showing that many common-pool resources are governed successfully by non-state provision units and some government and private arrangements also succeed. No simple governance system has been shown to be successful in all settings (Dietz *et al.*, 2003).

Many of the robust resource governance systems documented in the above-cited research do not resemble the textbook versions of either a government or a strictly private-for-profit firm, especially when participants have constituted self-governing units. Scholars who draw on traditional conceptions of "the market" and "the state" have not recognized these self-organized systems as potentially viable forms of organization and have either called for their removal or ignored their existence. It is a bit ironic that many vibrant, self-governed institutions have been misclassified or ignored in an era that many observers consider to be one of ever greater democratization.

One of the key findings of empirical field research on collective action and common-pool resources is the multiplicity of specific rules-in-use found in successful common-pool resource regimes around the world. One of the most important types of rules is *boundary* rules that determine who has rights and responsibilities and what territory is covered by a particular governance unit. Many different boundary rules are used successfully to control common-pool resources around the world (Ostrom, 1999), but the important aspect of these rules is the match between the organization of users and the resource rather than the specific rule used.

Some governance units face considerable biophysical constraints in dealing with a natural common-pool resource such as a groundwater basin, a river, or an air shed. Such resources have their own geographic boundaries, and creating a match between the boundary of those who are authorized users and the resource itself is a challenge. On the other hand, the biophysical world does not have as strong an impact on the efficacy of using diverse boundaries for governing and managing forest resources. More important is the agreement of those involved on who is included and the appropriate physical boundaries. Rules specifying duties as well as rules for sharing benefits are also crucial. No resource system functions well over time if all that users do is harvest from it with no investment to increase the productivity of the resource itself. Once basic rules, defining who is a legitimate beneficiary, who must contribute to the maintenance of the resource, and the actions that must or may be taken or are forbidden, have been accepted as legitimate by the users, many users will follow the rules so long as they believe others are also following these rules.

An essential attribute of rules is that to be effective they must be generally known and understood, considered relatively legitimate, generally followed, and enforced (Ostrom and Nagendra, 2006). Written legislation or contract provisions that are not common knowledge do not affect the structure of incentives unless someone involved in the situation invokes the rule and finds someone to enforce it. One of the problems in doing empirical research on the effect of diverse institutions on deforestation is trying to sort out the rules that exist on paper but are not used by participants as contrasted to unwritten rules that are, however, common knowledge of the users and enforced locally, but are not part of the formal legal structure (see Sproule-Jones, 1993, for a discussion of rules-in-form and rules-in-use). The crucial point is that rules affect the structure of the situation under analysis, and one should expect to see differences in the incentives and likely behavior when one configuration of rules is used versus another.

Many attributes of a community are also likely to affect the success of a local resource governance unit, including the size of the group affected, the homogeneity or heterogeneity of interests, the patterns of migration into or out of a community, and the discount rate used by individuals in ongoing situations. The range of specific attributes of a community that could potentially affect incentives and results is very large. For institutional analysis, however, addressing the following questions is key in any effort to understand why some systems succeed and others fail:

- Is there general agreement on the rules related to who is included as a member with both harvesting rights as well as responsibilities?
- Do the members have a shared understanding of what their mutual responsibilities are as well as the formulae used for distribution of benefits?
- Are these rules considered legitimate and fair?
- How are the rules transmitted from one generation to the next or to those who migrate into the group?

It takes time for a community to try out various rules related to the use of a resource and decide which works best in their particular ecology and culture. Large groups of users usually need ways of breaking themselves into smaller groups for many purposes to retain some elements of face-to-face communication and subtle monitoring of each other's actions. Heterogeneity may or may not be a major problem, depending on whether members of a community can develop trust and reciprocity and the rules-in-use are perceived by most members of a community to be fair and enforced (Varughese and Ostrom, 2001).

THREATS TO SUCCESSFUL COLLECTIVE ACTION: BLUEPRINT THINKING AND EASY ACCESS TO EXTERNAL FUNDS

In addition to finding a wide diversity of self-organized, robust resource regimes, recent research also has identified some of the threats to the creation or sustainability of user participation in the governance of common-pool resources (Ostrom, 2005). Two threats to long-term sustainable participation by users are blueprint thinking and easy access to funds. Blueprint thinking occurs whenever policymakers, donors, citizens, or scholars propose uniform solutions to a wide variety of problems that are clustered under a single name based on one or more successful exemplars. Korten (1980) called this the "blueprint approach" and made a devastating critique of its prevalence in development work at the end of the 1970s.

The response by donors to recent findings that users may self-organize and manage a common-pool resource successfully over time has been, unfortunately, to confuse the possibility of active participation with a universal likelihood of users participating actively only if given a chance. Little recognition has been given to the time needed by a community to develop some of the essential elements of achieving a self-governed or co-managed resource system. Projects or programs rely on some formula, rather than learning the specifics of a particular setting and enabling participants to experiment and learn from their own experience and that of others. In searching for the "holy grail," efforts to design homegrown solutions to unique ecological conditions are stymied while policymakers switch policies rapidly trying to copy whatever is considered the latest and best approach (Acuña and Tommasi, 2000). Pritchett and Woolcock (2003) bemoan the problem of trying to find solutions when "the" problem is actually the blueprint solution recommended by donors and national governments for solving a problem.

Tragically, advocates of community governance and local participation have sometimes fallen into this trap. Some of the projects that are called "participatory" or "community driven" turn out to be quick investments in something that can be counted. One evaluation of such "community-driven development projects" found very little impact beyond the initial investment (World Bank, 2002). Negative evaluations have not, however, had much impact on the fervent advocates of this new panacea (Platteau, 2004). Major risks exist in such programs of elite capture and fraud (Platteau and Gaspart, 2003).

The availability of funds from donors or from national government budgets that make no requirements for contributions from recipients can also undermine local institutions. Processes that encourage looking to external sources of funding make it difficult to build upon indigenous knowledge and institutions. A central part of the message asking for external funds is that local efforts have failed and massive external technical knowledge and funds are needed to achieve "development."

In some cases, no recognition is made at all of prior institutional arrangements. This has three adverse consequences: (1) property rights that resource users had slowly achieved under earlier regimes are swept away, and the poor lose substantial assets; (2) those who have lost prior investments are less willing to venture further investments; and (3) a general downgrading of the status of indigenous knowledge and institutions. In the light of their own analysis of a failed effort to use external funds to create an effective community forestry project, Morrow and Hull (1996) provide a good summary of the problems resulting from externally driven funding and priorities: "This case, along with the experience of other community forestry enterprises in Latin America, suggests that donor-driven projects often fail to analyze in sufficient depth the factors outlined by the design principles, particularly the issues of institutional and technological appropriateness and the impact of the larger political economy" (1996, p. 1655).

COLLECTIVE ACTION AND FOREST MANAGEMENT IN NEPAL

Nepal provides a particularly effective context to examine some of these theoretical puzzles in an empirical context. Contrary to predictions of the mid-1970s that all accessible forests in Nepal would disappear by the year 2000, Nepal is now recognized internationally as one of the more successful proponents of community forestry. Well-known examples exist in Nepal of traditional institutions that have managed the environment effectively for centuries, as well as new, experimental institutional structures of co-management and community management of forests.

Like other developing countries in the region, the landscape of Nepal represents a complex interface between people and nature, and there has been considerable discussion about the trajectory of forest change in the country. In the mid-1970s, an idea that was subsequently termed the "theory of Himalayan degradation" received international publicity (Ives and Messerli, 1989). This "theory" hypothesized that deforestation in the Himalayas was rapidly reaching alarming proportions because of population increase, with catastrophic consequences in store, including biodiversity loss, soil degradation, and downstream flooding. This model is now believed to be overly simplistic and exaggerated. This simple model of unidirectional land-cover change from forest to agriculture, caused by population increase, failed to describe adequately the complexity of land-cover change in this highly varied, socioculturally and biophysically complex mountain region. It also failed to take into account the diverse array of institutional

responses that mountain communities are capable of when faced with such change.

The rich social, ecological, physical, and cultural heterogeneity in the region makes it difficult, if not impossible, to generalize trends across the entire region, to identify a single unidirectional trend of forest change, or to highlight population increase as a single dominant factor that always leads to deforestation. We now know that the process of large-scale forest clearing had begun in much of Nepal as far back as the 18th century by the Gorkha rulers, who encouraged the conversion of forest land to agriculture in the late 18th and early 19th centuries in order to increase their tax base. In comparison to the substantial clearing of forests that happened at that time, there has been little reduction in forest cover since the early to mid-20th century, although there has been considerable deterioration in the quality of forest habitat, and in wildlife populations (Gautam and Watanabe, 2004; Ives, 2004).

In contrast, the major losses in forest cover that have occurred in recent decades have been largely concentrated in the Terai lowlands. Several studies have demonstrated that there has actually been a net regrowth of forests in the hills, and recently in the Terai lowlands, following the response of local communities to forest degradation (Gautam et al., 2002; Nagendra et al., 2004, 2005a). Much of this regrowth is scale dependent, and mostly confined to smaller patches of reforestation occurring in substantial parts of the landscape.

Traditional and indigenous practices of forest management have been prevalent through much of Nepal even prior to the 1950s (Messerschmidt, 1987; Thapa and Weber, 1995). Since population sizes were small, and forest resources relatively large, the pressures on the forests were nowhere near the levels that exist today. After the fall of the ruling Rana dynasty in 1950, these traditional land and forest holding rights were officially abolished. The Nationalization Act of 1957 and successive legislation over the next decade brought all forest land, as well as all trees planted on private land, under government ownership (Shrestha, 1998; Varughese, 1999). Thus, local control over forest resources was replaced by a central governance system. Lack of forest ownership resulted in a lack of incentives for people to control the forest. The state control over the public forest lands was, however, rather weak and ineffective in regulating forest access and, in some cases, even corrupt and implicated in illegal extraction (Chaudhary, 2000; Neupane, 2000).

In Nepal, nationalization of forests is believed to have been a major factor resulting in the alienation of local communities from the forest (Bajracharya, 1983; Neupane, 2000). During this process, traditional systems of forest management were taken over and replaced by state control of forests, and traditional access and harvest rights were altered, controlled, and denied at countrywide scales.

These communities no longer had any rights over the land, or incentives to manage it sustainably. The state had limited manpower and resources to safeguard these large stretches of forests that were now public property. As in many other developing countries, nationalization created open-access resources where previously limited- access resources had existed (Ostrom, 1990).

Awareness of the negative impact of nationalization on forest cover, a growing appreciation for the capacity of local communities to manage common-property institutions, and increasing donor pressure have encouraged the Nepalese state to begin to attempt to reverse this process since the early 1970s. After unsuccessful early experiments with turning some forests over to local governments in Nepal, extensive attempts to engage the communities with forest management have been made in the form of community forestry, leasehold forestry, and park buffer zone projects. The Community Forestry Act of 1993 was the starting point for most of these initiatives, and was established with the objective of handing over all accessible forests to user groups, which would be provided with the right to manage and protect the forests and the right to all forest produce and income derived from these forests (Thapa and Weber, 1995; Varughese, 1999). However, the existing models of community involvement have been conceptualized by the state and the formal expansion of the programs largely funded through external donor agencies. Local actors have had limited control over the planning or implementation of these efforts (Agrawal and Ostrom, 2001; Britt, 2002).

Community forestry has made substantial progress in terms of handing over of forests since its inception in 1993. By 1999, over 620,000 hectares of forest area had been handed over to 8500 forest-user groups in the Nepal hills and plains (Chaudhary, 2000). Several studies in the hills, where the worst crises were predicted, have demonstrated that there has actually been a net regrowth of forests following the response of local communities to forest degradation. Nepal is now internationally recognized as one of the most progressive proponents of community forestry. The initial positive experiences associated with community forestry encouraged state officials and international donors, and have sparked a range of programs aimed at community involvement from leasehold forestry to community forestry and park buffer zone programs.

How successful have these programs been in Nepal? Over the past 15 years we have collected a large database of studies on collective action in Nepal's forests, which is very useful in helping us to evaluate their effectiveness under different conditions. Under which conditions do these institutions function to enhance or to detract from collective action? There is considerable difference of opinion on the outcome of these programs, whether in terms

of institutional issues such as stability and heterogeneity, on social and economic issues relating to equity and participation, or in terms of forest conservation. Our research over the past decade has demonstrated that there is significant potential for these programs, as well as several instances of failure. We discuss these findings in greater detail below, as our results have wide-ranging implications that help us understand the interactions between collective action and formally crafted institutional policies in greater depth. The studies described below help to throw light on this crucial question.

RESEARCH METHODS: THE INTERNATIONAL FORESTRY RESOURCES AND INSTITUTIONS PROGRAM

We draw extensively on data collected from the International Forestry Resources and Institutions (IFRI) research program, an interdisciplinary research network of thirteen collaborating research centers in eleven countries (see Figure 39.1 for centers in Asia). This program was designed to further the study of collective action in the management of forest resources by developing a long-term database of the factors affecting forests and the communities that use them (Ostrom, 1998; Gibson *et al.*, 2000; Poteete and Ostrom, 2004). The interdisciplinary methodology developed for this purpose documents biophysical measures of forest and environmental conditions, demographic and economic information, and data about institutions that impact forest resources. This approach allows assessments of hypothesized relationships among demographic, economic, institutional, and biophysical variables. These features make IFRI an attractive resource for the study of relationships between attributes such as group size, heterogeneity, rules, and collective action for forest management. In Nepal, the IFRI research program is coordinated and conducted by a team of social science and natural science researchers from the Nepal Forest Resources and Institutions (NFRI) research program, located in Kathmandu. NFRI has so far conducted one-time visits and repeat visits in research locations carefully selected to maximize the coverage of biophysical and institutional variation in the country. Further details about the IFRI research program are available at http://www.indiana.edu/~ifri.

KEY FINDINGS

Blueprint thinking versus flexible adaptation

Institutions need to have the flexibility to be able to modify rules based on changing local environments and circumstance. Often this is not the case,

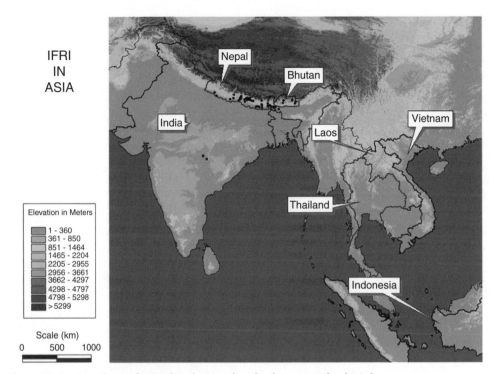

Figure 39.1   Locations of IFRI sites in Nepal and other countries in Asia

with national governments creating relatively inflexible one-size-fits-all rules, and limiting the capacity of local communities to adapt to change – a factor that is essential for long-term sustainability. In a mountain country like Nepal, topography and distance from roads dictate accessibility, and therefore forest condition. Often, neighboring communities can face dissimilar pressures on their forests due to biophysical differences in forest location, necessitating the adoption of different institutional rules for effective management (Schweik, 1998, 2000). In a detailed analysis of fourteen buffer zone user groups in the Chitwan district, we found that these groups were asked to function according to a rather restrictive set of management guidelines, in which they had limited flexibility to modify according to local circumstance (Nagendra *et al.*, 2004, 2005a,b). These management guidelines not only specify aspects related to forest management, they also dictate the manner in which income earned by user groups is to be spent, and specify the proportion of this income that is to be spent on forest management activities versus community development.

These restrictions have understandably created a sense of lack of ownership in the communities,

and a greater reliance on technical and management inputs provided by the state and international aid agencies such as the United Nations Development Programme and the Biodiversity Conservation Network. In contrast, the nearby community forestry user groups have had limited technical and financial support from external agencies, but also more flexibility to adapt and modify management practices to local needs. Although they have had initial problems, these community forestry groups have experimented and learned from their initial attempts, and are now putting better systems in place. The buffer zone user groups on the other hand are likely to face problems after a couple of years, once the term of international funding is complete, and they are left to rely on their own means.

Similar findings from the middle hills indicate that community groups do better when given the flexibility to modify rules according to local social or ecological circumstances. This is especially clear in the case of leasehold forestry. The leasehold forestry initiative was designed to target low-income families, by providing them with access to small, degraded patches of forest for agroforestry initiatives. This well-intentioned program did not take into account the conflict that was generated due to the refusal of adjacent, economically better off, and socially more powerful community forestry user groups to "buy in" to the program and support it. We found that some of these groups have developed innovative ways to deal with this conflict by entering into modifications that are outside the scope of current policies (Karmacharya *et al.*, 2003; Nagendra *et al.*, 2005b). One group has dealt with intra-group dissension over the distribution of benefits by converting leasehold forestry into a *de facto* private setup where each family is parceled out a plot of land for management (allocated by a random "lottery"). Another group, faced with tremendous social conflict by adjacent, socially and economically dominant community forestry users, has dealt with the challenge by expanding the group to involve these users, thus entering into a tradeoff

whereby they lose some share of the benefits but gain legitimacy and control over the forest.

Income: incentive or source of conflict?

Financial benefits can serve as a powerful incentive for effective management, but also can have negative consequences due to increased inequality, conflict, and dissension. In Nepal's Royal Chitwan National Park, buffer zone communities have been encouraged to reforest areas adjacent to the park (see Figure 39.2). These user groups have the potential capacity to earn significant income from tourists who visit the park, and this is projected as a strong incentive for forest protection. We find that only a few communities, those located near the park's entrance, can actually earn significant incomes from tourism, since these forests are

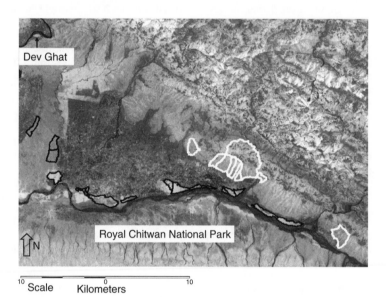

Figure 39.2   The Chitwan District of south-central Nepal provides an example of a landscape containing a mosaic of ownership forms, including a religious forest (Dev Ghat), a government-owned and -managed national park (Royal Chitwan National Park), co-managed Buffer Zone Forests located adjacent to the park boundary (outlined in black), and community forests (outlined in white). The Rapti River separates the large Royal Chitwan National Park, which extends farther south and west below the image, from the rest of the valley. This figure is a black and white image depicting the distribution of the Normalized Difference Vegetation Index (NDVI), which provides an indication of vegetation density in the landscape. The NDVI was produced from an Enhanced Thematic Mapper Plus (ETM+) Landsat image of March 2000. Bright areas indicate greater vegetation. The figure illustrates that under each form of ownership, well-protected forest boundaries or degraded forest may result. A key factor associated with stable forests is that they all have well-demarcated and legitimate boundaries and effective rule enforcement. When these requirements are not present we see major deforestation. This landscape is more fully described in Schweik *et al.* (2003) and Nagendra *et al.* (2004, 2005a).

more likely to be visited by tourists. These are also the forests that are better protected, contrary to the forests located at a distance from the park entrance, which earn lower incomes and have less protection. The higher economic potential from tourism in park buffer zone forests appears to be a powerful driver that can act as an incentive to maintain forest cover in the park's outer boundary. It can also be a powerful source of conflict.

In other parts of the Terai, the high-value *Shorea robusta* deciduous forests that have flourished under successful community protection have also generated conflict between the national government and the users of community forests. The government, which is now cognizant of the significant financial potential provided by these forests, has initiated policies partially to control the financial income of these user groups, and to collect 40% of the income. The long-term implications of such policy changes are unclear, but they have already led to much dissension and conflict between the state and local communities.

Population and collective action

Scale is an important factor that determines the relationship one observes between population and collective action. The relationship between the size of a group and the likelihood of successful collective action appears curvilinear (Agrawal and Goyal, 2001). Up to a certain point, it is useful to have more people involved with monitoring and management activities in a forest, as this leads to more effective protection. Beyond a certain point, however, having too many people dependent on a finite resource such as a forest is obviously detrimental to the sustainability of the resource over the longer term. Gautam (2002) found that very large user groups with over 300 households were not able to successfully manage forests, as indicated by the diameter of trees and the number of species in the forest. However, small- and medium-sized groups were better able to manage forests. Varughese (1999) found an overall negative (but nonsignificant) relationship between group size and the strength of collective action. However, on further examination, a majority of the user groups with high levels of collective action had a relatively large number of households (over 100).

In the Terai, it is more common to find large forest patches that are managed by very large user groups, consisting of over 1000 families. In the Chitwan district, a study of twenty-three forest user groups indicated that there is much greater regrowth in the buffer zones, where the people: forest ratio is higher (Poteete and Ostrom, 2004; Nagendra *et al.*, 2005a). Some of these buffer zone communities have over 1200 households, and yet manage to maintain high levels of reforestation. In contrast, there is much more deforestation in

the community forests, which are significantly larger than forests in the buffer zones and are managed by fewer households (less than 100 in some cases), and where the people:forest ratio is significantly lower than in the buffer zone.

The challenge of heterogeneity

Heterogeneity is often viewed as an obstacle for effective collective action. Increased ethnic, social, and economic heterogeneity within communities can, in theory, pose a challenge for the effective organization of people to deal with forest management. By leading to an increase in intra-group inequity, heterogeneity can give rise to greater dissension and conflict, decreasing the likelihood of successful collective action. In Nepal, the increased ethnic heterogeneity in the Terai is often cited as a cause for the difficulties related to community forestry in this region (Chakraborty, 2001). Several studies using the IFRI database have, however, indicated that the relationship between heterogeneity and collective action is not as straightforward as originally believed (Poteete and Ostrom, 2004). Not all forms of heterogeneity are associated with less collective action (Varughese and Ostrom, 2001; Gautam, 2002). To make matters more complicated, in addition to impacting institutional success, heterogeneity is in turn impacted by institutions (Gibson and Koontz, 1998).

In a study of eighteen forest communities in the middle hills, Varughese (1999) divided them into three groups based on the heterogeneity of ethnic and caste composition. His study did not find any association between heterogeneity and the level of collective action. In another study of eight forest communities in the Kabrepalanchowk middle hills district, Gautam (2002) found no association between sociocultural heterogeneity and effectiveness of forest management. In fact, biophysical conditions like slope, aspect, and elevation proved to be significantly better predictors of forest condition.

In a further study, Varughese (1999; see also Varughese and Ostrom, 2001) compared partitioned user groups according to whether they were homogeneous or heterogeneous with respect to distribution from the forest. Some user groups were located close to the forest, such that all families were located at roughly the same distance from the forest, and able to participate equally in monitoring activities. Other groups were heterogeneous in their distribution, with some families located very close to the forest, and others located at a distance. Since the families located at a distance from the forest had a substantially higher cost of participation due to the additional distance they had to walk to participate in forest management or monitoring activities, Varughese anticipated that these groups would have lower levels of collective action. However, results demonstrated the opposite to be true, with five of the groups with the greatest heterogeneity actually proving to have high levels of collective action. User groups had developed interesting and innovative ways to deal with the problem of an inequitable cost–benefit ratio by developing a two-tier system of membership. Households that were located at a distance from the forest compensated for their reduced contribution to forest monitoring and management by paying additional membership fees. These fees were then used to hire additional labor to patrol the forest.

Although these actors must overcome coordination problems and distributional inequities, these findings clearly indicate that communities can, and do, find innovative ways of dealing with these challenges under a variety of challenging circumstances. When groups are provided with the flexibility to innovate and adapt, heterogeneity can coexist with effective collective action, instead of being a negative factor.

The impact of accessibility and biophysical variability on forests and institutions

The biophysical environment is a powerful force that impacts the capacity of forests to regenerate and shapes the condition of forests. We find more forests at higher elevations and in areas that are relatively inaccessible for clearing, and Gautam *et al.* (2002) find that biophysical factors such as slope, aspect, and elevation significantly impact forest condition. These "external' conditions need to be taken into account while framing policies.

CONCLUSIONS

Instead of focusing on the *drivers* of environmental change as if they always have the same momentum and direction, scholars seriously interested in environmental policies need to pay much more attention to the need for adaptive crafting of institutions that fit the ecological system of interest. Biophysical scientists have long recognized that ecological systems vary dramatically according to elevation, slope, mean temperature, rainfall, soil types, native vegetation, and fauna that make their homes in a particular location. All forests are not the same.

Policy scientists need to recognize a similar diversity in the institutions that can assist human users to devise arrangements for sustainable management of a resource. The temptation to call for a single type of institution to solve all problems of resource allocation and management needs to be resisted strongly. It is impossible to formulate a one-size-fits-all policy that will allow for sufficient adaptation to changing local circumstance.

The key is to develop institutional policies that will encourage conditions favorable for the success of collective action, and encourage learning and adaptation. We have illustrated this general lesson with research findings from Nepal, where efforts to establish uniform policies, such as the nationalization of forests undertaken in the 1950s, had the opposite effect from the proposed outcome. We also showed that flexible, new institutions provide an opportunity for some communities to rehabilitate local forests and to overcome some of the increased challenges of working with larger groups and groups characterized by greater heterogeneity.

Acknowledgments

This research was supported by the MacArthur Foundation, the Ford Foundation, the National Science Foundation (grant SBR9521918), and the Society in Science (Branco Weiss Fellowship to Harini Nagendra). The authors are grateful to Arun Agrawal, Julie England, Ambika Gautam, Robin Humphrey, Birendra Karna, Mukunda Karmacharya, Charles Schweik, and George Varughese for their invaluable assistance with the IFRI program in Nepal. Last, but certainly not least, we thank Joanna Broderick for her careful editing.

REFERENCES

Abernathy, C. L. and Sally, H. (2000). Experiments of some government-sponsored organizations of irrigators in Niger and Burkina Faso, West Africa. *Journal of Applied Irrigation Studies* 35(2): 177–205.

Acuña, C. H. and Tommasi, M. (2000). Some reflections on the institutional reforms required for Latin America. In *Proceedings from the conference on institutional reforms, growth and human development in Latin America* (pp. 357–400). New Haven, CT: Yale Center for International and Area Studies, Yale University.

Agrawal, A. and Goyal, S. (2001). Group size and collective action: Third party monitoring in common-pool resources. *Comparative Political Studies* 34(1): 63–93.

Agrawal, A. and Ostrom, E. (2001). Collective action, property rights, and decentralization in resource use in India and Nepal. *Politics and Society* 29: 485–514.

Allen, J. C. and Barnes, D. F. (1985). The causes of deforestation in developing countries. *Annals of the Association of American Geographers* 75: 163–184.

Arnold, J. E. M. and Campbell, J. G. (1986). Collective management of hill forests in Nepal: The community forestry development project. In National Research Council, *Proceedings of the conference on common property resource management* (pp. 425–454). Washington, DC: National Academy Press.

Ascher, W. (1995). *Communities and Sustainable Forestry in Developing Countries.* San Francisco: ICS Press.

Bajracharya, D. (1983). Deforestation in the food/fuel context: Historical and political perspectives from Nepal. *Mountain Research and Development* 3: 227–240.

Bardhan, P. (1999). Water community: An empirical analysis of cooperation on irrigation in South India. Working paper. Berkeley: University of California, Department of Economics.

Baumol, W. J. and Oates, W. E. (1988). *The Theory of Environmental Policy*. Cambridge: Cambridge University Press.

Bray, D. B., Ellis, E. A., Armijo-Canto, A. and Beck, C. T. (2004). The institutional drivers of sustainable landscapes: A case study of the 'Mayan Zone' in Quintana Roo, Mexico. *Land Use Policy* 21: 333–346.

Britt, C. (2002). Changing the boundaries of forest politics: Community forestry, social mobilization, and federation-building in Nepal viewed through the lens of environmental sociology and PAR. Ph.D. diss., Department of Development Sociology, Cornell University, Ithaca, NY.

Brown, G. M. (2000). Renewable natural resource management and use without markets. *Journal of Economic Literature* 38: 875–914.

Chakraborty, R. N. (2001). Stability and outcomes of common property institutions in forestry: Evidence from the Terai region of Nepal. *Ecological Economics* 36: 341–353.

Chaudhary, R. P. (2000). Forest conservation and environmental management in Nepal: A review. *Biodiversity and Conservation* 9: 1235–1260.

Clark, C. W. (1976). *Mathematical Bioeconomics*. New York: Wiley.

Dayton-Johnson, J. (2000). Choosing rules to govern the commons: A model with evidence from Mexico. *Journal of Economic Behavior and Organization* 62: 19–41.

Dietz, T., Ostrom, E. and Stern, P. (2003). The struggle to govern the commons. *Science* 302 (special issue, December 12): 1907–1912.

Ehrlich, P. R. and Ehrlich, A. H. (1990). *The Population Explosion*. New York: Simon and Schuster.

Gautam, A. P. (2002). Forest land use dynamics and community-based institutions in a mountain watershed in Nepal: Implications for forest governance and management. Ph.D. diss., Asian Institute of Technology, Bangkok.

Gautam, A. P., Webb, E. L. and Eiumnoh, A. (2002). GIS assessment of land use/land cover changes associated with community forestry implementation in the Middle Hills of Nepal. *Mountain Research and Development* 22: 63–69.

Gautam, C. M. and Watanabe, T. (2004). Reliability of land use/land cover assessment in montane Nepal: A case study in the Kanchenjunga Conservation Area (KCA). *Mountain Research and Development* 24: 35–43.

Geist, H. J. and Lambin, E. F. (2001). What drives tropical deforestation? A meta-analysis of proximate and underlying causes of deforestation based on sub-national case study evidence. LUCC Report Series, no. 4., Louvain-la-Neuve: LUCC International Project Office, University of Louvain.

—— (2003). Regional differences in tropical deforestation. *Environment* 45: 22–36.

Gibson, C. and Koontz, T. (1998). When 'community' is not enough: Communities and forests in southern Indiana. *Human Ecology* 26(4): 621–647.

Gibson, C. C., McKean, M. and Ostrom, E. (2000). *People and Forests. Communities, Institutions and Governance*. Cambridge, MA: MIT Press.

Gordon, H. S. (1954). The economic theory of a common property resource: The fishery. *Journal of Political Economy* 62: 124–142.

Guillet, D. W. (1992). *Covering Ground: Communal Water Management and the State in the Peruvian Highlands*. Ann Arbor: University of Michigan Press.

Hackett, S. C. (1998). *Environmental and Natural Resource Economics: Theory, Policy, and the Sustainable Society*. London: M. E. Sharpe.

Hardin, G. (1968). The tragedy of the commons. *Science* 162 (December): 1243–1248.

Hayami, Y. (1998). Norms and rationality in the evolution of economic systems: A view from Asian villages. *The Japanese Economic Review* 49(1): 36–53.

Ives, J. D. (2004). *Himalayan Perceptions: Environmental Change and the Well-being of Mountain Peoples*. London and New York: Routledge and Taylor and Francis Group.

Ives, J. D. and Messerli, B. (1989). *The Himalayan Dilemma*. London and New York: Routledge and United Nations University Press.

Kaimowitz, D. and Angelsen, A. (1998). *Economic Models of Tropical Deforestation: A Review*. Bogor, Indonesia: Center for International Forestry Research.

Karmacharya, M., Karna, B. and Ostrom, E. (2003). *Rules, incentives and enforcement: Livelihood strategies of community forestry and leasehold forestry users in Nepal*. Paper presented at the conference on Rural Livelihoods, Forests and Biodiversity, Bonn, Germany, May 19–23.

Korten, D. C. (1980). Community organization and rural development: A learning process approach. *Public Administration Review* 40(5): 480–511.

Mather, A. S and Needle, C. L. (2000). The relationships of population and forest trends. *Geographical Journal* 166: 2–13.

McCay, B. and Acheson, J. (1987). *The Question of the Commons*. Tucson: University of Arizona Press.

Messerschmidt, D. A. (1987). Conservation and society in Nepal: Traditional forest management and innovative development. In P. D. Little and M. M. Horowitz, with A. E. Nyerges (eds) *Land at Risk in the Third World: Local Level Perspectives*. Boulder, CO: Westview, pp. 373–397.

Meyer, W. B. and Turner, B. L. II (1992). Human population growth and global landuse/land-cover change. *Annual Review of Ecology and Systematics* 23: 39–61.

Moor, M. de, Warde, P. and Shaw-Taylor, L. (eds) (2002). *The Management of Common Land in Northwest Europe, c. 1500–1850*. Comparative Rural History of the North Sea Area (CORN) Series, No. 8. Turnhout, Belgium: BREPOLS Publishers.

Morrow, C. E. and Hull, R. W. (1996). Donor-initiated common pool resource institutions: The case of the Yanesha Forestry Cooperative. *World Development* 24(10): 1641–1657.

Myers, N. (1993). Tropical forests: The main deforestation fronts. *Environmental Conservation* 20: 9–16.

Nagendra, H., Southworth, J., Tucker, C. M., Carlson, L. A., Karmacharya, M. and Karna, B. (2004). Remote sensing for policy evaluation: Monitoring parks in Nepal and Honduras. *Environmental Management* [online issue]. URL: http://www.springerlink.com/media/7PAB6QGBWQC-QRL5E8RWY/Contributions/9/5/C/L/95CL9VC6LTDMX1PQ_html/fulltext.html.

Nagendra, H., Karmacharya, M. and Karna, B. (2005a). Evaluating forest management in Nepal: Views across space and time. *Ecology and Society* 10(1):art.24 [online]. URL: http://www.ecologyandsociety.org/vol10/iss1/art24/

Nagendra, H., Karna, B. and Karmacharya, M. (2005b). Examining institutional change: Social conflict in Nepal's leasehold forestry programme. *Conservation and Society* 3(1): 72–91. URL: http://www.conservationandsociety.org/cs-3-1_5_enagendraetal.pdf

National Research Council (NRC). (1986) *Proceedings of the Conference on Common Property Management*. Washington, DC: National Academy Press.

—— (2002). *The Drama of the Commons*. Committee on the Human Dimensions of Global Change. E. Ostrom, T. Dietz, N. Dolšak, P. Stern, S. Stonich and E. Weber (eds). Washington, DC: National Academy Press.

Neupane, H. P. (2000). A grassroots organizer with a commitment to gender-balanced participation: An interview with Hari Prasad Neupane, Chairperson, Federation of Community Forestry Users in Nepal. *Mountain Research and Development* 20: 316–319.

Olson, M. (1965). *The Logic of Collective Action: Public Goods and the Theory of Groups*. Cambridge, MA: Harvard University Press.

Ostrom, E. (1990). *Governing the Commons*. Cambridge: Cambridge University Press.

—— (1998). The International Forestry Resources and Institutions program: A methodology for relating human incentives and actions on forest cover and biodiversity. In F. Dallmeier and J. A. Comiskey (eds) *Forest Biodiversity in North, Central and South America, and the Caribbean: Research and Monitoring*. Man and the Biosphere Series, ed. J. N. R. Jeffers (vol. 21, pp. 1–28). Paris: UNESCO; New York: Parthenon.

—— (1999). Coping with tragedies of the commons. *Annual Review of Political Science* 2: 493–535.

—— (2005). *Understanding Institutional Diversity*. Princeton NJ: Princeton University Press.

Ostrom, E. and Nagendra, H. (2006). Insights on linking forests, trees, and people from the air, on the ground, and in the laboratory. *Proceedings of the National Academy of Sciences of the United States of America* 103(51): 19224–19231.

Ostrom, V., Feeny, D. and Picht, H. (eds) (1988). *Rethinking Institutional Analysis and Development: Issues, Alternatives, and Choices*. San Francisco: ICS Press.

Platteau, J.-P. (2004). Monitoring elite capture in community-driven development. *Development and Change* 35(2): 223–246.

Platteau, J.-P. and Gaspart, F. (2003). The risk of resource misappropriation in community-driven development. *World Development* 31(10): 1687–1703.

Poteete, A. R. and Ostrom, E. (2004). Heterogeneity, group size and collective action: The role of institutions in forest management. *Development and Change* 35(3): 435–461. URL: http://www.blackwell-synergy.com/links/toc/dech/35/3

Pritchett, L. and Woolcock, M. (2003). Solutions when the solution is the problem: Arraying the disarray in development. *World Development* 35(3): 435–461.

Rudel, T. K. (2002). A tropical forest transition? Agricultural change, out-migration, and secondary forests in the Ecuadorian Amazon. *Annals of the Association of American Geographers* 92: 87–102.

Schweik, C. M. (1998). *The spatial and temporal analysis of forest resources and institutions*. CIPEC Dissertation Series, No. 2., Bloomington, IN: Center for the Study of Institutions, Population, and Environmental Change (CIPEC), Indiana University.

—— (2000). Optimal foraging, institutions, and forest change: A case from Nepal. In C. C. Gibson, M. A. McKean and E. Ostrom (eds) *People and Forests: Communities, Institutions, and Governance*. Cambridge, MA: MIT Press, pp. 99–134.

Schweik, C. M., Nagendra, H. and Sinha, D. R. (2003). Using satellite imagery to locate innovative forest management practices in Nepal. *Ambio* 32(4): 312–319. Available at http://www.bioone.org/pdfserv/i0044-7447-032-04-0312.pdf (subscription to BioOne required).

Scott, A. D. (1955). The fishery: The objectives of sole ownership. *Journal of Political Economy* 63 (April): 116–124.

Shreshtha, B. (1998). Changing forest policies and institutional innovations: User group approach in community forestry of Nepal. In *Proceedings of the international workshop on community-based natural resource management (CBNRM), Washington DC, May 10–14*. Washington, DC: World Bank Institute, World Bank.

Sproule-Jones, M. (1993). *Governments at Work: Canadian Parliamentary Federalism and its Public Policy Effects*. Toronto: University of Toronto Press.

Thapa, G. B. and Weber, K. W. (1995). Natural resource degradation in a small watershed in Nepal: Complex causes and remedial measures. *Natural Resources Forum* 19: 285–296.

Varughese, G. (1999). Villagers, bureaucrats, and forests in Nepal: Designing governance for a complex resource. Ph.D. thesis. Workshop in Political Theory and Policy Analysis, Indiana University. Bloomington, IN.

Varughese, G. and Ostrom, E. (2001). The contested role of heterogeneity in collective action: Some evidence from community forestry in Nepal. *World Development* 29(5): 747–765.

Wade, R. (1988). *Village Republics: Economic Conditions for Collective Action in South India*. New York: Cambridge University Press.

Weinstein, M. S. (2000). Pieces of the puzzle: Solutions for community-based fisheries management from native Canadians, Japanese cooperatives, and common property researchers. *Georgetown International Environmental Law Review* 12(2): 375–412.

World Bank (2002). *Social funds–Assessing effectiveness*. Washington, DC: World Bank Operations Evaluation Department.

World Resources Institute (1992). *Global biodiversity strategy: Guidelines for action to save, study and use Earth's biotic wealth sustainably and equitably*. Washington, DC: World Resources Institute, IUCN-The World Conservation Union, and United Nations Environment Programme, in consultation with the Food and Agricultural Organization of the United Nations and the United Nations Education, Scientific and Cultural Organization.

40

The Precautionary Principle in Environmental Policies

Albert Weale

INTRODUCTION

Environmental policies reveal an interesting conflict. On the one hand, there is widespread agreement on the need to protect biodiversity, environmental resources and the bases of sustainable development. On the other hand, there is considerable controversy across a wide range of particular decisions as to what to do in order to attain these consensual goals. Does the use of nuclear power provide a way of reducing greenhouse gas emissions? Should pest-resistant genetically modified crops be encouraged as a way of raising the productivity of agriculture in developing countries? How serious a threat to human health is long-term low-dose exposure to certain chemicals? How stringent should fishing quotas be in order to protect fish stocks? Should there be a cull of badgers to reduce the transmission of tuberculosis in cattle?

These questions have the characteristic that there are large uncertainties surrounding the effects of different measures, uncertainties that cannot be resolved by reference to the evidence available. For example, whilst it is possible to obtain evidence on exposure to particular chemical pollutants, it is not possible to say whether there are synergistic effects of which we are currently unaware. Similarly, whilst we can know in general on Darwinian grounds that pests will evolve in response to pest-resistant crops, we cannot say how easy or difficult it will be to deal with those resistant strains. In short, these policy problems have to be addressed by decisions where the evidential support of the premises of those decisions is weak or inadequate.

Sometimes it is thought that this problem of uncertainty is peculiar to modern environmental problems, but this is far from true. Legend has it that John Snow ended the London cholera epidemic of 1854 by removing the handle from the Broad Street water pump, on the clear demonstration that cholera was a water-borne disease. In fact, as Sir Austen Bradford-Hill (1962) showed a number of years ago, Snow's action on the Broad Street pump was not the turning point of the epidemic, and his demonstration of the water-borne nature of the disease required careful statistical work of the supply of two water companies in south London. Similarly, there was no clear line of action from the London smog of 1952 to the 1956 Clean Air Act, as is sometimes presented, but instead the action required energetic political action by the National Smoke Abatement Society to make people realise that smog was a health hazard to human beings (Parker, 1975). However, although one can find important and interesting cases in the past where uncertainty was an intrinsic element in the decision problem, it probably is true that the range, extent and depth of problems where uncertainty is a central element is a feature of contemporary environmental policy.

In relation to these problems, a precautionary approach is often urged. On the grounds that it is better to be safe than sorry, advocates of different policy solutions urge that their case is justified by reference to the precautionary principle. The argument here is that uncertainty should not be allowed to hold up effective action to deal with important problems. Against this view has been the argument that, in the absence of adequate scientific understanding of environmental problems, effective action cannot be taken, for effective action requires a science-based approach. Over the last twenty-five years, the dispute between these two positions have raged in a number of cases. Instances include the dispute between the UK and its European neighbours over whether or not retrofitted sulphur dioxide controls were needed on large combustion plants (a dispute that also occurred in the USA among different states), the setting of standards for automobile emissions, the co-disposal to landfill of different streams of waste and the dumping of sewage sludge in the North Sea. However, probably at present the major issue on which a precautionary approach has been advocated or resisted has been global climate change, with early proponents of action arguing a precautionary case and opponents suggesting the need to collect more evidence and to act more slowly.

The second main category of issues in respect of which precaution has been urged and disputed involves the control of innovative technologies. Thus, many opponents of the commercial use of genetically modified crops have sought to support their position by appeal to the precautionary principle, saying that unless there could be reassurance that adverse consequences, for example, pesticide-resistant strains, would not occur, then the technology should not be allowed. Similarly, many of those opposed to the use of animal organs in human transplantation have urged a precautionary approach, given the possibility, however remote, that dangerous pathogens could be transferred across species boundaries.

The invocation of the precautionary principle in relation to these issues is supported in part by the fact that the principle has received official recognition in policy statements, the legislation of various countries, as well as international treaties and agreements. Its use and development in international environmental policies and agreements is particularly significant, being cited as a guide to policy in important treaties and conventions. For example, Principle 15 of the 1992 Rio Declaration on Environment and Development reads as follows:

In order to protect the environment, the precautionary approach shall be widely applied by States according to their capabilities. Where there are threats of serious or irreversible damage, lack of full scientific certainty shall not be used as a reason for postponing cost-effective measures to prevent environmental degradation.

Earlier, representatives to the 1987 London Ministerial Declaration of the Second International Conference on the Protection of the North Sea agreed that:

... in order to protect the North Sea from possibly damaging effects of the most dangerous substances, a precautionary approach is necessary.

Similarly, the principle of precaution was formally introduced into European Union environmental policy by the 1992 Maastricht Treaty and it is explicitly stated in the Union's draft Constitution as follows:

Union policy on the environment shall aim at a high level of protection taking into account the diversity of situations in the various regions of the Union. It shall be based on the precautionary principle and on the principles of that preventive action should be taken, that environmental damage should as a priority be rectified at source and that the polluter should pay.

The range of these examples brings out the widespread acknowledgement of the precautionary principle in international agreements and illustrates the extent to which a precautionary approach is supposed to guide and shape policy in the setting of standards for pollutants and the protection of ecosystems and environmental resources.

The international acknowledgement of the precautionary principle reflects developments from the late 1960s and early 1970s in the thinking of national governments. These developments include a mention of the precautionary principle in the Swedish Environmental Protection Act of 1969 (Sunstein, 2005, p. 16) and the German government's assertion of the principle in its 1976 environmental policy statement, which read as follows:

precautionary environmental policy requires ... that natural resources are protected and demands on them made with care rather than just the warding off of imminent hazards and the elimination of damage that has occurred (*Umweltbericht' 76*, p. 26, as translated in von Moltke, 1988, p. 58)

Partly as a result of the controversies between the UK and Germany in the 1980s over acid precipitation and the disposal of waste to the North Sea, and partly as a result of the domestic development of thinking, the UK government formally adopted the precautionary principle in its 1990 policy strategy, published as *This Common Inheritance*, though its acceptance was framed in such a way that the government could be seen to be cautious about precaution:

Where there are significant risks of damage to the environment, the Government will be prepared to take precautionary action to limit the use of potentially dangerous materials or the spread of potentially dangerous pollutants, even where scientific knowledge is not conclusive, if the balance of likely costs and benefits justifies it. This precautionary principle applies particularly where there are good grounds for judging either that action taken promptly at comparatively low cost may avoid more costly damage later, or that irreversible effects may follow if action is delayed (*This Common Inheritance*, 1990, p. 11).

In short, acceptance of the precautionary principle in one form or another is widespread both at the domestic and the international levels, and it is taken as the basis upon which arguments may be mounted for measures of pollution control and nature protection.

Despite this history, the precautionary principle has not been without its critics. For example, Wilfred Beckerman (1995, pp. 2–3) has argued that there is nothing new in the precautionary principle, and that it is recognised in daily life simply as the maxim that we need to take measures to guard ourselves against foreseeable risks. He goes on to argue that interpretations of the precautionary principle that rest upon a one-sided emphasis of the risks to human health and the environment mean that the costs of measures taken in the name of precaution might actually be excessive in terms of their costs. Wildavsky (1995) offers a similar critique saying that the gains from ultra-cautious health and environmental regulation may be very small, whilst the disadvantage is that other values – including freedom, justice and excellence – may be sacrificed. Moreover, according to Wildavsky, to the extent to which rising prosperity is a necessary condition for improvements in human health action taken under the influence of a strong version of the precautionary principle is likely to be counterproductive, undermining the basis of prosperity upon which effective measures can be taken. In short, according to these critics, where the precautionary principle is unobjectionable, it is simply common sense; where the principle goes beyond common sense, it is objectionable both in theory and in practice.

To the view that the principle is too demanding on policymakers, we can add the criticisms from some sections of the green movement that the principle is not adopted in a demanding enough form. For example, it is sometimes argued that any possibility that a process, technology, chemical substance or product may be harmful to the environment or human beings is a reason to ban or prohibit the source of the risk in question. For example, a spokesperson from Greenpeace offers the following version of the principle:

The precautionary principle is a principle that puts the burden of proof onto the polluter rather than the environment. If a polluter cannot prove that what he is discharging will not damage the environment and will not harm the environment then he simply isn't allowed to discard that sort of waste. (Greenpeace spokesperson on BBC2 *Nature*, cited in North, 1995, p. 256).

In this version of the principle, caution in the face of environmental risks ceases to be a matter of common sense that is involved in taking measures that may cause harm and becomes a demand for the radical restructuring of production and the economy.

Why should there be these disagreements over the principle of precaution? If it is so widely cited and used as the basis for policy measures, we might expect there to be a clear understanding of its meaning, scope and implications. Instead, we find governments disputing its formulation and contesting its applications, whilst policy commentators, researchers and activists are divided as to whether the principle is valuable or not, and whether policy should or should not be based on the principle. To understand these agreements, it is useful to see discussions of the precautionary principle as turning on three sets of interrelated questions. First, there is the issue of how the principle should be defined, and what claims are being asserted as following from it. Second, there are questions about how policymakers should deal with the inevitable uncertainties about cause and effect relationships that environmental issues throw up. Finally, there are questions about how the values protected by application of the principle of precaution stand relative to other values. The remainder of this chapter looks at these three issues in turn.

DEFINING THE PRECAUTIONARY PRINCIPLE

In many ways we should not be surprised that there is controversy over exactly what claims are being made when the precautionary principle is asserted as a guide to policy. Even in the relatively closed context of a national policy community, there can be vagueness and ambiguity about what situations and problems the precautionary principle is supposedly addressing and what practical recommendations it might imply. Thus, one leading German legal scholar, writing in 1988, identified no less than eleven different senses in which in German law and policy the precautionary principle had been invoked (Rehbinder, 1988).

1   Preventing future damage that arises indirectly from certain sources.

2 Environmental quality improvements that allow a margin of protection for environmental restoration.
3 The reduction of risks that cause knowable, but very unlikely damage.
4 The reduction of risks where the damage is unknown.
5 Minimising environmental stress, even when there is no identified hazard or possible risk.
6 Choosing the best possible environmental option.
7 Stopping things getting worse.
8 Preventing contamination of the environment even when there is no proof that it is polluting.
9 Enforcing zero emission strategies.
10 Avoiding inconsistency in preventive strategies.
11 Forbidding the permanent use of certain environmental resources.

In other words, the principle of precaution was invoked as the basis and justification for a wide range of policies, some of which might even be inconsistent with one another. For example, choosing the best possible environmental option might involve using the assimilative capacity of the environment to absorb some pollutants, whereas enforcing zero emission strategies involves denying that the environment should be used as a disposal route. So, if within the circle of policy makers well used to thinking about and applying policy principles, there is disagreement as to what the precautionary principle means and implies, it is not surprising if it stimulates more extensive disagreement when international agreements are being sought or activists and commentators are arguing over its merits.

However, although there are important and substantive disagreements about the definition and implications of the precautionary principle, any formulation of the principle has to deal with an unavoidable problem in environmental policy, namely, how to set a standard for the control of pollution or a technology in a situation in which there is uncertainty about what cause and effect relationships are involved and in a situation in which action has costs, either in the form of economic costs or in terms of other important social and political values. In other words, any version of the precautionary principle has to involve reference to three questions. What should be the standard of control for pollution or a technology? (This includes the possibility that the standard may be a complete ban.) How should policymakers treat the scientific uncertainties that accompany the setting of such a standard? And, what benefits or values is it worth giving up in order to achieve precaution? Various formulations of the precautionary principle can be seen as offering different answers to these questions. But all three issues need to be addressed before we can have a discussion about how precautionary

environmental policy should be formulated and implemented.

To appreciate this point, consider the way in which environmental problems might be dealt with. For example, no matter how important the issue, if there is no way in which a standard can be set, then no policy is relevant whether precautionary or not. We cannot take a precautionary attitude towards the laws of thermodynamics, for example. It only makes sense to think we have a policy choice, if there are measures that policymakers can undertake that would prevent environmental damage. Equally, if there were no costs to adopting a highly precautionary attitude, then debates about the precautionary principle would be purely hypothetical. A strict standard could be adopted just as easily as a lax standard. If we can be strict as readily as we can be lax, then it would obviously make sense to set a strict standard, since even if the danger is remote and highly uncertain, we should want to guard against it. Of course, in practice, extra safety usually comes at a cost, and the point of asserting the precautionary principle is to say that we should be prepared to incur the cost in order to obtain the extra benefit of the added safety.

So the precautionary principle recommends a policy that guards against risk where there is uncertainty about how, whether and to what degree those risks will materialise. Understood in this very general way, the precautionary principle is an extension of well-established practices in the face of risk and uncertainty. Chairs are designed to take a weight that is much heavier than normal human weight, buildings are designed to withstand stresses from the weather that they are unlikely ever to encounter, planes are designed to function with only one engine working, sea defences are built to withstand wave conditions that may only occur once in a large number of years and shatter-proof glass is installed in cars and buses. All such examples are covered by the principle that there should be a margin of safety in circumstances where the exact stress that is to be placed on the system is unknown. Baroness Platt (1991, p. 3) gives the example of air transport in its early days when the passengers were weighed in addition to the baggage, in case the heaviest passengers were all seated at the after end of the aircraft, increasing the chance that the plane might go into a spin. These are all examples where under normal circumstances there is not a problem, but where it is impossible to foretell in every case what combination of conditions will obtain and where a particular combination of conditions, though extremely unlikely, might cause a serious untoward event.

At this point, it is useful to use a logical distinction familiar to philosophers, namely the distinction between concept and conception. A concept offers a general account of a particular idea, along the lines I have just set out in relation to the precautionary principle. The general concept of precaution is that policymakers should guard against risks to the environment and human health, even though there is some cost of doing so. Particular conceptions of the precautionary principle spell out this general concept in different ways. To illustrate this, consider the two contrasting versions of the precautionary principle I offered earlier from the UK government and Greenpeace. The UK government says that precaution is justified when scientific evidence of harm is not conclusive (so that there has to be some evidence suggesting harm, perhaps even requiring a balance of probabilities to that effect) and when the cost–benefit ratio is favourable or when irreversible damage is threatened. Its conception is a relatively mild version of the precautionary principle. It would justify action in some cases where a strict requirement to prove causation could not be met, but it would not licence huge investments where evidence of benefit was low. By contrast, the Greenpeace conception places the burden of proof upon the potential polluter to show that there is no evidence of harm, and it would seem that no benefit, no matter how large, could outweigh some slight evidence of risk. This highly risk-averse version of the precautionary principle is therefore strong in force and character.

These different conceptions of how precautionary policymakers should be in the face of threats represent only two possibilities, and there are clearly many more. These different versions will vary depending on how the relevant uncertainty is understood, and in particular how much evidence of the extent or likelihood of harm is required before public action can be justified. For example, if a great deal of evidence is required, particularly to justify a claim to serious or irreversible damage, then the corresponding precautionary principle will be weak. By contrast, if only indicative evidence is required, then the corresponding principle will be strong. However, in addition to the problem of uncertainty, the strength of the principle will also depend on how the costs are valued. For example, if the threshold of expenditure is set high, so that a lot of evidence of harm is needed in order to spend only modest amounts, then the

implied precautionary principle is weak. On the other hand, if large expenditures are willingly incurred for remote threats, then the precautionary stance is strong.

Table 40.1 illustrates this line of argument, although in a highly simplified form. On one axis it puts the degree of evidence that is available to the policymaker; on the other, the degree of environmental damage that may be at issue. Where evidence is high for a low or medium risk, then there is a clear argument for action. Where the degree of evidence corresponds to the degree of threat, we have cases where decision-makers will take action proportionate to the risk of damage. However, from the precautionary point of view, the interesting cells are the ones where the evidence may be low, but the prospect of damage high. In these cells, we have cases of decision-making that are precautionary or highly precautionary.

Distinguishing between the concept of precaution and varying conceptions of precaution enables us to see that there is not one precautionary principle. Rather, we should think of a precautionary attitude, characterised by a willingness to act on threats, even when the risk is unclear or unlikely, but to different degrees depending on how the threats and the costs are evaluated. Proponents of a strong conception of the precautionary principle will act with less evidence than those who hold to a weaker conception of the principle.

THE PROBLEM OF UNCERTAINTY

A strictly risk-proportionate approach to environmental policy decision-making would make the willingness to undertake costly decisions proportionate to the evidence of damage. If we have evidence that allowing a technology or permitting a level of pollution would cause environmental damage, then we should be prepared to take more costly measures than if there was less evidence. There are two underlying reasons for this approach. First, environmental damage is itself costly, both in terms of its direct environmental effects and in terms of the consequent economic effects that environmental damage might have. Thus, if the discharge of acidifying

Table 40.1 Evidence and likely risk

	Likely Risk Low	Likely Risk Medium	Likely Risk High
High	Clear case for action	Clear case for action	Proportionate case for action
Medium	Clear case for action	Proportionate case for action	Precautionary case for action
Low	Proportionate case for action	Precautionary case for action	Precautionary case for action

substances is permitted from large combustion plants, this cannot only cause direct environmental damage but can also lead to clean-up and repair costs. Hence, if we undertake a cost–benefit analysis in such cases, then we will find on a number of occasions that the cost of action is justified by the benefits that it brings. The more certain we can be of these beneficial effects, the more reason we have to believe that the expenditure will not be wasted.

The second reason why proportionality of evidence is important is that we should be more prepared to take action when the evidence is clear or it is more reliable evidence, because we have more reason to believe that our action will be effective. Indeed, in some cases, unless we have reliable evidence about the cause and effect relationship, it is possible that public action will not only be ineffective but also counter-productive, so that it will actually frustrate the end that it was designed to serve. For example, recent studies of the role of badgers in the transmission of tuberculosis in cattle suggest that destroying sets where badgers are suspected of being the cause of transmission can worsen the problem, because survivors of the cull range more widely than they otherwise would have done. Even if the effects are not nearly as counter-productive as this example suggests, they may cause side effects that are environmentally damaging. For example, the building of tall chimneys to deal with urban air pollution in the 1950s and 1960s was the cause of the dispersal of polluting substances that were then associated with acid rain.

What happens, however, when our evidence is uncertain, but the consequences of not acting or acting inadequately could be serious? If the evidence for global climate change is not strong or is disputed, as it has been in various stages of assessment of the issue, we may still feel compelled to take public measures, because the consequences of not acting are so widespread and serious that inaction is unjustified. In other words, the willingness to take action ought not only to depend on the strength of the evidence for the presumed damage, but also on the scale and seriousness of the consequences that are involved. Some evidence for a serious effect may be as much justification as we need to act as would strong evidence for a mild or not very harmful effect.

There can be various sorts of uncertainty involved in the assessment of effects. One type relates to whether or not there is an effect arising from a technology or process at all. It can be a matter of difficult scientific research to establish whether or not an effect is present. Do mobile phones have adverse health effects for those who use them? Will biodiversity be reduced by the introduction of genetically modified crops?

Is BSE transmissible to human beings via the food chain? These and many other questions have been at the centre of regulatory controversies in recent years, and the issues have turned as much on the question of whether there is a cause and effect relationship as to what the scale of that relationship might be. Moreover, even those who are prepared to admit that there are effects will sometimes disagree as to the seriousness of those effects for the environment. For example, those who defended the UK practice of dumping sewage sludge in the North Sea were willing to argue that not only did it fall within the assimilative capacity of the environment, but also that some of its effects were beneficial because it provided nutrients for bottom feeders.

However, even where a cause and effect relationship is accepted, there can still be important controversies over the scale and importance of the effects, difficulties that may be relevant in policy terms. In other words, the relevant question might be for a substance or process not 'does it cause damage?' but 'how much damage does it cause?'. For example, in setting safety standards for exposure to chemicals, the problem of estimating the size of the effects for any given dose is central. Moreover, issues of the existence of an effect and issues of the scale of an effect can come together when there is a need to consider whether there is likely to be a threshold below which no detectable effect occurs. In these circumstances, where it is difficult to establish whether or not there is such a threshold, the policy question is to decide whether one can assume that there is or whether the responsible course of action is to assume that there is not. With such issues of low-dose exposure the question of whether or not there is an effect merges into the question of whether an effect can or cannot be measured.

The problems of uncertainty do not end here, however. Questions can be raised not only about how weak or strong the evidence is for a particular claim of risk, but also about what constitutes reliable evidence and what sort of evidence is acceptable in making a practical decision on standards of protection. The UK's Royal Commission on Environmental Pollution (1998) in its report on standard setting draws attention to a scheme of classification proposed by Funtowicz and Ravetz (1990), which is reproduced in Table 40.2. This is a scheme of research pedigree, aiming to indicate the reliability that can be placed on empirical claims about environmental and related processes in the light of the research process that generated those claims. The classification turns on two dimensions of importance. The first is the nature of the research process that generates the relevant putative findings. How far does the process rely on informed

Table 40.2 Research-pedigree matrix

Rank	Theoretical structures	Data input	Peer acceptance	Colleague consensus
4	Established theory	Experimental data	Total	All but cranks
3	Theoretically based model	Historic or field data	High	All but rebels
2	Computational model	Calculated data	Medium	Competing schools
1	Statistical processing	Educated guesses	Low	Embryonic field
0	Definitions	Uneducated guesses	None	No opinion

Sources: Funtowicz and Ravetz (1998) as reproduced by Royal Commission on Environmental Protection (1998)

guesswork as distinct from rigorous testing, preferably of an experimental kind? The second aspect depends on the social processes of science. To what extent have claims been widely accepted through processes of peer review? To what extent have the claims secured warranted assertability?

The implication of this approach is that the more important and significant the policy decision being made, the more useful it would be to have highly reliable research findings that support the decision premises that underlie the decision. For example, if a proposal is being made to ban a polluting substance or a technology, thereby depriving some members of society from its benefits, it would be desirable to have evidence that supported the decision that itself had a high degree of warrant. This is important not only for the reasons mentioned earlier (that decisions need to reflect considerations of efficiency and effectiveness), but also because in a democratic society there is a norm of accountability, according to which citizens are entitled to know the grounds upon which decisions are taken that affect them. If a decision is taken, for example, to ban a chemical used as a fire-retardant, then manufacturers and consumers need the confidence to know that the decision has been taken on good grounds. Conversely, if a decision is taken to permit a process, for example, waste incineration, members of the public equally are entitled to know that there is sufficient evidence of a warranted kind to justify the decision that is made.

The problem for policymakers, however, is that whatever the merits of the research-pedigree approach in theory (and they are considerable), decisions need to be taken without having to wait for the results of scientific research. The paradox is that even a decision not to decide constitutes a form a decision. Thus, the moratorium on the commercial planting of genetically modified crops which has operated in recent years in the UK in effect means that there is a temporary ban on the technology and a ban that might be continued indefinitely. This situation has implications for investment decisions, research and commercial development generally. Inaction, no matter how justified it might be in terms of the need to deal with uncertainty, will have consequences.

Moreover, with highly complex or large-scale environmental processes, it can be intrinsically difficult to achieve the standards of evidence that the Funtowicz and Ravetz scheme requires. Perhaps with exposure to chemical substances we can identify a reasonable chain of evidential support that would run from guesswork to well-established findings, and as the knowledge of genetics increases the exact mechanisms involved in susceptibility to injury from chemical exposure will become clearer. Even in these cases, however, the need to make decisions will not wait upon full scientific understanding. With large-scale processes such as global climate change or the introduction of a major technology, there is no chance of securing the sort of reliability of evidence that the top rank of the Funtowicz and Ravetz scheme requires. As the Royal Commission (1998, pp. 31–33) notes, the need in such circumstances is to be open about the levels of uncertainty that obtain in relation to specific issues if shirking an important decision is to be avoided.

If action could be based on a purely risk-proportionate attitude with measures being undertaken in proportion to evidence of risk, this would be a moderately precautionary approach. Public decisions could be made on a cost–benefit analysis with evidence that was sufficient to justify the measures taken. This does not mean that there would not be any mistakes. Sometimes regulation would be over-stringent and sometimes it would be too lax, but, over time, the mistakes would cancel one another out. The difficulty is that such a situation is far from the pressures and needs of the actual world within which public policy decisions have to be made. Evidence is seldom adequate to the questions at hand, and in any case there may be disputes about what constitutes the relevant evidence. In this situation, we need to consider arguments for just how precautionary public policy should be. To those issues we now turn.

PRECAUTION AND OTHER VALUES

The argument so far has been that environmental policy cannot be purely evidence based, but has to deal with inevitable uncertainties, because sometimes there is a need to make a decision when the relevant evidence is simply not available, or where there is doubt about the quality of the evidence that is advanced. This is not to say that evidence can be ignored. It is simply to say that there is seldom likely to be an unambiguous or non-contentious chain of reasoning from the evidence to a particular set of policy measures. In these circumstances, what are the arguments for adopting a precautionary approach? The earlier discussion has also shown that a precautionary approach spans a wide range from the stringent version of the precautionary principle advanced by organisations like Greenpeace to the more cautious version favoured by the UK government. So the question we have to consider is not simply what the arguments are for a precautionary approach but how far along the spectrum of stringency should a precautionary policy be. Virtually everyone will sign up to weak precaution, because weak precaution is simply an extension of the existing practice of leaving a margin of safety when designing products or testing processes and technologies; few will sign up to the strongest version, which requires evidence that no harm will occur, because that would be too demanding a test for otherwise useful commodities. Since a precautionary approach will involve some cost in terms of other values, and not just economic values, there needs to be a balancing of different considerations. Somewhere along the spectrum, a balance needs to be found between an approach that is both precautionary and sensitive to other considerations that are relevant to the making of policy.

Corresponding to this spectrum from weak to strong there are a series of arguments in favour of a precautionary approach. Some of these arguments will imply only a weak version of the principle; others will imply a stronger version. We have already seen, for example, that existing practices of design and production operate on the assumption that it is prudent to leave a suitable margin for protection. For example, if we are considering pesticide residues in food, it will typically make sense to require standards that are

below the level at which there is any detectable health effect among those who are to consume the food. This is a commonsense position, not only because the costs of the requirement are not especially high, but also because mistakes happen and operating at tight safety margins is a way of ensuring that even if mistakes happen, they are not likely to be serious. Moreover, firms have an interest in their reputations, so that high safety standards can be seen as protection for those reputations.

Similarly, there are many cases where prevention is cheaper than cure, and taking a precautionary attitude in the design of products is likely to be economically sensible as well as environmentally prudent. For example, a 'no regrets' policy towards climate change will lead to investments in energy conservation on the grounds that this is not only good for the environment but also will lead to economic savings in the future. In this sort of case, there will be extra costs to be incurred in the development of a product, and usually in its cost to consumers, but these initial costs are more than recovered within a reasonable amount of time. As with the argument from the margin of safety, this is a relatively easy argument for a precautionary attitude and it is unlikely to be controversial.

Difficulties are more likely to arise where the margin of safety is extended very widely or where the expected environmental benefits are difficult to identify or estimate and the anticipated damage is more speculative. Critics of a precautionary approach argue in these cases that too strict a set of safety or environmental requirements is too costly in economic terms or counter-productive, even in relation to environmental goals themselves. This is the essence of the case that Wildavsky (1995) made against too strong an interpretation of the precautionary principle. He argued, for example, that the US public had been rendered over-anxious by media scare stories about toxic residues in food, sometimes leading to cases where farmers have lost revenue through consumers avoiding products that in all likelihood are perfectly safe. He also argued that the economic loss that unrealistically high standards involve undermine the prosperity which is the condition for dealing with environmental problems.

Some critics of a strong version of the precautionary principle go even further. In an attempt to reduce a strong version to absurdity forty scientists were polled for a meeting held at the Royal Institution in 2003 on the innovations that would not have taken place if the precautionary principle were applied rigidly. Their list included no heart surgery, antibiotics and hardly any pharmaceutical products at all. According to the same poll, there would have been no aeroplanes, bicycles, high-voltage power grids, pasteurisation, pesticides or biotechnology (Tudge, 2003). The point about this list is that all innovation involves some unforeseeable risk, and often some foreseeable risk. If the view is taken that no innovation is allowed unless there can be a guarantee that no risk occurs, then there will never be any innovation. Moreover, since existing technologies involve risk, it can be argued that a strict application of a precautionary approach in respect of innovation is simply self-defeating. In the search for lower risks, citizens will end up maintaining higher risks because lower-risk technologies can never be shown to be completely safe.

Thus, the arguments against too strong a version of the precautionary approach are that a strict attitude of precaution implies zero emission limits for certain toxic substances, which is simply unrealistic in economic or physical terms. Moreover, applied to new technologies, the requirement to show 'no harm' is excessive, when existing technologies may show as much harm. Proving a negative is an impossible burden of proof. A strict interpretation of the principle thus can only be held by those who assume what needs to be shown, namely that a 'precautionary' approach is better for health and the environment than the status quo. If this critical view is taken, then the strength of the precautionary approach is weakened. It can only be applied at the margins, determining increased safety standards, but not standards that go significantly beyond what the evidence suggests and what makes economic sense over the medium to long term. Is this the conclusion that we should draw? Not quite. There are at least four reasons why we should think that there is something distinctive and worthwhile in a modestly strong precautionary approach.

First, and contrary to the Royal Institution scientists (see Tudge, 2003), it is not clear that a precautionary approach in the face of innovation says that no risky innovation should be allowed. To be sure, some activists and commentators have argued along the lines, the Greenpeace spokesperson quoted earlier that if a polluter cannot prove that what he is discharging will not damage the environment then he is simply not allowed to discard that sort of waste. Similarly, they have argued that if the promoters of a technology, like the genetic modification of crops, cannot show that no harm will arise from commercial development, then no development should take place. However, this account of the precautionary principle suffers from the obvious problem of ignoring the risks that may be inherent in the status quo. Underlying the precautionary approach is a concern to reduce health and environmental risk, so that it makes no sense to ignore sources of risk that are already occurring. For example, the use of genetically modified cotton in China and

South Africa has reduced the exposure of agricultural workers to chemicals used as pesticides as well as raised production levels (Nuffield Council on Bioethics, 2004, pp. 30–34). This does not mean that the technology is without risks, including the emergence of resistant strains of pests and possibly less employment for farm workers. It does mean, however, that a precautionary appraisal of the technology is consistent with saying that it should be developed, even allowing for these risks (which are, after all, not unique to the technology of genetic modification).

If we take this view, we have to formulate the precautionary principle to allow for technological risks, whilst acknowledging that a simple demand that all innovations be absolutely safe is impossible in practice and wrong in principle. One way of putting this is to say that the precautionary approach should not lead to the prohibiting of a technology where health or environmental benefits are in principle demonstrable, or where the ban is solely on the grounds that there is a purely hypothetical chance that there may be an adverse risk. This is not to say that the benefits potentially obtainable by the technology have to be clear and obvious. All technologies may need time and experience to demonstrate their benefits. It is to say that technologies should not be precluded from being developed to the point where they can demonstrate their benefits because there is the mere possibility that they might cause significant damage.

This is a somewhat different case from the second consideration that should lead us to favour a strengthened conception of the precautionary approach, namely the need to deal with 'low probability/high consequence' events and technologies. Consider the case of nuclear power. It has often been pointed out that investment in safety standards for nuclear power stations is far higher per statistical death avoided than is true for other forms of electricity generation or for many everyday accidents like those arising from traffic on the roads. This 'excessive' investment in nuclear power safety might seem to be a perverse effect of the application of a precautionary approach. However, things are not quite so simple. Statistically speaking one can find a similar ratio between expenditure made and death or injury avoided in the case where there is a low probability of many great people dying and a relatively high probability of a small number of people dying. This does not mean, however, that the social and political significance of the two sets of events should be regarded as the same. Where there is an accident in a major plant causing injury or loss of life, questions are inevitably raised about the justice of concentrated risks, the balance between those who are exposed to the risk and those who benefit and the culture of those operating the plant. There are traditional principles of good employment practice as well as good neighbourliness that are involved in operating large plants that are not involved in recurrent and dispersed hazards, whatever the relative frequencies involved. In addition to these features that are inherent in technologies producing low-probability/high-consequence events, there are also important questions of public perception, questions that cannot be ignored in a democratic society. In narrow statistical terms, public anxiety over the safety of nuclear power plants and the like may be irrational, but they are in some societies an accompaniment to the use of the technology, and high standards are a way of underlining the fact that decisions are being made on a responsible basis.

Similar issues arise in the third consideration that argues for a relatively strict precautionary approach, namely issues to do with potentially irreversible environmental damage. In this case, it may be necessary to take action before an understanding of cause and effect relationships is established. There are of course major disputes about what are the values that are involved in environmental protection, with some accounts ranging on the human-centred side and some on the ecocentric side. However, whichever position is taken on that general question, an important element for any account of the value of environmental protection is going to lie in the uniqueness and distinctiveness of goods in question. Replica Grand Canyons would not be the same. If there is a threat that would damage or destroy something unique, then the values that are involved will be more significant than is easily captured in a conventional cost–benefit calculus.

There is a fourth and final argument for a relatively precautionary stance in environmental policy. In a democratic society governments act as the trustees for the body of citizens. As with all trustee relationships, political representatives in government are obliged to take a more cautious attitude than would be the case if they were acting on their own account. The most familiar form of trustee relationship is where someone administers the proceeds of an estate on behalf of a minor or some other incapacitated person. Trustees are expected to exercise these responsibilities carefully and cautiously. If you hear a good tip for a horse race in the pub, it is perfectly acceptable for you to spend a large amount of your own money on a bet, if you think the risk is worth the potential benefit. This is not so if you stand in a trustee relationship, where potentially large gains are foregone because the risks they involve are too great. The difference between the two cases is that trustees are supposed to be acting in the interests of others, and acquire responsibilities thereby that they would not otherwise have had.

For similar reasons, governments, acting on behalf of a collectivity and as agents for citizens, should take a more cautious attitude towards risk than individual citizens do. Critics of 'over-regulation' on precautionary grounds sometimes point out that often individual citizens choose to run higher risks than is permitted by government control over health and environment, and that this inconsistency is an embarrassment for precautionary regulation. If individuals choose to ski, hang-glide, take their cars when they could go by train, smoke, eat fatty foods and drink alcohol, why should governments seek to regulate pesticide residues in their food, the quality of bathing waters they swim in, the fuel efficiency of the products they buy and so on? Yet, there are obvious reasons why we should distinguish between individual risk-taking by citizens and collective risk-taking by governments. By and large it is individuals who bear the consequences of their own risky behaviour, knowing their own circumstances and conditions. When governments make decisions involving risks they do so on behalf of millions of anonymous others, in circumstances where they cannot reliably predict what the consequences will be for particular individuals. A more risk-averse attitude is justified by this need to make choices for anonymous others when in a trustee position. Moreover, governments are in a better position to understand the consequences of risk-taking than are individuals, who often learn by trial and error. Trial and error learning is very effective, but it does rely on the consequences of errors being allowed to lie where they fall, which is not an option for governments making collective decisions.

For these four reasons, there are grounds for governments sometimes occupying the precautionary cells in the matrix of decision-making that I have identified. This is not to say that governments should be ultra-precautionary, for there are good reasons why sometimes risky technologies have to be allowed and uncertain effects accepted. It is rather that, whereas risk-proportionality may be acceptable for individuals, the same is not true for governments.

CONCLUSION

It might seem that there is a paradox at the heart of the politics of precaution. On the one hand, the precautionary principle is widely accepted and invoked in international and domestic policy. On the other hand, there is pervasive contestation as to its meaning, scope, implications and importance relative to other values, particularly economic values. Yet, while this contrast holds, there is no reason to think it paradoxical. Indeed, in political terms, the more widely accepted a principle is in the abstract, the more liable it is to lead to contestation in particular cases, for the price of acceptance is commitment to action, and sometimes the action involved may be difficult or uncertain for governments. In these circumstances they need to offer an account of their policy stance that involves showing how that stance is consistent with acceptance of the precautionary principle but rejects the implications for policy that others have drawn.

An important reason for a relatively strong version of the precautionary principle is the trustee relationship that democratic governments have with citizens. Yet, in many ways, this justification raises as many problems as it solves. A reasonable precautionary stance requires consideration of many contingencies. It is not simply a matter of banning any technology or process that might cause a risk; nor is it a matter of permitting any technology or process where the risk of damage is uncertain. Instead, a judgement has to be made about the quality of the evidence for the risk and the seriousness of the damage that might occur if action is not taken. These are judgements that involve a great deal of deliberation and weighing of evidence. Despite the experiments in citizens' juries, consensus conferences, deliberative polls and other innovative forms of public consultation, it is difficult to open up to public discussion such deliberation and assessment of evidence. Whether we can democratise decisions on precaution remains an open question. That democracies need precaution is not.

REFERENCES

Beckerman, W. (1995) *Small is Stupid: Blowing the Whistle on the Greens*. London: Duckworth.

Bradford-Hill, Sir Austen (1962) *Statistical Methods in Clinical and Preventive Medicine*. Edinburgh and London: Livingston.

Funtowitz, S.O. and Ravetz, J.R. (1990) *Uncertainty and Quality in Science for Policy*. Dordrecht: Kluwer Academic Publishers.

Moltke, K. von (1988) 'The Vorsorgeprinzip in West German Environmental Policy' in Royal Commission on Environmental Pollution, *Twelfth Report, Best Practicable Environmental Option*. London: HMSO, Cm 310, Appendix 3, pp. 57–70.

North, R.D. (1995) *Life on a Modern Planet*. Manchester: Manchester University Press.

Nuffield Council on Bioethics (2004) *The Use of Genetically Modified Crops in Developing Countries*. London: Nuffield Council on Bioethics, available at www.nuffieldbioethics.org

Parker, R. (1975) 'The Struggle for Clean Air'. In P. Hall, H. Land, R. Parker and A. Webb (eds.) *Change, Choice and Conflict in Social Policy*. London: Heinemann, pp. 371–409.

Platt of Writtle, Baroness (1991) 'The Public Acceptance of Risk and Innovation'. In L. Roberts and A. Weale (eds.) *Innovation and Environmental Risk*. London and New York: Belhaven Press, pp. 1–12.

Rehbinder, E. (1988) 'Vorsorgeprinzip im Umweltrecht und präventive Umweltpolitik,' in Simonis, U.E. (ed.) *Präventive Umweltpolitik*. Frankfurt/NewYork: Campus Verlag, pp. 129–141.

Royal Commission on Environmental Pollution (1998), *Twenty-First Report: Setting Environmental Standards*. London: The Stationery Office, Cm 4053.

Sunstein, C.R. (2005) *Laws of Fear: Beyond the Precautionary Principle*. Cambridge: Cambridge University Press.

HMSO (1990) *This Common Inheritance: Britain's Environmental Strategy*. London: HMSO, 1990, Cm 1200.

Tudge, C. (2003) 'Where to Draw the Line,' *New Scientist* 178, (2395) (17 May): 23.

Wildavsky, A. (1995) *But Is It True? A Citizen's Guide to Health and Safety Issues*. Cambridge, MA: Harvard University Press. Conclusion: Rejecting the Precautionary Principle.

41

Environmental Risks and Public Perceptions

Ulrich Beck and Cordula Kropp

INTRODUCTION: ENVIRONMENTAL RISKS, MANUFACTURED UNCERTAINTIES AND THE THEORY OF RISK SOCIETY

The title "environmental risks and public perception" may suggest just another sociological analysis of the social perception of factual risks, at best in the footsteps of social constructionism or post-modern criticism. Such an approach focuses on different and contested social *risk constructions* which are strongly shaped by variations in individual and collective understandings. However, it programmatically abstains from all questions of the *constitution and the social impacts of risk*. But risks can teach us about possibilities of reflexive learning in political institutions which in times of pure administrative rationality tend to be fixed and immutable. So, the question of institutions and policies in mind, our contribution aims at a discussion of both, new risks, their characteristics and perceptions, and the interactions of institutions, public policies and environments in world risk society.

The term "environmental risk" misleads us to a conception which fundamentally distinguishes between risks in an (abused) nature outside and man in (an ignorant) society inside. Contrary to this view, it is exactly this distinction that is at risk to produce risks and to regard them as external whereas risky transformations go on – "inside" and "outside" society and its diverse environments (cf. Beck, 1992; Latour, 2004a). Moreover, the understanding to treat only those hazards and threads "risky to environment" as *environmental risks*, but not "risky environment" as such, is

becoming ever more obsolete by the implosion of its funding distinction.

Some social scientists, overwhelmed and shocked by the destructive force of the Tsunami in December 2004, blamed risk studies in this spirit for their hidden techno-averse agenda resulting from an inappropriate picture of endangered nature. They pointed to the fact that nature itself is a constant threat to humankind which only scientific–technical development can control. However, in Indian newspapers the seismic sea wave was not only interpreted as nature coming over us, but also as being baneful and deadly as a result of human high-density buildings on the coastline, the ecological destruction of mangrove woods and a decreasing awareness of coastal risks – all as a consequence of Western mass tourism and international competition. Obviously, in India even the Tsunami catastrophe has been perceived as socialized nature (*vergesellschaftete Natur*), as a risk presuming techno-economic modernization, decisions and considerations of utility. Today, natural disasters also differ from pre-industrial hazards by their (partial) origin in decision-making. Furthermore, if some populations can be warned but not others, one may ask whether the Tsunami has really been a "stroke of fate," raining down on people from "outside." As vulnerability to disaster is an "uneven matter" (see also Smith and Goldblatt, 2000), the problem of social accountability and responsibility irrevocably arises and scrutinizes the distinction between "risks" being related to decision-making and "hazards" to "strokes of fate" – both in the techno-scientifically produced world risk society.

The perils of global climate change, BSE, nuclear power, asbestos, and so on, tell us, that risk, risk perception and the social meaning of risk cannot be discussed separately. Nevertheless, this is what happens time and again in risk assessment, in risk management and in sociological risk studies as well. To broach the issue of risks without discussing society, civilization and the continuous process of manufacturing uncertainties on a global scale misses the sociological outlook and the societal self-reflection as well.

In order to seize the consequences of the emerging world risk society, however, it is necessary to develop a broader understanding of environmental risks and their public perception as embedded in the framework of (a) manufactured, interdependent uncertainties and (b) global risk constructions and deconstructions no longer in the hands of experts or (national) authorities alone. Both the increasingly perceived interconnectedness of risks resulting in a potential to connect and transform entire rule systems and landscapes of legitimacy and responsible authorities *and* the new (global) scope of negotiations and controversies about these risks urge us to move our attention from an isolated discussion of "environmental risks" to a comprehensive consideration of risks and side-effects, their consequences and secondary consequences in society and sociology as well.

This is why we will discuss environmental risks in the following sections against the backdrop of the theory of "world risk society" (cf. Beck 1996, 1999 and see pp. 5). We will first give a short outline of the characteristics of a specific type of risks which in sociological literature is referred to as "new risks" (Lau, 1999). Our aim is to identify not only the qualities of such new risks but also, more importantly, their impact on societies in an age of global uncertainties and global public awareness. We will go on to have a look at the social construction of environmental risks and the role of scientific experts, politicians, media, and the global public, and the interplay of all these actors in creating the world risk society. Our attention will focus on the challenge for the existing institutions concerned with risk regulation and management. Central social institutions in our health system, in the systems of attribution, liability and compensation, and in the administration of nature, technical development and political decision-making are still founded on more or less universal understandings of risk, harm and *force majeure*, respectively, in responsibility versus fate. These fundamental distinctions in constructing understandings of risk which restrict and regiment human behavior are fading away. It is important to acknowledge that, in the end, given distinctions, new social norms dealing with and deciding about risks have to be negotiated and established for reliable and legitimate problem solving (Beck and Lau, 2004). Finally, we will point to the relationship between generalized risk perception, global public debate and emerging options of reflexive governance.

NEW GLOBAL RISKS AND CHANGING STRUCTURES OF RISK PERCEPTION IN SOCIETY

Let us first explore the new landscape of manufactured uncertainties emerging in the course of successful modernization and especially as a result of the techno-economic progress. Then we will ask in which way these new risks call for a different self-reflection in and about society.

Landscape of manufactured uncertainties

A review of the historical concept of "risk" points out that the term, most of the time, referred to potential dangers consciously taken by individuals seizing economic or biographical opportunities (Beck, 1995; Bonß, 1995; Adam and van Loon, 2000). It applied to a world determined by laws of probability and more and more by the insurance principle (Ewald, 1991). At its core stood the probabilistic assessment of potential hazards in relation to expected benefits. The risk was measured by the extent of damage and its calculable probability of occurrence. Even though this calculus of risk is persisting in most of society's institutions of risk management, the risks we are speaking of cannot be calculated. They are not limited in space, time and probability anymore. Climate change, air pollution and holes in the ozone layer are risks that transgress the boundaries of nation-states, individual life spans or responsibilities.

These dangers to civilization arising from civilization – the risks of the risk society – go far beyond such limited concepts and, instead, set off complex and fluid landscapes of manufactured uncertainties with up to now unseen risk mobilities, "full of unexpected and irreversible time–space movements" (Urry, 2004, p. 97; see also Law, 2004). Even less they do result from external, natural forces or objectionable behavior. Rather, modern risks result from the societal, usually technology-based pursuit of highly valued goals and successes in the process of industrial modernization; in short they are *manufactured uncertainties*. As manufactured uncertainties emerge in a complex network of heterogeneous areas and involved rationalities, in an ever-expanding context of technical, social, financial and economic opportunities and decisions, they have also been called "systemic risks" (Renn and Klinke, 2004; Klinke and Renn, 2006). At the crossroads between events and developments classified as natural (even though more or less altered by human action), economic (but always embedded in cultural logics), social (or political) and technological (more scientific), both at the domestic and the international level (Klinke and Renn, 2006), complex and per se unlimited risk phenomena and perceptions arise in disregard of all societal boundary making. It is their unlimited scope, their embeddedness in different though interrelated fields, their indeterminacy and multi-causal nature, and, most importantly, the irreducible uncertainty which shape emerging global risks today. These can be qualified by six main characteristics:

1. *Manufactured uncertainties may produce irreversible consequences, unlimited in time and space.* Measuring "risk" in probabilistic terms presupposes a concept of "accidents" as things that happen at a particular time and in a particular place to a particular group. But none of these tacit assumptions hold for the accident that occurred at Chernobyl. Twenty years afterwards, some of the victims may not have been born yet and some of the victims in Germany or Sweden may not know about the connection. Now, one might argue that while a 10,000-year time horizon might be necessary in the case of nuclear accidents, it does not apply to genetically modified (GM) organisms. But the opposite is the case. Once released into the environment, some GM crops or foods may cause accidents just as unbounded as a nuclear accident, because the successful industrial universalization of food will *abolish all spatial, temporal and social boundaries.* Everything that is celebrated as the triumph of gene technology – like its universal applicability, and its power to increase productivity – will have the effect of spreading it that much faster throughout the worldwide food chain. If genetically modified organisms are allowed to spread out, they may last until the end of time.
2. *The problem of long latency periods.* Many environmental problems are characterized by a delay between initial event and hazardous outcome that may go unnoticed for a long time. The clear example for this problem is asbestos. Everyone knows it was used extensively for decades before it was finally unmasked as a health hazard. Less familiar are the losses, which have totalled $17 billion so far, with final figures predicting some $40 billion. The danger lies in the speed of hidden change and its extent. Another case in point is climate change. Neither can we ultimately know when and whether the "cattle

culture" (Rifkin, 1992) will set off climate change nor when and whether the changing climate will cause floods in this or devastation of landmasses in this region.

3. *Known and unknown unawareness.* The unknown outweighs by far what is known. Clearly, scientists today know much more about BSE than when the crisis started. But even now, 20 years after the disease's discovery, its origins, its host range, its means of transmission, the nature of the infectious agent and its relation to its human counterpart nvCJD (new variant Creutzfeldt–Jakob disease) remain largely unknown. New systemic risk may offer no narrative closure, no ending by which the truth is recovered and the boundaries stabilized. This is why new ways of dealing with these risks and the controversies about them are needed – politically (Beck, 1997; Latour, 2004a) and epistemologically (Wehling, 2003). Today, the public and the decision-maker are in need for instruments to enable broad and reflexive negotiations with all relevant knowledge claims where data from different sources are integrated to a holistic approach identifying hazard – and risk deliberation (Böschen *et al.*, 2007). The challenges to integrate deliberation and heterogeneous expertise call for new and legitimated distinctions and decision rules while the old ones are becoming obsolete and implausible (Beck and Lau, 2004).
4. *The dominance of public perception.* Risk acceptability depends on whether or not those who carry the potential losses will also receive the benefits. If this is not the case, or if such an attribution is systematically impossible, the risk will be unacceptable to those affected, to sum up the findings of the psychometric paradigm and of system theory as well (Fischoff *et al.*, 1978; Slovic *et al.*, 1980; Luhmann, 1993). If even the benefit is in some dispute – as is the case with GM food – it is not enough to demonstrate that the "residual risk" is, statistically speaking, highly improbable. A risk cannot be considered in and of itself. It is always framed by the criteria used in evaluating it, and colored by the cultural assumptions that surround it (see below). Or to put it another way risks are as big as they appear. This comprehension of early risk studies is particularly true in the case of manufactured uncertainties.
5. *Virtual risks.* In the context of manufactured uncertainties, the subjunctive has replaced the indicative. Possibility is accorded the same significance as existence. This is in large part because our past has been so thoroughly rewritten. Many things that were once considered universally certain and safe, such as agricultural fertilizers for instance, and vouched for by every conceivable authority, turned out to be deadly. Applying this knowledge to the present and the

future devalues the certainties of today. This is the soil which nurtures the fear of conceivable threats. Virtual risks no longer need to exist in order to be perceived as facts. These risks may be criticized as phantom risks, but this does not matter economically. Perceived as risks, they cause enormous financial and social losses by effects of spill over and functional amplification. Thus, the distinction between "real" risks and "hysterical" perception no longer holds. At least *economically* it makes no difference.

6. *Global risks.* By definition, global environmental risks cannot be regulated at a national level. BSE and FMD (food and mouth disease) painfully remind us how easily national risks become global ones. Both hazards confronted nation states with their limits to predict, manage or control risks in a chaotically interacting world, for example, the cross-national diversity and tension of regulatory standards resulting from conflicting EU regulation, from different decision-making processes, and so on. FMD, which seemed to be a beef disease spreading apart from human action first, finally had to be regarded as the result of political decision-making and non-vaccination policy that reacted to the trading rules of global markets. But we can see in many other cases too, nuclear energy or genetically modified crops for instance, that modern science has left the laboratory and the safety of contained and monitored experiments behind, now using the whole world as an experimental playground for risky research and experimentation. Risks are now intensely mobile, fluid, easily spreading beyond national and social borders and quickly expanding across the globe (Bankoff *et al.*, 2004). Nevertheless, we witness heterogeneous spaces of problematization, and, ever more noteworthy, contingent "assemblages" of technology, politics and norms defining risk management in "zones of technological regulation" with highly variable effects in rich and poor countries (Ong, 2004).

Environmental risks in the world risk society

The result of the outlined six main characteristics of risks today is a *crisis of global interdependence*, perceived as the "world risk society" (Beck, 1999). Therein at least three types of (interrelated) risks may be identified: global financial risks, which initially appear as individual and national, the threat from terrorist networks supported by particular states and – well known in the meantime – ecological risks that easily set up a global dynamic of risk definition and risk avoidance conflicts. Although for all three types growing risk awareness has the potential to cause border-crossing conflicts, environmental risks have their particular logic: first, although decisions are at their origin, ecological risks are the non-intended results of decision-making and must be conceptualized as the side-effects of successful modernization, as the accumulation of "bads" that goes together with the production of "goods" (similar to risks stemming from global financial markets and in sharp contrast to the deliberately planned effects from global terrorist networks).

Second, most of the time environmental risks emerge after long periods of latency and have to pass scientific, media and public attention to come into existence. Thus, risk debates and contending perceptions are their first typical outcome. Third, contrary to economic risks it is hard to individualize and nationalize ecological risks which generally spread in, through, over and under national borders. Global environmental risks highlight the new historical realization that no nation can deal with its problem alone. Fourth, in contrast to financial risks and risks resulting from terrorism, ecological risks become only sociologically important by their "externalities." They affect social structure only indirectly. Nevertheless, sociologists should distinguish between wealth-related ecological destruction such as the ozone hole or the greenhouse effect and poverty-related destruction such as the loss of rainforest. The perceived interdependency in the world risk society can result in conflicts and in cooperation – it will always create transnational commonalities. These will alter social interactions on the macro-, meso- and micro-level.

On the meso- and micro-level, the growth of global risk potentials enters everyday life and public discourse. Thereby, the public perception of unleashed risks and uncertainties is strongly influenced by mass media-dominated perceptions. The more ubiquitous a threat is represented in the mass media, the greater its political power to explode all business-as-usual answers up to the macro level (Adam *et al.*, 2000). We therefore argue global risk expectations constitute a transition from one epoch into another. Nowadays, increasing portions of everybody's normal course of life, from the food we eat to the entire lifestyle we choose, become significantly conditioned by risk recognition and responses that try to deal with uncertainty and assess personal impacts. At the same time the space of everyday experience has become a space of cosmopolitan interdependence. Global risks symbolize global interdependence and they force, in turn, the pace of further dependencies. Sociologically speaking, this means an alteration in spatial and temporal relations; what was far away is drawn near, in the dimension of time as well as of space (Harvey, 1996). In contrast to the global chains of commodity or food production, which need not to become objects of perception or awareness, the explosiveness of global risks and their attendant propensity to shatter all political boundaries ensures that they break as a rule into public awareness. One of their peculiarities consists in the combination they present of global interdependence and actual consciousness of global interdependence, that is, in their "reflexive globality" (Beck, 2005). Hence, risk-regulating institutions are concerned ever more as they do not dispose of the same infrastructure and mobility as the risks themselves.

The transformation of society as secondary consequences of risks and of global public awareness

Last but not least, modern society has become a "risk society" in the sense that it is increasingly occupied with debating, preventing and managing risks that it has to attribute to its own intentional decisions and developments. Consequences and side-effects of manufactured uncertainties cause radical changes in social structure, politics and cultural experience. As a secondary consequence of risks or, more precisely, as a secondary consequence of their global perception and negotiation, the optimistic idea of "progress" had to give way to greater social awareness of endangering risks and longstanding pollution, so that "today 'risk', by its very pervasiveness, seems to be the defining marker of our own less sanguine historical moment" (Jasanoff, 1999, p. 136). We witness the emergence of an increasingly "self-conscious" risk society. Here, the self-imperilling process of environmental destruction and risky technological development gives rise to a risk-sensitive public across the globe which is reflexive about ultimate uncertainties as well as about the narrowness of national management systems.

A qualitative transformation in the public perception of society, its functions and promises is the immediate result. At best, a global public sphere might appear as a side-effect of the unintended side-effects. If the international system of second modernity is subject to transnational imperilments (ozone depletion, climate change) which are undermining the stability of the system as it emerged during the era of first modernity, the public, as citizens, as employees and as consumers, feels to be pushed in a new global age of uncertainty. In second modernity, the global perception of global risks might perhaps represent the last source of new cosmopolitan bindings and interaction networks.

The global perceptions of interdependence throw also social science and political theory into crisis: the enormous shift cannot be grasped in categories stemming from the early industrial centuries.

This is why global risk society poses new challenges, not only to the politician and the man on the street but also to the practitioner of the social sciences. The social sciences must not only recognize that social interaction is no longer so clearly defined spatially and temporally as this interaction has been assumed to be in the old nation-state paradigm (Beck, 2006b). Furthermore, global risk society suggests a model of global socialization at odds with the traditional picture of *positive* social integration on the basis of shared norms and values and which is resting, as it does, upon the conflicts around *negative* values (risks, crises, dangers of annihilation). Here, the issue is not so much the multiplication of uncontrollable risks of a sort apt to generate a global interdependence as their liberation from all formerly decisive boundaries – and this all at once at the spatial, the temporal and the social plane.

GLOBAL PUBLIC PERCEPTION AS POLITICAL CHALLENGE

For any sociological consideration of environmental risks, it is today almost trivial to state that risk is a social construction – and a contested one. In consequence, one must clearly distinguish between, on the one hand, the (physical) event of factional and predicted catastrophe (or ongoing process of destruction) and, on the other, the global risk as the increasingly boundless expectation of such catastrophes. The more apparently it becomes that global risks divest scientific methods of their predictive capacity, the greater is the influence acquired by *global risk perception*. Whether a possible destructive event counts as a global risk or not does not here depend only on the number of the dead and injured or the extent of the actual damage done to nature. This perception and denotation is rather the expression of a career of social recognition.

In particular, concern for environmental issues was perceived for a long time as a quirk of the Germans; this is no longer the case since the Rio de Janeiro Conference of 1992. In other words, the ongoing process of destruction and the career of recognition which falls to the lot of a global risk are in no sense to be conceived of in terms of a cause–effect relation; rather, process and risk have their places in opposite contexts and systems of meaning and must, as regards their actual causes and effects, be examined and understood quite separately from one another.

We mentioned above that constructions of risk have their own part in secondary consequences: they might establish a public sphere, which by the way does not extend so far as the mass consciousness staged by the mass media. One of the effects of global risks which we can reliably expect to follow from them as an unintended consequence is their contributory role in regard to the public and political aspect of modern existence. In Western cultures we have just passed in the last decades an epoch of flux, uncertainty and rapid social change. At the beginning of the 21st century, the process of globalization continue to amplify these feelings: "Economic convergence, political fluctuation and national insecurity have become the motifs of the age. We are living in a 'runaway world' stippled by ominous dangers, military conflicts and environmental hazards" (Mythen, 2004, p. 1).

In this risk-sensitive global sphere, where we had once seen national integration on the basis of shared values and wealth distribution, risk has become the omnipresent issue and we see now a new kind of dialectic cross-border conflict because of the negative logic of risk and forced cooperation which is likely to intermingle social groups from very different spheres and levels: leading national politicians side by side with representatives of global economic players, with indigenous activists, epistemic communities, NGOs, mass media professionals, social movements and alerted citizens as well. Generally, what calls the public into being is not the state in itself or its practice of decision-making, but the consequences of decisions, insofar as citizens feel to be concerned and the decisions are perceived to be problematical or even to be dangerous. Matters of concern (Latour, 2004b) have always been in the beginning of the public (Dewey, 1954). Decisions as such are a matter of indifference and even tend to render social subjects indifferent as decision-making is just giving evidence of the (disencumbering) existence and functioning of responsible authorities and advising experts. What tears the normally somnolent subjects out of their general indifference – first by annoying them and dismaying them, but, in so doing, also animating them and making them concerned – is perception or just suspicion of the problematical consequences of decisions and communicating to one another of such perceptions and suspicions. It is in this way that a commonality and communities of trans- and post-national public spaces of action might be gradually established. But when public spheres stretch across borders and boundaries, the heterogeneity of contending understandings of risk is not likely to implode. Instead, a stirring up of negotiations and controversies, and thus of politics and processes of more or less reflexive reinvention of politics, can be expected when public debates about manufactured uncertainties spread across the globe.

So we want shortly to attract your attention to the enormous significance of different risk perceptions between individuals and social groups (Lupton, 1999), to the general meaning of "social rationality" (Krimsky and Plaugh, 1988; Beck, 1992) and to the institutional challenge in handling global environmental risks and global risk perception.

Uncertainty lends power to perception

All matters of concern may provoke huge losses of money and credibility and a lot of institutional irritation once they are perceived as risks. These secondary consequences of potential risks increase when the risk in question is characterized by high complexity and not well understood by science and regulation – as it is typical for environmental risks in the moment of first problematization. If systematic and legitimated knowledge about the distribution of consequences is essential but missing, the unmasked uncertainty lends power to perception. The more obvious it becomes that global risks are insusceptible of being calculated or precisely predicted by scientific method, the more influence accrues to the perception of risk. The distinction diminishes and gradually vanishes between real risks and the perception of risk. The perspective of Cultural Theory launched by Mary Douglas and Aaron Wildavsky (1982) has been crucial in the scientific discussion on social risk perception and heterogeneous risk interpretations. As risk is all about thoughts, beliefs and "ways of life," framed by language, social context, public discourses and storylines inherent to certain practices, standpoints and institutions, a person's estimate of risk here may be very different from another one's risk judgments there (Wildavsky and Dake, 1990; Adams, 1995; Hajer, 1995; Palmer, 1996; Rippl, 2002; Tulloch and Lupton, 2003). People choose what to fear and how much to fear it. But what is perceived as dangerous is not only a function of cultural and social contexts but also of an issue's career of media representation and social recognition.

Perceptions of risk are socially formed as a result not only of individual subjectivities or socially embedded perspectives, but even more arising out of individual attitudes and risk estimations in perpetual interplay with institutional discourses. The latter enable or restrict a person's capability of sense-making and judgment against the background of surrounding cultural belief systems. Social factors of risk awareness such as gender, age or ethnicity are likely to be eclipsed by careers of risk recognition in the media. Due to the media's weight, in the end it may often be the mediated public perception that ultimately defines the likelihood of political risk decisions to be broadly accepted or rebuffed. In world risk societies, the central question of power is a question of definitional authority.

So more important than all the ingenious probability scenarios of the experts becomes the question of who believes there to be a risk, and why. In other words: central importance accrues to the social-scientific investigation of cultural risk perception (Renn and Rohrmann, 2000). Global risks mean the start of a meta-power-game involving the deconstruction and reconstruction of boundaries, rules, responsibilities, "them and us" identities, spaces of action and action priorities. Paradoxical in all this is the fact that the ruling uncertainty, the indeterminability of the risk also goes to create new certainties – for example, the good conscience of being in possession of the right global risk, while the risk consciousness of the others is plainly paranoid, irrational, highly questionable and misleading as to their own real interests. The choice between different risks is also a choice between different visions of the world. Essentially involved in it are the grander questions of who is guilty and who innocent, whose star is in the ascendant and whose in decline – military force or human rights, the logic of war or the logic of peaceful accord. Varying the famous phrase of Samuel P. Huntington (1993), we might say that we are concerned with a *clash of risk cultures* in the recent trans-Atlantic political disagreements: the USA accuses Europe of suffering from ecological hysteria, while Europe accuses the USA of having succumbed to a hysterical fear of terrorism.

The loss of science-oriented dispute-settlement mechanisms and the dominance of cultural perceptions have two main implications. They increase and enforce the cross-national diversity of regulatory standards. And this diversity can cause enormous tensions not only domestically, but also in global, regional and bilateral trading systems. Even existing supranational democratic institutions have difficulties in reaching decisions. For example, in the EU, which has probably made the greatest progress in establishing transnational decision-making bodies, member states have long accepted or rejected the clearance certificates for British beef according to their own rights.

All rationality is positioned

No environmental risk is considered in a social vacuum. Risk assessment is framed by the criteria and standards used in its judgment, and colored by the cultural assumptions that surround all assessment processes. Often, authorities and citizens refer for risk evaluation to contradictory certainties (Schwarz and Thompson, 1990), to divergent epistemic discourses (von Schomberg, 1993), underlying assumptions or criteria of valuation (O'Neill, 1997). Especially where scientific rationality refers to dominant technical discourses used by scientific experts, social rationality stems from cultural evaluations convened through everyday lived experience (Mythen, 2004, p. 56). It is against this background that technical experts perceive the populations that surround them as irrational or hysterical, either because they seem to be making bad calculations of personal risk – as when smokers protest against nuclear energy – or because they express themselves with lurid images – as when Great Britain, seemingly invaded by German angst, demonized their genetically modified wonders as "Frankenstein food." It is true – it's a striking phrase. And it did serve as something of an ultimate weapon in the war of words against GM food. But it contained the important insight that even "objective" risks contain implicit judgments about what is right. Moreover, the expansion of information by the new media and information technologies has contributed to conflicts over the meaning and impacts of risk amongst competing interest groups. The ubiquity and fluidity of information and knowledge claims from different angles of legitimate speaking have enhanced channels of public communication and added fuel to visible debates between stakeholders (Strydom, 2003). Moreover, because there is no definitive authority on risk, small, self-proclaimed alternative experts in the risks of modernizations are mushrooming with highly divergent truth claims. So the various advances in knowledge production have failed to result either in a more secure social climate or in establishing new authorities in risk matters.

In the end, technical experts have lost their monopoly on rationality in the original sense: they no longer dictate the proportions by which judgment is measured. Statements of risk are based on cultural standards, technically expressed, about what is *still* and what is *no longer* acceptable. When scientists say that an event has a low probability of occurring, and hence is a negligible risk, they are necessarily encoding their judgment about relative payoffs. So it is wrong to regard social and cultural judgments as things that can only distort the perception of risk. Without social and cultural judgments, there *are* no risks. Those judgments *constitute* risk, although often in contradictory and often in hidden ways.

All of this makes clear that the socially transformative risk that concerns us here – non-probabilistic, incalculable environmental risk – is more than ever a social construction. And as such, it depends on the resources necessary to define it, and on access to those resources. This is what is meant by the "power relations of risk definition" (Beck, 1995). Once we define risk conflicts in these terms, each conflict reveals a microstructure of subsidiary struggles over the same set of questions which repeatedly recur: Who has to prove what?

Who has the burden of proof? What constitutes proof under conditions of uncertainty? What norms of accountability are being used? Who is responsible morally? And who is responsible for paying the costs? And this is true both nationally and transnationally, including along the north–south divide. When the politics of risk are explicated along these lines, they cast a rare light on shifts in epistemology and their relation to political strategy. And this in turn gives concreteness to the idea of social evolution. Changing power relations of definition are closely connected to changes in some of society's central self-definitions. And to the extent that power in risk conflicts has changed to favor social movements and NGOs, it shifts the whole context of risk conflict into a more reflexive constellation. The claimed democratization of expertise is a hopeful beginning of this process of transformation, but the proliferation of conspiracy theories is party too (Latour, 2004b).

Manufactured uncertainties and the institutional apparatus of risk regulation

The incapacity of the extant relations of definition to regulate against risk is highlighted by the continued production of ecological destruction: a symbol for the systemic failure of responsible institutions for "organized irresponsibility" (Beck, 1995). With this as a starting point, the thesis is that the resistance of modern society towards such things as genetically manipulated foodstuffs is not in essence a matter of understanding or misunderstanding calculable risks. Instead, it renders visible that the basis of power and legitimacy has been changed.

In all such cases, rather than particular technologies or the decisions made about them, it is the *unforeseeability of the consequences* that has become the source of politics. The risk profile of new, controversial technologies is determined by uneasy dissent in terms of risk perception rather than by agreed consent among stakeholders concerning opportunities. The question, therefore, is not whether a given technology is dangerous, but whether it is *perceived* as being dangerous. Genetic engineering is one of the prime examples. Some call them "phantom risks" or "virtual risks." Such theorists are inadvertently highlighting an important fact, that in the case of manufactured uncertainties, most cause-and-effect relationships are controversial, and they often remain controversial. What they miss is that this controversiality is itself a risk – an economic and political risk.

Forecasting the error zone about potentially hazardous outcomes of industrial developments and, more generally, every single claim about potential environmental risks puts the feared failure of national and international rule systems on the public agenda. The phrase global risk "society" must be understood in a specifically post-social sense, since there exist neither in national nor in international politics and society rules and institutions such as to provide fixed procedures for dealing with these risks, for ranking them in terms of urgency and for developing political and military strategies to counter them (cf. van Loon, 2002). To this extent, there tends to be played out, in each specific conflict between different perceptions of risk, also a meta-power-game concerning what rules must be adopted in future with regard to such indeterminate and illimitable risks.

The results are immense challenges of uncertainty. Global risks compel us to a new politics of uncertainty. They make imperative a distinction between things which lie by their nature beyond all possible control and things which happen in fact presently not to be under control. These are not "dangers" in the pre-modern sense, because they rest on decisions and therefore raise questions regarding the attribution of responsibility and the just distribution of blame and of costs. Political countermeasures are in every case seen to be imperative. Neither national nor international political authorities – and multinational companies are often today in the same case – can point to the fatal uncontrollability of the modern world to absolve themselves of the obligation to take action; rather, they are placed by the discourse of risk conducted throughout the global public sphere under an absolutely irreducible pressure to justify themselves by action. They are damned to countermeasures. This intense expectation of counteraction alone suffices to lend life to the contrafactual hypothesis of controllability, even when all available models of response prove inadequate. Not to take action in the face of recognized risks is *politically* out of the question – regardless of whether the measures taken do in fact minimize the risk, increase it or have no effect at all.

CONCLUSION: THE TRANSFORMATIVE POWER OF GLOBAL ENVIRONMENTAL RISKS

The consciousness of the unpredictability of ultimate consequences has given rise to a world public that is "ultimate-risk sensitive." Global risk responsiveness in public debate may engender something like "reflexive governance." But the other part of a risk-sensitive public is the unpredictable consumer (Gabriel and Lang, 1995), among whom chain reactions can be triggered by the merest hint of plausible evidence. Since uncontested scientific evidence is increasingly rare, public perception becomes the determining element in such scenarios. And because of its political weight, it is public perception that ultimately defines the likelihood of product bans or the success of liability claims. What is true for product politics is likely to apply also for other cases depending on public approval.

Citizenship is conceived in the West in terms of nation-bound rights and duties, and this is the framework that regulates the risks that anyone living within a national territory may face. But the globalization of risk has created huge difficulties for the nation state in its effort to manage risks in a world of global flows and networks, especially when nobody is accountable for the outcomes. Environmental risks such as dioxin and CFCs are explosive reminders of the inability of nation states to predict, manage and control risk in an interdependent world of politically hybrid forms. Politicians claim not to be in charge, since the best they can do is to set the regulatory framework for the market. Scientific experts say they merely create technological opportunities but that they do not decide whether and how these are implemented. Businesses say they are simply responding to consumer demand. Society has become a laboratory with nobody responsible for the outcome of the experiment.

Thus, the inability to manage manufactured uncertainties, both nationally and globally, could become one of the main counter-forces to neo-liberalism. It could leave those bitterly behind who have put their hopes in market solutions for consumer safety problems. Recent consumer protection and product liability legislation has shown a clear tendency towards anticipating potential losses rather than being geared to losses actually sustained. Furthermore, the burden of proof seems to be shifting from the consumer to the producer in a number of fields, including genetic engineering. This opens up the field for coalitions of international law firms and consumer NGOs to try and raid the treasure chests of multinational corporations.

The nature and extent of this fundamental transformation we recognize in new key concepts – new semantics of conflict – are currently in the process of obscuring, subverting and rendering open new political possibilities in the old shared languages and self-evidences in the politics of the nation state. One of these universal "buzz words" is "sustainable development." Noteworthy here is how the demand for "sustainability" has eclipsed or entirely displaced both the discourse of technical-economic "progress" and that of "Nature" and "the destruction of Nature." This concept of "sustainable development," accusatory and exhortatory and ever more omnipresent, is an indication not only that the old shared self-evidences regarding "economic growth" and "technological progress" have ceased to be perfect and immediate self-evidences but also that their proponents now find themselves very much in defense, forced to argue for these erstwhile axioms, and against alternatives to them, at every level of the process of industrial modernization.

The heat, moreover, of the debate about what "sustainable development" actually is – or what it should be, what it implies and what it does not imply – is evidence of the degree to which the so-called environmental problems have in fact long ceased to be seen and treated as problems merely of the "world around" us and have become integrated into the social world itself, breaking forth, *in* and indeed *in almost all* social institutions (from the traffic system through architecture to the system of consumption), as political (ethical, economic, legal) conflicts. This is all the more the case as the very term "sustainable development" entails a potentially litigious contradiction: sustention *and* development, which is to say, development and non-development. Which of these two mutually exclusive demands actually imposes itself (or which one of the two imposes itself to the greater, which one to the lesser, degree) remains – as we are taught by a fundamental theorem of sociology – *qua* institutionalized contradiction a question to be decided in the arena of political power.

Global risk society is no option which might have, or might still be, chosen or rejected in a process of political debate. This societal form is rather one which arises due to the autonomous dynamic of processes of modernization which have acquired an impetus of their own and which are quite blind to consequences and quite deaf to warnings of danger. These processes tend, taken as a whole, latently to engender various self-imperilments which go to delete, to transform and to politicize the foundations of that first modernity associated with industrial society.

Conflicts over civilization risks, as just described, emerge, for example, where opinions diverge as to how far the industrialized countries have a right to demand that developing countries protect such important global resources as the rainforests, given that the former countries arrogate to themselves the lion's share of energy resources. A certain reasoning sees, indeed, in precisely such differences of opinion a reason not to speak here of a form of global socialization. But this line of thought commits the mistake of equating "society" and "consensus." In fact, such conflicts already have an integrative function on their own, inasmuch as they make it clear that the solutions found will have to be cosmopolitan ones. Still, such solutions are hardly conceivable except on the basis of new global institutions and parameters – and thereby also of a diminishing gap between the industrial and developing world. This alone is a key feature of the idea of reflexive governance: the long-term consequences, by their nature transgressive of all borders and boundaries, of that constitutively unexpected to which, nonetheless, there can be no other response than the development of uncertain expectations can provide the spark and the fundament for transnational risk communities – *Folgen-Öffentlichkeiten*, or public spheres emerging from and sustained by the necessity to deal with commonly suffered consequences – which might in turn lead to an (indeed involuntary) politicization and thus, should the circumstances be right, to a *reinvention of politics on a transnational or global level* (Beck, 2006a).

REFERENCES

Adam, B. and van Loon, J. (2000) Repositioning Risk: The Challenge for Social Theory. In: Adam, B., Beck, U. and van Loon, J. (eds) *The Risk Society and Beyond*. London: Sage, pp. 1–33.

Adam, B., Beck, U. and van Loon, J. (eds) (2000) *The Risk Society and Beyond*. London: Sage.

Adams, J. (1995) *Risk*. London: UCL Press.

Bankoff, G., Frerks, G. and Hilhorst, D. (eds) (2004) *Mapping Vulnerability*. London: Earthscan.

Beck, U. (1992) *Risk Society: Towards a New Modernity*. London: Sage.

Beck, U. (1995) *Ecological Politics in an Age of Risk*. Cambridge: Polity Press.

Beck, U. (1996) World risk society as cosmopolitan society? Ecological questions in a framework of manufactured uncertainties. *Theory, Culture & Society*, 13(4):1–32.

Beck, U. (1997) *The Re-Invention of Politics*. Cambridge: Polity Press.

Beck, U. (1999) *World Risk Society*. Cambridge: Polity.

Beck, U. (2006a) *Power in the Global Age*. Cambridge: Polity Press.

Beck, U. (2006b) *The Cosmopolitan Vision*. Cambridge: Polity Press.

Beck, U. (2006c) Reflexive Governance. Politics in the Global Risk Society. In: Voß, J.-P., Bauknecht, D. and Kemp, R. (eds) *Reflexive Governance for Sustainable Development*. Edinburgh: Edward Elgar, pp. 31–56.

Beck, U. and Lau, C. (2004) *Entgrenzung und Entscheidung. Was ist neu an der Theorie reflexiver Modernisierung?* Frankfurt a.M.: Suhrkamp.

Bonß, W. (1995) *Vom Risiko. Unsicherheit und Ungewissheit in der Moderne*. Hamburg: Hamburger Edition.

Böschen, S. Kropp, C. and Soentgen, J. (2007) Gesellschaftliche Selbstberatung: Visualisierung von Risikokonflikten als Chance für Gestaltungsöffentlichkeiten. In: Leggewie, C. (ed) *Von der Politik-zur Gesellschaftsberatung. Neue Wege öffentlicher Konsultation*. Frankfurt: Campus, pp. S223–246.

Dewey, J. (1954) [1927] *The Public and its Problems*. Athens, OH: Swallow Press/Ohio University Press.

Douglas, M. and Wildavsky, A. (1982) *Risk and Culture*. Berkeley, Los Angeles, London: University of California Press.

Ewald, F. (1991) Insurance and Risk. In: Burchell, G., Gordon, C. and P. Miller (eds) *The Foucault Effect: Studies in Governmentality*. London: Harvester Wheatsheaf, pp. 197–210.

Gabriel, Y. and Lang, T. (1995) *The Unmanageable Consumer*. London: Sage.

Hajer, M. (1995) *The Politics of Environmental Discourse. Ecological Modernization and the Policy Process*. Oxford: Clarendon Press.

Harvey, D. (1996) *Justice, Nature and the Geography of Difference*. Oxford: Blackwell Publishers.

Huntington, S.P. (1993) The Clash of Civilizations. *Foreign Affairs*, 72(3): 22–49.

Fischoff, B., Slovic, P., Lichtenstein, S. *et al*. (1978) How safe is safe enough? A psychometric study of attitudes towards technological risks and benefits. *Policy Studies* 9: 127–152.

Jasanoff, S. (1999) The songlines of risk. *Environmental Politics*, 9(2): 135–153.

Klinke, A. and Renn, O. (2006) Systemic risks as challenge for policy making in risk governance. Forum: Qualitative Social Research 7(1), Art. 33. Available at http:www.qualitative-research.net/fqs-texte/1-06/06-1-33-e.htm

Krimsky, S. and Plough, A. (eds) (1988) *Environmental Hazards: Communicating Risks as a Social Process*. Dover: Auburn House.

Lau, Chr. (1999) Neue Risiken und gesellschaftliche Konflikte. In: Beck, U., Hajer, M. and Kesselring, S. (eds) *Der unscharfe Ort der Politik*. Opladen: Leske + Budrich, pp. 248–266.

Latour, B. (2004a) *Politics of Nature. How to Bring the Sciences into Democracy*. Cambridge, MA: Harvard University Press.

Latour, B. (2004b) Why has critique run out of steam? From matters of fact to matters of concern. *Critical Inquiry* 30: 225–248.

Law, J. (2004) *Disaster in Agriculture: or Foot and Mouth Mobilities*. Published by the Centre for Science Studies, Lancaster University at www.comp.lancs.ac.uk/sociology/papers/law-disaster-mobilities-foot-and-mouth.pdf

Luhmann, N. (1993) *Risk: A Sociological Theory*. New York: Aldine/de Gruyter.

Lupton, D. (1999) *Risk*. London: Routledge.

Mythen, G. (2004) *Ulrich Beck. A Critical Introduction to the Risk Society*. London: Pluto Press.

Palmer, C.G.S. (1996) Risk perception: an empirical study of the relationship between worldview and the risk construct. *Risk Analysis* 16: 717–723.

O'Neill, J. (1997) Value Pluralism, Incommensurability and Institutions. In: Foster, J. (ed) *Valuing Nature? Economics, Ethics and Environment*. London: Routledge, pp. 75–88.

Ong, A. (2004) Assembling around SARS: Technology, Body Heat, and Political Fever in Risk Society. In: Poferl, A.and Sznaider, N. (eds) Ulrich Becks kosmopolitisches Projekt. Auf dem Weg in eine andere Soziologie. Baden-Baden: Nomos Verlagsgesellschaft. pp. S.81–89.

Renn, O. and Klinke, A. (2004) Systemic risks: a new challenge for risk management. *EMBO Rep.* 2004 October 5 (Suppl 1): S41–S46.

Renn, O. and Rohrmann, B. (eds) (2000) *Cross-Cultural Risk Perception. A Survey of Empirical Studies.* (*Technology, Risk and Society*, 13). Berlin and New York: Springer.

Rifkin, J. (1992) *Beyond Beef: The Rise and Fall of the Cattle Culture.* New York: Penguin.

Rippl, S. (2002) Cultural theory and risk perception: a proposal for a better measurement. *Journal of Risk Research*, 5: 147–165.

Schwarz, M. and Thompson, M. (1990) *Divided We Stand – Redefining Politics, Technology and Social Choice.* Philadelphia: University of Pennsylvania Press.

Slovic, P., Lichtenstein, S. and Fischoff, B. (1980) Facts and Fears: Understanding Perceived Risk. In: Schwing R. C. and Albers, W. A. (eds) *Societal Risk Assessment: How Safe is Safe Enough?* New York: Plenum.

Smith, B. and Goldblatt, D. (2000) Whose Health Is It Anyway? In: Hinchcliffe, S. and Woodward, K. (eds) *The Natural and the Social: Uncertainty, Risk, Change.* London: Sage, pp. 43–77.

Strydom, P. (2003) *Risk, Environment and Society.* Buckingham: Open University Press.

Tulloch, J. and Lupton, D. (2003) *Risk and Everyday Life.* London: Sage.

Urry, J. (2004) Risks and Mobilities. In: Poferl, A. and Sznaider, N. (eds) Ulrich Becks kosmopolitisches Projekt. Auf dem Weg in eine andere Soziologie. Baden-Baden: Nomos Verlagsgesellschaft. S. 90–97.

Van Loon, J. (2002) Risk and Technological Culture. Towards a Sociology of Virulence. London and New York: Routledge.

von Schomberg, R. (1993) Science, Politics and Morality. Scientific Uncertainty and Decision Making. Dordrecht: Kluwer Academic Publishers.

Wehling, P. (2003) Reflexive Wissenspolitik: das Aufbrechen tradierter Wissensordnungen der Moderne. Anmerkungen zu Werner Rammerts "Zwei Paradoxien einer innovationsorientierten Wissenspolitik. *Soziale Welt* 54: 509–518.

Wildavsky, A. and Dake, K. (1990) Theories of risk perception: Who fears what and why? *Deadalus* 119, 41–60.

Index

b indicates a box, f indicates a figure, m indicates a map, t indicates a table